PLANT HORMONES

Physiology, Biochemistry and Molecular Biology

PLANT HORMONES
Physiology, Biochemistry
and Molecular Biology

Edited by

PETER J. DAVIES

Section of Plant Biology,
Division of Biological Sciences,
Cornell University, Ithaca, New York, U.S.A.

KLUWER ACADEMIC PUBLISHERS
DORDRECHT / BOSTON / LONDON

Library of Congress Cataloging-in-Publication Data

ISBN 0-7923-2984-8 (HB)
ISBN 0-7923-2985-6 (PB)

Published by Kluwer Academic Publishers,
P.O. Box 17, 3300 AA Dordrecht, The Netherlands.

Kluwer Academic Publishers incorporates
the publishing programmes of
D. Reidel, Martinus Nijhoff, Dr W. Junk and MTP Press.

Sold and distributed in the U.S.A. and Canada
by Kluwer Academic Publishers,
101 Philip Drive, Norwell, MA 02061, U.S.A.

In all other countries, sold and distributed
by Kluwer Academic Publishers Group,
P.O. Box 322, 3300 AH Dordrecht, The Netherlands.

The camera ready text was   ꞁ ₒ ₒ ₒ 𝟓 𝟑𝟓𝟑𝟗𝟗
prepared by the Editor

Printed on acid-free paper

Contents

Preface ix

A. INTRODUCTION

1 The plant hormones: Their nature, occurrence and functions
  P.J. Davies 1
2 The plant hormone concept: Concentration, sensitivity and transport
  P.J. Davies 13

B. HORMONE SYNTHESIS AND METABOLISM

1 Auxin biosynthesis and metabolism
  R.S. Bandurski, J. D. Cohen, J. Slovin and D.M. Reinecke 39
2 Gibberellin biosynthesis and metabolism
  V.M. Sponsel 66
3 Cytokinin biosynthesis and metabolism
  B.A. McGaw 98
4 Biosynthesis and metabolism of ethylene
  T.A. McKeon, J.C. Fernández-Maculet and S.F. Yang 118
5 Abscisic acid biosynthesis and metabolism
  D.C. Walton and Y. Li 140

C. OTHER HORMONAL COMPOUNDS

1 Polyamines as endogenous growth regulators
  A.W. Galston and R. Kaur-Sawhney 158
2 Jasmonates, salicylic acid and brassinolides
  a) Jasmonate activity in plants: P.E. Staswick 179
  b) Salicylic Acid: I. Raskin 188
  c) Brassinosteriods: R.N. Arteca 206

D. HOW HORMONES WORK

1 Auxin and cell elongation
  R.E. Cleland 214
2 The control of gene expression by auxin
  G. Hagen 228

3   Gibberellin action in germinated cereal grains
    J.V. Jacobsen, F. Gubler and P.M. Chandler 246
4   Hormone binding and signal transduction
    K.R. Libbenga and A.M. Mennes 272
5   Calcium and plant hormone action
    P.C. Bethke, S. Gilroy and R.L. Jones 298

E. MOLECULAR ASPECTS OF HORMONE
SYNTHESIS AND ACTION

1   Genes specifying auxin and cytokinin biosynthesis in prokaryotes
    R.O. Morris 318
2   Transgenic plants in hormone biology
    H.J. Klee and M.B. Lanahan 340
3   Molecular approaches to the study of the mechanism of
        action of auxins
    J. Schell, K. Palme and R. Walden 354
4   Ethylene genes and fruit ripening
    S. Picton, J.E. Gray and D. Grierson 372
5   The role of hormones in gene activation in response to wounding
    H. Peña-Cortés and L. Willmitzer 395

F. HORMONE ANALYSIS

1   Instrumental methods of plant hormone analysis
    R. Horgan 415
2   Immunoassay methods of plant hormone analysis
    J.L. Caruso, V.C. Pence and L.A. Leverone 433

G. THE FUNCTIONING OF HORMONES IN
PLANT GROWTH AND DEVELOPMENT

1   Hormone mutants and plant development
    J.B. Reid and S.H. Howell 448
2   Ethylene in plant growth, development, and senescence
    M.S. Reid 486
3   Auxin transport
    T.L. Lomax, G.K.Muday and P.H. Rubery 509
4   The induction of vascular tissues by auxin and cytokinin
    R. Aloni 531
5   Hormones and the orientation of growth
    P.B. Kaufman, L-L. Wu, T.G. Brock and D. Kim 547
6   Hormonal regulation of apical dominance
    I.A. Tamas 572

7   Hormones as regulators of water balance
    T.A. Mansfield and M.R. McAinsh 598
8   Hormones and reproductive development
    J.D. Metzger 617
9   The role of hormones in photosynthate partitioning and seed filling
    M.L. Brenner and N. Cheikh 649
10 The role of hormones during seed development
    C.D. Rock and R.S. Quatrano 671
11 The role of hormones in potato (*Solanum tuberosum* L.)
    tuberization
    E.E. Ewing 698
12 Postharvest hormone changes in vegetables and fruit
    P.M. Ludford 725
13 Natural and synthetic growth regulators and their use in
    horticultural and agronomic crops
    T.J. Gianfagna 751
14 Hormones in tissue culture and micropropagation
    A.D. Krikorian 774

INDEX 797

PREFACE

Plant hormones play a crucial role in controlling the way in which plants grow and develop. While metabolism provides the power and building blocks for plant life, it is the hormones that regulate the speed of growth of the individual parts and integrate these parts to produce the form that we recognize as a plant. In addition, they play a controlling role in the processes of reproduction. This book is a description of these natural chemicals: how they are synthesized and metabolized; how they work; what we know of their molecular biology; how we measure them; and a description of some of the roles they play in regulating plant growth and development. Emphasis has also been placed on the new findings on plant hormones deriving from the expanding use of molecular biology as a tool to understand these fascinating regulatory molecules. Even at the present time, when the role of genes in regulating all aspects of growth and development is considered of prime importance, it is still clear that the path of development is nonetheless very much under hormonal control, either via changes in hormone levels in response to changes in gene transcription, or with the hormones themselves as regulators of gene transcription.

This is not a conference proceedings, but a selected collection of newly written, integrated, illustrated reviews describing our knowledge of plant hormones, and the experimental work that is the foundation of this knowledge. This volume forms the second edition of a book originally published in 1987 under the title *Plant Hormones and Their Role in Plant Growth and Development*. The title has been changed in order to reflect the changing nature of the field of plant hormones, namely an increased understanding of hormone biosynthesis and action deriving from the advances in molecular biology that have taken place since the first edition was published. Almost every chapter from the first edition has been extensively revised and rewritten, and several new chapters have been added to cover recently emerging areas.

The information in these pages is directed at advanced students and professionals in the plant sciences: molecular biologists, botanists, biochemists, or those in the horticultural, agricultural and forestry sciences. It should also form an invaluable reference to molecular biologists from other disciplines who have become aware of the fact that plants form an exiting class of organisms for the study of development, and who need information on the regulators of development that are exclusive to plants. It is intended that the book should serve as a text and guide to the literature for graduate level courses in the plant hormones, or as a part of courses in plant, comparative, or molecular aspects of development. Scientists in other disciplines who wish to know more about the plant hormones and their role in plants should also find this volume a valuable resource. I hope that anyone with a reasonable scientific background can find valuable information in this book expounded in an understandable fashion.

The subject matter ranges from basic biochemistry and molecular biology to the use of natural and synthetic plant growth regulators in agriculture and horticulture. Recent findings deriving from the use of the tools of molecular biology are emphasized throughout. As far as possible chapters are grouped according to the area of the topic: Introduction; Biosynthesis; Mode of action; Molecular Aspects; and Roles in Plant Growth and Development, ordered approximately in a developmental plant life cycle. As, however, many chapters span two or more areas, their presence in one location rather than another is largely a matter of editorial choice. Thus for example the chapter on the role of abscisic acid in seed development and dormancy is located in the section on development, though from its extensive description of the molecular mechanisms of ABA action it could equally well be located in the molecular aspects section.

It is most noticeable as one progresses through the chapters that while we know a lot, though certainly not everything, about the action of plant hormones at the molecular and cellular level, our knowledge of hormones in the whole plant functions of importance to agriculture and horticulture, such in flowering, tuberization and dormancy, is still at a superficial level. Only when such systems are fully understood can we hope to manipulate plant growth to human advantage. However, the pace of acquisition of knowledge in this area is increasing and practical applications are on the horizon. One very notable area of success over the last few years has been the elucidation of knowledge on the molecular aspects of ethylene production in relation to fruit ripening. This has enabled the modification of ripening, and the production of genetically engineered tomatoes in which ripening can be regulated to fit the requirements of the market. At the present time such tomatoes are rapidly progressing to commercial introduction.

Gone are the days when one person could write a comprehensive book in an area such as plant hormones. I have thus drawn together a team of sixty-four experts who have individually or jointly written about their own area. At my direction they have attempted to tell a story in a way that will be both informative and interesting. Their styles and approaches vary, because they each have a tale to tell from their own perspective. The choice of topics has been my own. Within each topic the coverage and approach have been decided by the authors. While the opinions expressed by the authors are their own, they are, in general, also mine, because I knew their perspective before I invited them to join the project.

Where appropriate, the reader will find cross references between chapters. In addition the extensive, sub-divided index at the end of the volume should allow this book to be used as a reference to find individual pieces of information. Sometimes the same information can be found in more than one location, though usually from a different perspective. Rather than edit out such duplication, I have chosen to let it remain so that the complete story on any topic can be obtained without having to excessively transfer between chapters.

A volume such as this cannot be encyclopedic. Nevertheless, we have covered the majority of topics in which active research is taking place. The author of each chapter has provided a set of references that will guide the reader to more detailed recent reviews, as well as classical papers and the latest advances in that area. The number of references has, however, been limited so as not to disrupt the narrative excessively. Because of this, a reference citation at any point may not necessarily be to the paper describing that particular finding, but may instead be to a later paper in a series, or to a review covering the material, from which the original citation(s) can be obtained if desired.

I would like to thank all the authors who made this volume possible, and produced their chapters not only in a timely fashion, but in line with the many restrictions placed on them by this editor; Imre Tamas and Maureen Kelly for marking the index entries; Tom Silva for translating many of the contributors discs; and Gilles Jonker of Kluwer Academic Publishers for continuing support and cooperation. Finally I would like to thank my family, Linda, Kenneth and Caryn, and my students, who put up with my many hours of absence during which I edited and produced this book.

Peter J. Davies
Ithaca, New York
February 1995

A. INTRODUCTION

A1. The Plant Hormones: Their Nature, Occurrence, and Functions

Peter J. Davies
Section of Plant Biology, Cornell University, Ithaca, New York 14853, USA.

INTRODUCTION

The Meaning of a Plant Hormone

Plant hormones are a group of naturally occurring, organic substances which influence physiological processes at low concentrations. The processes influenced consist mainly of growth, differentiation and development, though other processes, such as stomatal movement, may also be affected. Plant hormones have also been referred to as 'phytohormones' though this term is seldom used.

The term "hormone" comes originally from the Greek and is used in animal physiology to denote a chemical messenger. Its original use in plant physiology was derived from the mammalian concept of a hormone. This involves a localized site of synthesis, transport in the bloodstream to a target tissue, and the control of a physiological response in the target tissue via the concentration of the hormone (14). Auxin was similarly thought to produce a growth response at a distance from its site of synthesis, and thus to fit the definition of a *transported* chemical messenger. It is now clear that plant hormones do not fulfil the requirements of a hormone in the mammalian sense. In a controversial article that has caused a rethinking of much dogma in the area, Trewavas (14) has argued that plant physiologists have been so concerned with making plant hormones fit the same characteristics as animal hormones that we have overlooked their unique characteristics in controlling growth and development. The synthesis of plant hormones may be localized (as occurs for animal hormones), but may also occur in a wide range of tissues, or cells within tissues. While they may be transported and have their action at a distance this is not always the case. They may also act in the tissue in which they are synthesized or even within the same cell. At one extreme we find the transport of cytokinins from roots to leaves where they prevent senescence and maintain metabolic activity, while at the other extreme the production of the gas ethylene may bring about changes within the same tissue, or within the same cell, where it is synthesized. Thus, we

1

P. J. Davies (ed.), Plant Hormones, 1–12.
© 1995 *Kluwer Academic Publishers. Printed in the Netherlands.*

must abandon the concept of *transport* as being an *essential* property of a plant hormone. Trewavas has also strongly argued that control is not by concentration but by a change in sensitivity of the tissue to the compound. This issue has been more hotly debated and will be discussed later (see Chapter A2). However, as a result of the lack of direct rigid parallels with animal hormones he has argued that we should abandon the term "plant hormone". In a rebuttal Cleland (15) has characterized this argument as "semantic quibbling". Trewavas suggests we should replace the term hormone with "plant growth substance"[1]. The disadvantage of this name, besides being cumbersome and containing the rather vague term "substance," is that it does not describe fully what these natural regulators do. Growth is only one of the many processes influenced. Changing the name to "plant growth and development substances" becomes even more awkward and still may not cover all cases. While the term plant growth regulator is a little more precise this term has been (unfortunately?) usurped by the agrichemical industry to denote synthetic plant growth regulators (see Chapter G13) as distinct from endogenous growth regulators. Thus the term plant growth regulator is largely unused in reference to endogenous regulators of growth and differentiation. One could invent an entirely new single word name for this group of compounds but, rather like the international language Esperanto, that would seem to be unlikely to catch on in the face of over fifty years of habit. Thus we are left with the imperfect term "plant hormone" and all that it implies. We must break with the characteristics expected of animal hormones: plant hormones are a unique set of compounds, with unique metabolism and properties, that form the subject of this book. Their only universal characteristics are that they are natural compounds in plants with an ability to affect physiological processes at concentrations far below those where either nutrients or vitamins would affect these processes. In fact, this notion of a plant hormone is much closer to the meaning of the Greek origin of the word (*to set in motion* or *to stimulate*) than is the current meaning of the word hormone used in the context of animal physiology. The Greek origin of the word does not have in it the implicit idea of either transport or action at a distance. The term *plant hormones* has, therefore, been retained in this book.

The Discovery, Identification and Quantitation of Plant Hormones.

The concept of plant hormones derives from Darwin's experiments on the phototropism of coleoptiles, which indicated the presence of a transported signal. This led to Went's elucidation of auxin in 1928 and its subsequent identification as indoleacetic acid (IAA) (5). Other lines of investigation led to the discovery of the other hormones: research in plant pathogenesis led to gibberellins (GA); efforts to culture tissues led to cytokinins (CK); the control

---

[1] The international Society for the study of plant hormones is named the "International Plant Growth Substance Association" (IPGSA)

of abscission and dormancy led to abscisic acid (ABA); and the effects of illuminating gas and smoke led to ethylene. These accounts are told in virtually every elementary plant physiology textbook, and further elaborated in either personal accounts (5,13) or advanced treatises devoted to individual hormones (2,10) so that they need not be repeated here. More recently other compounds, namely polyamines (Chapter C1), jasmonates (Chapter C2a) (and its derivative tuberonic acid; Chapter G12), salicylic acid (Chapter C2b), brassinosteroids (Chapter C2c) and the peptide systemin (Chapters C2a and E5) are being added to the list of (potential?) plant hormones, though whether these have universal effects or act in just a few special cases has not yet been fully determined.

It is interesting to note that, of all the main group of plant hormones, only the chemical identification of abscisic acid was made from higher plant tissue. The original identification of the others came from extracts that produced hormone-like effects in plants: auxin from urine, gibberellins from fungal culture filtrates, cytokinins from autoclaved herring sperm DNA, and ethylene from illuminating gas. Today we have at our disposal methods of purification (such as high performance liquid chromatography: HPLC) and characterization (gas chromatography-mass spectrometry: GC-MS, and high performance liquid chromatography-mass spectrometry: HPLC-MS) that can operate at levels undreamed of by early investigators (see Chapter F1). Thus while early purifications from plant material utilized tens or even hundreds of kilograms of tissues, modern analyses can be performed on a gram or less of tissue, making the characterization of hormone levels in individual leaves, buds, or even from tissues within the organs much more feasible. The advent of immunoassay (Chapter F2) promises to enable the localization of the hormones within cells. Only when the exact level and location of the hormones within the tissues are known will their precise roles and modes of action in a process be fully elucidated.

THE NATURE, OCCURRENCE, AND EFFECTS OF THE PLANT HORMONES

Before we become involved in the various subsequent chapters covering aspects of hormone biochemistry and action it is necessary to review what hormones do. In subsequent chapters some or most of these effects will be described in more detail, whereas others will not be referred to again. It is impossible to give detailed coverage of every hormonal effect. The choice of topics for subsequent chapters has been determined largely by whether there is active research in progress in that area. Thus, while the mode of action of auxin and some effects of auxin are subjects of subsequent chapters, most of the current work on cytokinins is on their synthesis and metabolism (which is described), but so little progress has been made on how they act that there is little to add to the information below. The effects produced by each hormone have been elucidated largely from exogenous applications. In some cases, using correlations between hormone levels and growth of defined genotypes (Chapter G1) or transgenic plants (Chapter E2), we have evidence that the endogenous hormone also fulfills that role. In other cases it has not been conclusively proved that the endogenous hormone functions in the same manner. The nature, occurrence, transport and effects of each hormone (or hormone group) are given below. It should, however, be emphasized that hormones do not act alone but in conjunction, or in opposition, to each other such that the final condition of growth or development represents the net effect of a hormonal balance (7).

Auxin

Nature

Indole-3-acetic acid (IAA) is the main auxin in most plants.

INDOLEACETIC ACID

Compounds which serve as IAA precursors may also have auxin activity (e.g., indoleacetaldehyde). Some plants contain other compounds that display weak auxin activity (e.g., phenylacetic acid) (16). IAA may also be present as various conjugates such as indoleacetyl aspartate (Chapter B1)). 4-chloro-IAA has also been reported in several species (11), though it is not clear to what extent the endogenous auxin activity in plants can be accounted for by 4-Cl-IAA. Several synthetic auxins are also used in commercial applications (Chapter G13).

Sites of biosynthesis

IAA is synthesized from tryptophan or indole (Chapter B1) primarily in leaf primordia and young leaves, and in developing seeds.

Transport

IAA transport is cell to cell (Chapters G3 and G4). Transport to the root probably also involves the phloem.

Effects

• Cell enlargement - auxin stimulates cell enlargement and stem growth (Chapter D1).

- Cell division - auxin stimulates cell division in the cambium and, in combination with cytokinin, in tissue culture (Chapters G4 and G14).
- Vascular tissue differentiation - auxin stimulates differentiation of phloem and xylem (Chapter G4).
- Root initiation - auxin stimulates root initiation on stem cuttings, and also the development of branch roots and the differentiation of roots in tissue culture (Chapter G14).
- Tropistic responses - auxin mediates the tropistic (bending) response of shoots and roots to gravity and light (Chapters G5 and G3).
- Apical dominance - the auxin supply from the apical bud represses the growth of lateral buds (Chapter G6).
- Leaf senescence - auxin delays leaf senescence.
- Leaf and fruit abscission - auxin may inhibit or promote (via ethylene) leaf and fruit abscission depending on the timing and position of the source (Chapters G2, G6 and G13).
- Fruit setting and growth - auxin induces these processes in some fruit (Chapter G13)
- Assimilate partitioning - assimilate movement is enhanced towards an auxin source possibly by an effect on phloem transport (Chapter G9).
- Fruit ripening - auxin delays ripening (Chapters G2 and G12).
- Flowering - auxin promotes flowering in Bromeliads (Chapter G8).
- Growth of flower parts - stimulated by auxin (Chapter G2).
- Promotes femaleness in dioecious flowers (via ethylene) (Chapters G2 and G8).

In several systems (e.g., root growth) auxin, particularly at high concentrations, is inhibitory. Almost invariably this has been shown to be mediated by auxin-produced ethylene (3, 6) (Chapter G2). If the ethylene synthesis is prevented by various ethylene synthesis inhibitors, the ethylene removed by hypobaric conditions, or the action of ethylene opposed by silver salts (Ag+), then auxin is no longer inhibitory.

Gibberellins (GAs)

Nature

The gibberellins (GAs) are a family of compounds based on the *ent*-gibberellane structure (Chapter B2). While the most widely available compound is GA_3 or gibberellic acid, which is a fungal product, the most important GA in plants is GA_1, which is the GA primarily responsible for stem elongation (Chapters A2, B2, and G1). Many of the other GAs are precursors of the growth-active GA_1.

GIBBERELLIN A₁ or GA₁

Sites of biosynthesis.

GAs are synthesized from mevalonic acid (Chapter B2) in young tissues of the shoot (exact location uncertain) and developing seed. It is uncertain whether synthesis also occurs in roots (12).

Transport

GAs are probably transported in the phloem and xylem.

Effects
- Stem growth - GA_1 causes hyperelongation of stems by stimulating both cell division and cell elongation (Chapters A2 and G1). This produces tall, as opposed to dwarf, plants.
- Bolting in long day plants - GAs cause stem elongation in response to long days (Chapter G8).
- Induction of seed germination - GAs can cause seed germination in some seeds that normally require cold (stratification) or light to induce germination.
- Enzyme production during germination - GA stimulates the production of numerous enzymes, notably α-amylase, in germinating cereal grains (Chapter D3).
- Fruit setting and growth - This can be induced by exogenous applications in some fruit (e.g., grapes) (Chapter G13). The endogenous role is uncertain.
- Induction of maleness in dioecious flowers (Chapter G8).

Cytokinins (CKs)

Nature

ZEATIN

CKs are adenine derivatives characterized by an ability to induce cell division in tissue culture (in the presence of auxin). The most common cytokinin base in plants is zeatin. Cytokinins also occur as ribosides and ribotides (Chapter B3).

Sites of biosynthesis

CK biosynthesis is through the biochemical modification of adenine (Chapter B3). It occurs in root tips and developing seeds.

Transport

CK transport is via the xylem from roots to shoots.

Effects
- Cell division - exogenous applications of CKs induce cell division in tissue culture in the presence of auxin (Chapter G14). This also occurs endogenously in crown gall tumors on plants (Chapter E1). The presence of CKs in tissues with actively dividing cells (e.g., fruits, shoot tips) indicates that CKs may naturally perform this function in the plant.
- Morphogenesis - in tissue culture (Chapter G14) and crown gall (Chapter E1) CKs promote shoot initiation. In moss, CKs induce bud formation (Chapters G1 and G6).
- Growth of lateral buds - CK applications, or the increase in CK levels in transgenic plants with genes for enhanced CK synthesis, can cause the release of lateral buds from apical dominance (Chapters E2 and G6).
- Leaf expansion (8) - resulting solely from cell enlargement. This is probably the mechanism by which the total leaf area is adjusted to compensate for the extent of root growth, as the amount of CKs reaching the shoot will reflect the extent of the root system. However, this has not been observed in transgenic plants with genes for increased CK biosynthesis (Chapter E2).
- CKs delay leaf senescence (Chapter E2).

- CKs may enhance stomatal opening in some species (Chapter G7).
- Chloroplast development - the application of CK leads to an accumulation of chlorophyll and promotes the conversion of etioplasts into chloroplasts (9).

Mode of Action

CKs are the only hormone in this book for which there is no chapter on the mode of action in any system. This is because the action of CKs is still poorly understood and insufficient evidence exists to conclusively identify any biochemical point of action (4).

Ethylene

Nature

The gas ethylene (C_2H_4) is synthesized from methionine (Chapter B4) in many tissues in response to stress. It does not seem to be essential for normal vegetative growth (Chapter E2). It is the only hydrocarbon with a pronounced effect on plants.

Sites of synthesis

Ethylene is synthesized by most tissues in response to stress. In particular, it is synthesized in tissues undergoing senescence or ripening (Chapters E4 and G2).

Transport

Being a gas, ethylene moves by diffusion from its site of synthesis. A crucial intermediate in its production, 1-aminocyclopropane-1-carboxylic acid (ACC) can, however, be transported and may account for ethylene effects at a distance from the causal stimulus (Chapter G2).

Effects

The effects of ethylene are fully described in Chapter G2. They include:
- Release from dormancy.
- Shoot and root growth and differentiation.
- Adventitious root formation.
- Leaf and fruit abscission.
- Flower induction in some plants (Chapter G8).
- Induction of femaleness in dioecious flowers (Chapter G8).
- Flower opening.
- Flower and leaf senescence.
- Fruit ripening (Chapter E4).

Abscisic acid (ABA)

Nature

Abscisic acid is a single compound with the following formula:

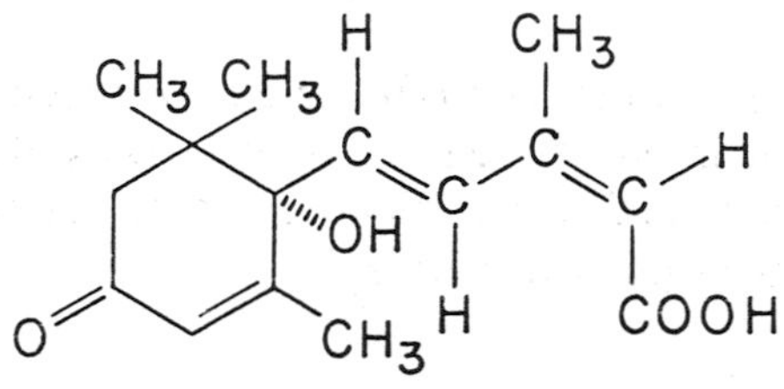

ABSCISIC ACID

Its name is rather unfortunate. The first name given was "abscisin II" because it was thought to control the abscission of cotton bolls. At almost the same time another group named it "dormin" for a purported role in bud dormancy. By a compromise the name abscisic acid was coined (1). It now appears to have little role in either abscission (Chapters G2) or bud dormancy, but we are stuck with this name. As a result of the original association with abscission and dormancy, ABA has become thought of as an inhibitor. While exogenous applications can inhibit growth in the plant, ABA appears to act as much as a promoter (e.g., storage protein synthesis in seeds - Chapter G10) as an inhibitor, and a more open attitude towards its overall role in plant development is warranted.

Sites of synthesis

ABA is synthesized from mevalonic acid (Chapter B5) in roots and mature leaves, particularly in response to water stress (Chapters G7 and G9). Seeds are also rich in ABA which may be imported from the leaves or synthesized in situ (Chapter G9).

Transport

ABA is exported from roots in the xylem and from leaves in the phloem. There is some evidence that ABA may circulate to the roots in the phloem and then return to the shoots in the xylem (Chapters A2, G7 and G9).

Effects
- Stomatal closure - water shortage brings about an increase in ABA which leads to stomatal closure (Chapter G7).
- ABA inhibits shoot growth (but has less effect on, or may promote, root growth). This may represent a response to water stress (Chapters A2 and G7).
- ABA induces storage protein synthesis in seeds (Chapter G10).

- ABA counteracts the effect of gibberellin on α-amylase synthesis in germinating cereal grains (Chapter D3).
- ABA affects the induction and maintenance of some aspects of dormancy in seeds. It does not, however, appear to be the controlling factor in 'true dormancy' or 'rest,' which is dormancy that needs to be broken by low temperature or light (Chapters G1 and G10).
- Increase in ABA in response to wounding induces gene transcription, notably for proteinase inhibitors, so it may be involved in defense against insect attack (Chapter E5).

Polyamines

There is some controversy as to whether these compounds (fully described in Chapter C1) should be classified as hormones, even within our rather broad current definition. They were first tentatively accepted by their inclusion (in the form of a specific session) at the International Conference on Plant Growth Substances in 1982. Galston (personal communication) justifies their classification as hormones on the following grounds:

- They are widespread in all cells and can exert regulatory control over growth and development at micromolar concentrations.
- In plants where the content of polyamines is genetically altered, development is affected. (E.g., in tissue cultures of carrot or *Vigna*, when the polyamine level is low only callus growth occurs; when polyamines are high, embryoid formation occurs. In tobacco plants, which are overproducers of spermidine, anthers are produced in place of ovaries.)

Such developmental control is more characteristic of hormonal compounds than nutrients such as amino acids or vitamins.

Polyamines have a wide range of effects on plants and appear to be essential for plant growth, particularly cell division and normal morphologies. At present it is not possible to make an easy, distinct list of their effects as for the other hormones. A variety of cellular and organismal effects is discussed in Chapter C1. It appears that polyamines are present in all cells rather than having a specific site of synthesis.

Jasmonates

Jasmonates (Chapter C2a) are represented by jasmonic acid and its methyl ester. They are named after the jasmine plant in which the methyl ester is an important scent component. As such they have been known for some time in the perfume industry. There is also a related hydroxylated compound that has been named tuberonic acid which, with its methyl ester and glycosides, induces potato tuberization. Jasmonic acid is synthesized from linolenic acid, while jasmonic acid is most likely the precursor of tuberonic acid.

Effects

Jasmonates inhibit many plant processes such as growth and seed germination. They also promote senescence, abscission, tuber formation, fruit ripening, pigment formation and tendril coiling. An important role appears to be in plant defense, where jasmonates induce the synthesis of proteinase inhibitors which deter insect feeding, and, in this regard, act as intermediates in the response pathway induced by the peptide systemin.

Salicylic Acid

Salicylates have been known for a long time to be present in willow bark, but have only recently been recognized as potential regulatory compounds. Salicylic acid is biosynthesized from the amino acid phenylalanine.

Effects

Salicylic acid (Chapter C2b) is the calorigenic substance that causes thermogenesis in *Arum* flowers. It may have a more general role in the resistance to pathogens by inducing the production of 'pathogenesis-related proteins.' It has also been reported to enhance flower longevity, inhibit ethylene biosynthesis and seed germination, block the wound response, and reverse the effects of ABA. Whether these are endogenous roles remains to be determined.

Brassinosteroids

Brassinosteroids (Chapter C2c) are a range of over 60 steroidal compounds, typified by the compound brassinolide that was first isolated from *Brassica* pollen. At first they were regarded as somewhat of an oddity but they appear to be widely distributed. As they produce effects on growth and development at very low concentrations it is possible that they play a role in the endogenous regulation of these processes.

Effects

Brassinosteroids promote stem elongation, inhibit root growth and development, and promote ethylene biosynthesis and epinasty.

Whether in the long run we classify polyamines and other compounds as plant hormones or not is irrelevant. Hormones are a human classification and organisms care naught for human classifications. Natural chemical compounds affect growth and development in various ways, or they do not do so. Polyamines are present in all organisms and clearly fall under the first of the two groups. Whether the other compounds should be regarded as plant hormones will depend on whether, in the long run, these compounds are shown to be endogenous regulators of growth and development in plants in general.

References

1. Addicott, F.T., Carns, H.R., Cornforth, J.W., Lyon, J.L., Milborrow, B.V., Ohkuma, K., Ryback, G., Smith, O.E., Thiessen, W.E., Wareing, P.F. (1968) Abscisic acid: a proposal for the redesignation of abscisin II (dormin). *In*: Biochemistry and Physiology of Plant Growth Substances, pp. 1527-1529, Wightman, F., Setterfield, G., ed. Runge Press, Ottawa.
2. Addicott, F.T., Carns, H.R. (1983) History and Introduction. *In*: Abscisic acid, pp. 1-21, Addicott, F.T., ed. Praeger, New York.
3. Burg, S.P., Burg, E.A. (1966) Interaction between auxin and ethylene and its role in plant growth. Proc. Natl. Acad. Sci. USA. 55, 262-269.
4. Horgan, R. (1984) Cytokinins. *In*: Advanced Plant Physiology, pp. 53-75, Wilkins, M.B., ed. Pitman, London.
5. Jacobs, W.P. (1979) Plant hormones and plant development. Cambridge University Press, Cambridge, New York. 339 pp.
6. Mulkey, T.J., Kuzmanoff, K.M., Evans, M.L. (1982) Promotion of growth and hydrogen ion efflux by auxin in roots of maize pretreated with ethylene biosynthesis inhibitors. Plant Physiol. 70, 186-188.
7. Leopold, A.C. (1980) Hormonal regulating systems in plants. *In*: Recent Developments in Plant Sciences, pp. 33-41, Sen S.P., ed. Today and Tomorrow Publishers, New Delhi.
8. Letham, D.S. (1971) Regulators of cell division in plant tissues. XII. A cytokinin bioassay using excised radish cotyledons. Physiol. Plant 25, 391-396.
9. Parthier, B. (1979) Phytohormones and chloroplast development. Biochem. Physiol. Pflanzen 174, 173-214.
10. Phinney, B.O. (1983) The history of gibberellins. *In*: The Biochemistry and Physiology of Gibberellins, Vol 1, pp. 19-52, Crozier. A., ed. Praeger, New York.
11. Pless, T., Bottger, M., Hedden, P., Graebe, J. (1984) Occurrence of 4-Cl-indoleacetic acid in broad beans and correlation of its levels with seed development. Plant Physiol. 74, 320-323.
12. Stoddart, J.L. (1983) Sites of gibberellin biosynthesis and action. *In*: The Biochemistry and Physiology of Gibberellins, Vol. 2, pp. 1- 55, Crozier, A., ed. Praeger, New York.
13. Thimann, K.V. (1977) Hormone action in the whole life of plants. Univ. of Massachusetts Press, Amherst. 448 pp.
14. Trewavas, A. (1981) How do plant growth substances act? Plant Cell Environment 4, 203-228.
15. Trewavas, A.J., Cleland, R.E. (1983) Is plant development regulated by changes in the concentration of growth substances or by changes in the sensitivity to growth substances? Trends in Biochem. Sci. 8, 354-357.
16. Wightman, F., Lighty, D.G. (1982) Identification of phenylacetic acid as a natural auxin in the shoots of higher plants. Physiol. Plant 55, 17-24.

A2. The Plant Hormone Concept: Concentration, Sensitivity and Transport

Peter J. Davies

Section of Plant Biology, Cornell University, Ithaca, New York   14853, USA.

CONCENTRATION VERSUS SENSITIVITY AS THE CONTROLLING ASPECT OF HORMONE ACTION.

The concept of control by changing concentrations is crucial to the original concept of hormones in mammals. A few years ago, a great stir was created amongst biologists working with plant hormones by the suggestion of Trewavas (56) that there is no evidence that plant hormones act via changes in the amount or concentration of the hormone, and that all change in response must be attributed to changes in the sensitivity of the tissue. The reason for this suggestion is the frequent lack of correlation between hormone concentrations measured in a tissue and the response of the tissue.   In addition, in most plants, growth is proportional to the logarithm of the applied hormone concentration, such that there may be an increasing response over three orders of magnitude in concentration. However, changes in the endogenous concentration in tissues are usually far smaller than would be expected to produce the vast changes in growth or development observed; (plant hormone workers tend to regard a doubling in concentration as a large change!). If concentration changes cannot account for the differences in growth and development then something else must be responsible, and changing tissue sensitivity is the only other logical alternative. While our knowledge on factors influencing sensitivity (or responsiveness) to hormones is still elementary, documentation of differential sensitivities, elucidation of hormone binding (Chapters D4 and E3), and understanding of the mechanisms of regulation of gene expression are steadily increasing, so that the potential mechanisms for the regulation of hormone sensitivity are becoming more evident. Our best examples of changing sensitivity are to ethylene. Immature flower or fruit tissue show no response to ethylene, but mature tissue responds to ethylene with ripening or senescence (see Chapter G2). However, an important question is what we mean by the rather vague term "sensitivity". As Firn (11) has pointed out, a change in sensitivity simply refers to an observation that the response to a given amount of hormone has changed. This could be caused by a change in the number of receptors (*receptivity*), a change in receptor affinity (*affinity*), or a change in

13

P. J. Davies (ed.), Plant Hormones, 13–38.

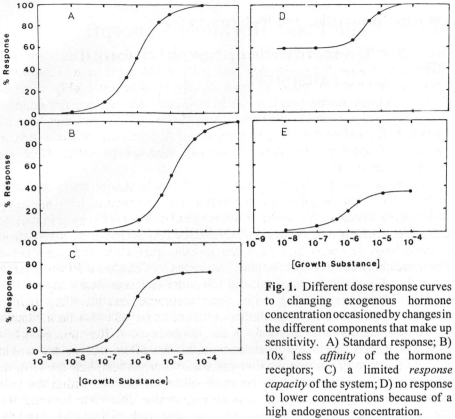

Fig. 1. Different dose response curves to changing exogenous hormone concentration occasioned by changes in the different components that make up sensitivity. A) Standard response; B) 10x less *affinity* of the hormone receptors; C) a limited *response capacity* of the system; D) no response to lower concentrations because of a high endogenous concentration. The response only occurs when the endogenous concentration is exceeded; E) A decrease in the number of hormone receptors (*receptivity*). Note change in slope of main part of response curve. From (11).

the subsequent chain of events (*response capacity*). A change in the response to a given amount of hormone could also be caused by a change in the level of other endogenous substances that enhance, or inhibit, the response to the hormone. When dealing with exogenously applied hormones, uptake efficiency must also be taken into account. A change in each of these would give a very different response curve to changing hormone concentrations (Fig. 1) (11).

Hormone concentration clearly has a major role to play in hormonal regulation. Even if sensitivity changed, there would still be a response to a change in concentration, though of a different magnitude (11). In addition we should ask what are we measuring when we calculate "concentration," and where we are measuring it, in relation to the tissues or cells that respond? If the accuracy of measurement or localization is vague then so is our knowledge of hormone concentration at the active site. Below I will list some of the pitfalls that can occur in measurement of these parameters.

Hormone Quantitation and Its Interpretation

One of the difficulties in correlating endogenous hormone concentration with differences in growth and development is the problem of what is measured when the hormone is assayed, even if this is done by highly accurate physico-chemical means (see Chapter F1). Quantitation is normally done by measuring total extractable hormone. Often this is of little relevance because the total hormonal amount tells us little about the hormonal concentration in the tissue in question, let alone in the cell, cell compartment, or at the hormone receptor site.

   Many hormonal extractions are done of whole shoots, roots or fruits. While this may enable a first approximation, it can be equated with analyzing the hormonal content of, say, the pituitary gland of a mammal by grinding up and assaying the whole animal. Studies investigating hormone metabolism in seeds have indicated that there are specific qualitative and quantitative differences between adjacent tissues. The embryo of pea has a different ABA and gibberellin content and metabolism than the seed coat (see Chapter G9). Applied GA_{20} is metabolized to GA_{29} in the embryo, and this GA_{29} is then further metabolized to GA_{29} catabolite in the seed coat (63). We do not know what further subdivisions may exist within the embryo. Differences probably also exist within seemingly uniform tissues. We often analyze the whole tissue in the case of phototropism or geotropism, or, at best, half the stem or coleoptile. However, there may be more subtle differences within the half tissue, even though there is only a small measurable difference between the two halves, and this may lead to the observed differential growth. Experiments on the gradient between the upper and lower epidermis of dicotyledonous shoots indicate that the difference may be two to four times the gradient recorded between the upper and lower halves. In maize coleoptiles a two fold difference between the IAA content of the two sides is sufficient to account for the observed phototropic curvature (2). More accurate measurements of auxin, such as the asymmetric promotion of gene activation, which will be described below, enable us to visualize a more precise localization of auxin redistribution in tropistically-stimulated tissues than is possible with assaying surgically isolated parts.

   In general, we have little idea of cell to cell variation. We know even less of the differences in hormonal contents within cells. It was noted many years ago that growth of stem or coleoptile sections correlates with the amount of auxin that will diffuse out of the tissue rather than the total extractable auxin (50). This tells us that much of the hormone in the tissue is probably compartmentalized away from the growth active site, possibly within an organelle. A demonstration of this compartmentation can be seen if a stem segment is loaded with radiolabelled auxin, and then washed over an extended period to remove the auxin effluxing from the tissue. Growth initially increases in response to the applied IAA, but then declines to a very low rate over 3- 4 hours, even though about 70% of the radiolabel is still in

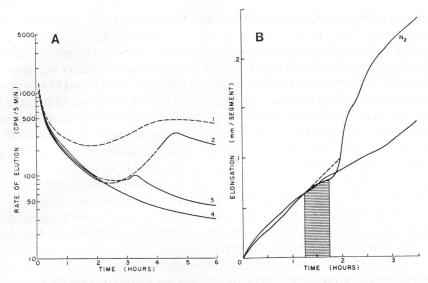

Fig. 2. A) The elution of label from [^{14}C]IAA-treated pea stem segments as influenced by anaerobiosis and subsequent oxygenation. Under aerobic condition (bubbled with oxygen-- solid line) the rate of efflux steadily decreases (curve 4), but if the sections become anaerobic (bubbled with nitrogen), as indicated by the dashed line (curves 2 and 3), the rate of efflux increases. The increase can subsequently be reversed by the return of aerobic conditions. Segments in curve 1 were anaerobic throughout. B) In segments that become anaerobic for a short period, and then return to aerobic conditions, growth is stimulated beyond that of those that remain in aerobic conditions. From (31).

the tissue, and can be chromatographically identified as IAA. If the tissue is then put in an anaerobic environment, under which growth ceases, the rate of auxin efflux increases. A return to aerobic conditions is accompanied by a burst in the growth rate such that the total growth more than makes up for the loss of growth during anaerobiosis (termed *emergent growth*), and the rate of auxin efflux gradually returns to its previous condition (31) (Fig. 2). We interpret this as an indication that the IAA is sequestered within an inner membrane-bound compartment, from which it leaks at a slower rate than when it leaves the cell through the plasmalemma. The growth-active site, with which auxin interacts, is exterior to the cellular compartment in which it is contained. Once the auxin originally in the cytoplasm leaves the cell, the auxin concentration at the growth active site remains very low (under normal aerobic conditions) because the auxin exits through the plasmamembrane faster than it leaves the internal compartment. Under anaerobic conditions the compartment membrane becomes leaky, possibly because an inwardly-directed, ATP-requiring IAA carrier becomes inactivated, allowing outward passive diffusion to predominate. Alternatively the compartment interior becomes more acid, thus increasing the diffusibility of the IAA, because a greater proportion of the IAA becomes protonated. This allows more IAA to leak into the rest of the cell and to the exterior. On the return of oxygen the high concentration of auxin at the growth active site allows

ATP-requiring growth to take place, while the compartment membrane ceases being leaky. We do not know the location of the IAA-storage compartment, nor the location of the growth active site (see Chapters C1 and C4 for the latter). One possibility for the storage compartment is the chloroplast (or proplastid) which, because of its relatively high pH, would tend to trap acidic molecules (see Chapter G3). In fact 30-40% of the cellular IAA has been reported to be localized in the chloroplast (47). The vacuole seems unlikely because its acidic pH would dictate an auxin concentration only about 1/10th that in the cytoplasm, though its vast bulk in parenchymatous cells might make it a hormone reservoir. (The vacuole could possibly function as a reservoir of hormone conjugates that might be later hydrolyzed to release free hormone.) The development of hormone immunoassay (Chapter F2) promises a method by which we may be able to localize hormones within tissues and cells, provided a way can be found to immobilize the highly soluble hormone molecules within the cells.

Correlations of Growth with Hormonal Concentration

One of Trewavas's principal examples to support his contention that hormones did not exert their influence via changes in concentration was the seeming lack (at that time) of any correlation between the endogenous GA concentration and tallness (in tall versus dwarf plants), despite the fact that applications of GA_3 to dwarf plants produced tall phenotypes (56). This has since been completely refuted by the work of Reid and co-workers in peas (Chapter G1) (19), and by Phinney and co-workers in maize (53). They have shown that tall plants have GA_1, while in dwarf plants GA_1 is absent or present at a very low level. It has subsequently been shown that GA_1 is the main growth-active GA in most plants, while the other GAs are not active in their own right, but only after conversion to GA_1. In peas, the tallness gene, *Le*, controls the conversion (by 3β-hydroxylation) of GA_{20} to GA_1. In the presence of *le* GA_{20} builds up and is not converted to GA_1 (19). If the gibberellin content is measured by bioassay, without prior chromatographic separation, then tall and dwarf plants are found to contain the same amount of "gibberellin activity." This is because the vast majority of bioassays will show no difference between GA_{20} and GA_1, since the bioassay plant has the ability to convert inactive GA_{20} to the active GA_1. (One newer bioassay, Waito-C rice, has the ability to distinguish 3C-OH GAs, and on such a bioassay the difference can be seen.) It should also be noted that the GA_1 is only present in the youngest

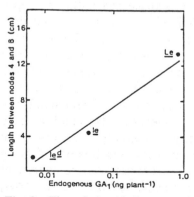

Fig. 3. The relationship between the endogenous GA_1 content and internode length in pea plants with different alleles of the *Le* gene. From (42).

internodes of the stem. Often the extracts used in comparing tall and dwarf plants have come from whole shoots, mature internodes, or even seeds, which do not display any difference between tall and dwarf. When the levels of GA_1 are compared in genotypes possessing different alleles of the *Le/le* gene that result in a range of heights, it can be seen that there is a reasonable relationship between internode length in the different genotypes and the log of the GA concentration (Fig. 3) (42). In sorghum genotypes, the genotype with the greatest height and weight also showed a 2-6 fold higher GA_1 concentration over two shorter, related genotypes (3).

The level of GA_1 also correlates with bolting in spinach under the influence of photoperiod. In response to increasing daylength spinach plants start to elongate after about 14 long days. The levels of all the GAs of the 13-hydroxylated GA pathway (see Chapter B2) start to increase after about 4 days (Fig. 4). Whilst the level of GA_{20} increases most (16 fold) over the first 12 days, it is the increasing level of GA_1 (5 fold) which brings about the increase in stem growth. This has been shown by the use of different inhibitors of GA synthesis and metabolism (62). The inhibitors AMO1618 and BX112 (see Chapter B2) both prevent bolting. The effect of AMO1618, which blocks GA biosynthesis prior to GA_{12}-aldehyde, can be overcome by

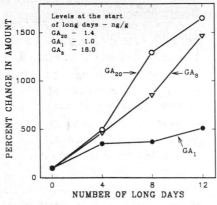

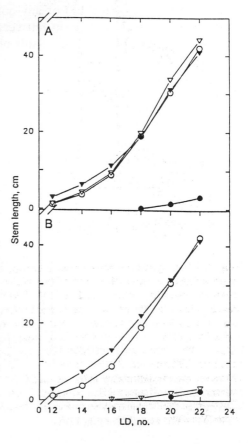

Fig. 4. (Above) Changes in the GA levels in spinach following exposure to an increasing number of long days. Bolting starts in about 14 days (Fig. 5). GA_8 is a biologically inactive GA. From data in (62).

Fig. 5. (Right) The effect of growth retardants (GA biosynthesis and metabolism inhibitors) on stem growth in spinach, induced by transfer to long-day conditions, and the reversal of the effects of the growth retardants by GA_{20} and GA_1. From (62). A) AMO1618; B) BX112. ○ = control; ● = inhibitor; ▽ = inhibitor + GA_{20}; ▼ = inhibitor + GA_1.

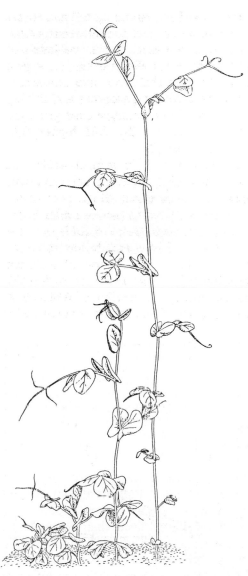

applications of GA_{20}, but the effect of BX112, which blocks the production of GA_1 from GA_{20}, can only be overcome by GA_1 itself (Fig. 5). This thus demonstrates that it is the rise in GA_1 which is the crucial factor in regulating spinach stem growth.

The relationship between stem length and GA content beaks down, however, when one considers the 'slender' peas (Fig. 6). These plants, which possess the alleles *la cry^s*, are ultratall regardless of their GA content, and look as if they are GA-saturated. However, the concentration of IAA shows a good relationship to tallness across a wide range of tallness genotypes, including slender (Fig. 7).

Early work seemed to indicate that IAA caused elongation in stem segments, but not in intact plants. Recent work with light grown pea seedlings has, however, shown that the continuous application of auxin solutions to the surface of the growing regions of the stem results in a considerable promotion of growth. The reason for the previous inability to demonstrate growth promotion in intact plants derives from the fact that auxin was usually applied on a single occasion either in aqueous solution or lanolin. Continuous recording with modern high sensitivity position transducers (Fig. 8) has shown that this tended to result in a very short-term growth promotion which rapidly changed to a growth inhibition, so that when measured after a period of several hours there is no net

Fig. 6. Four genetic lines of peas differing in internode length, and gibberellin and IAA content in their vegetative tissue. Left to right: *Nana* (*na*): an ultradwarf containing no detectable GAs and a very low IAA content; Dwarf (*Na le*) containing GA_{20}, but very little GA_1, and low IAA; Tall (*Na Le*) containing GA_1, and a medium IAA content; Slender (*na Le la cry^s*): an ultratall with no detectable GAs, but a high IAA content. See Chapter G1. Drawing from a photograph by the author, taken when the plants were the same age of about 4 weeks.

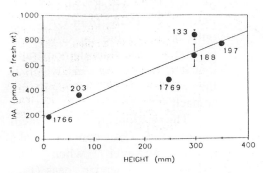

Fig. 7. The IAA levels of the stem elongation zone of a range of pea genotypes of varying heights (1766 = *nana* (*na*); 203 = dwarf (*le*); 1769 = tall (*Le*); 188, 133, 197 = slender (*la cry*^{s})), with differing GA$_1$ contents: 188 - no GAs (*na*); 133 - low GA$_1$ (*le*); 197 - high GA$_1$ (*Le*)). The heights given are above node 2 at the time that the plants had 5-6 nodes. From (21).

Fig. 8. Photograph of a pea seedling superimposed with the positions of the transducer arms (T1 and T2) used to measure growth following the application of IAA. IAA solutions were applied via a cotton wick wrapped round the apical bud and/or the uppermost one or two internodes before being diverted to waste. The closest stipule at the second node below the apical bud has been removed to show the apical bud and uppermost internode. From (61).

growth increase. Dark grown seedlings, often a favorite of experimenters because of their more rapid growth, are particularly sensitive to this inhibition. By contrast, continuous application of auxin solutions via a cotton wick wrapped around the upper stem of light-grown pea seedlings results in a 6 fold increase in growth rate in dwarf plants, and a 2 fold increase in growth rate in tall plants (61). This difference is most likely due to the differing hormonal content of the two height types: tall plants already contain more auxin (21) and GA$_1$ (19) than do dwarf plants. Even at saturating concentrations of applied auxin, tall plants still grow faster than dwarf plants, probably because of their enhanced content of GA$_1$. The application of auxin to intact plants induces a growth increase with similar kinetics to that seen in isolated segments (Chapter D1) i.e., a lag of about 15 min followed by a rise in the growth rate to a maximum, a drop in growth rate, and then a rise to a more steady rate slightly lower than the initial growth rate peak (Fig. 9). Over time the growth rate measured at any particular point slowly declines over 1-2 days as the tissues age and move out of the growing zone. Growth, of course, continues in the younger tissues that are being continually produced. The two peaks in growth rate appear to be related to two responses to auxin: an initial, short term, rapid growth response (IGR), followed by a more prolonged growth response (PGR) with a longer induction time. The IAA application data and the correlation of IAA content

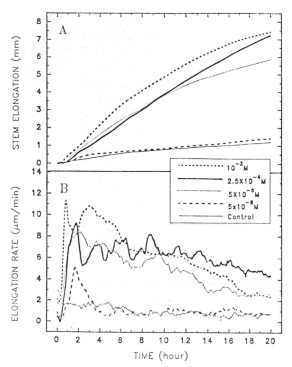

Fig. 9. The promotion of growth of dwarf pea seedlings by IAA continuously applied via a cotton wick to the two uppermost internodes of the plants. The IAA application started at time zero. Note that the initial growth response is more sensitive to IAA (responding to 5×10^{-6} M) than the prolonged growth response. From (61).

with tallness clearly show that stem elongation is governed both by IAA and GA_1, and that plant height has a clear relationship to hormonal concentration.

While it is relatively hard to find cases where precise variations in concentration correlate with parallel variations in growth or development, there are numerous cases where the presence of the hormone correlates with a distinct change in growth and development (57). Amongst these could be cited the effect of auxin coming from a leaf primordium on vascular development (Chapter G4) or the hyperelongation of the submerged stem of deep-water rice caused by the build-up of ethylene (Chapter G2). A developmental continuum from crown gall tumors with roots, to undifferentiated tumors, to tumors with shoots is shown to be correlated with increases in the cytokinin/auxin ratio upon infection with different strains of *Agrobacterium tumefaciens* bacteria (Chapter E1). Here we can see a distinct relationship with hormone concentration; sensitivity cannot be a factor as the infected tissue was the same in all cases.

The recent generation of transgenic plants with enhanced cytokinin or IAA synthesis, derived from the incorporation of the genes for their synthesis from *Agrobacterium*, enables the effects of increased hormone content on the growth and development of the whole plant to be more clearly demonstrated (see Chapter E2). For example the incorporation of the gene for cytokinin synthesis into tobacco plants shows the effect of cytokinin concentration on lateral branching and leaf senescence. The transformed plants, which contained an 3-23 fold increase in zeatin riboside level, displayed increased axillary branching, an underdeveloped root system and delayed leaf senescence (28). The relationship between IAA content and stem height in transgenic plants is less clear, possibly related to the morphological distortions observed when the IAA content is elevated throughout the entire plant. (Ethylene may play a part in this altered morphology.) Future work with more precisely controlled sites and levels of IAA synthesis may cast more light on this situation.

There has been considerable disagreement over whether tropistic responses to light or gravity are mediated by changes in IAA concentration. This has now been answered in the affirmative by an elegant technique. The promoter of an auxin-responsive gene was fused to the GUS (β-glucuronidase) gene and used to transform tobacco plants. This construct results in the production of a blue color in the presence of auxin when tissues are exposed to the GUS substrate. If the transformed tobacco seedlings were laid on their side, a shift in the position of the blue color from an even peripheral distribution in an upright stem to a more intense color localized only on the lower side of the stem was evident prior to the appearance of any gravitropic curvature (24). As the hypocotyl grew back into an upright position, the even distribution of the auxin, as detected by the blue color, returned. This is illustrated in Chapter D2.

A response to a wide (logarithmic) concentration range is not typical of all tissues. In maize coleoptiles the range is more linear, and over only about a two fold range of applied IAA concentration (2) (Fig. 10). It has also been pointed out (2) that the results indicating a response to the logarithm of applied IAA concentration have been obtained with isolated segments incubated in the solution, a possibly artifactual situation.

THE SENSITIVITY FACTOR.

There is little doubt that changes in hormone concentration at the hormone-responsive active site can account for many of the hormone-related changes in growth and development in plants. Nonetheless, it is likely that changes in sensitivity to the hormone also play a role in at least some cases. The increasing responsiveness of flowers and fruit to ethylene as they mature (Chapter G2) may represent either: a) a shift in the hormonal balance so that an inhibitor of ethylene action, such as IAA (59), may decrease with advancing maturity; b) a genuine change in the number or affinity of the ethylene responding sites, or c) the lack of inducibility of the subsequent enzymatic changes in immature tissue for reasons unknown. Only when hormone binding and receptor sites are fully understood can we hope to fully elucidate which of the above mechanisms might be operating.

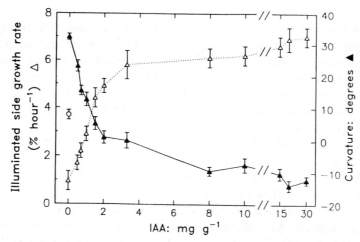

Fig. 10. The growth stimulation (△) and curvature repression (▲) induced by the application of different concentrations of IAA in lanolin (50 μg spot) to the irradiated side of phototropically stimulated maize coleoptiles. The single open circle shows the growth rate of the controls. Note that the growth response to IAA is approximately linear over the range of 0.5-2 mg/g and auxin totally overcomes the curvature at 2 mg/g of lanolin (a higher concentration than in aqueous solutions because of the limited diffusion in lanolin and the limited total quantity). The endogenous growth rate corresponds to that obtained by an application of IAA of about 1 mg/g or 50% of the concentration giving the maximum growth response. From (2).

Genetic Regulation of Sensitivity

Over the last few years many more cases of developmental regulation by changing sensitivity have been recorded. Our best opportunities for studying hormone sensitivity lie in genetic mutants (Chapter G1) which lack sensitivity to the hormone, or appear to already be "turned on". In addition, developments in molecular biology have provided data on the structure of hormone receptors and gene regulatory sequences that influence the responsiveness of the system to the presence of hormone molecules. This is beginning to lead to an improved understanding of factors that regulate hormone responsiveness.

In peas two stem-length mutants with abnormal responsiveness to GA have been described. The slender mutant, determined by the gene combination *la cry^s*, has extremely long internodes (35). It looks and behaves as if it is saturated with GA, yet the GA level is either normal, or deficient, depending on the genetic background. In the presence of the *na* allele the plants are free of detectable gibberellin, yet the plants still grow ultratall. Either the gibberellin receptor, or a subsequent receptor-controlled step, is turned on in an abnormal fashion. Alternatively it could mean that auxin, which, as described above, is abnormally high in all slender pea lines, plays an overriding role in stem elongation. If so, this would seem surprising, as it is currently not possible to get a dwarf plant to grow as fast as a tall plant by IAA applications alone. On the other side of the coin, dwarf pea mutants containing the alleles *1k*, *1ka* or *1kb* contain normal GA levels but are non-responsive to applied GA (22,41). Since *lk* is largely epistatic to (inhibits the expression of) the *la cry*^s gene combination, its action would appear to be after that of *la cry*^s. We will discuss this further below.

Some mutants are excessively sensitive while others are resistant. Elongation growth of auxin-hypersensitive mutants is inhibited, or, at high concentrations, the plants are killed by IAA, whereas wild type plant, are scarcely affected. The mutants are also more sensitive to ethylene, though whether the effects of auxin can be attributed entirely to auxin-induced ethylene synthesis is uncertain (52). On the other side, there are numerous resistant mutants including *Arabidopsis* mutants resistant to ethylene, auxin, ABA or cytokinins (see Chapter G1). The decreased response to auxin in auxin-resistant *Arabidopsis* mutants appears to have detrimental effects on plant growth and development, including reduced fertility, which becomes increasingly severe as the response to auxin decreases across a range of mutant lines (25). It is interesting to note that the ability of the *Arabidopsis* seedlings to emerge from the soil is directly proportional to their response to ethylene, indicating that the effect of ethylene in causing stem thickening and the hooking of the hypocotyl tip is of importance to seedling emergence, and therefore survival (17).

One interesting case of altered sensitivity to auxin is in the membrane hyperpolarization induced by auxin in tobacco protoplasts. In response to auxin the membrane potential becomes more negative with a sharp maximum effect at 5 x 10^{-5}M NAA. Increasing the content of auxin binding protein (ABP1) in the membrane increases the auxin sensitivity, whereas treating protoplasts with an antibody to ABP1 leads to a decrease in sensitivity (1) (Fig. 11). Protoplasts from tobacco plants transformed with the *rolB* gene from *Agrobacterium rhizogenes* also have increased sensitivity to auxin, in this case by up to 100 fold (27).

Sensitivity Within Genotypes

Not only are variations in hormone sensitivity dependent upon genotype, but

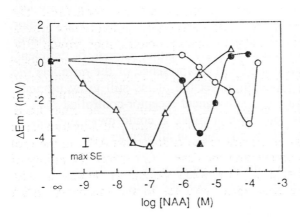

Fig. 11. The presence of auxin (NAA) (●) enhances the plasmamembrane electrical potential of tobacco protoplasts, but only at a very specific concentration (arrow) (5 x 10^{-6} M). This optimum concentration is affected by several factors. If protoplasts are treated with auxin-binding protein (ABP1) (△) the sensitivity is increased, whereas if they are treated with anti-ABP1 antibodies (○) the sensitivity is decreased. From (1).

sensitivity can vary with the process, the tissue, the age and developmental stage of the plant, the physiological conditions, and most notably the presence or absence of other hormones.

Differential Sensitivity of Different Process

It was noted above that *rolB* increased the sensitivity of membrane hyperpolarization to auxin. By contrast no effect of *rolB* was found on the optimal NAA concentration for cell division, indicating that changes in sensitivity may be for a single process and not necessarily a general phenomenon associated with a particular plant genotype (1). Indeed two parts of the same process may show differential hormonal responsiveness. The IGR and PGR stem elongation responses to auxin display different auxin responsiveness and can be separated by application and removal of different IAA concentrations at different times (61). Low IAA concentrations (5 x 10^{-6}M) induced only the initial growth response, showing that this is more sensitive to IAA than the reactions induced by the longer presence of auxin. Removal of the treating IAA solution results in a decline in the growth rate with a rate time of 25-30 min following a lag of about 25 min. Reapplication after 5 h produced a growth stimulation similar to the first application, but reapplication prior to the cessation of the growth stimulation resulted in a reappearance of the PGR, whereas the IGR to such reapplied IAA was barely visible. By contrast if IAA is supplied over only a short time (up to 10-15 min.) to previously untreated stems, then only the IGR is seen, possibly with a very truncated PGR. Thus the two components of auxin induced growth respond differently to both concentration and duration of IAA exposure.

Differences in Tissue Responsiveness

Addition of auxin can usually stimulate shoot elongation under appropriate conditions, though some very young stem tissues are inhibited by exogenous auxin (McGucken, unpublished results). In general roots are more sensitive to auxin than shoots: the addition of auxin to roots only promotes growth at very low concentrations (10^{-9}-10^{-10}M), being inhibitory at higher concentrations. In this context it is notable that auxin binding protein from roots had a lower dissociation constant (K_D) than that from shoots (36). The response to ABA is the inverse to IAA: ABA invariably inhibits shoot growth though it has much less effect on, and sometimes promotes, root growth (Fig. 12) (6, 45). This differential ABA sensitivity is of value in an ecological context. As the soil dries, ABA is synthesized in the root and is transported to the shoot in the xylem where it provides the warning of impending water shortage and shuts down shoot growth. Root growth, on the other hand, continues, so that the ability of the plant to access further soil moisture is not impeded. The responsiveness of growth to the endogenous ABA in the root and mesocotyl of maize also varies with the tissue. In the root the ability of ABA to protect cell expansion at low water potentials decreases with increasing distance from the root apex, whereas in the mesocotyl ABA

25

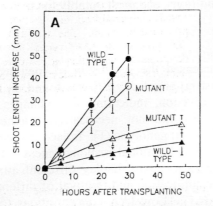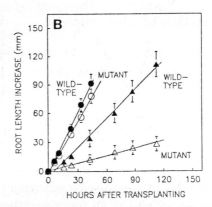

HOURS AFTER TRANSPLANTING   HOURS AFTER TRANSPLANTING

Fig. 12. Elongation of A) shoots and B) roots of normal (wild-type) and viviparous (ABA deficient) mutant plants growing in high (-0.03 MPa, circles) of low (-0.3 MPa in A and -1.6 MPa in B, triangles) water potential vermiculite. Note that while water stress depresses the growth of both roots and shoots, the presence of ABA (in the wild-type) promotes the growth of roots, whereas it depresses the growth of the shoots as compared to the ABA deficient plant. From (45).

becomes increasingly inhibitory to expansion as cells are further displaced from the meristemmatic region (44).

In carnation flowers ethylene responsiveness varies between the parts of the petal. Autocatalytic ethylene production is associated only with the basal parts of the petal, and is preceded by an increase in the mRNAs for the enzymes of ethylene biosynthesis. By contrast both basal and upper portions of the petals respond to ethylene by production of senescence related mRNAs, showing that responsiveness to ethylene in this case is specific to the production of individual enzymes and not simply to overall differences (9).

Effects of Tissue Age

Flower, fruit and abscission zone cells display an increasing responsiveness to ethylene as they age. Ethylene only induces ripening in mature fruit; in immature fruit ethylene has no effect or induces only some of the changes that are induced in the mature fruit. The same applies in the induction of flower senescence or leaf abscission. In certain flowers the dosage of exogenous ethylene needed to induce expression of senescence-related mRNAs decreases as the petals age. (23). The time or developmental stage at which responsiveness to ethylene occurs varies with the genotype, such that different flower longevity in varieties of petunia, for example, can be traced to differential sensitivity to ethylene (34).

Sensitivity Changes with Developmental Stage

I have already described the auxin-induced membrane hyperpolarization in protoplasts. There is also an auxin stimulation of proton-translocation

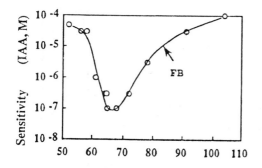

Fig. 13. The translocation of protons across the membrane of plasma membrane vesicles prepared from tobacco plants is promoted by auxin. The sensitivity of this process to IAA varies through development, increasing at the time of floral initiation. The time of flower bud appearance on the plants is shown by the arrow. From (43).

across the plasma membrane, which corresponds with an increase in the apparent affinity of ATPase for ATP (48). The optimal auxin concentration for the stimulation of the proton translocation has been shown to change during plant development. In both tobacco and petunia a transient but dramatic decrease in the optimal auxin concentration was observed at a time approximately corresponding to the time of floral induction (Fig. 13) (13, 43). Plants transformed with the *rolA* gene had the same developmental pattern, but at a level 100 times more sensitive (58). An increase in the amount of plasma membrane ATPase was also noted, starting on the same day as the maximum sensitivity to auxin. It has been proposed that these changes are part of the floral induction process, though the nature of involvement is currently unknown.

The Influence of Physical Environment

Responsiveness to plant hormones is also affected by the physical environment inside and surrounding the responding tissue. High temperatures inhibit tomato fruit ripening, in part because of reduced sensitivity to ethylene (60). Rye (*Secale cereale*) dwarfs with reduced GA responsiveness are more responsive at 10°C than at 20°C (4). In GA-deficient mutants of *Arabidopsis*, GA is required for germination. However, less GA is needed in the light, indicating that light increases sensitivity to GAs. Light was also found to enhance GA synthesis in these seeds (8). By contrast light induces a reduction in GA sensitivity in rice mesocotyl (30) or pea stem elongation (Fig. 14) (38).

The stomatal sensitivity to xylem-born ABA, which is produced in response to decreased soil water potential and causes stomatal closure, increases as the leaf water potential falls. It has been suggested that the ABA from the root would provide the plant with a means to sense the availability of water on a daily timescale, while the short term stomatal response to this signal would depend on the evaporative demand (55). However, the *in-vitro* response of stomata in detached leaves to ABA showed no effect of short term (on the order of minutes) water stress. An increase in sensitivity (as judged by the speed of a response) occurred after 1-10 days of stress,

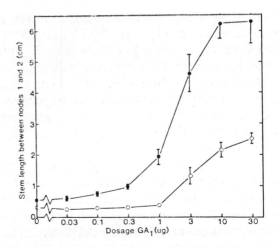

Fig. 14. The length between nodes 1 and 2 of *nana* (*na*) pea seedlings resulting from the application of differing amounts of GA₁ to plants grown in light (○) or darkness (●). From (38).

followed by a desensitization when exposed to longer periods of stress. This can be interpreted as being due to a change in the number of receptors or the effeciency with which the hormone-receptor complex produces the response (32). A computer model predicts that drought stress would cause a redistribution of ABA in the leaf in favor of the guard cell wall because of compartmental pH shifts (51). If the ABA receptor faces the apoplast then this might provide a mechanism for the modulation of ABA sensitivity by leaf water potential. Osmotic stress (0.4 M NaCl) has also been found to enhance the ABA-responsiveness of transcription of the Em storage protein gene. Either ABA or osmotic stress will induce the production of Em mRNA, but in the presence of salt, ABA, at levels which alone have no effect, will induce an increase in Em transcription.

The Influence of Other Hormones

The effect of any one hormone is often dependent upon the presence of one or more other hormones. A notable case in point is stem elongation. GA application to dwarf plants results in the production of the tall phenotype. However, it is now equally clear that the appropriate treatment of dwarf plants with IAA also increases the growth rate, though not to the same rate as is found in tall plants. Tall plants not only have a higher content of GA₁, but a high content of IAA. This all points towards both GA and IAA being needed for stem elongation.

Pea genotypes with alleles *lk*, *lka* and *lkb* have been characterized as GA-non-responders. They have a dwarf phenotype combined with several morphological abnormalities (See Chapter G1). When treated with GA, they show little response. It was originally thought that they lacked some part of the GA-transduction pathway. However, when intact *lkb* plants are treated with a continuous supply of auxin, they show a pronounced response, and will respond further to GA₃ (Yang, Davies and Reid, unpublished). Likewise stem segments of *lka*, which show no response to GA₃ in the absence of auxin, show a pronounced response to auxin, and a response to GA_3 in the presence of auxin (Fig. 15) (39). These plants thus behave as if 1) auxin is needed for the response to GA_3, and 2) they are deficient in IAA. This has in fact been shown to be the case, with *lkb* plants having about 1/3 the level of IAA present in the normal *Lkb* plants (39). In the GA-deficient dwarf plant containing the allele *ls*, which responds well to GA_3, the GA-response is enhanced in the presence of IAA and vice versa. Thus the level one hormone can influence the sensitivity of the plant to changing levels of another hormone.

Other examples of such synergistic hormone interactions can also be found in other systems. Cytokinin increases the sensitivity of vascular cambial cells to auxin stimulation (see Chapter G4). Ethylene (or its precursor, 1-aminocyclopropane-1-carboxylic acid, ACC) and IAA both enhanced adventitious root formation in sunflower hypocotyl. The effectiveness of ethylene depends on the presence of IAA, and ACC enhances the response to IAA. In this case it is thought that the primary controller of adventitious root formation was auxin, with the effect of ethylene mediated by auxin (26). In intact plants the effect of ethylene on leaf abscission increases with leaf age. This is associated with a decline in the leaf IAA content as the leaves age, and applied auxins 2,4-D or NAA block abscission while the auxin transport inhibitor naphthylphthalamic acid (NPA) enhances abscission. It therefore seems likely that the decreased effect of ethylene in younger leaves is because the action of ethylene is opposed by auxin (54). Ethylene also promotes the growth of internodal tissue of deepwater rice, which responds to flooding by a rapid elongation induced by ethylene formation. Ethylene promotes growth in part by increasing the responsiveness of the internodal tissue to GA_1, and appears to do so by causing a reduction in the endogenous levels of ABA, so that the growth rate is determined by the relative levels of endogenous GA_1 and ABA (18). In the reverse direction GA_3 reduces the sensitivity of carnation flowers to ethylene,

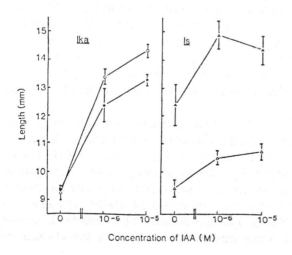

Length (mm)

Concentration of IAA (M)

Fig. 15. The response of 8 mm segments of etiolated *lka* and *ls* plants to IAA in the presence or absence of 10 μM GA_3, after 24 h. Note that *lka* segments, which are IAA-deficient, respond strongly to IAA and do not respond to GA unless IAA is present, whereas segments from dwarf *ls* plants respond to both IAA and GA, with IAA simply enhancing the GA effect. From (39).

29

so delaying their senescence (46).

Pretreatment with the same hormone can influence sensitivity. Petiole explants treated with ethylene during the 'aging' phase show enhanced sensitivity to ethylene during the abscission phase. The movement of auxin enhances its own transport (see Chapters G3 and G4), and the auxin responsiveness of H^+ translocation across the plasmamembrane is reported to be enhanced 100-fold by pretreatment with auxin of the tobacco plants from which the plasmamembrane vesicles are made (49), possibly providing a mechanism for the observed auxin stimulation of auxin transport. The auxin pretreatment has been associated with the accumulation of several polypeptides in the plasmamembrane, which may be the means by which the subsequent auxin sensitivity is increased.

Sensitivity Regulation by Changes in Detection or Transduction

In cases where the altered responsiveness to a hormone is not due to effects of other hormones, the likely reasons are a change in the hormone binding to its receptor, the subsequent transduction chain, including gene expression, or a change in the capacity of the system to respond. In many cases, the transduction chain is unknown. Our understanding of regulation at this level is limited. Molecular techniques have, however, allowed the isolation of genes that respond to the presence of hormones, together with their regulatory regions. Deletion analyses then pinpoint those promoter sequences essential for the hormonal response. For example, the promoter region of the ABA-inducible maize gene, *rab28*, which is related to water stress and embryo development, has an ABA responsive element CCACGTGG that is also found in other ABA responsive genes. A 134 base pair (bp) fragment between -194 to -60 bp upstream of the start signal is sufficient to convey ABA-responsiveness upon a fused GUS (ß-glucuronidase) reporter gene (33). Nuclear protein extracts from both embryos and water stressed leaves bind to this DNA fragment (see Chapter G10). GA and ABA responsive elements on the α-amylase gene in barley are located between -174 and -108 bp upstream from the transcription start site (16) (see Chapter D3).

In tobacco ethylene promotes the accumulation of the pathogenesis-related (PR) *PBR-1b* gene. GUS expression of a fused gene behind the *PBR-1b* promoter was regulated by a sequence -213 to -141 bp upstream of the transcription start site. This region also contained protein binding regions as detected by gel shift assays (29). Ethylene application also induces a rapid protein phosphorylation in tobacco leaves. Kinase inhibitors block both this phosphorylation and the expression of PR genes, whereas a phosphatase inhibitor by itself induced protein phosphorylation and PR protein accumulation (37). It therefore seems likely that the transduction pathway between ethylene and gene expression involves phosphorylated intermediates whose levels are regulated by specific kinases and phosphatases. In these and similar cases the presence of the hormone probably alters the level or nature of a nuclear DNA-binding protein, thereby regulating the expression of the responsive gene. However, the likely way in which sensitivity could directly be altered is if another transacting factor (possibly a protein) influences the binding of the hormone-induced DNA-binding protein to the gene promoter. Mutations in the hormone-responsive DNA element would also alter the hormonal response. However, *Arabidopsis* mutants with a decreased response to ABA showed a different mechanism. A protein encoded by the *ABI3* gene was identified. In the mutant possessing the most severe *abi3* allele this protein has been reduced in size by 40% due to the presence of an abnormal stop codon (14). The function of the protein is uncertain, but its modification provides a likely mechanism for the reduced response to ABA.

THE ROLE OF TRANSPORT AND REDISTRIBUTION IN PLANT HORMONE FUNCTION.

A prime function of hormones in plants, which lack a nervous system, is to convey information from one part of the plant to another. The idea that transport was an essential part of the role of plant hormones originally came from experiments on the phototropic control of coleoptile growth. The hypothesis was that the IAA was synthesized in the tip, transported basipetally and was then redistributed laterally to give differential growth and bending. We now know that most of the IAA coming from the coleoptile tip is not synthesized in situ, but comes from an IAA-inositol source in the endosperm of the grain (see Chapter B1), and is transported, as free IAA or as the IAA-conjugate, to the tip of the coleoptile where conjugate hydrolysis occurs. While so far we have only a minor deviation from the original concept, some studies have failed to show a redistribution of IAA in response to tropistic stimuli. At least in some cases, growth begins all along the stem at the same time, and faster than auxin can be transported from the tip (12). Thus, if auxin is involved in tropistic responses of stems it must already be in, or synthesized in, the responding tissue rather than being redirected in transit from the tip. If a stem is cut longitudinally in half and then laid on its side, the bottom half will grow faster than the top half, so that redistribution across the entire stem is not necessary for differential growth. This does not preclude the involvement of auxin, but indicates that if redistribution is important, then it must be on a smaller scale than the entire stem.

Auxin transport occurs in a very specific manner in a basipetal direction in stems (Chapter G3). The auxin undergoing transport is clearly involved in vascular differentiation (Chapter G4), lateral root initiation, and the regulation of stem elongation. The fact that transported auxin does regulate growth can be seen from the results of IAA applications to different parts of the stem of growing pea plants. When application is directly to the growing internodes, the classical growth response results. If, however, the IAA is applied to the apical bud, located above the growing internodes, then two or three initial peaks in the growth rate are observed before the PGR plateau occurs (Fig. 16) (61). This in fact represents the auxin being transported down the stem. When a transducer is also located at the tip of the second internode it can be seen that this internode responds later than the uppermost internode, corresponding to the time the basipetally-transported IAA reaches that internode. The multiple growth rate peaks observed by the upper transducer in fact correspond to the summed growth of the two or three internodes in which growth is taking place (in dwarf or tall plants respectively).

A reinvestigation as to whether transported auxin has a role in the phototropic curvature of maize coleoptiles has, however, seemingly answered the question in the affirmative, as there is a basipetal migration of the response from the tip, and this occurs at the same rate as the growth stimulation caused by exogenous auxin (2). Thus at the present time there is good evidence for a role of transported auxin in growth.

A most notable case of interorgan communication via long-distance hormone transport is root to shoot signaling via ABA. When the soil dries

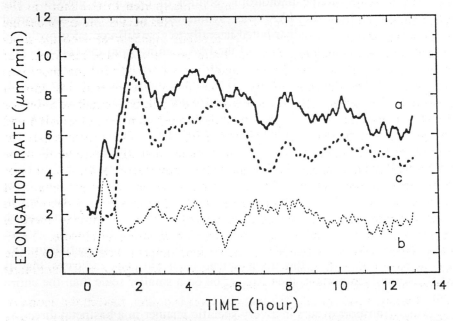

Fig. 16. The distribution of elongation of dwarf pea seedlings induced by 10^{-4} M IAA applied to the uppermost internode, starting at time zero, as detected by transducers positioned at nodes 1 (a) or 2 (c) below the apical bud (see Fig. 6). Line (a) represents the total growth; (c) the growth of the second internode (internodes below the second have ceased growth); and (b) represents the growth of the uppermost internode as determined by the difference between lines (a) and (c). Note that the growth of internode two lags behind that of node one, representing the time taken for the applied IAA to transport from the first to the second internode. Thus IAA in transport from the apical bud promotes growth as it moves down the stem. From (61).

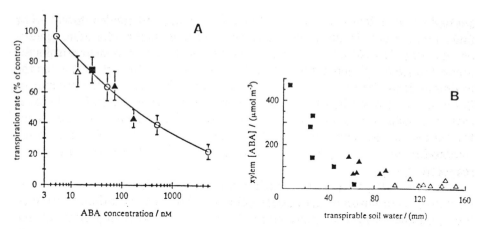

Fig. 17. A) The transpiration of detached wheat leaves as determined by the concentration of ABA in the xylem sap in which they were placed. The various symbols represent different sources of xylem sap. B) The ABA in xylem sap of field grown maize plants as a function of the predawn transpirable soil water. The symbols show different treatments. The correlation during the day is less obvious because it is affected by the flow rate through the xylem. From (15)

out ABA is synthesized in the root and is transported to the shoot in the xylem stream, causing stomatal closure (7,15). This occurs without a change in the water status of the shoot. For example, apple trees with the roots divided into two containers, one moist and one dry, showed restricted leaf expansion and leaf initiation, even though the water status of the shoot was unaffected. When the roots in the dry container were severed, leaf growth recovered. The xylem ABA level is usually found to be a sensitive indicator of water status, and there is a good relationship between the xylem ABA and the stomatal conductance of the leaves (Fig. 17). The leaves appear to respond to the amount of ABA arriving rather than the concentration, as during the day the concentration in the xylem may fluctuate with the flux of water through the xylem. Other factors, such as water potential and temperature, affect the sensitivity to the ABA. For example, a decrease in water potential leads to increased sensitivity. Thus at midday there is increased stomatal sensitivity to ABA, leading to stomatal closure. From solely physical (water potential) considerations stomata should reopen in the late afternoon though this is seldom observed in drying soil conditions because of the ABA reaching the leaves.

Trewavas has argued (56) that transport is unimportant in the action of "plant growth substances". As a prime example he claimed that grafts between tall plants and short plants show that tallness is not transmitted. We now know, from the elegant work of Reid and co-workers on peas (Chapter G1), that the results obtained depend on the tissue. The control of tallness resides in GA_1, which is found only in the youngest internodes and thus will not be transmitted through a graft. GA_1 is, however, the final product of a synthetic pathway which produces the biologically active compound, and the transmission of tallness can be seen if the correct system is chosen. The genotype *Na le* is dwarf as it has the ability to produce GA_{20} (through the gene *Na*), but it lacks the ability to convert GA_{20} to GA_1, because it lacks the dominant gene *Le*. The genotype *na Le* is ultradwarf (*nana*) as it lacks the ability to synthesize GAs (because of *na*), but it has the ability to convert GA_{20} to GA_1 (because of *Le*), though normally it does not do so because GA_{20} is lacking. Now if a *nana* scion is grafted onto a dwarf stock the resulting plant is tall. The stock synthesizes the GA_{20} and passes it to the scion, which converts the GA_{20} to GA_1, giving tall growth (40). This shows that a thorough knowledge of the system in question is necessary before conclusions can be accurately drawn.

In other cases plant hormones operate in or near the tissue in which they are produced. Ethylene is the prime example here. In almost all ethylene controlled phenomena, the ethylene is produced in the responding tissue (see Chapter G2). Ethylene is a plant hormone by all working definitions[1].

Although transport is not an *essential* part of the definition of a plant hormone, this does not mean that transport plays no part in hormone functioning. Indeed, we find that transport is important in the role of plant hormones in most, but not all, systems.

ON THE IMPORTANCE AND ROLE OF PLANT HORMONES

Plant hormones consist of only a few simple molecules. Canny has argued that there must be other controllers as the "hormones" are not specific or numerous enough to satisfy "Ashby's law of requisite variety: A situation can only be controlled by a controller that matches the variety of the situation" (5). He claims that, as plant hormones are simple compounds, they cannot match the variety of developmental directions. By contrast, animal hormones, being compounds like proteins, have a high information content and can match the required variety of controlled reactions. In plants we do have the variability provided by concentration and interaction between different hormones. Canny suggests, however, that these features are still not enough and that either the plant body is autonomous, or that other messages await discovery. In rebuttal to Canny, Firn notes that hormones do not carry information because they don't have to (10). They only provide a "turn on" or "turn off" signal with the information being provided by the cell. Some vertebrate hormones are complex because they need the specificity supplied by polypeptides. Concentration dependence could then provide some degree of control for the magnitude of the response. By comparison Ca^{2+} appears to be a very important regulator of a wide variety of processes, yet calcium is very simple; the specificity for the effects of Ca^{2+} depend on the cell. In fact

---

[1] "Whether or not we regard ethylene as a plant hormone is unimportant; bananas do..." Carl A. Price, in *Molecular Approaches to Plant Physiology*

there is evidence that Ca^{2+} may be an intermediate in some hormonal responses (see Chapter D5).

In general we can determine whether a designated plant hormone is active in controlling any process by invoking the PESIGS rules provided by Jacobs (20). These state that:

- Presence - the chemical must be present in an organism and parallel variation should exist between the amount of the chemical and the relative activation of the process. (We should, however, modify the latter requirement such that: a) the hormone level should be measured in the exact tissue, cells, or even subcellular compartment where the response is occurring; and b) the possible changes in sensitivity that may occur during development should be taken into account.)
- Excision - removal of the organ that is the source of the hormone should lead to a cessation of the process.
- Substitution - the substitution of the pure chemical for the source organ should lead to the restoration of the process.
- Isolation - when as much as possible of the reacting system is isolated then the effect of the chemical is the same as in the less isolated system.
- Generality - the chemical should be involved in all similar situations.
- Specificity - the chemical should be specific.

    Despite being formulated over thirty-five years ago these rules still provide a good set of guidelines for determining hormonal involvement. We should, however, add one more proviso now possible with modern genetics, and that is the principal of genetic control:

- A correlation should be shown between the presence, absence, or level of a process, and the corresponding presence, absence, or level of a hormone both in genetic lines of plants differing in the process purportedly controlled by the hormone, and in plants transformed with genes that regulate either the process in question or the hormone purported to influence that process.

There can be no doubt that hormones are important control agents in plants. The subsequent chapters provide more detail of our knowledge of this group of compounds and their effects.

Acknowledgments

I would like to thank Natalie McGucken and Jim Koch for constructive criticism, and Francine White for typing some of this chapter.

References

1. Barbier-Brygoo, H., Maurel, C., Ephritikhine, G.Guern, J. (1992) Differential sensitivity of protoplasts to auxin. *In*: Progress in Plant Growth Regulation, pp. 194-201, Karssen, C.M., Van Loon, L.C., Vreugdenhil, D., eds. Kluwer Academic Publishers, Dordrecht, The Netherlands.
2. Baskin, T.I., Briggs, W.R., Iino, M. (1986) Can lateral redistribution of auxin account for phototropism of maize coleoptiles? Plant Physiol. 81, 306-309.

3. Beall, F.D., Morgan, P.W., Mander, L.N., Miller, F.R., Babb, K.H. (1991) Genetic regulation of development of *Sorghum bicolor*. V. The *ma-3* allele results in gibberellin enrichment. Plant Physiol. 95, 116-125.

4. Boerner, A., Gale, M.D., Appleford, N.E.J., Lenton, J.R. (1993) Gibberellin status and responsiveness in shoots of tall and dwarf genotypes of diploid rye (*Secale cereale*). Physiol. Plant. 89, 309-314.

5. Canny, M.J. (1985) Ashby's law and the pursuit of plant hormones: A critique of accepted dogmas, using the concept of variety. Aust. J. Plant Physiol. 12, 1-7.

6. Creelman, R.A., Mason, H.S., Bensen, R.J., Boyer, J.S., Mullet, J.E. (1990) Water deficit and abscisic acid cause differential inhibition of shoot versus root growth, in soybean seedlings: Analysis of growth, sugar accumulation, and gene expression. Plant Physiol. 92, 205-214.

7. Davies, W.J., Tardieu, F., Tejo, C.L. (1994) How do chemical signals work in plants that grow in drying soil? Plant Physiol. 104, 309-314.

8. Derkx, M.P.M., Karssen, C.M. (1993) Effects of light and temperature on seed dormancy and gibberellin-stimulated germination in *Arabidopsis thaliana*: studies with gibberellin-deficient and -insensitive mutants. Physiol. Plant. 89, 360-368.

9. Drory, A., Mayak, S., Woodson, W.R. (1993) Expression of ethylene biosynthetic pathway mRNAs is spatially regulated within carnation flower petals. J. Plant Physiol. 141, 663-667.

10. Firn, R.D. (1985) Ashby's law of requisite variety and its applicability to hormonal control. Aust. J. Plant Physiol. 12, 685-687.

11. Firn, R.D. (1986) Growth substance sensitivity: The need for clearer ideas, precise terms and purposeful experiments. Physiol. Plant. 67, 267-272.

12. Firn, R.D., Digby, J. (1980) The establishment of tropic curvatures in plants. Ann. Rev. Plant Physiol. 31, 131-148.

13. Francois, J.M., Berville, A., Rossignol, M. (1992) Development and line dependent variations of petunia plasma membrane proton ATPase sensitivity to auxin. Plant Sci. 87, 19-27.

14. Giraudat, J., Hauge, B.M., Valon, C., Smalle, J., Parcy, F., Goodman, H.M. (1992) Isolation of the *Arabidopsis ABI3* gene by positional cloning. Plant Cell 4, 1251-1261.

15. Gowing, D.J.G., Davies, W.J., Trejo, C.L., Jones, H.G. (1993) Xylem-transported chemical signals and the regulation of plant growth and physiology. Philos. Trans. Roy. Soc. Lond. B. Biol. Sci. 341, 41-47.

16. Gubler, F., Jacobsen, J.V. (1992) Gibberellin-responsive elements in the promoter of a barley high-pI α-amylase gene. Plant Cell 4, 1435-1441.

17. Harpham, N.V.J., Berry, A.W., Knee, E.M., Roveda, H.G., Raskin, I., Sanders, I.O., Smith, A.R., Wood, C.K., Hall, M.A. (1991) The effect of ethylene on the growth and development of wild-type and mutant *Arabidopsis thaliana* (L.) Heynh. Ann. Bot. 68, 55-62.

18. Hoffmann-Benning, S., Kende, H. (1992) On the role of abscisic acid and gibberellin in the regulation of growth in rice. Plant Physiol. 99, 1156-1161.

19. Ingram, T.J., Reid, J.B., Murfet, I.C., Gaskin, P., Willis, C.L. (1984) Internode length in *Pisum*. The *Le* gene controls the 3-β-hydroxylation of gibberellin A_{20} to gibberellin A_1. Planta 160, 455-463.

20. Jacobs, W.P. (1979) Plant hormones and plant development, Cambridge University Press, Cambridge, New York. 339 pp.

21. Law, D.M., Davies, P.J. (1990) Comparative IAA levels in the slender pea and other pea phenotypes. Plant Physiol. 93, 1539-1543.

22. Lawrence, N.L., Ross, J.J., Mander, L.N., Reid, J.B. (1992) Internode length in Pisum. Mutants *lk, lka* and *lkb* do not accumulate gibberellins. J. Plant Growth Regul. 11, 35-37.

23. Lawton, K.A., Raghothama, K.G., Goldsbrough, P.B., Woodson, W.R. (1990) Regulation of senescence-related gene expression in carnation flower petals by ethylene. Plant Physiol. 93, 1370-1375.

24. Li, Y., Hagen, G., Guilfoyle, T.J. (1991) An auxin-responsive promoter is differentially induced by auxin gradients during tropisms. Plant Cell 3, 1167-1176.

25. Lincoln, C., Britton, J.H., Estelle, M. (1990) Growth and development of the *axr1* mutants of *Arabidopsis*. Plant Cell 2, 1071-1080.

26. Liu, J.H., Reid, D.M. (1992) Auxin and ethylene-stimulated adventitious rooting in relation to tissue sensitivity to auxin and ethylene production in sunflower hypocotyls. J. Exp. Bot. 43, 1191-1198.

27. Maurel, C., Barbier-Brygoo, H., Spena, A., Tempe, J., Guern, J. (1991) Single *rol* genes from the *Agrobacterium rhizogenes* T_L-DNA alter some of the cellular responses to auxin in *Nicotiana tabacum*. Plant Physiol. 97, 212-216.

28. Medford, J.I., Horgan, R., El-Sawi, Z., Klee, H.J. (1989) Alterations of endogenous cytokinins in transgenic plants using a chimeric isopentenyl transferase gene. Plant Cell 1, 403-413.

29. Meller, Y., Sessa, G., Eyal, Y., Fluhr, R. (1993) DNA protein interactions on a *cis*-DNA element essential for ethylene regulation. Plant Molecular Biology. 23, 453-463.

30. Nick, P., Furuya, M. (1993) Phytochrome dependent decrease of gibberellin-sensitivity. A case study of cell extension growth in the mesocotyl of japonica and indica type rice cultivars. Plant Growth Regul. 12, 195-206.

31. Parrish, D.J., Davies, P.J. (1977) Emergent growth - an auxin-mediated response. Plant Physiol. 59, 745-749.

32. Peng, Z-Y., Weyers, J.D.B. (1994) Stomatal sensitivity to abscisic acid following water deficit stress. J. Exp. Bot. 45, 835-845.

33. Pla, M., Vilardell, J., Guiltinan, M.J., Marcotte, W.R., Niogret, M.F., Quatrano, R.S., Pages, M. (1993) The *cis*-regulatory element CCACGTGG is involved in ABA and water-stress responses of the maize gene *rab28*. Plant Mol. Biol. 21, 259-266.

34. Porat, R., Reuveny, Y., Borochov, A., Halevy, A.H. (1993) Petunia flower longevity: the role of sensitivity to ethylene. Physiol. Plant. 89, 291-294.

35. Potts, W.C., Reid, J.B., Murfet, L.C. (1985) Internode length in *Pisum*. Gibberellins and the slender phenotype. Physiol. Plant. 63, 357-364.

36. Rademacher, E., Klaembt, D. (1993) Auxin dependent growth and auxin-binding proteins in primary roots and root hairs of corn (*Zea mays* L.). J. Plant Physiol. 141, 698-703.

37. Raz, V., Fluhr, R. (1993) Ethylene signal is transduced via protein phosphorylation events in plants. Plant Cell 5, 523-530.

38. Reid, J.B. (1988) Internode length in *Pisum*. Comparison of genotypes in the light and dark. Physiol. Plant. 74, 83-88.

39. Reid, J.B. Davies, P.J. (1992) The genetics and physiology of gibberellin sensitivity mutants in peas. *In*: Progress in Plant Growth Regulation, pp. 214-225, Karssen, C.M., Van Loon, L.C., Vreugdenhil, D., eds. Kluwer Academic Publishers, Dordrecht, The Netherlands.

40. Reid, J.B., Murfet, I.C., Potts, W.C. (1983) Internode length in *Pisum*. II. Additional information on the relationship and action of loci *Le La Cry Na* and *Lm*. J. Exp. Bot. 34, 349-364.

41. Reid, J.B., Potts, W.C. (1986) Internode length in *Pisum*. Two further mutants, *lh* and *ls*, with reduced gibberellin synthesis, and a gibberellin insensitive mutant, *lk*. Physiol. Plant. 66, 417-426.

42. Ross, J.J., Reid, J.B., Gaskin, P., Macmillan, J. (1989) Internode length in *Pisum*. Estimation of GA_1 levels in genotypes *Le, le* and *le*^{d}. Physiol. Plant. 76, 173-176.

43. Rossignol, M., Santoni, V., Francois, J.M. Vansuyt, G. (1992) Changing membrane sensitivity to auxin during plant development. *In*: Progress in Plant Growth Regulation,

pp. 207-213, Karssen, C.M., Van Loon, L.C., Vreugdenhil, D., eds. Kluwer Academic Publishers, Dordrecht, The Netherlands.

44. Saab, I.N., Sharp, R.E., Pritchard, J. (1992) Effect of inhibition of abscisic acid accumulation on the spatial distribution of elongation in the primary root and mesocotyl of maize at low water potentials. Plant Physiol. 99, 26-33.

45. Saab, I.N., Sharp, R.E., Pritchard, J., Voetberg, G.S. (1990) Increased endogenous abscisic acid maintains primary root growth and inhibits shoot growth of maize seedlings at low water potentials. Plant Physiol. 93, 1329-1336.

46. Saks, Y., Van, S.J. (1992) The role of gibberellic acid in the senescence of carnation flowers. J. Plant Physiol. 139, 485-488.

47. Sandberg, G., Gardestrom, P., Sitbon, F., Olsson, O. (1990) Presence of IAA in chloroplasts of *Nicotiana tabacum* and *Pinus sylvestris*. Planta 180, 562-568.

48. Santoni, V., Vansuyt, G., Rossignol, M. (1991) The changing sensitivity to auxin of the plasma-membrane proton ATPase relationship between plant development and ATPase content of membranes. Planta 185, 227-232.

49. Santoni, V., Vansuyt, G., Rossignol, M. (1993) Indoleacetic acid pretreatment of tobacco plants in-vivo increases the in-vitro sensitivity to auxin of the plasma membrane H$^+$-ATPase from leaves and modifies the polypeptide composition of the membrane. FEBS Letters 326, 17-20.

50. Scott, T.K., Briggs, W.R. (1962) Recovery of native and applied auxin from the light grown Alaska pea seedlings. Amer. J. Bot. 49, 1056-1083.

51. Slovik, S., Hartung, W. (1992) Compartmental distribution and redistribution of abscisic acid in intact leaves analysis of the stress-signal chain. Planta 187, 37-47.

52. Souza, L.D., King, P.J. (1991) Mutants of *Nicotiana plumbaginifolia* with increased sensitivity to auxin. Molec. Gen. Genet. 231, 65-75.

53. Spray, C., Phinney, B.O., Gaskin, P., Gilmour, S.I., Macmillan, J. (1984) Internode length in *Zea mays* L. The dwarf-1 mutant controls the 3β-hydroxylation of gibberellin A$_{20}$ to gibberellin A$_1$. Planta 160, 464-468.

54. Suttle, J.C., Hultstrand, J.F. (1991) Ethylene-induced leaf abscission in cotton seedlings: The physiological bases for age-dependent differences in sensitivity. Plant Physiol. 95, 29-33.

55. Tardieu, F., Davies, W.J. (1993) Integration of hydraulic and chemical signalling in the control of stomatal conductance and water status of droughted plants. Plant Cell Envir. 16, 341-349.

56. Trewavas, A. (1981) How do plant growth substances act? Plant Cell Envir. 4, 203-228.

57. Trewavas, A.J., Cleland, R.E. (1983) Is plant development regulated by changes in the concentration of growth substances or by changes in the sensitivity to growth substances? Trends in Biochem. Sci. 8, 354-357.

58. Vansuyt, G., Vilaine, F., Tepfer, M., Rossignol, M. (1992) *Rol* A modulates the sensitivity to auxin of the proton translocation catalyzed by the plasma membrane H$^+$-ATPase in transformed tobacco. FEBS Letters 298, 89-92.

59. Vendrell, M. (1985) Dual effect of 2,4-D on ethylene production and ripening of tomato fruit tissue. Physiol. Plant. 64, 559-563.

60. Yang, R.F., Cheng, T.S., Shewfelt, R.L. (1990) The effect of high temperature and ethylene treatment on the ripening of tomatoes. J. Plant Physiol. 136, 368-372.

61. Yang, T., Law, D.M., Davies, P.J. (1993) Magnitude and kinetics of stem elongation induced by exogenous indole-3-acetic acid in intact light-grown pea seedlings. Plant Physiol. 102, 717-724.

62. Zeevaart, J.A.D., Gage, D.A., Talon, M. (1993) Gibberellin A$_1$ is required for stem elongation in spinach. Proc. Natl Acad. Sci. USA. 90, 7401-7405.

63. Zhu, Y.-X., Davies, P.J., Halinska, A. (1991) Metabolism of gibberellin A$_{12}$ and A$_{12}$-aldehyde in developing seeds of *Pisum sativum* L. Plant Physiol. 97, 26-33.

B. HORMONE BIOSYNTHESIS AND METABOLISM

B1. Auxin Biosynthesis and Metabolism

Robert S. Bandurski[a], Jerry D. Cohen[b], Janet Pernise Slovin[c] and Dennis M. Reinecke[d]

[a]Department of Botany and Plant Pathology, Michigan State University, East Lansing, Michigan 48824, USA, [b]Horticultural Crops Quality and [c]Climate Stress Laboratories, Beltsville Agricultural Research Center, Agricultural Research Service, U.S. Department of Agriculture, Beltsville, MD 20705-2350, USA, and [d]Department of Plant Science, University of Alberta, Edmonton, AB, T6G 2P5 Canada.

INTRODUCTION

Hormones regulate growth and development in plants, but how does a plant manage to have hormone available in some specific amount at the right time and place? In this chapter, we attempt to bring together information relevant to this question for the hormone auxin. Of the several compounds having auxin activity we will focus most of our attention on indole-3-acetic acid (IAA)[1] since it is for IAA that the greatest body of knowledge exists. Other naturally occurring auxins such as IBA and 4-Cl-IAA will also be considered. Information on synthetic auxins and other compounds with auxin-like activity may be found in other reviews (14, 29, 48, 57, 75).

There is an enormous amount known concerning the response of plants to IAA application. However, there is substantially less known concerning the actual amounts of IAA in plant tissues. There is even less information available about how the amount of IAA in the tissue is regulated and what

---

[1] The following abbreviations are used throughout:
diOxIAA = dioxindole-3-acetic acid (3-hydroxy-2-indolone-3-acetic acid);
GC-MS = gas chromatography-mass spectrometry; **IAA** = indole-3-acetic acid;
IAAsp = indole-3-acetyl-L-aspartate; **IAAla** = indole-3-acetyl-L-alanine;
IAGlu = indole-3-acetyl-L-glutamate; **IAGluc** = indole-3-acetyl-glucose;
IAInos = indole-3-acetyl-*myo*-inositol;
IAInos-gal = indole-3-acetyl-*myo*-inositol-5'-galactoside;
IAInos-arab = indole-3-acetyl-*myo*-inositol-5-arabinoside; **IALys** = indole-3-acetyl-ε-L-lysine;
IAPhe = indole-3-acetyl-L-phenylalanine; **IBA** = indole-3-butyric acid;
7-OH-OxIAA = 7-hydroxy-oxindole-3-acetic acid; **OxIAA** = oxindole-3-acetic acid.

P. J. Davies (ed.), Plant Hormones, 39–65.
© 1995 *Kluwer Academic Publishers. Printed in the Netherlands.*

the relationship is between the amount of endogenous IAA and the plant response. There is now some evidence for a correlation between plant response and the amount of IAA in a responding tissue (12), but only recently have we been able to directly measure the turnover of IAA, an important parameter in regulating IAA levels. Equally obscure is our understanding of the relationship between hormone levels and the numbers and activity of IAA binding sites (48 and Chapter D4)

In the hope of shedding some light on the answers to these questions this chapter focuses on the "inputs to" and "outputs from" the IAA pool. Most of the knowledge is qualitative, and indicates only the existence of a given pathway, but there is also a growing body of knowledge concerning the enzymes involved in IAA synthesis and metabolism, the pool sizes of IAA and its metabolites in specific tissues, and the turnover of these pools. Ultimately, we should be able to account for the steady state level of IAA, and to understand how environmental stimuli, such as a tropic stimulus, alters the level of IAA in the tissue. In addition to the data presented here, several other reviews provide a more detailed analysis of particular aspects of auxin synthesis and metabolism (14, 29, 57, 75).

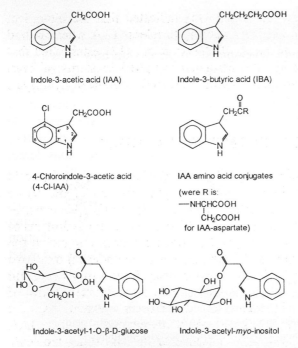

Indole-3-acetic acid (IAA)

Indole-3-butyric acid (IBA)

4-Chloroindole-3-acetic acid (4-Cl-IAA)

IAA amino acid conjugates

(were R is:

—NHCHCOOH
    |
    CH₂COOH
for IAA-aspartate)

Indole-3-acetyl-1-O-β-D-glucose

Indole-3-acetyl-*myo*-inositol

CHEMICAL FORMS OF AUXINS IN PLANTS

Several naturally occurring auxins are found in plants, including IAA, its halogenated derivative, 4-Cl-IAA, and IBA. In addition to these indolic auxins, various phenolic acids (such as phenylacetic acid) and other compounds in plants have low auxin activity. A physiological role for such non-indolic compounds in auxin regulation has not been established. Auxins are found in plants both as the free acid (which is thought to be the primary "active form") and in conjugated forms, one function of which may be to provide a

Fig. 1. Naturally occurring auxins and some examples of the lower molecular weight IAA conjugates found in plants. Related low molecular weight conjugates (such as IAInos-arabinose and IAInos-galactose) and higher molecular weight conjugates (such as an IAA-glucan, IAA-peptides, and an IAA-glycoprotein) have also been described following extraction from plant material.

readily accessible and easily regulatable source of free IAA without *de novo* synthesis. One type of conjugated form is linked through carbon-oxygen-carbon bridges and these compounds are referred to generically as "esters", although some 1-*O* sugar conjugates such as 1-*O*-IAGluc are actually linked by acyl alkyl acetal bonds. True esters include compounds such as 6-*O*-IAGluc and IAInos. The other type of conjugates are linked through carbon-nitrogen-carbon amide bonds, as in the IAA-amino acid and peptide conjugates (Fig. 1). All native auxins are found in both free forms and conjugated forms. However, in most tissues it is the conjugated forms which predominate.

Various conjugates of IAA, both ester and amide, have been used as "slow release" forms of IAA for tissue cultures (34) and for rooting of cuttings. IAA conjugates, each differing in ease of hydrolysis by the plant's enzymes, and having conjugating moieties of varying degrees of lipophilicity, could be used to "target" the IAA to a particular tissue or particular cell organelle with delivery of the hormone at the required rate. The conjugating moiety might thus be used as a "zip code," (7, 19) to bring the IAA to the desired location, with simultaneous protection against peroxidative attack (18, 67).

The studies of Hangarter and Good (34) indicated that the biological effect of the conjugate was related to its ability to release IAA when applied to the tissue. Additionally, the work of Bialek *et al.*, (12) established that differential growth induced by IAA-conjugates applied to stems of bean (*Phaseolus vulgaris*) was quantitatively correlated to the degree of hydrolysis of the conjugate by the tissue.

INPUTS TO AND OUTPUTS FROM THE IAA POOL

The known inputs to and outputs from the IAA pool are indicated by solid arrows in Fig. 2. Inputs to the IAA pool in a given tissue include: A) *de novo* synthesis, whether from tryptophan, from other indolylic precursors, or even total aromatic synthesis from non-indolylic precursors; B) hydrolysis of both amide and ester IAA conjugates; and C) transport from one site in the plant to another site. The outputs from the IAA pool include: D) oxidative catabolism; E) conjugate synthesis; F) transport away from a given site; and G) "use" of IAA in growth. Output G may include a special mechanism of IAA destruction that is identical to, or closely related to, the growth-promoting act. Such a linkage would assure that the same IAA molecule is not repeatedly used, as persistence of the signalling molecule would seriously limit a cell's capability to control the response reaction. Outputs "D" and "G" may therefore be closely related.

Fig. 3 presents an example of the usefulness of knowledge of the structures and amounts of IAA and its conjugates. These experiments were performed in the late 1970's, and aside from quantifying the amounts and

41

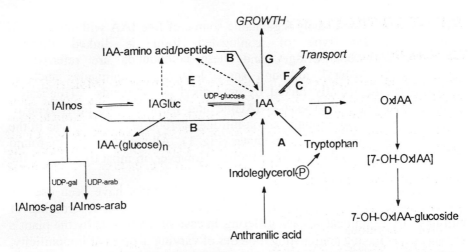

Fig. 2. Diagram of the metabolic transformations that determine the steady state level of IAA in plant tissues. Inputs to the IAA pool include: A) *de novo* synthesis from tryptophan or by the non-tryptophan pathway; B) ester and amide conjugate hydrolysis; and C) transport. Outputs from the IAA pool include: D) catabolic oxidation; E) ester and amide conjugate synthesis; F) transport; and G) IAA "use" during the growth process.

turnover of compounds related to tryptophan, they clearly demonstrated that there was an unknown major pathway involved in IAA degradation (see '?' in Fig. 3). These experiments still represent the only comprehensive picture of hormone economics in plants. One should keep in mind that this is a static picture and represents only one stage in the growth of this plant, when *de novo* synthesis represents a minor component in the kernel and there is no *de novo* synthesis occurring in the growing shoot.

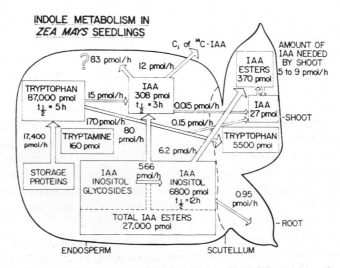

Fig. 3. Pool sizes and rates of turnover of the indolic compounds in the kernels of *Zea mays* after 4 days of germination. Fig. adapted from (31).

INPUTS TO THE IAA POOL

De Novo Synthesis (Fig. 2, A)

De novo synthesis implies the synthesis of the heterocyclic indole ring from non-aromatic precursors. In bacteria, fungi, and plants, this means some variation of the shikimic acid pathway in which anthranilic acid is made from sugar precursors (Fig. 4). This pathway is the only known source of all benzene ring compounds found in nature. One of these compounds, the abundant amino acid tryptophan, contains all the carbon and structural features necessary to make IAA.

IAA Synthesis from tryptophan

An enormous body of knowledge demonstrates that IAA can be synthesized from tryptophan; a summary of some of the known reactions is presented in Fig. 5. Data supporting these pathways have been the topic of several reviews (17, 29, 75). For most of the last half century, research on biosynthesis of IAA has focused on different possible routes for the conversion of tryptophan into IAA. However, concerns about the low rate of labeling from tryptophan of the IAA pool (31, 81) were largely overwhelmed by the preponderance of studies on tryptophan conversion in the literature. Recently it has been established that in some plants the quantitative importance of tryptophan conversion relative to other possible sources is minor (1) and that plants that cannot make tryptophan at all are, nevertheless, able to make IAA *de novo* (64, 81).

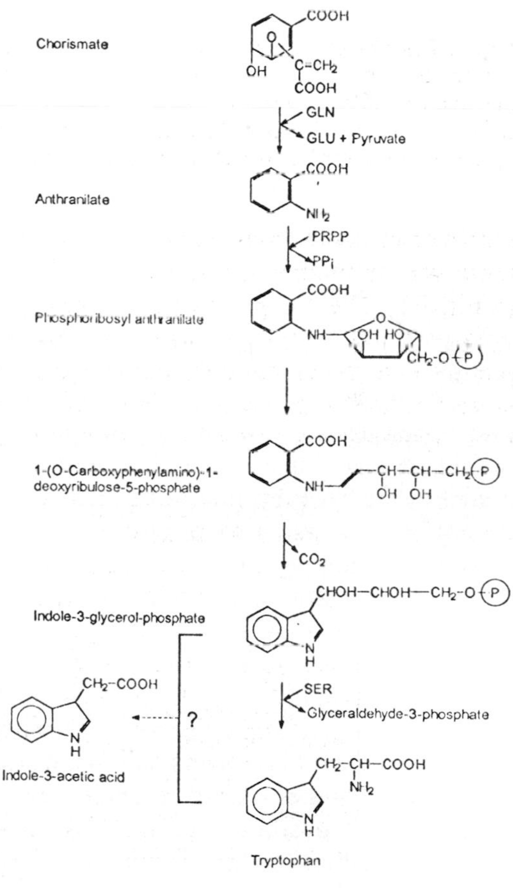

There are several potential pitfalls in this field, not the least of which includes the enormous disparity in the amounts of

Fig. 4. *De novo* biosynthesis of IAA from precursors in the indolic pathway. Intermediates in the non-tryptophan pathway have not been identified. However, mutant analysis studies have demonstrated that the branch point is likely to be at indole-glycerolphosphate (87).

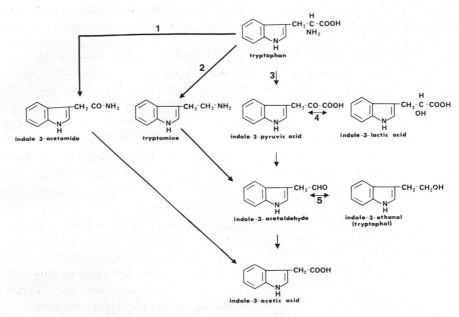

Fig. 5. Pathways for the synthesis of IAA from tryptophan: 1) indole-3-acetamide pathway of crown gall infected plants, *Pseudomonas savastanoi*, and other pathogenic bacteria; 2) the tryptamine pathway; and 3) the pathway through indole-3-pyruvic acid. Pathways 2 and 3 are related in that both involve sequential oxidative decarboxylation and deamination to produce indole-3-acetaldehyde. Indole-3-lactic acid (pathway 4), and indole-3-ethanol (pathway 5) may be side branches of pathways 2 and 3 involved in the regulation of IAA biosynthesis.

tryptophan and IAA in tissues. The pool size of tryptophan is three, or more, orders of magnitude larger than that of IAA (24). Tryptophan is readily converted to IAA by micro-organisms (41, 44, 56, 76) and even a minute microbial conversion of tryptophan to IAA would produce the picomolar amounts commonly found in plants. Radiolabeled tryptophan can be non-enzymatically converted to IAA, presumably by radio-chemical decomposition. Epstein et al., described the conversion of 5-[^{3}H]-tryptophan to labeled IAA in almost 30% yield during drying in vacuo in a glass tube (24). This oxidative conversion is mediated by peroxides and free radicals which accumulate in the radioactive tryptophan solution. Precautions can be taken to prevent this conversion (24), however they were not applied in many of the early studies. Until recently, few of the published studies attempted to calculate the amount of IAA synthesized. And finally, there are no published studies for higher plants showing that a mutant plant deficient in its capacity to produce IAA from tryptophan is, in fact, IAA deficient.

Considerations regarding the pool size of tryptophan

Law and Hamilton (50, 51) described a system in which the enormous tryptophan pool may be isolated biochemically from the IAA pool. They

44

observed that, in pea (*Pisum sativum*), L-tryptophan plus gibberellin A_3, or D-tryptophan alone, were as effective as IAA in promoting elongation of stem segments. They postulated that D-tryptophan was the immediate precursor of IAA and that the racemization of L to D tryptophan was GA controlled. These experiments were difficult to evaluate because of the disparity in pool sizes of D and L tryptophan but, the mechanism could provide a means of regulating the synthesis of the small IAA pool by separation of the precursor from the huge tryptophan pool. This theory was given support by a report that 4-Cl-tryptophan, the expected precursor to 4-Cl-IAA found in pea, also occurred in the D-form (57).

Baldi *et al.* (1) tested the hypothesis that D-tryptophan is the IAA precursor using the aquatic monocot *Lemna gibba* as a model system, but they could find no evidence for this pathway. With the *Lemna* system, the experiments could be done under sterile conditions and uptake of both D- and L-forms of tryptophan from the medium occurred rapidly. It was found that even after several days, the ^{15}N-D-tryptophan taken up from the medium was not converted into ^{15}N-IAA, although they measured a several hundred fold enrichment of the D-tryptophan pool. They also reported finding only low levels of L-tryptophan conversion, and this ^{15}N-L-tryptophan to ^{15}N-IAA labeling occurred without detectable labeling of the D-tryptophan pool.

Conversion of *N*-malonyltryptophan, a compound thought to be found *in vivo* predominantly in the D-tryptophan form, to indole-3-acetaldoxime and then to IAA has been proposed as another route to IAA (69). However, Ludwig-Muller and Hilgenberg (54) showed that while *N*-malonyltryptophan was converted to IAA, *N*-malonyl-L-tryptophan was the substrate for this reaction, not the D-tryptophan form. Recent publications from Marumo's laboratory (73) on the occurrence of 4-Cl-tryptophan and malonyl-tryptophan in pea, report that only about 2% of the 4-Cl-tryptophan is in the D-form and the bulk of 4-Cl-tryptophan is the L-isomer. They also present evidence that, contrary to the initial reports, the malonyl tryptophan in plants is predominantly in the L-form.

De novo synthesis of IAA not involving tryptophan

Although early studies equated IAA synthesis from tryptophan with *de novo* biosynthesis, recent more exacting and critical evaluations of when in a plant's life it begins or stops making IAA from early precursors have yielded surprising results. A particularly useful technique for measurement of *de novo* synthesis of aromatic ring compounds is to allow plants to grow in the presence of water enriched in deuterium oxide, or so-called "heavy water." Under these conditions any newly formed aromatic rings have deuterium locked into non-exchangeable positions on the ring during their biosynthesis. Such labeling techniques provide at least two advantages. First, since the "labeled precursor" is water, such an approach does not require exact knowledge of precursors or pathways in order to accurately ascertain the extent of *de novo* synthesis. Second, since all cell compartments are freely permeable to water, problems of compartmentation and uptake are not an issue.

Experiments from several laboratories in which young plants of *Zea mays*, *Arabidopsis*, or pea, and cell cultures of carrot were grown on 30% deuterium oxide demonstrate that IAA is made by a route resulting in the incorporation of deuterium molecules into non-exchangeable positions of the indole ring of IAA to a greater extent than that found in tryptophan (20, 59, 64, 68, 81).

Other experiments (e.g., 2) employing the deuterium incorporation technique, indicated that tryptophan synthesis begins before IAA biosynthesis in germinating *Zea mays* kernels. Thus, tryptophan and IAA synthesis appear to occur independently, and there is not *necessarily* an incorporation of deuterium from tryptophan into IAA. These results differ somewhat from the earlier experiments with maize (68) where *de novo* synthesis of IAA was not detected in shoot tissue. As more data is being obtained, it is becoming more and more apparent that the time of initiation of IAA biosynthesis is determined by the strain and type of plant and may also be determined by the availability of sufficient IAA conjugates from the endosperm.

Perhaps the most striking of these isotopic labeling studies is the report on the *orange pericarp* (*orp*) mutant of maize. This plant carries a double recessive trait caused by mutation of both genes in maize that encode the protein for tryptophan synthase b. Despite this metabolic block in the terminal step for tryptophan biosynthesis, the *orange pericarp* mutant produces IAA *de novo* and, in fact, accumulates up to 50 times the level of IAA as do non-mutant seedlings. Labeling studies established that the *orp* mutants are able to convert ^{15}N-anthranilate to IAA, but do not convert it to tryptophan. Neither *orp* seedlings nor control seedlings convert tryptophan to IAA in significant amounts even when the *orp* seedlings are fed levels of stable isotope labeled tryptophan high enough to reverse the lethal effects of the mutation (81). These results established that non-tryptophan biosynthesis of IAA does occur, and suggested that the non-tryptophan pathway actually predominates over the tryptophan pathway.

Despite the demonstration and now wide acceptance that IAA biosynthesis can occur without the amino acid tryptophan as an intermediate, the exact pathway for the production of IAA by such a route is not yet known. *In vivo* labeling techniques using *Arabidopsis* mutants (64) have extended the findings from the *orange pericarp* maize study, and suggest that the branch point for IAA production is probably at the point of indole (following tryptophan synthase a) or its precursor, indole-3-glycerol phosphate (the conversion of indoleglycerol phosphate to indole is a reversible reaction). Most current evidence favors the condensation of indole with a two carbon unit with a nitrogen at the terminal carbon, followed by conversion to the carboxylic acid. Until the pathway can be established using both *in vivo* and *in vitro* techniques, the data is only useful as a guide for further investigations. Rekoslavskaya *et al.* (70) reported obtaining an *in vitro* system from maize endosperm capable of converting radioactive indole into IAA by a reaction which is not inhibited by the addition of unlabeled tryptophan. The availability of this *in vitro* system should now make it possible to establish the biochemistry of the conversion.

These recent developments have certainly changed our concepts of IAA biogenesis from what we knew only a few years ago. It is important to keep in mind, however, that the establishment of the existence of a non-tryptophan pathway does not change the fact that many plant species have been shown to convert tryptophan to IAA, and in some cases this conversion is clearly at rates that make it important for the auxin economy of the plant. For example, in the bean seedling, *de novo* IAA biosynthesis begins even before the stored conjugates are fully used up (11, 13), and this biosynthesis comes primarily from tryptophan conversion. Similarly, Michalczuk et al. (59) showed that in embryogenic carrot suspension cultures, the conversion of tryptophan to IAA is the primary route. However, when 2,4-D was removed from the medium, which induces these carrot cells to form embryos, the conversion of tryptophan to IAA was no longer the primary route, and the non-tryptophan pathway appeared to predominate. These metabolic interactions, and the regulation of these pathways in relation to particular developmental programs, are clearly exciting topics for further detailed studies at a molecular level.

4-Chloroindole-3-acetic acid and indole-3-butyric acid in plants

Although IAA was the first auxin isolated from plants, and is widely considered the major plant auxin, other compounds with auxin activity are also found. In general most of these compounds are active only at much higher concentrations than IAA and their role in plant growth remain largely unknown. Two indolic auxins other than IAA, which may have regulatory roles for certain processes, have been isolated from plants. Several early studies reported finding a compound in plant extracts that appeared to have properties similar to that of IBA based on simple chromatographic evidence. IBA has recently been positively identified in plants by GC-MS (25). The exact role for IBA in plants is unknown although several authors have speculated that it could be involved in root formation (25, 53). IBA has been used commercially for plant propagation for decades because of its efficacy in the stimulation of adventitious roots (14). The interconversion of IAA to IBA and IBA to IAA has been shown to occur in plants (26, 53), so that a biochemical relationship between IAA and IBA appears to be likely.

A highly active halogenated indole auxin, 4-Cl-IAA, has been identified in a number of plants, mainly members of the *Fabaceae* (29, 57), but also in pine seeds (27). 4-Cl-IAA has been shown to have up to ten times the biological activity of IAA in bioassays (57). Most of the 4-Cl-IAA occurs as the methyl ester in many of the plants examined, however, 4-Cl-IAAsp and its monomethyl ester have also been described. As for IBA, the physiological role for 4-Cl-IAA is also not well established, however the recent report of its activity in the stimulation of pod growth in deseeded pea (where other auxins appear weakly or not active) and its presence in seeds and pod tissue suggest a function in pod development (66).

Microbial pathways for IAA biosynthesis

IAA production by plant pathogenic bacteria has been an important aspect of research on IAA metabolism. In particular, work on the crown gall forming bacteria *Agrobacterium tumefaciens*, which transfers genetic material (a fragment of DNA, termed T-DNA, born on the tumor inducing or Ti plasmid) from the infecting bacteria into the host dicotyledonous plant, has shown that encoded within the transferred DNA are genes with eukaryotic promoters for the enzymes tryptophan monooxygenase and indoleacetamide hydrolase (41). These two enzymes carry out the conversion of tryptophan to indoleacetamide and the hydrolysis of the indoleacetamide to IAA (Fig. 5). Several research groups have now used this system to ascertain the effect of increased IAA production on plants using engineered Ti plasmids that lack other genes involved in tumor induction (42, 76).

Another gall forming bacterium, *Pseudomonas savastanoi* uses the identical pathway for IAA production (44), however, in this case no genetic material is transferred and the bacteria themselves produce high levels of IAA. *Pseudomonas* also has the capacity to form the novel conjugate of IAA, IALys, as well as its a-N-acetyl derivative (28, 44). Although it has been shown that IAA-Lys formation can reduce the pool of free IAA produced by the bacteria by about 30% (30), the exact role of these conjugates in gall formation has not been established. Another gall former is *Erwinia herbicola*, some strains of which cause crown and root galls on host plants of *Gypsophila paniculata*. Strains of *Erwinia* that are pathogenic and form such galls have the capacity to make most of their IAA by the same indoleacetamide pathway as *Pseudomonas* and T-DNA transformed plant cells. However, *Erwinia* strains that do not form galls make IAA using the indolepyruvate pathway from tryptophan, and lack the indoleacetamide route (56, see Fig.5, pathway 3). The nonpathogenic strains of *Erwinia* are saprophytic epiphytic bacteria that are widespread in nature. Their presence in plants grown in non-sterile conditions is a potential complication for IAA biosynthesis studies because bacterial IAA production by the endophyte could be considerably greater than plant IAA production rates.

Hydrolysis of IAA Conjugates (Fig. 2, B)

Stored Auxins and Their Utilization

Only a small number of developmental systems have been studied in any detail relative to the utilization of stored conjugates for growth. The best studied system is the maize kernel and the subsequent germination of the kernel to form a seedling. The ester conjugate IAInos is one of the major conjugates in both the kernel and the shoot of corn (15). Five-[^{3}H]-IAInos applied to the kernel is transported to the shoot and is, in part, hydrolyzed to yield free 5-[^{3}H]-IAA (65). Thus, it is believed that conjugates from the kernel are the source of free IAA for the growing shoot. However, in a quantitative sense, there is uncertainty as to how much of the IAA of the shoot is derived from conjugates in a plant's seed or kernel, and this appears to be a function of kernel composition (2, 68) and plant species (13). Studies with 5-[^{3}H]-IAInos indicated that between 2 to 6 pmol/hr of IAA can be supplied to a maize shoot from the kernel and this would be a major source of the estimated 10 pmol/hr required by each shoot (4, 62, 65). There are difficulties, however, in determining how much the 5-[^{3}H]-IAInos is diluted by endogenous IAInos (24). Thus, quantitative aspects of the source of IAA for shoot growth require additional study.

Maize kernels contain about half of their conjugated IAA as high molecular weight glycans and about half as low molecular inositol glycosides such as IAInos-gal and IAInos-arab. 5-[^{3}H]-indole-3-acety-l-myo-inositol-[^{14}C]-galactoside applied to the endosperm is also transported to the shoot, there to yield free IAA (43). In these experiments, the double label compound was used to show that hydrolysis of the galactose from the conjugate occurred after the conjugate left the endosperm, but before it entered the shoot, presumably in the scutellum.

The major conjugate class in bean seeds is a series of small proteins/peptides where IAA is attached in amide linkage (9). These peptides accumulate during seed development essentially in parallel with other storage products during the late maturation stage of bean seed development (10). Unlike the situation in maize, these conjugated forms decrease dramatically only during the first few hours following imbibition. After only one or two days of growth, *de novo* IAA and IAA-peptide biosynthesis begins in the growing axes while conjugate hydrolysis continues in the cotyledonary tissue (11, 13).

The metabolism of auxins in carrot cell cultures has recently received attention because auxin removal from carrot cell cultures induces the formation of embryos. Removal from the medium of the synthetic, auxin-like growth regulator, 2,4-D, results in a decreased production of IAA and, as described in the section on IAA biosynthesis, a change in the metabolic route to IAA. Also, the rate of conjugate utilization goes up dramatically during this period of time so that order of magnitude decreases in the amide conjugate pool occur over a period of only a few days (58). This increase in conjugate utilization is accompanied by a 3-4 times increase in the activity of an IAA-amino acid hydrolase (47).

Enzymatic Hydrolysis of IAA-Conjugates

In 1935, Cholodny (19) showed that application of a water-moistened piece of endosperm to a seedling led to an auxin response, whereas alcohol-moistened endosperm elicited no response. Thus, Cholodny was the first person to observe enzyme catalyzed hydrolysis of the IAA conjugates stored in the endosperm. A few years later, Skoog (19) made the important observation that a "seed auxin precursor" moved from the seed to the shoot and that this precursor could be converted by the plant to an active auxin. Hamilton *et al.* (33) then showed that extraction of corn shoots or roots with ether for 3 hr at 4°C yielded IAA, whereas an extraction with 80% ethanol yielded no IAA. Cold ether does not, in general, inactivate enzymes, whereas 80% ethanol does. From these results it was concluded that the tissue was autolyzing in the wet ether, with consequent enzymatic hydrolysis of IAA conjugates to yield free IAA. Subsequent work established that the "bound" IAA of the seed could also be released by mild alkaline hydrolysis (19), and that the IAA was not the product of hydrolysis of tryptophan or storage proteins. At that time, nothing was known of the chemical structure of the "bound auxins" except that some were thought to be of high molecular weight and to be proteins.

While conjugate hydrolysis in ether-induced autolyzing tissues occurs fairly readily, it has been difficult to obtain *in vitro* hydrolysis of IAA conjugates with purified enzymes. Many commercial proteases and esterases that might be expected to hydrolyze IAA-amino acid conjugates fail to do so (19). *In vitro* hydrolysis has been reported (8, 32) but the hydrolytic enzymes have proven difficult to extract and purify. Recent work on an IAA amino acid hydrolase from carrot has resulted in its partial purification and characterization (47). This enzyme has an apparent molecular mass of over 200 kDa and is active toward IAA-Ala, IAA-Phe and several other monocarboxylic amino acid conjugates, but not IAAsp. Town *et al.* (79) reported on a cloning strategy for IAA-amino acid hydrolases using amino acid auxotrophic yeast mutants complemented by *Arabidopsis*. This approach has allowed the tentative identification of the gene for an enzyme with IAA-amino acid hydrolase activity toward both IAA-Ala and IAAsp.

Kowalczyk and Bandurski have reported on the co-fractionation of IAGlu synthetase and two enzymes of IAGlu hydrolysis, a 1-*O*-IAGluc hydrolase and a 6-*O*-IAGluc hydrolase (45, 46), and suggested the existence of a hormone-metabolizing complex. Membrane localization of such a hormone-metabolizing complex, capable of conjugate synthesis on one side and conjugate hydrolysis on the other side of a membrane could provide a transport mechanism similar to the vectorial PEP-sugar transport mechanism of bacteria. A summary of the synthetic and hydrolytic capabilities of this enzyme complex is shown in Fig. 6.

Success in purifying conjugate-hydrolyzing enzymes is of interest in designing experiments to understand the control of the amount of free IAA, and should also be of practical value in developing synthetic conjugates of varying degrees of ease of hydrolysis.

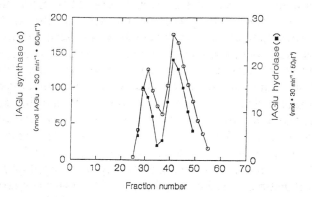

Fig. 6. Chromatography of an IAA-glucose synthase enzyme from *Zea mays* on amine ion exchange column. On this column and on other separation systems the synthase shows two peaks of activity. The protein in the leading peak catalyzes IAGluc synthesis but also catalyzes hydrolysis of 1-O-IAGluc. The protein in the second peak catalyzes 1-O-IAGluc synthesis and catalyzes the hydrolysis of 6-O-IAGluc. The association of IAGluc synthesis activity with IAGluc hydrolyzing activity has been referred to as a hormone metabolizing complex (61).

Transport of IAA from One Site to Another Site within the Plant (Fig. 2, C, F)

Although not involving a net increase in IAA, or its conjugates, transport from one site to another within the plant results changes in the levels of IAA within a given tissue or organ. An example is the transport from seed to growing shoot or root as illustrated in Fig. 3. A second case, most probably involved in asymmetric growth following a tropic stimulus, involves movement of IAA from the vascular stele to the surrounding cortical tissues.

Transport from seed to shoot and from coleoptile tip to basal growing regions
From the limited studies on transport from seed to shoot (discussed above), and determinations of IAA, IAInos and other esters in the fruit tissue (37, 43, 65), it may be concluded that IAInos is the chemical form in which IAA is transported from seed to shoot in seedlings of *Zea mays*, and that free IAA is almost certainly the form in which IAA is transported from the coleoptile tip to the more basal growing region (48). But is there some special zone in the tip where hydrolysis of the upward transported IAInos occurs so that downward transport of free IAA may occur? Epel and colleagues have delineated transport domains within the plant including a vascular domain in the stele for upward transport and a cortical domain separated from the stele by apoplastic barriers where presumably downward transport occurs (21, 22). Based on these recent findings, we conclude that a reexamination of the enzymology, chemistry, morphology and genetics of IAA transport within the plant should be rewarding.

Asymmetric transport of IAA during tropic stimulation

The earliest studies of IAA transport were concerned with how an asymmetric (tropic) stimulus could induce an asymmetric distribution of IAA. The asymmetric IAA distribution led to asymmetric growth resulting in the plant shoot bending towards the light or upwards from the earth, and there is now good molecular evidence that the asymmetric distribution of IAA leads to changes in gene expression (48, and see Chapter D2). This phenomenon of asymmetric IAA distribution was described, but not explained, by the phrase, "lateral transport", as used in the Went-Cholodny theory for tropic responses (4, 6).

Several recent findings help to explain how such lateral transport may occur. First, there is the finding that a geotropic stimulus results in an asymmetric distribution of *both* free and ester IAA in the mesocotyl of *Zea mays* seedlings (2, 5, 6). Thus the tropic response cannot be due *solely* to changes in the ratio of free to conjugated IAA. The ratio of free to conjugated IAA does change (4, 6), but since both free and ester IAA increased on the lower side of the gravity-stimulated mesocotyl, and since *de novo* IAA biosynthesis does not occur in young *Zea* seedlings (e.g., 2) there must also be selective transport of IAA and/or its esters into the rapidly growing cortical cells (5,6).

The sequence of events following a tropic stimulus is that following perception of the stimulus, a bioelectric perturbation occurs within seconds; IAA and Ca^{++} asymmetric distribution occurs in less than 5 minutes; and asymmetric growth occurs in about 5 minutes (4). This knowledge of the time sequence of events following the stimulus led to the formulation of a working theory for the transport of IAA in the maize mesocotyl. The postulates of the theory are that a) IAInos, the putative seed auxin precursor, moves upwards from seed to shoot in the stele; b) the IAA released from conjugates by hydrolysis then moves out of the stele and into the cortical cells through the plasmodesmata; c) movement of IAA or IAInos through the plasmodesmata from stele to cortex can be regulated by the plant; d) the plasmodesmatal connections between tissues are voltage "gated" in a manner analogous to the gating of animal gap junction cells and it is this which permits the gating; and e) the gravity stimulus alters the bioelectric potential of the plasmodesmata, thus bringing about asymmetric movement of IAA from the central stele into the upper and lower cortical cells (5, 6).

The implications of the voltage-gating theory are considerable in that the theory implies that the plant is able to regulate endogenous local concentrations of IAA by regulating the transport of IAA from one site to another site within the plant. Thus, selective movement of IAA within the plant and even from the plant into seeds or fruit could be controlled by metabolic gating of the plasmodesmatal connections of the plant symplast.

OUTPUTS FROM THE IAA POOL

Oxidative Catabolism (Fig. 2, D)

IAA oxidative catabolism is the chemical modification of the indole nucleus or side chain resulting in loss of auxin activity, and it is the only irreversible output regulating IAA levels. This discussion of IAA catabolism will include the oxidative decarboxylation of the side chain, and oxidation at the 2, 3, and 7 positions of the indole nucleus without decarboxylation.

The interest in IAA catabolism predates identification of IAA as a ubiquitous auxin in plants. In 1934 (78) it was observed that water extracts of leaves inactivate the "growth promoting substance" of *Avena*. Later work showed that peroxidase catalyzes the oxidative decarboxylation of IAA (75), and this reaction was generally assumed to be the physiological route for oxidative catabolism. More recently a new pathway of IAA catabolism with retention of the carboxyl side chain and oxidation of the indole nucleus has been identified (27, 72) and shown to occur in *Pinus sylvestris, Vicia faba,* and *Zea mays.*

Decarboxylation Pathway

The decarboxylation pathway is catalyzed by peroxidases from numerous plant species (75), and often by several peroxidase isozymes per plant species. The *in vitro* reaction may be monitored by $^{14}CO_2$ evolution from 1-[^{14}C]-IAA, manometrically by O_2 uptake, colorimetrically by loss of Salkowski color, or by UV absorbance changes (Fig. 7). In the literature, "IAA oxidase" and "peroxidative oxidation" of IAA with decarboxylation have been synonymous. This definition of IAA oxidation is now, however, too narrow, due to the discovery of oxidative non-decarboxylation pathways in several plant species.

The peroxidative decarboxylation of IAA occurs without added cofactors with purified horseradish peroxidase, the best studied "IAA oxidase," although Mn^{++} and monophenols increase the reaction rate, and H_2O_2 reduces the lag period for IAA oxidation (75). Tomato peroxidases are dependent on H_2O_2 for the oxidation of IAA. The main products of peroxidase oxidation of IAA

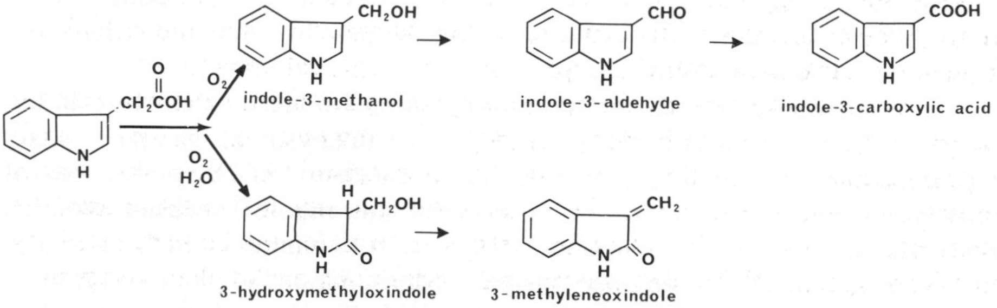

Fig. 7. The oxidative decarboxylation process catalyzed by peroxidase. This process is initiated by peroxidases contained in extracts of many plant species. The natural occurrence of some of these catabolites has been demonstrated.

are 3-methylene-oxindole, 3-hydroxymethyloxindole (oxindole-3-carbinol), indole-3-aldehyde, and indole-3-methanol (Fig. 7). The ratio of the various products depends on the enzyme/substrate ratio, cofactors, and pH of the reaction (75). A high enzyme/substrate ratio favors formation of indole-3-aldehyde and indole-3-methanol, and low amounts of the oxindoles. The addition of Mn^{++}, and 2,4-dichlorophenol also stimulates the formation of indole-3-methanol production at the expense of the other oxidation products. Indole-3-methanol is a precursor of indole-3-aldehyde since indole-3-methanol is converted by peroxidase to indole-3-aldehyde (35, 75). Most of the oxygen incorporation into indole-3-methanol is from molecular oxygen, and not water. Three-hydroxymethyl-oxindole is the immediate precursor of 3-methylene- oxindole via a non-enzymatic dehydration (35). The fact that labeled indole-3-methanol is not metabolized by horseradish peroxidase to 3-hydroxy-methyl-oxindole demonstrates that oxindole and indole formation are separate branches of the peroxidase pathway. It is now thought that peroxidase initially reacts with IAA by forming an IAA free radical which is subsequently attacked by oxygen (35, 75). Following decarboxylation of the IAA, the reaction proceeds either by the indole-3-aldehyde route or the 3-methylene-oxindole route.

Cofactors for IAA oxidation include Mn^{++} and monophenols. Mn^{+++} will non-enzymatically oxidize IAA, with subsequent decarboxylation, and it is thought that peroxidase plus monophenol oxidizes Mn^{++} to Mn^{+++}. Monophenols and *m*-diphenols stimulate IAA oxidation, while *p*-diphenols, *o*-diphenols, coumarins, and polyphenols inhibit the enzyme reaction (75). Although *in vivo* regulatory functions for these compounds in peroxidase oxidation of IAA have been suggested, the importance of the peroxidative oxidation of IAA *in vivo* itself is in question.

Oxindole-3-acetic acid/Dioxindole-3-acetic acid pathway

This oxidation pathway has been observed in several plant species including rice, corn, and broad bean. Based on simple colorimetric assays, the first product in this pathway (Fig. 8), OxIAA, also occurs in germinating seeds of *Brassica rapa*, and developing seeds of *Ribes rubrum* (40). *Oryza sativa*, rice, is interesting in that it is the only plant known to contain both OxIAA and DiOxIAA (39),and was also shown to have the 5-hydroxy analogs of OxIAA and DiOxIAA.

The standard methods for monitoring the peroxidative decarboxylation of IAA may not distinguish between the two pathways in an *in vitro* system. For example, IAA oxidation to OxIAA would result in O_2 uptake, loss of Salkowski color, and reduction in the 280 nm indole spectrum with an increase in the 247 nm oxindole spectrum, changes which are also characteristic of the peroxidase system. Measurement of the evolution of $^{14}CO_2$ from feeding 1-[^{14}C]-IAA is a clear indication of the peroxidase oxidative decarboxylation pathway. OxIAA has been shown not to be an intermediate or a substrate for the peroxidase pathway (35) so the pathways are independent. Careful chromatographic isolation and physicochemical identification of the catabolites is the best method of identifying the IAA catabolic pathways occurring in a particular plant.

The first report of IAA oxidation to oxindole-3-acetic acid (OxIAA) was in the basidiomycete *Hygrophorus conicus* (74). The conversion of tryptamine to IAA and then to OxIAA was unique to Hygrophorus conicus of the 12 basidiomycetes tested. OxIAA and DiOxIAA were found to be synthesized by *Zea mays* and *Vicia faba* respectively following feeding 1-[^{14}C]-IAA (72, 80) (Fig. 8). Isotope dilution experiments (72) showed that OxIAA was a naturally occurring compound in *Zea mays* endosperm and shoot tissues, occurring in amounts of 357 pmol per endosperm and 47 pmol per shoot, about the level of free IAA in these tissues. In *Vicia faba*, DiOxIAA was estimated by UV measurements to be 1 μmole kg^{-1} fresh weight in root tissues. In both of these experiments, labeled IAA was fed to etiolated seedlings via endosperm tissue or cotyledons.

In corn, OxIAA is further metabolized by hydroxylation at the 7 position, and by glucose addition to form 7-OH-OxIAA-glucoside (60). Isotope dilution assays have shown that the 7-OH-OxIAA-glucoside is a naturally occurring compound in corn in amounts of 62 pmol per shoot and 4.8 nmol per endosperm. *Vicia faba* is also reported to form a glucose derivative of DiOx-IAA-aspartate (e.g. 80).

The enzymology of the OxIAA and DiOxIAA pathways is much less understood than the peroxidase pathway owing to its more recent discovery,

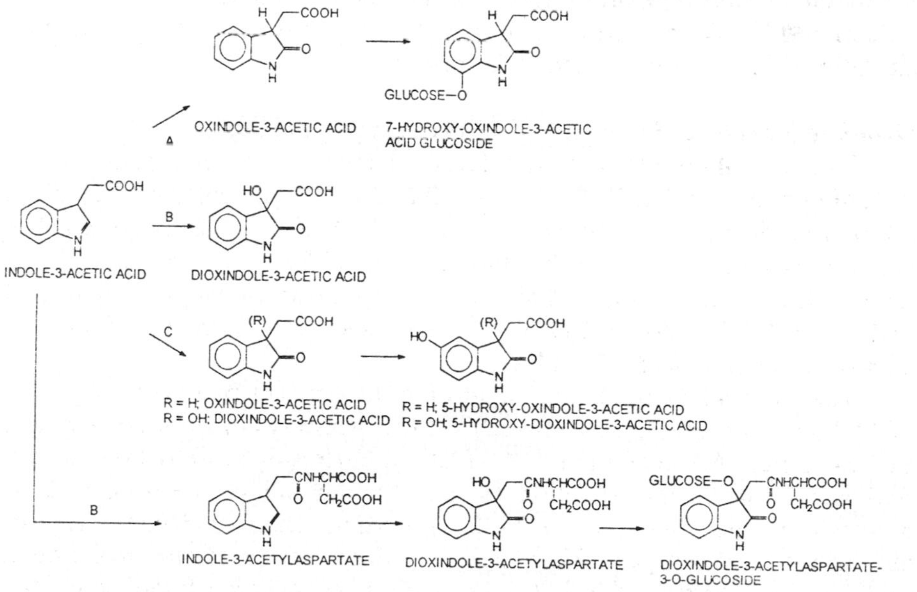

Fig. 8. The non-decarboxylation oxidative pathways from higher plants: Data has shown these catabolic transformations occur in A) *Pinus sylvestris* and *Zea mays*, B) *Vicia faba* and C) *Oryza sativa*.

and to the limited availability of enzymes and substrates. In *Zea mays*, the rate of oxidation of IAA to OxIAA has been measured in shoot, root, scutellar, and endosperm tissues at 1-10 pmol h^{-1} mg protein^{-1}. The enzyme is soluble, of high molecular weight and clearly different from lipoxygenase or peroxidase (72). Enzyme activity was reduced by 90 per cent when assayed under argon, indicating an oxygen requirement for the reaction. However, labeling studies to identify the source of oxygen have not been done. Enzyme activity was stimulated up to ten fold by addition of an ionic detergent extract of corn tissue. A heat stable component of the Triton X-100 extract increased enzyme activity when added to buffer extracted enzyme. This heat stable, detergent-extractable, corn tissue co-factor may be replaced by linolenic, linoleic, or arachidonic acid (72). Cofactors of mono-oxygenation reactions as well as peroxidase cofactors, are inactive in stimulating OxIAA formation.

7-OH-OxIAA has been identified as a catabolite of IAA in germinating kernels of *Zea mays*. It was found to be present at 3.1 nmol kernel^{-1} and has been shown to be an intermediate in the synthesis of 7-OH-OxIAA-7'-0-glucoside (63). The glucoside is present in much higher amounts than IAA, OxIAA, or 7-OH-OxIAA, and thus is hypothesized to accumulate in vacuoles (52). The further metabolism of 7-OH-OxIAA has not been studied except that 5-[^{3}H]-7-OH-OxIAA loses tritium to water upon further enzymatic oxidation (52). This implies a second oxidation of the benzenoid ring leading to a highly unstable dioxindole.

Physiological occurrence of catabolic pathways

The catabolic route for a plant hormone must be determined by feeding isotopically-labeled compounds in an appropriate manner not involving cut cell surfaces, followed by identification of the product(s), measurement of pool sizes and turnover, and determination of the rate at which the product is formed. The biological activity of the catabolites should be measured, and in addition, care must be taken with IAA metabolic studies because IAA may easily be non-enzymatically degraded by light, acid, silica gel, etc. (24, 75).

Despite all the research on the *in vitro* peroxidase-catalyzed decarboxylation of IAA, only a few reports on the isolation and physicochemical identification of endogenous peroxidase catabolites have been reported. Indole-3-methanol, and indole-3-carboxylic acid have been identified in pine (27), and indole-3-aldehyde, indole-3-methanol, and indole-3-carboxylic acid were reported to occur in pea sections fed IAA (55). In pine needles there were 2.3 ng per g fresh weight indole-3-carboxylic acid or about ten per cent of the level of free IAA (34). In etiolated pine shoots there were 19.7 ng per g fresh weight of indole-3-methanol. Whether indole-3-carboxylic acid is a natural metabolite of IAA is in question because in *Brassica* its occurrence can be an artifact of glucosinolate breakdown (64), and in *Pinus* [^{14}C]-IAA was not metabolized to indole-3-carboxylate (27).

3-Hydroxymethyl-oxindole and 3-methyleneoxindole are unstable compounds. However 3-hydroxymethyloxindole has been reported to occur after feeding labeled IAA to sections (27, 75). Indole-3-methanol glycoside, and indole-3-carboxylic acid were also reported to be catabolites of IAA. Generally, catabolites from radiolabel feeding studies have only been identified by thin-layer chromatography in several solvent systems and further substantiation by physicochemical methods is needed.

The occurrence in plants of very small amounts of IAA decarboxylation products shows that the reaction does occur *in vivo* (27) but it is unlikely to be a major pathway during normal development. For example, ~75% of the "IAA oxidase" can be lost from plant segments by a brief buffer wash, and "IAA oxidase" activity has shown to be proportional to the number of pieces into which the sections were cut (72, 75). Also, an *in vitro* corn peroxidase system decarboxylated IAA to several compounds (72, 75), but feeding labeled IAA to either corn endosperm or root segments resulted in rapid metabolism of IAA with minimal decarboxylation (24, 61). Thus, it is possible that the decarboxylation pathway is overemphasized by the manner of presenting IAA to the plant preparation. These findings, together with the fact that over 70% of the total enzyme activity was wall localized, may indicate a wound response role for peroxidase-catalyzed oxidation of IAA. The physiological meaning of the IAA oxidase activity of plant peroxidases has recently been subjected to a further, very serious, challenge by the finding that transgenic tobacco plants expressing either a ten fold excess of peroxidase or a ten fold reduction in peroxidase all have the same endogenous IAA content (49).

OxIAA and diOxIAA acid have been demonstrated to be endogenous compounds in three plant species, although the precursor-product relationship has been shown only for broad bean and corn. However, the occurrence of the carboxyl-retention pathway of IAA catabolism in both a monocotyledonous and a dicotyledonous plant indicates that the pathway may be widely distributed. Since Radiolabeled peroxidase metabolites and OxIAA/DiOxIAA metabolites can be synthesized chemically or enzymatically (27, 72, 80), the natural occurrence of these pathways in other plants may now be examined by an isotope dilution assay. A comprehensive determination of the biological activity of all the IAA catabolites from the peroxidase, and OxIAA/DiOxIAA pathways in several auxin bioassays, has not yet been performed. However, those catabolites tested have been found to be inactive in promoting growth in auxin bioassays (75).

The answer to whether IAA is catabolized *in vivo* by the relatively non-specific peroxidase isozyme system, and/or specifically by another enzyme will clarify how and where IAA catabolism is involved in IAA mediated growth. Attempts have been made to correlate peroxidase levels and age of tissue with responsiveness of tissues to IAA. In some studies there was a positive correlation, while other studies showed no correlation (75).

In summary, there are two pathways of IAA catabolism: the peroxidase catalyzed oxidative decarboxylation of IAA, and oxidation without decarboxylation to OxIAA or DiOxIAA. The peroxidase mechanism of oxidation is well understood with purified enzymes, but its physiological significance remains to be elucidated. The OxIAA/DiOxIAA pathway has been identified in at least three species; its occurrence in other species must be examined, and the mechanism of enzyme catalysis must be studied further. Future research will determine if IAA catabolism occurs in tissues non-responsive to IAA, or only in actively growing tissues where the oxidase reaction would destroy the hormone following the "growth promoting act". In either case, the elucidation of how and where IAA is catabolized in plants will help clarify how IAA levels are regulated during growth and development.

IAAsp oxidation

Early studies of the oxidation of IAA-conjugates focused on the attack by peroxidase (the classical "IAA-oxidase") on individual conjugates. The initial findings indicated that peroxidase did not attack the ester and amide conjugates tested (18). However, a more extensive study of amino acid conjugates by Park and Park (67) showed that less polar conjugates could be substrates for peroxidase. A more specific route of IAA-conjugate oxidation is shown, however, by the work of Tsurumi and Wada (80), who showed that in *Vicia* seedlings, IAAsp is oxidized to di-Ox-IAA-aspartate without prior hydrolysis to the free acid. As with IAA oxidation, once IAAsp oxidation occurs, the product is glycosylated to form the 3-(O-β-glucosyl) derivative. Tsurumi and Wada (80) also showed that IAAsp could be oxidized by peroxidase only when peroxide was added to the reaction mixture. The product of this oxidation was 2-OH-Ox-IAAsp, thus, the reaction with peroxidase and H_2O_2 yields a different product from that isolated from the plant.

Conjugate Synthesis (Fig. 2, E)

Synthesis of conjugates from endogenous IAA occurs in developing seeds and also under conditions leading to a change in growth rate (e.g., 4, 19). In addition, the application of IAA leads to conjugate formation (19) and the application of one conjugate, for example IAInos, can lead to the formation of other conjugates (65) and free IAA. Thus, IAA metabolism is a regulated, homeostatic, system involving "storage" of IAA in conjugated form with the possibility of subsequent hydrolysis to obtain free IAA. In the case of the *Pseudomonas* conjugation system, the gene for synthesis of IAA-lysine has been cloned, and for *Zea mays* the gene for synthesis of 1-O-IAGluc has been sequenced and cloned (Szerszen and Bandurski, unpublished data).

Enzymatic synthesis of
IAA-conjugates in Zea mays

The conjugate-synthesizing reactions of *Zea mays* that have been demonstrated *in vitro* are shown in Fig. 9 (2, 3, 45, 46). The first step in the synthesis of the IAA-*myo*-inositol family of conjugates is the synthesis of 1-*O*-ß-D-IAGluc from IAA and UDPG. The IAA is then transacylated to *myo*-inositol to form IAInos (3, 38). IAInos may then be glycosylated to form IAA-*myo*-inositol-galactoside or IAInos-arab by reaction with the appropriate uridine diphosphosugar (3). All of the low molecular weight IAA conjugates of *Zea mays* have been synthesized *in vitro* using crude enzyme extracts (2,3).

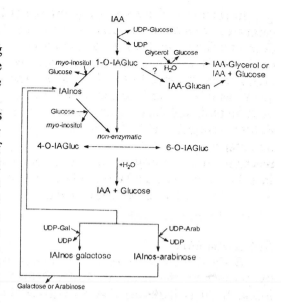

Fig. 9. Metabolic transformations between IAA and ester pools in *Zea mays*. The biosynthetic pathways, hydrolytic routes and the interconversions between ester pools are shown. Transacylation from IAInos back to glucose to form 4-O or 6-O-IAGluc and their subsequent hydrolysis is a reversible system for conjugate synthesis and hydrolysis.

Enzymatic synthesis of IAA-conjugates in Zea mays

The enzyme catalyzing the synthesis of 1-*O*-IAGluc was purified to homogeneity and polyclonal antibodies to the enzyme prepared (2). The enzyme was unstable except in the simultaneous presence of dithiothreitol and glycerol. It has an apparent molecular weight of about 50 kDa as estimated by sepharose chromatography and acrylamide gel electrophoresis. The sequence of the cDNA coding for the enzyme indicates a molecular weight of 50.6 kDa (Szerszen and Bandurski, unpublished data).

Synthesis of IAA-Amino Acid Conjugates

IAA-L-aspartate (Fig. 1) was the first IAA amide conjugate to be chemically characterized (18, 16). Its formation is also widespread in nature (18) and it is known to be an endogenous component of soybean seeds (16). Further, it is readily formed when IAA is applied to most plant tissues (18, 77) by what appears to be an inducible enzyme system. IAA-glutamate has also been demonstrated to occur naturally in some legumes (23) and cases are known where IAA is linked in amide linkage to a protein or peptide (9, 10, 11). Mysteriously, and despite numerous attempts, IAAsp and IAGlu have never been synthesized by an *in vitro* enzymatic system. This inability to obtain an *in vitro* enzymatic system for synthesis of IAAsp has resulted in a serious gap in our knowledge. Such an *in vitro* system would substantially contribute to a molecular understanding of how plants respond to exogenous growth regulators.

Synthesis of Indole-3-acetyl-ε-L-lysine

In some strains of *Pseudomonas savastanoi* (44) the IAA produced from tryptophan represents only a transitory intermediate since the IAA is then rapidly conjugated to form IALys (36). The synthesis of IALys requires L-lysine and ATP, and the IALys formed is then further metabolized to it α-N-acetyl derivative (36, 44). The gene for the IALys forming enzyme has been cloned and the IALys locus is some 2 kb upstream from the IAA operon (44, 71).

Use of IAA in Growth

IAA promotes a multitude of reactions that may interact to result in growth and differentiation. These include rapid induction of mRNAs, membrane phenomenon such as permeability changes, media pH changes and enzyme modification (48 and Chapters D1, D2 and E3). An argument can be made, based on kinetics, that IAA is somehow destroyed or deactivated immediately following, or concomitant with, the growth promoting act (18). This degradation or deactivation must be coupled with the tightly regulated, rapid turnover, usually associated with regulatory molecules like hormones (e.g., 17).

It is of interest that the enzyme oxidizing IAA requires an unsaturated fatty acid as cosubstrate so that the products of IAA oxidation are OxIAA and a prostaglandin (72). The physiological significance of prostaglandin formation in plants is not known. In animals, prostaglandins act as modulators of the response of other hormones, as well as participating in the contraction of smooth muscle, possibly by changing the cytosolic concentrations of secondary messengers such as Ca^{++}.

SUMMARY

The steady state levels of IAA are regulated by the "inputs to" the IAA pool including synthesis, conjugate hydrolysis and transport, and by the "outputs from" the IAA pool including oxidative catabolism, conjugate synthesis, transport of IAA away from the point of interest, and IAA "use" during growth. Work on the enzymes of IAA synthesis and metabolism is far from complete. Although the activity of many of the enzymes has been demonstrated *in vitro*, we still do not even know the primary pathways for *de novo* synthesis. It still remains necessary to establish the magnitude of the various routes of synthesis and metabolism. For higher plants, only the enzyme catalyzing the synthesis of IAGlu has been purified to homogeneity and sequenced. There remains the challenge of characterizing all of the enzymes involved in hormone homeostasis sufficiently to permit genetic and chemical manipulation of IAA levels within the plant.

Fig. 1 illustrates that key points for control of hormone levels could include: a) enzymes involved in the *de novo* synthesis of IAA; b) IAGluc synthetase since it is the first enzyme in the series of reactions leading to the IAA conjugates; c) 1-0, and 6-0, IAGluc hydrolases; d) IAInos synthetase since this enzyme shifts the equilibrium towards conjugate formation; e) the enzyme which oxidizes IAA to OxIAA since OxIAA itself is inactive as a growth regulator; and f) enzymes (carriers) involved in IAA or IAInos transport since these can affect local concentrations of IAA. Genetic and chemical knowledge of how these key reactions are controlled could lead to important agricultural applications for control of plant growth.

We predict that the agriculture of the future will use genetic and chemical manipulation of endogenous hormone levels to attain desirable levels of growth and differentiation. Ultimately it may be possible to regulate plant growth without reliance upon application of possibly hazardous growth-regulating chemicals.

Acknowledgments

Work in the laboratory of RSB at MSU was supported by the U.S. National Science Foundation; studies in the laboratory of JPS were supported by the U.S. National Science Foundation and USDA-ARS; and research in the laboratory of JDC was supported by USDA-ARS, USDA-NRI and US-Israel BARD.

References

1. Baldi, B.G., Maher, B.R., Slovin, J.P., Cohen, J.D. (1991) Stable isotope labeling *in vivo* of D- and L-tryptophan pools in *Lemna gibba* and the low incorporation of label into indole-3-acetic acid. Plant Physiol. 95, 1203-1208.
2. Bandurski, R.S., Desrosiers, M.F., Jensen, P., Pawlak, M., Schulze, A. (1992) Genetics, chemistry, and biochemical physiology in the study of hormonal homeostasis. *In*: Progress in Plant Growth Regulation, pp. 1-12, Karssen, C.M., Van Loon, L.C., Vreugdenhil, D., eds. Kluwer, Dordrecht, The Netherlands.
3. Bandurski, R.S., Nonhebel, H.M. (1984) Auxins. *In*: Advanced Plant Physiology, pp. 1-20, Wilkins, M.B., ed. Pitman, London.
4. Bandurski, R.S., Schulze, A., Desrosiers, M., Jensen, P., Epel, B., Reinecke, D. (1990) Relationship between stimuli, IAA, and growth. *In*: Plant Growth Substances 1988, pp. 341-352, Pharis, R.P., Rood, S.B., eds. Springer-Verlag, Heidelberg, Berlin.
5. Bandurski, R.S., Schulze, A., Desrosiers, M., Jensen, P., Reinecke, D., Epel, B. (1990) Voltage-gated channels as transducers of environmental stimuli. *In*: Inositol metabolism in plants, pp. 289-300, Morre, J., Boss, W., Loewus, F., eds. Wiley, New York.
6. Bandurski, R.S., Schulze, A., Jensen, P., Desrosiers, M., Epel, B., Kowalczyk, S. (1992) The mechanism by which an asymmetric distribution of plant growth hormone is attained. Adv. Space Res. 12, 203-210
7. Bandurski, R.S., Schulze, A., Leznicki, A., Reinecke, D.M., Jensen, P., Desrosiers, M., Epel, B. (1988) Regulation of the amount of IAA in seedling plants. *In*: Physiology and Biochemistry of Auxins in Plants, pp. 21-32, Kutacek, M., Bandurski, R.S., Krekule, R., eds. SPB Academic Publishing, The Hague.
8. Bialek, K., Cohen, J.D. (1984) Hydrolysis of an indole-3-acetic acid amino acid conjugate by an enzyme preparation from *Phaseolus vulgaris*. Plant Physiol. 75(S), 108.

9. Bialek, K., Cohen, J.D. (1986) Isolation and partial characterization of the major amide-linked conjugate of indole-3-acetic acid from *Phaseolus vulgaris* L. Plant Physiol. 80, 99-104.

10. Bialek, K., Cohen, J.D. (1989) Free and conjugated indole-3-acetic acid in developing bean seeds. Plant Physiol. 90, 398-400.

11. Bialek, K., Cohen, J.D. (1992) Amide-linked indoleacetic acid conjugates may control the levels of indoleacetic acid in germinating seedlings of *Phaseolus vulgaris*. Plant Physiol. 100, 2002-2007.

12. Bialek, K., Meudt, W.J., Cohen, J.D. (1983) Indole-3-acetic acid (IAA) and IAA conjugates applied to bean stem sections. IAA content and the growth response. Plant Physiol. 73, 130-134.

13. Bialek, K., Michalczuk, L., Cohen, J.D. (1992) Auxin biosynthesis during seed germination in *Phaseolus vulgaris*. Plant Physiol. 100, 509-517.

14. Blazich, F.A. (1988) Chemicals and formulations used to promote adventitious rooting. *In*: Adventitious Root Formation in Cuttings, pp 132-149. Davis, T.D., Haissig B.E., Sankhla, N., eds, Dioscorides Press, Portland, USA.

15. Chisnell, J.R. (1984) Myo-inositol esters of indole-3- acetic acid are endogenous components of *Zea mays* L. shoot tissue. Plant Physiol. 74, 278-283.

16. Cohen, J.D. (1982) Identification and quantitative analysis of indole-3-acetyl-L-aspartate from seeds of *Glycine max*. Plant Physiol. 70, 749-753.

17. Cohen, J.D. (1983) Metabolism of indole-3-acetic acid. What's New in Plant Physiol. 14, 41-44.

18. Cohen, J.D., Bandurski, R.S. (1978) The bound auxins: protection of indole-3-acetic acid from peroxidase-catalyzed oxidation. Planta 139, 203-208.

19. Cohen, J.D., Bandurski, R.S. (1982) Chemistry and physiology of the bound auxins. Ann. Rev. Plant Physiol. 33, 403-430.

20. Cooney, T.P., Nonhebel, H.M. (1991) Biosynthesis of indole-3-acetic acid in tomato shoots: measurement, mass-spectral identification and incorporation of ^{2}H from ^{2}H$_2$O into indole-3-acetic acid, D- and L-tryptophan, indole-3-pyruvate and tryptamine. Planta 184, 368-376.

21. Epel, B.L., Warmbrodt, R.P., Bandurski, R.S. (1992) Studies on the longitudinal and lateral transport of IAA in the shoots of etiolated corn seedlings. J. Plant Physiol. 140, 310-318

22. Epel, B.L., Bandurski, R.S. (1990) Tissue to tissue symplastic communication in the shoots of etiolated corn seedlings. Physiol. Plant. 79, 604-609.

23. Epstein, E., Cohen, J.D., Baldi, B.G. (1986) Identification of indole-3-acetylglutamate from seeds of *Glycine max* L. Plant Physiol. 80, 256-258.

24. Epstein, E., Cohen, J.D., Bandurski, R.S. (1980) Concentration and metabolic turnover of indoles in germinating kernels of *Zea mays* L. Plant Physiol. 65, 415-421.

25. Epstein, E., Cohen, J.D., Chen, K-H. (1989) Identification of indole-3-butyric acid as an endogenous constituent of maize kernels and leaves. Plant Growth Regulation 8, 215-223.

26. Epstein, E., Lavee, S. (1984) Conversion of indole-3-butyric acid to indole-3-acetic acid by cuttings of grapevine (*Vitis vinifera*) and olive (*Olea europea*). Plant Cell Physiol. 25, 697-703.

27. Ernstsen, A., Sandberg, G. (1988) Metabolism of indole-3-acetic acid in conifers. *In*: Physiology and Biochemistry of Auxins in Plants, pp. 47-55, Kutacek, M., Bandurski, R.S., Krekule, R., eds. SPB Academic Publishing, The Hague.

28. Evidente, A., Surico, G., Iacobellis N.S., Randazzo, G. (1986) α-N-Acetyl-indole-3-acetyl-ε-L-lysine: a metabolite of indole-3-acetic acid from *Pseudomonas syringae* pv. *savastanoi*. Phytochemistry 25, 125-128.

29. Gaspar, T., Hofinger, M. (1988) Auxin metabolism during adventitious rooting. *In*: Adventitious Root Formation in Cuttings, pp. 117-131, Davis, T.D., Haissig B.E., Sankhla, N., eds, Dioscorides Press, Portland, USA.

30. Glass, N.L., Kosuge, T. (1988) Role of indoleacetic acid-lysine synthetase in regulation of indoleacetic acid pool size and virulence of *Pseudomonas syringae*. subsp. *savastanoi*. J. Bacteriol. 170, 2367-2373

31. Greenberg, J.B., Galston, A.W., Shaw, K.N.F., Armstrong, M.D. (1957) Formation and auxin activity of indole-3-glycolic acid. Science 125, 992-993.

32. Hall, P.J., Bandurski, R.S. (1986) [³H]Indole-3-acetyl-*myo*-inositol hydrolysis by extracts of *Zea mays* L. vegetative tissue. Plant Physiol. 80, 374-377.

33. Hamilton, R.H., Bandurski, R.S., Grigsby, B.H. (1961) Isolation of indole-3-acetic acid from corn kernels and etiolated corn seedlings. Plant Physiol. 36, 354-359.

34. Hangarter, R.P., Good, N.E. (1981) Evidence that IAA conjugates are slow-release sources of free IAA in plant tissues. Plant Physiol. 68, 1424-1427.

35. Hinman, R.L., Lang, J. (1965) Peroxidase-catalyzed oxidation of indole-3-acetic acid. Biochem. 4, 144-158.

36. Hutzinger, O., Kosuge, T. (1968) 3-Indole-acetyl-ε-L-lysine, a new conjugate of 3-indoleacetic acid produced by *Pseudomonas savastanoi*. *In*: Biochemistry and Physiology of Plant Growth Substances, pp. 183-194, Wightman F., Setterfield, G., eds, Runge Press, Ottawa.

37. Jackson, D.L., McWha, J.A. (1983) Translocation and metabolism of endosperm applied 2-[¹⁴C] indoleacetic in etiolated *Avena sativa* L. seedlings. Plant Physiol. 73, 316-323.

38. Kesy, J.M., Bandurski., R.S. (1990) Partial purification and characterization of indol-3-ylacetylglucose:*myo*-inositol indol-3-ylacetyl transferase (indoleacetic acid-inositol synthase) Plant Physiol. 94, 1598-1604.

39. Kinashi, H., Suzuki, Y., Takeuchi, S., Kawarada, A. (1976) Possible metabolic intermediates from IAA to B-Acid in rice bran. Agr. Biol. Chem. 40, 2465-2470.

40. Klambt, H.D. (1959) Die 2-hydroxy-indol-3-essigsaure, ein pflanzliches indolderivat. Naturwiss 46, 649.

41. Klee, H., Estelle, M. (1991) Molecular genetic approaches to plant hormone biology. Ann. Rev. Plant Physiol. Plant Mol. Biol. 42, 529-551.

42. Klee, H., Horsch, R.B., Hinchee, M.A., Hein, M.B., Hoffman, N.L. (1987) The effects of overproduction of two *Agrobacterium tumefaciens* T-DNA auxin biosynthetic gene products in transgenic petunia plants. Genes Devel. 1, 86-96.

43. Komoszynski, M., Bandurski, R.S. (1986) Transport and metabolism of indole-3-acetyl-*myo*-inositol-galactoside in seedlings of *Zea mays*. Plant Physiol. 80, 961-964.

44. Kosuge, T., Sanger, M. (1986) Indoleacetic acid, its synthesis and regulation: A basis for tumorigenicity in plant disease. *In*: The Shikimic Acid Pathway (Recent Advances in Phytochemistry, Vol. 20) pp. 147-161, Conn, E.E., ed, Plenum Press, NY.

45. Kowalczyk, S., Bandurski R.S. (1990) Isomerization of 1-0-indol-3-ylacetyl-ß-D- glucose. Enzymatic hydrolysis of 1-0, 4-0, and 6-0-indol-3-ylacetyl-ß-D-glucose and the enzymatic synthesis of indole-3-acetyl glycerol by a hormone metabolizing complex. Plant Physiol. 94, 4-12.

46. Kowalczyk, S., Bandurski R.S. (1991) Enzymatic synthesis of 1-0-(indol-3-ylacetyl)-ß-D-glucose. Purification of the enzyme from *Zea mays* and preparation of antibodies to the enzyme. Biochem. J. 279, 509-514.

47. Kuleck G.A., Cohen J.D. (1992) The partial purification and characterization of IAA-alanine hydrolase from *Daucus carota*. Plant Physiol. 99(S), 18.

48. Kutacek, M., Elliott, M.C., Machackova, I. (1990) Molecular Aspects of Hormonal Regulation of Plant Development. SPB Academic Publishing, The Hague. 251pp.

49. Lagrimini, M.L. (1991) Peroxidase, IAA oxidase and auxin metabolism in transformed tobacco plants. Plant Physiol. 96(S), 1.

50. Law, D.M. (1987) Gibberellin-enhanced indole-3-acetic acid biosynthesis: D-tryptophan as the precursor of indole-3-acetic acid. Physiol. Plant. 70, 626-632.

51. Law, D.M., Hamilton, R.H. (1984) Effects of gibberellic acid on endogenous indole-3-acetic acid levels in dwarf pea. Plant Physiol. 75, 255-256.
52. Lewer, P., Bandurski, R.S. (1987) Occurrence and metabolism of 7-hydroxy-2-indolinone-3-acetic acid in *Zea mays*. Phytochemistry 26, 1247-1250.
53. Ludwig-Muller, J., Epstein, E. (1991) Occurrence and *in vivo* biosynthesis of indole-3-butyric acid in corn (*Zea mays* L.). Plant Physiol. 97, 765-770.
54. Ludwig-Muller, J., Hilgenberg, W. (1989) *N*-Malonyltryptophan metabolism by seedlings of Chinese cabbage. Phytochemistry 28, 2571-2575.
55. Magnus V., Iskric, S., Kveder, S. (1971) Indole-3-methanol: A metabolite of indole-3-acetic acid in pea seedlings. Planta 97, 116-125.
56. Manulis, S., Valinski, L., Gafni, Y., Hershenhorn, J. (1991) Indole-3-acetic acid biosynthetic pathways in *Erwinia herbicola* in relation to pathogenicity on *Gypsophila paniculata*. Physiol. Molec. Plant Pathol. 39, 161-171.
57. Marumo, S. (1986) Auxins. *In*: Chemistry of Plant Hormones, pp 9-56, Takahashi, N., ed. CRC Press, Boca Raton, FL.
58. Michalczuk, L., Cooke, T.J., Cohen , J.D. (1992) Auxin levels at different stages of carrot somatic embryogenesis. Phytochemistry 31, 1097-1103.
59. Michalczuk, L., Ribnicky, D.M., Cooke, T.J., Cohen, J.D. (1992) Regulation of indole-3-acetic biosynthetic pathways in carrot cell cultures. Plant Physiol. 100, 1346-1353.
60. Nonhebel H., Bandurski, R.S. (1984) Oxidation of indole-3-acetic acid and oxindole-3-acetic acid to 2,3-dihydro-7-hydroxy-2-oxo-1H indole-3-acetic acid 7-O-β-D-glucopyranoside in *Zea mays* seedlings. Plant Physiol. 76, 979-983.
61. Nonhebel, H.M., Hillman, J.R., Crozier, A., Wilkins, M.B. (1985) Metabolism of [^{14}C] indole-3-acetic acid by coleoptiles of *Zea mays* L. J. Exp. Bot. 36, 99-109.
62. Nonhebel, A.M. (1986) Measurement of the rate of oxindole-3-acetic acid turnover, and indole-3-acetic acid oxidation in *Zea mays* seedlings. J. Exp. Bot. 37, 1691-1697.
63. Nonhebel, H.M., Kruse, L.I., Bandurski, R.S. (1985) Indole-3-acetic acid catabolism in *Zea mays* seedlings. Metabolic conversion of oxindole-3-acetic acid to 7-hydroxy-2-oxindole-3-acetic acid-7'-O-ß-D-glucopyranoside. J. Biol. Chem. 260, 12685-12689.
64. Normanly, J., Cohen, J.D., Fink, G.R. (1993) A tryptophan auxotroph reveals two indole-3-acetic acid biosynthetic pathways in *Arabidopsis thaliana*. Proc. Natl. Acad. Sci. USA 90, 10355-10359.
65. Nowacki, J., Bandurski, R.S. (1980) Myo-inositol esters of indole-3-acetic acid as seed auxin precursors of *Zea mays* L. Plant Physiol. 65, 422-427.
66. Ozga, J.A., Reinecke, D.M., Brenner, M.L. (1993) Quantitation of 4-Cl-IAA and IAA in 6 DAA pea seeds and pericarp. Plant Physiol. 102(S), 7.
67. Park, R.D., Park, C.K. (1987) Oxidation of indole-3-acetic acid-amino acid conjugates by horseradish peroxidase. Plant Physiol. 84, 826-829.
68. Pengelly, W.L., Bandurski, R.S. (1983) Analysis of indole-3-acetic acid metabolism in *Zea mays* using deuterium oxide as tracer. Plant Physiol. 73, 445-449.
69. Rekoslavskaya, N.I. (1986) Possible role of N-malonyl-D-tryptophan as an auxin precursor. Biol. Plant. 28, 62-67.
70. Rekoslavskaya, N.I., Jensen, P.J., Bandurski, R.S. (1992) IAA biosynthesis in maturing corn endosperm. Plant Physiol. 99(S), 17.
71. Roberto, F.F., Klee, H., White, F., Nordeen, R., Kosuge, T. (1990) Expression and fine structure of the gene encoding N-ε-(indole-3-acetyl)-L-lysine synthase from *Pseudomonas savastanoi*. Proc. Natl. Acad. Sci. USA 87, 5797-5801.
72. Reinecke, D. M. (1990) The oxindole-3-acetic acid pathway in *Zea mays*. *In*: Plant Growth Substances 1988, pp. 367-373, Pharis R.P., Rood, S.B., eds. Springer-Verlag, Berlin.

73. Sakagami, Y., Manabe, K., Marumo, S. (1993) Assignment of the L-configuration to 4-Cl-trp and trp of their N-malonyl conjugates, and biosynthetic studies of 4-Cl-IAA from L- or D-4-Cl-trp. Abstracts (#4166), XV International Botanical Congress, Yokohama, Japan, August 28-September 3, 1993, pp 390.

74. Schuytema, E.C., Hargie, M.P., Merits, I., Schenck, J.R., Siehr, D.J., Smith, M.S., Varner, E.L. (1966) Isolation, characterization, and growth of basidiomycetes. Biotech. & Bioeng. 8, 275-286.

75. Sembdner, G., Gross, D., Liebisch, H-W., Schneider, G. (1981) Biosynthesis and metabolism of plant hormones. *In*: Hormonal Regulation of Development. I. Molecular Aspects of Plant Hormones (Encyclopedia of Plant Physiology Vol. 9), pp. 281-444, MacMillan, J., ed. Springer, Berlin.

76. Sitbon, F., Sundberg, B., Olsson, O., Sandberg, G. (1991) Free and conjugated indoleacetic acid (IAA) contents in transgenic tobacco plants expressing the *iaaM* and *iaaH* IAA biosynthesis genes from *Agrobacterium tumefaciens*. Plant Physiol. 95, 480-486.

77. Slovin, J.P., Cohen, J.D. (1993) Auxin metabolism in relation to fruit ripening. Acta Hort. 329, 84-89.

78. Thimann, K.V. (1934) Studies on the growth hormone of plants. VI. The distribution of the growth substance in plant tissues. J. Gen. Phys. 18, 23-34.

79. Town, C.D., Campell, B.R., Persinger, S.M., Campanella, J.J. (1993) Use of radiation-induced tumors to study growth control in *Arabidopsis* (Abstract). *In Vitro* Cellular & Developmental Biology - Animal 29A, 36A.

80. Tsurumi, S., Wada, S. (1990) Oxidation of indole-3-acetylaspartic acid in *Vicia*. *In*: Plant Growth Substances 1988, pp. 353-359, Pharis R.P., Rood, S.B. eds. Springer-Verlag, Berlin.

81. Wright A.D., Sampson, M.B., Neuffer, M.G., Michalczuk, L, Slovin, J.P., Cohen, J.D. (1991) Indole-3-acetic acid biosynthesis in the mutant maize *orange pericarp*, a tryptophan auxotroph. Science 254, 998-1000

B2. The Biosynthesis and Metabolism of Gibberellins in Higher Plants

Valerie M. Sponsel
Biology Department, Indiana University, Bloomington, IN 47405

INTRODUCTION

The gibberellins (GAs) are all tetracyclic diterpenoid acids with the *ent -* gibberellane ring system (structure 1). There are two main types of GAs, the C_{20}-GAs which have a full complement of 20 carbon atoms (structure 2), and the C_{19}-GAs in which the twentieth carbon atom has been lost by metabolism (structure 3). In almost all of the C_{19}-GAs, the carboxylic acid at carbon-19 bonds to carbon-10 to give a lactone bridge.

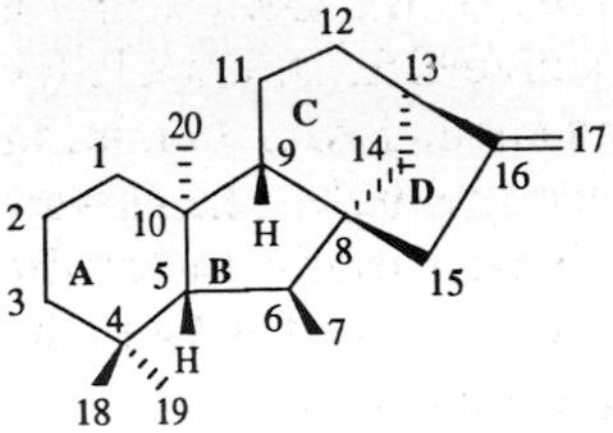

(1) 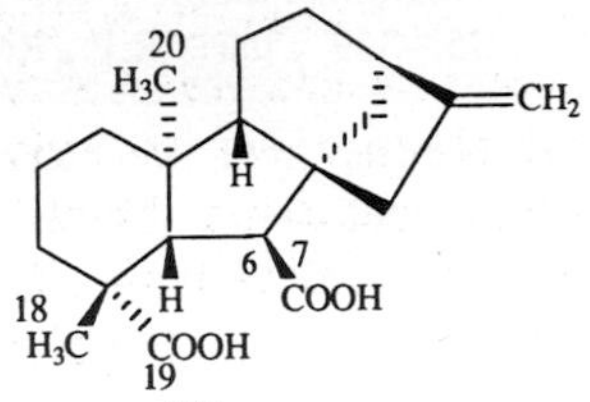 (2) 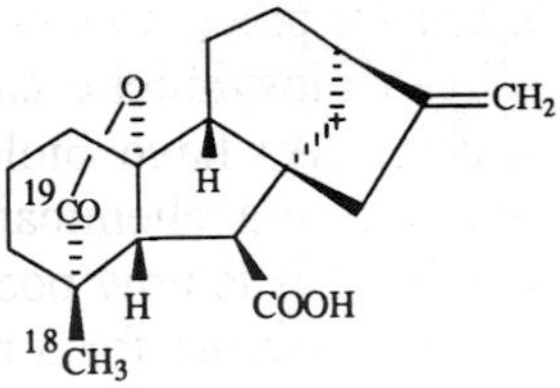 (3)

Many structural modifications can be made to the *ent*-gibberellane ring system. This diversity accounts for the large number of known GAs. Each different GA which is found to be naturally occurring and whose structure has been chemically characterized is allocated an "A number" (36). Eighty nine GAs are known to date, and these are numbered GA_1 through GA_{89} in approximate order of their discovery. Their structures are shown on pages 93-97.

Variations in GA structure arises in several ways. For instance, different oxidative states of carbon-20, namely methyl (CH_3), hydroxymethyl (CH_2OH), aldehyde (CHO) or carboxylic acid (COOH) occur. Additional functional groups can also be added to the *ent*-gibberellane skeleton, especially to the C_{19}-GAs. Insertion into the ring system of hydroxyl (OH) groups occurs frequently, and less common substituents are epoxide (>O) or ketone (=O) functions. The position and/or stereochemistry of these functional groups is very important. For example, the presence of a hydroxyl group in α- or β-stereochemistry (designated by ⸱⸱⸱ıllıl or ◄ bonds, respectively) can have quite different biochemical significance.

66

P. J. Davies (ed.), Plant Hormones, 66–97.
© 1995 *Kluwer Academic Publishers. Printed in the Netherlands.*

Gibberellins were first isolated from the fungus *Gibberella fujikuroi* in which they occur in large quantities as secondary metabolites. They are now known to be present in several other species of fungus, in some ferns, and in many gymno- and angiosperms. Of the 89 known GAs, 64 have been identified only in higher plants, 12 are present only in *Gibberella,* and 13 are present in both. Like *Gibberella,* individual angiosperms can contain many different GAs. For instance, immature seeds of apple (*Malus domestica*) contain 24 known GAs, together with several additional GA-like compounds which have not yet been fully characterized (27).

Gibberellins can also exist as conjugates. The most common naturally occurring GA-conjugates are those in which the GA is linked to a molecule of glucose, either by an ether or an ester linkage (42). Conjugates may be formed in order to inactivate a GA, either temporarily or permanently.

Gibberellic acid (GA_3), which is the end-product of GA metabolism in *G. fujikuroi,* has been commercially available for many years. It has high biological activity, and its application to dwarf or rosette plants, dormant buds, or dormant and germinating seeds can result in dramatic and diverse effects on growth (see Chapter A1).

Not all GAs have high biological activity. Many of the GAs within a plant are precursors or deactivation products of the active GA. A knowledge of GA biosynthetic and metabolic pathways is fundamental to determining which GAs have biological activity *per se.* By using single gene dwarf mutants and chemical growth retardants which inhibit specific metabolic reactions it is now possible to determine which GA(s) in a plant is (are) the active hormone for a particular growth or developmental event.

SITES OF GA BIOSYNTHESIS

All growing, differentiated tissues are potential sites of GA biosynthesis. There is incontrovertible evidence that developing fruits and seeds are sites of GA biosynthesis, for they contain enzymes which can convert mevalonic acid to C_{19}-GAs (18, 29). In immature seeds there are two main phases of GA biosynthesis. The first phase occurs shortly after anthesis and is correlated with fruit growth. At this stage of development seeds are very small, and contain GAs which are qualitatively and quantitatively similar to those in vegetative tissues. The second phase of GA biosynthesis occurs as the maturing seeds are increasing in size, and it results in a large accumulation of GAs. In most species chosen for study, the seeds at this stage of development are large enough to handle with ease and can readily be separated into constituent parts. Most studies on GA biosynthesis have utilized cell-free preparations. Their use obviates problems associated with substrate penetration into tissue, and in addition it provides an opportunity to study the biochemical properties of the enzymes involved. Preparations derived from the liquid endosperm of developing cucurbit seeds and from cotyledons of legumes are particularly active.

Definitive evidence that GA biosynthesis occurs in vegetative tissues is more difficult to obtain because cell-free enzyme preparations derived from vegetative tissues frequently have low GA-biosynthetic capability. However, as several GA metabolic sequences can be demonstrated in elongating internodes and petioles, expanding leaves and stem apical regions of several plants (15, 52) it is generally accepted that these immature organs of the shoot are sites of GA biosynthesis.

GAs have been identified in root extracts. Nevertheless, evidence for GA biosynthesis in roots is tenuous, although GAs have been identified in tomato roots which had been subcultured enough times to preclude carry-over from the initial inoculum (7).

THE GIBBERELLIN BIOSYNTHESIS PATHWAY

The following sections provide a brief outline of the GA biosynthetic pathway from mevalonic acid to the first formed GA which is GA_{12}-aldehyde. This part of the pathway is the same in *G. fujikuroi* and all higher plants examined so far. The characterization of individual stages in the pathway is well covered in review articles (8, 16, 24). The metabolism after GA_{12}-aldehyde is discussed in more detail in subsequent sections.

The Pathway from Mevalonic Acid to Geranylgeranyl Pyrophosphate (GGPP) (Fig. 1)

In the terpenoid pathway C_5-building blocks are linked head-to-tail to give branched polymers of different chain length, which can then undergo cyclization and other changes. In this way natural products such as monoterpenes (C_{10}), sesquiterpenes (C_{15}, from the Latin *sesqui* = 1.5), diterpenes (C_{20}), triterpenes (C_{30}) etc. are formed.

The diterpenoid nature of GAs was recognized by Cross and associates (9). Early steps in the terpenoid pathway have been characterized in plant, animal and bacterial systems (4). Birch et al. (5) were the first to show that [^{14}C]-labelled acetate was converted to [^{14}C]GA_3 by cultures of *G. fujikuroi*, and mevalonic acid (MVA) was found to be a key intermediate (see Fig. 1). MVA is formed by the reduction of hydroxymethylglutaryl-coenzyme A (HMGCoA). Both HMGCoA synthase and reductase are important enzymes which are known in mammals to be highly regulated by feedback control by MVA and steroids. HMGCoA reductase has been studied in several plant systems (3) and has been shown to be tissue-specific and developmentally regulated.

MVA is converted to mevalonate-5-pyrophosphate (MVAPP) in two steps which are catalysed by mevalonic kinases. Next, MVAPP is decarboxylated to isopentenyl pyrophosphate (IPP). Reversible isomerization

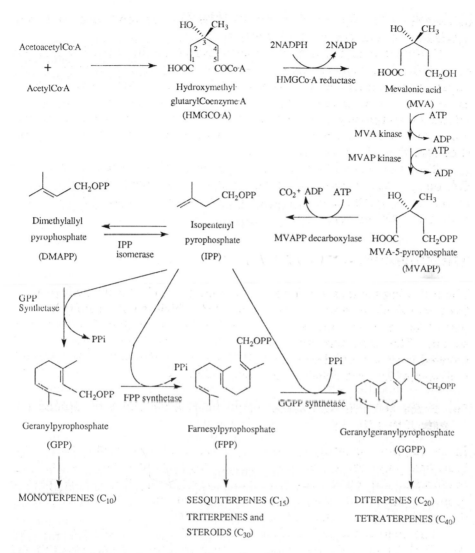

Fig. 1. The pathway to geranylgeranyl pyrophosphate.

of IPP gives dimethylallyl pyrophosphate (DMAPP). DMAPP is the starter unit for terpene biosynthesis. It condenses in a head-to-tail fashion with a molecule of IPP to give the C_{10}-intermediate geranyl pyrophosphate (GPP). This in turn condenses with another molecule of IPP to give farnesyl pyrophosphate (FPP). Further condensation of FPP with a third molecule of IPP gives the C_{20}-intermediate geranylgeranyl pyrophosphate (GGPP), which is the parent compound for all diterpenes. The three condensation reactions are catalysed by the same enzyme, namely GGPP synthetase, which belongs to a class of enzymes named prenyl transferases. Branch points on the pathway which lead to mono- and triterpenes are indicated in Fig. 1.

Cyclization of GGPP to *ent*-Kaurene (Fig. 2)

At GGPP the terpenoid pathway branches to give linear diterpenes (e.g., phytol), cyclic diterpenes (e.g., kaurenoids and gibberellins) and tetraterpenes (e.g., carotenoids). The cyclization of GGPP to the gibberellin precursor, *ent*-kaurene, is a two-stage reaction catalysed by the soluble enzymes, *ent*-kaurene synthetase A and *ent*-kaurene synthetase B. The A activity catalyses the conversion of GGPP to a bicyclic intermediate, copalyl pyrophosphate (CPP), whereas the B activity catalyses the further conversion of CPP to the tetracyclic diterpene, *ent*-kaurene[1] (Fig. 2). The mechanism of these cyclization reactions has been discussed in detail (24).

The *ent*-kaurene synthetase enzymes have been studied extensively. The A and B activities have proved difficult to separate physically, although it has long been known that they possess different pH optima and different divalent cation requirements. Evidence suggests that the two proteins associate during *ent*-kaurene synthesis (13).

The first cyclization reaction in the pathway, catalysed by *ent*-kaurene synthetase A, is the first committed step in the synthesis of cyclic diterpenes and as such is a potential site for regulation of the pathway. Numerous attempts have therefore been made to isolate the A activity of *ent*-kaurene synthetase but these attempts have been largely unsuccessful due to its instability and susceptibility to natural and synthetic inhibitors. However, the availability of mutant lines of several genera in which the synthesis of *ent*-kaurene is modified offers an alternative and attractive means to study this part of the GA pathway, especially in those genera such as *Arabidopsis thaliana* which are amenable to molecular genetic work.

In the *ga1* and *ga2* dwarf mutants of *Arabidopsis* GA biosynthesis is thought to be blocked before *ent*-kaurene, because only compounds beyond *ent*-kaurene in the pathway elicit a growth response when applied to mutant plants. Furthermore, because both mutants possess normal pigmentation the pathway up to GGPP is assumed to be intact. On the basis of this evidence the lesion in both mutants is inferred to be at *ent*-kaurene synthetase (53). However, although plants of both genotypes were shown to contain very reduced levels of native GAs, cell free preparations from siliques of the *ga1* mutant possessed normal *ent*-kaurene synthetase A and B activity (53). The *GA1* locus may therefore code for a regulatory factor, or for a carrier protein, rather than for the enzyme itself. Alternatively, GA biosynthesis in shoots and siliques may be controlled independently. The recent cloning of the *GA1* locus in *Arabidopsis* by genomic subtraction (50) will resolve these questions, and allow the regulation of *ent*-kaurene synthetase to be studied directly.

---

[1] Using IUPAC nomenclature the precursor of GAs is the enantiomeric form of kaurene, designated *ent*-kaurene. By convention α-substituents are designated *ent*-ß, and ß-substituents are designated *ent*-α.

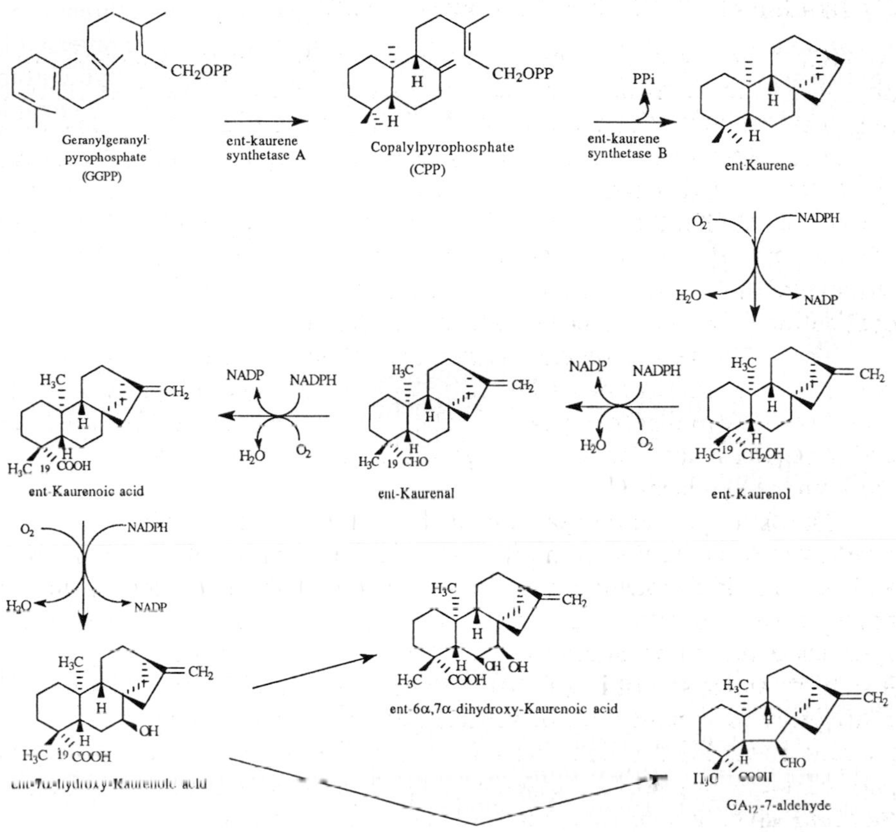

Fig. 2. The pathway from geranylgeranyl pyrophosphate to GA_{12}-aldehyde.

ent-**Kaurene to** GA_{12}-**aldehyde**

In the next part of the pathway, also shown in Fig.2, the methyl group at carbon-19 of *ent*-kaurene is oxidized in the sequence $CH_3 \rightarrow CH_2OH \rightarrow CHO \rightarrow COOH$, to give *ent*-kaurenol, *ent*-kaurenal and *ent*-kaurenoic acid. This sequence of oxidations is actually a series of successive hydroxylations so that all the reactions may be catalysed by a single enzyme active site. Evidence for and against this idea has been reviewed (24), leading to the conclusion that there are probably separate but similar catalytic sites for each substrate.

The enzymes involved in this series of oxidations have been studied in detail in several plants (22, 23). The enzymes are microsomal and each oxidative step requires molecular oxygen and a reduced pyridine nucleotide (e.g., NADPH) for activity. These properties are characteristic of the mixed-function oxidase or mono-oxygenase enzymes. This type of enzyme catalyses the insertion of one oxygen atom from molecular oxygen into an organic substrate: the second oxygen atom is reduced to water by electrons derived from a second substrate (e.g., NADPH). In the cucurbit, *Marah macrocarpus*, inhibition of the oxidation of *ent*-kaurene to *ent*-kaurenol, and of *ent*-kaurenal to *ent*-kaurenoic acid by carbon monoxide can be reversed by light of wavelength around 450 nm. Thus in the *Marah* system cytochrome P-450 is implicated as the electron acceptor at the active site of the enzyme.

Plant cytochrome P-450 enzymes catalyse the oxidative metabolism of many important plant components. In addition to catalysing several steps in GA biosynthesis, they have been implicated in the biosynthesis of lignin phenolics, membrane sterols, phytoalexins, monoterpenes and indole alkaloids (12). They also catalyse the detoxification of several herbicides. The sequences of many cytochrome P-450 enzymes from microorganisms and vertebrates are known (12). At the present time less information is available for enzymes from plant origin, although several have been sequenced. One cytochrome P450 from avocado (6) functions as a demethylase in herbicide metabolism, although its natural role may be as a hydroxylase in the terpenoid pathway. Work on the solubilization and purification of GA monooxygenases, particularly *ent*-kaurene oxidase, is proceeding in several laboratories with the goal of cloning the gene(s) for these important enzymes.

The GA biosynthetic pathway proceeds with hydroxylation of *ent*-kaurenoic acid at the 7β-position to give *ent*-7α-hydroxykaurenoic acid. (Alternatively *ent*-kaurenoic acid can be directed towards the synthesis of kaurenolide derivatives which accumulate in *G. fujikuroi* and seeds of some higher plants) (16).

At *ent*-7α-hydroxykaurenoic acid the pathway diverges once again. This branch-point is of critical importance. One of the products, GA_{12}-aldehyde, is the first-formed GA in all systems. It is formed by contraction of the B ring with extrusion of carbon-7. In contrast, the other product, *ent*-6α,7α-dihydroxykaurenoic acid, cannot be converted to GAs, nor has it any known function in plants. The enzymology of this reaction has been studied in detail by Graebe and co-workers, leading to the conclusion that both GA_{12}-aldehyde and *ent*-6α, 7α-dihydroxykaurenoic acid are probably formed from a single common intermediate (16).

Pathways from GA_{12}-aldehyde (Fig. 3)

The biosynthetic pathway up to GA_{12}-aldehyde appears to be the same in all plants. In contrast, the conversion for GA_{12}-aldehyde to other GAs can vary from genus to genus, consequently giving rise to several alternate pathways from GA_{12}-aldehyde. There is, however, a basic sequence of reactions starting from GA_{12}-aldehyde which is common to all pathways. This sequence, which is outlined below, involves the successive oxidation of carbon-20 leading to its elimination from the molecule as CO_2 and the formation of C_{19}-GAs.

Gibberellin A_{12}-aldehyde is first oxidized at carbon-7 to give the dicarboxylic acid, GA_{12} (see Fig. 3). A carboxyl group at C-7 is a feature of all GAs, and it is essential for biological activity. Next comes the sequential

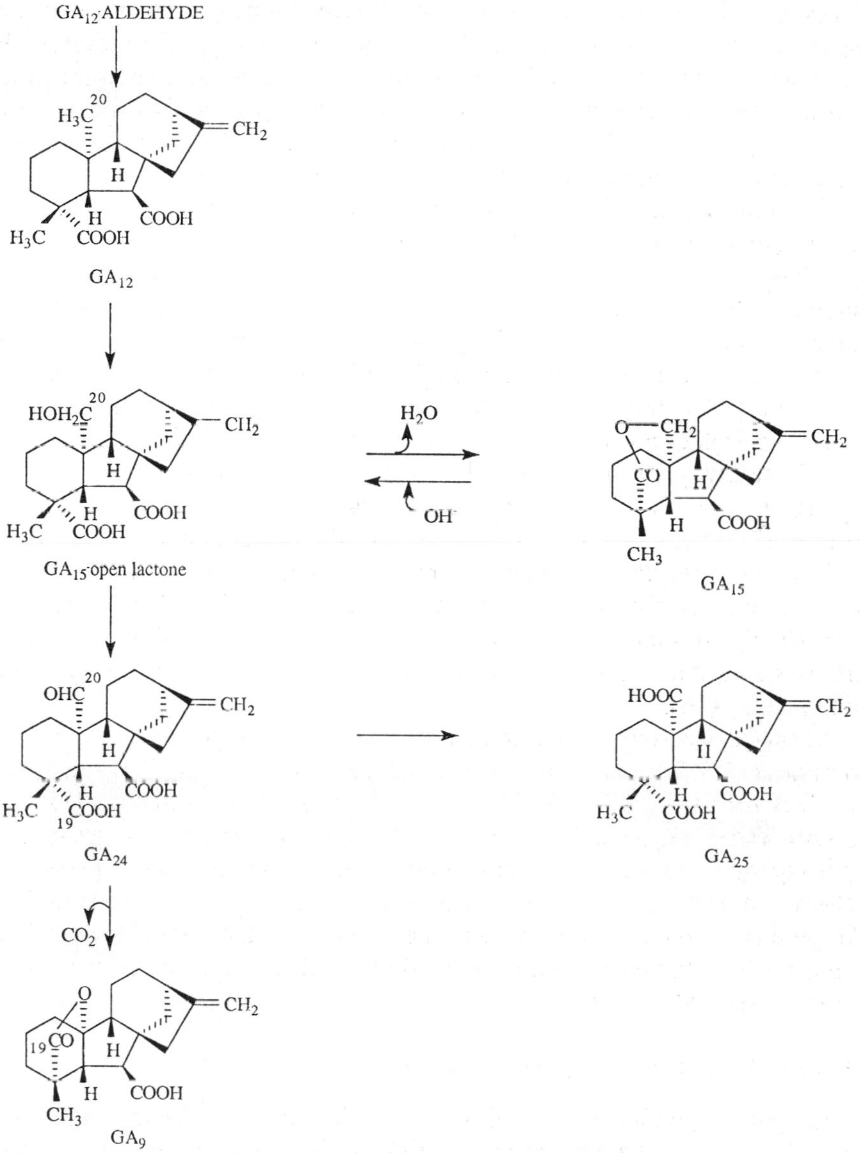

Fig. 3. The pathway from GA_{12}-aldehyde to the C_{19}-GA, GA_9.

oxidation of carbon-20. First the C-20 methyl group of GA_{12} is oxidized to a hydroxymethyl group (CH_2OH), which lactonises on extraction and work-up to yield GA_{15}. The true intermediate is probably the open-lactone form. In the next step of the pathway this intermediate is oxidized to the C-20 aldehyde, GA_{24}. At GA_{24} there is another branch-point in the pathway. Carbon-20 can be oxidized to the acid, giving GA_{25}. Alternately, and more importantly, carbon-20 can be eliminated from the molecule as CO_2, resulting

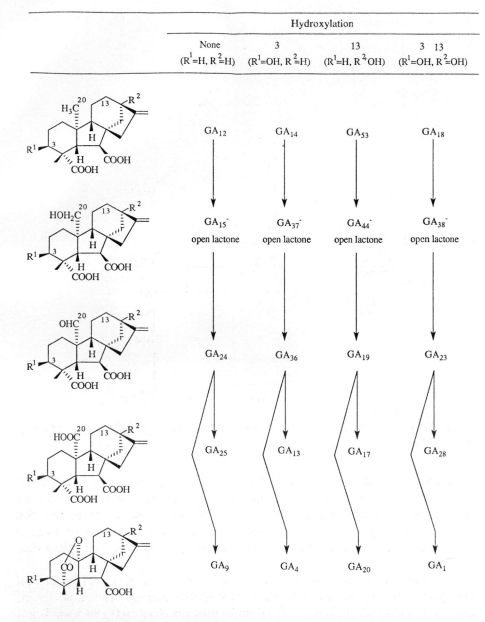

	Hydroxylation			
	None $(R^1=H, R^2=H)$	3 $(R^1=OH, R^2=H)$	13 $(R^1=H, R^2OH)$	3 13 $(R^1=OH, R^2=OH)$
	GA_{12}	GA_{14}	GA_{53}	GA_{18}
	GA_{15}^- open lactone	GA_{37}^- open lactone	GA_{44}^- open lactone	GA_{38}^- open lactone
	GA_{24}	GA_{36}	GA_{19}	GA_{23}
	GA_{25}	GA_{13}	GA_{17}	GA_{28}
	GA_9	GA_4	GA_{20}	GA_1

Fig. 4. Metabolic pathways to C_{19}-GAs.

in the formation a C_{19}-GA, GA_9 (11, 31). It is the C_{19}-GAs which have biological activity (see later).

The introduction of additional functional groups into the GA molecule can occur at any stage in this sequence of reactions. It is the position and order of insertion of all these additional substituents which distinguishes the metabolic pathways in different genera. Thus if GA_{12} were to be

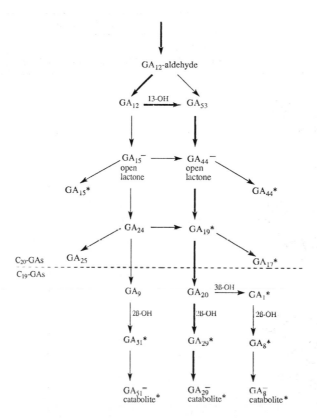

Fig. 5. GA metabolism in *Pisum sativum* L (pea). Thickened arrows refer to major pathway. 3β-Hydroxylation occurs only in shoots. (* = endogenous GA.)

hydroxylated at carbon-3, or at carbon-13, or at both the 3 and 13 positions then the GAs shown in Fig. 4 are formed on three separate pathways comparable to the route from $GA_{12} \rightarrow GA_9$ discussed above. In reality, the observed pathways are not always separate. They may converge or diverge to form a metabolic grid depending on the timing of hydroxylation (see Fig. 7).

The following discussion of GA metabolism is selective, and centres on two genera for which most information is available, namely pea and pumpkin, and on *Arabidopsis thaliana*. The data for pea are most straightforward and are considered first. Most recent work has concerned the purification of enzyme activities in order to characterise the enzymes, sequence the proteins and clone the corresponding genes. Progress towards these goals is described.

Pathways in *Pisum sativum*

There are two parallel GA metabolic pathways in pea seeds. One route, identical to the pathway shown in Fig. 4, leads to GA_9 whilst the second

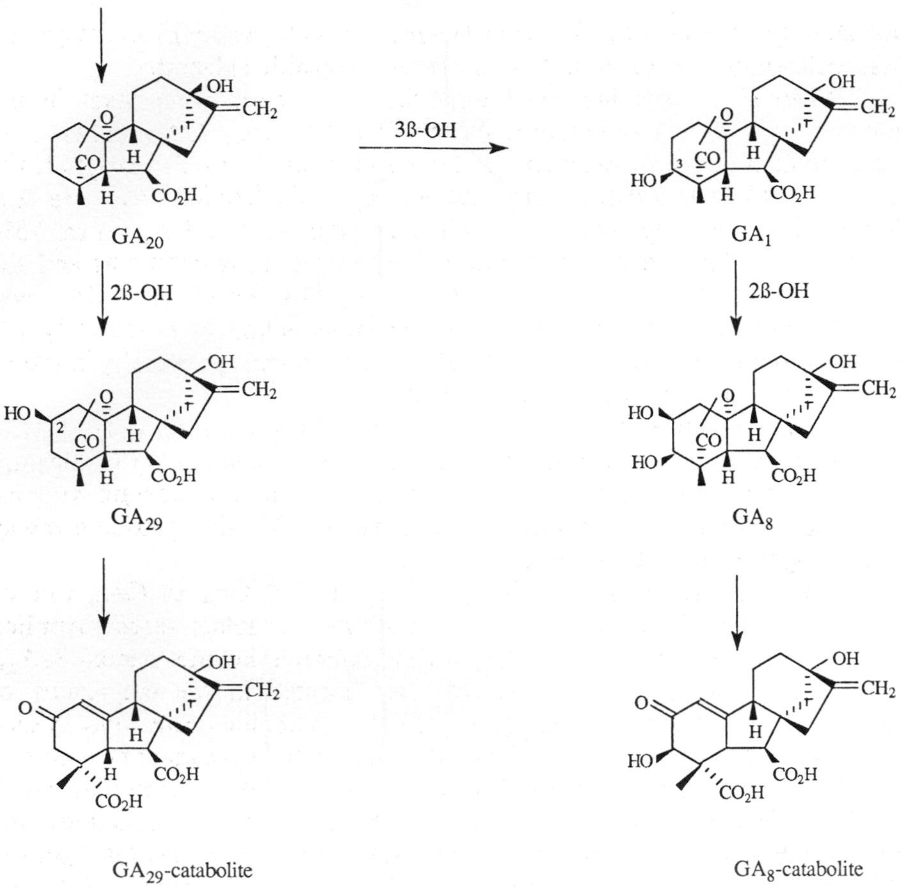

Fig. 6. The metabolism of GA_{20} in shoots of *Pisum sativum*.

pathway leads to the 13-hydroxylated GA, GA_{20} (see Fig. 5). The pathways were established by comprehensive feeding and refeeding studies using cell-free preparations, coupled with investigations into the identity of native GAs in pea seeds. The work is described here in some detail.

A high speed supernatant preparation from 18-day old pea embryos, together with Fe^{2+} and either NADPH or ascorbate, converted radio-labelled GA_{12} to GA_{15}, GA_{24}, GA_9 and GA_{51} (Fig. 3) (29). Gibberellin A_{15} was not metabolized but the open-lactone form, prepared by alkaline treatment of GA_{15}, was readily converted to GA_{24}, GA_9 and traces of GA_{51}. In turn GA_{24} was converted to GA_9 and some GA_{51}. Subsequently when GA_9 was incubated with a high speed supernatant from older embryos it was 2β-hydroxylated in good yield to give GA_{51}. Thus the pathway $GA_{12} \rightarrow GA_{15}$-open lactone $\rightarrow GA_{24} \rightarrow GA_9 \rightarrow GA_{51}$ was established (Fig. 5).

Alternatively, a low speed supernatant or washed microsomal pellet 13-hydroxylated either GA_{12}-aldehyde or GA_{12} to give GA_{53} (Fig. 5) (29). GA_{12} was the better substrate. Both NADPH and molecular oxygen are essential for activity. Gibberellin A_{15}-open lactone and GA_{24} were 13-hydroxylated less efficiently. Gibberellin A_9 was not an acceptable substrate.

When GA_{53} was incubated with the high speed supernatant in the presence of Fe^{2+} and ascorbate it was converted to GA_{44}, GA_{19} and GA_{20} in an analogous series of reactions to those shown in Figure 3, except all the GAs are 13-hydroxylated (Figs. 5 and 6). Gibberellin A_{20} was 2β-hydroxylated by preparations from older embryos to give GA_{29} in ca. 75% yield. Individual steps were confirmed by refeeding experiments and the major pathway $GA_{53} \rightarrow GA_{44}$-open lactone $\rightarrow GA_{19} \rightarrow GA_{20} \rightarrow GA_{29}$ was established (Figs. 5 and 6) (29). This pathways is known as the early 13-hydroxylation pathway and is a particularly common one, especially in shoots of many mono- and dicotyledonous plants.

Seven GAs are known to be native to developing pea seeds (48). All but one were observed as metabolites in cell-free preparations (29) confirming that the two pathways demonstrated *in vitro* do indeed operate *in vivo*. Quantitation of native GAs confirms that the early 13-hydroxylation pathway is the major route to C_{19}-GAs in pea seeds.

The 2β-hydroxylation of GA_{20} to GA_{29} and of GA_9 to GA_{51} can be observed *in vivo* by injecting labelled substrate into intact seeds. Applied GA_{29} is further converted to an α,β-unsaturated ketone named GA_{29}-catabolite. Although GA_{20} and GA_{29} are located in the cotyledons of maturing seeds, the conversion of GA_{29} to GA_{29}-catabolite occurs only in seed coats (47). This observation explains why GA catabolism cannot be observed in cell-free preparations derived from pea embryos (29). Gibberellins which are 2β-hydroxylated (e.g. GA_{29} and GA_{51}) have low biological activity and their catabolism may constitute a means for their removal and/or disposal. An alternative method for the removal of 2β-hydroxylated GAs by conjugation to glucose appears to be of minor importance in pea seeds.

The soluble enzymes involved in GA metabolism have the properties of 2-oxoglutarate dependent dioxygenases (26, 29). These enzymes have an absolute requirement for 2-oxoglutarate, which acts as a co-substrate, and for molecular oxygen and Fe^{2+}. Ascorbate is also required. In the overall reaction catalysed by this type of dioxygenase one oxygen atom oxidises 2-oxoglutarate to succinate and CO_2 while the other oxygen atom hydroxylates the primary substrate (i.e., in this case the GA). 2-Oxoglutarate-dependent-dioxygenases play a central role in primary and secondary metabolism in plants, animals and micro-organisms (40).

A catalytic mechanism for prolyl 4-hydroxylase, a 2-oxoglutarate-dependent-dioxygenase which catalyses the formation of 4-hydroxyproline in collagen, was proposed by Hanauske-Abel and Gunzler (21). Ferrous iron (Fe^{2+}) is bound at the enzyme active site. The formation of an iron-oxygen complex is essential for oxidative decarboxylation of 2-oxoglutarate to give succinate. This leads to the generation of a highly reactive ferryl ion which hydroxylates the primary substrate. Divalent iron remains bound to the enzyme's active site after the release of hydroxylated substrate, succinic acid and CO_2. Some decarboxylation of 2-oxoglutarate which is uncoupled from substrate hydroxylation can also occur because of the generation of Fe^{3+} which blocks the enzyme's active site. Ascorbate, in its role as a cofactor, functions to reduce this enzyme-bound Fe^{3+} back to Fe^{2+}, consequently allowing the coupled decarboxylation/hydroxylation to proceed.

A 20-oxidase, which catalyses the conversion of GA_{53} to GA_{44} has been partially purified from 20 day old pea embryos (32). The purified preparation also catalysed the conversion of GA_{44} to GA_{19}. The relative proportions of GA_{44} and GA_{19} remained constant throughout the purification procedure raising the possibility that a single enzyme catalyses the two reactions. An alternate explanation, namely that two enzymes were being co-purified, cannot be discounted as even the final enzyme preparation consisted of several proteins. The properties of the purified 20-oxidase were consistent with the known properties of 2-oxoglutarate-dependent dioxygenases. The relative molecular mass of 44,000 is similar to that of other dioxygenases and putative dioxygenases including a GA 2β-hydroxylase which has also been partially purified from pea embryos (45).

The early 13-hydroxylation pathway observed in pea seeds appears to operate in pea vegetative tissues too. 13-Hydroxylated C_{19}-GAs are present in seedlings, shoots and leaves of pea (10), although the levels are several orders of magnitude lower than in seeds. An additional metabolic conversion namely the 3β-hydroxylation of GA_{20} to GA_1 is observed in shoots but not seeds of pea (Figs. 5, 6) (28). Like GA_{20}, GA_1 can also be 2β-hydroxylated giving GA_8, which in turn is oxidized to GA_8-catabolite (Figs. 5, 6).

Gibberellin A_1 has high biological activity, and it has been implicated in the control of internode extension (28) (see Chapter G1). 3β-Hydroxylation is therefore a very important metabolic conversion which has profound effects on plant growth. The 3β-hydroxylase from pea stems has not, to date, been purified or characterized. The ontogenetic- and tissue-specific regulation of this enzyme will be particularly interesting to study since it has clearly been shown to be confined to immature tissues of the shoot.

Pea seeds cannot be used as a plentiful source of the 3β-hydroxylase, since this metabolic conversion does not occur in seeds of this genus. Instead, a 3β-hydroxylase which converts GA_{20} to GA_1 (and GA_5 and GA_{29}, see later) has been isolated and purified from *Phaseolus vulgaris* (bean) seeds (46). Whether there will be enough homology between the bean and pea proteins to allow the use of the bean enzyme as a probe for the pea 3β-hydroxylase remains to be determined.

The 3β-hydroxylase from bean seeds has several biochemical similarities to the 2β-hydroxylase from pea seeds (45, 46). Both are acid-labile, hydrophobic proteins of similar size, whose activity is dependent on the presence of 2-oxoglutarate, oxygen, Fe^{2+} and ascorbate. Despite these apparent characteristics of 2-oxoglutarate dependent dioxygenases, Smith and MacMillan (46) were not able to demonstrate that succinate was a product of the catalytic conversion of GA_{20} to GA_1. Instead they suggest that ascorbic acid, which is oxidatively degraded to threonic acid in incubation mixtures, may be the oxygen acceptor, and that 2-oxoglutarate may be a cofactor, rather than a cosubstrate. It should be emphasised that the 3β-hydroxylase described above has been isolated from bean seeds and is not necessarily identical to the 3β-hydroxylase in pea stems.

Despite some remaining uncertainty as to the catalytic mechanism of these oxidative enzymes, one of the most useful advances in this field over the past five years has been the synthesis of acylcyclohexanedione plant growth retardants which have some structural similarities to 2-oxoglutarate (38, 42). These retardants, which inhibit GA oxidation, are thought to exert their effect by competing with 2-oxoglutarate at the reaction center of the enzyme (19) (see later).

Pathways in *Cucurbita maxima*

Cell free preparations from pumpkin endosperm have been used for the past two decades to pioneer the study of GA biosynthesis and metabolism in higher plants. During that time the system has been modified and refined allowing the identification of new metabolic steps and additional pathways. Parts of all four pathways shown in Fig. 4 can be demonstrated in endosperm preparations, with additional hydroxylation at carbon-12 also occurring. Unlike the pea system, the formation of C_{19}-GAs in pumpkin preparations is low, and instead several polyhydroxylated C_{20}-GAs accumulate. The following discussion draws largely from two recent papers (33, 34), which describe the latest, most comprehensive feeding studies.

GA_{12} and GA_{12}-aldehyde are metabolized to similar products by a gel-filtered high speed supernatant from a pumpkin endosperm preparation, though GA_{12} is the more efficient precursor. When incubated with increasing concentrations of protein and appropriate cofactors for 2-oxoglutarate-dependent dioxygenases, labeled GA_{12} is converted in sequence to GA_{15}, GA_{24}, GA_{25} and the 3β-hydroxylated C_{20}-GAs, GA_{13} and GA_{43} (Fig. 7). At highest protein concentrations GA_{43} and GA_{39} (both dihydroxylated tricarboxylic acids) and $12\alpha OH$-GA_{43} accumulate. The 3β-hydroxylated C_{19}-GA, GA_4, was only a minor product and GA_9 is not formed. Individual metabolic steps were established by refeeding, giving the pathway shown on the left-hand side of Fig. 7.

When the gel-filtered high speed supernatant was incubated with the appropriate cofactors for NADPH-dependent mono-oxygenases no metabolism of GA_{12} was observed. In contrast, the microsomal preparation (washed pellet from the high speed centrifugation) displayed both mono- and di-oxygenase activity (33). When incubated in the presence of NADPH, GA_{12} was 13-hydroxylated to give GA_{53} (mono-oxygenase activity). In the presence of 2-oxoglutarate etc. mono-oxygenase activity was suppressed and GA_{12} was converted to the same products that were formed with high speed supernatant

79

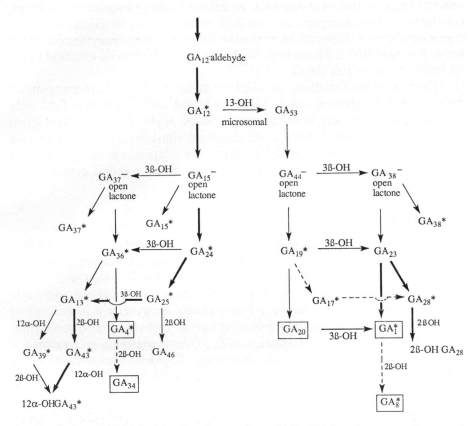

Fig. 7. GA metabolism in *Cucurbita maxima* (pumpkin). Thickened arrows refer to main pathway in endosperm. ----➤ refers to metabolic conversions occurring in embryos, and not in endosperm (* = endogenous GA). C_{19}-GAs are shown in boxes.

incubations (dioxygenase activity).

GA_{53}, incubated with the high speed supernatant in the presence of dioxygenase cofactors, was converted to GA_{44}, GA_{23}, and GA_1 (right hand side, Fig. 7) (33). GA_{38}, GA_{19} and GA_{20} were minor products. The presence of GA_{23} suggests that in pumpkin 3β-hydroxylation of 13-OH GAs occurs at the C_{20}-level ($GA_{19} \rightarrow GA_{23}$), in contrast to pea stems in which it occurs only at the C_{19}-level ($GA_{20} \rightarrow GA_1$). Therefore, unlike the situation in pea stems, in pumpkin endosperm GA_{23}, not GA_{20}, appears to be the immediate precursor of GA_1.

Reinvestigating the endogenous GAs in pumpkin endosperm, Lange et al (33) were able to identify 13 GAs in total. Two of the major native GAs, GA_{13} and GA_{43}, are major products in pumpkin endosperm preparations. However, the metabolic origin of two 3β,12α-hydroxylated C_{19}-GAs, GA_{58} and its 2β-hydroxylated derivative GA_{49}, which are major native GAs but which were not detected in cell free incubations, remains obscure.

Table 1. Enzymes catalysing reactions in the GA biosynthetic pathway--a summary of their properties.

Reaction or sequence of reactions	Enzyme Properties	Cosubstrates/ Cofactors, etc.	Ref.
MVA $\rightarrow\rightarrow$ *ent*-kaurene	soluble	ATP, Mn^{2+}/Mg^{2+}	4
ent-kaurene $\rightarrow\rightarrow$ GA_{12}-aldehyde	microsomal	NADPH, O_2	23, 24
GA_{12}-aldehyde $\rightarrow$ GA_{12}	(1) microsomal	NADPH, O_2	33
	(2) soluble	2-oxoglutarate, Fe^{2+}, O_2	33
sequential oxidation of carbon-20	soluble	2-oxoglutarate, Fe^{2+}, O_2	26
2β-hydroxylation	soluble	2-oxoglutarate, Fe^{2+}, O_2	45
3β-hydroxylation	soluble	2-oxoglutarate/ ascorbate, Fe^{2+}, O_2	46
12α-hydroxylation	(1) microsomal	NADPH, O_2	33
	(2) soluble	2-oxoglutarate Fe^{2+}, O_2	33
13-hydroxylation	(1) microsomal	NADPH, O_2	29, 33
	(2) soluble	2-oxoglutarate Fe^{2+}, O_2	33

There are some differences between pathways in pumpkin endosperm and embryo preparations (33, 34) . One of these differences concerns 2β-hydroxylation. In endosperm preparations 2β-hydroxylation occurs only at the C_{20}-level, whereas in embryos only C_{19}-GAs are 2β-hydroxylated. Thus only endosperm preparations yield the 2β-hydroxylated C_{20}-GAs, GA_{43} and 12α-GA_{43}, and only embryo preparations give GA_8 and GA_{34} (see Fig. 7).

Of particular interest with the pumpkin system are the observations that alternate enzyme activities exist for the same metabolic event. For example, the oxidation of GA_{12}-aldehyde to GA_{12} in pumpkin endosperm can be catalysed by both microsomal and soluble oxidases. Similarly, 12α-hydroxylation can be catalysed by microsomal and soluble enzyme activities. The two 12α-hydroxylating activities appear to have different substrates specificities, with the microsomal enzyme having GA_{12}-aldehyde as preferred substrate, and the soluble enzyme hydroxylating the more polar tricarboxylic acids.

A 20-oxidase has recently been purified from pumpkin endosperm (35) and has similar properties to the 20-oxidase in pea cotyledons (17, 32). The pumpkin 20-oxidase has now been cloned from immature cotyledons (35). *In vitro* translation of mRNA from cotyledons and endosperm produced measurable 20-oxidase activity, suggesting very high levels of expression in these tissues. An antibody, raised against a peptide from a tryptic digest of purified enzyme, was used to identify a clone in a λgt11 expression library derived from cotyledons. The recombinant enzyme expressed in *E. coli* catalysed the three-step oxidation of GA_{12} to GA_{25} and, with lower efficiency, of GA_{25} to GA_{17}. C_{19}-GAs were also produced by the enzyme in low (~1%) yield. Thus a single enzyme can catalyse all oxidation steps at C-20. The nucleotide sequence of this pumpkin 20-oxidase shows it to have 20-30% amino acid identity with members of a group of soluble Fe-containing dioxygenases. This group (40) includes 1-aminocyclopropane-1-carboxylic acid oxidase (ACC oxidase, previously known as the ethylene forming enzyme), which in contrast to most members, does not utilise 2-oxoglutarate as a co-substrate. There are several other dioxygenases such as prolyl 4-hydroxylase and lysyl hydroxylase that utilise 2-oxoglutarate but have little amino acid similarity to the GA 20-oxidase. This interesting group of enzymes may require some reclassification as additional information becomes available on enzyme mechanism, crystal structure, and sequence (40). A brief summary of these and other enzymes of GA biosynthesis and metabolism is given in Table 1.

Pathways in *Arabidopsis thaliana*

The features of *Arabidopsis* which make it so ideally suited to molecular genetic research have been repeatedly described (37). The identification of GAs, elucidation of GA biosynthetic pathways, and characterization of new dwarf mutants in *Arabidopsis* can therefore be expected to be particularly rewarding, opening up the possibilities for cloning genes involved in GA metabolism by chromosome walking, genomic subtraction and other techniques which are appropriate for *Arabidopsis* but would be unsuitable for pea or pumpkin. The cloning of the *GA1* locus, the recessive allele of which blocks the synthesis of *ent*-kaurene in *Arabidopsis* shoots, has recently been achieved (50).

The routes for GA biosynthesis in *Arabidopsis* shoots have been inferred from a knowledge of the native GAs in *Arabidopsis* shoots (51, 53). The twenty known GAs are components of three pathways, the 3β-hydroxylation pathway leading to GA_4, the early 13-hydroxylation pathway leading to GA_{20} and the 3,13-dihydroxylation pathway leading to GA_1 (Fig. 4, Fig. 8). The individual pathways seem to be connected in a grid, because 3β-hydroxylation can occur at several levels (Fig. 8). The metabolic conversions of C_{19}-GAs in *Arabidopsis* are particularly interesting as they differ from those in pea and pumpkin. From feeding studies with C_{19}-GAs it has been demonstrated that GA_9 can either be 3β-hydroxylated to give GA_4, or 13-hydroxylated to give GA_{20}. In turn, GA_4 can be 13-hydroxylated, and GA_{20} can be 3β-hydroxylated giving two alternate routes to GA_1(51, 53). However, in *Arabidopsis* GA_1 is not thought to be the active GA, instead GA_4 appears to fulfil the role of active hormone (see later).

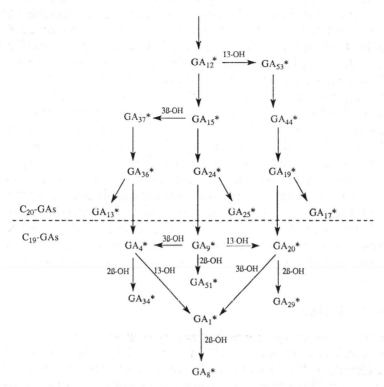

Fig. 8. GA metabolism in *Arabidopsis thaliana*. (* = endogenous GA).

The functionalization of Ring A in *Phaseolus, Zea and Marah*

Although the GA pathways in pea, pumpkin and *Arabidopsis* illustrate many of the important steps in GA biosynthesis, some metabolic conversion cannot be demonstrated in any of these genera. These conversions include the insertion of a 2,3 double bond into ring A as in the formation of GA_5, insertion of a 3β-hydroxyl group together with a 1,2 double bond as in the formation of GA_3 and GA_7, and of a 2,3 epoxidation as in GA_6. These metabolic steps are considered briefly here.

In cell-free systems from *Phaseolus vulgaris* the enzyme which converts GA_{20} to GA_1 also gives GA_5 and GA_{29} as additional products (2) (Fig. 9). All three appear to be formed by one enzyme since they always co-occur in incubations, and in the same relative proportions (2). The enzyme has the properties and cofactor requirements of a dioxygenase, although it may have ascorbate rather than 2-oxoglutarate as its cosubstrate (46). The conversion of GA_{20} to GA_5 does not involve the incorporation of oxygen into the molecule yet requires the presence of oxygen. This observation reinforces the proposition that the conversion of GA_{20} to GA_5 is mechanistically linked to the hydroxylation of GA_{20}.

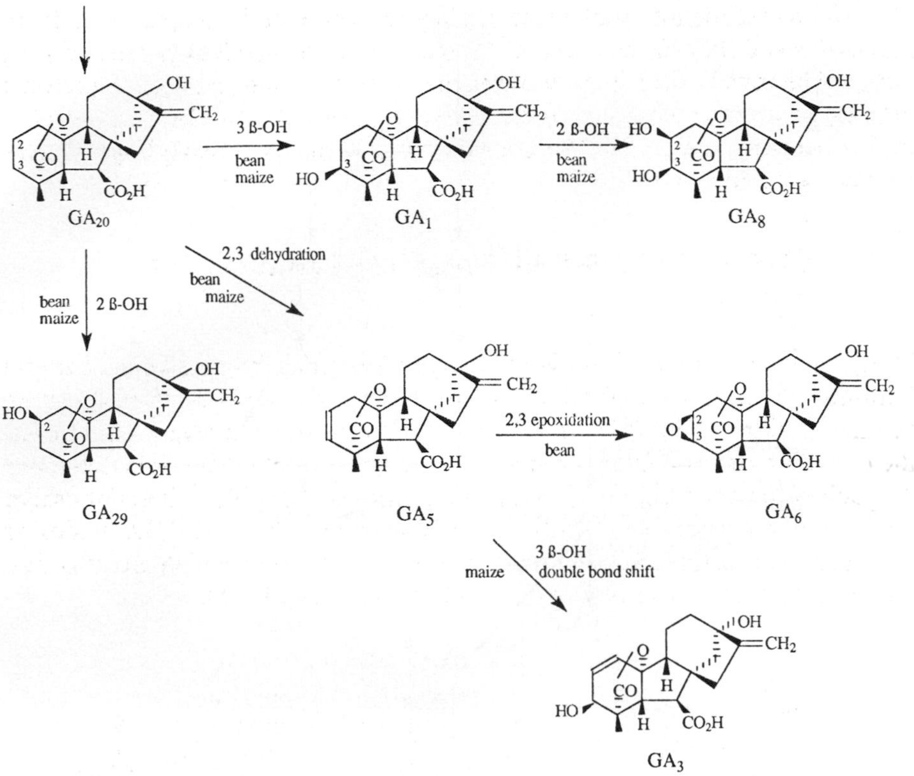

Fig. 9. The metabolism of GA_{20} in *Phaseolus vulgaris* (bean) and *Zea mays* (maize).

Kamiya and Kwak (30), working with a partially purified 3β-hydroxylase enzyme from bean seeds, also reported the epoxidation of GA_5 to GA_6. Although the cofactor requirements were the same for epoxidation and 3β-hydroxylation, it is not certain yet whether the conversions are catalysed by a single enzyme, or by two enzymes which are difficult to separate.

In *in vivo* feeds to maize, GA_{20}, in addition to being converted to GA_1, is also converted to GA_5 (14). Furthermore, GA_5 is converted to GA_3, establishing conclusively the biosynthetic origin of GA_3 in a higher plant (Fig. 9). Although GA_1 is purported to be the primary GA with hormonal function in maize stems (39), the presence of GA_3 in some maize mutants, and the verification of GA_3 as a true metabolic product suggests that it may also be important in the control of internode growth in *Zea*. GA_3 is not a substrate for 2β-hydroxylation and would be expected to have more persistent bioactivity than GA_1. Unlike the bean system, there is no evidence for the conversion of GA_5 to GA_6 in maize.

In a set of biosynthetic reactions analogous to the conversion of GA_{20} → GA_5 → GA_3, the non-13-hydroxylated GA, GA_9 is converted to GA_7 via 2,3-dehydro GA_9 in cell free systems from the Cucurbit, *Marah macrocarpus* (1).

The stereochemistry of these A ring substitutions in preparations from *Phaseolus* and *Marah* has been studied using stereospecifically labeled [^{2}H] GA$_{20}$. The metabolism in each instance occurred with loss of the β protons: 2β and 3β protons were lost in the conversion of GA$_{20}$ to GA$_5$, and of GA$_9$ to 2,3-dehydro GA$_9$. The 1β proton was lost in the metabolism of 2,3-dehydro GA$_9$ to GA$_7$ (1).

SYNTHETIC INHIBITORS OF GA BIOSYNTHESIS AND METABOLISM.

Many chemical inhibitors are known which block GA biosynthesis. Several inhibitors, e.g., AMO-1618[2] and cycocel, block the A activity of *ent*-kaurene synthesis. Other inhibitors block the pathway at the next stage, i.e., during the oxidation of *ent*-kaurene. These include ancymidol, tetcyclacis (structure 4), paclobutrazol (structure 5), and uniconazole (structure 6). These inhibitors are known to perturb the activity of P-450 mono-oxygenases. The modes of action and comparable activities of these inhibitors, all of which are effective growth retardants, have been extensively reviewed (20, 41).

(4) Tetcyclacis   (5) Paclobutrazol   (6) Uniconazole

Recently a series of acylcyclohexanedione compounds have been developed which act at even later stages in the pathway, namely at those reactions which are catalysed by 2-oxoglutarate dependent dioxygenases (38, 42). These compounds correspond closely with the structure of 2-oxoglutarate (Fig. 10) (19). Two of the compounds most commonly used are BX-112 (structure 7) (or its free acid, prohexadione), and LAB 198 999 (structure 8). At lower concentrations they act as competitive inhibitors, presumably competing with 2-oxoglutarate at the enzyme active site. For

---

[2] The chemical names of the inhibitors referred to in this section are as follows:
AMO-1618 = 2'-isopropyl-4'-(trimethylammonium chloride)-5'-methylphenylpiperidine-1-carboxylate;
cycocel = 2-chloroethyltrimethyl-ammonium chloride;
tetcyclacis = 5-(4-chlorophenyl)-3,4,5,9,10-pentaaza-tetracyclo [5,4,1,0^{2,6},0^{8,11}]-dodeca-3,9-diene;
paclobutrazol = 1-(4-chloro-phenyl)-4,4-dimethyl-2-(1*H*-1,2,4-triazol-1-yl)-pentan-3-ol;
uniconazole = (E)-1-(4-chlorophenyl)-4,4-dimethyl-2-(1*H*-1,2,4,triazol-1-yl)-1-penten-3-ol;
BX-112 (prohexadione-calcium) = calcium 3,5-dioxo-4-propionylcyclohexanecarboxylic acid;
LAB 198 999 = 4 (n propyl-α-hydroxymethylene)-3,5,-dioxocyclohexanecarboxylic acid ethyl ester.

some unknown but highly fortuitous reason, some dioxygense-catalysed reactions in the GA pathway seem to be more sensitive to these acylcyclohexanediones than are other reactions. The 3β-hydroxylation of GA$_{20}$ to GA$_1$ is particularly susceptible to the inhibitor (42). These inhibitors have been used to ascertain which GAs in a sequence have biological activity *per se* (see below).

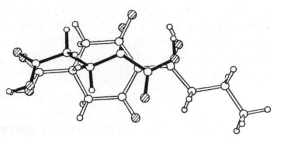

Fig. 10. Ball and stick representation of 2-oxoglutaric acid (filled bonds) superimposed on LAB 198 999 in its free acid form (open bonds) showing the structural similarities between the two compounds. Hatched circles represent oxygen atoms. (From (19) with modifications).

(7) BX-112

(8) LAB 198 999

PHYSIOLOGICAL CONSIDERATIONS: CONTROL OF BIO-SYNTHETIC AND DEGRADATIVE PATHWAYS.

The objective of GA biosynthesis and metabolism is to provide carefully regulated amounts of active GA to the appropriate tissue at the required time. Now that a considerable amount of information has been collected on GA pathways the focus of recent work is to ascertain which are the active GAs, and how are their levels regulated within the plant and during ontogeny. Some answers to these questions are now available.

In vegetative tissues the native GAs are less structurally diverse than in reproductive tissues, and are formed on pathways which appear to have been highly conserved (see Fig. 11). For instance in almost all species studied to date (e.g., maize, pea, rice, spinach) the major or only pathway in shoot tissues is the early 13-hydroxylated pathway. In this pathway the first formed C$_{19}$-GA is GA$_{20}$, and it is converted to GA$_1$ by 3β-hydroxylation. In pea and maize the *le* and *d1* mutations, respectively, block the conversion of GA$_{20}$ to GA$_1$ and cause a dwarf phenotype (28, 49). The evidence is therefore very strong that GA$_1$ is the active GA controlling internode elongation in these genera. In wildtype plants GA$_{20}$ and other members of the early 13-hydroxylation pathway are bioactive only because they are converted *in situ* to GA_1. The acylcyclohexanediones, which preferentially inhibit 3β-hydroxylation, have also been used in several other genera for which suitable mutants are unavailable. These experiments verify that 3-deoxy GAs have little activity *per se*, and that in almost all species studied to date GA_1 is the active hormone in shoot elongation.

GA_3, GA_4 and GA_7 are 3β-hydroxylated C_{19}-GAs like GA_1, and each of them is potentially as bioactive as GA_1. In *Arabidopsis thaliana* and *Cucumis sativus* applied GA_4 is actually very much more active than GA_1 (53). Although both genera do contain GA_1 in addition to GA_4, and applied GA_4 is metabolised to GA_1, the evidence is strong that in both genera GA_4 is the primary GA for stem growth. Thus, GA_9 has activity in both cucumber and *Arabidopsis*, yet is virtually inactive when applied in the presence of an acylcyclohexanedione which would prevent its conversion to GA_4 (53). Furthermore, synthetic GAs which are substituted at carbon-13 have reduced activity relative to the nonsubstituted parent compound in cucumber and *Arabidopsis*, indicating that in these genera 13-substitution actually inhibits bioactivity. One has to assume from these results that the GA receptors in shoots of cucumber and *Arabidopsis* have less tolerance for structural alterations on the C/D ring than do receptors from most other genera.

Now that it has been established that the active GAs for stem elongation are 3β-hydroxylated C_{19}-GAs, the next task is to determine what controls the levels of these types of GAs within the target tissue of the plant. In general terms it is a combination of rates of synthesis, degradation and transport. The control of synthesis and degradation are considered briefly.

Certain steps in the GA metabolic pathway are known to be rate limiting. Various factors can stimulate one or more of these steps, increasing the flux through the pathway. Photoperiod enhances the oxidation of $GA_{53} \rightarrow GA_{44}$ and of $GA_{19} \rightarrow GA_{20}$ in spinach (15), though how this is achieved at the molecular level is not known. The conversion of GA_{19} to GA_{20} is of particular interest. By comparing the relative levels of GA_{19}, GA_{20} and GA_1 in insensitive dwarfs, in wildtype plants and in slenders (in which the GA_1 response is constitutive) Hedden and Croker (25) have concluded that GA_1 exerts negative control on the conversion of $GA_{19} \rightarrow GA_{20}$ as a consequence of its action. Thus in situations in which GA_1 has a growth response or in which the response is constitutively expressed the levels of GA_{19} are high. Conversely if GA_1 is inactive because of a mutation conferring insensitivity then GA_{19} levels are low and GA_{20} and GA_1 accumulate. Scott (44) has suggested that this type of feed back control could occur if GA_1 induced the synthesis of a protein which down-regulated the transcription of mRNA for the GA_{19} 20-oxidase.

The amount of GA_1 produced can also be finely tuned by balancing the amount of GA_{20} which is 2β-hydroxylated to give the inactive GA_{29} versus the amount which is 3β-hydroxylated to give GA_1. The 2β-hydroxylation of GA_1 to form the inactive GA_8 is also an important factor in determining the steady state level of GA_1. Although GA hydroxylases from stems have not yet been isolated and purified, Smith and MacMillan (45, 46) have studied the kinetic parameters of 2β- and 3β-hydroxylases from bean seeds and a 2β-hydroxylase from pea. In bean seeds it appears that 2β-hydroxylation of GA_{20} to GA_{29} is less important than the 2β-hydroxylation of GA_1 to GA_8 in controlling the steady state level of GA_1. In pea, for which no data for 3β-hydroxylation is available, circumstantial evidence suggests that at high GA_{20} levels 2β-hydroxylation of GA_{20} is an important factor. Additional information on the regulation of these important enzymes is eagerly awaited.

The GA status of seeds is much more complex than that of shoots. In seeds the GAs are more abundant and more structurally diverse. Multiple hydroxylations, many observed only in seeds, lead to the accumulation of polyhydroxylated GAs at specific developmental stages. The appearance of these hydroxylating enzymes seems to be genetically programmed -- there is no evidence for them being inducible. To date there is no firm evidence that these polyhydroxylated GAs have a role in fruit maturation or seed development, although at earlier developmental stages GAs may control fruit growth.

CONCLUSIONS

The main pathways for GA biosynthesis and metabolism in higher plants are well defined. The four pathways shown in Figs 4 and 11 are the main routes to C_{19}-GAs. Not all pathways can be demonstrated in all plants, instead most genera studied have one or two of the pathways functioning in stems, with some additional structural elaborations to the basic pathways occurring specifically in seed tissues. Since the first edition of this book appeared seven years ago several new metabolic conversions have been studied, including the conversion of $GA_{20} \rightarrow GA_5 \rightarrow GA_3$. Substantial progress has been made with cell-free systems from pea and pumpkin, particularly with the further purification and characterization of the oxidative enzymes involved in GA metabolism. This work has recently come to fruition with the cloning of the 20-oxidase from pumpkin. Cloning of the *GA1* locus in *Arabidopsis* which is responsible for *ent*-kaurene synthesis has also been achieved. The continued use of single gene dwarf mutants and the new acylcyclohexanedione growth retardants which specifically inhibit GA dioxygenases has confirmed that 3β-hydroxylated C_{19}-GAs, primarily GA_1 (but probably GA_4 in *Arabidopsis* and cucumber) is the active hormone in stem elongation. Future work must now address the mechanism for regulating the levels of active GA. Of particular interest will be a study of tissue-specific and developmental regulation of the enzymes involved in GA metabolism.

Acknowledgements.
I would like to thank Drs. Mike Beale and Peter Hedden for their useful comments on the manuscript, and Serena Smith and Julie Blackwell for preparing the Figures.

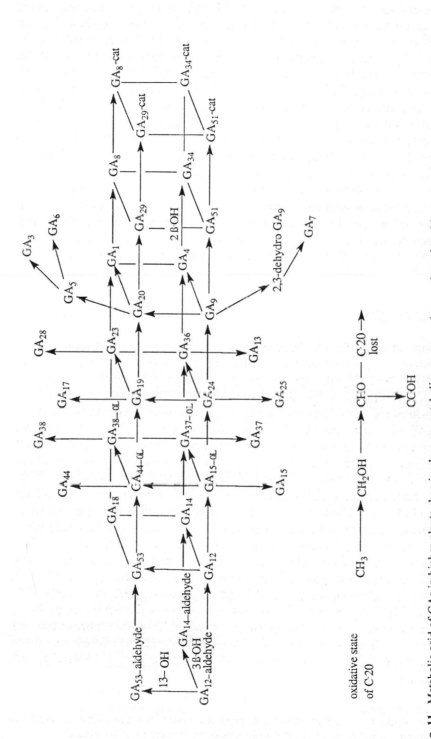

Fig. 11. Metabolic grid of GAs in higher plants showing known metabolic sequences (——▶) produced by successive oxidation of C-20 followed by its removal, combined with hydroxylation at C-13, C-3, and/or C-2. Not all reactions operate in all plants. Potential sequences which have not been confirmed are depicted (——). Some other reactions are known but not shown. (O.L. = open lactone; cat = catabolite (α,β-unsaturated ketone)).

References

1. Albone, K.S., Gaskin, P., MacMillan, J., Phinney, B.O., Willis, C. (1990) Biosynthetic origin of gibberellins A_3 and A_7 in cell-free preparations from seeds of *Marah macrocarpus* and *Malus domestica*. Plant Physiol. 94, 132-142.

2. Albone, K.S., Gaskin, P., MacMillan, J., Smith, V., Weir, J. (1989) Enzymes form seeds of Phaseolus vulgaris L.: Hydroxylation of gibberellins A_{20} and A_1 and 2,3-dehydrogenation of gibberellin A20. Planta 177, 108-115.

3. Aoyagi, K., Beyou, A., Moon, K., Fang, L., Ulrich, T (1993) Isolation and characterization of cDNAs encoding wheat 3-hydroxy-3-methylglutaryl coenzyme A reductase. Plant Physiol. 102, 623-628.

4. Beytia, E.D., Porter, J.W. (1976) Biochemistry of polyisoprenoid biosynthesis. Ann. Rev. Biochem. 45, 113-42.

5. Birch, A.J., Richards, R.W., Smith, H., Harris, A., Whalley, W.B. (1959) Studies in relation to biosynthesis-XXI. Rosenolactone and gibberellic acid. Tetrahedron 7, 241-51.

6. Bozak, K.R., Yu, H., Sirevag, R., Christoffersen, R.E., (1990) Sequence analysis of ripening-related cytochrome P-450 cDNAs from avocado fruit. Proc. Natl. Acad. Sci. USA 87: 3904-3908.

7. Butcher, D.N., Appleford, N.E.J., Hedden, P., Lenton, J.R. (1987) Plant growth substances in root cultures of *Lycopersicon esculentum*. Phytochem 27, 1575-1577.

8. Coolbaugh, R.C. (1983) Early stages of gibberellin biosynthesis. *In*: The biochemistry and physiology of gibberellins. Vol. 1, pp. 53-98. Crozier, A. ed., Praeger, New York.

9. Cross, B.E., Grove, J.F., MacMillan, J., Mulholland, T.P.C. (1956) Gibberellic acid, Part IV. The structures of gibberic and *allo* gibberic acids and possible structures for gibberellic acid. Chem. and Ind. 1956, 954-55.

10. Davies, P.J., Emshwiller, E., Gianfagna, T.J., Proebsting, W.M., Noma, M., Pharis, R.P. (1982) The endogenous gibberellins of vegetative and reproductive tissue of G2 peas. Planta 154, 266-72.

11. Dockerill, R.C., Hanson, J.R. (1978) Fate of C-20 in C_{19}-gibberellin biosynthesis. Phytochem. 11, 317-26.

12. Donaldson, R.P., Luster, D.G., (1991) Multiple forms of plant cytochromes P-450. Plant Physiol 96, 669-674.

13. Duncan, J.D., West, C.A. (1981) Properties of kaurene synthetase from *Marah macrocarpus* endosperm. Evidence for the participation of separate but interacting enzymes. Plant Physiol. 68, 1128-34.

14. Fujioka, S., Yamane, H., Spray, C.R., Phinney, B.O., Gaskin, P., MacMillan, J., Takahashi, N. (1990) Gibberellin A_3 is biosynthesized from gibberellin A_{20} via gibberellin A5 in shoots of *Zea mays* L. Plant Physiol. 94, 127-131

15. Gilmour, S.J., Zeevaart, J.A.D., Schwenen, L., Graebe, J.E. (1986) Gibberellin metabolism in cell-free extracts from spinach leaves in relation to photoperiod. Plant Physiol. 82, 190-195.

16. Graebe, J.E. (1987) Gibberellin biosynthesis and control. Ann. Rev. Plant Physiol. 38, 419-465.

17. Graebe, J.E., Bose, G., Grosselindemann, E., Hedden, P., Aach, H., Schweimer, A., Sydow, S., Lange, T., (1992) The biosynthesis of ent-Kaurene in germinating seeds and the function of 2-oxoglutarate in gibberellin biosynthesis. *In*: Progress in plant growth regulation, pp. 545-554, Karssen, C.M., van Loon, L.C., Vreugdenhil, D., eds. Kluwer Academic Publishers, Netherlands.

18. Graebe, J.E., Hedden, P., Gaskin, P., MacMillan, J. (1974) The biosynthesis of a C_{19}-gibberellin from mevalonic acid in a cell-free system from a higher plant. Planta 120, 307-309.

19.   Griggs, D.L., Hedden, P., Temple-Smith, K.E., Rademacher, W. (1991) Inhibition of gibberellin 2β-hydroxylases by acylcyclohexanedione derivatives. Phytochem. 30, 2513-2517.

20.   Grossmann, K. (1990) Plant growth retardants as tools in physiological research. Phys. Plant. 78, 640-648.

21.   Hanauske-Abel, H.M., Gunzler,V. (1982) A stereochemical concept for the catalytic mechanism of prolyhydroxylase, applicability to classification and design of inhibitors. J. Theor. Biol. 94, 421-455.

22.   Hasson, E.P., West, C.A. (1976) Properties of the system for the microsomes of immature seeds of *Marah macrocarpus*. Cofactor requirements. Plant Physiol. 58, 473-78

23.   Hasson, E.P., West, C.A. (1976) Properties of the system for the mixed function oxidation of kaurene and kaurene derivatives in microsomes of immature seeds of *Marah macrocarpus*. Electron transfer components. Plant Physiol 58, 479-484.

24.   Hedden, P. (1983) *In vitro* metabolism of gibberellins. *In*: The biochemistry and physiology of gibberellins, Vol. 1, pp. 99-150, Crozier, A. ed. Praeger, New York.

25.   Hedden, P., Croker, S.J., (1992) Regulation of gibberellin biosynthesis in maize seedlings. *In*: Progress in plant growth regulation, pp. 571-577, Karssen, C.M., vanLoon, L., Vreugdenhil, D. ed., Kluwer Academic Publishers, Dordrecht.

26.   Hedden, P., Graebe, J.E. (1982) The cofactor requirements for the soluble oxidases in the metabolism of the C_{20}-gibberellins. J. Plant Growth Regulation 1, 105-116.

27.   Hedden, P., Hoad, G.V., Gaskin, P., Lewis, M.J., Green, J.R., Furber, M., Mander, L.N. (1993) Kaurenoids and gibberellins, including the newly characterized gibberellin A_{88}, in developing apple seeds. Phytochem. 32, 231-237.

28.   Ingram, T.J., Reid, J.B., Murfet, I.C., Gaskin, P., Willis, C.L., MacMillan, J. (1984) Internode length in *Pisum*. The *Le* gene controls the 3β-hydroxylation of gibberellin A_{20} to gibberellin A_1. Planta 160, 455-63.

29.   Kamiya, Y., Graebe, J.E. (1983) The biosynthesis of all major pea gibberellins in a cell-free system from *Pisum sativum*. Phytochem. 22, 681-90.

30.   Kamiya, Y., Kwak, S.-S. (1991) Partial characterization of the gibberellin 3β-hydroxylase from immature seeds of *Phaseolus vulgaris*. *In*: Gibberellins, pp. 72-82, Takahashi, N., MacMillan, J., Phinney, B.O. eds. Springer-Verlag, New York.

31.   Kamiya, Y., Takahashi, N., Graebe, J.E. (1986) The loss of the C-20 carbon atom in C_{19}-gibberellin biosynthesis in a cell-free system from *Pisum sativum* L. Planta 169, 154-158.

32.   Lange, T., Graebe, J.E. (1989) The partial purification and characterization of a gibberellin C-20 hydroxylase from immature *Pisum sativum* L. seeds. Planta 179, 211-221.

33.   Lange, T., Hedden, P., Graebe, J.E. (1993a) Biosynthesis of 12α- and 13-hydroxylated gibberellins in a cell-free system from *Cucurbita maxima* endosperm and the identification of new endogenous gibberellins. Planta 189, 340-349.

34.   Lange, T., Hedden, P., Graebe, J.E. (1993b) Gibberellin biosynthesis in cell-free extracts from developing *Cucurbita maxima* embryos and the identification of new endogenous gibberellins. Planta 189, 350-358.

35.   Lange, T., Hedden, P., Graebe, J.E. (1994) Molecular cloning and heterologous expression of a cDNA encoding a gibberellin 20-oxidase from developing cotyledons of pumpkin (*Cucurbita maxima*). Proc. Natl. Acad. Sci. (in press).

36.   MacMillan, J., Takahashi, N. (1968) Proposed procedure for the allocation of trivial names to the gibberellins. Nature 217, 170-171.

37.   Meyerowitz, E.M. (1987) *Arabidopsis thaliana*. Ann. Rev. Genet. 21, 93-111.

38.   Nakayama, I., Miyazawa, T., Kobayashi, M., Kamiya, Y., Abe, H., Sakurai, A. (1990) Effects of a new plant growth regulator prohexadione calcium gibberellins in rice (*Oryza sativa* L.) seedlings. Plant Cell Physiol. 31, 195-200.

39. Phinney, B.O. (1984) Gibberellin A_1, dwarfism and the control of shoot elongation in higher plants. *In*: The biosynthesis and metabolism of plant hormones, pp. 17-41, Crozier, A., Hillman, J.R. eds. Soc. Exp. Biol. Seminar 23. Cambridge U.P., Cambridge.

40. Prescott, A.G. (1993) A dilemma of dioxygenases (or Where biochemistry and molecular biology fail to meet). J. Exp. Bot. 44, 849-861.

41. Rademacher, W. (1992) Biochemical effects of plant growth retardants. *In*: Plant biochemical regulators, pp. 169-199, Gausmann, H.W. ed. Marcel Dekker, Inc., New York.

42. Rademacher, W., Temple-Smith, K.E., Griggs, D.L., Hedden, P. (1992) The mode of action of acylcyclohexanediones - a new type of growth retardant. *In*: Progress in plant growth regulation, pp.571-577. Karssen,C.M., van Loon, L.C., Vreugdenhil, D. ed., Kluwer Academic Publishers, Dordrecht.

43. Schneider, G. (1983) Gibberellin conjugates. *In*: The biochemistry and physiology of gibberellins. Vol 1, pp. 389-456. Crozier, A. ed., Praeger, New York.

44. Scott, I.M. (1990) Plant hormone response mutants. Physiol. Plant. 78, 147-152.

45. Smith, V.A., MacMillan J. (1986) The partial purification and characterization of a gibberellin 2β-hydroxylases from seeds of *Pisum sativum*. Planta 167, 9-18

46. Smith, V.A., Gaskin, P., MacMillan, J. (1990) partial purification and characterization of the gibberellin A_{20} 3β-hydroxylase from seeds of *Phaseolus vulgaris*. Plant Physiol. 94, 1390-1401.

47. Sponsel, V.M. (1983) The localization, metabolism and biological activity of gibberellins in maturing and germinating seeds of *Pisum sativum* cv. Progress No. 9. Planta 159, 454-468

48. Sponsel, V.M. (1985) Gibberellins in *Pisum sativum*--their nature, distribution and involvement in growth and development of the plant. Plant Physiol. 65, 533-38.

49. Spray, C.R., Phinney, B.O., Gaskin, P., Gilmour, S.J., MacMillan, J. (1984) Internode length in *Zea mays* L. The dwarf-1 mutation controls the 3β-hydroxylation of gibberellin A_{20} to gibberellin A_1. Planta 160, 464-68.

50. Sun, T., Goodman, H.M., Ausubel, F. (1992) Cloning the *Arabidopsis* GA_1 locus by genomic subtraction. Plant Cell 4, 119-128.

51. Talon, M., Koornneef, M., Zeevaart, J.A.D. (1990) Endogenous gibberellins in *Arabidopsis thaliana* and possible steps blocked in the biosynthetic pathways of the semidwarf *ga4* and *ga5* mutants. Proc. Natl. Acad. Sci. USA 87, 7983-7987.

52. Zeevaart, J.A.D., Gage, D.A. (1993) *ent*-Kaurene biosynthesis is enhanced by long photoperiods in the long-day plants *Spinacia oleracea* L. and *Agrostemma githago* L. Plant Physiol. 101, 25-29.

53. Zeevaart, J.A.D., Talon, M. (1992) Gibberellin mutants in *Arabidopsis thaliana*. *In*: Progress in Plant Growth Regulation, pp. 34-41, Karssen, C.M., van Loon, L.C., Vreugdenhil, D. ed., Kluwer Academic Publishers, Dordrecht.

Appendix - Gibberellin structures

(F denotes fungal origin, P denotes plant origin)

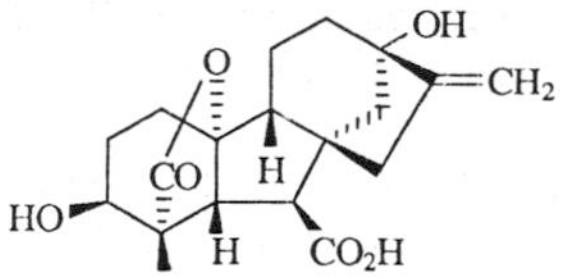

GA$_1$ (F, P)

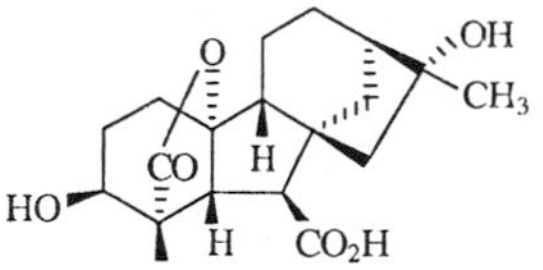

GA$_2$ (F)

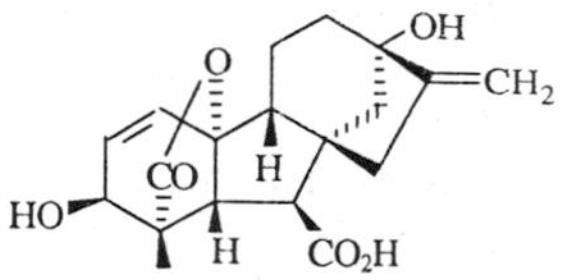

GA$_3$ (F, P)

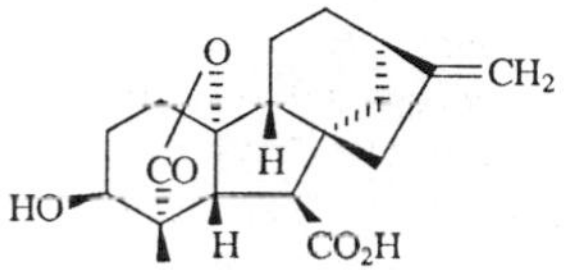

GA$_4$ (F, P)

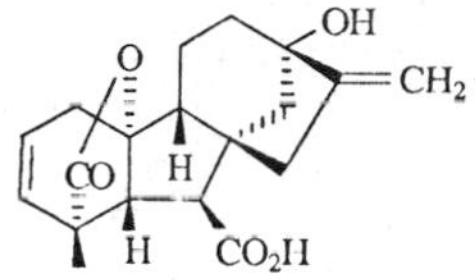

GA$_5$ (P)

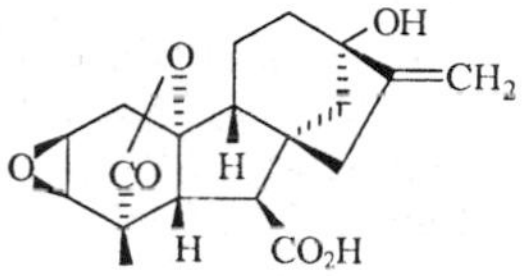

GA$_6$ (P)

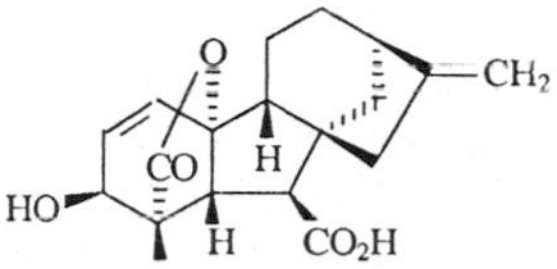

GA$_7$ (F, P)

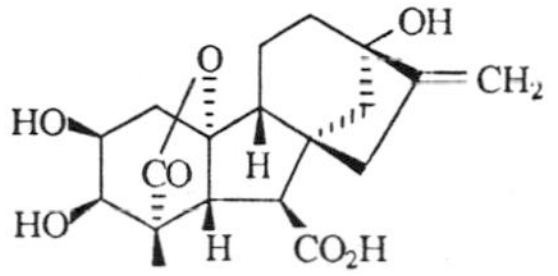

GA$_8$ (P)

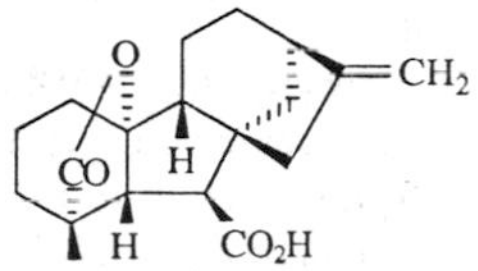

GA$_9$ (F, P)

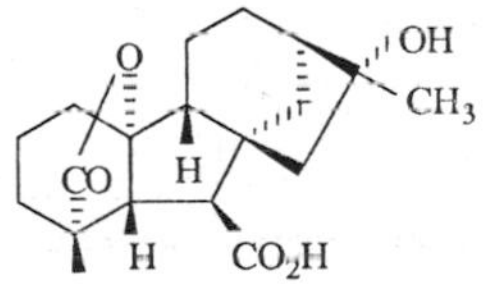

GA$_{10}$ (F)

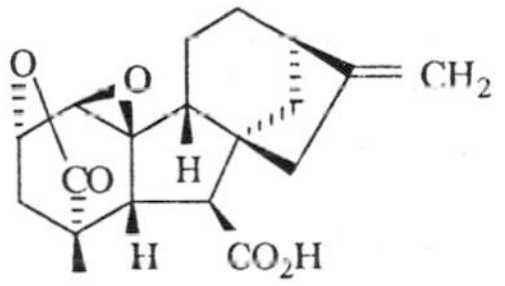

GA$_{11}$ (P)

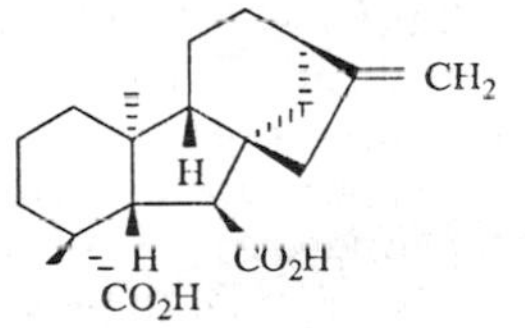

GA$_{12}$ (F)

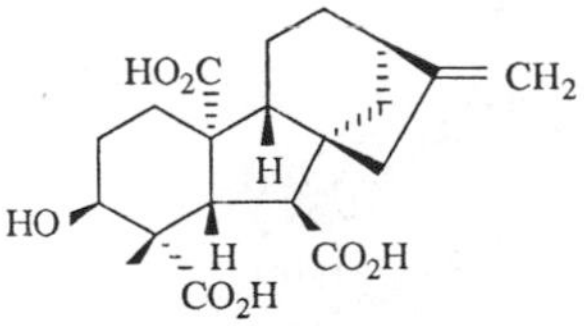

GA$_{13}$ (F, P)

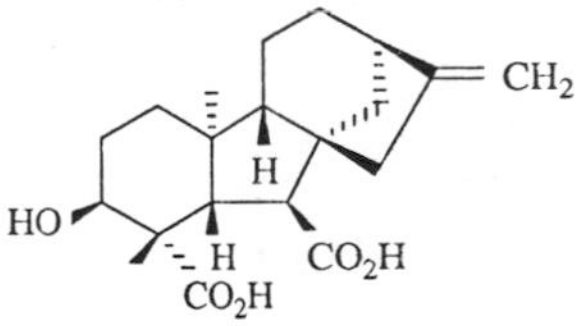

GA$_{14}$ (F)

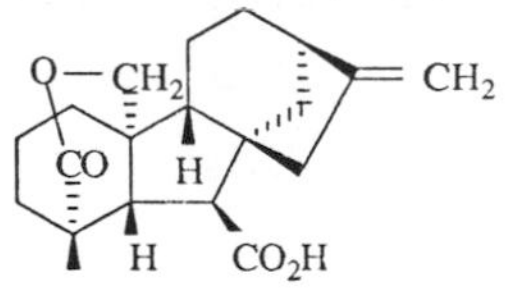

GA$_{15}$ (F, P)

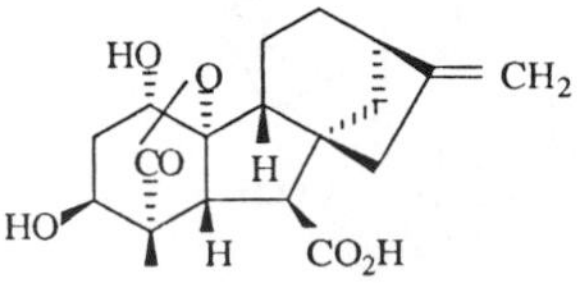

GA$_{16}$ (F, P)

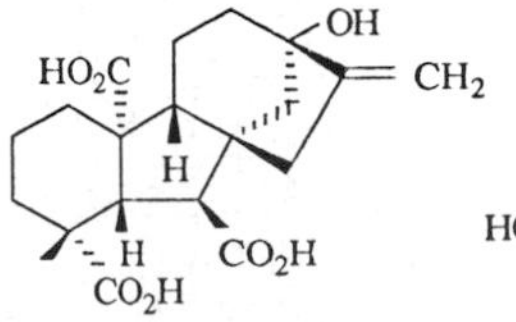

GA$_{17}$ (P)

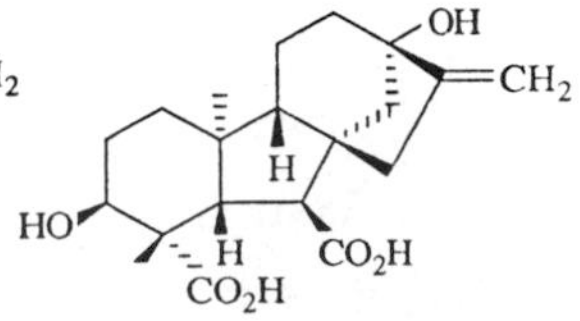

GA$_{18}$ (P)

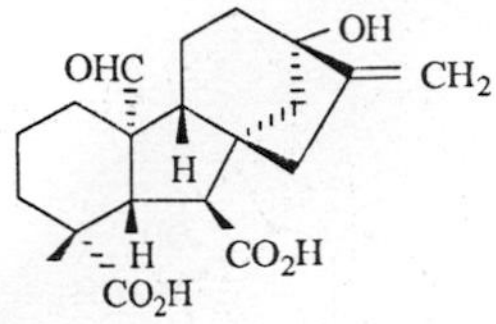

GA$_{19}$ (P)

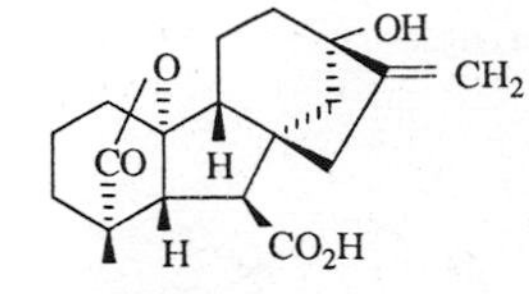

GA$_{20}$ (P)

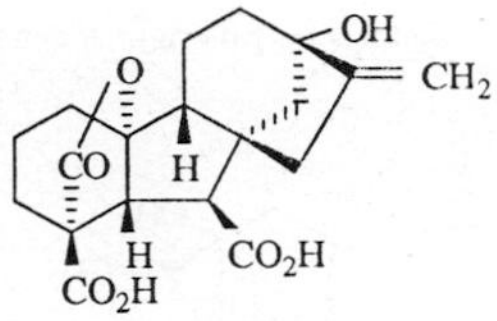

GA$_{21}$ (P)

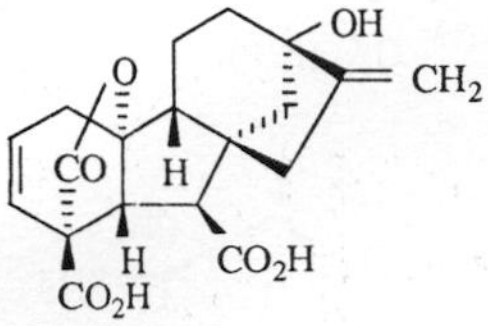

GA$_{22}$ (P)

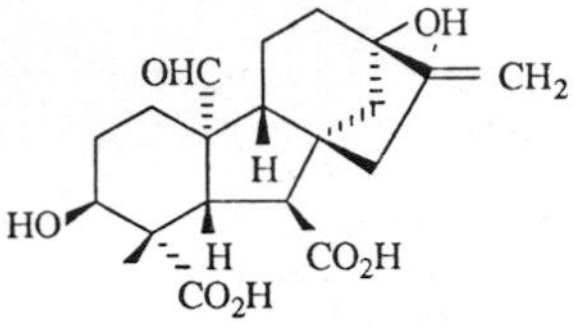

GA$_{23}$ (P)

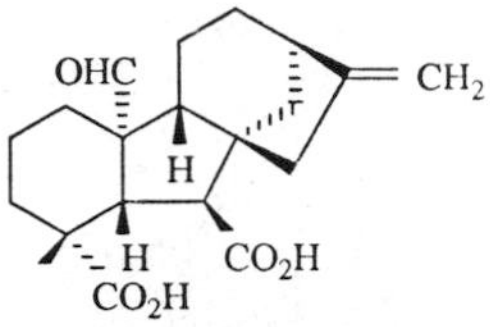

GA$_{24}$ (F, P)

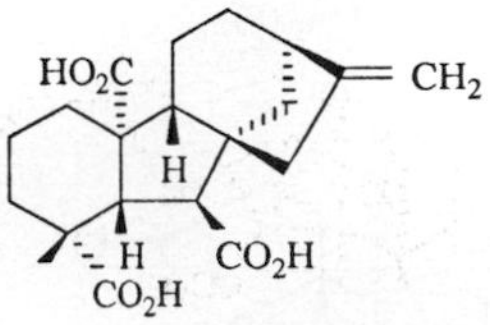

GA$_{25}$ (F, P)

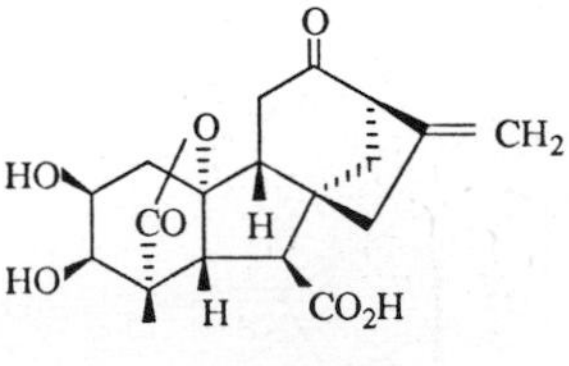

GA$_{26}$ (P)

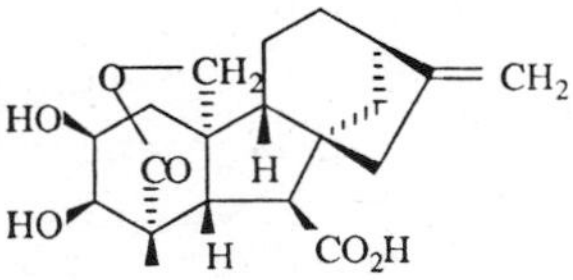

GA$_{27}$ (P)

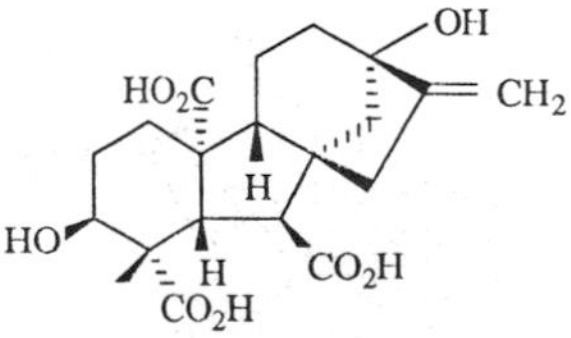

GA$_{28}$ (P)

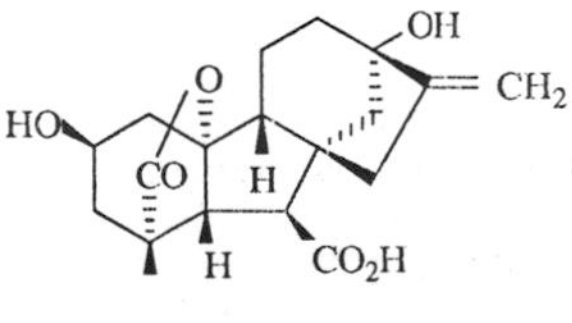

GA$_{29}$ (P)

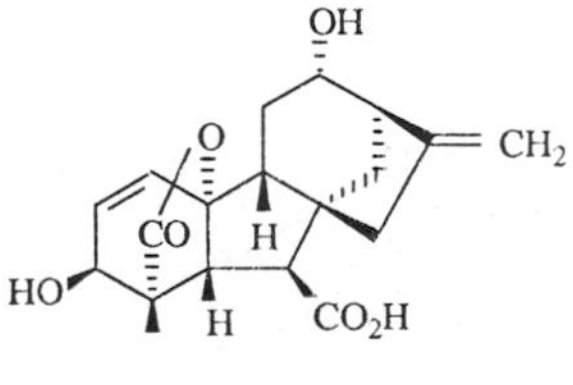

GA$_{30}$ (P)

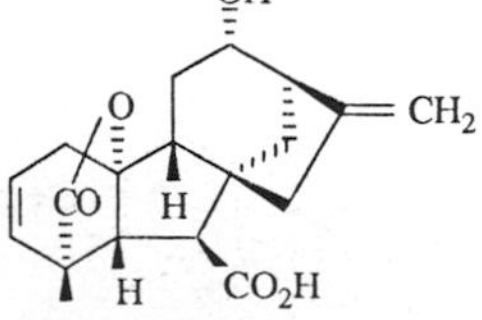

GA$_{31}$ (P)

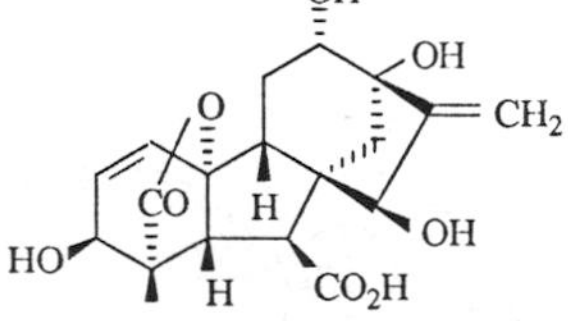

GA$_{32}$ (P)

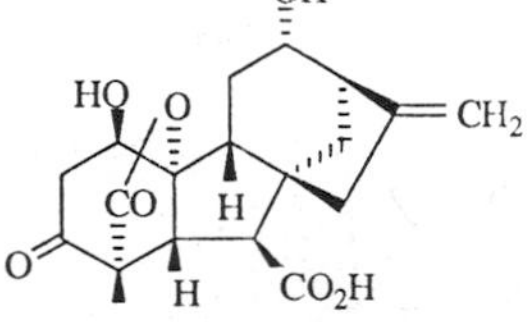

GA$_{33}$ (P)

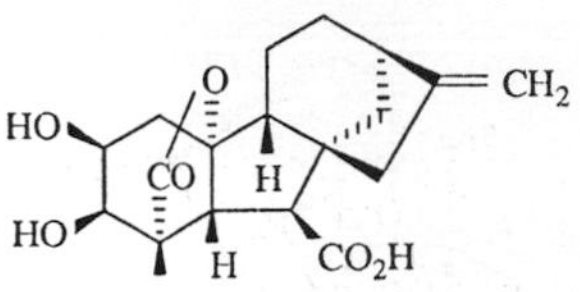

GA$_{34}$ (P)

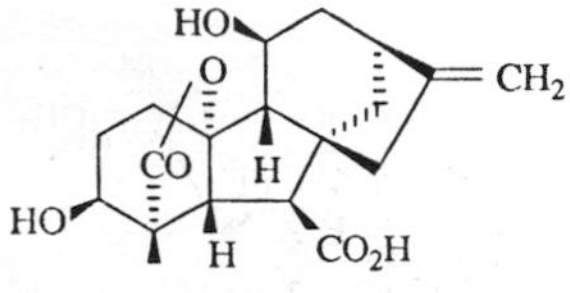

GA$_{35}$ (P)

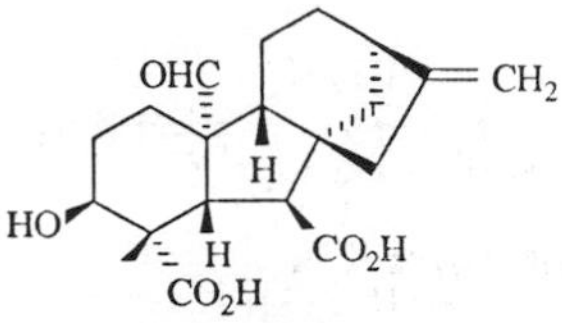

GA$_{36}$ (F)

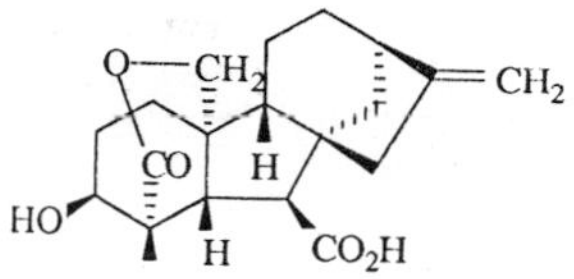

GA$_{37}$ (F,P)

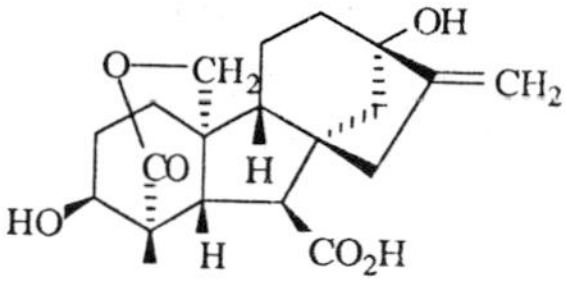

GA$_{38}$ (P)

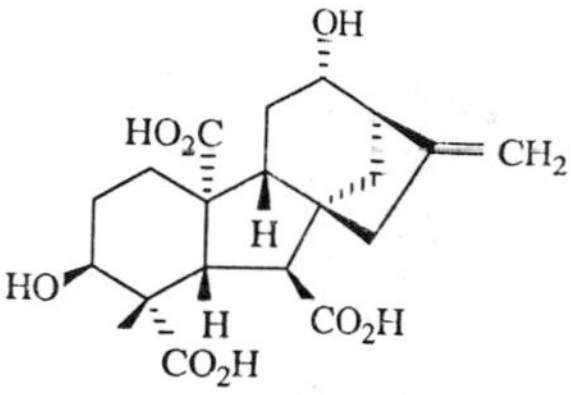

GA$_{39}$ (P)

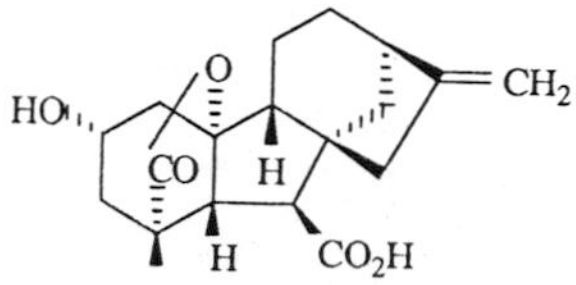

GA$_{40}$ (F)

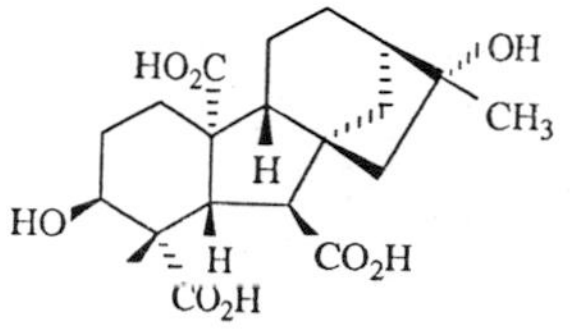

GA$_{41}$ (F)

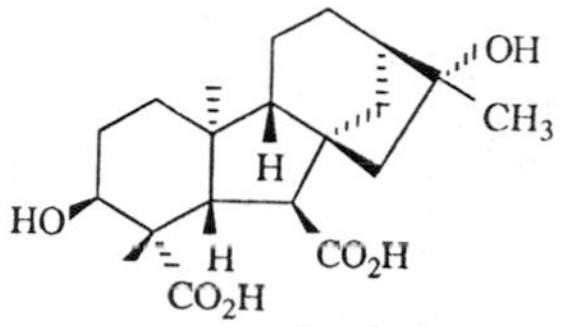

GA$_{42}$ (F)

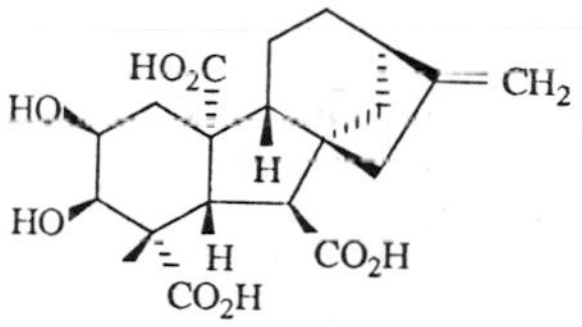

GA$_{43}$ (P)

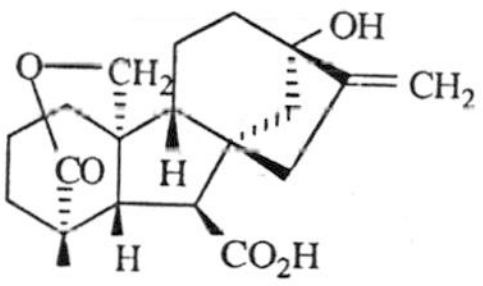

GA$_{44}$ (P)

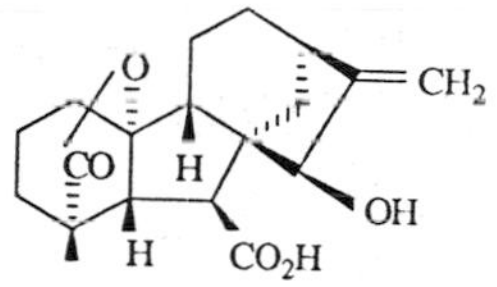

GA$_{45}$ (P)

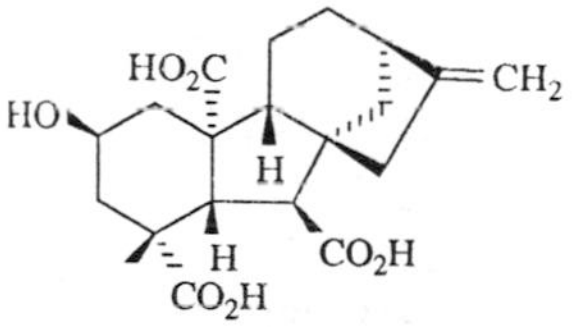

GA$_{46}$ (P)

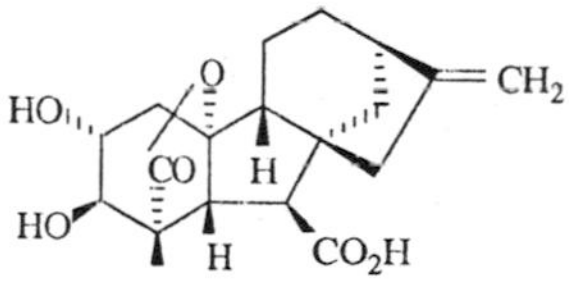

GA$_{47}$ (F)

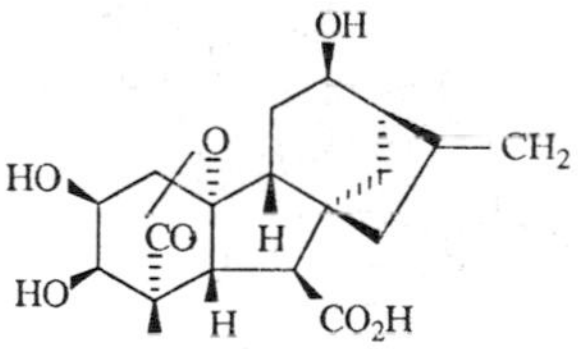

GA$_{48}$ (P)

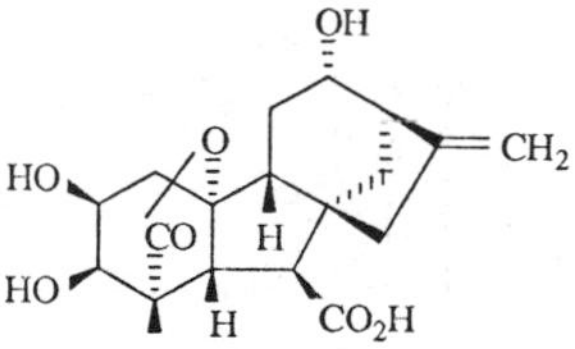

GA$_{49}$ (P)

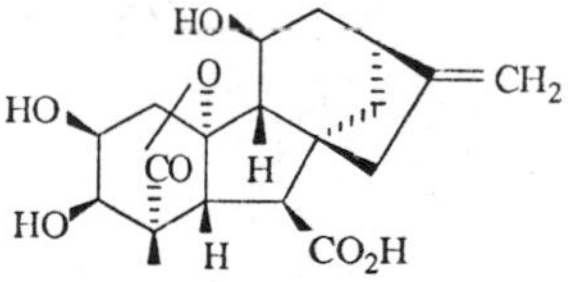

GA$_{50}$ (P)

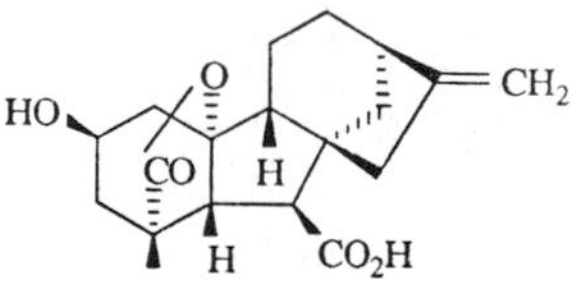

GA$_{51}$ (P)

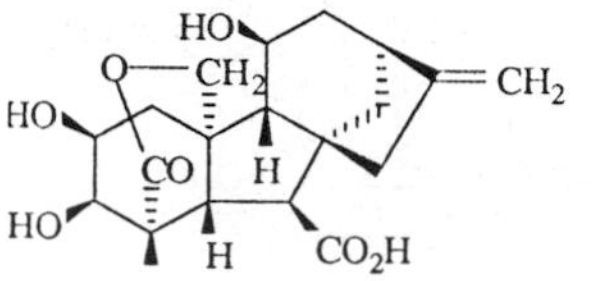

GA$_{52}$ (P)

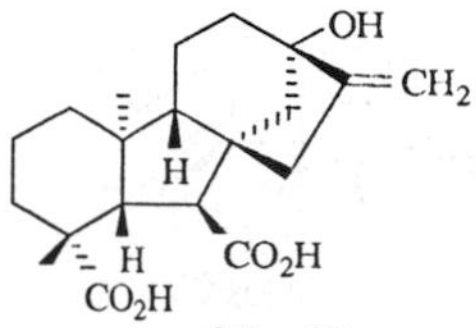

GA$_{53}$ (P)

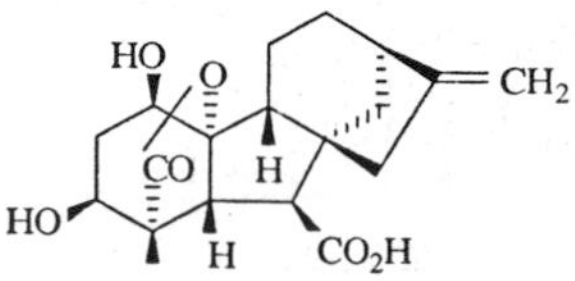

GA$_{54}$ (F, P)

95

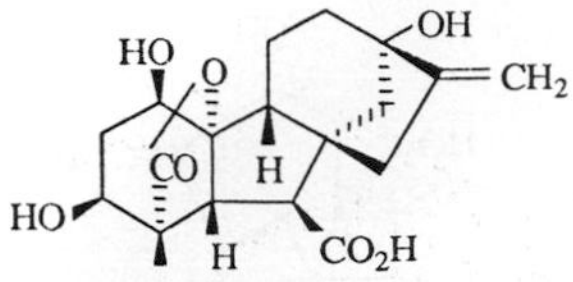

GA$_{55}$ (F, P)

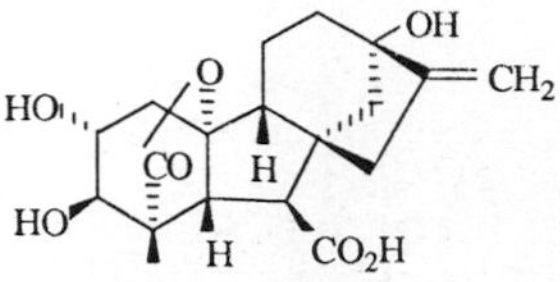

GA$_{56}$ (F)

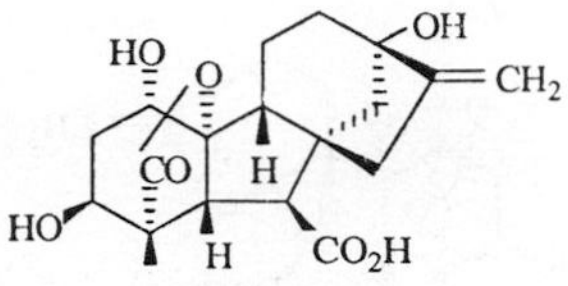

GA$_{57}$ (F)

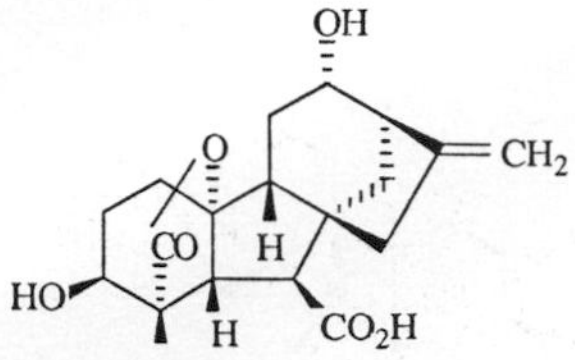

GA$_{58}$ (P)

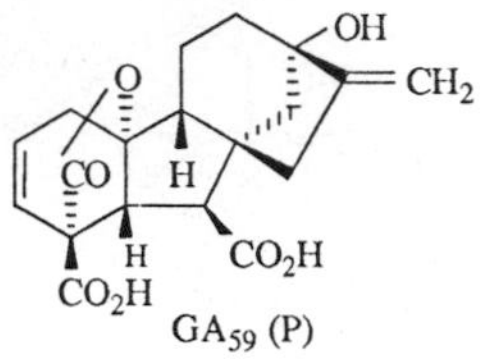

GA$_{59}$ (P)

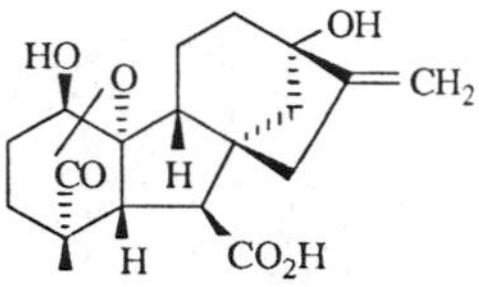

GA$_{60}$ (P)

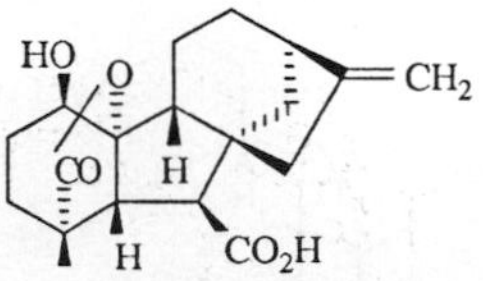

GA$_{61}$ (P)

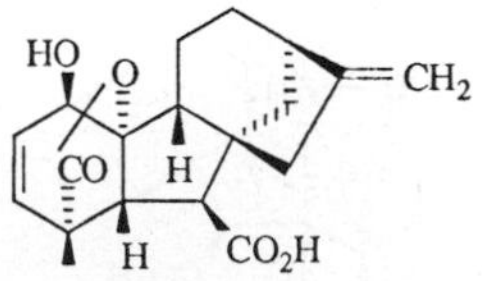

GA$_{62}$ (P)

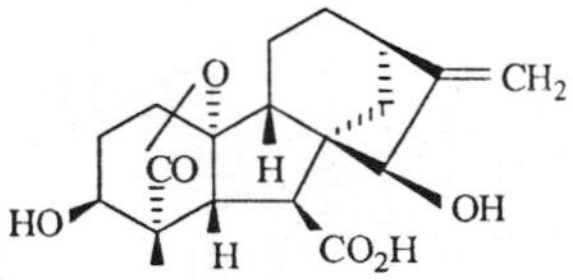

GA$_{63}$ (P)

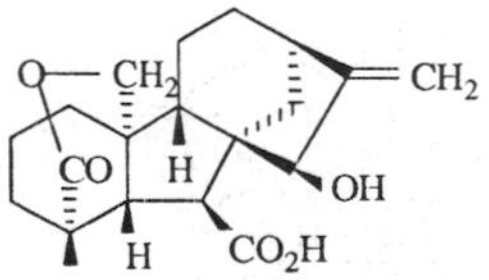

GA$_{64}$ (P)

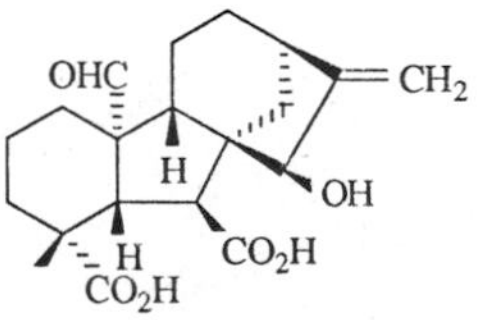

GA$_{65}$ (P)

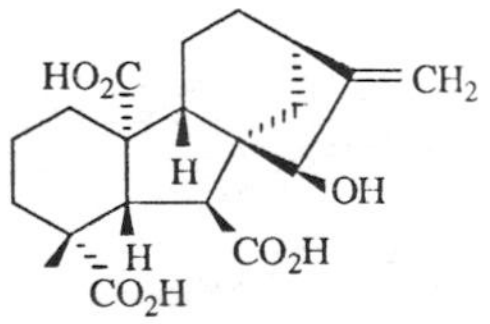

GA$_{66}$ (P)

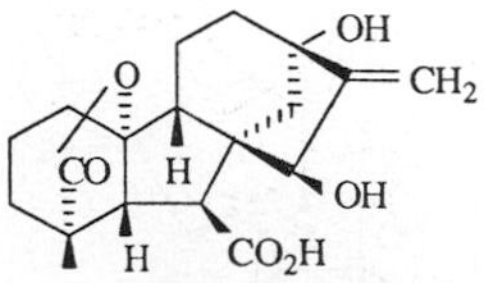

GA$_{67}$ (P)

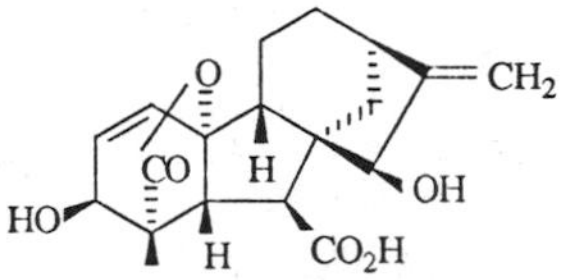

GA$_{68}$ (P)

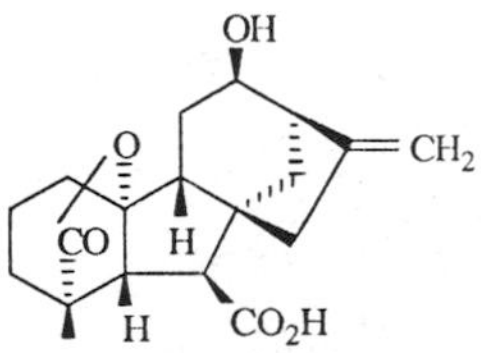

GA$_{69}$ (P)

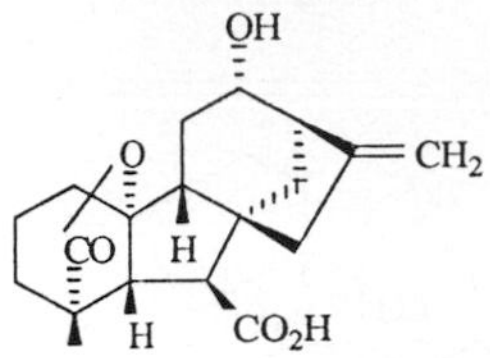

GA$_{70}$ (P)

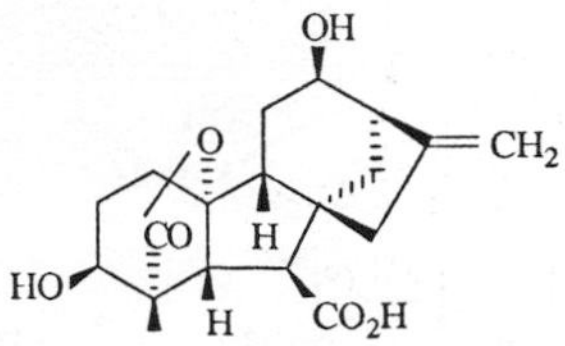

GA$_{71}$ (P)

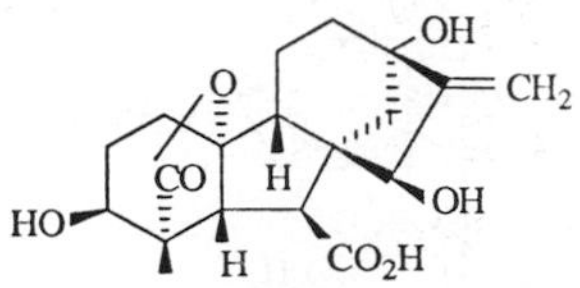

GA$_{72}$ (P)

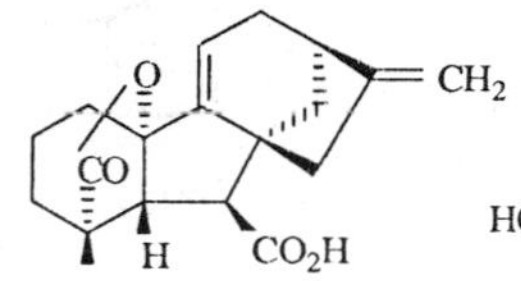

GA₇₃ (P)

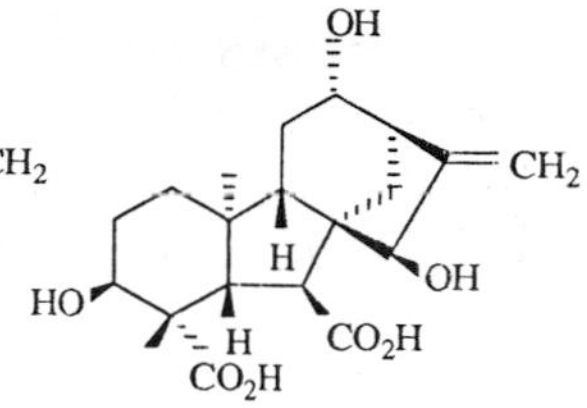

GA₇₄ (P)

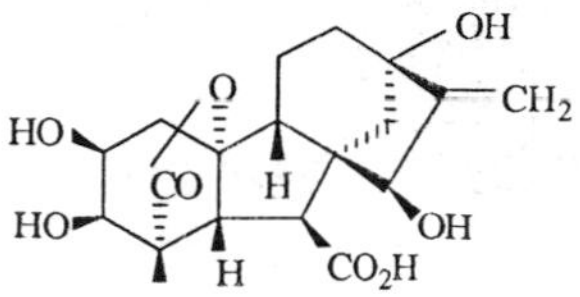

GA₇₅ (P)

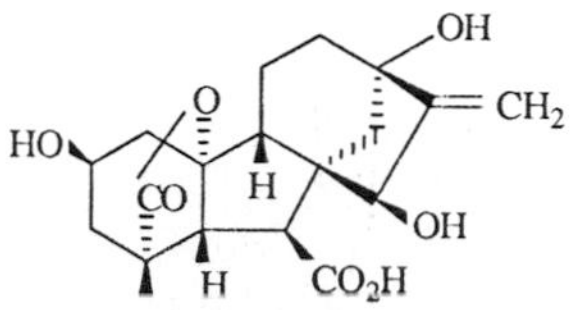

GA₇₆ (P)

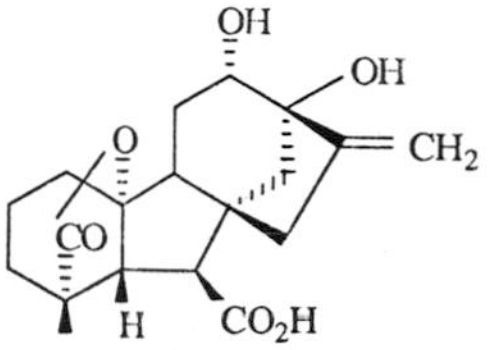

GA₇₇ (P)

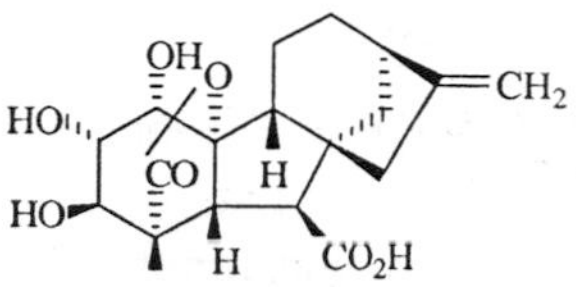

GA₇₈ (F)

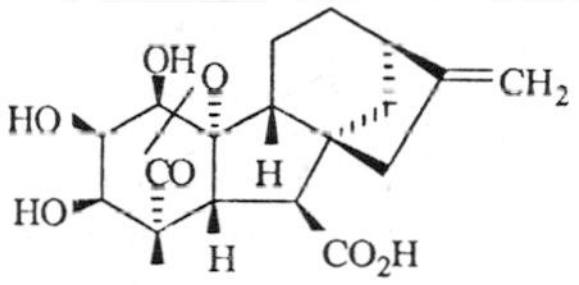

GA₇₉ (P)

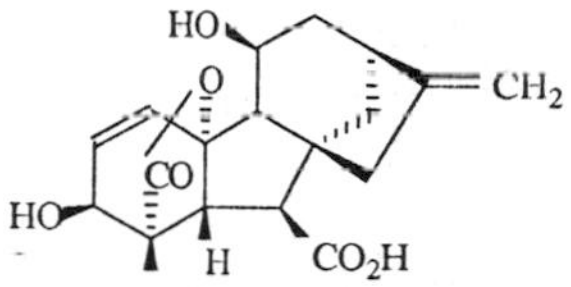

GA₈₀ (P)

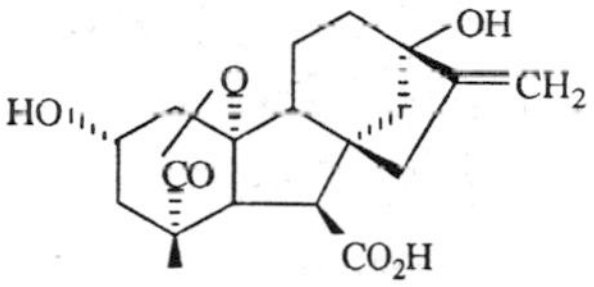

GA₈₁ (P)

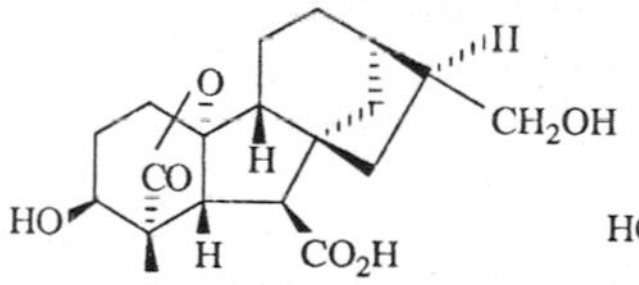

GA₈₂ (P)

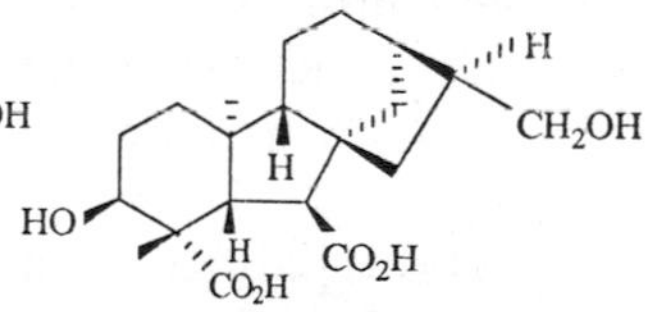

GA₈₃ (P)

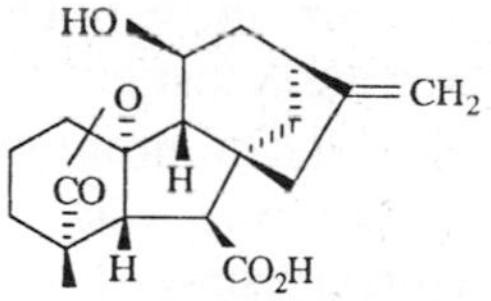

GA₈₄ (P)

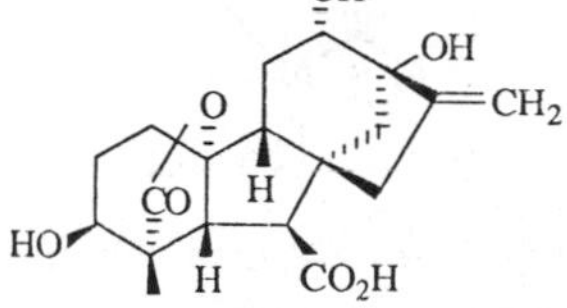

GA₈₅ (P)

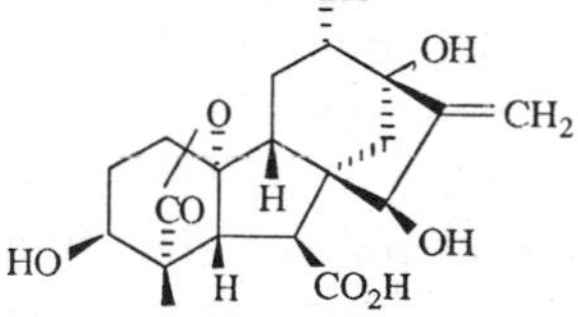

GA₈₆ (P)

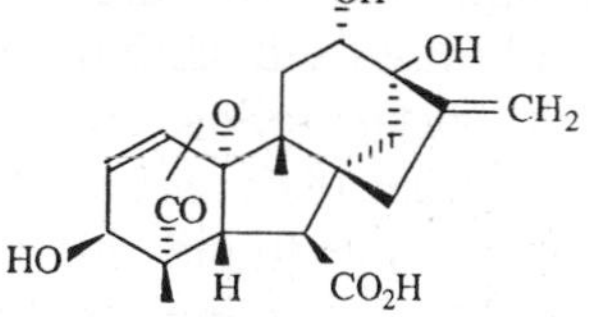

GA₈₇ (P)

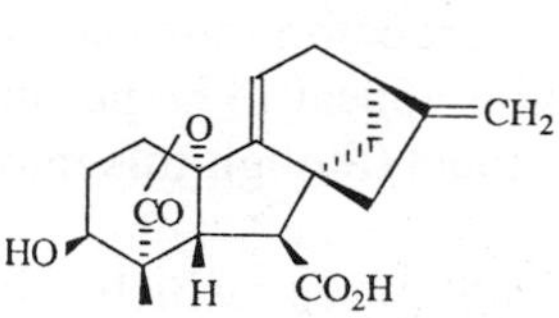

GA₈₈ (P)

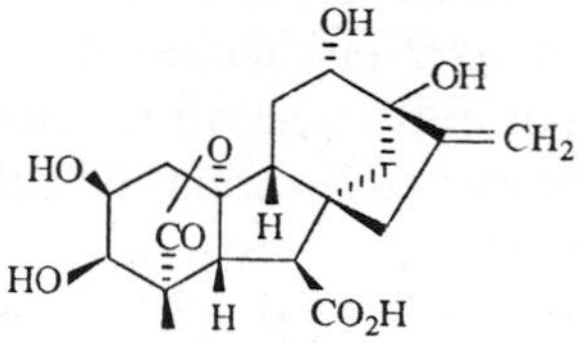

GA₈₉ (P)

B3. Cytokinin Biosynthesis and Metabolism

Brian A. McGaw and Lindsay R. Burch

School of Applied Sciences, The Robert Gordon University, St Andrew Street, Aberdeen, Scotland AB1 1HG and Department of Cellular and Environmental Physiology, Scottish Crop Research Institute, Invergowrie, Dundee DD2 5DA, Scotland.

INTRODUCTION

The view that plant cell division is chemically controlled is not new, indeed it can be traced back to the last century, but it was Haberlandt in 1913 (31) who provided the first experimental evidence for this concept. He showed that phloem diffusates could stimulate parenchymatous potato tuber cells to revert to a meristematic state. However the identity of the active substance/s in these diffusates was not established.

In the late 1940s and 1950s Folke Skoog and co-workers began a series of investigations into the nutritional requirements of tissue cultures derived from tobacco stem pith. On defined media in the presence of auxin the pith tissue cells enlarged, but failed to divide. 'Normal' cell division was restored, however, on the addition of several complex and undefined materials, most notably: coconut milk, vascular tissue extracts, autoclaved DNA and yeast extracts (58). The conclusive identification of the first active cell division promotor (or cytokinin as they are now known) was achieved in 1959 (57) when 6-(furfurylamino) purine (Fig. 1) was purified from autoclaved herring sperm DNA. This compound, though one of the most biologically active cytokinins (with activity recorded at levels of 10^{-9}M in tobacco callus growth bioassays (72)) is an artefactual rearrangement product of heated DNA and is not found in plant tissues. The first naturally occurring cytokinin was purified in 1963 by Letham (40) from immature kernels of *Zea mays* and identified as

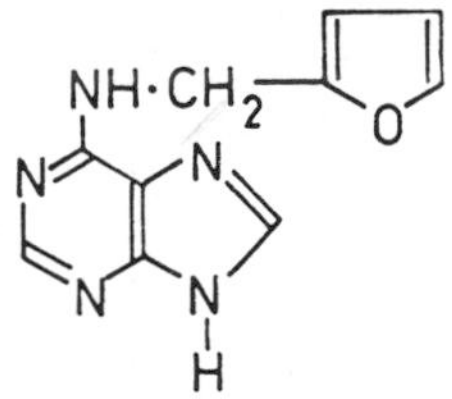

Fig. 1. 6-(furfurylamino) purine or kinetin

6-(4-hydroxy-3-methylbut -trans-2-enylamino) purine, more commonly known as zeatin. Virtually all the naturally occurring cytokinins appear to be purine derivatives (Fig. 2) with a branched 5 carbon N^6 substituent (though this may be lengthened by conjugation).

Cytokinins have been defined (80) "as substances which, in combination with auxin, stimulate cell division in plants and which interact with auxin in determining the direction which differentiation of cells takes". The term is now loosely used to cover all purine compounds that possess the

P. J. Davies (ed.), Plant Hormones, 98–117.
© 1995 *Kluwer Academic Publishers. Printed in the Netherlands.*

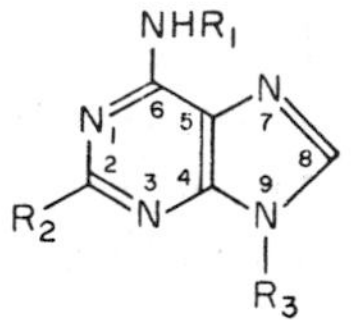

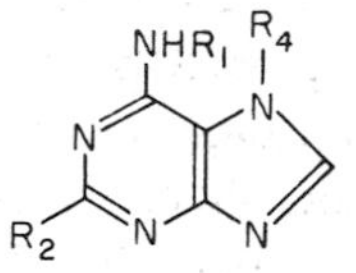

R_1	R_2	R_3	R_4	Trivial name	Abbreviation
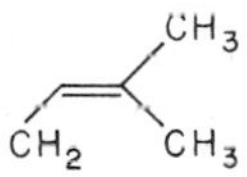 (isopentenyl: CH_2=, CH_3, CH_3)	H	H	-	$N^6(\Delta^2$-isopentenyl) adenine	iP
	H	ribosyl	-	$N^6(\Delta^2$-isopentenyl) adenosine	[9R]iP
	CH_3S	ribosyl	-	2 methylthio N^6 (Δ^2-iso-pentenyl) adenosine	[2MeS9R]iP
	H	ribotide	-	$N^6(\Delta^2$-isopentenyl) adeno-sine-5'- monophosphate	[9R-5'P]iP
	H	-	glucosyl	N^6 (Δ^2-isopentenyl) adenine-7-glucoside	[7G]iP
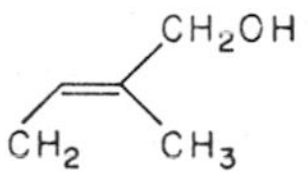 (zeatin: CH_2OH, CH_2=, CH_3)	H	H	-	*trans*-zeatin	Z
	H	ribosyl	-	*t*-zeatin riboside	[9R]Z
	H	glucosyl	-	*t*-zeatin-9-glucoside	[9G]Z
	H	-	glucosyl	*t*-zeatin-7-glucoside	[7G]Z
	H	alanyl	-	lupinic acid	[9Ala]Z
	H	ribotide	-	*t*-zeatin riboside-5'-monophosphate	[9R-5'P]Z
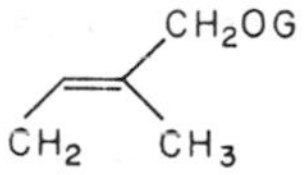 (CH_2OG, CH_2=, CH_3)	H	H	-	*t*-zeatin-O-glucoside	(OG)Z
	H	ribosyl	-	*t*-zeatin riboside-O-glucoside	(OG)[9R]Z
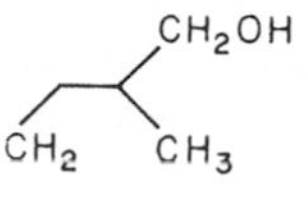 (dihydrozeatin: CH_2OH, CH_2, CH_3)	H	H	-	dihydrozeatin	(diH)Z
	H	ribosyl	-	dihydrozeatin riboside	(diH)[9R]Z
	H	glucosyl	-	dihydrozeatin-9-glucoside	(diH)[9G]Z
	H	-	glucosyl	dihydrozeatin-7-glucoside	(diH)[7G]Z
	H	alanyl	-	dihydrolupinic acid	(diH)[9Ala]Z
	H	ribotide	-	dihydrozeatin riboside-5'-monophosphate	(diH)[9R-5'P]Z
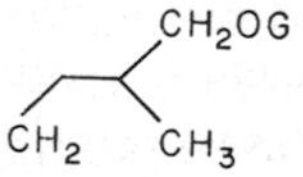 (CH_2OG, CH_2, CH_3)	H	H	-	dihydrozeatin-O-glucoside	(diHOG)Z
	H	ribosyl	-	dihydrozeatin riboside-O-glucoside	(diHOG)[9R]Z
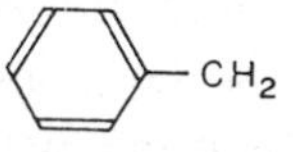 (benzyl: C_6H_5-CH_2)	H	H	-	N^6 (benzyl) adenine	BAP
	H	ribosyl	-	N^6 (benzyl) adenoside	[9R]BAP

Fig. 2. Free and tRNA cytokinin structures, nomenclature and abbreviations. *cis*-[9R]Z is a common cytokinin riboside in tRNA in which the -CH_2OH and -CH_3 in R_1 are reversed. Glucosyl [G] and ribosyl [R] refer to the ß-D-glucopyranosyl and ß-D-ribofuranosyl group.

necessary 5 carbon N^6 substituent, regardless of whether they exhibit cytokinin activity. Thus the biologically inactive (42) conjugates (i.e., the 7- and 9-D-glucosyl and 9-alanyl conjugates of zeatin and dihydrozeatin) are referred to as 'cytokinins' although they are probably the products of deactivation or control mechanisms designed to regulate levels of cytokinin activity in the plant (52).

There are now many bioassays available for the estimation of cytokinin activity. Some of these (i.e., tobacco pith callus and soybean callus growth, radish cotyledon expansion etc.) are directly related to the role of cytokinins in cell division, while other bioassay systems exploit their involvement in more specific metabolic processes (i.e., β-cyanin synthesis in *Amaranthus* seedlings, chlorophyll retention in oat leaves etc.). The structure/activity relationships of the various synthetic and naturally occurring cytokinins has been extensively studied in a number of different bioassay systems (42, 45). There is broad agreement, in these systems, on the activity of the cytokinin bases and their various conjugates. Whilst iP, Z, (diH)Z, (see Fig. 2 for abbreviations), benzyladenine (BAP) and their 9-ribosyl (and in the case of Z and (diH)Z their O-glucosyl derivatives), are generally very active, cytokinin activity is markedly reduced in the 7- and 9-glucosyl and 9-alanyl conjugates.

OCCURRENCE, NOMENCLATURE AND MODES OF CYTOKININ ACTION

Figure 2 shows the structures of the major naturally occurring cytokinins, giving the trivial names for these compounds and a list of abbreviations. There are several currently used abbreviation systems in use, but for simplicity that proposed by Letham and Palni (41) (see Fig. 2) should perhaps be generally adopted.

Several of the cytokinins in Fig. 2 occur as components of tRNA, namely: [9R]iP, *cis-* and *trans-*[9R]Z, [2MeS 9R]iP, *cis-* and *trans-*[2MeS 9R]Z. The biologically active 2-methylthio compounds are exclusive to tRNA in higher plants, but [9R]iP, *trans-*[9R]Z and, occasionally the biologically inactive *cis-*[9R]Z have also been identified as free compounds (41). These cytokinins constitute only a small proportion of the 30 or so unusual bases that are known to occur in tRNA (44), but unlike the simple methylated bases (i.e., 6-methyl adenosine, 2-methyl adenosine etc.) which are found in several regions of the tRNA, they appear to be exclusively located adjacent to the 3' end of the anti-codon (30). Recent work has shown that these cytokinins always occur as the middle residue of a triple A sequence and in tRNA species corresponding to codons with the initial letter U (i.e., isoacceptors for Cys, Leu, Phe, Ser, Trp and Tyr) (44).

Since cytokinins (and indeed other classes of plant growth regulator) appear to be important in so many aspects of plant growth and development

(34, 73) a role in plant primary metabolic processes would go a long way towards explaining their modes of action. The occurrence of cytokinins in tRNA has inevitably led to the hypothesis that these compounds are involved in the control of protein biosynthesis. That these cytokinins increase the binding affinity of the aminoacyl tRNA to the ribosomes and facilitate codon recognition is a commonly cited mechanistic explanation of their mode of action.

Unfortunately this hypothesis fails to explain certain points. First, most of the biologically active naturally occurring cytokinins do not occur as constituents of tRNA. Second, when the highly active cytokinin benzyladenine (BAP) was externally applied to tobacco callus tissue its incorporation into tRNA was extremely low and non-specific in nature (i.e., not confined to the 3' end of the anti-codon) (2). Third, the tRNA of cytokinin-requiring tobacco callus grown on BAP contains the usual complement of naturally occurring cytokinins (12).

The tRNA cytokinins may be of great importance in protein synthesis at the translational level, but it is clear that free cytokinin activity is not mediated *via* tRNA. There is strong evidence that externally applied cytokinins do stimulate protein synthesis in plants. The mechanism by which this occurs is not fully understood, but there are several possibilities. Firstly, cytokinins appear to cause an increased rate of RNA synthesis (including tRNA, rRNA and mRNA) perhaps by activation of chromatin-bound RNA polymerase (36). Secondly, cytokinins may act at the post-transcriptional level by the stimulation of polysome formation and/or the activation of polysomes in such a manner that increased recruitment of untranslated mRNA occurs. However, despite much work in these areas we remain a long way from understanding the exact mode of cytokinin action.

CYTOKININ ANALYSIS

The study of cytokinin biochemistry has benefited immensely from the development of modern analytical techniques (especially high-performance liquid chromatography and gas chromatography/mass spectrometry). For a detailed discussion of these aspects the reader is referred to Chapter E1 on instrumental methods in this volume.

Unfortunately many workers continue to rely on bioassay and co-chromatography for the quantitation and identification of cytokinins. It is nearly thirty years since the first isolation of Z from corn kernels (40) yet this compound has only been conclusively identified in a few plant species. There are less than ten plant tissues for which the application of rigorous analytical techniques have been applied to the construction of a total quantitative and qualitative picture (52). The data for some of these tissues are given in Table 1. These data are essential to our understanding of the roles these compounds play in plant development. Coupled with an

Table 1. Cytokinin levels in selected species and organs (expressed in nmoles 100 g.fwt^{-1}).

Cytokinin	V. rosea Crown gall	Tobacco Crown gall	Radish Seed	L. luteus Developing Seed	L. luteus Mature Seed	L. angustifolius Seed	L. angustifolius pod wall
Z	6.0	3.4	0.0	8.4	6.2		
[9R]Z	135.0	13.0	0.0	112.0	10.8	68.4	0.0
(diH)Z		0.7		188.7	0.01		
(diH)[9R]Z		2.2			42.5	56.7	45.3
[7G]Z		20.3	178.5				
[9G]Z	37.0	0.0	0.0				
(OG)Z	13.0	1.6		0.5	0.5	1.8	1.6
(OG)[9R]Z	64.0	2.4		10.5	5.9	7.2	7.6
(diHOG)Z				4.6	1.1	4.4	21.2
(diHOG)[9R]Z				76.1	39.1	71.8	213.6
[9R-5'P]Z		78.9					
(diH)[9R-5'P]Z		3.7					

understanding of the sites of cytokinin action and the systems designed to control the levels or expression of cytokinin activity (i.e., biosynthesis, metabolism, transport and compartmentation) we can go some way towards ascribing functions to the different cytokinin structures. Unfortunately we know very little about the sites of cytokinin action and their cellular compartmentation. There is, however, a considerable amount of data relating to the biosynthesis and metabolism of these compounds in plant tissues and an attempt is made, in the remainder of this chapter, to relate these data to our understanding of the control of cytokinin activity and the possible roles of the various cytokinin structures.

CYTOKININ BIOSYNTHESIS

The Biosynthesis of tRNA Cytokinins

The biosynthesis of tRNA cytokinins, and indeed other hypermodified tRNA bases (i.e., the methylated purines), are known to occur at the polymer level during post-transcriptional processing (32). The branched 5 carbon N^6 substituent of these cytokinins is derived from mevalonic acid pyrophosphate (16) which undergoes decarboxylation, dehydration and isomerisation to give 2-isopentenyl pyrophosphate (iPP) (Fig. 3). The latter then condenses with the relevant adenosine residue in the tRNA to give the [9R]iP moiety. It is not known whether the hydroxylation of the terminal methyl group (usually the *cis*-, but sometimes the *trans*-methyl) occurs in the 2-isopentenyl pyrophosphate or in the [9R]iP residues of the tRNA polymer.

Consistent with the above model a Δ^2-iPP:tRNA-Δ^2-isopentenyl transferase has been partially purified from *Escherichia coli* which can utilise tRNA, but not oligoadenylic acids, adenosine-5-monophosphate (AMP) or adenosine, as substrates (5). On the other hand, Holtz and Klambt (33) have purified a more catholic enzyme from *Zea mays* which was able to isopentenylate tRNA, oligo and polyadenylic acids, and adenosine. This work opens the interesting possibility that one enzyme may be responsible for the formation of both free and tRNA [9R]iP (subsequent modification of this cytokinin being dependent on enzymes which can discriminate between free and tRNA cytokinins).

The biosynthesis of the 2-methylthio derivatives has been extensively studied in *E. coli*, and occurs at the polymer level after isopentenylation. Thiolation (cysteine being the source of the sulphur atom (29)) is followed by methylation (1). S-Adenosyl methionine is the donor of the methyl group for these compounds and also for the hypermodified methylated purines and pyrimidines found in tRNA (29).

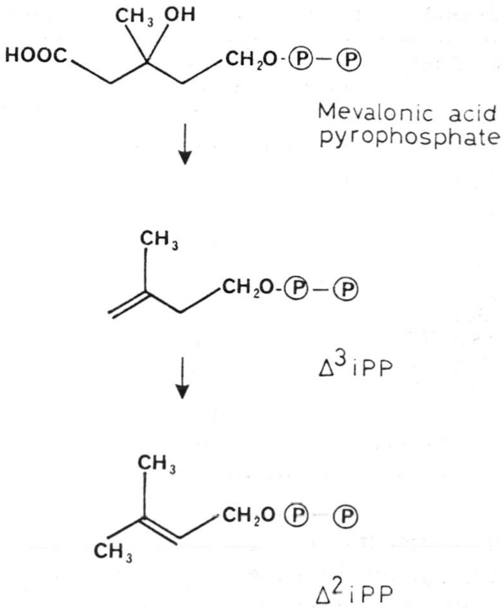

Fig. 3. The biosynthesis of Δ^2-isopentenyl pyrophosphate.

The Biosynthesis of the Free Cytokinins

Via tRNA

Since tRNA contains cytokinins, biosynthesis *via* the hydrolysis of tRNA to its constituent mononucleotides is a possibility. In certain circumstances (i.e., *Agrobacterium tumefaciens* and *Corynebacterium fascians*, the causative organisms of crown gall and fasciation diseases of plants) the cytokinins present as constituent bases within tRNA (2-methylthio cytokinins, *cis*-[9R]Z, [9R]iP and indeed the hypermodified methylated purines) are found as free compounds (40, 41). It is possible that tRNA hydrolysis may account for all the free cytokinins in these bacteria. In plant tissues, however, with the exception of [9R]iP and to a lesser extent *cis*- and *trans*-[9R]Z, which are common to both, the free and tRNA cytokinins are structurally distinct (e.g., free Z is mainly the *trans* isomer while Z present in tRNA is mainly the *cis* isomer). This remains one of the principal objections to the view that tRNA hydrolysis contributes to the pool of free cytokinins in plants. There is also further evidence against this view: first, the tRNA of cytokinin-requiring plant tissue cultures contains cytokinins (16), and secondly, tRNA 'turnover' rate is apparently not rapid enough in certain non-cytokinin-requiring tissues to account for the levels of endogenous cytokinins (32). In addition, the rate of incorporation of externally applied [^{14}C]-adenine into free cytokinins is too high to be accounted for by tRNA 'turnover' (76).

Counter to these arguments are the possibilities that there is selective 'turnover' of tRNA subpopulations rich in [9R]iP or *trans*-[9R]Z (there is evidence of this in animal tumour tissues (6)) or that a *cis-trans* isomerase system exists which can convert tRNA derived cis-[9R]Z into its trans isomer. In studies on the 'turnover' of labelled RNA, and the kinetics of its conversion to labelled cytokinins, it has been claimed (46) that cytokinin biosynthesis can only be accounted for in terms of an indirect pathway. This view is based, however, on gross tRNA and oligonucleotide turnover. Since the levels of cytokinins made available by this turnover is not known it is not possible to assess its contribution to the free cytokinin pool (another difficult parameter to measure when metabolism and other biosynthetic routes are considered). Though most workers in this field would accept the possibility that tRNA hydrolysis may contribute to the free cytokinin pool there are few who would support the view that this is the principal or sole route by which free cytokinins in plants are derived.

The situation is further confused by recent studies on cytokinin biosynthesis in crown gall tumour tissues. Crown gall is a neoplastic disease of dicotyledonous plants caused by the incorporation of a section of DNA (known as T-DNA) from the tumour inducing or Ti plasmid of the causative organism *Agrobacterium tumefaciens* into the DNA of the host cell. These tumour tissues have greatly enhanced cytokinin levels (71) and have been used by several laboratories as model systems for the study of cytokinin biosynthesis and metabolism. The free cytokinins in these tissues (being mainly *trans*-Z derivatives) have little in common with those in the bacterial tRNA or culture media where the Z compounds are all *cis* isomers. This indicated that the T-DNA genes are involved in the *de novo* (direct) synthesis of cytokinins (probably of [9R-5'P]iP [see next section below] which is then stereospecifically hydroxylated by the plant to give *trans zeatin* compounds). The bacterial enzymes responsible for the synthesis of *cis zeatin* derivatives (either by tRNA hydrolysis or stereospecific *cis* hydroxylation of free [9R-5'P]iP) are presumably not coded for by any of the genes transferred on the T-DNA. However, the most abundant cytokinin in the tRNA of *Vinca rosea* crown gall tissue is *trans*-[9R]Z (64), while cis-[9R]Z and its 2-methylthio derivatives are the predominant cytokinins in the tRNA of untransformed *Vinca* callus and the cultured bacteria, respectively. This is circumstantial evidence for an indirect pathway *via* tRNA, but new and important evidence on the function of the genes in the T-DNA has conclusively demonstrated that one of the genes (*ipt*; gene4) codes for a Δ^2-isopentenyl pyrophosphate : AMP-2-isopentenyl transferase (4) (for the biosynthesis of [9R-5'P]iP see next section below). Therefore, even in these tumour tissues, which provided perhaps the best evidence for a tRNA turnover pathway there seems little doubt that their enhanced cytokinin levels are due to the insertion of a *de novo* synthesising capability on the T-DNA.

De Novo (Direct) Cytokinin Biosynthesis

The above transferase enzyme from the T-DNA of *A. tumefaciens* is not the first enzyme to be investigated that could be responsible for *de novo* biosynthesis. Indeed an enzyme has been characterised in cell-free preparations of the slime mould *Dictyostelium discoideum* that catalyses the synthesis [9R-5'P]iP from AMP and 2-iPP (78). More recently an AMP-Δ^2-isopentenyl transferase has been partially purified from cytokinin-autonomous tobacco callus tissue (14) and its substrate requirements have been studied in detail. The reaction does not occur at the base or nucleoside level (i.e., adenine or adenosine are not substrates) and the side chain must be the correct isopentenyl isomer and contain a pyrophosphate group (14) (Fig. 4). The enzyme has not been studied with tRNA, oligoadenylic acids, polyadenylic acids or other potential polymeric substrates.

Infection of plants with *Agrobacterium tumefaciens*, which contains a tumour inducing (Ti) plasmid with genes coding for the biosynthesis of both auxins (genes 1 & 2) and cytokinins (*ipt* gene; gene 4), causes overproduction of auxin and cytokinin when transferred from the bacterium to the host genome, with the resulting characteristic tumour morphology.

The role of *ipt* in the *de novo* synthesis of endogenous cytokinins (as opposed to cytokinins supplied exogenously) has been shown in a number of experiments where plants have been transformed with the gene for *ipt*, and where a number of phenotypic changes were observed (50, 53, 61, 70, 74).

Tobacco and *Arabidopsis thaliana* transformed with the *ipt* gene from *A. tumefaciens* under the control of a heat shock (hsp) 70 promoter, resulted in an observable accumulation of *ipt* mRNA and an increased level of Z (52 fold), [9R]Z (23 fold) and [9R-5'P]Z (2 fold), after heat induction. A marked developmental effect was seen on the release of axillary buds, reduced stem and leaf area and an under-developed root system. It is interesting to note

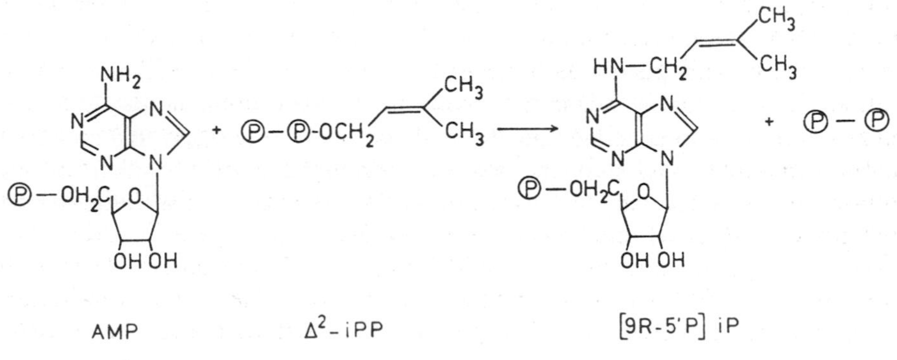

AMP Δ^2-iPP   [9R-5'P] iP

Fig. 4. *De novo* cytokinin biosynthesis.

105

that the observed large differences were in the *trans*-Z form of cytokinin. The stereospecific hydroxylation of the product of the reaction catalysed by *ipt*, [9R-5'P]iP, must be extremely rapid, as was the interconversion from nucleotide to the base, since this was the predominant product observed in both transgenic plants. In fact this hydroxylation step must normally occur with some rapidity in plants, since [9R-5'P]iP, [9R]iP and iP are rarely found as free compounds in most plants. Similarly when [^{14}C]-adenine was fed to *Vinca rosea* crown gall tissue labelled Z derivatives were readily recovered, but no radioactivity was detected in HPLC fractions corresponding to the elution volume of iP cytokinins (66, 76). From feeding studies the indication is that the hydroxylation step occurs at the level of the nucleotide; with [^{14}C]-adenine application to *Vinca rosea* crown gall tissue the peak of radioactivity in AMP preceded that of [9R-5'P]Z, but unlike the transgenic plants transformed with *ipt*, the amount of radioactivity in [9R-5'P]Z always exceeded that of the [9R]Z (76). A microsomal cytochrome P-450 preparation has been shown to be involved in the hydroxylation of iP and [9R]iP to Z and [9R]Z. The activity required NADPH and was inhibited by CO and metyrapone (a cytochrome P-450 inhibitor) (19). Where [^{14}C]-iP has been supplied to plant tissues it is always stereospecifically *trans* hydroxylated to give labelled Z derivatives (65). Very little iP or [9R-5'P]iP was recovered in this study.

The results of these experiments and the enzyme/cell free studies indicate that a *de novo* biosynthetic route is a strong possibility in plant tissues. However, there are great difficulties in studying the biosynthesis of compounds that occur at extremely low levels whose putative precursors (i.e., AMP and Δ^2-iPP) are major products of plant metabolism. Very careful consideration should be exercised in interpreting these results. For instance, the 'high' incorporations of adenine into the cytokinins is taken as circumstantial evidence for a *de novo* route, but it is possible that aberrant synthesis is occurring as a result of exogenous application. Second, it is not possible to accurately measure specific activities of cytokinin metabolites that are probably not pure (i.e., co-chromatography of radioactivity with a cytokinin metabolite on HPLC is no guarantee that we are dealing with a radioactive cytokinin). In our experience quite high levels of 'incorporation' of [^{14}C]-adenine into cytokinins can be observed in crude preparative HPLC runs. On further purification with different HPLC systems most of the radioactivity could be separated from the cytokinin fractions. We were unable to purify to constant specific activity. Future work in this area must attempt to adopt those standards set by biosynthetic studies in other areas of natural product chemistry. Convincing evidence for either biosynthetic route will not emerge until this approach is adopted.

CYTOKININ METABOLISM

The formation of the free cytokinins is presumed to begin with the biosynthesis of [9R-5'P]iP. This compound can be considered the precursor of the remaining twenty or so free cytokinins illustrated in Fig. 2 and quantified, for certain tissues, in Table 1. As already discussed it appears that the [9R-5'P]iP is rapidly and stereospecifically hydroxylated to give Z derivatives. This hydroxylation, therefore, is the first important metabolic event. From this point on many metabolic reactions occur with the consequent generation of aglycones, glucosides, ribosides, ribotides, amino acid conjugates, reduction and oxidation products (Fig. 5). These metabolic events can be categorised under four broad headings, namely: conjugation, hydrolysis, reduction and oxidation. The occurrence of these events is discussed in relation to our knowledge of their enzymology and an attempt is made to assign a role to the various metabolites in the regulation and expression of cytokinin activity.

Conjugation

Ribosides and Ribotides

The cytokinin ribosides and their 5' mono-, di- and tri-phosphates are probably the most abundant naturally occurring cytokinins (42, 52, 71). Ribosyl conjugation is always confined to the 9 position of the purine ring. When ^{14}C labelled zeatin is externally applied [9R]Z and [9R-5'P]Z are usually important metabolic products (37). Where time course studies have been performed it has been found that the nucleotides are the dominant metabolites in the early stages following the feed and that rapid hydrolysis of these compounds then occurs to give more 'stable' or less biologically active products (i.e., N-glucosides or products of oxidative side chain cleavage). The metabolism of externally applied [9R]Z and [9R]iP have also been studied and again phosphorylation was followed by hydrolysis and further metabolism (14, 43). The process involved in metabolic interconversion of cytokinin bases, ribosides and nucleotides are extremely important in all plant tissues studied, and there is evidence that they may be catalysed by the same enzyme system that metabolises the corresponding adenine base, riboside and nucleotide. However, the enzyme systems involved have been much less well defined than either the biosynthetic processes discussed above or the degradative processes discussed in the succeeding section.

The reactions catalysed by 5'-nucleotidase and nucleosidase have been shown to occur in wheat germ extracts (17, 18) and in tomato leaf and root extracts (9, 10), but in each case the enzyme showed lower affinities for the N^6-substituted purine than for the purine analog. That cytokinin interconversions involve purine metabolising enzymes has also been supported by feeding studies *in vivo* (38). Even so, we still understand very little concerning the control of these processes *in vivo*, particularly as the

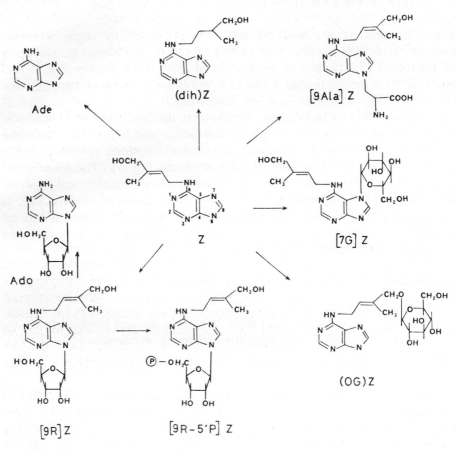

Fig. 5. Various metabolites of zeatin (Z). Ade and Ado refer to adenine and adenosine, respectively.

concentrations of adenine, adenosine and AMP in cells are normally several orders of magnitude higher than those of cytokinins. However, it has been suggested that even those enzymes which are substantially inhibited may be catalysing reactions at rates fast enough to account for the low rates, in absolute terms, of interconversion of cytokinins (75).

The enzyme adenine phosphoribosyl transferase, probably the main salvage route in plants for converting adenine to AMP, has been shown to catalyse the phosphoribosylation of cytokinin bases from a number of plant sources, including wheat germ (20) tomato (11), *A. thaliana* (55), and *Acer pseudoplatanus* (23, 69). Again the affinity of the enzyme for the cytokinin base was lower than that for adenine.

Uptake of externally applied cytokinin bases into cells has been associated with rapid formation of the corresponding nucleotide (3) involving phosphoribosylation by a cell wall bound adenine phosphoribosyl transferase

(APRT). However, the importance of APRT in the uptake of externally supplied cytokinin bases has been questioned in work carried out on APRT deficient mutants of *A. thaliana* (54), in which BAP is effectively metabolised to the riboside and glucosides even when conversion to BAMP is strongly reduced (55). This may of course be an alternative pathway which of necessity is utilised by the mutant plant, with the pathway using APRT predominating under normal circumstances.

An alternative pathway for conversation of the cytokinin base to the nucleotide would be by a two step reaction catalysed by adenosine phosphorylase and adenosine kinase. Both enzymes have been partially purified from wheat germ extracts and shown to ribosylate the cytokinin base (adenosine phosphorylase) (21) and phosphorylate the cytokinin riboside to give the corresponding nucleotide (adenosine kinase) (15). Again the affinity of both enzymes for the N^6-substituted purine was less than for the adenine analogue.

In the APRT-deficient *A. thaliana* (55) there would appear to be a significant activity of an adenosine phosphorylase, giving rise to the observed riboside formation. An active adenosine kinase, which would phosphorylate [9R]BAP to the corresponding nucleotide, would appear to be either reduced or absent in APRT deficient mutants, since there was little [9R]BAP formed. Although the action of an APRT may predominate in the uptake of cytokinins, the work with APRT mutant shows that there are alternative pathways that may also be responsible for cytokinin uptake, which would act in concert with APRT.

The ribosides and their aglycones (i.e., Z and iP) are extremely active in bioassay (42). Whether or not one or all of these compounds are the 'active' molecules is not known because of the extensive interconversions that they undergo when externally supplied. The same applies to the readily hydrolysable nucleotides.

Glucosides

Unlike the ribosyl conjugates glucosylation is not confined to the 9 position of the purine ring (Fig. 2). The 7- and 9-glucosides and side chain O-glucosides are major cytokinins in certain tissues (Table 1) and 3-glucosides (43, 47) have also been detected as metabolites of externally applied cytokinins. N-glucosylation has been studied in great detail in radish tissues (where [7G]Z is the predominant free cytokinin). Two glucosyltransferases, which separated on DEAE columns, have been partially purified from radish cotyledons (24). Both enzymes catalyse the formation of 7- and 9-glucosides of benzyladenine using UDPG or TDPG as the glucose source. However, the ratio of their products ([7G]BA/[9G]BA) was very different, being approximately 10 and 1.5. The latter enzyme has now been studied in relation to a large number of different naturally occurring and synthetic cytokinins (25). In all cases the rate of glucosylation and the [7G]/[9G] product ratio were determined. Interestingly, some cytokinins (e.g., *cis*-Z and *trans*-Z and BA) gave appreciable quantities of 9-glucosides, but others (e.g., (diH)Z and (OG)Z) gave only traces of 9-glucosyl products. There was little difference in the overall rates of N-glucosylation for these cytokinins.

In radish tissues N-glucosylation is the predominant fate of externally applied cytokinin. However, in contrast to the cell free work, [9G]Z was not identified as a metabolite when labelled zeatin was externally applied (51) and (diH)Z gave considerable quantities of its 3- and 9-glucosides. It is possible that other, as yet unidentified, enzyme systems are also involved in the formation of cytokinin glucosides.

N-Glucosyl conjugation is considered to be important in the regulation of cytokinin activity levels. The 7- and 9-glucosides are biologically inactive (42) and are extremely stable (i.e., are not hydrolysed to their active aglycones) in the tissues in which they are formed (67). However, cytokinin N-glucosides do appear to be available as a source of active cytokinins in plant cells transformed with the Ri plasmid of *A. rhizogenes*, the Ri plasmid containing specific genes (*rol C*) for β-glucosidases which are capable of hydrolysing cytokinin N-glucosides (27). Some plant tissues do not appear to form N-glucosides (either as endogenous compounds or on exogenous application of other cytokinins) and in these cases other methods of inactivation are adopted (i.e., oxidative side chain cleavage and/or the formation of amino acid conjugates).

In contrast, the O-glucosides appear to be candidates for cytokinin storage forms rather than as a means of inactivating cytokinins, as with the N-glucosides. They do seem to be biologically very active, either as the glucoside itself or as the products of β-glucosidase activity. Indeed several metabolic studies (51, 52) with labelled (OG)Z indicate that it is readily hydrolysed to its aglycone. However, in several other studies it has been noted that exogenous application of labelled Z can lead to the formation of large amounts of O-glucosyl derivatives which remain unmetabolised over long periods (68). These findings have lead to the proposal that O-glucosides may be cytokinin storage forms, being stable and yet readily metabolised under certain conditions to yield biologically active cytokinins when required. Further evidence for this view has been found in work on the endogenous cytokinins of *Phaseolus vulgaris* during different stages of development (62). Decapitation of these plants leads to a rapid rise in the amount of (diHOG)Z in their leaves. On lateral bud development the levels of this compound fall dramatically.

Amino Acid Conjugates

[9Ala]Z (lupinic acid) and (diH)[9Ala]Z (dihydrolupinic acid) are minor endogenous cytokinins in the immature pod walls and root nodules of *Lupinus luteus* (77). The 9-alanyl conjugates of Z and BAP have been identified when these cytokinins were exogenously supplied to *Lupinus spp.* (68), immature apple seeds and derooted *Phaseolus* seedlings (41).

An enzyme, a β-(6-allylaminopurine-9-yl) adenine synthase, has been characterised (26) from developing seeds of *L. luteus*. This enzyme utilises O-acetyl serine as the donor of the alanine residue, but is capable of conjugating a large number of different purine substrates (though the presence of an N^6 substituent greatly increases the rate of conjugation).

The role of these conjugates is probably similar to that of the N-glucosyl cytokinins. They are biologically inactive (42) and extremely stable compounds (68). As with glucosylation, the formation of amino acid conjugates is a common response of plant tissues to xenobiotic material and presumably, by rendering them more water soluble, facilitates their deposition in the vacuole.

Hydrolysis

The hydrolysis of cytokinin ribosides and ribotides is a major component in the metabolism of externally applied cytokinins, especially in the early stages of experiments associated with uptake. Two 5'-ribonucleotidase enzymes (differing in molecular weight) which catalyse the hydrolysis of [9R-5'P]iP to [9R]iP (17) and an adenine nucleosidase (18) which converts [9R]iP to iP have been characterised in wheat germ. Unlike the conjugating enzymes from wheat germ, N^6 substitution made little difference to the enzyme affinity (i.e., AMP or adenosine were equally good substrates as [9R-5'P]iP or [9R]iP).

The N-glucoside and N-alanyl conjugates are extremely stable (67, 68) in the tissues in which they are synthesised and therefore no enzymes capable of hydrolysing these compounds have been investigated. In certain circumstances, however, the O-glucosides are readily hydrolysed. Almond β-glucosidase (emulsin) preparations can cleave the O-glucosyl group in these compounds (though it is not capable of hydrolysing N-glucosyl conjugates), but this important enzyme has not been studied in any of the tissues where the endogenous cytokinins have been accurately quantified or their metabolism studied. Since the O-glucosides may occupy a key role as cytokinin storage forms this is an important area where future research could focus in an attempt to manipulate plant growth/development by controlling the size of the active cytokinin pool.

Reduction

Dihydrozeatin derivatives are commonly found in plant tissues and are frequent metabolites of applied Z (usually conjugated as their 9-ribosides, 9-ribotides or O-glucosides) (41, 52). In bioassay (diH)Z and its conjugates are equally as active as their zeatin analogues (42). In studies where (diH)Z has been externally supplied to plants it appears to be more 'stable' than Z (63). This may be because it is not a substrate for cytokinin oxidase (an enzyme which cleaves the N^6 side chain - see next section). As a more 'protected' species (diH)Z may be important in the maintenance of cytokinin activity levels in an oxidative environment. An enzyme which converts Z to (diH)Z has been partially purified from *Phaseolus* embryos and appears to be an NADPH-dependent Z reductase. The enzyme was specific for Z while the affinity of the enzyme for iP and [9R)Z was negligible and would support the central role of the free base in cytokinin action (56).

Oxidation

Oxidative side chain cleavage of externally applied Z, iP, [9R]Z and [9R]iP to give adenine, adenosine and adenine nucleotides is the major fate of these cytokinins in many tissues (52) (Fig. 6). Like the formation of N-glucosyl or N-alanyl conjugates, side chain cleavage leads to the irreversible loss of cytokinin activity and may be important in the regulation of cytokinin activity levels. An enzyme, cytokinin oxidase, has been partially purified from tobacco tissue (81), corn kernels and *Vinca rosea* crown gall tissue (49) wheat (37), *Phaseolus, vulgaris* (13) and purified from kernels Z. *mays* (8). In some of these tissues the specificity of the enzymes has been investigated with a large number of naturally occurring and synthetic cytokinins. Z, [9R]Z, iP, [9R]iP, [7G]Z, [9G]Z and [9Ala]Z all served as substrates for the enzymes, but side chain reduction (i.e., (diH)Z derivatives), the relocation of the Δ^2 double bond to Δ^3, the substitution of other functionalities (i.e., benzyladenine, kinetin) and the presence of an O-glucosyl group rendered the cytokinin resistant to oxidation. Some tissues (notably radish) seem to be unable to cleave the N^6 side chain. In these cases N-glucosylation is the mechanism by which cytokinin activity is controlled. In other tissues, for instance the *Lupinus spp.*, both methods of inactivation are utilised (52).

The mechanism of the oxidative cleavage is not fully understood, but 3-methyl-2-butenal (7) and adenine (48) have been unambiguously identified by GC/MS as the products of cytokinin oxidase action on iP (Fig. 6). It has been postulated (81) that an unstable imine intermediate is formed and this appears to have been confirmed by physico-chemical studies (37).

Work with *Phaseolus spp.* (13, 35), suggests that oxidase levels can be

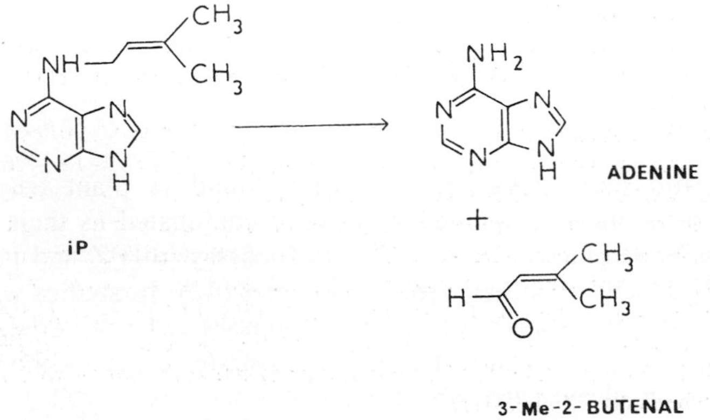

Fig. 6. The products of cytokinin oxidase action on iP.

regulated by the supply of cytokinin substrate, after the supply of exogenous cytokinin a rapid increase in oxidase activity was observed in these tissues.

Much information has been accrued concerning the biochemistry of oxidase (now known to be a copper containing amine oxidase (8) in Z. *mays*), but we are still unclear about its role *in vitro*; especially its compartmentation in tissue with both high levels of cytokinin and high apparent enzyme activity (e.g., Z. *mays* kernels and *Vinca rosea* crown gall tissue (49)). Immunohistochemistry using antibodies to cytokinin oxidase may help to resolve some of these questions.

CONCLUSION

The purification and identification of cytokinin bases, ribosides and nucleotides is now a relatively straightforward matter. With the development of selected ion monitoring/isotope dilution mass spectrometric quantitation and radio-immunoassay we are now able to accurately quantify extremely low levels of cytokinins. For the first time, therefore, it is possible to study the endogenous metabolism of cytokinins without recourse to the artificial application of labelled material. More recently genetically engineered mutants with and without genes responsible for cytokinin biosynthesis have been produced. These are now yielding important information on cytokinin biochemistry/physiology. It is hoped that these developments will eventually lead to an understanding of the mechanism and site/s of cytokinin action. If the quantitation of cytokinins at the cellular level can also be achieved we should then be in a position to explain how endogenous cytokinin activity levels are controlled.

References

1. Agris, P.F., Armstrong, D.J., Schafer, K.P., Soll, D. (1975) Maturation of a hypermodified nucleoside in transfer RNA. Nucleic Acid Res. 2, 691-698.
2. Armstrong, D.J., Murai, N., Taller, B.J., Skoog, F. (1976) Incorporation of cytokinin N^6-benzyladenine into tobacco callus transfer ribonucleic acid and ribosomal ribonucleic acid preparations. Plant Physiol. 57, 15-22.
3. Auer, C.A., Cohen, J.D., Laloue, M., Cooke, T.J. (1992) Comparison of benzyladenine metabolism in two *Petunia hybrida* lines differing in shoot organogenesis. Plant Physiol. 98, 1035-1041.
4. Barry, G.F., Rogen, S.G., Fraley, R.T., Brand, L. (1984) Identification of a cloned cytokinin biosynthetic gene. Proc. Natl. Acad. Sci. USA 81, 4776-4780.
5. Bartz, J.K., Soll, D. (1972) N^6-Δ^2-(isopentenyl) adenosine: biosynthesis in vitro in transfer RNA by an enzyme purified from *Escherichia coli*. Biochemie 54, 31-39.
6. Borek, E., Baliga, B.S., Gehrke, C.W., Kuo, G.W., Belman, S., Troll, W., Waalkes, T.P. (1977) High turnover rate of transfer RNA in tumor tissue. Cancer Res. 37, 3362-3366.
7. Brownlee, B.G., Hall, R.H., Whitty, C.D. (1975) 3-Methyl-2-butenal: an enzymatic product of the cytokinin, N^6-(Δ^2-isopentenyl) adenine. Can. J. Biochem. 53, 37-41.
8. Burch, L., Horgan, R. (1989) The purification of cytokinin oxidase from Z. *mays* kernels. Phytochemistry 28, 1313-1319.

9. Burch, L.R., Stuchbury, T. (1986) Metabolism of purine nucleotides in the tomato plant. Phytochemistry, 25, 2445-2449.
10. Burch, L.R., Stuchbury, T. (1986) Purification and properties of adenosine nucleosidases from tomato *(Lycopersicon esculentum)* roots and leaves. J. Plant Physiol. 125, 267-273.
11. Burch, L.R., Stuchbury, T. (1987) Activity and distribution of enzymes that interconvert purine bases, ribosides for cytokinin metabolism. Physiol. Plant. 69, 283-288.
12. Burrows, W.J., Skoog, F., Leonard, N.J. (1971) Isolation and identification of cytokinins located in the transfer ribonucleic acid of tobacco callus grown in the presence of 6-benzylaminopurine. Biochemistry 10, 2189-2194.
13. Chatfield, J.M., Armstrong, D.J. (1986) Regulation of cytokinin oxidase activity in callus tissue of *Phaseolus vulgaris* L. cv Great Northern, Plant Physiol. 80, 493-499.
14. Chen, C-M. (1982) Cytokinin biosynthesis in cell-free systems. *In*: Plant growth substances 1982, pp. 155-164, Wareing, P.F., ed. Academic Press, London New York.
15. Chen, C-M., Eckert, R.L. (1977) Phosphorylation of cytokinin by adenosine kinase from wheat germ. Plant Physiol. 59, 443-447.
16. Chen, C-M., Hall, R.H. (1969) Biosynthesis of N^6-Δ^2-(isopentenyl) adenosine in the transfer ribonucleic acid of cultured tobacco pith tissue. Phytochemistry 8, 1687-1695.
17. Chen, C-M., Kristopeit, S.M. (1981) Metabolism of cytokinin: dephosphorylation of cytokinin ribonucleotide by 5'-nucleotidase from wheat germ cytosol. Plant Physiol. 67, 494-498.
18. Chen, C-M., Kristopeit, S.M. (1981) Metabolism of cytokinin: deribosylation of cytokinin ribonucleoside by adenine nucleosidase from wheat germ cells. Plant Physiol. 68, 1020-1023.
19. Chen, C-M., Leisner, S.M. (1984) Modification of cytokinins by cauliflower microsomal enzymes. Plant Physiol. 75, 442-446.
20. Chen, C-M., Melitz, D.K. Clough, F.W. (1982) Metabolism of cytokinin: phosphoribosylation of cytokinin bases by adenine phosphoribosyltransferase from wheat germ. Archives of Biochemistry and Biophysics. 214, 634-641.
21. Chen, C-M., Petschow, B. (1978) Metabolism of cytokinin: ribosylation of cytokinin bases by adenine phosphorylase from wheat germ. Plant Physiol. 62, 871-874.
22. Dixon, S.C., Martin, R.C., Mok, M.C., Shaw, G., Mok, D.W.S. (1989) Zeatin glucosylation enzymes in *Phaseolus*. Isolation of 0-glucosyltransferase from *P. lunatus* and comparison to 0-xylosyltransferase from *P. vulgaris*. Plant Physiol. 90, 1316-1321.
23. Doree, M., Guern, J. (1973) Short term metabolism of some exogenous cytokinins in *Acer pseudoplatanus* cells. Biochim. Biophys. Acta 304, 611-622.
24. Entsch, B., Letham, D.S. (1979) Enzymatic glucosylation of the cytokinin, 6-benzylaminopurine. Plant Sci. Lett. 14, 205-212.
25. Entsch, B., Letham, D.S., Parker, C.W., Summons, R.E. (1979) Preparation and characterisation, using high performance liquid chromatography, of an enzyme forming glucosides of cytokinins. Biochim. Biophys. Acta 570, 124-139.
26. Entsch, B., Letham, D.S., Parker, C.W., Summons, R.E., Gollnow, B.E. (1979) Metabolism of cytokinins. *In*: Plant Growth Regulation 1979, pp. 109-115, Skoog, F., ed. Springer, Berlin, Heidelberg, New York.
27. Estruch, J.J., Chriqui, D., Grossmann, K., Schell, J., Spena, A. (1991) The plant oncogene rolC is responsible for the release of cytokinins from glucose conjugates. EMBO Journal. 10, 2889-2895.
28. Fittler, F., Hall, R.H. (1966) Selective modification of yeast seryl-tRNA and its effects on acceptance and binding functions. Biochem. Biophys. Res. Comm. 25, 441-446.
29. Geftner, M.L. (1969) The in vitro synthesis of 2'-O-methylguanosine and 2-methylthio N^6-(γ,γ-dimethylallyl) adenosine in transfer RNA of Escherichia coli. *Biochem*. Biophys. Res. Comm. 36, 435-441.
30. Geftner, M.L., Russell, R.L. (1969) Role of modifications in tyrosine transfer RNA: a modified base affecting ribosome binding. J. Mol. Biol. 39, 145-157.

31. Haberlandt, G. (1913) Zur physiologie der zellteilungen. Sitzungsber. K. Preuss. Akad. Wiss., 318-345.

32. Hall, R.H. (1973) Cytokinin as a probe of developmental processes. Ann. Rev. Plant Physiol. 24, 425-444.

33. Holtz, J., Klambt, D. (1978) tRNA isopentenyltransferase from *Zea mays* L. Characterisation of the isopentenylation reaction of tRNA, oligo(A) and other nucleic acids. Hoppe-Seylers Z. Physiol. Chem. 359, 89-101.

34. Horgan, R. (1984) Cytokinins. *In*: Advanced Plant Physiology, pp. 90-116, Wilkins, M.B., ed. Pitman, London.

35. Kaminek, M., Armstrong, D.J. (1990) Genotypic variation in cytokinin oxidase from *Phaseolus* callus cultures. Plant Physiol. 93, 1530-1538.

36. Kulaeva, O.N. (1981) Cytokinin action on transcription and translation in plants. *In*: Metabolism and Molecular Activities of Cytokinins, pp. 218-227, Guern, J., Peaud-Lenoel, C. eds. Springer, Berlin, Heidelberg, New York.

37. Laloue, M., Fox, J.E. (1989) Cytokinin oxidase from wheat: Partial purification and general properties. Plant Physiol. 90, 899-906.

38. Laloue, M., Pethe, C. (1982) Dynamics of cytokinin metabolism in tobacco cells. *In*: Plant Growth Substances 1982, pp. 185-195, Wareing P.F., ed. Academic Press, New York.

39. Laloue, M., Pethe-Terrine, C., Guern, J. (1981) Uptake and metabolism of cytokinins in tobacco cells: studies in relation to the expression of their biological activities. *In*: Metabolism and Molecular Activities of Cytokinins, pp. 80-96, Guern, J., Peaud-Lenoel, C. eds. Springer, Berlin, Heidelberg, New York.

40. Letham, D.S. (1963) Zeatin, a factor inducing cell division from Zea mays. Life Sci. 8, 569-573.

41. Letham, D.S., Palni, L.M.S. (1983) The biosynthesis and metabolism of cytokinins. *Ann. Rev.* Plant Physiol. 34, 163-197.

42. Letham, D.S., Palni, L.M.S., Tao, G.Q., Gollnow, B.I., Bates, C.M. (1983) Regulators of cell division in plant tissues. XXIX. The activities of cytokinin glucosides and alanine conjugates in cytokinin bioassay. J. Plant Growth Regulation 2, 103-115.

43. Letham, D.S., Tao, G.Q., Parker, C.W. (1982) An overview of cytokinin metabolism. *In*: Plant growth substances 1982, pp. 143-153, Wareing, P.F., ed. Academic Press, London.

44. Letham, D.S., Wettenhall, R.E.H. (1977) Transfer RNA and cytokinins. *In*: The Ribonucleic Acids, pp. 129-193, Stewart, P.R., Letham, D.S., eds. Springer, Berlin, Heidelberg, New York.

45. Matsubara, S. (1980) Structure-activity relationships of cytokinins. Phytochemistry 19, 2239-2253.

46. Maaβe, H., Klambt, D. (1981) On the biogenesis of cytokinins in roots of *Phaseolus vulgaris*. Planta 151, 353-358.

47. McGaw, B.A., Heald, J.K., Horgan, R. (1984) Dihydrozeatin metabolism in radish seedlings. Phytochemistry 23, 1373-1377.

48. McGaw, B.A., Horgan, R. (1983) Cytokinin catabolism and cytokinin oxidase. Phytochemistry 22, 1103-1105.

49. McGaw, B.A., Horgan, R. (1983) Cytokinin oxidase from *Zea mays* kernels and *Vinca rosea* crown gall tissue. Planta 159, 30-37.

50. McGaw, B.A., Horgan, R., Heald, J.K., Wullems, G., Schilperoort, R. (1988) Mass-spectrometric quantitation of cytokinins in tobacco crown-gall tumours induced by mutated octopine Ti plasmids of *Agrobacterium tumefaciens*. Planta 176, 230-234.

51. McGaw, B.A., Horgan, R., Heald, J.K. (1985) Cytokinin metabolism and the modification of cytokinin activity in radish. Phytochemistry 24, 9-13.

52. McGaw, B.A., Scott, I.M., Horgan, R. (1984) Cytokinin biosynthesis and metabolism. *In*: The Biosynthesis and Metabolism of Plant Hormones, pp. 105-133, Crozier, A., Hillman, J.R., eds. Cambridge University Press, Cambridge.

53. Medford, J.I., Horgan, R., El-Sawi, Z., Klee, H.J. (1989) Alterations of endogenous cytokinins in transgenic plants using a chimeric isopentenyl transferase gene. The Plant Cell 1, 403-413.

54. Moffatt, B., Somerville, C. (1988) Positive selection for male-sterile mutants of *Arabidopsis* lacking adenine phosphoribosyl transferase activity. Plant Physiol. 86, 1150-1154.

55. Moffatt, B., Pethe, C., Laloue, M. (1991) Metabolism of benzyladenine is impaired in a mutant of *Arabidopsis thaliana* lacking adenine phosphoribosyltransferase activity. Plant Physiol. 95, 900-908.

56. Mok, D.W.S., Mod, M.C., Shaw, G., Dixon, S.C., Martin, R.C. (1990) Genetic differences in the enzymatic regulation of zeatin metabolism in *Phaseolus* embryos. *In*: Plant Growth Substances 1988, pp. 267-274, Pharis R.P., Rood S.B., eds. Springer-Verlag, Berlin.

57. Miller, C.O., Skoog, F., Okomura, F.S., Saltza, M.H.von, Strong, F.M. (1956) Isolation, structure and synthesis of kinetin, a substance promoting cell division. J. Amer. Chem. Soc. 78, 1345-1350.

58. Miller, C.O., Skoog, F., Saltza, M.H.von, Strong, F.M. (1955) Kinetin, a cell division factor from deoxyribonucleic acid. J. Amer. Chem. Soc. 77, 1329-1334.

59. Morris, R.O., Regier, D.A., MacDonald, E.M.S. (1981) Analytical procedures for cytokinins: application to *Agrobacterium tumefaciens*. *In*: Metabolism and Molecular Activities of Cytokinins, pp. 3-16, Guern, J., Peaud-Lenoel, C., eds. Springer, Berlin, Heidelberg, New York.

60. Murai, N. (1981) Cytokinin biosynthesis and its relationship to the presence of plasmids in a strain of *Corynebacterium fascians*. *In*: Metabolism and Molecular Activities of Cytokinins, pp. 17-26, Guern, J., Peaud-Lenoel, C., eds. Springer, Berlin, Heidelberg, New York.

61. Ooms, G., Risiott, R., Kendall, A., Keys, A., Lawlor, D., Smith, S., Turner, J. and Young, A. (1991) Phenotypic changes in T-*cyt*-transformed potato plants are consistent with enhanced sensitivity of specific cell types to normal regulation by root-derived cytokinin. Plant Mol. Biol. 17, 727-743.

62. Palmer, M.V., Horgan, R., Wareing, P.F. (1981) Cytokinin metabolism in *Phaseolus vulgaris* L. I. Variation in cytokinin levels in leaves of decapitated plants in relation to bud outgrowth. J. Exp. Bot. 32, 1231-1241.

63. Palmer, M.V., Scott, I.M., Horgan, R. (1981) Cytokinin metabolism in *Phaseolus vulgaris* L. II. Comparative metabolism of exogenous cytokinins by detached leaves. Plant Sci. Lett. 22, 187-195.

64. Palni, L.M.S., Horgan, R. (1983) Cytokinins in transfer RNA of normal and crown-gall tissue of *Vinca rosea*. Planta 159, 178-181.

65. Palni, L.M.S.,Horgan, R. (1983) Cytokinin biosynthesis in crown gall tissue of V*inca rosea*: metabolism of isopentenyladenine. Phytochemistry 22, 1597-1601.

66. Palni, L.M.S., Horgan, R., Darrall, N.M., Stuchbury, T., Wareing, P.F. (1983) Cytokinin biosynthesis in crown gall tissue of *Vinca rosea*. Planta 159, 50-59.

67. Parker, C.W., Letham, D.S. (1973) Regulators of cell division in plant tissues. XVI. Metabolism of zeatin by radish cotyledons and hypocotyls. Planta, 114 199-218.

68. Parker, C.W., Letham, D.S., Gollnow, B.I., Summons, R.E., Duke, C.C., MacLeod, J.K. (1978) Regulators of cell division in plant tissues. XXV. Metabolism of zeatin in lupin seedlings. Planta 142, 239-251.

69. Sadorge, P., Doree, M., Terrine, C., Guern, J. (1970) Absorption and utilisation of exogenous adenine by plant tissue. I. Technique for the measurement of adenine pyrophosphoribosyl transferase activity. Physiol. Veg. 8, 499-514.

70. Schmulling, T., Beinsberger, S., DeGreef, J., Schell, J., Van Onckelen, H., Spena, A. (1989) Construction of a heat inducible chimeric gene to increase the cytokinin content in transgenic plant tissue. FEBS Lett . 249, 401-406.

71. Scott, I.M., Horgan, R. (1984) Mass spectrometric quantification of cytokinin nucleotides and glycosides in tobacco crown gall tissue. Planta 161, 345-354.

72. Skoog, F., Hamzi, H.Q., Szweykowska, M., Leonard, N.J., Carraway, K.L., Fugii, T., Hegelson, J.P., Leoppky, R.W. (1967) Cytokinins: structure/activity relationships. Phytochemistry 6, 1169-1192.

73. Skoog, F., Schmitz, R.Y. (1979) Biochemistry and physiology of cytokinins. *In:* Biochemical Actions of Hormones, Vol. VI pp. 335-413, Litwack, G. ed. Academic Press, London.

74. Smart, C.M., Schofield, S.R., Bevan, M.W., Dyer, T.A. (1991) Delayed leaf senescence in tobacco plants transformed with *tmr*, a gene for cytokinin production in Agrobacterium. The Plant Cell 3, 647-565.

75. Stuchbury, T., Burch, L.R. (1987) Enzymology of cytokinin and purine metabolism. *In:* Cytokinins - plant hormones in search of a role (monograph 14), pp. 19-24, Horgan, R., Jeffcoat, B., eds. British Plant Growth Regulator Group, Bristol.

76. Stuchbury, T., Palni, L.M.S., Horgan, R., Wareing, P.F. (1979) The biosynthesis of cytokinins in crown-gall tissue of *Vinca rosea*. Planta 147, 97-102.

77. Summons, R.E., Letham, D.S., Gollnow, B.I., Parker, C.W., Entsch, B., Johnson, L.P., MacLeod, J.K., Rolfe, B.G. (1981) Cytokinin translocation and metabolism in species of the Leguminoseae: studies in relation to shoot and nodule development. *In:* Metabolism and Molecular Activities of Cytokinins, pp. 69-80, Guern, J., Peaud-Lenoel, C. eds. Springer, Berlin, Heidelberg, New York.

78. Taya, Y., Tanaka, Y., Nishimura, S. (1978) 5'-AMP is a direct precursor of cytokinin in Dictyostelium discoideum. Nature 271, 545-547.

79. Tepfer, D.A., Fosket, D.E. (1978) Hormone mediated translational control of protein synthesis in cultured cells of *Glycine max*. Develop. Biol. 62, 486-497.

80. Whitty, C.D., Hall, R.H. (1974) A cytokinin oxidase in *Zea mays*. Can. J. Biochem. 52, 781-799.

B4. Biosynthesis and Metabolism of Ethylene

Thomas A. McKeon[a], Juan C. Fernández-Maculet[b] and Shang-Fa Yang[b]

[a] Western Regional Research Center, Agricultural Research Service, Albany, CA 94710, USA, and [b] Department of Vegetable Crops, University of California, Davis CA, 95616, USA.

INTRODUCTION

Ethylene is a plant hormone that is involved in the regulation of many physiological responses (2, 39). In addition to its recognition as a "ripening hormone", ethylene is involved in other developmental processes from germination of seeds to senescence of various organs and in many responses to environmental stresses.

In many ways, ethylene is the ideal plant hormone to investigate. As a simple gaseous hydrocarbon, it is readily isolated from plant material and is easily quantified at concentrations well below the physiologically active level. Recent progress has increased the understanding of enzymes involved in ethylene production and elucidated their genetic control, leading to the development of several ways to manipulate ethylene production by genetic alteration of plants (29).

Ethylene was recognized as a plant-produced hormone over 50 years ago, yet the biosynthetic pathway of ethylene in plants remained elusive until the key intermediate ACC[1] was shown to be the immediate precursor of ethylene. In the time elapsed since the first edition of this volume, considerable progress has been made in understanding ethylene biosynthesis and action, yet there are challenges remaining. The purpose of this chapter is to describe both the progress in ethylene biochemistry and avenues for future research.

---

[1] **Abbreviations: ACC** = l-Aminocyclopropane-l-carboxylic acid;
AdoMet = S-Adenosyl-L-methionine;
AEC = l-Amino-2-ethylcyclopropane-1-carboxylic acid;
AOA = amino-oxy acetic acid;
AVG = Aminoethoxy-vinylglycine;
EFE = Ethylene forming enzyme (ACC oxidase);
KMB = α-Keto-γ-methylthiobutyric acid;
MACC = N-Malonylaminocyclopropane-l-carboxylic acid;
MTA = 5'-Methylthioadenosine;
MTR = 5-Methylthioribose;
MTR-l-P = 5-Methylthioribose-l-phosphate;
SDS-PAGE = Sodium dodecyl sulfate-polyacrylamide gel electrophoresis

P. J. Davies (ed.), Plant Hormones, 118–139.
© 1995 *Kluwer Academic Publishers. Printed in the Netherlands.*

ELUCIDATION OF THE ETHYLENE BIOSYNTHETIC PATHWAY

The pathway for ethylene biosynthesis is shown below. Although a number of ethylene precursors were proposed after testing in plant tissue or in model systems, it was eventually shown that methionine was rapidly converted to ethylene in a chemical model system consisting of Cu^{2+} and ascorbic acid (35). Following up this work on the model system, Lieberman and coworkers showed that L-methionine labeled at the C-3,4 positions was readily converted by apple fruit tissue to labeled ethylene (35). Later, AdoMet was inferred as an ethylene precursor because the conversion of methionine to ethylene was inhibited by oxidative phosphorylation inhibitors, implying an energy (ATP) dependent step in the biosynthesis of ethylene from methionine. Adams and Yang (4) confirmed this proposal by demonstrating that the labeled [^{35}S]-methionine and [^{3}H-methyl] methionine released labeled MTA and its hydrolysis product MTR upon its conversion to ethylene in apple tissue. Thus, they deduced that methionine must be converted into AdoMet before ethylene is released.

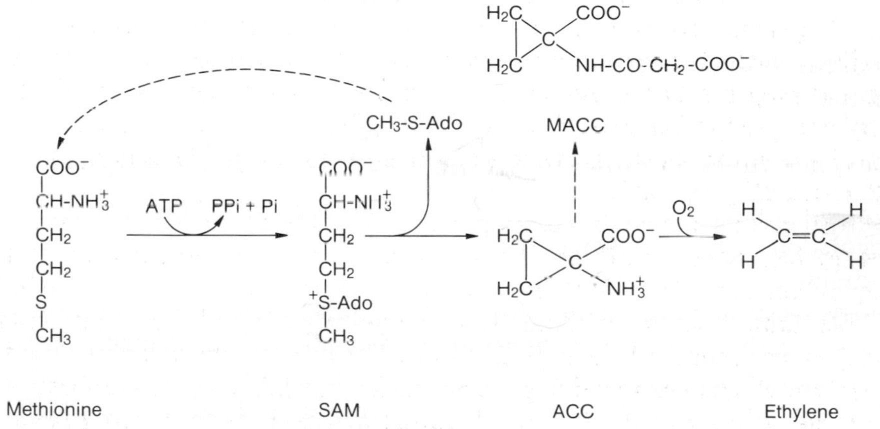

| Methionine | | SAM | | ACC | | Ethylene |

The next step in the pathway is the conversion of AdoMet to ACC. Adams and Yang (5) identified ACC, MTA and MTR as the labeled products which accumulated when L-[U^{14}C]methionine was incubated with apple tissue under anaerobic conditions which block ethylene production. Subsequent incubation of the tissue in air resulted in the production of labeled ethylene from the accumulated labeled ACC. Coinciding with these findings, Lürssen *et al.* (38), while screening a number of compounds as possible plant growth regulators, demonstrated that ACC dramatically stimulated ethylene production in plant tissues. By analogy to the chemical synthesis of ACC from a sulfonium intermediate, they deduced that ACC would be derived from AdoMet and were thus able to propose the correct biosynthetic pathway for ethylene.

ACC was first isolated in 1957 from ripe cider apples and perry pears (13) and was postulated to be involved in ripening. However, interest in this unusual, non-protein amino acid was not sparked until its recognition as an ethylene precursor.

In addition to its conversion to ethylene, ACC can be metabolized to N-malonyl-ACC (MACC). This conjugate was identified independently by two separate groups. Amrhein *et al.* (6) sought alternate pathways for ACC metabolism based on the fact that some bacteria and yeast metabolize ACC to products other than ethylene. They identified MACC as a conjugate of ACC formed in buckwheat hypocotyls. Hoffman *et al.* (28) sought an alternate pathway for ACC metabolism on finding that wilted wheat leaves lost more ACC than could be accounted for through conversion to ethylene. They subsequently demonstrated that this tissue metabolizes ACC to MACC.

An important consideration in ethylene biosynthesis is the limited amount of methionine present in plants. It was recognized that in order to maintain a high rate of ethylene production in apple fruit, the sulfur of methionine must be recycled back to methionine. It was first demonstrated that the 5'-methyl group of MTA is readily recycled to methionine (72), and then, using dual-labeled [^{35}S, ^{14}C-methyl]-MTA, it was shown that the CH_3S-group of MTA is converted as a unit to re-form methionine (4). Later work showed that the ribose moiety of MTA provides the carbon-skeleton for the 2-aminobutyrate portion of methionine (75).

Enzymes Involved in the Biosynthesis and Regulation of Ethylene

Knowing how ethylene is synthesized is essential to understanding its regulation. The following section is subdivided under headings of the individual enzymes or enzymatic reactions associated with ethylene biosynthesis and regulation. Although the first studies were carried out in vivo or with crude enzyme preparations, they allowed considerable progress toward understanding of the biochemical regulation of ethylene biosynthesis. Recently, important advances have been achieved in the characterization of ACC synthase and ACC oxidase at the biochemical and genetic levels (29).

ACC Synthase

ACC synthase (EC 4.4.1.4), which catalyzes the conversion of AdoMet to ACC and MTA, plays a key role in regulating ethylene production. Levels of ACC synthase are affected by changes in the growth environment, by changes in hormone levels and by physiological and developmental events (29, 72).

ACC synthase was first characterized and partially purified from tomato slices by ammonium sulfate fractionation and hydrophobic chromatography (10, 74). The enzyme requires pyridoxal phosphate for activity and is sensitive to pyridoxal phosphate inhibitors, especially AVG (K_i = 0.2 μM) and AOA (K_i = 0.8 μM). These inhibitors have proven invaluable in studying the regulation of ethylene production by distinguishing effects on ACC synthase from effects on ACC conversion to ethylene (5). The preferred substrate for the ACC synthase is (-)-S-adenosyl-L-methionine (which has the S-configuration at the sulfonium position), the naturally occurring isomer of AdoMet with a K_m of 20 μM, whereas (+)-AdoMet is an effective inhibitor, (I_{50}= 15 μM). It has also been shown that the ACC synthase is inactivated upon incubation with AdoMet (60). The authors proposed that AdoMet, when activated by the ACC synthase, can irreversibly modify the enzyme in a "suicide-inactivation". The substrate-induced inactivation of ACC synthase has been extensively studied by Satoh and Yang (61) who, by specifically radiolabeling ACC synthase with [3,4-^{14}C]AdoMet, demonstrated that the inactivation involves covalent linkage of a fragment of the AdoMet molecule to the active site of the enzyme. This AdoMet-dependent radiolabeling of the enzyme proved especially useful as a tool for confirmation of the identity of ACC synthase on SDS-PAGE. The interaction of ACC synthase with its substrate AdoMet is the basis for the rapid inactivation (turnover) of the enzyme observed both in vitro and in vivo (32).

Because of the very low levels of ACC synthase present with respect to other proteins, and its instability, progress in purification was slow. Since ACC synthase activity was known to be greatly induced under various conditions, the enzyme was isolated from various plant tissues following induction by IAA application, wounding, LiCl stress and the ripening of climacteric fruits. Several research groups have purified ACC synthase from wounded and ripe tomato fruits (8, 67), wounded winter squash fruit (46), wounded and IAA-induced zucchini fruit (58) and ripe apple fruit (20, 73). In most cases, partially purified enzyme preparations were suitable for preparation of antibodies to ACC synthase, and these antibodies subsequently provided the final, immunopurification step for the enzyme. This approach produced homogeneous ACC synthase that allowed determination of partial amino acid sequences.

Several different laboratories independently cloned ACC synthase cDNAs from zucchini fruit (59), winter squash fruit (45), tomato fruit (47, 68) and apple fruit (19) using either oligonucleotide probes deduced from peptide sequences or antibodies for immunoscreening expression cDNA libraries. Comparison of the deduced amino acid sequences of ACC synthases from the above species reveals about 40% identity and about 80% overall similarity with seven highly conserved regions (Fig. 1). Recently several more ACC synthase cDNAs have been reported from IAA-induced winter squash fruit (44), tomato fruit (55), carnation flowers (48) and mung bean hypocotyls (11, 31). Several observations indicate that ACC synthase exists in isoforms which are derived from a multigene family and that different members are expressed preferentially in response to stimuli such as wounding, IAA application, ripening and various stresses (19, 44, 47, 55).

ACC synthase plays a major role in regulating ethylene biosynthesis. Increased ethylene production is involved in developmental processes

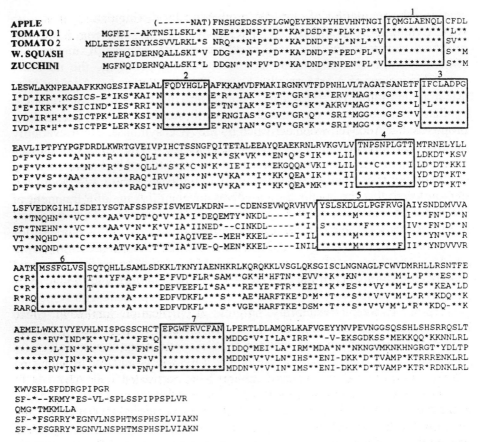

```
                                                                              1
APPLE              (------NAT)FNSHGEDSSYFLGWQEYEKNPYHEVHNTNGI IQMGLAENQL CFDL
TOMATO 1    MGFEI--AKTNSILSKL**  NEE***N*P**D**KA*DSD*F*PLK*P**V ********** *L**
TOMATO 2    MDLETSEISNYKSSVVLRKL*S NRQ***N*P**D**KA*DND*F*L*N*L**V ********** SV**
W.SQUASH    MEFHQIDERNQALLSKI*V DDG***N*P**D**KA*DND*F*PED*PL*V ********** S**M
ZUCCHINI    MGFNQIDERNQALLSKI*L DDGN**N*PV*D**KA*DND*FNPEN*PL*V ********** S**M
                              2                                            3
LESWLAKNPEAAAFKKNGESIFAELAL FQDYHGLP AFKKAMVDFMAKIRGNKVTFDPNHLVLTAGATSANETF IFCLADPG
I*D*IKR**KGSICS-E*IKS*KAI*N ******** E*R**IAK**E*T**GR*R***ERV*MAG***G****I ********
I*E*IKR**K*SICIND*IES*RRI*N ******** E*TN*IAK**E*T**G**K**AKRV*MAG***G****L *L******
IVD*IR*H***SICTPK*LER*KSI*N ******** E*RNGIAS**G*V**GR*Q***SRI*MGG***G*S**V ********
IVD*IR*H***SICTPE*LER*KSI*N ******** E*RN*IAN**G*V**GR*K***SRI*MGG***G*S**V ********
                                                             4
EAVLIPTPYYPGFDRDLKWRTGVEIVPIHCTSSNGFQITETALEEAYQEAEKRNLRVKGVLV TNPSNPLGTT MTRNELYLL
D*F*V*S****A*N***R*****QLI****E***N*K**SK*VK***EN*Q*S*IK***LIL ********** LDKDT*KSV
D*F*V*********N***R**S**QLL**S*K*C*N*K**IE*I****EKGQQA*VKI**LIL ***C*****I LD*DT*KKI
D*F*V*S**AA********RAQ*IRV**N***N**V*KA***I**KK*QEA*IK****II ********** YD*DT*KT*
D*F*V*S**A*********RAQ*IRV**NG**N**V*KA***I**KK*QEA*MK****II ********** YD*DT*KT*
                                                       5
LSFVEDKGIHLISDEIYSGTAFSSPSFISVMEVLKDRN---CDENSEVWQRVHVV YSLSKDLGLPGFRVG AIYSNDDMVVA
***TNQHN***VC****AA*V*DT*Q*V*IA*I*DEQEMTY*NKDL-----**I* ******M******** I***FN*D**N
ST*TNEHN***VC****AA*V*N**K*V*IA*IINED*--CINKDL-----**I* S******F****** IV**FN*D**N
VT**NQHD****C*****A*V*KA*T***IAQIVEE--MEH*KKEL-----I*IL ******M******** I***YN*V**R
VT**NQND****C*****ATV*KA*T*T*IA*IVE-Q-MEN*KKEL-----INIL ******M*******F II**YNDVVVR
                   6
AATK MSSFGLVS SQTQHLLSAMLSDKKLTKNYIAENHKRLKQRQKKLVSGLQKSGISCLNGNAGLFCWVDMRHLLRSNTFE
C*R* ******** T***YF*A**P**E*FVD*FLR*SAM**GK*H*HFTN**EVV**K**KN******M*L*P***ES**D
C*R* ******** T******AF****DEFVEEFLI*SA***RE*YE*FTR**EEI**K**ES***VY**M*L*S**KEA*LD
R*RQ ******** ******A*****EDFVDKFL***S***AE*HARFTKE*D*M**T***S***V*V*M*L*R**KDQ**K
RARQ ******** ******A*****EDFVDKFL***S**VGE*HARFTKE*DSM**T***S**V*V*M*L*R**KDQ-**K
                   7
AEMELWKKIVYEVHLNISPGSSCHCT EPGWFRVCFAN LPERTLDLAMQRLKAFVGEYYNVPEVNGGSQSSHLSHSRRQSLT
S**S**RV*IND*K**V*L***FE*Q ********** MDDG*V*I*LA*IRR***-V-EKSGDKSS*MEKKQQ*KKNNLRL
***S***L*IN**K**V*****FN*S V********** IDDQ*MEI*LA*IRM*MDA*N**NKNGVMKNKHNGRGT*YDLTP
******RV*IN**K**V*****F*V* ********** MDDN*V*V*LN*IHS**ENI-DKK*D*TVAMP*KTRRRENKLRL
******RV*IN**K**V*****FNV* ********** MDDN*V*V*IN*IMS**ENI-DKK*D*TVAMP*KTR*RDNKLRL

KWVSRLSFDDRGPIPGR
SF-*--KRMY*ES-VL-SPLSSPIPPSPLVR
QMG*TMKMLLA
SF-*FSGRRY*EGNVLNSPHTMSPHSPLVIAKN
SF-*FSGRRY*EGNVLNSPHTMSPHSPLVIAKN
```

Fig. 1. Comparison of deduced amino acid sequences of ACC synthases from apple (revised from (19); Genbank accession number U03294), tomato 1 (68), tomato 2 (47), winter squash (45) and zucchini (58). * = amino acid residues identical among all species; - = sequence gap. The seven highly conserved regions among all species are boxed in; the active site resides in Region 5 (73). (A = alanine; C = cysteine; D = aspartic acid; E = glutamic acid; F = phenylalanine; G = glycine; H = histidine; I = isoleucine; K = lysine; L = leucine; M = methionine; N = asparagine; P = proline; Q = glutamine; R = arginine; S = serine; T = threonine; V = valine; W = tryptophan; Y = tyrosine).

including germination, ripening, and senescence, and in stress responses to wounding, drought, water logging, chilling, toxic agents, infection or insect infestation (35, 72). In all these cases, it has been shown that the higher levels of ethylene are accompanied by increased ACC production, due to induction or activation of ACC synthase (72). This stimulation has been shown by direct measurement of increased ACC levels and by the ability of AVG to block or diminish the increase in ACC synthesis and the consequent increase in ethylene production. A true induction of ACC synthase has been inferred since cycloheximide, an inhibitor of protein synthesis, effectively blocks the increased ethylene production as do inhibitors of RNA synthesis

(72). With the availability of cDNA probes for ACC synthase, it is now known, in all cases examined, that the increases are due to a true induction, based on increased transcription of ACC synthase gene(s) (19, 44, 46, 47, 55).

Ethylene production rates are influenced by ethylene and other plant hormones. Auxin, cytokinins, abscisic acid and ethylene all regulate ethylene production at the level of ACC synthesis, although they may exert their effects by different biochemical mechanisms.

Auxin (IAA) promotes ethylene production by inducing the synthesis of ACC synthase, resulting in higher levels of ACC; the increase in ethylene production parallels the increase in ACC, and treatment with AVG blocks the IAA-induced ethylene increase (72). This auxin induction of ACC synthase is inhibited not only by protein synthesis inhibitors, but also by inhibitors of RNA synthesis. These data suggest that the induction of ACC synthase by auxin occurs at the transcriptional level, and Northern blotting confirms that auxin application results in increased mRNA for ACC synthase (44). Moreover, only a limited number of ACC synthase isoforms are induced in response to auxin (44, 47).

ABA effectively reduces wilting-induced ethylene production in wheat leaves (72). Pretreatment of wheat leaves with ABA inhibits ACC accumulation during the subsequent wilting treatment but does not significantly affect ACC levels in unwilted turgid leaves (72). In addition, when IAA-stimulated mung bean hypocotyls are co-treated with ABA, ACC accumulation is blocked as a result of reduced ACC synthase activity (72). It appears that ABA does not repress ACC synthase in uninduced tissue but may inhibit the induction of the enzyme caused by wilting or by IAA treatment. In citrus, tomato and some other plant tissues, ABA appears to promote ethylene production by stimulation of ACC levels (53).

Conversely, cytokinins stimulate ethylene production in conjunction with other treatments that increase ethylene synthesis. Thus, ethylene production from IAA-treated or water-stressed tissues rises in response to application of benzyladenine as a result of greater ACC synthase activity and the consequently increased ACC accumulation (72). Because cytokinins alone do not markedly affect ACC levels, their effect on ACC synthase must be through some other factor which affects the level of induction. Since cytokinins and ABA affect ACC synthase levels only under induced conditions, the induction of ACC synthase represents an ideal system for understanding the effect of these plant hormones on gene expression.

Depending on the tissue, ethylene can either promote ethylene production (autocatalysis) or inhibit ethylene production (autoinhibition). During autocatalysis in ripening fruits, ethylene initially affects ethylene production by promoting the conversion of ACC to ethylene, although a massive increase in ACC synthesis occurs later (36). ACC synthase is also the principal target of ethylene during autoinhibition. Excised grapefruit flavedo tissue treated with ethylene for 10 hours evolved ethylene at 6% of that treated with air (control), and this autoinhibitory effect of ethylene resulted from inhibited ACC synthesis (72).

One difficulty in interpreting hormone studies is that there is rarely a determination of how much of the hormone is taken up, translocated and metabolized by the tissue; this is the case for the studies described. Nevertheless, considerable progress has been made in understanding the role of hormones in ethylene biosynthesis. As analysis of plant hormones and their metabolism becomes more generally available, a clearer view of plant hormone interactions and their influence on ACC and ethylene production will result.

ACC oxidase

The conversion of ACC to ethylene is carried out by an oxidative enzyme that is now known as ACC oxidase. The enzyme was formerly named as ethylene-forming enzyme (EFE) by Yang and his colleagues (72) because the reaction mechanism was not known at that time. The ACC oxidase activity was first described by Adams and Yang (5), who trapped ACC in apple tissue incubated in a nitrogen atmosphere and demonstrated its conversion to ethylene under aerobic conditions. For more than a decade, authentic ACC oxidase activity could not be demonstrated in vitro. Hence, work on the characterization of ACC oxidase was carried out using an in vivo assay, in which ACC-dependent ethylene production was measured after uptake of exogenous ACC. Authentic ACC oxidase was ultimately isolated as a result of the expression of ACC oxidase activity in yeast and *Xenopus* transformed with a tomato ripening-specific cDNA clone (17, 25, 65).

While the conversion of ACC to ethylene by plants was initially considered to be a highly unusual reaction, a number of plant enzymes were shown to be capable of oxidizing ACC to ethylene in the presence of various cofactors. This includes IAA oxidase, peroxidase, lipoxygenase and H_2O_2-generating oxidases (72). Although these systems could generate ethylene from ACC, their K_m's for ACC were very high relative to observed levels of ACC in vivo. This lack of specificity for substrate led to the conclusion that these enzymes did not represent authentic ACC oxidase. Thus, one of the early problems in demonstrating ACC oxidase activity in vitro was proving that it was authentic.

The in vivo conversion of ACC to ethylene provided suitable conditions for validating putative ACC oxidase systems. The criteria developed are fundamental characteristics of enzymes: substrate specificity, stereospecificity, and competitive inhibition of activity.

When one of the ring hydrogens of ACC is substituted with an ethyl group, four stereoisomers of l-amino-2-ethylcyclopropane-1-carboxylic acid (AEC) are generated as shown at the top of the next page:

(1R,2S)	(1S,2R)	(1S,2S)	(1R,2R)
100	1.2	0.5	0.5

Hoffman *et al.* (27) showed that in apple, cantaloupe fruit and etiolated mung bean hypocotyls the ACC oxidase preferentially utilized one of the stereoisomers, (1R, 2S)-AEC, for the synthesis of the ethylene analogue 1-butene. Both ACC and AEC appear to be degraded by the same enzyme, ACC oxidase, since both reactions are inhibited to the same extent under varying levels of oxygen depletion and Co^{2+} concentrations, and since, when both substrates are present simultaneously, they are mutually inhibitory (27, 41). The substrate stereospecificity of ACC oxidase remains a key to validate authentic activity (22, 33, 40, 41, 70).

In addition to stereoselectivity among AEC isomers, the ACC oxidase displays high affinity of the ACC as its substrate. This was first demonstrated in vivo, based on the dependence of ethylene production rate upon the internal ACC concentration in pea epicotyl segments (41). The estimated K_m of 66 μM for ACC found in vivo compares favorably with the K_m of 6.4 to 85 μM for the isolated ACC oxidase (22, 33, 40, 70). Other important characteristics of authentic ACC oxidase based on properties determined in vivo are its dependence on O_2 (5), inhibition by Co^{2+} when applied in the range of 10 to 100 μM (38, 72), and competitive inhibition by α-aminoisobutyrate (37).

Unlike ACC synthase, ACC oxidase was first identified by its cDNA (see Chapter E4). While studying gene expression during ripening, Davies and Grierson (17) isolated a cDNA, pTOM13, by differential cloning techniques. Based on the observation that ACC oxidase activity in tomato plants was greatly reduced with a pTOM13 antisense gene, Hamilton *et al.* (26) suggested that the pTOM13 gene product is related to ACC oxidase. Later work confirmed that pTOM13 and its homologous sequence pHTOM5 confer ACC oxidase activity when expressed in yeast (25) or *Xenopus* oocytes (65). These cDNA sequences are homologous to cDNA for ACC oxidase from other species (Fig. 2). The deduced amino acid sequence of pTOM13 shows homology to those of flavanone-3-hydroxylase and other known hydroxylases (26). Although Yang (72) suggested as early as 1981 that ACC oxidase might be an ACC hydroxylase, attempts to assay ACC oxidase as a hydroxylase were not pursued. Ververidis and John (70) first demonstrated authentic ACC oxidase activity in vitro by extracting melon fruit and assaying

```
pTOM13   MEN---FPIINLEKLN---GDERANTMEMIKDACENWGFFELVNHGIPHE      44
pAVOe3   *D---S**VINMEKLE---*Q**AATMKL*N***E*****EL***SIPV*       44
pSR120   *ANIVN**IIDMEKLNNYN*V**SLVLDQ*K***H*****QV***SLSH*       50
pAE12    *AT---**VVDLSLVN---*E**AATLEK*N***E*****EL***GMST*       44

VMDTVEKMTKGHTKKCMEQRFKELVASKGLEAVQAEVTDLDWESTFFLRHLPTSNISQV     103
LM*E**RLT*E*Y**CM**R**ELMASK-VEGAVVDAN*M******FI**L*V**L*EI     102
LM*K**RMT*E*Y**FR**K**DMVQTKGLVSAESQVN*I******YL**R*T**I*EV     109
LL*T**KMT*D*Y**TM**R**EMVAAKGLDDVQSEIH*L******FL**L*S**I*EI     103

PDLDEEYRE-VMRDFAKRLKLEKLAEELLDLLCENLGLEKGYLKNAFYGSKG--PNFGT    159
***TDEH*KNV*KE**EK--L*K*A*QV***********G***MA*A*TTTGL*T***    159
***DDQY*K-L*KE**--AQI*R*S*QL***********A***NA*Y*ANG--*T***    163
***EEEY*K-T*KE**--VEL*K*A*KL***********G***KV*Y*SKG--*N***    157

KVSNYPPCPKPDLIKGLRAHTDAGGIILLFQDDKVSGLQLLKDEQWIDVPPMRHSIVVN    218
*********R*E*F******T***L*******R*A******GE*V****N****I*     218
*********K*D*I******T***I*******K*S******GH*V****K****V*     222
*********K*D*I******S***I*******K*S******GE*V****H****I*     216

LGDQLEVITNGKYKSVIHRVIAQTDGTRMSLASFYNPGSDAVIYPAKTLVEKEAEESTQ    277
****V***********M***V****N***L*******S*AV*F**PALV**EA*EKKE     277
****L***********M***I****N***I*******S*AV*Y**PTLV**E-*EKCR     280
****I***********M***I****T***I*******N*SF*S**PAVL**KT*DAP-     274

VYPKFVFDDYMKLYAGLKFQAKEPRFEAMKAMESD-----PIASA                 317
V*****-E***N**AG****A******V**MKAVETANLS*I-TT                 320
A*****FE***N**LK****E********AMETT----*G*IPTA                 321
T*****FD***K**SG****A********AKEST-----*VATA                   314
```

Fig. 2. Comparison of the amino acid sequences of ACC oxidase deduced from pTOM13 (tomato) (25), pHTOM5 (tomato) (65), pAVOe3 (avocado) (40), pSR120 (carnation) (71), pAE12 (apple) (18), and PCH313 (peach) (14). * = amino acid residues identical among all species; - = sequence gap.

the enzyme under the conditions for flavanone-3-hydroxylase. Their enzyme preparation requires Fe^{2+} and ascorbate for full activity, and meets all the criteria described above for ACC oxidase in vivo. Others have since detected ACC oxidase activity in homogenates of apple (22, 33) and avocado (40) fruits. All these enzyme preparations are soluble and require Fe^{2+} and ascorbate for activity. The failure of previous attempts to demonstrate ACC oxidase activity in vitro is now attributed to the loss of these factors during the enzyme extraction (22).

Recently, ACC oxidase has been purified to near homogeneity from apple fruits (18, 21). The enzyme is a monomer with a molecular mass of 35 kDa. Using the purified ACC oxidase, Dong *et al*. (18) determined the stoichiometry of the reaction catalyzed by the enzyme as follows (where AH_2 and A stand for ascorbate and dehydroascorbate, respectively):

$$ACC\ (C_4NO_2H_7) + O_2 + AH_2 \xrightarrow{Fe^{2+},\ CO_2} C_2H_4 + CO_2 + HCN + A + H_2O$$

Ascorbate serves as a co-substrate, and is concurrently oxidized to dehydroascorbate in an amount equivalent to that of ethylene produced. The stoichiometry agrees with the results obtained for ACC oxidase in vivo by Peiser *et al.* (49). The exact mechanism of the reaction is still unknown, but if it proceeds via the formation of N-hydroxyl-ACC, ACC oxidase can be referred to as ACC N-hydroxylase.

Study of the reaction products of ACC oxidase have provided insights in elucidating the enzyme's mechanism. Noting that 1-phenylcyclopropylamine is oxidized chemically by various oxidants to ethylene and benzonitrile via the intermediacy of the nitrenium ion, it was proposed (72) that ACC is oxidized by ACC oxidase to form N-hydroxy-ACC which could be considered equivalent to the nitrenium ion, similar to the proposed mechanism for the ACC chemical assay (72). This intermediate would then be degraded into ethylene (derived from C-2,3 of ACC) and cyanoformic acid (derived from carboxyl and C-1 of ACC), the latter being further degraded spontaneously into HCN (derived from C-1 of ACC) and CO_2 (derived from the carboxyl group of ACC). Support for the proposed reaction products was provided by Peiser *et al.* (49), who showed that the carboxyl group of ACC is liberated as CO_2, whereas C-1 of ACC yields HCN, which is then rapidly metabolized to yield β-cyanoalanine and asparagine:

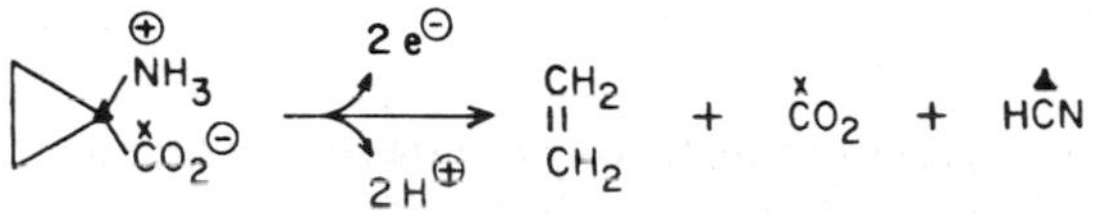

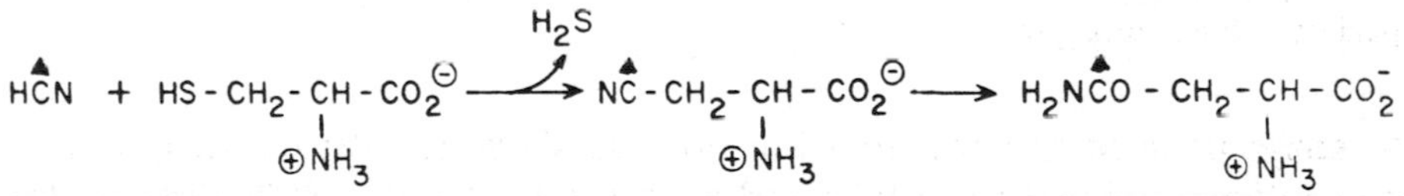

Additional information on the mechanism can be derived from the stereochemistry of ACC oxidation. When cis- and trans-2,3-dideutero-ACC are incubated with apples slices, the label is scrambled and equal amounts of cis- and trans-dideutero-ethylene are produced. When each of these dideutero-ACC's are oxidized with hypochlorite (72), the label is not scrambled, confirming that this reaction occurs by a concerted mechanism, probably via nitrenium ion (72).

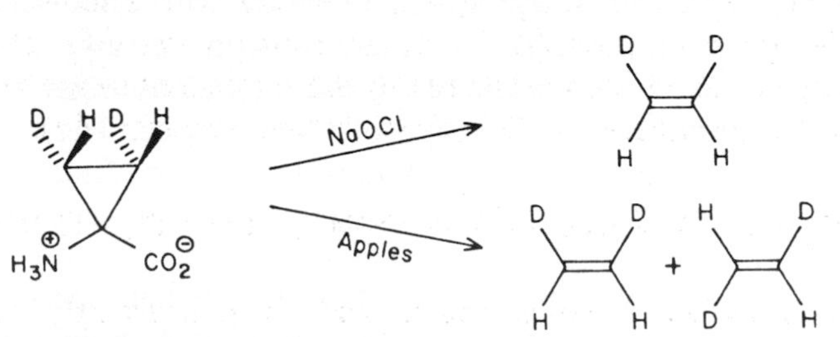

These observations indicate that the ACC oxidase reaction does not proceed in a concerted manner, but occurs by a stepwise mechanism involving an intermediate that allows scrambling of the ring hydrogens. With the availability of cloned ACC oxidase, it is now possible to generate large amounts of pure enzyme and use chemical and spectroscopic means to test putative mechanisms for this intriguing enzyme.

Kao and Yang (72) proposed in 1982 that CO_2 stimulates ethylene production in vivo by direct modulation of ACC oxidase activity. Using purified ACC oxidase, Dong *et al.* (18) showed that CO_2 and bicarbonate stimulate ACC oxidase activity markedly. Upon removal of CO_2, ACC oxidase activity was abolished, indicating that CO_2 is an essential activator of the enzyme. The concentration of CO_2 in the gas phase giving half maximal activity is 0.5%. The mechanism of this activation is not known, but is clearly reminiscent of ribulose-bis-phosphate carboxylase activation by CO_2. Recently Fernández-Maculet *et al.* (23) identified CO_2 as the species responsible for the activation of ACC oxidase and presented evidence that CO_2 forms a carbamate with an amino group of the enzyme.

ACC oxidase, as measured by ethylene production in the presence of a saturating concentration of ACC, is present in most tissues of higher plants (72). However, under some stress conditions, in response to ethylene, or during certain development stages (such as fruit ripening), the level of ACC oxidase increases markedly and effectively regulates ethylene production (72). Intact preclimacteric (unripe) cantaloupe and tomato fruits have low levels of ACC oxidase that increase markedly following treatment with ethylene (36). It is now known that the senescence and ripening induced increase in ACC oxidase is a result of increased transcription (17, 25, 71). ACC oxidase also responds to environmental stress, increasing up to eight-fold within one hour of wilting treatment and decreasing upon rehydration (72). In response to high temperature (>35°C) ACC oxidase activity drops and at 40°C is lost (72). The availability of cDNA and antisense technology to control the expression of ACC oxidase should allow the molecular dissection of ethylene-regulated development.

ACC N-Malonyltransferase

Because endogenous levels of ACC can increase during development or in response to stress, it is logical that the plant would require some means to sequester ACC to prevent overproduction of ethylene. The N-malonylation of ACC serves this purpose, as ethylene production in vivo is reduced by malonylation of ACC and promoted by blocking malonylation (66,72). The transferase is present in a wide range of plant tissues. The expression of malonyltransferase activity in preclimacteric tomato fruit is markedly promoted by ethylene treatment (66), thereby providing an autoregulatory mechanism for limiting ethylene production.

The ACC N-malonyltransferase has been isolated and partially purified from mung bean hypocotyls. The malonyl donor is malonyl-CoA, with a K_m

of 0.25 mM; at concentrations greater than 0.75 mM, malonyl-CoA inhibits the transferase. The K_m for ACC is 0.15 mM; AEC, non-polar D-amino acids (D-methionine, D-phenylalanine, and D-alanine) and α-aminoisobutyric acid can also be malonylated (37, 66). Based upon several observations it is thought that D-amino acid malonyltransferase and ACC malonyltransferase are the same enzyme (66). These observations include (a) enzyme preparations malonylate both D-amino acids and ACC, (b) the K_m values of those amino acids serving as substrates of malonyltransferase agree with their corresponding K_i values when the same amino acids act as competitive inhibitors of ACC malonyltransferase, and (c) the (lR,2S)- and (lR,2R)-AEC isomers, which have a D-amino acid configuration, are more effective substrates and inhibitors of malonyl-transferase than the (lS,2R)- and (lS,2S)-AEC isomers, which have an L-configuration.

What is the physiological role of MACC? It was initially hypothesized that MACC serves as a means for storing ACC in an unreactive form that could be hydrolyzed to ACC when needed for ethylene production. Such a system might, for example, be suitable for ethylene production in a germinating seed. Yet in germinating peanut seed, which contains a high level of MACC (50 to 100 nmol/g tissue), ethylene is derived almost exclusively from ACC produced *de novo* and MACC is converted to ethylene at less than 2% of the rate that ACC is (72). Previously, Satoh and Esashi (72) have reported that application of D-amino acids increases ACC content and promotes ethylene production in cocklebur cotyledons. When mung bean hypocotyls are fed D-amino acids, MACC formation from ACC is inhibited, leading to higher endogenous levels of ACC, a concomitant increase in ethylene production, and malonylation of the D-amino acid (66). The D-amino acids competitively inhibit malonylation of ACC, resulting in a higher ACC level and, thereby, a higher ethylene production rate. Malonyl-ACC is not an in vivo source of ACC; rather, it serves as a sink that allows depletion of ACC levels and thereby reduces ethylene production.

Recycling of 5'-Methylthioadenosine to S-Adenosyl-methionine

As described earlier, the recycling of the methylthio-group from AdoMet is important to the maintenance of ethylene production due to the limited amounts of sulfur present in plants. The known products and intermediates of the MTA-recycling pathway in plants (Fig. 3) and animals are the same with one exception: animals use MTA phosphorylase to metabolize MTA to MTR-1-P in one step, but plants utilize MTA nucleosidase, which converts MTA to MTR, and in turn, MTR kinase, which converts MTR to MTR-1-P (41). In both systems the MTR-1-P is then converted to methionine through several enzymatic steps which have been partially characterized (Fig. 3). The MTR-1-P is ultimately converted to KMB and equivalent amounts of formate (42) and the overall reaction represents a 4-electron oxidation. Since a requirement for molecular oxygen could not be demonstrated, it is probable that dehydrogenases rather than oxidases are involved in this oxidation. To

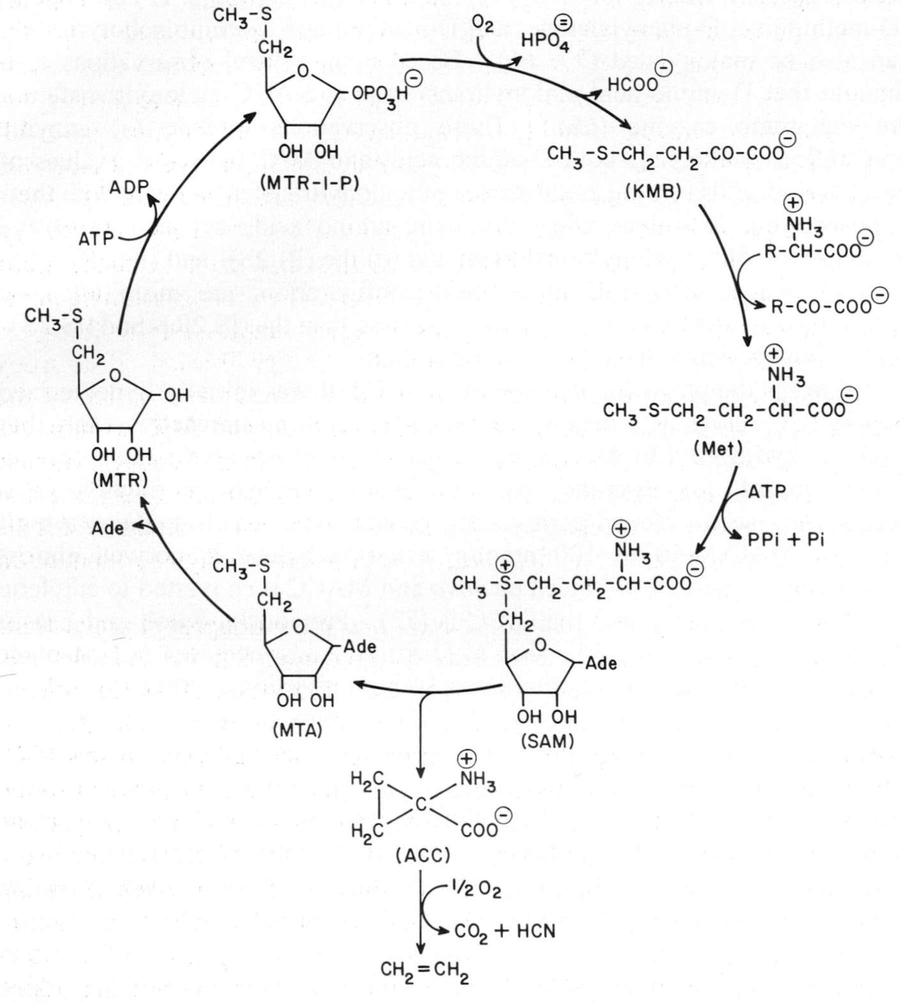

Fig. 3. Methionine cycle in relation to ethylene biosynthesis.

complete the cycle, KMB is transaminated to methionine by a specific transaminase (42), for which L-glutamine is the most effective amino donor, and finally the methionine formed is adenosylated to form AdoMet. The overall result of this cycle is that the ribose moiety of ATP furnishes the 4-carbon moiety of methionine from which ACC is derived; the methylthio-group of methionine is, however, conserved for continued regeneration of methionine (72). At this time, little is known about the regulation of this recycling pathway, although the conversion of MTR to methionine seems to be ample and does not appear to be a limiting factor of ethylene production (43).

In addition to its role in ethylene production, AdoMet is involved in the biosynthesis of polyamines and in methylation reactions. The enzyme methionine adenosyltransferase (AdoMet synthase) is responsible for the conversion of methionine to AdoMet. The partially purified methionine adenosyltransferase from pea seedling (1) appears to be similar to the enzyme from non-plant sources: in addition to similar K_m's for methionine (0.4 mM) and ATP (0.3 mM), the enzyme is also inhibited by high levels of AdoMet. Moreover, the transferase is inhibited by AMP and stimulated by ADP, suggesting possible regulation of the enzyme by adenylate energy charge (1). A gene for AdoMet synthase has been cloned and used to probe gene expression in *Arabidopsis* (50). Although described as a housekeeping gene, it is expressed differentially. The highest levels of expression for this gene are in vascular tissues, which require AdoMet for lignification. One study with senescing carnation petal demonstrated diminished transcription of the AdoMet synthase gene while those for ACC synthase and ACC oxidase are greatly increased (71). These results suggest that levels of AdoMet synthase activity are adequate to allow for the autocatalytic production of ethylene that occurs during flower senescence. It remains to be seen whether AdoMet synthase expression might be involved in regulation of ethylene production in any other plant system.

The polyamines and ACC each incorporate the aminopropyl group from AdoMet, and under certain conditions their competition for AdoMet can restrict biosynthesis of ethylene or polyamines. Inhibition of ACC synthesis by AOA results in increased polyamine production while inhibition of polyamine biosynthesis results in increased ACC and ethylene levels (54). These results indicate stimulation of one AdoMet-dependent pathway when the other is blocked, leading to the inference that ACC and polyamine biosynthesis are mutually inhibitory. Consistent with this interpretation is the delayed senescence and reduced ethylene production observed when senescing tissue is treated with polyamines (43) and decreased ethylene production concomitant with increased putrescine levels during ripening of a slow-ripening tomato cultivar (56). High levels of putrescine, whether added exogenously or produced endogenously, can interfere with ACC biosynthesis. When there is no competition for the availability of AdoMet due to low demand for polyamine or ACC, or due to compensation by increased levels of the MTR-recycling enzymes, there is no interaction between ethylene and polyamine production (34). It remains an open question whether ACC/polyamine interference represents a generally relevant means for controlling ethylene production.

ETHYLENE ANALOGUES AND ETHYLENE ANTAGONISTS

Burg and Burg (12) tested the ability of a number of ethylene analogues for ethylene-like action in the pea straight-growth test and found that the effectiveness of olefins that exert ethylene-like biological activity correlated with their ability to form a complex with Ag^+. They have proposed that the ethylene receptor site contains a metal ion. Among ethylene analogues, propylene and acetylene were found to require 100 and 2800 times, respectively, the concentration of ethylene to give half-maximal response (Table 1); alkanes are, however, inactive with no detectable response to ethane at 300,000 μl/l (12). Sisler (62) extended the Burgs' idea by postulating that ethylene acts via the "trans" effect on ligand-metal ion coordination, altering the geometry of the metal-binding site on the receptor protein when ethylene binds. He demonstrated that a number of π-acceptors (e.g., carbon monoxide, isocyanide, phosphorus trifluoride), with chemical structures quite different from ethylene, were effective in eliciting ethylene-like responses without any induction of ethylene synthesis.

There are three known types of antagonists that can be applied exogenously to inhibit ethylene action. These inhibitors have been used as diagnostic tests for ethylene action. The first, CO_2, prevents or delays many ethylene responses when ethylene concentration is low (1 μl/l or below). The mechanism of action is not known, but CO_2 has been suggested to be a competitive inhibitor of ethylene action presumably by competing with ethylene for the binding site, with a K_i of 1.5% (gas phase concentration) (12). CO_2 is used commercially in controlled atmosphere storage of fruits where high CO_2 levels help to delay the ripening action of ethylene. Interestingly, CO_2 is required for the conversion of ACC to ethylene by the ACC oxidase. The second, silver ion, inhibits ethylene action in a wide variety of ethylene-induced responses (7) and is used commercially to extend the shelf-life of cut carnations. It is thought that Ag^+ blocks ethylene action by interfering with ethylene binding. Finally, 2,5-norbornadiene, trans-cyclooctene, several other cyclic olefins, and cis-butene inhibit ethylene action competitively. It is assumed that they compete with ethylene for the binding site to form olefin-receptor complexes that do not induce an ethylene-like response but do block ethylene binding (64). Since these olefins are volatile and can readily diffuse away, they have proven useful for the elucidation of ethylene action and for characterization of the putative ethylene receptor. An interesting new analogue is diazo-cyclopentadiene (DACP), which also blocks

Table 1. A Comparison of ethylene analogues for inducing ethylene-like response and for inhibiting ethylene binding (7, 12).

Compounds	Relative Concentration for 1/2 Maximal Response	Relative Concentration for Inhibiting Ethylene Binding
Ethylene	1	1
Propylene	100	128
Carbon Monoxide	2,700	1,068
Acetylene	2,800	1,1013
1-Butene	270,000	601,277

ethylene action competitively. Additionally, it appears to be a photoaffinity label for the ethylene receptor: plants treated with the compound and then irradiated with fluorescent light irreversibly lose most sensitivity to ethylene (64). DACP thus provides a powerful new tool for detecting potential ethylene receptor proteins.

MODE OF ETHYLENE ACTION

Ethylene has been shown to induce the synthesis of new mRNA's and proteins (16). Experimental evidence points to ethylene binding, as opposed to ethylene metabolism, as the general mediator of ethylene action. First, the products of ethylene metabolism do not evoke ethylene-like responses; ethylene oxide has no ethylene-like effect (57). Secondly ethylene effects can be mimicked by some hydrocarbons (12) and counteracted by certain olefin antagonists (64) at levels which correlate to their ability to compete with ethylene binding to silver ion. Third, there are several inorganic compounds, such as phosphorus trifluoride, which cannot be metabolized to products related to any of the metabolic products of ethylene, yet they evoke ethylene effects (62, 64). Finally, CS_2, a potent inhibitor of ethylene oxidation, has no effect on ethylene action (3). It is now generally accepted that ethylene action is mediated by a receptor (9, 57, 64).

Possible ethylene receptors have been isolated and partially purified from several plant sources including tobacco, beans and tomatoes (63, 64). The binding of ethylene to these sites is saturable at physiological levels of ethylene and the K_D for ethylene in the gas phase is 0.1 to 0.3 $\mu l/l$ (or approximately 10^{-10} M in the liquid phase), in agreement with the value of 0.1 to 1 $\mu l/l$ ethylene causing a half-maximal response in a number of plant systems (2, 12). In all cases, the ethylene-binding component is membrane bound and detergent soluble. It appears to be a protein, based on heat-sensitivity, protease sensitivity, solubility and chromatographic behavior, and sensitivity to sulfhydryl agents (63,64). The isolated receptor displays properties similar to those expected from studies of ethylene action in vivo. Ethylene binding to the isolated receptor is competitively inhibited by propylene at 128 times and by acetylene at 1013 times the concentration of ethylene (gas phase), corresponding to the in vivo activity of these analogues (Table 1). Moreover, the ethylene antagonist 2,5-norbornadiene is similarly effective in blocking binding of ethylene to the receptor (64).

There is an important question remaining: What does the ethylene-binding protein do after ethylene binds? It is known that ethylene elicits many physiological responses (2). Furthermore, it has been shown that ethylene induces specific changes in genetic expression (16). Presumably, these changes would be mediated by the ethylene-binding protein. Increasingly, the study of ethylene perception is yielding to analysis of mutants. By screening *Arabidopsis* seedlings for elongation in the presence of ethylene, Bleecker *et al.* (9) first obtained a class of ethylene insensitive mutants, *etr*, which are lacking in a number of ethylene responses. The *etr* mutants are dominant and have reduced ethylene binding. Other ethylene insensitive mutants have been isolated, *ein1* and *ein2* (24) and *ain1* (69); The *ein1* maps near *etr*, has a similar phenotype and may therefore be identical. The *ain1* mutant is recessive, selected by resistance to ACC and is also resistant to ethylene. Based on chromosomal mapping, *ein2* and *ain1* are distinct from the other ethylene resistant mutants. Another class, the *ctr1*, is a recessive that displays the triple response in the absence of exogenous and endogenous ethylene (30). The *ctr1* mutation results in constitutive expression of ethylene regulated genes and, based on the gene sequence of the *ctr1* locus, it encodes a protein homologous to a family of eukaryotic protein kinases. Moreover, the carboxy-terminal half of the *etr* gene has conserved regions homologous to the prokaryotic signal transduction system (15). These homologies suggest that the ethylene response is transmitted by a multicomponent protein phosphorylation pathway (30), which could explain the involvement of calcium ion in the ethylene transduction pathway (52).

The question of how the ethylene-receptor complex induces the known physiological changes is proceeding from two complementary approaches, biochemical and genetic. Purification of ethylene-binding proteins will continue until a putative receptor can be assigned on the basis of amino acid sequence to an allele, such as *etr*, that is involved in ethylene perception.

ETHYLENE METABOLISM

Although ethylene metabolism by plants was considered for some time to be an artifact (2), it is now clear that in some plant tissues ethylene is oxidized (OX) to CO_2, in others it is incorporated in tissue (TI) by conversion to ethylene oxide and ethylene glycol and, in certain plants, both processes occur (57). The diversity of plants involved in ethylene metabolism suggests that it is a general phenomenon. In most instances, the rate of metabolism of ethylene is nearly first order with respect to ethylene, even at fairly high levels of ethylene (> 40 ml/l). These results indicate a very high K_m for ethylene metabolism, suggesting a chemical reaction as opposed to a controlled physiological process. In the pea, the concentration of ethylene giving a half-maximal rate for ethylene metabolism is approximately 1000 times the concentration necessary for half-maximal response in the pea growth test (7). However, the K_m for ethylene in TI by *Vicia faba* corresponds closely to the levels evoking physiological response (57). Although it was first demonstrated that oxygen deprivation and Ag^+ treatment similarly inhibited ethylene metabolism and action (7), later work (51, 57) demonstrated that the physiological effects of ethylene analogues and antagonists do not correlate with their metabolism or their effect on ethylene metabolism. Ethylene metabolism thus has no role in effecting ethylene action.

It is likely that ethylene metabolism is a nonessential consequence of ethylene action, resulting from high levels of ethylene production and an endogenous hydrocarbon oxidation system. It remains to be seen whether the TI and OX systems have any role in vivo.

CONCLUSION

Considerable progress has been made in understanding the biosynthesis of ethylene and its regulation. First, both ACC synthase and ACC oxidase have been isolated and purified from plant tissues. This will allow a systematic characterization of the enzymes and their catalytic mechanisms. Second, the cloning of the genes involved in ethylene biosynthesis during fruit ripening has permitted scientists to use biotechnology to genetically engineer crop plants with limited ethylene production. It is now possible to alter the timing and the extent of ethylene production by manipulating the expression of ACC synthase or ACC oxidase genes. This is an excellent illustration of the importance of fundamental research, initially elucidating the regulation and role of ethylene in plant development, but ultimately leading to applications for improving the quality of crops by extending shelf-life and reducing spoilage.

The mode of action of ethylene is not yet understood. It is known that ethylene elicits many physiological responses and induces specific changes in genetic expression. All of these changes must somehow be mediated by ethylene receptors. The means to explicate the ethylene response system are now available: ethylene response mutants and genes, the ability to genetically manipulate ethylene production *in planta*, and ethylene analogues and antagonists that can be used to control ethylene responses. With these tools, the understanding of the plant hormone ethylene will soon approach, and perhaps surpass, the level of sophistication that exists for knowledge of steroid hormone action in mammalian systems.

References

1. Aarnes, H. (1977) Partial purification and characterization of methionine adenosyltransferase from pea seedlings. Plant Sci. Lett. 10, 381-390.
2. Abeles, F.B., Morgan, P.W., Saltveit, M.E. Jr. (1992) Ethylene in plant biology, second edition. Academic Press, New York. 414 pp.
3. Abeles, F.B. (1984) A comparative study of ethylene oxidation in *Vicia faba* and *Mycobacterium paraffinicum*. J. Plant Growth Regul. 3, 85-95.
4. Adams, D.O., Yang, S.F. (1977) Methionine metabolism in apple tissue: implication of S-adenosylmethionine as an intermediate in the conversion of methionine to ethylene. Plant Physiol. 60, 892-896.
5. Adams, D.O., Yang, S.F. (1979) Ethylene biosynthesis: Identification of 1-aminocyclopropane-1-carboxylic acid as an intermediate in the conversion of methionine to ethylene. Proc. Natl. Acad. Sci. USA 76, 170-174.

6. Amrhein, N.D., Schneebeck, D., Skorupka, H., Tophof, S., Stöckigt, J. (1981) Identification of a major metabolite of the ethylene precursor 1-aminocyclopropane-1-carboxylic acid in higher plants. Naturwissenschaften 67, 619-620.

7. Beyer, E.M., Morgan, P.W., Yang, S.F. (1984) Ethylene. *In*: Advanced plant physiology, pp 111-126, Wilkins, M.B., ed. Pitman, London.

8. Bleecker, A.B., Herner, R.C., Kende, H. (1986) Use of monoclonal antibodies in the purification and characterization of 1-aminocyclopropane-1-carboxylate synthase, an enzyme in ethylene biosynthesis. Proc. Natl. Acad. Sci. USA 83, 7755-7759.

9. Bleecker, A.B., Estelle, M.A., Somerville, C., Kende, H. (1988) Insensitivity to ethylene conferred by a dominant mutation in *Arabidopsis thaliana*. Science 241, 1086-1089.

10. Boller, T., Herner, R.C., Kende, H. (1979) Assay for and enzymatic formation of an ethylene precursor, 1-aminocyclopropane-1-carboxylic acid. Planta 145, 293-303.

11. Botella, J.R., Schlagnhaufer, C.D., Arteca, R.N., Phillips, A.T. (1992) Identification and characterization of three putative genes for 1-aminocyclopropane-1-carboxylate synthase from etiolated mung bean hypocotyl segments. Plant Mol. Biol. 18, 793-797.

12. Burg, S.P., Burg, E.A. (1967) Molecular requirements for the biological activity of ethylene. Plant Physiol. 42, 144-152.

13. Burroughs, L.F. (1957) 1-Aminocyclopropane-1-carboxylic acid: A new amino acid in perry pears and cider apples. Nature 179, 360-361.

14. Callahan, A.M., Morgens, P.H., Wright, P., Nichols, K.E. Jr. (1992) Comparison of Pch313 (pTOM13 homolog) RNA accumulation during fruit softening and wounding of two phenotypically different peach cultivars. Plant Physiol. 100, 482-488.

15. Chang, C., Kwok, S.F., Bleecker, A.B., Meyerowitz, E.M. (1993) *Arabidopsis* ethylene-response gene *ETR1*: similarity of product to two-component regulators. Science 262, 539-544.

16. Christoffersen, R.E., Laties, G.G. (1982) Ethylene regulation of gene expression in carrots. Proc. Natl. Acad. Sci. USA 79, 4060-4063.

17. Davies, K.M., Grierson, D. (1989) Identification of cDNA clones for tomato (*Lycopersicon esculentum* Mill) mRNA that accumulate during ripening and leaf senescence in response to ethylene. Planta 179, 73-80.

18. Dong, J.G., Fernández-Maculet, J.C., Yang, S.F. (1992) Purification and characterization of 1-aminocyclopropane-1-carboxylic acid oxidase from ripe apple fruit. Proc. Natl. Acad. Sci. USA 89, 9789-9793.

19. Dong, J.G., Kim, W.T., Yip, W.K., Thompson, G.A., Li, L., Bennett, A.B., Yang, S.F. (1991) Cloning of a cDNA encoding 1-aminocyclopropane-1-carboxylate synthase and expression of its mRNA in ripening apple fruit. Planta 185, 38-45.

20. Dong, J.G., Yip, W.K., Yang, S.F. (1991) Monoclonal antibodies against apple 1-aminocyclopropane-1-carboxylate synthase. Plant Cell Physiol. 32, 25-31.

21. Dupille, E., Rombaldi, C., Lelièvre, J.M., Cleyet-Marel, J.C., Pech, J.C., Latché, A. (1993) Purification, properties and partial amino-acid sequence of 1-aminocyclopropane-1-carboxylic acid oxidase from apple fruits. Planta 190, 65-70.

22. Fernández-Maculet, J.C., Yang, S.F. (1992) Extraction and partial characterization of the ethylene-forming enzyme from apple fruit. Plant Physiol. 99, 751-754.

23. Fernández-Maculet, J.C., Dong, J.G., Yang, S.F. (1993) Activation of 1-aminocyclopropane-1-carboxylate oxidase by carbon dioxide. Biochem. Biophys. Res. Commun. 193, 1168-1173.

24. Guzmán, P., Ecker, J.R. (1990) Exploiting the triple response of *Arabidopsis* to identify ethylene-related mutants. Plant Cell 2, 513-523.

25. Hamilton, A.J., Bouzayen, M., Grierson, D. (1991) Identification of a tomato gene for the ethylene-forming enzyme by expression in yeast. Proc. Natl. Acad. Sci USA 88, 7434-7437.

26. Hamilton, A.J., Lycett, G.W., Grierson, D. (1990) Antisense gene that inhibits synthesis of the hormone ethylene in transgenic plants. Nature 346, 284-287.

27. Hoffman, N.E., Yang, S.F., Ichihara, A., Sakamura, S. (1982) Stereospecific conversion of 1-aminocyclopropane-1-carboxylic acid to ethylene by plant tissues. Conversion of stereoisomers of 1-amino-2-ethylcyclopropane-1-carboxylic acid to 1-butene. Plant Physiol. 70, 195-199.

28. Hoffman, N.E., Yang, S.F., McKeon, T. (1982) Identification of 1- (malonylamino) cyclopropane-1-carboxylic acid as major conjugate of 1-aminocyclopropane-1-carboxylic acid, an ethylene precursor in higher plants. Biochem. Biophys. Res. Commun. 104, 765-770.

29. Kende, H. (1993) Ethylene biosynthesis. Ann. Rev. Plant Physiol. Plant Mol. Biol. 44, 283-307.

30. Kieber, J.J., Rothenberg, M., Roman, G., Feldman, K.A., Ecker, J.R. (1993) *CTR1*, a negative regulator of the ethylene response pathway in *Arabidopsis*, encodes a member of the Raf family of protein kinases. Cell 72, 427-441.

31. Kim, W.T., Silverstone, A., Yip, W.K., Dong, J.G., Yang, S.F. (1992) Induction of 1-aminocyclopropane-1-carboxylate synthase mRNA by auxin in mung bean hypocotyls and cultured apple shoots. Plant Physiol. 98, 465-471.

32. Kim, W.T., Yang, S.F. (1992) Turnover of 1-aminocyclopropane-1-carboxylic acid synthase in wounded tomato tissues. Plant Physiol. 100, 1126-1130.

33. Kuai, J., Dilley, D.R. (1992) Extraction, partial purification and characterization of 1-aminocyclopropane-1-carboxylic acid oxidase from apple (*Malus domestica* Borkh.) fruit. Postharvest Biol. Technol. 1, 203-211.

34. Kushad, M.M., Yelenosky, G., Knight, R. (1988) Interrelationship of polyamine and ethylene biosynthesis during avocado fruit development and ripening. Plant Physiol. 87, 463-467.

35. Lieberman, M. (1979) Biosynthesis and action of ethylene. Ann. Rev. Plant Physiol. 30, 533-591.

36. Liu, Y., Hoffman, N.E., Yang, S.F. (1985) Promotion by ethylene of the capability to convert 1-aminocyclopropane-1-carboxylic acid to ethylene in preclimacteric tomato and cantaloupe fruit. Plant Physiol. 77, 407-411.

37. Liu, Y., Su, L.Y., Yang, S.F. (1984) Metabolism of α-aminoisobutyric acid in mung bean hypocotyls in relation to metabolism of 1-aminocyclopropane-1-carboxylic acid. Planta 161, 439-443.

38. Lürssen, K., Naumann, K., Schröder, R. (1979) 1-Aminocyclopropane-1-carboxylic acid. An intermediate of the ethylene biosynthesis in higher plants. Z. Pflanzenphysiol. 92, 285-294.

39. Mattoo, A.K., Suttle, J.C., eds. (1991) The Plant Hormone Ethylene. CRC Press, Boca Raton, FL. 337 p.

40. McGarvey, D.J., Christoffersen, R.E. (1992) Characterization and kinetic parameters of ethylene forming enzyme from avocado fruit. J. Biol. Chem. 267, 5964-5967.

41. McKeon, T.A., Yang, S.F. (1984) A comparison of the conversion of 1-amino-2-ethylcyclopropane-1-carboxylic acid stereoisomers to 1-butene by pea epicotyls and by a cell-free system. Planta 160, 84-87.

42. Miyazaki, J.H., Yang, S.F. (1987) Metabolism of 5-methylthioribose to methionine. Plant Physiol. 79, 277-281.

43. Miyazaki, J.H., Yang, S.F. (1987) The methionine pathway in relation to ethylene and polyamine biosynthesis. Physiol. Plant. 69, 366-370.

44. Nakagawa, N., Mori, H., Yamazaki, K., Imaseki H. (1991) Cloning of complementary DNA for auxin-induced 1-aminocyclopropane-1-carboxylate synthase and differential expression of the gene by auxin and wounding. Plant Cell Physiol. 32, 1153-1163.

45. Nakajima, N., Mori, H., Yamazaki, K., Imaseki, H. (1990) Molecular cloning and sequence of a complementary DNA encoding 1-aminocyclopropane-1-carboxylate synthase induced by tissue wounding. Plant Cell Physiol. 31, 1021-1029.

46. Nakajima, N., Nakagawa, N., Imaseki, H. (1988) Molecular size of wound induced 1-aminocyclopropane-1-carboxylate synthase from *Cucurbita maxima* Duch. and change of translatable mRNA of the enzyme after wounding. Plant Cell Physiol. 29, 989-998.

47. Olson, D.C., White, J.A., Edelman, L., Harkins, R.N., Kende, H. (1991) Differential expression of two genes from 1-aminocyclopropane-1-carboxylate synthase in tomato fruits. Proc. Natl. Acad. Sci. USA 88, 5340-5344.

48. Park, K.Y., Drory, A., Woodson, W.R. (1992) Molecular cloning of a 1-aminocyclopropane-1-carboxylate synthase from senescing carnation flower petals. Plant Mol. Biol. 18, 377-386.

49. Peiser, G.D., Wang, T.T., Hofmann, N.E., Yang, S.F., Liu, H.W., Walsh, C.T. (1984) Formation of cyanide from carbon-1 of 1-aminocyclopropane-1-carboxylic acid during its conversion to ethylene. Proc. Natl. Acad. Sci. USA 81, 3059-3063.

50. Peleman, J., Boerjan, W., Engler, G., Seurinck, J., Botterman, J., Alliotte, T., Van Montagu, M., Inze, D. (1989) Strong cellular preference in the expression of a housekeeping gene of *Arabidopsis thaliana* encoding S-adenosylmethionine synthase. Plant Cell 1, 81-93.

51. Raskin, I., Beyer, E.M. (1989) Role of ethylene metabolism in *Amaranthus retroflexus*. Plant Physiol. 90, 1-5.

52. Raz, V., Fluhr, R. (1992) Calcium requirement for ethylene-dependent responses. Plant Cell 4, 1123-1130.

53. Riov, J., Dagan, E., Goren, R., Yang, S.F. (1990) Characterization of abscisic acid-induced ethylene production in citrus leaf and tomato fruit tissues. Plant Physiol. 92, 48-53.

54. Roberts, D.R., Walker, M.A., Thompson, J.E., Dumbroff, E.B. (1984) The effects of inhibitors of polyamine and ethylene biosynthesis on senescence, ethylene production and polyamine levels in cut carnation flowers. Plant Cell Physiol. 25, 315-322.

55. Rottmann, W.H., Peter, G.F., Oeller, P.W., Keller, J.A., Shen, N.F., Nagy, B.P., Taylor, L.P., Campbell, A.D., Theologis, A. (1991) 1-Aminocyclopropane-1-carboxylate synthase in tomato is encoded by a multigene family whose transcription is induced during fruit and floral senescence. J. Mol. Biol. 222, 937-961.

56. Saftner, R.A., Baldi, B.G. (1990) Polyamine levels and tomato fruit development: Possible interaction with ethylene. Plant Physiol. 92, 547-550.

57. Sanders, I.O., Smith, A.R., Hall, M.A. (1989) Ethylene metabolism in *Pisum sativum* L. Planta 179, 104-114.

58. Sato, T., Oeller, P.W., Theologis, A. (1991) The 1-aminocyclopropane-1-carboxylate synthase of *Cucurbita*. Purification, properties, expression in *Escherichia coli*, and primary structure determination by DNA sequence analysis. J. Biol. Chem. 266, 3752-3759.

59. Sato, T., Theologis, A. (1989) Cloning the mRNA encoding 1-aminocyclopropane-1-carboxylate synthase, the key enzyme for ethylene biosynthesis in plants. Proc. Natl. Acad. Sci. USA 86, 6621-6625.

60. Satoh, S., Esashi, Y. (1986) Inactivation of 1-aminocyclopropane-1-carboxylic acid synthase of etiolated mung bean hypocotyl segments by its substrate, S-adenosylmethionine. Plant Cell Physiol. 27, 285-291.

61. Satoh, S., Yang, S.F. (1988) S-Adenosylmethionine-dependent inactivation and radiolabeling of 1-aminocyclopropane-1-carboxylate synthase isolated from tomato fruits. Plant Physiol. 88, 109-114.

62. Sisler, E.C. (1977) Ethylene activity of some π-acceptor compounds. Tobacco Sci. 21, 43-45.

63. Sisler, E.C. (1982) Ethylene-binding properties of a Triton X-100 extract of mung bean sprouts. J. Plant Growth Regul. 1, 211-218.

64. Sisler, E.C., Blankenship, S.M. (1993) Diazocyclopentadiene (DACP), a light sensitive reagent for the ethylene receptor in plants. Plant Growth Regul. 12, 125-132.

65. Spanu, P., Reinhardt, D., Boller, T. (1991) Analysis and cloning of the ethylene-forming enzyme from tomato by functional expression of its mRNA in *Xenopus laevis* oocytes. EMBO J. 10, 2007-2013.

66. Su, L.Y., Liu, Y., Yang, S.F. (1985) Relationship between 1-aminocyclopropane-1-carboxylate malonyltransferase and D-amino acid malonyltransferase. Phytochemistry 24, 1141-1145.

67. Van der Straeten, D., Van Wiemeersch, L., Goodman, H. M., Van Montagu, M. (1989) Purification and partial characterization of 1-aminocyclopropane-1-carboxylate synthase from tomato pericarp. Eur. J. Biochem. 182, 639-647.

68. Van der Straeten, D., Van Wiemeersch, L., Goodman, H.M., Van Montagu, M.V. (1990) Cloning and sequencing of two different cDNAs encoding 1-aminocyclopropane-1-carboxylate synthase in tomato. Proc. Natl. Acad. Sci. USA 87, 4859-4863.

69. Van der Straeten, D., Djudzman, A., Van Caeneghem, W., Smalle, J., Van Montagu, M. (1993) Genetic and physiological analysis of a new locus in *Arabidopsis* that confers resistance to 1-aminocyclopropane-1-carboxylic acid and ethylene and specifically affects the ethylene signal transduction pathway. Plant Physiol. 102, 401-408.

70. Ververidis, P., John, P. (1991) Complete recovery *in vitro* of ethylene-forming enzyme activity. Phytochemistry 30, 725-727.

71. Woodson, W.R., Park, K.Y., Drory, A., Larsen, P.B., Wang H. (1992) Expression of ethylene biosynthetic pathway transcripts in senescing carnation flowers. Plant Physiol. 99, 526-532.

72. Yang, S.F., Hoffman, N.E. (1984) Ethylene biosynthesis and its regulation in higher plants. Ann. Rev. Plant Physiol. 35, 155-189.

73. Yip, W.K., Dong, J.G., Yang, S.F. (1991) Purification and characterization of 1-aminocyclopropane-1-carboxylate synthase from apple fruits. Plant Physiol. 95, 251-257.

74. Yu, Y.B., Adams, D.O., Yang, S.F. (1979) 1-Aminocyclopropane-1-carboxylate synthase, a key enzyme in ethylene biosynthesis. Arch. Biochem. Biophys. 198, 280-286.

75. Yung, K.H., Yang, S.F., Schlenk, F. (1982) Methionine synthesis from 5-methylthioribose in apple tissue. Biochem. Biophys. Res. Commun. 104, 771-777.

B5. Abscisic Acid Biosynthesis and Metabolism

Daniel C. Walton[1] and Yi Li[2]

[1]Department of Environmental and Forest Biology and Forestry, SUNY College of Environmental Science and Forestry, Syracuse, NY 13210, USA, and [2]Division of Biology, Kansas State University , Manhattan KS 66506, USA.

INTRODUCTION

One of the questions that plant physiologists ask about a hormone is how its cellular levels are regulated. The concentration of a hormone, or of any other cellular constituent, will depend on its rate of synthesis and metabolism and on its rate of import into and export from the cell. Abscisic acid (ABA, Fig. 1) is a particularly interesting hormone with regard to regulation of its levels, since they rise and fall dramatically in several kinds of tissues in response to environmental and developmental changes. When leaves of mesophytic plants are water stressed, ABA levels can rise from 10- to 50-fold within 4 to 8 hours, apparently due to a greatly increased rate of biosynthesis. When the plants are rewatered, the ABA levels drop to pre-stress levels within 4 to 8 hours. The drop in concentration is due to a reduced biosynthetic rate, a vigorous metabolism and possibly export from the leaves. In developing seeds of various plants, ABA levels can rise a hundred-fold within a few days and then decline to low levels as the seeds mature and desiccate. Synthesis and metabolism, as well as import, are involved in changing the ABA levels. Dormant buds and seeds of woody plants accumulate high levels of ABA

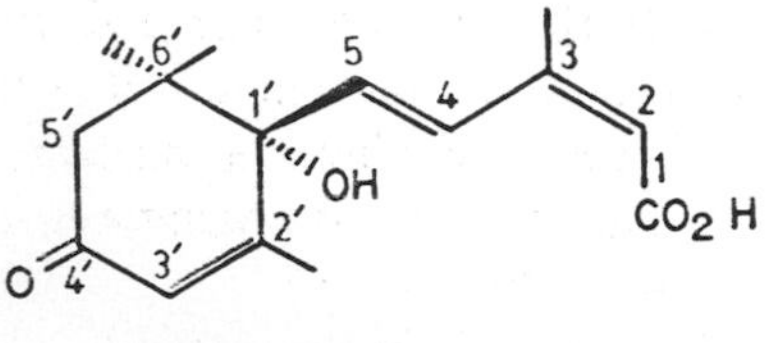

(S) – ABA

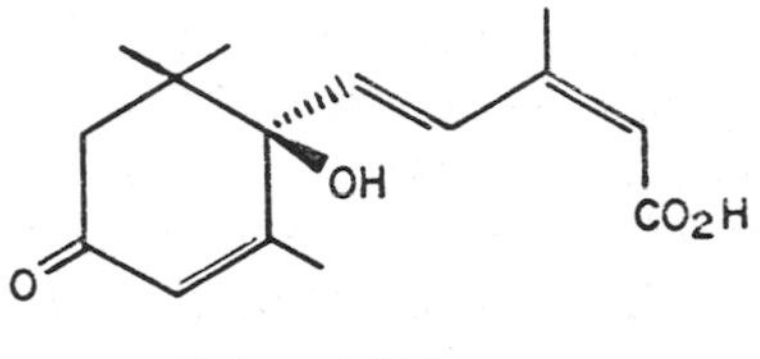

(R) – ABA

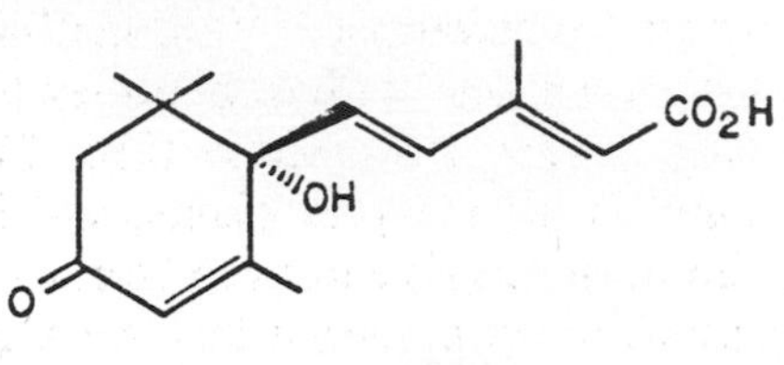

(S) 2-t- ABA

Fig. 1. Naturally occurring S-ABA and its R enantiomer.

140

P. J. Davies (ed.), Plant Hormones, 140–157.

which then decrease when the tissues are exposed to low temperatures. A combination of synthesis, metabolism, import and export is probably involved in determining the ABA levels in these issues.

In order to understand how ABA levels are regulated by biosynthesis and metabolism, it is first necessary to know the identity of the compounds which are involved in the pathways. Enzymes which interconvert these compounds must be characterized and the enzyme co-factors identified. Only then is it possible to begin to determine how the pathways are regulated. In the case of ABA, we have only recently unravelled many of the constituents of the pathways, particularly for biosynthesis. We have little knowledge of the enzymes involved. Consequently, our understanding of how these pathways are regulated will require considerably more information than we currently possess.

A number of more detailed reviews of the subject matter included in this chapter have been written during the past decade (1, 8, 22, 29, 34, 44).

BIOSYNTHESIS

ABA is a sesquiterpene which like other sesquiterpenes has been shown to be derived from mevalonic acid (MVA). Although ABA is a relatively simple molecule, most of the details of its biosynthesis have remained obscure until recently. We now know that in higher plants ABA is derived from the cleavage product of a xanthophyll (Fig. 2). Why was progress so slow in working out the ABA biosynthetic pathway since ABA was first described in 1965? As is the case for other hormones, ABA is usually present in the plant in very low concentrations. In most tissues the levels are from 10 to 50 ng/g fresh wt ($4x10^{-8}$M to $2x10^{-7}$M). Only in water-stressed leaves, developing and mature seeds and dormant buds are levels higher than 10^{-6}M. A second problem in studying ABA biosynthesis has been the poor incorporation of presumed precursors into ABA when they are applied in radioactive form. Even in water-stressed leaves in which ABA levels rise dramatically due to increased rates of synthesis, ^{14}C-MVA and ^{14}CO2 are poorly incorporated into ABA. Thus, one of the important tools used to study metabolic pathways in tissues has been of little use in studying ABA biosynthesis.

Although there had been suggestions that several C-15 compounds are intermediates on a direct ABA biosynthetic pathway, no convincing evidence was produced that they are involved. The suggestion that ABA may be derived from a xanthophyll came from several observations. Violaxanthin can be photo-oxidized to xanthoxin, a C-15 compound with similarities to ABA in its carbon skeleton and in its ring oxygen substitution (Fig. 2). Xanthoxin is a naturally-occurring compound in a variety of plants, and it was shown to be converted to ABA when fed in radioactive form to bean and tomato plants (37). Despite these observations direct evidence that ABA is derived from xanthoxin, was still lacking. Even that the xanthoxin found in

141

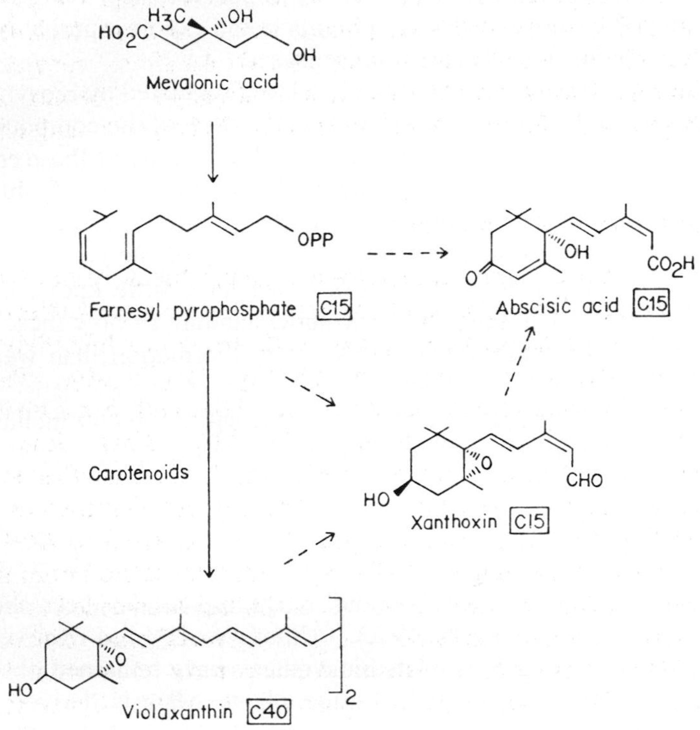

Fig. 2. Biosynthesis of ABA; C-15 vs C-40 pathways.

plants originated from a carotenoid was still not certain, since as indicated in Figure 2, xanthoxin could be derived directly from farnesyl pyrophosphate (FPP). The inability to unequivocally demonstrate ABA intermediates, coupled with the poor conversions of presumptive precursors to ABA, brought work on biosynthesis in plants to a virtual halt by the mid 1970s.

The discovery in 1977 that a rose pathogen, the fungus *Cercospora rosicola*, produces and excretes relatively large quantities of the naturally occurring enantiomer of ABA into its growth medium initiated work on the biosynthetic pathway in that organism (2). The hope was that the pathway in *C. rosicola* would be similar to, or identical with, the pathway in higher plants, and that the relatively large production of ABA by the fungus would allow its pathway to be determined more readily than can be done in higher plants. The discovery also stimulated investigators to look for ABA production in other fungi, with the result that at least a dozen fungi have been reported to produce ABA (6). Since only a relatively few fungi have been investigated for ABA production, it seems likely that the list will grow longer. The intriguing possibility that ABA may play a role in pathogenesis is still to be determined.

While studies proceeded with *C. rosicola,* investigators were restimulated to look at ABA biosynthesis in plants, using different techniques than previously. The results of these studies have shown almost conclusively that ABA is produced from the cleavage of a xanthophyll. The next 2 sections of this chapter will describe work done with *Cercospora* and with higher plants.

ABA Biosynthesis in *Cercospora*

The discovery that *C. rosicola* secretes a considerable amount of ABA into its growth medium led to the rapid development of a simple defined medium in which cell suspension cultures grow well, and into which they begin to secrete ABA after about 5 days growth (Fig. 3) (12, 30). The fungus converts both ^{14}C-acetate and ^{14}C-MVA into ABA with a reasonable yield, unlike the results obtained with plants (3, 29). When either of these compounds was fed in radioactive form to the fungus, another radioactive compound was isolated from the growth medium and identified as 1'-deoxy ABA (3 in Fig. 4a) (28). This compound was converted to ABA in good yield when refed to the fungus. 1'-Deoxy ABA was isolated from the fungal medium when neither ^{14}C-acetate nor ^{14}C-MVA had been added to the growth medium, so that it is a naturally occurring fungal metabolite. Since ABA and 1'-deoxy ABA differ by only a hydroxyl group at the 1' position, it is likely that 1'-deoxy ABA is an immediate precursor to ABA. Other compounds related to 1'-deoxy ABA, such as α-ionylidene acetic acid (1 in Fig. 4a) and 4'-hydroxy-α-ionylidene acetic acid (2 in Fig. 4a) are converted by the fungus to both 1'deoxy ABA and to ABA (29). Figure 4a shows one possible route for the later stages of ABA biosynthesis in *C. rosicola* based on these studies. The sequence of side-chain and ring oxidations shown in Figure 4 is not meant to imply a unique pathway, since it is possible that they could occur in one or more different sequences. Work with the related fungus, *C. cruenta,* has led to the suggestion that a variation of the apparent *C. rosicola* pathway may operate in this organism (31) (Fig. 4b). Whether

Fig. 3. Dry weight and ABA accumulation in cell suspension cultures of *C. rosicola.*

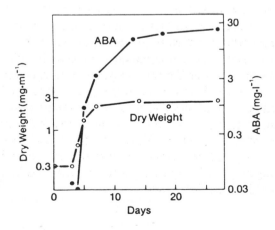

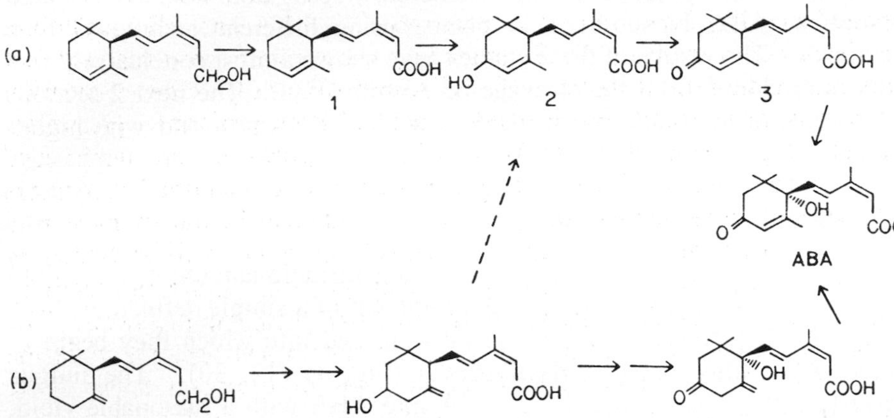

Fig. 4. Later stages of possible ABA biosynthetic pathways in (a) *C. rosicola* and (b) *C. cruenta.*

the two organisms really have the same pathways, or whether they vary slightly remains to be determined. Inhibitors of carotenoid biosynthesis did not affect the accumulation of ABA by *C. cruenta* so it is assumed that a direct pathway from FPP is involved (31). In neither fungus, nor in the others studied, has the presumptive first cyclic intermediate been isolated. The steps immediately after the formation of FPP are also still unknown (4).

After the discovery that α-ionylidene acetic acid and 1'-deoxy ABA may be intermediates on the ABA biosynthetic pathway in *C. rosicola*, these compounds were fed in radioactive form to several plant tissues (29). The results differed depending on the tissues fed. In the case of bean leaves, immature bean seeds, and avocado fruit, these compounds were not converted to ABA. α-Ionylidene acetic acid was converted to 1'-deoxy ABA and various polar compounds were formed from both compounds. These results were similar to those previously obtained with barley plants (18). In the case of *Vicia faba* leaves, however, convincing evidence was obtained for the conversion of both α-ionylidene acetic acid and 1'-deoxy ABA to ABA. Although these results appear to be genuine it seems likely that they were anomalous and due to the presence of an enzyme which is not universally distributed in higher plants. Thus, the hoped for similarity between the fungal and plant pathways was not supported by the evidence.

ABA Biosynthesis in Plants

As indicated earlier in this chapter, a major problem in studying the ABA biosynthetic pathway in higher plants is the poor incorporation of radioactive precursors such as MVA and CO_2 into ABA. The poor incorporation of MVA led Milborrow to propose that the chloroplast is the site of ABA synthesis on the assumption that transport of ^{14}C-MVA across the chloroplastic membrane was the rate-limiting step in incorporation. He reported that lysed chloroplasts incorporated ^{14}C-MVA into ABA, although the yields were very low (24). Subsequently, other investigators reported that chloroplasts did not incorporate ^{14}C-MVA into ABA (13). One explanation for both of these results is that ABA is derived from preformed precursors present at high levels relative to ABA, and that the precursors are synthesized at low rates in mature leaves. The consequences of preformed precursors with such attributes is that radioactive compounds will be incorporated into them very slowly and the radioactivity that is incorporated will be diluted by the high concentrations of precursor already present. Little radioactivity would appear in the ABA subsequently produced from the precursor. The major leaf xanthophylls such as lutein, violaxanthin and neoxanthin fit the requirements for such a precursor (Fig. 5). They are present at levels more than 10^3 times greater than ABA in unstressed leaves and their rates of synthesis in mature leaves are low.

Although the work with *C. rosicola* and *C. cruenta* suggested a direct C-15 pathway to ABA, evidence began to accumulate that an indirect pathway from the carotenoids is involved in higher plants. For example, several corn mutants were described which lack the ability to synthesize carotenoids due to specific defects in their biosynthetic pathway. These mutants also have a reduced ability to accumulate ABA in their leaves and roots (27). In addition, inhibitors of carotenoid synthesis, such as fluridone and norflurazon, also inhibit the accumulation of ABA under some conditions (15, 26).

A crucial experiment of a different type by Creelman and Zeevaart (6) also pointed to the possible role for xanthophylls as ABA precursors. These investigators water-stressed bean and *Xanthium* leaves in the presence of $^{18}O_2$, a heavy isotope of oxygen. Their rationale was that if 1-'deoxy ABA is the immediate precursor to ABA in plants, as it apparently is in *C. rosicola*, then one ^{18}O atom should be incorporated into the ABA ring at the 1' position. This assumes that the hydroxyl oxygen is derived from O_2 as normally would be expected, and as is the case for ABA made in *C. rosicola*. Their results showed, however, that one atom of ^{18}O was incorporated into the ABA, but it was in the carboxyl group and not in the ring. The results suggested that (a) 1'-deoxy ABA is not the immediate precursor to ABA, at least in water-stressed leaves and (b) ABA formed in the water stressed leaves was derived from a preformed precursor containing the oxygens which would become the 1' and 4' oxygens of ABA. One explanation for these results is that a xanthophyll, such as violaxanthin, was cleaved by oxygen to form an aldehyde containing ^{18}O. The aldehyde in the case of violaxanthin cleavage would be xanthoxin. If the xanthoxin were oxidized by dehydrogenases when converted to ABA, there would be only one atom of ^{18}O in the carboxyl group. The second oxygen atom would have been obtained from water which

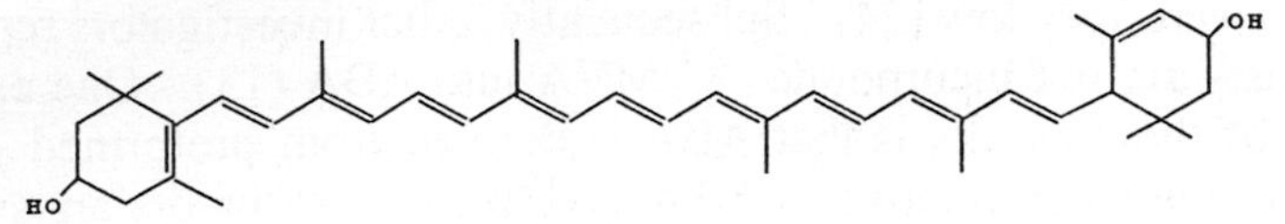

Lutein

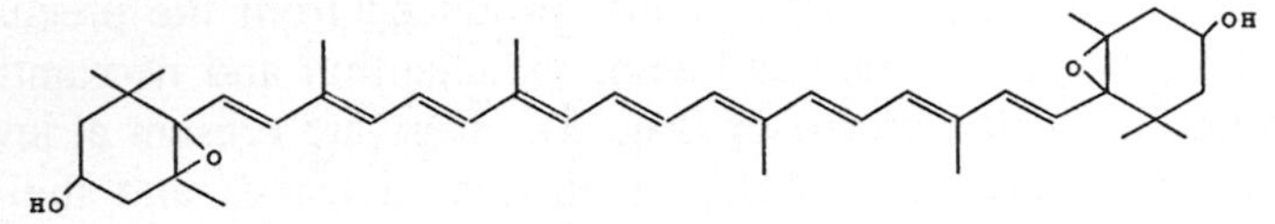

Violaxanthin

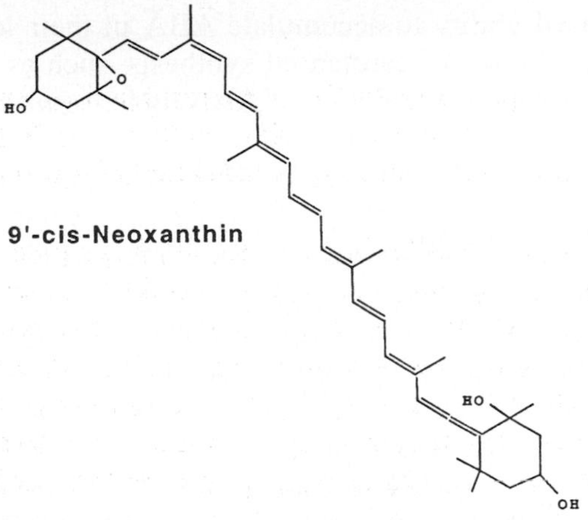

9'-cis-Neoxanthin

Fig. 5. Structures of several leaf xanthophylls.

would not contain ^{18}O. These conversions are summarized in Figure 6.

Although the experiments described in the previous paragraphs suggested that xanthophylls are ABA precursors, they did not give any information about the identities of the xanthophylls. In order to test whether violaxanthin is an ABA precursor, intact leaves were treated so that the epoxide oxygens were partially replaced by ^{18}O (20). The leaves were then water-stressed and both violaxanthin and ABA were isolated and analyzed by mass spectrometry. If the ABA produced during the water stress period had been derived from violaxanthin containing ^{18}O in its epoxide group, one would expect to observe

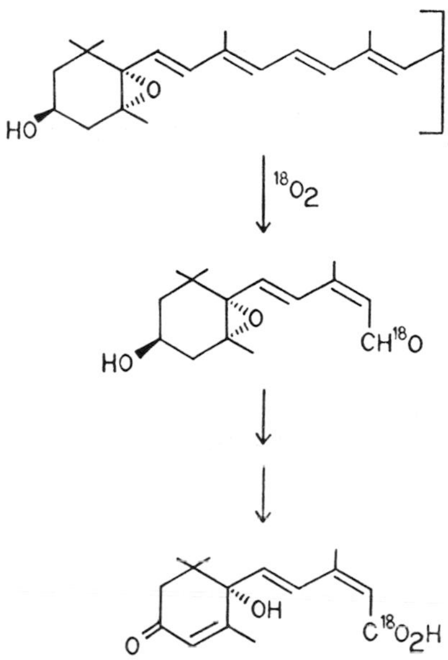

Fig. 6. Hypothetical cleavage of violaxanthin to xanthoxin by $^{18}O_2$ with subsequent conversion to ABA by dehydrogenases.

^{18}O at the ABA 1'-hydroxyl group (Fig. 7). The results suggested that a portion of the ABA was derived from the violaxanthin which had been labelled with ^{18}O, but that this violaxanthin accounted for only about 25% of the ABA produced.

As indicated above, one of the difficulties in establishing a precursor role for the xanthophylls in ABA biosynthesis is their high cellular concentrations compared with ABA. An attempt was made to reduce this problem by treating dark-grown barley seedlings with fluridone (10). Although the xanthophyll levels were reduced more than 95% by the fluridone treatment, ABA accumulation as a result of water stress still occurred albeit to a reduced level. Under these conditions there was also an indication that violaxanthin and neoxanthin levels dropped during water stress. These results were extended by Li and Walton with etiolated bean seedlings (21). In the etiolated bean tissue xanthophyll levels were greatly reduced, but ABA production during water stress was at least as great as it was in green seedlings. Xanthophyll levels, as well as ABA and its metabolites, were analyzed prior to and during a short water stress. It was observed that there was a 1:1 stoichiometry between the loss of violaxanthin, 9'-*cis*-neoxanthin and 9-*cis*-violaxanthin and the increase in ABA and its two major metabolites PA and DPA (Table 1). Treatments which inhibited ABA production also inhibited the loss of the xanthophylls. The results suggested that all-*trans* violaxanthin is a major source of ABA and that 9'-*cis*-neoxanthin, derived from all-*trans* violaxanthin, is cleaved to form one C-15 compound which

147

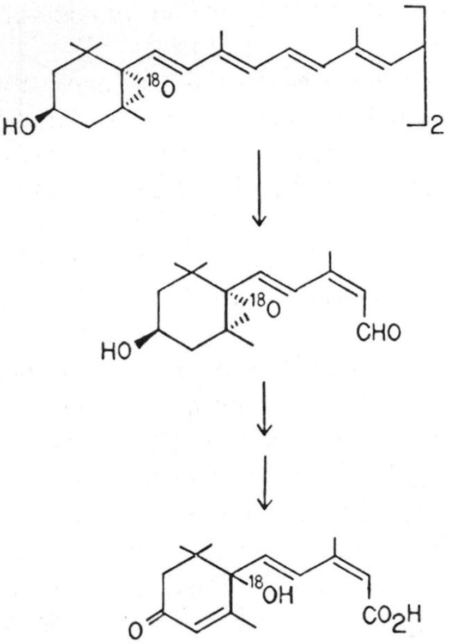

Fig. 7. Conversion of violaxanthin containing [18]O in the epoxide to ABA containing [18]O at the 1'-hydroxyl position.

is then converted to ABA. The finding that 9'-*cis*-neoxanthin is the apparent cleavage substrate is particularly significant since the stereochemistry at its 9' position is the same as that occurring in the side chain of ABA. The participation of 9'-cis-neoxanthin also helps to explain why there was only an apparent 25% incorporation of [18]O from trans-violaxanthin into ABA as described above (20). A portion of the ABA would have been formed directly from the pool of unlabelled 9'-cis-neoxanthin thus diluting the incorporation of [18]O.

These results were confirmed and extended to tomato roots by Parry et al (33). Further evidence that these xanthophylls are ABA precursors has been obtained from work with *Arabidopsis* ABA-deficient mutants which

Table 1. Kinetics of xanthophyll reduction and increase in ABA and its metabolites after water stress. (Adapted from 21)

Stress Time min	9'-cis -Nx[a]	9-cis -Vx	Vx	total Xt	ABA	PA	DPA	ABA+ PA+DPA
	decrease nmol. g fresh wt[-1]				increase nmol. g fresh wt[-1]			
40	-0.3	0.0	0.7	0.4	0.9	0.0	0.0	0.9
90	2.6	0.7	8.8	12.1	6.7	0.7	0.0	7.4
270	3.6	0.9	17.6	22.1	5.8	13.4	2.0	21.2

[a] Nx = neoxanthin; Vx = violaxanthin; Xt = xanthophylls; PA = phaseic acid; DPA = dihydrophaseic acid.

have a greatly reduced capacity to epoxidate zeaxanthin to violaxanthin and consequently to produce 9'-*cis*-neoxanthin (9, 35).

As previously indicated, xanthoxin had been implicated as an ABA precursor almost from the beginning of the work on ABA biosynthesis, but direct evidence for its involvement was lacking. Evidence that xanthoxin is a precursor has now been obtained with ABA-deficient mutants which convert xanthoxin to ABA very poorly (32, 39). Further work has shown that the lesion occurs in the enzyme which oxidizes ABA aldehyde, the intermediate between xanthoxin and ABA, to ABA (36, 38). Instead, the ABA aldehyde is converted to *t*-ABA alcohol which accumulates (39). A number of ABA-deficient mutants have now been described which are unable to convert ABA aldehyde to ABA (34). These results taken together indicate that 9'-*cis*-neoxanthin, derived from all-*trans* violaxanthin, is cleaved to form xanthoxin which is converted to ABA via ABA aldehyde (Fig. 8). The figure shows the two possible ways in which violaxanthin could be converted to 9'-*cis*-neoxanthin. While the work described above has now given us an understanding of the basic ABA biosynthetic pathway there are still a variety of questions to be answered. There is indirect evidence to suggest that the rate of ABA synthesis in leaves is controlled by the activity of the 9'-*cis* - neoxanthin cleavage enzyme, but since this enzyme activity has not been directly demonstrated we cannot be sure that it is rate controlling. The localization of the enzymes on the pathway are unknown although xanthoxin conversion to ABA appears to be cytoplasmic. Although the xanthophylls are localized within the chloroplast we do not know whether 9'-*cis* -neoxanthin cleavage occurs within the chloroplast, at the chloroplastic membrane, or outside the chloroplast. We also lack sufficient evidence about the conversion of all-*trans* violaxanthin to 9'-*cis*-neoxanthin and whether this conversion plays a role in the regulation of ABA biosynthesis.

METABOLISM

Research into ABA metabolism, which began shortly after its discovery, has been successful in identifying ABA metabolites. This success was due in part to the early availability of ^{14}C-ABA, and also to the rapid and extensive metabolism of ^{14}C-ABA when it is applied to plant tissues.

In discussing the metabolism of a compound which is administered to a plant, we must be careful to distinguish between those metabolites which are later shown to occur naturally in the plant and those which are not. This distinction is particularly important in the case of ABA in which most feeding experiments have been done with a racemic mixture of the naturally occurring *S*-enantiomer and the unnatural *R*-enantiomer (Fig. 1). There is evidence that the two enantiomers are metabolized not only at different rates, but in some instances to different compounds. Care must be taken to determine that a metabolite has in fact been derived from the *S*-enantiomer. Ultimately, the

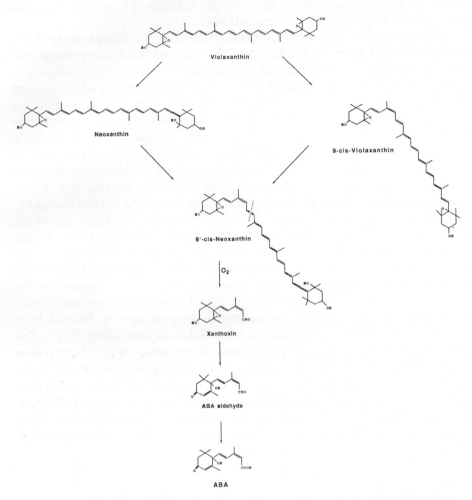

Fig. 8. The putative biosynthetic pathway from all-*trans* violaxanthin to ABA. Modified from (21).

isolation of the metabolite as a naturally occurring plant constituent is required in order to demonstrate that the compound is not an artifact of feeding. A number of ABA metabolites have been shown to occur naturally, while others have not been isolated from plants which have not been fed [14]C-*RS* ABA.

The initial ABA metabolites described were ABA glucose ester (ABA-GE, 6, Fig. 10), phaseic acid (PA, 3, Fig. 9) and 6'-hydroxymethyl ABA (HM-ABA, 2, Fig. 9). ABA glucose ester was first isolated as a naturally occurring compound from *Lupinus luteus* fruit and then shown to be produced when ^{14}C-ABA was fed to several plants (17). Although this compound is a naturally occurring ABA metabolite, feeding experiments with [14]C-*RS* ABA can exaggerate its importance since the *R*-enantiomer is often converted

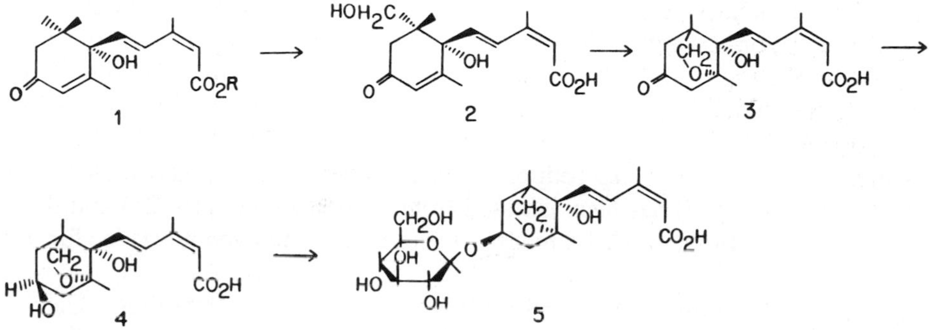

Fig. 9. Metabolic pathways for the conversion of ABA to DPA-GS.

preferentially to the glucose ester. HM-ABA was isolated from tomato shoots which had been fed ^{14}C-ABA. This compound has been isolated only once, apparently because of the ease with which it rearranges to PA. PA had actually been isolated prior to the discovery of ABA, but its correct structure was only determined as the rearrangement product of HM-ABA (23). The observation that HM-ABA so readily rearranged to PA in the absence of

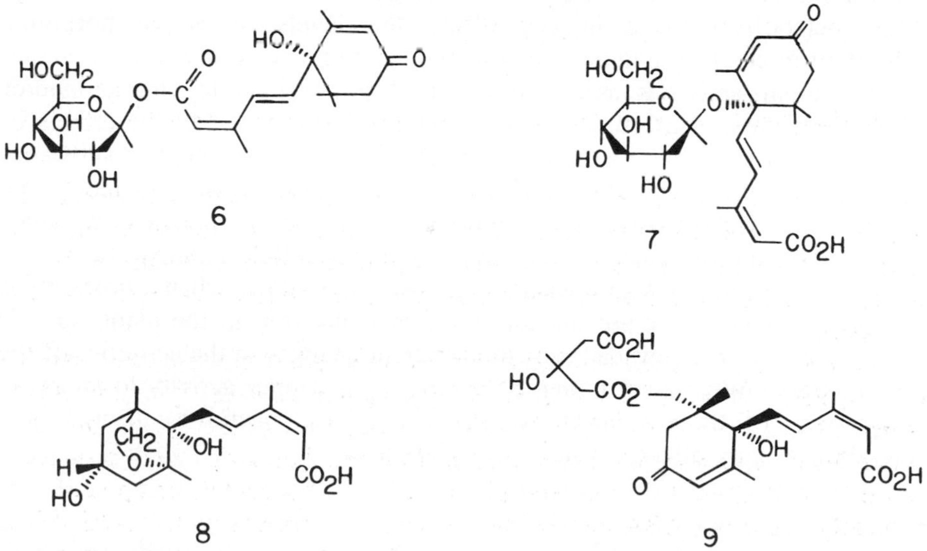

Fig. 10. Naturally occurring ABA metabolites not shown in Fig. 9.

enzymes, made it unclear whether PA was a natural ABA metabolite or an artifact of isolation. The discovery that an apparent PA reduction product dihydrophaseic acid (DPA, 4, Fig. 9) was present in very high concentrations in bean seeds suggested that PA is a naturally occurring ABA metabolite subject to further metabolism (42). Since the initial descriptions of PA and DPA, these compounds have been found in a wide variety of plants, and appear to be on the major pathway of ABA metabolism. An epimer of DPA, epi-DPA (8, Fig. 10) has also been shown to be a naturally occurring ABA metabolite although it usually occurs in lower concentrations than DPA (43). Although DPA exists in high concentrations in some tissues, it is clear that it is not necessarily the endproduct of metabolism. The 4'-glucoside of DPA (DPA GS, 5, Fig. 9) has been isolated from several tissues (16, 25) and there are indications from ^{14}C-ABA feeding experiments that even it may be further metabolized.

Figure 10 shows 2 other naturally occurring ABA metabolites. The 1'-glycoside of ABA (ABA-GS, 6) may be widespread in plants, although probably in low levels, and 3-hydroxy-3-methyl glutaryl HM-ABA (9, Fig. 10) which has so far been isolated only from the seeds of *Robinia pseudoacacia* (22). Feeding experiments with ^{14}C-*RS* ABA show that other metabolites can be formed, such as base-labile conjugates of PA, DPA and epi-DPA. These compounds have not yet been characterized, nor have they been shown to occur naturally (22).

Effects of Metabolism on Physiological Activities

The identification of metabolites and the elucidation of metabolic pathways are intellectually challenging and interesting in their own right. In the case of a hormone, however, the plant physiologist is more interested in the role that metabolism plays in controlling the levels of active hormone. Metabolism can play a number of roles, including activation and inactivation of hormonal activities and conversion of a hormone to storage and/or transport forms. Unfortunately, it is often difficult to assess the effects of metabolism on hormone activity. The effectiveness of any applied substance will depend on factors besides its intrinsic activity. These will include its rate of uptake into the tissue, its metabolism to more or less active compounds after uptake, and its rate of entry into the proper cellular compartment.

We often do not know whether the results observed when a hormone is applied to plant tissue are indicative of its actual role in the plant, so it is hard to assess the significance of the apparent changes in the activities of the metabolites. We may not even be aware of the proper activity to measure. Regardless of these caveats, we think it is still useful to discuss how metabolism may affect physiological activities.

Inactivation

When analogues of ABA are compared with ABA in their ability to affect various physiological processes, it has been found that almost any change in the ABA molecule reduces the apparent activity. The activity of ABA in the full range of bioassays tested appears to depend on the presence of a free carboxyl group, a 2-*cis*,4-*trans*-pentadienoic side chain, a 4'-ketone and a double bond in the cyclohexane ring (41) although a few synthetic compounds do show activity without all of these attributes. Many of the metabolites described so far lack one or more of these functional groups. ABA-GE lacks the free carboxyl, PA the ring double bond, and DPA and its further metabolites lack both the ring double bond and the 4'-keto group. HM-ABA, its 3-hydroxy-3-methyl glutaryl derivative and ABA-1'-glycoside all appear to have the necessary functional groups of ABA intact. Neither PA nor DPA appear to inhibit cell elongation in various tissues which indicates that the presence of the 4'-ketone and the ring double bond are both necessary for this activity. ABA glucose ester has been reported to inhibit cell elongation, but it seems likely that the activity depends on its hydrolysis to ABA as has been shown to be the case for other ABA esters. The 3-hydroxy-3-methyl conjugate of HM-ABA has been reported to inhibit cell elongation. Whether the activity requires the intact molecule was not determined. PA has been reported to have only about 10% of the activity of ABA in an abscission bioassay, and has reduced activity in stomatal closure bioassays in several plants, although apparently no activity in closing the stomata of *V. faba*. DPA has no activity in any of the bioassays in which it has been tested. It seems reasonable to conclude that DPA synthesis, and that of its further metabolites, is a primary mode of ABA inactivation,

Although PA is inactive, or has a greatly reduced activity, in several bioassays compared with ABA, it does appear to be as active as ABA in reducing the GA_3-stimulated synthesis of α-amylase in barley aleurone layers (40). DPA, however, has no activity in this assay, so that the 4'-keto group, but not the ring double bond, is necessary for the results observed with PA and ABA. One explanation for PA activity in this assay is that both ABA and PA are able to bind to and activate the necessary receptor equally well. A second explanation is that PA is the active molecule and that the apparent ABA activity is derived from its conversion to PA, which does occur readily in barley aleurone layers.

HM-ABA contains all of the ABA functional groups and has in addition a hydroxyl group at the 6'-methyl position. One intriguing possibility is that it is HM-ABA, rather than ABA, which is the active molecule in many of the ABA bioassays. Since all of the tissues which have been fed ^{14}C-ABA produced PA, and therefore HM-ABA, it is not possible to rule out an active role for HM-ABA. It is also not possible to test HM-ABA's activity directly. Even if it could again be isolated, it would be difficult to keep it from isomerizing to PA during any bioassay.

Storage and Transport Forms

One requirement that a metabolite must meet in order to be a storage or transport form of ABA is that it must be reconvertable to ABA. Of the various metabolites that have been described, only ABA-GE and ABA-GS seem likely to be able to meet this requirement, since esterase and glycosidase activity could release ABA from ABA-GE and ABA-GS, respectively. There is no evidence that either PA or DPA can be reconverted to ABA. ABA was shown to be released when ABA-GE was fed to tomato plants. It is difficult, however, to determine whether hydrolysis of endogenous ABA-GE occurs. The evidence obtained so far suggests that ABA-GE is not a major source of ABA, but is sequestered in the vacuole (14). When plants were subjected to a series of stress and rewatering cycles, ABA-GE continued to increase in concentration (5). ABA-GE has been shown to occur in the phloem and xylem. If it were hydrolyzed when it reached the sink tissue, it could be considered a transport form of ABA. ABA has also been shown to occur in the xylem and phloem, so it seems unlikely that ABA-GE is necessary as a long distance transport form of ABA. There is no evidence about the fate of ABA-GS. Since it appears to be a minor metabolite, it seems unlikely that it plays a major role as an ABA storage and/or transport form (22).

LOCALIZATION AND REGULATION OF METABOLISM

The capacity to metabolize applied ABA is widespread in plant tissues. ^{14}C-ABA has been shown to be metabolized by leaves, stems, roots, seeds and fruit, and ABA metabolites have been isolated from all of these tissues (22). However, since ABA metabolites have also been found in the xylem and phloem, the occurrence of metabolites in a particular tissue does not necessarily indicate that it was formed there. There is less evidence about the subcellular localization of ABA metabolism. The conversion of ABA to HM-ABA by a cell-free system obtained from the liquid endosperm of *Echinocystis lobata* was suggested to involve a mixed-function oxygenase (11). The report that the oxygen inserted into ABA to form HM-ABA is derived from O_2 is further evidence for the participation of an oxygenase (6). Since mixed function oxygenase activity is generally associated with the endoplasmic reticulum, these results suggest a cytoplasmic location for the initial step of ABA metabolism. The conversion of PA to DPA in the *E. lobata* system appeared to involve a soluble enzyme, presumably also of cytoplasmic origin. The conversion of ABA by spinach mesophyll protoplasts to PA and DPA was shown to occur in the extraplastidic fraction, indicating that chloroplasts do not metabolize ABA (14).

An enzyme which catalyzes the transfer of glucose from UDP-glucose to form ABA-GE has been isolated from cell suspension cultures of *Macleaya microcarpa* (19). Presumably the enzyme is of cytoplasmic origin.

Because little is known about the ABA metabolic enzymes, our knowledge of the regulation of metabolism is also limited. There is an indication, however, of possible regulation in barley aleurone layers. If layers are pretreated with 10^{-5}M ABA for several hours, the subsequent metabolism of exogenous ABA is from 2 to 5 times greater than it would have been without the pretreatment (40). The increased metabolic rate can be eliminated by inhibitors of protein and nucleic acid synthesis. These inhibitors also eliminate several new proteins which appear after ABA pretreatment. The suggestion has been made that ABA induces the synthesis of the ABA hydroxylating enzyme in this tissue since PA does not cause the new proteins to appear, nor does it increase ABA metabolism. The isolation of the hydroxylating enzyme from the aleurone layers has not been successful, so it has not been possible to demonstrate an increased level of the enzyme directly.

References

1. Addicott, F.T. (1983) Abscisic Acid. Praeger, New York, 607pp.
2. Asante, G., Merlini, Nasini, G. (1977) (+)-Abscisic acid, a metabolite of the fungus *Cercospora rosicola*. Experientia 33, 1556.
3. Bennett, R.D., Norman, S.M., Maier, V.P. (1981) The biosynthesis of abscisic acid from (1, 2-^{13}C)acetate in *Cercospora rosicola*. Phytochemistry 20, 2343-2444.
4. Bennett, R.D., Norman, S.M., Maier, V.P. (1990) Intermediate steps in the biosynthesis of abscisic acid from farnesyl pyrophosphate in *Cercospora rosicola*. Phytochemistry 27, 3473-3477.
5. Boyer, G.L., Zeevaart, J.A.D, (1982) Isolation and quantitation of β D glucopyranosyl abscisate from leaves of Xanthium and spinach. Plant Physiol. 70, 227-231.
6. Creelman, R.A., Zeevaart, J.A.D. (1984) Incorporation of oxygen into abscisic acid and phaseic acid from molecular oxygen. Plant Physiol. 75, 166-169.
7. Crocoll, C., Kettner, J., Dorffling, K. 1991. Abscisic acid in saprophytic and parasitic species of fungi. Phytochemistry 30, 1059-1060
8. Davies, W.J., Jones, H.G. Abscisic Acid: Physiology and Biochemistry. Bios Scientific Publishers, Oxford, 1991
9. Duckham, S.C., Linforth, R.S.T., Taylor, I.B. (1991) Abscisic-acid-deficient mutants at the *aba* gene locus of *Arabidopsis thaliana* are impaired in the epoxidation of zeaxanthin. Plant, Cell and Envir. 14, 601-606
10. Gamble, P.E., Mullet, J. (1986) Inhibition of carotenoid accumulation and abscisic acid biosynthesis in fluridone-treated dark-grown barley. Eur. J. Biochem. 160, 117-121.
11.. Gillard, D.F., Walton, D.C. (1976) Abscisic acid metabolism by a cell-free preparation from *Echinocystis lobata* liquid endosperm. Plant Physiol. 58, 790-795.
12.. Griffin, D.H., Walton, D.C. (1982) Regulation of abscisic acid formation in *Mycosphaerella (Cercospora) rosicola* by phosphate. Mycologia 74, 614-618.
13. Hartung, W., Heilmann, B. and Gimmler, H. (1981) Do chloroplasts play a role in abscisic acid synthesis? Plant Sci. Letters, 2, 235-242.
14. Hartung, W., Gimmler, H. (1982) The compartmentation of abscisic acid (ABA), of ABA-biosynthesis, ABA-metabolism and ABA-conjugation. *In*: Plant Growth Substances 1982, pp. 324-333, Wareing, P.F., ed. Academic Press, London.
15. Henson, I.E. (1984) Inhibition of abscisic acid accumulation in seedling shoots of pearl millet (*Pennisetum americanum* L.) following induction of chlorosis by norflurazon. Z. Pflanzenphysiol. 114, 35-43.

16. Hirai, N. and Koshimizu, K. (1983) A new conjugate of dihydrophaseic acid from avocado fruit. Agric. Biol. Chem. 47, 365-371.

17. Koshimizu, K., Inui, M., Fukui, H., Mitsui, T. (1968) Isolation of (+)-abscisyl-β-D-glucopyranoside from immature fruit of *Lupinus luteus*. Agric. Biol. Chem. 32, 789-791.

18. Lehmann, H., Schutte, H.R. (1976) Biochemistry of phytoeffectors. Biochem. Physiol. Pflanzen 169, 55-61.

19. Lehmann, H., Schutte, H.R. (1980) Purification and characterization of an abscisic acid glucosylating enzyme from cell suspension cultures of *Macleaya microcarpa*. Z. Pflanzenphysiol., 96, 277-280.

20. Li, Y., Walton, D.C. (1987) Xanthophylls and abscisic acid biosynthesis in water-stressed bean leaves. Plant Physiol. 85, 910-915.

21. Li, Y., Walton., D.C. (1990) Violaxanthin is an abscisic acid precursor in water-stressed dark-grown bean leaves. Plant Physiol. 92, 551-559.

22. Loveys, B.R., Milborrow, B.V. (1984) Metabolism of abscisic acid. *In*: The Biosynthesis and Metabolism of Hormones, pp. 71-104, Crozier A., Hillman, J.R.,ed. Cambridge Univ. Press, Cambridge.

23. Milborrow, B.V. (1969) Identification of 'metabolite C' from abscisic acid and a new structure for phaseic acid. Chem. Commun. pp. 966-967.

24. Milborrow, B.V. (1974) Biosynthesis of abscisic acid by a cell-free system. Phytochemistry 13, 131-136.

25. Milborrow, B.V. and Vaughan, G.T. (1982) Characterization of dihydrophaseic acid 4'-O-β-D-glucopyranoside as a major metabolite of abscisic acid. Australian Journal of Plant Physiology, 9, 361-372.

26. Moore, R., Smith, J.D. (1984) Growth, graviresponsiveness and abscisic acid content of *Zea mays* seedlings treated with fluridone. Planta 162, 342-344.

27. Moore, R., Smith, J.D. (1985) Graviresponsiveness and abscisic acid content of roots of carotenoid-deficient mutants of *Zea mays*. Planta 164, 126-128.

28. Neill, S.J., Horgan, R., Lee, T.S., Walton, D.C. (1981) 3-methyl-5-(4'-oxo-2', 6', 6'-trimethylcyclohex-2'-en-yl)-2, 4-pentadienoic acid, a putative precursor of abscisic acid from *Cercospora rosicola*. FEBS Lett. 128, 30-32.

29. Neill, S.J., Horgan, R., Walton, D.C. (1984) Biosynthesis of abscisic acid. *In*: The Biosynthesis and Metabolism of Hormones, pp. 43-70, Crozier A., Hillman, J.R., ed. Cambridge Univ. Press, Cambridge, U.K.

30. Norman, S.M., Maier, V.P., Echols, L.C. (1981) Development of a defined medium for growth of *Cercospora rosicola* Passerini. Appl. Environ. Microbiol. 41:981-85.

31. Oritani, T., Yamashita, K. (1985) Biosynthesis of (+)abscisic acid in *Cercospora cruenta*. Ag. Biol. Chem. 49, 245-249.

32. Parry, A.D., Neill, S., Horgan, R. (1988) Xanthoxin levels and metabolism in the wild type and wilty mutants of tomato. Planta 173, 397-404

33. Parry, A.D., Griffiths, A., Horgan, R. (1992) Abscisic acid biosynthesis in roots. II. The effects of water-stress in wild-type and abscisic-acid-deficient mutant (*notabilis*) plants of *Lycopersicon esculentum* Mill. Planta 187, 192-197

34. Parry, A.D, Horgan, R. (1992) Abscisic acid biosynthesis in higher plants. *In*: Progress in Plant Growth Regulation, pp. 160-172, Karssen, C.M. Van Loon, L.C., Vreugdenhil, D., eds. Kluwer Academic Pub, Netherlands.

35. Rock, C.D., Zeevaart, J.A.D. (1991) The *aba* mutant of *Arabidopsis thaliana* is impaired in epoxy-carotenoid biosynthesis. Proc. Natl. Acad. Sci. USA 88, 7496-7499

36. Sindhu, R.K., Griffin, D.H., Walton, D.C. (1990) Abscisic aldehyde is an intermediate in the enzymatic conversion of xanthoxin to abscisic acid in *Phaseolus vulgaris* leaves. Plant Physiol, 93, 689-694

37. Taylor, H.F., Burden, R.S. (1973) Preparation and metabolism of 2-^{14}C-*cis*, *trans* xanthoxin. J. Exptl. Bot. 24, 873-880.

38. Taylor, I.B. (1987) ABA-deficient tomato mutants. *In*: Developmental Mutants in Higher Plants, Thomas, H., Grierson, D., eds. S.E.B. Seminar Series, Cambridge University Press, Cambridge.

39. Taylor, I.B., Linforth, R.S.T., Al-Naieb, R.J., Bowman, W.R., Marples, B.A. (1988). The wilty tomato mutants *flacca* and *sitiens* are impaired in the oxidation of ABA-aldehyde to ABA. Plant, Cell and Environment 11, 739-745

40. Uknes, S J, .Ho, T.H.D. (1984) Mode of action of abscisic acid in barley aleurone layers: abscisic acid induces its own conversion to phaseic acid. Plant Physiol. 75:1126-1132

41. Walton, D.C. (1983) Structure-activity relationships of abscisic acid analogs and metabolites. *In*: Abscisic Acid, pp. 113-146, Addicott, F.T, ed. Praeger, New York.

42. Walton, D.C., Dorn, B., Fey, J. (1973) The isolation of an abscisic acid metabolite, 4'-dihydrophaseic acid, from non-imbibed Phaseolus vulgaris seed. Planta, 112, 87-90.

43. Zeevaart, J.A.D., Milborrow, B.V. (1976) Metabolism of abscisic acid and the occurrence of epi-dihydrophaseic acid in *Phaseolus vulgaris*. Phytochemistry, 15, 493-500

44. Zeevaart, J.A.D., Creelman, RA. (1988) Metabolism and physiology of abscisic acid. Plant Physiol. Plant Mol. Biol 39, 439-473.

C. OTHER HORMONAL COMPOUNDS

C1. Polyamines as Endogenous Growth Regulators

Arthur W. Galston and Ravindar Kaur-Sawhney
Department of Biology, Yale University, New Haven, Connecticut 06511, USA.

POLYAMINES AS ESSENTIAL CELLULAR COMPONENTS

It is probable that all cells contain the diamine putrescine (Put; 1,4-diaminobutane) and the triamine spermidine (Spd), while eukaryotic cells contain the tetramine spermine (Spm) as well (3, 8). In both prokaryotes and eukaryotes (54), including higher plants (13), mutants lacking the ability to biosynthesize polyamines (PAs) are unable to grow and develop normally (13, 54). Since the addition of PAs to these mutants generally restores normal growth and development, it is reasonable to conclude that PAs are essential to all cells. This conclusion is reinforced by the demonstrable effects of "suicide inhibitors" of the main PA-biosynthetic enzymes, ornithine decarboxylase (ODC) and arginine decarboxylase (ADC: Fig. 1). These compounds, α-difluoromethylornithine (DFMO) and α-difluoromethylarginine (DFMA), specifically and irreversibly bind to and inhibit ODC and ADC, respectively. The ensuing decline in cellular PA titers is accompanied by a diminution or cessation of growth and development, which are restored upon the addition of the relevant PA.

Given the biological ubiquity and apparent indispensability of PAs, we need, as plant physiologists, to ask at least the following basic questions:

1. How are PAs biosynthesized and metabolized in plants?
2. What is their location in the cell?
3. What is their probable function?
4. Do PAs normally regulate growth and development of "normal" plants?
5. Can changes in PA titer help explain the action of physical (light, temperature, stress) and chemical (nutrients, hormones) agents affecting plant growth?
6. Are PAs translocated?
7. What is the effect of the administration of exogenous PAs or PA

158

P. J. Davies (ed.), Plant Hormones, 158–178.

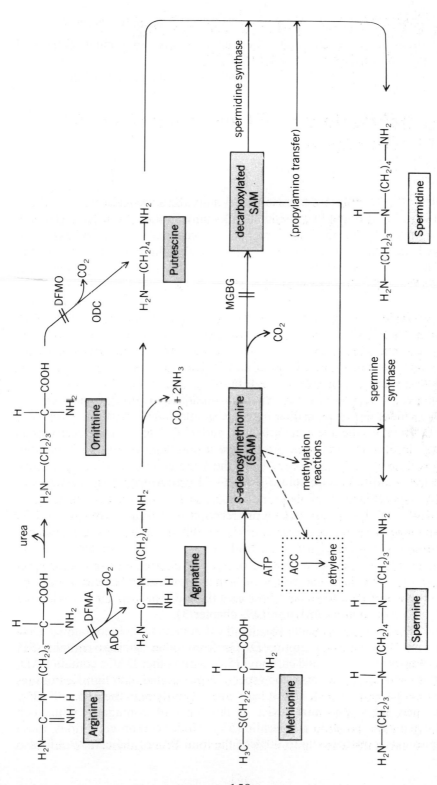

Fig. 1. Pathways for the biosynthesis of putrescine, spermidine and spermine in plants. ADC = arginine decarboxylase; DFMA = α-difluoromethylarginine; DFMO = α-difluoromethylornithine; MGBG = methylglyoxal bis-(guanylhydrazone). From (17).

analogs on entire plants, excised organs, individual cells and protoplasts?

8. Are any of these effects potentially important in agriculture?

THE BIOSYNTHESIS AND METABOLISM OF POLYAMINES IN PLANTS

Evidence supporting the biosynthetic scheme in Fig. 1 for plants has recently been extensively reviewed (13, 16, 19, 49, 53). While much of the original evidence came from microorganisms and animals, a recent spate of activity in the plant field has led to considerable documentation of similar systems in plants, especially in the angiosperms. Considerable interest has been displayed in the participation of S-adenosylmethionine (SAM) in the biosynthesis of Spd and Spm, since SAM is also the source of the plant hormone ethylene, by way of the intermediate 1-aminocyclopropane-1-carboxylic acid (ACC). Ethylene is well known as a senescence-inducer and PAs have antisenescence activity, particularly with excised plant parts (24). Thus, any system that controls the flow of carbon atoms through SAM could control the developmental fate of the cell, tissue or organ involved. It is noteworthy that many stress conditions will lead to an increased production of both ethylene and Put (see below), but not other PAs.

Put is oxidized by a diamine oxidase (DAO), yielding 4-aminobutyraldehyde, which then cyclizes spontaneously to form pyrroline: other products of the reaction are NH_3 and H_2O_2 (53). DAO activity is especially high in the *Leguminosae*, where it may represent up to 3% of the total protein of the cell. In some plants, the 4-aminobutyraldehyde is further oxidized to 4-aminobutyric acid (GABA) and then to sugars, glutamate and other organic acids (44). Polyamine oxidase (PAO), especially abundant in the cereals, catalyzes an analogous reaction, yielding pyrroline, 1, 3-diaminopropane (Dap) and H_2O_2 (53). When Spm is the substrate, aminopropylpyrroline is formed instead of pyrroline. In tomato fruit, the action of PAO leads to the formation of Put and β-alanine (44). Since PAO in cereals is localized in the cell wall, it has been suggested that it may be involved in lignification, since it produces the H_2O_2 needed for peroxidative conversion of precursors to lignin (49, chapter 3).

The DAO of pea contains two identical subunits, each containing Cu^{2+} and sharing one carbonyl group; DAOs from other sources contain -SH groups in addition to Cu^{2+} and carbonyl (53), while other DAOs contain FAD. PAO in several cereals seems to be especially associated with lignified tissues (25, 53) or guard cells (25), and at least some activity is in the cell wall (25).

Put and other PAs may exist in the form of conjugates with such phenolic acids as cinnamic and ferulic (53). Under certain conditions, these conjugates may constitute up to 90% of the total PAs of the cell. Put is also the source of carbon atoms for the formation of the 5-membered pyrrolidine ring of nicotine and related alkaloids. The fate of Put in alkaloid synthesizing cells appears to be determined by auxin: thus, at low levels of NAA (c. 1μM) tobacco callus grows little but makes lots of alkaloid, while at higher levels (c. 10μM), growth is stimulated and alkaloid synthesis falls, the Put being diverted into conjugate formation (55).

Cadaverine (Cad: 1, 5-diaminopentane) is formed from lysine in some plants (53). It is found free in actively-growing leguminous roots and is converted, via oxidation and cyclization, to the 6-membered piperidine ring of anabasine and other alkaloids. A lysine oxidase is apparently responsible for the initial oxidation.

LOCALIZATION OF POLYAMINES AND THEIR BIOSYNTHETIC ENZYMES

Since PAs are relatively small, soluble, diffusible molecules at cellular pH's, their immobilization in the cell for localization purposes is difficult to achieve. Disruption of the cell to achieve particle separation may result in many artifacts, especially if pH changes, affecting protonation of PAs, cause altered distribution patterns. The fact that endogenous and exogenous arginine may have entirely different metabolic fates *en route* to PAs and that endogenous and exogenous PAs are metabolized differently probably indicates some compartmentation of enzymes and substrates in the metabolic pathways related to PAs.

Several approaches (49, chapter 6) have been used to study the localization of PAs and their biosynthetic enzymes, including (1) subcellular fractionation (2) cytochemical and immunocytochemical staining methods and (3) autoradiographic localization of labelled PAs and their biosynthetic enzymes. Because of various limitations of the above techniques, information about localization of PAs and their biosynthetic enzymes is still rather limited. There are reports that a large portion of the intracellular PA pool in plants is sequestered within the vacuole, but PAs have also been found in isolated mitochondria and chloroplasts. As for cell walls, whose usually negative charge would be expected to absorb protonated PAs, studies comparing uptake of radiolabeled PAs by carrot cells versus protoplasts suggest that approximately two thirds of the total PA is bound to the cell wall. These results agree with earlier reports that PAs in mature plant tissues were associated primarily with the cell wall.

Little is also known about the localization of PA biosynthetic enzymes (49, chapter 6). The availability of DFMO and DFMA, which undergo irreversible covalent bond formation with ODC and ADC, respectively, affords a means of localizing these enzymes within tissues, but so far not conclusively within the structure of the cell. Labeling these inhibitors with ^{14}C or ^{3}H permits autoradiographic techniques to be employed, as it has been for ODC in animals (41). In plants, it appears by the use of ^{3}H-DFMO that ODC is localized in the nuclei of tobacco ovules (49, chapter 6), a conclusion in agreement with cell fractionation studies based on centrifugal separations. On the same basis, ADC seems to be in the cytosol, while Spd synthase, once reported to occur in chloroplasts, can now only be said to be particulate (49). PAO is probably wall-localized, at least in the cereals (25).

Most of the enzymes involved in plant PA metabolism have been at least partially purified and characterized and therefore specific antibodies for these enzymes should be available for localization studies in the near future. It would thus appear that immunocytochemistry has the best potential to provide unambiguous information about the tissue and subcellular localization of these enzymes. However, the localization of PAs and the enzymes of their metabolism within the plant cell is still largely *terra incognita*, and much remains to be done.

FUNCTIONS OF POLYAMINES

At cellular pH's, PAs are polycations (3, 8) and thus bind readily to such important cellular polyanions as DNA, RNA, phospholipids and acidic protein residues and cell wall components. Through such binding, PAs could affect the synthesis of macromolecules, the activity of macromolecules, membrane permeability, and partial processes of mitosis and meiosis. This has been shown convincingly *in vitro*, but the *in vivo* evidence is still scanty.

Membrane Structure and Function

Effects on the cellular plasma membrane are perhaps easiest to demonstrate. PAs can stabilize otherwise labile oat protoplasts against lysis, decrease the leakage of betacyanin from wounded root tissue and preserve thylakoid structure in excised barley leaves. The activity of membrane-localized enzymes in both animals and plants can be affected by PA-mediated changes in membrane fluidity and fine structure. One important example is the ethylene-generating system, whose activity is regulated in part by temperature-dependent inhibitory effects of Ca^{2+} and Spm. Similarly, PAs can counteract hormone-induced changes in membrane permeability and can affect membrane-localized proton-secreting systems, one of the probable targets for auxin (17, 51, 53, 54).

Interactions with Nucleic Acids

Spm-DNA complexes have a regular conformation, stabilizing the DNAs against thermal denaturation *in vitro*. Spd and Spm also facilitate conformational changes, such as the B $\rightarrow$ Z transition in methylated polynucleotides (4), and together with other basic cellular components such as histones, may control DNA conformation important in nucleosome assembly and gene expression. Depletion of intracellular PAs sensitizes DNA to alkylating agents, perhaps by exposing previously protected groups on DNA. These studies have been extensively reviewed (1, 51, 53, 54).

The structure and function of microbial tRNA's is affected by PAs (47), a fact that has led investigators with plant tRNA and rRNA to postulate similar action of PAs bound to those entities. PAs are also found as integral components of a plant virus and a bacteriophage, affect the organization of DNA in bacterial nucleoids (14) and control chromosome condensation and nuclear membrane dissolution during pre-prophase. Specific PAs are apparently required for maintenance of several yeast double-stranded RNA plasmids (51, 54).

Control of Protein Structure and Enzyme Activity

PAs have been reported to control the phosphorylation of a nucleolar protein in *Physarum polycephalum*, a slime mold (29) and in pea nuclei (11). This protein was identified as ODC, and the authors suggest that PAs control their own synthesis through phosphorylative inactivation of their main biosynthetic enzyme. Another worker suggests that ODC inactivation is brought about by direct linkage to Put. PAs have also been reported to stimulate various kinases in animals and plants, probably by virtue of their size and charge. Other enzymes whose activities are affected by PAs include an NADPH oxidizing enzyme and fructose 1, 6-bisphosphatase. While most of the effects are produced by ionic binding of PAs to the macromolecule, there is evidence of covalent binding of PAs to proteins, possibly mediated by transglutaminase (17, 51, 54).

Effects on the Synthesis of Macromolecules

In many types of cells, correlations have been reported between PA biosynthesis and titer on the one hand and cellular proliferation on the other (1). This is consistent with the finding that various systems for the *in vitro* synthesis of proteins are stimulated by the addition of PAs. Inhibition of PA biosynthesis retards growth in microbial (54), animal (32) and plant cells (18). In some instances, these inhibitions can be reversed by the addition of PAs (7, 51, 54). The suicide inhibitor of ODC, DFMO, was in fact synthesized in order to control the growth of malignant cells in cancer patients (37). MGBG, an inhibitor of SAM decarboxylase, is also able to inhibit the growth of some types of malignant cells, especially in concert with DFMO or other drugs.

When DFMO is added to rat hepatoma tissue culture cells, Spd titer declines, followed by a decline in polysome content, decreased ^{3}H-leucine incorporation into protein, and a prolongation of the G_1 phase of the cell cycle. It appears that a high titer of Spd or Spm is required for entry into the S phase, when DNA synthesis occurs. Similar events occur in a wide variety of prokaryotic and eukaryotic cells, both plant and animal (51, 54).

Buffering of Cellular pH

In plants showing Crassulacean acid metabolism (CAM), Put is synthesized along with secretion of malate into the cytoplasm (39). Put is also increased by increased external acidity, by SO_2, or by excess NH_4^+ (49, chapter 15), all of which produce an acid stress on the exterior of the plasma membrane. This raises the possibility that the reversible protonation of the multiple amino groups of PAs may thus serve to buffer cellular pH. Quantitatively, this seems possible, since the titer of Put in stressed cells may reach 0.4mM. Other stress stimuli that also result in the accumulation of Put (see below) may possibly act through a common effect on pH.

POLYAMINES IN PLANT GROWTH AND DIFFERENTIATION

Through the development of a series of mutants in *Escherichia coli* and *Saccharomyces cerevisiae* (54), PAs have been shown to be obligate growth factors for both prokaryotic and eukaryotic cells. In yeast, Put is required for attainment of optimal growth and development, while the higher PAs are required for sporulation. Mammalian cells, especially in tissue culture, have also been shown to require PAs for normal growth and development (3, 8).

In plants, PAs have been linked to a variety of growth and developmental processes, including cell division, vascular differentiation, embryoid formation in tissue culture, root initiation, adventitious shoot formation, flower initiation and development and control of fruit ripening and senescence (13, 16, 19, 49).

PA titer and plant growth rate are positively correlated in a wide variety of plant growth systems (18, 51, 53), and the interruption of PA biosynthesis by inhibitors (19, 49, 53) or mutation results in altered growth patterns (13), some of which can be reversed by the application of particular PAs. In a few cases, the application of exogenous PAs to protoplasts or cells in tissue culture has resulted in a temporary or sustained increase in cell division (13).

The ratio of diamines to higher polyamines (Put/Spd + Spm) is generally directly correlated with elongation rate, especially in seedling organs (Fig. 2, 48). It appears that the Put ➛ Spd transformation is especially important in controlling the rate of cell division. Put appears to be injurious to, while high Spd (and Spm) seem essential for the G_1 ➛ S transition (49, chapter 11).

Polyamines and Flowering

Martin-Tanguy and colleagues, while studying the conjugated PAs of tobacco and other plants (34) found, in general, that hydroxycinnamoyl acid amides, formed by the coupling of phenolic acids and PAs were absent from young plants, but accumulated progressively in apical leaves and then in large quantities in various organs of the flower. Clearly, at the time of flowering these amides are either translocated from apical leaves to the floral buds or are metabolically converted while the floral apex synthesizes large quantities of conjugated PAs *de novo*.

The different sex organs also have different spectra of PA-containing conjugates (6), indicating a possible relation to sexual differentiation. In corn, a male-sterile line has a very low titer of PA conjugates and a complete absence of feruloyl-Put conjugates (34).

A role for PAs in flowering has also been made probable by Malmberg and his colleagues (13). Presumptive PA tobacco mutants obtained by regenerating plants from mesophyll protoplasts that survived exposure to high levels of PA-biosynthetic inhibitors exhibited aberrant flowering

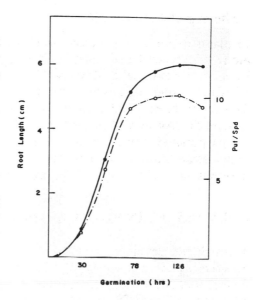

Fig. 2. Relation of Put/Spd ratio (----) to the length of root (——) during germination and growth of Alaska pea seedlings germination. From (45).

behavior, including altered patterns of floral organ morphogenesis. For example, in one MGBG resistant mutant (Mgr³), the resulting high titers of Spd and Spm are correlated with the development of anther-like ovules. In contrast, in a temperature sensitive revertant (Rt1), the altered PA metabolism showing substantially low titers of Spd and Spm produced flowers with a second row of petals in place of anthers. Since stamens and carpels are the spore-producing organs of the angiosperms, and since PA deprivation had been shown to affect sporulation in fungi, we decided to investigate the effects of PA deprivation and supplementation on patterns of flowering in tobacco, using a thin layer tissue culture technique (27). Cultures programmed to produce flowers showed high endogenous titers of Spd and progressively converted to the vegetative state as the Spd titer was decreased by increasing concentrations of cyclohexylamine, an inhibitor of Spd synthase. Contrariwise, cultures programmed to produce only vegetative buds showed low endogenous Spd titer, and were induced to produce some flowers with the addition of 0.5 to 5.0 mM Spd (Fig. 3) (27). While Spd is apparently specific for inducing such flowering in Wisconsin-38 tobacco, Flores and co-workers found that spermine is more effective with the variety Xanthi. Several other investigators have, however, found no involvement of PAs in the control of flowering in tobacco-thin layer cultures (13, 16, 19).

More convincing correlations between PAs and flowering have been observed in whole plants. Our studies on the relation between photoperiodic induction and PA titer in a well characterized short day plant, *Xanthium*

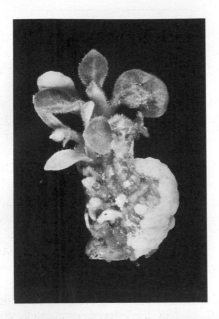 

Fig. 3. Effect of spermidine on bud differentiation in thin-layer explants of tobacco. Explants were cultured in the presence (right) or absence (left) of 5 mM spermidine. Photographs were taken when the cultures were one month old and buds fully developed (x 2.5). From (26).

strumarium L. revealed that exposure to one or two consecutive induction long nights greatly increases the titer of conjugates of the higher PAs in leaves, and later in buds (19). This investigation has also been extended to a long day plant, *Sinapis alba* L. Results on leaf exudates show that free Put increases quite early and dramatically during the floral transition, coinciding closely with movement of the floral stimulus out of the induced leaves. Recently, Kushad and colleagues (21) have also reported a close involvement of PAs in flowering in the long day plant *Rudbeckia hirta*. A rise in free PAs was linked to important cytological events during floral initiation and after 4 long days, the PA level was consistently higher in photoinduced plants than in non-induced controls. Studies from the same lab showed that 80% of the total PA content in fully developed citrus flowers is localized in the reproductive organs and only 20% in corolla and calyx.

The development of the ovary and ovules during maturation seems highly sensitive to PA titer, especially in tomato and other solanaceous plants (7, 50). Application of DFMO to young tomato pistils immediately after pollination blocks the subsequent growth and development of the ovary (7). This inhibition can be overcome in part by the subsequent application of Put or Spd, indicating that reduction of PA titer below some critical value inhibits the prolific cellular division that prepares the way for fruit development.

Polyamines and Embryogenesis

The involvement of PAs in embryogenesis was first observed in carrot cell cultures (13, 16, 19). Studies on somatic embryogenesis in carrot reveals three very important facts: (1) Embryogenic growth is suppressed by the presence of an auxin in the medium (2). Increased synthesis of PAs accompanies, and is required for, embryogenesis upon removal of auxin. (3) Ethylene suppresses embryogenesis. Just prior to embryoid formation, ADC activity increases rapidly, accompanied by elevated Spd titer. Application of DFMA to the cultures inhibits embryogenesis, and the inhibition is reversed by addition of Spd. In a non-embryogenic line of carrot, the rise of ADC activity and Spd titer do not occur. Similar effects have been noted in protoplast-derived cultures of *Vigna*. Recently Minocha and his colleagues (46) have shown that carrot-cells grown in the presence of auxin produced more ethylene and less PAs than those grown on minus auxin medium. DFMO promoted PA biosynthesis, inhibited ethylene production and permitted somatic embryogenesis even in the presence of auxin. Based upon the above observations and the fact that auxins generally promote ethylene biosynthesis in plant tissues, it was hypothesized that a competitive interaction between PAs and ethylene biosynthesis may be an important regulatory factor controlling embryogenesis in carrot cell culture.

Polyamines and Senescence

Senescence in many plant organs, both *in situ* and upon their excision from the plant, is correlated with a decline in PA titer (49, chapter 14). Since the addition of exogenous PAs in the millimolar concentration range delays or prevents such senescence-related processes as the decline of chlorophyll, protein and RNA content (measured by decreases in protease and RNase activities) in a wide range of monocot and dicot leaves (Fig. 4) (24), it has been proposed that PA titer controls senescence, at least in leaves (24). ADC activity is usually well correlated with leaf well-being (26). Thus inhibitors or other agents conducing to a decline in ADC activity generally promote senescence (26), while light, hormones and other treatments that inhibit senescence lead to an increase in ADC activity (26). Exogenous PAs seem to interact with a Ca^{2+}-specific site on the outer membrane (24), and only a brief contact between leaf and PA (a few minutes in a 48 h period) suffices to prevent senescence. PAs act through an inhibition of ethylene formation or action (49, chapter 5).

The apparent inverse relation between PAs and senescence has been tested in ripening fruits (49, chapter 5). Alcobaca and other varieties of tomato with retarded senescence and extended shelf lives have higher than usual levels of endogenous Put, which increases rather than showing the usual decline during ripening. Tomato over-ripening has also been delayed by infusion of exogenous Put via the peduncle (30). While at higher

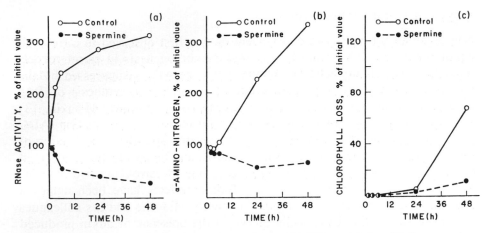

Fig. 4. Kinetics of the effect of 1 mM spermine on (a) RNase activity; (b) free α-amino nitrogen content a (a measure of protease activity) and; (c) chlorophyll loss in excised peeled oat leaf segments incubated in the dark. From (23).

concentrations, PAs inhibit both ripening and ethylene formation, at lower concentrations they still affect ripening with no effect on ethylene formation. This indicates that PAs may have their own independent action on the ripening process (19, 49).

Spermidine Binding To Proteins

One possible mechanism through which PAs might act as growth regulators involves their binding to specific regulatory proteins which have been observed in many organisms. We reported the binding of exogenous labelled Spd to a specific 18kD protein in thin-layer tobacco tissue cultures (2) and to a larger protein in oat protoplasts (38). Mehta *et al* (36), using two rice cell cultures, were also able to demonstrate stable incorporation of label from (^{3}H) Spd into an 18 kD protein, which has subsequently been identified as the eukaryotic translation initiation factor eIF-5A. The binding of exogenous labeled Spd increased in the presence of inhibitors of endogenous PA biosynthesis, suggesting that exogenous and endogenous PAs bind to the same site. More recently we have shown a close correlation between PA binding proteins and mitotic activity in a comparison of young and mature tissues of tobacco leaves, internodes and ovaries (23).

REGULATION OF POLYAMINE TITER IN THE PLANT

The synthesis of PAs is sensitive to several externally-manipulable variables, including light, hormones, excision and stress. Control seems to be exerted through ADC, ODC and Spd synthase in different plants.

Light

In excised oat leaves, ADC activity and PA titer fall in darkness and rise in white light (26). In etiolated peas, the $P_r \rightarrow P_{fr}$ phytochrome transition increases ADC activity in buds, while decreasing it in epicotyls (9). Thus, P_{fr} effects on ADC in each organ parallel P_{fr} effects in growth of the organ. This may be the only known case of simultaneous photoinduced induction and repression of the same enzyme in different organs. Changes in PA titer reflect the altered ADC activity and growth, and the altered ADC activity is not simply a consequence of altered growth rate, as shown by subsequent kinetic studies and surgical procedures.

Hormones

Either PAs or auxin will stimulate dormant *Helianthus tuberosus* explants to grow *in vitro*. Since auxin application results in an increase in PA titer and macromolecular synthesis, it has been proposed that auxins act through PAs to induce growth in that tissue. A similar situation exists in the tomato ovary, where auxin application that induces parthenocarpy requires active ODC. Auxin-induced rooting in mung bean seedlings is also reduced in the presence of MGBG, an inhibitor of SAM decarboxylation, and the inhibition is reversed by the application of arginine or ornithine (17, 51).

Auxin effects on PA biosynthesis have recently been reviewed (49, chapter 13). Auxin treatment was shown to induce a four fold increase in ODC activity during barley grain germination. This was correlated with a decrease in the titer of an ODC "antizyme" by auxin treatment. IAA also promoted ADC activity in rice embryos. In a few cases, auxins have been shown to have an inhibitory effect on PA levels and their biosynthesis. The reasons for these discrepancies are not clear.

The gibberellin-induced elongation of dwarf pea internodes in the light is accompanied by a rise in ADC activity and PA titer, while application of DFMA partially prevents these effects (49, chapter 13). A similar GA-PA interaction exists in lettuce hypocotyls. In the GA-induced elongation of pea internodes, which involves both cell division and cell elongation, PAs appear to play a role in cell division and not cell elongation (52). The GA-induced increase in α-amylase in barley aleurone is inhibited by MGBG, but there is no change in PA titer following GA administration. GA and PA both induce a large increase in ODC activity in this same barley system (17, 49, 51).

Cytokinins increase PA biosynthesis and titer in lettuce and cucumber cotyledons and in red-irradiated etiolated pea buds. In cucumber, cytokinins can also reverse an inhibitory effect of abscisic acid on PA biosynthesis (49, chapter 13). In a recent report (33) the growth stimulation of suspension culture of *Oryza sativa* L. var IR 20 by conditioned medium containing 2, 4-D and kinetin could be mimicked by Spd, but not by Put or Spm.

We have already mentioned the quantitative interplay of PAs and ethylene, possibly resulting in part from competition for SAM, a common precursor. PAs and ethylene also inhibit each other's biosynthesis and action (49, chapter 5). Thus, exogenous PAs reduce auxin-induced ethylene production in petals, leaves and fruits, and in senescing orange peel. PAs apparently block ethylene biosynthesis at the ACC → ethylene step, known also to be Ca^{2+}-sensitive, as well as at the prior ACC synthase step. When ACC synthesis is blocked by PAs, there is apparently an increased flow of carbons from SAM into PAs. Conversely, when PA synthesis is blocked by MGBG or DFMA, ethylene synthesis is promoted. In etiolated pea seedlings, exogenous application of ethylene inhibits ADC activity, while depletion of endogenous PAs increases ADC activity. Thus fluctuations in the comparative rates of ethylene and PA production could produce consequent changes in ADC activity.

Knowledge of the uptake and transport of PAs is still scant, mainly because at cellular pHs, PAs are polycations and thus bind readily to important cellular polyanions. Recent reports from Bagni's lab in Bologna details the pH dependent uptake of Put into *Saintpaulia* petals. Results from the Altman and Martin-Tanguy labs indicate the intertissue and interorgan translocation of polyamines in several plant species (49, chapter 7).

Stress

Physical and chemical

More than thirty years ago, K^+-deficiency was found to increase Put levels in plants, and this effect has been widely confirmed (53, 60). The increased Put "replaces" 30% of the cationic loss represented by K^+ deficiency. K^+ deficiency rapidly and reversibly induces higher ADC activities in young oat seedlings grown in sand culture on modified Hoagland's solution (60). The increase required *de novo* protein synthesis (60), and is correlated with an increased stress-induced incorporation of ^{35}S-methionine into a 39 kD band on a denaturing polyacrylamide gel.

Put accumulation and ADC increase also result from osmotic stress (15, Fig. 5), acid feeding (59, Fig. 6), high NH_4^+ feeding, or exposure to the atmospheric pollutants SO_2 or Cd^{2+} (13, 16, 19, 49). The response to osmotic stress in excised oat leaves occurs within minutes, as shown by timed application of cycloheximide to stressed systems. There is evidence that Put accumulation results not only from ADC activation, but also from a decrease in Spd synthase activity (57). This may also be true with acid feeding and the Put/Spd ratio increases with increase in acidity (Fig. 7).

The stress-increased Put could represent (a) the cause of the injury syndrome (b) the plant's defense against injury, or (c) a metabolic side-effect unrelated to stress. Recent experiments with simultaneous application of stress and DFMA indicate that the first possibility is closest to the truth (56).

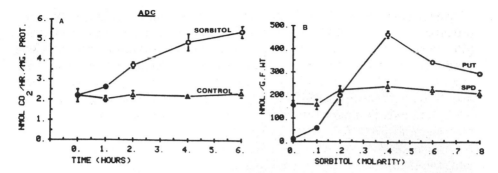

Fig. 5. A). Effect of osmotic stress on arginine decarboxylase activity of peeled oat leaf segments, which were floated on buffer (△) or buffer +0.4 M sorbitol (○) in the light. B). Effect of osmotic concentration on polyamine titer in peeled oat leaf segments. Polyamine titer determined after 4 h incubation in osmoticum or control buffer. From (14).

Chilling

Injury caused by chilling results in significant increases in Put levels in a variety of fruits and vegetables (28, 35, 58). The results do not indicate whether the increase in Put is a protective response or whether Put itself is the cause of the stress-induced injury. However, the results of others indicate that PAs protect plants from chilling injury. For example the cold hardening of several plant species correlates with increases in PAs; also, the reduction of chilling injury in squash, Chinese cabbage (58) and apples (28) by controlled atmosphere storage is coupled with increased PA titers. These results are consistent with the suggestion that PAs preserve membrane integrity, resulting in increased cell viability during chilling.

Heat and Drought

Plant cells can be induced to synthesize novel PAs when subjected to high temperature or drought. These PAs were first detected in the

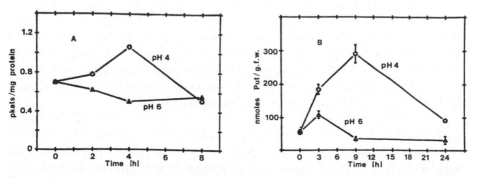

Fig. 6. Time-course of changes in A) arginine decarboxylase activity B) putrescine titer in peeled oat leaf segments incubated in buffer at pH 4.0 or 6.0. ○ = pH 4.0; △ = pH 6.0. Enzyme activity is expressed as pkat/mg protein (1 katal = 1 mol/s). From (59).

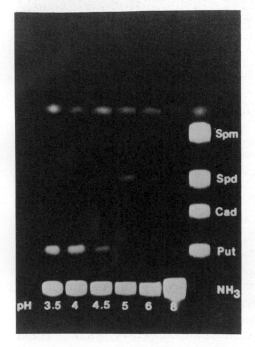

Fig. 7. Effect of external pH on polyamine titers in peeled oat leaf segments. Oat leaf tissue was incubated for 8 hours at various pH values. Perchloric acid extracts were dansylated and separated by thin-layer chromatography along with known polyamine standards. The thin-layer chromatograph was photographed in ultraviolet light to reveal dansylated polyamines. Note the rise in putrescine and decline in spermidine as the pH decreases. From (58).

thermophilic bacteria that usually contain an abundance of Spd when grown under more usual physiological conditions (25 to 35°C). However, a series of mostly symmetrical analogs of Spd and Spm appear when the temperature is increased above 50°C (40). The most common induced "thermopolyamines", norspermidine and norspermine, are apparently essential for continued protein synthesis at high temperature, both *in vitro* and *in vivo*. It is only recently that these "thermopolyamines" have been detected in plants. When cell cultures of drought-tolerant alfalfa lines are exposed to water deficit, significant amounts of norspermidine and norspermine accumulate (49, chapter 8). Pollen and cell suspension cultures of heat-tolerant cotton genotypes also accumulate these uncommon PAs when exposed to high temperature. The biosynthesis of these "thermopolyamines" may be related to stress responses in drought or heat tolerant plants.

Biological stress

Yeast and other fungi have only the ODC pathway for Put biosynthesis (54), while higher plants have both ADC and ODC. This led us to inquire whether phytopathogenic fungi might be selectively inhibited by DFMO. In petri dish culture on Czapek's medium, both DFMO and DFMA produced

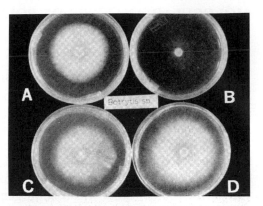

Fig. 8. The inhibition of mycelial growth of *Botrytis sp.* by DFMO and its reversal by putrescine. Upper left, control; upper right, 0.5 mM DFMO; lower left, 0.5 mM DFMO + 0.1 mM Put; lower right, 0.5 mM DFMO + 1.0 mM Put. From (41).

marked growth inhibitions that were reversible with applied Put or Spd (42, Fig. 8). The DFMA inhibition turned out to be trivial, since DFMA is converted to DFMO by arginase, in an analog of the arginine → ornithine conversion. When DFMO was sprayed on a unifoliate leaf of bean infected with uredospores of *Uromyces phaseoli*, partial or complete protection against infection was achieved, depending on the concentration and timing (Fig. 9, 43). Both pre- and post-infection sprays were effective, and it appeared that some protective effect moved from sprayed to unsprayed tissues. It appears crucial to inhibit fungal growth with DFMO before hyphal penetration of leaf cells, where the uninhibited leaf ADC could furnish Put to the fungus (43). Since several other diseases of plants are also benefitted by DFMO application, this differential and specific inhibition of PA biosynthesis could turn out be an important new method of disease control.

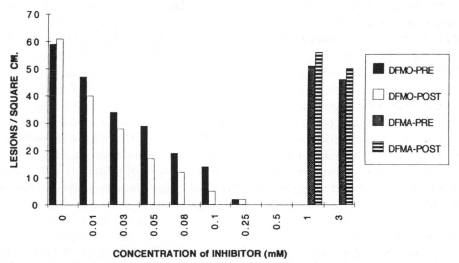

Fig. 9. The effect of pre- and post-infection application of DFMO and DFMA on lesion formation on bean leaves inoculated with *Uromyces phaseoli*. Twenty-four hours intervened between treatments with fungal spores and either DFMO or DFMA. At 0.5 and 1.0 mM DFMO, no lesions were produced. From (42).

173

MOLECULAR ANALYSIS OF POLYAMINE BIOSYNTHESIS AND ACTION

The genes for ADC, ODC, SAMDC, spermidine synthase and spermine synthase have been isolated from *E. coli*, yeast and mammalian tissues (22, 54) and cloned. Restriction maps and coding sequences for these genes have been worked out in considerable detail, and regulatory mechanisms have been studied, both in the original organism and in heterologous tissue. By contrast, studies with corresponding genes from plants are in their infancy. Malmberg and colleagues (5) have studied the ADC system of oats because of its induction by various stresses (15, 59, 60). The amino acid sequence of the purified oat enzyme was used to generate a cDNA for oat ADC, which resembled that of *E. coli* (54) in many respects. The open reading frame (ORF) encoded a 66 kD protein, but only a 24 kD ADC polypeptide could be isolated, from the C-terminus of the ORF. Part of this cDNA was expressed in *E. coli*, and a polyclonal antibody developed against the expressed polypeptide. The antibody detected only 34 and 24 kD polypeptides (5). Later work (31) indicated that oat ODC consists of two polypeptides, 42 and 24 kDs in mass, clipped from a precursor and reassembled, probably through disulfide linkages. The full length precursor could not be detected in plants, but showed up in *E. coli*, along with the 42 and 24 kD fragments. The nature of the clipping and reassembling mechanism, as well as its effect on enzyme activity, are currently under investigation. More recently Rastogi *et al.* (45) have isolated and characterized the ADC gene from tomato during the ripening of normal and Alcobaca (*alc*) fruit. The gene contains an open reading frame encoding a polypeptide of 502 amino acids and a molecular mass of about 55 kD. The predicted amino acid sequence shows 47% and 38% identity to oat (5) and *E. coli* (54) ADCs respectively. The tomato ADC is encoded by a single gene and is expressed as a transcript of *ca.* 2.2 kilobases in the fruit pericarp. The amount of the ADC transcript appeared to peak at the breaker stage of fruit ripening. Although *alc* fruit contain elevated Put and ADC activity at the ripe stage no significant differences were seen when steady-state ADC mRNA levels were compared between normal versus long-keeping alc fruit. This lack of correlation suggests a translational and/or post-transcription regulation of ADC gene expression during tomato fruit ripening.

Studies with transgenic plants have been made by Hamill *et al.* (20), who inserted the yeast ODC gene, linked to the cauliflower mosaic virus 35S promoter, into tobacco, where it was frequently over-expressed. The higher ODC expression in tobacco depended on the stage of the growth cycle, sometimes resulting in a trebling of ODC activity and a doubling of the content of nicotine in roots, where the alkaloid is synthesized from the putrescine presumably produced by ODC activity. DeScenzo and Minocha (12) introduced full length or truncated mouse ODC gene, under CaMV 35S promoter control, into strains of *Agrobacterium tumefaciens*, which were then

174

used to transform tobacco plants. Transformed plants contained a unique 50 kD polypeptide and higher ODC activity, shown to be mouse ODC by pH studies which distinguished it from the plant enzyme. Cellular putrescine content was augmented 4-12 fold. No information is yet available on any phenotypic effects of the extra putrescine.

Several research groups have reported that various polyamines can act like calcium in promoting the activity of kinases. Daniels *et al.* (10) reported that Spd and Spm can activate an enzyme from nuclei and nucleoli of the slime mold *Physarum polycephalum* that transfers labeled phosphate from ATP to a non-histone 70 kD protein. Datta *et al.* (11) showed that spermine stimulates the phosphorylation of a 47 kD nuclear protein from pea plumules, as well as a cyclic AMP-independent casein kinase. In producing the latter effect, Spm acted by lowering the Mg^{2+} requirement of the kinase. It is thus possible that polyamines contribute to regulation of the cell cycle by controlling the phosphorylation of proteins essential to the process.

CONCLUSIONS

While the precise physiological role of the PAs remains unclear, we are compelled to consider them as candidates for active regulators of plant growth (17), since they are present in all cells, and essential to normal growth and development in at least some organisms. Their approximately millimolar titer is responsive to such physiological controls as light, hormones, injury and stress, and external application can affect important physiological processes. While their endogenous levels are about two orders of magnitude higher than those of the traditional plant hormones, they clearly have a regulatory role distinct from a simple nutritional requirement. Effects of added PAs and inhibitors of their biosynthesis, both *in vivo* and *in vitro*, are impressive. The next few years should yield much exciting new information, including that from transgenic plants, permitting an evaluation of the true role of PAs in the control of plant growth.

References

1. Abraham, K.A., Alexander, P. (1981) Role of polyamines in macromolecular synthesis. Trends in Biochemical Sciences 64, 106-107.
2. Apelbaum, A., Canellakis, Z.N., Applewhite, P.B., Kaur-Sawhney, R., Galston, A.W. (1988) Binding of spermidine to a unique protein in thin-layer tobacco tissue culture. Plant Physiol. 88, 996-998.
3. Bachrach, U. (1973) Function of Naturally Occurring Polyamines, Academic Press, New York.
4. Behe, M., Felsenfeld, G. (1981) Effects of methylation on a synthetic polynucleotide: The B-Z transition in poly (dG-M^5dC)-poly(dG-M^5dC). Proc. Natl. Acad. Sci. USA 78, 1619-1623.
5. Bell, E., Malmberg, R.L. (1990) Analysis of a cDNA encoding arginine decarboxylase from oat reveals similarity to the *Escherichia coli* arginine decarboxylase and evidence of protein processing. Mol. Gen. Genet. 224, 431-436.

6. Cabanne, F., Martin-Tanguy, J., Martin, C. (1977) Phénolamines associées à l'induction floral et à l'état reproducteur du *Nicotiana tabacum* var. Xanthi n.c. Physiologie Veg. 15, 429-443.

7. Cohen, E., Arad, S.M., Heimer, Y.M., Mizrahi, Y. (1982) Participation of ornithine decarboxylase in early stages of tomato fruit development. Plant Physiol. 70, 540-543.

8. Cohen, S.S. (1971) Introduction to the Polyamines. Prentice-Hall, Englewood Cliffs, N.J.

9. Dai, Y.-R., Galston, A.W. (1981) Simultaneous phytochrome controlled promotion and inhibition of arginine decarboxylase activity in buds and epicotyls of etiolated peas. Plant Physiol. 67, 266-269.

10. Daniels, G.R., Atmar, V.J., Kuehn, G.D. (1981) Polyamine-activated protein kinase reaction from nuclei and nucleoli of *Physarum polycephalum* which phosphorylates a unique Mr 70000 nonhistone protein. Biochemistry, 20, 2525-2532.

11. Datta, N., Schell, M.B., Roux, S.J. (1987) Spermine stimulation of a nuclear NII kinase from pea plumules and its role in the phosphorylation of a nuclear polypeptide. Plant Physiol. 84, 1397-1401.

12. De Scenzo, R.A., Minocha, S.C. (1993) Modulation of cellular polyamines in tobacco by transfer and expression of mouse ornithine decarboxylase cDNA. Plant Mol. Biol. 22, 113-127.

13. Evans, P.T., Malmberg, R.L. (1989) Do polyamines have roles in plant development? Ann. Rev. Plant Physiol. Plant Mol. Biol. 40, 235-269.

14. Flink, L., Pettijohn, D.E. (1975) Polyamines stabilize DNA folds. Nature 253, 62-63.

15. Flores, H., Galston, A.W. (1982) Polyamines and plant stress: activation of putrescine biosynthesis by osmotic shock. Science 217, 1259-1261.

16. Flores, H.E., Protacio, C.M., Signs, M.W. (1989) Primary and secondary metabolism of polyamines in plants. *In*: Recent Advances in Phytochemistry, Vol. 23, pp. 329-393. Poulton, J.E., Romeo, J.Y., Conn, E.E., eds. Plenum Press, N.Y.

17. Galston, A.W. (1983) Polyamines as modulators of plant development. Bioscience 33, 382-88.

18. Galston, A.W. (1986) Plant morphogenesis. *In*: McGraw-Hill Yearbook of Science and Technology pp. 351-354.

19. Galston, A.W., Kaur-Sawhney, R. (1990) Polyamines in plant physiology. Plant Physiol. 94, 406-410.

20. Hammill, J.D., Robins, R.J., Parr, A.J., Evans, D.M., Furze, J.M., Rhodes, M.J.C. (1990) Over-expressing a yeast ornithine decarboxylase gene in transgenic roots of *Nicotiana rustica* can lead to enhanced nicotine accumulation. Plant Mol. Biol., 15, 27-38.

21. Harkess, R.L., Lyons, R.E., Kushad, M.M. (1992) Floral morphogenesis in *Rudbeckia hirta* in relation to polyamine concentration. Physiol. Plant. 86, 575-582.

22. Kahana, C. (1989) Molecular genetics of mammalian ornithine decarboxylase. *In*: The Physiology of Polyamines, Vol 1, pp. 281-295. Bachrach, U., Heimer, Y.M. eds. CRC Press, Boca Raton, Florida.

23. Kaur-Sawhney, R., Applewhite, P.B. (1993) Endogenous protein-bound polyamines: correlation with regions of cell division in tobacco leaves, internodes and ovaries. Plant Growth Regulation, 12, 223-228.

24. Kaur-Sawhney, R., Galston, A.W. (1979) Interaction of polyamines and light on biochemical processes involved in leaf senescence. Plant Cell Environ. 2, 189-196.

25. Kaur-Sawhney, R., Flores, H.E., Galston, A.W. (1981) Polyamine oxidase in oat leaves: a cell wall localized enzyme. Plant Physiol. 68, 494-498.

26. Kaur-Sawhney, R., Shih, L.M., Flores, H.E., Galston, A.W. (1982) Relation of polyamine synthesis and titer to ageing and senescence in oat leaves. Plant Physiol. 69, 405-410.

27. Kaur-Sawhney, R., Tiburcio, A.F., Galston, A.W. (1988) Spermidine and floral bud differentiation in thin-layer explants of tobacco. Planta, 173, 282-284.

28. Kramer, G.F., Wang, C.Y., Conway, W.S. (1989) Correlation of reduced softening and increased polyamine levels during low oxygen storage and McIntosh apples. J. Amer. Soc. Hort. Sci., 114, 942-946.

29. Keuhn, G.D., Affolter, H.-U., Atmar, V.J., Seebeck, T., Gubler, U., Braun, R. (1979) Polyamine-mediated phosphorylation of a nucleolar protein from *Physarum polycephalum* that stimulates r RNA synthesis. Proc. Natl. Acad. Sci. USA 76, 2541-2545.

30. Law, D.M., Davies, P.J., Mutschler, M.A. (1991) Polyamine-induced prolongation of storage in tomato fruits. Plant Growth Regulation 10, 283-290.

31. Malmberg, R.L. Smith, K.E., Bell, E., Cellino, M.L. (1992) Arginine decarboxylase of oats is clipped from a precursor into two polypeptides found in the soluble enzyme. Plant Physiol. 100, 146-152.

32. Mamont, P.S., Duchesne, M.C., Grove, J., Bey, P. (1978) Anti-proliferative properties of DL-α-difluoromethyl ornithine in cultured cells. A consequence of the irreversible inhibition of ornithine decarboxylase. Biochem. Biophys. Res. Commun. 81, 58-66.

33. Manoharan, K., Gnanam, A. (1992) Growth stimulation by conditioned medium and spermidine in low-density suspension cultures of rice. Plant Cell Physiol. 33, 1243-1246.

34. Martin-Tanguy, J. (1985) The occurrence and possible function of hydroxycinnamoyl acid amides in plants. Plant Growth Regul. 3, 381-399.

35. McDonald, R.E., Kushad, M.M. (1986) Accumulation of putrescine during chilling injury of fruits. Plant Physiol. 82, 324-326.

36. Mehta, A.M., Saftner, R.A., Schaffer, G.W., Mattoo, A.K. (1991) Translational modification of an 18 kilodalton polypeptide by spermidine in rice cell suspension cultures. Plant Physiol. 95, 1294-1297.

37. Metcalf, B.W., Bey, P., Danzin, C., Jung, M.J., Casara, P., Vevert, J.P. (1978) Catalytic irreversible inhibition of mammalian ornithine decarboxylase (E.C.4.1.1.17) by substrate and product analogues. J. Am. Chem. Soc. 100, 2551-2553.

38. Mizrahi, Y., Applewhite, P.B., Galston, A.W. (1989) Polyamine binding to proteins in oat and petunia. Plant Physiol. 91, 738-743.

39. Morel, C., Villanueva, V.R., Queiroz, O. (1980) Are polyamines involved in the induction and regulation of the crassulacean acid metabolism? Planta 149, 440-444.

40. Oshima, T. (1988) Polyamines in thermophiles. *In*: The Physiology of Polyamines Vol. 11, pp. 35-46, Bachrach, U., Heimer, Y.M. eds, CRC Press, Boca Raton, Florida.

41. Pegg, A.E., Seeley, J., Zagon, I.S. (1982) Autoradiographic identification of ornithine decarboxylase in mouse kidney by means of α-(5^{14}C) difluoromethyl ornithine. Science 217, 68-70.

42. Rajam, M.V., Galston, A.W. (1985) The effects of some polyamine biosynthetic inhibitors on growth and morphology of phytopathogenic fungi. Plant Cell Physiol. 26, 683-692.

43. Rajam, M.V., Weinstein, L.H., Galston, A.W. (1985) Prevention of a plant disease by specific inhibition of fungal polyamine biosynthesis. Proc. Natl. Acad. Sci. USA. 82, 6874-6878.

44. Rastogi, R., Davies, P.J. (1990) Polyamine metabolism in ripening tomato fruit. Plant Physiol. 94, 1449-1455.

45. Rastogi, R., Dulson, J. Rothstein, S.J. (1993) Cloning of tomato arginine decarboxylase gene and its expression during fruit ripening. Plant Physiol. 103, 829-834.

46. Robie, C.A., Minocha, S.C. (1989) Polyamines and somatic embryogenesis in carrot. The effects of difluoromethylornithine and difluoromethylarginine. Plant Sci. 65, 45-54.

47. Sakai, T.T., Cohen, S.S. (1976) Effects of polyamines on the structure and reactivity of tRNA. Prog. Nucl. Acid Res. 17, 15-42.

48. Shen, H. Galston, A.W. (1985) Correlations between polyamine ratios and growth patterns in seedling roots. Plant Growth Regulation 3, 353-363.

49. Slocum, R.D., Flores, H.E., eds. (1991) Biochemistry and Physiology of Polyamines in Plants. CRC Press, Boca Raton, FL.

50. Slocum, R.D., Galston, A.W. (1985) Changes in polyamine biosynthesis associated with post-fertilization growth and development in tobacco ovary tissues. Plant Physiol. 79, 336-343.
51. Slocum, R.D., Kaur-Sawhney, R., Galston, A.W. (1984) The physiology and biochemistry of polyamines in plants. Arch. Biochem. Biophys. 235, 283-303.
52. Smith, M.A., Davies, P.J., Reid, J.B. (1985) Role of polyamines in gibberellin-induced internode growth in peas. Plant Physiol. 78, 92-99.
53. Smith, T.A. (1985) Polyamines. Ann. Rev. Plant Physiol. 36, 117-143.
54. Tabor, C.W., Tabor, H. (1984) Polyamines. Ann. Rev. Biochem. 53, 749-790.
55. Tiburcio, A.F., Kaur-Sawhney, R., Galston, A.W. (1985) Correlation between polyamines and pyrrolidine alkaloids in developing tobacco callus. Plant Physiol. 78, 323-326.
56. Tiburcio, A.F., Kaur-Sawhney, R., Galston, A.W. (1986) Polyamine metabolism and osmotic stress II. Improvement of oat protoplasts by an inhibitor of arginine decarboxylase. Plant Physiol. 82, 375-378.
57. Tiburcio, A.F., Kaur-Sawhney, R., Galston, A.W. (1993) Spermidine biosynthesis as affected by osmotic stress in oat leaves. Plant Growth Regulation 13, 103-109.
58. Wang, C.Y., Ji, Z.L. (1989) Effect of low oxygen storage on chilling injury and polyamines in zucchini squash. Sci. Hortic. 39, 1-7.
59. Young, N.D., Galston, A.W. (1983) Putrescine and acid stress: induction of arginine decarboxylase activity and putrescine accumulation by low pH. Plant Physiol. 71, 767-771.
60. Young, N.D., Galston, A.W. (1984) Physiological control of arginine decarboxylase activity in K-deficient oat shoots. Plant Physiol. 76, 331-335.

C2. Jasmonates, Salicylic acid and Brassinolides

C2a. Jasmonate Activity in Plants

Paul E. Staswick
Department of Agronomy, University of Nebraska-Lincoln, Lincoln, Nebraska 68583-0915, USA.

INTRODUCTION

Until recently jasmonic acid (JA) and its fragrant methyl ester, methyl jasmonate (MeJA), had been studied only modestly since their discovery in plants over 30 years ago. Early research focused primarily on their potential role in plant growth and development. However, after jasmonate was shown to increase the expression of genes involved in plant defense, there was a surge in activity aimed at clarifying the function of these potentially important signaling molecules. Although considerable work remains, increasing evidence supports the hypothesis that jasmonate is involved in signaling stress responses in plants.

In this discussion references to the jasmonate literature will be limited to some of the more significant and most recent. Work on the biosynthetic pathway for JA was summarized in earlier reviews describing lipoxygenase-dependent metabolism of fatty acids in plants (1, 20). Structural and possible functional similarities between jasmonate, other related plant compounds, and animal eicosanoids are also discussed therein. General reviews of jasmonate action in plants have also appeared recently (15, 18).

Both JA and MeJA are widespread in the plant kingdom and both exhibit biological activity. Unless a specific compound is intended, I will refer to them collectively as jasmonate, since it is not known whether one or the other plays a more significant role in plant signaling.

THE OCCURRENCE OF JASMONIC ACID IN PLANTS

JA (3-oxo-2-(2'-cis-pentenyl)-cyclopentane-1-acetic acid) is one of several plant fatty acid derivatives that may regulate growth and development, as well as signal stress responses. Jasmonate biosynthesis involves a lipoxygenase-mediated oxygenation of α-linolenic acid followed by several steps, including a cyclization to form the cyclopentanone ring, and ß-oxidations to shorten the resulting side chain. A simplified scheme for JA

179

P. J. Davies (ed.), Plant Hormones, 179–187.
© 1995 *Kluwer Academic Publishers. Printed in the Netherlands.*

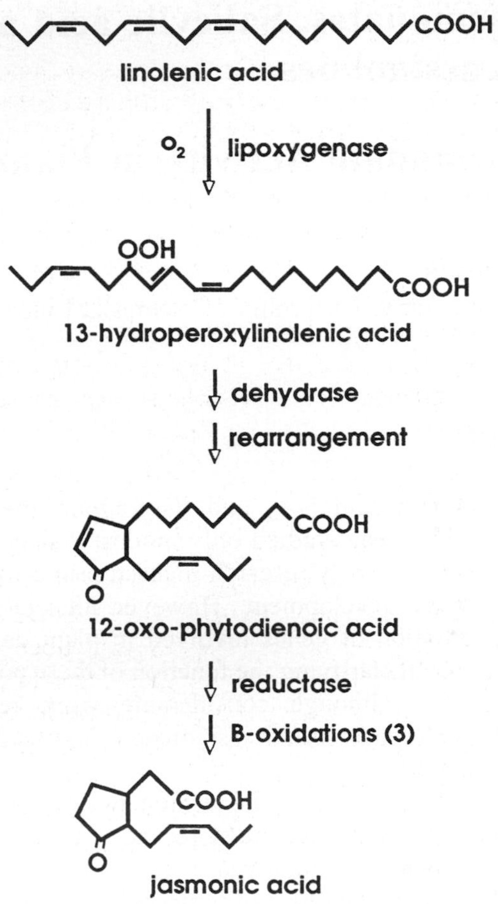

Fig. 1. Pathway for jasmonic acid biosynthesis in plants.

biosynthesis is illustrated in Figure 1; more detailed descriptions can be found in the references mentioned earlier. Interest in lipoxygenase has increased recently, as this enzyme is potentially an important regulatory step in JA biosynthesis. In contrast to the seed lipoxygenases, relatively little is known about this enzyme in vegetative tissues or about the other enzymes involved in JA biosynthesis.

Jasmonate is found in a variety of plant species and apparently in most if not all plant organs. The quantities that have been estimated by various methods are typical of other plant hormones, ranging from about 10 ng to as much 3 μg per gram fresh weight (14). Although the acid seems to be more prevalent, MeJA is usually more active than JA when applied exogenously. This may result from the volatility of the ester, and the fact that it is lipophilic.

MeJA is a fragrant oil which contributes to the distinctive aroma of certain fruits and flowers (e.g., jasmine). These characteristics have led to its synthesis for the perfume industry. Unfortunately the commercial product, which has been widely used by plant researchers, is not optically pure. Four stereoisomers of jasmonate are possible. As noted by others (20) evaluating the efficacy of jasmonate is complicated by the fact that (-)-jasmonate is easily epimerized to the (+) isomer, which is reported in some studies to have lower biological activity.

The picture is further confused by an array of closely related compounds which have been isolated from both plants and fungi. Several hydroxylated cyclopentanones occur, as do various amino acid conjugates and glucosyl esters of JA. A glucoside derivative induces tuberization in potato while growth inhibitor activity is associated with cucurbic acid. The latter is identical to JA except that a hydroxyl replaces the ketone function (1). The role of these compounds in plants has not been established.

JASMONATE ELICITS A VARIETY OF PLANT RESPONSES

The first reported effect of exogenously applied jasmonate was the inhibition of plant growth. Subsequent studies have described a wide variety of other activities. Among the stimulatory responses recently cataloged by Parthier (15) are those promoting senescence, abscission, tuber formation, tendril coiling, fruit ripening and pigment formation. Diverse processes are also inhibited by exogenous jasmonate including seed germination, callus growth, root growth, photosynthesis, and biosynthesis of ribulosebisphosphate carboxylase. It should be noted that these are probably not all hormone-like responses. Some may be more general stress responses brought on by the relatively high concentration of jasmonate applied in some studies (1).

JASMONATE AND PLANT DEFENSE

The list of genes or gene products (Table I) influenced by jasmonate continues to grow rapidly since the first report that proteins of unknown function were induced in many plant species (15). Many of these genes are implicated in plant defense of one sort or another. Whether this reflects the primary endogenous function of jasmonate or is an artifact of the limited kinds of genes that have been investigated is not yet clear.

Induction by jasmonate of protease inhibitors, which appear to be targeted primarily against certain insects, has been noted in several species. Enzymes of the phenylpropanoid pathway (e.g., phenylalanine ammonia lyase, chalcone synthase) on the other hand, are more important for defense against microorganisms. The phenylpropanoid pathway has been extensively studied in plants, especially regarding phytoalexin biosynthesis in response to pathogens and fungal elicitors. Recent evidence suggests, that jasmonate plays a role in this response (10). Barley thionins are also toxic to plant pathogenic fungi (3), but it is not known whether their induction in plants is

Table I. Genes and gene products induced by jasmonate

Gene or Protein	Plant, Organ	Reference, Year Reported*
Vegetative storage protein	Soybean, leaf and suspension cells	(2), 1989
Proteinase inhibitor	Tomato, Tobacco, Alfalfa leaf	(8), 1990
Napin and cruciferin seed storage proteins	*Brassica napus*, seed	(21), 1991
Lipoxygenase	Soybean, leaf and suspension cells	(4), 1991
Trypsin inhibitor Leucine aminopeptidase Threonine deaminase	Potato, leaf	(11), 1992
Chalcone synthase Proline-rich cell wall protein	Soybean, suspension cells	(5), 1992
Phenylalanine ammonia lyase	Soybean, suspension cells	(10), 1992
Thionin	Barley, leaf	(3), 1992
Late embryogenesis abundant-related	Barley, leaf Cotton, cotyledon	(16), 1992
Soybean VSP-related	*Arabidopsis thaliana*, leaf	(19), 1992
Lipoxygenase	*Arabidolsis thaliana*, root	(13), 1993
Kunitz-type proteinase inhibitor	Potato, tuber	(22), 1993

*List is not exhaustive. Only the first of multiple reports for a species is cited.

for defensive purposes.

The induction of lipoxygenase, a potential regulatory step in jasmonate biosynthesis, raises the interesting possibility of a signal amplification mechanism. Localized jasmonate synthesis or release of sequestered jasmonate in response to wounding might stimulate lipoxygenase and other enzymes of the jasmonate biosynthetic pathway, leading to further jasmonate production. This enhanced signal may then stimulate other defense pathways more effectively. However, it is not presently known whether the lipoxygenases induced by jasmonate play a direct role in jasmonate biosynthesis, or whether other lipoxygenases are involved.

Some of the genes induced by jasmonate, such as the developmentally regulated seed storage proteins of *Brassica napus* (21) are unlikely to be involved in plant defense. This suggests that jasmonate may also be an important signal controlling gene expression during plant growth. Even genes induced by wounding via jasmonate, need not be directly involved in plant defense. Some of these may assist in altering plant metabolism during times of stress. Under adverse conditions metabolites normally used for growth and development might be stored, possibly in an insoluble form (e.g., protein), for reuse when growth resumes. This appears to be the role of soybean vegetative storage proteins (VSPs), for which a defensive role is unknown. It has been postulated that the normal developmental control of VSPs in response to source-sink status may be partly controlled by an inter-organ flux of jasmonate in the vascular stream (18).

In addition to its putative roles in development and in signaling plant defense responses, jasmonate may be more directly involved as an antifungal agent. Jasmonate is produced by at least some fungi and limited evidence suggests fungi development is influenced by jasmonate (K. Nickerson, personal communication). Oxidation products of lipoxygenase also affect fungal growth phase and some lipoxygenase inhibitors are anti-fungal agents. The latter also convert *Ceratocystis ulmi*, the causative agent in Dutch Elm disease, from the mycelial to the yeast form (12).

The presence of jasmonate in fungi also raises the interesting possibility that fungal-derived jasmonate may be perceived by plants and elicit localized defense responses. Evidence that jasmonate signaling is more widespread in biology comes from the pheromone properties of MeJA in Oriental fruit moths. The pleasant fragrance of MeJA and its consequent widespread use in perfumes notwithstanding, it is perhaps pertinent examine whether jasmonate, which is related to important eicosanoid signaling molecules in animals, elicits additional biological responses in animals

SIGNALING BY GASEOUS METHYL JASMONATE

The volatile nature of MeJA led Anderson (1) to suggest that, like ethylene, MeJA may be effective as a gaseous hormone in plants. This hypothesis was subsequently confirmed by others, who demonstrated that airborne MeJA stimulates tomato proteinase inhibitor gene expression (8) and induces tendril coiling in *Bryonia dioica* (6). The latter is a developmental response normally triggered by mechanical stimulation of tendrils. The concentration of atmospheric MeJA effective in these responses was estimated to be around 80 nM, indicating that plants are quite sensitive to jasmonate. Farmer and Ryan (8) also concluded that MeJA volatilized from one plant signals a defensive response in another, when both are enclosed together in a small chamber. The efficacy of jasmonate as an interplant signal remains unproven in nature, however.

JASMONATE AS AN ENDOGENOUS REGULATOR

A variety of evidence now supports the idea that endogenous jasmonate plays a role in plant gene regulation. In general, the concentration of JA and/or

MeJA in plant organs is in the range found for ABA and thus appears sufficient for a physiological role (5, 14). To some extent at least, variation in the amount of jasmonate in organs during their development also correlates with developmental changes in the expression of certain genes. For example, soybean VSP genes are most active in young tissues, which is where highest levels of jasmonate and the enzymes for its biosynthesis occur. Jasmonate levels are also now known to increase coordinately with VSP gene expression in soybean seedlings following wounding (5), and with phytoalexin biosynthesis in suspension cell cultures of various plant species in response to fungal elicitors (10).

Further evidence for a regulatory role of endogenous jasmonate comes from studies involving the feeding of precursors of jasmonate or of putative inhibitors of jasmonate biosynthesis to leaves. Tomato proteinase inhibitors were induced by intermediates of jasmonate biosynthesis, but closely related compounds that are not JA precursors were ineffective (9). The implication is that the authentic intermediates were metabolized to jasmonate, which was active in gene induction.

On the other hand, induction of the soybean VSP genes by wounding was repressed by pretreatment with various inhibitors of lipoxygenase activity, while exogenously applied MeJA remained fully effective (18). Although lipoxygenase inhibitors have not been shown to inhibit jasmonate accumulation in wounded tissue, these results are consistent with the hypothesis that *de novo* synthesis of jasmonate is necessary for signal transduction in some wound response pathways.

An additional result supporting jasmonate's role in gene expression involves the restricted expression of soybean VSP genes in specific cells associated with leaf vasculature. The limited pattern of gene expression can be overcome by exogenously applied MeJA, which induces the VSP genes in all leaf cell types (18). MeJA is reportedly most abundant in the vascular regions of soybean pericarp tissue, also a site of relatively high VSP gene expression. One interpretation of these results is that endogenous jasmonate concentration and/or a cell's sensitivity to jasmonate conditions cell-specific expression of the VSP genes. Taken together, the evidence is consistent with the hypothesis that endogenous plant jasmonate is involved in both developmentally and environmentally modulated gene expression.

SYSTEMIC SIGNALING

The nature of systemic signaling in plants has remained enigmatic despite much effort in this area. Jasmonate is one of several possible transported chemical signals in plants. Although protease inhibitors are induced in distant untreated leaves following exposure of single leaves to jasmonate (9) or by wounding, it is not known whether jasmonate itself is mobilized to signal a response in untreated leaves in these cases. However, MeJA in aqueous solution is readily assimilated through leaf petioles and then distributed to leaves where gene expression is induced in leaf explants. Further study of the movement of jasmonate in plants, both naturally and in response to stimuli such as wounding, would provide needed information about whether jasmonate is an important systemic signal in plants.

Several other transported compounds including systemin, a novel 18-amino acid peptide from tomato leaves and a powerful inducer of the protease inhibitors, have been proposed as systemic plant signals (17). It is also possible that entirely different mechanisms are also involved, such as electrical signaling. The diversity of stimuli that are effective towards a number of stress-responding genes suggests the signaling pathway, and the role of individual components such as jasmonate, will be complex and challenging to understand.

JASMONATE RESPONSE MUTANTS

Mutants affecting plant hormone response have been a valuable resource to study hormone action and potentially for the isolation of the genes involved in these signal transduction pathways. An *Arabidopsis thaliana* mutant with impaired response to MeJA was recently characterized (19). Not only was the ability of MeJA to inhibit root growth diminished, also suppressed was the induction of two polypeptides which were detected immunologically with soybean VSP antisera. Response to abscisic acid (ABA) was not diminished in the mutant, nor was the response to cytokinin or auxin (Staswick and Stasinopolous, unpublished results). These results suggest the mutation affects a fundamental step in a signaling pathway that is specific to jasmonate.

Further study of this and additional mutants should help to define the mechanism for jasmonate action. As suggested for other plant hormones, jasmonate signaling may occur through specific receptors that bind jasmonate. Response mutants may define defects in binding ability, or subsequent steps in the signaling pathway. Other mutants deficient in JA biosynthesis or metabolism would also be useful for investigating the role of jasmonate in the great variety of plant responses in which it has been implicated.

RELATIONSHIP TO ABSCISIC ACID

Similarities in structure, physical properties, and activities for ABA and JA have been highlighted by others (1, 20). Both can inhibit growth, inhibit seed germination and promote senescence. Potato proteinase inhibitors, as well as seed storage and oilbody proteins in *Brassica* (2), are also induced by both ABA and MeJA. The latter result suggested an overlapping function for ABA and MeJA, which may account for the fact that ABA-deficient mutants still produce seed storage proteins. There is also evidence for synergism

185

between ABA and MeJA (19).

Other studies show there are distinct differences between ABA and jasmonate. For example, only MeJA induced the soybean VSP genes while ABA alone inhibited *Arabidopsis* seed germination (19). The isolation of *Arabidopsis* mutants that are deficient in their response to either jasmonate or ABA, not both, also indicates that to at least some degree the signaling pathways are independent. However, this does not preclude the possibility that the pathways merge at a later point, beyond what is defined by currently described mutants. Hopefully, these and other approaches will allow the genetic dissection of jasmonate signaling in plants and eventually, isolation of the genes involved.

The fact that many plant responses are influenced by multiple growth regulators suggests these responses result from a complex interaction of endogenous signaling molecules. This makes the mechanisms controlling plant growth, development, and response to the environment challenging to understand. Jasmonate is emerging as an important new piece to the puzzle and further clarification of its activity will assist in assembling a more complete picture of regulation in plants.

References

1. Anderson, J.M. (1989) Membrane derived fatty acids as precursors to second messengers. *In*: Second Messengers in Plant Growth and Development, pp. 181-212, Boss, W., Morre, J.D., eds. Alan R. Liss, New York.
2. Anderson, J.M., Spilatro, S.R., Klauer, S.F., Franceschi, V.R. (1989) Jasmonic acid-dependent increase in the level of vegetative storage proteins in soybean. Plant Science 62, 45-52.
3. Andresen, I., Becker, W., Schluter, K., Burges, J., Parthier, B., Apel, K. (1992) The identification of leaf thionin as one of the main jasmonate-induced proteins of barley (*Hordeum vulgare*). Plant Mol. Biol. 19, 193-204.
4. Bell, E., Mullet, J.E. (1991) Lipoxygenase gene expression is modulated in plants by water deficit, wounding, and methyl jasmonate. Mol. Gen. Genet. 230, 456-462.
5. Creelman, R.A., Tierney, M.L., Mullet, J.E. (1992) Jasmonic acid/methyl jasmonate accumulate in wounded soybean hypocotyls and modulate wound gene expression. Proc. Natl. Acad. Sci. USA 89, 4938-4941.
6. Falkenstein, E., Groth, B., Mithofer, A., Weiler, E.W. (1991) Methyl jasmonate and α-linolenic acid are potent inducers of tendril coiling. Planta 185, 316-322.
7. Farmer, E.E., Johnson, R.R., Ryan, C.A. (1992) Regulation of expression of proteinase inhibitor genes by methyl jasmonate and jasmonic acid. Plant Physiol. 98, 995-1002.
8. Farmer, E.E., Ryan, C.A. (1990) Interplant communication: Airborne methyl jasmonate induces synthesis of proteinase inhibitors in plant leaves. Proc. Natl. Acad. Sci. USA 87, 7713-7716.
9. Farmer, E.E., Ryan, C.A. (1992) Octadecanoid precursors of jasmonic acid activate the synthesis of wound-inducible proteinase inhibitors. Plant Cell 4, 129-134.
10. Gundlach, H., Muller, M.J., Kutchan, T.M., Zenk, M.H. (1992) Jasmonic acid is a signal transducer in elicitor-induced plant cell cultures. Proc. Natl. Acad. Sci. USA 89, 2389-2393.
11. Hildman, T., Ebneth, M., Pena-Cortes, H., Sanchez-Serrano, J.J., Willmitzer, L., Prat, S. (1992) General roles of abscisic and jasmonic acids in gene activation as a result of mechanical wounding. Plant Cell 4, 1157-1170.

12. Jensen, E.C., Ogg, C., Nickerson, K.W. (1992) Lipoxygenase inhibitors shift the yeast/mycelium dimorphism in *Ceratocystis ulmi*. Appl. Environ. Microbiol. 58, 2505-2508.
13. Melan, M.A., Dong, X., Endara, M.E., Davis, K.R., Ausubel, F.M., Peterman, T.K. (1993) An *Arabidopsis thaliana* lipoxygenase gene can be induced by pathogens, abscisic acid, and methyl jasmonate. Plant Physiol. 101, 441-450.
14. Meyer, A., Miersch, O., Buttner, C., Dathe, W., Sembdner, G. (1984) Occurrence of the plant growth regulator jasmonic acid in plants. J. Plant Growth Regul. 3, 1-8.
15. Parthier, B. (1991) Jasmonates, new regulators of plant growth and development: Many facts and few hypothesis on their actions. Botanica Acta 104, 405-464.
16. Reinbothe, S., Machmudova, C., Wasternak, C., Reinbothe, C., Parthier, B. (1992) Jasmonate-induced proteins in cotton: Immunological relationship to the respective barley proteins and homology of transcripts to late embryogenesis abundant (Lea) mRNAs. J. Plant Growth Regul. 11, 7-14.
17. Ryan, C.A. (1992) The search for the proteinase inhibitor-inducing factor, PIIF. Plant Mol. Biol. 19, 123-133.
18. Staswick, P.E. (1992) Jasmonate, genes, and fragrant signals. Plant Physiol. 99, 804-807.
19. Staswick, P.E., Su, W., Howell, S.H. (1992) Methyl jasmonate inhibition of root growth and induction of a leaf protein are decreased in an *Arabidopsis thaliana* mutant. Proc. Natl. Acad. Sci. USA 89, 6837-6840.
20. Vick, B.A., Zimmerman, D.C. (1987) Oxidative systems for the modification of fatty acids. *In*: The Biochemistry of Plants, Lipids, vol. 9, pp 53-90, Stumpf, P., Conn, E., eds. Academic, New York.
21. Wilen, R.W., van Rooijen, G.J., Pearce, D.W., Pharis, R.P., Holbrook, L.A., Moloney, M.M. (1991) Effects of jasmonic acid on embryo-specific processes in *Brassica* and *Linum* oilseeds. Plant Physiol. 95, 399-405.
22. Yamagishi, K., Mitsumori, C., Takahashi, K., Fujino, K., Kodo, Y., Kikuta, Y. (1993) Jasmonic acid-inducible gene expression of a Kunitz-type proteinase inhibitor in potato tuber disks. Plant Mol. Biol. 21, 539-541.

C2b. Salicylic Acid

Ilya Raskin

AgBiotech Center, Rutgers University, Cook College, New Brunswick, NJ 08903-0231, USA.

HISTORY AND PROPERTIES OF SALICYLIC ACID

Centuries ago, the American Indians and Ancient Greeks independently discovered that the leaves and bark of the willow tree cured aches and fevers. It was not until 1828 that Johann Buchner, working in Munich, successfully isolated a tiny amount of salicin - the glucoside of salicyl alcohol, which was the major salicylate in willow bark (56). The name salicylic acid (SA), from the Latin word *Salix* for the willow tree, was given to this active ingredient of willow bark by Raffaele Piria in 1838. The first commercial production of synthetic SA began in Germany in 1874. Aspirin, a trade name for acetylsalicylic acid, which is not a natural plant product, was introduced by the Bayer Company in 1898 and the compound rapidly became one of world's best-selling drugs. In spite of the fact that the mode of medicinal action of salicylates is a subject of a continual debate, they are being used to treat human diseases ranging from the common cold, to heart attacks. Since even in aqueous solutions, aspirin undergoes spontaneous hydrolysis to SA, both compounds have similar effects in plants and will be treated together in this review.

Several recent reviews summarize various aspects of SA biology in plants (34, 37, 40, 41). This compound belongs to an extraordinary diverse group of plant phenolics which are usually defined as substances which possess an aromatic ring bearing a hydroxyl group or its functional derivative. In the past, plant phenolics have often been referred to as secondary metabolites. The term "secondary" implies that such compounds are only of minor importance to the plant and were sometimes previously equated with waste products. However, in recent years this opinion has been gradually replaced by the view that many phenolic compounds play an essential role in the regulation of plant growth, development, and interaction with other organisms (18). For example, phenolics are essential for the biosynthesis of lignin, an important structural component in plant cell walls. Furthermore, phenolics, most notably phytoalexins (17), have been associated with the chemical defenses of plants against microbes, insects and herbivores (37). Several phenolics function as allelopathic compounds influencing germination and growth of neighboring plants (10). Phenolic molecules produced by plant roots are essential for germination, haustorium formation, and host attachment in parasitic *Striga* species (31). Experimental evidence increasingly suggests

P. J. Davies (ed.), Plant Hormones, 188–205.
© 1995 *Kluwer Academic Publishers. Printed in the Netherlands.*

that phenolics function as signals in plant-microbe interactions. *Agrobacterium tumefaciens* virulence gene expression was shown to be activated specifically by the plant-produced phenolic compounds acetosyringone and α-hydroxyacetosyringone (51). Species-specific flavonoids exuded from legume roots and seeds are essential for the induction of the *nod* genes of *Bradyrhizobium* and *Rhizobium* species (30).

According to a recently developed mathematical model (19, 25) the physical properties of SA (*pKa* = 2.98, log K_{ow} (octanol/water partitioning coefficient) = 2.26) are nearly ideal for long distance transport in the phloem. Therefore, unless free SA is actively transported, metabolized, or conjugated, it should translocate rapidly from the point of initial application or synthesis to distant tissues.

Salicylic Acid Levels in Plants

A comprehensive survey of SA levels in the leaves and reproductive structures of agronomically important species (43) and in foods derived from plants (52) has confirmed the ubiquitous distribution of this compound in plants. Rice leaves contained the highest levels of SA, as much as 30 $\mu g \ g^{-1}$ fresh weight. Unusually high levels of SA were also recorded in the inflorescences of thermogenic plants and in plants infected with necrotizing pathogens (see below).

Salicylic Acid and Flowering

Most people learn of the effects of salicylic acid on flowering from the finding that a tablet of aspirin dissolved in water will make cut flowers last longer. The origin of this observation could not be traced to the scientific literature. However, some indications of the mechanisms by which SA may increase flower longevity can be found in the discovery that SA or aspirin inhibits ethylene biosynthesis in pear cell suspension culture by blocking the conversion of 1-aminocyclopropane-1-carboxylic acid to ethylene (28). SA also prevented the accumulation of wound-inducible ACC synthase transcripts in tomato (29). It is unlikely that the endogenous levels of SA present in the floral tissues are sufficiently high to affect ethylene formation in the flower tissue. In addition, ethylene is not always involved in flower senescence. While aspirin will delay the senescence of roses, this effect is likely due to the acidification of the medium used to feed cut flowers and can be duplicated with other organic acids (D. Kuiper, unpublished information).

The first indication of the flower-inducing effects of SA was obtained in an organogenic tobacco tissue culture supplemented with kinetin and indole acetic acid (26). All monohydroxybenzoic acids were found to promote flower bud formation from tobacco callus, with SA being active even at 4 μM concentration. These observations have never attracted much attention since a number of different molecules were found to be active in inducing flower bud formation in tobacco cell cultures (9). The first suggestion that SA may be involved in the regulation of flowering in plants came from experiments in which aphids were allowed to feed on vegetative and reproductive plants of the short-day plant *Xanthium strumarium.* It was hypothesized that a phloem-transmissible factor responsible for the induction of flowering could be found in the honeydew excreted by aphids. Different fractions of honeydew were tested in a bioassay system using *Lemna gibba* strain G3, a long-day plant, kept under a non-photoinductive light cycle. Flower-inducing as well as flower-inhibiting components were identified in the collected honeydew (4). The regulatory substances were thought to be of plant origin since the honeydew produced by aphids feeding on a synthetic diet lacked any flower-inducing activity. The flower inducing substance from *X. strumarium* was identified as SA which at 5.6 μM caused a maximal induction of *L. gibba* flowering (5). SA accelerated flower induction in *Lemna*, while having little effect on the rate of subsequent flower development. The stimulatory effects of SA on flowering were demonstrated in other species of *Lemna*, both short and long-day (6). In addition, SA, aspirin, and related phenolics triggered flowering under non-inductive photoperiods in *Spirodella* and *Wolffia* species (22, 23) which belong to other genera of *Lemnaceae*. The reports of the florigenic effects of exogenous SA in *Lemnaceae* were soon followed by demonstrations of its ability to induce or promote flowering in various plants belonging to different families. However, just as in *Lemnaceae*, the flower-inducing effect of SA in these plants was not very specific.

The possibility that SA functions as the endogenous regulator of flowering in *Xanthium* and *Lemnaceae* was diminished by the fact that SA did not induce flowering in *X. strumarium* and that the levels of SA were not different in honeydew collected from vegetative and flowering plants. Also no changes in the endogenous levels of SA in vegetative or flowering *Lemna* have been reported. In addition, the SA effect on flowering is not specific: a variety of other chemicals also stimulate Lemna flowering. The mechanism by which SA may induce flowering in plants is not known. One hypothesis suggests that SA induces flowering by acting as a chelating agent, because the free *o*-hydroxyl group confers metal chelating activity on benzoic acids. This view is supported by the fact that chelating agents, can induce flowering in *Lemnaceae* (39).

Other Effects of Exogenous Salicylic Acid

Application of SA to plants was shown to elicit a plethora of responses. The most prominent effects of salicylic acid and its close analog, aspirin include: inhibiting ethylene biosynthesis and seed germination; blocking wound responses; interfering with membrane ion transport and absorption in roots; induction of rapid membrane depolarization and collapse of the transmembrane electrochemical potential; affecting nastic leaf movements; reducing transpiration in leaves and epidermal strips; reversing of ABA-induced stomatal closure, leaf abscission, and growth inhibition (34, 41). Other effects of SA on plant development include inducing anthocyanin production in maize seedlings (20) increasing the pod number and yield in mung beans and increasing the height and grain number of cheena millet. In combination with indoleacetic acid (IAA), SA stimulated adventitious root initiation in mung beans. SA also increased the *in vivo* activity of nitrate reductase in maize seedlings. Aspirin and related hydroxy-benzoic acids blocked the wounding response of tomato plants (8).

THERMOGENIC PLANTS AND THE SEARCH FOR CALORIGEN

Thermogenicity (heat production) in plants, first described by Lamarck in 1778 for the genus *Arum*, is now known to occur in the male reproductive structures of cycads and in the flowers or inflorescences of some Angiosperm species belonging to the families *Annonaceae, Araceae, Aristolochiaceae, Cyclanthaceae, Nymphaeaceae*, and *Palmae* (35). The heating is believed to be associated with a large increase in the cyanide-insensitive non-phosphorylating electron transport pathway unique to plant mitochondria. The increase in this so-called alternative respiratory pathway is so dramatic that oxygen consumption in the inflorescences of *Arum* lilies at the peak of heat production is as high as that of a hummingbird in flight. In addition to the activation of the alternative oxidase, thermogenicity involves activation of the glycolytic and Krebs cycle enzymes which provide substrates for this remarkable metabolic explosion.

In one of the *Arum* lilies, *Sauromatum guttatum* Schott (voodoo lily), the inflorescence develops from a large corm, and can reach 80 cm in height (Fig. 1). Early on the day of anthesis, a large bract (spathe) which surrounds the central column of the inflorescence (spadix) unfolds to expose the upper part of the spadix known as the appendix. Soon thereafter, the appendix starts to generate heat, which facilitates the volatilization of foul-smelling amines and indoles attractive to insect pollinators. By early afternoon the temperature of the appendix can increase by 14°C above ambient, but it returns to ambient in the evening. The second thermogenic episode in the lower spadix starts late at night and ends the following morning after maximum temperature increases of more than 10°C. In 1937 Van Herk (54) suggested that the burst of metabolic activity in the appendix of the voodoo lily is triggered by "calorigen", a water-soluble substance produced in the male (staminate) flower primordia located just below the appendix. Van Herk believed that calorigen began to enter the appendix on the day preceding the day of anthesis. Van Herk's ideas encountered some skepticism, partially because attempts to isolate and characterize calorigen were not successful until recently.

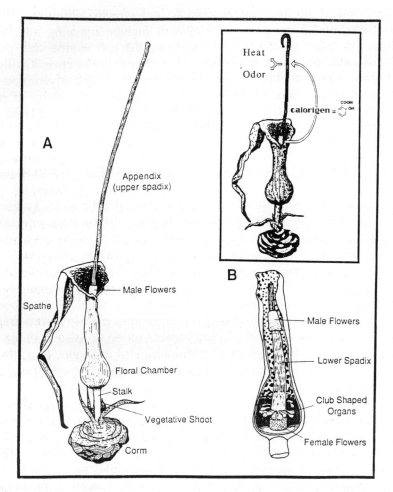

Fig. 1. The inflorescence of *Sauromatum guttatum* on the day of blooming and heat production. (A) Entire inflorescence. (B) Longitudinal cross section of the floral chamber. The drawing is not entirely to scale. Inset: Proposed action of salicylic acid in thermogenesis. On the day before blooming calorigen, identified as SA, moves from the male flowers to the appendix. There it induces heat and the production of odor attractive to insect pollinators. The heat is a product of cyanide-insensitive respiration which, along with the enzymes of Krebs cycle and glycolysis, is induced by SA.

Salicylic Acid as a Natural Inducer of Thermogenesis in Arum Lilies

In 1987 an attempt to identify the elusive calorigen ended in success. Mass spectroscopic analysis of highly purified calorigen extracted from the male flowers of voodoo lily indicated the presence of SA (42). Application of salicylic acid at 0.13 μg g^{-1} fresh weight to sections of the immature appendix, led to temperature increases of as much as 12°C. These increases duplicated the temperature increases produced by the crude calorigen extract both in magnitude and timing, indicating that SA is the calorigen (Fig. 2).

The sensitivity of appendix tissue to salicylic acid increased daily with the approach of anthesis and was controlled by the photoperiod. On the day preceding the day of blooming the levels of SA in the appendix of the voodoo lily increased almost 100-fold reaching the level of 1 μg g^{-1} fresh weight (44). The level of SA in the appendix began to rise in the afternoon and reached its maximum late in the evening, while the maximum accumulation of SA in the lower spadix occurred late at night. The concentration of SA in both thermogenic tissues promptly returned to basal, pre-blooming, levels at the end of the thermogenic periods. The observed kinetics of SA accumulation in the appendix was consistent with the original suggestion by Van Herk that calorigen is made in the male flowers and moves to the appendix during the day preceding the day of blooming. Of 33 analogs of SA tested, only 2,6-dihydroxybenzoic acid and aspirin were thermogenic. The activity of 2,6-dihydroxybenzoic acid exceeded that of SA. SA, 2,6-dihydroxybenzoic acid and aspirin also induce the production of large quantities of amines and indoles on the first day of blooming (44). The thermogenic effect of SA could not be separated from its odor-producing effect, suggesting that the transduction pathways for these processes are closely linked.

The levels of SA determined during heat production in thermogenic inflorescences of five other aroid species, and in male cones of four thermogenic cycads exceeded 1 μg g^{-1} fresh weight (43). However, SA was not detected in the thermogenic flowers of the water lily, *Victoria regia* Lindle (*Nymphaeaceae*), and *Bactris major* Facq (*Palmae*).

The nuclear gene from *Sauromatum guttatum* encoding the alternative oxidase protein with the calculated molecular mass of 38.9 kD was recently isolated and characterized (46). Both calorigen extract (11) and SA (12) cause the induction of the alternative oxidase gene, providing additional confirmation of the chemical identity of calorigen. Application of SA to the immature appendix of *Sauromatum guttatum* caused an increase in alternative

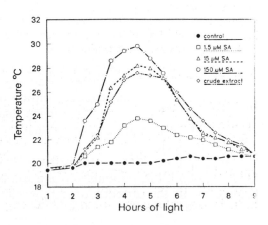

Fig. 2. Dose response of *Sauromatum guttatum* appendix tissue to salicylic acid and calorigen extract. Crude calorigen extract or salicylic acid solution (200 μl) was placed on top of the 3-cm-long appendix sections excised 2 days before blooming. From (42).

193

pathway capacity and a dramatic accumulation of the alternative oxidase proteins (47). SA can also induce alternative respiration in non-overtly thermogenic plants. For example, addition of 20 μM SA to young tobacco cell suspensions increased cyanide-resistant respiration and associated heat production (21). SA was also shown to induce the activity of alternative oxidase in *Chlamydomonas* (15). While the mechanism involved in SA induction of alternative respiration is being unraveled, the mechanism by which SA induces glycolysis, Krebs cycle, and odor production during the thermogenic syndrome still remains a mystery.

The discovery of the role of SA in the flowering of thermogenic plants was the first demonstration of an important regulatory role played by endogenous SA. The study ended a 50-year-long search for calorigen and laid the foundation for ongoing investigations of other processes in plants which may be regulated by SA. This discovery also moved SA research from the stage of phenomenological observations to serious attempts to understand the mechanisms of its action and the search for other plant processes which might be regulated by SA.

SALICYLIC ACID AND DISEASE RESISTANCE

Plants face a constant barrage of potentially pathogenic microorganisms, but infection and disease rarely develop from these contacts. In addition to chemical and physical barriers such as the cuticle, cell wall, and antimicrobial chemicals that are constitutively present, plants frequently restrict the spread of fungal, bacterial, or viral pathogens to a small area around the point of initial penetration where a necrotic lesion appears. This protective cell suicide is referred to as the hypersensitive reaction (HR). The HR may lead to systemic acquired resistance (SAR) defined as a resistance to subsequent pathogen attack which develops after the initial inoculation in the uninfected, pathogen-free parts of the plants (48). SAR develops following plant interactions with lesion-forming pathogens, is detected several days after the initial infection, can last for several weeks and is effective against a broad range of pathogens which may be unrelated to the inducing organism. SAR is sometimes referred to as plant immunization.

Associated with HR and SAR is the systemic synthesis of at least five families of serologically distinct, pathogenesis-related (PR) proteins. Chitinases and β-1,3-glucanases are among the best studied proteins belonging to this group. The localization, timing of appearance and defense-related functions of at least some PR proteins suggest their possible involvement in SAR. However, definitive proof that the induction of PR proteins causes SAR is still lacking.

It is well established that resistance to pathogens and the production of most, if not all, PR, proteins in plants can be induced by SA or acetylsalicylic acid, even in the absence of pathogenic organisms. The discovery of a protective function of salicylates was made in 1979, in tobacco (*Nicotiana tabacum* cv. Xanthi-nc) (57). Xanthi-nc tobacco contains the *N* gene from *N. glutinosa* and confers HR response to tobacco mosaic virus (TMV). Salicylate treatments also resulted in the induction of PR proteins in treated leaves. The level of PR protein induction and TMV protection increased with increasing aspirin concentrations. A recent comprehensive study utilizing modern molecular approaches showed that nine classes of PR protein mRNAs that are induced during the development of systemic acquired resistance to TMV in tobacco can be induced by SA to a similar degree (55).

In TMV-susceptible *Nicotiana tabacum* containing the recessive *n* allele, TMV does not trigger the induction of PR proteins and HR. Instead, the virus spreads systemically causing a characteristic mosaic in younger leaves. However, aspirin induced PR proteins in *n* tobacco, and simultaneously reduced the spread and total accumulation of TMV (58). The extent to which SA-induced resistance is based on the induction of PR proteins is still unknown. It is certainly possible that SA activates other resistance mechanisms.

Since SAR can be induced systemically by localized infections, the existence of a systemic signal that activates PR proteins and/or other resistance mechanisms has been hypothesized for at least 25 years (49). Evidence from stem girdling and grafting experiments suggests that the putative signal moves through the phloem tissue of the vascular system of the plant (16).

The observations that exogenous SA applications induce resistance and PR proteins in plants, that SA is an important endogenous messenger in thermogenic plants, together with the development of analytical methods to quantify its endogenous levels in plant tissues, prepared the way to test the possibility that SA is an endogenous messenger which activates important elements of host resistance to pathogens. The single-gene inheritance of TMV resistance in tobacco, provided a suitable experimental system to investigate this possibility.

A new chapter in SA research started from the observation that SA levels in TMV-resistant (Xanthi-nc), but not susceptible, (Xanthi) tobacco increase almost 50-fold in TMV-inoculated leaves, and at least 10-fold in uninfected leaves of the same plant (32). The highest concentration of free SA, almost 20 μg g^{-1} fresh weight, representing a 400-fold increase over basal levels was observed in and around hypersensitive lesions, with relatively lower increases observed in other tissues (14) (Fig. 3). Induction of PR1 genes in virus-inoculated and systemically protected leaves always paralleled the rise in SA levels (32). While TMV induced PR proteins only in Xanthi-nc tobacco, SA was effective in both Xanthi *n* and Xanthi-nc *N* plants (32). By feeding SA to excised leaves of Xanthi-nc (*NN*) tobacco it was shown that the observed increase in endogenous SA levels is sufficient for the systemic induction of PR-1 proteins (60) and increased resistance to TMV (14). At 32°C TMV infection becomes systemic and Xanthi-nc plants fail to accumulate PR-1 proteins. This loss of HR at high temperature was associated with an inability to accumulate SA. However, spraying leaves with SA induced PR-1 proteins at both 24°C and 32°C (60). Different components of TMV were compared for their ability to induce SA accumulation and exudation: three different aggregate states of coat protein failed to induce SA, but unencapsidated viral RNA elicited SA accumulation in leaves and phloem (60). It is important to note that SA does not activate such wound-related responses as ethylene synthesis and production of proteinase inhibitors. In addition, mechanical leaf injury did not stimulate SA production and exudation.

SA was also exported from the primary site of infection to the uninfected tissues. When leaves of Xanthi-nc tobacco were excised 24 hr after TMV inoculation and exudates from the cut petioles collected, the increase in endogenous SA in TMV-inoculated leaves paralleled SA levels in exudates (60). Exudation and leaf accumulation of SA were proportional to TMV concentration. In cucumber a fluorescent metabolite, identified as SA, increased dramatically in the phloem of plants inoculated with tobacco necrosis

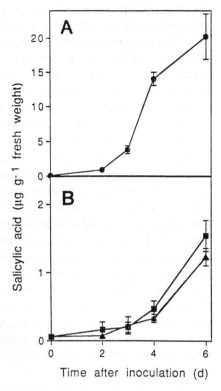

Fig. 3. Accumulation of salicylic acid in TMV-tobacco. (A) Salicylic acid levels in inoculated tobacco leaf tissue expressing a hypersensitive response, and (B) in the basal uninoculated portion of an inoculated leaf (squares) and in the untreated leaf immediately above the inoculated leaf (triangles). Plants were inoculated at zero time.

virus (TNV) or the fungal pathogen *Colletotrichum lagenarium* (36) (Fig. 4A). Levels of SA increased transiently after inoculation, and reached a peak before SAR was detected (Fig. 4B). However, analysis of phloem exudate from cucumber leaves demonstrated that the earliest detectable increase in SA occurred 8 hours after inoculation with *Pseudomonas syringae* pv. *syringae* (45). The systemic accumulation of SA was observed even when the inoculated leaf remained attached to the plant for only 4 hr. While supporting the role of SA as a component of the transduction pathway leading to resistance, these results suggest that another chemical signal may be required for the systemic accumulation of SA in cucumber.

If SA is the natural signal for induced resistance, then treatments that inhibit or enhance SA accumulation should have a corresponding effect on levels of resistance. In fact it was demonstrated that environmental (see above), genetic, and developmental changes in pathogen resistance and PR protein expression correlate with SA levels. Other evidence for the involvement of SA in plant disease resistance comes from an amphidiploid hybrid generated from *Nicotiana glutinosa* and *N. debneyi*. This hybrid is known for high levels of constitutive expression of PR proteins and high resistance to viral, bacterial, and fungal pathogens when compared to its parental species or *N. tabacum* cv. Xanthi-nc (1). Healthy hybrid plants have levels of SA 30-times greater than less resistant Xanthi-nc plants (61) supporting an involvement of SA in resistance. Finally, PR proteins and increased TMV-resistance can be detected in the leaves of healthy, flowering Xanthi-nc tobacco. These developmentally-induced increases in resistance are also correlated with elevated leaf levels of SA (62). Changes in SA levels upon infection with necrotizing pathogens were not specific to the Xanthi-nc cultivar of *Nicotiana tabaccum*. Inoculating different tobacco cultivars and species with necrotizing viral, bacterial, and fungal pathogens inevitably resulted in increased levels of SA (50).

Little is known about the transduction pathway preceding and following the increases in salicylic acid during development of SAR. However it was established that ethylene, a stress

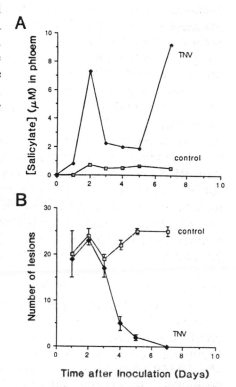

Fig. 4. Time course of the appearance of salicylic acid in the cucumber phloem (A) in relation to the induction of resistance against *Colletotrichim lagenarium* after the initial infection with TNV (B). At time zero the first leaves of cucumber plants were inoculated with TNV. At each point, two cucumber plants were cut with a razor blade through the stem above the first and below the second leaf and the phloem exudate collected. At the same times the other group of inoculated plants were challenged with a secondary inoculation of *C. lagenarium*. Adapted from (36)

hormone commonly produced during pathogenesis, is not likely to be a part of the SA signal transduction pathway leading to SAR (50). Recently a SA-binding protein with a physiologically relevant K_d for SA, was identified in tobacco leaves (3). However, the role of this protein in SA action remains to be elucidated. It was also shown that SA is not involved in the transduction pathway by which cultured rose cells respond to UV-C radiation and *Phytophthora* cell-wall elicitor (38). Another observation suggests that SA may be involved in rapid defense responses. One such response mimicked by SA, involves stimulation of oxidative cross-linking of cell wall structural proteins which leads to strengthening of the cell wall (2).

BIOSYNTHESIS OF SALICYLIC ACID

Biochemical logic and some of the previously published reports (40) suggest that, in plants, SA is likely synthesized from *t*-cinnamic acid, an intermediate of the phenylpropanoid pathway which yields a variety of phenolics with structural and defense-related functions. The formation of SA from *t*-cinnamic acid may occur by a chain-shortening reaction followed by 2-hydroxylation or vice versa. Feeding both healthy and TMV-inoculated tobacco leaf tissue with different putative precursors showed that only benzoic acid was capable of increasing tissue levels of SA. On the other hand, feeding [^{14}C]-labeled cinnamic acid resulted in formation of labeled benzoic acid and SA (62). No radioactive *o*-coumaric was formed from cinnamic acid. Feeding leaf tissue with labeled benzoic acid resulted in the formation of SA with specific radioactivity almost equal to that initially supplied as benzoic acid, suggesting that most of the SA in tobacco is formed from benzoic acid. Figure 5 shows the time course of accumulation of SA and its possible precursors in TMV-inoculated tobacco leaves following the temperature shift described previously. Large increases in the levels of free benzoic and SA were detectable by 7.5 h. By 10.5 h, TMV-inoculated leaves contained 15 μg benzoic and 19 μg SA per g fresh weight compared to less than 2 μg benzoic and 0.7 μg SA at time 0. Levels of free cinnamic and *o*-coumaric acid did not change significantly. The data suggests that benzoic acid is a direct precursor of SA in tobacco.

The monooxygenase activity which catalyzes 2-hydroxylation of benzoic acid to SA was recently isolated from tobacco leaves. Benzoic acid 2-hydroxylase (BA2H), required NAD(P)H or reduced methyl viologen as an electron donor (27). BA2H activity was detected in healthy tobacco leaf extracts, but increased significantly following inoculation with TMV. This increase paralleled the levels of free SA in the leaves and was particularly strong following the transfer of inoculated plants from 32°C to 24°C (Fig. 6). The effect of TMV-inoculation on BA2H could be

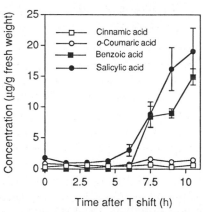

Fig. 5. Accumulation of SA and its putative precursors in TMV-inoculated tobacco leaves following temperature shift from 32°C to 24°C. Seedlings were inoculated with TMV on one leaf and kept at 32°C for 96 h. At time = 0 h, the incubation temperature was lowered to 24°C. Unhydrolyzed leaf extracts were examined. Each point is the mean of triplicate samples ($\pm$ SE). From (62)

duplicated by infiltrating leaf discs of healthy plants with benzoic acid at the levels observed *in vivo* after virus inoculation. BA2H appears to be a pathogen-inducible protein with an important role in the SA accumulation during the development of induced resistance to TMV in tobacco. However, the observed induction of this enzyme by increased benzoic acid levels suggests that BA2H is not the primary regulator of SA production, and that the rate limiting step may be the formation of benzoic acid from cinnamic acid. Further identification of the intermediate(s) in the formation of benzoic acid from cinnamic acid and characterization of the rate-limiting enzymes of SA biosynthesis will be important for designing strategies for increasing the resistance of plants to pathogens.

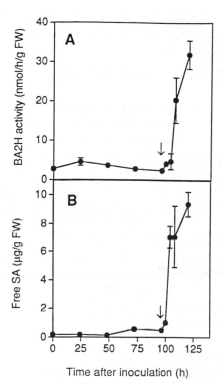

Fig. 6. BA2H activity (A) and free SA content (B) in TMV-inoculated leaves of Xanthi-nc tobacco incubated at 32°C (0-96 h) and then transferred to 24°C. The arrows indicate the time of temperature shift. Each value is the mean of three replicates ±SE. From (27)

METABOLISM OF SALICYLIC ACID

Various hydroxybenzoic acid glucosides have been reported to occur in higher plants. For example, sunflower hypocotyls incubated with carboxy-labeled benzoic acid formed trace amounts of SA and larger amounts of glucosyl-SA (GSA) (24). Leaves of Xanthi-nc tobacco rapidly metabolize exogenously supplied, or endogenously produced, SA to β-O-D-glucosyl-SA (14). More interestingly, endogenous SA produced in the TMV-inoculated leaves was also rapidly metabolized to the same conjugate. Actually, most of the SA in TMV-inoculated leaves of tobacco is present in the form of β-O-D-glucosyl-SA (14, 33) which could be converted back to SA following the *in vitro* enzymatic digestion with β-glucosidase. Large amounts of β-O-D-glucosyl-SA was found only in leaves that exhibited HR with the highest levels present in and around lesions. Phloem sap and pathogen-free leaves of TMV-inoculated tobacco did not contain significant levels of β-O-D-glucosyl-SA (14), indicating that only free SA can move in the plant.

SA-inducible UDP-glucose:SA-glucosyltransferase (GTase), an enzyme which can catalyze the glucosylation of SA to β-O-D-glucosyl-SA, has been partially purified from cell suspension cultures of *Mallotus japonicus* (53) and from oat roots (59). It was soon discovered that healthy tobacco tissues also have constitutive SA-glucosyl-transferase activity of 0.076 mU g^{-1} fresh weight (13). This activity started to increase 48 h after TMV-inoculation reaching its maximum (6.7-fold induction over the basal levels) by 72 h (Fig. 7A). This increase in SA-glucosyltransferase activity coincided with the accumulation of free SA and β-O-D-glucosyl-SA (the difference between total SA released after hydrolysis with β-glucosidase and initially present free SA) in the inoculated leaf (Fig. 7B,C). No significant β-O-D-glucosyl-SA accumulation or elevated SA-glucosyl-transferase activity could be detected in the healthy leaf immediately above the TMV-inoculated leaf. The effect of TMV-inoculation on the SA-glucosyltransferase and β-O-D-glucosyl-SA accumulation could be duplicated by infiltrating tobacco leaf discs with SA at the levels naturally produced in TMV-inoculated leaves (13). Of 12 analogs of SA tested only 2,6-dihydroxy-benzoic acid induced β-GTase activity. The ability of SA to induce SA-glucosyl-transferase activity may serve as an effective mechanism for the feed-back regulation of SA levels in plant tissues. This regulation of tissue SA may be an important control point in the signal transduction pathway leading to the activation of disease resistance mechanisms in plants. The available data demonstrate that β-GTase is one of the many

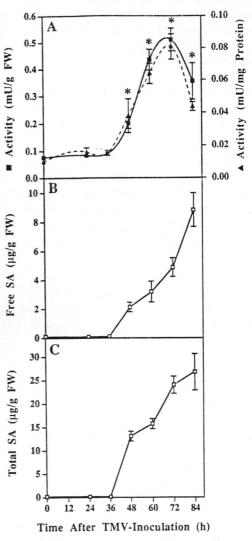

Fig. 7. β-GTase activity expressed as (n) mU g^{-1} FW and (s) mU mg^{-1} protein (A), free SA content (B), and total SA (SA + β-O-D-glucosyl-SA) content (C) in TMV-inoculated leaves of Xanthi-nc tobacco at various times after inoculation. Vertical bars denote ± SE. (*) denotes a significant level of β-GTase induction at the $p \leq 0.05$ level. GSA content is derived by subtraction of (B) from (C); for example, GSA is 67% of the total SA at 84 h after inoculation. From (13)

proteins induced during the HR response. Therefore, it may be appropriate to refer to this enzyme as a PR protein.

CONCLUSIONS

Centuries have passed since the healing substance from willow bark was shown to have value not only for humans but for the plants which synthesize it. However, only in the last six years has rapid progress in understanding SA biology, biosynthesis and metabolism in plants has been made. Fig. 8 summarizes current knowledge of SA biosynthesis and metabolism and shows the steps which are activated during pathogenesis. Surprisingly, some of the effects of SA in plants are also associated with reduction of disease symptoms. Unfortunately, we still do not know whether there are any connections between the therapeutic effects of salicylates in plants and animals. In addition, we still do not understand the biochemical link between the action of SA in plant disease resistance and its thermogenic and odor-producing effects in *Arum* lilies. It is also important to identify genes involved in SA biosynthesis, metabolism and reception. Furthermore, the molecular components of the SA signal transduction pathway(s) should be elucidated, and other possible regulatory functions for SA in plants investigated.

The growing appreciation of the role of SA in plants may bring some practical applications. For example, manipulating the level of SA in plants may be a promising area for the application of biotechnology to crop protection. Increases in endogenous SA may be achieved via enhancing transcription and translation of the genes for SA biosynthesis or by blocking the expression of genes involved in SA metabolism. Engineering transgenic plants with elevated SA levels may be the first step in the creation of crops with increased resistance to agronomically important pathogens.

The commonly used definition of a plant hormone, simply states that it is a "natural compound in plants with an ability to affect physiological processes at concentrations far below those where either nutrients or vitamins would affect these processes" (7 and Chapter A1). All the information on the role of SA in thermogenesis and disease resistance suggests that SA meets these qualifying criteria for a plant hormone.

References

1. Ahl Goy, P., Felix, G., Metraux, J.-P., Meins, F. (1992) Resistance to disease in the hybrid *Nicotiana glutinosa* x *Nicotiana debneyi* is associated with high constitutive levels of β-1,3-glucanase, chitinase, peroxidase and polyphenoloxidase. Physiol. Molec. Plant Pathol. 41, 11-21.
2. Bradley, D.J., Kjellbom, P., Lamb, C.J. (1992) Elicitor- and wound-induced oxidative cross-linking of a proline-rich plant cell wall protein: a novel, rapid defense Response. Cell 70, 21-30.

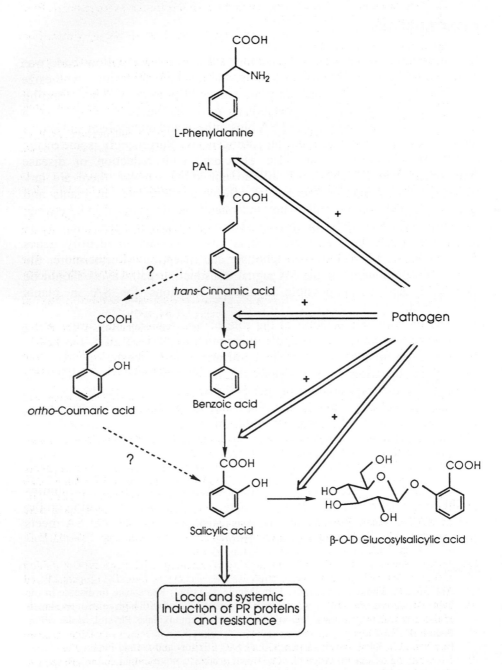

Fig. 8. Salicylic acid biosynthesis and metabolism in plants. Plus signs indicate steps induced in the vicinity of hypersensitive response lesions. From (63)

3.   Chen, Z., Klessig, D.F. (1991) Identification of a soluble salicylic acid-binding protein that may function in signal transduction in the plant disease-resistance response. Proc. Natl. Acad. Sci. USA 88, 8179-8183.

4.   Cleland, C.F. (1974) Isolation of flower-inducing and flower-inhibiting factors from aphid honeydew. Plant Physiol. 54,899-903.

5.   Cleland, C.F., Ajami, A. (1974) Identification of the flower-inducing factor isolated from aphid honeydew as being salicylic acid. Plant Physiol. 54, 904-06.

6.   Cleland, C.F., Tanaka, O. (1979) Effect of daylength on the ability of salicylic acid to induce flowering in the long-day plant *Lemna gibba* G3 and the short-day plant *Lemna paucicostata* 6746. Plant Physiol. 64, 421-424.

7.   Davies, P.J. (1988) The plant hormones: their nature, occurrence, and functions. *In*: Plant Hormones and Their Role in Plant Growth and Development, pp. 1-11, Davies P.J., ed. Kluwer, Dordrecht.

8.   Doherty, H.M., Selvendran, R.R., Bowles, D.J. (1988) The wound response of tomato plants can be inhibited by aspirin and related hydroxy-benzoic acids. Physiol. Mol. Plant Pathol. 33, 377-384.

9.   Eberhard, S., Doubrava, N., Marta, V., Mohnen, D., Southwick, A., Darvill, A., Albersheim, P. (1989) Pectic cell wall fragments regulate tobacco thin-cell-layer explant morphogenesis. Plant Cell 1, 747-755.

10.   Einhellig, F.A. (1986) Mechanisms and modes of action of allelochemicals. *In*: The Science of Allelopathy, p. 317, Putnam, A.R., Tang, C.S., eds. John Wiley & Sons, New York.

11.   Elthon, T.E., McIntosh, L. (1987) Identification of the alternative terminal oxidase of higher plant mitochondria. Proc. Natl. Acad. Sci. USA 84, 8399-8403.

12.   Elthon, T.E., Nickels, R.L., McIntosh, L. (1989) Mitochondrial events during development of thermogenesis in *Sauromatum guttatum* (Schott). Planta 180, 82-89.

13.   Enyedi, A.J., Raskin, I. (1993) Induction of UDP-Glucose: salicylic acid glucosyltransferase activity in tobacco mosaic virus-inoculated tobacco (*Nicotiana tabacum*) leaves. Plant Physiol. 101, 1375-1380.

14.   Enyedi A.J., Yalpani N., Silverman P., Raskin I. (1992) Localization, conjugation, and function of salicylic acid in tobacco during hypersensitive reaction to tobacco mosaic virus. Proc Natl Acad Sci. USA 89, 2480-2484.

15.   Goyal, A., Tolbert N.E. (1989) Variations in the alternative oxidase in *Chlamydomonas* grown in air or high CO_2. Plant Physiol. 89, 958-62.

16.   Guedes M.E.M., Richmond S., Kuc J. (1980) Induced systemic resistance to anthracnose in cucumber as influenced by the location of the inducer inoculation with *Colletotrichum lagenarium* and the onset of flowering and fruiting. Physiol. Plant Pathol. 17, 229-233.

17.   Hahlbrock, K., Scheel, D. (1989) Physiology and molecular biology of phenylpropanoid metabolism. Annu. Rev. Plant Physiol. Plant Molec. Biol. 40, 347-369.

18.   Harborne, J.B., (1980) Plant phenolics. *In*: Secondary Plant Products, pp. 329-402, Bell, E.A., Charlwood, B.V., eds. Springer Verlag, Berlin.

19.   Hsu, F., Kleier, D.M. (1990) Phloem mobility of xenobiotics. III. Sensitivity of unified model to plant parameters and application to patented chemical hybridizing agents. Weed Sci. 38, 315-23.

20.   Jain, A., Srivastava, H.S. (1984) Effect of phenolic acids on anthocyanin content in maize roots. Biologia Plant. 26, 241-245.

21.   Kapulnik, Y., Yalpani, N., Raskin, I. (1992) Salicylic acid induces cyanide-resistant respiration in tobacco cell-suspension cultures. Plant Physiol. 100, 1921-1926.

22.   Khurana, J.P., Maheshwari, S.C. (1980) Some effects of salicylic acid on growth and flowering in *Spirodela polyrrhiza* SP_{20}. Plant Cell Physiol. 21, 923-27.

23. Khurana, J.P., Maheshwari, S.C. (1987) Floral induction in *Wolffia microscopica* by non-inductive long days. Plant Cell Physiol. 24, 907-12.

24. Klambt H.D. (1962) Conversion in plants of benzoic acid to salicylic acid and its β-*D*-glucoside. Nature 196, 491.

25. Kleier, D.A. (1988) Phloem mobility of xenobiotics. I. Mathematical model unifying the weak acid and intermediate permeability theories. Plant Physiol. 86, 803-10.

26. Lee T.T., Skoog F. (1965) Effect of substituted phenols on bud formation and growth of tobacco tissue culture. Physiol. Plant. 18, 386-402.

27. Leon J., Yalpani, N., Raskin, I., Lawton, M. A. (1993) Induction of benzoic acid 2-hydroxylase in virus-inoculated tobacco. Plant Physiol. 103, 323-328.

28. Leslie, C.A., Romani R.J. (1988) Inhibition of ethylene biosynthesis by salicylic acid. Plant Physiol. 88, 833-37.

29. Li, N., Parsons, B.L., Liu, D., Mattoo, A.K. (1992) Accumulation of wound-inducible ACC synthase transcript in tomato fruit is inhibited by salicylic acid and polyamines. Plant Mol. Biol. 18, 477-487.

30. Long, S.R. (1989) *Rhizobium*-legume nodulation: life together in the underground. Cell 56, 203-214.

31. Lynn D.G., Chang, M. (1990) Phenolic signals in cohabitation: implications for plant development. Annu. Rev. Plant Physiol. Plant Mol. Biol. 41, 497-526.

32. Malamy J., Carr J.P., Klessig D.F., Raskin I. (1990) Salicylic acid: a likely endogenous signal in the resistance response of tobacco to viral infection. Science 250, 1002-1004.

33. Malamy J., Hennig J., Klessing D.F. (1992) Temperature-dependent induction of salicylic acid conjugates during the resistance response to tobacco mosaic virus infection. Plant Cell 4, 359-366.

34. Malamy, J., Klessig, D.F. (1992) Salicylic acid and plant disease resistance. The Plant Journal 2(5), 643-654.

35. Meeuse, B.J.D., Raskin, I. (1988) Sexual reproduction in the Arum lily family, with emphasis on thermogenicity. Sex. Plant Reprod. 1, 3-15.

36. Metraux J.P., Signer H., Ryals J., Ward E., Wyss-Benz M., Gaudin J., Raschdorf K., Schmid E., Blum W., Inverardi B. (1990) Increase in salicylic acid at the onset of systemic acquired resistance in cucumber. Science 250, 1004-1006.

37. Metraux, J.P., Raskin, I. (1992) Role of phenolics in plant disease resistance. *In*: Application of Biotechnology in Plant Pathology, pp. 191-209, Chet, I., ed. Wiley, New York.

38. Murphy, T.M., Raskin, I., Enyedi, A.J. (1993) Plasma membrane effects of salicylic acid treatment on cultured rose cells. Environ. Expt. Botany 33, 267-272.

39. Oota, Y. (1972) The response of *Lemna gibba* G3 to a single long day in the presence of EDTA. Plant Cell Physiol. 13, 575-580.

40. Raskin, I. (1992) Role of salicylic acid in plants. Annu. Rev. Plant Physiol. Plant Mol. Biol. 43, 439-463.

41. Raskin, I. (1992) Salicylate, a new plant hormone. Plant Physiol. 99, 799-803.

42. Raskin, I., Ehmann, A., Melander, W. R., Meeuse, B.J.D. (1987) Salicylic acid - a natural inducer of heat production in Arum lilies. Science 237, 1545-1556.

43. Raskin, I., Skubatz, H., Tang, W., Meeuse, B.J.D. (1990) Salicylic acid levels in thermogenic and non-thermogenic plants. Ann. Bot. 66, 376-373.

44. Raskin, I., Turner, I.M., Melander, W.R. (1989) Regulation of heat production in the inflorescences of an Arum lily by endogenous salicylic acid. Proc. Natl. Acad. Sci. USA 86, 2214-2218.

45. Rasmussen J.B., Hammerschmidt R., Zook M.N. (1991) Systemic induction of salicylic acid accumulation in cucumber after inoculation with *Pseudomonas syringae* pv. *syringae*. Plant Physiol. 97, 1342-1347.

46. Rhoads, D.M., McIntosh, L. (1991) Isolation and characterization of a cDNA clone encoding an alternative oxidase protein of *Sauromatum guttatum* (Schott). Proc. Natl. Acad. Sci. USA 88, 2122-2126.
47. Rhoads, D.M., McIntosh, L. (1992) Salicylic acid regulation of respiration in higher plants: alternative oxidase expression. Plant Cell 4, 1131-1139.
48. Ross, A.F. (1961) Systemic acquired resistance induced by localized virus infections in plants. Virology 13, 340-358.
49. Ross, A.F. (1966) Systemic effects of local lesion formation. *In:* Viruses of Plants, pp. 127-150, Beemst, A.B.R., Bijkstra J., eds. North Holland, Amsterdam.
50. Silverman, P., Nuckles, E., Ye, X.S., Kuc, J., Raskin, I. (1993) Salicylic acid, ethylene, and pathogen resistance in tobacco. Mol. Plant Micro. Int. 6, 775-781.
51. Stachel, S.E., Messens, E., Van Montagu, M., Zambryski, P. (1985) Identification of the signal molecules produced by wounded plant cells that activate T-DNA transfer in *Agrobacterium tumefaciens*. Nature 318, 624-629.

C2c. Brassinosteroids

Richard N. Arteca

Department of Horticulture, The Pennsylvania State University, University Park, PA 16802, USA.

INTRODUCTION

In the early 70's John Mitchell and co-workers at the USDA's Department of Agriculture Research Center began screening pollens in search of new plant hormones. It had been known for many years that pollen is a rich source of plant growth regulating substances, thereby, making it a logical choice for screening. Nearly 60 species of plants were screened and about half caused increased growth in the bean second internode bioassay. The greatest growth increases were obtained from alder tree (*Alnus glutinosa* L.) and rape plant pollen (*Brassica napus* L.). Extracts from these two pollens caused such rapid growth in the bean second internode bioassay that the stem would split above the second pair of leaves. The USDA workers proposed that this was a new class of lipoidal hormones, which they termed brassins (defined as a crude lipoidal extract from rape pollen). In 1972 Mitchell and Gregory (18) showed that brassins could enhance crop yield, crop efficiency and seed vigor. Milborrow and Pryce (17) believed that brassins was a crude extract containing gibberellins and other compounds rather than endogenous lipids. In an effort to isolate the active components of brassins 500 pounds of bee-collected rape pollen, which was more readily available then alder pollen, was extracted and purified resulting in 10 mg of active crystalline material. In 1979 Grove and co-workers (10) identified brassinolide (Fig. 1) as the active component in brassins. Shortly after the identification of brassinolide by the USDA group it was also identified in *Distylium* extracts (16). Much of this and subsequent work has been outlined in reviews (1, 9, 15, 21).

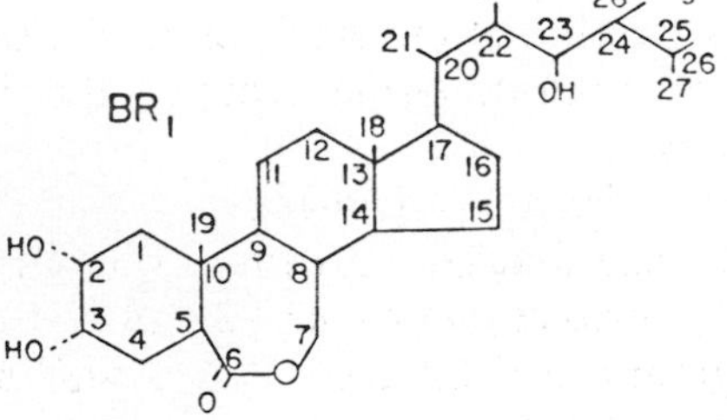

Fig. 1. The structure of brassinolide (2α, 3α, 22α, 23α-tetrahydroxy-24α-methyl-B-homo-7-oxa-5α-cholestan-6-one).

P. J. Davies (ed.), Plant Hormones, 206–213.
© 1995 *Kluwer Academic Publishers. Printed in the Netherlands*

Brassinolide is the first plant hormone shown to have a steroidal structure and the only naturally occurring steroid which has a seven-membered lactone ring as part of a fused ring system (15). Since the discovery of brassinolide, sequential numerical suffix-designated brassinosteroids (BR_x) have been used to describe a class of compounds having activity similar to brassinolide (BR_1) in the bean second internode bioassay. At the present time over sixty kinds of brassinosteroids have been found. Thirty-one are fully characterized, including 29 free compounds and 2 conjugates (14). Many are ubiquitous in the plant kingdom (14) and thought to be another class of plant hormones.

BRASSINOSTEROIDS AND THEIR DISTRIBUTION IN PLANTS

Brassinosteroids (BR) have been found in a wide range of plants, including dicots, monocots, gymnosperms and algae (14), and in various plant parts such as pollen, leaves, flowers, seeds, shoots, galls and stems. Although plant roots have not yet been investigated, it is likely that they will also contain brassinosteroids (22). Among the naturally occurring brassinosteroids, brassinolide and castasterone are considered to be the most important, because of their wide distribution, as well as their potent biological activity.

BIOSYNTHESIS OF BRASSINOLIDE

At the present time the biosynthesis of brassinolide has not yet been thoroughly investigated, though a proposed pathway for the biosynthesis of brassinolide has been outlined (15, 21, 28).

STRUCTURE/ACTIVITY RELATIONSHIPS

All naturally-occurring brassinosteroids are known to be derivatives of 5-α-cholestan. Variation of kinds and orientation on this skeleton have been shown to have an effect on activity. It has been shown, using the bean first and second internode bioassay (25, 26), BR-induced ethylene production in mung bean (3, 4) and radish and tomato hypocotyl elongation assays (22), that in order to have brassinosteroid activity the following structural requirements must be meet:
1. trans A/B ring system (5α-hydrogen);
2. 6-ketone or a 7-oxa-6-ketone system in ring B;
3. cis α-oriented hydroxyl groups at C-2 and C-3 positions;
4. cis hydroxy groups at C-22 and C-23 as well as a methyl or ethyl group at C-24;

5. α-orientation at C-22, C-23 and C-24 are more active then β-oriented compounds.

TRANSPORT AND METABOLISM OF BRASSINOSTEROIDS

At the present time the mode of brassinosteroids transport is not known. When brassinosteroids are applied to the roots of tomato plants there is a stimulation in ethylene biosynthesis resulting in epinasty (23). Prior to this work indirect evidence was presented by several workers indicating that BR could be transported from the roots to the shoots of plants. It was shown that when BR was applied to the roots little or no ACC was found in the xylem sap, indicating that there was a signal (presumably BR) from the roots which stimulated ACC synthesis in leaf tissue. Others have shown that when BR is applied to the roots of tomato and radish plants there was an increase in petiole and hypocotyl elongation, and when applied to the base of mung bean cuttings it promoted elongation of the epicotyls (22).

The application of [³H]BR to the roots of tomato plants for 12 h led to the production of two unknown metabolites. When the plants were returned to a solution minus BR, the ACC content in these tissues decreased after 24 h, and there was a large increase in the two BR metabolites, suggesting that the plant metabolizes BR to inactive forms resulting in a decrease in ethylene production (24). The feeding of radiolabeled castasterone or brassinolide to mung bean or rice seedlings also led to an increase in polar metabolites (28).

PHYSIOLOGICAL EFFECTS OF BRASSINOSTEROIDS

Comparisons with other hormones in different bioassays

Since the discovery of brassinolide its biological activity in bioassay systems designed for auxins, gibberellins and cytokinins has been investigated. One of the main effects of brassinolide appears to be the close relationship between IAA and BR. Typically they show a synergistic relationship. Although in most cases brassinolide acts in a similar manner to auxins, gibberellins or cytokinins, bioassays based on root formation, including those on the mung bean hypocotyls, lateral bud growth in decapitated pea shoot, and cress seedling root elongation, BR and IAA act differently. In the dock leaf disc senescence bioassay BR promotes, whereas gibberellins delay senescence. Cytokinins and BR also act differently in the dwarf pea apical hook and tip expansion, pigweed betacyanin formation, and cocklebur leaf disc senescence bioassays (29).

Promotion of ethylene biosynthesis and epinasty

In etiolated mung bean hypocotyl segments BR increases ethylene biosynthesis by stimulating ACC synthase activity. BR-induced ethylene can be inhibited by AOA, Co^{2+}, fusicoccin (a fungal toxin) and the auxin transport inhibitors 2,3,5-triiodobenzoic acid and 2-(p-chlorophenoxy)-2-methylpropionic acid. In ethylene production, BR acts synergistically with active auxins and calcium, whereas it has an additive effect when used in combination with cytokinins. Light has been shown to inhibit BR-induced ethylene production while having little effect on ethylene produced in response to IAA (2). BR applications to the roots of hydroponically-grown tomato plants have been shown to promote dramatic increases in ACC, ethylene and petiole bending (Fig. 2) (23).

Fig. 2. Epinasty resulting from BR (1 µM) applied to the roots of tomato plants grown hydroponically, picture show response 24 hours following treatment.

Shoot elongation

Brassinosteroids have been shown to promote elongation of vegetative tissue in a wide variety of plants at very low concentrations. The promotive effects of BR on elongation have clearly been shown under white, green or weak red light conditions. However, little or no effects have been found in complete darkness suggesting that brassinolide action may result by overcoming the inhibitory effects of light (13). In a recent report by Wang and co-workers (27) it was shown that brassinosteroid can stimulate hypocotyl elongation in Pakchoi by increasing wall relaxation without a concomitant change in wall mechanical properties.

Root growth and development

Brassinosteroids are powerful inhibitors of root growth and development. BR and IAA effects are generally similar, and a synergism between the two is typically reported. However, in the case of root elongation they act quite differently, IAA stimulating and BR having an inhibitory effect. The possible reasons for these differences may be that BR acts independently of IAA in roots, or it acts as an antagonist of IAA. As ethylene has an inhibitory effect on root growth (20) and BR stimulates ethylene, it is possible that the inhibition of root growth is due to BR-induced ethylene production.

Plant tissue culture

24-epibrassinolide has been shown to mimic culture conditioning factors and to synergize with these factors in promoting carrot cell growth (6). However, in transformed tobacco cells brassinosteroids have been shown to significantly inhibit cell growth at concentrations as low as 10^{-8} M (5).

Antiecdysteroid effects in insects

Structurally brassinosteroids are very similar to ecdysteroids which are molting hormones of insects and other arthropods. Brassinosteroids have been shown to interfere with ecdysteroids at their site of action, and are the first true antiecdysteroids observed thus far. Since brassinosteroids are natural products they are good candidates for safer insect pest control. However, their cost must be reduced before becoming economically feasible (19).

Other biological effects

Brassinosteroids have been shown to induce changes in plasmalemma energization and transport, assimilate uptake, enhancement of xylem differentiation, enhance resistance to: chilling, disease, herbicides and salt stress, promote germination, and decrease fruit abortion and drop (9, 11).

EFFECTS OF BRASSINOSTEROIDS ON NUCLEIC ACID AND PROTEIN SYNTHESIS

When bean plants were treated with BR there was a significant increase in RNA and DNA polymerase activities and synthesis of RNA, DNA and protein (12). Putative inhibitors of RNA and protein synthesis have since been shown to interfere with BR-induced epicotyl elongation indicating that the growth effects induced by BR depend on the synthesis of nucleic acids and proteins (15). In a two dimensional gel analysis of *in vitro*-translated mRNA produced during BR-stimulated elongation of soybean epicotyls, it was found that gene expression patterns were altered by BR either plus or minus IAA, indicating that BR was having its effect alone (8). However, the possibility still exists that it could be acting with endogenous auxin. In order to take this work one step further the effects of BR on several known auxin regulated genes was evaluated. This work indicated that the molecular mechanism of BR-induced elongation is different than auxin-induced elongation in this system (8).

There is a synergistic relationship in the stimulation of ACC synthase when BR and IAA are used in combination in etiolated mung bean hypocotyl sections. Recently, a full length cDNA (pAIM-1) for IAA-induced ACC synthase was identified and characterized in this tissue (7). Using this cDNA as a probe it was shown that BR could turn on the same gene for ACC synthase.

BR can protect cereal leaf cells from heat shock and salt stress. Pretreatment with both 22S, 23S-homobrassinolide and 24-epibrassinolide activated total protein synthesis and *de novo* synthesis of different polypeptides, when wheat leaves were heat shocked by subjecting them to 40°C as well as at normal temperatures. In addition, 22S, 23S-homobrassinolide stimulated heat shock granules in the cytoplasm and increased thermotolerance of total protein synthesis under heat shock. 24-epibrassinolide protected the leaf cell ultrastructure in leaves under salt stress and also prevented nuclei and chloroplast degradation.

PRACTICAL APPLICATIONS OF BRASSINOSTEROIDS

In the early 80's USDA scientists showed that BR could increase yields of radish, lettuce, bean, pepper and potatoes. However, subsequent results under field conditions were disappointing because inconsistent results were obtained. As a result testing was phased out in the United States. More recently large scale field trials in China and Japan over a six year period have shown that 24-epibrassinolide, an alternative to brassinolide, increased the production of agronomic and horticultural crops (including wheat, corn, tobacco, watermelon and cucumber). However, once again, depending on cultural conditions, method of application and other factors, the results sometimes were striking, while other times they were marginal. Further improvements in the formulation, application method, timing, and investigation of the effects of environmental conditions and other factors need to be undertaken in order to identify the reason for these variable results (9).

ARE BRASSINOSTEROIDS A NEW CLASS OF PLANT HORMONES?

Brassinosteroids are thought by some to be a new class of plant hormones (22). However, there are gaps in our knowledge in some areas which allow for a degree of skepticism. The first line of evidence supporting the idea that brassinosteroids are a new class of plant hormones is that they are widely distributed in the plant kingdom. Second, they have an effect at extremely low concentrations, both in bioassays and whole plants. Third, they have a range of effects which are different from the other classes of plant hormones, and there are strict structural requirements for a brassinosteroid to be active in promoting a physiological response. Fourth, they can be applied to one part of the plant and transported to another where in very low amounts elicit a biological response. At the present time the actual mechanism of BR action remains unclear. Recent studies using molecular technology suggest that BR have the ability to regulate gene expression resulting in elongation (8) and ethylene production. However, before definitive conclusions on the mechanism of action can be made more work is necessary.

211

References

1. Adam, G., Marquardt, V. (1986) Brassinosteroids. Phytochem. 25, 1787-1799.
2. Arteca, R.N. (1990) Hormonal stimulation of ethylene biosynthesis. *In*: Polyamines and ethylene: Biochemistry, physiology and interactions, pp. 216-223, Flores, H.E., Arteca, R.N., Shannon, J.C., ed. American Society of Plant Physiologists, Rockville, MD.
3. Arteca, R.N., Bachman, J.M., Mandava, N.B. (1988) Effects of indole-3-acetic acid and brassinosteroid on ethylene biosynthesis in etiolated mung bean hypocotyl segments. J. Plant Physiol. 133, 430-435.
4. Arteca, R.N., Bachman, J.M., Yopp, J.H., Mandava, N.B. (1985) Relationship of steroidal structure to ethylene production by etiolated mung bean segments. Physiol. Plant. 64, 13-16.
5. Bach, T.J., Roth, P.S., Thompson, M.J. (1991) Brassinosteroids specifically inhibit growth of tobacco tumor cells. *In*: Brassinosteroids. Chemistry, bioactivity, and applications, pp. 176-188, Cutler, H.G., Tokota, T., Adam, G., ed. American Chemical Society, Washington, DC.
6. Bellincampi, D., Morpurgo, G. (1991) Stimulation of growth induced by brassinosteroid and conditioning factors in plant-cell cultures. *In*: Brassinosteroids. Chemistry, bioactivity, and applications, pp. 189-199, Cutler, H.G., Yokota, T., Adam, G., ed. American Chemical Society, Washington, DC.
7. Botella, J.R., Arteca, J.M., Schlagnhaufer, C.D., Arteca, R.N., Phillips, A.T. (1992) Identification and characterization of a full-length cDNA encoding for an auxin-induced 1-aminocyclopropane-1-carboxylate synthase from etiolated mung bean hypocotyl segments and expression of its mRNA in response to indole-3-acetic acid. Plant Mol. Biol. 20, 425-436.
8. Clouse, S.D., Zurek, D.M., McMorris, T.C., Baker, M.E. (1992) Effect of brassinolide on gene expression in elongating soybean epicotyls. Plant Physiol. 100, 1377-1383.
9. Cutler, H.G., Yokota, T., Adam, G. (1991) Brassinosteroids. Chemistry, bioactivity and applications. American Chemical Society, Washington, D.C. 375 pp.
10. Grove, M.D., Spencer, F.G., Rohwedder, W.K., Mandava, N.B., Worley, J.F. (1979) A unique plant growth promoting steroid from *Brassica napus* pollen. Nature 281, 216-217.
11. Iwahori, S., Tominaga, S., Higuchi, S. (1990) Retardation of abscission of citrus leaf and fruitlet explants by brassinolide. Plant Growth Reg. 9, 119-125.
12. Kalinich, J.F., Mandava, N.B., Todhunter, J.A. (1985) Relationship of nucleic acid metabolism to brassinolide-induced responses in beans. J. Plant Physiol. 120, 207-214.
13. Kamuro, Y., Inada, K. (1991) The effect of brassinolide on the light-induced growth inhibition in mung bean epicotyl. Plant Growth Reg. 10, 37-43.
14. Kim, S-K (1991) Natural occurrences of brassinosteroids. *In*: Brassinosteroids. Chemistry, bioactivity, and applications, pp. 26-35, Cutler, H.G., Yokota, T., Adam, G., ed. American Chemical Society, Washington, DC.
15. Mandava, N.B. (1988) Plant growth-promoting brassinosteroids. Ann. Rev. Plant Physiol. Plant Mol. Biol. 39, 23-52.
16. Marumo, S., Hattori, H., Nanoyama, Y., Munakata, K. (1968) The presence of novel plant growth regulators in leaves of *Distylium racemosum* Sieb et Zucc. Agric. Biol. Chem. 32, 528-529.
17. Milborrow, B.V., Pryce, R.J. (1973) The brassins. Nature 243, 46.
18. Mitchell, J.W., Gregory, L.E. (1972) Enhancement of overall growth, a new response to brassins. Nature 239, 254.
19. Richter, K., Koolman, J. (1991) Antiecdysteroid effects of brassinosteroids in insects. *In*: Brassinosteroids. Chemistry, bioactivity, and applications, pp. 265-279, Cutler, H.G., Tokota, T., Adam, G., ed. American Chemical Society, Washington, DC.
20. Roddick, J.G., Guan, M. (1991) Brassinosteroids and root development. *In*: Brassinosteroids. Chemistry, bioactivity, and applications, pp. 231-245, Cutler, H.G., Tokota, T., Adam, G., ed. American Chemical Society, Washington, DC.

21. Sakurai, A., Fujioka, S. (1993) The current status of physiology and biochemistry of brassinosteroids. Plant Growth Reg. 13, 147-159.
22. Sasse, J.M. (1991) Brassinosteroids - are they endogenous plant hormones? Plant Growth Reg. Soc. Amer. Quarterly 19, 1-18.
23. Schlagnhaufer, C., Arteca, R.N. (1985) Brassinosteroid-induced epinasty in tomato plants. Plant Physiol. 78, 300-303.
24. Schlagnhaufer, C.D., Arteca, R.N. (1991) The uptake and metabolism of brassinosteroid by tomato (*Lycopersicon esculentum*) plants. J. Plant Physiol. 138, 191-194.
25. Thompson, M.J., Mandava, N.B., Meudt, W.J., Lusby, W.R., Spaulding, D.W. (1981) Synthesis and biological activity of brassinolide and its 22, 23-isomer. Novel plant growth promoting steroids. Steroids 38, 567-580.
26. Thompson, M.J., Meudt, W.J., Mandava, N.B., Dutky, S.R., Lusby, W.R., Spaulding, D.W. (1982) Synthesis of brassinosteroids and relationship of structure to plant growth-promoting effects. Steroids 39, 89-105.
27. Wang, T., Cosgrove, D.J., Arteca, R.N. (1993) Brassinosteroid stimulation of hypocotyl elongation and wall relaxation in pakchoi (*Brassica chinensis* cv Lei-Choi). Plant Physiol. 101, 965-968.
28. Yokota, T., Ogino, Y., Suzuki, H., Takahashi, N., Saimoto, H., Fujioka, S., Sakurai, A. (1991) Metabolism and biosynthesis of brassinosteroids. *In*: Brassinosteroids. Chemistry, bioactivity, and applications, pp. 86-96, Cutler, H.G., Yokota, T., Adam, G., ed. American Chemical Society, Washington, DC.
29. Yopp, J.H., Mandava, N.B., Thompson, M.J., Sasse, J.M. (1981) Brassinosteroids in selected bioassays. *In*: Proc. Plant Growth Reg. Soc. Amer., pp. 110-126, St. Petersburg, FL.

D1. Auxin and Cell Elongation

Robert E. Cleland

Department of Botany, University of Washington, Seattle, Washington 98195, USA.

INTRODUCTION

One of the most dramatic and rapid hormone responses in plants is the induction by auxin of rapid cell elongation in isolated stem and coleoptile sections. The response begins within 10 minutes after addition of auxin, results in a 5-10 fold increase in the growth rate, and persists for hours or even days (18). It is hardly surprising that this may be the most studied hormonal response in plants.

How do auxins produce this response? To answer this, the process of cell enlargement must first be considered. Cell enlargement consists of two interrelated processes; osmotic uptake of water, driven by a water potential gradient across the plasma membrane, and extension of the existing wall area, driven by the turgor-generated stress within the wall. The process of cell enlargement can be described (32) by two equivalent equations:

$$dV/dt = Lp.\Delta\psi \qquad \text{Equation 1}$$

$$dV/dt = m(P\text{-}Y) \qquad \text{Equation 2}$$

where dV/dt is the rate of increase in cell volume, Lp is the hydraulic conductivity, $\Delta\psi$ is the water potential gradient across the plasma membrane, m is the wall extensibility, P is the turgor pressure and Y is the wall yield threshold (the turgor that must be exceeded for wall extension to occur). These two equations can be combined into a third equation:

$$dV/dt = \frac{m.Lp}{m+Lp} (\psi_a - \pi - Y) \qquad \text{Equation 3}$$

where ψ_a is the apoplastic water potential and π is the osmotic potential of the cell. In the absence of auxin, the growth rate is low because of a low m, low Lp, low P or high Y (or a combination of these). When auxin initiates rapid cell enlargement, it must do so by increasing m, Lp, ψ_a or P, or by decreasing π or Y. No matter what the initial effect of auxin is (e.g., gene activation, ATPase activation, or change in membrane permeability), increased cell enlargement can be initiated *only* if the ultimate effect is a change in one of these cellular growth parameters.

P. J. Davies (ed.), Plant Hormones, 214–227.

Characteristics of Auxin-induced Cell Elongation

Auxin-induced cell elongation has a number of distinct characteristics (Table 1). The first is its time course (Fig. 1). Upon addition of auxin there is a lag of at least 8 minutes before the growth rate begins to increase. Then the rate rises until a maximum is reached after 30-60 minutes. The length of the lag can be increased from 8-10 minutes by lowering the temperature or by using suboptimal auxin concentrations, but it cannot be decreased below 8 minutes by raising the temperature, using superoptimal auxin or by removing the cuticle surrounding the tissue (18). Thus we must conclude that the lag is not the time needed for auxin simply to penetrate to its site of action. This is further indicated by the fact that in some, but not all cases, the growth rate actually decreases immediately after addition of auxin (18). The lag is independent of the rate of protein synthesis; thus it is unlikely to reflect the time required to synthesize some new protein (19). The rapidity with which the maximum growth rate is achieved, and the actual rate, are dependent on the tissue and can be influenced by the past history of the tissue. For example, freshly cut maize coleoptile sections respond slowly and rather poorly to auxin, while sections "aged" in water for 3 hours are far more responsive (51). The maximum growth rate in this tissue is 8-10% per hour.

In contrast, oat coleoptiles show no such change in sensitivity to auxin, but have a maximum growth rate of only 4-6% per hour (7). Once a maximum rate is achieved, it can be maintained for up to 18 hours in oat coleoptiles, as long as auxin and absorbable solutes are both present (7). If auxin is removed the growth rate soon declines to the control level (22). In most dicot stem sections there is a decrease in the growth rate

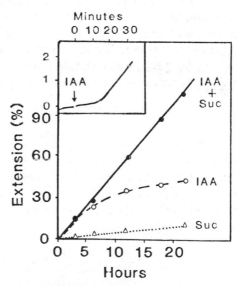

Fig. 1. Time course of auxin-induced growth of *Avena* coleoptile sections. Section incubated in 10 mM phosphate buffer, pH 6.0, ± 2% sucrose and with 10 μM IAA added at time zero. Rapid elongation is initiated after an 8-10 minute lag. The initial rate is independent of the presence or absence of absorbable solutes (sucrose), but absorbable solutes are required for continual rapid elongation.

Table 1. Characteristics of the initial growth response to auxin.

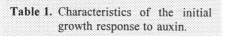

1) Minimum lag 8-10 minutes
2) Growth rate = log(IAA)
3) Requires:
   a) ATP synthesis
   b) Active ATPase
   c) Continual protein synthesis
   d) Turgor in excess of yield threshold
4) Not required:
   a) Exogenous sugars
   b) Exogenous K^+ or Ca^{++}

215

after the first maximum is reached, followed by a rise to a second, lower steady-state rate (Fig. 2). Unlike coleoptile sections, dicot stem sections rarely maintain a constant growth rate for more than 2-3 hours; thereafter the growth rate falls continuously so that by 16-20 hours growth has ceased (23).

In most tissues, the growth rate over the first 2-4 hours is proportional to the log of the external auxin concentration over a range of about 2 1/2 decades, but the concentration range varies from tissue to tissue (Fig. 3). Thus for oat coleoptiles the range is from 3 nM to about 1 μM (7), while for light-grown peas stems it is 300 nM to 50 μM (23). This initial rate of auxin-induced growth is not increased by the addition of any particular ion or sugar; auxin, alone, is sufficient to initiate this response (8). This makes it unlikely that the uptake of any specific solute is directly involved in the initiation of cell enlargement.

Auxin-induced cell enlargement is an energy-requiring process. All inhibitors of ATP synthesis (e.g. KCN, 2,4-dinitrophenol, azide) or ATPase activity (vanadate, N,N'-dicyclohexyl-carbodiimide [DCCD], diethylstilbesterol [DES]) block auxin-induced growth within minutes (18). These data indicate that the energy of ATP drives some critical step and that an ATPase must be involved. Inhibitors of protein synthesis also inhibit auxin-induced growth within minutes after they inhibit protein synthesis (16). It has been suggested that some short-lived protein is required (6), but the identity of the protein has not been determined. RNA synthesis antagonists inhibit growth

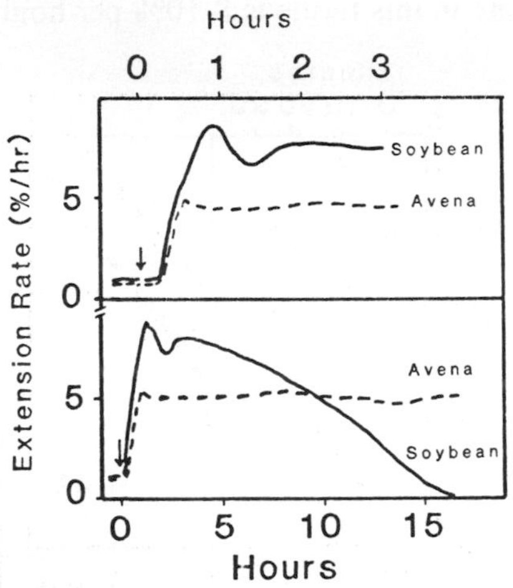

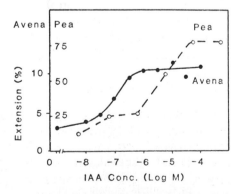

Fig. 2. Comparison of the growth kinetics for *Avena* coleoptile and soybean hypocotyl sections, incubated with IAA (10 μM) and sucrose (2%). The growth rate is enhanced after a lag of 10-12 minutes for both. The rate for *Avena* sections reaches a maximum after 30-45 minutes and then remains nearly constant for 18 hours. The rate for soybean hypocotyls increases to a first maximum after 45-60 minutes, then declines before climbing to a second maximum, after which it steadily falls over the next 16 hours as turgor falls due to dilution of the osmotic solutes during growth.

Fig. 3. A comparison of the auxin concentration curves for *Avena* coleoptile (8) and green pea stem sections (23). For both, the growth rate is proportional to the log(IAA) over a 300-fold range, but the range differs between the two tissues.

with a slightly longer lag (16).

Another requirement for auxin action is a sufficient cell turgor. Addition of osmoticum sufficient to reduce $(\psi_a\text{-}\pi\text{-}Y)$ to zero inhibits elongation by eliminating the driving force for wall extension, but it also largely blocks the ability of auxin to change any of the cellular parameters. This is shown (Fig. 4) by the fact that when turgor is restored after a period of time with auxin at reduced turgor, there is no burst of growth, as would be expected if auxin had been acting normally on the cellular parameters during the period of low turgor (43).

It must be remembered that auxin-induced growth is a tissue response; because the cells are linked together by their common cell walls, all cells must elongate or not elongate together. Individual cells, however, may differ in their response to auxin. In dicot stem sections, the outer cell layers (epidermis and collenchyma) appear to be the major targets of auxin (34), although the inner cells can respond to auxin under certain conditions. Studies on the cellular nature of auxin-induced cell elongation in dicots should focus on these outer cell layers. In coleoptiles, on the other hand, both epidermal and mesophyll cells are capable of responding to auxin (10).

In conclusion, then, we can say that the initiation of auxin-induced growth requires the continued presence of auxin, a continued supply of ATP and active ATPases, protein synthesis and turgor in excess of Y. Any mechanism to explain auxin-induced elongation must take into account these requirements, as well as the 8-10 minute lag which always occurs.

THE INITIATION OF CELL ENLARGEMENT BY AUXIN

The Cellular Parameters

Auxin must initiate cell elongation by changing one or more of the cellular growth parameters. Which one? Direct measurement of cell turgor, P, using a micro-pressure probe (13), has shown that auxin causes no increase in P in pea stem cells. On the other hand, measurement of wall extensibility (m) by any of several techniques

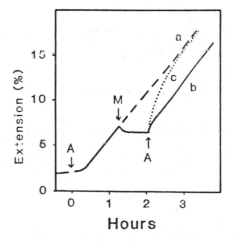

Fig. 4. Demonstration that turgor is required for auxin-induced wall loosening in *Avena* coleoptile sections (lack of stored growth). Sections were treated in water with IAA (10 μM) at A, then transferred to IAA + 0.2 M mannitol at M. Curve **a** is a projection of the growth if the section had remained in A. Upon return to A, rapid growth resumes (**b**). If auxin-induced wall loosening had occurred during the period of low turgor, the extension would have followed curve **c**, since the growth potential would have been stored up.

217

including Instron stress-strain analysis, stress-relaxation, creep, or a pressure-block technique always gives the same answer: auxin causes a large and rapid increase in wall extensibility (5, 12, 49). Any condition other than a reduction in turgor that inhibits auxin-induced elongation also blocks the auxin-induced increase in wall extensibility.

The situation with the wall yield threshold is less clear. When measured by a pressure-jump technique, the P of *Vigna* hypocotyls appears to be only slightly greater than Y, and an effect of auxin is to cause a significant decrease in Y (38). On the other hand, if the steady-state growth of *Avena* coleoptile sections is measured as a function of the solution osmotic potential, Y appears to be considerably below P and auxin does not affect Y (8). This confusion about Y reflects the fact that the actual physical basis for a yield threshold is unknown.

The following picture has emerged as to how a cell elongates (Fig. 5). The walls are extended elastically by the force of turgor until the turgor is sufficient to reduce $\Delta\psi$ to zero. The wall is kept from further extension by load-bearing bonds; by crosslinks between wall polymers or by entanglements between the polymers. The first step in cell elongation is breakage of load-bearing bonds, with a consequent rearrangement of wall polymers. This reduces tension in the wall and cell turgor, permitting additional water uptake and thus additional elastic extension. In essence, elastic extension is converted to irreversible extension, and elastic extension is then regenerated; this is a form of viscoelastic extension.

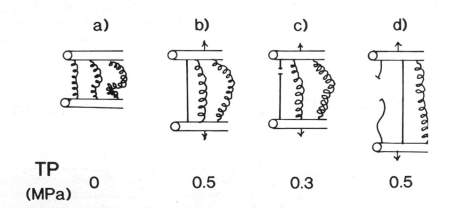

Fig. 5. Model of wall extension. In a cell at incipient plasmolysis (**a**), cellulose microfibrils are crosslinked by polymers which are not under tension or elastically extended. Water uptake results in elastic stretching of the crosslinks until the wall stress causes turgor to rise sufficiently to bring the cell to equilibrium with the apoplastic solution (**b**). The crosslinks are not under equal tension. Cleavage of the crosslink under the most tension (**c**) results in a reduction in stress in the wall and therefore turgor. This also converts the elastic extension into irreversible extension. The cell then takes up water again, resulting in additional elastic extension until water equilibrium is again achieved (**d**).

Two major questions arise from this picture. First, how does auxin bring about cleavage of load-bearing bonds? Secondly, what bonds are cleaved? Each of these questions will be discussed in turn.

The Wall-loosening Factor Concept

In order to act, auxin must first attach to a receptor. This receptor may be on the outside of the plasma membrane or at some internal site in the cell (42), but it certainly is not in the wall itself (30). And yet it is the wall which undergoes the biochemical wall loosening. This means that there must be some communication between the cell, when stimulated by auxin, and the cell wall. This must occur by means of one or more chemicals, which can be called *wall-loosening factors* (WLF, Fig. 6).

To date, only one WLF has been positively identified: protons. In 1970-71, Hager et al. (26) and Rayle and Cleland (42) independently suggested that when coleoptile or stem cells are stimulated by auxin, they excrete protons into the apoplastic solution, where the lowered pH activates wall loosening enzymes (Fig. 6). This "acid-growth theory" can be tested by means of four predictions. If it is correct, it should be possible to show: 1) auxin causes growing cells to excrete protons, 2) addition of acid to tissues should substitute for auxin and induce rapid cell enlargement, as long as the acid can penetrate into the walls, 3) neutral buffers infiltrated into the walls should prevent auxin-induced growth by preventing the decline in wall pH, and 4) any other agent which induced proton excretion should also induce rapid cell elongation. These predictions have been tested for only a few tissues, but in each case, the predictions have been confirmed qualitatively (44), although doubts have persisted about some details (31). While acid-induced wall loosening may explain the initiation of growth, it is not sufficient to explain

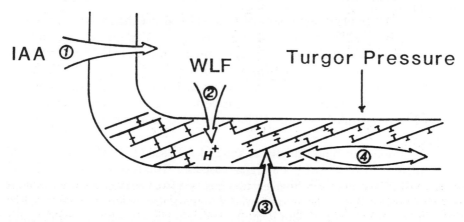

Fig. 6. The wall-loosening factor (WLF) concept and the acid-growth theory. Auxin enters the cell (1) and interacts with a receptor. A WLF is then exported to the wall (2), where it induces wall loosening (3). In acid-growth, the WLF is H+, and the resulting lowered apoplastic pH activates wall polysaccharidases which cleave load-bearing bonds in the wall (3), permitting turgor-driven wall expansion (4).

long-term auxin-induced cell elongation (11). There must be other WLFs, but whether they are excreted wall-loosening enzymes, substrates for wall synthesis or some other agent is totally unknown.

The Mechanism of Auxin-induced Proton Excretion

Auxin-induced acidification of the wall is due to excretion of protons, not organic acids. Hager et al. (26) originally suggested that auxin activates a proton ATPase located in the plasma membrane (PM). Such an enzyme exists in the PM of all plant cells, causing an electrogenic export of protons into the apoplastic solution (47). The fact that ATP is required for auxin-induced proton excretion, and that the H^+-excretion is blocked by inhibitors of ATPase activity strongly supports this idea. In addition, auxin can cause the expected hyperpolarization of the membrane potential, with a time-course which matches that of the proton excretion; i.e., both have a lag of about 6-8 minutes (2).

The auxin-induced increase in proton excretion could be due to an enhanced amount of PM ATPase, or to greater activity of existing ATPases. Hager et al. (24) have found that the amount of ATPase protein in the plasma membranes of *Zea mays* coleoptiles, as determined by antibody binding, increased rapidly, starting about 5 minutes after addition of auxin, and nearly doubled by 40 minutes. After addition of cycloheximide, the ATPase protein level in the plasma membrane decreased to the control level within an hour. The PM ATPase, then, might be the labile protein that must be synthesized following addition of auxin (16). Although it is not known whether ATPase activity showed parallel changes, these results suggest that auxin- induced increases in ATPase activity could be due to greater amounts of ATPase. This mechanism is similar to the "bucket-brigade" mechanism proposed by Ray (41).

On the other hand, auxins have been shown to enhance the activity of preexisting PM ATPases. For example, both the ATPase activity and the ATP-driven proton transport of plasma membrane vesicles isolated from tobacco leaves was enhanced up to 50% by auxin (45). Since no physiological auxin response is known to occur in most of these leaf cells, the significance of the increased activity is uncertain. A comparable response has not been demonstrated yet for ATPases isolated from an auxin-responsive tissue.

Auxin does not bind directly to the PM ATPase (15), which means that auxin must exert its effect via a signal cascade. The receptor involved may be the 20-22 kDa auxin-binding protein isolated from maize, since antibodies to this protein block the auxin-induced hyperpolarization of tobacco leaf mesophyll protoplasts (1). But what happens after auxin binds to the receptor is uncertain.

One possibility is that the critical step is a reduction in cytoplasmic pH (pH_c). The pH optimum of the PM ATPase is 6.5, but the pH_c of most cells is 7.3- 7.6, at which pH the activity of the ATPase is definitely suboptimal (Fig 7). An auxin-induced decrease in pH_c of 0.2 units has been recorded by two different methods (20, 24); such a decrease in pH should cause a large increase in ATPase activity. But how does auxin cause this cytoplasmic acidification? One possibility is that proton channels on the plasma membrane or the tonoplast are opened by auxin. A problem with this idea is that the required proton channels have not been shown to exist. Auxin does alter the characteristics of *Vicia faba* guard cell PM chloride channels (33); this would not alter the cytoplasmic pH, but may be the cause of the initial membrane depolarization that is induced by auxin.

A second possibility is that the activity of the PM ATPase is enhanced by phosphorylation of the ATPase, mediated by an auxin-sensitive protein kinase cascade. This cascade might involve the IP_3 (L-α-phosphatidylinositol 4,5- bisphosphate) cycle, with an increase in diacylglycerol or cytoplasmic Ca^{2+}, resulting in an activated protein kinase. A rapid increase in cytoplasmic calcium occurs in maize coleoptiles after addition of auxin (20, 24), and effects of auxins on IP_3 formation have been reported (17). Alternatively, an auxin-activated G- protein might be the cause of the activated protein kinase. Although there is some evidence for all parts of this mechanism, a problem is that it has not been shown that the PM ATPase is phosphorylated differently in the presence and absence of auxin.

A third possibility is that auxin activates a phospholipase, resulting in the formation of lysophospholipids (LPL) in the PM (46). LPLs have been shown to increase the activity of the PM ATPase, and alter its pH profile so that the activity at pHs more alkaline than its pH optimum are greatly increased (39).

There is no lack, then, or ways in which auxin might lead to activation of the PM ATPase. What is needed is concrete evidence which supports or rejects each of the possible steps. It must also be explained why a lag of 5-10 minutes occurs between binding of auxin to its receptor and the enhancement of ATPase activity.

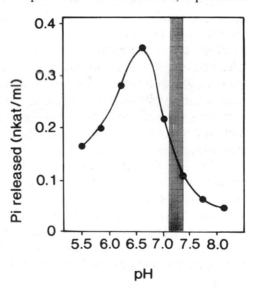

Fig. 7. Effect of pH_c on the PM ATPase activity from maize coleoptiles. Plasma membrane vesicles were purified and ATPase activity was assayed in solutions of pH between 5.5 and 8.1. Hatched region denotes the proposed pH_c change that occurs in response to auxin; the right margin is the pH in the absence of auxin, while the left margin is the pH after auxin. A decrease in pH_c of 0.2 units results in a doubling in ATPase activity.

What is needed is concrete evidence which supports or rejects each of the possible steps. It must also be explained why a lag of 5-10 minutes occurs between binding of auxin to its receptor and the enhancement of ATPase activity.

The Mechanism of Auxin-induced Wall Loosening

A result of the export of wall-loosening factors in response to auxin, whether they be H^+ or something else, is the cleavage of load-bearing cell-wall crosslinks. The identity of these crosslinks has been a matter of some controversy. The wall is a complex mixture of polymers, with at least three distinct networks coexisting; cellulose crosslinked by hemicelluloses, pectic chains crosslinked by calcium, and structural proteins (4). It is unlikely that calcium crosslinks between pectic chains are major load-bearing crosslinks, as removal of up to 95% of soybean hypocotyl cell wall calcium with Quin-2 (2-[(2-bis-[carboxymethyl]-amino-5-methylphenoxy)methyl]-6-methoxy-8-bis[carboxymethyl]aminoquinoline) resulted in little loosening of these cell walls (52). Likewise, treatment of isolated walls with proteases failed to cause measurable wall loosening (43). Most of the attention has focused on the hemicellulose components of the wall as the site of auxin- induced wall loosening.

In dicot walls, one of the major hemicelluloses, the xyloglucans (XG), seem to undergo bond breakage in response to auxin. For example, their molecular weight has been shown to decrease after auxin treatment of azuki bean epicotyls (27), and XG is lost from pea epicotyl walls upon auxin treatment (50). Treatment of azuki bean epicotyl sections with a lectin that binds the fucose of XG inhibited auxin-induced wall loosening (27). These data are consistent with xyloglucans as load-bearing crosslinks that are cleaved during auxin-induced growth.

The enzyme which may be responsible for degradation of XG is xyloglucan endotransglycosylase (21), which reduces XG chainlength by transferring large blocks of XG to small xyloglucan acceptors. This enzyme will reduce the molecular weight of XG, and is localized in the growing region of pea epicotyls. The ability of a small oligomer of XG (XG9) to promote auxin-induced elongation in pea stem sections would be compatible with such an idea (35). However, purified xyloglucan endoglycoslase failed to cause any loosening of isolated cucumber walls (37), raising real doubts as to whether this enzyme can be a wall-loosening enzyme.

The cell walls of grass coleoptiles are rich in (1-3,1-4)-β-glucans rather than in xyloglucans (4). During growth of coleoptiles this mixed-linked glucan is extensively degraded. Antibodies to either the mixed-linked glucan (28), or to a β-glucanase that degrades this polymer (29) block auxin-induced growth of maize coleoptiles. The possibility exists, therefore, that these glucans are load- bearing crosslinks, and that their degradation is a mechanism of wall loosening in coleoptiles.

enzymes involved in acid-mediated wall loosening[1]. They do not appear to be glycosylases, as no sugars or small oligomers were released during wall loosening. Thus they are unlikely to be the proteins involved in XG breakdown. Cosgrove and Li have shown that it is not the level of this protein, which they call "expansin", that controls the rate of cell elongation, but rather the capacity of the walls to undergo loosening in the presence of this enzyme and an acidic wall pH. The importance of this capacity was pointed out earlier by Cleland (1976).

In conclusion, breakdown of xyloglucans in dicot walls and mixed-linked glucans in coleoptile walls occurs during auxin-induced growth, and inhibition of these degradations by lectins or antibodies prevents the growth. But the only proteins yet isolated from walls which are capable of causing walls to become loosened in a manner consistent with normal extension do not seem to cause degradation of either of these polymers. Thus the identity of the load-bearing crosslinks which are cleaved during auxin-induced wall loosening remains a mystery.

THE MAINTENANCE OF AUXIN-INDUCED GROWTH

An analysis of long-term auxin-induced growth shows that it is more complex than the initial growth response. While the primary effect of auxin during the initial growth response may be to cause wall acidification, during the prolonged growth other unknown wall-loosening factors appear to be involved (11). In addition, at least two additional processes are required for prolonged growth. The first is osmoregulation. As the cells start to enlarge and take up water, the dilution of the osmotic solutes will reduce the effective turgor unless additional solutes are taken up or are manufactured in the cells. Thus when *Avena* coleoptile sections are incubated with auxin but without absorbable solutes, the growth rate declines after the first hour in parallel with the decline in osmotic concentration, while in the presence of absorbable solutes such as sucrose or KCl, both the growth rate and the osmotic concentration remain nearly constant for hours (48). While the enhanced rate of uptake of solutes into *Avena* coleoptile sections in the presence of auxin is actually in response to the cell enlargement rather than to auxin itself (48), in intact stems auxin may play a direct role in facilitating movement of solutes through the phloem to the growing zone (40).

A second important process is the maintenance of the ability of cell walls to undergo auxin-induced wall loosening. When cell walls are isolated by freezing- thawing tissues, then placed under tension and given acidic

---

[1] Editors note: Cosgrove and coworkers have claimed that expansin has no glycanase activity, but causes paper to fall apart under acid conditions, presumably by breaking the hydrogen bonds between the cellulose microfibrils. If so, this would provide a possible mechanism for the increase in wall extensibility induced by acidification of the cell wall. However, as these results have not been fully published at the time of going to press, it is impossible to critically examine their significance.

in intact stems auxin may play a direct role in facilitating movement of solutes through the phloem to the growing zone (40).

A second important process is the maintenance of the ability of cell walls to undergo auxin-induced wall loosening. When cell walls are isolated by freezing/thawing tissues, then placed under tension and given acidic solutions, they undergo acid-mediated wall extension (42, 43). The capacity for this *in vitro* extension depends on the past history of the tissue (Fig. 8). If coleoptile or hypocotyl sections are incubated in water, this capacity is slowly lost (9). On the other hand, if sections are given auxin, this capacity increases for at least 6-8 hours (9). It would appear that one of the effects of auxin is to regenerate the capacity of walls to undergo this acid-mediated extension. This might involve renewal of the wall-loosening enzymes. Alternatively, it might involve synthesis of new cell wall polymers that are susceptible to the wall-loosening enzymes. Synthesis of new wall cannot lead directly to wall loosening, but if wall components are intercalated in after cleavage of load-bearing bonds, it may facilitate further cleavage. The orientation of the wall polysaccharides is another important factor. In order for walls to extend the cellulose must be in a transverse orientation. In maize coleoptiles, auxin causes the newly synthesized cellulose microfibrils in the outer epidermal wall, which are longitudinal in the absence of auxin, to assume a transverse orientation (3).

CONCLUSIONS

Auxin-induced elongation is initiated when auxin binds to a receptor, probably located on the outside of the plasma membrane, and sets into motion a cascade of events leading to enhanced proton excretion by the PM ATPase.

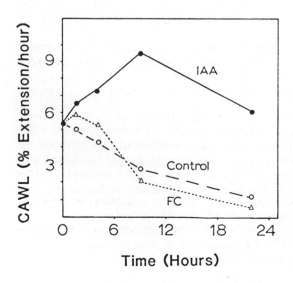

Fig. 8. The effect of auxin on the capacity of *Avena* coleoptile cell walls to undergo loosening in response to acid (CAWL) depends on the past history of the tissue. Sections were incubated 0-20 hours in phosphate buffer without (control) or with 10 μM IAA or fusicoccin (FC). The sections were then frozen-thawed, and the rate of extension in response to 20 g tension at pH 3.0 was determined, one hour after addition of acid. Auxin, but not fusicoccin has increased the capacity (CAWL).

This may involve an increase in amount of ATPase in the PM, or an activation of pre-existing ATPase by a lowering of cytoplasmic pH or by phosphorylation of the enzyme. During the first critical few minutes there are changes in a number of components of possible signal cascades, such as IP_3 and cytoplasmic Ca^{2+}.

The resulting acidification of the apoplast activates proteins which cause cleavage of load-bearing wall crosslinks. Two proteins isolated from cell walls appear to loosen walls without the release of wall fragments, but other evidence suggests that degradation and solubilization of xyloglucans in dicot walls or mixed-linked glucans in coleoptiles are an integral part of wall loosening.

After the initiation of elongation, prolonged auxin-induced growth probably involves release of additional but yet unknown wall loosening factors. In addition, osmoregulation is required to maintain sufficient turgor, and regeneration of the ability of walls to undergo loosening must also occur. If these processes are all integrated, elongation of stem or coleoptile cells in response to auxin can persist for considerable periods of time.

Acknowledgements

I wish to thank the NASA Space Biology Program and the Division of Energy Biosciences, Dept. of Energy, for support of research described here.

References

1. Barbier-Brygoo, H., Ephritikhine, G., Klämbt, D., Ghislain, M., Guern, J. (1989) Functional evidence for an auxin receptor at the plasmalemma of tobacco mesophyll protoplasts. Proc. Natl. Acad. Sci. U.S. 86, 891-895.
2. Bates, G.W., Goldsmith, M.H.M. (1983) Rapid response of the plasmamembrane potential in oat coleoptiles to auxin and other weak acids. Planta 159, 231- 237.
3. Bergfeld, R., Speth, V., Schopfer, P. (1988) Reorientation of microfibrils and microtubules at the outer epidermal wall of maize coleoptiles during auxin-mediated growth. Bot. Acta 101, 31-41.
4. Carpita, N.C., Gibeaut, D.M. (1993) Structural models of primary cell walls in flowering plants: consistency of molecular structure with the physical properties of walls during growth. Plant J. 3, 1-30.
5. Cleland, R.E. (1971) Cell wall extension. Annu. Rev. Plant Physiol. 22, 197- 222.
6. Cleland, R.E. (1971) Instability of the growth-limiting proteins of the *Avena* coleoptile and their pool size in relation to auxin. Planta 99, 1-11.
7. Cleland, R.E. (1972) The dosage response curve for auxin-induced cell elongation: a reevaluation. Planta 104, 1-9.
8. Cleland, R.E. (1977) The control of cell enlargement. In: Integration of activity in the higher plant, pp. 101-115, Jenning, D.H. ed. Cambridge Press, Cambridge.
9. Cleland, R.E. (1983) The capacity for acid-induced wall loosening as a factor in the control of *Avena* coleoptile cell elongation. J. Exp. Bot. 34, 676-680. 10. Cleland, R.E. (1991) The outer epidermis of *Avena* and maize coleoptiles is not a unique target for auxin in elongation growth. Planta 186, 75-80.
11. Cleland, R.E. (1992) Auxin-induced growth of *Avena* coleoptiles involves two mechanisms with different pH optima. Plant Physiol. 99, 1556-1561.
12. Cosgrove, D.J. (1993) Wall extensibility: its nature, measurement and relationship to plant cell growth. New Phytol. 124, 1-23.

13. Cosgrove, D.J., Cleland, R.E. (1983) Osmotic properties of pea internodes in relation to growth and auxin action. Plant Physiol. 72, 332-338.
14. Cosgrove, D.J., Li, Z-C. (1993) Role of expansin in cell enlargement of oat coleoptiles. Plant Physiol. 103, 1321-1328.
15. Cross, J.W., Briggs, W.R., Dohrmann, U.C., Ray, P.M. (1978) Auxin receptors of maize coleoptile membranes do not have ATPase activity. Plant Physiol. 61, 581-584.
16. Edelmann, H., Schopfer, P. (1989) Role of protein and RNA synthesis in the initiation of auxin-mediated growth in coleoptiles of *Zea mays* L. Planta 179, 475-485.
17. Ettlinger, C., Lehle, L. (1988) Auxin induces rapid changes in phosphatidyl inositol metabolites. Nature 33, 176-178.
18. Evans, M.L. (1985) The action of auxin on plant cell elongation. Critical Rev. Plant Sci. 2, 317-365.
19. Evans, M.L., Ray, P.M. (1969) Timing of the auxin response in coleoptiles and its implications regarding auxin action. J. Gen. Physiol. 53, 1-20.
20. Felle, H. (1988) Auxin causes oscillations of cytosolic free calcium and pH in *Zea mays* coleoptiles. Planta 174, 495-499.
21. Fry, S.C., Smith, R.C., Renwick, K.F., Martin, D.J., Hodge, S.K., Matthews, K.J. (1992) Xyloglucan endotransglycosylase, a new wall-loosening enzyme activity from plants. Biochem. J. 282, 281-828.
22. Fuente, R.K. dela, Leopold, A.C. (1970) Time course of auxin stimulations of growth. Plant Physiol. 46, 186-189.
23. Galston, A.W., Baker, R.S. (1951) Studies on the physiology of light action. IV. Light enhancement of auxin-induced growth in green peas. Plant Physiol. 26, 311-317.
24. Gehring, C.A., Irving, H.R., Parish, R.W. (1990) Effects of auxin and abscisic acid on cytosolic calcium and pH in plant cells. Proc. Natl. Acad. Sci. U.S. 87, 9645-9649.
24. Hager, A., Debus, G, Edel, H-G., Stransky, H., Serrano, R. (1991) Auxin induces exocytosis and the rapid synthesis of a high-turnover pool of plasma-membrane H$^+$-ATPase. Planta 185, 527-537.
26. Hager, A., Menzel, H., Krauss, A. (1971) Versuche und Hypothese zur Primar wirkung des Auxins beim Streckungswachstum. Planta 100, 47-75.
27. Hoson, T., Masuda, Y. (1991) Inhibition of auxin-induced elongation and xyloglucan breakdown in azuki bean epicotyl segments by fucose-binding lectins. Physiol. Plant. 82, 41-47.
28. Hoson, T., Nevins, D.J. (1989) β-D-Glucan antibodies inhibit auxin-induced cell elongation and changes in the cell wall of *Zea* coleoptile segments. Plant Physiol. 90, 1353-1358.
29. Inouhe, M., Nevins, D.J. (1991) Inhibition of auxin-induced cell elongation of maize coleoptiles by antibodies specific for cell wall glucanases. Plant Physiol. 96, 426-431.
30. Jones, A.M. (1990) Do we have the auxin receptor yet? Physiol. Plant. 80, 154-158.
31. Kutschera, U., Schopfer, P. (1985) Evidence against the acid growth theory of auxin action. Planta 163, 483-493.
32. Lockhart, J.A. (1965) An analysis of irreversible plant cell elongation. J. Theor. Biol. 8, 264-275.
33. Marten, I., Lohse, G., Hedrich, R. (1991) Plant growth hormones control voltage-dependent activity of anion channels in plasma membrane of guard cells. Nature 353, 758-762.
34. Masuda, Y., Yamamoto, R. (1972) Control of auxin-induced stem elongation by the epidermis. Physiol. Plant. 27, 109-115.
35. McDougall, G.J., Fry, S.C. (1990) Xyloglucan oligosaccharides promote growth and activate cellulase: evidence for a role of cellulase in cell expansion. Plant Physiol. 93, 1042-1048.
36. McQueen-Mason, S.J., Durachko, D.M., Cosgrove, D.J. (1992) Endogenous proteins that induce cell wall expansion in plants. Plant Cell 4, 1425-1433.

37. McQueen-Mason, S.J., Fry, S.C., Durachko, D.M., Cosgrove, D.J. (1993) The relationship between xyloglucan endotransglycosylase and in-vitro cell wall extension in cucumber hypocotyls. Planta 190, 327-331.

38. Okamoto, H., Lue, Q., Nakahori, K, Katou, K. (1989) A pressure-jump method as a new tool in growth physiology for monitoring physiological wall extensibility and effective turgor. Plant & Cell Physiol. 30, 979-985.

39. Palmgren, M.G., Sommarin, M., Ulvskov, P., Jorgensen, P.L. (1988) Modulation of plasma membrane H^+-ATPase from oat roots by lysophosphotidylcholine, free fatty acids and phospholipase A_2. Physiol. Plant. 74, 11-19.

40. Patrick, J.W. (1982) Hormone control of assimilate transport. In: Plant growth substances 1982, pp. 669-678, Wareing, P.F., ed. Academic Press, New York.

41. Ray, P.M. (1977) Auxin-binding sites of maize coleoptiles are localized on membranes of the endoplasmic reticulum. Plant Physiol. 59, 594-599.

42. Rayle, D.L., Cleland, R.E. (1970) Enhancement of wall loosening and elongation by acid solutions. Plant Physiol. 46, 250-253.

43. Rayle, D.L., Cleland, R.E. (1972) The in-vitro acid-growth response: relation to in-vivo growth responses and auxin action. Planta 104, 282-296.

44. Rayle, D.L., Cleland, R.E. (1992) The acid growth theory of auxin-induced cell elongation is alive and well. Plant Physiol. 99, 1271-1274.

45. Santoni, V., Vansuyt, G., Rossignol, M. (1990) Differential auxin sensitivity of proton translocation by plasma membrane H^+-ATPase from tobacco leaves. Plant Sci. 68, 33-38.

46. Scherer, G.F.E., Andre, B. (1989) A rapid response to a plant hormone: auxin stimulates phospholipase A_2 in vivo and in vitro. Biochem. Biophys. Res. Comm. 163, 111-117.

47. Serrano, R. (1989) Structure and function of plasma membrane ATPase. Annu. Rev. Plant Physiol. Plant Mol. Biol. 40, 61-94.

48. Stevenson, T.T., Cleland, R.E. (1981) Osmoregulation in the *Avena* coleoptile in relation to growth. Plant Physiol. 67, 749-753.

49. Taiz, L. (1984) Plant cell expansion: regulation of cell wall mechanical properties. Annu. Rev. Plant Physiol. 35, 585-657.

50. Talbott, L.D., Ray, P.M. (1992) Changes in molecular size of previously deposited and newly synthesized pea cell wall matrix polysaccharides. Plant Physiol. 98, 369-379.

51. Vesper, M.J., Evans, M.L. (1978) Time-dependent changes in the auxin sensitivity of coleoptile segments. Apparent sensory adaptation. Plant Physiol. 61, 204-208.

52. Virk, S.S., Cleland, R.E. (1988) Calcium and the mechanical properties of soybean hypocotyl cell walls: possible role of calcium and protons in wall loosening. Planta 176, 60-67.

D2. The Control of Gene Expression by Auxin

Gretchen Hagen

Biochemistry Department, University of Missouri, Columbia, MO 65211, USA.

INTRODUCTION

Auxin has been shown to be involved in a variety of diverse plant growth and developmental responses, including cell division, cell elongation and cell differentiation. The mechanism(s) by which auxin affects such diverse processes is unknown, but the suggestion that auxin-mediated changes in growth are accompanied by alterations in nucleic acid metabolism was made 40 years ago (37). In the ensuing years, the relationship between auxin action and gene expression has been pursued in a number of laboratories. Major advances in identifying specific, auxin-induced changes in gene expression have come from the application of high-resolution biochemical techniques, and the use of recombinant DNA technologies. Taken together, the results of these molecular studies clearly indicate that when auxins are applied to intact plants, excised plant organs and tissue culture cells from a variety of monocot and dicot species, rapid and specific changes in gene expression can be detected. As will be discussed below, a number of auxin-responsive sequences have been characterized. They represent potentially valuable probes to define macromolecules involved in auxin-regulated growth processes and to dissect the auxin signal transduction pathway.

Following a brief review of some of the plant systems that have been used and molecular approaches that have been taken to study auxin effects on gene expression, this chapter will summarize selected information on auxin-responsive sequences that have been characterized. The reader is directed to a number of reviews on auxin-regulated gene expression, which provide additional information, historical perspectives and an extensive reference list (13, 16, 21, 22, 23, 44).

PLANT SYSTEMS

The effects of auxin on cell division responses have been examined in several, different plant systems. In dicot seedlings, such as in three-day old etiolated soybean seedlings, auxin application at high doses (most frequently, the synthetic auxin 2,4-dichlorophenoxyacetic acid or 2,4-D at 1-2mM) causes the cessation of cell division in the apical meristem and cell elongation in the elongation zone directly below the apical hook (21). In general, at these

P. J. Davies (ed.), Plant Hormones, 228–245.

levels of applied auxin, normal growth of intact plants is inhibited. Within 12-24 hours after auxin application, the basal or mature region of the hypocotyl (a normally quiescent region below the elongation zone) begins to enlarge radially due to massive cell proliferation. Early studies demonstrated that prior to or concomitant with these morphological changes, there are dramatic increases in DNA, RNA and protein content. Further molecular characterization has shown that changes in gene expression include increases in RNA polymerase I (the enzyme involved in the transcription of ribosomal RNA genes) amount and activity (13), ribosomal RNA levels (21), messenger RNA (mRNA) levels for ribosomal proteins (10, 11) and polyribosome pools (21).

Two additional systems that have been used to study auxin effects on cell division are tobacco cell suspension cultures (48, 49) and cultured tobacco mesophyll protoplasts (38). Cell division in both systems is auxin-dependent. In suspension cultures, division can be detected within 10-12 hours after auxin addition to auxin-deprived cultures; in cultured mesophyll protoplasts, division can be observed after 24 hours. As will be discussed below, rapid changes in gene expression have been detected in auxin-treated tobacco cells and protoplasts, prior to the cell division response.

Perhaps the best characterized auxin-regulated growth response is that of cell elongation. This may be due in part to the fact that, in many systems, elongation growth occurs rapidly in response to auxin and this growth can be easily measured. In excised, incubated monocot and dicot organs that are capable of elongation, such as maize coleoptile, pea epicotyl or soybean elongating hypocotyl, the growth response to auxin addition can be observed after a short lag (10-15 minutes; 21, 23). Results of early studies indicated that sustained elongation in these systems requires continued RNA and protein synthesis (21), suggesting that alterations in gene expression are important in this growth response. Excised, elongating organ systems offer many technical advantages, and have been used extensively in molecular studies of auxin action.

MOLECULAR APPROACHES

Early attempts to detect specific auxin-induced changes in gene expression involved the resolution of polypeptides that were newly synthesized following auxin application (13). For these studies, excised organs were incubated in solutions containing radioactive amino acid precursors, in the presence or absence of auxin. When *in vivo* labeled polypeptides were isolated and separated using one-dimensional or high-resolution, two-dimensional (2D) polyacrylamide gel electrophoresis, a limited number of auxin-induced changes were observed (13, 57). This technical approach has a major disadvantage in that the analysis is dependent upon the time that is required to achieve sufficient radiospecific activity of the polypeptides. In some cases, labeling times of up to several hours were required (57), which precluded an analysis of rapid auxin-induced changes in protein profiles.

An alternative method that has been used to study changes in protein profiles involves *in vitro* translation of purified mRNA in cell-free extracts (such as wheat germ extracts or rabbit reticulocyte lysates). With this technique, it is possible to isolate mRNA from untreated organs or organs that have received a brief exposure to auxin, translate the mRNA *in vitro* in the presence of a labeled amino acid precursor and resolve hundreds of labeled translation products on 2D gels. Early studies of *in vitro* translation products revealed that up to 40 translation products undergo either an upward or downward shift in relative levels over a period of 5 hours following auxin treatment of intact soybean seedlings (4). An analysis of the patterns of *in vitro* translation products on 2D gels revealed that a similar, specific set of 10-12 products was induced by auxin treatment of either elongating sections or basal sections of the soybean hypocotyl. Most of the translation products showed no change between auxin-treated and untreated samples (58, 59). Based on a comparison of 2D gel patterns, the same set of products was also induced in intact, auxin-sprayed seedlings (Fig. 1). These data were of particular interest, as they suggested that although there are differences in growth potential (elongation verses division) within different regions of the hypocotyl, there may be common, primary gene targets in auxin action (59).

Some of the changes in *in vitro* translation products occurred very rapidly (within minutes) after auxin application. In soybean, one translation product was induced in elongating sections within 15 minutes following auxin treatment, and the remaining products were induced within 30 minutes (58). Rapid, auxin-induced changes in translation products have also been detected within 15-20 minutes of treatment of excised pea epicotyl sections (46), and within 10 minutes of treatment of elongating maize coleoptile sections (60).

The rapid modulation of specific *in vitro* translation products following auxin treatment could result from the regulation of gene expression at several levels. For example, a change in mRNA levels for specific polypeptides could be controlled at the level of transcription, mRNA processing and/or mRNA stability. Alternatively, auxin treatment could affect the efficiency of *in vitro* translation of specific mRNAs. A molecular strategy that has been taken to define the level(s) at which auxin regulates gene expression has focused on the isolation and characterization of mRNAs and genes that are specifically modulated by auxin. This has been accomplished using a wide range of molecular and recombinant DNA techniques (for specific details on methodologies, see 36). The general approach has been to construct complementary DNA (cDNA) libraries and to select auxin-responsive sequences by differential hybridization screening techniques. This approach has resulted in the identification of numerous auxin-responsive cDNA clones.

Characterization of the different auxin-responsive cDNA clones has revealed that they can be grouped into 2 classes, based on the kinetics of induction. One class represents sequences that are induced after one hour (or

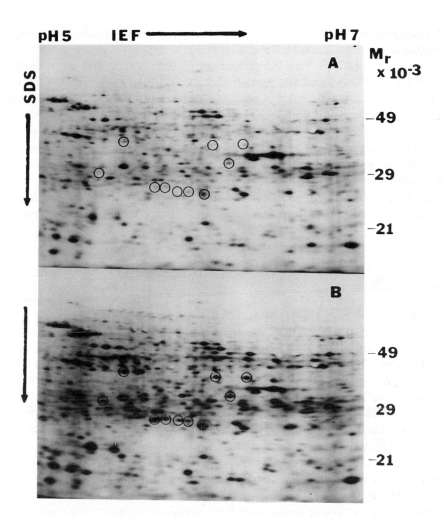

Fig. 1. Two-dimensional polyacrylamide gel analysis of *in vitro* translation products from poly(A)+RNA isolated from intact soybean hypocotyl of seedlings that were untreated (A) or (B) treated with 2,4-D for 2 hr. Several of the *in vitro* translation products that are induced with auxin treatment are indicated.

more) of auxin exposure ("long term" responses). The other class of clones represents sequences that are modulated within minutes after auxin exposure (short term responses). Sequences exhibiting the "long term" response to auxin have been isolated from *Arabidopsis* peduncles (2), strawberry receptacles (35) and mung bean hypocotyls (55, 56). The auxin-responsive cDNA clones that have been characterized most extensively are those representing rapidly-induced mRNAs. To date, these include sequences isolated from soybean (18, 30, 50), pea (45) and tobacco (38, 40, 48). These clones are potentially important molecular probes to study auxin action at the level of gene expression, as well as to study auxin action on growth. The rapid induction kinetics, for example, suggest that the genes encoding these mRNAs may be primary targets for auxin action. Thus, elucidation of sequence elements within the promoters of these genes that are required for auxin inducibility, and the protein factors that interact with these elements, have become primary goals of current studies. Additionally, the polypeptides encoded by the auxin-inducible mRNAs may represent critical regulatory or structural macromolecules involved in auxin-regulated growth. In a broader sense, the information gained from studying the rapidly induced sequences may be of value in linking the components of the auxin signal transduction pathway.

The following section will focus primarily on summarizing the information obtained from studies of the rapidly-induced, auxin-responsive sequences.

CHARACTERIZATION OF AUXIN-RESPONSIVE SEQUENCES

RNA Studies

Three groups of rapidly induced, auxin-responsive mRNAs from soybean have been described. Each group was selected using different isolation strategies. The cDNA clones pJCW1 and pJCW2 were isolated from mRNAs of elongating hypocotyl sections of untreated, intact seedlings (50, 51). The pGH series (pGH1, pGH2/4, pGH3) of cDNA clones were selected from 2,4-D treated, intact seedling mRNAs (18). The SAURs (Small Auxin Up-regulated RNAs) were isolated from a library of cDNA clones made from elongating hypocotyl sections that were incubated in the presence of auxin (30). In pea, 2 cDNA clones (pIAA4/5, pIAA6) from auxin-treated, third internode epicotyl sections have been described (45). Several auxin-responsive cDNA clones have been characterized from tobacco tissue culture cells. These cells were first depleted of auxin and subsequently treated with 2,4-D (48). Recently, one of these clones (pCNT103) has been characterized further (47). From auxin-treated tobacco mesophyll protoplasts, several cDNA clones (par A, 38; par B, 40; par C, 41) have been isolated. The cDNA clones from all plant systems have been used as probes in RNA blot hybridization analyses to study the induction/accumulation processes in response to auxin application.

For many of these sequences, the mRNA levels in intact plants (Fig. 2) or organ sections (Fig. 3A) not treated with auxin range from undetectable to moderately high. When organ sections are excised and incubated in the absence of auxin for up to 24 hours, the mRNA levels generally decline, often to undetectable levels. Addition of auxin at concentrations ranging from 10^{-8} to 10^{-5}M causes a rapid (within 10-30 minutes) increase in mRNA levels. In most systems, the mRNA levels increase over a period of hours, with maximum induction ranging from 2-50 fold over untreated controls;

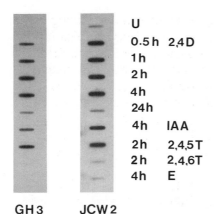

U		
0.5h	2,4D	
1h		
2h		
4h		
24h		
4h	IAA	
2h	2,4,5T	
2h	2,4,6T	
4h	E	

GH3   JCW2

Fig. 2. RNA blot analysis of auxin-responsive clones. Three-day old etiolated soybean seedlings were untreated (U) or sprayed once with 2,4-dichlorophenoxyacetic acid (2,4-D), indole 3-acetic acid (IAA), 2,4,5-trichlorophenoxyacetic acid (2,4,5-T), 2,4,6-trichlorophenoxyacetic acid (2,4,6-T) or Ethephon (E, an ethylene generating compound) for the times indicated, and RNA was extracted. Two micrograms of poly(A)+ mRNA were "slot blotted" and hybridized to labeled cDNA plasmids GH3(18) and JCW2(47). For details, see (18). The cDNA clone JCW2 was a gift of J.C. Walker and J.L. Key.

mRNA levels generally decline by 24 hours of incubation (Fig. 2). A notable exception to the rapid kinetics observed with most of these sequences is the very rapid accumulation kinetics of the SAURs (30). The levels of these mRNAs increase within 2-5 minutes after auxin treatment (Fig. 3B), with the accumulation rate reaching half-maximal by 10 minutes and steady state by 30 minutes. For most sequences, a linear increase in mRNA levels is observed over several orders of log increase in auxin concentration ranging from 10-7 to 10-3M. In some cases, saturation of the induction response is not observed even at millimolar concentrations (Fig. 3C; 30).

The rapid increase in mRNA levels could be controlled at the level of gene transcription and/or by post-transcriptional events such as mRNA processing, transport and stability. Using *in vitro* nuclear "run-on" transcription assays, it has been shown that the auxin-induced increase in mRNA levels of 6 soybean sequences (17, 33) and 6 tobacco sequences (48) is, at least partially, the result of increased transcription rates on the corresponding genes. The transcriptional response to auxin is rapid; for the soybean GH3 sequence, transcription rates increase within 5 minutes of auxin exposure (Fig. 4A; 17). Interestingly, increased transcription rates of the SAURs are observed only after 5-10 minutes of auxin treatment (33), which contrasts with the observed steady state accumulation kinetics of 2-5 minutes. This observation has led to the suggestion that auxin may affect SAUR mRNA stability, as well as transcription.

In soybean, transcriptional induction in response to auxin has been shown to occur in the presence of protein synthesis inhibitors such as cycloheximide (9, 17). This result suggests that the transcriptional induction is a direct response to auxin, in that protein synthesis is not required to induce transcription on auxin-responsive genes. Additionally, protein synthesis inhibitors do not affect auxin-induced mRNA accumulation of the pea (45) and tobacco (48) sequences. In the presence of cycloheximide alone, however, mRNA levels for the 2 pea (45), several tobacco (48) and the SAUR sequences (9) are elevated. Using *in vitro* nuclear "run-on"

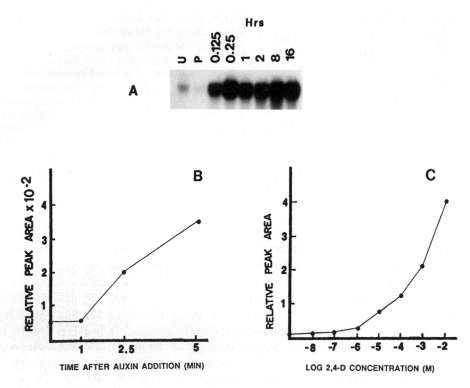

Fig. 3. Characterization of the auxin-responsive SAUR 15 sequence in soybean elongating hypocotyl sections (EHS). A). Autoradiogram of SAUR 15 mRNA accumulation. EHS were excised (U), preincubated (P, 4h) in the absence of auxin, and incubated in 2,4-D for the times indicated. Poly(A)+ mRNA was isolated, separated in gels, blotted and hybridized to labeled cDNA clone SAUR 15. B). Short term kinetics of RNA accumulation in response to added auxin. EHS were preincubated and incubated in 2,4-D for the times indicated, mRNA was isolated, blotted and hybridized as in A. Autoradiograms were scanned by a densitometer and relative areas of each peak were determined and plotted. C). 2,4-D dose response curve. EHS were preincubated and incubated for 2h at the 2,4-D concentration indicated. mRNA was isolated, blotted and hybridized as in A. Autoradiograms were scanned by a densitometer, relative areas determined and plotted. Data from (29).

transcription assays, it has been shown that, unlike auxin, cycloheximide does not activate transcription on SAUR genes. These results suggest that in the presence of protein synthesis inhibitors, SAUR mRNA levels increase due to a stabilization of SAUR mRNAs (9).

Studies designed to determine whether the observed mRNA accumulation or transcriptional induction is specific to added auxins have indicated that, with several exceptions, the response is specific to both naturally occurring and synthetic auxins (Figs. 2 & 4B). The magnitude of the response differs, however, depending on the auxin and the plant system being used. Nonauxin analogues, other plant growth regulators and environmental stresses such as heat shock, cold shock and anaerobiosis generally do not induce mRNA levels. Clones that fail to exhibit strict auxin specificity are the soybean

GH2/4 (17), several of the tobacco CNT clones (47, 48) and the tobacco parA clone (39). The GH2/4 mRNA has been shown to be induced by several nonauxin analogues and some heavy metals (17, 20). The tobacco pCNT sequences are most strongly induced by auxins, but are also induced by salicylic acid at high concentrations and slightly induced by some heavy metals (47). The tobacco parA sequence is induced by cadmium (39). As will be discussed below, these sequences appear to belong to a related family of polypeptides, based on deduced amino acid sequence comparisons.

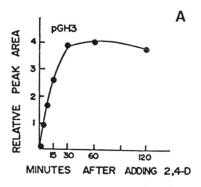

Tissue and organ expression patterns of the auxin-responsive mRNAs have been analyzed, and they show some interesting differences. RNA blot analysis has revealed that some of the sequences (pJCW1, pJCW2 and SAURs from soybean and pIAA4/5 and pIAA6 from pea) are primarily expressed in organ regions undergoing elongation growth (30, 45, 50). In contrast, pGH1 and pGH2/4 show little organ-specific expression, while pGH3 is difficult to detect in any organ unless treated with auxin (18). The techniques of tissue printing (32) and *in situ* hybridization (12) have further localized the expression of SAURs to the epidermal and cortical cells, while GH3 mRNAs are expressed at low levels in soybean roots, and transiently expressed in soybean floral organs. In these organs, GH3 transcripts

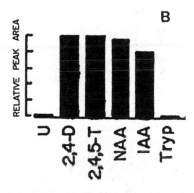

Fig. 4. Characterization of the expression of GH3 using *in vitro* nuclear run-on transcription assays in isolated soybean plumule nuclei. A. Kinetics of induction by 2,4-D (10-4M). B. Specificity of induction. Plumules were untreated (U) or treated with 2,4-D, 2,4,5-T, α-naphthaleneacetic acid (NAA), IAA or tryptophan (Tryp). For details, see (17).

were found predominantly associated with the vascular system. A striking difference between the SAURs and GH3 is that when intact soybean seedlings or excised organs are treated with auxin, induction of SAUR transcripts remains confined to cortical and epidermal cells of elongation regions, while GH3 transcripts are induced in virtually all tissues and organs examined.

Results of the RNA studies clearly demonstrate that auxin can rapidly and specifically modulate mRNA levels. The observation that the induction of transcription by auxin is rapid, specific to auxin and direct (not dependent on protein synthesis) suggests that these transcriptional events may be close to the primary site of auxin action. The studies of tissue localization of

SAUR and GH3 transcripts (12) indicate that the auxin response pathway(s) may be complex. These studies showed that most, if not all, tissues within a soybean organ (both external and internal tissues) are capable of responding rapidly to auxin at the gene expression level. Therefore, these tissues must possess the requisite components of an auxin signal transduction pathway. However, results indicating tissue-specific expression of SAUR and GH3 transcripts suggest that there may be tissue-specific auxin receptors and/or response pathways. Further, the observation that GH3 transcripts are induced in multiple tissues by auxin application suggests that some tissues may possess multiple auxin receptors or signal transduction pathways.

DNA Analysis

Auxin-responsive Genes

Genomic and cDNA sequence information has recently been reported for most of the rapidly induced, auxin-responsive clones. Based on deduced amino acid sequence, some clones appear to be related (Fig. 5). The genes for pJCW1 and pJCW2 (Aux 28 and Aux 22, respectively) are homologous in that they share 5 regions of amino acid identity or similarity (1). Four of the same regions of amino acid homology are also found in 2 *Arabidopsis* genes (which were selected by homology to Aux28 and Aux22; 5), in 2 mung bean cDNAs selected as auxin-responsive (56), and in the soybean GH1 gene (Hagen, unpublished).

A second group of related, auxin-induced sequences includes the genes for GH2/4, CNT103 and parA. The DNA sequence of GH2/4 was found to be identical to the soybean Gmhsp 26A gene (6, 20), which was originally isolated as a heat shock-inducible sequence (6). Characterization of the Gmhsp 26A gene has shown that a variety of compounds (2,4-D, ABA, sodium arsenite and heavy metals) will cause an increase in mRNA levels (6). The tobacco pCNT 103-like clones show a high degree (58%) of amino acid identity to Gmhsp 26A (47), and 42% identity to the tobacco parA gene

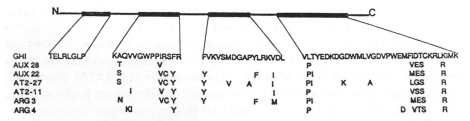

Fig. 5. Comparison of conserved amino acids within polypeptides encoded by auxin-responsive sequences. Deduced amino acid sequences were derived from cDNA and gene sequence analysis. Four "islands" of homology (black boxes) have been identified and sequences within these "islands" are compared. Only amino acid differences are listed. Data for GH1 from soybean (Hagen, unpublished); AUX 28 & AUX 22 from soybean (1); At 2-27 & At 2-11 from *Arabidopsis* (5); ARG 3 & ARG 4 from mung bean (56).

product (47); the parA sequence is also homologous (37%) to Gmhsp 26A (39). Related members of this gene family have been shown to be induced in other plant systems. They include a transcript that was induced after treatment of potato leaves with fungal elicitor (43) and a transcript that was cytokinin-responsive (7). Based on deduced amino acid sequence, computer searches of protein databases revealed homology to an *E.coli* stringent starvation protein (7, 14, 39) and glutathione S-transferases (GST; 34). Additionally, the products of the auxin-responsive tobacco cDNAs par B (40) and CNT 103 (8) and the soybean cDNA GH2/4 (14), when expressed in *E. coli*, were shown to have *in vitro* GST activity.

The GH3 gene in soybean is part of a small multigene family and encodes a polypeptide of 70 kD (19). A computer search has failed to identify any homologous sequences in the nucleotide or amino acid sequence databases. Two highly homologous *Arabidopsis* GH3-like genes have been sequenced (G. Hagen and T. Ulmasov, unpublished).

The genes encoding the soybean SAURs possess several striking features (33). Five SAUR genes have been found to be clustered within about 7 kilobase (kb) pairs. They are transcribed in opposite orientations giving rise to small (0.5-0.6 kb) transcripts, the open reading frames of these genes contain no introns, and the genes encode small (9-10 kD) polypeptides that share a high degree of homology. Sequences homologous to the SAURs have been identified in pea (Hagen and McClure, unpublished), *Arabidopsis* (Hagen and Hong, unpublished) and mung bean (56).

Auxin-responsive Promoters

Elucidation of the sequence of auxin-responsive genes has focused research efforts on determining how these genes are regulated by auxin. The initial focus has been on defining regulatory *cis* elements within the promoter of these genes. From sequence comparisons alone, it appears that there may not be a simple, single DNA sequence element that functions as an auxin-response element. A comparison of all the auxin-responsive soybean genes, for example, shows little conservation of sequence elements in the promoters. Within gene families, conserved promoter sequences have been identified, such as the DUE-NDE element found in soybean SAUR genes (33). However, while common promoter sequence elements may be shared between homologous family members, these elements may not be found in homologous genes of another plant species. For example, there are several sequences that are found in both the soybean Aux28 and Aux22 promoters, but these same elements are not found in the *Arabidopsis* homologues (5; Fig. 6).

To define the auxin response elements more precisely, promoter fragments are being fused to reporter genes such as the *E. coli uidA* gene which encodes β-glucuronidase (GUS), the *lacZ* gene which encodes β-galactosidase, or the *cat* gene which encodes chloramphenicol acetyltransferase. Promoter activity can be assayed following delivery of

237

```
AUX 22   TGATAAAAG
AUX 28   TGATAAAAG

   22    GGCAGCATGCA
   28    GGCAGCATGCA

AtAUX 2-11   GACTATGAATATGTT
AtAUX 2-27   GACCATGAATATGTT

   2-11   GAGAGAGA
   2-27   GAGAAAGA

   2-11   AAACAATCTTCTTCACAAAGCC
   2-27   AAACAAGCTTGTCTCAAACGCC

SAUR 15   CTTAAGAAAGTCCTCTAAGACA
SAUR  6   CTTGATAAAGTCCTCCAAGGCA
SAUR 10   CTTGA AAGACCTCTGAGACA

   15   CCATATGCCA TGTCTCTCATTTGGT   CCCAT
    6   CCATATGCCCCTGTCTCT GTCGGT   CCCAT
   10   CCATGTGATCCTTTCTCCCTCAGTAGACCCCAT
```

Fig. 6. Comparison of conserved promoter sequence elements found in auxin-responsive gene families. Gene families are AUX 22 & AUX 28 from soybean (1), AtAUX 2-11 & AtAUX 2-27 from *Arabidopsis* (5) and SAUR 15, 6 & 10 from soybean (33).

constructs into protoplasts (transient expression studies), or after stably incorporating constructs into plant genomes (using *Agrobacterium*-mediated transformation, for example). Promoter activity can be detected by measuring reporter enzyme activity and, depending on the reporter gene used, monitored *in planta* using histochemical stains. An additional advantage of using histochemical stains is that expression patterns in specific organs, tissues and cell types can be examined throughout plant development in the absence of exogenous auxin.

Promoter fragments with auxin-responsive activity have been identified from several soybean genes (19, 26), a tobacco gene (47) and an *Arabidopsis* gene (54). In these studies, large promoter fragments (ranging from 0.6 to 4.5 kbp) were fused to a reporter gene. Results of these studies confirm and extend the information obtained using RNA blot, tissue print and *in situ* hybridization analyses. To define minimal auxin-inducible elements within promoter fragments, reporter gene constructs containing promoter deletions (or mutations) have been used. In the tobacco parA gene promoter, a 111 bp region has been shown to be important for auxin inducibility in tobacco protoplasts (42). Promoter deletion constructs have also been used to identify sequences involved in auxin induction of 2 bacterial plasmid genes. A 10 bp sequence in the promoter of the nopaline synthase gene (3) found in the *Agrobacterium tumefaciens* Ti plasmid, and a 90 bp region in the Ti plasmid gene 5 promoter (25) have been shown to be important for auxin inducibility. Alignment of the 90 bp region of gene 5 with promoter sequences of other auxin-responsive genes revealed that certain groups of bases are shared in some genes (25). To date, however, there is no clear understanding of what constitutes an auxin-response element.

AUXIN-REGULATED GENES AS PROBES TO STUDY GROWTH

The auxin-responsive genes are providing new tools to study growth and developmental processes. Tropistic growth, such as gravitropism and phototropism, involves asymmetric elongation growth in response to a stimulus such as gravity or light. It has been proposed that the stimulus causes a directional transport of auxin, creating an asymmetric distribution of auxin. This auxin gradient would differentially affect rapid elongation growth, resulting in the observed asymmetric growth or curvature of an organ (52). The data indicating that the SAURs are specifically induced by auxins in tissues capable of rapid elongation led McClure and Guilfoyle (31) to examine SAUR expression in gravistimulated soybean hypocotyls. Using tissue print hybridization, it was shown that in a vertically grown seedling, SAURs are symmetrically distributed in the cortex and epidermis of the elongating region of the hypocotyl. During gravistimulation, SAURs rapidly disappear from the upper part of the hypocotyl and accumulate on the lower, more rapidly elongating side of the hypocotyl (Fig. 7). This asymmetric SAUR expression pattern is detected well before elongation growth of the hypocotyl is observed. These results suggest that there is a redistribution of internal auxin during gravitropism, which would activate transcription of the SAUR genes and cause the asymmetric accumulation of SAURs (31). Support for auxin redistribution during tropisms has come from studies of transgenic plants expressing fusion constructs containing auxin-responsive promoters fused to reporter genes. Gravistimulation of transgenic tobacco plants expressing either the SAUR promoter-GUS fusion (26; Table 1) or the GH3 promoter-GUS fusion (28) and transgenic *Arabidopsis* plants expressing the AtAux 2-11 promoter-*lacZ* fusion (54) resulted in higher detectable levels of reporter gene activity associated with the more rapidly elongating side of gravistimulated stems. Studies of transgenic tobacco plants expressing the SAUR promoter-GUS construct (26) showed that during phototropism, GUS activity increased on the non-illuminated side of tobacco stems, although the kinetics of this response were slower than that observed during gravitropism. Additional studies with these plants revealed that auxin transport inhibitors and manipulations that reduce endogenous auxin supply affect the differential GUS expression, as well as asymmetric growth.

In tobacco plants expressing the GH3 promoter-GUS construct, there is little endogenous reporter gene activity except for a low level in mature roots and a transient detectable level in developing flowers (19). The promoter can be activated in almost every tissue, however, by exogenous auxin application. Recently, it was reported that this promoter is activated in the basal end of cut transgenic tobacco stems within several hours after cutting (28). This was thought not to be due to wounding, as the promoter was not activated in the region of the stem below the cut. It was suggested that the accumulation of auxin at the cut basal end was responsible for activation of the GH3 promoter. The GH3 promoter was also shown to be activated at an early

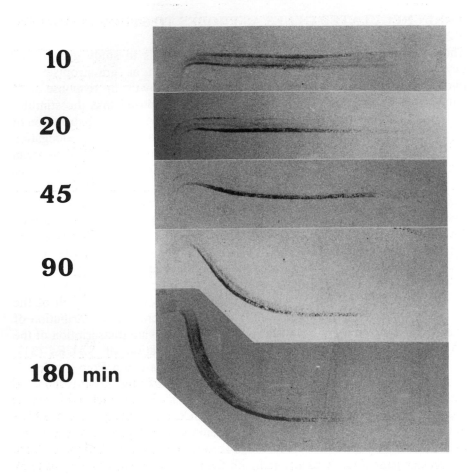

Fig. 7. Tissue print hybridization autoradiograms of gravistimulated hypocotyls of 3-day old etiolated soybean seedlings. Hypocotyls were sectioned longitudinally at the different times indicated after being reoriented from a vertical to a horizontal position. The cut surface was blotted onto a nylon membrane and hybridized with a mixed SAUR 6, 10A, and 15 antisense 35S-labeled RNA probe. From top to bottom, the time of horizontal displacement is 10, 20, 45, 90, and 180 minutes. Figure from (15), reprinted with permission.

stage in adventitious root formation. The studies reported to date would suggest that auxin-responsive promoter-reporter gene fusion constructs may be useful as probes to study growth processes that are controlled, at least in part, by auxins. Further, these constructs might be used to monitor qualitative changes in auxin concentration that occur during tropisms and other auxin-regulated growth responses (14).

Auxin-responsive gene promoters have also been used to drive the expression of hormone biosynthetic and conjugating enzyme genes in transgenic plants (14, 27). These constructs are being used to gain new insights into hormone action by manipulating internal hormone concentrations (24; see Chapter E2). Auxin-responsive gene promoters provide a means to

Table 1. Shoot bending and GUS expression during gravitropism in transgenic tobacco plants expressing a SAUR-GUS construct (26).

	Curvature (degree)	GUS Activity (% in lower side)
KINETICS		
0	0	50.4
1 hr	0.3	55.6
2 hr	20.5	66.5
3 hr	34.3	72.7
5 hr	45.9	83.3
TREATMENTS		
TIBA (1mM, 5hr)	2.3	53.1
NPA (1mM, 5hr)	3.5	52.6
Control, 4hr	43.7	76.8
Removal of leaves	28.3	67.0
Removal of apex	12.5	56.3
Removal of leaves & apex	3.6	52.5
Removal of leaves & apex; addition of IAA (50µM) to tip	20.3	68.5

alter hormone levels in a tissue-, organ- and, in some cases, developmental stage-specific manner. Initial results indicate that such constructs will provide valuable information. Expression of a construct consisting of the SAUR promoter fused to the *Agrobacterium tumefaciens* iptZ gene in tobacco, for example, resulted in some dramatic alterations in morphology and physiology, which were correlated with elevated cytokinin levels in specific organs (27).

CURRENT RESEARCH CHALLENGES

Research efforts on auxin-regulated gene expression are being focused in several areas. One major challenge is to determine the function of the polypeptides encoded by auxin-responsive genes and to define the role they play in auxin-regulated growth. As mentioned earlier, based on amino acid sequence homology and some biochemical studies, it appears that one group of sequences is related to glutathione S-transferases. The evidence that these genes are induced by a variety of chemical and nonchemical stresses, in addition to auxin, suggests that the gene products may be induced as part of a general stress response, and may function solely as detoxifying enzymes (34). On the other hand, some members of this family of sequences may be important to auxin action. To date, their role as auxin-induced gene products has not been determined. For the other groups of auxin-responsive genes, deduced amino acid sequence information gives no clue as to function.

Current approaches being used to probe for function include the use of antibodies raised against the auxin-responsive gene products (53) and transgenic plants that either overproduce the polypeptides or that express antisense constructs to deplete the polypeptide levels.

The other major research effort continues to focus on defining promoter sequence elements that are important for induction by auxins. This information is essential in order to achieve the goal of isolating trans-acting factors that specifically interact with the auxin-response elements. This goal represents another major challenge, especially in view of the evidence that there may be multiple auxin-response elements. Ultimately, the information gained from these studies should contribute in defining the components of the auxin signal transduction pathway.

Acknowledgments
I want to thank Bruce McClure and Tom Guilfoyle for comments on the manuscript, and Linda Zurfluh for providing the data presented in Fig. 1.

References

1. Ainley, W.M., Walker, J.C., Nagao, R.T., Key, J.L. (1988) Sequence and characterization of two auxin-regulated genes from soybean. J. Biol. Chem. 263, 10658-10666.
2. Alliotte, T., Tire, C., Engler, G., Peleman, J., Caplan, A., Van Montague, M., Inze, D. (1989) An auxin-regulated gene of *Arabidopsis thaliana* encodes a DNA-binding protein. Plant Physiol. 89, 743-752.
3. An, G., Costa, M.A., Ha, S-B. (1990) Nopaline synthase promoter is wound inducible and auxin inducible. Plant Cell 2, 225-233.
4. Baulcombe, D.C., Key, J.L. (1980) Polyadenylated RNA sequences which are reduced in concentration following auxin treatment of soybean hypocotyls. J. Biol. Chem. 255, 8907-8913.
5. Conner, T.W., Goekjian, V.H., LaFayette, P.R., Key, J.L. (1990) Structure and expression of two auxin-inducible genes from *Arabidopsis*. Plant Mol. Biol. 15, 623-632.
6. Czarnecka, E., Nagao, R.T., Key, J.L., Gurley, W.B. (1988) Characterization of Gmhsp 26-A, a stress gene encoding a divergent heat shock protein of soybean: Heavy-metal-induced inhibition of intron processing. Mol. Cell. Biol. 8, 1113-1122.
7. Dominov, J.A., Stenzler, L., Lee, S., Schwarz, J.J., Leisner, S., Howell, S.H. (1992) Cytokinin and auxin control the expression of a gene in *Nicotiana plumbaginifolia* by feedback regulation. Plant Cell 4, 451-461.
8. Droog, F.N.J., Hooykaas, P.J.J., Libbenga, K.R., van der Zaal, E.J. (1993) Proteins encoded by an auxin-regulated gene family of tobacco share limited but significant homology with glutathione S-transferases and one member indeed shows *in vitro* GST activity. Plant Mol. Biol. 21, 965-972.
9. Franco, A.R., Gee, M.A., Guilfoyle, T.J. (1990) Induction and superinduction of auxin-responsive mRNAs with auxin and protein synthesis inhibitors. J. Biol. Chem. 265, 15845-15849.
10. Gantt, J.S., Key, J.L. (1983) Auxin-induced changes in the levels of translatable ribosomal protein messenger ribonucleic acids in soybean hypocotyl. Biochem. 22, 4131-4139.
11. Gantt, J.S., Key, J.L. (1985) Coordinate expression of ribosomal protein mRNAs following auxin treatment of soybean hypocotyls. J. Biol. Chem. 260, 6175-6181.

12. Gee, M.A., Hagen, G., Guilfoyle, T.J. (1991) Tissue-specific and organ-specific expression of soybean auxin-responsive transcripts GH3 and SAURs. Plant Cell 3, 419-430.

13. Guilfoyle, T.J. (1986) Auxin-regulated gene expression in higher plants. CRC Crit. Rev. Plant Sci. 4, 247-277.

14. Guilfoyle, T.J., Hagen, G., Li, Y., Ulmasov, T., Liu, Z., Strabala, T., Gee, M. (1993) Auxin-regulated transcription. Aust. J. Plant Physiol. In press.

15. Guilfoyle, T.J., McClure, B.A., Hagen, G., Brown, C., Gee, M., Franco, A. (1990) Regulation of plant gene expression by auxins. *In*: Gene manipulation in plant improvement II, pp. 401-418, Gustafson, J.P., ed. Plenum Press, New York.

16. Hagen, G. (1989) Molecular approaches to understanding auxin action. New Biol. 1, 19-23.

17. Hagen, G., Guilfoyle, T.J. (1985) Rapid induction of selective transcription by auxins. Mol. Cell. Biol. 5, 1197-1203.

18. Hagen, G., Kleinschmidt, A., Guilfoyle, T.J. (1984) Auxin-regulated gene expression in intact soybean hypocotyl and excised hypocotyl sections. Planta 162, 147-153.

19. Hagen, G., Martin, G., Li, Y., Guilfoyle, T.J. (1991) Auxin-induced expression of the soybean GH3 promoter in transgenic tobacco plants. Plant Mol. Biol. 17, 567-579.

20. Hagen, G., Uhrhammer, N., Guilfoyle, T.J. (1988) Regulation of expression of an auxin-induced soybean sequence by cadmium. J. Biol. Chem. 263, 6442-6446.

21. Key, J.L. (1969) Hormones and nucleic acid metabolism. Ann. Rev. Plant Physiol. 20, 449-474.

22. Key, J.L. (1986) Auxin-regulated gene expression: A historical perspective and current status. *In*: Molecular biology of plant growth control, pp. 1-21, Fox, J.E., Jacobs, M., eds. Liss, New York.

23. Key, J.L. (1989) Modulation of gene expression by auxin. BioEssays 11, 52-57.

24. Klee, H., Estelle, M. (1991) Molecular genetic approaches to plant hormone biology. Annu. Rev. Plant Physiol. Plant Mol. Biol. 42, 529-551.

25. Korber, H., Strizhov, N., Staiger, D., Feldwisch, J., Olsson, O., Sandberg, G., Palme, K., Schell, J., Koncz, C. (1991) T-DNA gene 5 of *Agrobacterium* modulates auxin response by autoregulated synthesis of a growth hormone antagonist in plants. EMBO J. 10, 3983-3991.

26. Li, Y., Hagen, G., Guilfoyle, T.J. (1991) An auxin-responsive promoter is differentially induced by auxin gradients during tropisms. Plant Cell 3, 1167-1175.

27. Li, Y., Hagen, G., Guilfoyle, T.J. (1992) Altered morphology in transgenic tobacco plants that overproduce cytokinins in specific tissues and organs. Develop. Biol. 153, 386-395.

28. Li, Y., Strabala, T.J., Hagen, G., Guilfoyle, T.J. (1993) Auxin physiology and expression of the GH3 promoter-GUS fusion gene. Plant Physiol. Suppl. 102, 25.

29. McClure, B.A. (1987) Characterization of auxin-regulated polyadenylated RNAs from soybean. PhD dissertation, University of Minnesota.

30. McClure, B.A., Guilfoyle, T. (1987) Characterization of a class of small auxin-inducible soybean polyadenylated RNAs. Plant Molec. Biol. 9, 611-623.

31. McClure, B.A., Guilfoyle, T.J. (1989) Rapid redistribution of auxin-regulated RNAs during gravitropism. Science 243, 91-93.

32. McClure, B.A., Guilfoyle, T.J. (1989) Tissue print hybridization. A simple technique for detecting organ-and tissue-specific gene expression. Plant Molec. Biol. 12, 517-524.

33. McClure, B.A., Hagen, G., Brown, C.S., Gee, M.A., Guilfoyle, T.J. (1989) Transcription, organization and sequence of an auxin-regulated gene cluster in soybean. Plant Cell 1, 229-239.

34. Pickett, C.B., Lu, A.Y.H. (1989) Glutathione S-transferases: gene structure, regulation and biological function. Annu. Rev. Biochem. 58, 743-764.

35. Reddy, A.S.N., Jena, P.K., Mukherjee, S.K., Poovaiah, B.W. (1990) Molecular cloning of cDNAs for auxin-induced mRNAs and developmental expression of the auxin-inducible genes. Plant Molec. Biol. 14, 643-653.

36. Sambrook, J., Fritsch, E.F., Maniatis, T. (1989) Molecular Cloning. Cold Spring Harbor Laboratory Press, New York.

37. Silberger, J., Skoog, F. (1953) Changes induced by indoleacetic acid in nucleic acid content and growth of tobacco pith tissue. Science 118, 443-444.

38. Takahashi, Y., Kuroda, H., Tanaka, T., Machida, Y., Takebe, I., Nagata, T. (1989) Isolation of an auxin-regulated gene cDNA expressed during the transition from G0 to S phase in tobacco mesophyll protoplasts. Proc. Natl. Acad. Sci. USA 86, 9279-9283.

39. Takahashi, Y., Kusaba, M., Hiraoka, Y., Nagata, T. (1991) Characterization of the auxin-regulated par gene from tobacco mesophyll protoplasts. Plant J. 1, 327-332.

40. Takahashi, Y., Nagata, T. (1992) par B: An auxin-regulated gene encoding glutathione S-transferase. Proc. Natl. Acad. Sci. USA 89, 56-59.

41. Takahashi, Y., Nagata, T. (1992) Differential expression of an auxin-regulated gene, parC, and a novel related gene C-7, from tobacco mesophyll protoplasts in response to external stimuli and in plant tissues. Plant Cell Physiol. 33, 779-787.

42. Takahashi, Y., Niwa, Y., Machida, Y., Nagata, T. (1990) Location of the cis acting auxin-responsive region in the promoter of the par gene from tobacco mesophyll protoplasts. Proc. Natl. Acad. Sci. USA 87, 8013-8016.

43. Taylor, J.L., Fritzemeier, K.-H., Hauser, I., Kombrink, E., Rohwer, F., Schroder, M., Strittmatter, G., Hahlbrock, K. (1990) Structural analysis and activation by fungal infection of a gene encoding a pathogenesis-related protein in potato. Molec. Plant-Microbe Interact. 3, 72-77.

44. Theologis, A. (1986) Rapid gene regulation by auxin. Ann. Rev. Plant Physiol. 37, 407-438.

45. Theologis, A., Huynh, T.V., Davis, R.W. (1985) Rapid induction of specific mRNAs by auxin in pea epicotyl tissue. J. Mol. Biol. 183, 53-68.

46. Theologis, A., Ray, P.M. (1982) Early auxin regulated polyadenylated mRNA sequences in pea stem tissue. Proc. Natl. Acad. Sci. USA 79, 418-421.

47. van der Zaal, E.J., Droog, F.N.J., Boot, C.J.M., Hensgens, L.A.M., Hoge, J.H.C., Schilperoort, R.A., Libbenga, K.R. (1991) Promoters of auxin-induced genes from tobacco can lead to auxin-inducible and root tip-specific expression. Plant Molec. Biol. 16, 983-998.

48. van der Zaal, E.J., Memelink, J., Mennes, A.M., Quint, A., Libbenga, K.R. (1987) Auxin-induced mRNA species in tobacco cell cultures. Plant Molec. Biol. 10, 145-157.

49. van der Zaal, E.J., Mennes, A.M., Libbenga, K.R. (1987) Auxin-induced rapid changes in translatable mRNAs in tobacco cell suspensions. Planta 172, 514-519.

50. Walker, J.C., Key, J.L. (1982) Isolation of cloned cDNAs to auxin-responsive poly(A)+ RNAs of elongating soybean hypocotyl. Proc. Natl. Acad. Sci. USA 79, 7185-7189.

51. Walker, J.C., Legocka, J., Edelman, L., Key, J.L. (1985) An analysis of growth regulator interactions and gene expression during auxin-induced cell elongation using cloned complementary DNAs to auxin-responsive messenger RNAs. Plant Physiol. 77, 847-850.

52. Wilkins, M.B. (1984) Gravitropism. In: Advanced plant physiology, pp. 163-185, Wilkins, M.B., ed. Pitman Publishing, London.

53. Wright, R.M., Hagen, G., Guilfoyle, T. (1987) An auxin-induced polypeptide in dicotyledonous plants. Plant Molec. Biol. 9, 625-634.

54. Wyatt, R.E., Ainley, W.M., Nagao, R.T., Conner, T.W., Key, J.L. (1993) Expression of the *Arabidopsis AtAux2-11* auxin-responsive gene in transgenic plants. Plant Mol. Biol. 22, 731-749.

55. Yamamoto, K.T., Mori, H., Imaseki, H. (1992) Novel mRNA sequences induced by indole-3-acetic acid in sections of elongating hypocotyls of mung bean (*Vigna radiata*). Plant Cell Physiol. 33, 13-20.

56. Yamamoto, K.T., Mori, H., Imaseki, H. (1992) cDNA cloning of indole-3-acetic acid-regulated genes: Aux 22 and SAUR from mung bean (*Vigna radiata*) hypocotyl tissue. Plant Cell Physiol. 33, 93-97.
57. Zurfluh, L.L., Guilfoyle, T.J. (1980) Auxin-induced changes in the patterns of protein synthesis in soybean hypocotyl. Proc. Natl. Acad. Sci. USA 77, 357-361.
58. Zurfluh, L.L., Guilfoyle, T.J. (1982) Auxin-induced changes in the population of translatable messenger RNAs in elongating sections of soybean hypocotyl. Plant Physiol. 69, 332-337.
59. Zurfluh, L.L., Guilfoyle, T.J. (1982) Auxin-and ethylene-induced changes in the population of translatable messenger RNA in basal sections and intact soybean hypocotyl. Plant Physiol. 69, 338-340.
60. Zurfluh, L.L., Guilfoyle, T.J. (1982) Auxin-induced changes in the population of translatable messenger RNA in elongating maize coleoptile sections. Planta 156, 525-527.

D3. Gibberellin Action in Germinated Cereal Grains

John V. Jacobsen[1], Frank Gubler[2] and Peter M. Chandler[1]
[1]CSIRO Division of Plant Industry, PO Box 1600, Canberra, ACT 2601, Australia, and [2]Co-operative Research Center for Plant science, PO Box 475, Canberra, ACT 2601, Australia.

INTRODUCTION

The study of the response of cereal aleurone to gibberellin and abscisic acid (GA and ABA, respectively), particularly with reference to α-amylase synthesis, has made a significant contribution to our understanding of GA action in plant cells, especially as it relates to the control of protein synthesis, and also to the function of the endosperm during germination. While much of the work has been carried out using isolated aleurone from a single cultivar of barley ("Himalaya"), it seems so far that the principles which have emerged from this system can be applied to *in vivo* behaviour of barley and other cereal grains.

The incubation of isolated aleurone layers in media containing specified concentrations of GA_3 and/or ABA, has been extensively studied as a model for hormone action in plants. The main observation is that GA_3 can stimulate aleurone cells to secrete a range of hydrolytic enzymes, the major one being α-amylase (Fig. 1). These enzymes are responsible for the mobilisation of stored endosperm reserves which provide the growing seedling with a supply of fixed carbon, reduced nitrogen and other nutrients. The interest in ABA lies in the observation that it can prevent the action of GA_3 if present in excess, and it also induces its own set of proteins in isolated aleurone.

Fig. 1. Schematic diagram illustrating some of the principal features associated with reserve mobilization in a wheat or barley grain (and probably other cereals as well) following germination. Gibberellin produced by the embryo stimulates cells of the aleurone layer to synthesise and secrete α-amylase and other hydrolases which degrade starch and other polymeric reserves in the endosperm, providing nutrients for the developing seedling.

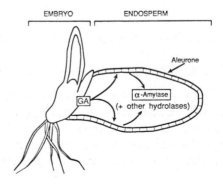

246

P. J. Davies (ed.), Plant Hormones, 246–271.
© 1995 *Kluwer Academic Publishers. Printed in the Netherlands.*

This article is a synopsis of our understanding of mainly GA_3 action in germinating cereal grains, focusing on aleurone, but also ABA in the same system. It very briefly discusses the role of GA and ABA in grain development, and the influence of environment and genotype of the developing grain on hormone responsiveness of mature aleurone. The major part of our attention is given to endosperm degradation, the response of aleurone to hormones and to our knowledge of the molecular mechanisms of hormone action which has progressed considerably since the first edition of this chapter was published in 1987 (37). Background information is provided in previous reviews (2, 13, 40).

HORMONES IN DEVELOPING GRAIN

While the responses of mature isolated aleurone to applied GA or ABA are fairly well understood, quantitative aspects of the response can vary between different harvests of the same cultivar. This observation suggests that environmental conditions during grain development can influence the eventual hormone responsiveness of aleurone. For example, it has been shown that aleurone from barley grown in relatively high temperatures produces high levels of α-amylase in the absence of added GA (57). In normal development, cereal grains accumulate significant levels of both GA and ABA. The activity of these hormones during grain development, or their residual levels in the dry seed, may influence subsequent aleurone (and embryo) behaviour during and following germination. It is unknown whether high temperature during development causes accumulation of abnormally high amounts of GA resulting in the production of high levels of α-amylase in germinated grain. Low temperature during development strengthens dormancy in barley grain. Perhaps such conditions cause accumulation of high levels of ABA and this produces stronger dormancy. The accumulation and role of GA and ABA in developing grain has been discussed previously (37). The responses of aleurone from mature grains to these hormones is discussed below.

EMBRYONIC CONTROL OF ENDOSPERM FUNCTION

Gibberellin Production During Germination

The high levels of GAs present in developing grain usually decrease during maturation so that the dry grain normally contains very low levels. Following germination, GA levels again rise with GA_1 being the predominant species in barley although low amounts of other GAs (GA_3, GA_8, GA_{17}, GA_{19}, GA_{20}, GA_{29}, GA_{34} and GA_{48}) are also present (5, 12, 14). Many of these are products of the early 13-hydroxylation GA synthesis pathway which is the major pathway in vegetative tissues of many higher plants. Similar changes in GAs would appear to occur in wheat (46).

In barley, if the embryo is removed from the endosperm earlier than about 24 h after germination, there is no increase in GA in the endosperm and very little subsequent production of α-amylase. This indicates that it is GA from the embryo which stimulates the production of α-amylase in the aleurone (Fig. 1). However, inhibition of GA biosynthesis does not affect α-amylase production indicating that although GA is produced by the embryo, the supply of GA to the endosperm during germination may not arise directly by new synthesis but from a pre-formed GA pool (17). The alternative interpretation of these results is that GA is not the signal for α-amylase production although there is a large amount of circumstantial evidence which argues against such an interpretation. GA_1 (plus perhaps GA_3) is probably the major factor controlling shoot elongation as well as aleurone response (5).

The pathway by which GA leaves the scutellum and reaches the aleurone is not well understood although there is evidence that there is asymmetric transport of GA towards the apex of the scutellum from which it is released into the aleurone layer (60). GA movement has usually been thought of as occurring by diffusion in free space, but movement within the scutellum could occur via the vascular system, and symplastic movement between aleurone cells via plasmodesmata has also been suggested.

Patterns of Endosperm Breakdown During Germination

Morphological evidence (based on scanning electron microscopy) showed that initial starch degradation in barley occurred adjacent to the scutellar epithelium (52) and proceeded symmetrically along the grain towards the distal end (Fig. 2). This indicated that the scutellum was the major source of

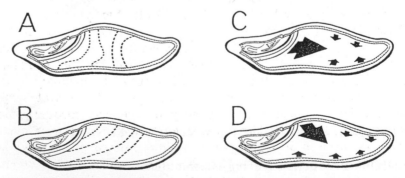

Fig. 2. Theories of endosperm hydrolysis in germinated grains. A, The "symmetrical" pattern of endosperm hydrolysis. Dashed lines represent approximate areas of modified endosperm after 2, 3, 4 and 5 days of hydration at 15 °C. B, The "assymetric" pattern of modification. C, The pattern of enzyme movement suggested to produce the modification outline in A. The size of the arrows indicate the relative importance of scutellum and aleurone in hydrolase production. D, The pattern of enzyme movement suggested to produce the modification outlined in B. From (55).

hydrolases involved in endosperm degradation (Fig. 2A and C), particularly at early stages of endosperm hydrolysis. However, there is also evidence that the hydrolysis of endosperm begins at the apex of the scutellum (Fig. 2B and D), primarily under the control of hydrolytic enzymes released asymmetrically from the aleurone. This enzyme release is under the control of GA that accumulates in (and is released from) from the scutellar apex (60). Hydrolysis probably begins near the embryo and proceeds distally along the grain on a front, which is at first approximately parallel to the scutellar surface and moves faster immediately beneath the aleurone layer on the dorsal surface of the grain.

The relative contributions of scutellum and aleurone to the hydrolytic enzyme complement of the endosperm has been a major focus of research. It now seems that in barley, scutellum accounts for only a minor part (5-10%) of total α-amylase and that scutellum produces α-amylase in 1-2 days whereas aleurone production peaks after 3-4 days. In germinating grain, the sites of endosperm hydrolysis are not necessarily adjacent to the sites of enzyme production and thus endosperm hydrolysis may not be a reliable indicator of the localisation of the enzyme production. It is now possible to assess relative mRNA levels for particular hydrolases in scutellum versus aleurone by tissue section (or *in situ*) hybridisation. By hybridising radioactive cloned DNA molecules for (1-3, 1-4)-β-glucanase to longitudinal sections of germinating grains, it was shown that over the first 2-3 days, message accumulated in the scutellum (confined to the epithelial cells) and that subsequently, while scutellar mRNA decreased, levels increased strongly in the aleurone progressing distally along the layer (48). Studies using Northern analysis of tissue specific RNA preparations have confirmed these results (70). While the relative mRNA levels may not necessarily reflect levels of hydrolase production (and subsequent endosperm hydrolysis), in the case of (1-3, 1-4)-β-glucanase, quantitative and temporal aspects of mRNA appearance are in general accord with the results of enzyme assays (13). *In situ* hybridisation studies examining α-amylase production in germinated rice grains have produced results which are very similar to those for β-glucanase and in this case it was also shown that there was differential expression of different isoforms in aleurone and scutellum (61). In summary, it is now clear that initiation of endosperm hydrolysis adjacent to the scutellum early in germination is accomplished by enzymes (perhaps specific isoforms) produced and secreted by the scutellar epithelium. Studies of numerous enzymes involved in the hydrolysis of β-glucan, starch, protein and nucleic acids in the starchy endosperm (13) show that a proportion of the enzyme activity (or perhaps specific isozymes) of most or all of these enzymes originates in the scutellum. As germination progresses, enzyme production by the scutellum appears to decline and the aleurone becomes the major source. The situation in some of the sub-tropical cereals, e. g. sorghum, maize, millet and rice may be different in that the scutellum might be a more significant source of α-amylase and other enzymes (1).

Although there is abundant evidence that hydrolase synthesis in aleurone is under hormonal control, the question of whether hydrolase synthesis in scutellum is similarly controlled has been more difficult to answer. In the main, experiments involving isolated embryos of barley as well as the sub-tropical cereals have shown that scutellar enzyme synthesis is largely insensitive to applied GA indicating either that hydrolase genes were not hormonally controlled in embryonic tissue, or that there were already high endogenous levels of GA. Using a GA-deficient mutant of barley, it has now been shown that high-pI α-amylase gene expression is also induced by GA in the scutellum (10).

ALEURONE RESPONSE

Reserve Degradation

During early seedling development, the stored reserves (protein, carbohydrate, lipid, nucleic acid and mineral complexes) in the aleurone and starchy endosperm are degraded and mobilised via the scutellum to the developing seedling. The battery of enzymes required for eventual hydrolysis and solubilization of the stored reserves (13) originates mainly in the aleurone layer, but some are also produced in the starchy endosperm and the scutellum, and many of these are also under GA control.

The progression of aleurone cell and endosperm hydrolysis has been examined mainly by light and electron microscopy. Changes in the aleurone include vacuolation of the cells and digestion of the cell walls (Fig. 3). In the sub-aleurone and starchy endosperm cells, cell wall and starch grain hydrolysis and general disorganisation of the cells can be seen. Aleurone cell walls appear to be impermeable to proteins so that hydrolysis of aleurone (and probably also endosperm) cell walls is probably necessary for efficient release of hydrolytic enzymes from the aleurone layer and for maximal access of the hydrolytic enzymes to the starch grains and surrounding protein matrix in the endosperm. α-amylase is distributed evenly throughout the cell (endoplasmic reticulum (ER), Golgi and other locations?) and it is initially released into the cell wall mostly on the inner (starchy endosperm) side of the cell (Fig. 3A). Figure 3B shows the tunnelling of walls which occurs in the early stages of aleurone cell wall hydrolysis. Digestion is originally localised around plasmodesmata but ultimately involves the whole wall. The gold grains over the digested areas in the micrograph show the localisation of α-amylase in the digested areas. α-amylase does not appear to exist in undigested wall in agreement with their apparent impermeability. Hydrolysis of aleurone cell walls leaves an innermost layer of wall material (IW) (Fig. 3B) which is apparently resistant to degradation. This layer could be similar to cell wall connections which persist around plasmodesmata connecting digested aleurone cells, and it may contribute to mechanical stability of the aleurone layer as a whole as wall digestion progresses.

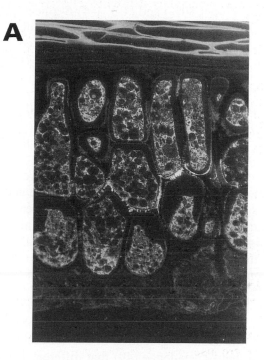

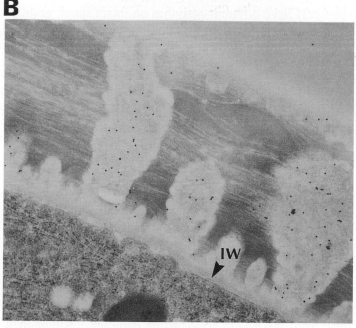

Fig. 3. Localization of α-amylase in aleurone cells and in digested pockets within the cell wall of GA-treated tissue as determined by immunofluorescence (A) and by immunogold labelling (B). From (18).

Arabinoxylans (about 80%) and (1-3, 1-4)-β-glucan (about 20%) are the major cell wall polysaccharides of barley aleurone. There are five arabinoxylan degrading enzymes which are released from aleurone in response to GA$_3$, and of these endo-(1-4)-β-xylanase is the one most likely to be involved in the initial cell wall hydrolysis. However there is a discrepancy (13) between the relatively late appearance of endoxylanase activity and the relatively early beginning of cell wall hydrolysis, which at present is unresolved. The other major polysaccharide is believed to be degraded by two endo-(1-3, 1-4)-β-glucanases which have been extensively characterised (13).

The starch deposits of wheat and barley exist in two size classes of starch granules embedded in a protein matrix throughout the endosperm. The two glucose polymers forming the starch are amylose, a linear (1-4)-α-glucan and amylopectin, a highly branched form of amylose containing (1-6)-α-branches. Following germination of barley the small starch granules are first degraded, primarily from the surface. Large starch granules frequently suffer initial surface pitting, but hydrolysis may then occur inside the granule and proceed through a hollow shell stage.

Starch hydrolysis proceeds through the action of four different enzymes, collectively known as diastase. α-Amylase, an endohydrolase is probably very important in the initial degradation of amylose and amylopectin within starch grains, and there is evidence (49) that different isozyme groups of α-amylase (see later) have different activities on intact starch grains, low-pI α-amylase being able to digest starch grains more efficiently than high-pI α-amylase both in barley and wheat. Although it was once thought that α-amylase was the only enzyme capable of attacking whole starch grains, it is now evident that α-glucosidase, an exohydrolase liberating glucose, has the same ability and that together, the two enzymes are strong synergistic (73). Thus it would appear that these two enzymes are responsible for the initial attack on starch grains *in vivo* and that the other diastatic enzymes, β-amylase and limit dextrinase, are involved in reducing the size of the initial products of hydrolysis. The latter, known also as debranching enzyme, hydrolyses α-1, 6 links in amylopectin.

α-amylase and α-glucosidase are synthesised in the aleurone layer during germination. β-amylase accumulates in the starchy endosperm during development and at maturity, it exists in the endosperm of the dry grain mostly in an inactive form, disulphide bonded to a component of protein bodies. The increase in β-amylase activity which follows germination results from the activity of GA-induced cysteine endopeptidases which cleave the β-amylase-protein complex, releasing active β-amylase (20). These endopeptidases are synthesised in the aleurone in response to GA$_3$. The situation with limit dextrinase seems to be complex. Although it is newly synthesised by aleurone cells, it becomes bound in starchy endosperm during germination and, as with β-amylase, proteolysis is required for its activation in the later stages (47).

Although relatively little is known about enzymes causing storage protein hydrolysis compared to starch and β-glucan, it is clear that protein is reduced to amino acids and peptides by a number of exo- and endopeptidases (13). Carboxypeptidases appear to be the main exopeptidases involved. Aminopeptidases appear to play no part.

Other reserves which survive in mature starchy endosperm in small quantities are also hydrolysed during germination (13). Nucleic acids appear to be hydrolysed by nuclease 1, ribonuclease and nucleosidases and (1-3)-β-glucan which occurs extracellularly mainly in the sub-aleurone cells, is hydrolysed by GA-induced (1-3)-β-glucanase.

What controls reserve degradation? Is it limited only by the availability of hydrolases? In the early stages of germination and seedling growth, it would seem that endosperm hydrolysis is initiated by hydrolases which are newly synthesised and secreted by the scutellum and aleurone. However, as seedling growth progresses, the products of reserve hydrolysis accumulate in the starchy endosperm cavity up to 570 milliosmolar and the concentrations of some of the component products are high enough to inhibit further hydrolase synthesis. Thus several days after germination, osmotic controls may come into play. Are there other controls? One possibility is that pH within the starchy endosperm may exercise some limitation. Many endosperm hydrolases have pH optima in the vicinity of pH 5, and thus would function best in an acidic environment. Such an environment is established in the endosperms of germinating barley and wheat by protons apparently released by the aleurone layer. Barley endosperm is acidified during the latter stages of development as well as during germination (53), but while mature wheat aleurone can acidify (21), developing wheat does not (J.V. Jacobsen, unpublished). Gradients of pH in the endosperm may favour hydrolysis of reserves by providing an optimum environment for the enzymes to function or by increasing accessibility of enzymes to substrates which become more soluble at acid pH (21). However, little is known about endosperm pH and its control. Reduced pH in developing barley endosperm is associated with the release of hydrogen ions (as malic acid) from the aleurone cells.

Enzyme Response

The observation that isolated aleurone layers of barley respond to exogenously applied GA_3 in a manner similar to intact aleurone layers following seed germination has made this a favourite system for studying the effect of GA_3 on the activity of the hydrolases associated with mobilisation of seed reserves. From a theoretical viewpoint we can envisage GA having either a direct effect on the amount, cellular location, or activity of an enzyme, or an indirect effect as found for the appearance of β-amylase and limit dextrinase activity in the endosperm (see preceding section). GA_3 has been found to cause increased activity of about 29 enzymes in the aleurone layer (40) Most of the secreted enzymes are hydrolytic and are involved in cell wall and reserve hydrolysis, while enzymes which are not secreted are mostly metabolic enzymes. Therefore, in aleurone cells, GA_3 appears to regulate not only the production of hydrolytic enzymes, but also the metabolic machinery. Numerous studies of GA_3-induced changes in ultrastructure of aleurone cells in the early 1970's, mainly by R. L. Jones and colleagues, are consistent with major metabolic changes during hydrolase synthesis. More recently, it has been shown that GA_3 and ABA regulate other processes such as calcium transport into the endoplasmic reticulum, and this is associated with the synthesis, stabilisation and secretion of α-amylase (39). ER-associated proteins including calmodulin, a calcium transporting ATPase, and a BiP-related protein (calcium binding protein), increase simultaneously (15). It would seem likely that as other GA_3-controlled processes are studied in greater detail more hormonally-regulated proteins will be described.

The responses to GA_3 of about 18 aleurone hydrolases have been classified into four groups (35). The first group is characterised by a rapid change in activity of the enzyme, with, presumably, little change in amount. Enzymes involved in phospholipid metabolism fall into this group, and at present there is little information concerning the mechanism of their regulation.   The second group includes enzymes whose synthesis and secretion are stimulated by GA_3 such as α-amylase (Fig. 4) and protease. Many aspects of hormonal control of gene expression in aleurone have been studied with particular reference to α-amylase, and this topic will dominate the remainder of this chapter. The third group includes β-glucanase, acid phosphatase and ribonuclease, which increase in activity by new synthesis in the absence of added GA_3, but which may show an additional increase in activity if GA_3 is added. Finally, there are some enzymes (xylopyranosidase and arabinofuranosidase) whose activity is constant in the absence of GA_3 but increases in the presence of GA_3 and also show secreted activity. For enzymes of the last two groups, the precise relationship between isozymes formed in the absence or presence of GA_3 is frequently not known, so it is difficult to assess whether GA_3 is stimulating levels of existing isozymic forms, or causing *de novo* accumulation of new isozymes.

Much of the research on hormone-induced enzyme changes has been

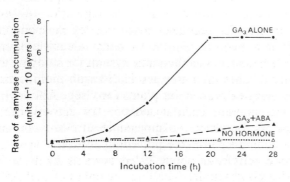

Fig. 4. Rates of α-amylase accumulation in isolated barley aleurone (plus surrounding medium) incubated for the indicated time without hormone, in the presence of 1 μM GA_3, or in the presence of 1 uM GA_3 plus 25 μM ABA. From (23).

done on a total activity basis, but, as is often the case, examination of enzyme production at the isozyme level has produced interesting results. For example, studies of (1-3, 1-4)-β-glucanase (13, 48, 70) indicate that the early enzyme increase in germinating grain occurs in the scutellum and consists of only isozyme E1 whereas later, most of the enzyme is produced in the aleurone and consists of both E1 and E11. Study of α-amylase at the isozyme level has also been rewarding. A particularly interesting result is that the kinetics of production of low- and high-pI α-amylases in aleurone of germinating grains and in isolated aleurone layers is different. In whole grains, high-pI isozymes dominate early in germination while low-pI forms appear later. In isolated layers, α-amylases appear to accumulate co-ordinately, but whereas high-pI forms decrease, low-pI forms persist (9). Such results highlight the fact that different isozymes can be controlled differentially by hormones, and that the full story of the aleurone response to GA and ABA is unlikely to unfold unless individual isozymes are studied. Studies of the aleurone response to hormones has been dominated by up-regulation of enzymes, but some enzymes are also down-regulated (40). That GA$_3$ caused both up- and down-regulation of synthesis of proteins was evident from the very early studies of protein synthesis in aleurone. Studies of the molecular mechanisms of hormone action are also dominated by up-regulation (see later), and little is known about down-regulation.

ACTION OF GIBBERELLIN AND ABSCISIC ACID

Hormone Perception

Although a considerable amount of information about transcriptional and post-transcriptional events in α-amylase induction has accrued in recent times, little is known about hormone perception and the events which lead to regulation of gene transcription and other hormone-associated changes. For a cell to respond to a hormone, the cell must perceive the hormone. As in the case of the steroid hormone receptor proteins in animal cells, hormone perception in plants is thought to involve binding proteins, and the result of this recognition would be to set in motion a sequence of processes, commonly referred to as a signal transduction pathway, leading to regulation of gene expression or some other hormonally regulated response. Such a binding protein would then qualify as a receptor (56), but clearly not all hormone binding proteins (for example hormone metabolising enzymes) are receptors.

In general, despite a considerable amount of work, there is no strong case for soluble (cytosolic?) GA receptor proteins in plants (56, 71). Although a number of GA-binding proteins have been described, few have had the desired characteristics of specific, saturable and reversible hormone binding, and none have been shown to be involved in mediating hormone responses. Similarly, studies of aleurone cells have not revealed proteins likely to be receptors. Early studies using photoaffinity labelling for both GA and ABA were hampered by high levels of non-specific binding of the probes to many proteins, but refinement of the procedures may now be paying dividends. *In vivo* labelling of oat aleurone layers with photoaffinity labelled GA has identified a 60 kD protein which appears to bind biologically active GA's specifically (27). Such a protein must be characterised extensively and shown to be involved in the signal transduction pathway.

The possibility that GA perception might occur by interaction with membrane components has been considered for some time, but recently the notion has received some experimental support. One approach involves the use of anti-idiotype antibodies. An anti-idiotype antibody is an antibody to the idiotype or antigen-binding region of another antibody, and, as such, may resemble some of the configuration of the antigen against which the first antibody was raised. If the antigen was a hormone, then the anti-idiotype antibody may recognise the hormone receptor, and either mimic the hormone action or compete with the hormone. An anti-idiotype antiserum raised against antibodies to GA was found to block GA action in oat aleurone protoplasts (26). Because antibodies are not thought to enter intact cells, and because the anti-idiotype antiserum agglutinated the oat protoplasts, this indicates that the receptor is on the outer membrane (26). However, anti-idiotype antibodies may also recognise GA-uptake or metabolising proteins, so any moiety recognised by the probe is not necessarily a receptor. The full potential of the anti-idiotype approach does not appear to have been exploited yet. Another approach facilitated by the development of hormone sensitive aleurone protoplasts has also provided evidence that GA does not have to enter the cell to be effective. GA immobilised on agarose beads was found to be able to induce α-amylase synthesis in protoplasts, indicating that perception occurred at the protoplast surface and that GA receptors may exist in the plasmalemma (25). However, efforts to isolate GA-binding proteins from solubilised aleurone membrane proteins using affinity chromatography with immobilised GA have not yet been achieved. Although there would appear to be a long way to go before a GA receptor is isolated from aleurone cells and its role in α-amylase induction proven, the approaches described above offer the hope that such a result is achievable. It is hoped that elucidation of the steps downstream of hormone perception and the steps upstream of gene regulation, will soon arrive at the same point. However, the possibility that the GA receptor in some mobilised or modified form may be a DNA-binding factor which interacts with the promoters of GA-regulated genes, seems unlikely at this time, and it appears more likely that there is a much more complex signal transduction pathway.

No studies of ABA perception in aleurone cells have yet been reported even although all of the above approaches are as applicable to the study of ABA action as they are to GA.

When probes for GA receptors and signal transduction pathway components (are they proteins?) become available, there exists the possibility for studying hormone perception in mutants with altered hormone sensitivity to GA and ABA which are potentially receptor or signal transduction pathway mutants. In wheat and corn respectively, the Rht 3 and D8 dwarf mutations are insensitive to GA, and in barley, *"slender"* appears to be a GA "full-on" mutation (7, 43). These mutations are also reflected in aleurone response. Although there is little information on GA-insensitive mutants in barley, some have been isolated and are in the process of being characterised (8).

α-Amylases and Their Genes

Cereal α-amylases are highly conserved proteins (29, 30, 63). They have 428-440 amino acids, there are large tracts of homology, and they diverge at three locations, the carboxy terminus, the signal peptide, and in a region of about 20 amino acids centered around amino acid 280, termed the α-amylase signature region (Fig. 5). One of the highly conserved features relates to the arrangement of introns and exons in the genes. All cereal α-amylase genes so far examined have 2 or 3 introns similarly placed within the genes. Introns 1 and 2 occur close together at the 5' end of the gene, and intron 3 occurs at the 3' end of the gene. Introns 1 and 3 occur in all high- and low-pI α-amylase genes, whereas intron 2 occurs only in the low-pI forms. The same occurs in wheat, but there is also a third group of α-amylases (on chromosome 5) with 2 introns. In rice, α-amylase genes have 2 or 3 introns, with a distribution across all of the 5 groups. These comparisons indicate that cereal genes arose from a common ancestor with three introns. Sequence comparisons by Rodriguez and colleagues have led to an understanding of the phylogenetic relationships between a number of cereal and dicot α-amylases (29).

The α-amylases of barley have been studied extensively and many of the characteristics have been described previously (13, 35, 37, 40). The isozymes separate into two isoelectric point (pI) groups (Fig. 6), high-pI and low-pI, and these two groups of isozymes differ in many ways while isozymes within groups are more alike. Proteolytic fingerprinting studies, cell free translation

Fig. 5. The placement of introns and exons in α-amylase genes from barley (B) and rice (R). The letters below the horizontal lines correspond to the one letter codes for amino acids occurring before and after the intron splicing sites. The slash (/) indicates the intron splice sites in the DNA sequence. I = intron; E = exon; from (30).

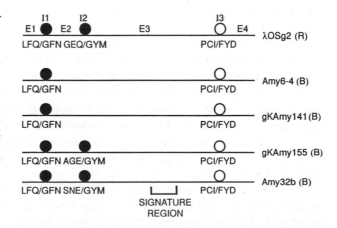

257

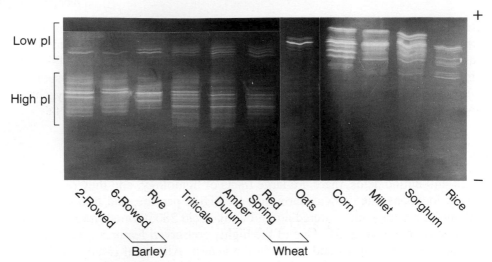

Fig. 6. Zymogram of α-amylases from germinated cereal grains. The enzymes were separated by isoelectric focusing on a pH 3.5-9.5 gradient. From (50).

studies, and the demonstration that the genes for the low- and high-pI isozymes occur on different chromosomes (chromosomes 1 and 6 respectively) indicate that there are genetic differences, not only between groups of isozymes, but also between isozymes within groups. Nucleotide sequence heterogeneity has been demonstrated for various cDNA clones, and the clones also fall into two groups corresponding to the two pI groups (13, 40). Nucleotide sequence homology between clones of the two groups is about 75%, while within groups it is about 90-95%.

The two enzyme groups also have different enzymatic characteristics. The specific activity of the high-pI group is 2.3 times that of the low-pI group (51), and low-pI enzyme is more efficient in attacking starch grains than high-pI (49). The groups are also differentially regulated by GA_3 (see later). It seems likely that there is still a good deal to be learnt about heterogeneity among the isozymes, their behaviour within the starchy endosperm, and their interaction with a heterogeneous population of starch grains.

Is each isoform encoded by a different gene? Analyses have been done at the mRNA, cDNA and gene levels (40). Primer extension analysis of mRNAs has demonstrated that there are two or three low-pI products (mRNAs), and three to five high-pI products (Fig. 7) (9). At least five different high-pI cDNAs have been described. Southern blots of genomic DNA using group specific probes have demonstrated about seven or eight DNA fragments containing α-amylase gene sequences, three of them falling into the low-pI group, and five into the high-pI group. Studies of wheat-barley addition lines demonstrate that about three genes occur on chromosome 1, and about six on chromosome 6. These numbers are in substantial agreement with the numbers of enzyme isoforms indicating that probably

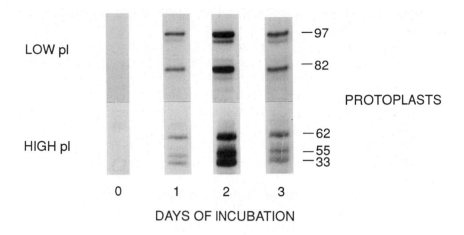

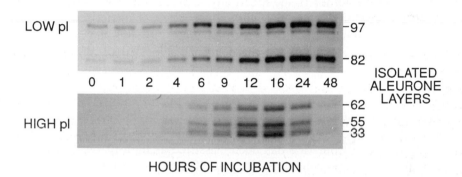

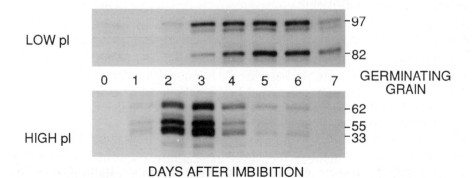

Fig. 7. Primer extensions on RNA from aleurone protoplasts, isolated aleurone layers and germinated grain of barley. All incubations occurred in the presence of 1 μM GA$_3$. Primers specific for either low-pI or high-pI mRNAs were used. Extended primers were visualised by autoradiography after electrophoresis. The numbers define individual primer bands according to the length of the 5' untranslated region of the mRNA from which each primer was extended. From (9).

259

most or all of the genes are functional and not pseudogenes, and that most of the isoforms are encoded by different genes.

The situation in wheat is very similar to that in barley (33). Wheat isozymes also fall mainly into two pI groups (Fig. 6) (50) and although there are many more isozymes in wheat (about 25) due to the presence of the A, B and D genomes, most of them have been located on chromosomes 6 and 7 (45), which correspond to chromosomes 6 and 1 of barley respectively. The kinetics of synthesis also resemble those of barley. The rates at which the low- and high-pI groups digest starch grains differ as in barley (68). A third class of genes which is expressed only in developing wheat grain has been mapped to chromosome 5 (3). The α-amylases in rice are considerably more complex than in wheat or barley, but they all seem to have low isoelectric points (Fig. 6). Ten genes have been cloned and they have been divided into five subfamilies occurring spread out over five chromosomes (1, 2, 6, 8, and 9) (62). These genes are expressed widely in the rice plant (28) with a strong tendency for tissue specificity, especially between aleurone and embryo (63). The only α-amylase gene cluster to be characterised in cereals is in rice. Three genes have been shown to occur in 28 kb of DNA (72). It is also of interest to note that within this cluster, two of the genes are expressed primarily in aleurone and the other primarily in embryo-derived callus tissue.

Not all cereals have high- and low-pI groups of α-amylase isozymes. Figure 6 shows that while wheat and barley have two groups, oat, corn, millet, sorghum and rice appear to have only one group with low-pI values. Although one gene from rice has been classified as high-pI on the basis of deduced amino acid sequence, its isozyme (if there is one) has not yet been detected.

Control of gene expression

It is envisaged that plant hormones act solely or in part by controlling transcription of genes, and thus levels of mRNA, which would, in turn, regulate rates of synthesis of specific hormone-induced proteins. It follows therefore that if this were the case for the hormonal regulation of α-amylase synthesis in aleurone cells, one would expect the level of α-amylase mRNA to accumulate in the presence of GA. There is an abundance of evidence indicating that this occurs.

The first specific assay for α-amylase mRNA was developed in 1976 (24), and involved indirect (cell-free translation) assay of the message. Results of this and other similar studies agreed that GA_3 caused translatable mRNA for α-amylase to accumulate (35). These studies did not discriminate between different α-amylase mRNAs. However, using two dimensional gel electrophoresis, it was demonstrated that *in vitro* synthesised α-amylase had components corresponding to both the high- and low-pI α-amylase groups, thus indicating that GA_3 caused mRNA for both α-amylase groups to accumulate.

Cell-free mRNA translation studies presented the opportunity of observing the effects of GA_3 and ABA on mRNA species other than α-amylase mRNA (37). GA_3 causes many changes in the translatable mRNA population, some increasing, some decreasing and some remaining constant, whereas ABA inhibits the accumulation of translatable α-amylase mRNA. At the same time, ABA causes increased levels of some mRNAs and has no effect on others. It was also shown that ABA-induced mRNAs were prevented from appearing by GA_3. In principle, GA_3 and ABA act in similar but opposing ways at the level of mRNA. Both cause increased levels of some mRNAs and decreased levels or no change in others, and each hormone appears to antagonise the other's action. The effects of GA and ABA on mRNA levels in aleurone are usually reflected in their effects on the pattern of protein synthesis (35). Together, these results indicate that many of the effects of both hormones on protein synthesis are likely to be mediated by regulation of mRNA levels, as in the case of α-amylase. However, the mechanism involved in the down regulation of expression of many genes remains obscure. There is essentially no indication whether it occurs at the protein, mRNA or gene levels (or some combination of them).

The availability of α-amylase cDNA clones, made from isolated α-amylase mRNA, presented the opportunity for direct analysis of mRNA, and the possibility of assaying total α-amylase gene transcripts (as opposed to translatable transcripts which perhaps did not reflect the total). By using the dot blot or Northern blot techniques, hybridisation analysis of RNA from control and hormone treated aleurone has provided evidence which supports the cell-free translation studies. Results of several studies on barley indicate that total α-amylase mRNA accumulates in the presence of GA_3, but not in its absence, and that ABA inhibits α-amylase mRNA accumulation (Fig. 8). Hybridisation analyses of the accumulation of α-amylase mRNA in isolated aleurone layers, using group specific cDNA probes (31, 58, 64) as well as primer extension analysis of individual isozyme mRNAs (Fig. 8) (9, 64), have shown that mRNA for the low-pI α-amylase is already present in hydrated aleurone. It increases gradually to about 10-20 fold over 24 h in response to GA_3, and responds maximally to a low level (about $10^{-9}M$ or below) of GA_3. In contrast, mRNA for the high-pI α-amylase is present in very low amounts

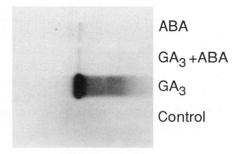

Fig. 8. Hybridisation of an α-amylase cDNA (pHV 19) clone to size fractionated RNA isolated from aleurone layers which were incubated with and without (control) hormones for 24h. From (11).

ABA

GA_3 +ABA

GA_3

Control

in hydrated tissue, increases by 50-100 fold over 12-16 h and then decreases, and responds maximally to a high (10^{-6} M or above) level of GA_3. In general, all of these results are in accord with the responses of the low and high-pI groups of α-amylase isozymes to GA_3 (37), and they reinforce the notion that the two groups of α-amylase genes are differentially regulated.

The studies reported above were done with isolated aleurone layers, but there is evidence that the timing of synthesis of isozyme groups can change depending on the aleurone system used (9). Primer extension studies (Fig. 7) demonstrated that in protoplasts, the induction of low- and high-pI mRNAs was co-ordinated for all mRNAs both within and between isozyme groups. For isolated aleurone layers, the major differences were that high-pI mRNA reached a maximum at 16 h and then declined, whereas low-pI mRNA increased to a maximum at 24 h and remained high. In germinated grain, high-pI mRNAs dominated at 2-3 days and then decreased, whereas low-pI isozymes began to accumulate after 3 days and decreased only after 7 days. Thus synthesis of high-pI isozyme mRNAs in germinated grains occurs 2 days ahead of low-pI, whereas, in the other systems, it occurs either at the same time or after low-pI, indicating that expression of the two groups of α-amylase genes can be regulated differentially, and that perhaps different mechanisms are involved.

The results described above demonstrate clearly that GA_3 and ABA regulate levels of mRNAs for α-amylase and other proteins. How does this mRNA arise? Is it by an increased rate of transcription or by increased mRNA stability, or both? Inhibitors of transcription prevent α-amylase induction, indicating that new or continued RNA synthesis is necessary, but such a result is consistent with both models. More convincing evidence for the induction of new synthesis came with the development of procedures for producing hormonally responsive protoplasts from aleurone of oat and barley, which permitted the isolation of transcriptionally active nuclei in good yield. Detection of the synthesis of specific gene transcripts in such nuclei showed that nuclei from GA_3-treated barley cells made less total RNA, more α-amylase mRNA, but less rRNA than nuclei from control cells (Fig. 9) (36).

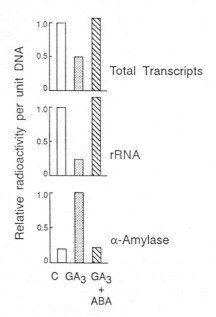

Fig. 9. The effects of GA_3 and ABA on the accumulation of total and two specific RNA transcripts (rRNA and α-amylase) in nuclei isolated from barley aleurone cell protoplasts incubated for 24 h. C = control nuclei; GA_3 = nuclei from GA_3-treated protoplasts; GA_3 + ABA = nuclei from protoplasts treated with both GA_3 and ABA. Data from (36).

If ABA was present during incubation of protoplasts with GA$_3$ at levels which were inhibitory to α-amylase synthesis, all of the effects of GA$_3$ on transcription were prevented. Studies using oat nuclei (76) have given essentially similar results, and together the results indicate that both GA$_3$ and ABA control α-amylase synthesis at least in part by controlling events within the nucleus, probably through the transcription of the α-amylase genes. These effects of GA$_3$ and ABA on α-amylase gene transcription are consistent with the effects of the hormones on levels of α-amylase mRNA described above. The studies of transcription described above employed cDNA probes and hybridisation conditions which probably favoured assay of transcripts from all α-amylase genes. By using appropriate hybridisation conditions and more selective cDNA probes, it would be possible to measure the transcription of each major group of α-amylase genes, and perhaps even individual genes.

Is there a role for hormonal control of translation? The studies relating mRNA levels to synthesis of α-amylase have shown that the accumulation of α-amylase mRNA is correlated with the rate of α-amylase synthesis under a number of experimental conditions (see above). This indicates that mRNA is probably translated without delay, and that α-amylase synthesis is regulated mainly by the level of newly synthesised mRNA. On the quantitative level, it has been estimated, using cell-free translation techniques, that α-amylase mRNA accumulates to a maximum of about 20% of the total translatable mRNA, while *in vivo* protein labelling studies indicate that α-amylase constitutes about 30% of total protein synthesis (11). Thus, there would seem to be little reason to invoke control of α-amylase synthesis at post-transcriptional levels. The case for ABA acting on translation of α-amylase mRNA is also not strong (35).

Promoter studies: *Cis-* and *trans-*acting factors

Results indicating that α-amylase synthesis is controlled, at least in part, at the level of transcription, led to the belief that control was probably exerted through hormone response elements located in the gene promoter (*cis*-acting), and that proteinaceous transcription factors (*trans*-acting) were involved. This belief was strengthened when chimeric genes containing promoters of wheat and barley α-amylase genes fused to reporter genes were introduced into barley or oat aleurone protoplasts, and the expression of the reporter genes was shown to be regulated by GA$_3$ and ABA in a dose-dependent manner (16, 32, 38, 67). More recently, similar results have been obtained by using the particle gun (biolistics) to introduce DNA into intact aleurone, as an alternative transient expression system to protoplasts (44).

Deletion analysis of wheat, barley and rice α-amylase promoters showed that the hormone response elements were within the proximal 200-300 bp upstream of the transcription start (32, 38, 42, 44). This region of each of the promoters contains three short sequence motifs, C/TCTTTTC/T, called the pyrimidine box, and TAACAA/GA and TATCCAC/T boxes, which are highly conserved among promoters of GA$_3$-responsive α-amylase genes of wheat, barley and rice (28). The placement, order and orientation of the sequences also appears to be highly conserved in these promoters. The functional roles of these conserved elements have been tested in the two different expression systems above. The results based on expression analyses of a barley high-pI α-amylase promoter (Amy6-4) in barley aleurone protoplasts indicate that the TAACAAA box can act alone as a gibberellin-response element (GARE) (69). Gain-of-function experiments showed that six copies of the -148 to -128 sequence (GGCCGATAACAAACTCCGGCC) from a barley α-amylase promoter conferred GA$_3$-regulated expression on a reporter gene. Removal of the TAACAAA sequence by site-directed mutagenesis virtually abolished the GA$_3$ response of the barley α-amylase promoter/reporter gene constructs in aleurone protoplasts (19). Mutagenesis of the TATCCAC box resulted in a loss in GA$_3$-induced expression, but less so than for TAACAAA. In contrast, mutation of the pyrimidine box had little effect on GA$_3$-regulated expression. These results indicated that both the TAACAAA and TATCCAC boxes act co-operatively and that together they may form a gibberellin-response complex (GARC). This was supported by evidence from functional analyses of a wheat α-amylase promoter in oat aleurone protoplasts (34). Partial deletion of both the TAACAGA and TATCCAT boxes virtually abolished GA$_3$-regulated expression.

Expression analysis in whole aleurone tissue has revealed both qualitative and quantitative differences in the roles of two of the conserved boxes in GA$_3$ regulation of gene expression as compared with data obtained with aleurone protoplasts. Mutation of the pyrimidine box in a low-pI barley α-amylase (Amy32b) promoter resulted in decreased expression from the promoter in response to GA$_3$ (44). The importance of the pyrimidine box has been confirmed (Gubler and Jacobsen, unpublished) in a study of a different (high-pI) barley α-amylase promoter of the gene corresponding to the cDNA clone, pHV19. Thus it seems that all three conserved boxes are required for a full GA$_3$-response. Mutation of the individual conserved boxes caused large decreases in the levels of expression in GA$_3$ treatments, but failed to abolish the GA$_3$ response. This evidence argues against the TAACAAA box acting as an autonomous GARE as found in protoplasts, but instead indicates that the three conserved boxes comprise a gibberellin response complex. Further evidence for this was obtained by gain-of-function experiments with individual conserved boxes in front of a minimal α-amylase promoter. Tandem copies of oligos containing either the pyrimidine box, or the TAACAAA or TATCCAC boxes, failed to confer GA$_3$-responsiveness onto the minimal α-amylase promoter.

These differences between the two expression systems appear to reflect differences in the control of gene expression depending on the state of aleurone cells. It would seem likely that the transcriptional controls present in whole tissue can be disturbed during preparation of protoplasts. The absence of a cell wall following treatment with crude mixtures of cell wall hydrolases may change the way aleurone cells respond to GA_3 and the results highlight the need to select transient expression systems carefully and to interpret results cautiously. Until efficient stable transformation procedures for wheat and barley are available, use of biolistics and intact aleurone would seem to be the best expression system for analysing GA_3-regulation of expression of α-amylase promoter/reporter gene constructs. Procedures for the stable transformation of rice have already been developed.

There are interesting differences emerging between high- and low-pI α-amylase gene promoters in the composition of *cis*-acting elements involved with GA action . Low-pI promoters of both barley (44) and wheat (34) have a number of *cis*-acting elements upstream of the GARC (as defined here) which are required for a full GA_3 response. Upstream of the pyrimidine box in a low-pI α-amylase promoter (Amy32b) there is an Opaque-2 like element and another sequence (from -192 to -158), both of which play an important role in GA_3-regulation of gene expression (44). Deletion of an Opaque-2 like element and the other sequence virtually abolished GA_3 induction of transcription. However there do not appear to be similar elements in a barley high-pI α-amylase promoter (19, 38). Functional analysis of a wheat low-pI promoter has also identified two regions upstream of the pyrimidine box which are required for high level of expression in response to GA_3 (34).

Although there has been considerable progress made towards understanding molecular mechanisms involved with GA_3-regulation of α-amylase gene expression, functional analysis of promoters of other GA_3-regulated genes has only recently been initiated. Two genes encoding barley (1-3, 1-4)-β-glucanase isoenzymes have been isolated, one of which has been shown to be under GA_3-regulation in barley aleurone cells (70, 75) Functional analysis of the promoter of isoenzyme II (EII) has shown that the promoter confers GA_3-responsiveness onto a reporter gene in barley aleurone protoplasts (75). Deletion analysis indicates that the hormone response elements lie downstream of -310. This region contains two motifs, TAACAAC and TAACCTC, which closely resemble the TAACAA/A and TATCCAC/T motifs present in the GARC in α-amylase promoters. There are also several putative pyrimidine boxes present upstream of the TAACAAC sequence. Analysis of another GA_3-regulated gene, a cathepsin-like gene from wheat, has shown that GA_3-responsive region of the promoter lies downstream of -276 (6). Sequence comparison of this downstream region failed to identify any TAACAA/GA homologue. Further functional analysis is required to identify the hormone response elements within this region. Sequence analysis of a promoter of GA_3-regulated rice carboxypeptidase has revealed a number of pyrimidine boxes in the region from -171 to -135 upstream of the transcription start (74). No obvious TAACAA/GA and TATCCAC/T motifs appear to be present downstream of these pyrimidine boxes. These results indicate that the occurrence of all three conserved elements is associated with strong GA_3 regulation, but perhaps a somewhat reduced strength of expression can be obtained from fewer elements, or from elements which more or less resemble the conserved gene sequences in α-amylase genes. A number of other GA$_3$-regulated genes from both monocots and dicots have elements similar to the α-amylase gene elements discussed above. No doubt, as time passes, there will be more functional analysis of these genes and it will be possible to test the hypothesis that there is a single basic type of GARC.

A number of promoter studies have also investigated whether ABA antagonism of GA$_3$-regulation of α-amylase gene expression is mediated via an autonomous ABA response element, or via components of the GARC. Evidence to date indicates that ABA action occurs via elements in the GARC, probably the TAACAAA box (18, 69). Whether ABA acts at the level of the regulated gene, perhaps directly on transcription factors which bind to the GARC, or at sites further up the GA-signal transduction chain, remains to be determined.

Hormonal regulation of transcription is likely to involve *trans*-acting factors which interact with *cis*-acting elements in the GARC. Regulation of *trans*-acting factors may occur at the level of synthesis, or by activating pre-existing forms. There is some evidence to support the former. Cycloheximide, a protein synthesis inhibitor, inhibits GA$_3$-induced transcription of α-amylase genes in barley aleurone layers, indicating that *de novo* synthesis of at least one protein is required to activate transcription (54). DNase 1 footprinting experiments, which detect zones of DNA that bind protein, have shown that rice and wheat α-amylase gene promoters have multiple sites of DNA-protein interaction with nuclear proteins (62, 66, 82), and no clear protein-binding pattern has yet emerged. Also, whether or not all of these interactions play roles in GA regulation of gene expression is yet to be determined.

How *cis*- and *trans*-acting factors interact to regulate gene transcription is unknown, however, two possibilities for a high-pI α-amylase gene are depicted in figure 10. Proteins may bind to the GARC elements in such a way as to stabilize a transcription initiating complex involving RNA polymerase. Alternatively, protein binding may give rise to a "hair-pin" structure in which GARC elements would be relocated adjacent to the transcription initiating site. In this position, the elements would be more directly involved in formation of the transcription initiating complex. Possible protein/DNA interactions have also been proposed to explain operation of a low-pI α-amylase gene promoter (65).

Sequence analysis of the GARC of a barley high-pI α-amylase promoter, indicates that the sequence T<u>AACA</u><u>AACTCCG</u> has two possible MYB-binding sites (underlined) (4). MYBs are transcription factors which regulate growth and development. Preliminary evidence based on Northern analysis indicates that expression of *myb*-gene-transcripts is under GA$_3$-regulation in barley aleurone layers (Gubler, unpublished data). The expression of two *myb*-gene transcript size classes is down-regulated by GA$_3$

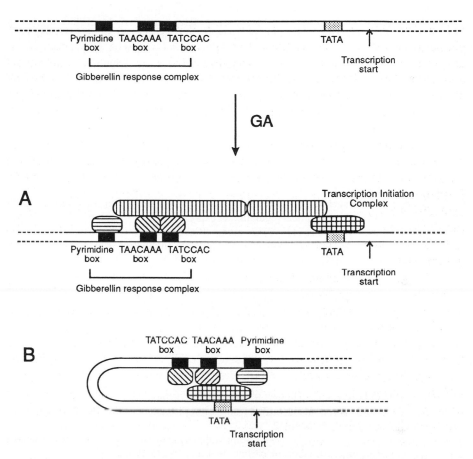

Fig. 10. Diagram of two hypothetical structures depicting the role of the gibberellin response complex in transcription initiation of a high-pI α-amylase gene.

while a third is induced by GA₃. These results lead to speculation that a GA₃-induced MYB protein(s) is responsible for *trans*-activation of α-amylase gene expression. It is relevant to note that ABA has also been shown to regulate *myb* gene expression in maize (22). It will be of interest to determine if the antagonistic action of GA₃ and ABA at the level of gene expression is mediated by GA₃- and ABA-regulated *myb* genes.

References

1. Aisien, A.O., Palmer, G.H., Stark, J.R. (1986) The ultrastructure of germinating sorghum amd millet grains. J. Inst. Brew. 92, 162-167.
2. Akazawa, T., Mitsui, T., Hayashi, M. (1988) Recent progress in α-amylase biosynthesis. *In*: The Biochemistry of Plants: A Comprehensive Treatise, Vol 14, pp. 465-492, Preiss, J., ed. Academic, New York.
3. Baulcombe, D.C., Huttly, A.K., Martienssen, R.A., Barker, R.F., Jarvis, M.G. (1987) A novel wheat α-amylase gene (Amy 3) Mol. Gen. Genet. 209, 33-40.

4.   Biedenkapp, H., Borgmeyer, U., Sippel, A.E., Klempnauer, K.-H. (1988) Viral *myb* oncogene encodes a sequence-specific DNA-binding activity. Nature 335, 835-837.

5.   Boother, G.M., Gale, M.D., Gaskin, P., McMillan, J., Spansel, V.M. (1991) Gibberellins in shoots of Hordeum vulgare. A comparison between cv. Triumph and two dwarf mutants which differ in their response to gibberellin. Physiologia Plantarum 81: 385-392.

6.   Cejudo, F.J., Ghose, T.K., Stabel, P., Baulcombe, D.C. (1992) Analysis of the gibberellin-responsive promoter of a cathepsin B-like gene from wheat. Plant Molec. Biol. 20, 849-856.

7.   Chandler, P.M. (1988) Hormonal regulation of gene expression in the "slender" mutant of barley (Hordeum vulgare L.). Planta 175, 115-120.

8.   Chandler, P.M. (1992) Gibberellin responses in barley. *In*: Barley: Genetics, Biochemistry, Molecular Biology and Biotechnology, pp. 403-411, Shewry, P.R., ed. Alden Press, Oxford.

9.   Chandler, P.M., Jacobsen, J.V. (1991) Primer extension studies on α-amylase mRNAs in barley aleurone II. Hormonal regulation of expression. Plant Molec. Biol. 16, 637-645.

10. Chandler, P.M., Mosleth, E. (1990) Do gibberellins play an *in vivo* role in controlling α-amylase gene expression? *In*: Proceedings of the 5th International Symposium on Pre-Harvest Sprouting in Cereals, pp. 100-109, Ringlund, K., Mosleth, E., Mares, D.J. eds. Westview Press, Boulder, Colorado.

11. Chandler, P.M., Zwar, J.A., Jacobsen, J.V., Higgins, T.J.V., Inglis, A.S. (1984) The effects of gibberellic acid and abscisic acid on α-amylase mRNA levels in barley aleurone layers: Studies using an α-amylase cDNA clone. Plant Molec. Biol. 3, 407-418.

12. Croker, S.J., Hedden, P., Lenton, J.R., Stoddart, J.L. (1990) Comparison of gibberellins in normal and slender barley seedlings. Plant Physiol. 94: 194-200.

13. Fincher, G.B. (1989) Molecular and cellular biology associated with endosperm mobilisation in germinating cereal grains. Ann. Rev. Plant Physiol. Plant Molec. Biol. 40, 305-346.

14. Gaskin, P., Gilmour, S.J., Lenton, J.R., MacMillan, J., Sponsel, V.M. (1984) Endogenous gibberellins and kaurenoids identified from developing and germinating barley grains. J. Plant Growth Regul. 2, 229-242.

15. Gilroy, S., Jones, R.L. (1993) Calmodulin stimulation of unidirectional calcium uptake by the endoplasmic reticulum of barley aleurone. Planta 190, 289-296.

16. Gopalalkrishnan, B., Sonthayanon, B., Rahmatullah, R., Muthukrishnan, S., (1991) Barley aleurone layer cell protoplasts as a transient expression system. Plant Molec. Biol. 16, 463-467.

17. Grosselindemann, E., Graebe, J.E., Stöckll, D., Hedden, P. (1991) ent-Kaurene biosynthesis in germinating barley (Hordeum vulgare L., cv Himalaya) caryopsis and its relating to α-amylase production. Plant Physiol. 96, 1099-1104.

18. Gubler, F., Ashford, A.E., Jacobsen, J.V. (1987) The release of α-amylase through gibberellin-treated barley aleurone cell walls. Planta 172, 155-161.

19. Gubler, F., Jacobsen, J.V., (1992) Gibberellin-responsive elements in the promoter of a barley high-pI α-amylase gene. Plant Cell 4, 1435-1441.

20. Guerin, J.R., Lance, R.C.M., Wallace W. (1992) Release and activation of barley β-amylase by malt endopeptidases. J. Cer. Sci. 15, 5-14.

21. Hamabata, A., Garcia-Maya, M., Romero, T., Bernal-Lugo I. (1988) Kinetics of the acidification capacity of the aleurone layer and its effect upon solubilisation of reserve substances from starchy endosperm of wheat. Plant Physiol. 86, 643-644.

22. Hattori, T., Vasili, V., Rosenkrans, L., Hannah, C., McCarty, D.R., Vasil, I.K. (1992) The *viviparous-1* gene and abscisic acid activate the *C1* regulatory gene for anthocyanin biosynthesis during seed maturation in maize. Genes Dev. 6, 609-618.

23. Higgins, T.J.V., Jacobsen, J.V., Zwar, J.A. (1982) Gibberellic acid and abscisic acid modulate protein synthesis and mRNA levels in barley aleurone layers. Plant Molec. Biol. 1 191-215.

24. Higgins, T.J.V., Zwar, J.A., Jacobsen, J.V. (1976) Gibberellic acid enhances the level of translatable mRNA for α-amylase in barley aleurone layers. Nature 260, 166-169.
25. Hooley, R., Beale, M.H., Smith, S.J. (1991) Gibberellin perception at the plasma membrane of Avena fatua aleurone protoplasts. Planta 183, 274-280.
26. Hooley, R., Beale, M.H., Smith, S.J., MacMillan, J. (1990) Novel affinity probes for gibberellin receptors in aleurone protoplasts of *Avena fatua*. *In*: Plant Growth Substances 1988, pp. 145-153, Pharis, R.P., Rood, S.B., eds. Springer-Verlag, Berlin.
27. Hooley, R., Smith, S.J., Beale, M.H., Walker, R.P. (1993) Photoaffinity labelling of gibberellin-binding proteins in Avena fatua aleurone.
28. Huang, N., Koizumi, N., Reinl, S., Rodriguez, R.L. (1990) Structural organisation and differential expression of rice α-amylase genes. Nucleic Acids Res. 18, 7007-7014.
29. Huang, N., Stebbins, L., Rodriguez, R.L. (1992) Classification and evolution of α-amylase genes in plants. Proc. Natl. Acad. Sci. 89, 7526-7530.
30. Huang, N., Sutliff, T.D., Litts, J.C., Rodiguez, R.L. (1990) Classification and characterization of the rice α-amylase multigene family. Plant Molec. Biol. 14, 655-668.
31. Huang, J., Swegel, M., Dandekar, A.M., Muthukrishnan, S. (1984) Expression and regulation of α-amylase gene family in barley aleurones. J. Molec. Appl. Genet. 2, 579-588.
32. Huttly, A.K., Baulcombe, D.C., (1989) A wheat α-Amy2 promoter is regulated by gibberellin in transformed oat aleurone protoplasts. EMBO 8, 1907-1913.
33. Huttly, A.K., Baulcombe, D.C., (1990) Hormonal control of wheat α-amylase genes. *In*: Genetic Engineering of Crop Plants, pp. 171-189, Lycett, G.W., Grierson, D., eds. Butterworths, London.
34. Huttly, A.K., Phillips, A.L., Tregear, J.W. (1992) Localisation of *cis* elements in the promoter of a wheat α-amy2 gene. Plant Molec. Biol. 19: 903-911.
35. Jacobsen, J.V. (1983) Regulation of protein synthesis in aleurone cells by gibberellin and abscisic acid. *In*: The biochemistry and physiology of gibberellins, Vol 2 pp. 159-187, Crozier, A., ed. Praeger, New York.
36. Jacobsen, J.V., Beach, L.R. (1985) Evidence for control of transcription of α-amylase and ribosomal RNA genes in barley aleurone protoplasts by gibberellic acid and abscisic acid. Nature 316, 275-277.
37. Jacobsen, J.V., Chandler, P.M. (1987) Gibberellin and abscisic acid in germinating cereals. *In*: Plant Hormones and Their Role in Plant Growth and Development, pp. 164-193, Davies, P.J., ed. Martinus Nijhoff, Dordrecht, The Netherlands.
38. Jacobsen, J.V., Close, T.J., (1991) Control of transient expression of chimaeric genes by gibberellic acid and abscisic acid in protoplasts prepared from mature barley aleurone layers. Plant Molec. Biol. 16, 713-724.
39. Jones, R.L., Bush, D.S. (1991) Gibberellic acid regulates the level of BiP cognate in the endoplasmic reticulum of barley aleurone cells. Plant Physiol. 97, 456-459.
40. Jones, R.L., Jacobsen, J.V. (1991) Regulation of synthesis and transport of secreted proteins in cereal aleurone. Int. Rev. Cytol. 126, 49-88.
41. Khursheed, B., Rogers, J.C. (1988) Barley α-amylase genes. J. Biol. Chem. 263, 18953-18960.
42. Kim, J.K., Cao, J., Wu, R. (1992) Regulation and interaction of multiple protein factors with proximal promoter regions of a rice high pI α-amylase gene. Mol. Gen. Genet. 232, 383-393.
43. Lanahan, M.B., Ho, T-H.D. (1988) Slender barley: A constitutive gibberellin-response mutant. Planta 175, 107-114.
44. Lanahan, M.B., Ho, T-H.D., Rogers, S.W., Rogers, J.C. (1992) A gibberellic response complex in cereal α-amylase gene promoters. Plant Cell 4, 203-211.
45. Lazarus, C.M., Baulcombe, D.C., Martienssen, R.A. (1985) α-amylase genes of wheat are two multigene families which are differentially expressed. Plant Molec. Biol. 5, 13-24.

46. Lenton, J.R., Appleford, N.E.J. (1991) Gibberellin production and action during germination of wheat. *In*: Gibberellins, pp. 125-135, Takahashi N., Phinney, B.O., MacMillan, J., eds. Springer-Verlag, Berlin.

47. Longstaff, M.A., Bryce, J.H. (1993) Development of limit dextrinase in germinating barley (*Hordeum vulgare* L) - Evidence of proteolytic activation. Plant Physiol. 101, 881-889.

48. McFadden, G.I., Ahluwalia, B., Clarke, A.E., Fincher, G.B. (1988) Expression sites and developmental regulation of genes encoding (1-3, 1-4)-β-glucanase in germinated barley. Planta 173, 500-508.

49. MacGregor, A.W. (1980) Action of malt α-amylases on barley starch granules. MBAA Technical Quarterly 17, 215-221.

50. MacGregor, A.W., Marchylo, B.A., Kruger, J.E. (1988) Multiple α-amylase components in germinated cereal grains determined by isoelectric focusing and chromatofocusing. Cereal Chem. 65, 326-333.

51. MacGregor, A.W., Morgan, J.E. (1992) Determination of specific activities of malt α-amylases. J. Cereal Sci. 16, 267-277.

52. MacGregor, A.W., Matsuo, R.R. (1982) Starch degradation in endosperms of barley and wheat kernels during initial stages of germination. Cereal Chem. 59, 210-216.

53. Macnicol, P.K., Jacobsen, J.V. (1992) Endosperm acidification and related metabolic changes in the developing barley grain. Plant Physiol. 98, 1098-1104.

54. Muthukrishnan, S., Chandra, G.R., Maxwell, E.S., (1983) Hormonal control of α-amylase gene expression in barley: studies using a cloned cDNA probe. J. Biol. Chem. 258, 2370-2375.

55. Mundy, J., Munck, L. (1985) Synthesis and regulation of hydrolytic enzymes in germinating barley. *In*: New Approaches to Research on Cereal Carbohydrates. Progress in Biotechnology, Vol 1, pp. 139-148, Hill, R.D., Munck, L., eds. Elsevier, Amsterdam.

56. Napier, R.M., Venis, M.A. (1990) Receptors for plant growth regulators: Recent advances. J. Plant Growth Regulation 9, 113-126.

57. Nicholls, P.B. (1982) Influence of temperature during grain growth and ripening of barley on the subsequent response to exogenous gibberellic acid. Aust. J. Plant. Physiol. 9, 373-383.

58. Nolan, R.C., Ho, T-H.D. (1988) Hormonal regulation of gene expression in barley aleurone layers. Planta 174, 551-560.

59. Ou-Lee, T-M., Turgeon, R., Wu, R. (1988) Interaction of a gibberellin-induced factor with the upstream region of an α-amylase gene in rice aleurone tissue. Proc. Natl. Acad. Sci. USA 85, 6366-6369.

60. Palmer, G.H., Shirakashi, T., Sanusi, L.A. (1989) Physiology of germination. EBC Congress Proceedings pp. 63-74.

61. Ranjhan, S., Karrer, E.E., Rodriguez, R.L. (1992) Localizing α-amylase gene expression in germinated rice grains. Plant Cell Physiol. 33, 73-79.

62. Ranjhan, S., Litts, J.C., Foolad, M.R., Rodriguez, R.L. (1991) Chromosomal localization and genomic organization of α-amylase genes in rice (*Oryza sativa* L.). Theor. Appl. Genet. 82, 481-488.

63. Rodriguez, R.L., Huang, N., Sutliff, T.D., Ranjhan, S., Karrer, E.E., Litts, J.C. (1990) Organization, structure and expression of the rice α-amylase multigene family. Proc. 2nd International Rice Genetics Symposium. Los Banos, Phillipines. pp. 1-16.

64. Rogers, J.C. (1985) Two barley alpha amylase gene families are regulated differently in aleurone cells. J. Biol. Chem. 260 3731-3738.

65. Rogers, J.C., Lanahan, M.B., Rogers, S.W., Mundy, J. (1992) The gibberellin response element: A DNA sequence in cereal α-amylase gene promoters that mediates GA and ABA effects. *In*: Progress in Plant Growth Regulation, pp. 136-146, Karssen, C.M., van Loon, L.C., Vreugdenhil, D., eds. Kluwer, Dortrecht.

270

66. Rushton, P.J., Hooley, R., Lazarus, C.M. (1992) Aleurone nuclear proteins bind to similar elements in the promoter regions of two gibberellin-regulated α-amylase genes. Plant Molec. Biol. 19, 891-901.
67. Salmenkallio, M., Hannus, R., Teeri, T.H., Kauppinen, V., (1990) Regulation of α-amylase promoter by gibberellic acid and abscisic acid in barley protoplasts transformed by electroporation. Plant Cell Rep. 9, 352-355.
68. Sargeant, J.G. (1980) α-amylase isoenzymes and starch degradation. Cereal Res. Commun. 8, 77-86.
69. Skriver, K., Olsen, F.L., Rogers, J.C., Mundy, J. (1991) *Cis*-acting DNA elements responsive to gibberellin and its antagonist abscisic acid. Proc. Natl. Acad. Sci. USA 88, 7266-7270.
70. Slakeski, N., Fincher, G.B. (1992b) Developmental regulation of (1-3, 1-4)-β-glucanase gene expression in barley. Plant Physiol. 99, 1226-1231.
71. Srivastava, L., Sechley, K.A. (1991) In search of a gibberellin receptor. Jour. Iowa Acad. Sci. 98, 51-62.
72. Sutliff, TD., Huang, N., Litts, J.C., Rodriguez, R.L. (1991) Characterization of an α-amylase multigene cluster in rice. Plant Molec. Biol. 16, 579-591.
73. Sun, Z., Henson, C.A. (1991) A quantitative assessment of the importance of barley seed α-amylase, β-amylase, debranching enzyme and α-glucosidase in starch degradation. Arch. Biochem. Biophys. 284, 298-305.
74. Washio, K., Ishikawa, K. (1992) Structure and expression during the germination of rice seeds of the gene for a carboxypeptidase. Plant Molec. Biol. 19, 631-640.
75. Wolf, N., (1992) Structure of the genes encoding Hordeum vulgare (1-3, 1-4)-β-glucanase isoenzymes I and II and functional analysis of their promoters in barley aleurone protoplasts. Mol. Gen. Genet. 234, 33-42.
76. Zwar, J.A., Hooley, R. (1986) Hormonal regulation of α-amylase gene transcription in wild oat (Avena fatua L.) aleurone protoplasts. Plant Physiol. 80, 459-463.

D4. Hormone Binding and Signal Transduction

Kees R. Libbenga and Albert M. Mennes

Institute of Molecular Plant Sciences, Clusius Laboratory, University of Leiden, 2333 AL Leiden, The Netherlands.

INTRODUCTION

Communication between cells in multicellular organisms is required to regulate their differentiation and organization into tissues, to control their growth and division and to regulate their diverse activities. When the organism becomes more complex during its development, communication between its different parts requires signaling systems which operate over relatively long distances. Hormonal systems belong to such long-range communication systems. Both plants and animals make use of hormonal systems, but in higher plants, which do not possess a nervous system, long-range communication is largely dependent upon a complex hormonal system.

In a hormonal system, cells of the different tissues and organs not only transmit signals, but they are also capable of detecting signals which they receive from other parts and responding to those signals in their own characteristic way. In this chapter we will discuss the molecular mechanisms by which target cells for plant hormones translate the signals into a specific response. Assuming that those mechanisms are largely unknown, which is very close to reality, how should one proceed to unravel them? Of course, the first thing to do is to study what we know about such mechanisms from investigations with other organisms and to try to elucidate a few general principles. As a working hypothesis one might then assume that these principles are, *mutatis mutandis*, also valid for higher plants. It is along this line that we will discuss the mechanism of primary action of plant hormones.

We know from other organisms that target cells are equipped with a distinctive set of receptors for detecting a complementary set of chemical signals. Receptors are (glyco)proteins which specifically and reversibly bind chemical signals but, unlike enzymes, do not convert them chemically. Upon binding, the receptor molecules are, through a conformational change, transformed into an activated state. This causes the initiation of a molecular program that ultimately leads to the characteristic response. Thus, receptor proteins act both as signal detectors and transducers.

Hormones often have pleiotropic effects, i.e., different types of target cell all respond to the same set of signals, but in a different way. In many cases these types of target cell have similar perception-and-transduction mechanisms, but the molecular programs which are elicited by these mechanisms are different. This is illustrated in a very simplified way by Fig.

P. J. Davies (ed.), Plant Hormones, 272–297.
© 1995 *Kluwer Academic Publishers. Printed in the Netherlands.*

1. In this example, receptor activation triggers a molecular program which is simply a direct activation of a distinct set of enzymes. If the set of responsive enzymes is different in another type of target cell, then the same signal elicits a different response via a similar perception-and-transduction chain.

Of course, hormones not only modulate enzyme activities in target cells. In general, most chemical signals ultimately influence target cells either by altering the properties (activities) or rates of synthesis of existing proteins or by altering the synthesis of new ones. Moreover, perception-and-transduction chains may be more complex than the ones shown in Fig. 1. This can be

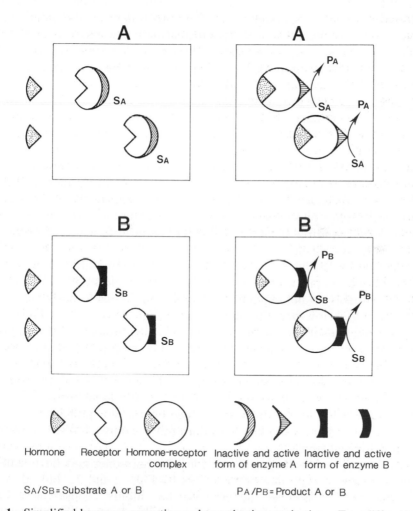

SA/SB = Substrate A or B   PA/PB = Product A or B

Fig. 1. Simplified hormone perception-and-transduction mechanism. Two different target cells, A and B respectively, contain the same receptor protein for a certain hormone. In cell A enzyme A, and in cell B enzyme B is transformed into an active state upon binding of the hormone. In this way the same signal triggers a different molecular programme in different target cells.

inferred from current models on perception and transduction of animal hormones. These models also show that, although each hormone is detected by a specific receptor, transduction follows only a limited number of pathways. Figs. 2 and 3 represent two major pathways of hormone transduction in animal systems. In one pathway (Fig. 2) the receptors are localized at the plasmamembrane, with the hormone-binding moiety facing outside the cell. These receptors function as sensory systems for external hormone levels and transduce the signal into intracellular signals, either via activation of adenylate cyclase which converts ATP into cyclic AMP and/or activation of phospholipase C which converts the membrane lipid phosphatidylinositol-4,5-biphosphate into inositol triphosphate (IP3) and diacylglycerol (DG). These signals activate c-AMP-dependent protein kinase, liberate Ca^{2+} from internal sources thus increasing the levels of free Ca^{2+}, and activate protein kinase C, respectively. These processes somehow trigger the cell's response, which may include alterations in gene expression. Many peptide hormones, for example, follow such transduction routes.

Fig. 3 represents a transduction pathway for hormones, like steroids, which are readily taken up by the target cells by simple diffusion through the plasmamembrane. These target cells are equipped with internal receptors detecting intracellular hormone levels. These receptors are regulatory proteins which may directly interact with target-cell-specific non-histone proteins and DNA sequences of the chromatin. This interaction results in increased rates and/or altered patterns of gene transcription. Target cells for steroid hormones also have plasmamembrane-bound receptors which detect external hormone levels and might be responsible for steroid-induced rapid responses in Na^+/H^+ fluxes over the plasmamembrane and in c-AMP and Ca^{2+} levels.

Transduction chains as shown by Figs. 2 and 3 are multistep regulatory circuits, of which many details are still poorly understood. The above mentioned models do not explain the complex responses to hormones, such as cell division. However, they give us an insight into the first process of perception and transduction of the signals producing these complex responses. The aim of plant-hormone receptor research is to unravel these initial events in plant-hormone action. We do not pretend that perception and transduction of plant hormones are necessarily identical with those of animal hormones. However, we want to propose the following working hypothesis:
   a. Plant hormones are detected by specific receptor proteins
   b. The transduction of plant hormones follows only a limited number of major pathways.
   c. The major transduction chains for plant hormones may have features in common with those represented by Figs. 2 and 3. Indeed, some regulatory systems appear to be highly conservative, the Ca^{2+}/calmodulin system being an example. For a critical review see (62).

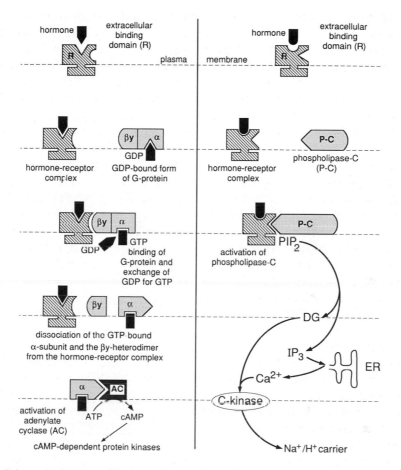

Fig. 2. Schematic diagram of the perception-and-transduction mechanisms for animal peptide hormone. In the left part of the picture a cell-surface receptor is shown. It consists of an extracellular specific binding domain coupled to a transmembrane domain. Upon binding of the hormone the receptor domain exposes a binding site for the inactive GDP-form of a G-protein. Binding results in a conformational change in the G-protein, thus exchanging GDP, bound to the α subunit, for GTP and the subsequent dissociation of the α-GTP complex from the ßγ heterodimer. The α-GTP complex binds to adenylate cyclase which is activated to produce the second messenger cAMP from ATP. This second messenger stimulates regulatory phosphorylations.

In the right part of the picture the cell-surface receptor, a transmembrane tyrosine kinase-type, upon activation by a hormone, stimulates a membrane phospholipase-C. In case the receptors are no tyrosine kinases, but seven membrane-spanning receptors, they act, upon hormone binding, via G-proteins as shown in the left part of the figure, but the effector is phospholipase-C instead of adenylate cyclase. The activated phospholipase-C hydrolyzes inositol-containing phospholipids like phosphatidylinositol 4,5-biphosphate (PIP2) present in the membrane, to inositol triphosphate (IP3) and diacylglycerol (DG). IP3 acts as a second messenger in the cytoplasm in mobilizing intracellular calcium from the endoplasmic reticulum. The second regulatory signal DG operates within the plane of the membrane to activate, in concert with the elevated concentration of calcium ions in the cytoplasm, a protein kinase C. One of its functions which has been proposed is the activation of the Na^+/H^+ exchange carrier to increase the cytoplasmic pH.

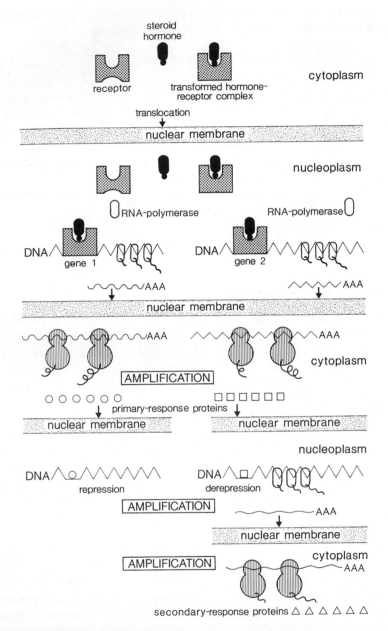

Fig. 3. Schematic diagram of the early primary response and the delayed secondary response to a steroid hormone. Steroid hormones bind to their receptor in the cytoplasm thus inducing a transformation resulting in a translocation of the complex into the nucleus. Many steroid receptors are always present in the nucleus and therefore do not show translocation. The hormone-receptor complex binds specifically and with high affinity to regulatory proteins and DNA sequences of the chromatin. The result is an increase in RNA-polymerase II activity producing gene-specific mRNAs which are translated in the cytoplasm into primary-response proteins, thus amplifying the signal. Each of these proteins may in turn activate or repress other genes thus producing an amplified secondary response.

PERCEPTION AND TRANSDUCTION OF PLANT HORMONES: AUXINS

In the preceding section we have shown that in order to understand plant-hormone action at the molecular level we have to think in terms of signal perception, transduction and response. Therefore, we have to search for receptors, which by definition are the signal-perceiving molecules, and then describe how they are involved in the transduction of the signal.

Auxins, often in combination with other hormones, influence many processes in higher plants. In this section we want to discuss auxin-receptor research on two examples of auxin action:

- Initiation and maintenance of tissue proliferation and formation of adventitious roots or shoots (regeneration).
- Stimulation of cell elongation.

For complete reviews on auxin receptors the reader is referred to (24, 27, 34, 39, 46, 66, 67).

Proliferation and Regeneration

A classical model system for the study of hormonal control of proliferation and regeneration is cultured stem-pith tissue from *Nicotiana tabacum*. In 1957 it was demonstrated for the first time that proliferation and regeneration in this tissue can be controlled by exogenous auxins and cytokinins (57). These two classes of plant hormones are both required for proliferation; at relatively high auxin concentrations ($[IAA] \approx 10^{-5}M$) and low cytokinin concentrations ($[kinetin] \approx 10^{-7}M$) the proliferating tissues produce roots, whereas at relatively low auxin concentrations ($\approx 10^{-7}M$) and high cytokinin concentrations ($\approx 5 \times 10^{-6}M$) they produce shoots. At intermediate auxin/cytokinin ratios (ca. $10^{-5}M$ IAA and $10^{-6}M$ kinetin) only proliferation occurs. Although this discovery initiated a vast amount of fundamental and applied research on the hormonal control of proliferation and regeneration, we still know very little about how auxins and cytokinins work and interact at the molecular level.

About twenty years ago it was decided in the authors' laboratory to approach this problem by unraveling the perception-and-transduction chains of these hormones. We started with a search for high-affinity auxin-binding sites[1] in 3-week old callus tissues, derived from freshly isolated tobacco stem pith. The results of extensive binding experiments with various cell fractions have been summarized in Table 1. The callus tissue contains three classes of auxin-binding proteins, which can be distinguished by their binding behaviour and their location. Two of these classes of proteins are membrane-bound and probably localized on the plasmalemma. One has a high affinity for the auxin-transport inhibitor naphthylphthalamic acid (NPA) and binding takes

---

[1] For a mathmatical treatment of the measurement of hormone binding the reader is refered to the first edition of this book.

place rapidly at 0°C at an optimal pH of 4; the natural auxin IAA shows low-affinity binding, which can only be demonstrated in competition experiments with radiolabeled NPA. Because of its high affinity for NPA this class is called NPA-binding proteins.

The other membrane-bound auxin-binding protein can be distinguished from the NPA-binding protein by its temperature- and time-dependent binding behaviour. With 10^{-7}M IAA maximal binding is reached after 60 min at 25°C at an optimal pH of 5, and it does not bind NPA. This binding protein exhibits complex binding and its concentration is much higher than that of the NPA-binding protein (40). In solubilized fractions of homogenized tissue and in salt extracts of isolated nuclei a third binding protein is present in apparently very low concentrations. The binding of active auxins to this protein is also temperature- and time-dependent. At 2.5 nM IAA, maximum binding is reached after 45 min at an optimal pH of 7.5. The affinity of this binding protein for IAA is much higher than that of the membrane-bound auxin-binding protein. This protein is called the cytoplasmic/nuclear auxin-binding protein (35).

Cytoplasmic/Nuclear Auxin-binding Proteins

Auxin is readily taken up by the cells and one class of specific high-affinity auxin-binding proteins is present in both the cytoplasm and the nucleus. Therefore, as a working hypothesis it was assumed that a major pathway of auxin perception and transduction roughly follows the scheme as established for steroid hormones, i.e., auxin directly controls transcriptional activity in nuclei via coupling to a cytoplasmic/nuclear receptor (Fig. 3). In order to verify this working hypothesis the influence of partially purified receptor (R) fractions on the transcriptional activity of isolated nuclei was studied. It was found that addition of R-fractions resulted in a reproducible auxin-dependent stimulation of RNA-polymerase-II activity, provided that a minimum concentration of specific binding protein is present (Fig. 4; 34, 63). A similar stimulation of transcription has been reported for mung bean hypocotyl nuclei (30) which, in the presence of auxin-binding proteins (ABP-I and ABP-II), showed an IAA-dependent initiation of RNA synthesis, thus giving rise to some specific translation products. Concerning our

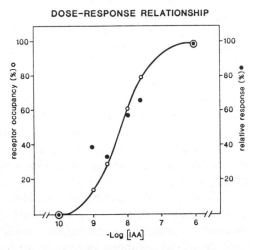

Fig. 4. Dose-response relationship of the stimulation of transcription in isolated nuclei by a high-affinity binding protein fraction from the cytosol of tobacco callus. From (63).

results (Fig.4) it is not clear whether this was an overall stimulation or that it was a stimulation of the transcription of specific sets of genes. Comparable results were obtained with a 2,4-D-dependent cell suspension line from *Nicotiana tabacum* (34). Moreover, it was found that in early-stationary phase cells most binding activity is present in the cytosol, whereas in rapidly dividing log-phase cells most specific auxin-binding is present in salt extracts of isolated nuclei. This observation is extremely interesting in that it resembles the apparent translocation of occupied steroid receptors to the nuclei in hormone-activated target tissues.

During the growth cycle of a soybean cell suspension, a similar decline of two cytoplasmic auxin-binding sites was reported (15). Interestingly, an increase in the binding affinity during the growth of the cells for both binding sites was observed, while the number of one of the two sites increased at the onset of rapid cell division thus suggesting a possible causal correlation. However, one should be very careful in drawing conclusions on receptor concentrations from *in vitro* binding experiments. Experience with the cytoplasmic/nuclear binding proteins tells us that even with highly standardized procedures, the number of binding sites per mg of protein varies considerably, and often is below the detection level of the binding assay. This indicates that unknown factors might influence the recovery of the high-affinity binding sites. For example, we have found that the amount of specific IAA-binding in crude protein extracts can be significantly increased by adding MgATP and/or excess artificial substrate for phosphatases (p-nitrophenylphosphate) to the binding-assay medium (63). These observations suggest that the auxin-binding protein might be liable to affinity modulation by ATP-dependent phosphorylation and by dephosphorylation, transforming the binding proteins into a high- or a low-affinity form respectively. The phosphatase activity in the preparations - which may vary from experiment to experiment - was probably a strong modulating factor in the recovery of high-affinity auxin-binding in the original experiments. Such an affinity modulation has also been proposed for steroid receptors (42). This putative auxin-receptor protein could also be detected in various kinds of proliferating tissue from *Nicotiana tabacum* (Table 1). An examination of different tobacco shoot tissues revealed that this binding protein is present in shoot tips, but apparently not in mature tissue like stem pith (63). A reexamination of the latter tissue is required because we cannot exclude the possibility that the binding proteins are present after all, but predominantly in their low-affinity form.

We have tried to purify the cytoplasmic / nuclear binding sites by affinity chromatography after activation of the binding sites with MgATP in the presence of p-nitrophenylphosphate. Addition of a fraction, which was eluted from the affinity column by IAA, to isolated nuclei has resulted in a significant stimulation of transcription on several occasions (35).

As already mentioned, the number of cytoplasmic/nuclear binding sites in crude preparations varies considerably thus giving rise to often non-

279

Table 1. General characteristics of auxin receptors in tobacco tissues. From (63).

General characteristics	Receptors		
	Membrane bound		Soluble
	Auxin recep.	NPA* recep.	Auxin recep.
Presence in tissues:			
Stem pith	+	+	-
Stem-pith callus	+	+	+
Leaves	+	+	nd[†]
Protoplasts (2 days)	+	+	-
Shoot tips	nd[†]	nd[†]	+
Cell-suspension cul.	+/-	+	+
Location in the cell	Plasmalemma	Plasmalemma	Cytoplasm Nucleus
pH optimum of binding	5.0	4.0	7.5
Specificity:			
Ka for IAA (M^{-1})	6×10^4	5×10^3	1.6×10^8
1-NAA	1×10^7	2×10^4	1×10^8
2-NAA	1×10^6	1×10^5	1×10^7
2,4-D	3×10^5	5×10^5	3×10^7
TIBA*	1.5×10^4	8.9×10^4	1×10^8
NPA*	$<1 \times 10^3$	$3\text{-}5 \times 10^8$	$<1 \times 10^5$
Fusicoccin	$<1 \times 10^3$	1.1×10^4	nd[†]
Tryptophan	$<1 \times 10^3$	$<1 \times 10^3$	$<1 \times 10^5$
Concentration (pM/mg protein)	50	1.6	0-0.2

† nd = not determined.

* NPA = naphtylphthalamic } auxin transport inhibitors
  TIBA = triiodobenzoic

reproducible results. Therefore, most investigators, including ourselves, have focussed on the specific gene responses evoked by the hormone binding and the consecutive transduction of the signal (29).

In tobacco we have isolated 7 cDNA clones of mRNA that were specifically transcribed after addition of auxin to an auxin-starved cell suspension (65). The induction of genes corresponding to cDNA clone, pCNT103, showed a good correlation with cell division and the corresponding mRNAs were found to accumulate transiently prior to the cell division response after auxin treatment (64, 65). The fusion of the promoters of two 103-like genes to the 5' end of the coding sequence of the ß-D-glucuronidase (GUS) (64) provides us with a system for probing the function of cell-surface receptors in auxin-induced gene expression, using protoplasts derived from transgenic cell cultures. Moreover, *in vitro* experiments with nuclei isolated from auxin-treated cells showed that the 103-gene was clearly expressed, but not in nuclei from untreated control cells (65). These control nuclei may, therefore, be used in reconstitution experiments to show a hormone- and auxin-binding protein-dependent induction of specific gene transcripts like the

103-mRNAs. Since our nuclear system was shown still to possess transcription initiation factors (41), a clear identification and isolation of putative soluble auxin receptors is desperately needed. Maybe the recently published strategy of using anti-idiotypic antibodies will prove to be useful (49).

Membrane-bound Auxin-binding Proteins

In order to establish a receptor function for membrane-bound hormone-binding proteins, one first needs to know on which membrane system they are localized. Indirect evidence indicates that the membrane-bound auxin-binding proteins from tobacco tissues are localized on the plasmamembrane. This evidence stems from experiments with tobacco leaf-mesophyll protoplasts. The binding proteins can be detected in microsome fractions from leaf-mesophyll tissue, but they are missing, or at least cannot be detected, in freshly isolated protoplasts from that tissue. Upon culturing these protoplasts the binding proteins can be detected again just after the onset of cell divisions. It was found that the binding proteins are not destroyed by osmotic stress, but most probably by proteolytic enzymes that are present in the cellulase and macerozyme preparations used for protoplast isolation (34). This is further supported by the observation that pure protease preparations have the same effect as the enzymatic preparation of protoplasts. If these enzymes do not penetrate the plasmalemma of the protoplasts - and up to now there are no reports describing endocytosis of proteins into protoplasts - then only the proteins in the cell wall and at the external face of the plasmalemma will be degraded. Furthermore, it is known from animal systems that proteins taken up by endocytosis are usually transferred to lysosomes and destroyed. We can exclude a possible localization in the cell wall, because it was found that cell-wall fragments from tobacco callus are pelleted at low-speed centrifugation, while binding proteins are not. Moreover, the density, in linear sucrose gradients, of particles containing the binding proteins is much lower than that of cell-wall fragments. Hence, an obvious explanation for the disappearance of binding proteins by exogenous proteases is that they are located at the external face of the plasmalemma. Such a location of the binding sites is in agreement with the pH optimum of 5 for auxin binding (34).

As distinct from NAA + kinetin-requiring cell-suspension lines from tobacco, 2,4-D-dependent cell lines from the same source do not have detectable amounts of membrane-bound auxin-binding proteins. These cell lines seem to have lost their ability to regenerate roots. However, after subculture on medium with NAA + kinetin the binding proteins reappear and the ability to regenerate roots is restored (Fig. 5; 39). The 2,4-D-dependent cell lines still possess the cytoplasmic/nuclear auxin-binding proteins and the NPA-binding proteins. On the basis of these results it is tempting to assume that the membrane-bound auxin-binding proteins might have some function in auxin-induced root regeneration. The high concentrations of IAA which are generally required for root regeneration are in agreement with the apparently low affinity of this binding protein for natural auxins. Based on this assumption the levels of membrane-bound auxin-binding proteins and, as a marker, a root-specific peroxidase were studied in several tobacco cell lines, including an auxin-resistant variant lacking the ability to differentiate roots (44). It was shown that in the cell lines there exists a correlation between the presence of membrane-bound auxin-binding protein and the root-specific peroxidase, while in the variant neither the binding protein nor the peroxidase could be detected.

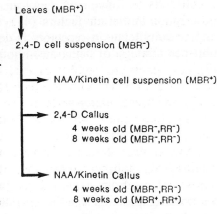

Fig. 5. Sequence of cultures derived from tobacco leaves. MBR$^+$/ MBR$^-$=presence and absence of specific membrane-bound NAA binding; RR$^+$/RR$^-$=appearance and non-appearance of roots in the culture within two weeks after transfer to solid medium containing 10^{-4} M IAA and 10^{-6} M kinetin, respectively.

Solubilization of the membrane-bound auxin-binding protein from tobacco cell suspension cultures showed that the binding protein clearly differs from that of *Zea mays* (see below), and that the membrane-bound and soluble binding proteins present in tobacco are two distinct auxin-binding proteins (43).

Much information on the membrane-bound and cytoplasmic/nuclear auxin-binding proteins from tobacco was obtained from binding experiments, but a biochemical characterization, as seen in studies on the auxin-binding protein (ABP) in maize (see below), was still missing. We, therefore, studied the effect of a specific antibody, raised against a part of the auxin-binding site of the maize ABP (68), on the auxin-induced expression of one member of the 103-gene family in the 2,4-D-dependent cell suspension. Combined with additional experiments on auxin-affected transmembrane potentials (1, 9; see below), we have obtained substantial evidence that, in plasmamembranes of tobacco protoplasts, an auxin-binding protein is present which is immunologically related to the maize ABP, and which is involved in the induction of a specific gene (submitted).

Concluding Remarks

The evidence obtained thus far indicates that tobacco cells are equipped with genetic information coding for at least two classes of auxin-binding proteins and a class of putative auxin-export carriers.

The cytoplasmic/nuclear binding proteins are probably receptors which are capable of detecting *relatively low intracellular* concentrations of auxin. The occupied receptors presumably stimulate the transcription of genes that are either directly and/or indirectly involved in the cell's response to auxin, i.e. cell division. Some of these genes are now known and can be used to study the auxin transduction pathway. Up to now the data obtained for the cytoplasmic nuclear auxin-receptor agree well with the model for steroid hormones (Fig. 3). The membrane-bound auxin-binding proteins may be receptors which are capable of detecting only *relatively high extracellular* auxin concentrations. The occupied receptors somehow transduce the auxin-signal into intracellular signals, which trigger the response, i.e., possibly pro-gramming the cells in such a way that the proliferating tissue acquires the potential to regenerate roots.

If these two classes of auxin-binding proteins constitute the receptor system by which auxin controls proliferation and root formation, then, since cytokinins do not interfere with the binding of auxin to these receptors, the cells must also be equipped with a distinct set of cytokinin receptors. The interaction between auxins and cytokinins occurs probably at the level of signal transduction. In this connection it is interesting to mention that, in cell lines in which there is a correlation between the presence or absence of membrane-bound auxin-binding protein and a root-specific peroxidase, both activities were expressed if kinetin was supplied (44). Unfortunately, at present nothing is known about cytokinin receptors in tobacco tissue-cultured cells (see also the section on cytokinin receptors).

Whether the maize ABP-related auxin-binding protein, present in the plasmamembranes of tobacco protoplasts and involved in auxin-regulated gene transcription, represents a third class of auxin-binding proteins is not yet clear.

Cell Elongation

The experimental systems most widely used in the study of the mechanism of auxin action are those in which auxin seems to be a major limiting factor, i.e., the classical auxin bioassays. In auxin bioassays (coleoptiles and stem segments from etiolated seedlings) exogenous auxin stimulates growth by cell elongation in a concentration-dependent way. In auxin-receptor research coleoptiles from *Zea mays* have been studied most extensively. Therefore the results obtained with this system illustrate the progress that has been made in the isolation and characterization of auxin receptors involved in cell elongation.

Up to now, three auxin-binding proteins in maize coleoptiles have been detected in microsome fractions, indicated as Site I, II, and III, respectively (34). Most reports agree that *in vitro* specific binding of auxin occurs rapidly at 0°C at an optimal pH of 5.5. With the aid of gradient-centrifugation techniques and the use of membrane-marker enzymes it was found that most binding sites are located on the endoplasmic reticulum (ER, Site I). However, besides the ER, the plasmamembrane (Site III) and the tonoplast (Site II) possibly contain high-affinity auxin-binding sites as well (34). The binding proteins can easily be solubilized by either acetone or Triton X-100 treatment of the microsome preparations (34, 66). In the first reported extensive purification, solubilized binding proteins from crude microsome fractions were purified to a high degree by a combination of ligand-affinity chromatography and immunological methods (36). This purified auxin-binding protein (ABP) has a sharp binding optimum at pH 5.5. It seems to be a 40-kDa dimer in its native form, and it has a higher affinity for NAA (K_d $5.7 \times 10^{-8}M$) than is found for NAA binding to microsome fractions (K_d $4 \times 10^{-7}M$). Scatchard analysis of NAA binding in the purified binding-protein preparations reveals only one class of binding sites, showing the characteristics of Site I. With monospecific antibodies against the purified binding protein (IgGanti-ABP), and using indirect immunofluorescence labelling of microscopic preparations of fixed coleoptile segments, it could be shown that the binding proteins are localized within the outer epidermal cells. IgGanti-ABP at $10^{-8}M$ specifically inhibits auxin-induced growth in coleoptile segments and it strongly reduces the auxin response of split coleoptile sections. Two conclusions are drawn from these observations. 1) The binding protein is involved in auxin-induced elongation growth; 2) The binding protein has to be located at the external face of the plasmamembrane, because it seems highly unlikely that the IgGanti-ABP reaches the cytoplasm of living cells (37).

If these auxin-binding proteins are indeed receptors involved in elongation growth then their relatively high concentration within the outer epidermal cells raises the interesting question of whether this tissue is the main auxin-sensitive part of the coleoptile. The answer to this question might be affirmative, since removal of the outer epidermis seems to make the coleoptile sections substantially less sensitive to auxin (48). Moreover, it was also found that the specific activity of auxin binding by a major 21-kDa subunit, purified from the membrane fraction of maize shoots via a NAA-affinity column, was lower in abraded coleoptiles in which many epidermal cells are destroyed (55, 56).

Further circumstantial evidence for a receptor function of membrane-bound auxin-binding sites in cell elongation is provided by results from maize mesocotyls. Microsome fractions from maize mesocotyls contain an abundant class of Site I-high-affinity auxin-binding proteins localized on the ER. Mesocotyls from etiolated and red-light-treated maize seedlings show a large difference in the amounts of auxin-binding proteins, either expressed per mg of membrane protein or per g fresh weight. Red light considerably reduces the amount of binding proteins (30-50%) and reduces the sensitivity to auxin-induced elongation. This treatment does not alter the affinity but only the number of binding proteins. Auxin-dose-response curves obtained from mesocotyl segments isolated from either irradiated or etiolated seedlings do not show a difference in the auxin concentration at which half-maximum response is obtained, but only in the maximum-response value. This can be expected if the response is proportional to the amount of occupied receptors (70). Comparison of some properties of these crude Site I-binding sites with properties of a purified maize 43-kDa auxin-binding protein having 22-kDa subunits (36, 47, 55), showed that they share the spatial distribution in maize mesocotyls, the effects of light on its abundance, and its subcellular location (26).

Since the first report on large purification of ABP (36), many others have identified ABP's in maize. Thus, by ion exchange and affinity chromatography a Site I, 21-kDa subunit ABP was obtained (56). Applying anion exchange, gel filtration, and FPLC Mono Q, preparations over 50% purity were reported which were used to raise monoclonal antibodies (47). Native polyacrylamide gel electrophoresis (PAGE) yielded a 44-kDa ABP which consisted of a glycosylated homodimer of two 22-kDa subunits.

With the specific photoaffinity labeling agent, 5-azidoindole-3-acetic acid, two peptides with a subunit molecular mass of 24 and 22 kDa, which saturably and specifically bind the hormone, were tagged in purified extracts of maize membrane proteins (28). The native molecular masses of each ABP were between 40 and 50 kDa. Using the same photoaffinity labeling technique a polypeptide doublet of 40 and 42 kDa was specifically labelled in plasma membrane-enriched fractions of tomato (18) and in zucchini (*Cucurbita pepo*) (17). The *dgt* mutation of tomato, assumed to be an auxin receptor mutant, lacked the photoaffinity labeling of the polypeptides, but only in the stems, not in the roots (18).

Diverse polyclonal and monoclonal antibodies have been raised against ABP at different steps of purification (47). Two monoclonal antibodies specifically recognized the major 22-kDa binding protein and could be mapped to epitopes within a C-terminal 1-kDa region. This region appeared to be involved in a conformational change which is induced when auxin binds (45). Recently, it was found that antiserum (D16), raised against a synthetic peptide that is expected to form part of the auxin-binding site of the maize ABP, hyperpolarizes protoplast transmembrane potential in an auxin-like manner (68), thus acting as an auxin agonist.

Oligonucleotide probes constructed on the basis of the NH$_2$-terminal sequence of the purified ABP (56) were used to screen a cDNA library derived from shoots of 3-day-old maize seedlings (22). By the same screening strategy the cDNA encoding the 22-kDa ABP has also been cloned at the same time by a second group (16). A third group used a highly purified ABP-antibody as probe to select an ABP cDNA clone from a λgt 11 cDNA library from maize coleoptiles (61). All three laboratories have obtained very similar amino acid sequence data. The predicted precursor for the binding protein contains 201 amino acid residues and has a molecular weight of 22 kDa. The sequence indicated a 38 amino acid signal peptide as had been calculated from *in vitro* translation experiments (38). The hydrophobic N-terminal leader sequence could represent a signal for translocation of the ABP to the ER. The mature ABP contains a potential N-glycosylation site as predicted before (38, 47), and, at the C-terminus, a tetrapeptide sequence, KDEL (lys-asp-glu-leu), known as a common signal for proteins that are retained within the lumen of the ER. The presence of a KDEL sequence is in accordance with the finding by two groups that the 22-kDa ABP subunit and Site I activity comigrate on sucrose gradient density fractionation predominantly with the ER-marker, cytochrome-c reductase (26, 55).

Auxin Receptors and the Acid-Growth Theory

For about the last 25 years the study of auxin action in cell elongation has been dominated by the so-called acid-growth theory. According to this theory auxin induces acidification of the free space in the cell wall, presumably by the activation of plasmamembrane-bound proton pumps (see Chapter D1). The increase in proton concentration brings about an increase in the plasticity of the cell wall, thus causing a rapid increase in elongation rate of the tissues.

The acid-growth theory suggests that proton pumps might be targets for auxin receptors. A rather direct approach to verify this hypothesis is to isolate receptors and pumps and to reconstitute a functional system by incorporating these putative components of the auxin-transduction chain into artificial membranes. Only one report has appeared describing such an approach (60). The auxin-binding proteins from maize coleoptiles were solubilized and partially purified. The purified fractions exhibited both high-affinity auxin binding and ATPase activity. These proteins were incorporated into artificial membranes in which transmembrane currents were measured. ATP, binding proteins + ATPases and NAA were added to the membranes in different sequences. Only in the experiments in which NAA was added as the last component was a convincing, immediate increase in current observed. Unfortunately no experiment was done in which binding protein + ATPase and NAA were added before the addition of ATP. The optimum pH for stimulation (pH 5.3) is in good agreement with the optimum pH of binding (pH 5.5). Finally, the K_d of the binding, ca 10^{-7}M, is close to the lower detection limit of an NAA effect (ca 10^{-7} M). These very interesting results have never been repeated by any other laboratory.

Functional evidence for an auxin receptor localised at the plasmamembrane was reported for tobacco mesophyll protoplasts (Table 2; 1). Auxin-induced hyperpolarization of membrane potential (Em) can be measured by a microelectrode technique (9), and this functional test was used to show the effects of polyclonal antibodies raised against the ABP of maize (37). Although these antibodies were derived from maize, they blocked the auxin-induced hyperpolarization in the tobacco leaf protoplasts, but not that induced by fusicoccin (FC). Antibodies directed against a plasmalemma H^+-ATPase, however, blocked both, auxin-induced and FC-induced hyperpolarization (1). These results demonstrate 1) that an auxin receptor is present in tobacco plasmamembranes and that it shares some homology with the maize ABP, and 2) that the auxin- and FC-receptors are distinct, but both act via a H^+-ATPase. Of particular interest is the finding that addition of

Table 2. Effects of anti-ABP IgG and anti-ATPase IgG on the variation in membrane potential difference in tobacco protoplasts induced by auxin or fusicoccin.

Addition to medium	Em, mV	NAA- or FC-induced Em, mV
None	-5.3 ± 0.3	—
NAA (5μM)	-12.7 ± 0.3	- 7.4
Anti-ABP IgG (0.4μM)	-5.8 ± 0.2	—
Anti-ABP IgG (0.4μM) + NAA (5μM)	-6.0 ± 0.4	- 0.7
None	-6.1 ± 0.3	—
NAA (5μM)	-12.5 ± 0.4	- 6.4
Anti-ATPase IgG (0.4μM)	+0.8 ± 0.6	—
Anti-ATPase IgG (0.4μM) + NAA (5μM)	-0.2 ± 0.4	- 1.0
FC (1μM)	-10.8 ± 0.4	- 4.7
Anti-ATPase IgG (4μM) + FC (1μM)	+0.9 ± 0.6	+0.1

Combined data from (1).

purified maize ABP increases the sensitivity of the protoplasts towards NAA (Fig. 6; 2).

Impermeant auxin analogues (69) are capable of inducing membrane hyperpolarization and extension growth, though only when the cell walls were removed, or the epicotyls were abraded to perforate the cuticle, respectively.

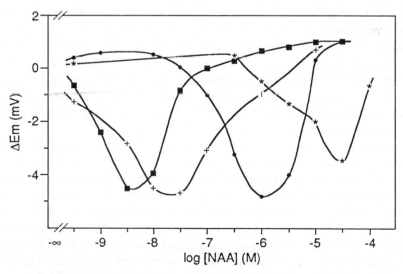

Fig. 6. Dose-response curves for Em of tobacco protoplasts in the presence of NAA. Mesophyll protoplasts were isolated from tobacco leaves of wild-type plants (●) or *Agrobacterium rhizogenes* transformed plants (■). Also shown are the effects of ABP and anti-ABP IgG on the sensitivity of the protoplasts towards NAA. Before the addition of NAA, protoplasts isolated from wild-type plants were treated with either 10^{-10}M ABP (+) or with 4.10^{-9}M anti-ABP IgG (*). Combined data from (2) and (3).

This shows that the functional auxin receptors are localized on the outside of the plasmamembrane. It also indicates that auxin can produce both, rapid and long-term responses, without entering the cell.

When protoplasts from an auxin-resistant tobacco mutant were used, it was demonstrated, by immunotitration with antibodies against maize ABP (37), that the mutant had fewer receptors. Similarly, protoplasts isolated from root tips of plants transformed by *Agrobacterium rhizogenes* exhibited a 100 to 1000 fold increase in their sensitivity to auxin and a corresponding increase in the concentration of receptors (Fig. 6; 3). In these transformants the auxin sensitivity of the cell division response was not different from the wild type, but there was an apparent correlation between the abundance of membrane receptors and the ability of plants to undergo rooting (40, 44).

If the conclusions drawn from all these experiments with IgGanti-ABP are correct, then the auxin-responsive cells are equipped with receptors at the external face of the plasmamembrane that can detect auxin-signals outside the plasmamembrane and translate them into a response. In our opinion, an attractive possibility is that these auxin receptors are the first elements in a transduction chain that might be comparable to the IP3 and DG generating system as discovered in animal cells (see Introduction). This would mean that auxin transduction is a subtle and complicated system which, among other things, regulates cytoplasmic Ca^{2+} levels and plasmamembrane-bound protein complexes, such as particular classes of proton pumps, probably by phosphorylation processes.

There is strong evidence, however, that the bulk of the 22-kDa ABP is located in the lumen of the ER (16, 22, 26, 55, 61), whereas the site of its action is at the outer face of the plasmamembrane. This so-called ER paradox (24) is difficult to explain, but some models have been proposed. One model assumes that the Ca^{2+} level of the cell is perturbed from its resting level and that this perturbation causes the movement of the auxin receptor from the ER via the normal secretory pathway to the plasmamembrane. The action of a specific carboxypeptidase removing the ER retention signal was supposed (24). Based on the finding that with antibodies against KDEL only one of two purified ABP bands from *in vitro* ABPcDNA translation products (61) was detected, the second model also favours the action of a carboxypeptidase (31). The ABP secreted via the endomembrane system to the plasmamembrane may be primarily bound to acceptor sites. Because there are inconsistencies with the acid-growth theory (see Chapter C1), the model proposes two different membrane-integrated or docking proteins for the ABP-auxin complex; one specific for H^+ secretion and the other receptor for cell elongation. In this model the ABP is presented as an apo-proteohormone, which on auxin binding reaches the state of a holohormone (31).

In an attempt to harmonize several conflicting data it was proposed that ER-auxin-binding proteins might cycle through the cell. In response to endogenous auxin, ABP-auxin complexes are formed in the ER making the KDEL tail inaccessible for binding to its receptor. These complexes are released from the lumen of the ER, following a secretory pathway to the cell surface and possibly serving as a molecular chaperone for cell wall precursors. At the cell surface, ABP-auxin would activate the proton pump and thereby increase the membrane potential. After secretion, ABP would interact with membrane proteins (specific ABP-receptors?) at the cell surface and become subject to endocytosis. Once in the membrane system KDEL receptors would return ABP back to the ER (8).

In accordance with this model are the recently published immunocytochemical results (25) showing that the auxin-binding protein present in the ER is secreted to plasmamembrane and into the cell wall without loss of the carboxy-terminal KDEL sequence. When the cells of the studied maize cell line were starved for 2,4-D less ABP was retained in the cell and more was secreted into the medium. Possibly the latter finding presents a good opportunity to apply other methods for identification of ABP, like binding assays or even amino acid sequencing after purification of the secreted protein.

Although these models give some useful ideas to explain the ER-paradox we still have to solve the problem of auxin signal transduction leading to specific gene expression.

Concluding Remarks

There is now substantial evidence that maize coleoptiles and mesocotyls contain membrane-bound auxin-receptor proteins involved in elongation growth. It seems that functional receptors are located at the external face of the plasmamembrane and it is assumed that the ER-bound auxin-binding sites may represent precursors of functional receptors which undergo a maturation process to a final form in the plasmamembrane (8, 24, 31). With monospecific antibodies it has recently been possible to establish the intracellular localization of the auxin receptors more directly by means of electronmicroscopical immunolabelling techniques (25). In those cases where localization studies have been performed the plasmalemma and the ER are tentatively indicated as the membrane system which contain the binding sites. Most of the binding sites bind the ligand rapidly at 0°C at a pH optimum varying between pH 4 to 6.5. All these binding sites need further characterization and it remains to be proven that they have a receptor function. In unravelling the auxin transduction chain in cell elongation, one must be alert not to be limited by the acid- growth theory, especially since it is shown that the induction of hyperpolarization and the stimulation of proton extrusion is not IAA-specific. Also growth-inactive structural analogues, like 2-NAA and 3,5-D, were found to be active (10). An open mind for alternative concepts and research strategies is required.

OTHER PLANT HORMONES

To an even greater extent than in auxin-receptor re search, the study of transduction chains of the other plant hormones has predominantly stuck to the characterization of high-affinity binding sites. For this reason we shall discuss these hormones only briefly.

Cytokinins

The most studied cytokinin-binding proteins are the ones present in wheat germ. This system contains a binding protein (CBF-l=wheat germ cytokinin-binding protein) which is highly specific for 6-substituted purine cytokinins. This protein appears to be loosely bound to ribosomes (reviewed in 5). A similar, if not identical, protein and a 30-kDa high-affinity cytokinin-binding protein (CBF-2) are present in an apparently free form in the cytosol. The CBF-l protein from the ribosomal fraction has been purified to a very high degree by ligand-affinity chromatography. The purified CBF-l appears to consist of 4 subunits but later work has conclusively demonstrated that the native protein could be obtained as a 160-kDa holo-protein composed of 3 54-kDa polypeptides (4). By using antiserum to CBF-1 it was shown that the binding protein is localized in membrane-bound protein bodies, found only in cells of tissues surrounding the embryonic axis (5), and that it is immunologically related to CBF-1 in other cereals. It was speculated that the protein is not a cytokinin receptor but functions by sequestering cytokinins, thus acting as a regulator of cytokinin availability (6).

Comparable to our own data with auxins (63), an *in vitro* cytokinin-dependent stimulation of nuclear transcription by a cytokinin-binding protein, isolated from excised barley leaves, was reported (32). This protein was 12000-fold purified by means of anti-idiotype antibodies (33). In the presence of 6-benzyladenine the protein activates *in vitro* rRNA synthesis in the transcription elongation system, containing chromatin-bound RNA-polymerase I from barley leaves.

In maize shoots a soluble cytokinin-binding protein has been extensively purified and characterized (51). In terms of molecular size (46 kDa) and binding affinity this protein shows to be similar to the cytokinin-binding protein from green barley leaves (50).

Gibberellins

The gibberellin (GA)-induced initiation and control of hydrolytic enzymes in the cereal aleurone tissue is one of the better known hormonal responses in plants. GA induces a considerable increase in the rates of transcription of α-amylase genes and it was suggested that the hormone perception in the aleurone layer involves an interaction between the ligand and specific receptors. But, although in elongating tissues from pea and cucumber soluble GA-binding proteins show some characteristics expected of a receptor (58), it was not possible to detect any GA receptor in aleurone tissues. An exciting result was obtained by applying impermeant GA_4 to isolated aleurone protoplasts. High levels of α-amylase mRNAs were induced, thus indicating that this response was due to an interaction between the immobilised ligand and the protoplast surface (19). Support for the concept that GA is perceived at the aleurone plasmamembrane was given by an immunological approach, where anti-idiotypic sera acted as antagonists of GA-action in aleurone protoplasts (20). All these experiments present evidence that GA does not really act in a manner analogous to that of steroid hormones (cf. Fig. 3). Therefore, the role of Ca^{2+} acting as a second messenger in the GA-transduction pathway (cf. Fig. 2), has now become an important part of the GA-research (7).

The aleurone system is not only highly attractive for GA-receptor research because the molecular responses to the hormone are relatively well known, but also because sensitivity mutants are available (11), thus providing potent tools for the unravelling of GA-perception and transduction.

An interesting report has appeared providing evidence that GA (and ABA) controls α-amylase and ribosomal RNA synthesis by regulating gene transcription (23). These results were obtained in run-off transcription studies with nuclei isolated from aleurone cell protoplasts, incubated for 24 h with and without GA (and ABA). It, therefore, seems worthwhile to analyze these protoplasts for the presence of high-affinity GA-binding proteins, and to see if these results can be reproduced in reconstitution experiments with nuclei isolated from untreated aleurone cells.

Abscisic acid

There is only one report that provides strong evidence for the presence of ABA receptors in a particular class of ABA-target cells, i.e., guard cells (21). ABA affects guard cells by stimulating efflux of K^+, their major osmotically active constituent, thus inducing stomata closure. High-affinity binding of $[^3H]cis(+)ABA$, the physiologically active enantiomer, to guard-cell protoplasts from *Vicia faba*, could be demonstrated by photoaffinity labelling. The binding proteins have an apparent K_d for ABA of $3\text{-}4 \times 10^{-9}M$, which corresponds well with the ABA concentration ($5 \times 10^{-9}M$) at which half-maximum response is obtained in stomatal closure bioassays. Upon sodium dodecyl sulphate polyacrylamide gel electrophoresis the binding proteins resolve into three species: A (Mw 20.2 kDa), B (Mw 19.3 kDa) and C (Mw 24.3 kDa). At an alkaline pH in the medium ABA binds preferentially to A, whereas at an acidic pH most ABA binds to B and C. In such a system binding is largely independent of apoplastic pH and this corresponds well with the observation that ABA induces stomatal closure at alkaline as well as at acidic pH. There is a close correlation between physiological activity and the ability to displace $[^3H]ABA$ from its high-affinity binding sites, for a range of ABA analogues. Mild tryptic treatment of guard-cell protoplasts

291

before incubation with ABA completely eliminates the ABA-binding sites, but leaves the protoplasts intact. These results indicate that the binding sites are proteins that are located at the plasmalemma of guard cells, with the ABA-binding moiety facing the apoplastic space. As the reader will recall, a similar location has been assumed for membrane-bound auxin-binding proteins in tobacco cells and maize coleoptiles, and recently for GA-binding proteins in the aleurone cells.

Although the guard-cell system looks very promising for further receptor research, there has been no follow-up. Research is now mainly focussed on the possible role of second messengers, like Ca^{2+} and IP3 (cf. Fig. 2), in the transduction of the ABA signal towards initiation of stomatal closure.

Ethylene

Specific high-affinity membrane-bound ethylene-binding sites (EBP) have been described for various tissues (reviewed in 14). The binding kinetics differ from those of binding sites for other plant hormones by low association and dissociation rates. The EBP from *Phaseolus vulgaris* is a highly hydrophobic integral membrane protein but has now been purified to homogeneity and N-terminal and internal sequences are being obtained (14). Using isokinetic sucrose gradients and gel permeation chromatography, it was suggested that the native protein is a heterotrimer or heterotetramer of subunits of 12 to 14 kDa (59).

When they are acting as receptors controlling rapidly responding systems, it would be expected that EBPs have high rate constants of association and dissociation. By developing techniques for the *in vivo* and *in vitro* measurement of EBPs having high rate constants (see 53), it was shown that in pea epicotyls there are two classes of binding sites. One class shows very high rate constants of association and dissociation, while the other class has very low rate constants similar to those in *Phaseolus* and mungbean (54). The same pattern was discovered in other systems including rice, tomato, and *Arabidopsis* (39).

Polyclonal antibodies have been raised to the EBP from *Phaseolus* and these antibodies also recognize homologous proteins in pea, rice, and *Arabidopsis*. Immunogold localisation studies on abscission zones and petioles of *Phaseolus*, epicotyls of peas, and hypocotyls of *Arabidopsis*, showed that staining in all these tissues was consistent with results obtained from direct ethylene-binding assays (39).

In ethylene-insensitive mutants of *Arabidopsis* the concentration of binding sites is much lower than in the wild type, thus indicating that these sites might be putative receptors (52). When genes for the fast associating sites are available, it will be possible to confer sensitivity in mutants by transformation assays.

Now it has been reported that it is possible to obtain viable protoplasts from abscission zones and petioles of *Phaseolus,* that respond to ethylene in the same way as intact cells (13), the effect of anti-EBP antibodies can be tested in a functional assay, comparable to the approaches used in the studies with both auxin- and gibberellin-receptors.

Recent work (12) has shown that ethylene can effect phosphorylation of EBP and that the degree of phosphorylation has a direct effect on the ethylene-binding characteristics of the protein.

Considering all data, it seems that useful tools are available for the elucidation of the ethylene perception-and- transduction mechanisms.

SUMMARY AND CONCLUSIONS

- Specific high-affinity binding proteins have been described for all major classes of plant hormones. There is good, albeit circumstantial evidence that at least the membrane-bound binding proteins in maize coleoptiles (up till now the best characterized), tobacco cells, *Avena* aleurone cells and broad bean guard cells have a receptor function in auxin-controlled elongation, auxin-induced root regeneration, GA-induced α-amylase synthesis and abscisic acid-controlled stomatal closure, respectively. These putative receptors appear to be localized at the plasmamembrane with the binding moiety facing the apoplast, thus constituting sensory systems for hormone levels outside the cells.
- The fact that hormones can easily be taken up by plant cells suggests that these cells might also be equipped with intracellular receptors. It is not unlikely that the cytoplasmic/nuclear auxin-binding proteins in tobacco cells and perhaps the cytoplasmic cytokinin-binding sites in wheat germ and in barley leaves represent such receptors.
- At present no conclusive evidence as to a biochemical function of the plant hormone-binding sites exists, although all evidence obtained thus far indicates that the cytoplasmic/nuclear binding proteins in tobacco cells and the cytokinin-binding proteins in barley leaves are directly involved in transcriptional activity in the nuclei. In this respect it is important to note that over the past years it has been found that auxin is capable of stimulating the transcription of specific genes with response times measured in minutes (29) (see Chapter C2). Although it is not certain that auxin-induced acidification is a primary trigger of the rapid auxin response in cell elongation, regulation of apoplastic pH by modu-lation of proton pumps may be part of the auxin-transduction chain.
- One may wonder how many pathways for plant-hormone perception, and transduction do exist. Perhaps there are only a few major pathways, possibly based on very conservative principles of hormone perception and transduction in higher eukaryotes. As a working hypothesis we will propose at least two major pathways:
  1. Intracellular receptor proteins which are directly involved in gene expression either on the transcriptional and/or translational level.

2. Plasmamembrane-bound receptor proteins which function as sensory systems for external hormone levels and which transduce intracellular signals, possibly via the phosphatidylinositol pathway. The secondary signals control the cell's activity via modulation of cytoplasmic Ca^{2+} levels and protein kinase activity. Some evidence supporting this view has recently been discussed (62).

References

1. Barbier-Brygoo, H., Ephritikhine, G., Klämbt, D., Ghislain, M., Guern, J. (1989) Functional evidence for an auxin receptor at the plasmalemma of tobacco mesophyll protoplasts. Proc. Natl. Acad. Sci. USA 86, 891-895.
2. Barbier-Brygoo, H., Ephritikhine, G., Shen, W.H., Delbarre, A., Klämbt, D., Guern, J. (1990) Characterization and modulation of the sensitivity of plant protoplasts to auxin. *In*: Transducing Pathways: Activation and Desensitisation, pp. 231-244, Konijn, T.M., ed. Springer-Verlag, New York.
3. Barbier-Brygoo, H., Maurel, C., Shen, W.H., Ephritikhine, G., Delbarre, A., Guern, J. (1990) Use of mutants and transformed plants to study the action of auxins. *In*: Hormone Perception and Signal Transduction in Animals and Plants, SEB Symposia 44, pp. 67-77, Kirk, C., Roberts, J., Venis, M.A., eds. The Company of Biologists, Cambridge.
4. Brinegar, A.C., Fox, J.E. (1985) Resolution of the subunit composition of a cytokinin-binding protein from wheat embryos. Biol. Plant. 27, 100-104.
5. Brinegar, A.C., Fox, J.E. (1987) Immunocytological localization of a wheat embryo cytokinin binding protein and its homology with proteins in other cereals. *In*: NATO ASI Series H10, Plant Hormone Receptors, pp. 177-184, Klämbt, D., ed. Springer-Verlag, Berlin, Heidelberg.
6. Brinegar, A.C., Stevens, A., Fox, J.E. (1985) Biosynthesis and degradation of a wheat embryo cytokinin-binding protein during embryogenesis and germination. Plant Physiol. 79, 706-710.
7. Bush, D.S. (1992) The role of Ca^{2+} in the action of GA in the barley aleurone. *In*: Progress in Plant Growth Regulation, pp. 96-104, Karssen, C.M., Van Loon, L.C., Vreugdenhil, D., eds. Kluwer Academic Publishers, Dordrecht, The Netherlands.
8. Cross, J.W. (1991) Cycling of auxin-binding protein through the plant cell: pathways in auxin signal transduction. New Biologist 3, 813-819.
9. Ephritikhine, G., Barbier-brygoo, H., Muller, J.F., Guern, J. (1987) Auxin effect on the transmembrane potential difference of wild-type and mutant tobacco protoplasts exhibiting a differential sensitivity to auxin. Plant Physiol. 83, 801-804.
10. Felle, H., Peters, W., Palme, K. (1991) The electrical response of maize to auxins. Biochim. Biophys. Acta 1064, 199-204.
11. Fujioka, S., Yamane, H., Spray, C.R., Katsumi, M., Phinney, B,O., Gaskin, P., MacMillan, J., Takahashi, N. (1988) The dominant non-gibberellin-responding dwarf mutant (*D8*) of maize accumulates native gibberellins. Proc. Natl. Acad. Sci. USA 85, 9031-9035.
12. Gilroy, S., Fricker, M.D., Read, N.D., Trewavas, A.J. (1991) Role of calcium in signal transduction of *Commelina* guard cells. The Plant Cell 3, 333-344.
13. Hall, M.A., Berry, A.W., Cowan, D.S., Evans, J.S., Harpham, N.V.J., Moshkov, I., Novikova, G., Raskin, I., Smith, A.R., Turner, R.J., Zhang X. (1991) Ethylene receptors. *In*: Proc. Int. Symposium on "Biochemical Mechanisms involved in Growth Regulation", Milan, in press, Smith, C.J., Gallon, J., eds. Oxford University Press.
14. Hall, M.A., Connern, C.P.K., Harpham, N.V.J., Ishizawa, K., Roveda-Hoyos, G., Raskin, I., Sanders, I.O., Smith, A.R., Turner, R., Wood, C.K. (1990) Ethylene: receptors and action. *In*: Hormone Perception and Signal Transduction in Animals and Plants, SEB

Symposia 44, pp. 87-110, Kirk, C., Roberts, J., Venis, M.A, eds. The Company of Biologists, Cambridge.

15. Herber, B., Ulbrich, B., Jacobsen, H-J. (1988) Modulation of soluble auxin-binding proteins in soybean cell suspensions. Plant Cell Reports 7, 178-181.

16. Hesse, T., Feldwisch, J., Balshüsemann, D., Bauw, G., Puype, M., Vandekerckhove, J., Löbler, M., Klämbt, D., Schell, J., Palme, K. (1989) Molecular cloning and structural analysis of a gene from *Zea mays* (L.) coding for a putative receptor for the plant hormone auxin. EMBO J. 8, 2453-2461.

17. Hicks, G.R., Rayle, D.L., Jones, A.M., Lomax, T.L. (1989) Specific photo-affinity labeling of two plasma membrane polypeptides with an azido auxin. Proc. Natl. Acad. Sci. USA 86, 4948-4952.

18. Hicks, G.R., Rayle, D.L., Lomax, T.L. (1989) The *Diageotropica* mutant of tomato lacks high specific activity auxin binding sites. Science 245, 52-54.

19. Hooley, R., Beale, M.H., Smith, S.J. (1990) Gibberellin perception in the *Avena fatua* aleurone. *In*: Hormone Perception and Signal Transduction in Animals and Plants, SEB Symposia 44, pp. 79-86, Kirk, C., Roberts, J., Venis, M.A., eds. The Company of Biologists, Cambridge.

20. Hooley, R., Beale, M.H., Smith, S.J., MacMillan, J. (1990) Novel affinity probes for gibberellin receptors in aleurone protoplasts of *Avena fatua*. *In*: Plant Growth Substances 1988, pp. 145-153, Pharis, R.P., Rood, S.B., eds. Springer-Verlag, Heidelberg.

21. Hornberg, C., Weiler, E.W. (1984) High-affinity binding sites for abscisic acid on the plasmalemma of *Vicia faba* guard cells. Nature 310, 321-324.

22. Inohara, N., Shimomura, S., Fukui, T., Futai, M. (1989) Auxin-binding protein located in the endoplasmic reticulum of maize shoots: Molecular cloning and complete primary structure. Proc. Natl. Acad. Sci. USA 86, 3564-3568.

23. Jacobsen, J.V., Beach, L.R. (1985) Control of transcription of α-amylase and rRNA genes in barley aleurone protoplasts by gibberellin and abscisic acid. Nature 316, 275-277.

24. Jones, A.M. (1990) Do we have the auxin receptor yet? Physiol Plant 80, 154-158.

25. Jones, A.M., Herman, E.M. (1993) KDEL-containing auxin-binding protein is secreted to the plasma membrane and cell wall. Plant Physiol. 101, 595-606.

26. Jones, A.M., Lamerson, P., Venis, M.A. (1989) Comparison of Site 1 auxin binding and a 22-kilodalton auxin-binding protein in maize. Planta 179, 409-413.

27. Jones, A.M., Prasad, P.V. (1992) Auxin-binding proteins and their possible roles in auxin-mediated plant cell growth. BioEssays 14, 43-48.

28. Jones, A.M., Venis, M.A. (1989) Photoaffinity labeling of indole-3-acetic acid-binding proteins in maize. Proc. Natl. Acad. Sci. USA 86, 6153-6156.

29. Key, J.L. (1989) Modulation of gene expression by auxin. BioEssays 11, 52-28.

30. Kikuchi, M., Imaseki, H., Sakai, S. (1989) Modulation of gene expression in isolated nuclei by auxin-binding proteins. Plant Cell Physiol. 30, 765-773.

31. Klämbt, D. (1990) A view about the function of auxin-binding proteins at plasma membranes. Plant Mol. Biol. 14, 1045-1050.

32. Kulaeva, O.N. (1985) Hormonal regulation of transcription and translation in plants. *In*: Proc. 16th FEBS meeting at Moscow, Part C, pp. 391-396, Ovchinnikov, Y.A., ed., VNU Science Press.

33. Kulaeva, O.N., Karavaiko, N.N., Moshkov, I.E., Selivankina, S.Ya., Novikova, G.V. (1990) Isolation of a protein with cytokinin-receptor properties by means of anti-idiotype antibodies. FEBS Lett. 261, 410-412.

34. Libbenga, K.R., Maan, A.C., Van der Linde, P.C.G., Mennes, A.M. (1985) Auxin receptors. *In*: Hormones, Receptors and Cellular Interactions in Plants, pp. 1-68, Chadwick, C.M., Garrod, D.R., eds. Cambridge University Press.

35. Libbenga, K.R., Van Telgen, H.J., Mennes, A.M., Van der Linde, P.C.G., Van der Zaal, E.J. (1987) Characterization and function analysis of a high-affinity cytoplasmic auxin-

binding protein. *In*: Molecular Biology of Plant Growth Control, pp. 229-243, Fox, J.E., Jacobs, M., eds. A.R. Liss, NY.

36. Löbler, M., Klämbt, D. (1985) Auxin-binding protein from coleoptile membranes of corn (*Zea mays* L.). I. Purification by immunological methods and characterization. J. Biol. Chem. 260, 9848-9853 .

37. Löbler, M., Klämbt, D. (1985) Auxin-binding protein from coleoptile membranes of corn (*Zea mays* L.). II. Localization of a putative receptor. J. Biol. Chem. 260, 9854- 9859.

38. Löbler, M., Simon, K., Hesse, T., Klämbt, D. (1987) Auxin receptors in target tissue. *In*: Molecular Biology of Plant Growth Control, pp. 279-288, Fox, J.E., Jacobs, M., eds. A.R. Liss, NY.

39. Mennes, A.M., Maan, A.C., Hall, M.A. (1991) Plant hormone receptors. *In*: NATO ASI Series, Vol. H51, Cell to Cell Signals in Plants and Animals, pp. 301-314, Neuhoff, V., Friend, J., eds. Springer-Verlag, Berlin, Heidelberg.

40. Mennes, A.M., Nakamura, C., Van der Linde, P.C.G., Van der Zaal, E.J., Van Telgen, H.J., Quint, A., Libbenga, K.R., (1987) Cytosolic and membrane-bound high-affinity auxin-binding proteins in tobacco. *In*: NATO ASI Series H10, Plant Hormone Receptors, pp. 51-62, Klämbt, D., ed. Springer-Verlag, Berlin, Heidelberg.

41. Mennes, A.M., Quint, A., Gribnau, J.H., Boot, C.J.M., Van der Zaal, E.J., Maan, A.C., Libbenga, K.R. (1992) Specific transcription and reinitiation of 2,4-D-induced genes in tobacco nuclei. Plant Mol. Biol. 18, 109-117.

42. Moudgil, V.K. (1990) Phosphorylation of steroid hormone receptors. Biochim. Biophys. Acta 1055, 243-258.

43. Nakamura, C., Ono, H. (1988) Solubilization and characterization of a membrane-bound auxin-binding protein from cell suspension cultures of *Nicotiana tabacum*. Plant Physiol. 88, 685-689.

44. Nakamura, C., Van Telgen, H.J., Mennes, A.M., Ono, H., Libbenga, K.R. (1988) Correlation between auxin resistance and the lack of a membrane-bound auxin-binding protein and a root-specific peroxidase in *Nicotiana tabacum*. Plant Physiol. 88, 845-849.

45. Napier, R.M., Venis, M.A. (1990) Monoclonal antibodies detect an auxin-induced conformational change in the maize auxin-binding protein. Planta 182, 313-318.

46. Napier, R.M., Venis, M.A. (1990) Receptors for plant growth regulators: recent advances. J. Plant Growth Reg. 9, 113-126.

47. Napier, R.M., Venis, M.A., Bolton, M.A., Richardson, L.I., Butcher, G.W. (1988) Preparation and characterisation of monoclonal and polyclonal antibodies to maize membrane auxin-binding protein. Planta 176, 519-526.

48. Pope, D.G. (1982) Effect of peeling on IAA-induced growth in *Avena* coleoptiles. Ann. Bot. 49, 495-501.

49. Prasad, P., Jones, A.M. (1991) Putative receptor for the plant growth hormone auxin identified and characterized by anti-idiotypic antibodies. Proc. Natl. Acad. Sci. USA 88, 5479-5483.

50. Romanov, G.A., Taran, V.Ya., Chvojka, L., Kulaeva, O.N., (1988) Receptor-like cytokinin-binding protein(s) from barley leaves. J. Plant Growth Reg. 7, 1-17.

51. Romanov, G.A., Taran, V.Ya., Venis, M.A. (1990) Cytokinin-binding protein from maize shoots. J. Plant Physiol. 136, 208-212.

52. Sanders,I.O., Harpham, N.V.J., Raskin, I., Smith, A.R., Hall, M.A. (1991) Ethylene binding in wild type and mutant *Arabidopsis thaliana*(L.) Heynh. Ann. Bot. 68, 97-103.

53. Sanders, I.O., Smith, A.R., Hall, M.A. (1989) The measurement of ethylene binding and metabolism in plant tissue. Planta 179, 97-103.

54. Sanders, I.O., Smith, A.R., Hall, M.A. (1991) Ethylene binding in epicotyls of *Pisum sativum* L. cv. Alaska. Planta 183, 209-217.

55. Shimomura, S., Inohara, N., Fukui, T., Futai, M. (1988) Different properties of two types of auxin-binding sites in membranes from maize coleoptiles. Planta 175, 558-566.

56. Shimomura, S., Sotobayashi, T., Futai, M., Fukui, T. (1986) Purification and properties of an auxin-binding protein from maize shoot membranes. J. Biochem. 99, 1513-1524.

57. Skoog, F., Miller, C.O. (1957) Chemical regulation of growth and organ formation in plant tissues cultured *in vitro*. Symp. Soc. Exp. Biol. 11, 118-131.

58. Srivastava, L.M. (1987) The gibberellin receptor. *In*: NATO ASI Series H10, Plant Hormone Receptors, pp. 199-228, Klämbt, D., ed. Springer-Verlag, Berlin, Heidelberg.

59. Thomas, C.J.R., Smith, A.R., Hall, M.A. (1985) Partial purification of an ethylene-binding site from *Phaseolus vulgaris* L. cotyledons. Planta 164, 272-277.

60. Thompson, M., Krull, U.J., Venis, M.A. (1983) .A chemo-receptive bilayer lipid membrane based on an auxin-receptor ATPase electrogenic pump. Biochem. Biophys. Res. Comm. 110, 300-304.

61. Tillmann, U., Viola, G., Kayser, B., Siemeister, G., Hesse, T., Palme, K., Löbler, M., Klämbt, D. (1989) cDNA clones of the auxin-binding protein from corn coleoptiles (*Zea mays* L.): isolation and characterization by immunological methods. EMBO J. 8, 2463-2467.

62. Trewavas, A., Gilroy, S. (1991) Signal transduction in plant cells. Trends in Genetics 7, 356-361.

63. Van der Linde, P.C.G., Maan, A.C., Mennes, A.M., Libbenga, K.R. (1985) Auxin receptors in tobacco. *In*: Proc. 16th FEBS meeting at Moscow, Part C, pp. 397-403, Ovchinnikov, Y.A., ed., VNU Science Press.

64. Van der Zaal, E.J., Droog, F.N.J., Boot, C.J.M., Hensgens, L.A.M., Hoge, J.H.C., Schilperoort, R.A., Libbenga, K.R. (1991) Promoters of auxin-induced genes from tobacco can lead to auxin-inducible and root tip-specific expression. Plant Mol. Biol. 16, 983-998.

65. Van der Zaal, E.J., Memelink, J., Mennes, A.M., Quint, A., Libbenga, K.R. (1987) Auxin-induced mRNA species in tobacco cell cultures. Plant Mol. Biol. 10, 145-157.

66. Venis, M.A. (1985) Hormone binding in plants. Longman Inc., New York-London.

67. Venis, M.A., Napier, R.M. (1991) Auxin receptors: recent developments. J. Plant Growth Reg. 10, 329-340.

68. Venis, M.A., Napier, R.M., Barbier-Brygoo, H., Maurel, C., Perrot-Rechenmann, C., Guern, J. (1992) Antibodies to a peptide from the maize auxin-binding protein have auxin agonist activity. Proc. Natl. Acad. Sci. USA 89, 7208-7212.

69. Venis, M.A., Thomas, E.W., Barbier-Brygoo, H., Ephritikhine, G., Guern, J. (1990) Impermeant auxin analogues have auxin activity. Planta 182, 232-235.

70. Walton, J.D., Ray, P.M. (1981) Evidence for receptor function of auxin-binding sites in maize. Plant Physiol. 68, 1334-1338.

D5. Calcium and Plant Hormone Action

Paul C. Bethke, Simon Gilroy and Russell L. Jones
Department of Plant Biology, University of California, Berkeley, CA 94720,
USA.

INTRODUCTION

It is widely accepted that a complex system of signal-transduction pathways
links the responses of cells to external stimuli. In plants, hormones and
environmental stimuli such as light, temperature and touch are transduced into
a cellular response by a set of reactions that are only beginning to be
understood. Compounds that act as second messengers have a central role in
the initial transduction steps leading from these stimuli. Rather than forming
simple, linear signal-transduction chains, however, signaling molecules are
thought to form complex webs of interconnected pathways (Fig. 1). The
interactions between signal-transduction chains, the so-called cross-talk, can
be a source of signal amplification as well as signal complexity. The
existence of multiple, interconnected signaling pathways may help to explain
how the small number of plant hormones can elicit such a wide variety of
cellular responses. After interacting with a receptor, the hormonal signal may
be transduced within the cell via one of a number of alternative pathways, the

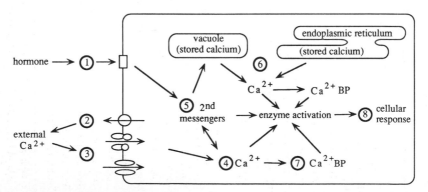

Fig. 1. Elements of calcium homeostasis and transport linking a hormonal signal to a cellular
response in an idealized plant cell. Phytohormones (1) are perceived by a receptor, possibly
on the outer surface of the PM. Hormone perception may then change the activity of PM
Ca^{2+}-ATPases (2) or PM Ca^{2+} channels and other transporters (3), resulting in a change, usually
an increase, in $[Ca^{2+}]_i$ (4). Alternatively, the hormone may cause a second messenger (5) other
than $[Ca^{2+}]_i$, such as IP_3, to release Ca^{2+} from intracellular stores (6). Cross-talk between
$[Ca^{2+}]_i$ and other second messengers is expected. Increased $[Ca^{2+}]_i$ activates Ca^{2+}-binding
proteins (Ca^{2+} BP) (7), and these proteins activate enzymes which begin the chain of events
leading to a specific cellular response (8).

P. J. Davies (ed.), Plant Hormones, 298–317.
© 1995 Kluwer Academic Publishers. Printed in the Netherlands.

Table 1. Stimuli that cause changes in cytosolic Ca²⁺ in plants.

Stimulus[1]	Cell type	Physiological Response[2]	Change in $[Ca^{2+}]$[3]	Reference
Hormonal				
ABA	Guard cell	Stomatal closure	TI	28, 36, 52, 53, 73
	Maize coleoptile & root	Growth?	TI	25
	Parsley hypocotyl & root	Growth?	TI	25
	Aleurone cell	Inhibition of secretion?	TD	83
	Aleurone cell	Inhibition of secretion	SD	29
Auxin	Maize epidermal cell	Growth?	OSC	23
	Maize coleoptile & root	Growth?	SI	25
	Parsley hypocotyl & root	Growth?	SI	25
CK	*Funaria*	Bud initiation	SI	32
GA	Aleurone cell	Stimulation of secretion	SI	10, 13, 14, 29
Non hormonal				
Cold shock	Tobacco seedling	Cold acclimation?	TI	43
Gravity	Maize coleoptile	Negative gravitropism	SI	26
Light	Maize coleoptile	Positive phototropism	SI	26
	Oat protoplast	Phytochrome?	SD	15
	Wheat leaf protoplast	Phytochrome mediated protoplast swelling	TI	70
	Nitellopsis	Regulation of photosynthesis?	SD	54
	Mougeotia scalaris	Chloroplast rotation	I	65
Salinity	Maize root protoplast	Salt tolerance?	I	49
Touch	Tobacco seedling	Mechanoperception	TI	43, 44
Wind	Tobacco seedling	Mechanoperception	TI	43
Yeast elicitors	Tobacco seedling	Pathogen resistance	TI	43

[1] ABA, abscisic acid; CK, cytokinin; GA, gibberellic acid.
[2] Physiological response thought to be regulated by the change in $[Ca^{2+}]_i$.
[3] I, increase; OSC, oscillations; SD, sustained decrease; SI, sustained increase; TD, transient decrease; TI, transient increase.

response of the cell being dependent on the pathway or combination of pathways employed. Alternatively, some cells may not possess all the elements of every signaling pathway. Thus, the response of differentiated cells to hormones may be determined by the presence of tissue- or cell-type-specific signal-transduction molecules.

The calcium ion plays a prominent role in eukaryotic signal transduction, and it has been implicated in almost every response of plant cells to hormones (Table 1; 60, 69), with the notable exception of ethylene. Calcium is ideally suited to its role as a second messenger. The physico-chemical properties of the Ca^{2+} ion allow for highly specific interactions between it and other molecules (35). Because the concentration of free Ca^{2+} in the cytosol ($[Ca^{2+}]_i$) of resting cells is low, around 100 nM, it can be readily modulated. The opening of Ca^{2+} channels in the membranes that separate the cytosol from external or internal stores of Ca^{2+} can rapidly raise $[Ca^{2+}]_i$ (Figs. 1 and 2; 5). Calcium pumps in the plasma membrane (PM) and endomembrane system, on the other hand, maintain the low level of $[Ca^{2+}]_i$ and can quickly restore $[Ca^{2+}]_i$ to basal levels following an increase (8). Calcium acts via Ca^{2+}-binding regulatory proteins, such as calmodulin (CaM) (64). These Ca^{2+}-binding proteins respond to changes in $[Ca^{2+}]_i$ by modulating their activities. In this chapter we shall discuss the evidence that links changes in Ca^{2+} and Ca^{2+}-binding regulatory proteins to the action of plant hormones.

CALCIUM HOMEOSTASIS AND TRANSPORT

Because the synthesis of metabolic energy in the form of ATP depends on a large pool of free phosphate ions, the $[Ca^{2+}]_i$ of living cells must be kept low to avoid the formation of insoluble calcium phosphate salts. There can be no doubt that plant cells maintain $[Ca^{2+}]_i$ within a very narrow range, rarely exceeding 1 μM. Indeed, using a variety of methods for measuring $[Ca^{2+}]_i$, resting levels of this ion are generally found to be 100 nM to 200 nM (61). A system of ion carriers, pumps and channels maintains $[Ca^{2+}]_i$ within this narrow range (Fig. 2). An understanding of how the activities of these transporters are regulated is crucial to an understanding of how Ca^{2+} functions as a signaling molecule in cells.

Calcium Channels

Calcium channels at the PM and tonoplast (TP) exploit the large electrochemical gradient that favors the passive influx of Ca^{2+} into the cytoplasm of all cells. Calcium channels have been identified at the PM and TP of plant cells but not yet on the endoplasmic reticulum (ER) (38). At least two types of Ca^{2+} channel have been identified at the PM based on their sensitivity to channel blockers, and two channels can be distinguished in the TP based on their response to the regulatory ligand inositol trisphosphate (IP$_3$) (5, 38). A large number of factors affect the gating of Ca^{2+} channels in

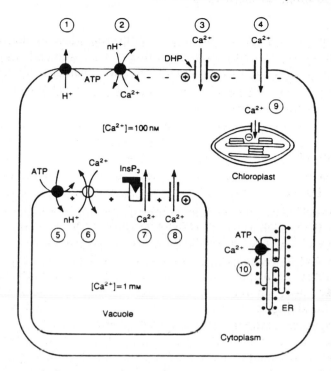

Fig. 2. The identity and location of transport systems involved in cytosolic Ca^{2+} homeostasis. Transport systems at the PM are as follows: 1, Primary electrogenic H^+-ATPase which generates a highly negative membrane potential (typically -150 mV); 2, Primary Ca^{2+}-ATPase which energizes export of cytosolic Ca^{2+} in exchange for extracellular H^+; 3, Dihydroypryidine (DHP)-sensitive Ca^{2+} channel activated by depolarizing voltages (+); and 4, Other classes of Ca^{2+}-permeable channels. Transport systems at the TP include: 5, Primary electrogenic H^+-ATPase which generates a vacuole positive membrane potential between +20 and +50 mV and an inside-acid pH gradient of 2 units; 6, Ca^{2+}/H^+ exchanger driven by the chemical and electrical component of the proton motive force; 7, IP_3-gated Ca^{2+} channel; and 8, Voltage-operated Ca^{2+} channel opened by positive shifts in trans-TP electrical potential (+). Ca^{2+} transport at other membranes: 9, Ca^{2+} channel in the chloroplast envelope facilitating Ca^{2+} uptake in response to an inside-negative membrane potential; 10, Primary Ca^{2+}-ATPase at the ER. From (38).

higher plants and algae including voltage, stretch and Ca^{2+} (5).

Experimental evidence suggests that Ca^{2+} channels at the PM and TP are used to raise $[Ca^{2+}]_i$ in plants, and supports a role for external Ca^{2+} in modulating $[Ca^{2+}]_i$. Many researchers have observed that changes in external Ca^{2+} concentration result in very rapid changes in $[Ca^{2+}]_i$. For example, using Ca^{2+}-sensitive microelectrodes Felle (24) has shown that in maize roots an increase in external Ca^{2+} from 100 μM to 10 mM caused a three-fold increase in $[Ca^{2+}]_i$ within seconds. Similar observations have been made with stomatal guard cells (28) and wheat aleurone tissue (10). High external Ca^{2+} concentrations have also been shown to be required for the responses of plant cells to abscisic acid (ABA) (27), auxins (23, 25, 51), cytokinins (69) and gibberellins (GAs) (14, 29). Furthermore, a rise in $[Ca^{2+}]_i$ has been shown to be associated with responses to these hormones. The evidence that Ca^{2+} channels in the PM play a role in the influx of Ca^{2+} into cells as cited above is indirect, but the inhibitory effects of Ca^{2+} channel blockers and the demonstration of inward Ca^{2+} currents associated with elevated cytosolic Ca^{2+} strongly implicate channels as a pathway for Ca^{2+} entry and regulation of $[Ca^{2+}]_i$.

Calcium Carriers and Pumps

Energy-dependent Ca^{2+} transporters at the PM, ER and TP play an equally important role in Ca^{2+} homeostasis by lowering $[Ca^{2+}]_i$ (Fig. 2). These transporters are of two types: Ca^{2+} pumps that couple the hydrolysis of ATP directly to the transport of Ca^{2+}, and H^+/Ca^{2+} antiporters that exploit H^+ gradients established by H^+-ATPase or H^+-translocating pyrophosphatases (PPases) to couple energetically favorable H^+ movement to Ca^{2+} movement (Fig. 2; 8, 63). Calcium pumps are located at the PM and ER, whereas H^+/Ca^{2+} antiporters have been found in the PM and TP. As in animal cells, the activities of the ER and PM Ca^{2+} pumps in plants are regulated by CaM (22). Since CaM activation is dependent on Ca^{2+} (64), CaM stimulation of Ca^{2+} pump activity provides a feedback mechanism for the lowering of $[Ca^{2+}]_i$. Much less is known about regulation of the H^+/Ca^{2+} antiporter at the PM and TP. The H^+/Ca^{2+} antiporter, however, is sensitive to the transmembrane proton gradient, and the activity of this Ca^{2+} transporter may reflect regulation imposed via the PM or TP ATPase (75) or the PPase (63). It should be noted that the activities of both PM ATPase and the TP PPase are regulated by Ca^{2+} concentrations that are within the accepted range of $[Ca^{2+}]_i$ (63, 64).

Calcium Stores in the Symplast and Apoplast

Reservoirs of Ca in the cell wall and in organelles are important for Ca^{2+} homeostasis in plants, although the evidence favoring a role for internal Ca^{2+} stores in cytosolic Ca^{2+} homeostasis is circumstantial. The cell wall contains millimolar concentrations of Ca, and depending on the pH and charge properties of the wall, the concentration of Ca^{2+} in the apoplast of growing cells is estimated to be on the order of 100 μM to 200 μM (18). The vacuole and ER also contain relatively high Ca concentrations (38). The vacuole by default has been assumed to be the most important internal store of Ca^{2+} for cytosolic Ca^{2+} homeostasis, in part because of its relatively large size. Vacuolar Ca^{2+} is thought to be in the 100's of micromolar to the millimolar range. Few accurate estimates exist of vacuolar free Ca^{2+} because this compartment also contains high levels of oxalic, phytic and phosphoric acids, which lead to insolubilization of Ca^{2+}. There is evidence, however, that in storage organs Ca insolubilized as phytate can be mobilized (41).

The ER, in barley aleurone cells for example, contains millimolar levels of calcium, with free Ca^{2+} reaching concentrations of at least 5 μM (11, 12).

The role of the ER Ca^{2+} store in cytosolic Ca^{2+} homeostasis in plants remains enigmatic. The ER contains an abundant supply of Ca^{2+}, and the surface area of the ER greatly exceeds that of the vacuole or PM, but there is as yet no evidence that this Ca^{2+} store is used to regulate cytosolic Ca^{2+}. Indeed, attempts to cause gating of putative ER Ca^{2+} channels with signaling molecules such as IP_3 have not been successful, suggesting that in plants, the IP_3-sensitive Ca^{2+} channel is confined to the TP (38). An interesting aspect of the ER Ca^{2+} store concerns the role of Ca^{2+}-binding proteins. A family of Ca^{2+}-binding proteins, referred to as reticuloplasmins, has been identified in the lumen of the ER of animal cells (45, 66), and many of these proteins are found in the ER of plant cells (39). The reticuloplasmins are highly acidic proteins and they possess the ER retention signal XDEL (59). In addition to acting as molecular chaperones, the reticuloplasmins are also Ca^{2+}-binding proteins which may be involved in buffering ER Ca^{2+}. The ER luminal protein BiP (binding protein), for example, is thought to participate in the translocation of newly synthesized proteins into the ER (67), but can also bind as many as 30 mol Ca^{2+}/mol BiP (66). BiP appears to be widely distributed in plants, and its distribution is correlated with the synthesis of proteins at the ER (39). Interestingly, GA elevates BiP levels in the ER of aleurone tissue that is actively synthesizing secretory proteins (40).

Chloroplasts and mitochondria also contain millimolar levels of Ca, and it has been argued that these two organelles play important roles in Ca^{2+} homeostasis (9, 17, 20, 47). The mechanism of Ca^{2+} uptake by mitochondria is controversial. A low affinity Ca^{2+} pump located on the mitochondrial membrane has been described, although the K_m for Ca^{2+} (ca. 100 μM) of this transporter suggests that it does not play a role in cytosolic Ca^{2+} homeostasis under physiological conditions (20). Furthermore, since Ca that accumulates within the mitochondrial matrix is thought to be present largely as insoluble salts, the mitochondrial pool of Ca may not be available for replenishment of cytosolic Ca^{2+}. An inwardly-rectifying Ca^{2+} channel (Fig. 2) and an unidirectional Ca^{2+} uniporter (42) are thought to be located in the chloroplast envelope (46). Light plays an important role in driving the uptake of Ca^{2+} into the chloroplast, and light-dependent Ca^{2+} uptake by chloroplasts is associated with a lowering of $[Ca^{2+}]_i$ (54). Although the significance of light-dependent cytosolic Ca^{2+} regulation is not fully understood, it is widely accepted that Ca^{2+} plays an important role in regulating the biochemical activities of the chloroplast (47). For example, the activity of fructose 1, 6 bisphosphate is regulated by light in a Ca^{2+}-dependent manner. Chloroplasts also contain CaM and enzymes whose activities are regulated by CaM (78).

Hormones and Cytosolic Ca^{2+} Homeostasis

Calcium acts as a signaling molecule via changes in its cytosolic concentration. Change in $[Ca^{2+}]_i$ in response to phytohormones is characterized by a slow resetting of Ca^{2+} concentration, generally to a higher

level. Using Ca^{2+}-sensitive
fluorescent dyes it has been shown
that in stomatal guard cells ABA
brings about a two- to three-fold
increase in [Ca^{2+}]$_i$ from about 100
nM to about 200-300 nM (Fig. 3C).
Similar changes are found in the
[Ca^{2+}]$_i$ of barley aleurone cells
following exposure to gibberellic
acid (GA) (Fig. 3A). These
observations of hormone-induced
change in [Ca^{2+}]$_i$ have been
confirmed for auxins by Felle using
microelectrodes to monitor [Ca^{2+}]$_i$
(Fig. 3B). Interestingly, Felle (24)
reports that he has never observed
spikes in [Ca^{2+}]$_i$ in response to
stimuli such as auxin. Such spikes
are observed when plants are
perturbed by touch or temperature
shock, with [Ca^{2+}]$_i$ rising rapidly to
low micromolar concentrations then
rapidly declining to resting levels
(Fig. 4; 43, 44). Felle (24) argues
that plant cells respond to hormones
with subtle changes in [Ca^{2+}]$_i$ (Fig.
3). These changes act to modulate
cellular reactions rather than to
amplify signals as in most animal
cells. CaM and other Ca^{2+}-binding
proteins would be appropriate
candidates for modulating cellular
reactions in response to [Ca^{2+}]$_i$.

Scanning laser confocal
microscopy has allowed the mapping
of cytosolic Ca^{2+} in cells using Ca^{2+}-
specific fluorescent dyes. The
picture that emerges from confocal
microscopy is that Ca^{2+}
concentrations within cells are not
always uniform. In barley aleurone
cells exposed to GA, for example,
elevated [Ca^{2+}]$_i$ is localized in areas
just inside the PM and in the region
of the cytosol that is rich in ER

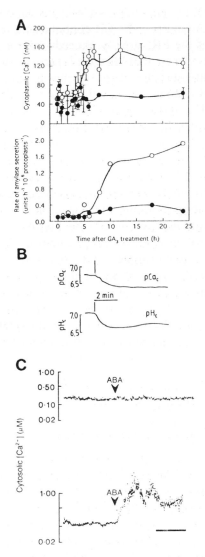

Fig. 3. The effects of hormones on [Ca^{2+}]$_i$. A,
the effects of incubating barley aleurone
protoplasts in the presence (○) or absence (●)
of GA on α-amylase secretion and [Ca^{2+}]$_i$
measured with Indo-1. From (29). B, The
effect of IAA on cytosolic Ca^{2+} concentration
(pCac) and cytosolic pH (pHc) as measured
with microelectrodes in corn coleoptiles. From
(24). C, The effect of ABA on [Ca^{2+}]$_i$ as
measured with Indo-1 in guard cells of
Commelina communis. Although stomata
always close in response to ABA, [Ca^{2+}]$_i$ does
not always increase, as shown in the top trace.
From (28).

(29). Note that in the aleurone cell, high external Ca^{2+} concentrations are required for the response of the cell to GA, and the ER of the aleurone cell accumulates high Ca^{2+} concentrations (11, 12). In stomatal guard cells, on the other hand, the rise in $[Ca^{2+}]_i$ following ABA treatment or exposure to high external Ca^{2+} or K^+, all stimuli that cause stomatal closure, is localized to the cytosol around the vacuoles and the ER-rich perinuclear cytoplasm, suggesting that elevated $[Ca^{2+}]_i$ results from efflux out of intracellular stores (28). Clearly the changes in $[Ca^{2+}]_i$ brought about by hormones are complex. In addition to the quantitative information conveyed by changes in bulk $[Ca^{2+}]_i$, localized gradients of Ca^{2+} could dictate position-specific responses within cells.

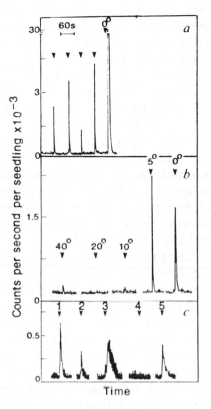

Fig. 4. The effects of touch (a), temperature shock (b) and fungal elicitors (c) on Ca^{2+}-dependent aequorin luminescence in transgenic tobacco seedlings. From (43).

CALCIUM AND SIGNAL TRANSDUCTION

Calcium-binding Proteins

Calcium, although uniquely suited to be a signaling molecule, is unsuitable as an effector molecule. As noted above, changes in $[Ca^{2+}]_i$ are often an integral part of plant responses to hormonal or environmental signals. Yet to produce the varied physiological responses that follow a change in $[Ca^{2+}]_i$, responses that may depend on cell type, stage of development, or environmental history, requires effector molecules downstream of $[Ca^{2+}]_i$. These effector molecules are often proteins which, upon binding one or more calcium ions, change conformation and become activated. It is through these molecules that Ca^{2+}-based signals are propagated, and it is they that create specificity of response.

Calmodulin

Of the Ca^{2+}-binding proteins, one of the most important seems to be CaM. CaM is a small polypeptide containing four Ca^{2+}-binding pockets called EF-hands. One pair of EF-hands is linked to the other by a flexible a-helical region, giving the molecule its dumbbell shape (64). Throughout evolution, the CaM molecule has been highly conserved. Animal and plant

CaMs are >90% similar at the amino acid level (64). Nevertheless, multiple alleles of the CaM gene are often present within single species. Once activated by Ca^{2+} binding, CaM itself binds to numerous other proteins, causing their activation. The value of a multi-functional activator like CaM may be that it allows for simultaneous, coordinated control of numerous biochemical processes.

The activity of CaM itself can be regulated on at least three levels. The first is at the level of Ca^{2+} concentration. At resting levels of $[Ca^{2+}]_i$ (ca. 100 nM), CaM exists largely in its inactive, Ca^{2+}-free form. Only after $[Ca^{2+}]_i$ rises does appreciable Ca^{2+} binding take place. CaM activity can also be regulated at the level of protein concentration. Although little is known about the dynamics of CaM concentration within plant cells, CaM levels in barley aleurone protoplasts have been shown to increase approximately 50% or more following GA_3 stimulation (30); strawberry fruit whose achenes have been removed show higher levels of CaM mRNA when treated with auxin (37); and CaM transcript levels increase more than tenfold in *Arabidopsis thaliana* 10 min after stimulation by touch (7). Finally, intracellular CaM activity can be regulated by localizing the molecule to discrete regions within the cell. To date, little evidence for this exists in plants, although CaM has been shown to be associated with the mitotic apparatus of endosperm cells (82). CaM localization has been elegantly demonstrated in fibroblasts using MeroCaM, a fluorescent dye which reports regions of Ca^{2+}-activated CaM (33).

Numerous proteins are activated by Ca^{2+}-CaM in both plants and animals. In plants, NAD kinase, NTP kinase, quinate: NAD^+ oxidoreductase, an ER and PM Ca^{2+}-ATPase, and a TP ion channel have all been shown to be stimulated by CaM (64, 84). There is also evidence for a CaM-stimulated protein kinase similar to CaM-dependent protein kinase II (61).

Calcium-dependent Protein Kinase

Of the other Ca^{2+}-binding proteins that may serve as effector molecules in plants, the best characterized are the calcium-dependent protein kinases (CDPKs) (64). A member of this protein family was first identified in soybean, and others have since been found in numerous species, tissues and intracellular locations. Multiple isoforms of CDPK exist within single species, and both soluble and membrane-bound forms have been reported. Ranging in size from 40 kDa to 90 kDa, CDPKs comprise a kinase domain, a CaM-like domain, and an auto-inhibitory domain. Calcium binds to the four EF-hands within the CaM-like domain to cause a change in protein conformation (64). This in turn displaces the autoinhibitory domain from the kinase domain and activates the kinase. The K_{max} for Ca^{2+} is typically 1-10 μM, similar to that for CaM. In vitro many proteins are phosphorylated by CDPKs on serine and threonine residues. The in vivo substrates required for further propagation of the Ca^{2+} signal are currently being sought. One such substrate may be an oat root PM H⁺-ATPase which has been shown to be phosphorylated by a Ca^{2+}-stimulated kinase (64, 71).

Other Ca^{2+}-binding Proteins

Additional Ca^{2+}-binding proteins or Ca^{2+}-regulated proteins that may be important in signaling include other ion transporters and cytoskeletal elements. Calcium has been shown to be required for the opening of K⁺ efflux channels and closing of K⁺ influx channels in the PM of stomatal guard cells, presumably by direct interaction of Ca^{2+} with the channel. The fast vacuolar channel seems to be similarly activated. Microtubule disassembly is mediated by Ca^{2+}-CaM (76), and polymerization of actin, as well as movement of actin along myosin filaments is Ca^{2+}-dependent. Finally, in animal systems $[Ca^{2+}]_i$ has been tied to changes in transcription. This is thought to occur through Ca^{2+}-dependent phosphorylation of transcription factors that interact with Ca^{2+}-responsive or cAMP-responsive elements in the promoter region of genes such as *c-fos* (77).

The CaM-activated proteins, CDPKs, and other Ca^{2+}-binding proteins are additional links in the signal transduction chains utilizing changes in $[Ca^{2+}]_i$. Some, such as protein kinases and ion transporters, have the potential to serve as branch points, propagating the signal in several directions. Dissecting out complete signaling pathways from signal to physiological response will be a challenge for biologists in the years to come, yet some of the methodologies needed to approach this complex problem are already at hand. In *Arabidopsis thaliana*, for example, Northern analysis has shown that transcription of CaM and CaM-related genes is rapidly increased following physical stimuli, perhaps linking CaM to thigmotropism (7). In etiolated tomato hypocotyls, microinjection of Ca^{2+}-CaM into living cells resulted in expression of a chlorophyll a/b-binding protein-GUS (Cab-GUS) reporter gene and chloroplast development, establishing CaM as an intermediate in the transduction chain downstream of phytochrome (55). Experiments with yeast having a conditional-lethal mutation in the gene for CaM have shown that CaM is required for nuclear division (1). And mammalian cells expressing an anti-sense CaM gene were arrested in growth at both G1 and M phases of the cell cycle, further suggesting that one of the signaling pathways utilizing Ca^{2+}-CaM regulates the cell cycle (1). Transgenic tobacco plants that overexpress the CaM gene up to 50-fold or have reduced CaM mRNA resulting from an anti-sense CaM gene have wild-type levels of CaM protein, and nearly wild-type growth. This suggests that in plants the amount of CaM protein may be regulated at the translational or post-translational level (61).

Other Signaling Pathways

While many signaling systems in plants use $[Ca^{2+}]_i$ as a second messenger, other second messengers and signals reflecting the metabolic status of the cell are used as well. A full understanding of how $[Ca^{2+}]_i$ is utilized to achieve a desired physiological response requires an understanding of how these other signals interact with and influence Ca^{2+}-mediated signaling pathways. Cross-talk between the second messengers inositol 1, 4, 5-trisphosphate (IP_3) and [H^+] or pH, as well as a metabolic parameter, membrane potential, are briefly discussed below. Other second messengers that may function in plants are listed in Table 2.

Inositol Phosphates

Perhaps the most direct link between Ca^{2+} and another second messenger is that between Ca^{2+} and IP_3. In animals, numerous environmental or hormonal signals, acting through receptors at the PM, activate phospholipase C. This activation is either direct, as in the case of tyrosine-kinase-linked receptors, or requires an intervening GTP-binding protein (G protein). Once activated, phospholipase C hydrolyzes phosphatidyl-inositol 4, 5-bisphosphate (PIP_2) to produce the second messenger IP_3. IP_3 stimulates the release of stored Ca^{2+} from the lumen of the ER or sarcoplasmic reticulum into the cytosol. It is by regulating $[Ca^{2+}]_i$ that IP_3 affects numerous physiological processes (3).

Many of the key players in this signaling system have been shown to exist in plants (21). The most notable exception being a receptor capable of activating phospholipase C either directly or indirectly. Accumulating evidence suggests that IP_3 has a signaling role in plants, and that it is involved in regulating $[Ca^{2+}]_i$ (19). Microinjection and release of caged IP_3 into living stomatal guard cells, for example, results in an increase in $[Ca^{2+}]_i$ and causes stomatal closure (31). IP_3 also stimulates Ca^{2+} release from vacuolar membrane vesicles (74) and from intact vacuoles (62). This suggests that in plants the vacuole may be a source of IP_3-mobilizable Ca^{2+}.

An intriguing feature of IP_3-stimulated Ca^{2+} release in animal cells is that the process is regulated by Ca^{2+}. The activity of the IP_3 receptor, a Ca^{2+} channel, increases as $[Ca^{2+}]_i$ increases, up to approximately 300 nM. Higher $[Ca^{2+}]_i$ are inhibitory to Ca^{2+} release (3). Although such Ca^{2+}-induced Ca^{2+}

Table 2. Second messengers in plants.

2nd Messenger	Activity	References
Inositol Phosphates	Release of Ca^{2+} from internal stores	19, 21, 31, 62, 74
Cytoplasmic pH ([H^+])	Unknown	25, 36
G-Proteins	Ion channel regulation Cell cycle regulation? Vesicle-mediated transport?	81
Diacylglycerols	Ion pump activation	48
cAMP	Unknown	56, but see 79
cGMP	Unknown	57

release has not been directly demonstrated in plants, it is suggested in guard cells (31) and serves to illustrate the complexities possible in the interactions between signaling pathways.

Cytoplasmic pH

The role of [H$^+$] as a second messenger and the cross-talk between cytoplasmic pH (pH$_i$) and [Ca^{2+}]$_i$ have not been extensively studied in plant cells. Yet recent findings suggest that in plants, as in animals, there is a strong interaction between these two signaling molecules, and that they may be co-regulatory. When individual, living maize coleoptile cells were impaled with double-barreled microelectrodes to simultaneously measure [Ca^{2+}]$_i$ and pH$_i$, an increase in [Ca^{2+}]$_i$ and an acidification of the cytoplasm were detected following application of IAA (Fig. 3B and ref. 23). It was also noted that IAA induced oscillations in membrane potential, [Ca^{2+}]$_i$ and pH$_i$. By simultaneously loading corn coleoptiles with two fluorescent dyes, one reporting [Ca^{2+}]$_i$ and another reporting pH$_i$, Gehring et al. (25) demonstrated that [Ca^{2+}]$_i$ increased and pH$_i$ decreased within 4 min following application of the synthetic auxin 2, 4-D. Changes in pH were 0.1-0.2 pH unit. ABA, while still raising [Ca^{2+}]$_i$, caused an increase in pH$_i$ of 0.05-0.1 pH unit (25). Using similar methodology, Irving et al. (36) showed that guard cell [Ca^{2+}]$_i$ and pH$_i$ responded qualitatively similarly to those parameters in corn coleoptiles when treated with auxin or ABA. Kinetin also caused an increase in [Ca^{2+}]$_i$ and a decrease in pH$_i$. It was suggested that both [Ca^{2+}]$_i$ and pH$_i$ are important in regulating guard cell movement, since changes in concentration preceded hormone-induced stomatal opening or closure (36).

Membrane Potential

The environmental and developmental history of plant cells may be reflected in their metabolic status. One parameter that may mirror this status is membrane potential. For example, auxin has been shown to hyperpolarize the PM of tobacco protoplasts (2), and anion channels in the PM of guard cells may respond to auxin by producing a transient depolarization (51). How such metabolic parameters interact with Ca^{2+}-signaling systems is only beginning to be known. What is clear, however, is that there is an enormous potential for cross-talk between them. A few examples follow. Voltage-dependent Ca^{2+} channels are well known in animal cells, and at least one voltage-dependent Ca^{2+}-permeable channel has been reported in higher plants (58). This channel in the TP of sugar beet opens when E$_{cytoplasm}$- E$_{vacuole}$ is positive. Its function is not fully understood. A voltage-sensitive H$^+$/Ca^{2+} antiport in the TP of beets and maize is thought to sequester Ca into the vacuole (4, 16). Membrane potential also influences the activity of several Ca^{2+}-activated channels. Within the stomatal guard cell, for example, depolarization of the PM results in activation of two Ca^{2+}-dependent, voltage-gated anion channels. The R-type anion channel rapidly activates and then inactivates during prolonged stimulation, while the S-type may regulate long term anion flux and membrane depolarization (72).

PARADIGMS

Having indicated the cellular mechanisms of calcium homeostasis and signalling in plant cells, we now describe three experimental systems, stomatal closure, moss bud development and enzyme synthesis and secretion by cereal aleurone cells, where models of the transduction of a hormonal signal by Ca^{2+} can be constructed.

Stomatal Guard Cell Function

Perhaps our most complete understanding of Ca^{2+}-based signal transduction in plants comes from the stomatal guard cell. Gaseous exchange at the leaf surface occurs through stomatal pores in the epidermis. Control of stomatal aperture provides the mechanism by which plants regulate their rate of water loss and uptake of CO_2 for photosynthesis. The aperture of the stomate is controlled by the turgor of the two guard cells surrounding the pore; when the guard cells are fully turgid the pore is maximally open. Many physical and chemical stimuli are known to influence stomatal aperture, including light quantity and quality, CO_2 concentration, water stress and plant hormones (50). ABA has recently been shown to cause a rapid rise in $[Ca^{2+}]_i$ that precedes ABA-induced stomatal closure (28, 36, 52, 53). Stomatal closure and the ABA-induced increase in $[Ca^{2+}]_i$ were slowed in the presence of the Ca^{2+}-channel blocker La^{3+}, but an increase still occurred with external EGTA pretreatment, suggesting that Ca^{2+} may be mobilized from internal stores in response to ABA. Indeed, imaging of the increase in $[Ca^{2+}]_i$ in guard cells of *Commelina communis* revealed hot spots of Ca^{2+} in the ER-enriched cytoplasm surrounding the nucleus and vacuole (28, 53), which may reflect Ca^{2+} release from these internal stores. Schroeder and Hagiwara (73), however, using simultaneous patch clamping of the guard cell PM and fluorescent imaging of $[Ca^{2+}]_i$ showed that ABA induced repetitive opening of nonspecific, Ca^{2+} permeable cation channels in the PM. Increases in $[Ca^{2+}]_i$ occurred as these channels opened. These results suggest the PM is a site of ABA-induced Ca^{2+} influx.

To directly test the role of Ca^{2+} in triggering stomatal closure, Gilroy et al. (31) microinjected caged Ca^{2+} into guard cells. Release of Ca^{2+} from its cage triggered stomatal closure, providing $[Ca^{2+}]_i$ increased to above a threshold of approximately 600 nM. Similar experiments using caged-IP_3 photolysis showed that release of IP_3 induced an increase in $[Ca^{2+}]_i$ (31), inhibition of K^+ channel activity (6) and subsequent stomatal closure . These experiments reinforce the hypothesis that an increase in cytosolic Ca^{2+}, at least in part from intracellular sources, triggers stomatal closure (28).

This extensive work on guard cells and our knowledge of guard cell electrophysiology has led to a plausible model for Ca^{2+}-transduced, ABA-induced stomatal closure. In this model, ABA opens non-specific cation channels in the guard cell PM. This allows for Ca^{2+} influx and a consequent increase in $[Ca^{2+}]_i$ (73). Increased $[Ca^{2+}]_i$ triggers Ca^{2+} release from internal stores and/or IP_3 release, which induces Ca^{2+} release from internal stores (31). Elevated $[Ca^{2+}]_i$ activates voltage-dependent anion channels (34) tending to further depolarize the membrane. Depolarization closes inward K^+ channels. Outward K^+ channels may then be activated by changes in pH. The net effect would be an increase in K^+ efflux from the cell, subsequent loss of turgor and stomatal closure (50, 72).

This model describes a mechanism whereby ABA and Ca^{2+} could lead to stomatal movements. However, the transduction of the ABA signal may involve more than a simple increase in $[Ca^{2+}]_i$. Gilroy et al. (28) observed that although guard cells always closed in response to ABA, Ca^{2+} levels were elevated in only 40% of the cells tested. Schroeder and Hagiwara (73) also found that one third of the guard cells in epidermal strips of *Vicia faba* closed in response to ABA, and only 37% showed an ABA-induced rise in $[Ca^{2+}]_i$. These results suggest that ABA-induced stomatal closure may occur through Ca^{2+}-dependent and Ca^{2+}-independent pathways, although this is controversial (53).

Auxin also affects stomatal responses, eliciting stomatal opening. Martin et al. (51) have used patch clamping to show that auxin may interact directly with the external face of ion channels in the guard cell PM. Active auxins (IAA and 1-naphthyacetic acid), but not inactive auxin analogs, ABA, cytokinins or GA, cause anion channel activation potential to move toward the resting potential of the cell, promoting transient channel opening and thus the uptake of ions that leads to stomatal opening (51). Thus, the receptor for auxin action in the guard cell may be an ion channel itself and, unlike the effects of ABA in guard cells, the auxin signal may not require a second-messenger-based transduction system to regulate ion channel activity (51). Blatt and Thiel (6) disagree with this interpretation of the effects of IAA on guard cell membrane depolarization. They argue that the rapid effects of IAA on membrane depolarization are brought about by the electrophoretic co-transport of H^+ which brings about depolarization, not by an interaction of auxin directly with the channel (6).

Bud Formation in *Funaria hygrometrica*

In response to cytokinins, target cells of the moss *Funaria hygrometrica* undergo asymmetrical cell division, a process thought to involve both spatial and temporal changes in $[Ca^{2+}]_i$ (35, 69). Cytokinin (CK) induces a localized swelling of the target cell (caulonemal cell) and nuclear migration to this region. This cell then undergoes asymmetrical cell division and the side branch cells divide to form buds. These CK-induced events can be disrupted with Ca^{2+} channel antagonists such as Verapamil and La^{3+} or by removal of extracellular Ca^{2+}. The Ca^{2+} ionophore A23187 in the presence of Ca^{2+} can mimic the effect of CK addition (35). These data suggest that Ca^{2+} influx at the PM is involved in the transduction of the CK signal. Visualization of membrane-associated Ca^{2+} using the fluorescent indicator chlortetracycline showed an accumulation of membrane-associated Ca^{2+} at the site of nuclear migration, lateral wall swelling and bud cell division. In a direct test of this model, Hahm and Saunders (32) have loaded *Funaria* cells with the fluorescent Ca^{2+} indicator Indo-1 to monitor $[Ca^{2+}]_i$ changes in response to CK treatment. Ca^{2+} in quiescent, i.e. non-target, cells remained at 250 nM irrespective of CK treatment. However, CK induced a three-fold increase in Ca^{2+} in target caulonemal cells.. This increase was dependent on extracellular Ca^{2+}, and when ion fluxes were monitored with the vibrating microprobe, current influx (partly carried by Ca^{2+}?) predicted the site of bud initiation (68). However, Ca^{2+} cannot entirely account for the response of these cells, as non-target caulonemal and tip-growing cells, which do not go on to divide asymmetrically, also responded with an increase in $[Ca^{2+}]_i$. Thus an elevation in $[Ca^{2+}]_i$ in conjunction with more specialized response elements further downstream of Ca^{2+}, or interaction with other signaling pathways, may account for the induction of cell division in caulonemal cells of *Funaria*.

Enzyme Synthesis and Secretion by the Cereal Aleurone

The aleurone layer of cereal grains is a digestive tissue that secretes a spectrum of hydrolytic enzymes, principally a-amylases, that mobilizes storage polymers in the endosperm for use in germination and seedling growth (41). Gibberellins stimulate enzyme synthesis and secretion from the aleurone cell and this is reversed by ABA. Barley aleurone protoplasts show a slow (4 h) rise in cytosolic Ca^{2+} from 100 to 300 nM upon incubation with GA. This increase precedes the onset of a-amylase synthesis and secretion (13, 14, 29). Subsequent treatment with ABA lowers $[Ca^{2+}]_i$, and secretion ceases. Ratio and confocal imaging of individual protoplasts loaded with fluorescent Ca^{2+}-indicators revealed the increase in Ca^{2+} to be localized to the cytoplasm just below the PM (29). Removal of extracellular Ca^{2+} abolished both secretion of a-amylase and the rise in $[Ca^{2+}]_i$. This suggests that in the aleurone cell the increase in $[Ca^{2+}]_i$ results from influx at the PM. Similarly, in the isolated aleurone layer of wheat, GA causes cytosolic Ca^{2+} levels to rise within minutes (10). This increase is blocked by the Ca^{2+} channel antagonist Nifedipine, reinforcing the idea that GA modulates Ca^{2+} fluxes at the PM in aleurone cells. GA has also been shown to increase the level of CaM and the activity of a CaM-dependent Ca^{2+}-ATPase on the ER in barley aleurone (10, 12, 30). This provides a mechanism for increasing the transport of Ca^{2+} into the ER, Ca^{2+} which is needed for the synthesis of the Ca^{2+}-containing a-amylase holoenzyme (41). Ca^{2+} and CaM may also regulate other activities in the aleurone cell, such as the fusion of secretory vesicles, a well-known

Ca^{2+}-dependent event (80). Thus changes in $[Ca^{2+}]_i$ and CaM levels provide one mechanism to integrate and coordinate the complex events that change as GA and ABA regulate secretory activity in the aleurone cell.

CONCLUSIONS AND FUTURE DIRECTIONS

It is now clear that many hormonal signals are transduced by changes in $[Ca^{2+}]_i$. These changes are often characterized by a resetting of $[Ca^{2+}]_i$ to a new, usually higher level. Heterogeneity of $[Ca^{2+}]_i$ within individual cells may provide a spatial component to the Ca^{2+} signal. There is an increasing awareness that other second messengers may play a role in amplifying or modifying the Ca^{2+} signal. The full extent of such cross-talk is not yet known, but its potential for creating specific cellular responses from more generic hormonal signals is tremendous. The cellular machinery required to provide complicated Ca^{2+} dynamics is also beginning to be understood. Calcium pumps have been well characterized. Our knowledge of Ca^{2+}-channels and their regulation is expanding rapidly. Calcium-binding proteins and Ca^{2+}-regulated proteins are being used to trace Ca^{2+}-based signaling pathways from the Ca^{2+} signal to the final response. Yet for those seeking an understanding of how Ca^{2+} is used by plants to transduce hormonal signals many challenges lie ahead. Much more remains to be learned about how Ca^{2+}-transport through the PM and organellar membranes is regulated in vivo. Although some Ca^{2+}-binding proteins have been identified, others almost certainly await discovery. Even for the most extensively studied Ca^{2+}-binding proteins, CaM and the CDPKs, little is known about their activity in vivo. Until a more complete list of the proteins with which Ca^{2+}-binding proteins interact is in hand, and until such interactions have been quantified with regard to binding affinity and kinetics, we can only begin to guess at how the responses to plant hormones that are mediated by Ca^{2+} are brought about. Finally, although at present the number of elements known to be links in Ca^{2+}-based signal transduction chains is small, it is bound to increase. It is the complex nature of signaling within plant cells that makes this field of study at once so daunting and so exciting. The challenge for the future is to identify missing pieces of the signal transduction puzzle, to establish the linkages between all the pieces, and to interpret the multi-dimensional picture that emerges. At present we see only a crude sketch, notable more for its omissions than for its completeness. Yet detail is being added at a rapid rate, and the pace is likely to quicken.

References

1. Anraku, Y., Ohya, Y., Iida, H. (1991) Cell cycle control by calcium and calmodulin in *Saccharomyces cerevisiae*. Biochim. Biophys. Acta 1093, 169-177.
2. Barbier-Brygoo, H., Ephritikhine, G., Klämbt, D., Murel, C., Palme, K., Schell, J., Guern, J. (1991) Perception of the auxin signal at the plasma membrane of tobacco mesophyll protoplasts. Plant J. 1, 83-93.

3. Berridge, M.J. (1993) Inositol trisphosphate and calcium signalling. Nature 361, 315-325.
4. Blackford S., Rea P.A., Sanders P.A. (1990) Voltage sensitivity of H^+/Ca^{2+} antiport in higher plant tonoplast suggests a role in vacuolar calcium accumulation. J Biol. Chem. 265, 9617-20.
5. Blatt, M.R., Thiel, G. (1993) Hormonal control of ion channel gating. Annu. Rev. Plant Physiol. Plant Mol. Biol. 44, 543-568.
6. Blatt, M.R., Thiel, G., Trentham, D.R. (1990) Reversible inactivation of K^+ channels of *Vicia* stomatal guard cells following the photolysis of caged inositol-1, 4, 5-trisphosphate. Nature 346, 766-769.
7. Braam, J. (1992) Regulation of expression of calmodulin and calmodulin-related genes by environmental stimuli in plants. Cell Calcium 13, 457-463.
8. Briskin, D.P. (1990) Ca^{2+}-translocating ATPase of the plant plasma membrane. Plant Physiol. 94, 397-400.
9. Burchert, M., Surek, B., Kreimer, G., Latzko, E. (1990) Calcium binding by chloroplast stroma proteins and functional implications. Am. Soc. Plant Physiol. Symp. Ser. 4, 17-25.
10. Bush, D.S. (1992) Hormonal regulation of Ca^{2+} transporters leads to rapid changes in steady-state levels of cytosolic Ca^{2+} in the cereal aleurone. Abs., Ninth International Workshop on Plant Membrane Biology, Monterey, CA.
11. Bush, D.S., Biswas, A.K., Jones, R.L. (1989) Gibberellic-acid-stimulated Ca^{2+} accumulation in endoplasmic reticulum of barley aleurone: Ca^{2+} transport and steady-state levels. Planta 178, 411-420.
12. Bush, D.S., Biswas, A.K., Jones, R.L. (1993) Hormonal regulation of Ca^{2+} transport in the endomembrane system of the barley aleurone. Planta 189, 507-515.
13. Bush, D.S., Jones, R.L. (1987) Measurement of cytoplasmic calcium in aleurone protoplasts using Indo-1 and Fura-2. Cell Calcium 8, 455-472.
14. Bush, D.S., Jones, R.L. (1998) Cytoplasmic calcium and a-amylase secretion from barley aleurone protoplasts. Eur. J. Cell Biol. 46, 466-469.
15. Chae, Q., Park, H.J., Hong, S.D. (1990) Loading Quin 2 into oat protoplasts and measurement of cytosolic calcium ion concentration changes by phytochrome action. Biochim. Biophys. Acta 1051, 115-122.
16. Chanson, A. (1991) A Ca^{2+}/H^+ antiport system driven by the tonoplast pyrophosphate-dependent proton pump from maize roots. J. Plant Physiol. 137, 471-476.
17. Clarkson, D.T., Hanson, J.B. (1980) The mineral nutrition of plants. Annu. Rev. Plant Physiol. 31, 239-298.
18. Cleland, R.E., Virk, S.S., Taylor, D., Bjorkman, T. (1990) Calcium, cell walls and growth. Am. Soc. Plant Physiol. Symp. Ser. 4, 9-16.
19. Cote, G.C., Crain, R.C. (1993) Biochemistry of phosphoinositides. Annu. Rev. Plant. Physiol. Plant Mol. Biol. 44, 333-356.
20. Dieter, P., Marme, D. (1980) Ca^{2+} transport in mitochondrial and microsomal fractions from higher plants. Planta 150, 1-8.
21. Drøbak, B.K. (1993) Plant phosphoinositides and intracellular signalling. Plant Physiol. 102, 705-709.
22. Evans, D.E., Theodoulan, F.L., Williamson, I.M., Boyce, J.M. (1994) The calcium pumps of plant cell membranes. *In*: Membrane transport in plants and fungi: molecular mechanisms and control. Blatt, M.R., Leigh, R., Sanders, D., eds. The Company of Biologists, Ltd., UK. in press.
23. Felle, H. (1988) Auxin causes oscillations of cytosolic free calcium and pH in *Zea mays* coleoptiles. Planta 174, 495-499.
24. Felle, H. (1991) The control of cytoplasmic levels of Ca^{2+} and H^+ in plants. *In*: Plant signalling, plasma membrane and change of state, pp. 79-104, Penel, C. and Greppin, H., eds. Universite de Geneve.
25. Gehring, C.A., Irving, H.R., Parish, R.W. (1990) Effects of auxin and abscisic acid on cytosolic calcium and pH in plant cells. Proc. Natl. Acad. Sci. USA 87, 9645-9649.

26. Gehring, C.A., Williams, P.A., Cody, S.H., Parrish, R.W. (1990) Phototropism and geotropism in maize coleoptiles are spatially correlated with increase in cytosolic free calcium. Nature 345, 528-530.

27. Gilroy, S., Bethke, P.B., Jones, R.L. (1993) Calcium homeostasis in plants. J. Cell Sci. 106, 453-462.

28. Gilroy, S., Fricker, M.D., Read, N.D., Trewavas, A.J. (1991) Role of calcium in signal transduction in *Commelina* guard cells. Plant Cell 3, 333-344.

29. Gilroy, S., Jones, R.L. (1992) Gibberellic acid and abscisic acid coordinately regulate cytoplasmic calcium and secretory activity in barley aleurone protoplasts. Proc. Natl. Acad. Sci. U.S.A. 89, 3591-3595.

30. Gilroy, S., Jones, R.L. (1993) Calmodulin stimulation of unidirectional calcium uptake by endoplasmic reticulum of barley aleurone. Planta 190, 289-296.

31. Gilroy, S., Read, N.D., Trewavas, A.J. (1990) Elevation of cytoplasmic calcium by caged calcium or caged inositol trisphosphate initiates stomatal closure. Nature 346, 769-771.

32. Hahm, S.H., Saunders, M.J. (1991) Cytokinin increases intracellular calcium in *Funaria*: detection with Indo-1. Cell Calcium 12, 675-681.

33. Hahn, K., DeBiasio, R., Taylor, D.L. (1992) Patterns of elevated free calcium and calmodulin activation in living cells. Nature 359, 736-738.

34. Hedrich, R., Busch, H., Raschke, K. (1990) Ca^{2+} and nucleotide dependent regulation of voltage dependent anion channels in the plasma membrane of guard cells. EMBO J. 9, 3889-3892.

35. Hepler, P.K., Wayne, R.O. (1985) Calcium and plant development. Annu. Rev. Plant Physiol. 36, 397-439.

36. Irving, H.R., Gehring, C.A., Parish, R.W. (1992) Changes in cytosolic pH and calcium of guard cells precede stomatal movements. Proc. Natl. Acad. Sci. USA 89, 1790-1794.

37. Jena, P.K., Reddy, A.S.N., Poovaiah, B.W. (1989) Molecular cloning and sequencing of cDNA for plant calmodulin: signal-induced changes in the expression of calmodulin. Proc. Natl. Acad. Sci. USA 86, 3644-48.

38. Johannes, E., Brosnan, J.M., Sanders, D. (1991) Calcium channels and signal transduction in plant cells. BioEssays 13, 331-336.

39. Jones R.L. Synthesis and secretion of hydrolytic enzymes during germination. *In*: Seed development and germination. Galili, G. Kiegel, J., eds. Marcel Dekker, New York. In press.

40. Jones, R.L., Bush D.S. (1991) Gibberellic acid regulates the level of a BiP cognate in the endoplasmic reticulum of barley aleurone cells. Plant Physiol. 97, 456-459.

41. Jones, R.L., Jacobsen, J.V. (1991) Regulation of synthesis and transport of secreted proteins in cereal aleurone. Int. Rev. Cytol. 126, 49-88.

42. Kauss, H. (1987) Some aspects of calcium-dependent regulation in plant metabolism. Annu. Rev. Plant Physiol. 38, 47-72.

43. Knight, M.R., Campbell, A.K., Smith, S.M., Trewavas, A.J. (1991) Transgenic plant aequorin reports the effects of touch, cold-shock and elicitors on cytoplasmic calcium. Nature 352, 524-526.

44. Knight, M.R., Read, N.D., Campbell, A.K., Trewavas, A.J. (1993) Imaging calcium dynamics in living plants using semi-synthetic recombinant aequorins. J. Cell Biol. 117, 111-125.

45. Koch, G.L.E. (1990) The endoplasmic reticulum and calcium storage. BioEssays 12, 527-531.

46. Kreimer, G., Melkonian, M., Holtum, J.A., Latzko, E. (1985) Characterization of calcium fluxes across the envelope of spinach chloroplasts. Planta 166, 515-523.

47. Kreimer, G., Melkonian, M., Holtum, J.A., Latzko, E. (1988) Stromal free calcium concentration and light-mediated activation of chloroplast fructose-1, 6-bisphosphatase. Plant Physiol 86, 423-428.

48. Lee, Y., Assmann, S.M. (1991) Diacylglycerols induce both ion pumping in patch-clamped guard-cell protoplasts and opening of intact stomata. Proc. Natl. Acad. Sci. USA 88, 2127-2131.
49. Lynch, J., Polito, V.S., Lauchli, A. (1989) Salinity stress increases cytoplasmic Ca^{2+} activity in maize root protoplasts. Plant Physiol. 90, 1271-1274.
50. MacRobbie, E.A.C. (1992) Calcium and ABA-induced stomatal closure. Phil. Trans. Roy. Soc. Lond. B 338, 5-18.
51. Martin, I., Lohse, G., Hedrich, R. (1991) Plant growth hormones control voltage-dependent activity of anion channels in plasma membrane of guard cells. Nature 353, 758-762.
52. McAinsh, M.R., Brownlee, C., Hetherington, A.M. (1990) Abscisic acid induced elevation of guard cell cytosolic free Ca^{2+} precedes stomatal closure. Nature 343, 186-188.
53. McAinsh, M.R., Brownlee, C., Hetherington, A. M. (1992) Visualizing changes in cytosolic free Ca^{2+} during the response of stomatal guard cells to abscisic acid. Plant Cell 4, 1113-1122.
54. Miller, A.J., Sanders, D. (1987) Depletion of cytosolic free calcium induced by photosynthesis. Nature 326, 769-400.
55. Neuhaus, G., Bowler, C., Kern, R., Chua, N.H. (1993) Calcium/calmodulin-dependent and -independent phytochrome signal transduction pathways. Cell 73, 937-952.
56. Newton, R.P., Brown, E.G. (1986) The biochemistry and physiology of cyclic AMP in higher plants. *In*: Hormones, receptors and cellular interactions in plants, pp. 115-153, Chadwick, C.M. Garrod, D.R., eds. Cambridge University Press.
57. Newton, R.P., Chiatante, D., Ghosh, D., Brenton, A.G., Walton, T.J., Harris, F.M., Brown, E.L. (1989) Identification of cyclic nucleotide constituents of meristematic and non-meristematic tissue of *Pisum sativum* roots. Phytochemistry 28, 2243-2254.
58. Pantoja, O., Gelli, A., Blumwald, E. (1992) Voltage-dependent calcium channels in plant vacuoles. Science 255, 1567-1570.
59. Pelham, H.R.B. (1989) Control of protein exit from the endoplasmic reticulum. Annu. Rev. Cell Biol. 5, 1-23.
60. Poovaiah, B.W., Reddy, A.S.N. (1987) Calcium messenger system in plants. CRC Crit. Rev. Plant Sci. 6, 47-103.
61. Poovaiah, B.W., Reddy, A.S.N. (1993) Calcium and signal transduction in plants. CRC Crit. Rev. Plant Sci. 12, 185-211.
62. Ranjeva, R., Carrasco, A., Boudet, A.M. (1988) Inositol trisphosphate stimulates the release of calcium from intact vacuoles isolated from *Acer* cells. FEBS Lett. 230, 137-141.
63. Rea, P.A., Poole, R.J. (1993) Vacuolar H^+-translocating pyrophosphatase. Annu. Rev. Plant Physiol. Plant Mol. Biol. 44, 157-180.
64. Roberts, D.M., Harmon, A.C. (1992) Calcium-modulated proteins: Targets of intracellular calcium signals in higher plants. Annu. Rev. Plant Physiol. Plant Mol. Biol. 43, 375-414.
65. Russ, U., Grolig, F., Wagner, G. (1991) Changes in cytoplasmic free Ca^{2+} in the green alga *Mougeotia scalaris* as monitored with Indo-1, and their effect on the velocity of chloroplast movements. Planta 184, 105-112.
66. Sambrook, J.F. (1990) The involvement of calcium in transport of secretory proteins from the endoplasmic reticulum. Cell 61, 197-199.
67. Sanders, S.L., Schekman, R. (1992) Polypeptide translocation across the endoplasmic reticulum membrane. J. Biol. Chem. 267, 13791-13794.
68. Saunders, M.J. (1986) Cytokinin activation and redistribution of plasma membrane ion channels in *Funaria*. Planta 167, 402-409.
69. Saunders, M.J. (1990) Calcium and plant hormone action. Proc. Soc. Exp. Biol. 44, 271-283.

70. Shacklock, P.S., Read, N.D., Trewavas, A.J. (1992) Cytosolic free calcium mediates red light-induced photomorphogenesis. Nature 358, 753-755.
71. Schaller, G.E., Sussman, M.R. (1988) Phosphorylation of the plasma membrane H⁺-ATPase of oat roots by a calcium-stimulated protein kinase. Planta 173, 509-518.
72. Schroeder, J.I. (1992) Plasma membrane ion channel regulation during abscisic acid-induced closing of stomata. Phil. Trans. Roy. Soc. Lond. B. 338, 83-89.
73. Schroeder, J.I., Hagiwara, S. (1990) Repetitive increases in cytosolic Ca^{2+} of guard cells by abscisic acid activation of nonselective Ca^{2+} permeable channels. Proc. Natl. Acad. Sci. U.S.A. 87, 9305-9309.
74. Schumaker, K.S., Sze, H. (1987) Inositol(1,4,5)trisphosphate releases Ca^{2+} from vacuolar membrane vesicles of oat roots. J. Biol. Chem. 262, 3944-3946.
75. Serrano, R. (1989) Structure and function of plasma membrane ATPase. Annu. Rev. Plant Physiol. Plant Mol. Biol. 40, 19-38.
76. Shelanski, M.L. (1989) Intracellular ionic calcium and the cytoskeleton in living cells. Ann. New York Acad. Sci. 568, 121-124.
77. Sheng, M., McFadden, G., Greenberg, M.E. (1990) Membrane depolarization and calcium induce *c-fos* transcription via phosphorylation of the transcription factor CREB. Neuron 4, 571-582.
78. Simon, P., Bonzon, M., Greppin, H., Marme, D. (1984). Subchloroplastic localization of NAD kinase activity: evidence for a Ca^{2+}, calmodulin-dependent activity at the envelope and for a Ca^{2+}, calmodulin independent activity in the stroma of pea chloroplasts. FEBS Lett. 167, 332-338.
79. Spiteri, A., Viratelle, O.M., Raymond, P., Rancillac, M., Labouesse, J., Pradet, A. (1989) Artifactual origins of cyclic AMP in higher plant tissues. Plant Physiol. 91, 624-628.
80. Steer, M.J. (1988) The role of calcium in exocytosis and endocytosis in plant cells. Physiol. Plant. 72, 213-220.
81. Terryn, N., Van Montagu, M., Inzé, D. (1993) GTP-binding proteins in plants. Plant Mol. Biol. 22, 143-152.
82. Vantard, M., Lambert, A.-M., Mey, J.D., Picquot, P., Eldik, L.J.V. (1985) Characterization and immunocytochemical distribution of calmodulin in higher plant endosperm cells: localization in the mitotic apparatus. J. Cell Biol. 101, 488-499.
83. Wang, M., Van Duijn, B., Schram, A.W. (1991) Abscisic acid induces a cytosolic calcium decrease in barley aleurone protoplasts. FEBS Lett. 278, 69-74.
84. Weiser, T., Blum, W., Bentrup, F.W. (1991) Calmodulin regulates the Ca^{2+}-dependent slow-vacuolar ion channel in the tonoplast of *Chenopodium rubrum* suspension cells. Planta 185, 440-442.

E. MOLECULAR ASPECTS OF HORMONE SYNTHESIS AND ACTION

E1. Genes Specifying Auxin and Cytokinin Biosynthesis in Prokaryotes

Roy O. Morris
Biochemistry Department, University of Missouri, Columbia, Missouri 65211, USA.

INTRODUCTION

Much effort has been expended to understand the biosynthesis and mode of action of the auxins and cytokinins, with only moderate success. For example, although cytokinin structures, internal distribution and transport within plants are documented (64,37), cytokinin catabolism is well understood (47), and the morphological and physiological consequences of cytokinin application have been described (37), we still do not know fully how they are synthesized. Putative plant cytokinin biosynthetic enzymes have been observed (14,13), but none has been purified to homogeneity. Further, we do not know when and where such enzymes are expressed in the plant or even if they are responsible for plant cytokinin synthesis *in vivo*. A longstanding debate continues (7; see Chapter B3) as to whether cytokinins are synthesized directly or via an indirect pathway involving tRNA catabolism. Plant genes that encode biosynthetic enzymes for both pathways remain to be isolated. A similar situation obtains for auxins.

Recently, light has been shed from an unexpected source on the biosynthesis of both auxins and cytokinins. Several phytopathogenic, symbiotic, or free-living soil bacteria have been shown to synthesize one or both hormones (Table 1).

The phytopathogens listed in Table 1 generally form galls on their plant hosts, suggesting that they alter the endogenous phytohormone content of infected tissues. Biochemical and molecular studies have confirmed that this does happen. Tumors incited by *Agrobacterium tumefaciens* and *Pseudomonas savastanoi* have elevated auxin and cytokinin levels and the bacteria themselves contain genes for auxin and cytokinin (1, 6, 36, 39, 60, 80) biosynthesis.

P. J. Davies (ed.), Plant Hormones, 318–339.
© 1995 *Kluwer Academic Publishers. Printed in the Netherlands.*

Table 1. Bacteria known to produce phytohormones

Bacteria	Location/ Function	Auxins	Cytokinins
Agrobacterium rhizogenes (3)	Hairy root disease	IAA	Z
Agrobacterium tumefaciens (45, 62)	Crown gall tumors	IAA	Z, [9R]Z, iP, [9R]iP
Azospirillum brasilense (84)	Rhizosphere associated	IAA	
Azotobacter chroococcum (54)	Rhizosphere associated		cZ, Z, [9R]Z, [9R]dHZ, iP, [9R]iP
Azotobacter vinelandii (72)	Rhizosphere associated		Z, iP, [9R]iP
Bradyrhizobium japonicum (68)	Symbiont		2ms[9R]Z, [9R]iP, 2ms[9R]iP
Erwinia herbicola pv gypsophilae (42)	Gypsophila gall		Z, [9R]Z, iP, [9R]iP
Frankia sp. (66)	Symbiont		[9R]iP
Pseudomonas amygdali (31)	Almond canker	IAA	[9deoxyR]Z, Z, dHZ, iP
Pseudomonas solanacearum (3)	Bacterial wilt		Z
Pseudomonas savastanoi (69)	Olive galls		Z, [9R]Z, 1'MeZ, 1"Mc[9R]Z
Rhizobium sp. (4)	Symbiont nodules		Z, [9R]Z, iP, [9R]iP
Rhizobium sp. (71)	Symbiont		Z
Rhizobium sp. (IC3442) (74)	Symbiont		Z, iP
Rhodococcus fascians (49)	Witches Broom disease		cZ, iP, [9R]iP
Vibrio sp. (44)	Marine, free living		iP, [9R]iP

In this chapter, the mechanisms of tumorigenesis by *A. tumefaciens* and *P. savastanoi* are outlined. Evidence relating to tumor phytohormone status is presented and details of the isolation, structure and function of the auxin and cytokinin biosynthetic genes from *A. tumefaciens* and *P. savastanoi* are described. Recent studies of the existence of cytokinin biosynthetic genes in other pathogens and of the relative importance of tRNA-mediated vs. direct cytokinin synthesis in *A. tumefaciens* are also presented.

Mechanisms of plant tumor formation

At least four classes of phytopathogenic bacteria, *A. tumefaciens*, *Erwinia herbicola*, *P. savastanoi* and *Rhodococcus fascians* (formerly *Corynebacterium fascians*) produce neoplastic or hyperplastic diseases in plants (53, 17). The resulting galls display either completely unorganized growth or incomplete organogenesis with production of abnormal shoots or roots (Fig. 1).

Although galls incited by all four classes of bacteria appear similar, there is a fundamental difference between those incited by *A. tumefaciens* and those incited by the other pathogens. Galls resulting from A. *tumefaciens* are transformed and no longer need the inciting bacteria to grow. Galls incited by *E. herbicola*, *P. savastanoi* and *R. fascians* need the continued presence of bacteria (or a supply of auxin and cytokinin) to grow in culture (17).

Fig. 1. Tumors and galls incited by phytopathogenic bacteria. (A) Undifferentiated crown gall tumor incited on *Kalanchoë* stem by *A. tumefaciens* octopine strain A6NC. Roots arise not from the tumor but from the stem below the point of infection. (B) *Nicotiana tabacum* shoot-bearing teratoma in culture. Incited by *A. tumefaciens* nopaline strain T37. (C) Hairy root tumor incited on *Kalanchoë* by *Agrobacterium rhizogenes* strain A4. (D) Witches broom disease incited on Shasta Daisy by *Rhodococcus fascians* (photograph courtesy of Don Cooksey).

A. tumefaciens incites unorganized crown gall tumors on most dicotyledonous plant species (Fig. 1A) although on some hosts some strains of *A. tumefaciens* incite shooty teratomas (Fig. 1B). The closely related *A. rhizogenes* causes hairy root disease, an overproduction of grossly abnormal roots (Fig. 1C). Crown gall and hairy root tumors are hormone-autotrophic in culture. They contain elevated levels of auxins and cytokinins (47) which result from the expression of bacterial auxin and cytokinin biosynthetic genes which have been transferred from the *Agrobacterium* genome to the plant genome. The process of tumorigenesis is therefore a natural genetic engineering event, and *A. tumefaciens* and *A. rhizogenes* may be classed as transforming pathogens.

In contrast to *A. tumefaciens*, the three other classes of bacteria (*E. herbicola*, *P. savastanoi* and *R. fascians*) produce galls not by transformation but by virtue of close association with the host. They may properly be termed associative pathogens. *E. herbicola* pv *gypsophilae* induces shoot proliferation on *Gypsophila paniculata* (17). *P. savastanoi* produces unorganized galls on olives, oleanders and privet. *R. fascians* produces witches broom disease, a proliferation of shoots at the apex of herbaceous plants (Fig. 1D).

Molecular basis for crown gall tumor formation

Fig. 2 illustrates the molecular events that underlie crown gall tumor formation by *A. tumefaciens*. Full details may be found in recent reviews

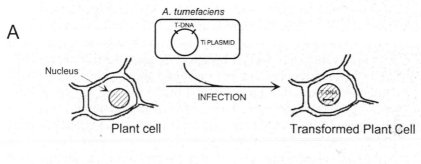

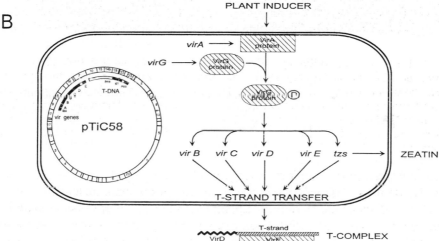

Fig. 2. Molecular mechanisms of crown gall tumorigenesis. Top: The T-DNA of the *A. tumefaciens* Ti plasmid is transferred to and integrated into the nuclear genome of the host plant. Bottom: Schematic of gene activation in *A. tumefaciens* during infection. The organism represented is a typical *A. tumefaciens*, nopaline strain, C58. A phenolic signal from the plant is sensed by the VirA protein which catalyzes the phosphorylation of VirG. Phosphorylated VirG activates transcription of all *vir* genes which then effect T-complex formation and transfer. Details from work summarized in (47, 82).

pTiC58: *A. tumefaciens* strain C58 Ti plasmid (*HindIII* restriction map)
vir A-G: Genetic loci controlling T-DNA transfer to plant
tms: T-DNA locus controlling tumor auxin content
ipt: T-DNA locus controlling tumor cytokinin content (*ipt = tmr*)
nos: T-DNA locus controlling tumor nopaline synthesis
VirA, VirD, VirE, VirG: Proteins encoded by the appropriate *vir* loci
Ⓟ : Phosphate

(52,53,82).   In 1974, Zaenen *et al.* (81) demonstrated that *A. tumefaciens* harbors a large (ca 200 kb) plasmid responsible for virulence.   This was termed the tumor-inducing (Ti) plasmid.   If a strain of *A. tumefaciens* is cured of its endogenous Ti plasmid by growth at elevated temperatures, it loses virulence.   If the Ti plasmid from the same strain or from a different strain is introduced into the cured bacterium, virulence is regained.

Although many different Ti plasmids have been characterized, they all have two regions in common.   These are the T-DNA, which is responsible for tumorigenesis, and the *virulence* or *vir* region that is responsible for the process of infection (Fig. 2).   In a sense, the T-DNA may be regarded as the ultimate pathogenic entity.   It is a region (approximately 10%) of the Ti plasmid, bounded by well-defined 23 bp direct oligonucleotide repeats (79). Between these boundaries lay genes encoding the synthesis of auxins, cytokinins and opines (unusual amino acids) (82).   During tumorigenesis a copy of the T-DNA is transferred from the Ti plasmid to the plant cell. There it is integrated into the nuclear genome where it functions to induce and maintain the tumorous condition.

T-DNA transfer from the Ti plasmid to the plant is controlled by *virulence* genes which   initiate T-DNA transfer but are not themselves transferred.   They cluster into at least six complementation groups (*virA, virB, virC, virD, virE and virG*) (32,26) and some details of their function are now understood.   The products of the *virD* and *virE* genes are responsible for excision and packaging T-DNA for transfer.   VirD2 is a strand-specific endonuclease which cuts the T-DNA border repeats, attaches covalently to one end (15) and initiates packaging of one strand of the T-DNA into a linear complex with one molecule of VirD2 and many molecules of VirE.   This linear structure, termed the T-complex (27), is moved to the plant cell through a pore in the bacterial wall formed by proteins encoded by the *virB* locus (63).

Interestingly, *Agrobacteria* are not constitutively competent to transfer T-DNA.   They can do so only in a plant wound site.   Competence is acquired on receipt of a wound-specific signal emanating from the plant.   This signal is sensed by the VirA protein which then activates the VirG protein by phosphorylation (33).   Phosphorylated VirG is a positive transcriptional activator that then stimulates transcription of the other *vir* genes. The process is thus a classical two component regulatory cascade similar to known homeostatic bacterial cascades (78).   It is illustrated in Fig. 2B.   Several plant phenolics that activate the cascade have been identified.   They include acetosyringone (65) and coniferyl alcohol (50).

Once the T-complex enters the plant cell, it moves to the nucleus under the direction of nuclear targeting sequences present within the VirD2 and VirE proteins (16).   It is then integrated into the plant nuclear genome by a process that is not yet clearly understood.   What is clear however, is that there can be different patterns of T-DNA integration depending on the origin of a particular Ti plasmid.   For example, in those Ti plasmids that specify the synthesis of the opine octopine, T-DNA tends to be integrated in two separate sections derived from the leftmost (TL-DNA) and rightmost (TR-DNA) segments of the T-DNA (see Fig. 6). Octopine TL-DNA is 12 kb long, encodes seven genes (73) and is alone responsible for the tumor phenotype. TR-DNA is not required. On the other hand, the T-DNA from *A. tumefaciens* strains which encode synthesis of the opine, nopaline, tend to integrate as a single tandem repeat. Nopaline T-DNA is larger (23 kb) than octopine TL-DNA and encodes thirteen genes (77). There is, however, extensive sequence homology between regions of the octopine and nopaline T-DNA (22), and it is within these regions that genes responsible for auxin and cytokinin biosynthesis are found.

INDOLEACETIC ACID PARTICIPATION IN CROWN GALL TUMOR GROWTH

An extensive body of evidence suggests that auxin production, encoded by T-DNA genes, contributes significantly to the maintenance of the transformed state. The first evidence that crown gall tumors have altered phytohormone status came from the pioneering work of Braun (10). Tumors exhibited auxin-like growth effects and could be grown in axenic culture lacking auxin (in contrast to untransformed tissue that could not). Braun suggested that growth was supported by the endogenous production of auxins and cell division factors. Subsequent studies, summarized in (52, 47) confirmed and extended his theory. The evidence falls into six categories:

- Tumor IAA levels are elevated over those of untransformed tissues in culture. Because the analytical techniques used for auxin determination have been various, because cloned and uncloned tumor lines (containing mixtures of transformed and untransformed cells) have been examined, and because tumors have been incited on different plant species, there is some conflict between the reports. However, most studies show that IAA levels are higher in tumors than in untransformed tissues.
- IAA accumulation rates (presumably reflecting endogenous synthesis rates) of tumors in culture are greater than those of untransformed tissue. Again there is considerable variation between studies. IAA levels have been reported to peak early in exponential growth, late in growth or not at all. But, in general, levels increase more rapidly and to a greater degree in tumors than in untransformed tissues.
- The process of conversion of tryptophan to IAA by tumor tissue differs from that of untransformed tissue. Both tissues can metabolize tryptophan to IAA but conversion is more efficient in crown gall.
- Some tumors incited by mutated Ti plasmids are IAA-auxotrophs whereas wild type tumors are not. For example, a mutant of the octopine strain Ach5 bearing the insertion element IS60 within its T-DNA was avirulent on *Kalanchoë* and produced only small tumors on

tomato (55). Tumor growth was restored by application of NAA. Other mutants (A66) exhibited a requirement for IAA for growth in culture (9).

- Tumor IAA levels decrease significantly if the T-DNA bears certain well-defined transposon insertions in the T-DNA (Fig. 3). Garfinkel *et al.* (23) used *Tn3* and *Tn5* transposon mutagenesis to define three T-DNA loci (*tms, tmr,* and *tml*) which control tumor morphology. Inactivation of the *tms* locus causes normally undifferentiated tumors to produce shoots and concomitantly decreases IAA levels (1). Because the *tms* locus encompasses two genes (*tms1* and *tms2*) (39) and because insertional mutagenesis of either causes reduction in IAA content, both genes must be involved in IAA metabolism.

- Some tumors incited by *tms* mutants contain excessively high levels of the putative IAA biosynthetic intermediate, indoleacetamide (76). Levels are 1000 times greater than in wild type tumors.

Auxin biosynthetic genes in *Agrobacterium tumefaciens*

Auxin participation in tumor formation was confirmed by experiments

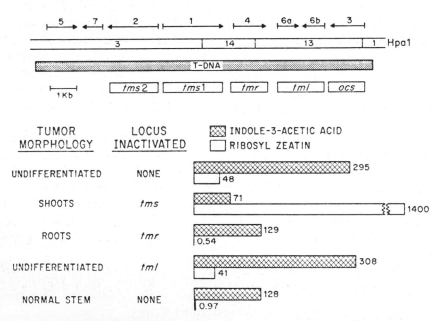

Fig. 3. Loci affecting phytohormone content of crown gall tumors. Horizontal arrows represent major T-DNA transcripts from a typical octopine T-DNA. Transcripts 1 and 2 correspond to the *tms1* and *tms2* genes encoding auxin biosynthetic enzymes; transcript 4 corresponds to the cytokinin biosynthetic locus, *tmr (ipt)*; and transcript 3 to the locus encoding octopine synthase *(ocs)*. The boundaries of the loci determining tumor morphology are indicated by the boxes below the T-DNA. Inactivation of *tms* gives shoot-bearing tumors, inactivation of *tmr* gives root-bearing tumors, and inactivation of *tml* gives large tumors. Horizontal bars represent the auxin and cytokinin levels present in tobacco tumors incited by strains bearing mutated loci. Adapted from (47, 53).

showing that the *tms1* and *tms2* genes of *A. tumefaciens* actually specify auxin biosynthetic enzymes. The genes were sequenced and found to be homologous to known auxin biosynthetic genes in *P. savastanoi*. They encode the two enzymes necessary for a complete auxin biosynthetic pathway (Fig. 4). The first enzyme is tryptophan-2-monooxygenase, encoded by *tms1*. The second is indoleacetamide hydrolase, encoded by *tms2*.

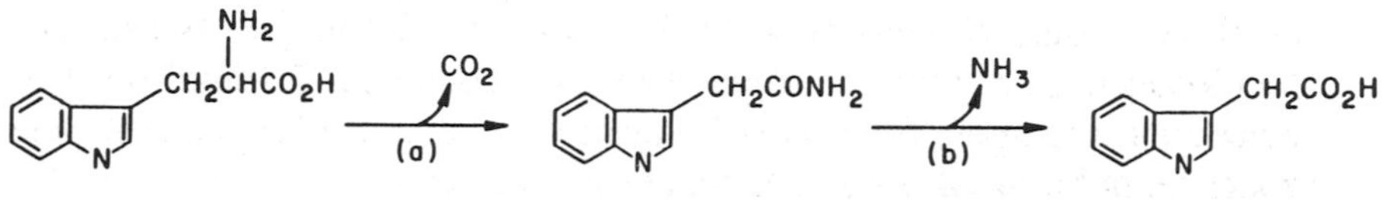

Fig. 4. Auxin biosynthetic pathway in *A. tumefaciens* and *P. savastanoi*. (a) tryptophan-2-monooxygenase encoded by *tms1* in *A. tumefaciens* and by *iaaM* in *P. savastanoi* (b) indole acetamide hydrolase encoded by *tms2* in *A. tumefaciens* and by *iaaH* in *P. savastanoi*.

Tms1 was sequenced (39) and shown to contain an open reading frame capable of specifying an 84 kd protein. The deduced amino acid sequence displayed significant homology with the FAD-dependent 4-hydroxybenzoate hydroxylase from *P. fluorescens*. Specifically, it contained a consensus FAD binding sequence indicating that *tms1* was a flavoprotein. Expression of *tms1* in *E. coli* conferred production of tryptophan-2-monooxygenase activity. If *tms1* was introduced alone into plants, the transformants displayed high tryptophan-2-monooxygenase activity, were able to convert tryptophan to indoleacetamide and contained excessively high levels of indoleacetamide (75).

The second enzyme of the pathway, indoleacetamide hydrolase, was also characterized by direct cloning. The *tms2* open reading frame encodes a 49 kd protein (61) which, when expressed in *E. coli* (60), confers the ability to catalyze conversion of indoleacetamide to IAA. The enzyme was partially purified (36) and found to hydrolyse indoleacetonitrile, IAA esters, naphthaleneacetamide, phenylacetamide and, of course, indoleacetamide. It cannot hydrolyze IAA-aspartate or the IAA conjugates of alanine, glutamic acid or glycine.

The *tms1* and *tms2* loci are now usually called *iaaM* and *iaaH* in conformity to their homologs in *P. savastanoi* and to reflect their function.

Auxin biosynthetic genes in *Pseudomonas savastanoi*

In contrast to crown gall tumors, galls produced by *P. savastanoi* or *E. herbicola* do not exhibit characteristics of permanent transformation. Galls senesce at the end of the growing season and do not resume growth in subsequent years. There is no direct experimental evidence for integration of *P. savastanoi* DNA into plant cells (T. Kosuge, unpublished data) nor is there evidence for transfer of *E. herbicola* DNA to host plant cells (17). All evidence suggests that these pathogens incite tumor growth by overproducing and secreting high levels of IAA and cytokinins.

Early studies of the galls formed on olives by *P. savastanoi* (see 53) indicated that IAA and other growth promoting substances were present. A series of elegant experiments by Kosuge and his coworkers established (18, 19, 20, 30) that there is an active IAA biosynthetic pathway in the bacteria.

The pathway is identical to that found in crown gall tumors incited by *A. tumefaciens* (Fig. 4). Initially, both tryptophan-2-monooxygenase and indoleacetamide hydrolase activities were demonstrated in semi-purified extracts of *P. savastanoi*. Subsequently, genes encoding these activities were identified and designated *iaaM* and *iaaH*. A clear connection between IAA production and virulence was established by examination of *P. savastanoi* strains resistant to tryptophan analogs. Some were defective in IAA production and had lost virulence. Others overproduced IAA and exhibited enhanced virulence (19, 20).

As in *A. tumefaciens*, the IAA biosynthetic genes of *P. savastanoi* are carried on plasmids (19). *P. savastanoi*, strain 2009, contains four medium size plasmids (58, 52, 41, and 34 Kb). Mutants that have lost the 52 kb plasmid (pIAA1) cannot synthesize IAA or indoleacetamide. Reintroduction of pIAA1 into IAA-minus mutants of *P. savastanoi* causes reacquisition of tryptophan-2-monooxygenase and indoleacetamide hydrolase activities and virulence.

Characterization of tryptophan-2-monooxygenase and indoleacetamide hydrolase was achieved by cloning and expression of *iaaM* and *iaaH* in *E. coli*. The *iaaM* gene was isolated after transformation of *E. coli* with restriction fragments of pIAA1. Transformants bearing the appropriate fragments produced indoleacetamide but not IAA and expressed tryptophan-2-monooxygenase but not indoleacetamide hydrolase. The gene encodes a 62 kd open reading frame that contains a consensus FAD binding sequence. The overexpressed enzyme contained one mole FAD per mole of protein and catalyzed conversion of tryptophan to indoleacetamide (29).

Similar studies (80) allowed characterization of *iaaH*. It is next to *iaaM* on pIAA1 and was cloned (with *iaaM*) on a 4 kb subfragment of pIAA1. Sequence data and *in vitro* transcription/translation experiments indicated that it specifies a 47 kd protein that has indoleacetamide hydrolase activity.

POSSIBLE CYTOKININ BIOSYNTHETIC PATHWAYS

The topic of plant cytokinin biosynthesis has been covered in detail in Chapter D3. Fig. 5 illustrates the two most likely pathways.

Direct cytokinin synthesis from AMP

A direct route is possible (Fig. 5A) via the condensation of dimethylallyl-pyrophosphate (DMAPP) and 5'-AMP to form isopentenyladenosine

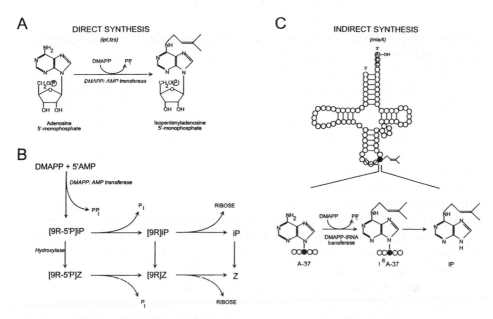

Fig. 5. Possible cytokinin biosynthetic pathways. (A) Direct synthesis catalyzed by DMAPP: AMP transferase encoded by the *A. tumefaciens* genes *ipt* or *tzs*. (B) Putative metabolic interconversions following direct isopentenyladenosine 5'-monophosphate synthesis. (C) Indirect synthesis via tRNA. Adenine residues adjacent to the anticodon triplet are isopentenylated by the DMAPP: tRNA prenyl transferase encoded by *miaA*. Subsequent cleavage of the nucleoside bond leads to the formation of free iP.
DMAPP = dimethylallylpyrophosphate; [9R-5'P]iP = isopentenyladenosine 5'-monophosphate, [9R iP] = isopentenyladenosine; iP = isopentenyladenine; [9R-5'P]Z = zeatin riboside 5'-monophosphate; [9R]Z = zeatin riboside; Z = zeatin.

5'-phosphate ([9R-5'P]iP). Subsequent dephosphorylation and deribosylation and/or side chain hydroxylation (Fig. 5B) could lead to the production of [9R]iP, iP, [9R-5'P]Z, [9R]Z and Z. The condensing enzyme is a prenyl transferase, dimethylallylpyrophosphate: 5'-AMP transferase (DMAPP: AMP transferase or isopentenyl transferase) first described in habituated tobacco callus by Chen and Melitz (14). The gene encoding this enzyme has not been isolated.

Indirect cytokinin synthesis mediated by TRNA

An alternative indirect biosynthetic route (Fig. 5C) is via isopentenylated TRNA (40). Almost all organisms isopentenylate some adenine residues in subpopulations of their TRNA. The enzyme responsible is a DMAPP: TRNA transferase encoded by the *miaA* gene (48). Isopentenylated tRNA is a potential source of free cytokinin by the excision process illustrated. The possibility that plant cytokinins arise by this latter mechanism has been advanced several times (7,38) but never confirmed.

EVIDENCE FOR CYTOKININ PARTICIPATION IN PLANT TUMOR GROWTH

The body of evidence implicating cytokinins in crown gall growth is smaller than that related to the auxins but is more internally consistent. Differences in cytokinin levels and metabolism between untransformed and crown gall tissues are greater than those observed for the auxins. Again, the pioneering work of Braun (10) laid the foundation with the finding that crown gall tumors in culture were autotrophic for cell division factors. Subsequently, Miller showed that tumors contained high levels of zeatin and zeatin riboside (45). The evidence for altered cytokinin levels and metabolism in tumors is reviewed in detail in (47) and, in outline, is as follows:

- Tumor cytokinin levels are elevated over those of untransformed tissues. The most precise measurements were made by Horgan and his coworkers (62, and references cited therein) using mass spectrometric techniques. They found high levels of zeatin, zeatin riboside, and zeatin side chain glucosides in crown gall tumors.
- Cytokinin accumulation rates are greater in tumors than in untransformed tissues in culture. Cytokinin concentrations peaked early in the exponential phase of tumor growth in culture and reached levels greater than those of untransformed tissues (25).
- Cloned octopine tobacco crown gall tumors in culture contain high levels of cytokinins and display DMAPP: AMP transferase activity. The activity is associated solely with the presence of TL-DNA and not with TR-DNA (Fig. 6), indicating that the gene responsible is carried within the TL-DNA.
- Crown gall lines bearing mutations in the TL-DNA have cytokinin levels that differ from those of wild type tumors. Inactivation of the *tmr* locus gives rise to "rooty" tumors (23) which have a cytokinin requirement for growth in culture. The cytokinin content of mutant tumors is much lower than that of wild-type tumors (1) (Fig. 3). For example, wild-type primary tobacco tumors contain 50 pmole/g zeatin riboside whereas *tmr* mutant tumors contain less than 1 pmole/g, not significantly different from that of untransformed stem tissue.
- If *tmr* is inactivated, tumor DMAPP: AMP transferase activity is abolished (46).

GENES SPECIFYING CYTOKININ BIOSYNTHESIS IN *AGROBACTERIUM TUMEFACIENS*

The tmr gene is responsible for tumor cytokinin biosynthesis

Although these data suggested that the *tmr* gene product might be directly responsible for tumor cytokinin biosynthesis, it was not until the gene was cloned and expressed in *E. coli* that this was found to be true. Barry *et al.* (6) were the first to find that *tmr* encodes a cytokinin biosynthetic prenyl

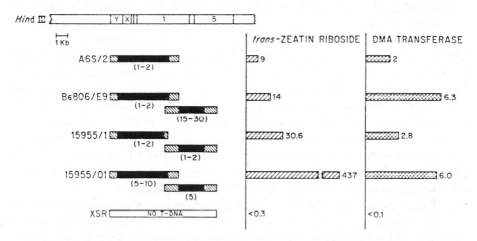

Fig. 6. Cytokinin levels and cytokinin prenyl transferase activity in cloned crown gall tumors. Cytokinin levels were determined by HPLC and RIA of extracts of cloned tumor lines grown in axenic culture. DMAPP: AMP transferase activity was measured *in vitro* as described in (1). HindIII cut sites of the t-region are shown. Horizontal bars represent the extent and copy number of TL-DNA and TR-DNA segments present in tumors (determined by Kwok, Gordon and Nester; personal communication). Hatched bars represent the cytokinin and DMA:AMP transferase levels (D. Akiyoshi, Thesis, Oregon State University).

transferase. The nucleotide sequence of an octopine T-DNA became available (5) and allowed identification of an open reading frame associated with *tmr*. It specifies a protein of about 27 kd. When *tmr* was cloned and expressed in *E. coli*, a protein of this size was produced and simultaneously DMAPP: AMP transferase activity was detected (6).

Table 2 shows that expression of *tmr* in *E. coli* results in DMAPP: AMP transferase activity. *Tmr* was cloned in the correct (pIP192) or inverted (pIP182) orientation behind the *lac* promoter on a multicopy plasmid (51). Although DMAPP: AMP transferase activity is expressed in both cases, it is higher when the gene is in the correct orientation. That the DMAPP: AMP transferase expressed in *E. coli* is identical to the enzyme contained in crown gall tumors is likely. Buchmann *et al.* (11) found that an antibody raised against a synthetic decapeptide, whose sequence was derived from the *tmr* coding region, crossreacted with both enzymes.

When expressed in crown gall tumors, the cytokinin prenyl transferase encoded by *tmr* results in elevated levels of many cytokinins and concomitant massive cell proliferation. Because *tmr* possesses a eukaryotic promoter, it is not expressed in free-living *A. tumefaciens*, but is expressed in the host plant, where it contributes to maintenance of the transformed state (6, 11). In light of its function, *tmr* is now usually called *ipt* (isopentenyl transferase).

A second cytokinin biosynthetic gene, tzs, is present in A. tumefaciens

The DMAPP: AMP transferase encoded by *ipt* is not the only cytokinin prenyl transferase present in *A. tumefaciens*. A second gene is present in some strains. The first indication of the existence of a second copy came from studies of cytokinin secretion by free-living *A. tumefaciens* (35). The nopaline strain C58, secreted zeatin in culture whereas octopine strains did not. Only wild-type nopaline strains or transconjugants containing nopaline Ti plasmids secrete zeatin (59) because the responsible locus (*tzs*, trans zeatin secretion) is present only on nopaline Ti plasmids. *Tzs* mapped outside the T-DNA, close to the *virA* locus (8). Production of zeatin by *E. coli* containing the cloned *tzs* gene from pTiC58 is illustrated in Fig. 7.

Since expression of *tzs* in *E. coli* results in secretion of zeatin, it first appeared that the gene might encode an enzyme responsible for hydroxylation of the cytokinin isoprenoid side chain (see Fig. 5B). However, the nucleotide sequence of *tzs* was found to exhibit extensive homology with *ipt* suggesting that the gene might express DMAPP: AMP transferase activity (8). This proved to be true (Table 2) (51) and it now appears that *tzs* encodes a cytokinin prenyl transferase similar in function to that encoded by *ipt*.

The *tzs*-encoded transferase has been purified to homogeneity from acetosyringone-induced *A. tumefaciens*, strain C58 (48). The K_m for DMAPP was 8.2 μM and for 5'-AMP it was 11.1 μM. The only effective acceptor was 5'-AMP. Neither adenine, adenosine nor ATP were prenylated. Expression of *tzs* is controlled by the same phenolic compounds that control expression of the *vir* cascade (34, 57). In the presence of acetosyringone, nopaline strains of *A. tumefaciens* secrete mg levels of zeatin into the culture filtrate. However, the biological function of *tzs* remains obscure. Octopine strains do not possess it yet are perfectly capable of inciting tumors. Deletion of *tzs* from nopaline strains impairs tumorigenicity on some hosts (12) but does not abolish it. It has been suggested that *tzs* plays a role in initiation of oncogenesis by stimulating cell division at the wound site thus making the plant more susceptible to transformation (83).

The role of the miaA gene

A. tumefaciens cured of the Ti plasmid, and therefore lacking the *ipt* and

Table 2. Cytokinin prenyl transferase activity expressed by cloned *ipt* and *tzs* genes

Gene	Strain	DMAPP: AMP transferase activity[a]
ipt (tmr)	HB101(pIP192)	5.1
ipt (tmr)	HB101(pIP182)	1.5
tzs	HB101(pTZ120)	6.5
none	HB101	n.d.

a) iP plus [9R]iP synthesized (fmole/hr/mg protein)

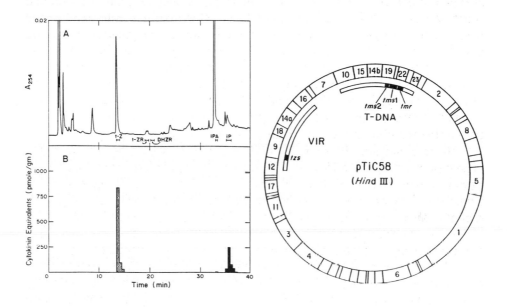

Fig. 7. Cytokinin production and location of *tzs*. Left: Cytokinin production by *tzs* cloned in *E. coli*. Production by the clone IIB101(pTZ121) was measured in mid-exponential phase growth by immunoaffinity chromatography followed by HPLC and RIA. RIA with anti-zeatinriboside antibody and anti-isopentenyladenine antibody. Right: Restriction map of pTiC58 showing *Hind*III cut sites and approximate locations of the *virulence* genes, the T-DNA and the genes specifying phytohormone biosynthesis: *ipt (tmr)* - cytokinin prenyl transferase (transferred from the T-DNA to the plant); *tms1* - tryptophan-2-monooxygenase; *tms2* - indole-3-acetamide hydrolase; *tzs* - cytokinin prenyl transferase (not transferred to plant).

tzs genes, still produces small but significant amounts of Ip in culture (59). This chromosomally-encoded free iP production could be due to the activity of a yet undiscovered *ipt* or *tzs* homolog or it could be tRNA-mediated and, if it is, the gene responsible for tRNA isopentenylation, *miaA*, may be involved. The *A. tumefaciens miaA* gene has been isolated recently (24). It encodes a DMAPP: tRNA transferase that displays significant amino-acid sequence homology with the DMAPP: AMP transferases (see Fig. 9).

The availability of *Tn5*-induced *miaA* mutants has the potential to allow a direct estimate to be made of the significance of tRNA-mediated cytokinin biosynthesis. Transfer RNA from such mutants should exhibit impaired isopentenylation and, if the indirect path is significant, the cells should secrete lower levels of cytokinin into the culture medium. A comparison of the isopentenylated nucleoside content of total tRNA from *A. tumefaciens* and a *miaA* mutant showed that tRNA isopentenylation was reduced significantly. However, this mutation was leaky since there was still some residual tRNA isopentenylation (48). Likewise, cytokinin secretion by the mutants was decreased but not totally eliminated. Complete deletion of *miaA* by marker exchange will be necessary for a final assessment of the role of the contribution of the indirect pathway to bacterial cytokinin synthesis. Current results suggest, however, that is not significant when compared to the direct route.

CYTOKININ BIOSYNTHETIC GENES IN
PSEUDOMONAS SAVASTANOI

Cytokinin production by *P. savastanoi* was first observed by Surico *et al.*, (70) who showed that culture filtrates contained iP and [9R]iP-like cytokinins. Subsequently, the presence of an unusual cytokinin, 1"-methyl-*trans*-zeatin riboside was reported (69). Examination of several wild-type *Pseudomonas* strains revealed considerable variation in the amounts and types of cytokinins produced. In general, levels of cytokinin in culture filtrates of *P. savastanoi* were very much greater than those present in cultures of *A. tumefaciens* (43). Stationary phase cultur*es of P. savastanoi* can contain as much as 10 μM cytokinin, most of which is zeatin and zeatin riboside. Some strains produce 1"-methyl-zeatin riboside whereas others do not. In two strains of *P. savastanoi* the existence of plasmid-born cytokinin biosynthetic genes has now been shown (43). They are necessary for pathogenicity (T. Kosuge, unpublished data) and are very similar to the genes from *A. tumefaciens*. The oleander-specific *P. savastanoi* strain 213 contains five plasmids ranging in size from 38 to 64 kb; the olive-specific strain 1006 has two, 84 kb and 105 kb. Cytokinin production by plasmid deletion mutants indicated that the gene was probably on the 42 kb plasmid in strain 213 and on the 105 kb plasmid of strain 1006 (51). Cytokinin production by strain 1006 and its dependence on plasmid status is illustrated in Fig. 8.

The isolation of the biosynthetic gene, *ptz*, was achieved by screening a *P. savastanoi* library in *E. coli* (58). On expression in *E. coli*, zeatin was produced and DMAPP: AMP transferase activity was detected. The gene was sequenced and shown to have substantial homology to *ipt* and *tzs*.

CYTOKININ PRODUCTION BY OTHER BACTERIA

Recently, *P. amygdali* has been shown to produce IAA and a variety of cytokinins, some with unusual structures such as zeatin 2'-deoxyriboside (31). Another pseudomonad, *P. solanacearum* contains a *tzs* homolog (2) and secretes cytokinins (3). *Azotobacter chroococcum* also secretes high levels of cytokinins (54). Recently, *Rhodococcus fascians* has been shown to secrete cytokinins into the culture medium (49) and to possess an *ipt*-like gene (21). Barash and collaborators (Lichter, Barash and Gray, unpublished data) have shown that the gall forming pathogen *E. herbicola* secretes high

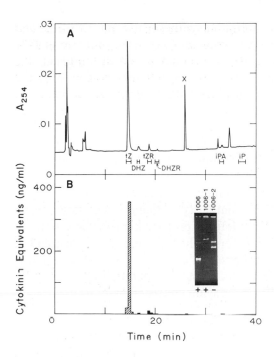

Fig. 8. Cytokinin production and plasmid status of *P. savastanoi* strain 1006. Cytokinins present in one ml of mid-exponential phase culture filtrate of *P. savastanoi* strain 1006, purified by immunoaffinity chromatography and analyzed by HPLC and RIA. (A) Absorbance at 254 nm. X is 1"-methyl zeatin riboside. (B) RIA with anti-zeatin riboside antibody and anti-isopentenyladenine antibody. *Inset:* Plasmid status and zeatin production: 1006: plasmids, 105 kb and 81 kb; 1006-1: plasmids, 105 kb; 1006-2: plasmids, 81 kb and 65 kb (deletion of 105 kb).

levels of cytokinins into the medium. Cytokinin secretion correlates with pathogenicity and the organism apparently possesses both IAA and cytokinin biosynthetic pathways. For both *R. fascians* and *E. herbicola*, the levels of cytokinins produced in free culture seem insufficient to produce the symptoms observed on the plant. It is possible that phytohormone gene expression may be controlled by molecules present in the host plant (21). Finally, there have been a number of reports of cytokinin production by *Rhizobium* sp. (4, 68, 71, 74), but the evidence is still contradictory (reviewed in 4). Some strains apparently secrete cytokinins whereas others do not. The only well-documented case is that of the tropical Rhizobium strain IC3442, which produces classical symptoms of cytokinin overproduction on its host (leaf hyponasty, deep green foliage etc.). It secretes cytokinins at reasonably high levels in culture (74) but as yet, no cytokinin biosynthetic gene has been isolated.

COMPARISON OF BACTERIAL PHYTOHORMONE BIOSYNTHETIC GENES

Kosuge and coworkers (80) have shown that the *P. savastanoi* auxin biosynthetic genes, *iaaM* and *iaaH*, maintain substantial sequence homology with their counterparts, *tms1* and *tms2* from *A. tumefaciens*. The *tms1* coding region is 54% homologous to that from *iaaM* while *tms2* is 38% homologous to *iaaH*. A comparison of deduced amino acid sequences from the coding regions of the genes specifying cytokinin biosynthesis by both the direct and indirect routes (Fig. 9) reveals similar homology (48). Overall there is 44% identity between *tzs* and *ipt*. Within the N-terminal region of the genes, the extent of homology is highest, at 58%. Little similarity is found at the C-terminus, suggesting that this region plays only a minor role in catalytic activity. Interestingly, the prenyl transferases encoded by *miaA* are related to the *ipt* group in that they possesses a consensus adenine nucleotide binding domain (GXXXXGK(S)T) and a conserved "GGS(T)"sequence similar to that found in the direct transferases. An evolutionary relationship between the two sets of genes seems possible.

Although substantial sequence homology is observed within the open reading frames of the *ipt* group, little similarity is found in the non-coding regions. Both *ptz* and *tzs* display well-defined prokaryotic ribosome binding sites and apparent transcription terminators. The *ptz* promoter is identical in sequence to the *E. coli* consensus promoter as expected for a gene which is expressed constitutively. The *ipt* promoter bears no identifiable prokaryotic transcription or translation control signals, again consistent with its expression only in the plant.

CONCLUSION

Several intriguing biological questions remain to be answered regarding the phytohormone biosynthetic genes present in these bacteria. First, what is the

```
ipt   ........MD LHLIFGPTCT GKTTTAIALA QQTGLPVLSL DRVQCCPQLS TGSGRPTVEE LKGTT.RLYL
tzs   ........ML LHLIYGPTCS GKTDMAIQIA QETGWPVVAL DRVQCCPQIA TGSGRPLESE LQSTR.RIYL
mia   MMKNLDQNFD AILITGPTAS GKSALALRLA RERNGVVINA DSMQVYDTLR VLTARPSDHE MEGVPHRLYG

ipt   DDRPL..... .......VEG IIAAKQAHHR LIEEVYNHEA NGGLILEGGS TSLLNCMARN SYWSADFRWH
tzs   DSRPL..... .......TEG ILDAESAHRR LIFEVDWRKS EEGLILEGGS ISLLNCMAKS PFWRSGFQWH
mia   HVPAGSAYST GEWLRDIS.G LLSDLRGEGR FP........ ....VIVGGT GLYFKALTGG LSDMPAIPDD

ipt   I...IRHKLP DQETFMKAAK ARVKQMLHPA AGH....... .......... .......... ........SI
tzs   V...KRLRLG DSDAFLTRAK QRVAEMFAIR EDR....... .......... .......... .......PSL
mia   LREGLRARLE EG.....AAK L.HAELVSRD PSMAQMLQPG DGQRIVRALE VLEATGKSIR DFQRASGPMI

ipt   TQELVYLWNE .PRLRPILKE IDGYRY.AML FASQNQITAD MLLQLDANME GKLIN.GIAQ EYFIHARQQE
tzs   LEELAELWNY .PAARPILED IDGYRC.AIR FARKHDLAIS QLPNIDAGRH VELIE.AIAN EYLEHALSQE
mia   IDPERAQKFI VLPERRVLHD RINRRFEAMM DSGAVEEVQA LLALNLAPDA TAMKAIGVAQ IADMLTGRMG

ipt   QKFPQVN.AA AFDGFEGHPF GMY
tzs   RDFPQWPEDG AGQPVCPVTL .......... .TRIR
mia   AAEVIEKSAA ATRQYAKRQM TWFRNQMGDD WTRIQP
```

Fig. 9. Homology between cytokinin prenyl transferases from *Agrobacterium tumefaciens*. The deduced amino acid sequences of the coding regions of *ipt*, *tzs* and *miaA* from *Agrobacterium tumefaciens* are illustrated. Areas of identity or close similarity are shaded. (A = alanine; C = cysteine; D = aspartic acid; E = glutamic acid; F = phenylalanine; G = glycine; H = histidine; I = isoleucine; K = lysine; L = leucine; M = methionine; N = asparagine; P = proline; Q = glutamine; R = arginine; S = serine; T = threonine; V = valine; W = tryptophan; Y = tyrosine).

origin of the genes? How are they related to native plant biosynthetic genes (if at all)? If plants are ultimately shown to have a direct cytokinin biosynthetic pathway, then bacterial genes may have been acquired from plants in the past, or (less likely) the reverse transfer may have occurred. Until the native plant genes are isolated, such questions remain open. Second, what are the biological functions of these genes? In the gall-forming pathogens, functions are readily rationalized on the basis that their expression stimulates formation of an environment *in planta* that is favorable to bacterial survival and nutrition. This rationale is not present for the non gall-forming pathogens such as *P. solanacearum,* or the rhizosphere bacteria such as *Azospirillum* or *Azotobacter.* Finally, the relationship between the *ipt* and *miaA* genes needs to be determined. Presumably, they once diverged from a common precursor. Any understanding of cytokinin biosynthesis in plants must consider the relationship between these sets of genes.

Acknowledgements

I wish to express sincere thanks to those colleagues who participated in the work described here and I especially acknowledge helpful discussions with and John Gray and Richard Meilan. Much of the work of the laboratory was supported by grants from the National Science Foundation and the U.S. Department of Agriculture

References

1. Akiyoshi, D.E., Morris, R.O., Hinz, R., Mischke, B.S., Kosuge, T., Garfield, D.J., Gordon, M.P., Nester, E.W. (1983) Cytokinin-auxin balance in crown gall tumors is regulated by specific loci in the T-DNA. Proc. Natl. Acad. Sci. USA 80, 407-11.
2. Akiyoshi, D.E., Regier, D.A., Gordon, M.P. (1989) Nucleotide sequence of the *tzs* gene from *Pseudomonas solanacearum* strain K60. Nucleic Acids Res. 17, 8886.
3. Akiyoshi, D.E., Regier, D.A., Gordon, M.P. (1987) Cytokinin production by *Agrobacterium* and *Pseudomonas* sp. J. Bacteriol. 169, 4242-8.
4. Badenoch-Jones, J., Parker, C.W., Letham, D.S. (1987) Phytohormones, *Rhizobium* mutants, and nodulation in legumes. VII. Identification and quantification of cytokinins in effective and ineffective pea root nodules using radioimmunoassay. J. Plant Growth Regul. 6, 97-111.
5. Barker, R.F., Idler, K.B., Thompson, D.V., Kemp, J.D. (1983) Nucleotide sequence of the T-DNA from *Agrobacterium tumefaciens* octopine Ti plasmid pTi15955. Plant Mol. Biol. 2, 335-50.
6. Barry, G.F., Rogers, S.G., Fraley, R.T., Brand, L. (1984) Identification of a cloned cytokinin biosynthetic gene. Proc. Natl. Acad. Sci. USA 81, 4776-80.
7. Bassil, N.V., Mok, D.W.S., Mok, M.C. (1993) Partial purification of a *cis-trans*-isomerase of zeatin from immature seed of *Phaseolus vulgaris* L. Plant Physiol. 102, 867-872.
8. Beaty, J.S., Powell, G.K., Lica, L., Regier, D.A., MacDonald, E.M.S., Hommes, N.G., Morris, R.O. (1986) *Tzs*, a nopaline Ti plasmid gene from *Agrobacterium tumefaciens* associated with trans-zeatin biosynthesis. Mol. Gen. Genet. 203: 274-280.
9. Binns, A.N., Sciaky, D., Wood, H.N. (1982) Variation in hormone autonomy and regenerative potential of cells transformed by strain A66 of *Agrobacterium tumefaciens*. Cell 31, 605-612.
10. Braun, A.C. (1958) A physiological basis for the autonomous growth of the crown gall tumor cell. Proc. Natl. Acad. Sci. USA 44, 344-349.

11. Buchmann, I., Marner, F.J., Schröder, G., Waffenschmidt S., Schröder, J. (1985) Tumor genes in plants: T-DNA encoded cytokinin biosynthesis. EMBO J. 4, 853-9.
12. Castle, L.A., Morris, R.O. (1993) Transient expression assays using GUS constructs and fluorometric detection for analysis of T-DNA Transfer. Plant Molecular Biology Manual. In press.
13. Chen, C.M., Leisner, S.M. (1984) Modification of cytokinins by cauliflower microsomal enzymes. Plant Physiol. 75, 442-6.
14. Chen, C.M., Melitz, D.K. (1979) Cytokinin biosynthesis in a cell- free system from cytokinin-autotrophic tobacco tissue cultures. FEBS Lett. 107, 15-20.
15. Citovsky, V., Wong, M.L., Zambryski, P. (1989) Cooperative interaction of Agrobacterium VirE2 protein with single-stranded DNA: implications for the T-DNA transfer process. Proc. Natl. Acad. Sci. USA 86, 1193-7.
16. Citovsky, V., Zupan, J., Warnick, D., Zambryski, P. (1992) Nuclear localization of *Agrobacterium* VirE2 protein in plant cells. Science 256, 1802-5.
17. Clark, E., Vigodsky-Haas, H., Gafni, Y. (1989) Characteristics in tissue culture of hyperplasias induced by *Erwinia herbicola* pathovar *gypsophilae*. Physiol. Mol. Plant Pathol. 35, 383-90.
18. Comai, L., Kosuge, T. (1980) Involvement of plasmid deoxyribonucleic acid in indoleacetic acid synthesis in *Pseudomonas savastanoi*. J. Bacteriol. 143, 950-57.
19. Comai, L., Kosuge T. (1982) Cloning and characterization of iaaM, a virulence determinant of *Pseudomonas savastanoi*. J. Bacteriol. 149, 40-46.
20. Comai, L., Surico, G., Kosuge, T. (1982) Relation of plasmid DNA to indoleacetic acid production in different strains of *Pseudomonas syringae* pv. *savastanoi*. J. Gen. Microbiol. 128, 2157-63.
21. Crespi, M., Messens, E., Caplan, A. B., Van Montagu, M., Desomar, J. (1992) Fasciation induction by the phytopathogen *Rhodococcus fascians* depends upon a linear plasmid encoding a cytokinin synthase gene. EMBO Journal 11, 795-804.
22. Engler, G., Depicker, A., Maenhaut, R., Villaroel-Mandiola, R., Van Montagu, M., Schell, J., Hernalsteens, J.P. (1981) Physical mapping of DNA base sequence homologies between an octopine and a nopaline Ti plasmid of *Agrobacterium tumefaciens*. J. Mol. Biol. 152, 183-208.
23. Garfinkel, D.J., Simpson, R.B., Ream, L.W., White, F.F., Gordon, M.P., Nester, E.W. (1981) Genetic analysis of crown gall: Fine structure map of the T-DNA by site-directed mutagenesis. Cell 27, 143-53.
24. Gray, J., Wang, J., Gelvin, S.B. (1992) Mutation of the *miaA* gene of *Agrobacterium tumefaciens* results in reduced *vir* gene expression. J. Bacteriol. 174, 1086-1098.
25. Hansen, C.E., Meins, F. Jr., Milani, A., (1985) Clonal and physiological variation in the cytokinin content of tobacco-cell lines differing in cytokinin requirement and capacity for neoplastic growth. Differentiation 29, 1-6.
26. Hille, J., Van Kan, J., Schilperoort, R. (1984) Trans-acting virulence functions of the octopine Ti plasmid from *Agrobacterium tumefaciens*. J. Bacteriol. 158, 754-6.
27. Howard, E., Citovsky, V. (1990) The emerging structure of the Agrobacterium T-DNA transfer complex. BioEssays 12, 103-8.
28. Howard, E.A., Winsor, B.A., De Vos, G., Zambryski, P. (1989) Activation of the T-DNA transfer process in *Agrobacterium* results in the generation of a T-strand-protein complex: tight association of VirD2 with the 5' ends of T-strands. Proc. Natl. Acad. Sci. USA 86, 4017-21.
29. Hutcheson, S., Kosuge, T. (1985) Regulation of 3-indoleacetic acid production in *Pseudomonas syringae* pv. *savastanoi*. Purification and properties of tryptophan-2-monooxygenase. J. Biol. Chem. 260, 6281-7.
30. Hutzinger, O., Kosuge, T. (1967) Microbial synthesis and degradation of indole-3-acetic acid. Biochim. Biophys. Acta 136, 389-391.

31. Iacobellis, N.S., Evidente, A., Surico, G., Sisto, A., Gammaldi, G. (1990) Production of phytohormones by *Pseudomonas amygdali* and their role in the hyperplastic bacterial canker of almond. J. Phytopathol. 129, 177-86.

32. Iyer, V.N., Klee, H.J., Nester, E.W. (1982) Units of genetic expression in the virulence region of a plant tumor-inducing plasmid of Agrobacterium tumefaciens. Mol. Gen. Genet. 188, 418-24.

33. Jin, S., Prusti, R.K., Roitsch, T., Ankenbauer, R.G., Nester, E.W. (1990) Phosphorylation of the VirG protein of *Agrobacterium tumefaciens* by the autophosphorylated VirA protein: essential role in biological activity of VirG. J. Bacteriol. 172, 4945-50.

34. John, M.C., Amasino, R.M. (1988) Expression of an Agrobacterium Ti-plasmid gene involved in cytokinin biosynthesis is regulated by virulence loci and induced by plant phenolic compounds. J. Bacteriol. 170, 790-795.

35. Kaiss-Chapman, R.W., Morris, R.O. (1977) Trans-zeatin in culture filtrates of *Agrobacterium tumefaciens*. Biochem. Biophys. Res. Comm. 76, 453-59.

36. Kemper, F., Waffenschmidt, S., Weiler E.W., Rausch, T., Schröder, J. (1985) T-DNA-encoded auxin formation in crown-gall cells. Planta 163, 257-262.

37. Kende, H. (1971) The cytokinins. Int. Rev. Cytol. 31, 301-38.

38. Klämbt, D. (1992) The biogenesis of cytokinins in higher plants: Our present knowledge. In: Physiology and Biochemistry of Cytokinins in Plants, Kaminek, M., Mok, D.W.S., Zazimalova F., eds. SPB Academic Publishing, The Hague.

39. Klee, H., Montoya, A., Horodyski, F., Lichtenstein , C., Garfinkel, D. *et al.* (1984) Nucleotide sequence of the *tms* genes of the pTiA6NC octopine Ti plasmid: 2 gene products involved in plant tumorigenesis. Proc. Natl. Acad. Sci. USA 81, 1728-1732.

40. Klemen F., Klämbt, D. (1974) Half life of sRNA from primary roots of Zea mays. A contribution to the cytokinin production. Physiol. Plant. 31, 186-188.

41. Letham, D.S., Palni, L.M.S. (1983) The biosynthesis and metabolism of cytokinins. Ann. Rev. Plant Physiol. 34, 163-97.

42. Lichter, A., Manulis, S., Sagee, O., Gafni, Y., Gray, J., Meilan, R., Morris, R , Barash, I. (1993) Identification of cytokinins produced by *Erwinia herbicola pv. gypsophilae* and isolation of a cytokinin biosynthetic gene. 6th Intl. Congress Plant Path., Montreal, Canada.

43. MacDonald, E.M.S., Powell, G.K., Regier, D.A., Glass, N.L., Roberto, F., Kosuge, T., Morris, R.O. (1986) Secretion of zeatin, ribosylzeatin and ribosyl-1"-methylzeatin by *Pseudomonas savastanoi*: plasmid-coded cytokinin biosynthesis. Plant Physiol. 82: 742-747.

44. Maruyama, A, Yamaguchi, I, Maeda, M, Simidu, U. (1988) Evidence of cytokinin production by a marine bacterium and its taxonomic characteristics. Can. J. Microbiol. 34, 829-33.

45. Miller, C.O. (1974) Ribosyl-trans-zeatin. a major cytokinin produced by crown gall tumor tissue. Proc. Natl. Acad. Sci. USA 71, 334-338.

46. Morris, R.O., Akiyoshi, D.E., MacDonald, E.M.S., Morris, J.W., Regier, D.A., Zaerr, J.B. (1982) Cytokinin metabolism in relation to tumor induction by *Agrobacterium tumefaciens*. In: Plant Growth Substances. Wareing, P.F., ed. Academic Press, London.

47. Morris, R.O. (1986) Genes specifying auxin and cytokinin biosynthesis in phytopathogens. Ann. Rev. Plant Physiol. 37, 509-38.

48. Morris R.O., Blevins D.G., Dietrich J.T., Durley R.C., Gelvin S.B., Gray J., Hommes N.G., Kaminek M., Matthews L.J., Meilan R., Reinbott T.M., Luis Sayavedra-Soto L. (1993) Cytokinins in plant pathogenic bacteria and developing cereal grains. Aus. J. Plant Physiol 20, 621-637.

49. Morris, R.O., Jameson, PE., Laloue, M, Morris, J.W. (1991) Rapid identification of cytokinins by an immunological method. Plant Physiol. 95, 1156-61.

50. Morris, J.W., Morris, R.O. (1990) Identification of an Agrobacterium tumefaciens virulence gene inducer from *Pseudotsuga menziesii*. Proc. Natl. Acad. Sci. USA 87: 3614-3618.

51. Morris, R.O., Powell, G.K., Beaty, J.S., Durley, R.C., Hommes, N.G., Lica, L., MacDonald, E.M.S. (1986) Cytokinin biosynthetic genes and enzymes from *Agrobacterium tumefaciens* and other plant associated prokaryotes. In: Plant Growth Substances 1985, pp. 185-196, Bopp M., ed. Springer-Verlag, Berlin.

52. Nester, E.W., Gordon, M.P., Amasino, R.M., Yanofsky, M.F. (1984) Crown gall: A molecular and physiological analysis. Ann. Rev. Plant Physiol. 35, 387-413.

53. Nester, E.W., Kosuge, T. (1981) Plasmids specifying plant hyperplasias. Ann. Rev. Microbiol. 35, 531-565.

54. Nieto, K.F., Frankenberger, W.T., Jr. (1989) Biosynthesis of cytokinins by *Azotobacter chroococcum*. Soil Biol. Biochem. 21, 967-72.

55. Ooms, G., Hooykaas, P.J.J., Moolenaar, G., Schilperoort, R.A. (1981) Crown gall plant tumors of abnormal morphology induced by *Agrobacterium tumefaciens* carrying mutated octopine Ti plasmids: analysis of T-region DNA functions. Gene 14, 33-50.

56. Palni, L.M.S., Horgan R., Darrall, N.M., Stuchbury, T., Wareing P.F. (1983) Cytokinin biosynthesis in crown gall tissue of *Vinca rosea* L. The significance of nucleotides. Planta 159, 50-59.

57. Powell, G.K., Hommes, N.G., Kuo. J., Castle, L.A., Morris, R.O. (1988) Inducible expression of cytokinin biosynthesis in Agrobacterium tumefaciens by plant phenolics. Molec. Plant-Microbe Interactions 1: 235-242.

58. Powell, G.K., Morris, R.O. (1986) Nucleotide sequence and expression of a *Pseudomonas savastanoi* cytokinin biosynthetic gene: Homology with *Agrobacterium tumefaciens tmr* and *tzs* loci. Nucleic Acids Res. 14, 2555-65.

59. Regier, D.A., Morris, R.O. (1982) Secretion of trans-zeatin by *Agrobacterium tumefaciens:* A function determined by the nopaline Ti plasmid. Biochem. Biophys. Res. Comm. 104, 1560-1566.

60. Schröder, G., Waffenschmidt, S., Weiler, E.W., Schröder, J. (1984) The T-region of Ti plasmids codes for an enzyme synthesizing indole-3-acetic acid. Eur. J. Biochem. 138, 387-91.

61. Sciaky, D., Thomashow, M.F., (1984) The sequence of the *tms* transcript 2 locus of the *A. tumefaciens* plasmid pTiA6 and characterization of the mutation in pTiA66 that is responsible for auxin attenuation. Nucleic Acids Res. 12, 1447-61.

62. Scott, I.M., Horgan, R. (1984) Mass spectrometric quantification of cytokinin nucleotides and glycosides in tobacco crown gall tissue. Planta 161, 345-54.

63. Shirasu, K., Morel, P., Kado, CI. (1990) Characterization of the *virB* operon of an *Agrobacterium tumefaciens* Ti plasmid: nucleotide sequence and protein analysis. Mol. Microbiol. 4, 1153-63.

64. Skoog, F., Armstrong D.J. (1970) Cytokinins. Ann. Rev. Plant Physiol. 21, 359-84.

65. Stachel, S.E., Messens, E., Van Montagu, M., Zambryski, P. (1985) Identification of the signal molecules produced by wounded plant cells that activate T-DNA transfer in *Agrobacterium tumefaciens* Nature 318, 624-9.

66. Stevens, G.A., Jr., Berry, A.M. (1988) Cytokinin secretion by Frankia sp. HFP ArI3 in defined medium. Plant Physiol. 87, 15-16.

67. Stuchbury, T., Palni, L.M., Horgan, R., Wareing, P.F. (1979) The biosynthesis of cytokinins in crown gall tissue of *Vinca rosea* L. Planta 147, 97-102.

68. Sturtevant, D.B., Taller, B.J. (1989) Cytokinin production by *Bradyrhizobium japonicum*. Plant Physiol. 89, 1247-52.

69. Surico, G., Evidente, A., Iacobellis, N., Randazzo, G. (1985) A new cytokinin from the culture filtrate of *Pseudomonas syringae* pv. *savastanoi*. Phytochemistry 24, 1499-1502.

70. Surico, G., Sparapano, L., Legario, P., Durbin, R.D., Iacobellis, N. (1975) Cytokinin-like activity in extracts from the culture filtrate of *Pseudomonas savastanoi*. Experientia 31, 929-30.

71. Taller, B.J., Sturtevant, D.B. (1991) Cytokinin production by rhizobia. Curr. Plant Sci. Biotechnol. Agric. 10, (Adv. Mol. Genet. Plant-Microbe Interact., Vol. 1), 215-21.

72. Taller, B.J., Wong, T.Y. (1989) Cytokinins in *Azotobacter vinelandii* culture medium. Appl. Environ. Microbiol. 55, 266-7.

73. Gelvin, S.B., Gordon, M.P., Nester, E.W., Aronson, A.I. (1981) Transcription of the *Agrobacterium* Ti plasmid in the bacterium and in crown gall tumors. Plasmid 6, 17-29.

74. Upadhyaya, N.M., Parker, C.W., Letham, D.S., Scott, K.F., Dart, P.J. (1991) Evidence for cytokinin involvement in *Rhizobium* (IC3342)-induced leaf curl syndrome of pigeonpea (*Cajanus cajan* Millsp.) Plant Physiol. 95, 1019-25.

75. Van Onckelen, H., Prinsen, E., Inze, D., Rudelsheim, P., Van Lijsebettens, M., Follin, A., Schell, J., Van Montagu, M., De Greef, J. (1986) *Agrobacterium* T-DNA gene 1 codes for tryptophan-2-monooxygenase activity in tobacco crown gall cells. FEBS Lett. 198, 357-60.

76. Van Onckelen, H., Rudelsheim, P., Inze, D., Follin, A., Messens, E., Hiremans, S., Schell, J., Van Montagu, M., De Greef, J. (1985) Tobacco plants transformed with the *Agrobacterium* T-DNA gene 1 contain high amounts of indole-3-acetamide. FEBS Lett. 181, 373-6.

77. Willmitzer, L., Dhaese, P., Schreier, P.H., Schmalenbach, W., Van Montagu, M., Schell, J. (1983) Size, location and polarity of T-DNA-encoded transcripts in nopaline crown gall tumors: common transcripts in octopine and nopaline tumors. Cell 32, 1045-56.

78. Winans, S.C., Ebert, P.R., Stachel, S.E., Gordon, M.P., Nester, E.W. (1986) A gene essential for Agrobacterium virulence is homologous to a family of positive regulatory loci. Proc. Natl. Acad. Sci. USA 83, 8278-82.

79. Yadav, N.S. Vanderleyden, J., Bennett, D.R., Barnes, W.M., Chilton, M.-D. (1982) Short direct repeats flank the T-DNA on a nopaline Ti plasmid. Proc. Natl. Acad. Sci. USA 79, 6322-26.

80. Yamada, T., Palm, C.J., Brookes, B., Kosuge, T. (1985) Nucleotide sequence of the *Pseudomonas savastanoi* indoleacetic acid genes show homology with *Agrobacterium tumefaciens* DNA. Proc. Natl. Acad. Sci. USA 82, 6522-6.

81. Zaenen, I., Van Larabeke, N., Teuchy, H., Van Montagu, M., Schell, J. (1974) Supercoiled circular DNA in crown gall inducing *Agrobacterium* strains. J. Mol. Biol. 86, 109-27.

82. Zambryski, P.C. (1992) Chronicles from the *Agrobacterium*-plant cell DNA transfer story. Ann. Rev. Plant Physiol. Plant Mol. Biol. 43, 465-90.

83. Zhan, X., Jones, D.A., Kerr, A. (1990) The pTiC58 *tzs* gene promotes high-efficiency root induction by agropine strain 1855 of *Agrobacterium rhizogenes*. Plant Mol. Biol. 14, 785-92.

84. Zimmer, W., Bothe, H. (1988) The phytohormonal interactions between *Azospirillum* and wheat. Plant Soil 110, 239-47.

E2. Transgenic Plants in Hormone Biology

Harry J. Klee and Michael B. Lanahan
Monsanto Company, 700 Chesterfield Village Parkway, Chesterfield, MO 63198, USA.

INTRODUCTION

Plant biology has been revolutionized by the development of genetic transformation systems. Genes are routinely engineered and reintroduced into a variety of plant species. The functions of cloned genes can be determined through the use of antisense gene expression and cosuppression. Transformation has further broken down species barriers, permitting transfer of genes from any source into plants. In the area of hormone biology, approaches that greatly extend technical feasibility are now possible. Experiments that previously required exogenous application of chemicals can be accomplished more precisely. While transgene technology has its limitations, it eliminates or greatly reduces problems associated with uptake, transport and metabolism of exogenously applied materials. It is now possible to direct hormone alterations to specific tissues or developmental stages. Application of the technology to the analysis of the biosynthesis and functioning of ethylene has been particularly useful. Transgenic plants have facilitated elucidation of the mechanism of biosynthesis as well as the role of ethylene in development, culminating in the control of fruit ripening. While the ethylene work is more advanced than that of other hormones, it illustrates the great opportunity presented to the hormone field.

EXISTING MUTANTS

Our understanding of the mechanisms of hormone action has been greatly aided by the availability of a variety of mutants that affect either metabolism or perception of hormones (17). For example, mutants that are insensitive to ethylene are facilitating a fine dissection of the transduction mechanism from hormone binding onward (11, 16). However, while there are numerous mutants that are insensitive to all of the major classes of phytohormones, there are few mutants altered in hormone metabolism. This deficiency is complemented by the use of transgenic plants. With transgenes, the levels of several hormones can be significantly increased or decreased. This can be done in a tissue or temporal specific manner, allowing a researcher to test a hypothesis or alter development in an advantageous way. In many cases, all that is limiting is the availability of appropriate transcriptional promoters.

P. J. Davies (ed.), Plant Hormones, 340–353.
© 1995 *Kluwer Academic Publishers. Printed in the Netherlands.*

While there are clearly limitations to what can be accomplished, the research that has been done to date with transgenic plants has greatly extended our knowledge of hormone action and led to some very practical applications.

HORMONE METABOLIC ENZYMES

A number of genes encoding enzymes capable of altering hormone levels have been cloned. These genes come from both bacteria and plants. Several of the genes were first isolated from *Agrobacterium tumefaciens,* and their discovery is described elsewhere in this book. The available genes are summarized according to the hormone that they affect.

Auxin

iaaM and *iaaH*

    The growth of *Agrobacterium* crown gall tumors is due in part to two enzymes, encoded by genes located in the transferred bacterial DNA (T-DNA), that together synthesize the auxin indole-3-acetic acid (IAA). This pathway for auxin synthesis is illustrated in Figure 1. The first enzyme, tryptophan monooxygenase (*iaaM*), converts tryptophan to indoleacetamide (IAM) (41). The IAM is then converted to IAA by the action of indoleacetamide hydrolase (*iaaH*) (34, 40). This pathway for synthesis of IAA appears not to be utilized by plants, and tissues expressing the genes contain higher IAA. When the *iaaM* and *iaaH* genes are expressed from their

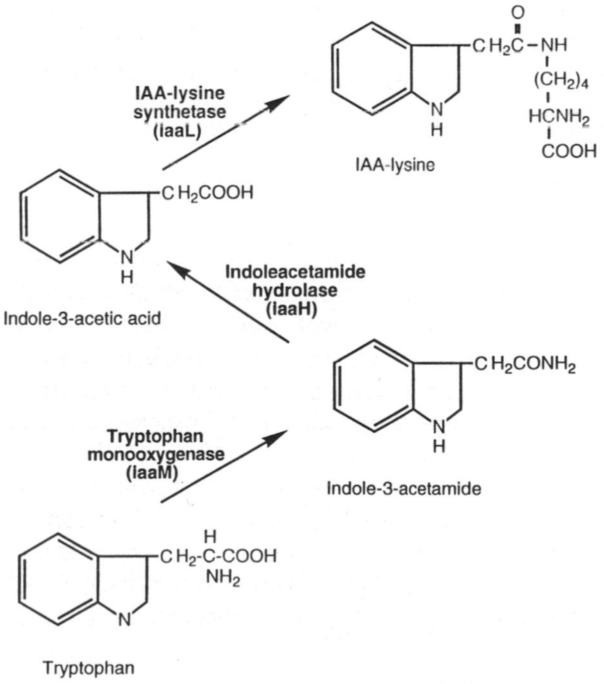

Fig. 1. Activities of enzymes encoded by auxin metabolic genes. Indole-3-acetic acid (IAA) is synthesized from tryptophan by the sequential action of tryptophan monooxygenase (*iaaM*) and indole-3-acetamide hydrolase (*iaaH*). The IAA can also be converted to the biologically inactive N^ε-(indole-3-acetyl)-L-lysine (IAA-lysine) by IAA-lysine synthetase (*iaaL*).

own transcriptional promoters, expression of both genes is required for increasing free IAA levels (15). However, elevated IAA can be accomplished by expressing only the *iaaM* gene under the control of a strong transcriptional promoter (20). Under these circumstances IAM accumulates in the plant tissue. Some of the IAM is converted to IAA. Whether this conversion is enzymatic or chemical has not been determined. Expression of the *iaaM* gene under the control of a strong plant promoter typically results in a four- to ten-fold increase in the level of IAA in plant tissues. Expression of the auxin biosynthetic genes also leads to increased auxin conjugate formation in tobacco, suggesting that the plant does respond to increased IAA by attempting to inactivate the excess (36).

iaaL

The plant pathogen *Pseudomonas savastanoi* causes galls on some plants by producing and secreting auxin. The bacterium also can modify IAA by conjugation to lysine (Fig. 1). The gene encoding IAA-lysine synthetase has been cloned (28). As is the case with other IAA-amino acid conjugates, the lysine-conjugated IAA is biologically inactive. Although the reversibility of IAA-amino acid conjugation in plants is still the subject of much speculation, lysine conjugates are not normally synthesized in plants and it appears that this reaction is not reversible. Thus, the effect of *iaaL* expression in plants is to reduce the pool of free IAA, and transgenic plants with up to a 20-fold reduction have been produced (30).

Cytokinin

ipt

A gene encoding the cytokinin biosynthetic enzyme, isopentenyl transferase (ipt), has also been isolated from the *Agrobacterium* T-DNA. This enzyme catalyzes the condensation of adenosine monophosphate and isopentenyl pyrophosphate to form isopentenyl adenosine monophosphate (IPA) (1, 4). Synthesis of IPA appears to be the rate limiting step for cytokinin synthesis in plants since IPA does not accumulate in transgenic plants expressing the ipt gene whereas large elevations in zeatin-related cytokinins are observed (2, 23). Construction of transgenic plants expressing the ipt gene has been extremely difficult since even weak expression of the gene during regeneration suppresses root formation. Thus, transgenic plants produced to date have used the ipt under the control of regulated transcriptional promoters or have activated the gene following plant regeneration.

Ethylene

The reactions related to ethylene metabolism for which genes are available are shown in Figure 2.

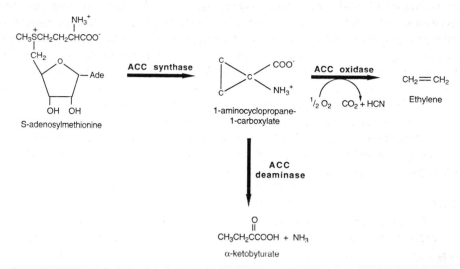

Fig. 2. Enzymes involved in ethylene biosynthesis. Ethylene is synthesized from S-adenosylmethionine via the intermediate 1-aminocyclopropane-1-carboxylate (ACC). Genes encoding the ACC synthase and ACC oxidase enzymes have been cloned from several plant species. Synthesis of ethylene can be inhibited by expression of the gene encoding ACC deaminase in transgenic plants. This enzyme shunts ACC to α-ketobutyric acid and ammonia, thus preventing its conversion to ethylene.

ACC synthase

The rate limiting step in the synthesis of ethylene is formation of 1-aminocyclopropane-1-carboxylate (ACC) (see Chapter B4). The gene encoding this enzyme was first cloned from zucchini (32) and has since been cloned from many different plant species. The ACC synthase gene has been used for manipulation of ethylene synthesis in plants; overexpression of the gene with a constitutive promoter can cause plants to produce in excess of 100 times normal levels (H. Klee and M. Lanahan, submitted) while antisense gene expression can cause up to a 99% reduction in ethylene synthesis (26). A difficulty with antisense gene shutoff is that ACC synthase is a divergent gene family in species such as tomato (31). Since different gene family members can be quite divergent in DNA sequence, the degree of ethylene inhibition can vary from tissue to tissue. Thus an antisense tomato gene construction can virtually shut off ethylene synthesis in ripening fruit (26) but not affect leaf ethylene production (H. Klee, unpublished). The ability to shut off different gene family members should permit a critical evaluation of the regulation and roles of individual genes in a plant.

ACC oxidase

The gene encoding ACC oxidase was first identified by its ability, in an antisense orientation, to prevent tomato fruit ripening (12). Only after this initial observation, was it demonstrated that ethylene is synthesized by an

343

iron-dependent oxidation mechanism (42). Thus, transgenic plants proved invaluable in directing subsequent biochemical analyses. Like ACC synthase, ACC oxidase is encoded by a gene family (13). Antisense gene expression has been reported to reduce ethylene synthesis in ripening fruit by 97% relative to controls (12, 27).

ACC deaminase

ACC deaminase degrades ACC to α-ketobutyric acid, thus effectively preventing its conversion to ethylene (14). The gene encoding this enzyme has been cloned from two different bacteria (19, 35) and is not a normal constituent of plants. Expression of the gene in plants greatly reduces ethylene synthesis in all tissues where the gene is expressed. In tomato, the gene has been used to reduce ethylene synthesis in leaves by more than 95% and in fruit by 90% (19).

TRANSGENIC PLANTS

Several species of transgenic plants containing one or more of the above genes have been produced. Some of the better model systems include *Arabidopsis*, tobacco and tomato. Factors that have influenced the choice of model systems include ease of transformation, availability of hormone-related mutants, background information on hormone levels, and physiological work on effects of hormones on growth and development. Each plant species has distinct advantages and disadvantages for doing phytohormone experimentation. For example, *Arabidopsis* has distinct genetic advantages such as an abundance of characterized hormone-related mutants but has the disadvantage of having very little information concerning hormone content and metabolism, as well as limited tissue availability and difficulty in manipulation due to its small size.

Perhaps the most remarkable aspect of introduction of genes altering hormone concentrations is that fertile plants can be recovered. In tobacco the range of IAA concentration between overproducers (*iaaM*) and underproducers (*iaaL*) is two hundred-fold. With cytokinins, increases of 100-200-fold are tolerated. In the case of ethylene, tomato plants with rates of biosynthesis covering three orders of magnitude have been produced. While there are a number of major effects on size and shape of cells in different organs, it is clear that many tissues and organs do not respond to alterations in hormone levels. For example, overproduction of auxin in maturing petunia embryos has no detectable effect (20). These observations indicate that perception, as well as metabolism of hormones is developmentally regulated and represents a major control point in modulating hormone responses.

A related aspect of the transgene experiments is the effect of expression on overall plant development. The hormonal perturbations have a major impact on rates of cell division and expansion. Thus, auxin overproduction causes downward curvature of leaves because of uneven expansion of cells on the abaxial and adaxial surfaces (20). In a similar manner, auxin overproduction and the consequent ethylene overproduction causes a decrease in internode length due to less stem cell elongation (see below). In some cases, cells can be forced to undergo a change in identity such as when high auxin leads to proliferation of adventitious roots (20). However, it has been somewhat surprising that alteration of the levels of such important molecules has not led to homeotic-type changes in cell identity. It cannot be concluded at this point that there is no role for hormones in determining cell and organ identity. The negative results to date may be a consequence of a lack of substantial alterations to the hormone gradients within a plant. This question can only be adequately addressed with tissue specific hormone modification.

Another significant conclusion of the transgenic plant experiments is that there are complex interactions between the various hormones. For example, the synthesis and activities of auxin, cytokinin and ethylene are intimately interrelated. Overproduction of IAA with the *A. tumefaciens* genes has major phenotypic effects on plants (discussed below) but significant increases in auxin stimulate ethylene production and it has been difficult to distinguish the "auxin" effects from ethylene effects. Further, some developmental processes are regulated by the ratio of auxin to cytokinin and the absolute levels of each are not critical.

Apical dominance

That auxin and cytokinin are capable of regulating apical dominance is well established (see Chapter G6). High auxin suppresses release of lateral buds from dormancy while cytokinin stimulates their growth. Thus, it is not surprising that transgenic plants with elevated IAA exhibit increased apical dominance (reduction of lateral branching) (20) and plants with elevated cytokinins exhibit reduced apical dominance (23). What is more surprising is that the apical dominance exhibited by auxin overproducing plants can be overcome by exogenous application of cytokinin to a dormant lateral bud (H. Klee, unpublished). This result suggests that absolute levels of auxin and cytokinin are not the determinants of lateral growth. Rather, it is the ratio of auxin to cytokinin that controls growth. Dormancy can be induced by raising the auxin level tenfold but can be relieved by further increasing the cytokinin in the bud. If this is the case, then reducing the effective auxin level should be equivalent to increasing the cytokinin level. This prediction was verified when plants overexpressing the *iaaL* gene were produced (30). These plants, containing 5- to 20-fold lower IAA levels than controls, showed reduced apical dominance similar to the *ipt* overproducing plants.

An additional question concerns the role of auxin in reducing lateral bud growth. It has been suggested that auxin acts by stimulating high ethylene production and that it is actually ethylene controlling apical dominance.

Ethylene does act to slow growth of many plant tissues. This can be seen in Figure 3. In this experiment, auxin overproduction (and its concomitant ethylene overproduction) causes a large reduction in internode elongation. When the ethylene synthesis is brought back to wild-type levels by inclusion of an ACC deaminase gene, much, though not all, of the internode length reduction is eliminated. Analysis of these same plants clearly rules out a role for ethylene in control of apical dominance (29). By crossing auxin overproducing plants with plants either reduced in ethylene synthesis or sensitivity, auxin and ethylene effects could be uncoupled. In tobacco, petunia and *Arabidopsis* there is no role for ethylene in the control of apical dominance. Thus, lateral growth appears to be regulated principally by the ratio of auxin to cytokinin.

FMV-ACCase   19S-iaaM,   19S-iaaM
                 FMV-ACCase

Fig. 3. The morphological effects of auxin and ethylene overproduction on transgenic tobacco plants. Left to right plants transgenic for ACC deaminase alone (Left), ACC deaminase and *iaaM* (Center), and *iaaM* alone (Right). The *iaaM* alone plant exhibits the typical reduced stature as a consequence of a 4.5x increase in IAA and a 5x increase in ethylene compared to a non-transgenic control. The center plant contains the same elevated level of auxin but wild-type levels of ethylene. The ACC deaminase alone plant produces approximately 50% less ethylene and is phenotypically indistinguishable from a non-transgenic control plant. From (30).

Vascular differentiation

Auxin is a major controlling factor of vascular differentiation in plants (see Chapter G4). Auxin is believed to affect xylem formation both quantitatively and qualitatively. Work with transgenic plants having altered levels of IAA generally confirms the direct relationship between auxin and the degree of differentiation of xylem (20, 30). Auxin overproducing petunia plants contain more xylem elements than control plants. The cells are, however, smaller and more lignified. Conversely, tobacco plants with lowered IAA levels contain fewer xylem elements of larger size and are less lignified (Fig. 4). These results support the idea that auxin stimulates cell division within the vascular cambium and that there is a direct relationship between auxin content and rate of cell division (3). Cell size is most likely affected because auxin stimulates secondary cell wall formation and lignin synthesis. The more rapidly the wall is synthesized, the less time there is for cell expansion.

Fruit ripening

The role of ethylene in promoting fruit ripening of climacteric fruits has been

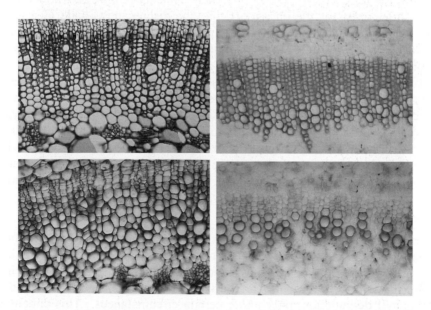

Fig. 4. Cross sections of stems from wild-type (top) and transgenic tobacco plants expressing IAA-lysine synthetase (bottom). Sections are taken from the fourth node from the bottom of mature plants and stained with either toluidine blue (left) or phloroglucinol (right). The transgenic plant contains fewer, larger xylem elements than the control. The transgenic plant is also considerably less lignified as can be seen by the phloroglucinol, which preferentially stains lignin. From (31).

examined in transgenic plants using several independent approaches to reduce its synthesis. Ethylene synthesis has been reduced using antisense gene constructs against ACC oxidase (12) or ACC synthase (26) and by expressing ACC deaminase (19). Since fruit ripening is covered in detail elsewhere in this book, it will be covered only briefly here. Analysis of transgenic plants indicates that there is a direct correlation between ethylene inhibition and the rate of ripening; the greater the inhibition, the slower the rate of ripening. Ethylene seems to act as a catalyst and coordinator of ripening. However, even in the absence of significant ethylene, aspects of ripening do occur. For example, chlorophyll breakdown occurs and fruit undergo significant softening over time relative to mature green fruit. These ethylene-independent aspects of ripening may be analogous to ripening in nonclimacteric fruits. What ethylene accomplishes is to coordinately induce a large number of genes that are responsible for rapid and uniform fruit ripening. The ability to modulate the rate of ripening of climacteric fruits presents a major economic opportunity and promises to significantly reduce losses of crops to spoilage.

One interesting aspect of ripening that is observed in ethylene inhibited transgenic lines is that fruit left attached to the plant ripen at a faster rate than detached fruit (12, 19, 27). In the most inhibited lines where detached fruit never fully ripen, attached fruits accumulate significantly more carotenoids and become measurably softer than their detached controls. At least two possibilities exist for the on vs. off the vine discrepancy. One idea is that there may be translocation of ripening modifying factor(s). Alternatively, the attached fruit may simply have a higher effective concentration of ethylene than fruit that have been detached. In fruit where ethylene has become rate-limiting due to ACC deaminase expression, a higher ethylene content would result in more rapid and complete ripening. This idea is supported by the observation that the skin of a tomato is relatively impermeable to ethylene and that 97% of ethylene released from detached fruit occurs through the stem scar (5). Attached fruit do contain higher internal levels of ethylene than similarly staged detached fruit (18). Analysis of an ethylene insensitive tomato mutant indicates no difference between on vs. off the vine ripening.

Senescence

Experiments with transgenic plants have both confirmed and cast doubt on current models for senescence. For example, cytokinin has long been promoted as an inhibitor of senescence (10). However, elevated cytokinins in transgenic tobacco plants can inhibit leaf senescence, as measured by chlorophyll degradation, only under certain circumstances. This conclusion is based on observations of plants containing the ipt gene under the control of heat shock promoters (23, 37, 38). Elevated levels of cytokinin throughout the entire plants did not lead to any significant delays in senescence relative to non-transgenic controls. But when leaves were detached and a localized heat shock applied, delays were observed in the heat shocked portion (37). When similar experiments were performed on transgenic plants containing an ipt gene fused to an auxin-inducible promoter, significant delays in senescence were observed only when plants were actively transpiring (22). In the absence of transpiration, leaves actually senesced more rapidly than controls. The patterns of ipt gene expression in these plants were more complex and localized than those observed with the hsp constructions. All of the results can be explained in terms of cytokinin gradients and the effects of localized cytokinins on nutrient mobilization. The results of Li et al. (22) indicate that locally elevated cytokinin causes accumulation of sucrose in that tissue. The source:sink relationships of tissues in plants with generally elevated cytokinins, i.e. the hsp/ipt plants, would not be expected to be altered in their distribution of nutrients and therefore not be altered in their pattern of senescence relative to controls.

Ethylene has been implicated as having a role in the senescence of many plant tissues. In fact, ethylene-mediated ripening of fruits can be viewed as a specialized case of senescence. In tomato, transgenic plants that both over- and underproduce ethylene are greatly altered in some aspects of senescence. For example, plants that contain an ACC synthase gene under the control of the Cauliflower Mosaic Virus 35S promoter synthesize up to 100 times as much ethylene as controls (M. Lanahan and H. Klee, submitted). These plants are almost infertile since most flowers abort prior to fertilization. The abortion is due to premature induction of the abscission zone in the pedicel. However, leaves of these plants do not exhibit accelerated senescence and vegetative tissue remains green for up to a year. Only when the plants set fruit do the leaves senesce, suggesting that the signal for vegetative senescence is probably due to establishment of a strong sink tissue. Transgenic tomato plants that synthesize reduced levels of ethylene, due to an antisense ACC oxidase gene, have been reported to show minor delays in leaf senescence, suggesting that there is a correlation between ethylene and the rate of senescence (27). These results have been confirmed with an ethylene insensitive mutant of tomato, *Never ripe* (*Nr*) (M. Lanahan et al., submitted). In the *Nr* mutant, leaf senescence occurs at a slightly slower rate than in the isogenic control. However, flower petals remain viable and abscission zone formation is delayed for four weeks or longer whether fertilization has occurred or not. Thus, it can be concluded that ethylene has a critical role in senescence of petals and the pedicellar abscission zone. However, ethylene alone is not itself sufficient to cause senescence of vegetative tissues in the same plant. As is the case for fruit ripening, ethylene accelerates the process when it is developmentally appropriate for a tissue to senesce.

THE FUTURE

As hormone levels have been manipulated, many unexpected results have appeared. For example, plants that synthesize very low levels of ethylene in all tissues or are ethylene insensitive do not appear to be morphologically abnormal. These results necessitate reassessment of the roles of the hormones in development. In the case of ethylene, a major role appears to be mediating response to stresses. Environmental challenges such as temperature, water or pathogen stress that a plant encounters lead to increased production of ethylene. This elevated ethylene in turn leads to a reduction in growth rate. Thus, ethylene reduces growth of a plant when it encounters adverse environmental conditions. It can still, however, be argued that ethylene levels have simply not been reduced sufficiently to uncover "essential" functions in development. Combinations of reduced ethylene synthesis and insensitivity should be able to address this outstanding issue conclusively.

One apparent limitation of altering hormone levels through transgenes relates to the lack of cell autonomous effects. While hormone synthesis or degradation can be manipulated in a cell-specific manner, diffusion of hormones to and from adjacent cells or tissues cannot be precisely controlled. A more precise approach to fine tuning effects to specific tissues will be possible with genes affecting hormonal sensitivity. For example, eliminating ethylene synthesis throughout a plant with a transgene may in certain circumstances be less desirable than modulating the ethylene responsiveness of specific tissues with the dominant Etr1 gene product.

Other hormone-related genes

There are several additional genes derived from *A. tumefaciens* and *A. rhizogenes* that may affect plant morphology by altering hormonal regulation. For example, it has been proposed that the rolC gene of *A. rhizogenes* encodes an enzymes capable of hydrolyzing certain cytokinin conjugates (6). While this activity has been demonstrated in vitro, morphologically abnormal plants are not significantly altered in their cytokinin content (25, 33). The *A. rhizogenes* rolB gene is also purported to cleave auxin conjugates, releasing free IAA (7). Recent transgenic plant experiments have cast doubt on this interpretation. These plants are not altered in their auxin or auxin conjugate levels or turnover (25).

In the area of hormone perception, rapid progress is being reported. One *Arabidopsis* auxin perception gene, AXR1, has been identified (21). This gene encodes an enzyme with homology to a ubiquitin activating enzyme. In the case of ethylene, two genes involved in signal transduction have been cloned from *Arabidopsis*. One gene, *ctr*, encodes a protein homologous to a Raf protein kinase (16). The second gene, *Etr1*, encodes a histidine kinase that exhibits homology to the sensory proteins of bacterial two component signal transduction systems (C. Chang and E. Meyerowitz, submitted).

Finally, a gene involved in sensitivity to ABA has been cloned (9). This gene encodes a protein with homology to the maize Vp1 protein.

At present, none of the genes involved in abscisic acid or gibberellin (GA) metabolism are available. However, this is likely to change. There is already one report of the cloning of a gene involved in GA biosynthesis (39). A number of mutants involved in GA and abscisic acid (ABA) metabolism and perception have been accurately mapped in *Arabidopsis* (17). Available technology makes the cloning of the genes feasible. The possibility of manipulating GA and ABA metabolism represents a significant opportunity.

Levels of salicylate, a signaling molecule involved in establishing systemic acquired resistance (SAR) after pathogen infection, can also be modulated by transgene technology. Recently Gaffney et al. (8) have shown that expression of the *Pseudomonas NahG* gene product, which hydroxylates salicylate, will inhibit salicylate accumulation in plants. Their results indicate that salicylate accumulation is necessary for induction of the SAR response. Thus, transgene technology has unequivocally demonstrated the role salicylate plays in mediating the SAR response.

Practical Applications

The work with delayed ripening tomatoes is the most advanced example of a practical application of hormone manipulation in plants. It illustrates the great potential of transgenic technology. Based on what is known about the functions of hormones it is likely that many new applications will occur in the near future. For example, the inhibition of ethylene synthesis or perception can be immediately applied to other climacteric fruits and some floricultural crops. Also, patterns of auxin and/or cytokinin synthesis may be modified in ways that affect source:sink relationships and ultimately affect yield. With proper targeting of hormone synthesis, it should be possible to alter the architecture of a plant, creating novel horticultural and agricultural crops. Finally, as transformation technology advances, hormone manipulation in a wider variety of plant species should be possible, greatly expanding the range of potential experimentation

References

1. Akiyoshi, D., Klee, H., Amasino, R., Nester, E.W., Gordon, M. (1984) T-DNA of Agrobacterium tumefaciens encodes an enzyme of cytokinin biosynthesis. Proc. Natl. Acad. Sci. USA 81, 5994-5998.
2. Akiyoshi, D.E., Morris, R.O., Hinz, R., Mischke, B.S., Kosuge, T., Garfinkel, D., Gordon, M., Nester, E.W. (1983) Cytokinin auxin balance in crown gall tumors is regulated by specific loci in the T-DNA. Proc. Natl. Acad. Sci. USA 80, 407-411.
3. Aloni, R., Zimmermann, M.H. (1983) The control of vessel size and density along the plant axis-a new hypothesis. Differentiation 24, 203-208.
4. Barry, G.F., Rogers, S.G., Fraley, R.T., Brand, L. (1984) Identification of a cloned cytokinin biosynthetic gene. Proc. Natl. Acad. Sci. USA 81, 4776-4780.
5. Cameron, A.C., Yang, S.F. (1982) A simple method for determination of resistance to gas diffusion in plant organs. Plant Physiol. 70, 21-23.

6.   Estruch, J.J., Chriqui, D., Grossmann, K., Schell, J., Spena, A. (1991) The plant oncogene *rolC* is responsible for the release of cytokinins from glucoside conjugates. EMBO J. 10, 2889-2895.
7.   Estruch, J.J., Schell, J., Spena, A. (1991) The protein encoded by the *rolB* plant oncogene hydrolyses indole glucosides. EMBO J. 10, 3125-3128.
8.   Gaffney, T., Friedrich, L., Vernooij, B., Negrotto, D., Nye, G., Uknes, S., Ward, E., Kessmann, H., Ryals, J. (1993) Requirement of salicylic acid for the induction of systemic acquired resistance. Science 261, 754-756.
9.   Giraudat, G., Hauge, B., Valon, C., Smalle, J., Parcy, F., Goodman, H. (1992) Isolation of the *Arabidopsis* ABI3 gene by positional cloning. Plant Cell 4, 1251-1261.
10. Goldthwaite, J. (1987) Hormones in plant senescence. *In*: Plant hormones and their role in plant growth and development, pp. 553-573, Davies, P., ed. Martinus Nijhoff, Dordrecht.
11. Guzman, P., Ecker, J. (1990) Exploiting the triple response of *Arabidopsis* to identify ethylene-related mutants. Plant Cell 2, 513-523.
12. Hamilton, A.J., Lycett, G.W., Grierson, D. (1990) Antisense gene that inhibits synthesis of the hormone ethylene in transgenic plants. Nature 346, 284-287.
13. Holdsworth, M., Bird, C., Ray, J., Schuh, W., Grierson, D. (1987) Structure and expression of an ethylene-related mRNA from tomato. Nucleic Acids Res. 15, 731-739.
14. Honma, M., Shimomura, T. (1978) Metabolism of 1-aminocyclopropane-1-carboxylic acid. Agric. Biol. Chem. 42, 1825-1831.
15. Inze, D., Follin, A., Van Lijsebettens, M., Simoens, C., Genetello, C., Van Montagu, M., Schell, J. (1984) Genetic analysis of the individual T-DNA genes of *Agrobacterium tumefaciens*; further evidence that two genes are involved in indole-3-acetic acid synthesis. Mol. Gen. Genet. 194, 265-274.
16. Kieber, J., Rothenberg, M., Roman, G., Feldmann, K., Ecker, J. (1993) CTR1, a negative regulator of the ethylene response pathway in *Arabidopsis*, encodes a member of the Raf family of protein kinases. Cell 72, 427-441.
17. Klee, H., Estelle, M. (1991) Molecular genetic approaches to plant hormone biology. Annu. Rev. Plant Physiol. Plant Mol. Biol. 42, 529-551.
18. Klee, H.J. (1993) Ripening physiology of fruit from transgenic tomato plants with reduced ethylene synthesis. Plant Physiol. 102, 911-916.
19. Klee, H.J., Hayford, M.B., Kretzmer, K.A., Barry, G.F., Kishore, G.M. (1991) Control of ethylene synthesis by expression of a bacterial enzyme in transgenic tomato plants. Plant Cell 3, 1187-1193.
20. Klee, H.J., Horsch, R.B., Hinchee, M.A., Hein, M.B., Hoffmann, N.L. (1987) The effects of overproduction of two *Agrobacterium tumefaciens* T-DNA auxin biosynthetic gene products in transgenic petunia plants. Genes & Dev. 1, 86-96.
21. Leyser, H.M.O., Lincoln, C., Timpte, C., Lammer, D., Turner, J., Estelle, M. (1993) *Arabidopsis* auxin-resistance gene *AXR1* encodes a protein related to ubiquitin-activating enzyme E1. Nature 364, 161-164.
22. Li, Y., Hagen, G., Guilfoyle, T. (1992) Altered morphology in transgenic tobacco plants that overproduce cytokinins in specific tissues and organs. Dev Bio 153, 386-395.
23. Medford, J.I., Horgan, R., El-Sawi, Z., Klee, H.J. (1989) Alterations of endogenous cytokinins in transgenic plants using a chimeric isopentenyl transferase gene. Plant Cell 4, 403-413.
24. Nilsson, O., Crozier, A., Schmulling, T., Sandberg, G., Olsson, O. (1993) Indole-3-acetic acid homeostasis in transgenic tobacco plants expressing the *Agrobacterium rhizogenes* rolB gene. Plant J. 3, 681-689.
25. Nilsson, O., Moritz, T., Imbault, N., Sandberg, G., Olsson, O. (1993) Hormonal characterization of transgenic tobacco plants expressing the rolC gene of *Agrobacterium rhizogenes* T_L-DNA. Plant Physiol. 102, 363-371.

26. Oeller, P.W., Min-Wong, L., Taylor, L.P., Pike, D.A., Theologis, A. (1991) Reversible inhibition of tomato fruit senescence by antisense RNA. Science 254, 437-439.

27. Picton, S., Barton, S., Bouzayen, M., Hamilton, A., Grierson, D. (1993) Altered fruit ripening and leaf senescence in tomatoes expressing an antisense ethylene-forming enzyme transgene. Plant J. 3, 469-481.

28. Roberto, F.F., Klee, H., White, F., Nordeen, R., Kosuge, T. (1990) Expression and fine structure of the gene encoding $N^{\text{Œ}}$-(indole-3-acetyl)-L-lysine synthetase from *Pseudomonas savastanoi*. Proc. Natl. Acad. Sci. USA 87, 5797-5801.

29. Romano, C.P., Cooper, M.L., Klee, H.J. (1993) Uncoupling auxin and ethylene effects in transgenic tobacco and Arabidopsis plants. Plant Cell 5, 181-189.

30. Romano, C.P., Hein, M.B., Klee, H.J. (1991) Inactivation of auxin in tobacco transformed with the indoleacetic acid-lysine synthetase gene of *Pseudomonas savastanoi*. Genes Dev. 5, 438-446.

31. Rottmann, W.H., Peter, G.F., Oeller, P.W., Keller, J.A., Shen, N.F., Nagy, B.P., Taylor, L.P., Campbell, A.D., Theologis, A. (1991) 1-aminocyclopropane-1-carboxylate synthase in tomato is encoded by a multigene family whose transcription is induced during fruit and floral senescence. J. Mol. Biol. 222, 937-961.

32. Sato, T., Theologis, A. (1989) Cloning the mRNA encoding 1-aminocyclopropane-1-carboxylate synthase, the key enzyme for ethylene synthesis in plants. Proc. Natl. Acad. Sci. USA 86, 6621-6625.

33. Schmulling, T., Fladung, M., Grossmann, K., Schell, J. (1993) Hormonal content and sensitivity of transgenic tobacco and potato plants expressing single rol genes of *Agrobacterium rhizogenes* T-DNA. Plant J. 3, 371-382.

34. Schröder, G., Waffenschmidt, S., Weiler, E.W., Schröder, J. (1984) the T-region of Ti plasmids codes for an enzyme synthesizing indole-3-acetic acid. Eur. J. Biochem. 138, 387-391.

35. Sheehy, R., Honma, M., Yamada, M., Sasaki, T., Martineau, B., Hiatt, W.R. (1991) Isolation, sequence, and expression in *E. coli* of the *Pseudomonas* sp strain ACP gene encoding 1-aminocyclopropane-1-carboxylate deaminase. J. Bacteriol. 173, 5260-5265.

36. Sitbon, F., Sundberg, B., Olsson, O., Sandberg, G. (1991) Free and conjugated indoleacetic acid (IAA) contents in transgenic tobacco plants expressing the *iaaM* and *iaaH* IAA Biosynthesis genes from *Agrobacterium tumefaciens*. Plant Physiol. 95, 480-485.

37. Smart, C., Scofield, S., Bevan, M., Dyer, T. (1991) Delayed leaf senescence in tobacco plants transformed with *tmr*, a gene for cytokinin production in *Agrobacterium*. Plant Cell 3, 647-656.

38. Smigocki, A.C. (1991) Cytokinin content and tissue distribution in plants transformed by a reconstructed isopentenyl transferase gene. Plant Mol. Biol. 16, 105-115.

39. Sun, T., Goodman, H., Ausubel, F. (1992) Cloning the *Arabidopsis* GA1 locus by genomic subtraction. Plant Cell 4, 119-128.

40. Thomashow, L.S., Reeves, S., Thomashow, M.F. (1984) crown gall oncogenesis: evidence that a T-DNA gene from the *Agrobacterium* Ti plasmid pTiA6 encodes an enzyme that catalyzes synthesis of indoleacetic acid. Proc. Natl. Acad. Sci. USA 81, 5071-5075.

41. Thomashow, M.F., Hugly, S., Buchholz, W.G., Thomashow, L.S. (1986) Molecular basis for the auxin-independent phenotype of crown gall tumor tissues. Science 231, 616-618.

42. Ververidis, P., John, P. (1991) Complete recovery *in vitro* of ethylene-forming enzyme activity. Phytochem. 30, 725-727.

E3. Molecular Approaches to the Study of the Mechanism of Action of Auxins

Jeff Schell, Klaus Palme and Rick Walden
Max-Planck-Institut für Züchtungsforschung, Carl-von-Linné-Weg 10, D-50829 Köln, F.R.G.

INTRODUCTION

Phytohormones play a crucial role in normal plant growth, but despite their fundamental influence on almost all aspects of plant development, very little is known on their mode of action at the molecular level. Here we describe three different approaches to addressing this question: the analysis of proteins which bind auxin with high specificity and as such might act as auxin receptors; the study of the effect of expression of bacterial genes whose products modify the intracellular levels of auxins and cytokinins, and the generation of gene tagged plant mutants altered in their response to auxins. While diverse in their approach, these strategies have been devised to dissect the auxin signal transduction pathway at the molecular level. The results that we will describe, while sharing a common thread, indicate that the plant cell has evolved a remarkable array of molecular pathways in response to plant growth substances.

AUXIN BINDING PROTEINS

Receptor-like proteins which bind auxin and transmit the auxin signal are central to most models of auxin action (20, 48, 66). Auxin binding proteins, thought to represent potential auxin receptors, have been identified by ligand binding studies in several monocotyledonous and dicotyledonous plants, and were detected in cellular fractions such as the endoplasmic reticulum (ER), the vacuole, and the plasma membrane (see Chapter D4) (10, 28, 42, 66). At least three auxin binding proteins are thought to be present in the plasma membrane: an auxin uptake carrier (21), an auxin efflux carrier (27) and a receptor that influences elongation growth (35).

To overcome problems associated with traditional auxin binding studies (e.g. low receptor protein concentration, instability of auxin binding under experimental conditions, denaturation or loss of the auxin binding proteins during purification) we have used auxin specific photoaffinity labeling techniques for identification of auxin binding proteins. Photoaffinity labeling techniques have contributed greatly to the identification and structural studies of animal receptor proteins. Photoaffinity ligands covalently label the ligand-

P. J. Davies (ed.), Plant Hormones, 354–371.
© 1995 *Kluwer Academic Publishers. Printed in the Netherlands.*

binding polypeptide and allow receptor molecules to be followed throughout purification under both denaturing and nondenaturing conditions. Initial photoaffinity labeling experiments were not successful as a result of signals being overwhelmed by background noise thus making discernment of radioactive peaks difficult (29). More recent improvements with this technique have resulted in a better signal to noise ratio (23). Special care to ensure the purity of the photoaffinity labeling agent, and to maximize the specificity of the assay has allowed us to identify and characterise several auxin binding proteins from maize and *Arabidopsis* (6, 7, 15, 74, 75). Photoaffinity labeling techniques have been applied to identify auxin binding proteins in plasma membrane vesicles from zucchini, tomato and maize (23, 24, 30). Using 5-azido-[7-^{3}H]IAA for photoaffinity labeling we were able to identify three proteins in plasma membranes from maize coleoptiles with molecular masses of 60 kDa (pm60), 58 kDa (pm58) and 23 kDa (pm23). Whereas pm60 and pm58 are not yet characterised, pm23 has been analyzed in detail (15). Binding of 5-azido-[7-^{3}H]IAA to pm23 was competed by auxins and functional analogues. A purification scheme allowing purification of pm23 was designed. Homogenous pm23 were obtained from coleoptile extracts after 7000 fold purification. Partial amino acid sequences were obtained for both proteins. The primary amino acid sequence information obtained allowed the synthesis of specific oligonucleotides and subsequent isolation of pm23 specific cDNAs. The proteins predicted from the deduced open reading frames were not found in data bases and represent novel proteins with unknown functions.

The NPA-Receptor

It is generally assumed that polar auxin transport by carrier proteins plays an important role in controlling auxin concentrations in various tissues. Little is known about the molecular elements required for auxin transport through the various tissues of a plant (53; see Chapter G3). Binding and inhibitor studies, using active auxins or auxin transport inhibitors, as well as electrophysiological evidence, point to the presence of auxin uptake and efflux carriers localized on the plasma membrane of responsive plant cells (16). Transport of auxin through shoot tissues probably involves a saturable, specific H$^+$/IAA$^-$ influx-carrier (3, 18, 21). Passive uptake of undissociated indole-3-acetic acid across the plasma membrane can probably also occur. Binding studies have indicated that a H$^+$/IAA$^-$ influx-carrier must be evenly distributed on the plasma membrane of maize coleoptiles or zucchini hypocotyl cells. In addition, there is evidence for an auxin efflux carrier (53). This efflux carrier was found to be inhibited by phytotropins such as 1-*N*-naphthylphthalamic acid (NPA) or 2,3,5-triiodobenzoic acid (TIBA). NPA, for example, inhibits polar auxin transport, affects root growth and abolishes gravitropic responses. These responses to NPA *in vivo* correlate with an *in vitro* inhibition of auxin efflux from PM vesicles isolated from zucchini hypocotyls or maize coleoptiles, resulting in an increase of net accumulation of IAA in these membrane vesicles (21).

Analysis of high-specific binding of [³H]NPA to maize microsomes or plasma membranes led to the identification of NPA binding sites, called NPA receptors. It has been proposed that the NPA binding protein could be either an auxin efflux carrier or a regulatory element of the auxin efflux carrier. Although NPA binding activities were detected in membrane vesicles by the use of highly-specific binding of radioactively-labeled NPA, such assays were not sensitive enough to allow isolation and molecular characterization of NPA proteins from *Avena sativa* coleoptiles or other plant tissues. We therefore aimed at developing a photoaffinity probe for auxin carrier proteins, and synthesized a tritiated and light sensitive NPA-analogue, 5'-azido-[3,6-³H₂]NPA ([³H₂]N₃NPA) (74, 75). The binding of [³H₂]N₃NPA to maize plasma membrane vesicles was competed by nonlabeled NPA, but not by benzoic acid. After incubation of plasma membrane vesicles with [³H₂]N₃NPA and exposure to UV light, we observed specific photoaffinity labeling of a protein with an apparent molecular weight of 23 kDa. Pretreatment of the plasma membrane vesicles with IAA or with the auxin transport inhibitor TIBA strongly reduced specific labeling of this protein. This 23 kDa protein was also labeled by addition of 5-azido-[7-³H]IAA to plasma membranes prior to exposure to UV light. The 23 kDa protein was solubilized from plasma membranes using 1% Triton X-100.

The Endoplasmic Reticulum-Auxin Binding Protein Family

A binding site (site I) from maize, initially thought to be membrane associated, has been most thoroughly studied. Historical arguments in favour of this site being a putative receptor were the specificity of binding of various auxin analogues, a correlation of changes in auxin binding activity with physiological events (67), and its apparent absence in several auxin-insensitive mutants (43, 51). Additionally, alterations in the degree of specific auxin binding correlated with inhibition of growth of maize mesocotyls by high-irradiance red light and with growth alterations in artichoke tubers and tobacco cell cultures (65, 68, 71).

The major ER-located auxin-binding protein from maize coleoptiles (Zm-ERabp1, for *Zea mays* ER located auxin binding protein) has an apparent molecular weight of 22 kDa. Its primary structure was independently deduced from different cDNAs isolated from maize (22, 25). Meanwhile additional members of this family were isolated from maize, as well as from *A. thaliana* (49, 57, 77). All these proteins contain an N-terminal hydrophobic signal sequence that appears to be responsible for the uptake of these proteins into microsomes. This was confirmed by *in vitro* studies which demonstrated that both the maize as well as the *Arabidopsis* ERabp's are translocated into ER-derived microsomes and cotranslationally glycosylated (8, 49). Biochemical experiments reported by Hesse et al. (22) and Palme et al. (47) established that this auxin binding protein is a luminal component of the ER. This is consistent with the absence of hydrophobic sequences that could be responsible for membrane insertion and by the presence of a C-terminal tetrapeptide sequence, -Lys-Asp-Glu-Leu (-KDEL) previously reported for other proteins to be a retention signal for the endosplasmic reticulum. The signal K/HDEL is recognized as being responsible for retrieval of proteins from a post-ER salvage compartment in several eukaryotic organisms, and has apparently a similar function in plants.

Although the primary sequences of the proteins encoded by this gene family do not fit the structural requirements of "animal" receptor proteins, Zm-ERabp1 is considered to be an "auxin receptor". The basis for this belief originally was a correlation of auxin-binding profiles with the cellular pattern of auxin-stimulated elongation growth. In addition, light-induced changes in cell elongation were related to a modulation of the number of these binding sites (71). Recent experiments using antibodies raised against Zm-ERabp1, demonstrated a specific inhibition of electrical responses at the plasma membrane of tobacco protoplasts (1, 2). This was further confirmed by analyzing the electrical response of maize protoplasts using the patch clamp technique (54). It was found that after a short lag phase, IAA or NPA induced an outwardly directed current that was dependent on the concentration of the auxin applied to the protoplasts. This current was inhibited by the application of antibodies directed against the Zm-ERabp1 protein. Furthermore, Venis et al. (67) demonstrated that antibodies to a peptide from Zm-ERabp1 that probably harbours the auxin binding site, have auxin agonist activity. The question, however, arises as to how a putative receptor located in the lumen of the ER could be involved in the perception of exogenous auxins and affect phenomena such as cell expansion and membrane hyperpolarization. A number of observations indeed imply that ERabp plays a role in these phenomena. These data indicate that the Zm-ERabp1 protein may be a site for auxin perception through which the activity of the plasma membrane H$^+$ATPase is activated and modulated. The answer may well be that although the ERabp encoded proteins can indeed be found to be associated with the ER, they can apparently, under circumstances that still await elucidation, go through the secretory pathway and appear on the plasma membrane. That such a remarkable movement of a luminal ER auxin-binding protein to the cell surface might indeed occur and have functional significance, was first suggested by experiments (2) showing that ERabp was involved in the perception of the auxin signal at the plasma membrane. Indeed, incubation of tobacco mesophyll protoplasts with exogenous Zm-ERabp1 increased their sensitivity to hyperpolarization by external auxins through several orders of magnitude, whereas incubation of the protoplasts with antibodies directed against Zm-ERabp1 decreased their sensitivity. This interpretation was recently supported in a publication by Jones and Herman (31) who used microscopic immunochemistry and immunochemical assays to study the subcellular distribution and secretion of an ABP in maize coleoptiles and suspension culture cells.

Hormonal Homeostasis by Conjugate Hydrolysis

Maize β-glucosidase

Plant cells must have a means to regulate the cellular activity of phytohormones such as auxins and cytokinins. The reversible inactivation of such phytohormones by conjugation and hydrolysis might be one way by which a fine tuning of intracellular phytohormone activity could be achieved (55). Cohen and Bandurski (9) have defined hormonal homeostasis as "the maintenance of a steady state concentration of the hormones in the receptive tissue appropriate to any fixed environmental condition".

While we shall see that the first genetic evidence that this was the case came from studying genes derived from bacteria which modify free auxin and cytokinin levels (see later), a plant gene coding for a glucosidase capable of hydrolyzing phytohormone conjugates has recently been isolated. Protein extracts from maize coleoptiles grown in the dark were photoaffinity labeled with 5'-azido-[7-^{3}H]indole-3-acetic acid ([^{3}H]N$_3$IAA). A 60-kDa protein, termed p60, was thus identified. This protein was initially detected in the post-ribosomal supernatant, indicating that it might be present in the cytosol of intact cells. p60 was also detected in protein extracts prepared after solubilization of microsomal fractions. In both cases, labeling of p60 was strong and no other protein present in the extract was labeled. To demonstrate the specificity of photoaffinity labeling of p60, competition studies were performed using various unlabeled auxin analogues. Physiologically active natural and synthetic auxins significantly reduced the incorporation of [^{3}H]N$_3$IAA into p60. Compounds specific for the indole ring, such as L-tryptophan, or the aromatic ring system, or radical scavengers, such as ρ-amino-benzoic acid, did not compete the labeling of p60.

To study the presence of p60 in different organs of the maize seedling, crude microsomal fractions from coleoptiles (including the node and the primary leaf), from the mesocotyl and from roots were isolated. Photoaffinity labeling of the corresponding protein extracts showed that p60 was mainly present in the coleoptile and root fractions. In addition an isoform of p60 (pm60) was found in plasma membranes (7).

Microsequencing studies were performed with purified p60 after proteolytic digestion. The primary amino acid sequence of p60 revealed similarities to ß-glucosidases (ßGlu). p60 was therefore analyzed for ßGlu activity. It was found that a fraction containing p60 indeed exhibited ßGlu activity towards general ßGlu substrates (e.g. p-nitrophenyl-glucopyranoside or 6-bromo-2-napthyl-ß-D-glucopyranoside/fast blue for activity staining in native polyacrylamide gels). Specific staining of the p60 with 6-bromo-2-naphthyl-ß-D-glucopyranoside/fast blue demonstrated that p60 has ß-D-glucoside glucohydrolase activity (E.C. 3.2.1.21). To define the substrate specificity of p60, different compounds that are commonly cleaved by a wide variety of glucosidases were tested. In contrast to other ßGlu enzymes, which hydrolyze a broad range of substrates, a distinct pattern of substrate specificity was found for p60.

The results of photoaffinity labeling discussed above support the view that p60 can bind auxins. In this context it is important to know whether the presence of auxins has any effect on the ßGlu activity of p60. Both IAA and 1-naphthylacetic acid, as well as the auxin transport inhibitor NPA, were found to inhibit p60 associated ßGlu activity in a competitive manner. In contrast, the presence of non-functional auxin analogues such as L-tryptophan or 5-hydroxy-IAA, or aromatic compounds such as benzoic acid, had no effect on the ßGlu activity of p60. These results suggest that IAA and related compounds are aglycones which can bind to the active site of p60. Further experiments demonstrated that p60 readily hydrolyzed indoxyl-O-glucoside, a synthetic compound structurally related to the natural auxin conjugate indole-3-acetyl-ß-D-glucose. This activity appears to be highly specific since p60 is not able to hydrolyze other IAA conjugates like IAA-*myo*-inositol or IAA-aspartate. Present data suggest that p60 might be involved *in vivo* in the hydrolysis of glucosidic phytohormone conjugates.

Extensive amino acid sequence analysis of this protein allowed the construction of several synthetic oligodeoxynucleotide probes which were used to isolate a cDNA clone coding for a protein related to p60. The cDNA, named Zm-p60.1, corresponded to a mRNA with a 3'-poly(A)$^+$ sequence with a single open reading frame. When the Zm-p60.1 primary sequence was compared with other amino acid sequences available in protein data bases, similarities were observed to other ßGlu enzymes from archaebacteria, eubacteria and eukaryotes. Amino acid sequence motifs showing similarity to the *rolC* protein from *Agrobacterium rhizogenes* were also found. The motifs shared between the *rolC* protein and Zm-p60.1 pointed to the possibility that both proteins could share common substrates.

Maize kernels are a rich source for phytohormone conjugates, compounds which accumulate in the endosperm during seed maturation and which are mobilized to other parts of the seedling during germination. Hydrolysis and transport of conjugates from the endosperm to the shoot and to the root could be of importance for controlling maize seedling development (see Chapter B1). Protein p60 could play a pivotal role in the germination process by controlling the release of free cytokinin. To test whether Zm-p60.1 expression was able to influence plant growth, tobacco protoplasts were transformed with Zm-p60.1 DNA. These protoplasts acquired the ability to use exogenous cytokinin glucosides to initiate division. Further immunocytochemical analysis of maize seedling roots localized Zm-p60.1 to meristematic cells suggesting that Zm-p60.1 is a glucosidase capable of supplying the developing embryo with biologically active cytokinins (5).

An attractive model explaining plant growth control could be based on the action of ßGlu. Particularly interesting is the simplicity by which developmental adaption to environmental cues could be provided. Auxin and cytokinin conjugates have been found to be broadly distributed in plants. The activity of phytohormone-specific ßGlu could easily be regulated by environmental as well as by endogenous factors. Thus, phytohormone-specific ßGlu might be a link between environmental stimuli and the activation of phytohormones in precise locations of the plant. Although these ideas are far from being proven, they open a promising area of research in plant development. We hope that future investigations will contribute to define more precisely the importance of phytohormone-specific ßGlu in the control of developmental processes in plants.

GENETIC APPROACHES

In contrast to progress made with other plant growth substances such as ethylene, for example, genetic analysis of auxin action has been limited. Two types of general approach have been adopted: first, the study of the phenotypic changes in transgenic plants engineered to contain genes from plant pathogenic bacteria which either encode enzymes involved in the synthesis of auxins, or play a role in their action; second the generation and analysis of mutants which either have a changed sensitivity to auxin, require auxin for growth, or can grow in tissue culture without the addition of auxins.

The study of genes which encode proteins that modify auxin biosynthetic pathways from pathogenic bacteria has grown from the study of the molecular basis of neoplastic growth promoted by the soil bacteria *Agrobacterium tumefaciens*, *A. rhizogenes* and *Pseudomonas syringae* (60; see Chapter E1). The causative agent of crown gall, *A. tumefaciens*, transfers a defined portion of DNA, the T-DNA, to the nuclear genome of the infected plant cell (73). The T-DNA contains genes which are active in the plant cell and encode genes which not only deregulate the synthesis of auxin (and cytokinin), but also fine tune the action of auxin by synthesizing an auxin antagonist (33). The *iaaM* and *iaaH* genes contained on the T-DNA encode enzymes which convert tryptophan to IAA (26) (see Chapter E1). Transfer of these genes, linked to a strong constitutive promoter, to petunia resulted in plants containing up to 10 times higher levels of free IAA (32) (see Chapter E2). The resultant plants were viable and displayed phenotypes, such as enhanced apical dominance and vascular development, resembling those obtained by external application of auxin. In contrast, reduction of the internal levels of auxin can be obtained by engineering plants to contain the *iaaL* gene of *P. syringae* subsp. *Savastanoi* (52, 61), which encodes IAA-lysine synthase, an enzyme which conjugates auxin to lysine and in so doing reduces its biological activity. In this case, after flowering, plants display reduced apical dominance and vascular growth, and suffer from defects in some respects similar to those seen in plants with increased levels of cytokinin. This work confirmed, by internal modification of the relative levels of active auxin and inactive conjugates, that conjugation is a process that can be used by the plant to internally modulate the intracellular levels of active growth substances (9).

Rol Genes

Further support for the notion that internal levels of active growth substances may be reversibly controlled by conjugation comes from the study of the action of the products of the *rol* genes from *Agrobacterium rhizogenes*. *A. rhizogenes* is the causative agent of "hairy root", a disease incited on many dicotyledonous plants which is characterised by the proliferation of root growth at the site of infection. Growth of hairy root is dependent on 3 genes transferred from the bacterium to the plant cell on T-DNA: *rolA, B* and *C*. In order to understand the mechanism of action of the *rol* gene products, studies have centered on transferring them, linked to their own promoters or the constitutive 35S RNA promoter of CaMV, to new plant hosts by transformation, and studying their effect on plant growth and development (60).

rolA

Tobacco plants containing the *rolA* gene linked to its own promoter are dwarfed, due to reduced growth and internode distance, and have wrinkled leaves. When the *rolA* gene is linked to a stronger promoter the dwarfing and leaf wrinkling is more pronounced, and the plants display delayed flowering as well as flower malformation. This delayed flowering is associated with a delayed and decreased accumulation of both free and conjugated polyamines (38). Protoplasts from *rolA*-containing plants are 1000 times more sensitive to auxin as judged by membrane depolarisation (39), although, intriguingly, in transient expression assays using an auxin-responsive promoter element, expression appears to be repressed in comparison with untransformed protoplasts (70).

Currently the function of the *rolA* product remains unresolved. It has been proposed that it may, directly or indirectly, alter polyamine metabolism (63), or be a nucleic acid binding protein (34). Interestingly, the whole plant phenotypes displayed by *rolA*-containing plants resemble those containing the *iaaL* gene of *P. savastanoi* which conjugates IAA to lysine and ornithine (52, 61). Indeed, the *iaaL* enzyme and the *rolA* peptide have a discrete region of homology (40). Taken together with the whole plant phenotype and the inhibition of auxin responsive gene expression in protoplasts one is tempted to think that the *rolA* product may have a negative effect on IAA content and/or activity.

rolB

Inoculation of Kalanchoë leaves with *A. tumefaciens* indicated that *rolB* alone is able to induce root growth (62). Transgenic tobacco containing *rolB* linked to its own promoter display altered leaf and flower morphology, heterostyly and increased formation of adventitous roots or stems (56). Linking *rolB* to a stronger promoter resulted in additional novel phenotypes, such as a more spherical leaf shape and increased leaf necrosis. Protoplasts from *rolB* plants appear more sensitive to auxin (39, 70).

The *rolB* protein possesses *in vitro* indole-ß-glucosidase activity (14), and this activity can be inhibited by anti-*rolB* antibody, indole and, to a lesser extent, IAA. However, how the *rolB* product acts to influence the intracellular activity of auxin remains unknown.

rolC

Transgenic plants containing the *rolC* gene linked to its own promoter are dwarfed, and the size of the corolla of the flowers is reduced (46). Linking the gene to a stronger promoter results in transgenic plants with greatly reduced apical dominance, pale green lanceolate leaves, and small male sterile flowers (56). The *rolC* product is cell autonomous and located in the cytosol (12a). This suggests that the *rolC* product does not produce a diffusible growth factor and is unlikely to be involved in *de novo* phytohormone synthesis.

The phenotypes of transgenic plants containing the *rolC* gene are suggestive that it has a cytokinin-like function. *In vitro* assays showed that the *rolC* product has a cytokinin-N-glucosidase activity (12), and it is likely that it acts by releasing active cytokinin from inactive conjugates.

While informative, the use of plants engineered to contain individual *rol* genes to study their effect on growth and development is not without complication. First, the promoters used to direct the expression may themselves respond to differing developmental and hormonal cues. Hence uniform levels of expression may not occur throughout the plant. Second, the plant itself may modify the effect produced by the *rol* gene product in a differential, tissue-specific manner. With the later point in mind it is worth remembering that *A. rhizogenes*, when inoculated on aerial plant tissue may, instead of forming hairy roots, form tumors. Such observations may explain the apparent discrepancy between the observations that the *rolC* gene codes for a glucosidase-hydrolyzing cytokinin-N-glucoside, and the *rolB* gene for a glucosidase-hydrolyzing indoxyl-glucoside (12, 14), and the reports by Nilsson et al. (44, 45) who did not detect increases in the free IAA or cytokinin pools in tissues of transgenic *rolB* or *rolC* tobacco plants respectively. However, since Nilsson et al.'s transgenic plants exhibited very marked growth abnormalities, it must be concluded that the overall levels of phytohormone pools cannot be responsible for the observed growth phenomena. Either relatively modest variations in the ratio of free to bound phytohormones, or changes in this ratio in specific cells, or even subcellular compartments, could perhaps explain the observed effects of *rolB* and *rolC* gene products in transgenic tobacco plants. That intracellular pools of IAA can be modified by conjugation was demonstrated by analysis of free- and conjugated forms of IAA in IAA-overproducing transgenic tobacco plants, in comparison with wild-type (58). Consistent with the proposed mechanism of action of *rol* genes is the recent observation (70) that whereas tobacco protoplasts expressing the *rolA* gene require auxin and cytokinin in the growth medium for callus formation, protoplasts expressing *rolB* and *C* form callus in the absence of exogenously applied auxin and cytokinin, respectively.

Mutant Analysis To Study Hormone Effects

It seems reasonable to suspect that there might be also other, molecular ways in controlling the activity of phytohormones. It is noteworthy that the phenotypic changes observed in plants transgenic for auxin biosynthetic genes result from internal modifications in the levels of accumulated auxins, and are not subject to variation and difficulties in interpretation which haunt studies involving the uptake of external applied growth substances. With the potential of using promoters whose expression can be controlled at will, this work raises the opportunity to modify, in a specific tissue at a specific time, the levels of phytohormones, and to assess what effect this might have on growth and development. This goal was reached by an approach in which a cytokinin-synthesizing gene was inactivated by the insertion of a transposable element in its untranslated leader sequence. Random genomic excision in various somatic tissues of tobacco plants transgenic for this interrupted gene, resulted in ectopic production of cytokinin in different tissues and at different times during development leading, for example, to the formation of viviparous leaves and other growth abnormalities (13).

There are apparently relatively few naturally occurring auxin mutants. This has led workers to suspect that such mutants might have a lethal character and hence not be observed routinely. One exception are the recessive *diageotropica* (*dgt*) mutants of tomato (76) and barley (64). The tomato mutant is characterised by a decrease in sensitivity to auxin in excised stem segments, diagravitropic shoot growth, changed leaf morphology and an inability to form lateral roots (76). Photoaffinity labeling with azido-IAA showed that the *dgt* mutant lacks an auxin binding protein considered to be a potential auxin receptor (24).

N. plumbaginifolia mutants that are more sensitive or resistant to auxin have been isolated (59). Each of these mutants display phenotypic changes that reflect modification in the plants response to auxins. Unfortunately, at the moment there is no easy means available for the isolation of the mutated gene sequences.

Arabidopsis, with its small and relatively well characterised genome, has been the target of EMS mutagenesis to generate a variety of mutants which are resistant to normally toxic levels of 2,4-dichlorophenoxyacetic acid (11, 36, 37) or IAA (36, 72). The rationale behind the generation of these mutants is that resistance to auxin might mean that the plants could have been mutated at some point along the system of auxin perception, for example at a receptor, so that the plant no longer is able to sense toxic levels of auxin. Mutants resistant to toxic concentrations of auxins display pleiotropic morphological changes in roots, flowers and leaves (11). For example, *axr1* displays reduced plant height, root gravitropism, hypocotyl elongation and fertility, and *axr2* is defective in the orientation of shoot and root growth. Such phenotypes have been used to infer that the mutations result in changes of the ability of the plants to sense auxin. Interestingly, some mutants appear to be cross resistant to toxic levels of other plant growth substances, and *axr1, axr2* and *aux1* are all resistant to one other plant growth substance (36, 37, 50, 72). While this raises the issue that if the plants are mutated in auxin perception, this pathway might be common for the perception of other non-auxin growth substances. It can also be argued that this may be a consequence of the method of mutant selection and that other steps in hormone response might have been affected (4).

T-DNA Tagging

The ultimate aim of the experiments involving *Arabidopsis* is the isolation of the mutated gene sequences by map-based cloning. However, while feasible with *Arabidopsis*, this technique is involved and time consuming (17).

In order to overcome some of the experimental limitations discussed above, a genetic approach was adopted to study the molecular basis of auxin action, based on T-DNA tagging (69). This capitalizes on the ability of the T-DNA to insert randomly into the plant genome and, in so doing, create an insertion mutation. Generally, such mutations are recessive, as a result of disruption of a functional sequence upon insertion. We decided to create a T-DNA tag that in contrast would produce dominant mutations. The creation of dominant mutations has the advantage that it allows direct selection for a particular phenotype at the level of the primary transformant. To do this a T-DNA tag was constructed that contained the transcriptional enhancer sequence of the 35S RNA-promoter of cauliflower mosaic virus, cloned as a tetramer near the right border sequence, in addition to a hygromycin-resistance selectable marker. The idea was that, following insertion of this T-DNA into the plant genome, flanking plant sequences would become overexpressed. In designing a selection scheme with which to use the vector we considered two aspects from the literature concerning plant growth substances: first, that isolated protoplasts require both auxin and cytokinin for cell division *in vitro* (41); second, that tissue that has been transformed with the T-DNA of wild-type *Agrobacterium tumefaciens* grows *in vitro* in the absence of growth substances. The first observation indicates the importance of growth substances in the cell cycle, whereas the second suggested to us that this importance could be overridden should mutations be generated in either the auxin biosynthetic or perception pathway. This T-DNA tag was therefore used as a mutagen in transformed protoplasts, and transgenic cells were selected *in vitro* for growth as calli in the absence of auxin in the culture media.

Tobacco protoplasts were co-cultivated with *Agrobacteria* containing the tagging vector and selected for growth of transgenic callus in the absence of auxin. The use of co-cultivation as a method of transformation has the advantage that a large number of individual transgenic cells can be generated with ease, and that a biochemical selection for a particular phenotype can be simply applied. In a typical experiment involving 30 million protoplasts, and a transformation frequency of approximately 20%, 12 calli were obtained that grew under selective conditions. Eleven of these could be regenerated into plants that flowered and set seed. Protoplasts derived from mesophyll cells of all the plants regenerated from mutant calli were able to grow and form calli on media devoid of auxin. Plants displaying obvious phenotypic changes showed a tendency to form side shoots, had reductions not only in the rate of root growth, but also in root initiation, and senesced earlier when compared with untransformed control plants. One of the regenerated plant lines, *axi 159*, has been studied in detail (19).

Axi 159 contains a single insert of T-DNA and displays no obvious phenotypic changes. The ability of protoplasts isolated from *axi 159* mesophyll cells to grow in culture in the absence of auxin, genetically co-segregates with the T-DNA insert. The T-DNA in *axi 159* is located on a single 17.5 kb EcoRI fragment of genomic DNA, and this has been rescued from the plant genome as plasmid pHH159. Once transfected into SR1 protoplasts, pHH159 confers the ability of the protoplasts to grow *in vitro* in the absence of auxin indicating that it contains the DNA responsible for this characteristic. This allows deletion analysis to define the position of the gene sequence which carries this out. Using this sequence as a hybridisation probe we isolated a full length cDNA from *axi 159* leaves corresponding to the region in question, which, upon subcloning into an expression vector and reintroduction into protoplasts, directed growth of callus in the absence of auxin. Comparison of the sequence of the cDNA and the genomic sequence indicates that the gene responsible for producing auxin-independent growth upon overexpression, *axi 1*, is approximately 4000 bp long and contains 9 introns. Sequence comparison with data bases reveals no obvious homology with previously characterised proteins.

Currently studies are underway to investigate how the overexpression of *axi 1* might lead to callus formation in the absence of auxin. Preliminary Northern analysis of the pattern of *axi 1* expression indicates that it is expressed primarily in root tissue, and that its expression in freshly isolated protoplasts requires auxin in the media. Our experiments at the moment are aimed at determining the location of the *axi 1* gene product in the cell and its activity.

Our studies have revealed that the overexpression of particular plant genes can lead to the growth of protoplasts in the absence of auxin. It is remarkable that in the majority of the cases studied thus far, this overexpression does not prevent regeneration to fertile plants. The *axi 1* gene normally requires auxin for expression, and possibly codes for a transacting transcription factor that stimulates expression of auxin-dependent target genes. Given that so little is known about the auxin biosynthetic pathway, or about the molecular basis of auxin perception, we anticipate that the further characterization of *axi* genes will provide interesting results.

DISCUSSION

It is possibly ironic that the molecular analysis of hormone action in plants was initiated by the analysis of pathogenic events: tumor or hairy root induction. The genes of the T-DNA of *Agrobacterium*, as well as those of *Pseudomonas*, have provided valuable insight into how plant cells respond to plant growth substances. That being said, as illustrated by the example of the *rol* genes, although the genes themselves have been characterised and the phenotypic changes caused by their action have been well documented, there remains dispute concerning their precise action. This in part may be a reflection of our inability to precisely measure, at a subcellular level, the amount of both active and inactive growth substances, and to relate this to changes in the growth and morphology of the whole plant.

Possibly one of the most significant points to emerge from the study of the action of the *rol* genes is the confirmation, at the molecular level, of the proposal that growth substance conjugation plays an important role in modulating subcellular levels of active growth substances. It would appear that the individual plant cell has a variety of checks and balances at its disposal to finely regulate its hormonal status and requirements. *A. rhizogenes* has exploited this as a tactic in how it induces hairy root disease, although it is intriguing to wonder how the controls of active hormone levels are apparently deactivated during establishment of hairy root. The simplest explanation for this is that other T-DNA genes may act to modify the normal control processes of active hormone regulation in the plant cell. With this in mind it is relevant to note that the product of gene 5 of the T-DNA of *A. tumefaciens* appears to act as an auxin antagonist (33).

Increasingly, we are beginning to understand more of the plant genes involved in hormone action. Proteins which could act as potential receptors are being characterised, and it is significant that one auxin binding protein appears to be an analogue of *rolC*, reinforcing the view that at least cytokinin conjugation and hydrolysis is a relevant process in normal plant development. Further mutant analysis of T-DNA tagged or EMS-mutated genes is likely to increase the trend of characterising plant genes involved in auxin action. While the means of generating mutations is in place, the limitation remains in devising adequate screening of selection procedures to isolate individuals mutated at a defined point in the hormonal signal transduction pathway.

Given the recent upsurge in plant molecular genetics it is to be expected that this field will expand rapidly in the near future.

References

1. Barbier-Brygoo, H., Ephritikhine, G., Klämbt, D., Gishlain, M., Guern, J. (1989) Functional evidence for an auxin receptor at the plasmalemma of tobacco mesophyll protoplasts. Proc. Natl. Acad. Sci. USA 86, 231-237.
2. Barbier-Brygoo, H., Ephritikhine, G., Klämbt, D., Maurel, C., Palme, K., Schell, J., Guern, J. (1991) Perception of the auxin signal at the plasma membrane of tobacco mesophyll protoplasts. Plant J. 1, 83-93.
3. Benning, C. (1986) Evidence supporting a model of voltage-dependent uptake of auxin into *Cucurbita* vesicles. Planta 169, 228-237.
4. Blonstein, A.D., Stirnberg, P., King, P. (1991) Mutants of *N. plumbaginifolia* with specific resistance to auxin. Mol. Gen. Genet. 228, 361-371.
5. Brzobohaty, B., Moore, I., Kristoffersen, P., Bako, L., Campos, N., Schell, J., Palme, J. (1993) Release of active cytokinin by a ß-glucosidase localized to the maize root meristem. Science, in press.
6. Campos, N., Feldwisch, J., Zettl, R, Boland, W., Schell, J., Palme, K. (1991) Identification of auxin-binding proteins using an improved assay for photoaffinity labeling with 5-N$_3$-(7-^{3}H)-indole-3-acetic acid. Technique 3, 69-75.
7. Campos, N., Bako, L., Feldwisch, J., Schell, J., Palme, J. (1992) A protein from maize labeled with azido-IAA has novel ß-glucosidase activity. Plant J. 2, 675-684.
8. Campos, N., Schell, J., Palme, K. (1993) *In vitro* uptake and processing of maize auxin-binding proteins by ER-derived microsomes. Plant Cell Physiol. 34, in press.
9. Cohen, J.D., Bandurski, R.S. (1982) Chemistry and physiology of the bound auxins. Ann. Rev. Plant Physiol. 34, 163-197.
10. Cross, J.W. (1985) Auxin action: The search for the receptor. Plant Cell Environ. 8, 351-359.
11. Estelle, M., Somerville, C. (1987) Auxin resistant mutants of *Arabidopsis thaliana* with altered morphology. Mol. Gen. Genet. 206, 200-206.
12. Estruch, J.J., Chriqui, D., Grossmann, K., Schell, J., Spena, A. (1991) The plant oncogene *rolC* is responsible for the release of cytokinins from glucoside conjugates. EMBO J. 10, 2889-2895.
12a. Estruch, J.J., Parets-Soler, T., Schmülling, T., Spena, A. (1991) Cystolic localisation in transgenic plants of the *rolC* peptide from *Agrobacterium rhizogenes*. Plant Mol. Biol. 17, 547-550.
13. Estruch, J.J., Prinsen, E., Van Onckelen, H., Schell, J., Spena, A. (1991) Viviparous leaves produced by somatic activation of an inactive cytokinin-synthesizing gene. Science 254, 1364-1367.
14. Estruch, J.J., Schell, J., Spena, A. (1991) The protein encoded by the *rolB* plant oncogene hydrolyses indole glucosides. EMBO J. 10, 3125-3128.
15. Feldwisch, J., Zettl, R., Hesse, F., Schell, J., Palme, K. (1992) An auxin-binding protein is localized to the plasma membrane of maize coleoptile cells: Identification by photoaffinity labeling and purification of a 23-kDa polypeptide. Proc. Natl. Acad. Sci. USA 89, 475-479.
16. Felle, H., Peters, W., Palme, K. (1991) The electrical response of maize to auxins. Biochim. Biophys. Acta 1064, 199-204.
17. Gibson, S.I., Somerville, C. (1992) Chromosome walking in *Arabidopsis thaliana* using yeast artificial chromosomes. *In*: Methods in *Arabidopsis* Research, pp 119-143, Koncz, C., Chua, N.-H., Schell, J., eds. World Scientific, Singapore.
18. Goldsmith, M.H.M. (1982) A saturable site responsible for polar transport of indole-3-acetic acid in sections of maize coleoptiles. Planta 155, 68-75.

19. Hayashi, H., Czaja, I., Schell, J., Walden, R. (1992) Activation of a plant gene by T-DNA tagging: auxin-independent growth *in vitro*. Science 258, 1350.
20. Hertel, R. (1987) Auxin transport: Binding of auxins and phytotropins to the carriers. Accumulation into and efflux from membrane vesicles. *In*: Plant Hormone Receptors. NATO ASI Series, Vol. H10, pp 81-92. Klämbt, D., ed. Springer Verlag, Berlin, Heidelberg.
21. Hertel, R., Lomax, T.L., Briggs, W.R. (1983) Auxin transport in membrane vesicles from *Cucurbita pepo* L. Planta 157, 193-201.
22. Hesse, T., Feldwisch, J., Balshüsemann, D., Bauw, G., Puype, M., Vandekerckhove, J., Löbler, M., Klämbt, D., Schell, J., Palme, K. (1989) Molecular cloning and structural analysis of a gene from *Zea mays* (L.) coding for a putative receptor for the plant hormone auxin. EMBO J. 8, 2453-2461.
23. Hicks, G.R., Rayle, D.L., Jones, A.M., Lomax, T.L. (1989) Specific photoaffinity labeling of two plasma membrane polypeptides with an azido auxin. Proc. Natl. Acad. Sci. USA 86, 4948-4952.
24. Hicks, G.R., Rayle, D.L., Lomax, T. (1989) The Diageotropica mutant of tomato lacks high specific activity auxin binding sites. Science 245, 52-54.
25. Inohara, N., Shimomura, S., Fukui, T., Futai, M. (1989) Auxin-binding protein located in the endoplasmic reticulum of maize shoots: Molecular cloning and complete primary structure. Proc. Natl. Acad. Sci. USA 86, 3564-3568.
26. Inzé, D., Follin, A., Van Lijsebettens, M., Simoens, C., Genetello, C., Van Montagu, M., Schell, J. (1984) Genetic analysis of the individual T-DNA genes of *Agrobacterium tumefaciens*; further evidence that two genes are involved in indole-3-acetic acid synthesis. Mol. Gen. Genet. 194, 265-274.
27. Jacobs, M., Gilbert, S.F. (1983) Basal localization of the presumptive auxin transport carrier in pea stem cells. Science 220, 1297-1300.
28. Jones, A.M. (1990) Do we have the auxin receptor yet? Physiol. Plant. 80, 154-158.
29. Jones, A.M., Melhado, L.L., Ho, T.-H., Leonhard, N.J. (1984) Azido auxins. Quantitative binding data in maize. Plant Physiol. 74, 295-301.
30. Jones, A.M., Venis, M.A. (1989) Photoaffinity labeling of indole-3-acetic acid-binding proteins in maize. Proc. Natl. Acad. Sci. USA 86, 6153-6156.
31. Jones, A.M., Herman, E.M. (1993) KDEL-containing auxin-binding protein is secreted to the plasma membrane and cell wall. Plant Physiol. 101, 595-606.
32. Klee, H., Horsch, R.B., Hinchee, M., Hein, M.B., Hoffmann, N.L. (1987) The effects of overproduction of two *Agrobacterium tumefaciens* T-DNA auxin biosynthetic gene products in transgenic petunia plants. Genes Dev. 1, 86-89.
33. Körber, H., Strizhov, N., Staiger, D., Feldwisch, J., Olsson, O., Sandberg, G., Palme, K., Schell, J., Koncz, C. (1991) T-DNA gene 5 of *Agrobacterium* modulates auxin response by autoregulated synthesis of a growth hormone antagonist in plants. EMBO J. 10, 3983-3991.
34. Levesque, H., Delepelaire, P., Rouzé, P., Slightom, J., Tepfer, D. (1988) Common evolutionary origin of the central portions of the Ri TL-DNA of *Agrobacterium rhizogenes* and the Ti T-DNA of *Agrobacterium tumefaciens*. Plant Mol. Biol. 11, 781-744.
35. Libbenga, K.R., Mennes, A.M. (1987) *In*: Plant Hormones and their Role in Plant Growth and Development, pp 194-221. Davies, P.J., ed. Martinus Nijhoff Publishers, Dordrecht.
36. Lincoln, C., Britton, J.H., Estelle, M. (1990) Growth and development of the *axr1* mutants of *Arabidopsis*. Plant Cell 2, 1071-1080.
37. Maher, E.P., Martindale, S.J.B. (1980) Mutants of *Arabidopsis* with altered responses to auxin and gravity. Biochem. Genet. 18, 1041-1053.
38. Martin-Tanguy, J., Tepfer, D., Paynot, M., Burtin, D., Heisler, L., Martin, C. (1990) Inverse relationship between polyamine levels and the degree of phenotypic alteration

induced by the root inducing left-hand transferred DNA from *Agrobacterium rhizogenes*. Plant Physiol. 92, 912-918.

39. Maurel, C., Barbier-Brygoo, H., Brevet, J., Spena, A., Tempé, J., Guern, J. (1991) *Agrobacterium rhizogenes* T-DNA genes and sensitivity of plant protoplasts to auxins. *In*: Advances in Molecular Genetics of Plant-Microbe Interactions, pp 343-351. Hennecke, H., Verma, D.P.S., eds. Kluwer Academic Publishers, Dordrecht.

40. Michael, T., Spena, A. (1993) The plant oncogenes *rolA, B* and *C* from *Agrobacterium rhizogenes*; effects on morphology, development and hormone metabolism. *In*: Agrobacterium Protocols, Methods in Molecular Biology Series, Gartland, K., Davey, M., eds. Humana Press, New Jersey (in press).

41. Nagata, T., Takebe, I. (1970) Cell wall regeneration and cell division in isolated tobacco mesophyll protoplasts. Planta 92, 301-308.

42. Napier, R.M., Venis, M. (1991) From auxin-binding protein to plant hormone receptor? Trends Biochem. Sci. 16, 72-75.

43. Narayanan, K.R., Mudge, K.W., Poovaiah, B.W. (1981) Demonstration of auxin binding to strawberry fruit membranes. Plant Physiol. 68, 1289-1293.

44. Nilsson, O., Crozier, A., Schmülling, T., Sandberg, G., Olsson, O. (1993) Indole-3-acetic acid homeostasis in transgenic tobacco plants expressing the *Agrobacterium rhizogenes rolB* gene. Plant J. 3, 681-689.

45. Nilsson, O., Moritz, T., Imbault, N., Sandberg, G., Olsson, O. (1993) Hormonal characterization of transgenic tobacco plants expressing the *rolC* gene of *Agrobacterium rhizogenes* T_L-DNA. Plant Physiol. 102, 363-371.

46. Oono, Y., Handa, T., Kanaya, K., Uchimiya, H. (1987) The TL-DNA gene of Ri plasmids responsible for dwarfness of tobacco plants. Jpn. J. Genet. 62, 501-505.

47. Palme, K., Feldwisch, J., Hesse, T., Bauw, G., Puype, M., Vandekerckhove, J., Schell, J. (1990) Auxin binding proteins from maize coleoptiles: Purification and molecular characterization. *In*: Hormone Perception and Signal Transduction in Animals and Plants, pp 299-313. Roberts, J., Kirk, C., Venis, M., eds. Society for Experimental Biology.

48. Palme, K., Hesse, T., Moore, I., Campos, N., Feldwisch, J., Garbers, C., Hesse, F., Schell, J. (1991) Hormonal modulation of plant growth: The role of auxin perception. Mech. Dev. 33, 97-106.

49. Palme, K., Hesse, T., Campos, N., Garbers, C., Yanofsky, M.F., Schell, J. (1992) Molecular analysis of an auxin binding protein gene located on chromosome 4 of *Arabidopsis*. Plant Cell 4, 193-201.

50. Pickett, F.B., Wilson, A.K., Estelle, M. (1990) The *aux1* mutation of *Arabidopsis* confers both auxin and ethylene resistance. Plant Physiol. 94, 1462-1466.

51. Poovaiah, B.W. (1982) Strawberry fruit as a model system to study target tissue specificity and the physiological relevance of auxin-binding. Plant Physiol. 69, Suppl., 151.

52. Romano, C.P., Hein, M.B., Klee, H.J. (1991) Inactivation of auxin in tobacco transformed with the indoleacetic acid-lysine synthetase gene of *Pseudomonas savastonoi*. Genes and Devel. 5, 438-446.

53. Rubery, P.H. (1990) *In*: Hormone Perception and Signal Transduction in Animals and Plants, pp 119-146. Roberts, J., Kirk, C., Venis, M., eds. Company of Biologists Limited, Cambridge.

54. Rück, A., Palme, K., Venis, M.A., Napier, R.M., Felle, H.H. (1993) Patch-clamp analysis establishes a role for an auxin binding protein in the auxin stimulation of plasma membrane current in *Zea mays* protoplasts. Plant J. 4, 41-46.

55. Schliemann, W. (1991) Zum Konzept der reversiblen Konjugation bei Phytohormonen. Naturwissenschaften 78, 392-401.

56. Schmülling, T., Schell, J., Spena, A. (1988) Single genes from *Agrobacterium rhizogenes* influence plant development. EMBO J. 7, 2621-2629.

57. Schwob, E., Choi, S.-Y., Simmons, C., Migliaccio, F., Ilag, L., Hesse, T., Palme, K., Söll, D. (1993) Molecular analysis of three maize 22 kDa auxin binding protein genes - transient promoter expression and regulatory regions. Plant J. 4, 423-432.
58. Sitbon, F., Östin, A., Sundberg, B., Olsson, O., Sandberg, G. (1993) Conjugation of indole-3-acetic acid (IAA) in wild-type and IAA-overproducing transgenic tobacco plants, and identification of the main conjugates by frit-fast atom bombardment liquid chromatography-mass spectrometry. Plant Physiol. 101, 313-320.
59. Souza de, L., King, P.J. (1991) Mutants of *Nicotiana plumbaginifolia* with increased sensitivity to auxin. Mol. Gen. Genet. 231, 65-75.
60. Spena, A., Estruch, J.J., Schell, J. (1992) On microbes and plants: new insights in phytohormonal research. Current Opinion Biotech. 3, 159-163.
61. Spena, A., Prinsen, E., Fladung, M., Schulze, S.C., Van Onckelen, H. (1991) The indoleacetic acid-lysine synthetase gene of *Pseudomonas syringae* subsp. *savastanoi* induces developmental alterations in transgenic tobacco and potato plants. Mol. Gen. Genet. 227, 205-212.
62. Spena, A., Schmülling, T., Koncz, C., Schell, J. (1987) Independent and synergistic activity of *rolA, B* and *C* loci in stimulating abnormal growth in plants. EMBO J. 6, 3891-3899.
63. Sun, L.-Y., Monneuse, M.-O., Martin-Tanguy, J., Tepfer, D. (1991) Changes in flowering and the accumulation of polyamines and hydroxycinnamic acid-polyamine conjugates in tobacco plants transformed by the *rolA* locus from the Ri TL-DNA of *Agrobacterium rhizogenes*. Plant Science 80, 145-156.
64. Tagliani, L., Nissen, S., Blake, T.K. (1986) Comparison of growth, exogenous auxin sensitivity and endogenous indole-3-acetic acid content in roots of *Hordeum vulgare* L. and an agravitropic mutant. Biochem. Genet. 24, 839-848.
65. Trewavas, A. (1980) An auxin induces the appearance of auxin-binding activity in artichoke tubers. Phytochemistry 1, 1303-1308.
66. Venis, M., ed. (1985) Hormone binding sites in plants. Longmann, New York.
67. Venis, M.A., Napier, R.M., Barbier-Brygoo, H., Maurel, C., Perrot-Rechenmann, C., Guern, J. (1992) Antibodies to a peptide from the maize auxin-binding protein have auxin agonist activity. Proc. Natl. Acad. Sci. USA 89, 7208-7212.
68. Vreugdenhil, D., Burgers, A., Harkes, P.A.A., Libbenga, K.R. (1981) Modulation of the number of membrane-bound auxin-binding sites during the growth of batch-cultured tobacco cells. Planta 152, 415-419.
69. Walden, R., Hayashi, H., Schell, J. (1991) T-DNA as a gene tag. Plant J. 1, 281-288.
70. Walden, R., Czaja, I., Schmülling, T., Schell, J. (1993) *Rol* genes alter hormonal requirements for protoplast growth and modify the expression of an auxin responsive promoter. Plant Cell Rep. 12, 551-554.
71. Walton, D.J., Ray, P.M. (1981) Evidence for receptor functions of auxin-binding sites in maize; red light inhibition of mesocotyl elongation and auxin binding. Plant Physiol. 68, 1334-1338.
72. Wilson, A., Pickett, F.B., Turner, J., Estelle, M. (1990) A dominant mutation in *Arabidopsis* confers resistance to auxin, ethylene and abscisic acid. Mol. Gen. Genet. 222, 377-383.
73. Zambryski, P. (1992) Chronicles from the Agrobacterium-plant cell transfer story. Ann. Rev. Plant Physiol. Plant. Mol. Biol. 43, 465-490.
74. Zettl, R., Campos, N., Feldwisch, J., Schell, J., Boland, W., Palme, K. (1991) Synthesis and application of 5'-azido-[3,6-^{3}H$_2$]naphthylphthalamic acid, a photo-activatable probe for auxin efflux carrier proteins. Technique 3, 151-158.
75. Zettl, R., Feldwisch, J., Boland, W., Schell, J., Palme, K. (1992) 5'-Azido-[3,6-^{3}H$_2$]-1-naphthylphthalamic acid, a photoactivatable probe for naphthylphthalamic acid receptor proteins from higher plants: Identification of a 23-kDa protein from maize coleoptile plasma membranes. Proc. Natl. Acad. Sci. USA 89, 480-484.

76. Zobel, R.W. (1973) Some physiological characteristics of the ethylene-requiring tomato mutant diageotropica. Plant Physiol. 52, 385-389.

77. Yu, L.-W., Laza Rus, C.M. (1991) Structure and sequence of an auxin-binding protein gene from maize (*Zea mays* L.). Plant Mol. Biol. 16, 925-930.

E4. Ethylene Genes and Fruit Ripening

Steve Picton, Julie E. Gray and Don Grierson.

AFRC Research Group in Plant Gene Regulation, Department of Physiology and Environmental Science, University of Nottingham, Sutton Bonington Campus, Loughborough, LE12 5RD, UK.

INTRODUCTION

Ethylene was identified as affecting plant growth at the start of this century (42), and is produced in at least trace amounts by almost all higher plants. It is involved in the control and coordination of a diverse range of plant growth and developmental processes including seed germination, root growth and development, vegetative growth, flower development, the ripening of fruits and the senescence and abscission of flower, leaf and fruit organs (Chapter G2). The hormone thus exerts its influence throughout the entire developmental progression of plants. Ethylene has also been implicated in the modulation of responses of plants to a wide range of biotic and abiotic stresses. The role of ethylene in regulating the ripening of tomato fruits has received particular attention, partly because of its intrinsic scientific interest, but also for reasons of experimental convenience and the economic importance of the tomato fruit as a major food crop.

Fruits are classified as climacteric or non-climacteric according to their respiratory output at the onset of the ripening process and the ability of ethylene or its analogue, propylene, to stimulate the autocatalytic production of ethylene in climacteric fruits (39). Classic climacteric fruits such as bananas, apples, pears and tomatoes show a clear increase in respiratory CO_2 at the onset of ripening concomitant with a dramatic increase in the rate of ethylene evolution. Non-climacteric fruits, such as the citrus family and strawberries, show no increase in respiratory activity and do not display an increase in ethylene evolution at the onset of ripening. Changes associated with the ripening process have been widely studied in both climacteric and non climacteric fruits. These generally, but not always, involve alteration in colour, flavour, aroma and texture (Fig. 1) that occur by a variety of different biochemical pathways in different species. The characterisation of ripening at the molecular level is, at present, most advanced in a small group of climacteric fruits such as tomatoes, avocados and melons.

That ethylene plays a pivotal role in the control and coordination of the ripening of climacteric fruits is in little doubt (6, 7, 16, 17). It is known that ethylene production from one fruit may advance the onset of ripening of neighbouring fruits and many tropical fruits show a highly coordinated pattern of ripening (1). Delay in the onset or progression of ripening may be

P. J. Davies (ed.), Plant Hormones, 372–394.

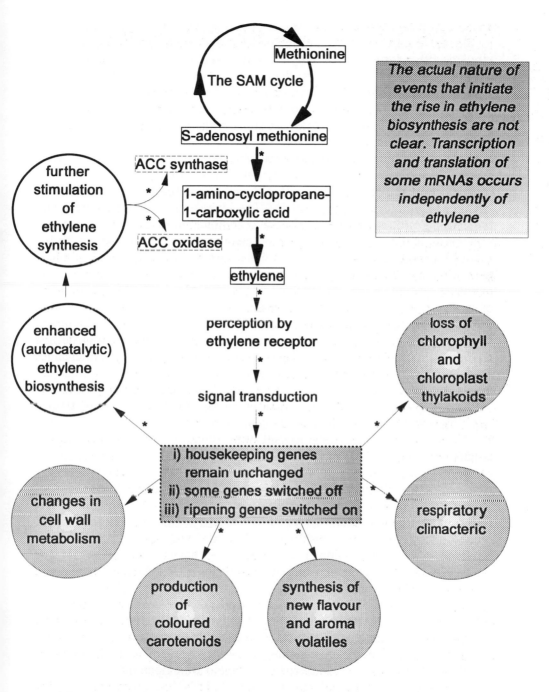

* Each of these pathways may be modified independently with antisense-RNA constructs or naturally occurring mutations.

Fig. 1. Illustration of the interactions between ethylene biosynthesis, tomato fruit ripening and the associated changes in gene expression.

achieved by constant ventilation under hypobaric pressures, a process believed to aid the escape of ethylene and reduce the sensitivity of the fruits to ethylene (9). Inhibitors of ethylene synthesis, such as aminoethoxyvinyl glycine (AVG), can be used to delay ripening (8) whilst treatment of fruit with inhibitors of ethylene action or perception, such as norbornadiene (35) or silver (11, 22, 53) can delay both the onset and progression of the ripening process. The treatment of climacteric fruits with silver at advanced stages of ripening impairs the further progression of the process (11) and indicates that the role of ethylene must be of a regulatory nature and not simply a switch for the ripening process. The widely held view is that ethylene is involved in regulation of fruit ripening, either directly or indirectly, by coordinating changes in gene expression (Fig. 1). The interaction between ethylene and the ripening process is often exploited in commercial horticulture to achieve uniform ripening or to initiate the ripening of mature harvested fruits. Examples include controlled atmosphere storage with decreased ethylene levels to slow the ripening process during the shipping of bananas and the exposure of mature fruit to ethylene in order to advance and coordinate the onset of the cascade of ripening events.

The tomato (*Lycopersicon esculentum)* has provided an ideal model system to study, at the molecular level, interactions between ethylene and developmental changes (Fig. 1) that occur during fruit ripening. All changes associated with the ripening process take place after cell division and expansion are completed. The tomato plant has a relatively small genome (7.1×10^5 kbp), an extensive genetic map and a number of well defined ripening mutants are available in isogenic backgrounds. The plant can be easily grown and fruit obtained all year round. Tomato explants are amenable to stable *Agrobacterium*-mediated transformation (2, 13, 15, 20, 33, 38, 43, 57) thus allowing *in planta* manipulation of gene expression. In the past decade the application of recombinant DNA technology to the field of fruit ripening has led to the construction of a number of tomato fruit cDNA libraries that have been screened to identify ripening clones whose homologous genes have been subsequently used to study interactions between ethylene and gene expression (15). These studies have indicated that ethylene is causally related to changes in gene expression at both transcriptional and post-transcriptional levels.

This chapter will concentrate specifically on the isolation, cloning and identification of genes involved in the ethylene biosynthetic pathway of tomato and how manipulation of these genes with antisense RNA technology has provided further understanding of the role of ethylene in the developmental changes characteristic of tomato fruit ripening.

THE ETHYLENE BIOSYNTHETIC PATHWAY OF HIGHER PLANTS

The ethylene biosynthetic pathway of higher plants has been extensively studied in the past two decades and ethylene synthesis has been shown to proceed from the S-adenosyl-L-methionine (SAM) cycle via the intermediate SAM (67). The pathway from SAM to ethylene is catalysed exclusively in higher plants by the enzymes 1-aminocyclopropane-1-carboxylic (ACC) synthase and ACC oxidase (formerly referred to as the ethylene-forming enzyme or EFE) (Figs. 1 and 7). A more detailed discussion of ethylene biosynthesis is provided in Chapter B4. Data obtained with excised tissues has led to the conclusion that ACC oxidase activity is constitutive and that ACC synthase therefore provides the rate limiting step for ethylene biosynthesis; however, small changes in ACC oxidase activity may provide fine tuning of ethylene evolution (67), evidenced by the induction of ACC oxidase activity by ethylene in immature tomato fruit (36). However, the dramatic induction of ACC oxidase mRNA at the onset of tomato fruit ripening (56) and the demonstration that ACC oxidase mRNA accumulation is induced very rapidly in tissues in response to a variety of wounding stimuli (25) means it may now be pertinent to re-examine some of the early data on ACC oxidase activity obtained with isolated, and thus wounded, tissues, since both ACC oxidase and ACC synthase enzymes may play a regulatory role in the evolution of ethylene from ripening tomato fruit (47). Recent interest in several laboratories around the world has centred on the biochemical purification, isolation and subsequent cloning of enzymes responsible for steps in ethylene synthesis and the utilization of antisense RNA both to identify the function of cloned mRNAs and to manipulate ethylene biosynthesis and ripening in transgenic plants.

THE USE OF ANTISENSE RNA CONSTRUCTS TO DOWN REGULATE ENDOGENOUS PLANT GENE EXPRESSION: AN OVERVIEW

Amongst the many advantages of the tomato plant for molecular analysis of developmental gene regulation is the availability of a routine method to achieve stable integration of recombinant genes using *Agrobacterium* mediated transformation. Transformation with antisense RNA transgenes has proved a powerful technique to selectively down-regulate the expression of a targeted endogenous gene (15, 29, 30, 57, 62) and has proved particularly useful in studies in ethylene synthesis and action in tomato fruits. Expression of a gene or part of a gene placed in an inverse orientation downstream of a constitutive promoter sequence leads to the production of an mRNA that is an "antisense RNA" complementary to the normal endogenous mRNA transcript (Fig. 2). The presence of such an antisense RNA molecule in

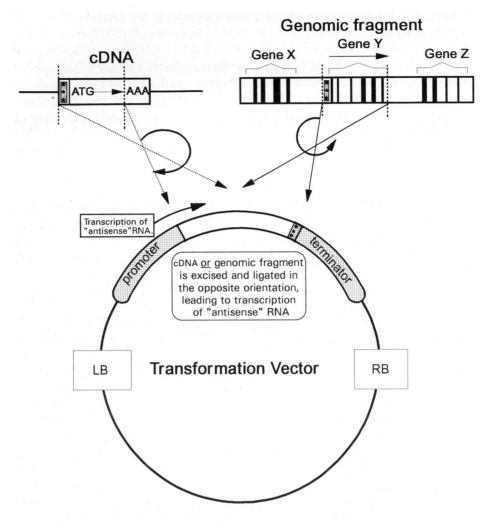

Fig. 2. Schematic construction of a vector to transcribe an antisense RNA *in planta*. A transformation vector containing a complete or partial cDNA or genomic sequence inserted in the "antisense" or reversed orientation is prepared. Following transformation of explants and stable integration of the sequence (contained between the left and right borders [LB and RB] of the vector) into the plant genome, transcription of the "antisense" RNA leads to reduction in accumulation of the homologous endogenous mRNA and thus disruption of expression of the endogenous homologous gene.

transgenic plants leads to a dramatic reduction in the level of the homologous endogenous mRNA, and therefore in the accumulation of its encoded polypeptide, thus leading to a reduction in the level of expression of the gene. Although the mechanism responsible for this reduction is as yet unclear, it is postulated that it involves specific nucleic acid base pairing at the DNA or RNA level which leads to the observed gene inactivation and thus creates a targeted mutation. Based on the results of many experimental reports, it is clear that the method is highly gene specific, leading to inactivation of only the endogenous gene or highly homologous members of a multigene family. Antisense RNA constructs have been particularly useful in the identification of genes for ACC oxidase (20, 21) and for regulating ethylene biosynthesis and thus altering fruit ripening (5, 20, 46, 47).

CLONING AND IDENTIFICATION OF ACC OXIDASE FROM TOMATO FRUIT

Cloning

The enzyme activity responsible for the terminal step in ethylene biosynthesis, ACC oxidase, has been studied *in vivo* by the application of its substrate, ACC, to plant tissue. More detailed study of the enzyme had been hampered by the lack of a cell free assay system. Activity of the enzyme was rapidly lost on tissue homogenisation (67), and although maintenance of enzyme activity was reported in isolated leaf mesophyll vacuoles (19) and membrane isolates from kiwi fruit extracts (41), this activity was very low (19, 41, 48). The observation that the loss of membrane integrity abolished the activity (19, 37) supported the hypothesis that ACC oxidase activity depended on membrane integrity (67) and may require the maintenance of membrane potentials (27). Cell-free systems have been isolated that display an ability to convert ACC to ethylene but the activity has been shown to be due to non-enzymic reactions. John (28) summed up the situation with study of the ACC oxidase enzyme stating "All in all, little progress was being made in characterising EFE. Until, that is, the molecular biologists entered the scene".

In early investigations on the molecular changes during tomato fruit ripening, differential screening of a ripe tomato fruit cDNA library with radioactive probes derived from unripe and ripe fruit mRNA identified 19 non-homologous groups of "ripening-related" clones for mRNAs that accumulated preferentially during ripening (55). Extensive characterisation of these and other clones led to the identification of mRNAs and genes involved in cell wall metabolism, carotenoid biosynthesis and ethylene synthesis and enabled their role in ripening and senescence to be more clearly defined (15). One clone, originally designated pTOM 13 (55), hybridized to an mRNA that accumulated at the onset of fruit ripening and was also rapidly induced following the wounding of unripe fruit and tomato leaves (56), both processes that are accompanied by a dramatic rise in ethylene synthesis. The TOM 13 cDNA hybrid-selected an mRNA that encoded a 35kD protein (55). Utilizing *in vivo* labelling studies, a protein of equivalent size was shown to be synthesised rapidly in wounded fruit tissue (56). Based on these observations, Smith *et al*. (56) suggested that the TOM 13 mRNA may encode one of the enzymes involved in ethylene biosynthesis.

The TOM13 cDNA was used to isolate homologous genomic sequences from a tomato genomic library and the TOM 13 mRNA was shown to originate from one of a small multigene family of 3 members (23, 24, 25) (Table 1). Sequencing of the TOM 13 cDNA (24), a related genomic sequence GTOM A (23) and other homologous genomic clones designated GTOM 17 (25) and ETH 1 (34) (Table 1) gave no clue as to the identity of the products or their biochemical function.

Identification

In 1990 it was reported that inhibition of endogenous TOM 13 mRNA accumulation by the stable integration and expression of an antisense TOM 13 RNA in transgenic tomato plants led to a transgene-dosage dependent reduction in ethylene synthesis in both wounded leaf tissue (Fig. 3A) and ripening fruit (Fig. 3B) obtained from the transgenic plants (20). These results clearly implicated the TOM 13 encoded product in playing a role in either ethylene production, perception or metabolism (20). Purification of ACC synthase from tomato fruits had resulted in estimations of its molecular weight as 50kD (3), 65-67kD (40) and 45kD (63) and thus the TOM 13 product, previously shown to be 35kD (55, 56), appeared to be too small to encode an ACC synthase enzyme. In order to test whether the TOM 13 product could encode ACC oxidase, its activity was assayed in wounded transgenic leaf tissue and was shown to be inhibited in an antisense gene-dosage dependent manner (20) (Fig. 3C). Since a proposed reaction mechanism for ACC oxidase involved the hydroxylation of ACC (67), sequence homology of the TOM 13 clone with a flavanone-3-hydroxylase from *Antirrhinum majus* (20) provided further support that the TOM 13 product played a role in the ACC oxidase system. Final verification of the function of the TOM 13 product came following the expression of a full length reconstructed TOM 13 cDNA in the sense orientation (pRC 13, Table 1) in *Saccharomyces cerevisiae* (21). Initial attempts to express the TOM 13 encoded product in both *E. coli* and yeast were unsuccessful due to cloning artifacts present in the 5' region of the cDNA. These cloning artifacts were identified by direct sequencing of the TOM 13 mRNA and a correct, reconstructed TOM 13 clone, pRC 13, was created by ligating a 5' genomic region of the corresponding gene, *ETH 1* (34), to a fragment of the original TOM 13 clone. Cultures of the transformed yeast containing the reconstructed TOM 13 sequence in an expression cassette were able to convert the ethylene precursor ACC to ethylene whilst control cultures were unable to catalyse the reaction; furthermore, the recombinant enzyme displayed many characteristics of the ACC oxidase enzyme found in plant tissues, being stereospecific in its ability to convert the ACC analogue, 1-amino-2-ethylcyclopropane-1-carboxylic acid, to ethylene (21, 67) and being severely inhibited by cobalt ions and 1,10-phenanthroline (21). This data verified that the TOM 13 product constituted the entire functional polypeptide of the elusive ACC

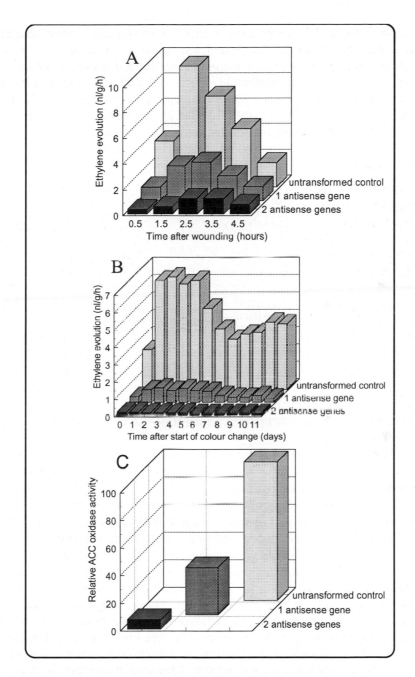

Fig. 3. Ethylene evolution and ACC oxidase activity in normal and ACC oxidase antisense tomato plants. A. Ethylene evolution from wounded leaves. Leaf discs were obtained from untransformed controls or plants containing one or two copies of an ACC oxidase antisense gene construct. B. Ethylene evolution during fruit ripening. Ethylene evolution from detached control and transformed fruit, as indicated, measured daily from the onset of colour change. C. Relative ACC oxidase activity measured in leaf discs obtained from control and transformed plants as indicated. (20).

oxidase, and not simply a part of the enzymatic system responsible for catalysis of ACC to ethylene. The enzymatic activity expressed in yeast was also shown to be stimulated by ascorbate and to be inhibited by Fe-chelating agents (5, 21), providing additional evidence that the enzyme catalysing this step belonged to a group of hydroxylase enzymes. Subsequently, a further TOM 13-like sequence (pHTOM 5, Table I) was cloned and its biochemical function verified by heterologous expression in *Xenopus* oocytes (59). The biochemical nature of ACC oxidase had previously been difficult to study since extraction methods had resulted in almost complete loss of activity, suggesting that the enzyme was membrane bound and required membrane integrity for activity. With the knowledge that ACC oxidase is a hydroxylase type enzyme, an extraction method preserving activity by the addition of ascorbate has now been successfully used to extract, completely solubilise and assay, *in vitro*, ACC oxidase activity from melon fruits (66). Subsequent analysis of the predicted amino acid sequence of ACC oxidase and comparison with other related enzymes indicated that it is one of a family of dioxygenases that require Fe and ascorbate (49). This series of experiments not only allowed the identification of the elusive ACC oxidase but also demonstrated the power of the "reverse genetics" approach in identifying cloned sequences of unknown biochemical function. A similar approach has been used successfully to identify a component of the tomato fruit carotenoid biosynthetic pathway (2).

CLONING AND IDENTIFICATION OF ACC SYNTHASE GENES FROM TOMATO

The second enzyme to be identified in the pathway to ethylene biosynthesis in tomato, ACC synthase, was cloned and identified by a more conventional molecular and biochemical approach. ACC synthase has been shown to have a short half life *in vivo* and proved very unstable following purification. The enzyme is present in tissues at only very low levels, estimated at less than 0.0005% of the total soluble protein of tomato pericarp extracts (3, 64). This hampered progress but led to a plethora of published extraction and purification methods including physical and chemical treatments to over-induce the enzyme (26). Several putative cDNAs encoding ACC synthase were originally isolated from a zucchini cDNA expression library by screening with antibodies (54). The identification of a clone encoding the ACC synthase enzyme and proof of its biochemical function as an ACC synthase was confirmed by expression in heterologous systems (54). The first published ACC synthase DNA sequences resulted from the 5000 fold purification of ACC synthase from tomato pericarp. The ACC synthase activity correlated with a 45kD protein from which peptide sequences were obtained. Degenerate oligonucleotide probes were synthesized utilizing the obtained amino acid sequence and used to probe a cDNA library prepared

Table 1 Tomato ACC synthase and ACC oxidase clones

Clone name	Insert size(bp)	Polypeptide size (kD)	Source	Reference
ACC Synthase. (Genes *A CS 1A*, *A CS1B*, *A CS 2*, *A CS 3*, *A CS 4* and *A CS 5* : Enzyme name, ACS)				
pcVV4A[a]	1846	54.7	LiCL/ Wounded Fruit pericarp cDNA	Van Der Staeten et al. [67].
pcVV4B[b]	420	N/A	LiCL/Wounded Fruit pericarp cDNA	
ptACC2[a]	1775	54.69	Ripe tomato fruit cDNA	Rottmann et al. [54].
ptACC4[b]	1616	53.5	Ripe tomato fruit cDNA	
L-ACC2[a]	N/A	54.6	Tomato genomic DNA	
L-ACC1A[d]	N/A	54.8	Tomato genomic DNA	
L-ACC1B[e]	N/A	54.5	Tomato genomic DNA	
L-ACC3[c]	N/A	53	Tomato genomic DNA	
L-ACC4[b]	N/A	53.5	Tomato genomic DNA	
pBEN11[b]	288	N/A	PCR from unwounded fruit pericarp cDNA	Olsen et al. [47].
ACCSYN1[a]		54.7		
ACCSYN2[h]	1652	53.5	Unwounded Tomato fruit pericarp cDNA	
pBEN17[a]	267	N/A	PCR product	
pBEN18[b]	209	N/A	PCR product	
pBTAS1[a]	268	N/A	PCR from fruit mRNA/cell suspension culture	Yip et al. [70].
pBTAS2[c]	271	N/A	PCR from Tomato cell suspension culture	
pBTAS3[f]	271	N/A	PCR from Tomato cell suspension culture	
pBTAS4[b]	246	N/A	PCR from Tomato fruit mRNA	

Following the recent suggestions of H.Kende [34], clones have been grouped and assigned to specific genes: [a], sequences derived from *A CS2*; [b], sequences derived from *A CS4*; [c], sequences derived from *A CS3*. [d]LE-ACC1A is derived from *A CS1A* and [e]LEACC1B is derived from *A CS1B*. [f]pBTAS3 is derived from *A CS5*.

Clone name	Insert size(bp)	Polypeptide size (kD)	Source	Reference
ACC oxidase (formerly the ethylene-forming enzyme. Genes, *A CO 1*, *A CO 2*, *A CO 3*: Enzyme name, ACO)				
pTOM 13[A]	1370	33.5	Ripe tomato fruit pericarp cDNA	Slater et al. [58] and Holdsworth et al. [24].
pRC 13[A]		33.5	Reconstructed pTOM 13	Hamilton et al. [20, 21].
pGTOMA[B]	3526	N/A	Tomato genomic DNA	Holdsworth et al. [23, 25].
pGTOMB[C]		N/A	Tomato genomic DNA	Holdsworth et al. [25].
pGTOM17[A]		N/A	Tomato genomic DNA	Holdsworth et al. [25].
pETH 1[A]	1800	N/A	PCR product from genomic DNA	Koch et al. [36].
pHTOM5[C]	1035	35.9	Elicitor treated Tomato suspension culture cDNA	Spanu et al. [62].

As above, clones have been grouped and assigned to specific genes: [A], sequences derived from *A CO1*(formally *ETH 1*); [B], sequences derived from *A CO2* (formally *ETH 2*); [C], sequences derived from *A CO 3* (formally *ETH 3*) from "induced" tomato fruit pericarp resulting in the isolation of two non-identical ACC synthase clones (pcVV4a and pcVV4b, Table 1) (64). The largest of the obtained clones contained an insert of 1.9 kbp (Table 1) with a single open reading frame coding for a protein of approximately 55kD (64). As with the zucchini clone, the entire coding region was expressed in *Escherichia coli* and the recombinant peptide was used to raise polyclonal antibodies. The polyclonal antibodies were shown to immunoprecipitate and immunoinhibit ACC synthase activity from a tomato extract, thus demonstrating the biochemical identity of the clone (64). Following these early reports there has been much interest in cloning further ACC synthase genes from other plants and specific tissues in order to study the differential regulation of the ACC synthase. Published data now shows that ACC synthase is encoded by a large and divergent multigene family (60). In tomato there are at least six ACC synthase genes that show differential expression, with at least two members of the gene family being expressed during the normal fruit ripening process (60, 61).

THE USE OF ANTISENSE RNA TO INHIBIT ENZYMES INVOLVED IN ETHYLENE BIOSYNTHESIS

With the isolation and functional identification of sequences encoding enzymes that catalyse the two key steps in ethylene biosynthesis, ACC synthase and ACC oxidase, much work has centred on the use of these sequences in a "reverse genetics" approach. The generation in transgenic plants of antisense RNA transcribed from an antisense gene based on a cDNA insert, or genomic fragment (Fig. 2) has been demonstrated to inhibit the expression of the homologous endogenous plant gene, greatly reducing the accumulation of its encoded protein (15, 29, 30, 57, 62). This approach has enabled the creation of transgenic tomato plants in which a single step in the ethylene biosynthetic pathway is inhibited, by down regulation of either the ACC oxidase genes (20) or ACC synthase (43) genes. These plants and further transgenic lines have aided in dissection and definition of the role of ethylene in the control and regulation of the fruit ripening process and provided a model system in which it may be possible to study other ethylene-mediated processes such as organ abscission (33) and foliar senescence (46, 47).

The phenotype of transgenic tomato plants expressing an ACC oxidase antisense RNA construct

Transgenic plants expressing an ACC oxidase antisense transgene (20) were propagated and self pollinated to yield a segregating population containing either zero, one or two copies of the stably integrated ACC oxidase antisense construct. Fruit obtained from plants that had not inherited an antisense gene produced ethylene at a similar level to non-transformed wild-type fruits. This demonstrated that the phenotypic effects of the transgene were the result of direct gene specific interaction between the integrated transgene and the wild-type ACC oxidase genes and that the function of the endogenous plant ACC oxidase genes was not permanently affected by the antisense construct. Plants in which two copies of the transgene had been inherited gave rise to fruit whose ethylene production during ripening was inhibited by approximately 95% (Fig. 3b). Following pollination, ACC oxidase antisense fruit developed normally and the onset of colour change, co-incident with the normal rise in ethylene evolution, also occurred at a similar chronological stage. The ACC oxidase antisense fruit were, however, reported to show reduced reddening and an increased resistance to over-ripening and shrivelling when stored at room temperature for extended periods (5, 20, 46, 47). Further examination of fruit, borne on plants homozygous for the ACC oxidase antisense gene, showed reduced accumulation of carotenoids, particularly lycopene, in the ripening fruit (5, 46, 47). It was also reported that detachment of the fruit from the vine prior to the onset of ripening increased the severity of the antisense phenotype with detached fruit turning slowly yellow and becoming pale orange after several weeks, with normal colour change being restored by supplying ethylene (5) and Fig. 5 in Gray *et al* (15).

A more detailed biochemical and molecular examination of the homozygous ACC oxidase antisense fruit extended these observations and provided further information about the role of ethylene and the possible contribution of other factors in fruit ripening (47). Lycopene accumulation was reduced in ACC oxidase antisense fruit ripened both on and off the plant (Fig. 4), with the severity of the detached phenotype being dependent upon the stage of fruit removal (Fig. 5). Other ripening parameters were also examined

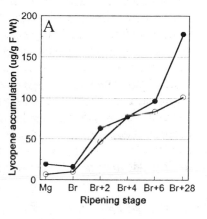

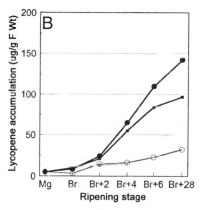

Fig. 4. Changes in lycopene accumulation in fruit pericarp ripened on and off the plant. A. Fruit ripened on the plant; lycopene accumulation was measured at mature green, Mg, (before onset of ripening), breaker, Br, (at the start of pigment change) and up to 28 days after the breaker stage in control fruit (●) and fruit containing 2 ACC oxidase antisense genes (○). B. Fruit ripened off the plant; lycopene accumulation was measured at ripening stages, as above, in control fruit (●) and fruit, containing 2 ACC oxidase antisense genes, incubated in air (○) or with the addition of ethylene (■). (47).

Wild-type (in air)	ACC oxidase antisense (in air)	ACC oxidase antisense (in air with ethylene)	

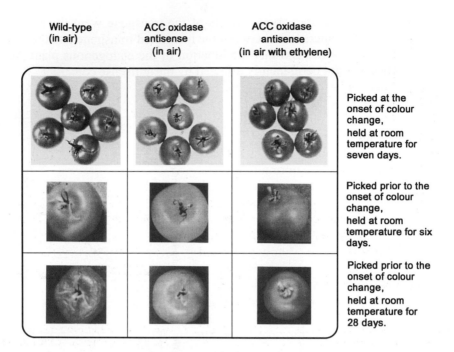

Picked at the onset of colour change,

held at room temperature for seven days.

Picked prior to the onset of colour change,

held at room temperature for six days.

Picked prior to the onset of colour change,

held at room temperature for 28 days.

Fig. 5. Phenotypic changes in ripening of detached fruit containing 2 ACC oxidase antisense genes. Wild-type and homozygous ACC oxidase fruit were detached from plants at the onset of colour change (top panel) and stored for seven days in air or air + 20 μl l^{-1} ethylene as indicated. Similar fruit from wild-type and homozygous ACC oxidase plants were detached **prior** to the onset of any colour change (bottom two panels) and were stored at room temperature in air or air + 20 μl l^{-1} ethylene for times as indicated. (47).

and shown to be more severely disrupted in detached fruits, including chlorophyll degradation and accumulation of a number of previously defined ripening-related mRNAs, including one shown to encode phytoene synthase, an enzyme involved in carotenoid biosynthesis (Fig. 6). A further difference was that the antisense fruit were far less susceptible to over-ripening and shrivelling compared to the wild-type controls (Fig. 5). It was shown that the application of ethylene to the detached ACC oxidase antisense fruit could only *partially* restore normal ripening. This was consistent with the idea that ethylene was intimately involved with ripening. However, the later stages of ripening, including continued production of lycopene (Fig. 4b) and the extensive softening and shrivelling associated with over-ripening of wild-type fruit (Fig. 5) were not fully restored. This implicated a further factor, associated with fruit attachment to the parent plant and working in concert with ethylene, that was required for fruit to undergo full ripening (47). The ACC oxidase antisense plants were also shown to have a second phenotypic characteristic of temporally delayed leaf senescence. The onset of foliar senescence of the lowest true leaves was delayed approximately two weeks although once senescence had begun the rate, judged by loss of chlorophyll,

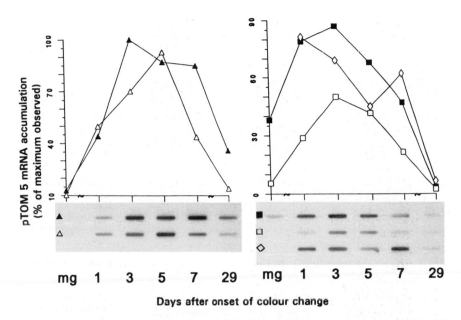

Days after onset of colour change

Fig. 6. Alterations of phytoene synthase (pTOM 5) mRNA accumulation in ACC oxidase antisense fruit. mRNA accumulation was measured at various ripening stages of fruit attached to the plant (Left panel, wild-type fruit ▲; homozygous ACC oxidase fruit Δ) or fruit removed from the plant prior to the onset of ripening (Right panel: Wild-type fruit in air ■; homozygous ACC oxidase fruit in air □ ; homozygous ACC oxidase fruit in air + 20 μl l⁻¹ ethylene ◇) Graphs are derived from densitometric scans of the mRNA accumulation patterns presented below. (46).

appeared to be the same as that observed in control tissues (47). The two sets of data taken together indicate the action of ethylene in developmental processes may be more complex than previously thought since the low ethylene background of the ACC oxidase antisense plants yielded fruit whose ripening initiation was not delayed temporally but appeared to proceed at a much reduced rate, whilst foliar senescence was temporally delayed but once initiated, proceeded at a rate equivalent to that observed for wild-type plants.

The phenotype of plants expressing an ACC Synthase antisense construct

A similar antisense approach was used to perturb ethylene biosynthesis by the stable integration and expression in tomato plants of an antisense RNA construct containing an isolated tomato ACC synthase clone, LE-ACC2 (43). Fruit from a homozygous ACC synthase antisense plant, estimated to contain 10 copies of the antisense gene, showed an extreme reduction in ethylene biosynthesis during ripening, to a level of less than 1% of that observed in wild-type controls. The transgenic fruit, either left on the plant or detached and stored in air, failed to develop normal colour as a result of delayed chlorophyll loss and a lack of lycopene accumulation. Further, the fruit failed to undergo a respiratory climacteric, did not show characteristic softening and did not attain a normal aroma (43). The authors found that application of ethylene, or its analogue propylene, restored the respiratory climacteric, reversed the mutant phenotype and gave rise to fruit that were indistinguishable from wild-type controls with respect to parameters of texture, pigmentation, aroma and fruit compressibility. Photographs of these fruit (43), show the apparent complete reversal of pigment accumulation changes; however, the quantitative measurements of lycopene accumulation are more consistent with the conclusions drawn from the ACC oxidase antisense fruit (47) that only partial restoration of lycopene accumulation is achieved by ethylene treatment of the low ethylene fruit. In addition to reversal of the antisense phenotype, it was shown that ethylene treatment was required for at least 6 days in order to overcome the inhibition of ripening. Shorter treatments were only partially effective (43), which supports the suggestion that ethylene plays a continuing role in regulating the process of ripening rather than simply being a switch for initiation of the process. Analysis of mRNA changes in the these ACC synthase antisense plants showed that both the polygalacturonase (PG) and ACC oxidase mRNAs accumulated to normal levels in the fruit, which led the authors to conclude that these mRNAs were not regulated by ethylene (43, 60, 61). The conclusion that PG mRNA accumulation does not require ethylene is at variance with the experiments of Davies *et al.* (11), who used silver to inhibit ethylene action and showed this prevented the rise in PG mRNA during tomato ripening. It remains to be elucidated if the PG gene is transcriptionally activated by even the low level of ethylene produced by the ACC synthase antisense plants or whether the inhibition of its accumulation by silver treatment of fruit is a secondary effect rather than a direct result of inhibition of ethylene perception or action. It has been subsequently reported (60) that although the PG mRNA is present in the ACC synthase antisense fruit, its polypeptide fails to accumulate, suggesting that ethylene also exerts an influence on gene expression at the post-transcriptional level, either controlling translation of the PG mRNA or dramatically altering the stability of the translated polypeptide.

An alternative approach in the manipulation of ethylene biosynthesis: Metabolism of the ethylene precursor, ACC

In a complementary approach to the expression of antisense RNA constructs in transgenic plants, Klee and co-workers (33) inhibited the production of ethylene by over-expression in transgenic tomato plants of a gene isolated from *Pseudomonas* sp., ACC deaminase, that metabolizes the ethylene precursor ACC to α-ketobutyric acid (Fig. 7). Fruit from homozygous ACC deaminase transgenic plants showed a 90-97% reduction in ethylene evolution during ripening. These fruit also displayed an altered ripening phenotype.

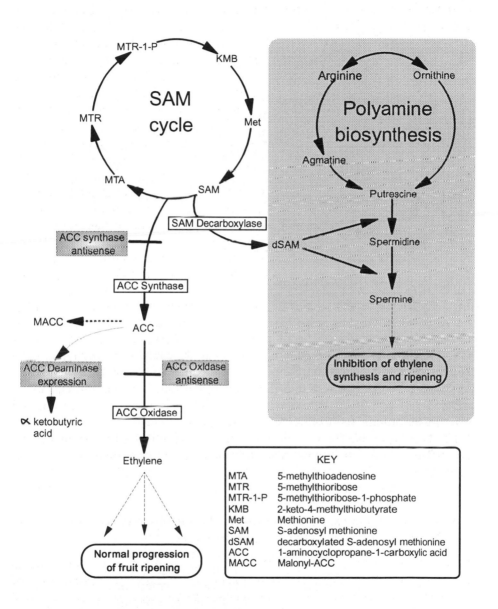

Fig. 7. Schematic representation of the interaction between ethylene and polyamine biosynthesis in ripening tomato fruit. The inter-relationship between polyamine and ethylene biosynthesis and the points of inhibition of ethylene biosynthesis by expression of antisense genes (ACC oxidase or ACC synthase) or expression of a non-plant enzyme (ACC deaminase) are also indicated.

Control fruit detached from the plant at breaker stage were reported to reach full red colour after 7 days and showed significant softening after 14 days whilst transgenic fruit, harvested at a similar stage, took 24 days to reach an equivalent full red stage and remained firm (33). Unfortunately, pigment comparison was only made visually so it is not possible to ascertain if transgenic fruit ripened to the full extent but over an extended period. This is important since the extent of lycopene accumulation (Fig. 4) during the later stages of ripening cannot be accurately assessed by eye (Fig. 5) (47). In direct comparison to the results of Hamilton *et al.* (20), Bouzayen *et al.* (5) and Picton *et al.* (46, 47), detached transgenic ACC deaminase fruit showed much reduced over-ripening and shrivelling. The authors further reported that transgenic fruit left attached to the plant showed a much less rapid rate of softening and also failed to abscise after 40 days, whilst control fruit abscised after 14 days (33). This latter observation also suggests, in parallel with the delay in onset of foliar senescence, (47), that inhibition of ethylene biosynthesis in transgenic plants may result in alterations to other developmental processes and that such transgenic plants may prove valuable in the dissection of other ethylene-mediated events.

Examination of other genes with a role in the biosynthesis of ethylene

In a further antisense experiment designed to identify the function of a previously cloned gene, E8 (35), an antisense E8 construct was transformed into tomato plants (45). The E8 gene has been shown to be transcriptionally activated at the onset of ripening (35) and its encoded product has been suggested, by sequence homology, to be a member of the dioxygenase family of enzymes which includes the ACC oxidase identified by Hamilton *et al* (20, 21). A full length E8 antisense cDNA was transformed into tomato plants and transformants analyzed. Two independently transformed plants showed a characteristic reduction in the endogenous E8 mRNA accumulation and were selfed to yield homozygous E8 antisense progeny that were further analyzed (45).

At the onset of fruit ripening, the E8 antisense fruit showed an equivalent rate of ethylene evolution to wild-type controls; however, two days later ethylene production by the transgenic fruit was approximately sixfold higher than the controls. The effect of the transgene appeared transitory, with ethylene production from transgenic fruit falling to equivalent wild-type levels after five days. There was no effect on ethylene evolution from unripe transgenic fruit, which correlates with the known pattern of expression of the endogenous E8 gene during fruit development and ripening. It was reported that detached fruit from one of the transgenic lines showed delayed ripening although the extent of the delay was variable (45); it has since been shown that the delayed ripening phenotype does not co-segregate with ethylene over-production and may have arisen as a result of somaclonal variation following tissue culture (R.L. Fischer, personal communication).

Clearly these results implicate a role for the E8 protein in the control of ethylene evolution after the onset of ripening, since inhibition of E8 protein accumulation results in substantial ethylene over-production in transgenic fruit after the fruit have initiated the ripening process. It has been suggested that the E8 polypeptide may play a role in a negative feedback mechanism responsible for regulation of ethylene biosynthesis during ripening perhaps by directly limiting the amount of ACC, or, indirectly by constituting part of the signal transduction pathway that regulates ethylene biosynthesis during ripening (45). The involvement of the E8 protein in ethylene perception was also discussed by Theologis (60). Drawing on evidence presented in recent papers (18, 45) it was suggested that the E8 protein may constitute a part of the proposed metalloprotein ethylene receptor. The observed over-production of ethylene by both antisense E8 fruit and wild-type plants treated with silver, which is thought to affect ethylene perception, is consistent with this idea. Thus, interfering with the ethylene receptor may be perceived by cells as "no ethylene present" and invoke over production of the hormone. The model does not, however, extend to an explanation of the transitory nature of ethylene over-production observed in the E8 antisense plants. Furthermore, since E8 encodes a member of a large group of dioxygenase enzymes with diverse functions (49), it is possible that it plays a more indirect role in modulating ethylene synthesis.

DISCUSSION

All approaches to inhibit ethylene synthesis in transgenic tomato plants demonstrated different degrees of inhibition of ethylene synthesis and produced fruit show varying degrees of ripening inhibition. The most extreme reduction in ethylene production has been achieved utilizing multiple insertion of an ACC synthase antisense construct (43). Fruit from these plants also display the most retarded ripening phenotype. Taken together, data obtained from the antisense ACC oxidase and ACC synthase fruits and results obtained from the expression of ACC deaminase in fruit prove that the synthesis of ethylene during tomato fruit ripening occurs solely from the SAM cycle (Figs. 1 and 7). The findings that the application of ethylene to the transgenic fruit can induce the climacteric respiratory rise in the absence of endogenous ethylene production and reverse the antisense phenotype (43) demonstrates that ethylene controls the climacteric rise in respiration in tomato fruit and that ethylene is therefore a causal agent of fruit ripening and not simply a by-product of the ripening process. Other data supports the hypothesis that ethylene *alone* is not the sole determinant of ripening, since application of ethylene to transgenic ACC oxidase antisense fruit fails to cause complete reversal of the antisense phenotype (46, 47). The visual difference observed in the phenotype of either ACC oxidase antisense or ACC deaminase fruit when left attached to the plant or detached and stored in air

(33, 47) supports the involvement of a further factor associated with attachment of the fruit to the plant, as proposed by Picton *et al* (47).

The requirement for continued ethylene treatment to overcome the ACC synthase antisense phenotype demonstrates that ethylene acts in the ripening process as a regulator or rheostat rather than simply a switch that activates the process at the onset of ripening. Furthermore, the failure of ethylene treatment to fully reverse the ACC oxidase antisense phenotype of detached fruit strongly suggests that ethylene acts in concert with a further factor(s) when fruit remain attached to the plant. The identity of this factor, either a fruit associated inhibitor which is metabolised, mobilised or inactivated at the onset of ripening or an enhancer imported from the plant remains to be elucidated.

Further studies on the role of ethylene in the accumulation of PG mRNA (43, 47, 60) and the PG polypeptide (60) are required to ascertain whether the very low levels of ethylene produced in the antisense fruit are sufficient to promote PG mRNA accumulation. The present indications are, however, that ethylene may affect the translation or stability of translated polypeptides in the ripening fruit (60, 61) and thus play a dual role in regulation of ripening gene expression.

It is possible that differences in the methods used to inhibit ethylene biosynthesis in transgenic plants may provide a plausible explanation for the differences observed in the results discussed in this chapter. There is a clear metabolic link between the pathways for biosynthesis of ethylene and polyamines (Fig. 7). In both the ACC oxidase antisense and ACC deaminase plants, no substantial increase in the concentration of SAM would be expected and thus no significant effect on the flux of SAM into the polyamine pathway. In ACC deaminase fruit, SAM would continue to be converted to ACC that is subsequently metabolized by the recombinant enzyme (Fig. 7). In contrast, however, in ACC oxidase antisense fruit, SAM also continues to be converted to ACC which accumulates to higher levels in the pericarp (47) and can be further metabolised to malonyl-ACC (Fig. 7). In ACC synthase antisense fruits there could be an increased flow of SAM into the polyamine pathway via SAM decarboxylase as a result of the inhibition of conversion of SAM to ACC; such a change in flux has been reported in aged orange peel slices where the conversion of SAM to ACC was blocked by AVG and led to increased incorporation of SAM into spermidine (10). If the hypothesis is correct, there may be increased levels of polyamines in the ACC synthase antisense fruit compared to the ACC oxidase antisense or ACC deaminase fruits (Fig. 7). This could be important, since polyamines have been demonstrated to have antisenescent properties (58) and the tomato variety Liberty and the mutant *alc* both contain increased levels of polyamines, show reduced ethylene biosynthesis during ripening and display a delayed over-ripening phenotype (12, 52). This raises the possibility that higher endogenous polyamine levels, in addition to severe inhibition of ethylene biosynthesis may, in part, contribute to the extreme phenotype of the ACC synthase antisense fruit. The possibility that increased synthesis of ethylene is part of a mechanism to reduce the level of polyamines at the onset of normal ripening and that they may constitute part of the fruit-associated ripening inhibitor (47) is intriguing (14).

Despite the purification, identification and cloning of two key enzymes involved in ethylene production, little is known about the underlying mechanisms responsible for the perception of the ethylene stimulus or how signals are transduced following perception in order to bring about specific alterations to gene expression. Steps to address this specific area have been taken recently using the model plant system *Arabidopsis*. Ethylene insensitive mutants have been identified (4, 18, 65) and more recently a mutant that constitutively activates ethylene mediated responses, the *ctr 1* mutant (31), has been isolated. The gene corresponding to the *ctr 1* mutation has been cloned and shown to encode a peptide that resembles the Raf family of serine/threonine protein kinases. The putative CTR 1 gene has been postulated to act as a negative regulator in the ethylene signal transduction pathway and the identification of three other loci involved in ethylene perception (18, 50) and cloning of one of the genes, EIN 3, cited in (31) may aid dissection of the signal transduction pathway responsible for gene regulation by ethylene.

Acknowledgements

Work in the authors laboratory was supported by a grant from the Agricultural and Food Research Council to Don Grierson. All work was carried out under Ministry of Agriculture, Food and Fisheries licences. We express our thanks to Lee Whotton for assistance in preparation of figures.

References

1. Abeles, F.B. (1992) Ethylene in plant biology. Academic Press, New York.
2. Bird, C.R., Ray, J.A., Fletcher, J.D., Boniwell, J.M., Bird, A.S., Teuheres, C., Blain, I., Bramley, P.M., Schuch, W. (1991) Using antisense RNA to study gene function: Inhibition of carotenoid biosynthesis in transgenic tomatoes. Bio/technology 9, 635-639.
3. Bleeker, A.B., Kenyon, W.H., Somerville, S.C., Kende, H. (1986) Use of monoclonal antibodies in the purification and characterisation of 1-aminocyclopropane-1-carboxylate synthase, an enzyme in ethylene biosynthesis. Proc. Natl. Acad. Sci. USA 83, 7755-7759.
4. Bleeker, A.B., Estelle, M.A., Somerville, C., Kende, H. (1988) Insensitivity of ethylene conferred to a dominant mutation in *Arabidopsis thaliana*. Science 241, 1086-1089.
5. Bouzayen, M., Hamilton, A.J., Picton, S., Barton, S., Grierson, D. (1992) Identification of genes for the ethylene-forming enzyme and inhibition of ethylene synthesis in transgenic plants using antisense genes. Biochem. Soc. Trans. 20, 76-79.
6. Brady, C.F. (1987) Fruit ripening. Ann. Rev. Plant Physiol. 38, 155-178.
7. Brady, C.F., Speirs, J. (1991) Ethylene in fruit ontogeny and abscission. *In:* The Plant Hormone Ethylene, pp. 235-258, Mattoo, A.K., Suttle, J.C., eds. CRC Press, Boston.
8. Bramlage, W.J., Greene, D.W., Autio, W.R., McLaughlin, J.M. (1980) Effects of aminoethoxyvinylglycine on internal ethylene concentrations and storage of apples. J. Am. Soc. Hort. Sci. 105, 847-851.
9. Burg, S.P., Burg, E.A. (1965) Ethylene action and the ripening of fruits. Science 148, 1190-1196.

10. Chen, Z.V., Mattoo, A.K., Goren, R. (1982) Inhibition of ethylene biosynthesis by aminoethoxyvinylglycine and polyamines shunts label from 3,4-[^{14}C]Methionine into spermidine in aged orange peel discs. Plant Physiol. 69, 385-388.

11. Davies, K.M., Hobson, G.E., Grierson, D. (1988) Silver ions inhibit the ethylene-stimulated production of ripening-related mRNAs in tomato. Plant, Cell Envir. 11, 729-738.

12. Davies, P.J., Rastogi, R., Law, D.M. (1990) Polyamines and their metabolism in ripening tomato fruit. *In*: Polyamines and Ethylene: Biochemistry, Physiology, and Interactions, pp. 112-125, Flores, H.E., Arteca, R.N., Shannon, J.C., eds. American Society of Plant Physiologists, Rockville, MD, USA.

13. Fray, R.G., Grierson, D. (1993) Identification and genetic analysis of normal and mutant phytoene synthase genes of tomato by sequencing, complementation and co-suppression. Plant Mol Biol. 22, 589-602.

14. Fray, R.G., Grierson, D. (1993) Molecular genetics of tomato fruit ripening. Trends in Genetics 9, 438-443.

15. Gray, J., Picton, S., Shabbeer, J., Schuch, W., Grierson, D. (1992) Molecular biology of fruit ripening and its manipulation with antisense genes. Plant Mol. Biol. 19, 69-87.

16. Grierson, D. (1985) Gene expression in ripening tomato fruit. Crit. Rev. Plant Sci. 3, 113-132.

17. Grierson, D. (1987) Senescence in fruits. HortScience 22, 859-862.

18. Gutzmán, P., Ecker, J.R. (1990) Exploiting the triple response of *Arabidopsis* to identify ethylene-related mutants. Plant Cell 2, 513-523.

19. Guy, M., Kende, H. (1984) Conversion of 1-aminocyclopropane-1-carboxylic acid to ethylene by isolated vacuoles of *Pisum sativum* L. Planta 160, 281-287.

20. Hamilton, A.J., Lycett, G.W., Grierson, D. (1990) Antisense gene that inhibits synthesis of the hormone ethylene in transgenic plants. Nature 346, 284-287.

21. Hamilton, A.J., Bouzayen, M., Grierson, D. (1991) Identification of a tomato gene for the ethylene forming enzyme by expression in yeast. Proc. Natl. Acad. Sci. USA 88, 7434-7437.

22. Hobson, G.E., Nichols, R., Davies, J., Atkey, P.T. (1984) The inhibition of tomato fruit ripening by silver. J. Plant Physiol. 116, 21-29.

23. Holdsworth, M.J., Schuch, W., Grierson, D. (1987) Nucleotide sequence of an ethylene-related gene from tomato. Nucl. Acids Res. 15, 10600.

24. Holdsworth, M.J., Bird, C.R., Ray, J., Schuch, W., Grierson, D. (1987) Structure and expression of an ethylene-related mRNA from tomato. Nucl. Acids Res. 15, 731-739.

25. Holdsworth, M.J., Schuch, W., Grierson, D. (1988) Organisation and expression of a wound/ripening-related small multigene family from tomato. Plant Mol. Biol. 11, 81-88.

26. Imaseki, H. (1991) The biochemistry of ethylene biosynthesis. *In*: The Plant Hormone Ethylene, pp. 1-20, Mattoo, A.K., Suttle, J.C., eds. CRC Press, Boston, USA.

27. John, P. (1983) The coupling of ethylene biosynthesis to a transmembrane, electrogenic proton flux. FEBS Lett. 152, 141-143.

28. John, P. (1991) How plant molecular biologists revealed a surprising relationship between two enzymes, which took an enzyme out of a membrane where it was not located, and put it into the soluble phase where it could be studied. Plant Mol. Biol. Rep. 9, 192-194.

29. Jorgensen, R. (1990) Altered gene expression in plants due to *trans* interactions between homologous genes. Trends Biotech. 8, 340-344.

30. Jorgensen, R. (1991) Silencing of plant genes by homologous transgenes. AgBiotech News Info. 4, 265-273.

31. Keiber, J.J., Rothenberg, R., Roman, G., Feldmann, K.A., Ecker, J.R. (1993) *CTR1*, a negative regulator of the ethylene response pathway in Arabidopsis, encodes a member of the Raf family of protein kinases. Cell 72, 427-441.

32. Kende, H. (1994) Standardizing the nomenclature of the ACC synthase and ACC oxidase gene families. Plant Molecular Biology Reporter, In press.

33. Klee, H.J., Hayford, M.B., Kretzner, K.A., Barry, G.F., Kishore, G.M. (1991) Control of ethylene synthesis by expression of a bacterial enzyme in transgenic tomato plants. Plant Cell 3, 1187-1193.

34. Köck, M., Hamilton, A.J., Grierson, D. (1991) *eth 1*, a gene involved in ethylene synthesis in tomato. Plant Mol. Biol. 17, 141-142.

35. Lincoln, J.E., Cordes, S., Read, E., Fischer, R.L. (1987) Regulation of gene expression by ethylene during *Lycopersicon esculentum* (tomato) fruit development. Proc. Natl. Acad. Sci. USA 84, 2793-2797.

36. Liu, Y., Hoffman, N.E., Yang, S.F. (1985) Promotion by ethylene of the capability to convert 1-aminocyclopropane-1-carboxylic acid to ethylene in preclimacteric tomato and cantaloupe fruits. Plant Physiol. 77, 407-411.

37. Mayne, R.G., Kende, H. (1986) Ethylene biosynthesis in isolated vacuoles of *Vicia faba* L.-requirement for membrane integrity. Planta 167, 159-165.

38. McCormick, S., Neidermeyer, J., Fry, J., Barnason, A., Horsch, A., Fraley, R. (1986) Leaf disc transformation of cultured tomato (*L. esculentum*) using *Agrobacterium tumefaciens*. Plant Cell Rep. 5, 81-84.

39. McMurchie, E.J., McGlasson, W.B., Eaks, I.L. (1972) Treatment of fruit with propylene gives information about the biogenesis of ethylene. Nature 237, 235-236.

40. Mehta, A.M., Jordan, R.L., Anderson, J.D., Mattoo, A.K. (1988) Identification of a unique isoform of 1-aminocyclopropane-1-carboxylic acid synthase by monoclonal antibody. Proc. Natl. Acad. Sci. USA 85, 8810-8814.

41. Mitchel, T., Porter, A.J.R, John, P. (1988) Authentic activity of the ethylene-forming enzyme observed in membranes obtained from kiwifruit (*Actinidia deliciosa*). New Phytol. 109, 313-319.

42. Neljubow, D. (1901) Üeber die horizontale nutation der stengel von *Pisum sativum* und einiger anderer. Pflanzen Beih. Bot. Zentralb. 10, 128-239.

43. Oeller, P.W., Wong, L.M., Taylor, L.P., Pike, D.A., Theologis, A. (1991) Reversible inhibition of tomato fruit senescence by antisense RNA. Science 254, 437-439.

44. Olsen, D.C., White, J.A., Edelman, L., Harkins, R.N., Kende, H. (1991) Differential expression of two genes for 1-aminocyclopropane-1-carboxylate synthase in tomato fruits. Proc. Natl. Acad. Sci. USA 88, 5340-5344.

45. Peñarrubia, L., Aguilar, M., Margossian, L., Fischer, R.L. (1992) An antisense gene stimulates ethylene hormone production during tomato fruit ripening. Plant Cell 4, 681-687.

46. Picton, S.J., Hamilton, A., Fray, R., Gray, J., Bouzayen, M., Smith, C., Watson, C.F., Evans, A., Barton, S., Smith, H., Turner, A., Grierson, D. (1992) Modifying gene expression and ripening in transgenic tomatoes. *In*: Profiles on Biotechnology, pp.645-656, Villa, T.G., Abalde, J., eds. Intercambio Cientifico, Universidade de Santiago, Spain.

47. Picton, S.J., Barton, S.L., Bouzayen, M., Hamilton, A.J., Grierson, D. (1993) Altered fruit ripening and leaf senescence in tomatoes expressing an antisense ethylene-forming enzyme transgene. Plant Journal 3, 469-481.

48. Porter, A.J.R., Borlakoglu, J.T., John, P. (1986) Activity of the ethylene-forming enzyme in relation to plant cell structure and organisation. J. Plant Physiol. 125, 207-216.

49. Prescott, A.G. (1993) A dilemma of dioxygenases (or where biochemistry and molecular biology fail to meet). J. Exp. Bot. 44, 849-861.

50. Rothenberg, M., Ecker, J.R. (1993) Mutant analysis as an experimental approach towards understanding plant hormone action. Sem. Dev. Biol. In Press.

51. Rottmann, W.H., Peter, G.F., Oeller, P.W., Keller, J.A., Shen, N.F., Nagy, B.P., Taylor, L.P., Campbell, A.D., Theologis, A. (1991) 1-Aminocyclopropane-1-carboxylate synthase in tomato is encoded by a multigene family whose transcription is induced during fruit and floral senescence. J. Mol. Biol. 222, 937-961.

52. Saftner, R.A., Baldi, B.G. (1990) Polyamine levels and tomato fruit development. Plant Physiol. 92, 547-550.
53. Saltveit, M.E., Bradford, K.J., Dilley, D.R. (1978) Silver ion inhibits ethylene synthesis and action in ripening fruits. J. Am. Soc. Hort. Sci. 103, 472-475.
54. Sato, T., Theologis, A. (1989) Cloning the mRNA encoding 1-aminocyclopropane-1-carboxylate synthase, the key enzyme for ethylene biosynthesis in plants. Proc. Natl. Acad. Sci. USA 86, 6621-6625.
55. Slater, A., Maunders, M.J., Edwards, K., Schuch, W., Grierson, D. (1985) Isolation and characterisation of cDNA clones for tomato polygalacturonase and other ripening-related proteins. Plant Mol. Biol. 5, 137-147.
56. Smith, C.J.S., Slater, A., Grierson, D. (1986) Rapid appearance of an mRNA correlated with ethylene synthesis encoding a protein of molecular weight 35,000. Planta 168, 94-100.
57. Smith, C.J.S., Watson, C.F., Ray, J., Bird, C.R., Morris, P.C., Schuch, W., Grierson, D. (1988) Antisense RNA inhibition of polygalacturonase gene expression in transgenic tomatoes. Nature 334, 724-726.
58. Smith, T.A. (1985) Polyamines. Ann. Rev. Plant Physiol. 36, 117-143.
59. Spanu, P., Reinhardt, D., Boller, T. (1991) Analysis and cloning of the ethylene forming enzyme from tomato by functional expression of its mRNA in *Xenopus laevis* oocytes. EMBO J. 10, 2007-2013.
60. Theologis, A. (1992) One rotten apple spoils the whole bushel: The role of ethylene in fruit ripening. Cell 70, 181-184.
61. Theologis, A., Zarembinski, T.I., Oeller, P.W., Liang, X., Abel, S. (1992) Modification of fruit ripening by suppressing gene expression. Plant Physiol. 100, 549-551.
62. van der Krol, A.R., Lenting, P.E., Veenstra, J., van der Meer, I.M., Koes, R.E., Gerats, A.G.M., Mol, J.N.M., Stuitje, A.R. (1988) An anti-sense chalcone synthase gene in transgenic plants inhibits flower pigmentation. Nature 333, 866-869.
63. van der Straeten, D., van Wiermeersch, L., Goodman, H.M., van Montague, M. (1989) Purification and partial characterisation of 1-aminocyclopropane-1-carboxylate synthase from tomato pericarp. Eur. J. Biochem. 182, 639-647.
64. van der Straeten, D., van Wiemeersch, L., Goodman, H.M., van Montague, M. (1990) Cloning and sequence of two different cDNAs encoding 1-aminocyclopropane-1-carboxylate synthase in tomato. Proc. Natl. Acad. Sci. USA 87, 4859-4863.
65. van der Straeten, D., Djudzman, A., Caeneghem, W.V., Smalle, J., Van Montague, M. (1993) Genetic and physiological analysis of a new locus in *Arabidopsis* that confers resistance to 1-aminocyclopropane-1-carboxylic acid and ethylene and specifically affects the ethylene signal transduction pathway. Plant Physiol. 102, 401-408.
66. Ververidis, P., John, P. (1991) Complete recovery in vitro of ethylene-forming enzyme activity. Phytochem. 30, 725-727.
67. Yang, S.F., Hoffman, N.E. (1984) Ethylene biosynthesis and its regulation in higher plants. Ann. Rev. Plant Physiol. 35, 155-189.
68. Yip, W-K., Moore, T., Yang, S.F. (1992) Differential accumulation of transcripts for four tomato 1-aminocyclopropane-1-carboxylate synthase homologs under various conditions. Proc. Natl. Acad. Sci. USA 89, 2475-2479.

E5. The Role of Hormones in Gene Activation in Response to Wounding

Hugo Peña-Cortés and Lothar Willmitzer

Institut für Genbiologische Forschung GmbH, Ihnestr. 63, 14195 Berlin, Germany.

INTRODUCTION

Plants are unable to move of their own volition from one place to another where a more comfortable environment for growth and reproduction might be available. The frequent exposure to severe physical, chemical and biological stresses has led higher plants to acquire and develop, during their evolution, self-defense mechanisms   protect themselves against the many kinds of environmental stress to which they are exposed. Thus the expression of many plant genes is influenced by environmental conditions. Under adverse conditions plant tissues are often damaged. Wounding can be an extremely severe stress which may result in the death of the plant. Injuries inflicted by herbivore feeding or other mechanical damage may also provide ready entrance points for pathogens, removing the protective layers which normally coat the entire plant. The gene expression pattern of the plant is drastically altered by wounding. Processes which are likely to require large amounts of energy, such as those related to plant growth, are brought to a stop. The expression of some genes is turned off; for example, the small subunit of the ribulose biphosphate carboxylase (30) and the 10 kd protein of the water splitting apparatus, both involved in photosynthesis, whilst the expression of others is triggered. Thus, plants respond to mechanical injury by inducing a defense response characterized by the expression of a set of proteins, mainly aimed at wound healing and prevention of pathogen invasion. These responses include reinforcement of the cell wall by deposition of callose, lignin, and hydroxyproline-rich glycoproteins; synthesis of the antimicrobial compounds such as phytoalexins; and production of proteinase inhibitors and lytic enzymes such as chitinases and glucanases. With the exception of callose production, which involves Ca^{2+} stimulation of preexisting callose synthase at the plasma membrane, induction of all known defense proteins involves transcriptional activation of the corresponding genes and, as a consequence, correlates with a substantial alteration of gene

---

Abbreviations: ABA=abscisic acid; JA=jasmonic acid; MeJA=methyljasmonate; LA=linolenic acid; LOX=lipoxygenase; 13HPLA=13-hydroxyperoxylinolenic acid; 12-oxo-PDA=12-oxo-phytodienoic acid; SHAM=salicylhydroxamic acid; ETYA=eicosatetraynoic acid.

P. J. Davies (ed.), Plant Hormones, 395–414.

expression in the host plant. This gene activation can be confined to the close vicinity of the wound site. Some genes, however, are not only induced locally but also in distal organs which are themselves not damaged (4).

The activation of wound-induced genes is most likely due to the generation of an activating signal at the site of the injury. The subsequent distribution of this putative "wound signal" might be directly responsible for the activation of different subsets of genes either locally or systemically. The local induction of genes might require a high concentration of the "wound-signal" which is only reached close to the site of the injury, whilst the induction threshold would be much lower for distantly located systemically induced genes. It would not be surprising therefore if some substances with short migration range were responsible for the local induction, whilst other substances able to migrate long distances through the plant would trigger the expression of the systemically induced genes. A number of different substances have been shown to modulate the activity of genes induced in the wounded tissue and thus have been put forward as putative wound signals. The aim of this chapter is to discuss the evidence for a role of abscisic acid (ABA) and jasmonic acid (JA) (see Chapter C2a) in the induction of plant gene expression upon wounding using the expression of the proteinase inhibitor II (Pin2) gene family as a model system.

THE PROTEINASE INHIBITOR II (PIN2) GENE EXPRESSION AS A MODEL SYSTEM FOR GENE ACTIVATION BY WOUNDING

Plants react to wounding and pathogen attack by activating a set of genes, most of them playing a role in wound healing and the prevention of subsequent pathogen invasion. Some of these genes are expressed in the vicinity of the wound site while others are also systemically activated in the non-damaged parts of the plant. The potato and tomato proteinase inhibitor II (Pin2) gene families are the best studied examples of genes which are systemically activated upon mechanical damage (4, 40). Potato Pin2 belongs to a small family consisting of some 3-5 members per haploid genome. Complementary DNAs (cDNAs) have been isolated from both potato and tomato plants and shown to share a high degree of similarity in their sequences (over 80%). The Pin2 gene family is constitutively expressed in potato tubers, where Pin2 protein can make up to 5% of the total tuber protein content. Additionally, Pin2 mRNA is also constitutively present in the early stages of floral development. Young floral buds accumulate Pin2 mRNA which is absent later in the organs of the fully developed potato flower. Tomato flowers also accumulate Pin2 mRNA but, in contrast to potato, adult flowers exhibit readily detectable levels of it in virtually every organ (33). In addition to its constitutive expression in tubers and flowers, Pin2 mRNA accumulates in the foliage of both potato and tomato plants following wounding, either mechanical or by herbivore feeding. This accumulation is due to transcriptional activation of the Pin2 genes upon wounding, and is not confined to the site of injury. Indeed, the non-damaged leaves of a wounded plant readily accumulate Pin2 mRNA, but with a short delay compared to the directly wounded ones, thus resulting in lower levels of Pin2 mRNA at a given time point in the systemically induced leaves as compared to the locally induced ones. This systemic induction is likely to be related to the synthesis or release of a wound signal at the site of the injury which migrates throughout the plant activating Pin2 genes in distal tissues. This signal is most likely transported via the phloem since non-wounded leaves both above and below the wound site accumulate Pin2 mRNA, indicating acropetal and basipetal distribution which are characteristic of phloem transport (30). Transgenic potato plants have been obtained with a Pin2 promoter driving the expression of a bacterial b glucuronidase (GUS) reporter gene (18). These transgenic plants exhibited constitutive high levels of GUS activity in tubers and floral buds, and high levels of GUS were also observed in leaves upon wounding. Consistent with the idea of a transport of the wound signal via the phloem, the Pin2 promoter activity in systemically induced leaves is highest in the tissue surrounding the vascular bundles. Every organ of the potato plant appears to be able to release the wound signal. Thus in addition to the already described effect of wounding leaves, tubers or roots wounded while still attached to the plant also trigger the systemic accumulation of Pin2 throughout the plant. However, not each tissue reacts to the wound signal by accumulating Pin2 mRNA. In the same set of experiments it was shown that whereas the Pin2 gene family was readily induced systemically in leaves and young stems, neither roots nor the lower part of the stem (close to the roots) expressed Pin2 as a reaction to the stimulus (30). Roots wounded by nematodes do not accumulate Pin2 mRNA which, however, accumulates systemically in the rest of the plant (14). In the case of the lower stem the absence of Pin2 accumulation upon wounding is not due to a general non-responsiveness of the tissue since it reacted to wounding by decreasing the levels of RNA of the small subunit of rubisco, suggesting that it has the ability to detect changes related to wound signals (30).

It thus appears that upon wounding of any organ of the plant a signal is released and distributed, triggering the activation of the Pin2 gene family in the aerial part of the plant, whereas in other organs, such as roots, different responses may occur.

DIFFERENT COMPOUNDS INDUCE Pin2 GENE ACTIVATION

Several different stimuli have been shown to stimulate Pin2 mRNA accumulation in leaves, and have therefore been suggested to play a role in the transduction of environmental or developmental cues to Pin2 expression. Thus, plant cell-wall derived oligosaccharides with different degrees of polymerization have been suggested to be the wound signal (39). Oligosaccharide fragments could be generated from the plant cell wall upon contact with endogenous polygalacturonases, released in the medium due to the loss of compartmentalization by mechanical disruption. It was reported that pectic fragments with different degrees of polymerization derived from the plant cell wall were able to induce Pin2 gene expression, and they were therefore assumed to be the proteinase-inhibitor inducing factor (PIIF) (2). The expression of the Pin2 gene family was also induced in detached leaves supplied with chitosan, a β-1-4-glucosamine homopolymer present in fungal cell walls (30). By using radiolabeled oligosaccharides, however, Baydon and Fry (1) have shown that molecules with a degree of polymerization greater than 6 do not travel long distances through the plant vascular system, which argues against them actually mediating the systemic activation of the Pin2 gene. These compounds are rather thought to be released from the wounded tissues as early signals in the pathway that ultimately leads to both localized and systemic wound-induced expression of Pin2 genes.

The phytohormone ethylene has been suggested to be involved in the transmission of the wound stimulus leading to the expression of plant-defense-related genes. Ethylene production is indeed increased upon injury as well as plant stresses such as freezing or drought, mainly as a result of increasing synthesis of ACC (1-aminocyclopropane-1-carboxylic acid), a precursor in the ethylene biosynthetic pathway. Several genes involved in processes of plant defense, such as some PR (pathogenesis-related) proteins, are thereby activated and this activation can be mimicked by ethylene treatment (3). Ethylene, however, appears to play no role in Pin2 induction upon wounding. Ethylene-forming compounds or precursors such as ethephon or ACC were unable to induce the Pin2 gene family over the level already reached in detached leaves incubated in water alone. In addition, known inhibitors of ethylene formation and action i.e. Co^{2+} and Ag^+ do not prevent the accumulation of Pin2 mRNA upon wounding (42). These data suggest that ethylene may participate in a wound stimulus transduction pathway other than the one leading to the activation of Pin2. Consistent with this, it has been shown that the induction pattern of the cell wall hydroxyproline-rich glycoprotein genes upon wounding is different from the induction by ethylene treatment (8). Moreover, only 15% of the proteins accumulating upon wounding of tomato fruit are affected by the presence of inhibitors of ethylene action (15). These results suggest the existence of at least two different pathways acting to transmit the wound stimulus.

On the other hand, the phytohormone auxin prevents the wound-induced Pin2 gene expression. Exogenously applied auxin can act as a repressor of the wound-inducible activation of a chimeric Pin2-CAT (chloramphenicol acetyltransferase) chimeric gene in transgenic tobacco callus and in whole plants (17). In addition, the endogenous levels of indole-3-acetic acid (IAA) declines two- to threefold within 6 hours after wounding. The kinetics of auxin decline are consistent with the kinetics of activation of the Pin2-CAT construction in the foliage of transgenic tobacco. These results suggest that endogenous levels of IAA in unwounded plant tissues are sufficient to maintain the inhibitor II gene system in a repressed state so that the genes are not expressed. However, following a wound, the levels of IAA in bulk tissues decline, allowing a derepression of the gene system with concomitant expression of the CAT protein (48).

Additional signalling mechanisms in the regulation of the wound response system have been reported and diverse theories have been advanced regarding the nature of the systemic signal transmitted from wound sites.. Thus, changes in membrane polarity are among the first effects to be detected in wounded plants. These alterations of the membrane polarity are also systemically transmitted but it still unclear if they are responsible for the systemic induction of genes or if they are just the result of the systemic changes occurring upon wounding (52). Phytohormones such as ABA (31) and jasmonic acid (JA) derivatives (9, 10, 32, 34) are able to induce Pin2 gene activation (the role of both ABA and JA is discussed in this chapter).Other substances including sucrose also induce Pin2 gene expression without wounding (34). In addition, a peptide mediating the systemic wound response of Pin1 and Pin2 genes in wounded tomato leaves has been reported (29). This molecule, termed systemin, is an 18-amino acid oligopeptide, rich in proline and basic amino acids, which has been shown to move systemically in the phloem to distant plant tissues. On the other hand, many of the stimuli that initiate systemic responses in plants are also known to cause simultaneous electrical activity, and an increasing number of plant species have been shown to be able to transmit action potentials and variation potentials (21). Thus, it has been recently reported that a propagated electrical signal may be the messenger for the systemic induction of Pin2 (53). In contrast to our results (20 minutes for systemic wound-induced gene activation), systemic accumulation of Pin2 mRNA was already detected five minutes after wounding. The signal would move from the wound site to the systemic tissue via some system which exclude the involvement of phloem tissue, previously blocked by cooling. The signal would also travel apparently unhindered through significant lengths of stem which consisted of alternating live and dead (at least temporally) regions. Electrical signal transmission from cell to cell would require the presence of functional membranes. Therefore, one crucial point will be to demonstrate the capacity of these forms of signal in propagating through nonfuntional tissues. More recently, Malone (22) proposed that hydraulic signal represent the mechanism of systemic signalling of wounding in tomato plants. He demonstrated that wound-induced mass flows can transport solutes extensively and rapidly from the wound site. The rate mass flow (ca. 10 mm s^{-1}) is sufficiently rapid to distribute elicitors throughout the plant within the shortest time observed for systemic wound-induced gene activation (20 min) (30).

THE INVOLVEMENT OF ABA IN WOUND-INDUCED Pin2 GENE EXPRESSION

The plant growth regulator abscisic acid (ABA) appears to play a predominant role in the conversion of environmental signals into changes in plant gene expression. A rise in its endogenous concentration precedes and is involved in the establishment of seed dormancy, with mature seeds not germinating in the presence of exogenous ABA (37). Moreover, viviparous mutants have been described in a number of plant species, with seeds which precociously germinate in the fruits. The absence of dormancy is due either to a deficiency in ABA synthesis, in which case external application reverts the seed to the wild-type phenotype or, alternatively, to a lack of responsiveness to endogenous ABA levels, and in this case exogenously applied ABA is not able to prevent premature seed germination. ABA treatment at normal temperature can also mimic the effect of cold temperatures by inducing the synthesis of the same set of proteins which are thought to increase the tolerance of plants to freezing. Similarly, drought and high salt concentration trigger a rise in the endogenous ABA level which may be responsible for the activation of most of the water- and salt-stress induced genes (43).

Recently strong evidence has accumulated for the involvement of ABA in the induction of the Pin2 gene expression upon wounding. In a series of experiments, potato plants were sprayed with 100 μM ABA, and Pin2 mRNA accumulated subsequently in the absence of any wounding (31). Tissues that were sprayed directly, as well as non-sprayed leaves, showed increased Pin2 mRNA levels. This accumulation was tissue-specific, being detected in leaves and stems but not in roots or the lower part of the stems. ABA sprayed to the leaves of a plant is thus able to trigger the systemic induction of the Pin2 gene with a pattern identical to the one described for wounded plants (30). In contrast, no activation of Pin2 could be detected in tomato or transgenic tobacco plants containing a wound-inducible potato Pin2 gene when these were sprayed with ABA. An enhancement of Pin2 mRNA accumulation was only seen in these plants upon incubation of detached leaves in an ABA solution, suggesting that differences in ABA absorption through the epidermis between potato and tomato or tobacco leaves might could be responsible for the contrasting results. Conclusive evidence for the involvement of ABA on wound-induced Pin2 activation was obtained by analyzing the wounding effect in mutant plants impaired in ABA synthesis. These plants provide the best required control for the experiments involving the external application of ABA. The potato mutant droopy and the *sitiens* mutant of tomato have a mutation blocking the last step in ABA biosynthesis, namely the conversion of the ABA-aldehyde to ABA (7, 47). The altered phenotype of these mutants results from their lower internal level of ABA (9-12%) and can be reversed by exogenous application of this hormone. Wound induction of Pin2 is not observed in mutants of potato and tomato deficient in the synthesis of

ABA. In these plants, however, treatment with ABA causes a reversion of the accumulation of Pin2 mRNA to the levels normally found in wild-type plants upon wounding (Fig. 1). Moreover, endogenous levels of ABA increase, both locally and systemically, in wild-type plants upon wounding but not in the droopy mutant (Fig. 2).

Both the local accumulation of Pin2 mRNA around the site of injury and the systemic activation of Pin2 transcription in the distal non-wounded tissue are affected by the ABA deficiency. In the mutant plants, very low levels of Pin2 mRNA can be detected in the tissue closest to the wounding site, whilst Pin2 mRNA concentration is below the limits of detection in the systemically induced foliage (33). This very low Pin2 mRNA accumulation in the vicinity of the injury is consistent with the low ABA levels present in the mutant plants (36, 47), and further suggests that ABA is involved in the release of a local wound signal or, alternatively, that it is itself the local signal. The lack of Pin2 accumulation in the tissue distal to the wound site indicates that ABA is also involved in the systemic induction of Pin2, either by preventing the formation of a signal at the site of the injury, which subsequently migrates to the distal tissue, or by being the systemic signal itself. Externally applied ABA is also able to induce the systemic expression of the Pin2 gene family. Potato plants sprayed with ABA in the lower part of the foliage accumulated Pin2 mRNA, not only in the tissue sprayed directly, but also in the upper part of the foliage which was not sprayed. By performing this experiment in the ABA-deficient droopy plants, it was possible to demonstrate that the ABA levels increase in the distal non-sprayed tissue to within levels normally found in the systemically induced wild-type plants (Fig. 3). Accumulation of Pin2 mRNA in the distal, non-sprayed tissue accompanied the rise in the hormone concentration (Fig. 4). Since

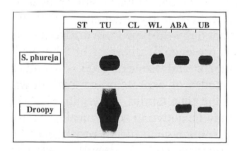

Fig. 1. Pin2 expression pattern in wild-type and ABA-deficient (droopy) *S. phureja* plants. RNA was isolated from stolons (ST), tubers (TU), non-wounded leaves (CL), wounded leaves (WL), leaves sprayed with ABA (ABA) and unripe floral buds (UB) from wild-type (*S. phureja*), or ABA-deficient mutant plants (droopy). The autoradiogram shows the result of an RNA gel blot hybridization of total RNA (20 μg per slot) against Pin2 cDNA probe.

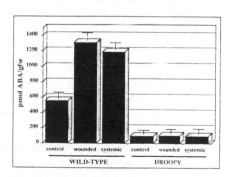

Fig. 2. Endogenous levels of ABA produced upon wounding in wild-type and ABA deficient (droopy) potato plants. ABA was determined in damaged (wounded) and non-wounded leaves (systemic) of wild-type (*S. phureja*) and ABA-deficient mutant plants as well as in leaves before wounding (control).

401

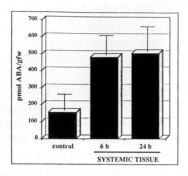

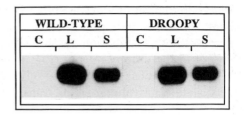

Fig. 4. Comparison of the Pin2 gene expression upon ABA spraying in wild-type and ABA-deficient (droopy) mutant plants. Potato wild-type and ABA-deficient droopy plants were sprayed with a 100 μM ABA solution. Both directly sprayed (L) and non-sprayed systemically (S) induced leaves were harvested after 24 h. The RNA gel blot was hybridized against a Pin2 cDNA probe. As a control, RNA was isolated from plants 24 h after spraying with water (lane C).

Fig. 3. ABA moves throughout in ABA-deficient mutant plants. Droopy plants were sprayed with a 100 μm ABA solution. The ABA concentration attained upon treatment in the systemically induced leaves at the time of harvesting was determined. The ABA concentration from a droopy plant 24 h after spraying with water was also determined (control).

droopy plants are not able to synthesize ABA de novo, the high ABA levels detected were most likely due to the migration of the exogenously applied ABA to the distal, non-sprayed tissue. The results demonstrate the ability of ABA to migrate throughout the plant and thereby trigger the expression of the Pin2 gene family in distal tissues in a similar manner to their systemic induction upon wounding.

Whilst ABA is involved at some stage during wound-induced gene activation, constitutive Pin2 expression in potato flowers and tubers is apparently not affected by the block in the ABA biosynthetic pathway (Fig. 1). Thus, the ABA-deficient droopy mutant has wild-type Pin2 mRNA levels in tubers and flower buds, suggesting the presence of different or additional factors for Pin2 gene activation in tubers and flowers (33). Another explanation could be that the different modes of expression of Pin2 might reflect differential activity of members of the gene family. Alternatively, one or several members of the gene family may be active in some or all of these situations. Transgenic potato plants containing a gene fusion consisting of a Pin2 promotor hooked to a β-glucuronidase (GUS) reporter showed constitutive GUS activity in tubers and floral buds, and wound-induced activity in leaves (18). Additionally, this construction endowed transgenic ABA-deficient droopy plants with constitutive GUS activity in tuber and floral buds, and ABA-induced activity in leaves (33). A single promoter element therefore mirrors the expression pattern of the whole gene family indicating the presence within this promoter of all cis-acting sequences responsible for the complex regulation of Pin2 by environmental and developmental factors. The activity of another highly similar Pin2 promoter has been shown to be also wound inducible in transgenic tobacco plants. Its activity in potato has not been so far tested, and it is not known yet if it is a general rule that Pin2 promoters are active in both tubers and wounded leaves.

Analysis of the Pin2 promoter (5′, 3′, and internal deletions) in transgenic tobacco and potato plants has revealed that it consists of at least two functionally distinct elements: an upstream quantitative enhancer, required for maximal levels of expression, whose activity is modulated by downstream regulatory elements (19). Within this latter element are located cis-regulatory sequences involved in inducible expression in wounded leaves and constitutive expression in tubers and flowers. Deletion analysis of the promoter did not distinguish wound-inducible from tuber-specific elements, with the promoter carrying internal deletions being active or inactive in tubers and damaged tissues in a parallel fashion. In contrast to this, promoter activity in flowers can be endowed by sequences located both within and outside the region where sequences involved in expression in tubers and wounded leaves have been mapped, and therefore has to be different from them (20). Thus, the maximal level of expression in flowers depends on a different promoter region as compared to the one needed for maximal expression in wounded leaves (from -624 to -504). Moreover, analysis of the effectiveness of ABA, JA, and sucrose in inducing the different Pin2 internal promoter deletions suggest that their action is mediated by the -624 to -405 promoter region. In this respect, ABA, JA, and sucrose cannot be distinguished from the stimuli responsible for expression in tubers or wounded leaves. However, in leaves, none of these compounds activated a promoter deletion which was constitutively expressed only in flowers, indicating that this tissue-specific expression has to involve other, so far unidentified, compounds.

WOUND RESPONSE AND THE INVOLVEMENT OF ABA

The *in vivo* involvement of ABA in the gene activation processes which follow mechanical damage of the plant tissue is supported by the fact that the endogenous ABA concentration rises three- to fivefold upon wounding. This increase is, moreover, not restricted to the tissue damaged directly but can also be detected in the non-wounded systemically induced tissue (31). This phenomenon is common to several plant species. ABA increases upon wounding have been detected in potato, tomato and tobacco leaves (42). Furthermore, in all three plant species a correlation could be established between the ABA increase and either the expression of the Pin2 gene family (in the case of potato and tomato), or the activity of an introduced Pin2 promoter (in the case of transgenic potato or tobacco plants) (31).

An increase of ABA levels after mechanical wounding of the foliage of other non-solanaceous species is likely to be a common feature of the plant response to wounding. For instance, hevein, a chitin-binding protein present in the laticifers of the rubber tree, has recently been shown to accumulate in the foliage of the tree, but not in the roots, upon mechanical wounding. ABA, as well ethylene treatment, leads to an increase in the hevein mRNA level, displaying the same organ specificity as shown upon wounding (5). In any case, it is also likely that the activation of other potato genes would be triggered by wounding along with the Pin2 gene family. The expression of a gene coding for a highly anionic peroxidase is induced during healing of potato tubers and tomato fruits. This gene is also responsive to ABA treatment; its mRNA accumulates in potato calli grown in 100 μM ABA (38). Similarly, the isolation by differential screening of four other wound- and ABA-responsive genes was recently reported (16). The distribution of the corresponding mRNA in the different organs of non-wounded, wounded, and ABA-sprayed wild-type and mutant potato plants showed that wounding or ABA treatment leads to a pattern of expression of these genes that is very similar to that of Pin2, thus supporting the direct involvement of this hormone in the signal transduction of mechanical damage.

The role of ABA in wound induction has proven to be more subtle than promoting gene transcription by directly triggering promoter activity. We can consider the expression of Pin2 in situations of plant water deficit as an example. Water deficits result in an increase in the endogenous ABA concentration, which in potato is some eight- to tenfold higher than the nonstressed levels. This leads to stomatal closure, thus preventing further water losses through the transpiration stream. Despite these high ABA levels, the Pin2 gene family is not transcribed in water-stressed potato or tomato plants. This inactivity is not due to toxic levels of ABA, as water-stressed plants transcribe and accumulate Pin2 when wounded (31). Moreover, a consensus ABA responsive element (ABRE) is found in the promotor region responsible for promoter activity in damaged leaves (-624 to -504) which might mediate the observed effect of ABA on wound-induced Pin2 expression. However, a mutation of this element to a sequence that did not allow binding of the cognate nuclear protein in the wheat Em promoter had no effect on Pin2 promoter activity (20). This suggest a more complex mode of ABA action on Pin2 transcription which must be consistent firstly with the fact that Pin2 is not active under water-stress conditions, despite a much higher rise in the endogenous ABA levels (31), and second with the absence of any activity in potato nuclear extracts binding to the ABRE consensus in this region (41). Most of the genes induced in plants subjected to water deficit also appear to be ABA-responsive (43), for example the late embryo-abundant (LEA) proteins, such as those coded for by the rice RAB 16 gene (27) and the Em gene of wheat (23). In addition, a maize seed protein has been described as ABA inducible which is also weakly induced upon wounding (12). However, neither the Pin2 gene family nor any of the other wound-inducible potato genes, such as cathepsin D inhibitor (Cdi) or threonine deaminase (Td) could be induced by water stress (16). Their mode of activation suggests that factors additional to the presence of ABA are necessary for the induction of this set of genes. Supporting this assumption, it was shown that the effect of ABA on Pin2 gene expression was not exerted directly. Detached potato leaves pre-treated with cycloheximide prevented Pin2 accumulation upon ABA treatment, suggesting the presence of a labile protein factor mediating the ABA action. In contrast to this, water-stress-responsive genes, such as RAB16 or the Em, appears to be independent of cytosolic protein synthesis. These results suggest a different induction pathway for Pin2 and the other wound-induced potato genes, as compared to the LEA genes as a result of ABA treatment. These results further indicate that whereas ABA would directly mediate responses to osmotic stress, a more complex signaling pathway might lead to transcriptional activation of the defense-related genes as the end result of the increased levels of ABA caused by wounding.

THE ROLE OF JASMONIC ACID IN WOUND-INDUCED Pin2 GENE EXPRESSION

The occurrence of JA and its derivatives is widespread in the plant kingdom and several different regulatory effects have been reproducibly associated with their presence. One of the most consistently observed effects is that JA treatment leads to plant responses similar to those caused by ABA treatment. Several ABA-induced proteins could be detected upon incubation of barley leaves in a JA solution, and these jasmonic acid-induced proteins (JIPS) were immunologically related to the proteins accumulating upon ABA treatment (51). It is interesting to note that both ABA and JA have been associated with wound-induced gene expression in diverse plant species (46). However, the expression of some of these genes is also affected by others factors showing the complexity of their regulation.

In many plant species wounding induces a systemic response in untreated leaves, including the induction of soybean vegetative storage proteins (VSPs) (25) and proteinase inhibitors (10, 34). Jasmonates also induces proteinase inhibitors in distant untreated leaves, and assimilation of methyl jasmonate (MeJA) through soybean petioles is an effective way to induce VSP gene expression in leaf plants. Wound induction of VSP is blocked by pretreatment with lipoxygenase inhibitors, whereas exogenous MeJA remains fully effective in VSP gene induction, suggesting a role of endogenous jasmonates during VSP induction upon mechanical damage (45). Jasmonic acid is synthesized from α-linolenic acid (LA) by a lipoxygenase (LOX)-mediated oxygenation leading to 13-hydroxyperoxylinolenic acid (13HPLA) which is the subsequently transformed to JA by the action of hydroxyperoxide dehydrase and additional modification steps (49). Both JA and its methyl ester are though to be significant components of the signalling pathway regulating wound-response in higher plants.

Airborne MeJA and JA (9), as well as intermediates of the JA biosynthetic pathway such as LA, 13HPLA and 12-oxo-phytodienoic acid (12-oxo-PDA) also lead to an accumulation of Pin2 mRNA in tomato leaves (11). Likewise, we have demonstrated that JA strongly induces the accumulation of Pin2 (32, 34) as well as other ABA-responsive/wound-induced genes in potato leaves (16). More interestingly, treatment of potato leaves with JA resulted in similar levels of mRNA accumulation in both wild-type and ABA-deficient mutant plants (Fig. 5). These data suggest that JA is involved in a step downstream of ABA in the pathway which links wounding to Pin2 gene activation. Additionally, JA could bypass the initial recognition events requiring ABA and thus trigger the induction of the genes even in the absence of the ABA. Nevertheless, we observed that JA-induced Pin2 gene activation, like ABA-induced Pin2 gene expression, can also be blocked by cycloheximide. This result suggests that some later step, necessary for Pin2 gene activation, depends on *de novo* protein biosynthesis.

Both the chemical structure as well as the biosynthetic pathway of JA resemble those of the mammalian eicosanoids (prostaglandins and leukotrienes) which are derived from LOX- and cyclooxygenase (COX)-mediated reactions (28, 49). To assess the role of endogenous JA in the wound-response, we treated tomato and potato plants with different mammalian LOX- and COX-inhibitors and analyzed the expression of Pin2 gene family as well as other wound-inducible genes (i.e. Cdi and Td) (35). The mammalian COX-inhibitor aspirin (acetyl salicylic acid) was one of the most effective inhibitors, blocking both local and systemic accumulation of wound-induced Pin2 gene activation (Fig. 6). This results suggest a common feature in both local and systemic activation of this wound-induced gene.

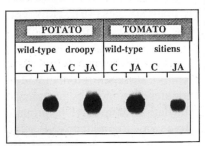

Fig. 5. Jasmonic acid induces Pin2 gene expression in both wild-type and ABA-deficient mutant plants. Wild-type and ABA-deficient potato (droopy) and tomato (*sitiens*) plants were sprayed with a 50 μM JA solution. The directly sprayed leaves (JA) were harvested 24 h after spraying. The RNA gel blot was hybridized against a Pin2 cDNA probe. As a control (C), RNA was isolated from a plant 24 h after spraying with water.

Mechanical wounding increases the endogenous level of JA in soybean (6). In tomato, wounding also leads to an increase of JA levels by 6 h. Thereafter, levels of JA decline, being at 24 h slightly higher than the control leaves. Tomato plants pretreated with aspirin and subsequently wounded showed the same low levels of JA as non-wounded plants (Fig. 7). These results strongly suggest that aspirin blocks some step of the JA biosynthetic pathway, thus preventing the synthesis of this compound upon wounding.

In order to obtain more information about the effect of aspirin on Pin2 gene expression, we investigated the influence of this substance on ABA- and JA-induced Pin2 mRNA accumulation. Additionally,

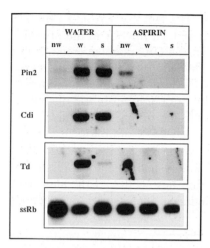

Fig. 6. Influence of aspirin on local and systemically induced Pin2 gene expression in tomato plants. Whole tomato plants were cut at the base of the stem and supplied with water alone or aspirin for 6 h and subsequently wounded. Both directly wounded (W) and non-wounded systemically induced leaves (S) were harvested after 20 h. The RNA gel blot was hybridized against wound inducible Pin2, Cdi, Td and the small subunit of ribulose biphosphate carboxylase (ssRB) (internal control) cDNA probes. As a control (NW) RNA was isolated from tomato leaves 6 h after supplying with water or aspirin, before wounding.

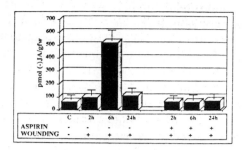

Fig. 7. Aspirin prevents the wound-induced increase of endogenous JA levels. Whole tomato plants were cut at the base of the stem and supplied with water alone or aspirin for 6 h and subsequently wounded. The endogenous levels of (-) JA was measured at different times (2, 6 and 24 h) after wounding in the presence (+) and absence (-) of aspirin.

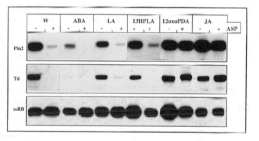

Fig. 8. Influence of ABA, LA, 13-HPLA, 12-oxo-PDA and JA on the inhibitory effect of aspirin. Detached tomato leaves were supplied either with water (W), abscisic acid (ABA), linolenic acid (LA), 13 hydroperoxylinolenic acid (13-HPLA), 12-oxo-phytodienoic acid (12-oxo-PDA) or jasmonic acid (JA), in the absence (-) or presence (+) of aspirin. The RNA gel blot was hybridized with Pin2, Td, and ssRB cDNA probes.

experiments using different intermediates of the JA biosynthetic pathway might allow to elucidate where aspirin is affecting this pathway. To this end, detached tomato leaves were pre-treated with aspirin (and others LOX and COX-inhibitors) and after 3 h supplied either with ABA, LA, 13HPLA, 12-oxo-PDA or JA (Fig. 8). Neither LA nor 13HPLA were able to complement the inhibition mediated by aspirin. Contrary to this, 12-oxo-PDA and JA which are the strongest inducers of Pin2 gene expression in tomato plants, were also able to overcome the inhibitory effect of aspirin. The effect of ABA on aspirin-mediated inhibition is difficult to assess due to the low levels of Pin2 gene expression, but it seems that this phytohormone is not able to overcome this inhibition. Essentially the same results were obtained upon analysis of other wound-inducible genes (Cdi and Td, Fig. 8). The fact that LA and 13HPLA do not restore the accumulation of Pin2, Cdi and Td mRNA in the presence of aspirin or propyl gallate, whereas 12-oxo-PDA does overcome this inhibition, strongly suggests that hydroperoxide dehydrase which mediates the formation of 12-oxo-PDA from 13HPLA is the target for these inhibitors.

Similar experiments were performed with LOX-inhibitors and the results obtained are summarized in Figure 9. As indicated in this schematic model, preincubation of tomato plants with mammalian LOX-inhibitors leads to a reduction of wound-induced gene expression by inhibiting 13HPLA formation. Aspirin prevents both the local and systemic gene expression of all three wound-inducible potato gene families (Pin2, Cdi and Td). Additionally, aspirin inhibits the wound-induced increase of endogenous

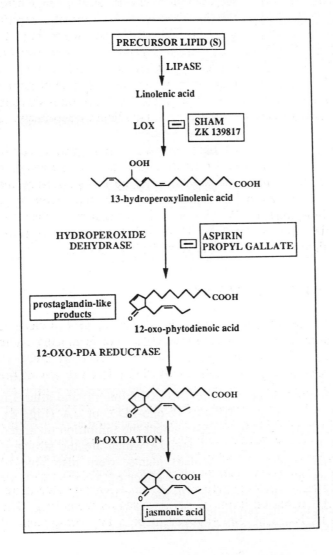

Fig. 9. Target-sites of different mammalian LOX- and COX-inhibitors on JA biosynthetic pathway in tomato leaves.

levels of JA by blocking the conversion of 13HPLA to 12-oxo-PDA, i. e., hydroperoxide dehydrase activity. Thus, these results suggest that a similar mechanism modulates the expression of three different wound inducible genes in tomato leaves. Moreover, the fact that ABA appears not to be able to overcome the inhibitory effect of aspirin in tomato, and that JA suppresses the inhibition by all inhibitors, strongly support the assumption that the step in the signal transduction chain in which JA is involved would be located downstream of ABA (35).

The same experiments were performed to study the role of endogenous JA in wound-induced gene expression in potato plants (our unpublished results). Interestingly, most of the results obtained in this system differ from those already described in tomato plants. As in tomato, only salicylhydroxamic acid (SHAM), 3,4,5-trihydrobenzoic acid propyl ester (propyl gallate) and aspirin block Pin2 activation. Additionally, aspirin prevents local and systemic Pin2 mRNA accumulation upon wounding. In contrast to tomato plants, LOX-inhibitors such as ibuprofen, eicosatetraynoic acid (ETYA) and ZK139817 do not affect wound-induced Pin2 activation. Assuming that LOX-inhibitors are taken up in potato petioles an explanation for this unexpected observation could be that potato LOX are insensitive to these substances or that an alternative pathway might mediate wound-response in this plants.

ABA leads to an accumulation of Pin2 mRNA in potato plants even in the presence of aspirin. This suggests either that ABA is able to overcome the inhibitory effect of this substance, which would be different from the situation in tomato, or that ABA can activate the gene upon wounding, independently of JA biosynthesis. We can not rule out that more than one mechanism may mediate the wound-response, and that there may be some differences between distinct though closely related plant species.

SIGNAL TRANSDUCTION PATHWAY

ABA appears not to be the only regulator involved in the control of changes in gene expression that occur in response to wounding. For instance, despite water stress promoting an increase of endogenous ABA levels by eight- to tenfold, this does not lead to any accumulation of Pin2 mRNA or any of the other wound-inducible genes from potato (16, 31). In agreement with these results, accumulation of water stress-responsive genes appears to be independent of *de novo* protein synthesis, whereas accumulation of Pin2 mRNA is not (31). These results indicate that different transduction mechanisms regulate these two ABA-mediated responses. Whereas ABA would directly mediate responses to osmotic stress, a more complex signalling pathway might lead to transcriptional activation of the defense-related genes as the final result of the increased levels of ABA caused by wounding. The fatty acid derivative JA has been hypothesized to be a key component of intracellular signaling in response to wounding or pathogen attack. MeJA was shown to stimulate the accumulation of wound-inducible vegetative storage proteins in soybean plants and suspension culture (24, 44). In addition JA and MeJA induce the expression of phenylalanine ammonia lyase (PAL) genes that are known to be involved in the chemical defense mechanism of plants against pathogens (13). As mentioned above, methyl jasmonate and intermediates of the JA biosynthetic pathway lead to an accumulation of Pin2 mRNA in both tomato and potato leaves (11, 32, 34) as well as all the isolated ABA-responsive/wound-induced genes (16). More interestingly, treatment of potato leaves with JA resulted in similar levels of mRNA accumulation in both wild-type and ABA-deficient mutant plants. These data suggest that the step where JA is involved is located downstream to the ABA requirement in the pathway which links wounding to Pin2 gene activation. That would be consistent with the association often found in plant responses to JA and ABA treatments. Additionally, JA could bypass the initial recognition events requiring ABA and thus trigger the induction of the genes even in the absence of ABA. The fact that ABA appears not to be able to overcome the inhibitory effect of aspirin in tomato, and that JA suppresses the inhibition by all inhibitors, further strongly indicates that the step in the signal transduction chain in which JA is involved is located downstream of ABA (35). However, JA-induced Pin2 gene activation can be also blocked by cycloheximide. This suggests that the *novo* protein biosynthesis is essential to produce the required factors involved in the most later steps mediating Pin2 gene activation.

JA is synthesized in plants from linolenic acid by an oxidative pathway similar to that leading to the synthesis of eicosanoids (prostaglandins and leukotrienes) in animals. Actually, the chemical structure of JA is very similar to that of prostaglandins (50). In mammals, eicosanoid synthesis is triggered by release of arachidonic acid from membranes into the cytoplasm, where it is metabolized into stress-related second messengers. In an analogous way, metabolites of linolenic acid might function as plant stress second messengers released during defense responses to wounding or pathogen attack (11). In this model, wounding is proposed to activate systemin, perhaps by releasing this potent 18-amino acid peptide from an inactive propeptide. Systemin could serve as a systemic signal that releases linolenic acid from membranes after binding to a plasma membrane receptor. Jasmonate synthesized from linolenic acid in turn might activate genes, possibly through another receptor.

Wounding induces Pin2 expression and leads to an increase of both endogenous ABA and JA levels in wild-type plants. Mechanical wounding does not induce the gene in ABA-deficient mutant plants nor increase endogenous ABA levels. Additionally, our preliminary results show that potato and tomato wild-type plants treated with ABA contain higher levels of endogenous JA than the control plants (untreated). On the other hand, ABA as well as JA are able to promote the accumulation of Pin2 and others wound-inducible genes in the ABA-deficient mutant plants (16, 31). According to the proposed model (11), increased levels of ABA as a result of tissue injury may lead to the activation of a lipase in the plasma membrane and the release of linolenic acid or, alternatively, to the activation of specific lipoxygenases that, when acting on linolenic acid, would produce a rapid accumulation of JA. Recently, it has been shown that a LOX gene from *Arabidopsis thaliana* can be induced by pathogens, ABA and MeJA (26). We have also observed that another enzyme mediating JA biosynthesis such as hydroperoxide dehydrase is accumulated either upon wounding or ABA treatment in both wild-type and ABA-deficient mutant plants (our unpublished results). Whatever the exact mechanism of ABA action is, it is important to note that high levels of this phytohormone are required to trigger the JA signaling pathway. Actually, elevated concentrations of ABA are required for expression of Pin2 and the other ABA/wound-responsive genes, and, accordingly, in ABA-deficient plants, only very low levels of the transcripts corresponding to these genes accumulate upon mechanical wounding (16). More interestingly, the polypeptide systemin is not able to induce Pin2 gene expression in ABA-deficient mutant plants (our unpublished results), supporting the importance of ABA for mediating the wound-response in higher plants. Further experiments will be required to elucidate the modes of action of ABA, systemin and electrical signal in order to aid in the understanding of how endogenous levels of JA are regulated. The isolation and characterization of genes encoding enzymes involved in JA biosynthesis as well as the creation of JA mutants could provide a more complete picture of the signal transduction pathway mediating wound-responses in higher plants.

References

1. Baydon, E.A., Fry, S.C. (1985) The immobility of pectic substances in injured tomato leaves and its bearing on the identity of the wound hormone. Planta, 165, 269-276.
2. Bishop, P.D., Pearce, G., Bryant, J.E., Ryan, C.A. (1984) Isolation and characterization of the proteinase inhibitor-inducing factor from tomato leaves. J. Biol. Chem. 259, 13172-13177.
3. Boller, T. (1988) Ethylene and the regulation of antifungal hydrolases in plants. Ox. Surveys Plant Mol. Cell Biol. 5, 145-174.
4. Bowles, D. (1990) Defense-related proteins in higher plants. Annu. Rev. Biochem. 59, 873-907.
5. Broekaert, W., Lee, H-I., Kush, A., Chua, N-H., Raikhel, N. (1990) Wound-induced accumulation of mRNA containing a hevein sequence in laticifers of rubber tree (*Hevea brasiliensis*). Proc. Natl. Acad. Sci. USA 87, 7633-7637.
6. Creelman, R.E., Tierney, M.L., Mullet, J.E. (1992) Jasmonic acid/methyl jasmonate accumulate in wounded soybean hypocotyls and modulate wound gene expression. Proc. Natl. Acad. Sci. USA 89, 4938-4941.
7. Duckham, S.C., Taylor, I.B., Linforth, R.S.T., Al-Naieb, R.J., Marples, B.A., Bowman, W.R. (1989) The metabolism of *cis*-ABA aldehyde by the wilty mutants of potato, pea and *Arabidopsis thaliana*. J. Exp. Bot. 40, 901-905.
8. Ecker, J.R., Davies, R.W. (1987) Plant defense genes are regulated by ethylene. Proc. Natl. Acad. Sci. USA 84, 5202-5206.

9. Farmer, E.E., Ryan, C.A. (1990) Interplant communication: Airborne methyl jasmonate induces synthesis of proteinase inhibitors in plant leaves. Proc. Natl. Acad. Sci. USA 87, 7713-7716.

10. Farmer, E.E., Ryan, C.A. (1992) Regulation of expression of proteinase inhibitor genes by methyl jasmonate and jasmonic acid. Plant Physiol. 98, 995-1002.

11. Farmer, E.E., Ryan, C.A. (1992b) Octadecanoid precursors of jasmonic acid activate the synthesis of wound-inducible proteinase inhibitors. Plant Cell 4, 129-134.

12. Gómez, J., Sanchez-Martinez, D., Stiefel, V., Rigau, J., Puigdomènech, P., Pagès, M. (1988) A gene induced by the plant hormone abscisic acid in response to water stress encodes a glycine-rich protein. Nature 334, 262-264.

13. Gundlach, H., Müller, M.J., Kutchan, T.M., Zenk, M.H. (1992) Jasmonic acid is a signal transducer in elicitor-induced plant cell cultures. Proc. Natl. Acad. Sci. USA 89, 2389-2393.

14. Hammond-Kosack, K.E., Atkinson, H.J., Bowles, D.J. (1990) Changes in abundance of translatable mRNA species in potato roots and leaves following root invasion by cyst-nematode *G. rostochiensis* pathtypes. Physiol. Mol. Plant Pathol. 37, 339-354.

15. Henstrand, J.M., Handa, A.K. (1989) Effect of ethylene action inhibitors upon wound-induced gene expression in tomato pericarp. Plant Physiol. 91, 157-162.

16. Hildmann, T., Ebneth, M., Peña-Cortés, H., Sanchez-Serrano, J., Willmitzer, L., Prat, S. (1992) General roles of abscisic and jasmonic acid in gene activation as a result of mechanical wounding. Plant Cell 4, 1157-1170.

17. Kernan, A., Thornburg, R.W. (1989) Auxin levels regulate the expression of a wound-inducible proteinase inhibitor II-chloramphenicol acetyl transferase gene fusion *in vitro* and *in vivo*. Plant Physiol. 91, 73-78.

18. Keil, M., Sanchez-Serrano, J., Willmitzer, L (1989) Both wound-inducible and tuber-specific expression are mediated by the promoter ogf a single member of the potato proteinase inhibitor II gene family. EMBO J. 8, 1323-1330.

19. Keil, M., Sanchez-Serrano, J., Schell, J., Willmitzer, L (1990) Localization of elements important for the wound-inducible expression of a chimeric potato proteinase inhibitor II-CAT gene in transgenic tobacco plants. Plant Cell 2, 61-70.

20. Lorberth, R., Damman, C., Ebneth, M., Amati, S., Sanchez-Serrano, J. (1992) Promoter elements involved in environmental and developmental control of potato proteinase inhibitor II expression. Plant J. 2, 477-486.

21. Malone, M., Stankovic, B. (1991) Surface potentials and hydraulic signals in wheat leaves following localized wounding by heat. Plant Cell Envir. 14, 431-436.

22. Malone, M. (1993) Hydraulic signals. Phil. Trans. R. Soc. Lond. B. 341, 33-39.

23. Marcotte, W.R., Bayley, C.C., Quatrano, R.S. (1988) Regulation of a wheat promoter by abscisic acid in rice protoplasts. Nature 335, 454-457.

24. Mason, H.D., Mullet, J.E. (1990) Expression of two vegetative storage protein genes during development and in response to water deficit, wounding and jasmonic acid. Plant Cell 2, 569-579.

25. Mason, H.D., DeWald, D.B., Creelman, R.A., Mullet, J.E. (1992) Coregulation of soybean vegetative storage protein gene expression by methyl jasmonate and soluble sugars. Plant Physiol. 98, 859-867.

26. Melan, M.A., Donng, X., Endara, M.E., Davis, K.R., Ausubel, F.M., Kaye Peterman, T. (1993) An Arabidopsis thaliana lipoxygenase gene can be induced by pathogens, abscisic acid, and methyl jasmonate. Plant Physiol. 101, 441-451.

27. Mundy, J., Chua, N-H. (1988) Abscisic acid and water-stress induce the expression of a novel rice gene. EMBO J. 7, 2279-2286.

28. Needleman, P., Turk, J., Jakschik, B.A., Morrison, A.R., Lefkowith, J.B. (1986) Arachidonic acid metabolism. Annu. Rev. Biochem. 55, 69-102.

29. Pearce, G., Strydom, D., Johnson, S., Ryan. C.A. (1991) A polypeptide from tomato leaves induces the synthesis of wound-inducible proteinase inhibitor proteins. Science 253, 895-898.

30. Peña-Cortés, H., Sanchez-Serrano, J., Rocha-Sosa, M., Willmitzer, L. (1988) Systemic induction of proteinase inhibitor II gene expression in potato plants by wounding. Planta 174, 84-89.

31. Peña-Cortés, H., Sanchez-Serrano, J., Mertens, R., Willmitzer, L., Prat, S. (1989) Abscisic acid is involved in the wound-induced expression of the proteinase inhibitor II gene in potato and tomato. Proc. Natl. Acad. Sci. USA 86, 9851-9855.

32. Peña-Cortés, H. (1990) Untersuchungen zur Charakterisierung der Expression von Proteinase-Inhibitor-II-Genen in Kartoffelpflanzen (*Solanum tuberosum* L.). Ph.D. Thesis, Freie Universität Berlin, Germany.

33. Peña-Cortés, H., Willmitzer, L., Sanchez-Serrano, J. (1991) Abscisic acid mediates wound induction but not developmental-specific expression of the proteinase inhibitor II gene family. Plant Cell 3, 963-972.

34. Peña-Cortés, H., Liu, X., Sanchez-Serrano, J., Schmid, R., Willmitzer, L. (1992) Factors affecting gene expression of patatin and proteinase inhibitor II gene families in detached potato leaves: Implications for their co-expression in developing tubers. Planta 186, 495-502.

35. Peña-Cortés, H., Albrecht, T., Prat, S., Weiler, E.W., Willmitzer L. (1993) Aspirin prevents wound-induced gene expression in tomato leaves by blocking jasmonic acid biosynthesis. Planta 191, 123-128.

36. Quarrie, S.A. (1982) Droopy: a wilty mutant of potato deficient in abscisic acid. Plant Cell Environ. 5, 23-26.

37. Quatrano, R.S. (1987) The Role of Hormones during seed development. *In*: Plant Hormones and Their in Plant Growth and development, pp. 494-514, Davies, P.J., ed. Kluwer Academic Publishers, Dordrecht, The Netherlands.

38. Roberts, E., Kolattukudy, P.E. (1989) Molecular cloning, nucleotide sequence, and abscisic acid induction of a suberization-associated highly anionic peroxidase. Mol. Gen. Genet. 217, 223-232.

39. Ryan, C.A. (1987) Oligosaccharide signalling in plants. Ann. Rev. Cell. Biol. 3, 295-317.

40. Ryan, C.A. (1990) Protease inhibitors in plants: Genes for improving defenses against insects and pathogens. Annu. Rev. Phytopath. 28, 425-449.

41. Sanchez-Serrano, J., Amati, S., Keil, M., Peña-Cortés, H., Prat, S., Recknagel, C., Willmitzer L. (1990) Promoter elements and hormonal regulation of the proteinase inhibitor II gene expression in potato. *In*: The Molecular and Cellular Biology of the Potato, pp. 57-69, Vayda M.E., Park W.E., eds. CAB International, Wallingford, UK.

42. Sanchez-Serrano, J., Amati, S., Ebneth, M., Hildmann, T., Mertens, R., Peña-Cortés, H., Prat, S., Willmitzer, L. (1991) The involvement of ABA in wound responses of plants. *In*: Abscisic acid: physiology and biochemistry, Davies, W.J., Jones H.G., eds. Bios Scientific Oxford.

43. Skriver, K., Mundy, J. (1990) Gene expression in response to abscisic acid and osmotic stress. Plant Cell 2, 503-512.

44. Staswick, P.E. (1990) Novel regulation of vegetative storage protein genes. Plant Cell 2, 1-6.

45. Staswick, P.E., Huang, J., Rhee, Y. (1991) Nitrogen and methyl jasmonate induction of soybean vegetative storage protein genes. Plant Physiol. 96, 130-136.

46. Staswick, P.E. (1992) Jasmonate, genes, and fragrant signals. Plant Physiol. 99, 804-807.

47. Taylor, I.B., Linforth, R.S.T., Al-Naieb, R.J., Bowman, W.R., Marples, B.A. (1988) The wilty tomato mutants flacca and sitiens are impaired in the oxidation of ABA-aldehyde to ABA. Plant Cell Environ. 11, 739-745.

48. Thornburg, R.W., Li, X. (1991) Wounding Nicotiana tabacum leaves causes a decline in endogenous indole 3-acetic acid. Plant Physiol. 96, 802-805.

49. Vick, B.A., Zimmermann, D.C. (1983) The biosynthesis of jasmonic acid: a physiological role for plant lipoxygenase. Biochem. Biophys. Res Commun. 111, 470-477.
50. Vick, B.A., Zimmermann, D.C. (1987) Oxidative systems for the modification of fatty acids. *In*: The Biochemistry of Plants, Vol 9, pp. 53-90, Stumpf, P.K., Conn, E.E., eds. Academic Press, New York.
51. Weidhase, R.A., Kramell, H-M., Lehman, J., Liebisch, H-W., Lerbs, W., Parthier, B. (1987) Methyl jasmonate-induced changes in the polypeptide pattern of senescing barley leaf segments. Plant Sci. Lett. 51, 177-186.
52. Wildon, D.C., Doherty, H.M., Eagles, G., Bowles, D.J., Thain, J.F. (1989) Systemic responses arising from localized heat stimuli in tomato plants. Ann. Bot. 64, 691-695.
53. Wildon, D.C., Thain, J.F., Minchin, P.E.H., Gubb, I.R., Reilly, A.J., Skipper, Y.D., Doherty, H.M., O'Donnell, P.J., Bowles, D. (1992) Electrical signalling and systemic proteinase inhibitor induction in the wounded plant. Nature 360, 62-65.

F. HORMONE ANALYSIS

F1. Instrumental Methods of Plant Hormone Analysis.

Roger Horgan

Department of Botany and Microbiology, University College of Wales, Aberystwyth, Dyfed, SY23 3DA, Wales, U.K.

INTRODUCTION

From information presented in previous chapters it will be clear to readers that plant hormones are, as a rule, present at very low levels in most plant tissues. Whilst relatively high levels of some hormones are found in immature seeds of certain species (e.g., GAs in developing pea seeds (10)) even these levels are low when compared with the levels of most plant secondary metabolites. Thus while many alkaloids, terpenoids and phenolics may be present at levels of mgs per gm dry weight of plant material, plant hormones are usually present at several hundred to several thousand fold lower levels. It is not suprising therefore that knowledge of the chemical identity of plant hormones has been limited by the techniques available for their isolation in a pure state and by the sensitivity of the spectroscopic techniques required to elucidate their chemical structure.

In the last fifteen years there have been spectacular improvements in the sensitivity of spectroscopic methods of structure determination and corresponding increases in performance in chromatographic techniques, principally via the development of high performance liquid chromatography (HPLC) and capillary column gas-liquid chromatography (GLC). This improvement in methodology can be clearly seen if one compares the isolation and identification of zeatin by Letham in 1963 (15), where 60 kg of plant material had to be extracted and purified by traditional chromatographic methods to yield the mg of material needed for spectroscopic studies, with the identification of 1'-deoxy ABA as a precursor of abscisic acid (ABA) in the fungus *Cercospora rosicola* (25), where, after purification by HPLC, identification was possible at the μg level.

This chapter is concerned with the application of modern instrumental techniques to the isolation, identification and quantitation of plant hormones. Clearly the theoretical background to these techniques is beyond the scope of this work and readers are referred to suitable textbooks for this information (e.g., 9, 28, 19). However, it is very important for a critical understanding of the methods used in the identification of plant hormones to appreciate the

P. J. Davies (ed.), Plant Hormones, 415–432.

inherent limitations of the various techniques and so these will be touched upon in the relevant sections. In particular it is necessary to appreciate the importance of sample purity to the interpretation of spectroscopic data. Even with the most sophisticated instrumentation, correct identifications can only be made if the spectroscopic data obtained is relevant to the hormone under investigation. Two examples of 'mistaken identities' provide informative reading on this point (33, 5).

This chapter is organised in the chronological order in which a real analysis of a plant hormone would probably proceed. First, the compound would have to be isolated in a sufficiently pure form, second, its structure would have to be determined by appropriate methods, and finally a strategy would need to be devised for its quantitative measurement.

ISOLATION AND PURIFICATION OF PLANT HORMONES

Extraction and Preliminary Purification

The methods of extraction and preliminary purification of plant hormones using traditional methods such as solvent partitioning, ion exchange chromatography and, paper and thin layer chromatography will not be discussed in this chapter as strictly speaking they fall outside the area of instrumental methods and many of these methods are being superseded by HPLC based methods. Nevertheless it is often necessary to revert to older methods particularly with plant extracts that are too large for the initial use of HPLC. The readers attention is drawn to the extremely comprehensive treatment of these methods by Yokota et al. (34).

Bioassays

Although the bulk of this chapter is concerned with the use of physical methods for the detection of plant hormones, it should be noted at this point that the primary detection of any novel plant hormones is dependent on bioassay. Bioassays are also necessary when studying the hormone content of novel plant materials, particularly with regard to gibberellins and cytokinins. Because of the trace nature of plant hormones in most extracts, direct physico-chemical detection is impossible during the early stages of purification. In the case of the gibberellins even detection at the latter stages of purification is difficult due to the low wavelength and low extinction coefficient of UV absorption by these compounds. In these situations bioassays have to be used to detect the compounds of interest and to monitor the purification process. The choice of suitable bioassays can be problematical. Ideally a bioassay should respond quantitatively to all the members of a certain group of plant hormones, be highly selective towards that particular group of compounds, have high sensitivity and not be inhibited by other compounds in the plant extract. These conditions are never met in practice. In the case of the gibberellins the most suitable assays are probably the dwarf maize assay (26) and the Tanginbozu dwarf rice assay[1] (22). Both respond well to a good range of GAs, and the former assay is particularly suitable for relatively crude plant extracts where the presence of inhibitors may interfere with other assays. In the case of the cytokinins, which by definition are promoters of cell division in plant callus cultures, the Tobacco callus (23) or the Soybean callus (20) bioassay is required to unambiguously reveal the presence of cytokinins in a plant extract. However these assays require about 21 days for growth and so the more rapid, but probably less specific, *Amaranthus* betacyanin bioassay (3) is frequently used for routine monitoring of cytokinin activity during purification processes.

It should be pointed out that although bioassays are necessary for the detection of plant hormones, they are now generally accepted to be unsuitable for quantitative work. The presence of inhibitors in most plant extracts and the logarithmic nature of the response of most assays makes for very inaccurate and imprecise measurements. To large extent realisation of the quantitative limitations of bioassays has provided the stimulus for the development of the physical methods of quantitation described in this chapter.

High Performance Liquid Chromatography

HPLC is distinguished from traditional chromatography by its high efficiency and resolution, and speed of separation. The first two improvements are achieved by the utilisation of small particle size stationary phase materials at the cost of relatively complex and expensive instrumentation. This results mainly from the high liquid pressures required to achieve fast flow rates through small particle size columns. The resolution of a system is largely dependent on the chemistry of the mobile and stationary phases. The relative importance of the chemical (separation factor α and capacity factor k') and physical (efficiency N) terms in governing the resolution (Rs) can be seen from the chromatographic equation:

$$R_s = \frac{1}{4}\left(\frac{\alpha-1}{\alpha}\right)\left(\frac{k'}{k'+1}\right)\sqrt{N}$$

Optimising α and k' can only be achieved through careful selection of mobile and stationary phases. The power of HPLC and high resolution capillary GLC stems from the huge increases in N possible in these systems. From the point of view of plant hormone analysis any increase in resolution of a chromatographic system is valuable since it increases the chances of separating the compound(s) of interest from other interfering materials.

---

[1] The use of prohexadione calcium, which blocks the 2β- and 3β-hydroxylation of GAs (e.g., GA_{20} to GA_1), combined with uniconazole, an inhibitor of GA biosynthesis, now enables the use of non-dwarf rice and increased sensitivity in this bioassay (Nishijima et al., Plant Growth Regulation 13, 241-247, 1993).

In assessing HPLC methods for plant hormone analysis it is important to appreciate the presence of two often conflicting factors. Firstly the need to separate the hormone from other compounds present in the extract, and secondly the ability of the system to separate closely related hormones. In general most HPLC systems for plant hormones have been developed using the second criterion. Whilst this is a useful indicator of the resolution of the system it should be borne in mind that when used preparatively on a relatively crude plant extract the desired compound may not be obtained in a degree of purity sufficient for unambiguous interpretation of subsequent spectroscopic data. The first attempt to develop a HPLC system for plant hormones was directed at improving the resolution of an open column partition chromatography system for gibberellins (27). Although this work illustrated the potential of HPLC it has not found widespread use due to the technical complexity of using aqueous stationary phases. The most important development in HPLC as far as plant hormones are concerned was the commercial production of so called 'bonded reverse phase materials'. In reverse phase chromatography the support material, usually silica, is coated with a liquid phase of very low polarity. In aqueous solution low polarity compounds will partition into the stationary phase and will only be eluted with solvents of lower polarity than water.

For plant hormone HPLC the most frequently used stationary phase is octadecylsilane (ODS or C18), which is covalently bonded to microparticulate or microspherical silica. Solvents are usually binary mixtures or gradients of water (weak solvent) and the lower alcohols or acetonitrile (strong solvents). Since the selectivity of reverse phase HPLC is at its best when the compounds being analysed are un-ionised, acidic aqueous phases are used for the analysis of acid plant hormones such as GAs, ABA and IAA, whereas neutral buffers are preferred to suppress the ionisation of basic compounds such as cytokinins. The great value of reverse phase HPLC for the isolation of plant hormones lies in the enormous range of compound polarities that can be accommodated on a single stationary phase if gradient elution is used. Under reversed phase conditions polar compounds will elute before non-polar compounds. Thus polar plant hormone conjugates and free compounds may be fractionated in a single HPLC step. Gradient elution reverse phase HPLC therefore provides a very valuable method for the preliminary purification of plant extracts. Indeed reverse phase HPLC on reasonably large size columns (10 to 20 mm ID) is gradually superseding the older methods of solvent partitioning, paper chromatography and open column chromatography as the first step in plant hormone purification. With columns of this size injection sizes of several mls can be used and if the initial solvent is of low polarity the sample will concentrate at the end of the column. With subsequent gradient elution surprisingly good resolution can be achieved with very crude samples. A typical chromatogram resulting from the use of this technique is shown in Fig. 1. An 80% methanol extract of bean leaves was chromatogaphed on a ODS reverse phase column (10 x 150mm) with a linear

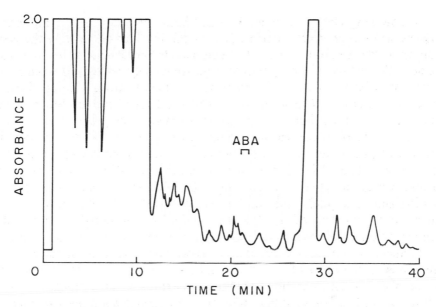

Fig. 1. Reverse phase HPLC of an extract of *Phaseolus vulgaris* leaves on a 150 x 10 mm column of Spherisorb ODS-2 using a linear gradient of 20% methanol in 0.1 M acetic acid to 100% methanol over 40 min at a flow rate of 5 ml min^{-1}. The fraction denoted by the bar, at the retention time of an ABA standard, was collected and analysed as described in the text.

gradient of 20% methanol in 0.1 M acetic acid to 100% methanol over 40 mins. The chromatogram shown in Fig. 1 is the absorbance trace obtained by monitoring the column eluate at 265 nm. The fraction corresponding to the elution time of ABA was collected, methylated with diazomethane and examined by GC-MS. Methyl ABA was clearly identified by its mass spectrum. The superior resolution of HPLC over conventional open column chromatography is illustrated in Fig. 2 which compares the reverse phase chromatography of a mixture of cytokinins on LH 20 Sephadex and on an octadecyl silica HPLC material. Details of reverse HPLC systems for the separation of gibberellins (14), cytokinins (12) and, ABA and IAA (7) have been published.

It should be pointed out that at the stage of initial purification the collection of the correct fraction(s) from a preparative HPLC column will be on the basis of the elution time of a standard compound if a known hormone

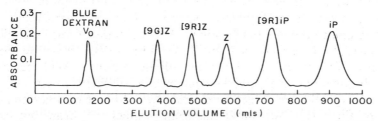

Fig. 2a. Separation of a series of cytokinins on a 80 x 2.5 cm column of LH20 Sephadex eluted with 35% ethanol at a flow rate of 30 ml hr^{-1}.

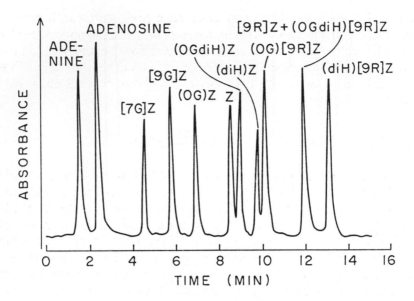

Fig. 2b. Separation of a series of cytokinins by HPLC on a 150 x 4.5 mm column of Spherisorb ODS-2 using a linear gradient of 5% acetonitrile in water (pH adjusted to 7 with triethylammonium bicarbonate) to 20% acetonitrile over 30 mins at a flow rate of 2 ml min^{-1}.

is being purified or on the basis of biological activity, resulting from bioassaying successive fractions, if a new compound is being isolated. With a crude extract the usual detectors, UV absorption or refractive index (RI), are of little value in locating the compound(s) of interest. This point will be discussed later in the context of the quantitative measurement of plant hormones.

Gas-Liquid Chromatography

GLC is usually used as the last chromatographic step of a hormone analysis, either utilising a specific detector on the GLC instrument or in combination with mass spectrometry (GC-MS). In all cases the specificity of the analysis is affected by the resolution of the GLC system itself. It is instructive to compare the efficiency term N in the chromatographic expression for the traditional packed chromatographic column, where the liquid phase is deposited on a particulate support, and a wall coated capillary column where the phase is coated as a very thin layer on the wall of the column. Although the value N, expressed as theoretical plates per meter is similar for the two types of column (i.e., about 1500), the practicality of using long capillary columns, typically 25 meters, means that these columns can exhibit efficiencies of say 50000 theoretical plates compared to about 5000 theoretical plates for 3 or 4 meter packed column. In seeking to identify and quantify plant hormones in complex extracts the increased resolution obtained with capillary columns makes a very valuable contribution to the analytical system. For most GC-MS analyses of plant hormones, capillary columns, particularly those in which the stationary phase is chemically bonded to the wall of the column, would be used. Fig. 3a shows the separation of a group of permethylated cytokinins on a 1.5 m packed GLC column. Fig. 3b shows the separation of the same compounds on a 12 m bonded phase capillary column.

IDENTIFICATION OF PLANT HORMONES

The problems of the chemical characterisation of plant hormones need to be considered at two levels of difficulty. At the easier level there is the problem of confirming the identity of known plant hormones in extracts of plant in which they have not previously been identified. At the much more difficult level there is the problem of identifying unknown compounds exhibiting biological activity corresponding to one of the known classes of plant hormone or possibly a previously undescribed form of biological activity. For confirming the identity of known plant hormones there is no doubt that mass spectrometry (MS) and in particular combined gas chromatography-mass spectrometry (GC-MS) is the technique 'par excellence'. Whilst GC-MS also

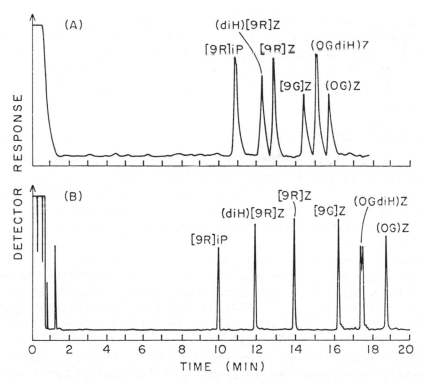

Fig 3. GLC separation of a series of permethyl cytokinins on (A) a conventional packed column (1.5 m x 4 mm 3% OV1) and (B) a 12 m x 0.3 mm bonded phase capillary column.

plays an important role in the tentative identification of novel compounds, complete identification at this level frequently requires the use of additional techniques and eventually requires confirmation of the proposed structure by unambiguous chemical synthesis.

GC-MS Identification of Plant Hormones

The coupling of the outlet of a gas chromatography column to the source of a mass spectrometer enables mass spectra to be obtained as individual components elute from the gas chromatograph. Since modern mass spectrometers can scan a decade of mass (e.g., 300-30 amu (atomic mass units)) in less than one second, several mass spectra may be obtained on a single GC peak. With the addition of computerised acquisition, storage and processing of the mass spectra, enormous amounts of information may be obtained about the chemical nature of complex plant extracts. Ideally the GLC system should be able to resolve all the components of the mixture under analysis. However, with suitable data processing software individual components can be identified even if they coelute from the GC column. Thus provided the mass spectrum of the plant hormone under investigation is available as a reference, and additionally if possible its retention time (Kovats retention index) on the GLC system being used, it is a relatively easy matter to confirm the presence of a known compound in a partially purified plant extract. Details of the theory and operation of GC-MS systems and associated data systems may be found in Rose and Johnson (28). In this section some of the features of the mass spectra of members of the main classes of plant hormones will be described in relation to the use of MS and GC-MS for their identification.

With the exception of ethylene, GLC of all plant hormones requires that they first be converted to volatile derivatives. In general carboxylic acids are converted to methyl esters by diazomethane, alcohol groups to trimethylsilyl (TMS) derivatives with a reagent such as bis-trimethylsilyl acetamide (BSA) or to methyl ethers with the methyl sulphonyl anion and methyl iodide (DMSO$^-$/CH$_3$I), and amino groups TMSed or methylated as above. Thus in identifying known compounds mass spectra are rarely recorded for the free compound, but usually via GC-MS of one or more derivatives.

Auxins

The mass spectra of a large number of IAA derivatives have been recorded (8, 32, 2) and thus provide information for identifying these compounds in plant extracts. The mass spectra of compounds related to IAA are usually very simple and are typified by the spectrum of IAA methyl ester and its TMS derivative shown in Fig. 4. The ion corresponding to the unfragmented molecule, the M$^+$ or molecular ion, is apparent in both spectra. The great stability of the indole nucleus is reflected in the base peaks which have the structures shown in Fig. 5. The presence of one of these spectra in

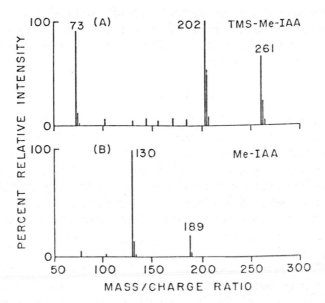

Fig. 4. Electron impact mass spectra of (A) trimethylsilyl-IAA methyl ester and (B) IAA methyl ester.

the GC-MS run of a suitably derivatised plant extract, at the correct retention time, may be taken as conclusive proof for the existence of IAA in the extract.

Gibberellins

GC-MS was first used for plant hormone analysis by MacMillan and his co-workers in identifying gibberellins (17). Although basically similar in structure, many of the gibberellins exhibit widely differing substitution patterns which are reflected in the mass spectra of their methyl-TMS derivatives (4). Whilst certain ions are very diagnostic of hydroxyl groups at certain positions in the molecule, e.g., 207/208 for 13-OH and 129 for 3-OH, the unambiguous identification of a known GA requires careful comparison of the mass spectrum with that of a standard compound, and

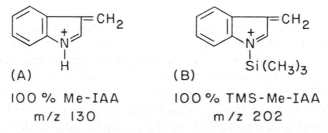

Fig. 5. Structures of the base peaks in the electron impact mass spectra of (A) IAA methyl ester and (B) trimethylsilyl-IAA methyl ester.

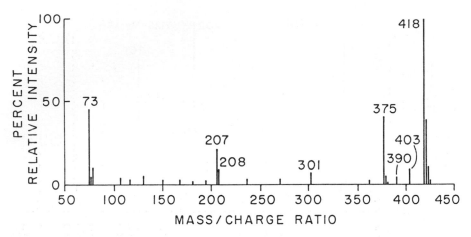

Fig. 6. Electron impact mass spectrum of trimethylsilyl GA_{20} methyl ester.

often requires the addition information provided by the Kovats retention index (1). Because of their cyclic structures, the molecular ions of many methyl/TMS GAs are very prominent and the spectra are very simple. This is illustrated by the spectrum of methyl/TMS GA_{20} shown in Fig. 6. The presence of the 13-OH group is clearly indicated by the prominent ions at m/z 207 and 208.

Cytokinins

Both TMS and permethyl derivatives of cytokinins exhibit good GLC properties and have mass spectra which provide excellent fingerprints for the identification of known compounds in plant extracts by GC-MS (21, 16, 31). From the point of view of interpretation of fragmentation patterns, the mass spectra of permethyl cytokinins are easier to understand than those of the TMS compounds. Maximum structural information is shown by the MS of underivatised cytokinins introduced directly into the source of the mass spectrometer using a heated probe. The mass spectrum of underivatised zeatin is shown in Fig. 7. The molecular ion at m/z 219 is clearly visible and suggests the presence of an odd number of nitrogen atoms on the molecule. The ions at m/z 202 and 188 arise by losses of fragments of 17 and 31 amu from the molecular ion and indicate the presence of a primary -OH group. The presence of strong ions at m/z 136, 135, 119 and 108 provide strong evidence that the molecule contains an adenine moiety. The location of the side chain on the 6-amino group is indicated by the presence of an ion at m/z 148 and the presence of an additional alkyl group in the side chain is indicated by an ion at m/z 160.

Readers should note that detailed analysis of the fragmentation of zeatin is largely retrospective and that mass spectra of unknown compounds are often very difficult to interpret. In the case of zeatin, mass spectrometry alone would not be sufficient to completely characterise the side chain.

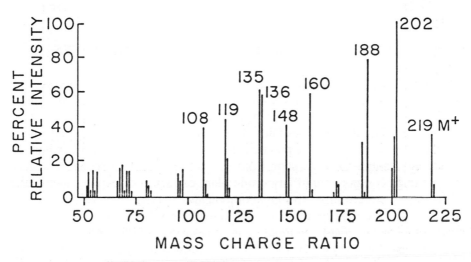

Fig. 7. Electron impact mass spectrum of zeatin.

Abscisic Acid and Related Compounds

The mass spectral fragmentation of methyl ABA, shown in Fig. 8, has been examined in considerable detail (11). Whilst the molecular ion is very weak there are a sufficient number of characteristic ions to enable the compound to be easily identified in plant extracts. It should be noted that methyl ABA and its 2-trans isomer have identical mass spectra. Thus identification of ABA in a plant extract should also involve the information provided by the GLC retention time.

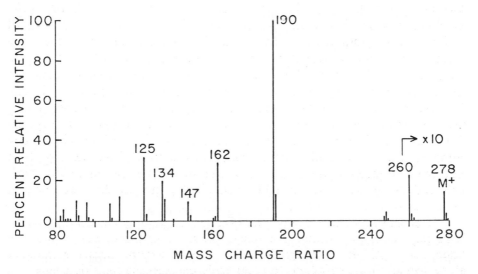

Fig. 8. Electron impact mass spectrum of methyl abscisate.

QUANTITATIVE MEASUREMENT OF PLANT HORMONES

The ultimate aim of most analyses of plant hormones is to measure precisely and accurately the level of the compound(s) of interest under certain known physiological conditions. Whilst theoretically any physico-chemical detector with a defined quantitative response to the particular compound(s) may appear suitable for this purpose, the practical problems of hormone measurement are enormous. In general these stem from the fact that plant hormones are nearly always trace components in any crude plant extract and are present together with many structurally similar compounds. Thus the selectivity and sensitivity of the measurement system is extremely important. All currently used physico-chemical methods for measuring plant hormones use a chromatographic system, HPLC or GLC, coupled to detectors with varying degrees of sensitivity and selectivity. Detectors with a low degree of selectivity, e.g., flame ionisation with GLC, or single wavelength UV with HPLC, necessitate the use of very pure extracts to ensure that the peak being measured contains only the compound of interest. For this reason these detectors are rarely used to measure hormones in plant extract, and results obtained with these detectors should be judged very critically. The recent advent of diode array UV/VIS detectors, which can record whole spectra on HPLC peaks, or even sections of HPLC peaks, as they pass through a cell, offers a way to selectively measure plant hormones directly in HPLC eluates.

An example of the use of a highly sensitive and specific detector may be seen in the use of an electron capture detector (ECD) for the GLC determination of ABA (29). The unsaturated carbonyl group of ABA renders it electron capturing and, in a suitably partially purified plant extract, ABA may often be the only compound at the correct GLC retention time to exhibit this property and therefore can be accurately measured. In a similar way the use of a nitrogen specific detector has been proposed as a method of measuring cytokinins although the method had not yet found widespread use.

IAA exhibits a very strong and characteristic fluorescence with an excitation maximum at 280nm and an emission maximum at 350nm. A sensitive spectrofluorimeter with a narrow band width for excitation and emission therefore makes a sensitive and selective HPLC detector for IAA. With the correct instrumentation IAA can be measured at the pg level in relatively crude plant extracts (6).

In general plant extracts need to be purified to varying degrees before measurements can be made. During the purification process losses of material occur and therefore these have to be accounted for in the final result. Thus methods for the reliable quantitative measurement of plant hormones require the use of a suitable internal standard. In principal, an internal standard may be any compound sufficiently similar in its chemical and physical properties to the compound of interest to pass through the extraction and purification procedures with the same degree of loss. However, it must be sufficiently different from the compound of interest to be measured independently in the final quantitative step. Thus by adding a known amount of internal standard at the initial extraction step and measuring it accurately at the final step its percentage recovery may be calculated. Provided the first criterion above has been met this will be identical to the recovery of the endogenous compound and thus the losses of this during workup may be corrected for.

The need to meet the two somewhat conflicting criteria mentioned above place several restrictions on types of internal standards suitable for plant hormone quantitation. Thus isomers and homologues are not usually suitable internal standards. For example 2-trans abscisic acid (2T-ABA) has different solubility in several organic solvents to ABA and may not be recovered to the same degree on solvent partitioning of plant extracts. It separates from ABA in a number of chromatographic systems, and although this makes it a suitable internal standard from the point of independent measurement at the last step, it introduces severe complications when purifying plant extracts by TLC or HPLC.

The most suitable internal standards for plant hormone determinations are isotopically labelled versions of the compounds themselves. Compounds may be labelled with radioactive and/or heavy isotopes. The techniques involved in using these different types of internal standard depend on the methods necessary to detect and measure the internal standard and the hormone itself. Radioactive internal standards may be measured by liquid scintillation counting. Thus the recovery of labelled hormone in the course of purification procedure is simply calculated as ratio of the DPM added at the initial extraction phase to the DPM present at the final measurement stage. If this is assumed to be the same as the recovery of the endogenous compound, then multiplication of the final measured amount of endogenous compound by this ratio will give the amount present in the original extract. Superficially this appears to be a very straightforward method limited only by the availability of suitable radioactive hormones. In practice, however, it is subject to certain limitations. Whilst the measurement of amount of internal standard is specific for the radioactive material the same is not true for the measurement of the endogenous compound. Measurement systems such as GLC or HPLC with UV or fluorescence detection will respond to both labelled and unlabelled molecules of hormone. Thus the amount of hormone measured will be the sum of the endogenous material plus the internal standard. Since the amount of internal standard, at this stage, will be know from the radioactivity present in the extract, it can be subtracted from the amount measured by the particular detection system in use to give the correct value for the endogenous hormone. If the specific activity of the internal standard is high (usually only possible with 3H-labelled materials) the amount that has to be added to an extract to give a measurable number of counts in the final analysis may be so small that its contribution to the endogenous compound may be neglected. On the other hand, if the specific activity of the internal standard is low it may effectively overwhelm the endogenous compound to such an extent that the amount cannot be accurately measured. It is also important that the internal standard remains radiochemically pure throughout the procedure and any labelled compounds formed by chemical or biochemical transformations during the workup are separated from it before the final counting. The need to measure both radioactivity and amount of hormone present means that two independent measurements need to be made, and this introduces a greater degree of error into the proceedings.

In general the best methods for the quantitation of plant hormones use heavy isotope-labelled internal standards. Selected examples of these are shown in Fig. 9. Since mass spectrometers can easily distinguish between isotopically labelled molecules GC-MS can be used to measure the relative proportions of an endogenous compound and its heavy isotope-labelled internal standard. Thus if the ratio of these in the final sample is know together with the amount of internal standard added at the initial extraction stage the simple expression:

Endogenous amount =
$$\frac{\text{Endogenous ion intensity}}{\text{Internal standard ion intensity}} \quad \text{x Amount of internal standard added}$$

can be used to calculate the original amount of endogenous hormone in the extract. Heavy isotope internal standards have been used to measure ABA (24), IAA (18), cytokinins (30) and gibberellins (13) in plant extracts. In practice the ratio of endogenous to heavy isotope-labelled compound may be measured from suitable ions in a full mass spectrum obtained from a GC-MS analysis. Because there is often slight separation of the labelled and

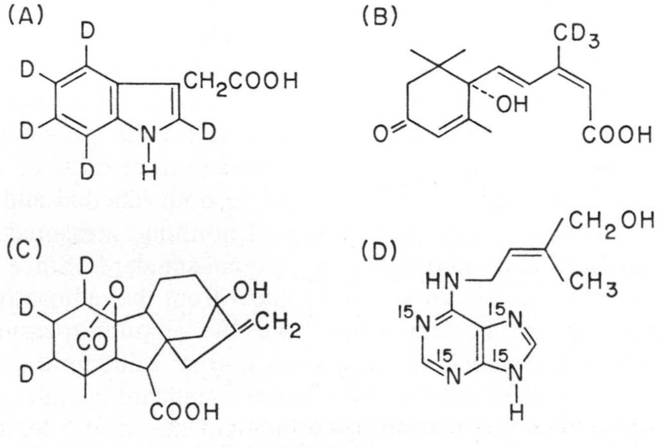

Fig. 9. Structures of some heavy isotope labelled plant hormones used as internal standards. (A) [^{2}H$_5$]IAA. (B) [^{2}H$_3$]ABA. (C) [^{2}H$_3$]GA$_{20}$. (D) [^{15}N$_4$]zeatin.

unlabelled compounds on the GC, a single mass spectrum taken at one point on a GC peak may give an erroneous value of the isotope ratio.

Measurements of this type are usually made with the mass spectrometer operating in the selected ion monitoring (SIM) mode. In this mode the mass spectrometer is adjusted to monitor one or two ions of the endogenous compound and the equivalent mass shifted ions of the internal standard. Because the ions are close together in mass, they can be selectively monitored by electrical rather than magnetic scanning. This allows them to be cyclically selected very rapidly and measured with a much improved signal to noise ratio. Thus many measurements of the isotope ratios may be made over a single GLC peak with much greater sensitivity than can be achieved from the full mass spectra method. The ratio of the areas under the relevant peaks can be used as measure of the relative amounts of endogenous compound to internal standard. Thus the amount of endogenous hormone originally present in the extract can be calculated as described above. A SIM trace from the GC-MS of extract of ABA from *Phaseolus vulgaris* leaves is shown in Fig. 10. The ions monitored are the m/z 190 ion from the endogenous methyl-ABA and the corresponding ion at m/z 193 from the $^2H_3$ internal standard shown in Fig. 9.

An obvious limitation of this method is the presence of interfering ions from other compounds with the same GC retention time as the hormone. Before the SIM method is used, full MS should be obtained from a sample of the hormone purified in an identical manner to that used for the quantitative determinations. If potentially interfering ions are observed, the purification method has to be modified to eliminate the interfering compound(s).

It can be seen from the foregoing account that the identification and measurement of plant hormones is a technically difficult problem. In this

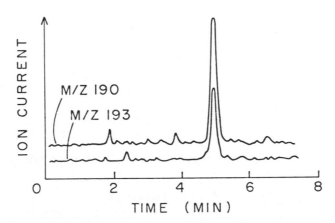

Fig. 10. SIM trace from the determination of ABA in an extract of *Phaseolus vulgaris* leaves utilising [$^2H_3$]ABA as an internal standard. The peak for the ion at m/z 190 is due to endogenous methyl abscisate and the peak at m/z 193 is due to the internal standard.

chapter it has only been possible to discuss briefly the procedures involved. Whilst the methods described are in general use in many other areas of analytical biochemistry, their application to plant hormone analysis is not necessarily as straightforward as this account may suggest. Because of the complex chemical nature of plant extracts and the low levels of many plant hormones analytical methods often have to be operated at their limits of sensitivity and resolution. In addition different plant species often pose different analytical problems due to qualitative and quantitative differences in interfering compounds. These factors have to be taken into account when applying the techniques discussed in this chapter.

Editors Note:

In contrast to all other chapters, this chapter has not been revised, except for a few small changes, due to the unfortunate illness of Roger Horgan. Over the past eight years there have been only incremental advances in analytical techniques, resulting primarily from improved commercial columns and analytical instruments. While the resulting resolution and sensitivity in analysis has increased, the basic principles outlined in this chapter remain the same. Notable instrumental advances have been the increased use of diode array absorbance detectors, which enable the absorbance spectra of compounds eluting from an HPLC column to be recorded, the downsizing of benchtop GC-MS instruments, and the appearance of HPLC-MS. On the techniques side, small cartridges of HPLC-type packing material for pre-HPLC purification purposes, and the routine availability of [^{2}H] gibberellins[2], [^{13}C]IAA[3], and [^{2}H]cytokinins[4] have increased the ease with which the techniques described in this chapter can be carried out. The recent volume *GC-MS of the Gibberellins and Related Compounds: Metholodogy and a Library of Spectra* by P. Gaskin and J. MacMillan (Cantocks Press, Bristol, UK)[5] is an essential reference for all those who wish to undertake the analysis of plant hormones by GC-MS. It includes mass spectra not only of GAs but all other hormones including derivatives, metabolites and possible contaminants in hormone extracts.

---

[2] Available for purchase from Dr. L. N. Mander, Research School of Chemistry, Australian National University, Canberra, Australia (Fax: 61-62-490750).

[3] Available from Cambridge Isotope Laboratories, Woburn, MA, USA (Fax: 1-508-749-2768).

[4] Available from Apex Organics Ltd., Honiton, Devon, UK (Fax: 44-404-47525). The mention of this and the other company above is given for the convenience of readers and does not constitute an endorsement of these companies. Other sources may be available.

[5] Available from Dr. P. Gaskin at Long Ashton Research Station, Long Ashton, Bristol UK.

References

General Reference:
   Principles and Practice of Plant Hormone Analysis, vol. 1 and 2, Rivier L., Crozier, A.,
   eds. Academic Press, London, 401 pp, (1987).

1.   Albone, K.S., Gaskin, P., MacMillan, J., Sponsel, V.M. (1984) Identification and
     localisation of gibberellins in maturing seeds of the cucurbit *Sechium edule*, and a
     comparison between this cucurbit and the legume *Phaseolus coccineus*. Planta 162,
     560-565.
2.   Badenoch-Jones, J., Summons, R.E., Entsch, B., Rolfe, B.G., Parker, C.W., Letham, D.S.
     (1982) Mass spectrometric identification of indole compounds produced by *Rhizobium*
     strains. Biomed. Mass Spectrometry 9, 430-436.
3.   Biddington, N.L., Thomas, T.H. (1973) A modified *Amaranthus* betacyanin bioassay for
     the rapid determination of cytokinins in plant tissues. Planta 111, 183-186.
4.   Binks, R., MacMillan, J., Pryce, R.J. (1969) Plant hormones VIII. Combined gas
     chromatography-mass spectrometry of the methyl esters of gibberellins A1 to A24 and
     their trimethylsilyl ethers. Phytochemistry 8, 271-284.
5.   Bowman, W.R., Linforth, R.S.T., Rossall, S., Taylor, I.B. (1984) Accumulation of an
     ABA analogue in the wilty tomato mutant, flacca. Biochemical Genetics 22, 369-378.
6.   Crozier, A., Loferski, J., Zaerr, J., Morris, R.O. (1980) Analysis of picogram quantities
     of indole-3-acetic acid by high performance liquid chromatography-fluorescence
     procedures. Planta 150, 366-370.
7.   Durley, R.C., Kannangara, T., Simpson, G.M. (1982) Leaf analysis for abscisic, phaseic
     and 3-indolylacetic acids by high performance liquid chromatography. J. Chromatography
     236, 181-188.
8.   Ehmann, A. (1974) Identification of 2-O-(indole-3-acetyl)-D-glucopyranose,
     4-O-(indole-3-acetyl)-D-glucopyranose and 6-O-(indole-3-acetyl)-D-glucopyranose from
     kernels of *Zea mays* by gas liquid chromatography-mass spectrometry. Carbohydrate
     Research 34, 99-114.
9.   Engelhart, H. (1978) High Performance Liquid Chromatography. Springer-Verlag, Berlin.
10. Frydman, V.M., Gaskin, P., MacMillan, J. (1974) Qualitative and quantitative analysis
     of gibberellins throughout seed maturation in *Pisum sativum* cv. Progress No. 9. Planta
     118, 123-132.
11. Gray, R.T., Mallaby, R., Ryback, G., Williams, V.P. (1974) Mass spectra of methyl
     abscisate and isotopically labelled analogues. J. Chem. Soc Perkin Trans. 2, 919-924.
12. Horgan, R., Kramers, M.R. (1979) High-performance liquid chromatography of
     cytokinins. J. Chromatography 173, 263-270.
13. Ingram, T.J., Reid, J.B., Murfet, I.C., Gaskin, P., Willis, C.L., MacMillan, J. (1984)
     Internode length in *Pisum*. The *Le* gene controls the 3-hydroxylation of gibberellin A_{20}
     to gibberellin A_1. Planta 160, 455-463.
14. Koshioka, M., Harada, J., Takeno, K., Noma, M., Sassa, T., Ogiyama, K., Taylor, J.S.,
     Rood, S.B., Legge, R.L., Pharis, R.P. (1983) Reverse phase C18 high-performance liquid
     chromatography of acidic and conjugated gibberellins. J. Chromatography 256, 101-115.
15. Letham, D.S., Shannon, J.C., MacDonald, I.R.C. (1964) The structure of zeatin, a
     (kinetin like) factor inducing cell division. Proc. Chem. Soc. London 230-231.
16. MacLeod, J.K., Summons, R.E., Letham, D.S. (1976) Mass spectrometry of cytokinin
     metabolites. Per(trimethylsilyl) and permethyl derivatives of glucosides of zeatin and
     6-benzylaminopurine. J. Org. Chem. 41, 3959-3967.
17. MacMillan, J., Pryce, R.J., Eglinton, G., McCormick, A. (1967) Identification of
     gibberellins in crude plant extracts by combined gas chromatography-mass spectrometry.
     Tetrahedron Lett. 2241-2243.

18. Magnus, V., Bandurski, R.S., Schulze, A. (1980) Synthesis of 4,5,6,7 and 2,4,5,6,7 deuterium labelled indole-3-acetic acid for use in mass spectrometric assays. Plant Physiol. 66, 775-781.
19. Millard, B.J. (1979) Quantitative mass spectrometry. Heyden, London.
20. Miller, C. O. (1963) Kinetin and kinetin-like compounds. *In*: Modern methods of plant analysis, Vol. VI, pp. 194-202, Linskins, H.F., Tracey, M.V., eds. Springer-Verlag, Berlin, Heidelberg, New York.
21. Morris, R.O. (1977) Mass spectrometric identification of cytokinins. Glucosyl zeatin and glucosyl ribosylzeatin from *Vinca rosea* crown gall. Plant Physiol. 59, 1029-1033.
22. Murakami, Y. (1968) The microdrop method, a new rice seedling test for gibberellins and its use for testing extracts of rice and morning glory. Bot. Mag. 79, 33-43.
23. Murashige, T. and Skoog, F. (1962) A revised medium for rapid growth and bioassays with tobacco tissue cultures. Physiol. Plant. 15, 473-497.
24. Neill, S.J., Horgan, R., Heald, J.K. (1983) Determination of the levels of abscisic acid-glucose ester in plants. Planta 157, 371-375.
25. Neill, S.J., Horgan, R., Lee, T.S., Walton, D.C. (1981) 3-methyl-5(4-oxo-2,6,6-trimethylcyclohex-2-en-1-yl)-2,4 pentadienoic acid from *Cercospora rosicola*. FEBS Letters 128, 30-32.
26. Phinney, B.O. (1956) Growth response of single-gene mutants of maize to gibberellic acid. Proc. Natl. Acad. Sci. USA. 42, 185-186.
27. Reeve, D.R., Crozier, A. (1978) The analysis of gibberellins by high performance liquid chromatography. *In*: Isolation of Plant Growth Substances, SEB Seminar Series, Vol. 4, pp. 41-77, Hillman, J.R., ed. Cambridge University Press.
28. Rose, M.E., Johnson, R.A.W. (1982) Mass spectrometry for chemists and biochemists. Cambridge University Press.
29. Saunders, P.F. (1978) The identification and quantitative analysis of abscisic acid in plant extracts. *In*: Isolation of Plant Growth Substances, SEB Seminar Series, Vol.4, pp. 115-134, Hillman, J.R., ed. Cambridge University Press.
30. Scott, I.M., Horgan, R. (1984) Mass spectrometric quantification of cytokinin nucleotides and glycosides in tobacco crown gall tissue. Planta 161, 345-354.
31. Scott, I.M., Horgan, R., McGaw, B.A. (1980) Zeatin-9-glucoside a major endogenous cytokinin of *Vinca rosea* L. crown gall tissue. Planta 149, 472-475.
32. Udea, M., Bandurski, R.S. (1974) Structure of indole-3-acetic acid myoinositol esters and pentamethyl-myoinositols. Phytochemistry 13, 243-253.
33. Van Staden, J., Drewes, S.E. (1974) Identification of cell division inducing compounds from coconut milk. Physiol. Plant. 32, 347-352.
34. Yokota, T., Murofushi, N., Takahashi, N. (1980) Extraction, purification and identification. *In*: Encyclopedia of Plant Physiology, Vol. 9, Molecular aspects of plant hormones, pp. 113-201, MacMillan J., ed. Springer-Verlag, Berlin.

F2. Immunoassay Methods of Plant Hormone Analysis

John L. Caruso[1], Valerie C. Pence[2] and Leslie A. Leverone[2]

[1]Department of Biological Sciences, University of Cincinnati, Cincinnati, OH, 45221, USA, and [2]Center for Reproduction of Endangered Wildlife, Cincinnati Zoo and Botanical Garden, Cincinnati, OH, 45220, USA.

INTRODUCTION

The use of immunoassays in the analysis of plant hormones has increased significantly since the founding work of Weiler and associates (58). Immunoassays are based on the ability of animals to produce proteins (*antibodies*) which recognize and bind to specific compounds (*antigens*) foreign to the animal. The specificity and sensitivity of antibodies make immunoassays attractive for the quantitation of plant hormones, which usually occur at low concentrations. Small amounts of plant tissue can thus be used in the assay, and, in addition, large numbers of samples can be processed in a brief period of time. Although some purification of extracts is generally required for the immunoassays, it is generally less than the more extensive clean-up needed in physico-chemical methods.

This chapter will review some of the major features of immunoassays as well as some of the problems in assaying for specific hormones. The use of antibodies in the localization of plant hormones will also be described.

ANTIBODY PRODUCTION

Plant hormones are *haptens*, or small molecules which on their own cannot elicit an antibody response. Thus they must be covalently linked to a large protein, such as an albumin (BSA, HSA, OA). Keyhole limpet hemocyanin has also been used as a carrier protein because of its reportedly high immunogenicity (40). A solution of the hormone-protein antigen, described below, is mixed with an oily adjuvant which slows the release of the antigen after injection. A series of injections are then made into an appropriate animal to stimulate antibody production.

Rabbits have been the most commonly used animal for polyclonal antibody (PAb), or antiserum, production. The animals are injected at several sites on the back and boosted for a series of weeks. Other injection sites, such as footpads, have been used, but are not recommended because of the discomfort to the animal. Bleedings can be made from the ear, and the

P. J. Davies (ed.), Plant Hormones, 433–447.

antibody-containing serum isolated. A number of spleen cells are involved in producing antibodies in response to the antigen, and these antibodies will differ in their recognition of the *epitope*, or portion of the antigen molecule, against which they are directed. As a result, some of the antibodies may be less specific than others, if they react against an epitope shared by related compounds. The PAb preparation is thus a mixture of more and less specific antibodies, as well as other, unrelated, antibodies.

The production of monoclonal antibodies (MAbs) allows for the isolation of individual antibodies and those which are most specific for the hormone can be selected (see 42, for review). Mice are often used, and undergo a series of injections similar to those used for rabbits. The mouse is then killed and the spleen removed and aseptically macerated in a sterile medium. The spleen cells which produce the antibodies cannot continue growth in culture on their own. Thus, they are fused with mouse tumor (myeloma) cells, which grow well in culture. The fusion products, or hybridoma cells, are capable of both growing in tissue culture and of producing antibodies. A series of dilution cultures isolates clones from single hybridoma cells, each of which synthesizes only one form of an antibody against the hormone.

Antibodies from a variety of clones can be tested and those which are most specific for the antigen are selected for the immunoassay. The hybridoma lines can be stored frozen, and as needed, grown for the production of antibodies, either in tissue culture or by injection into mice. In the latter case, the cells form tumors. Mice are killed, and tumor fluid (ascites) containing the antibodies is removed.

Dilutions of the rabbit serum or mouse ascites may be used directly in the immunoassays, or they can be partially purified by ammonium sulfate precipitation. In some cases, affinity purification is used to improve activity, particularly with PAbs (34). Both anion- (1) and cation- exchange chromatography have also been used (30).

The decision to use MAbs or PAbs must be based on a number of factors, including the greater initial cost and facilities needed for producing MAbs, as well as any improvement in specificity the MAbs may provide for measurement of a particular hormone. Some commercially produced antibodies are currently available, providing the researcher with the option of purchasing them rather than engaging in the labor-intensive production of these proteins[1].

THE IMMUNOASSAYS

Both radioimmunoassays (RIAs) and enzyme-immunoassays (EIAs) have been used in the quantitation of plant hormones. These techniques differ

---

[1] MAbs are currently available from Idetek Inc, Sunnyvale, CA 94089; MAb kits are obtainable from Sigma Chemical Company, St. Louis, MO 63178; ABA PAbs can be obtained from ICN Biomedicals, Costa Mesa, CA, 92626, and from Sigma.

essentially in the type of tracer or labeled hormone used. The EIA has the advantage of avoiding the use of radioisotopes, and thus is safer and is less expensive; it also requires less space and equipment. By utilizing enzymes, however, the EIA is subject to factors which affect enzyme activity, both in the synthesis of the hormone-enzyme tracer and in running the assay (27). The worker should be aware that the enzyme tracer may introduce into the assay the same bridge between hormone and protein which was present in the original antigen. When PAbs are used, antibodies may be present which recognize this bridge in preference to the hormone itself, causing preferential binding of the tracer over the free hormone, thus decreasing the sensitivity of the assay (58). In choosing between the two types of immunoassay, available equipment, expertise and antibodies must be considered. Whereas both EIAs and RIAs may be viable options for most laboratory use, the non-isotopic character of the former is probably a significant factor in its greater use.

There are a number of protocols for immunoassays, but Figure 1 outlines one which has been commonly used for plant hormones. Labeled and unlabeled hormones are mixed with the antibodies to compete for binding sites. The antibodies with their bound hormone are then separated from the unbound hormone and tracer. The antibody may be precipitated with ammonium sulfate, or it may be adsorbed onto a polystyrene substrate. Unbound materials are removed by washing. An EIA in which the antibody is immobilized by adsorption onto a solid substrate is called an enzyme-linked immunosorbent assay, or ELISA, and these have been the EIAs used in plant hormone analyses. An avidin-biotin complex can also be incorporated into the ELISA to enhance its sensitivity (28, 47).

The amount of tracer is measured, either as counts per minute (RIA) or as enzyme activity (EIA). Tritium is the most commonly used radiolabel, although ^{125}I has also been used. The activity of the most commonly used enzyme, alkaline phosphatase (AP), is measured colorimetrically following the conversion of substrate, *p*-nitrophenylphosphate, to *p*-nitrophenol. An AP-dependent conversion of NADP to NAD, with enzymic cycling of the latter, has also successfully increased the sensitivity of hormone detection (14). The amount of bound tracer is inversely proportional to the amount of unlabeled hormone with which it competes in the incubation medium; thus, the color produced in the above reaction is inversely proportional to the amount of unlabeled hormone present. Standard curves are often presented as the concentration of standard hormone plotted against

the logit $\dfrac{B}{B_0}$ (or $\log{}_n \dfrac{B/B_0}{1-B/B_0}$, where

B = binding of tracer to antibody in the presence of hormone,
B_0 = binding of same in the absence of hormone).

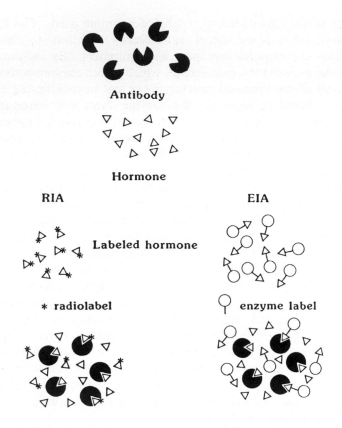

Bound hormones isolated from unbound hormones

Bound label measured

Fig. 1. Generalized scheme of immunoassays for plant hormones.

SPECIFIC HORMONE ANALYSIS

Both RIAs and EIAs have been developed for the four major classes of non-volatile plant hormones. The sensitivities of these assays frequently compare favorably with those of the physico-chemical methods of detection, ranging from 0.03 pmol in the RIA for zeatin riboside to as little as 0.001 pmol in EIAs for GAs (Table 1).

Table 1. Some reported sensitivities of immunoassays for plant hormones.

Hormone (p mol)	Assay	Sensitivity	Reference
IAA	EIA	0.02	59
	RIA	0.3	26
ABA	EIA	0.002	14
	RIA	0.02	38
ZR	EIA	0.015	33
	RIA	0.03	55
GAs	EIA	0.001	3
	RIA	0.005	60

Indoleacetic Acid

Two types of antigens have generally been used for producing antibodies to IAA. The first is an IAA protein conjugate synthesized by the Mannich formaldehyde reaction (34, 37), which couples the ring of IAA to the carrier protein, presumably at the indolic nitrogen of IAA (IAA-N). The second is coupling of IAA to protein via the carboxyl group (IAA-C1'), using the mixed anhydride (56) or carbodiimide reaction (51, 59). Because of the different epitopes presented by these two methods, the resulting antibodies differ in their cross-reactivities. Antibodies raised against the IAA-C1' antigen are poorly reactive with free IAA unless the charged carboxyl group is neutralized by methylation. In contrast, antibodies raised against the IAA-N antigen recognize free IAA, and methylation, which commonly employs highly toxic and explosive diazomethane, can be avoided. PAbs to IAA-N, however, have been reported as giving lower titers than IAA-C1' antibodies (37).

There have been several approaches to avoiding cross-reactivity with IAA conjugates when measuring free IAA. One of these has been the use of acidic partitioning (pH 3.0) of the extract with an organic solvent to separate free IAA from the conjugates. The production of MAbs has also been used as a way to improve detection of free IAA. IAA-C1' MAbs can be more specific than the PAbs, although there still may be some cross-reactivity with selected conjugates (29). The production of a MAb to IAA-N avoided the problem of low titers often seen with IAA-N PAbs (1). A C5-linked hapten has also been used to produce a free IAA- recognizing MAb. This MAb, however, showed a high degree of cross-reactivity for 4-Cl-IAA (49), which could pose a problem for those working with certain leguminous species. Another approach is the use of affinity purification to concentrate IAA-N Pabs, compensating for the low antibody titers. This is accomplished by passing the crude antibody preparation through an agarose column to which

Table 2. Comparison of crude and affinity purified IAA-N antibody preparations in the ELISA (34).

| Antibody | Coating Concentration (µg/well) | Reaction Time (min) | Absorbance in ELISA[1] | | %Drop With IAA |
			Without IAA	With IAA[2]	
Crude	10	45	0.551	0.416	25
Purified	0.8	38	1.871	0.249	87

[1] Absorbance is inversely proportional to free hormone concentration.
[2] Saturating concentration.

IAA has been covalently linked by the same reaction used to produce the antigen (34). After elution from the column, the antibody may show marked improvement in the ELISA (Table 2).

Abscisic Acid

Conjugation of ABA to a protein carrier for the production of the antigen can be accomplished at the carboxyl group, using either the mixed anhydride or carbodiimide reaction. The resulting antibodies, however, cross react with naturally occurring carboxyl-linked ABA conjugates posing a problem similar to that seen in IAA determinations. Antibodies to carboxyl-linked (C1) ABA often have a higher affinity for ABA methyl ester than for free ABA, although the difference is less than that seen between free and methylated IAA. By linking ABA to the carrier protein at the C4' keto group, antibodies are obtained which recognize free ABA (40, 50). In a study of potential receptor probes, MAbs were produced which recognize different biologically active analogs of ABA (52).

As with IAA-Cl', enzyme tracers for ABA ELISAs have been made with the carbodiimide reaction. An exception has been the conjugation of alkaline phosphatase to a previously synthesized ABA-BSA conjugate using glutaraldehyde to link the two proteins (9).

Another consideration in the synthesis of antibodies to ABA has been their differing reactivities with (S)-ABA and (R)-ABA, or (+)-ABA and (-)-ABA, respectively. In the first assays reported, PAbs were raised against antigen synthesized from the racemic mixture (R,S)-ABA. When tested against the individual enantiomers, these antibodies showed a preference for (R)-ABA, rather than the naturally occurring (S)-ABA, thus greatly reducing the effectiveness of the assay (53). PAbs raised against carboxyl-linked (S)-ABA, an expensive compound, showed little reactivity with (R)-ABA (57). In contrast, use of the C4'-linked (R,S)-ABA generated antibodies with a strong affinity for the (S)-ABA enantiomer (54). A MAb against C1-linked (S)-ABA showed exclusive recognition for (S)-ABA (38).

As with IAA-Cl', enzyme tracers for ABA ELISAs have been made with the carbodiimide reaction. An exception has been the conjugation of alkaline phosphatase to a previously synthesized ABA-BSA conjugate using glutaraldehyde to link the two proteins (9).

Cytokinins

Antigens for the production of antibodies and enzyme tracers for cytokinin ELISAs are produced using cytokinin-ribosides in a periodate reaction which opens the sugar ring and links dialdehydes to free amino groups (forming Schiff's base) on the carrier protein (25). PAbs have been used in RIAs (16, 17, 43) and ELISAs (39) for the determination of cytokinins. Avidin linked to an AP-conjugate of isopentenyl adenosine (iPA) had strong affinity for biotinylated PAb raised against the same hapten and the result was a sensitivity of 0.003 pmol (47). The specificity for a single family of cytokinins (e.g. zeatin and zeatin riboside) was improved by the production of a MAb (35). Using MAbs, a comparison was made between RIA and a fluorescence ELISA for the assay of ZR. The fluorescence ELISA showed a greater sensitivity at 0.03 pmol (48).

Gibberellins

The large number of GAs having similar chemical structures, and hence nondiscriminate epitopes, frequently results in antibody recognition of sets rather than specific hormones (30). For example, antibodies raised against GA_3 also recognized GA_7, GA_{20}, and to some extent GA_1 (2). Linking the hapten to the immunogenic carrier at the carboxylic acid (C7) generally gives less discrimination than forming a linkage at other positions. Furthermore, as with the ester-recognizing antibodies against ABA and IAA, samples must be methylated prior to analyses. An alternative method which used GA-protein conjugates with linkages at C3 (GA_1) and C17 (GA_4 and GA_9) produced MAbs which specifically recognized GA_1, GA_{20}, GA_4, and GA_9 (22).

Since only a few GAs are commercially available as standards, it is perhaps understandable why this class of hormones has not been as widely used in immunoassays as the other plant hormones.

VALIDATION OF IMMUNOASSAYS

Materials in plant extracts can either competitively or non-competitively inhibit the interaction between antibody and antigen. These interfering chemicals have been referred to as specific and non-specific, respectively (20). Such influences can vary with extracts of tissues from different species, with different experimental treatments of the same tissue and with different treatments of the same extract. Thus there is a frequent need for a comparison of the immunoassay with those physico-chemical methods which give the investigator more certitude. Combined gas chromatography-mass spectrometry (GC-MS) is such a method and it is becoming more frequently employed for validation. Internal radiolabeled standards have also been used

(8, 18). A less costly method, however, and one that is frequently encountered in the literature is the construction of dilution curves from the mixture of standard hormone and extract. An extract dilution curve which is not parallel to that of a standard hormone indicates the presence of interfering materials which must be removed before analysis. Interference of antibody generally gives a lower value of B_o and consequently an overestimation of endogenous hormone.

Whereas nonparallel dilutions indicate interference, parallel dilutions cannot assure the absence of interference. Spiking samples with known amounts of standard can indicate interfering interactions, even with extracts which exhibit parallelism (19). In addition, compounds with binding affinities similar to that of the antigen can give parallel dilutions, as has been shown for NAA in the IAA-N RIA (37). In such cases, purification of the sample is needed.

Again, the performance of an extract in the immunoassay depends on the particular hormone, the tissue being extracted and the treatment given. As an illustration, crude extracts gave parallel dilution curves in a ZR ELISA (33) whereas acidic ether partitioning was necessary to eliminate interference in extracts of crown gall tumors in an IAA-N ELISA (34; Fig. 2). Relatively simple modifications of partitioning improved an IAA-Cl' ELISA (51). Acidic and basic ether partitioning were also effective in the preparation of an IAA-N RIA, but base hydrolysis of the extract to release auxin from its conjugates produced interfering chemicals (36). Thus, more stringent purification, as provided by such methods as HPLC, is frequently necessary (7, 38, 39). An intensive comparison of crude and HPLC-purified extracts using validation of GC-MS revealed a 10-fold overestimation of IAA in crude

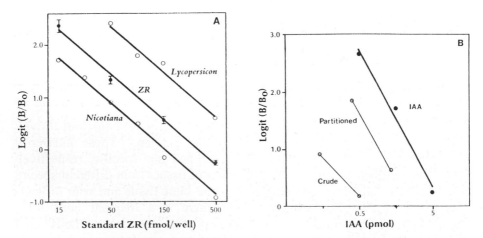

Fig. 2. a) Dilutions of tobacco (*Nicotiana tabacum*) and tomato (*Lycopersicon esculentum*) crown gall tumor tissue crude extracts compared with standard ZR dilutions in the ZR PAb ELISA. **b)** Dilution of tobacco crown gall tumor tissue extracts, both crude and acidic (pH 3.0) ether-partitioned, compared with standard IAA dilutions in the IAA N-PAb ELISA.

extracts of apple leaf, a five-fold overestimation in *Lemna* and a 60% overestimation in dry bean seeds, as determined by ELISA (8). An average of 16% overestimation of IAA was seen in willow buds, using RIA and comparing its results with GC-MS (5).

Similarly, immunoassays of ABA can be troublesome. The choice of buffer for dilution curves is important since HEPES gave false-positives in the ELISA readings of ABA (4). Significantly high false values for ABA were also obtained following the use of PVPP, polyamide and C_{18} Sep Pak columns in the clean up process (31). Furthermore, mannitol has been reported to contain up to 1.7 ng ABA per gram commercial product (4). Contaminating ABA was effectively removed by precipitating mannitol with methanol. It is also important to add that good agreement between ABA analyses by RIA and GC-MS has been observed using crude extracts of tomato leaf, bean leaf and the endosperm of *Sechium* (50).

Agreement has also been observed between RIA and GC-MS analyses of cytokinin in cultured anise cells (12) and minor differences were reported when the same methods were compared for several cytokinins of dry *Zea mays* seed (17). Cytokinin immunoassays, however, are not exempt from problems posed by contaminants. A fraction which was washed from a DEAE cellulose column used to purify cytokinins showed activity in a PAb-RIA for the zeatin family (17). The same fraction inhibited the *Amaranthus* bioassay but showed no bioassay activity of its own and did not have the partitioning characteristics of the cytokinins being assayed. Thus, here, as with IAA and ABA, stringent purification is sometimes necessary.

A difference of only 10% was observed between RIA and GC-MS analyses of GA_{29} extracted from pea seed cotyledons and 8.4% between similar analyses of GA_{20} in the liquid endosperm of *Marah macrocarpus* (45).

OTHER USES OF ANTIBODIES

Immunoaffinity Purification of Hormones

Immobilizing antibodies on immunoaffinity columns offers the possibility of sequestering the hormone of interest from a complex mixture of chemicals in plant extracts. When eluted, the purified and concentrated hormone can be quantitated by a variety of methods. Elution of hormone with an organic solvent may denature the antibody, although there have been reports of repeated methanol elutions from PAb-columns (10) and numerous cycles of acetone elution from a MAb-column (58). CNBr-activated Sepharose (45) and microcrystalline cellulose (25), as well as other supports, have been used to build an affinity matrix. The capacity of a column can be tested with radiolabeled antigen.

Immunocytochemical Localization of Hormones

Immunocytochemistry is a powerful tool in plant biology and the technique has been used to localize organ-specific antigens in a wide variety of plant tissues (21). Although the more easily localized proteins and polysaccharides have long been considered the target molecules of choice, success has also been achieved in using immunocytochemical techniques for the localization of small molecules such as plant hormones.

There are many variations in the preparatory steps for localizing an antigen, whether it be at the light microscope or ultrastructure level. In the simplest terms, embedded or non-embedded tissue, which has been fixed or frozen, is sectioned to insure contact between antibody and antigen as well as to visualize the location. A *primary antibody* (for example, antibody against mouse immunoglobulin G, or IgG MAb) with a well defined specificity for the hormone in question, is applied to the section. Following a sufficient number of washes to remove excess antibody, a *secondary antibody* (for example, goat anti-mouse IgG) is applied. The secondary antibody may be conjugated to a fluorophore (such as fluorescein or rhodamin) or an enzyme. The secondary antibody attaches to the primary antibody and, following washes, the preparation is examined with an appropriate microscope. The secondary antibody may instead be labelled with colloidal gold and applied to grids supporting ultrathin sections for examination with the electron microscope.

Nonspecific retention properties of either the primary or secondary antibody may pose a major problem in the interpretation of results. Primary antibodies must of course be specific and sensitive. Those which lack these properties may fail to bind selectively and in sufficient quantity to the target hormone.

In addition to antibody characterization, a series of controls should be used. These include, but are not limited to, the following:

1. binding of antigen to antibody before application to the specimen;
2. deletion of the primary or secondary antibody, or both;
3. substitution of a preimmune antiserum or control MAb such as P3-X63Ag8 (with no known antigen) for the primary PAb or MAb, respectively.

A cellular localization of cytokinins was accomplished using antisera against dihydrozeatin (DHZR) in corn root sections. Specimens were fixed in formaldehyde, fresh frozen or freeze substituted. Immunofluorescence and immunogold were both used to show the presence of cytokinin in meristem cells surrounding the quiescent center (63). A specific MAb against iP (isopentenyl adenine) localized this hormone (or a closely related substance) in protonemal cells of a moss mutant which overproduced cytokinin (11).

ABA was localized in *Chenopodium* by first coupling the hormone to cellular proteins with carbodiimide. Anti-ABA PAbs were applied to semi-thin sections and the light microscope was used to show an insoluble product of the peroxidase-antiperoxidase (PAP) complex (46). In this instance, water stress was used to the advantage of the investigators in that hormone levels increased prior to the tissues being fixed for sectioning. ABA was shown by the immunogold technique to vary in its distribution in tomato root tips, according to zones at the tip (6). In root cap cells, ABA appeared to accumulate in the apoplast, cytoplasmic vesicles and immediately surrounding the starch grains. In the meristematic cells, the hormone appeared to accumulate near the junction of the root cap. The carbodiimide typically used to link a carboxyl group of the hormone to an amino group of protein is l-ethyl-3 (3-dimethyl-amino-propyl) carbodiimide hydrochloride (EDC). It has been shown, however, that EDC has a much lower fixation efficiency compared with di-isopropyl carbodiimide (IPC) in linking ABA to cellular protein (44). It was also shown in the same study that the PAP staining method failed to detect ABA that was known to be at a level in excess of 1 pmol in pea cotyledons.

PAbs made against N-linked IAA localized auxin in chloroplasts and mitochondria of expanding peach leaves fixed in paraformaldehyde (32). This fixative was also used in this laboratory to link free IAA to proteins in root tips of four day-old *Zea mays* seedlings. Paraffin-embedded sections were dewaxed and incubated in IAA-N antiserum, washed and then exposed to a goat-anti-rabbit rhodamin conjugate. Sections viewed with epifluorescent optics revealed fluorescence associated with amyloplasts in columella cells (Fig. 3). In another study of root cap cells in corn, IAA was anchored at the carboxyl group with EDC. An MAb which recognized C1'-linked IAA was then used in conjunction with a goat anti-mouse colloidal gold conjugate. Gold particles were found in nuclei, vacuoles, mitochondria, dictyosomes and dictyosome-derived vesicles (41). The same type of MAb was used in localizing an IAA-zein conjugate in corn endosperm, which was confirmed by base hydrolysis and GC/MS analysis (24).

FUTURE PROSPECTS

The use of immunological techniques in plant hormone research is increasing, albeit slowly. Recent improvements in immunoassays have been refinements of an already useful tool. The electrochemical (EC) immunoassay is such an innovation which appears to be at the forefront of biosensor technology. EC has evolved since we first described it (35) to the point where a few thousand molecules of analyte can be detected (13, 62).

The development of the immunoassay has contributed significantly to multiple growth regulator analysis, within the context of a single experimental study (15, 23, 28, 61), and this approach will most likely increase in the years ahead. Furthermore, if the number of sources of commercial antibodies should increase, there will probably be a corresponding increase in the number of users of the immunoassay.

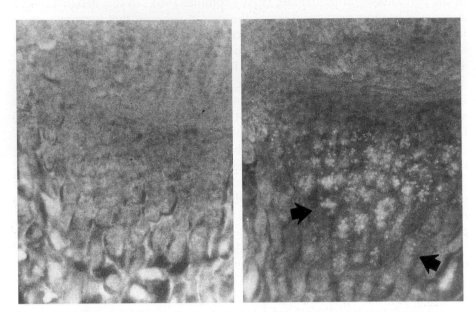

Fig. 3. Immunofluorescence of free IAA in the root cap of *Zea mays*. Root tips of four day-old seedlings were fixed in 4% paraformaldehyde. Control specimen (left, X120) for which primary antibody was omitted. Positive immunostaining (right, X 120) with goat anti-rabbit rhodamin conjugate, in columella cells of root cap (arrows).

Acknowledgements

The authors are grateful to Dr. J. Richard Hess for supplying several of the references used in this chapter.

References

1. Arteca, R.N., Arteca, J.M. (1989) Use of a monoclonal antibody for the determination of free indole-3-acetic acid. J. Plant Physiol. 135, 631-634.
2. Atzorn, R., Weiler, E.W. (1983) The immunoassay of gibberellins. I. Radioimmunoasssays for the gibberellins A_1, A_3, A_4, A_7, A_9 and A_{20}. Planta 159, 1-6.
3. Atzorn, R., Weiler, E.W. (1983) The immunoassay of gibberellins. II. Quantitation of GA_3, GA_4 and GA_7, by ultra-sensitive solid-phase enzyme immunoassays. Planta 159, 1-11.
4. Belefant, H., Fong, F. (1989) Abscisic acid ELISA: organic acid interference. Plant Physiol. 91, 1467-1470.
5. Bergman, L., Sandberg, G., von Arnold, S., Eriksson, T. (1986) Indole-3-acetic acid content in buds of five willow genotypes. J. Plant Physiol. 125, 485-489.
6. Bertrand, S., Benhamou, N., Nadeau, P., Dostaler, D., Gosselin, A. (1992) Immunogold localization of free abscisic acid in tomato root cells. Can. J. Bot. 70,1001-1011.
7. Cahill, D.M., Ward, E.W.B. (1989) An indirect enzyme-linked immunosorbent assay for measurement of abscisic acid in soybean inoculated with *Phytophthora megasperma* f. sp. *glycinea*. Phytopath. 79,1238-1242.
8. Cohen, J.D., Bausher, M.G., Bialek, K., Buta, J.G., Gocal, G.F.W., Janzen, L.M., Pharis, R.P., Reed, A.N., Slovin, J.P. (1987) Comparison of a commercial ELISA assay for indole-3-acetic acid at several stages of purification and analysis by gas chromatography-

selected ion monitoring-mass spectrometry using a $^{13}C_6$-labeled internal standard. Plant Physiol. 84, 982-986.

9. Daie, J., Wyse, R. (1982) Adaptation for the enzyme-linked immunosorbent assay (ELISA) to the quantitative analysis of abscisic acid. Anal. Biochem. 119, 365-371.

10. Davis, G.C., Hein, M.B., Chapman, D.A., Neely, B.C., Sharp, C.R., Durley, R.C., Biest, D.K., Heyde, B.R., Carnes, M.G. (1986) Immunoaffinity columns for the isolation and analysis of plant hormones. *In*: Plant growth substances 1985, pp. 44-51, Bopp, M., ed. Berlin, Springer-Verlag.

11. Eberle, J., Wang, T.L., Cook, S., Wells, B., Weiler, E.W. (1987) Immunoassay and ultrastructural localization of isopentenyladenine and related cytokinins using monoclonal antibodies. Planta 172, 289-297.

12. Ernst, D., Schafer, W., Oesterhelt, E. (1983) Isolation and quantitation of isopentenyladenosine in an anise cell culture by single-ion monitoring, radioimmunoassay and bioassay. Planta 159, 216-221.

13. Halsall, H.B., Heineman, W.R., Jenkins, S.H. (1988) Capillary immunoassay with electrochemical detection. Clinical Chem. 34, 1702.

14. Harris, M.J., Outlaw, W.H., Jr., Mertens, R., Weiler, E.W. (1988) Water-stress-induced changes in the abscisic acid content of guard cells and other cells of *Vicia faba* L. leaves as determined by enzyme-amplified immunoassay. Proc. Natl. Acad. Sci. USA 85, 2584-2588.

15. Hartung, W., Weiler, E.W., Radin, J.W. (1992) Auxin and cytokinin in the apoplastic solution of dehydrated cotton leaves. J. Plant Physiol. 140, 324-327.

16. Heylen, C., Vendrig, J.C., Onckelen, H.V. (1991) The accumulation and metabolism of plant growth regulators during organogenesis in cultures of thin cell layers of *Nicotiana tabacum*. Physiol. Plant. 83, 578-584.

17. Hocart, C.H., Badenoch-Jones, J., Parker, C.W., Letham, D.S., Summons, R.E. (1988) Cytokinins of dry *Zea mays* seed: quantification by radioimmunoassay and gas chromatography-mass spectrometry. J. Plant Growth Reg. 7, 179-196.

18. Hocher, V., Sotta, B., Maldiney, R., Bonnet, M., Miginiac, E. (1992) Changes in indole-3-acetic acid levels during tomato (*Lycopersicon esculentum* Mill.) seed development. Plant Cell Repts. 11, 253-256.

19. Jones, H.G. (1985) Correction for non-specific interference in hormone immunoassay. 12th International Conference on Plant Growth Substances, Heidelberg, Abstract no. 1105.

20. Jones, H.G. (1987) Correction for non-specific interference in competitive immunoassays. Physiol. Plant. 70, 146-154.

21. Knox, R.B. (1982) Immunology of plants. *In*: Antibody as a tool: the applications of immunochemistry, pp. 293-346, Marchalonis, J.J., Warr, G.W., eds. New York, John Wiley and Sons.

22. Knox, J.P., Beale, M.H., Butcher, G.W., MacMillan, J. (1987) Preparation and characterization of monoclonal antibodies which recognize different gibberellin epitopes. Planta 170, 86-91.

23. Lee, B., Martin, P., Bangerth, F. (1989) The effect of sucrose on the levels of abscisic acid, indoleacetic acid and zeatin/zeatin riboside in wheat ears growing in liquid culture. Physiol. Plant. 77, 73-80.

24. Leverone, L.A., Kossenjians, W., Jayasimihulu, K., Caruso, J.L. (1991) Evidence of zein-bound indoleacetic acid using gas chromatography-selected ion monitoring-mass spectrometry and analysis and immunogold labeling. Plant Physiol. 96, 1070-1075.

25. MacDonald, E.M.S., Morris, R.O. (1985) Isolation of cytokinins by immunoaffinity chromatography and analysis by high-performance liquid chromatography radioimmunoassay. *In*: Methods in Enzymology, vol. 110. Steroids and isoprenoids, Part A, pp. 347-358, Law, J.H., Rilling, H.C., eds. New York, Academic Press.

26. Madej, A., Jaggblom, P. (1985) Radioimmunoassay for determination of indole-3-acetic acid in fungi and plants. Physiol. Plant. 64, 389-392.
27. Maggio, E.T. (1980) Enzymes as immunochemical labels. *In*: Enzyme-immunoassay, pp. 53-70, Maggio, E.T., ed. Boca Raton, FL, CRC Press, Inc.
28. Maldiney, R., Leroux, B., Sabbagh, I., Sotta, B., Sossountzov, L., Miginiac, E. (1986) A biotin-avidin-based enzyme immunoassay to quantify three phytohormones: auxin, abscisic acid and zeatin-riboside. J. Immunol. Meth. 90, 151-158.
29. Mertens, R., Eberle, J., Arnscheidt, A., Ledebur, A., Weiler, E.W. (1985) Monoclonal antibodies to plant growth regulators. II. Indole-3-acetic acid. Planta 166, 389-393.
30. Nester-Hudson, J.E., Beale, M.H., MacMillan, J. (1992) Large-scale purification and preliminary structural studies of MAC 182, a gibberellin-binding monoclonal antibody. Phytochemistry 31, 3337-3339.
31. Norman, S.M., Poling, S.M., Maier, V. (1988) An indirect enzyme-linked immunosorbent assay for (+)-abscisic acid in *Citrus, Ricinus* and *Xanthium* leaves. J. Agric. Food Chem. 36, 225-231.
32. Ohmiya, A., Hayashi, T. (1992) Immuno-gold localization of IAA in leaf cells of *Prunus persica* at different stages of development. Physiol. Plant. 85, 439-445.
33. Pence, V.C., Caruso, J.L. (1986) Auxin and cytokinin levels in selected and temperature-induced morphologically distinct tissue lines of tobacco crown gall tumors. Plant Sci. 46, 233-237.
34. Pence, V.C., Caruso, J.L. (1987) ELISA determination of IAA using antibodies against ring-linked IAA. Phytochemistry 26, 1251-1255.
35. Pence, V.C., Caruso, J.L. (1988) Immunoassay methods of plant hormone analysis. *In*: Plant hormones and their role in plant growth and development, pp. 240-256, Davies, P.J., ed. Kluwer Academic Publishers, Dordrecht, The Netherlands.
36. Pengelly, W.L., Bandurski, R.S., Schulze, A. (1981) Validation of a radioimmunoassay for indole-3-acetic acid using gas chromatography-selected ion monitoring-mass spectrometry. Plant Physiol. 68, 96-98.
37. Pengelly, W.L., Meins, F., Jr. (1977) A specific radioimmunoassay for nanogram quantities of the auxin, indole-3-acetic acid. Planta 136, 173-180.
38. Perata, P., Vernieri, P., Armellini, D., Bugnoli, M., Presentini, R., Picciarelli, P., Alpi, A., Tognoni F. (1990) A monoclonal antibody for the detection of conjugated forms of abscisic acid in plant tissues. J. Plant Growth Reg. 9, 1-6.
39. Pilate, G., Sossountzov, L., Miginiac, E. (1989) Hormone levels and apical dominance in the aquatic fern *Marsilea drummondii* A. Br. Plant Physiol. 90, 907-912.
40. Quarrie, S.A., Galfre, G. (1985) Use of different hapten-protein conjugates immobilized on nitrocellulose to screen monoclonal antibodies to abscisic acid. Anal. Biochem. 151, 389-399.
41. Shi, L. (1992) Immunocytochemical localization of indoleacetic acid in primary roots of *Zea mays*. M.S. thesis, Wright State University, Dayton, OH USA.
42. Siddle, K. (1985) Properties and application of monoclonal antibodies. *In*: Alternative immunoassays, pp. 13-37, Collins, W.P., ed. New York, John Wiley & Sons.
43. Singh, S., Letham, D.S., Zhang, X., Palni, L.M.S. (1992) Cytokinin biochemistry in relation to leaf senescence. VI. Effect of nitrogenous nutrients on cytokinin levels and senescence of tobacco leaves. Physiol. Plant. 84, 262-268.
44. Skene, D.S., Browning, G., Jones, H.G. (1987) Model systems for the immunolocalization of cis, trans abscisic acid in plant tissues. Planta 172, 192-199.
45. Smith, V., MacMillan, J. (1989) An immunological approach to gibberellin purification and quantification. Plant Physiol. 90, 1148-1155.
46. Sotta, B., Sossountzov, L., Maldiney, R., Sabbagh, I., Tachon, P., Miginiac, E. (1985) Abscisic acid localization by light microscope immunochemistry in *Chenopodium polyspermum* L. J. Histochem. Cytochem. 33, 201-208.

47. Sotta, B., Pilate, G., Pelese, F., Sabbagh, I., Bonnet, M., Maldiney, R. (1987) An avidin-biotin solid phase ELISA for femtomole isopentenyladenine and isopentenyladenosine measurements in HPLC purified plant extracts. Plant Physiol. 84, 571-573.

48. Trione, E.J., Banowetz, G.M., Krygier, B.B., Kathrein. J.M., Sayavedra-Soto, L. (1987) A quantitative fluorescence enzyme immunoassay for plant cytokinins. Anal. Biochem. 162, 301-308.

49. Ulvskov, P., Marcussen, J., Seiden, P., Olsen, C.E. (1992) Immunoaffinity purification using monoclonal antibodies for the isolation of indole auxins from elongation zones of epicotyls of red-light-grown Alaska peas. Planta *188,* 182-189.

50. Vernieri, P., Perata, P., Armellini, D., Bugnoli, M., Presentini, R., Lorenzi, R., Ceccarelli, N., Alpi, A., Tognoni, F. (1989) Solid phase radioimmunoassay for the quantitation of abscisic acid in plant crude extracts using a new monoclonal antibody. J. Plant Physiol. 134, 441-446.

51. Veselov, S.Y., Kudoyarova, G.R., Egutkin, N.L., Gyuli-Zade, V.Z., Mustafina, A.R., Kof, E.M. (1992) Modified solvent partitioning scheme providing increased specificity and rapidity of immunoassay for indole-3-acetic acid. Physiol. Plant. 86, 93-96.

52. Walker-Simmons, M.K., Reaney, M.J.T., Quarrie, S.A., Perata, P., Vernieri, P., Abrams, S.R. (1991) Monoclonal antibody recognition of abscisic acid analogs. Plant Physiol. 95, 46-51.

53. Walton, D., Dashek, W., Galson, E. (1979) A radioimmunoassay for abscisic acid. Planta 146, 139-145.

54. Weiler, E.W. (1980) Radioimmunoassays for the differential and direct analysis of free and conjugated abscisic acid in plant extracts. Planta 148, 262-272.

55. Weiler, E.W. (1980) Radioimmunoassays for *trans*-zeatin and related cytokinins. Planta 149, 155-162.

56. Weiler, E.W. (1981) Radioimmunoassay for pmol-quantities of indole-3-acetic acid for use with highly stable(^{125}I)-and (^{3}H)IAA derivatives as radiotracers. Planta 153, 319-325.

57. Weiler, E.W. (1982) An enzyme-immunoassay for cis-(+)-abscisic acid. Physiol. Plant. 54, 510-514.

58. Weiler, E.W., Eberle, J., Mertens, R., Atzorn, R., Feyerabend, M., Jourdan, P.S., Arnscheidt, A., Wieczorek, U. (1986) Antisera-and monoclonal antibody-based immunoassay of plant hormones. *In*: Immunology in plant science, pp. 27-58, Wang, T.L., ed. Cambridge, Cambridge University Press.

59. Weiler, E.W., Jourdan, P.S., Conrad, W. (1981) Levels of indole-3-acetic acid in intact and decapitated coleoptiles as determined by a specific and highly sensitive solid-phase enzyme-immunoassay. Planta 153, 561-571.

60. Weiler, E.W., Wieczorek, U. (1981) Determination of femtomol quantities of gibberellic acid by radioimmunoassay. Planta 152, 159-167.

61. Wood, B.W., Payne, J.A. (1988) Growth regulators in chestnut shoot galls infected with oriental chestnut gall wasp, *Dryocosmus kuriphilus* (Hymenoptera: Cynipidae) Environ. Entomol. 17, 915-920.

62. Xu, Y., Halsall, H.B., Heineman, W. (1989) Solid-phase electrochemical enzyme immunoassay with attomole detection limit by flow injection analysis. J. Pharm. & Biomed. Anal. 7, 1301-1311.

63. Zavala, M.E., Brandon, D.L. (1983) Localization of a phytohormone using immunocytochemistry. J. Cell Biol. 97, 1235-1239.

G. THE FUNCTIONING OF HORMONES IN PLANT GROWTH AND DEVELOPMENT

G1. Hormone Mutants and Plant Development

James B. Reid and Stephen H. Howell
Botany Department, University of Tasmania, Hobart 7001, Australia, and
Boyce Thompson Institute, Cornell University, Ithaca NY 14853, USA.

INTRODUCTION

Genetics is a powerful tool for studying hormones and their impact on plant development. Mutants affecting hormone biosynthesis or responses have been obtained in various plant systems by predicting phenotypes based on physiological information, by assaying hormone levels in plants with unexplained developmental defects or by selecting for resistance to hormones or their biosynthesis inhibitors. These mutants have given us greater insight into the role of hormones in a wide variety of plant developmental processes. The genes responsible for some hormone mutants have been cloned, and in several cases, the sequences of the cloned genes have provided us with a key to understanding the mechanisms of hormone action at the molecular level.

Single gene mutations are particularly useful in unraveling complex biochemical or developmental pathways since they result from single genetic lesions. The use of various inhibitors or environmental treatments is often less helpful in understanding the effects of hormones on development simply because inhibitor action is usually less specific. Without mutants or inhibitors, the only recourse left is to examine the effects of applying a hormone or to correlate the endogenous level of the hormone with physiological responses. Single gene mutants in identical genetic backgrounds can be obtained when large numbers of progeny are screened following mutagenesis of a pure line (e.g., as in many studies with *Arabidopsis*). In other plant systems, where mutations or variants are found in different genetic lines, it is necessary to produce isogenic lines so that the effects of a mutant are not confused by the differences in the genetic

448

P. J. Davies (ed.), Plant Hormones, 448–485.
© 1995 *Kluwer Academic Publishers. Printed in the Netherlands.*

background.   Regardless of how they are obtained, single gene mutations affecting hormone synthesis or responses are sometimes not simple to interpret even though their defects can be traced back to a single primary lesion.   Plant hormones influence a multitude of processes and, therefore, mutants will frequently show a range of pleiotropic effects.   Part of the challenge in working with single gene mutants affecting hormone action is understanding the complex phenotypes produced.

NEW OPPORTUNITIES IN PLANT HORMONE RESEARCH: *ARABIDOPSIS* GENETICS

While mutants affecting hormone biosynthesis and responses have been isolated in many plants (e.g., maize, tomato, pea and rice), work on hormone mutants in *Arabidopsis thaliana* has proceeded at a particularly fast pace over the last few years.   Hormone mutants are relatively easy to identify and characterize in *Arabidopsis*, and more importantly, it is possible to isolate the genes responsible for various mutations in *Arabidopsis*. The small plant size and rapid life cycle allow large numbers of *Arabidopsis* plants to be screened easily for mutants affecting hormone production or response.   At standard rates of mutagenesis, the entire *Arabidopsis* genome can be scanned for loss-of-function mutations by examining a few thousand seedlings.   Because of this, mutants involving every major class of hormones have been identified in *Arabidopsis*.

Certainly, one of the most compelling reasons for studying hormone mutants in *Arabidopsis* is that the genes responsible for the mutant phenotypes can be cloned.   While only a few such genes have been cloned so far in *Arabidopsis*, undoubtedly the number will increase rapidly in the next few years.   Map-based cloning procedures are probably most useful to investigators working on hormone mutants, because map-based cloning can be used to track down the most common type of mutants;   EMS-generated point mutations.   Map-based cloning, once thought to be a formidable task, is now a reality in *Arabidopsis*.   It has become practicable because the small *Arabidopsis* genome is so densely marked that it is possible to find markers that lie within reasonable physical proximity to a gene of interest (37).   Map-based cloning involves mapping a gene of interest with respect to other DNA markers, e.g., restriction fragment length polymorphisms (RFLP) or randomly amplified polymorphic DNAs (RAPD) markers.   Then the nearest DNA markers are used to identify families of yeast artificial chromosomes (YACs) that contain large segments of the *Arabidopsis* genome.   Gibson and Somerville (24) have pointed out that if the nearest DNA marker lies within 1 centimorgan[1] of a gene of interest, there is a 40% probability that the YAC identified by the DNA marker will overlap with the gene of interest.   In that

---

[1] A centimorgan or cM is a unit of genetic distance, and markers that are separated by 1 cM undergo recombination at a frequency of 1%.

case, there is no need to undertake a "chromosome walk"[2] to find the gene of interest.

There has also been much interest in identifying genes in *Arabidopsis* by tagging them with T-DNA inserts. T-DNA is the DNA segment transferred to plants during *Agrobacterium* infection. T-DNA insertion mutagenesis[3] has been useful for tagging some hormone mutants, particularly those that survive as null mutants, i.e., mutants that totally eliminate the function of a gene. For mutants that appear infrequently, the problem is one of numbers. T-DNA insertion mutants are kept as individual lines (and in pools of lines) and must be planted out as lines or pools. Because of this, the numbers of mutants that can be screened is relatively small. The problem is further compounded by the fact that in T-DNA insertional mutagenesis, it is desirable to have few insertional events per genome (~1.4 random inserts per plant) in order to minimize the number of T-DNA copies per genome in any prospective mutant. At this mutation rate, over 100,000 independent lines must be screened in order to achieve a 95% probability of having an insert every 2 kb in the genome. Nonetheless, an insertion mutation was recently found that confers a constitutive ethylene response phenotype. Kieber *et al.* (34) took advantage of the tagged mutant to clone the *CTR1* gene from *Arabidopsis*. However, for most purposes, T-DNA insertion collections may not be the best source of material for finding subtle hormone mutants.

Another means of identifying genes responsible for mutations in *Arabidopsis* is to use genomic subtraction methods (95). Cloning by genomic subtraction may only be applicable to plant genomes as small as *Arabidopsis*, and even the *Arabidopsis* genome may be too large for routine use of this technique. Genomic subtraction involves the isolation of a DNA sequence that is missing in a deletion mutant. In genomic subtraction, a DNA sequence present in the wild-type genome, but not in a deletion mutant, is progressively enriched using DNA hybridization techniques. Although genomic subtraction has already led to the cloning of a hormone biosynthesis gene (97), the procedure requires a deletion and is very difficult to carry out successfully. Furthermore, the deletion must not be lethal in the homozygous state, because homozygotes are needed as a source of DNA for the genomic subtraction procedure. While mutants can be generated at high frequency with certain forms of mutagenesis, considerable effort is required to prove that a mutation is truly a deletion mutation. In the genomic subtraction procedure, DNA from wild-type plants is subjected to multiple rounds of hybridization with biotinylated DNA[4] from a homozygous deletion mutant. In each round, DNA from the wild type plant that hybridizes with DNA in

---

[2] A chromosome walk is a process in which one progresses along a chromosome by identifying YACs that overlap each other.

[3] T-DNA insertion mutagenesis results from the random incorporation of T-DNA into the DNA of the plant genome, which will on occasion disrupt the function of a gene.

[4] DNA synthesized from biotinylated nucleotides.

the deletion mutant is removed by binding the biotinylated hybrid DNA to avidin-coated beads. After the last round, linkers (small oligonucleotides) are attached to the remaining unbound DNA (enriched for the gene of interest that is deleted in the mutant), and used to amplify the DNA by the polymerase chain reaction (PCR).

A word of caution should be added about the selection of hormone mutants, in general, but particularly in *Arabidopsis*. Since so many seedlings can be screened in mutant searches, broad-based screens may find mutants that are totally unrelated to the interests of the investigator. Consequently, further screens may be essential to fully characterize the mutants obtained (for example, to separate hormone synthesis from hormone response types). Finally, the terminology regarding mutants can be confusing. Biosynthesis mutants may occasionally be referred to as hormone "sensitive" mutants, but in microbial genetics terms they are hormone auxotrophs or hormone-requiring mutants. Biosynthesis mutants may also be overproduction (87) or unregulated biosynthesis mutants. Response mutants are frequently "insensitive" mutants and selected because they appear to be insensitive to their own endogenous levels of hormone or resistant to toxic or growth-inhibiting levels of exogenous hormone. Unlike sensitive mutants, insensitive mutants cannot be rescued by adding hormones. Another group of response mutants are the constitutive response types. These are mutants that, on the face of it, appear to be hormone overproduction mutants, but are unaffected by hormone synthesis inhibitors (69).

General considerations

This is an exciting time for hormone research in *Arabidopsis*, because some of the genes responsible for the hormone response genes have been cloned (12, 25, 34, 61). From these reports, we have a first glimpse of the components that compose plant hormone signaling pathways. Response mutants have been selected in *Arabidopsis* for most of the major classes of hormones. Through an understanding of the physiology of plant hormones and their effects on *Arabidopsis* development, investigators have devised clever means to select for mutants. As described above, a number of hormone response mutants have been selected because they are resistant to growth-inhibiting doses of exogenous hormone or they have mutant phenotypes that cannot be rescued by adding hormones (Fig. 1).

Several interesting general observations can be made about hormone response mutants. First, it is surprising to be able to find such mutants at all, particularly mutants with defects in response to growth hormones, without having to resort to the selection for conditional mutants. This suggests that the plant hormones are not essential for survival, that the mutations are compensated for by some alternative response pathway or that the mutations are "leaky," i.e. not totally blocked in the hormone response. The discovery of various hormone response mutants in plants contrasts with the difficulty

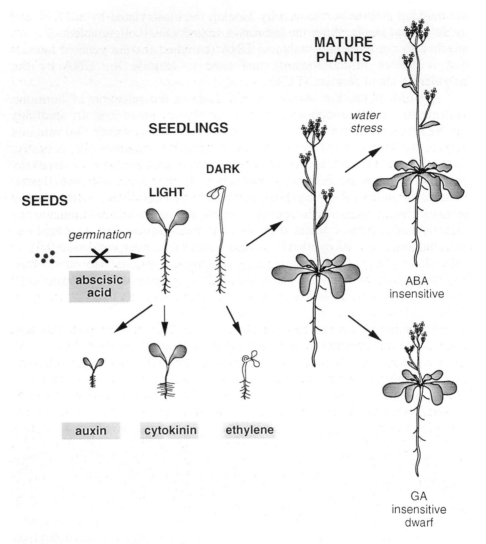

Fig. 1. Growth characteristics of hormone resistant or insensitive mutants of *Arabidopsis*. Various mutants have been selected based on the effects of hormones on growth at different developmental stages (seeds, seedlings, mature plants). Illustration shows the growth characteristics of wild type seedlings in the presence of various hormones and hormone inhibitors (gray boxes) and the growth characteristics of hormone insensitive mutants (ABA and GA insensitives).

in obtaining certain nutritional mutants, such as amino acid auxotrophs. The reason usually given for this difficulty in finding metabolic mutants in plants is that there may be multiple genes that encode any given gene product or multiple pathways that produce the same metabolic product (45).

Second, although response mutants have been identified for all the major classes of hormones, the number of genes in some of the hormone categories is small. Admittedly, the genome has not been saturated for mutants, and some mutants in certain hormone response pathways may be lethal. So far, there is only one locus, *ckr1*, that confers cytokinin resistance (96). Likewise only one locus conferring gibberellic acid insensitivity (*gai*) has been described (38). Most of the auxin resistant mutants represent two loci, *axr1* (47) and *aux* (49). The mutants at a third locus, called *axr2* (110), do not appear to be auxin-specific, and a fourth locus, *axr3*, is relatively uncharacterized. There are three loci at which mutations confer abscisic acid resistance, but one of these only affects certain ABA responses (42). The fact that there are few response mutants runs counter to the notion that there are many steps in hormone response pathways. If hormone response pathways are multi-step, then it should be possible to find many different mutants that affect different steps in the pathways, providing that the mutations are not lethal. At this time, the best example of a multi-step pathway appears to be the ethylene response pathway. Kieber *et al.* (34) have several mutations in the ethylene response pathway and have ordered the mutations by their epistatic[5] relationships (88). By determining whether mutations are epistatic to one another, one can order the events represented by the mutations (69, 73, 88).

Third, the results of cloning the first few hormone response mutants have been quite unexpected. So far, the genes that have been identified do not encode membrane-associated signal transduction pathway components such as G-protein subunits, inositol phospholipases, or calcium channels. However, in the ethylene response pathway, the genes that have been cloned appear to encode protein kinases that may be involved in a protein phosphorylation cascade (34). One of the components on the ethylene response pathway that has been recently cloned is a novel form of a procaryotic signaling pathway component, and the identification of this component in *Arabidopsis* has set a precedence for other eucaryotic organisms (12). Therefore, the hormone signal transduction pathways in plants may offer some real surprises for investigators.

ROOT AND SHOOT GROWTH: AUXINS AND CYTOKININS

Auxins

From the classical experiments of Skoog and Miller (91), the action of auxins and cytokinins have been closely tied to root and shoot growth. However, most of the auxin response mutants in *Arabidopsis* have been isolated on the basis of more general characteristics, such as auxin resistant germination or seedling growth (Fig. 1). The auxin mutants that have been isolated so far represent four different loci in the *Arabidopsis* genome, *aux* and *axr1* to *axr3*

---

[5] A mutation is epistatic to another when one mutation obscures the expression of the other mutation.

(21, 49, 110). Maher and Martindale (49) identified *aux* mutants with roots resistant to relatively low auxin concentrations (2 μM 2,4-D). All *aux* mutations map to a single locus on chromosome 2. Untreated *aux* mutants are morphologically normal, but the alleles at the *aux* locus differ in sensitivity to auxin. The *aux-1* alleles are most resistant to auxin and the roots are agravitropic. On the other hand, the *aux-2* alleles are less resistant, and roots are gravitropically normal (53).

Estelle and Somerville (21) isolated mutants in which the germination of seeds was resistant to moderate levels of auxin (5 μM 2,4-D). Most of these mutants were recessive mutants at the *axr1* locus on chromosome 1. These mutants were more severe than the *aux* mutants and had phenotypes that one might suspect for an auxin response mutants, such as reduced apical dominance and shorter stature (47). Recently *axr1* was cloned (61) using map-based cloning techniques. The authors narrowed in on the *axr1* site by analyzing RFLPs in recombinants between *axr1* and an outside marker. The authors used the most closely linked RFLPs to identify a corresponding YAC from a library of the *Arabidopsis* genome. Then, they subcloned DNA fragments from the YAC in a series of cosmid[6] clones that were individually tested for their ability to complement *axr1* in transgenic plants. Two complementing cosmid clones were identified and used as probes to identify RNA transcripts from *Arabidopsis* plants. One of two transcripts was expressed at different levels in the various *axr1* mutants except *axr1-3*, a weak allele, and the cDNA (complementary DNA) for that transcript was used to identify the *axr1* gene. Evidence verifying that the identified gene was, indeed, *axr1* was the fact that two *axr1* alleles showed alterations in the coding region of the gene.

The protein encoded by *axr1* is similar in sequence to the ubiquitin-activating enzyme E1 of human, yeast and wheat (61). However, it is not known whether the *axr1* product functions in ubiquitination. The E1 enzyme catalyzes the formation of a thiol ester bond between ubiquitin and the E1 protein. The ubiquitin moiety is then transferred to one member of a family of ubiquitin-conjugating E2 enzymes which then transfers the ubiquitin to target proteins. In support of the idea that the function of the *axr1* gene product might be related to ubiquitination is the finding that a conserved cysteine residue (at position 154 in *axr1*) required for E1 function is changed in the weak allele of *axr1* (*axr1-3*). On the other hand, AXR1, the *axr1* gene product, does not have a critical cysteine residue at a position equivalent to 626 in the E1 enzyme. This residue forms the thiol bond between E1 and ubiquitin and is essential for E1 function. Furthermore, AXR1 only has 22-27% similarity to E1 (excluding sequence gaps), whereas the E1 enzymes from humans, yeast and wheat are more highly conserved. Therefore, it is

---

[6] Cosmids are plasmids that can be packaged as bacteriophage. Cosmids have the advantage that they can be handled as easily as plasmids, yet they can support the genetic complexity of bacteriophage DNA libraries.

not known whether AXR1 has E1 activity or whether a domain from the ubiquitin E1 enzyme has been commandeered for other functions. Alternatively, AXR1 might not have anything to do with auxin signal transduction directly, but AXR1 might regulate the stability of a component in the auxin signaling pathway by regulating its turnover.

Wilson et al. (110) recovered a new dominant mutant at a locus called *axr2* in a selection for seedlings in which root growth was resistant to higher concentrations of auxin (50 μM IAA) The mutant was dark green, vigorous and had normal fertility, but had short internodes, smaller, rounder leaves and short petioles. In addition, roots in the mutant grew more slowly, lacked root hairs and did not respond normally to gravity. Most unusual was the fact that root growth in the mutant was also resistant to other hormones such as ethylene and abscisic acid.

Cytokinins

Although cytokinin action is usually associated with shoot development (91), *Arabidopsis* mutants have been isolated recently on the basis of cytokinin resistance in roots (96). It has been difficult to obtain mutants in which shoot growth is resistant to cytokinin, because high concentrations of cytokinin are required to inhibit shoot growth, and high concentrations of cytokinin produce general stress responses (51). Therefore, mutants in other plant systems that were selected for resistance to high concentrations of cytokinin did not have defects in specific cytokinin responses, rather they seem unable to mount general stress responses. For example, a *Nicotiana plumbaginifolia* mutant that was resistant to high concentrations of cytokinin was defective in a terminal step in ABA biosynthesis (8). To avoid selecting for general stress response mutants, Su and Howell (96) determined what aspect of seedling development in *Arabidopsis* was most sensitive to exogenous cytokinin, benzyladenine (BA). They found that root growth was most sensitive, and at relatively low concentrations of BA seedlings displayed a "cytokinin root syndrome" in which primary root growth was inhibited and root hair elongation was stimulated (Fig. 1). These observations were used to isolate mutants that did not exhibit the root syndrome in the presence of cytokinin. Despite the fact that a very general screen was used, the mutants all belonged to the same complementation group at a locus called *ckr1*. The mutants appeared to be response mutants rather than hormone uptake mutants, because the uptake and metabolism of BA in the mutants was comparable to wild type (Su, Auer and Howell, in preparation).

A surprising feature about the *ckr1* mutants was their phenotype when grown under standard lighting conditions; the plants were nearly normal in growth, stature, fertility, etc. (96). The only abnormalities in light-grown *ckr1* mutants were cup-shaped leaves in seedlings, larger but slightly more chlorotic leaves at maturity and longer roots. It was argued that *ckr1* mutants appeared normal because other responses, particularly light signals, compensated for the defects in the cytokinin response (Su and Howell, submitted for publication). As a consequence, many of the effects of the *ckr1* mutants on germination, hypocotyl elongation, time to flowering, etc., were masked by light. Hence, *ckr1* mutants may be useful for studying the interactions between cytokinin and light signaling pathways.

Biosynthesis mutants

Because auxin and cytokinin play such critical roles in plant development, the biosynthetic pathways of these hormones would seem to be obvious targets for mutant analysis. However, until recently very little genetics has been done to elucidate the biosynthetic pathways of these hormones. However, the matter is important because the normally accepted pathway for auxin biosynthesis in plants is in dispute, and the regulation of auxin biosynthesis is not understood. One reason why auxin and cytokinin biosynthesis mutants may not have been pursued aggressively is that auxin and cytokinin biosynthesis can be controlled and perturbed in transgenic plants by introducing hormone biosynthesis and metabolism genes from *Agrobacterium* and other sources (e.g., 36, 50, 82, 83). However, these transgenic studies have not lessened the need to understand the normal auxin or cytokinin biosynthetic pathway(s) through the analysis of mutants.

Hints that there were problems with the generally accepted idea that auxin is synthesized from tryptophan have come from studies of tryptophan mutants in *Arabidopsis* and maize. Orange pericarp mutants in maize are defective in terminal steps of tryptophan biosynthesis, yet they accumulate high levels of indole acetic acid (IAA) (112). *Arabidopsis* mutants that block the terminal steps of tryptophan biosynthesis, such as the *trp2*-1 mutants defective in tryptophan synthase b activity, do not have phenotypes predicted for auxin biosynthetic mutants (45). However, mutants blocked at earlier steps in the pathway appear to be auxin deficient (46). Furthermore, when *trp2*-1 mutants were grown under high light conditions (conditions in which the mutants require tryptophan for growth), there was a dramatic elevation in ester and amide-linked IAA levels (61). On the basis of feeding experiments, Normanly *et al.* argued that a compound earlier in the tryptophan biosynthetic pathway, possibly between anthranilate and indole, might be a more direct precursor for IAA, and that indole-3-acetonitrile (IAN) might be an intermediate product.

Auxin deficient mutants have not been found, possibly because such mutations might be lethal (21). However, transgenic tobacco plants can be regenerated and propagated bearing the *iaaL* gene from *Pseudomonas savastanoi* (83). The *iaaL* gene encodes indoleacetic acid-lysine synthetase that catalyzes the production of inactive conjugates of auxin. Transgenic plants expressing the *iaaL* gene are reported to have reduced levels of unconjugated IAA. These plants have reduced apical dominance, rooting and vascular development. Such plants can be reverted to near normal phenotypes by overproduction of IAA through expression of the *iaaM* gene. Therefore, it should be possible to recover *Arabidopsis* mutants that are specifically defective in auxin biosynthesis.

Recently, an *Arabidopsis* mutant, *amp1*, has been described by Chaudhury *et al.* that has elevated levels of the cytokinins, zeatin and dihydrozeatin (13). The mutant was obtained by predicting that a cytokinin overproducer would affect cotyledon leaf number and leaf enlargement in seedlings. In the dark, the *amp1* mutant displays morphogenic properties similar to light grown seedlings. The mutation also leads to faster vegetative growth and precocious flowering. It is not known whether the mutation directly affects a step in cytokinin biosynthesis or whether it has some indirect affect on cytokinin accumulation or breakdown.

ETHYLENE AND THE TRIPLE RESPONSE

Ethylene is involved in determining the shape and stature of germinating seedlings as they emerge from the soil. Ethylene mediates the so-called "triple response" in which hypocotyls do not elongate, but swell in girth, apical hooks do not fully open, and seedlings fail to show normal geotropic responses. In the presence of exogenous ethylene, seedlings show a constitutive triple response: root and hypocotyl elongation are inhibited, directional growth is abnormal, the apical hooks show exaggerated tightening and hypocotyls swell (Fig. 1).

Based on the triple response, *Arabidopsis* mutants with altered responses to ethylene have been isolated. The first ethylene resistant mutant was selected because it grew conspicuously taller than other seedlings in the presence of exogenously supplied ethylene (7) (see book cover). The mutant, conferred by a dominant mutation at a locus designated *etr*, was defective in other ethylene responses. For example, ethylene failed to stimulate germination in the mutant seeds, and leaf senescence in mutant plants. The mutant showed typical levels of ethylene, but lacked the normal feedback inhibition of ethylene production. The mutant had reduced capacities to bind ethylene and was thought to have a defect in an ethylene receptor.

The *ETR1* gene in *Arabidopsis* was cloned recently by Chang *et al* (12) using map-based cloning techniques. An RFLP marker was found that cosegregated with *etr1*, and the RFLP marker was used to initiate a chromosome walk among cosmid clones. The identity of the gene was ultimately confirmed by demonstrating that the gene was changed in various *etr1* mutants, and that the dominant mutant gene conferred ethylene insensitivity when introduced into wild-type *Arabidopsis* plants by *Agrobacterium*-mediated transformation.

The sequence of the *ETR1* gene yielded information that was quite exciting because the carboxyl-terminal portion of the gene product was similar to both components of two component signal transducing systems in bacteria (12). Two component systems are characterized by having one component that serves as a sensor with protein kinase activity and another component that acts as a response regulator. The stimulus results in activation of a protein kinase in the sensor component and autophosphorylation of a carboxyl-terminal histidine. The phosphate is transferred to an aspartate residue on the amino-terminal end of the response regulator. The response regulator then activates some effector such as a transcription factor.

The N-terminal region of *ETR1* is not similar to other proteins but consists of three hydrophobic or potential membrane spanning domains. The various *etr1* mutant alleles are all missense mutations (amino acid change mutations) in the three hydrophobic domains. It is not known why the *etr1* mutations are dominant although it was speculated by Chang *et al.* (12) that the mutations may produce a defective subunit which poisons the function of a multimeric protein or locks *ETR1* into a state where it is constitutively expressed. It is also not known whether *ETR1* is an ethylene receptor, although, as mentioned above, *etr1* mutants have reduced binding capacity for ethylene. Nonetheless, the reduced binding capacity could be an indirect effect of the *etr1* mutation.

Other ethylene insensitive mutants (*ein* mutants) were isolated by Guzman and Ecker (27). Both dominant and recessive mutations conferring ethylene resistance were obtained. The dominant mutation, *ein1*, maps to chromosome 1 and may be allelic to the *etr* mutant described by Bleecker *et al* (7). The recessive mutation, *ein2*, mapped to chromosome 4. Etiolated seedlings ordinarily produce a hook at the end of the shoot axis, and ethylene affects the formation of the hook. In addition to the *ein* mutants, Guzman and Ecker (27) identified two different genes affecting hook formation, *hls1* and *hls2*. These mutants appeared to produce normal levels of ethylene and, therefore, are thought to be mutants involved in the perception of ethylene. Interestingly, the roots and hypocotyls respond normally to ethylene in the *hls1* mutants, and, therefore, these mutants appear to uncouple the events associated with the triple response. These findings suggest that the *ein* mutants, which have more general effects, control earlier steps in the ethylene response pathway than the *hls* mutations.

Recently, Kieber *et al.* (34) defined another class of ethylene response mutants called constitutive response mutants. These mutants have the phenotype of an ethylene overproducer, that is, they show the ethylene triple response in the absence of ethylene. From a group of constitutive triple response mutants, the authors culled out the constitutive response mutants by their insensitivity to ethylene biosynthesis inhibitors. None of the insensitive mutants produced significantly more ethylene than the parents, and all belonged to the same complementation group at a single genetic locus designated *ctr1*. Double mutants were constructed to determine the epistatic relationship between *ctr1* and two *ein* mutants, *ein1* and *ein3*. It was found that *ctr1* was epistatic to *ein1*, but not to *ein3*. From this the authors argued that *ctr1* is downstream in a response pathway from *ein1*, but upstream from *ein3* (Fig. 2).

The *CTR1* gene was mapped to the top of chromosome 5, and a T-DNA insertion mutant was found that tagged the *CTR1* gene. The T-DNA and flanking sequences were rescued and used to select for corresponding l genomic clones, which in turn were used to identify cDNAs. The identification of the *CTR1* gene was confirmed by sequencing the gene from various chemically-induced mutants and showing that the mutants bore point mutations in the putative gene (34). The *CTR1* gene itself encodes a protein with similarity to the Raf-type, serine/threonine protein kinases. In animal systems, Raf-type protein kinases are thought to mediate the transmission of signals from a number of different growth-stimulating ligands. Raf is usually considered to be a protein kinase kinase that ferries signals in multi-step protein phosphorylation cascades ultimately to activate gene expression. From the genetic data, ETR1 is upstream from CTR1, but it has not yet been determined whether ETR1 interacts directly with and phosphorylates CTR1.

Recently, it has been reported that *ein3* and *etr* have been cloned. So far, the sequence of *ein3* has not provided much information about its function because it does not match other sequences in the database (Ecker, personal communication).

Guzman and Ecker (27) also selected for other mutants in which ethylene responses were constitutive in the absence of exogenously supplied ethylene. They recovered two mutants that were allelic at a locus they called *eto1*. The mutation *eto1-1* produces high levels of ethylene in etiolated seedlings, generating 40 times the level of wild-type seedlings. The phenotype of the mutations could be reversed by treatment with inhibitors of ACC synthase, such as aminoethoxyvinylglycine (AVG). Thus, the mutation was thought to be a defect in the regulation of ACC synthase, or a prior step (Fig. 2). No mutants of *Arabidopsis* have been found that underproduce ethylene, although they should survive, because Guzman and Ecker (27) found that the ethylene insensitive (*ein*) mutants that fail to perceive ethylene appear like normal plants. However, such mutants may not possess a unique phenotype, hence, ethylene deficient mutants would be difficult to isolate. Although the ethylene biosynthesis pathway is one of the best characterized hormone

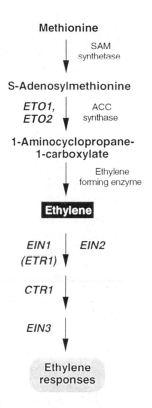

Fig. 2. Model of a pathway for ethylene biosynthesis and response in *Arabidopsis*. Steps in the response pathway are represented by mutants (bold) or by enzymatic activities (non-bold). The order of steps in the response pathway have been determined by epistasis analysis (after 88).

biosynthesis pathways from a biochemical point of view, there is a real need for ethylene biosynthesis mutants to study the regulation of the pathway.

CAULONEMATA AND GAMETOPHORE PRODUCTION IN MOSSES

Hormone mutants have been effectively applied to the analysis of plant development and cell differentiation in bryophytes (mosses and liverworts) (4). The advantages of studying development in mosses include a short life cycle, a relatively small genome, and a conspicuous haploid phase during which relatively simple changes in cell development occur. Some of the changes in cell development appear to be under the control of the auxins and cytokinins (4, 9). Mutants that regulate hormone levels or sensitivity have been isolated and used extensively to study the control of development (2, 4).

The development of the protonema in mosses such as *Funaria hygrometrica* or *Physcomitrella patens* commences with spore germination to produce the primary chloronemata. The chloronemata consist of irregular branching filaments of cylindrical cells which have cross walls perpendicular to the filament axis and contain a large number of round chloroplasts (4, 9). Subsequent development leads to the production of a second cell type, the caulonemata, which differs from chloronemata by virtue of the fact that they have cross walls oblique to the filament axis, a reduced number of spindle-shaped chloroplasts and a very regular pattern of branching.

Branching in the caulonemata may then give rise to secondary chloronemata or to buds that develop into leafy gametophores (2, 4). The transition from chloronemata to caulonemata is under strong environmental control and does not occur if the protonema are grown at low light intensities, low temperatures or in a continuous flow of liquid medium (16). The latter suggests that some substance normally responsible for caulonemata production is leached from the system. That substance appears to be an auxin, because IAA or α-NAA stimulates the transition from chloronemata to caulonemata (32). The production of buds and gametophores requires cytokinins. In *P. patens* auxins are also necessary for this development (16) and light may also be required (2, 3).

Cove, Ashton and coworkers isolated a range of developmental mutants from *P. patens* and divided them into a number of categories based upon their stage of developmental arrest (2, 3) (Table 1). The possible defects in the mutants have been explained by comparing phenotypes produced in wild-type plants treated in various ways (e.g., application of auxins and cytokinins, drip-cultures, etc.), and by observing the effect of auxins and cytokinins applied to the mutant plants (e.g., reversion of the mutant type to the wild-type form or insensitivity to applied hormones) (Table 1) (3, 4). Evidence of this type has underscored the importance of auxin and cytokinin to the normal development of the moss plant and confirmed that auxin, in addition

Table 1. Categories of developmental mutants in *Physcomitrella patens*, their phenotypes compared with the wild-type (normal), their response to applied auxins and cytokinins and their possible modes of action (3,4,16,106,107).

Mutant Category	Phenotype on Minimal Medium			Effect of Applied		Possible Mode of Action
	Chloronemata primary + secondary	Caulonemata	Gametophores	Auxin	Cytokinin	
1A	more	none	none	little	insensitive	auxin or cytokinin reception or transduction impaired
1B	more	few	none	little	little	auxin or cytokinin reception or transduction impaired
2A	more	none	none	little	normal growth	cytokinin deficient, non-leaky
2B	more	normal	none	little	normal growth	blockage of cytokinin production, leaky
3	normal	normal	normal	resistant	wild-type	possible defect in NAA uptake
4	slightly more	no normal caulonemata	none	normal growth	little	blockage of auxin production
5	many more	slightly fewer	few or none	normal growth	little	reduced auxin production
6	many more	slightly more	few or none	normal growth	little	reduced auxin production
7	more	normal or slightly fewer	few	normal growth	little	reduced auxin production
8	many more	more	normal	sensitive	resistant	unknown but not related to auxin or cytokinin levels
9 (OVE)	few	normal or slightly more	many, abnormal	-	-	cytokinin overproducers
10 (OVE)	normal	normal	more, slightly abnormal	-	-	cytokinin overproducers?

461

to being required for caulonemata production, is essential for gametophore production. Cytokinin appears to control the development of gametophores (3), and may also be necessary at lower levels for caulonemata production (4).

The division of mutants into categories based on phenotypic and not genetic grounds has yielded a larger number of mutant categories than appears to be warranted at the physiological and biochemical levels (Table 1). For example, categories 4, 5, 6 and 7 all appear to be caused by mutations which reduce the level of auxin production to some extent. The mutants are 'leaky' to varying degrees resulting in differing levels of phenotypic disturbance. The same applies for categories 2A and 2B. Such a range of mutants has shown that low levels of auxin are essential to caulonemata production while higher levels are necessary for gametophore production (3). The gametophore (cytokinin) over-producing mutants (categories 9 and possibly 10 in Table 1) contain 100-fold greater levels of N^6-(Δ^2-isopentenyl) adenine and zeatin in the culture medium compared to wild-type plants (107). The use of protoplast fusion to carry out somatic hybridization and complementation analysis (necessary since category 9 mutants are sterile) has elegantly shown that at least 3 loci are involved with the category 9 phenotype (106). Therefore this system is well-suited to studies on cytokinin biosynthesis.

In addition to mutations involved with auxin and cytokinin accumulation (categories 2A, 2B, 4, 5, 6, 7, 9, and 10), mutations which influence hormone sensitivity (categories 1A and 1B) have been isolated (3). The availability of mutations influencing many aspects of development, such as occurs in the moss *P. patens*, allows developmental processes to be taken apart step by step, and may yield a far greater understanding than in systems where fewer mutations are found. The production of mutants by insertional mutagenesis in *P. patens* should allow the tagging of genes involved in hormone-regulated developmental processes (4).

WILT, SEED DORMANCY AND ABSCISIC ACID

Mutants have been used with great success to understand the role of abscisic acid (ABA) in water stress (42, 71, 101, 103) and seed germination (33, 92). In tomato three non-allelic, recessive, wilty mutants, have been isolated, *flacca* (*flc*), *sitiens* (*sit*) and *notabilis* (*not*) (101). All three result in a tendency for the mutants to wilt if subjected to mild water stress. The mutants also possess much higher rates of transpiration than control varieties because their stomata open wider, and resist closure in the dark or in wilted leaves. Plants homozygous for the genes *flc*, *sit* and *not* treated with ABA behave like the normal control varieties, Rheinlands Ruhm (for *flc* and *sit*) and Lukullus (for *not*). The endogenous level of ABA was found to be reduced in all three mutants when measured by gas chromatography-mass

462

spectroscopy (GC-MS, 59). The reduction was greater in mutants with the more pronounced abnormalities, *sit* and *flc*, compared with the less extreme mutant *not*. This evidence has frequently been used to indicate the involvement of ABA in water stress.

These and other mutants have also played a crucial role in allowing the biosynthetic pathway for ABA to be determined. Two pathways had been proposed: a 'direct' pathway in which ABA is formed by the cyclization and oxidation of farnesyl pyrophosphate (C_{15} pathway) and secondly, an 'indirect' pathway via the oxidative cleavage of an epoxy carotenoid (C_{40} pathway). Recent evidence favors the C_{40} pathway (62) since mutants possessing inhibited carotenoid biosynthesis also possess reduced ABA levels and viviparous phenotypes (e.g., in maize, 54). ABA appears to arise from the asymmetric cleavage of 9'-cis-neoxanthin to yield xanthoxin plus a C25 apo-carotenoid (62, Fig. 3). This is further metabolized to ABA-aldehyde, and then ABA. The most likely control point appears to be the cleavage of 9'-cis-neoxanthin. The *sit* and *flc* mutants in tomato both appear to block the step from ABA-aldehyde to ABA (70, 117) whereas *not* blocks prior to cleavage of the 9'-cis-neoxanthin (62, Fig. 3). A series of double mutants were produced from the mutants *flc*, *sit* and *not* (62). The double mutants, *not flc* and *not sit*, gave a more extreme wilty phenotype than any of the single mutant types, while *flc sit* produced only a marginally more severe phenotype than the single homozygous mutant types *flc* and *sit* (103). These results support the notion that all three mutants are leaky (101, 62). However, the rank order in severity of the double mutant types (*not sit* > *not flc* > *flc sit*) was originally unexpected (103), given the severity of the single mutants (*sit* > *flc* > *not*; 101), but now that the biochemical site of action of these mutations is known (Fig. 3), this result is understandable.

Wilty mutants have also been found in potato (71), pea (105), corn (68), *Capsicum scabrous* (*diminutive*) (100) and *Arabidopsis thaliana* (41). In potato the *droopy* mutant results in excessive stomatal opening. Stomatal conductances were reduced by applied ABA, and the leaves from *droopy* plants accumulated less ABA than normal plants when water-stressed (71). These findings suggest *droopy* is similar to the wilty mutants described in tomato. Recent results suggest that droopy blocks the conversion of ABA-aldehyde to ABA, the same step as *sit* and *flc* (20). A wilty mutant in corn appears to be caused by an inadequate water supply due to a delay in the differentiation of metaxylem vessels (68), while in *Capsicum scabrous* (*diminutive*) (100) an increased concentration of ions in the guard cells may be responsible. In peas, a single recessive gene, *wil*, causes wilting and lowers percent water content, water potential and diffusive resistance (18). Grafting studies indicate that this is not attributable to the rootstock (105). Analysis of endogenous ABA levels by GCMS-MIM using a deuterated internal standard showed that leaves of water stressed *wil* plants possess less ABA than *Wil* types, and in this respect they are similar to the ABA deficient mutants in tomato and potato (105).

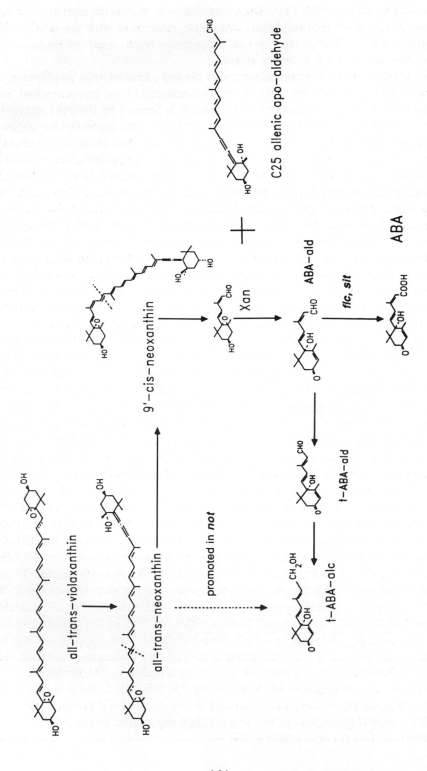

Fig. 3. The possible biosynthetic pathway for ABA showing the proposed sites of action of the wilty mutants in tomato, *flc*, *sit* and *not* (after 62).

The wilty-like mutants in *Arabidopsis* are of particular interest since both ABA insensitive and ABA deficient mutants have been isolated. The isolation of the ABA deficient mutants employed an ingenious screen that involved the use of gibberellin (GA) deficient dwarf mutants (41, 42). Koornneef and coworkers (43) identified 'non-germinating GA-dwarfs' in *Arabidopsis*. These plants failed to germinate unless provided with gibberellin (GA). As in other species, 'germinating GA-dwarfs' were also found. Five loci were identified, three (*ga1, ga2* and *ga3*) possessing both 'nongerminating' and 'germinating' mutant alleles. It was suggested that the mutations at all five loci control steps in the GA biosynthetic pathway. The difference in germination was attributed to the possibility that the GA requirement for germination was much lower than for stem elongation, and that some mutants were leakier than others (43). Since ABA has been implicated in seed dormancy, the selection for ABA mutants began by screening for germination revertants among M_2 and M_3 progenies from EMS treated, nongerminating *ga1* mutants. This was an elegant strategy since second site suppressors would appear as germinating seeds in a population of non-germinating seeds (i.e. the suppressor trait would be self-selecting) (41). From the range of suppressors produced, germinating GA-responsive extreme dwarfs were crossed to parental nongerminating *ga1* types (Table 2). The ability to restore germination was due to a recessive gene (called *aba*) which, on a wild-type (e.g., *GA1*) background, resulted in a smaller and weaker plant with a slightly yellow-brownish color and symptoms of withering. The withering symptoms can be reduced by applying ABA, and both seeds and leaves of *aba* types contain reduced ABA levels. The mutation *aba* appears to block the epoxidation of zeaxanthin, resulting in reduced levels of possible C_{40} ABA precursors including 9'-cis-neoxanthin (19, 81). Homozygous *aba* restores germination on a *ga1* background, suggesting that GA is required for germination only if ABA was present during seed development. Mutant *aba* also leads to reduced seed dormancy and increased rate of water loss compared with comparable wild-type plants (Table 2). A final point is that homozygous *aba* plants segregating from plants heterozygous at the *aba* locus lacked seed dormancy, indicating that the germination behavior of ripe seeds is determined by the embryo genotype and not by that of the maternal genotype (41). Karssen *et al.* (33) showed that although both maternal and embryonic ABA occurred in developing seeds, it was embryonic ABA that

Table 2. The genotypes of certain developmental phenotypes of *Arabidopsis*.

Phenotype	Genotype
Wild-type	*Ga1 Aba*
Nongerminating, extreme-dwarf	*ga1 Aba*
Germinating (revertant) extreme-dwarf	*ga1 aba*
Withering, reduced seed dormancy and ABA levels	*Ga1 aba*

controlled seed dormancy. The use of mutant types in this way is a powerful tool if it is necessary to determine whether a seed or fruit character is under the control of the maternal or embryo genotype.

ABA-insensitive mutants in *Arabidopsis* were selected by their good growth on a medium supplemented with 10 μM ABA (normal plants grow poorly on this medium) (42). Mutations at three loci were isolated, and these, like the *aba* mutants, possessed reduced seed dormancy. Mutations at two of the three loci also exhibited increased withering and water loss like the ABA deficient *aba* types. ABA levels in young seeds of the insensitive types were found to be higher than, or similar to, wild-type seeds. The three loci involved have been named *abi1*, *abi2* and *abi3*. Mutations *abi1* and *abi2* also influence water relations. This may suggest that *abi1* and *abi2* operate at an earlier step than *abi3*, thus providing a genetic tool to further explore the physiology of hormone action. Although seed development prior to imbibition appeared normal in *aba* and *abi3-1* seeds (42), seeds of genotype *aba abi3-1* or *abi3-3* fail to accumulate normal levels of the late embryogenic abundant (LEA) proteins (23, 39, 58).

Recently, *abi3* in *Arabidopsis* was cloned by map-based cloning (25). The mutation was found closely linked to another DNA marker, cosmid 4711. A DNA fragment from the cosmid was found to complement the mutant phenotype, and the location of a long open reading frame within the subfragment corresponded to an independently isolated cDNA. The open reading frame corresponding to *ABI3* encodes a protein (ABI3) that has discreet regions with up to 87% sequence similarity to the product of the *VP1* gene in maize. Koornneef and coworkers (42), in their early work on the mutant, pointed out the similarities of the *Arabidopsis abi3* mutant to the maize mutant *vp1,* which was identified on the basis of precocious germination. As Koornneef predicted, ABI3 may be a molecular equivalent to VP1, at least by the criterion of sequence similarity. Of course, it has not been demonstrated that ABI3 functions in the same way as does maize VP1. Indeed, what VP1 does in maize is still a matter of much discussion. Suffice it to say that VP1 is not a typical signal transduction chain component. McCarty et al. (48) have argued persuasively that VP1 is a transcription factor that mediates the hormone responsiveness and/or developmental expression of genes such as *Em* and *Cl* (48). The problem is that VP1 itself does not bind to DNA, as one might expect for a transcription factor. Instead it is thought to bind at the *Em* promoter site via other proteins. In any case, because it does not bind directly, it has been more difficult to demonstrate convincingly that VP1 is, indeed, a transcription factor.

With the function of VP1 in mind, it seems possible that ABI3 acts at the end of a signal transduction chain that activates ABA responsive genes. When the functions of ABI1 and ABI2 are determined, it will be interesting to relate them to ABI3, since these mutants are thought to be further upstream in the signal transduction chain because their effects are more pleiotropic than *abi3*. The striking effect of ABA mutants on dormancy in *Arabidopsis* also appears to occur in other species since the *sit* mutation in tomato also reduces seed dormancy (26). Further, both the *droopy* mutant in potato (89) and the *sit* mutant in tomato (41) promote precocious seed germination as does the double mutant *aba abi3-1* (39) or the more severe *abi3-3* allele (58) in *Arabidopsis*. The precocious germination of viviparous mutants in maize also appears to involve reduced ABA levels and sensitivity (92).

The work with wilty mutants across all species shows that a wide range of physiological causes are involved. These range from anatomical changes that limit water transport (68), to changes in ionic concentrations within the guard cells, which limit their function (100), to mutations that influence the level of, or response to, ABA (41, 42, 59, 71). The development of a broad range of mutations influencing different aspects of a particular developmental process is of importance since it allows the individual components to be examined in detail. Production of double or triple mutants may then allow the interaction and interdependence of these processes to be determined, so that the limiting or controlling steps may be identified. Unfortunately, such recombinants have not always been produced, limiting the full potential of the physiological-genetic approach. The work with ABA deficient and insensitive mutants outlined above is one of the key pieces of evidence that supports the role of ABA in the normal water relations of the plant and the development of seed dormancy. The mutants also help to clarify the role of ABA in the induction of desiccation tolerance during seed development (52).

INTERNODE LENGTH AND THE GIBBERELLINS

The elegant work of Phinney and coworkers (64) with maize, and of Brian and coworkers (10) with garden pea, was the first to link internode length genes with gibberellins (gibberellic acid, GA) and represents the beginning of the use of hormone mutants to understand plant development. Since then, similar studies on internode length have been carried out with many other species, including sweet pea (85), lettuce (108), wheat (94), tomato (6), *Arabidopsis thaliana* (43, 114) and *Brassica* (84). Internode length mutants have received this attention because they occur frequently, are easy to identify and are commercially important.

In maize over fifty mutants which influence plant height have been described (67). The early work by Phinney (64) demonstrated that the five non-allelic dwarf mutants, d_1, d_2, d_3, d_5, and an_1, became phenocopies of normal types after treatment with small amounts of GA_3. All contained less GA-like activity than normal types when examined using the d_5 maize bioassay. This led to the conclusion that all five mutations inhibit different steps in the GA biosynthetic pathway. In addition, another five non-allelic dwarf mutants, D_8, pe_1, mi_2, na_1 and na_2 showed little or no response to applied GA_3. The four GA-sensitive (synthesis) dwarfs, d_1, d_2, d_3 and d_5 have received considerable attention, and their sites of action in the GA

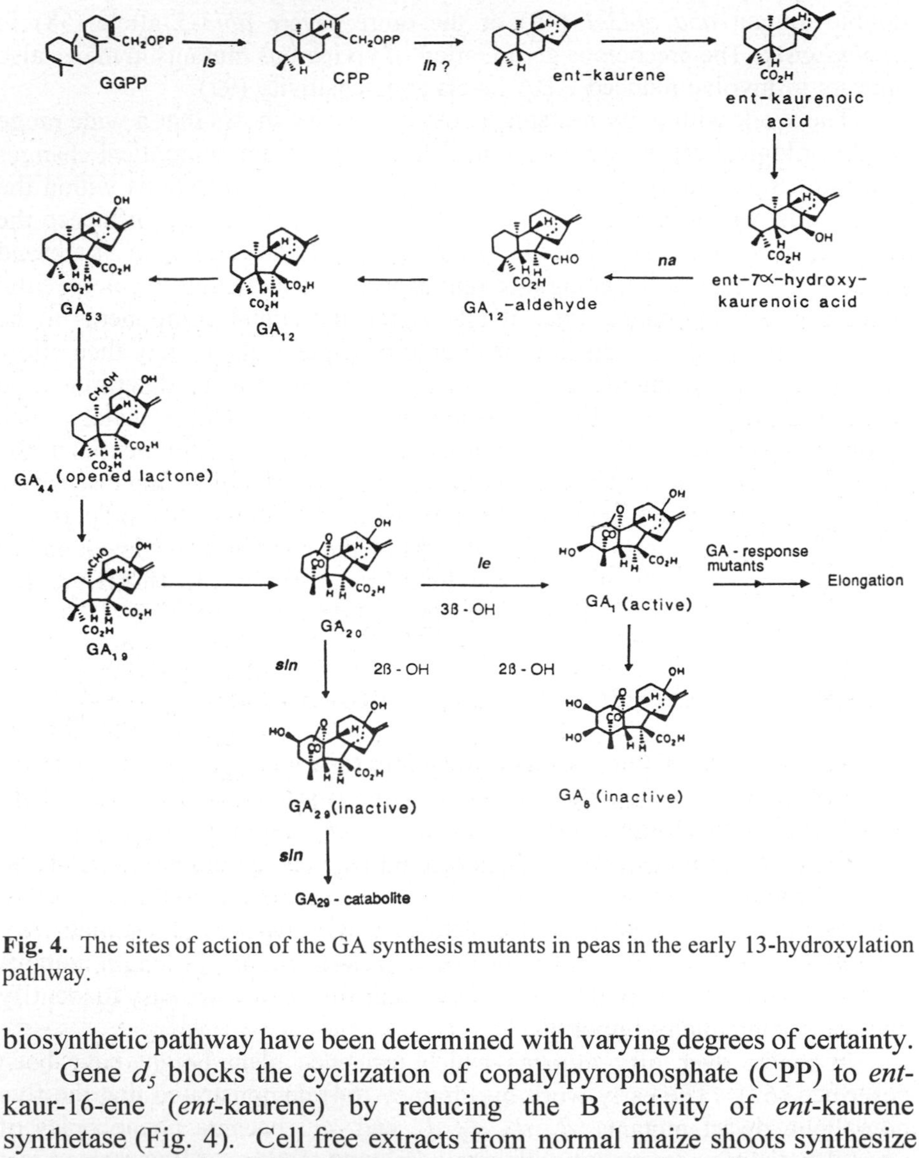

Fig. 4. The sites of action of the GA synthesis mutants in peas in the early 13-hydroxylation pathway.

biosynthetic pathway have been determined with varying degrees of certainty.

Gene d_5 blocks the cyclization of copalylpyrophosphate (CPP) to *ent*-kaur-16-ene (*ent*-kaurene) by reducing the B activity of *ent*-kaurene synthetase (Fig. 4). Cell free extracts from normal maize shoots synthesize *ent*-kaurene as the major diterpene. Extracts from d_5 seedlings showed a marked reduction in *ent*-kaurene synthesis regardless of whether mevalonic acid, geranylgeranylpyrophosphate or CPP was used as substrate. A concomitant increase in *ent*-kaur-15-ene (*ent*-isokaurene) synthesis was observed (28), a metabolite not in the pathway leading to the GAs.

Gene d_1 blocks the 3β-hydroxylation of GA_{20} to GA_1 since $[^{13}C,^3H]GA_{20}$ was metabolized to $[^{13}C,^3H]GA_{29}$-catabolite and $[^{13}C,^3H]GA_1$ by shoots of normal seedlings, whereas $[^{13}C,^3H]GA_{29}$ and $[^{13}C,^3H]GA_{29}$-catabolite were produced by shoots of d_l seedlings (93) (Fig. 4). The reduced levels of endogenous GA_1 and GA_8 in d_l plants and the build up of GA_{20} and GA_{29} levels confirm the position of this block (22). Application data and grafting results also support the site of action of d_5 and d_l (66). These results have been made possible by the determination of the native GAs in maize (22), the construction of a probable biosynthetic pathway (65-67) and the availability of appropriate intermediates. Biochemically, these mutations are amongst the best understood genes controlling plant hormone production.

In peas, an undisputed role for the involvement of certain internode length genes in GA biosynthesis has also been established (30, 80, Fig. 4). Skepticism about this relationship during the 1960-70s existed because extraction of GA-like substances could not establish a consistent difference between tall (conferred by the gene *Le* on an otherwise wild-type background) and Mendel's dwarf (*le*) varieties of peas. However, *le* was subsequently shown to block the 3β-hydroxylation of GA_{20} to GA_1 (Fig. 4) since immature shoot tissue of *Le* peas can metabolize $[^{13}C,^3H]GA_{20}$ to $[^{13}C,^3H]GA_1$, $[^{13}C,^3H]GA_8$ and $[^{13}C,^3H]GA_{29}$ while in *le* plants no $[^{13}C,^3H]GA_1$ and only a trace of $[^{13}C,^3H]GA_8$ was found (30). This was confirmed by recent work which has shown that the shoots of tall (*Le*) plants possess 10-20 times the level of GA_1 found in the shoots of light-grown dwarf (*le*) plants (86). As expected, GA_{20} levels are elevated in *le* plants (86). The *le* mutation is leaky since low levels of GA_1 are found in *le* plants and a more severe allele at this locus, le^d, has been identified. As would be expected, *Le* plants show a more or less equivalent growth response to the application of GA_1 or GA_{20}, while plants possessing the leaky gene *le* respond far more to GA_1 than to GA_{20} . The alleles at the *le* locus in peas control the same step as d_l in maize (93), *dy* in rice and *ga4* in *Arabidopsis* (114).

There were several reasons for the uncertainties in earlier experiments surrounding the action of the *Le* gene. The reasons include poor separation techniques, extraction of plants at an age before the tall (*Le*)/dwarf (*le*) difference was maximal, and extraction of inappropriate tissue. It would seem obvious that if the biochemical effect of a gene controlling a developmental response is to be examined, the tissue extracted should show the developmental difference and be of an age where this effect is maximal. However, this has often been overlooked because plants of the appropriate age have been too large and difficult to grow, or the hormone concentration has been greater in some other tissue (e.g., developing seeds) and thus easier to measure.

Fourteen non-allelic internode length genes, in addition to *le*, have now been examined in peas (78, 80). Three further dwarfing mutations, *na, lh* and *ls* (Fig. 5) block other steps in the GA biosynthetic pathway (80). Treatment with GA-precursors suggests that *na* may block the conversion of *ent*-7α-hydroxy kaurenoic acid to GA_{12}-aldehyde since *na* plants show no response to even 103 µg of *ent*-kaurene, *ent*-kaurenol, *ent*-kaurenal, *ent*-kaurenoic acid or *ent*-7α-hydroxy kaurenoic acid while wild-type plants do respond. The *na* plants do, however, respond normally to GA_{12}-aldehyde (29) (Fig. 4). Metabolism studies also show that, while *na* plants can convert [^{2}H]GA_{12}-aldehyde to C_{19} GAs such as GA_{20}, GA_{29}, GA_1 and GA_8, these plants do not metabolize [^{3}H]kaurenoic acid to substances co-eluting with GA_{20}, GA_1 or GA_8, even though *Na* plants do carry out this conversion (29).

Application studies show that *ls* and *lh* plants respond to *ent*-kaurene in a similar way to comparable wild-type plants (29). This may indicate that these mutant alleles block very early steps in GA-biosynthesis (prior to *ent*-kaurene) but probably after geranylgeranylpyrophosphate (GGPP) (Fig. 4). While the precise step blocked by *lh* is still unknown, *in vitro* metabolism studies using GGPP and CCP suggest *ls* blocks metabolism between these two compounds (Swain *et al.* unpub. results).

Excellent quantitative relationships between the log of the GA_1 concentration and elongation can be found using these mutants by examining endogenous GA_1 levels in multiple allelic series (98). Like *le*, the mutations *na*, *lh* and *ls* affect elongation in both the light and the dark (74) (Fig. 5) and are probably leaky since the homozygous double recessive plants (e.g., *na ls* and *na lh* plants) are shorter than comparable single homozygotes (73), and all the single mutants contain small but detectable levels of GA_1 (98).

Recently, a mutant containing elevated GA_1 levels has also been isolated in peas (87). This mutant, *sln*, results in a slender phenotype in which the lower internodes are elongated compared with wild-type. The threefold increase in internode length is associated with an eight fold increase in GA_1 levels in young shoots. The *sln* allele results in a blockage of the conversion of GA_{29} to GA_{29}-catabolite in the developing seed (Fig. 4; 87). This is associated with a 400-fold increase in the level of GA_{20} in the dry seed, which, on germination, appears to be converted to the biologically active

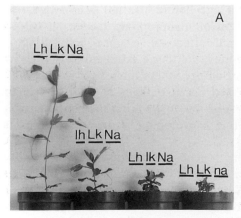
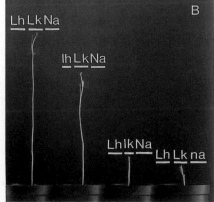

Fig. 5. The effect of the internode length mutants *lh*, *lk* and *na* on the growth of (A) 29 day-old light grown pea plants and (B) 14 day-old dark grown pea plants.

GA_1, resulting in elevated GA_1 levels in the shoot, and rapid elongation (87).

The following results summarize the evidence that GA_1 does directly regulate elongation in light grown wild-type peas. GA_1 is endogenous to wild-type peas (30) and is active *per se*. Precursors are not fully active on all GA-deficient mutants (e.g., GA_{20}) while metabolites such as GA_8 possess little activity. GA_1 levels do vary during ontogeny and correlate with changes in internode length (86). Since elevated GA_1 levels increase the length of the basal internodes in *sln* plants, GA_1 levels are not saturating for growth in the wild-type. Decreases in GA_1 levels result in reduced elongation, and the deficient mutants can be converted into phenocopies of wild-type plants by the addition of GA_1 (Fig. 6). Taken together these results define a key role for GA_1 in controlling internode elongation (80).

None of the GA biosynthesis genes in peas or maize have yet been cloned. However, in *Arabidopsis* one of the five identified GA biosynthesis mutations (114), *ga1*, has been cloned by genomic subtraction (97). To use genomic subtraction one must have a deletion that covers the gene of interest. There was evidence from Koornneef (43) that one of the *ga1* alleles, *ga1-3*, generated by fast neutron bombardment was a deletion mutation. It failed to recombine with several other alleles at the *GA1* locus. In the experiments by Sun *et al.* (97), one out of six clones resulting from the cloning of the PCR amplified DNA represented a prospective *GA1*-specific clone. The clone was verified to be from the *GA1* locus by complementing the *ga1* mutants, by RFLP mapping, and by demonstrating that the corresponding DNA in the *ga1-3* mutant bore a deletion. The action of the *GA1* gene at the molecular level is currently being analyzed by Sun and coworkers.

Many GA response mutants have been identified in various plants (e.g., ten in peas, 80). The mutants can be divided into two groups, short mutants that possess a reduced response to added GA_1 and slender mutants, in which the GA response is thought to be constitutive (11, 44, 69). In *Arabidopsis*, a semi-dominant mutant, *gai*, was isolated under non-selective conditions because it phenotypically resembled other GA sensitive dwarfs (i.e. dwarf, increased numbers of axillary shoots, narrower, darker green leaves, 38). The only morphological feature that distinguished *gai* from GA sensitive mutants was that the size of flower petals in *gai* was not reduced. However, the phenotype of *gai*, could not be rescued by repeated application of GA (GA_{4+7}). Furthermore, the GA activity in *gai*, determined in a bioassay, was indistinguishable from wild type. Therefore, *gai* is considered to be a GA response mutant (38) .

The *gai* mutation maps near *GA4* on chromosome 1. Since the mutation affects many different morphological features, the defect may be at a step that is common to GA responses in different organs, or the defect may be in an early step in the pathway that affects all organs. Recently, a novel procedure was devised for selecting for suppressors of *gai* (63). The suppressors were obtained by irradiating seeds homozygous for *gai* and selecting in the M1 generation for mutants which bolted normally. This was done to find somatic chimeras expressing secondary mutations with effects that are semi-dominant and that overcome the *gai* phenotype. Such suppressors were found, and, surprisingly, all of them appeared to be intragenic suppressors at the *gai* locus (63). Interestingly these mutations, called *gai-d* (derivative mutations), had no obvious phenotype when present in a homozygous condition.

In an effort to find other GA response mutants in *Arabidopsis*, Jacobsen and Olszewski (31) isolated mutants that germinate in the presence of the GA biosynthesis inhibitor paclobutrazol (Fig. 1). To eliminate ABA biosynthesis and/or response mutants from the screen, the mutants were further examined at the seedling stage for resistance to the dwarfing effects of paclobutrazol. The mutants obtained by this screen were all alleles at a locus called spindly (*spy*), and in the absence of the inhibitor the mutants resembled wild type plants that are repeatedly sprayed with GA_3. To demonstrate that the *spy* mutations conferred resistance to paclobutrazol by suppressing the effects of GA deficiency, rather than by ameliorating the action of the inhibitor, Jacobsen and Olszewski constructed double mutants between *ga1* and *spy-1*. They found that *spy-1* suppressed the requirement conferred by the *ga1* mutation for exogenous GA to germinate (31).

In peas, where 10 non-allelic GA-response mutants have been identified, the majority possess reduced stature. However examination of these mutants clearly demonstrates that they are a heterogeneous group of mutants influencing a wide range of processes, not just the perception of the GA_1 signal. For example, while the dwarf *lgr* mutant is a true phenocopy of the dwarf GA_1 deficient mutants (Murfet *et al.*, unpub. results) and may indeed influence the perception of the GA_1 signal, the other GA_1 response mutants possess a range of additional pleiotropic effects that include altered branching patterns, stem shape and leaf characters. This suggests that the mutants influence steps well down the transduction pathway leading from GA_1 perception to elongation. Some modify the chemorheological properties of the cell wall by reducing the wall yield threshold and also result in reduced IAA levels (e.g., *lka* and *lkb*, 5, Mackay *et al.*, unpub. results). Another mutation acts by modifying the photomorphogenic response (*lw*, 109). This may come about by enhancing the response to phytochrome which in turn reduces the GA_1-responsiveness of the stem tissue.

The GA constitutive behavior of the slender mutant, conferred by the gene combination *la cry^s*, appears to act close to the point of GA_1 perception (69). Consequently, reductions of endogenous GA_1 levels (e.g., in the triple recessive *na la cry^s*) do not alter elongation (i.e. it has a slender phenotype, or the *la cry^s* gene combination is epistatic to the GA_1-deficient mutant *na*). However, the short mutant, *lk* (Fig. 5), which acts further down the transduction pathway is epistatic to *la cry^s* (i.e. the genotype *lk la cry^s* is extremely short, 73). Finally, genotypes deficient in phytochrome B function (e.g., *lv*, 109) are elongated, not due to elevated levels of GA but due to an enhanced responsiveness to GA_1 (79). These GA-response mutations consequently can be used to explore the partial processes leading from GA_1 reception to the elongation response. They already indicate the complexity of this response pathway since they show that the levels of other hormones (e.g., IAA and ethylene) may be involved, that the physical properties of the cell wall are important, and that the pathway interacts with the control of photomorphogenesis by phytochrome.

FLOWER INDUCTION AND INITIATION

In most plants species, there is considerable genetic variation in time to flowering. Single gene differences have been isolated both from natural populations and mutagenesis programs (40, 55). However, unlike the developmental processes so far discussed, the chemical nature of the 'flowering signal(s) or hormone(s)' remains obscure even though there is considerable indirect evidence from grafting experiments to support their existence. The concept of a flowering signal or hormone is also important to link flower induction in the leaves with evocation and initiation events in the shoot apex. The best examined species are *Arabidopsis thaliana* (1, 40, 113) and *Pisum sativum* (peas; 55, 57) where over ten loci affecting floral induction have been identified. Peas possess the advantage that they are well suited to physiological and biochemical studies because of their morphology and size. Consequently, grafting and extractions from specific tissues are possible. *Arabidopsis* is superior for molecular studies for the reasons previously outlined. Work with both species demonstrates that even though the biochemical control of flowering has not been determined, flowering mutants can be isolated and divided into groups according to whether they influence the level of the 'hormone(s)' or response to the 'hormone(s),' or have indirect or secondary effects due to changes in plant growth rate, etc.

The genes influencing 'hormone' levels include those involved with the induction events in the leaves in photoperiod sensitive varieties, while the genes involved with 'hormone' response may include those operating in the apex and influencing evocation and determination in the apex. To illustrate that flowering mutants can be classified in terms of 'hormone' synthesis and response types, the genetic control of flower induction and initiation will be outlined for peas.

In peas the genes at least eleven major loci, *lf, e, sn, hr, dne, ppd, gi, det, dm, veg-1* and *veg-2* interact to control the flowering phenotype (55, 57) (Table 3). Phenotypes range from day-neutral types, through quantitative long day types, to types with an almost qualitative requirement for long days (55) (Fig. 6). A short-day mutant has even been isolated, as have mutants that never produce a flower (57). Certain genotypes possess large vernalization responses, while others show little or no response to vernalization (63). The change from vegetative to reproductive growth in

Table 3. The phenotype and possible site and mode of action of the flowering mutations in peas (55, 57).

Mutation	Phenotype	Site of Action	Mode of action
sn	Promotes flowering, day neutral	Leaves including cotyledons	Blocks production of a graft transmissible flower inhibitor
dne	(as above)	(as above)	(as above but leaky)
ppd	(as above)	(as above)	(as above for sn)
hr	Reduced quantitative LD photoperiod response	Leaves excluding cotyledons	Causes Sn Dne Ppd activity to be reduced at an early age
e	Delays flowering initiation	Primarily cotyledons	Increases Sn Dne Ppd activity in cotyledons
gi	Delays flowering	Leaves	Blocks production of graft transmissible flower promotor
fsd	Delays flowering, no flowering in LD	Leaves	(as above)
lf	Promotes flowering, reduces length of juvenile period	Shoot apex	Increases sensitivity to the floral signal
veg-1	Vegetative, no flowers produced	Shoot apex	Homeomeotic; inflorescence sites occupied by vegetative branches
veg-2	(as above)	(as above)	(as above)
det	Determinate	Shoot apex	Growth of apical meristem ceases soon after the onset of flowering

peas is thought to be controlled by the time at which the balance of a flower promoter to a flower inhibitor passes a threshold, even though the biochemical nature of these substances is not known (63).

The three complementary genes, *Sn*, *Dne* and *Ppd* confer the ability to respond to photoperiod since recessive alleles at any one of these loci results in an early day neutral phenotype. These genes appear to control steps leading to the synthesis of the flower inhibitor in the cotyledons and leaves (35, 55) since stocks homozygous for either *sn* or *dne* lack the ability to delay flowering in *Sn Dne* stocks when grafted to *sn Dne* or *Sn dne* scions. Since the delaying ability of the gene combination *Sn Dne Ppd* occurs only under short photoperiods, it seems likely that long photoperiods promote flowering by retarding activity of the pathway leading to inhibitor production (35, 55). This process is phytochrome controlled, with the size of the inhibitory effect depending on the length of the dark period when 25 h cycle lengths are used (76). The phytochrome B response mutant, *lv*, reduces the size of the photoperiod response (109) but does not influence flowering on a day-neutral background (e.g., *sn*, 79). This suggests that *Sn*, *Dne* and *Ppd* act downstream of phytochrome and its direct response pathway. This is supported by the fact that *sn*, *dne* and *ppd* plants do not possess any of the other pleiotropic effects of the phytochrome mutants. However, the product of the *Sn Dne Ppd* pathway, as well as influencing flower induction, and hence initiation, also influences a range of other developmental processes including flower development, yield, apical senescence (or more correctly cessation of apical growth), vegetative vigor and the production of basal

Fig. 6. The effect of the flowering genes controlling hormone levels, *E, Sn* and *Hr* on the flowering node (indicated by tape and arrows), development of pods, cessation of apical growth and senescence in 12 week old pea plants grown in a 10 h photoperiod. Note that *Sn* and *Hr* influence many facets of growth (A) even if the flowering node of the plants is similar due to the presence of *E* (B). All plants possessed gene *lf*.

laterals (55, 77). This may occur if the primary role of the *Sn Dne Ppd* product is to direct assimilate flow (55, 77). Consequently, if other genes mask the photoperiodic effect of the *Sn Dne Ppd* combination on flower initiation their presence may be traced using one or more of these pleiotropic characters.

The genes *E* and *Hr* modify the expression of the *Sn Dne Ppd* combination since the *E/e* and *Hr/hr* gene differences have little effect in homozygous *sn* plants (55) (Fig. 6). Gene *E* has its major influence in the cotyledons where it appears to reduce inhibitor production by the *Sn Dne Ppd* pathway resulting, in some circumstances, in early flower initiation (55). *Hr* exerts its major influence in the foliage leaves (72). The output of the *Sn Dne Ppd* pathway appears to decrease rapidly with age in *hr* plants, which then behave as quantitative long day plants. This decrease in activity is reduced in *Hr* plants, and they therefore behave as near qualitative long day plants (72) (Fig. 6). Genes *E* and *Hr* are therefore regulatory genes influencing hormone levels by regulating the ontogenetic stages and tissues in which the 'structural' genes *Sn, Dne* and *Ppd* are expressed (Table 3).

Mutations *gi* and *fsd* are also considered to be hormone synthesis mutants since they appear to operate by controlling steps in the pathway leading to the synthesis of the flower promoter. Grafting studies clearly show *gi* and *fsd* plants are deficient in some substance essential for flowering since leafy wild-type stocks are capable of promoting flowering in *gi* and *fsd* scions (57). Plants possessing *fsd* (and some *gi* plants) do not flower in long photoperiods (18-24 h) possessing normal levels of far red (FR) light, and thus can be classified as short day plants. However, flowering was delayed substantially relative to the progenitor varieties even in short days. The *gi* plants show a strong vernalization response which has been localized to the apical bud (Beveridge and Murfet, unpub. results), consistent with the view that vernalization may influence the sensitivity to the flowering signal (75). When mutations in both the flower promoter and flower inhibitor pathway were combined (e.g., *sn gi*) flowering was unstable. Some plants flowered earlier than wild-type plants while others remained vegetative indefinitely. Vegetative reversion was common, and the phenotype suggests that the hormonal situation in *sn gi* plants hovers close to the flowering threshold for a long period of time. Normal ontogenetic strengthening of the flowering signal with age may be absent.

Gene *Lf* operates at or near the shoot apex and influences the sensitivity of the plant to the putative flowering hormones (55) (Table 3). Four distinct alleles, *lf*^{a}, *lf*, *Lf* and *Lf*^{d} are known at the *lf* locus and respectively lead to a higher promoter to inhibitor ratio being required for flowering as determined by grafting experiments (Fig. 7). The *Lf* alleles consequently determine the length of the juvenile phase under highly promoting conditions resulting in minimum flowering nodes of 15, 11, 8 and 5 for *Lf*^{d}, *Lf*, *lf* and *lf*^{a} respectively. They disrupt the normal relationship between vegetative maturity as indicated by leaf form and the onset of flowering and thus provide an excellent example of heterochrony.

The remaining mutations in peas (e.g., *veg-1*, *veg-2*, *det*, *fds*, *pim*) appear to affect processes beyond the perception of the flowering signal in the apex. Some of these may control the identity of the floral meristem, or floral organogenesis, in a similar fashion to *lfy* in *Arabidopsis* or *flo* in *Antirrihinum* (14, 15) (e.g., *veg-1* and *veg-2*; 57). Consequently, while the growth pattern of the plant clearly indicates the flowering signal from the leaves has arrived and been perceived at the apex in these mutants, the *veg-1* and *veg-2* mutants do not produce any flowers regardless of the environmental conditions (57, 77). The *det* mutation on the other hand causes the shoot to cease growth after a small number of reproductive nodes (57), but does not greatly influence flower initiation itself. However, the double mutant *det veg-1* is reported to produce flowers (90). Therefore, while it is possible by grafting to define genes involved with induction in the leaves, it becomes difficult to define which mutants influence the ability of the apex to respond to this inductive signal, and which control subsequent steps leading to flower initiation and organogenesis. A detailed study of the interactions of these genes may, however, elucidate the steps leading from induction, through evocation to initiation and hence to flower development. While the later part

Fig. 7. The effect of the flowering alleles *lf^a*, *lf* and *Lf*, which control the sensitivity of the apex to the flowering hormones, on the growth of 7 week old, day-neutral (*sn hr*) pea plants. The flowering node for each plant is indicated by the arrows. Note the pleiotropic effects of the *lf* alleles on peduncle length, lateral outgrowth and pod development (55, 57). (Photograph provided by Dr. I.C. Murfet, Univ. of Tasmania).

of this sequence has been well defined by the work in *Arabidopsis* and *Antirrihinum* (15), the early steps near reception in the apex clearly deserve further examination.

Other genes in peas (e.g., the internode length genes *le*, *la* and *cry^s*, 55) have minor pleiotropic effects on flowering. These effects are probably secondary consequences of the altered growth rate in these genotypes. The involvement of gibberellins with the genes controlling flowering and apical senescence in peas has been investigated extensively using application data, metabolism of labeled gibberellins, and examination of the levels of endogenous gibberellins by both bioassays and GC-MS (17, 70). These studies provided inconclusive results regarding the direct involvement of any of the flowering genes with the gibberellins. However, genetic evidence shows that the expression of the flowering genes is not masked by the presence of altered gibberellin levels in mutants *le*, *na*, *ls* or *lh* (56), or in double mutants such as *lh ls* or *ls na*. Flower development is relatively normal even though expanding tissue may possess less than two percent of the normal level of GAs (30). However, the staminal filaments may be reduced in length and thus fertilization may be reduced. This result is contrary to the view for *Arabidopsis* where GA has been suggested to be essential for flowering (111). These results suggest the flowering genes do not operate by directly influencing gibberellin metabolism and demonstrate the usefulness of producing combinations of developmental mutants since they provide firm evidence where physiological and biochemical data have been inconclusive.

Although the nature of the 'flowering hormone(s)' remains obscure, an understanding of the genetic control of flowering allows the physiology of the partial processes to be examined and provides a firm basis for biochemical comparisons. The similarity of the categories of mutants involved in flowering to those shown for the biochemically better understood developmental responses (e.g., internode length) offers hope that resolution of the control of this fundamental transition may soon be possible. This will be helped if the molecular nature of the genes controlling the early steps in flowering can be determined.

CONCLUDING REMARKS

The preceding discussion shows the power of genetics in deciphering the role of hormones in different developmental processes in plants. In many cases, mutants affecting levels or responses to hormones provide the most direct evidence for the involvement of a specific class of hormones with a particular developmental process (30, 41, 65, 101). The use of genetics in plant hormone biology has drawn together hormone physiologists and developmental biologists. Over the last 5 years there has been an intense effort to select mutants specifically for the purpose of examining physiological processes in higher plants, especially in *Arabidopsis*. It is to be hoped that over the next few years, through the use of a combined genetic, physiological, biochemical and molecular approach, an understanding of the control of development in higher plants by plant hormones will be obtained.

A feature common to all the developmental processes examined (e.g., internode length, protonema development, and transpiration) is that hormone mutants fall into two clear categories, hormone synthesis and hormone response types. Hormone synthesis mutants are most easily recognized when the effects of hormone overproduction or underproduction can be predicted from physiological studies, such as in the case of the effects of ABA and GA on transpiration and internode elongation, respectively. The response mutants are a heterogeneous group which, on closer examination, may be further characterized and categorized. The work with the GA-response mutants influencing stem elongation clearly illustrates this complexity, and also demonstrates that such mutants may, in some cases, act primarily by modifying other developmental control systems (e.g., by influencing phytochrome responses or IAA levels). The rapid development of molecular techniques and the amenability of *Arabidopsis* to such studies appears to provide the opportunity in the near future to determine the elements in the response pathway for particular hormones. This has proven a major stumbling block using physiological and biochemical techniques alone.

It has been emphasized throughout this review that the hormonal control of developmental processes can be regulated by hormone levels or hormone responsiveness (sensitivity) (contrast with 104). So far, little attention has been paid to the question whether the mutants have different effects in various organs or at different times during development. However, where this has been investigated the mutants have shown precise ontogenetic and tissue specificity. For example, the *na* mutation in pea reduces GA levels in vegetative tissue and the pods, but does not reduce GA levels in the developing seeds (Swain unpub. results). The *lh* and *lh*^{i} alleles in peas have differential effects on GA production in seeds and shoot tissue resulting in contrasting phenotypes (99).

Further, discrete maternal and embryonic ABA pools have been identified in developing seeds of *Arabidopsis* using the *aba* mutation (33, 39). Some of these results are explained most simply if certain synthesis mutants (e.g., *na, sln, lh*^{i}) are the result of mutations of regulatory genes or elements rather than mutations of structural genes directly encoding biosynthetic enzymes. If this turns out to be true it will add a new dimension to our understanding of hormone biosynthetic pathways and their tissue specificity. Since hormones are almost certainly involved in controlling ontogenetic and tissue specific developmental changes, detailed examination of the present mutants and the selection of further mutants with ontogenetic or tissue specificity may be most beneficial.

Acknowledgments

Thanks to Ian Murfet, John Ross, Stephen Swain and Jim Weller for helpful comments on the manuscript and for reference to unpublished work and the Australian Research Council for financial support for J.B.R. and to the USDA/CSRS for partial support for S.H.H.

References

1. Araki, T., Komeda, Y. (1993) Analysis of the role of the late flowering locus, *GI*, in the flowering of *Arabidopsis thaliana*. Plant Journal 3, 231-239.
2. Ashton, N.W., Cove, D.J. (1990) Mutants as tools for the analytical dissection of cell differentiation in *Physcomitrella patens* gametophytes. *In*: Bryophyte development: physiology and biochemistry, pp. 17-31, Chopra, R.N., Bhatla, S.C., eds. CRC Press, Boca Raton, Florida.
3. Ashton, N.W., Cove, D.J., Featherstone, D.R. (1979) The isolation and physiological analysis of mutants of the moss, *Physcomitrella patens*, which over-produce gametophores. Planta 144, 437-442.
4. Ashton, N.W., Cove, D.J., Wang, T.L., Saunders, M.J. (1990) Developmental studies of *Physcomitrella patens* using auxin and cytokinin sensitivity mutants. *In*: Plant Growth Substances, 1988, pp. 57-64, Pharis, R.P., Rood, S.B., eds. Springer-Verlag, Berlin.
5. Behringer, F.J., Cosgrove, D.J., Reid, J.B., Davies, P.J. (1990) The physical basis for altered stem elongation rates in internode length mutants of *Pisum*. Plant Physiol. 94, 166-173.
6. Bensen, R.J., Zeevaart, J.A.D. (1990) Comparison of *ent*-kaurene synthetase A and B activities in cell-free extracts from young tomato fruits of wild-type and *gib-1, gib-2* and *gib-3* tomato plants. J. Plant Growth Regul. 9, 237-242.

7. Bleecker, A.B., Estelle, M.A., Somerville, C., Kende, H. (1988) Insensitivity to ethylene conferred by a dominant mutation in *Arabidopsis thaliana*. Science 241, 1086-1089.
8. Blonstein, A.D., Parry, A.D., Horgan, R., King, P.J. (1991) A cytokinin-resistant mutant of *Nicotiana plumbaginifolia* is wilty. Planta 183, 244-250.
9. Bopp, M. (1990) Hormones of the moss Protonema. *In*: Bryophyte development: physiology and biochemistry, pp. 55-77, Chopra, R.N., Bhatla, S.C., eds. CRC press, Boca Raton, Florida.
10. Brian, P.W. (1957) The effects of some microbial metabolic products on plant growth. Symp. Soc. Exp. Biol. 11, 166-181.
11. Chandler, P.M. (1988) Hormonal regulation of gene expression in the "slender" mutant of barley (*Hordeum vulgare* L.). Planta 175, 115-120.
12. Chang, C., Kwok, S.F., Bleecker, A.B., Meyerowitz, E.M. (1993) *Arabidopsis* ethylene-response gene *ETR1*: Similarity of product to two-component regulators. Science 262, 539-544.
13. Chaudhury, A. M., Letham, S., Craig, S., Dennis, E. S. (1993) *Amp1* - a mutant with high cytokinin levels and altered embryonic pattern, faster vegetative growth, constitutive photomorphogenesis and precocious flowering. Plant J. 4, 907-916.
14. Coen, E.S. (1991) The role of homeotic genes in flower development and evolution. Ann. Rev. Plant Physiol. Mol. Biol. 42, 241-279.
15. Coen, E.S., Meyerowitz, E.M. (1991) The war of the whorls: Genetic interactions controlling flower development. Nature 353, 31-37.
16. Cove, D.J., Ashton, N.W., Featherstone, D.R., Wang, T.L. (1980) The use of mutant strains in the study of hormone action and metabolism in the moss *Physcomitrella patens*. *In* The proceedings of the fourth John Innes Symposium, 1979, pp. 231-241, Davies, D.R., Hopwood, D.A., eds. John Innes, Norwich, UK.
17. Davies, P.J., Emshwiller, E., Gianfagna, T.J., Proebsting, W.M., Noma, M., Pharis, R.P. (1982) The endogenous gibberellins of vegetative and reproductive tissue of G2 peas. Planta 154, 266-272.
18. Donkin, M.E., Wang, T.L., Martin, E.S. (1983) An investigation into the stomatal behaviour of a wilty mutant in *Pisum sativum*. J. Exp. Bot. 34, 825-834.
19. Duckham, S.C., Linforth, R.S.T., Taylor, I.B. (1991) Abscisic-acid-deficient mutants at the *aba* gene locus of *Arabidopsis thaliana* are impaired in the epoxidation of zeaxanthin. Plant Cell Environ. 14, 601-606.
20. Duckham, S.C., Taylor, I.B., Linforth, R.S.T., Al-Naieb, R.J., Marples, B.A., Bowman, W.R. (1989) The metabolism of *cis*-ABA-aldehyde by the wilty mutants of potato, pea and *Arabidopsis thaliana*. J. Exp. Bot. 217, 901-909.
21. Estelle, M.A., Somerville, C. (1987) Auxin-resistant mutants of *Arabidopsis thaliana* with an altered morphology. Mol. Gen. Genet. 206, 200-206.
22. Fujioka, S., Yamane, H., Spray, C.R., Gaskin, P., MacMillan, J., Phinney, B.O., Takahashi, N. (1988) Qualitative and quantitative analyses of gibberellins in vegetative shoots of normal, *dwarf-1*, *dwarf-2*, and *dwarf-5* seedlings of *Zea mays* L. Plant Physiol. 88, 1367-1372.
23. Galau, G.A., Bijaisoradat, N., Hughes, D.W. (1987) Accumulation kinetics of cotton late embryogenesis abundant mRNAs and storage protein mRNAs: coordinate regulation during embryogenesis and the role of abscisic acid. Dev. Biol. 123, 198-212.
24. Gibson, S.I., Somerville, C. (1992) Chromosome walking in *Arabidopsis thaliana* using yeast artificial chromosomes. *In*: Methods in *Arabidopsis* research, pp. 119-143, Koncz, C., Chua, N.-H., Schell, J., eds. World Scientific, Singapore.
25. Giraudat, J., Hauge, B.M., Valon, C., Smalle, J., Parcy, F., Goodman, H.M. (1992) Isolation of the *Arabidopsis AB13* gene by positional cloning. Plant Cell 4, 1251-1261.
26. Groot, S.P.C., Karssen, C.M. (1992) Dormancy and germination of abscisic acid-deficient tomato seeds. Studies with the *sitiens* mutant. Plant Physiology 99, 952-958.

27. Guzman, P., Ecker, J.R. (1990) Exploiting the triple response of *Arabidopsis* to identify ethylene-related mutants. Plant Cell 2, 513-524.
28. Hedden, P., Phinney, B.O. (1979) Comparison of *ent*-kaurene and *ent*-isokaurene synthesis in cell-free systems for etiolated shoots of normal and *dwarf-5* maize seedlings. Phytochem. 18, 1475-1479.
29. Ingram, T.J., Reid, J.B. (1987) Internode length in *Pisum*: Gene *na* may block gibberellin synthesis between *ent*-7α-hydroxy-kaurenoic acid and gibberellin A_{12}-aldehyde. Plant Physiol.83, 1048-1053.
30. Ingram, T.J., Reid, J.B., Murfet, I.C., Gaskin, P., Willis, C.L., Macmillan, J. (1984) Internode length in *Pisum*. The *Le* gene controls the 3β-hydroxylation of gibberellin A_{20} to gibberellin A_1. Planta 160, 455-463.
31. Jacobsen, S.E., Olszewski, N.E. (1993) Mutations at the SPINDLY locus of Arabidopsis alter gibberellin signal transduction. Plant Cell 5, 887-896.
32. Johri, M.M., Desai, S. (1973) Auxin regulation of caulonema formation in moss protonema. Nature New Biol. 245, 223-224.
33. Karssen, C.M., Brinkhorst-van der Swan, D.L.C., Breekland, A.E., Koornneef, M. (1983) Induction of dormancy during seed development by endogenous abscisic acid: studies on abscisic acid deficient genotypes of *Arabidopsis thaliana* (L.). Heynh. Planta 157, 158-165.
34. Kieber, J.J., Rothenberg, M., Roman, G., Feldmann, K.A., Ecker, J.R. (1993) *CTR1*, a negative regulator of the ethylene response pathway in *Arabidopsis*, encodes a member of the RAF family of protein kinases. Cell 72, 1-20.
35. King, W., Murfet, I.C. (1985) Flowering in *Pisum*: a sixth locus, *Dne*. Ann. Bot. 56, 853-846.
36. Klee, H.J., Horsch, R.B., Hinchee, M.A., Hein, M.B., Hoffmann, N.L. (1987) The effects of overproduction of two *Agrobacterium tumefaciens* T-DNA auxin biosynthetic gene products in transgenic petunia plants. Genes and Devel. 1, 86-96.
37. Koncz, C., Chua, N.-H., Schell, J. (1992) Methods in *Arabidopsis* research. World Scientific Singapore.
38. Koornneef, M., Elgersma, A., Hanhart, C.J., van Loenen-Martinet, E.P., van Rijn, L., Zeevaart, J.A.D. (1985) A gibberellin insensitive mutant of *Arabidopsis thaliana*. Physiol. Plant. 65, 33-39.
39. Koornneef, M., Hanhart, C.J., Hilhorst, H.W.M., Karssen, C.M. (1989) In vivo inhibition of seed development and reserve protein accumulation in recombinants of abscisic acid biosynthesis and responsiveness mutants in *Arabidopsis thaliana*. Plant Physiol. 90, 463-469.
40. Koornneef, M., Hanhart, C.J., van der Veen, J.H. (1991) A genetic and physiological analysis of late flowering mutants in *Arabidopsis thaliana*. Mol. Gen. Genet. 229, 57-56.
41. Koornneef, M., Jorna, M.L., Brinkhorst-van der Swan, D.L.C., Karssen, C.M. (1982) The isolation of abscisic acid (ABA) deficient mutants by selection of induced revertants in nongerminating gibberellin sensitive lines of *Arabidopsis thaliana* (L.) Heynh. Theor. App. Genet. 61, 385-393.
42. Koornneef, M., Rueling, G., Karssen, C.M. (1984) The isolation and characterisation of abscisic acid-insensitive mutants of *Arabidopsis thaliana*. Physiol. Plant. 61, 377-383.
43. Koornneef, M., van der Veen, J.H. (1980) Induction and analysis of gibberellin sensitive mutants in *Arabidopsis thaliana* (L.) Heynh. Theor. Appl. Genet. 58, 257-263.
44. Lanahan, M.B., Ho, T.H.-D. (1988) Slender barley: A constitutive gibberellin-response mutant. Planta 175, 107-114.
45. Last, R.L., Bissinger, P.H., Mahoney, D.J., Radwanski, E.R., Fink, G.R. (1991) Tryptophan mutants in *Arabidopsis*: The consequences of duplicated tryptophan synthase b genes. Plant Cell 3, 305-310.
46. Last, R.L., Fink, G.R. (1988) Tryptophan-requiring mutants of the plant *Arabidopsis thaliana*. Science 240, 305-310.

47. Lincoln, C., Britton, J.H., Estelle, M. (1990) Growth and development of the *axr1* mutants of *Arabidopsis*. Plant Cell 2, 1071-1080.

48. McCarty, D.R., Hattori, T., Carson, C.B., Vasil, V., Lazar, M., Vasil, L.K. (1991) The *viviparous-1* development gene of maize encodes a novel transcription activator. Cell 66, 895-905.

49. Maher, E.P., Martindale, S.J.B. (1980) Mutants of *Arabidopsis thaliana* with altered responses to auxins and gravity. Biochem. Genet. 18, 1041-1052.

50. Medford, J.I., Horgan, R., El-Sawi, Z., Klee, H.J. (1989) Alterations of endogenous cytokinins in transgenic plants using a chimeric isopentenyl transferase gene. Plant Cell 1, 403-413.

51. Memelink, J., Hoge, J.H.C., Schilperoort, R.A. (1987) Cytokinin stress changes the developmental regulation of several defence-related genes in tobacco. EMBO J. 6, 3579-3583.

52. Meurs, C., Basra, A.S., Karssen, C.M., van Loon, L.C. (1992) Role of abscisic acid in the induction of desiccation tolerance in developing seeds of *Arabidopsis thaliana*. Plant Physiol. 98, 1484-1493.

53. Mirza, J.I., Olsen, G.M., Iversen, T.-H., Maher, E.P. (1984) The growth and gravitropic responses of wild-type and auxin-resistant mutants of *Arabidopsis thaliana*. Physiol. Plant. 60, 516-522.

54. Moore, R., Smith, J.D. (1985) Graviresponsiveness and abscisic-acid content of roots of carotenoid-deficient mutants of *Zea mays* L. Planta 164, 126-128.

55. Murfet, I.C. (1985) *Pisum sativum* L. *In* Handbook of flowering, vol. IV, pp. 97-1126, Halevy, A.H., ed. CRC Press, Boca Raton, Florida.

56. Murfet, I.C., Reid, J.B. (1987) Flowering in *Pisum*: gibberellins and the flowering genes. J. Plant Physiol. 127, 23-29.

57. Murfet, I.C., Reid, J.B. (1993) Developmental mutants. *In*: Peas - genetics, molecular biology and biotechnology, Davies, D.R., Casey, R., eds. (in press), CAB International, Wallingford, U.K.

58. Nambara, E., Satoshi, N., McCourt, P. (1992) A mutant of *Arabidopsis* which is defective in seed development and storage protein accumulation is a new *abi3* allele. Plant J. 2, 435-441.

59. Neill, S.J., Hogan, R. (1985) Abscisic acid production and water relations in wilty tomato mutants subjected to water deficiency. J. Exp. Bot. 36, 1222-1231.

60. Ottoline Leyser, H.M., Lincoln, C.A., Timpte, C., Lammer, D., Turner, J., Estelle, M. (1993) Arabidopsis auxin-resistance gene *AXR1* encodes a protein related to ubiquitin-activating enzyme E1. Nature 364, 161-164.

61. Normanly, J., Cohen, J.D., Fink, G.R. (1993) *Arabidopsis thaliana* auxotrophs reveal a typtophan-independent biosynthetic pathway for indole-3-acetic acid. Proc. Natl. Acad. Sci. USA 93, 10355-10359.

62. Parry, A.D., Horgan, R. (1992) Abscisic acid biosynthesis in higher plants. *In* Progress in plant growth regulation, pp. 160-172, Karssen, C.M., van Loon, L.C., Vreugdenhil, D., eds. Kluwer Academic Publishers, Netherlands.

63. Peng, J., Harberd, N.P. (1993) Derivative alleles of the *Arabidopsis* gibberellin-insensitive (*gai*) mutation confer a wild-type phenotype. Plant Cell 5, 351-360.

64. Phinney, B.O. (1961) Dwarfing genes in *Zea mays* and their relation to the gibberellins. *In*: Plant growth regulation, pp. 489-501, Klein, R.M., ed. Iowa State College Press, Ames, Iowa.

65. Phinney, B.O. (1984) Gibberellin A$_1$, dwarfism and the control of shoot elongation in higher plants. *In*: The biosynthesis and metabolism of plant hormones, Soc. Exp. Biol. Seminar Series 23, pp. 17-41, Crozier, A., Hillman, J.R., eds. Cambridge University Press, London.

66. Phinney, B.O., Spray, C. (1982) Chemical genetics and the gibberellin pathway in *Zea mays* L. *In*: Plant growth substances 1982, pp. 101-110, Wareing, P.F., ed. Academic Press, London.

67. Phinney, B.O., Spray, C.R. (1990) Dwarf mutants of maize - research tools for the analysis of growth. *In*: Plant growth substances 1988, pp. 65-73, Pharis, R.P., Rood, S.B., eds., Springer-Verlag, New York.

68. Postlethwait, S.N., Nelson, O.E. (1957) A chronically wilted mutant of maize. Amer. J. Bot. 44, 628-633.

69. Potts, W.C., Reid, J.B., Murfet I.C. (1985) Internode length in *Pisum*. Gibberellins and the slender phenotype. Physiol. Plant. 63, 357-364.

70. Proebsting, W.M., Davies, P.J., Marx, G.A. (1978) Photoperiod-induced changes in gibberellin metabolism in relation to apical growth and senescence in genetic lines of peas (*Pisum sativum* L.). Planta 141, 231-238.

71. Quarrie, S.A. (1982) Droopy: a wilty mutant of potato deficient in abscisic acid. Plant

72. Reid, J.B. (1979) Flowering in *Pisum*: the effect of age on the gene *Sn* and the site of action of gene *Hr*. Ann. Bot. 44, 163-173

73. Reid, J.B. (1986) Internode length in *Pisum*. Three further loci, *lh*, *ls* and *lk*. Ann. Bot. 57, 577-592.

74. Reid, J.B. (1988) Internode length in *Pisum*. Comparison of genotypes in the light and dark. Physiol. Plant. 74, 83-89.

75. Reid, J.B., Murfet, I.C. (1975) Flowering in *Pisum*: the sites and possible mechanisms of the vernalisation response. J. Exp. Bot. 26, 860-867.

76. Reid, J.B., Murfet, I.C. (1977) Flowering in *Pisum*: the effect of light quality on genotype *lf e Sn Hr*. J. Exp. Bot. 28, 1357-1364.

77. Reid, J.B., Murfet, I.C. (1984) Flowering in *Pisum*: a fifth locus, Veg. Ann. Bot. 53, 369-382.

78. Reid, J.B., Murfet, I.C., Potts, W.C. (1983) Internode length in *Pisum*. II. Additional information on the relationship and action of loci *Le, La, Cry, Na* and *Lm*. J. Exp. Pot. 34, 349-364.

79. Reid, J.B., Ross, J.J. (1988) Internode length in *Pisum*. A new gene, *lv*, conferring an enhanced response to gibberellin A_1. Physiol Plant 72, 595-604.

80. Reid, J.B., Ross, J.J. (1993) A mutant-based approach, using *Pisum sativum*, to understanding plant growth. Int. J. Plant Sci. 154, 22-34.

81. Rock, C.D., Zeevaart, J.A.D. (1991) The *aba* mutant of *Arabidopsis thaliana* is impaired in epoxy-carotenoid biosynthesis. Proc. Natl. Acad. Sci., USA 88, 7496-7499.

82. Romano, C.P., Cooper, M.L., Klee, H.J. (1993) Uncoupling auxin and ethylene effects in transgenic tobacco and *Arabidopsis* plants. Plant Cell 5, 181-189

83. Romano, C.P., Hein, M.B., Klee, H.J. (1991) Inactivation of auxin in tobacco transformed with the indoleacetic acid-lysine synthetase gene of *Pseudomonas savastanoi*. Gene Develop. 5, 438-446.

84. Rood, S.B., Pearce, D., Williams, P.N., Pharis, R.P. (1989) A gibberellin-deficient *Brassica* mutant - *rosette*. Planta 175, 107-114.

85. Ross, J.J., Davies, N.W., Reid, J.B., Murfet, I.C. (1990) Internode length in *Lathyrus odoratus*. Effects of mutants *l* and *lb* on gibberellin metabolism and levels. Physiol. Plant. 79, 453-458.

86. Ross, J.J., Reid, J.B., Dungey, H.S. (1992) Ontogenetic variation in levels of gibberellin A_1 in *Pisum*. Implications for the control of stem elongation. Planta 186, 166-171.

87. Ross, J.J., Reid, J.B., Swain, S.M. (1993) Control of stem elongation by gibberellin A_1: Evidence from genetic studies including the slender mutant, *sln*. Aust. J. Plant Physiol. 20 585-599.

88. Rothenberg, M., Ecker, J.R. (1993) Mutant analysis as an experimental approach towards understanding plant hormone action. Seminars in Developmental Biology 4, 3-13.

89. Simmonds, N.W. (1966) Linkage to the S-locus in diploid potatoes. Heredity 21, 473-479.

90. Singer, S.R., Maki, S.L. (1993) Interactions of meristem identity genes affecting floral architecture in pea. Plant Physiol. 102, S 121.

91. Skoog, F., Miller, C. O. (1989) Chemical regulation of growth and organ formation in plant tissue culture in vitro. Symp. Soc. Exp. Biol. 11, 118-131.

92. Smith, J.D., McDaniel, S., Lively, S. (1978) Regulation of embryo growth by abscisic acid in vitro. Maize Genet. Co-op. Newslet. 52, 107-108.

93. Spray, C., Phinney, B.O., Gaskin, P., Gilmour, S.J., MacMillan, J. (1984) Internode length in *Zea mays* L. The dwarf-1 mutant controls the 3β-hydroxylation of gibberellin A$_{20}$ to gibberellin A$_1$. Planta 160, 464-468.

94. Stoddart, J.L. (1984) Growth and gibberellin-A$_1$ metabolism in normal and gibberellin-insensitive (*Rht3*) wheat (*Triticum aestivum* L.) seedlings. Planta 161, 432-438.

95. Straus, D., Ausubel, F.M. (1990) Genomic subtraction for cloning DNA corresponding to deletion mutations. Proc. Natl. Acad. Sci. USA 87, 1889-1893.

96. Su, W., Howell, S.H. (1992) A single genetic locus, *ckr1*, defines *Arabidopsis mutants* in which root growth is resistant to low concentrations of cytokinin. Plant Physiol. 99, 1569-1574.

97. Sun, T.P., Goodman, H.M., and Ausubel, F.M. (1992) Cloning the *Arabidopsis GA1* locus by genomic subtraction. Plant Cell 4, 119-128.

98. Swain, S.M., Reid, J.B. (1992) Internode length in *Pisum*. A new allele at the *Lh* locus. Physiol. Plant 86, 124-130.

99. Swain, S.M., Reid, J.B., Ross, J.J. (1993) Seed development in *Pisum*. The *lh^i* allele reduces gibberellin levels in developing seeds, and increases seed abortion. Planta 191, 482-488.

100. Tal, M., Eshel, A., Witztum, A. (1976) Abnormal stomatal behaviour and ion imbalance in *Capsicum scabrous diminutive*. J. Exp. Bot. 27, 953-960.

101. Tal, M., Nevo, Y. (1973) Abnormal stomatal behaviour and root resistance, and hormonal imbalance in three wilty mutants of tomato. Biochem. Genet. 8, 291-300.

102. Taylor, I.B., Linforth, R.S.T., Al-Naieb, R.J., Bowman, W.R., Marples, B.A. (1988) The wilty mutants *flacca* and *sitiens* are impaired in the oxidation of ABA-aldehyde to ABA. Plant Cell Envir. 11, 739-745.

103. Taylor, I.B., Tarr, A.R. (1984) Phenotypic interactions between abscisic acid deficient tomato mutants. Theor. Appl. Genet. 68, 115-119.

104. Trewavas, A. (1981) How do plant growth substances work? Plant Cell Envir. 4, 203-228.

105. Wang, T.L., Donkin, M.E., Martin, E.S. (1984) The physiology of a wilty pea: abscisic acid production under water stress. J. Exp. Bot. 351, 1222-1232.

106. Wang, T.L., Futers, T.S., McGeary, F., Cove, D.J. (1984) Moss mutants and the analysis of cytokinin metabolism. *In*: The biosynthesis and metabolism of plant hormones, Soc. Exp. Biol. Seminar Series 23, pp. 135-164, Crozier, A., Hillman, J.R. eds. Cambridge University Press, London.

107. Wang, T.L., Horgan, R., Cove, D.J. (1981) Cytokinins from the moss *Physcomitrella patens*. Plant Physiol. 68, 735-738.

108. Waycott, W., Smith, V.A., Gaskin, P., MacMillan, J., Taiz, L. (1991) The endogenous gibberellins of dwarf mutants of lettuce. Plant Physiol. 95, 1169-1173.

109. Weller, J.L., Reid, J.B. (1993) Photoperiodism and photocontrol of stem elongation in two photomorphogenic mutants of *Pisum sativum* L. Planta 189, 15-23.

110. Wilson, A.K., Pickett, F.B., Turner, J.C., Estelle, M. (1990) A dominant mutation in *Arabidopsis* confers resistance to auxin, ethylene and abscisic acid. Mol. Gen. Genet. 222, 377-383.

111. Wilson, R.N., Heckman, J.W., Somerville, C.R. (1992) Gibberellin is required for flowering in *Arabidopsis thaliana* under short days. Plant Physiol. 100, 403-408.

112. Wright, A.D., Sampson, M.B., Neuffer, M.G., Michalczuk, L., Slovin, J.P., Cohen, J.D. (1991) Indole-3-acetic acid biosynthesis in the mutant maize orange pericarp tryptophan auxotroph. Science 254, 377-383.

113. Zagotta, M.T., Shannon, S., Jacobs, C., Meeks-Wagner, D.R. (1992) Early-flowering mutants of *Arabidopsis thaliana*. Aust. J. Plant Physiol. 19, 411-418.

114. Zeevaart, J.A.D., Talon, M. (1992) Gibberellin mutants in *Arabidopsis thaliana*. *In*: Progress in plant growth regulation, pp. 34-42, Karssen, C.M., van Loon, L.C., Vreugenhil, D., eds. Kluwer Academic Publishers, The Netherlands.

G2. Ethylene in Plant Growth, Development, and Senescence

Michael S. Reid

Department of Environmental Horticulture, University of California, Davis, California 95616, USA.

INTRODUCTION

Amongst hormones in both plant and animal kingdoms, ethylene, a gaseous hydrocarbon, is unique. Despite its chemical simplicity, it is a potent growth regulator, affecting the growth, differentiation, and senescence of plants, in concentrations as little as 0.01 µl/l. As recently as twenty-five years ago, plant physiologists were divided as to whether this gas, which had been shown to have a range of striking effects on plant tissues, could properly be called a hormone. Since then, the advent of gas chromatographic means of detecting and measuring ethylene, the elucidation of its biosynthetic pathway, and the discovery of potent regulators of its production and action, have provided powerful tools for physiologists to explore the role of ethylene in plant growth and development. Ethylene is now considered to be one of the important natural plant growth regulators, and the literature abounds with reports of its effects on almost every phase of the life of plants. Although the majority of studies have concentrated on particular processes, particularly fruit ripening, flower senescence, and abscission, many other reported responses of plants to ethylene may be important parts of normal growth and development. The proceedings of several symposia (7, 51) and a number of reviews (1, 18, 66) provide an excellent background in the subject; the pathway of ethylene biosynthesis and molecular aspects of the role of ethylene in fruit ripening are discussed in Chapters B4 and E4 respectively. The aim of this chapter is briefly to review present understanding of the role of ethylene in plant growth and development, highlighting effects of this gas that have received less attention in previous reviews.

What are the selective advantages of this gaseous compound that led to its widespread adoption by plants as a hormone? Although the normal pathway of ethylene biosynthesis in plants is a complex, carefully regulated series of enzymatic processes, ethylene is also a common product of oxidation of many types of organic molecules. In particular, peroxidation of unprotected long-chain fatty acids can generate substantial quantities of ethylene. It is conceivable that there was a selective advantage to primitive plants of the ability to recognize and respond to situations where stresses such as high temperature, high light, or attack by microorganisms caused the

P. J. Davies (ed.), Plant Hormones, 486–508.

generation of ethylene gas. One can imagine that once such a response system was selected, there would be a further advantage to plants of developing a controlled system of biosynthesis that would enable this new regulator to be produced in response to a variety of stimuli, not simply external oxidizing stresses.

The gaseous nature of ethylene may also explain its manifold hormonal actions in plants. Lacking the nervous system of the animal kingdom, plants appear poorly equipped to sense distant stimuli. The ready diffusion of ethylene through the intercellular spaces may, for example, act as a signal of damage, stress, or physical contact. Similarly the diffusion of ethylene may be an important part of the coordination of ripening of quite different tissues in ripening fruits. Another advantage of ethylene's diffusibility might be the fact that its concentrations in the tissue can be altered simply by changing the rate of synthesis. Whereas reducing the concentration of other hormones requires metabolism or detoxification of that already in the tissue, ethylene diffuses away, and the concentration in the tissue rapidly reaches the equilibrium between the rate of production and the rate of diffusion. One can imagine that rapid growth responses such as response to contact stimuli (6) and the inhibition or acceleration of shoot elongation (18) would benefit from such a rapid and sensitive response system.

In contrast to volatile insect pheromones which function as long-distance sex attractants, ethylene does not appear to be utilized by plants as a "long distance" hormone, despite its ability to be transported over long distances by diffusion or air movements. There is no evidence, for example, that ethylene levels in orchards are sufficiently augmented by the presence of some ripening fruits to initiate the ripening of others. Of course, the diffusibility of ethylene is a frequent problem for operators of greenhouses, cool stores, and transportation facilities, who must avoid exposing sensitive crops (such as lettuce and some cut flowers) to products of incomplete combustion, gases from ripening rooms, or ripening fruits.

TECHNIQUES

Because ethylene is a common environmental pollutant, plants are often accidentally exposed to it in physiologically active concentrations, sometimes with devastating results (Fig. 1). Such effects are often then studied in the laboratory, where ethylene can easily be applied, either as a gas, or by utilizing Ethephon (2-chloroethyl phosphonic acid), a compound that decomposes to release ethylene at physiological pH. Almost every phase of plant growth and development has been shown to be affected by ethylene in some plant or other. Of course, this does not mean that any particular response is a general effect on all plants, nor that ethylene is a natural regulator of the affected process. A satisfactory demonstration of a natural role for ethylene should also include evidence that ethylene production

Fig. 1. Plants grown in a greenhouse heated with unvented heaters. Ethylene concentrations in the greenhouse air were greater than 0.5 µl/l. (a) Defoliation in *Pisonia*; (b) discoloration of *Asplenium*; (c) yellowing and defoliation of *Schefflera*; (d) epinasty of *Syngonium*.

increases prior to, or concomitantly with, the process being studied, or that ethylene sensitivity of the tissue increases at that time. In addition, application of stimulators and inhibitors of ethylene production and action should, respectively, accelerate and retard the process.

Physiologists working with ethylene are particularly fortunate in the techniques that they have available for testing its role in plant growth and development. The gas chromatograph, which spurred the tremendous advances in our knowledge of ethylene effects during the last twenty-five years, has been refined by the addition of photo-ionization and photoacoustic detectors (63), which enable even more sensitive measurement of ethylene, permitting determination of resting production rates where before we had to content ourselves with that frustrating concentration, "trace amounts".

Modulators of Ethylene Biosynthesis and Action

The elucidation of the pathway of ethylene biosynthesis (67, Chapter B4) included the discovery of the immediate precursor to ethylene, 1-aminocyclopropane-1-carboxylic acid (ACC). This compound is an excellent exogenous precursor for ethylene, and enables application of "ethylene" in liquid media without the disturbing effects of pH which accompany the use of Ethephon (49), the previous best choice. In addition, we have some useful inhibitors of ethylene production. Aminoethoxyvinylglycine (AVG) and its analogues inhibit ACC synthase, the enzyme that converts S-adenosyl methionine (SAM) to ACC. These chemicals are effective at vanishingly low concentrations (2), which is fortunate, since they are expensive. Amino(oxyacetic) acid (AOA), another inhibitor of pyridoxal phosphate-requiring enzymes like ACC synthase, is also relatively effective (19). Although rather toxic, it is freely available, and inexpensive. Inhibitors of the terminal step in ethylene biosynthesis, ACC oxidase, include free-radical scavengers such as sodium benzoate, high temperature, uncouplers, low oxygen, and Co^{++} (67).

For years, ethylene physiologists used CO_2 as an inhibitor of ethylene action (11) in testing the involvement of ethylene in processes they were examining. Although many recognized the possible other explanations of CO_2 effects (e.g., inhibition of respiration, effects on a range of enzymes, on cellular pH, photosynthesis and stomatal aperture), it was all that was available. The discovery that Ag^+ was a potent and specific inhibitor of ethylene action (4), and formulation of this ion in the stable, mobile and non-phytotoxic thiosulfate complex (STS), provided an excellent tool for probing the involvement of ethylene in almost any phase of plant growth and development (58). It seems surprising that many workers studying ethylene effects fail to utilize this compound in exploring ethylene responses.

In recent years, a number of new inhibitors of ethylene action have been synthesized. The first in the series was 2,5-norbornadiene (NBD), a cyclic olefin which appears to compete with ethylene at its binding site (55). This chemical had the advantage over Ag^+ of being volatile, and could therefore be applied in experiments where Ag^+ could not be used, for example because of phytotoxicity, or because of the mass of the tissue under study. Its volatility also allowed experiments where it was desirable to treat with an inhibitor only temporarily. The disadvantages of NBD were that relatively high concentrations (ca. 2,000 µl/l) were required to overcome some ethylene responses, that it was difficult to design a protocol to maintain a steady treatment concentration (50), and that it had an offensive aroma. Subsequently, a compound designed as a photoaffinity label for the ethylene binding site (diazocyclopentadiene, DACP) was shown to have similar properties to those of NBD, with the advantages that it was active at lower concentrations, and inhibited ethylene action irreversibly if the tissue was illuminated, during treatment, with fluorescent light (54). Further research on these volatile inhibitors promises to provide a non-toxic and very effective gaseous inhibitor of ethylene action.

Being volatile itself, ethylene can also be applied for short periods, a property of this hormone of which advantage is all too rarely taken. Frequently, however, the "hypobaric" reduction of tissue ethylene concentrations has been used to test for ethylene involvement in plant responses. In this technique, reduction of the total pressure to 0.1 atm and supplying pure oxygen places the tissue in a normal oxygen partial pressure but where the diffusivity of ethylene is greatly increased. Hypobaric ethylene removal is relatively difficult, technically, because of the problems of providing appropriate vessels, of controlling pressures, and of maintaining humidity at low pressures.

ETHYLENE RESPONSES

The Release of Dormancy

The ability of many plants to grow in regions with adverse climates, such as freezing winters, or hot dry summers, is closely tied to their ability to suspend operations during adverse weather, and to recommence, often quite rapidly, when the weather improves. The timing and the mechanism of the reactivation of growth and development is still imperfectly understood, but in some cases it appears that ethylene may play some role.

The dormancy of many seeds can be broken by application of growth regulators. In a number of species, ethylene application stimulates the germination of dormant seeds (32, 57). A number of authors have demonstrated substantial production of ethylene by germinating seeds, but it is probable that this ethylene is the result of the stresses of germination, and therefore not associated with release of dormancy.

The flowering of many bulb crops is controlled by a proper schedule of storage at high and low temperatures to break dormancy and initiate flower development. For years it has been known that ethylene pollution of the storage area can cause flower malformations or abortion (termed "blasting") in these crops (17). It has been now shown (27) that exposure of dormant iris, narcissus, tulip, freesia and gladiolus propagules to smoke or to ethylene at the appropriate time will hasten shoot and root growth, shorten the time to flowering, and increase the number of small propagules which successfully flower (Fig. 2). The mechanism by which ethylene "breaks" the dormant state of these propagules is still unexplored. The pronounced effect of ethylene on sprouting of potatoes (52) is associated with increased respiration and mobilization of carbohydrates (25). Although it seemed possible that such

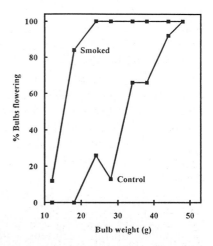

Fig. 2. Effect of ethylene exposure prior to forcing on the flowering of *Narcisssus tazetta* cv. Soleil d'Or. Cured bulbs were graded by weight and exposed to..."smoke from smouldering wood and fresh leaves for several hours on each of 10 consecutive days". The bulbs were then stored, and planted in the normal manner. (Redrawn from the data of Imanishi, H. (1963) Scientia Hortic. *21*, 173-80).

mobilization would be the basis for the shorter dormant period and improved flowering of bulbous crops treated with ethylene, measurement of soluble carbohydrate changes in the apex of dormant iris bulbs during ethylene treatment did not reveal any significant changes (26). In a study of *Brodiaea* inflorescences, it was shown that the flower-inducing effects of ethylene treatment were associated with rapid expansion of the apical meristem (22).

Shoot Growth and Differentiation

The period between imbibition of a seed and the unfolding of the first true leaf is the most perilous time in the life of any green plant. Even if the succulent new plant survives the attacks of pests and pathogens, its path to the sun may be frustrated by barriers in the soil. The dramatic effects of ethylene on the growth and development of etiolated seedlings were the basis for the discovery by Neljubov (41) of the physiological action of ethylene. The so-called "triple response" involves a reduction in elongation, swelling of the hypocotyl, and a change in the direction of growth. This response may well be the means by which seedlings grow around obstacles in the soil (20). It presumably results from a redistribution of auxin in response to the ethylene induced by the contact stress between the seedling and the obstacle.

In another "avoidance" mechanism, some aquatic plants, such as star-wort (28), and rice (31), grow much faster when exposed to ethylene. In the star-wort, the compressed internodes of the floating rosette that forms the surface portion of the plant start to elongate very rapidly following submergence, or treatment with ethylene (Fig. 3). Not only is the ethylene content of the aerial spaces in the submerged shoots high, but also the effect of submersion can be eliminated by a pretreatment with Ag$^+$. The data are consistent with the hypothesis that increased intercellular ethylene content is the signal that the plant has been flooded. Rapid diffusion into the air keeps intercellular ethylene low when the plants are above water. Flooding (and the sparing solubility of ethylene in water) results in a rapid rise in intercellular ethylene content, and a growth response directed to returning the leaves to an aerial environment.

One of the classic symptoms of ethylene exposure in plants is the downward growth of the petioles, termed

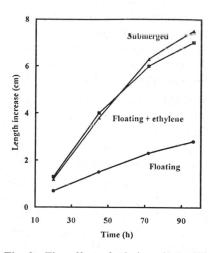

Fig. 3. The effect of ethylene (1.5 µl/l) or submergence on stem extension in the star-wort (*Callitriche platycarpa* L.). Detached rosettes were placed in half-strength Hoagland's solution and treated. Length of the plant was measured at intervals. (Redrawn from, Jackson, M.B. (1982) *In*: Plant Growth Substances, pp. 291-301, Wareing, P.F., ed. Academic Press, N.Y.)

491

"epinasty" (Fig. 4). This symptom is considered to result from a redistribution of auxin in response to the ethylene treatment; increased auxin in the upper part of the petioles causes increased growth there, and a consequent downward bending of the petiole. Epinasty is typically a symptom of inadvertent exposure of plants to ethylene, and is a symptom common to plants that are flooded, or stressed in other ways. The epinastic curvature of leaves of flooded plants is now known to be due to the transport of ACC from the roots to the petioles, and its conversion, there, to ethylene (9). It is possible that other epinastic responses to stress are also mediated via enhanced ethylene production in the petiole.

Stress ethylene may also modify shoot growth. The reduced growth and improved trunk diameter of unstaked shade trees is probably due to ethylene produced in response to flexing stress (40). Tree trunks treated for an extended period with ethylene are of markedly greater diameter than air-treated controls (Fig. 5). The site of production of the ethylene induced by flexing, and the way in which it exerts its effects have not yet been determined.

Low concentrations of ethylene applied to plants of the "diageotropica" tomato, a mutant which normally grows horizontally, partially restored the plants to a normal growth habit. Zobel (71) suggested that this mutant was unable to produce ethylene, and hypothesized that basal ethylene production is required for normal growth. Other workers have shown that the mutant can produce ethylene at normal rates (29), and suggest that its response may be the result of insensitivity to ethylene, or possibly auxin.

In our studies, we have found that early application of Ag^+ as STS to potted plants of various taxa has no perceptible effect on growth development, and flowering, while effectively inhibiting the eventual abscission of flower petals (12). Likewise the ethylene-resistant *etr1* mutant

Fig. 4. Epinasty in poinsettia (*Euphorbia pulcherrima*) plants treated with ethylene at different concentrations. The bracts and leaves were removed from the plants to demonstrate the curvature of the petioles.

Fig. 5. Effect of ethylene on the growth of pine tree trunks. The portion of the trunk between the arrows was treated with air (A) or 5 to 20 µl/l ethylene (B) for 99 days. (Photo, Neel).

of *Arabidopsis* has normal morphology (70). Since normal growth continues even when the action of ethylene is strongly inhibited, it seems that ethylene action is not essential for normal growth and development. The curious behavior and ethylene response of the *diageotropica* mutant must relate to modulation of the effects of growth regulators other than ethylene.

Root Growth and Differentiation

Roots, too, respond markedly to the ethylene content of their environment. This is of particular importance, because the soil atmosphere can contain quite high concentrations of ethylene (28). Typically, root growth is stimulated at low concentrations of ethylene, and inhibited at higher concentrations (29, Fig. 6). The anatomy of roots can also be affected by their ethylene content. A particularly good example of this response is the development of aerenchymatous roots in flooded maize plants (28). This adaptive response to increased intercellular ethylene can also be induced by ethylene treatment and prevented by Ag^+. Ethylene appears to play an important role in the penetration of dense soil media by roots. When seeds of normal tomato plants are placed on dense agar medium containing STS, the roots fail to enter the agar, and the developing plantlet develops a corkscrew appearance as it repeatedly fails to anchor itself (69). *Etr1* seedlings showed the same behavior when placed on water agar. It appears that ethylene produced by young roots in response to the obstruction of dense

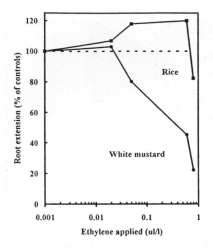

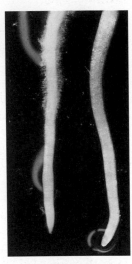

Fig. 6. Elongation of primary roots of rice and white mustard after 24 h treatment with ethylene at various concentrations. (Redrawn from, Jackson, M.B. (1982) *In*: Plant Growth Substances, pp. 291-301, Wareing, P.F., ed. Academic Press, N.Y.).

Fig. 7. Effect of treatment for 4 days with 1.5 µl/l ethylene (left) or air (right) on the formation of root hairs on mung bean roots (Photo, Robbins).

medium may be involved in a growth response that allows the roots to penetrate.

There has been little study of another interesting response of roots of a number of species to ethylene: the massive production of root hairs (Fig. 7). It is not yet known whether ethylene is involved in the normal development of root hairs. However, it is conceivable that in this phenomenon, too, ethylene acts as a signal, indicating that the root is well embedded in the soil.

Responses to Physical Stimuli

Plant physiologists have long been fascinated by the ability of plants to respond to physical stimuli. Although a number of these responses (for example, those to gravity and light) appear to be mediated through changes in the distribution of auxin, ethylene has been implicated as the active agent in some cases (6). Even quite gentle stimulation of plants will result in increased production of ethylene (23). Ethylene may therefore be a mediator in the response of some plants to tactile stimuli. In many climbing vines, roots attach the vine to supporting walls or other plants by modified root hairs. The stimulation of root hair development on roots of plants exposed to ethylene (Fig. 7) suggests that localized contact-induced ethylene may be the signal for the elaboration of these supports.

Adventitious Rooting

Ever since it was demonstrated that application of high concentrations of auxin will stimulate ethylene production, researchers have questioned the relative roles of auxin and ethylene in the formation of adventitious roots during auxin-stimulated rooting of cuttings. The possible involvement of ethylene is strengthened by reports that rooting of hard-to-root plants can sometimes be improved by scarification (wounding) of the base of the cuttings, a treatment that would induce production of wound ethylene. Studies of the effects of ethylene, ethylene producing compounds, and inhibitors of ethylene synthesis and action on rooting of cuttings have led to conflicting interpretations of ethylene's

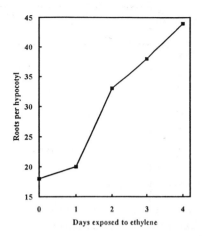

Fig. 8. Effect of basal application of dissolved ethylene (equilibrated with 6 µl/l in air) on the number of adventitious roots on the hypocotyls of mung bean cuttings after 4 days. Redrawn from (50).

role in the process. Using a rooting chamber which allowed us to apply ethylene to the base of mung bean cuttings and not to the upper portions, we showed (50) that ethylene does indeed stimulate adventitious rooting (Fig. 8). Since application of inhibitors of ethylene synthesis and action to similar cuttings reduced rooting, we concluded that ethylene plays a role in adventitious rooting of untreated mung bean cuttings. Its role in rooting of other cuttings is not yet known.

Abscission

Amongst the most spectacular and commercially important effects of ethylene is the stimulation of abscission. Physiologists are still divided about the role of ethylene in natural abscission, but the majority of the data on hormonal control of abscission are consistent with the following hypothetical scheme (45):
• A gradient of auxin from the subtended organ to the plant axis maintains the abscission zone in a nonsensitive state. This gradient is itself maintained by factors which inhibit senescence of the organ. Thus, auxins, cytokinins, light, and good nutrition tend to reduce or delay abscission.
• Reduction or reversal of the auxin gradient causes the abscission zone to become sensitive to ethylene. Application of auxin proximal to the abscission zone, removal of the leaf blade, or treatments which accelerate its senescence (shading, poor nutrition, ethylene) therefore hasten abscission. Ethylene, or stresses which enhance its production, also may hasten abscission by reducing auxin synthesis and/or interfering with its transport from the leaf. Where

ABA stimulates abscission, it may do so by stimulating ethylene production, or by interfering with the production, transport, or action of auxin.

• Once sensitized, the cells of the abscission zone respond to low concentrations of ethylene, whether exogenous or endogenous, by the rapid production and secretion of cellulase and other hydrolytic enzymes, and subsequent shedding of the subtended organ.

This scheme suggests a key role in abscission for ethylene, a view that is still somewhat controversial (53). The importance of ABA in abscission is widely questioned, although this hormone has been shown to be of primary importance in seed shedding in grasses (44).

Flower Induction

The induction of flowering in pineapple by ethylene or ethylene-producing chemicals is of considerable commercial importance. Other tropical fruits (mango, wax apple) also appear to respond to ethylene. In mango orchards it has long been the practice to hasten flowering by "..keeping smoky fires going for periods of six to seventeen days in such a way that a dense smoke is passing up through the trees" (14). This treatment ensures flowering in climates where trees do not experience drought or chilling stresses (which might be suggested to act through stimulation of stress ethylene production). No information is yet available on the mechanism by which ethylene induces flowering in these species. Application of high concentrations of ethylene during the inductive long night has been shown to prevent the induction of flowering in *Pharbitis nil* (56). We used this response as a probe to examine the molecular basis of the induction of flowering in this species (37).

Sex Expression

Not long after Ethephon became available commercially, it was demonstrated that application of this material to seedlings would dramatically change the ratio of male to female flowers in members of the Cucurbitaceae (1). Such applications were used to accelerate fruit production in lines of cucumber that naturally produced a number of male flowers before any female flowers appeared. It seems as though ethylene may be an important part of the natural determination of flower sex. Application of Ag^+ has been shown to induce male flowers on the gynoecious cucumber cultivars used to produce F1 hybrid seed, and is now used commercially to permit propagation of these lines.

Flower Opening

Ethylene may be involved in different ways in the opening of flowers. Low concentrations of ethylene may accelerate the opening of carnation buds (13); in roses ethylene has diverse effects, depending on the cultivar (46). Opening of some cultivars, such as 'Lovely Girl' is strongly inhibited by low concentrations of ethylene (Fig. 9). In others the process is unaffected ('Gold Rush'), or substantially accelerated ('Stirling Silver').

De Candolle (16) suggested that the ..."unusual movements (of the sexual organs of several flowers)...interest the physiologist on two counts, firstly as a proof of plant sensitivity, and secondly as an indication of the analogy existing between these organs and the corresponding organs of animals"! The rapid growth of the filaments in intact flowers of *Gaillardia* can also be seen in detached filaments treated with auxin (33). The hypothesis that auxin-induced ethylene plays a part in this rapid growth needs to be tested by exploring the

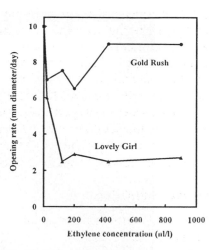

Fig. 9. Effect of ethylene concentrations on the rate of opening of two rose cultivars.

effects of some of the specific inhibitors' of ethylene synthesis and action that are now available.

Flower Senescence

Most research on the physiology of flower senescence has been carried out on a limited number of species; morning glory (*Ipomoea*) and carnation (*Dianthus*) have been the most thoroughly studied (21). In these flowers, senescence is associated with a burst of ethylene production, can be induced by treatment with exogenous ethylene or ACC and is prevented by pre-treatment with inhibitors of ethylene synthesis or action. In a pattern similar to that seen in many fruits, the onset of senescence is associated with a coordinated increase in the activities of ACC synthase and of ACC oxidase (10). Detailed study of ethylene production in aging carnation petals has revealed that the biosynthesis of ethylene is separated between the base and the upper portion of the petal (39). Basal portions of the petals produce little ethylene when treated with ACC. Upper portions produce little ethylene on their own. These results are interpreted to indicated that the majority of the ACC synthase is located in the lower portion of the petal, and that ACC oxidase is largely located in the upper portion. Studies of the distribution of mRNAs coding for the enzymes of ethylene biosynthesis in *Phaleonopsis* orchids (43) demonstrate a similar situation - although their petals contribute up to 26% of total flower ethylene and accumulate high levels of ACC oxidase mRNA and activity following pollination, no ACC synthase mRNA or activity was detected in the petals. The ACC is apparently synthesized in the ovary and transported to the petals where it is oxidized to ethylene.

Fig. 10. Effect of pretreatment with STS (4 mM Ag⁺, 10 min) on carnation flowers held in deionized water at 20°C for 14 days.

The inhibition of flower senescence by Ag^+ has formed the basis of an effective practical treatment for cut flowers (48). Commercial growers of carnations commonly treat the flowers with STS after harvest to inhibit ethylene action. This treatment can extend the life of the flowers as much as 4-fold (Fig. 10), but its continued commercial use is in question because of environmental concerns over the use of silver salts. Researchers have therefore continued to examine the biology of ethylene-induced flower senescence in order to develop alternative control methods. Substantial variation in sensitivity to ethylene has been shown among carnation varieties (66), suggesting that a classical or molecular genetic approach to preventing ethylene-induced flower senescence might be successful.

Woodson and his colleagues have examined molecular events associated with senescence in carnations (65, 35). Using differential hybridization to a cDNA library prepared from senescent carnation petals, they isolated three senescence-related cDNA clones (35). Expression of two of the cloned mRNAs in response to ethylene was floral specific (36), while the expression of another was induced in both flowers and leaves exposed to ethylene. Ethylene treatment increased expression of these mRNAs in petals (Fig. 11); a six hour treatment with 1 $\mu l.l^{-1}$ induced high expression of pSR5 and pSR12, but pSR8 was not fully induced unless the treatment concentration was 100 $\mu l.l^{-1}$. Abundance of some of the mRNAs decreased after removal of ethylene, indicating that continued perception of ethylene is required for their expression. Sequence analysis suggests that pSR12 codes for ACC oxidase (59), and that pSR8 codes for a glutathione s-transferase (38). In the near future we may expect the senescence of ethylene-sensitive flowers to be modified by molecular engineering, using these tools.

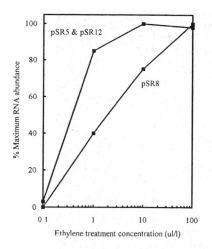

Fig. 11. Ethylene-induced changes in expression of mRNA binding to cDNAs isolated from a senescing carnation petal cDNA library. RNA was isolated from the petals of open flowers treated for 6 hours with different concentrations of ethylene. Redrawn from (36).

The spectacular effects of ethylene on carnations and some other flowers, and the benefits obtained by pretreating them with Ag^+ have given rise to the erroneous generalization that flower senescence is caused by ethylene. For many flowers this is not true. A study of the responsiveness of flowers from different families to ethylene (64) revealed that ethylene caused petal wilting in flowers from the Caryophyllaceae, Solanaceae, and Orchidaceae, but that in most other families, ethylene treatment does not curtail flower life. Nor will pretreatments with inhibitors of ethylene synthesis or action extend the life of such flowers. These observations suggest that the ethylene-mediated senescence of carnation and morning glory is only one type of flower senescence.

Pollination

The rapid senescence of some flowers following their pollination has provided an interesting experimental system for examining the control of flower senescence (42). Pollination causes a very rapid increase in ethylene production, first by the gynoecium, and then by the petals (Fig. 12).

A number of workers, intrigued by the rapidity of the pollination response, have sought the nature of the pollination stimulus. Auxin, which is present in high concentrations in pollen, has been suggested as a possible pollination stimulus, but it does not move in the stigmas of carnations (47). We demonstrated that the stimulus was not ethylene *per se* because pollinated flowers wilted just as rapidly when we placed a tube over the pollinated stigmas and aspirated the tube to remove ethylene produced by the gynoecium (61). The finding that pollen of many species contains substantial concentrations of ACC (61) led us to hypothesize that ACC might be the stimulus for the pollination response. The stigma of carnation flowers contains a much higher activity of ACC oxidase than the other tissues of the flower, and is therefore well prepared for rapid conversion of ACC to ethylene.

Although there is sufficient ACC in the pollen to explain the initial burst in ethylene production by the gynoecium, the role of pollen ACC in the early ethylene production by pollinated flowers has been questioned (24). Cosmos pollen, which contained very little ACC, greatly stimulated ethylene

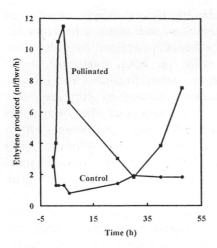

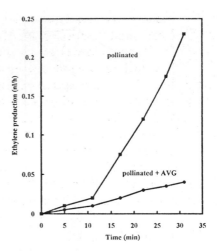

Fig. 12. Effect of pollination on ethylene production in *Petunia* flowers. Redrawn from (62).

Fig. 13. Effect of 2 µl droplets of AVG (2 nmol) or H_2O (control), applied to *Petunia* stigmas 3 h in advance of pollination, on early ethylene evolution from styles excised from freshly opened flowers. Redrawn from (24).

production by *Petunia* stigmas. Pretreatment of stigmas with AVG prevented the pollen-stimulated burst of ethylene production, suggesting that it comes from newly synthesized ACC (Fig. 13). These data suggest that the early effects of pollination are related to some other interaction between the pollen and the style.

Whatever the role of pollen ACC, there is increasing evidence that synthesis and transport of ACC plays an important part in post-pollination events. Increased ethylene production by the gynoecium following pollination or wounding indicates an increase in the activity of ACC synthase (since ACC oxidase is constitutive and in high activity in this organ) and we have demonstrated such an increase in pollinated Petunia flowers. Using radioactively labelled ACC, we showed that ACC applied to the stigma is transported as ACC to the petals (47), and have suggested that ACC may be the mobile pollination stimulus. The finding that pretreatment of the stigma with AVG delayed pollination-induced senescence (24) supports this suggestion.

Wound Responses

In many plant responses, ethylene appears to function as a wound hormone. Its production rises rapidly when plants are wounded or exposed to stress, and the responses of plants to ethylene often seem to be tailored to reducing stress or averting infection of wounds. Thus, for example, leaf abscission accompanying severe drought stress may be an adaptive response to ethylene which reduces the evaporative surface of the affected plant. Similarly, ethylene can elicit the production of phytoalexins--compounds which appear to combat fungal or bacterial infection in wounded plants.

A spectacular wound response of some plants to ethylene is the stimulation of secretory processes. In some species of *Prunus* application of Ethephon results in the stimulation of gum production--normally a response to infection or other stresses. In *Hevea*, the rubber tree, a similar response is valuable commercially. Ethephon treatment maintains the flow of latex from the trees, thereby considerably improving yields, and reducing labor costs. There are some indications that the vascular blockages that accompany "wilt" diseases, and are also found in the stems of some cut ornamentals, may also be an ethylene-mediated wound response.

Leaf senescence

In isolated systems (leaves or disks, typically held in the dark), the progress of leaf senescence shows some of the physiological characteristics of flower senescence and fruit ripening. In the few systems that have been examined, leaf senescence has been shown to be accompanied by increased respiration (a "climacteric" peak is sometimes seen), and increased ethylene production. The fact that ethylene treatment of intact plants or detached leaf tissue often accelerates leaf senescence has been taken as evidence that ethylene is a primary agent in the process. Leaves of ethylene-resistant mutants of *Arabidopsis* show reduced senescence rates when treated with ethylene (70) but senesce anyway in ethylene-free air, suggesting that senescence can occur in the absence of ethylene action. Molecular tools have permitted a comparison of the events of natural senescence on the plant, and senescence of detached leaves in the dark. Three senescence-related cDNA clones were isolated by subtractive differential screening of a cDNA library prepared from senescing detached leaves of barley *(Hordeum vulgare)* (3). Two of the three transcripts showed very similar expression patterns: in detached leaves they were induced by abscisic acid and inhibited by kinetin. They were also induced by wounding and osmotic stress, but could not be detected in naturally senescing leaves. The third mRNA, represented by only one of the six cDNA clones, behaved differently. It accumulated in senescing detached leaves, but its expression was not affected by ethylene application, wounding, or drought conditions. This third transcript did accumulate during natural senescence of barley flag leaves. These data call into question the relevance to natural leaf senescence of the many physiological and growth regulation studies that have been carried using detached leaves. They are consistent with a model where ethylene may be involved in stress-induced acceleration of leaf senescence, but is not involved in natural senescence.

Fruit Development and Ripening

The stimulation of fruit ripening was one of the earliest reported effects of ethylene gas, and is the basis for what is probably the most widespread use of ethylene and compounds which liberate it. Commercially, harvested bananas, tomatoes, avocados, melons, and kiwifruit are ripened with ethylene gas. Ethephon is sprayed in processing tomato fields late in the season to ensure uniform ripeness of the fruit. For many years the question of whether ethylene production by fruit rose before or after the other events of ripening (usually signalled by the onset of the respiration climacteric) was hotly debated. For most of the fruit that have been examined using single-fruit samples and high-sensitivity gas chromatography, it has been found that increased ethylene production is coincident with, or follows, the onset of the respiration climacteric (Fig. 14).

The melon, in which a substantial rise in ethylene production occurs *before* the onset of ripening, is not typical of climacteric fruits. Most workers studying the control of the onset of ripening now support the hypothesis that the concentrations of an unidentified "ripening inhibitor" must fall before endogenous ethylene can induce the onset of the "climacteric" patterns of respiration and ethylene production that accompany fruit ripening. Perhaps the best evidence for such an inhibitor comes from work with avocados. Some cultivars of avocado will not ripen while attached to the tree, even if

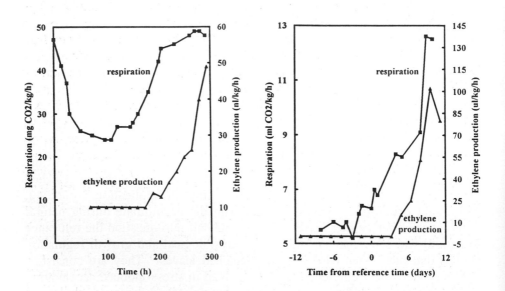

Fig. 14. Respiration and ethylene production of a feijoa fruit (a) and of 'Cox's Orange' apples (b) during the onset of ripening. The apple data are obtained from many different fruit sampled at different times - day 0 is the day for each apple on which ethylene production started to increase.

treated with high concentrations of ethylene. Shortly after removal they become very sensitive to applied ethylene, and will ripen in the absence of exogenous ethylene after a few days (5). Extraction and identification of the hypothetical inhibitor would provide an important practical tool for controlling fruit ripening that might have far-reaching practical implications. The molecular approaches described in Chapter E4 provide convincing evidence of the requirement for endogenous synthesis and action of ethylene in climacteric ripening.

Seed Dispersal

The fact that abscission of tissues within the fruit precedes explosive dispersal of seeds in some species suggest that here, too, ethylene may play an important role. In the "squirting cucumber", ethylene production by the fruits rises substantially in the period prior to the splitting of the fruit rind (29). This increased ethylene production is presumed to be responsible for inducing the abscission of the turgid fruit tissues which precedes seed ejection. A similar mechanism may be the basis of seed dispersal in many other species.

Mechanism of ethylene action.

Most of the physiological and biochemical evidence suggests that the manifold effects of ethylene are mediated through induction of new proteins and enzymes. For example, treatment of *Coleus* explants with inhibitors of transcription, message processing, and protein synthesis, substantially delays ethylene-induced senescence (45). Studies of ethylene binding in plants suggested a model for ethylene action where reversible association of ethylene with a membrane-bound receptor would result in activation or release of a "second message" (55). Ethylene-mediated changes in the transcription of mRNA would result from the activity of this second message. Bleecker et al. (8) used the response of etiolated seedlings to ethylene as a tool to isolate ethylene-resistant mutants of *Arabidopsis* (*etr1*). These mutants provided an experimental system for investigating the molecular basis of ethylene action since binding assays suggested that the mutation was in the hypothetical ethylene binding site. The *ETR1* gene has now been cloned by the method of chromosome walking (15). The role of the cloned gene in ethylene responsiveness was tested by transforming wild-type *Arabidopsis* plants with the dominant *ETR1-1* mutant gene. The transformed plants were resistant to ethylene. Although the sequence of the amino-terminal half of the deduced *ETR1-1* protein appears to be novel, the carboxyl-terminal half is similar in sequence to both components of the prokaryotic family of signal transducers known as the two-component systems (34). In these systems, attachment of a hormone results in activation of a kinase, which autophosphorylates (usually on a histidine residue), and then transfers the phosphate to a "response regulator" protein which binds to an output protein, such as a transcription factor. Since polypeptide regions with strong homology to both of these component proteins are present in the deduced amino acid sequence of the *ETR1-1* protein, it appears possible that an early step in transduction of the ethylene signal in plants may therefore involve transfer of phosphate as in prokaryotic two-component systems (34). A hypothetical model based on Sisler's ligand hypothesis (55) and the two-component systems is shown in Fig. 15. In this model, attachment of ethylene to the binding site in the *ETR1-1* polypeptide activates the kinase domain, which autophosphorylates then transfers phosphate to the response regulator domain. This phosphorylation could result in release of an activated ligand or activation of soluble factors that would then potentiate the derepression of genes involved in the effects of ethylene.

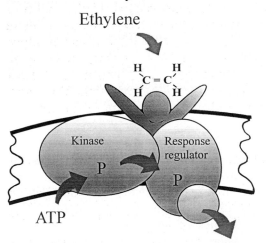

Fig. 15. Hypothetical model for two-component regulation of ethylene action. Binding of ethylene to the receptor activates the kinase domain, which autophosphorylates, then transfers the phosphate to the response regulator protein, resulting in release of a ligand, or activation of a transcription factor.

DISCUSSION

In reviewing the role of ethylene in plant growth and development it becomes clear that this role may be quite different from those of the other plant hormones. Although ethylene appears to be involved in a great many plant responses, it is almost always as a signal of environmental or physiological change. There is scant evidence that ethylene is an important component of the balance of growth regulators that maintains normal vegetative growth; indeed, plants grow quite normally under hypobaric conditions, or after application of sufficient Ag^+ to strongly inhibit the action of ethylene, and the *etr* mutant of *Arabidopsis* appears perfectly normal apart from having dramatically reduced responsiveness to external ethylene. Changes in plant growth are, however, often associated with increased production of or sensitivity to ethylene. Ethylene production increases in response to environmental stresses (physical stimuli, flooding); sensitivity to endogenous ethylene increases in organs reaching physiological maturity (maturing fruits and flowers, or aging leaves).

The induction of any change in plant behavior is the result of an interaction between tissue sensitivity and the concentration of the effective hormone. Ethylene does not act alone in affecting plant behavior; the action of ethylene can almost always be modulated by application of other plant hormones. It may be that the sensitivity of the tissue to ethylene is a function of the concentrations of other hormones.

It is also apparent that information on the response of plant tissues to ethylene, and its role in growth and development far outstrips our understanding of how these responses are mediated. We know much about the biochemical changes associated with ethylene-mediated abscission and fruit ripening, and several groups are now studying the molecular biology of these processes. In contrast, almost nothing is known of the mechanisms by which ethylene mediates such processes as dormancy release, aerenchyma development, growth inhibition or stimulation and stimulation of latex secretion. Elucidation of these mechanisms provides a rich field of research for those interested in the regulation of plant growth and development.

References

1. Abeles, F.B., Morgan, P.W., Saltveit, M.E. (1992) Ethylene in Plant Biology, 2nd Ed. Academic Press, New York.
2. Baker, J.E., Wang, C.Y., Lieberman, M., Hardenburg, R.E. (1977) Delay of senescence in carnations by a rhizobitoxine analog and sodium benzoate. HortScience 12, 38-39.
3. Becker, W., Apel, K. (1993) Differences in gene expression between natural and artificially induced leaf senescence. Planta, 189, 74-79.
4. Beyer, E.M.Jr. (1976) A potent inhibitor of ethylene action in plants. Plant Physiol. 58, 268-271.
5. Biale, J.B., Young, R.E. (1981) Respiration and ripening in fruits--retrospect and prospect. In: Recent Advances in the Biochemistry of Fruits and Vegetables, pp. 1-39, Friend, J., Rhodes, M.J.C., eds. Academic Press, London
6. Biro, R.L., Jaffe, M.J. (1984) Thigmomorphogenesis: Ethylene evolution and its role in the changes observed in mechanically perturbed bean plants. Plant Physiol. 62, 289-296.
7. Blanpied, G.D. (1985) Ethylene in postharvest biology and technology of horticultural crops. Symposium of the American Society for Horticultural Science. HortScience 20, 40-60
8. Bleecker, A.B., Estelle, M.A., Somerville, C., Kende, H. (1988) Insensitivity to ethylene conferred by a dominant mutation in *Arabidopsis thaliana.* Science 241, 1086-1089.
9. Bradford, K.J., Yang, S.F. (1980) Xylem transport of 1-aminocyclopropane-1-carboxylic acid, an ethylene precursor, in waterlogged tomato plants. Plant Physiol. 65, 322-326.
10. Bufler, G., Mor, Y., Reid, M.S., Yang, S.F. (1980) Changes in the 1-aminocyclopropane-1-carboxylic acid content of cut carnation flowers in relation to their senescence. Planta 150, 439-442.
11. Burg, S.P., Burg, E.A. (1967) Molecular requirements for the biological activity of ethylene. Plant Physiol. 42, 144-152.
12. Cameron, A.C., Reid, M.S. (1983) The use of silver thiosulfate to prevent flower abscission from potted plants. Scientia Hortic. 19, 373-378.
13. Camprubi, P., Nichols, R. (1979) Ethylene-induced growth of petals and styles in the immature carnation inflorescence. J. Hort. Sci. 54, 225-258.

14. Chandler, W.H. (1950) Evergreen Orchards. Lea & Febiger, Philadelphia. 452 pp.

15. Chang, C., Kwok, S.F., Bleecker, A.B., Meyerowitz, E.M. (1993) *Arabidopsis* ethylene-response gene *ETR1*: Similarity of product to two-component regulators. Science 262,539-544.

16. De Candolle, A.P. (1832) Physiologie Vegetale, ou Exposition des Forces et des Fonctions Vitales des Vegetaux. Bechet Jeune, Paris. Vol. 2, 517.

17. De Munk, W.J. (1975) Ethylene disorders in bulbous crops during storage and glasshouse cultivation. Acta Hort 51, 321-335.

18. Eisinger, W. (1983) Regulation of pea internode expansion by ethylene. Ann. Rev. Plant Physiol. 34, 225-240.

19. Fujino, D.W., Reid, M.S., Yang, S.F. (1981) Effects of aminooxyacetic acid on postharvest characteristics of carnation. Acta Hortic. 113, 59-64.

20. Goeschl, J.D., Rappaport, L., Pratt, H.K. (1966) Ethylene as a factor regulating the growth of pea epicotyls subjected to physical stress. Plant Physiol. 41, 877-884.

21. Halevy, A. H., Mayak, S. (1979) Senescence and postharvest physiology of cut flowers, Part I. *In*: Horticultural Reviews, Vol. 1, pp. 204-36, Janick, J., ed. AVI, Westport.

22. Han, S.S., Halevy, A.H., Sachs, R.M. Reid., M.S. (1990) Enhancement of growth and flowering of *Triteleia laxa* by ethylene. J. Am. Soc. Hortic. Sci. 115, 482-486.

23. Hiraki, K., Ota, Y. (1975) The relationship between growth inhibition and ethylene production by mechanical stimulation in *Lilium longiflorum*. Plant Cell Physiol. 16, 185-189.

24. Hoekstra, F.A., Weges, R. (1986) Lack of control by early pistillate ethylene of the accelerated wilting of *Petunia hybrida* flowers. Plant Physiol. 80, 403-408.

25. Huelin, F.E., Barker, J. (1939) The effect of ethylene on the respiration and carbohydrate metabolism of potatoes. New Phytol. 38, 85-104.

26. Imanishi, H., Halevy, A.H., Kofranek, A.M., Han, S., Reid, M.S. (1993). Respiratory and carbohydrate changes during ethylene-mediated flower induction in iris. Scientia Hortic. *In Press*.

27. Imanishi, H., Fortanier, E.J. (1982/83) Effects of exposing Freesia corms to ethylene or to smoke on dormancy-breaking and flowering. Scientia Hortic. 18, 381-389.

28. Jackson, M.B. (1985) Ethylene and responses of plants to soil waterlogging and submergence. Ann. Rev. Plant Physiol. 36, 145-174.

29. Jackson, M.B. (1979) Is the *diageotropica* tomato ethylene deficient? Physiol. Plant. 46, 347-351.

30. Jackson, M.B., Morrow, I.B., Osborne, D.J. (1972) Abscission and dehiscence in the squirting cucumber (*Ecballium elaterium*). Can. J. Bot. 50, 1465-1471.

31. Kende, H., Acaster, M.A., Jones, J.F., Metraux, J.P. (1982) On the mode of action of ethylene. *In*: Plant Growth Substances, 1982, pp. 269-277, Warding, P.F., ed. Academic Press, London.

32. Ketring, D.L. (1977) Ethylene and seed germination. *In*: The Physiology and Biochemistry of Seed Dormancy and Germination, pp. 157-178, Khan, ed. Elsevier, Amsterdam.

33. Koning, R.E. (1983) The roles of auxin, ethylene, and acid growth in filament elongation in *Gaillardia grandiflora (Asteraceae)*. Amer. J. Bot. 70, 602-610.

34. Koshland, D.E. Jr. (1993) The two-component pathway comes to eukaryotes. Science 262, 532.

35. Lawton, K.A., Huang, B., Goldsbrough, P.B., Woodson, W.R., (1989) Molecular cloning and characterization of senescence-related genes from carnation flower petals. Plant Physiol. 90, 690-696.

36. Lawton, K.A., Raghothama, K.G., Goldsbrough, P.B., Woodson, W.R. (1990) Regulation of senescence-related gene expression in carnation flowers petals by ethylene. Plant Physiol. 93, 1370-1375.

37. Lay-Yee, M., Sachs, R.M., Reid, M.S. (1987) Changes in cotyledon mRNA during ethylene inhibition of floral induction in *Pharbitis nil* strain violet. Plant Physiol. 84, 545-548.

38. Meyer, R.C., Goldsbrough, P.B., Woodson, W.R. (1991) An ethylene-responsive flower senescence-related gene from carnation encodes a protein homologous to glutathione s-transferases. Plant Mol. Biol. 17, 277-281.

39. Mor, Y., Halevy, A.H., Spiegelstein, H., Mayak, S. (1985) The site of 1-aminocyclopropane-1-carboxylic acid synthesis in senescing carnation petals. Physiol. Plant. 65, 196-202.

40. Neel, P.L., Harris, R.W. (1971) Factors influencing tree trunk growth. Arborist's News 36, 115-138.

41. Neljubov, D. (1901) Ueber die horizontale Nutation der Stengel von *Pisum sativum* und einiger anderen Pflanzen. Beih. Bot. Centralbl. 10, 128-139.

42. Nichols, R., Bufler, G., Mor, Y., Fujino, D.W., Reid, M.S. (1983) Changes in ethylene production and 1-aminocyclopropane-1-carboxylic acid content of pollinated carnation flowers. J. Plant Growth Regul. 2, 1-8.

43. O'Neill, S.D., Hadeau, J.A., Zhang, X.S., Bui, A.Q., Halevy, A.H. (1993). Interorgan regulation of ethylene biosynthetic genes by pollination. The Plant Cell 5,419-432.

44. Osborne, D.J. (1983) News Bulletin of the British Plant Growth Regulator Group 6, 8-11.

45. Reid, M.S. (1985) Ethylene and abscission. HortScience 20, 45-50.

46. Reid, M.S., Evans, R.Y., Dodge, L.L. (1989) Ethylene and silver thiosulfate influence opening of cut rose flowers. J. Amer. Soc. Hortic. Sci. 114,436-440.

47. Reid, M.S., Fujino, D.W., Hoffman, N.E., Whitehead, C.S. (1984) 1-Aminocyclopropane-1-carboxylic acid (ACC) - The transmitted stimulus in pollinated flowers? J. Plant Growth Regul. 3, 189-196.

48. Reid, M.S., Paul, J.L., Farhoomand, M.B., Kofranek, A.M., Staby, G.L. (1980) Pulse treatments with the silver thiosulfate complex extend the vase life of cut carnations. J. Amer. Soc. Hortic. Sci. 105, 25-27.

49. Reid, M.S., Paul, J.L., Young, R.E. (1980) Effects of ethephon and betacyanin leakage from beet root discs. Plant Physiol. 66, 1015-1016.

50. Robbins, J., Reid, M.S., Rost, T., Paul, J.L. (1985) The effect of ethylene in adventitious root formation of Mung bean (*Vigna radiata*) cuttings. J. Plant Growth Regul. 4, 147-157.

51. Roberts, J.A., Tucker, G.A. (1985) Ethylene and Plant Development. Butterworths, London. 116 pp.

52. Rylski, I., Rappaport, L., Pratt, H.K. (1974) Dual effects of ethylene on potato dormancy and sprout growth. Plant Physiol. 53, 658-662.

53. Sexton, R., Lewis, L.N., Trewavas, A.K., Kelly, P. (1985) Ethylene and abscission. In: Ethylene and Plant Development, pp. 173-96, Roberts, J.A., Tucker, G.A., eds. Butterworths, London.

54. Sisler, E.C., Blankenship, S.M. (1993) Diazocyclopentadiene, a light sensitive reagent for the ethylene receptor. Plant Growth Reg. 12,125-132.

55. Sisler, E.C., Yang, S.F. (1984) Ethylene, the gaseous plant hormone. BioSci. 34, 234-238.

56. Suge, H. (1972) Inhibition of photoperiodic floral induction in *Pharbitis nil* by ethylene. Plant Cell Physiol. 13, 1031-1038.

57. Taylorson, R.B. (1979) Response of weed seeds to ethylene and related hydrocarbons. Weed Sci. 27, 7-10.

58. Veen, H. (1983) Silver thiosulphate: an experimental tool in plant science. Scientia Hort 20, 211-224.

59. Wang, H., Woodson, W.R. (1991) A flower senescence-related mRNA from carnation shares sequence similarity with fruit ripening-related mRNAs involved in ethylene biosynthesis. Plant Physiol. 96, 1000-1001.

60. Whitehead, C.S., Fujino, D.W., Reid, M.S. (1983) The roles of pollen ACC and pollen tube growth in ethylene production by carnations. Acta Hortic. 141, 229-234.
61. Whitehead, C.S., Fujino, D.W., Reid, M.S. (1983) Identification of the ethylene precursor, 1-aminocyclopropane-1-carboxylic acid (ACC) in pollen. Scientia Hortic. 21, 291-297.
62. Whitehead, C.S., Halevy, A.H., Reid, M.S. (1984) Roles of ethylene and 1-aminocyclopropane-1-carboxylic acid in pollination and wound-induced senescence of *Petunia hybrida* flowers. Physiol. Plant. 61, 643-648.
63. Woltering E. J. (1990) Interrelationship between the different flower parts during emasculation-induced senescence in cymbidium flowers. J. Exp. Bot. 41, 1021-1029.
64. Woltering E.J., van Doorn, W.G. (1988) Role of ethylene in the senescence of petals - Morphological and taxonomical relationships. J. Exp. Bot. 39, 1605-1616.
65. Woodson, W.R., Lawton, K.A. (1988) Ethylene-induced gene expression in carnation petals. Plant Physiol. 87, 498-503.
66. Wu, M.J., van Doorn W.G., Reid M.S. (1991) Variation in the senescence of carnation (*Dianthus caryophyllus* L.) cultivars. 1. Comparison of flower life, respiration and ethylene biosynthesis. Scientia Hortic. 48, 99-107.
67. Yang, S.F. (1985) Biosynthesis and action of ethylene. HortScience 20, 41-45.
68. Yang, S.F., Hoffman, N.E. (1984) Ethylene biosynthesis and its regulation in higher plants. Ann. Rev. Plant Physiol. 35, 155-189.
69. Zacarias, L., Reid, M. (1992) Inhibition of ethylene action prevents root penetration through compressed media in tomato (*Lycopersicon esculentum*) seedlings. Physiol. Plant. 86,301-307.
70. Zacarias, L., Reid, M.S. (1990) Role of growth regulators in the senescence of *Arabidopsis thaliana* leaves. Physiol. Plant. 80,549-554.
71. Zobel, R.W. (1973) Some physiological characteristics of the ethylene-requiring tomato mutant *diageotropica*. Plant Physiol. 52, 385-389.

G3. Auxin Transport

Terri L. Lomax[1], Gloria K. Muday[2], and Philip H. Rubery[3]
[1]Department of Botany and Plant Pathology, Oregon State University, Corvallis, OR 97331-2902, USA, [2]Department of Biology, Wake Forest University, Winston-Salem, NC 27109, USA, and [3]Department of Biochemistry, University of Cambridge, UK.

INTRODUCTION

The concept of translocation of chemical messengers in higher plants was proposed in the nineteenth century by Charles Darwin. From experiments on the response of oat seedlings to light, Darwin suggested that the translocation of an unknown substance might regulate the phototropic response. The chemical messenger was later shown to be auxin, of which indole-3-acetic acid (IAA; Fig. 1, I) was found to be the predominant, naturally-occurring form. The polarity of auxin transport in cereal seedlings was established by the 1930s, and later found to be a widespread feature of shoot and root tissues. Excellent reviews of the historical development of auxin polar transport studies are available elsewhere (12, 25, 62).

Polar auxin transport has long been linked, at least in theory, to polar development, differentiation, and growth phenomena, such as apical dominance, vascular development, and tropisms. The focus here will be to re-examine these links in the context of recent analyses exploring the biochemistry and cell biology of auxin transport, with the goal of identifying molecular mechanisms for the regulation of plant growth and development through modulation of auxin transport.

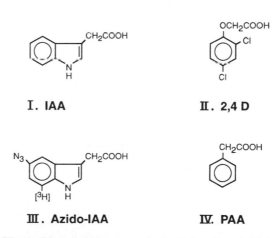

Fig. 1. Chemical structure of selected auxins. I. IAA = indole-3-acetic acid, the major native auxin; II. 2,4-D = 2,4-dichlorophenoxyacetic acid, a synthetic auxin widely used as a herbicide; III. azido-IAA = 7-[^{3}H]-5N$_3$-indole-3-acetic acid, a photoaffinity analog; IV. PAA = phenylacetic acid, a weakly active native auxin.

509

P. J. Davies (ed.), Plant Hormones, 509–530.
© 1995 *Kluwer Academic Publishers. Printed in the Netherlands.*

PHYSIOLOGICAL CHARACTERISTICS OF POLAR AUXIN TRANSPORT

Polar auxin transport has classically been detected and analyzed by applying either unlabeled or radiolabeled auxin (usually IAA) to one end of a tissue segment, either as a fixed source or as a pulse, and following movement of the hormone through the tissue. In early studies, movement of auxin was determined by bioassay. Currently, auxin transport is followed by quantitation of radiolabel in dissected sections of the tissue segment or by measurement of movement of the radiolabel into an agar 'receiver' block at the opposite end of the segment (Fig. 2). Recent analyses using physical or chemical techniques have demonstrated that, in many tissues, all or most of the polarly transported radiolabel remains in the form of IAA. The results of classical polar auxin transport studies have been extensively reviewed (12, 25) and will only be summarized here.

While polar auxin transport is the main theme of this Chapter, it should be emphasized that both endogenous and applied auxins can also move in the plant's vascular system. For example, auxin applied to mature leaves is translocated in the phloem like a photoassimilate, and has also been reported to move in the xylem transpiration stream. In the vascular system, transport occurs by mass flow. In contrast, the direction of polar auxin transport is determined by distribution of protein carriers rather than by gross potential gradients. Inactive auxin conjugates such as inositol esters may move in the

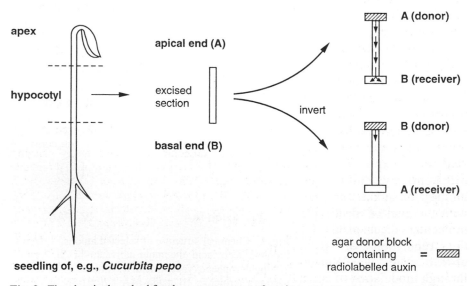

Fig. 2. The classical method for the measurement of auxin transport. Auxin transport through a seedling hypocotyl is basipetal. The polarity is independent of orientation with respect to gravity.

vascular system and be 'activated' by enzymatic hydrolysis at unloading zones, before being distributed by polar transport. An important question is how auxin, in theory a highly permeant molecule, is kept relatively confined and prevented from exchanging freely between the different transport systems (47).

Polar transport is specific for IAA and synthetic auxins, although 2,4-D (Fig. 1, II) in particular moves at only about 20% the rate of IAA. It is an energy-requiring process, as evidenced by the facts that an uphill auxin concentration gradient can be established along the tissue, and that anaerobiosis and metabolic poisons, such as cyanide and dinitrophenol, inhibit polar auxin transport. Auxin transport inhibitors, such as 1-N-naphthylphthalamic acid (NPA, Fig. 3, I) and 2,3,5-triiodobenzoic acid (TIBA, Fig. 3, II), have been identified based on their ability to specifically inhibit polar transport of radiolabeled IAA through segments of etiolated hypocotyls and coleoptiles (12, 26). The enhancement of radiolabeled auxin accumulation into stem segments or thin sections by addition of the auxin transport inhibitors indicates that these compounds act at the site of efflux (7, 59).

In shoot tissues the polarity of auxin transport is basipetal, i.e. auxin moves preferentially from the morphologically apical to more basal regions (Fig. 2). While in coleoptiles all cells may transport auxin, in most tissues polar auxin transport is confined to certain cells that do not include the mature vascular elements. In maize seedlings,

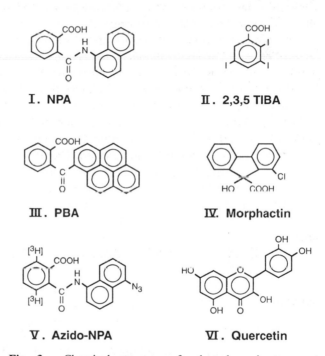

I. NPA

II. 2,3,5 TIBA

III. PBA

IV. Morphactin

V. Azido-NPA

VI. Quercetin

Fig. 3. Chemical structure of selected auxin transport inhibitors. I. NPA = 1-N-naphthylphthalamic acid, a synthetic phytotropin; II. 2,3,5-TIBA = 2,3,5,-triiodo-benzoic acid, a synthetic auxin transport inhibitor which does not bind to the phytotropin binding-protein and which is itself polarly transported. III. PBA = 1-pyrenoylbenzoic acid, the most active synthetic phytotropin; IV. morphactin = 2-chloro-9-hydroxyfluorene-9-carboxylic acid, also causes changes in morphology which phytotropins do not; V. azido-NPA = 3,6[3H2]-5'N3-naphthylphthalamic acid, a photoaffinity analog; VI. quercetin = one of the naturally-occurring flavonoids which have polar auxin transport inhibiting activity and compete for NPA binding.

transport may be through the epidermis (24), but the elongated amyloplast-containing parenchyma cells surrounding the vascular strands (starch sheath cells) appear to be the major transport pathway in dicot stems (39). In roots, the polarity of auxin transport is not as clear. Two distinct transport streams have been identified in roots, but in different tissues. Application of radiolabeled auxin to the tip of roots leads to basipetal transport of the auxin in bean (6) and maize (5). This basipetal transport is inhibited by TIBA, and microautoradiography indicates that this basipetal transport is through bean root epidermal tissues (65). In addition, application of radiolabeled auxin to the base of maize roots leads to acropetal auxin transport (57). This acropetal transport is also sensitive to inhibition by TIBA and microautoradiography indicates that this transport is through the central cylinder of bean roots (65).

Both flux and velocity of auxin transport can be estimated from the kinetics of auxin arrival at the receiver or by following the movement of a pulse of radiolabeled auxin through the tissue. The velocity measurements suggest that IAA transport occurs at between 5 and 20 mm/h in shoots and coleoptiles of a variety of plants. This is consistent with the basic mechanism of movement being diffusion rather than mass flow (12). Although isotropic diffusion does not have a velocity as such, the polar movement of auxin is a true transport process resulting from a biologically-specified asymmetric diffusion. Thus, molecules have a greater probability of basipetal than of acropetal movement, which is manifested as an overall velocity.

A number of criticisms have been directed toward the classical auxin transport experiments using stem or coleoptile segments. Wound responses occurring in response to excision of shoot segments may alter the transport capacity (56) and wounding has been demonstrated to abolish the phototropic response of maize coleoptiles (43). Also, early experiments performed with low specific activity radiolabeled auxin required large concentrations of auxin in order to allow transport measurements. These concentrations were sufficient to perturb growth responses and to alter auxin transport. Several recent experiments have overcome these problems by measuring auxin transport through application of tracer concentrations of radiolabeled IAA using intact plants. Auxin is transported through stems of intact light-grown pea plants at rates of 7-15 mm/h, similar to the rates found with etiolated samples and with segments of pea stems (23). In etiolated maize coleoptiles, the velocity of auxin transport is greater through intact coleoptiles than through tissue segments (43). The rate of transport is greatest at the tip (20 mm/h) and decreases toward the base (12 mm/h at the mesocotyl node). Transport in intact plants is saturable by high levels of IAA, suggesting that this auxin movement is protein-mediated. Furthermore, tracer quantities of IAA did not influence growth, and the radiolabel stayed in the form of IAA. Auxin transport inhibitors block polar transport in intact coleoptiles and hypocotyls, elegantly demonstrating their effect on auxin transport *in vivo* (23, 43).

THE CHEMIOSMOTIC HYPOTHESIS FOR POLAR AUXIN TRANSPORT

Polar transport of auxin appears to involve both active and passive components. Evidence for the existence of a passive diffusion component is provided by the fact that respiration inhibitors do not completely stop the uptake and efflux of auxin from individual cells (12). The participation of a transport component that requires energization is indicated by the inhibition of polar transport by anaerobiosis and by inhibitors of respiration. The specificity and saturability of the process also indicates mediation by protein components. In addition, specific inhibitors of auxin transport, such as NPA, have been used to identify proteins associated with the plasma membrane (PM). Polar auxin transport occurs in a cell-to-cell fashion, but not in all cells. This specific cell-to-cell polar transport requires protein carriers for auxin movement. Specific protein-mediated transport is a critical component of an auxin transport system that can be differentially regulated during plant growth, development, and response to the environment.

The current concept of polar auxin transport, known as the chemiosmotic hypothesis of polar auxin transport (59), was proposed independently by Rubery and Sheldrake in 1974 (51) and Raven in 1975 (46). As diagrammed in Figure 4, the chemiosmotic hypothesis proposes that the proton motive force across the PM provides the driving force for auxin transport. IAA is

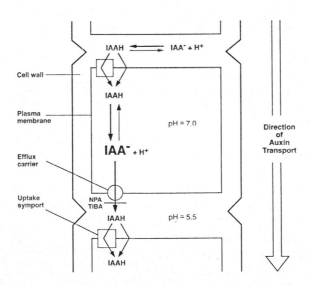

Fig. 4. Schematic model of the chemiosmotic hypothesis for polar auxin transport. Undissociated IAA in the cell wall space can enter the cell either by diffusion or via an uptake symport. Once in the more basic cytoplasm, the IAA dissociates and can exit only via an anion efflux carrier. Metabolically-maintained pH differences across the plasma membrane provide the driving force for transport, while the basal localization of the efflux carrier provides its polar nature.

a weak acid with a pK_a of 4.7. It has a carboxyl group that will be more protonated in the extracytoplasmic space (with a pH approximately equal to 5.5) than inside the cell (with a pH of 7). Undissociated lipophilic IAA in the cell wall can thus move across the cell membrane into the cytoplasm. The IAA then dissociates in the more basic environment of the cytoplasm, and accumulates as a result of the lower membrane permeability of the anionic form. In addition uptake is suggested to be partly via a saturable, specific H^+/IAA$^-$ symport carrier that is evenly distributed around the membrane. Efflux of IAA anions is thought to take place via a saturable, auxin-specific carrier that occurs primarily at the basal end of cells, and which is dependent upon IAA-concentration and electrical gradients for transport to take place. It is the basal location of this carrier which is thought to be responsible for the polar nature of auxin transport. According to the chemiosmotic theory, metabolic energy is expended in order to maintain the pH and electrical gradients and the polar permeability. The identification and characterization of the carriers involved in both IAA uptake and polar efflux is described next.

THE AUXIN UPTAKE CARRIER

An uptake carrier for auxin was originally proposed because of kinetic detection of a saturable component of auxin uptake by both cultured suspension cells and stem segments (7, 51) The carrier-mediated nature of IAA uptake has been substantiated by studies of the association of radiolabeled IAA with isolated membrane vesicles. Membranes isolated from zucchini (*Cucurbita pepo* L.) hypocotyls have been used to demonstrate uptake of ^{14}C-IAA into sealed microsomal vesicles prepared such that they mimic the transmembrane pH and electrical gradients which exist *in vivo* (14). The zucchini membrane vesicles have all of the essential components of *in vivo* polar auxin transport, including an efflux carrier that could be inhibited by NPA and TIBA (Fig. 5). NPA-sensitive IAA uptake was shown to be exclusively into the population of outside-out, sealed PM vesicles (14, 29, 31).

In vitro IAA uptake by vesicles is a saturable, carrier-mediated process which is specific to active auxins (29). The ability of the uptake carrier to function in an electrogenic manner was demonstrated by the addition of valinomycin, a specific K^+ ionophore, to membrane vesicles loaded with 25 mM KCl and resuspended in a low K^+ medium. This resulted in a large increase in ^{14}C-IAA accumulation, presumably due to the creation of a considerable diffusion potential across the membrane (14). Nigericin, an K^+/H^+ ionophore, and FCCP, a protonophore, both reduced IAA accumulation to the background level found with no pH gradient present (31). Using electron spin resonance (ESR) probes to measure the ΔpH, nigericin was shown to totally dissipate the pH gradient, while FCCP reduced the pH gradient only to the point where the proton motive force was reduced to zero by an equal but opposite electrical potential. This experiment indicates that IAA uptake depends on proton motive force rather than ΔpH. Using ESR to quantitate both the sealed vesicle volume and magnitude of the pH gradient, it was determined that IAA accumulation was considerably larger than predicted by diffusional equilibrium at the measured pH gradient. This led to the proposal that the IAA uptake carrier is electrogenic, with each protonated IAA molecule accompanied by an additional proton, thus providing additional driving force (31). Analysis of the uptake of radiolabeled probes as a measure of ΔpH and membrane potential measurements substantiated this hypothesis, and provided evidence that a 2H⁺/IAA⁻ symport is a more likely mechanism than IAAH/H⁺ (52).

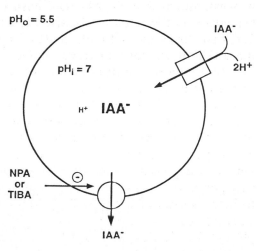

Fig. 5. Model of auxin transport in sealed, right-side out plasma membrane vesicles from shoot tissues. The vesicles are isolated such that the pH and electrical potential gradients are similar to those of intact cells. Such vesicles isolated from zucchini hypocotyls can maintain a pH gradient for many hours and studies with radiolabeled IAA have shown that they contain the essential components of the polar auxin transport system, including an electrogenic uptake symport which transports IAA along with two protons and an efflux carrier which can be inhibited by polar auxin transport inhibitors.

The role of an uptake carrier for IAA in the overall process of auxin transport and/or growth and development of the plant is perhaps less obvious than the role of the auxin efflux carrier in polar transport . Nevertheless, several potential functions are worth highlighting. The polarity of auxin transport could be achieved or enhanced by apical distribution of the uptake carrier rather than the basal localization of the efflux carrier, since it is the relative permeability that matters. The uptake symport may also optimize IAA accumulation when low auxin concentrations are present in the extracellular space or when the pH gradient across the PM is small.

Another possibility is that the IAA uptake carrier is more important in cells other than the few cell files where polar auxin transport takes place. Since the lateral transport of IAA is also important to the development and growth physiology of the plant, the uptake symport may play a major role in this phenomenon. The response to both gravitropic and phototropic stimuli requires a rapid change in the lateral movement of IAA to direct the change in growth pattern. There is evidence which suggests that not only auxin transport, but also sensitivity to auxin is altered during the gravitropic response of plants (53), and data have been presented which suggest that changes in uptake kinetics can alter tissue sensitivity to IAA (58). There is currently little information on the mechanism of this change in auxin transport, but some possibilities include; 1) increasing the driving forces for IAA transport, 2) changing the amount of a particular carrier, 3) altering the affinity or efficiency of a carrier, or 4) altering the distribution of a carrier. The isolation of the IAA uptake and efflux carriers should allow probes to be developed with which to study their distribution, and such studies will help distinguish between these possibilities.

Biochemical characterization of a potential auxin uptake carrier.

The isolation and characterization of proteins which bind IAA has recently been made possible by the simple but powerful technique of photoaffinity labeling. With this technique, auxin-binding proteins are covalently tagged with a tritiated azido-IAA ($7[^3H]$-5N$_3$-IAA; Fig. 1, III). Using this method, high specific activity labeling of two PM polypeptides (with molecular masses of approximately 40 and 42 kD) with azido-IAA has been observed in a variety of dicotyledonous and monocotyledonous plants, as well as in conifers (15, 30). The polypeptides are very low abundance, integral membrane proteins (16), which are detected in auxin-responsive tissues such as roots, stems, and fruits, but not in auxin non-responsive tissues such as mature leaves. The ability of different auxins and auxin analogs to inhibit azido-IAA labeling is similar to the specificity of those compounds for competition with both *in vitro* ^{14}C-IAA uptake into membrane vesicles, and *in vivo* IAA transport studies with hypocotyl segments (15). The specificity does not agree with that for auxin efflux, and auxin transport inhibitors do not affect azido-IAA labeling of the 40 and 42 kD polypeptides. Taken together, these characteristics are consistent with the 40 and 42 kD azido-IAA binding proteins being components of an auxin uptake carrier.

Surprisingly for proteins which behave as hydrophobic membrane proteins, the 40 and 42 kD azido-IAA binding polypeptides both partition into the hydrophilic aqueous phase during extraction with the non-ionic detergent Triton X-114 (16). Such partitioning behavior is, however, consistent with the behavior of known multimeric integral membrane proteins that aggregate to form water-filled channels through membranes, such as the acetylcholine receptor. The 40 and 42 kD polypeptides appear to aggregate under native conditions, forming an apparent dimer at ca. 80 kD and a multimeric complex at ca. 300 kD (16). These characteristics are consistent with the hypothesis that the 40 and 42 kD polypeptides are subunits of a multimeric integral membrane protein that has an auxin binding site, and which may possess a transporter or channel function. While this evidence is consistent with the notion that the 40 and 42 kD polypeptides are components of an auxin uptake carrier, it does not exclude the possibility that they represent either a PM-localized auxin receptor, an auxin-regulated channel protein, or some other protein that specifically binds IAA.

THE AUXIN EFFLUX CARRIER

Compounds which specifically inhibit polar auxin transport have served as valuable tools in the characterization of the auxin efflux carrier. In addition to NPA and TIBA, a number of synthetic inhibitors of auxin transport have been identified (26, 49). The most studied class of such inhibitors is the phytotropins, compounds that inhibit plant gravitropic and phototropic responses in addition to the polar transport of auxin, and which have the common structural theme of benzoic acid ortho-linked to a second aromatic ring system. NPA is a widely-used phytotropin, although 1-pyrenoylbenzoic acid (PBA, Fig. 3, III) is the most active synthetic compound (49). Not all polar auxin transport inhibitors are classed as phytotropins due to structural deviations from the definition: TIBA and the morphactins, such as 2-chloro-9-hydroxyfluorene-9-carboxylic acid (Fig. 3, IV), are the best established exceptions. As described above, the accumulation of auxin into stem segments and membrane vesicles is accentuated by inhibitors of auxin transport, with kinetics confirming that the auxin efflux carrier is the target of these inhibitors (7, 14). Auxin transport and binding studies indicate that all types of auxin transport inhibitors act at a site distinct from the auxin binding site on the efflux carrier; i.e., they are non-competitive with auxins.

Biochemical characterization of the auxin efflux carrier

Transport of molecules across membranes can be mediated either by single polypeptides or by complexes containing several polypeptides. Originally, it was assumed that both the IAA transport activity and the NPA binding activity of the auxin efflux carrier resided on the same polypeptide. Recent indirect evidence, however, has suggested that the auxin efflux carrier has at least two components, with distinct catalytic (transport) and regulatory activities. This evidence is based on procedures that can separate NPA binding activity from IAA efflux activity. Treating zucchini hypocotyl sections with inhibitors of protein synthesis does not significantly affect the ability of hypocotyl segments to accumulate radiolabeled IAA (38). Vesicles isolated from these segments also accumulate radiolabeled IAA and bind [^{3}H]NPA. The ability of NPA to elevate auxin accumulation is, however, greatly reduced by the inhibition of protein synthesis, in both segments and vesicles. This result suggests that the auxin transport site on the efflux carrier system and the receptor site for NPA may reside on separate proteins linked by a third, rapidly turned-over, transducing protein (38). Further investigations have indicated that the phytotropin effects on the auxin efflux carrier are not transduced by a GTP-binding protein (G-protein; 68).

Examination of the biochemical behavior of the NPA-binding protein also suggests that this protein is distinct from the IAA transport activity. Proteins which transport hydrophilic molecules across the PM must be integral to the membrane. Recent experiments have demonstrated that the NPA-binding protein is peripheral to the membrane. Treatments with sodium carbonate at pH 11 and potassium iodide are frequently used to dissociate the ionic interactions by which peripheral proteins associate with the membrane or membrane proteins. Both sodium carbonate and potassium iodide treatments release NPA binding activity from zucchini PM vesicles (Cox and Muday, unpublished data), demonstrating that this protein is peripheral. This substantiates the role of the NPA-binding protein as a distinct regulatory polypeptide.

The phytotropin-binding site

The binding of radiolabeled phytotropins, predominantly [³H]NPA, has now been characterized extensively (40, 49). Membranes isolated from most plant organs are able to bind NPA, with similar affinity for radiolabeled NPA reported in coleoptiles, leaves, stems, and roots of *Zea mays* (27). Isolation by aqueous two-phase partitioning leads to a purified PM fraction that will bind [³H]NPA with high affinity (K_d=15 nM) (2, 40). Although evidence has been presented for two NPA-binding sites in zucchini membrane preparations (35), a detailed analysis of [³H]NPA binding to zucchini microsomes and PM indicates that only a single [³H]NPA-binding site can be resolved on Scatchard, Hill, and Lineweaver-Burk plots using two different filtration assays (40). A single NPA-binding site has also been indicated by linear Scatchard plots in maize, pea, *Arabidopsis*, and sunflower membranes (2, 60, 61, 64). These analyses suggest that NPA and other auxin transport inhibitors bind not only with high affinity, but with high specificity to a single site associated with the PM. The two sites observed in zucchini (35) have several alternative explanations, including the possibility that they may represent both an NPA-binding site and the contributions of uptake of NPA, since isolation of zucchini membranes results in very tightly sealed vesicles which can maintain pH and electrical gradients for long periods of time.

Detailed comparisons of the structure and activity of auxin transport inhibitors in several physiological assays and binding assays have been performed (26, 49). Using this detailed information, molecular modeling of the NPA-binding site has been performed (3), resulting in a map of the site at which auxin transport inhibitors bind. Identification and purification of the NPA-binding protein will provide additional information on the biochemical character of this protein and its function. Several approaches have been used to identify candidate proteins, but purified NPA-binding protein has not yet been isolated. Monoclonal antibodies which prevent NPA binding to both intact microsomal membrane vesicles and proteins solubilized from these vesicles (20) have been used to identify a 77 kD protein in fractionated extracts of pea membranes (22), while a 23 kD protein has been identified (72) by photoaffinity labeling of maize coleoptile PM with 5'-azido-[3,6-^{3}H$_2$]-NPA (Fig. 3, V). Binding of the azido-NPA derivative was reduced by adding high concentrations of auxin transport inhibitors or auxins (1 mM) to the labeling reaction. It is not yet clear why the two potential candidates for the NPA-binding protein are so different in size, or if both are involved in NPA regulation of auxin efflux.

Naturally-occurring auxin transport inhibitors

The phytotropin-binding site associated with the auxin efflux carrier presumably has an *in vivo* function, and may regulate auxin transport in plants by binding a natural regulator of auxin transport. The first potential native regulator identified was phenylacetic acid (PAA; Fig. 1, IV), a weak native auxin which was shown to be an inhibitor of polar auxin transport in pea shoots (36). This compound is not itself polarly transported, but appears to block the movement of auxin, apparently at the site of efflux. Binding experiments, however, show that PAA does not inhibit NPA binding at concentrations as high as 1 mM, suggesting that this compound does not act at the same site as phytotropins (21).

Flavonoid compounds, such as quercetin (Fig. 3,VI), have been shown to block auxin efflux from hypocotyl segments, to compete for [^{3}H]NPA binding in membrane vesicles (21), and to stimulate accumulation of radiolabeled IAA by membrane vesicles (11). Consequently, these compounds have the biochemical activity of auxin transport inhibitors, although the concentration of flavonoid required for action is much higher than the concentrations required for the synthetic inhibitors. It has been more difficult to demonstrate directly that these flavonoid compounds prevent polar auxin transport in segments, perhaps due to the immobility of flavonoids (50). Other characteristics of flavonoids make them attractive candidates as endogenous regulators of auxin transport, as discussed below.

Subcellular localization of the NPA-binding protein

The basal localization of the auxin efflux carrier in stem cells has been proposed to drive the polarity of auxin transport (46, 51). Immunocytochemical studies using monoclonal antibodies against the NPA-binding protein supported a basal localization in pea stem cells (20). This exciting and potentially important result was the first experimental evidence suggesting the distribution of the NPA binding reflected the polarity of auxin transport. Unfortunately, those NPA-monospecific antibodies no longer exist and these intriguing findings can not be expanded. In other studies, when auxin-accumulating PM vesicles from maize and zucchini were fractionated on dextran density gradients, two membrane populations, which had distinctly different auxin transport characteristics, were isolated (32). Vesicles from the lower part of the gradient accumulated IAA and this accumulation could be strongly increased by addition of phytotropins. These vesicles also had high levels of NPA binding activity. In contrast, vesicles from the top part of the gradient showed saturable, specific accumulation of IAA with only a small stimulation by phytotropins, and with very few binding sites for NPA. The accumulation of IAA was reduced relative to the vesicles from the lower fraction. The authors suggested that these vesicle populations could have been derived from the basal and apical parts of the PM, respectively, and resulted from an asymmetric localization of the auxin efflux carrier. Recent evidence indicating that the NPA-binding protein is on a polypeptide separate from the efflux carrier suggests that these results could have two interpretations. Either the entire efflux carrier complex may be restricted to the base of stem cells, or the efflux carrier may be evenly distributed in the membrane of plant cells while the NPA-binding protein may be localized to the base of the membrane and associated with only a subset of the efflux carriers.

In order to understand the mechanisms by which auxin transport is regulated by the NPA-binding protein, it is necessary to determine whether this peripheral protein is associated with the extracytoplasmic or cytoplasmic face of the membrane. Several lines of evidence suggest that this protein may function on the cytoplasmic face of the membrane. Two membrane vesicle populations which bind NPA can be isolated from zucchini hypocotyls, and it has been proposed that they differ in the orientation of the vesicles (34). The population suggested to be outside-out was unable to bind the NPA anion, but bound the membrane permeant uncharged NPA species. The inside-out population was able to bind NPA in either ionization state. Furthermore, NPA binding activity increases as outside-out zucchini vesicles are converted to inside-out or open vesicles by multiple freeze thaw cycles or by removal of sucrose in the buffer (14). After hypocotyl segments are incubated with NPA or quercetin and the segments subsequently washed, the inhibitors trapped within the vesicles were still able to accentuate IAA accumulation over the control (50). Lastly, a modification of the flavonoid, quercetin, by addition of a sulfate group has been found to increase the affinity of that flavonoid for the NPA-binding site, but reduce the ability of that compound to block IAA accumulation into membrane vesicles (11). The authors suggest that an external site could be occupied non-productively, but that the binding of NPA to the external site would exclude binding to an internal site. An even simpler interpretation of their results is that the charge of the sulfated molecule made it membrane impermeant and that the site of action for this compound is at the internal face of the membrane.

AUXIN TRANSPORT IN PLANT GROWTH AND DEVELOPMENT

Polar auxin transport is required for the normal growth and development of plants. The transport of auxin is not a constant process, but changes throughout the plant life cycle. Changes in auxin transport can occur in response to the developmental program of plants, as well as in response to environmental stimuli such as light and gravity. In addition, modulation of auxin transport may control growth and development. Evidence for such control is given below. The role of auxin transport in embryogenesis has recently been summarized (4), so it will not be discussed further.

Role of auxin transport in gravitropism and phototropism

The role of auxin transport in the gravitropic and phototropic responses of plants has been suggested in a theory commonly referred to as the Cholodny-Went hypothesis. This hypothesis suggests that gravitropism occurs as a result of changes in the transport of auxins, i.e. when stems are reoriented relative to the gravity vector, lateral auxin transport occurs and the redistribution of auxin leads to the differential growth on the two sides of the stem, resulting in an orientation that is again normal with respect to the gravity vector. The effect is opposite in roots where the auxin which is reoriented to the lower side inhibits growth, so the root grows down (10, 45). Strong support for this hypothesis comes from the ability of auxin transport inhibitors to completely abolish the gravity response in both shoots and roots (26, 41). It should be noted that auxin transport is believed to be important in transduction of the gravity signal, not in it's perception.

The Cholodny-Went hypothesis has been difficult to prove because of the rapidity of the gravitropic response (10), and the limitations in the sensitivity of measurements of auxin redistribution (63). Changes in plant growth in response to gravistimulus are often visible in less than 5 minutes (45). Although the development of auxin asymmetries in response to gravistimulation has been observed by many investigators, it has been more difficult to demonstrate that the rapidity and magnitude of the changes in auxin distribution are sufficient to cause the reorientation of growth. Recent papers have reported that auxin asymmetries in response to both gravity and light precede the growth response in tomato shoots (13), maize coleoptiles (18, 44) and maize roots (71).

An alternative approach to examine auxin gradients in response to gravity and light stimuli utilized a gene fusion between an auxin-responsive promoter and the gene encoding the reporter enzyme β-glucuronidase (GUS) (28). The expression of this fusion protein in transgenic plants was differentially induced across stems during gravitropism and phototropism, supporting the hypothesis that redistribution of auxin is an important part of the gravitropic and phototropic response (see Chapter D2, Fig. 7). Using NPA and TIBA to block auxin transport, Li et al. showed that auxin transport was essential for the auxin-specific gene expression in response to both gravitropic and phototropic stimulus, further strengthening the relationship between auxin transport and tropistic responses. Clearly, auxin transport is required for gravitropism, since auxin transport inhibitors abolish this response, but evidence is accumulating which indicates that changes in auxin transport are not sufficient for gravitropism (10, 63). This subject is the basis of Chapter G5.

Role of auxin transport in elongation growth

An increase in growth in subapical regions of a dicot stem can be seen as they are reached by a pulse of applied auxin transported down from the apex (69) (see Chapter A2). This demonstrates that IAA transport has a regulatory role in elongation growth. Auxin transport inhibitors alter elongation of roots and shoots, but in opposite fashions, since the auxin sensitivity of these tissues are very different. Exogenous auxin stimulates shoot growth, while it inhibits root growth, suggesting that the physiological concentrations are sub- and supra-optimal, respectively. Not surprisingly, low levels of auxin transport inhibitors have been shown to increase elongation growth of maize coleoptile segments and to sustain IAA-induced elongation (66). These effects occur, presumably, by raising auxin concentration in tissues in which growth is stimulated by auxins. In contrast, auxin transport inhibitors reduce root growth and this growth inhibition is further promoted when auxins and auxin transport inhibitors are applied simultaneously (41), again, presumably, by raising auxin concentrations in tissues to supra-optimal levels.

Another link between auxin transport and elongation is apparent when the distribution of IAA is compared to the elongating and non-elongating regions of stems. The sub-apical region of hypocotyls elongates rapidly, and this tissue also has much higher levels of auxin (42), which also correlates with a greater auxin transport capacity (54, 61). In tissues of greater physiological age which are located further from the apex, the growth rate and auxin transport rate are reduced in parallel. IAA accumulation is also elevated in vesicles or hypocotyl segments of greater physiological age, suggesting that auxin efflux is reduced in these tissues. The fact that [^{3}H]NPA-binding capacity and the sensitivity of IAA accumulation to NPA are also decreased in vesicles of greater physiological age (61) suggests that changes in the capacity for auxin transport mediated by the auxin efflux carrier correlates with both auxin levels and elongation growth.

Role of auxin transport in lateral root development

Auxin transport inhibitors completely abolish lateral root development, while auxins stimulate lateral root growth (41). These results suggest that the effect of auxin transport inhibitors on lateral root growth is different from that on primary growth. This effect cannot be due to an elevation of auxin concentration to inhibitory levels, since auxins stimulate lateral root growth. An alternative hypothesis is that auxin transport inhibitors prevent the movement of auxin to tissues in which elevated concentrations are a requirement for normal lateral root development. When both IAA and NPA are applied simultaneously, the effect of auxin transport inhibitors on lateral root growth is dominant. In contrast, application of either IAA, NPA, or both compounds together inhibits elongation of the primary root, which supports the hypothesis that phytotropins reduce primary root elongation by raising auxin concentrations. It is possible that auxin transport inhibitors reduce primary root growth by raising auxin concentrations, but reduce lateral root growth by depleting auxin concentrations, if elongation growth and lateral growth are found in two spatially separated tissues with differing auxin sensitivities.

Auxin-inhibited and auxin-requiring growth responses in roots are spatially separated in both longitudinal and lateral directions. Lateral root growth occurs basal to the elongation region of roots. Consequently, it is possible that treatment with NPA leads to IAA accumulation in the elongation region of the primary root, where it inhibits elongation, but does not reach the region of lateral root initiation. This pattern of effects is consistent with the polarity of auxin transport in roots, being basipetal. This separation of tissues is also interesting in light of evidence suggesting that there are two distinct polar auxin transport streams in different tissues in roots, with basipetal transport occurring in the epidermal tissues and acropetal transport occurring in the central cylinder of roots. Localized application of auxin has indicated that the IAA elongation growth signal moves in a basipetal fashion (5, 33), suggesting that the basipetal, epidermal transport stream controls growth.

In contrast, the acropetal transport stream may provide the auxin to the tissues from which lateral roots emerge. Radiolabeled auxin applied to the apical bud of intact pea seedlings reached the root and accumulated in the lateral root primordia, although little radiolabel was detected in the apical region of the primary root (48). Thus, the concentration of auxin in the pericycle cells, from which lateral roots emerge, may be controlled by this acropetal transport. Furthermore, removal of the root tip leads to increase in number of lateral root primordia, with the largest number of primordia forming in the most apical segment of the decapitated roots, suggesting the accumulation of an acropetally-moving promoter at the cut surface (67). Removal of cotyledons leads to the loss of this acropetal signal. Together, these experiments indicate distinct functional significance for the two polar auxin transport streams in controlling root growth and development.

Although the elimination of lateral roots by auxin transport inhibitors indicates that auxin transport is required for lateral root growth (41), it does not demonstrate whether subtle variations in auxin transport can lead to variations in lateral root development. A clear correlation has been identified between lateral root growth and the abundance of the NPA binding protein in roots of a variety of dicotyledonous plant species (8). As shown in Table 1, plants which have a low amount of lateral root growth have a greater abundance of NPA binding activity, whereas plants with a high degree of lateral root growth have a lower NPA binding activity.

Table 1. Comparison of the abundance of NPA-binding protein activity with the ratio of primary and secondary root growth.

Plant	primary / secondary root growth	Root B_{max}[a] (pmol/mg)
Zucchini	0.35	0.45
Bean	0.47	0.59
Radish	0.79	0.63
Pea	1.5	1.4
Tomato	4.1	2.3

[a] Bmax is the maximum number of binding sites as calculated from the x-intercept of a Scatchard plot of ^{3}H-NPA binding

MECHANISMS FOR THE REGULATION OF AUXIN TRANSPORT

Changes in a natural regulator of auxin transport

Auxin transport can be reduced in living tissue by exogenous application of auxin transport inhibitors. As described above, reduction of auxin transport by synthetic inhibitors profoundly affects root and shoot growth, lateral root development, and ability of plants to respond to gravity and light. The flavonoid compounds mentioned above may regulate auxin transport *in vivo* through binding to the same site as phytotropins (21, 50). Other characteristics of flavonoids make them attractive candidates for endogenous regulators of transport. Flavonoids are ubiquitous in higher plants, where they are concentrated in specific regions that include the reported locations of polar auxin transport. The variety of flavonoids affords great potential for diverse patterns of auxin transport regulation in different species, tissues, and physiological circumstances (49). Furthermore, flavonoid biosynthesis is tightly regulated and responsive to stimuli which also alter polar auxin transport (21).

Rhizobial nodule initiation may be a system which allows demonstration of the regulation of auxin transport by flavonoids *in vivo*. Evidence suggests that an increase in the concentration of naturally occurring auxin transport regulators may control nodulation in roots of several species. Flavonoid aglycones have been shown to regulate rhizobial nodule initiation at the genetic level. Formation of 'empty' nodule-like structures on uninfected roots have been documented in response to application of the synthetic auxin transport inhibitors, TIBA and NPA. These pseudonodules express nodulation genes in correct temporal and spatial patterns (55), suggesting that an early step in nodulation may be inhibition of auxin transport. These nodules also contain a 23 kDa protein identified by azido-NPA and azido-IAA binding as a candidate for the NPA binding protein (19). In addition, it has recently been reported that a compound produced by *Rhizobium meliloti* strains is able to act as a competitive inhibitor of auxin transport (17). Addition of luteolin, a flavonoid compound that induces expression of nodulation genes, to the culture medium increased auxin transport inhibiting activity. Luteolin itself does not appear to act as a transport inhibitor, but a luteolin metabolite may be the active molecule. Although there presently is no direct evidence linking changes in concentration of an endogenous auxin transport inhibitor with changes in auxin transport, these results are suggestive of this type of regulation.

Changes in activity of the NPA-binding protein

Changes in abundance or catalytic activity of the NPA-binding protein could also control the amount of auxin transport and thereby influence plant development. Factors which affect the development of plants, including the physiological age of tissue, the concentrations of other hormones, and the calcium concentration, can also affect the amount of auxin transport through modulation of the levels of the NPA-binding protein. As mentioned above, analyses of *Helianthus* hypocotyls of different physiological ages demonstrated that the reduction of polar auxin transport was correlated with a decrease in the activity of the NPA-binding protein (61). In ethylene-treated tissues, the reduction in transport clearly was related to the decrease in the abundance of active NPA-binding protein (60). Additionally, depletion of endogenous auxin concentration in segments of maize coleoptiles by incubation in buffer leads to a reduction of auxin transport (70). The loss of basipetal transport is accompanied by a reduction in auxin efflux (indicated by an increase in auxin accumulation by segments), reduction in phytotropin sensitivity of this auxin accumulation, and a decrease in the abundance of NPA binding activity. These results suggest that auxin transport in maize coleoptiles is autoregulated by the endogenous auxin level through its influence on the auxin efflux carrier.

Changes in the spatial distribution or association of the NPA-binding protein

The differential growth response of plants to environmental factors such as light and gravity requires changes in auxin transport. One mechanism that could explain the reorientation of the auxin transport stream during gravitropism and phototropism is the spatial redistribution of either the auxin efflux carrier or the regulatory protein which controls transport. Several groups have examined the regulation of auxin transport and their evidence, albeit indirect, has indicated that redistribution of auxin transport proteins may be responsible for alterations in auxin transport. Auxin transport is reduced both in decapitated pea shoots and in secondary shoots under apical dominance as compared with intact dominant shoots, but polar transport capacity can be restored rapidly by treatments which re-established an apical auxin supply. Although these shoots exhibit reduced polar auxin transport, there is no significant change in either the net uptake of [^{14}C] IAA in stem segments or in the sensitivity of such IAA uptake to NPA (37). Morris and Johnson concluded that both subordinate shoots and decapitated stems have approximately the same densities of functional efflux carrier as the cells of normal shoots, and suggested that the loss of an apical source of auxin "leads to a randomization of the distribution of auxin-anion efflux carriers in the plane of the PM". It should be noted that these data cannot differentiate between randomization of the efflux carrier or the regulatory, NPA-binding protein.

Polar auxin transport has been reported to be dependent on cellular calcium concentrations or calcium movements (1, 9). Polar auxin transport is reduced in calcium-depleted plants, but this reduction is reversed by addition of calcium. In membranes isolated from calcium-depleted zucchini hypocotyls, NPA binding is similar to the control values, but the NPA-sensitivity of transport is diminished (1). This result could be interpreted to suggest that under conditions where calcium is limited, the NPA-binding protein is still present in cells, but unable to regulate the transport through the efflux carrier, perhaps due to an alteration in association with the carrier complex.

Summary of regulation of auxin transport

Experimental evidence has indicated at least three major mechanisms by which auxin transport can be regulated. Changes in the concentration of endogenous regulators of transport, changes in activity of the auxin efflux carrier proteins, and changes in the association or distribution of proteins associated with the efflux carrier have all been experimentally implicated. These results are intriguing, but molecular probes with which to examine the proteins involved in auxin transport are necessary before study of the mechanisms by which this regulation occurs will be approachable.

References

1. Allan, A.C., Rubery, P.H. (1991) Calcium deficiency and auxin transport in *Cucurbita pepo* L. seedlings. Planta 183, 604-612.
2. Brunn, S.A., Muday, G.K., Haworth, P. (1992) Auxin transport and the interaction of phytotropins. Plant Physiol. 98, 101-107.
3. Bures, M.G., Black-Schaefer, C., Gardner, G. (1991) The discovery of novel auxin transport inhibitors by molecular modeling and three-dimensional pattern analysis. J. Comput.-aided Molec. Design 5, 323-334.
4. Cooke, T.J., Racusen, R.H., Cohen, J.D. (1993) The role of auxin in plant embryogenesis. Plant Cell 5, 1494-1495.
5. Davies, P.J., Doro, J.A., Tarbox, A.W. (1976) The movement and physiological effect of indole-acetic acid following point applications to root tips of *Zea mays*. Physiol. Plant. 36, 333-337.
6. Davies, P.J., Mitchell, E.K. (1972) Transport of indoleacetic acid in intact roots of *Phaseolus coccineus*. Planta 105, 139-154.

7. Davies, P.J., Rubery, P.H. (1978) Components of auxin transport in stem segments of *Pisum sativum* L. Planta 142, 211-219.
8. Dela Fuente, R.K., Leopold, A.C. (1973) A role for calcium in auxin transport. Plant Physiol. 51, 845-847.
9. Donaldson, A.W., Muday, G.K. (1993) Auxin transport and root development in different plant species: Is there a correlation? Plant Physiol. 102(S), 323.
10. Evans, M.L. (1991) Gravitropism: Interaction of sensitivity, modulation, and effector redistribution. Plant Physiol. 95, 1-5.
11. Faulkner, I.J., Rubery, P.H. (1992) Flavonoids and flavonoid sulphates as probes of auxin-transport regulation in *Cucurbita pepo* hypocotyl segments and vesicles. Planta 186, 618-625.
12. Goldsmith, M.H.M. (1977) The polar transport of auxin. Ann. Rev. Plant Physiol. 28, 439-478.
13. Harrison, M.A., Pickard, B.G. (1989) Auxin asymmetry during gravitropism by tomato hypocotyls. Plant Physiol. 89, 890-894.
14. Hertel, R., Lomax, T.L., Briggs, W.R. (1983) Auxin transport in membrane vesicles from *Cucurbita pepo* L. Planta 157, 193-201.
15. Hicks, G.R., Rayle, D.L., Jones, A.M., Lomax, T.L. (1989) Specific photoaffinity labeling of two plasma membrane polypeptides with an azido auxin. Proc. Natl. Acad. Sci. USA 86, 4948-4952.
16. Hicks, G.R., Rice, M.S., Lomax, T.L. (1993) Characterization of auxin-binding proteins from zucchini plasma membrane. Planta 189, 83-90.
17. Hirsch, A.M., Jacobs, M. (1993) NPA binding activity in *Rhizobium meliloti* culture filtrates may be due to breakdown of luteolin, the NOD gene inducer. Plant Physiol. 102(S), 620.
18. Iino, M. (1991) Mediation of tropisms by lateral translocation of endogenous indole-3-acetic acid in maize coleoptiles. Plant Cell Environ. 14, 279-286.
19. Jacobi, A., Zettl, R., Palme, K., Werner, D. (1993) An auxin binding protein is localized in the symbiosome membrane. Z. Naturforsch. 48, 35-40.
20. Jacobs, M., Gilbert, S.F. (1983) Basal localization of the presumptive auxin transport carrier in pea stem cells. Science 220, 1297-1300.
21. Jacobs, M., Rubery, P.H. (1988) Naturally occurring auxin transport regulators. Science 241, 346-349.
22. Jacobs, M., Short, T.W. (1986) Further characterization of the presumptive auxin transport carrier using monoclonal antibodies. *In*: Plant Growth Substances 1985, pp. 218-226, Bopp, M., ed. Springer Verlag, Berlin.
23. Johnson, C.F., Morris, D.A. (1989) Applicability of the chemiosmotic polar diffusion theory to the transport of indol-3yl-acetic acid in the intact pea (*Pisum sativum* L.). Planta 178, 242-248.
24. Jones, A.M. (1990) Location of transported auxin in etiolated maize shoots using 5-azidoindole-3-acetic acid. Plant Physiol. 93, 1154-1161.
25. Kaldeway, H. (1984) Transport and other modes of movement of hormones (mainly auxins). *In*: Hormonal Regulation of Development II, pp. 80-148, Scott, T.K., ed. Springer Verlag, Berlin.
26. Katekar, G.F., Geissler, A.E. (1977) Auxin transport inhibitors III. Chemical requirements of a class of auxin transport inhibitors. Plant Physiol. 60, 826-829.
27. Katekar, G.F., Geissler, A.E. (1989) The distribution of the receptor for 1-N-naphthylphthalamic acid in different tissues of maize. Physiol. Plant. 76, 183-186.
28. Li, Y., Hagen, G., Guilfoyle, T.J. (1991) An auxin-responsive promoter is differentially induced by auxin gradients during tropisms. The Plant Cell 3, 1167-1175.
29. Lomax, T.L. (1986) Active auxin uptake by specific plasma membrane carriers. *In*: Plant Growth Substances 1985, pp. 209-213, Bopp, M., ed. Springer Verlag, Berlin.

30. Lomax, T.L., Hicks, G.R. (1992) Specific auxin-binding proteins in the plasma membrane: Receptors or transporters? Biochem. Soc. Trans. 20, 64-69.

31. Lomax, T.L., Mehlhorn, R.J., Briggs, W.R. (1985) Active auxin uptake by zucchini membrane vesicles: Quantitation using ESR volume and ΔpH determinations. Proc. Natl. Acad. Sci., USA 82, 6541-6545.

32. Lützelschwab, M., Asard, H., Ingold, U., Hertel, R. (1989) Heterogeneity of auxin-accumulating membrane vesicles from *Cucurbita* and *Zea*: A possible reflection of cell polarity. Planta 177, 304-311.

33. Meuwly, P., Pilet, P.-E. (1991) Local treatment with indole-3-acetic acid influences differential growth responses in *Zea mays* L. roots. Planta 185, 58-64.

34. Michalke, W. (1982) pH-shift dependent kinetics of NPA-binding in two particulate fractions from corn coleoptile homogenates. *In*: Plasmalemma and Tonoplast: Their Functions in the Plant Cell, pp. 129-135, Marme, D., Marre, E., Hertel, R., eds. Elsevier Biomedical Press, Berlin.

35. Michalke, W., Katekar, G.F., Geissler, A.E. (1992) Phytotropin-binding sites and auxin transport in *Cucurbita pepo*: Evidence for two recognition sites. Planta 187, 254-260.

36. Morris, D.A., Johnson, C.F. (1987) Regulation of auxin transport in pea (*Pisum sativum* L.) by phenylacetic acid: Inhibition of polar auxin transport in intact plants and stem segments. Planta 172, 408-416.

37. Morris, D.A., Johnson, C.F. (1990) The role of auxin efflux carriers in the reversible loss of polar auxin transport in the pea (*Pisum sativum* L.) stem. Planta 181, 117-124.

38. Morris, D.A., Rubery, P.H., Jarman, J., Sabater, M. (1991) Effects of inhibitors of protein synthesis on transmembrane auxin transport in *Cucurbita pepo* L hypocotyl segments. J. Exp. Bot. 42, 773-783.

39. Morris, D.A., Thomas (1978) A microautoradiographic study of auxin transport in the stem of intact pea seedlings (*Pisum sativum* L.) J. Exp. Bot. 29, 147-157.

40. Muday, G.K., Brunn, S.A., Haworth, P., Subramanian, M. (1993) Evidence for a single naphthylphthalamic acid binding site on the zucchini plasma membrane. Plant Physiol. 103, 449-456.

41. Muday, G.K., Haworth, P. (1994) Tomato root growth, gravitropism, and lateral development: Correlation with auxin transport. Plant Physiol. Biochem., in press.

42. Ortuño, A., Sánchez-Bravo, J., Moral, J.R., Acosta, M., Sabater, F. (1990) Changes in the concentration of indole-3-acetic acid during the growth of etiolated lupin hypocotyls. Physiol. Plant. 78, 211-217.

43. Parker, K.E., Briggs, W.R. (1990a) Transport of indoleacetic acid in intact corn coleoptiles. Plant Physiol. 94, 417-423.

44. Parker, K.E., Briggs, W.R. (1990b) Transport of indole-3-acetic acid during gravitropism in intact maize coleoptiles. Plant Physiol. 94, 1763-1769.

45. Pickard, B.G. (1985) Role of hormones, protons, and calcium in geotropism. *In*: Hormonal Regulation of Development III, Role of Environmental Factors, Encyclopedia of Plant Physiology, pp. 193-281, Pharis, R.P., Reid, D.M., eds. Springer Verlag, New York.

46. Raven, J.A. (1975) Transport of indoleacetic acid in plant cells in relation to pH and electrical potential gradients, and its significance for polar IAA transport. New Phytol. 74, 163-172.

47. Raven, J.A., Rubery, P.H. (1982) Coordination of development: Hormone receptors, hormone action, and hormone transport. In: Molecular Biology of Plant Development, pp. 28-48, Smith, H., Grierson, D., eds. Blackwell Scientific, Oxford.

48. Rowntree, R.A., Morris, D.A. (1979) Accumulation of ^{14}C from exogenous labelled auxin in lateral root primordia of intact pea seedlings (*Pisum sativum* L.). Planta 144, 463-466.

49. Rubery, P.H. (1990) Phytotropins: Receptors and endogenous ligands. Soc. Exp. Biol. Symp. 44, 119-146.

50. Rubery, P.H., Jacobs, M. (1990) Auxin transport and its regulation by flavonoids. In: Plant Growth Substances 1988, pp. 428-440, Pharis, R.P., Rood, S.B., eds. Springer Verlag, Berlin.

51. Rubery, P.H., Sheldrake, A.R. (1974) Carrier-mediated auxin transport. Planta 188, 101-121.

52. Sabater, M., Rubery P.H. (1987) Auxin carriers in *Cucurbita* vesicles. Planta 171, 507-513.

53. Salisbury, F.B. (1993) Gravitropism: Changing ideas. Hort. Rev. 15, 233-278.

54. Sánchez-Bravo, J., Ortuño, A., Botia, J.M., Acosta, M., Sabater, F. (1991) Lateral diffusion of polarly transported indoleacetic acid and its role in the growth of *Lupinus albus* L. hypocotyls. Planta 185, 391-396.

55. Scheres, B., McKhann, H.I., Zalensky, A., Löbler, M., Bisseling, T., Hirsch, A.M. (1992) The PsENOD12 gene is expressed at two different sites in Afghanistan pea pseudonodules induced by auxin transport inhibitors. Plant Physiol. 100, 1649-1655.

56. Schwark, A., Schierle, J. (1992) Interaction of ethylene and auxin in the regulation of hook growth I. The role of auxin in different growing regions of the hypocotyl hook of *Phaseolus vulgaris*. J. Plant Physiol. 140, 562-570.

57. Scott, T.K., Wilkins, M.B. (1968) Auxin transport in roots II. Polar flux of IAA in *Zea* roots. Planta 83, 323-334.

58. Shinkle, J.R., Briggs, W.R. (1984) Indole-3-acetic acid sensitization of phytochrome-controlled growth of coleoptile sections. Proc. Natl. Acad. Sci. USA 81, 3742-3746.

59. Sussman, M.R., Goldsmith, M.H.M. (1981) Auxin uptake and action of N-1-naphthylphthalamic acid in corn coleoptiles. Planta 151, 15-25.

60. Suttle, J.C. (1988b) Effect of ethylene treatment on polar IAA transport, net IAA uptake and specific binding of N-1-naphthylphthalamic acid in tissues and microsomes isolated form etiolated pea epicotyls. Plant Physiol. 88, 795-799.

61. Suttle, J.C. (1991) Biochemical bases for the loss of basipetal IAA transport with advancing physiological age in etiolated *Helianthus* hypocotyls. Plant Physiol. 96, 875-880.

62. Thimann, K.V. (1988) A history of the knowledge of auxin. *In*: Physiology and Biochemistry of Auxins in Plants, pp. 3-20, Kutacek, M., Bandurski, R.S., Kreukele, J., eds. SPB Academic Publishing, Prague.

63. Trewavas, A.J. (1992) FORUM: What remains of the Cholodny-Went theory? Plant Cell Environ. 15, 759-794.

64. Trillmich, K., Michalke, W. (1979) Kinetic characterization of N-1-naphthylphthalamic acid binding sites from maize coleoptile homogenates. Planta 145, 119-127.

65. Tsurumi, S., Ohwaki, Y. (1978) Transport of ^{14}C-labeled indoleacetic acid in *Vicia* root segments. Plant Cell Physiol. 19, 1195-1206.

66. Vesper, M.J., Kuss, C.L. (1990) Physiological evidence that the primary site of auxin action in maize coleoptiles is an intracellular site. Planta 182, 486-491.

67. Wightman, F., Thimann, K.V. (1980) Hormonal factors controlling the initiation and development of lateral roots. Physiol. Plant. 49, 13-20.

68. Wilkinson, S., Morris, D.A. (1993) Effects of G-protein probes on interactions between auxin efflux carriers and phytotropin receptors in zucchini (*Cucurbita pepo* L.) microsomal membranes. Plant Growth Reg. 13, 213-220.

69. Yang, T., Law, D.M., Davies, P.J. (1993) Magnitude and kinetics of stem elongation induced by exogenous indole-3-acetic acid in intact light grown pea seedlings. Plant Physiol. 102, 717-724.

70. Yoon, I.S., Kang, B.G. (1992) Autoregulation of auxin transport in corn coleoptile segments. J. Plant Physiol. 140, 441-446.

71. Young, L.M., Evans, M.L., Hertel, R. (1990) Correlations between gravitropic curvature and auxin movement across gravistimulated roots of *Zea mays*. Plant Physiol. 92, 792-796.

72. Zettl, R., Feldwisch, J., Boland, W., Schell, J., Palme, K. (1992) 5'-azido-[3,6-^{3}H$_2$]-1-naphthylphthalamic acid, a photoactivatable probe for naphthylphthalamic acid receptor proteins from higher plants: Identification of a 23-kDa protein from maize coleoptile plasma membranes. Proc. Natl. Acad. Sci. USA 89, 480-484.

G4. The Induction of Vascular Tissues by Auxin and Cytokinin

Roni Aloni
Department of Botany, The George S. Wise Faculty of Life Sciences, Tel Aviv University, Tel Aviv 69978, Israel

INTRODUCTION

The vascular system of the plant connects the leaves and other parts of the shoot, with the roots, and enables efficient long-distance transport between the organs. In higher plants it is composed of two kinds of conducting tissues: the *phloem*, through which organic materials are transported and the *xylem*, which is the pathway for water and soil nutrients. In angiosperms, the functional conduits of the *phloem* are the *sieve tubes*; and those of the *xylem* are the *vessels* (4, 37). Vascular development in the plant is an open type of differentiation, it continues as long as the plant grows from apical and lateral meristems. The continuous development of new vascular tissues enables regeneration of the plant and its adaptation to changes in the environment. This differentiation of vascular tissues along the plant is induced and controlled by longitudinal streams of inductive signals (4, 42). In spite of the complexity of structure and development of the vascular tissues (37), there is evidence that the differentiation of both the sieve tubes and the vessels is induced by two hormonal signals, namely: (i) auxin, indole-3-acetic acid (IAA), produced mainly by young leaves (4, 6, 26, 27, 42), and (ii) cytokinin produced by root apices (8, 9, 18). This fact raises the question how these two hormonal signals control the differentiation of complex patterns of phloem and xylem? Nevertheless, it should be emphasized that additional growth regulators, like gibberellin (1) and ethylene (7, 45), may also be involved in vascular differentiation. They are beyond the scope of this article and the reader is directed to reviews on the topic (4, 5, 6, 24, 39, 42, 46).

A major problem in studying vascular differentiation is the difficulty of observing the phloem (4, 39). Therefore, it is not surprising that reports on phloem differentiation are confusing and often contradictory. Some of these contradictions have been clarified by using a clearing technique which enables to study the three-dimensional structure of both the phloem and xylem in thick preparations (15).

The aim of this chapter is to present a summary of current thoughts and evidence on the role of both auxin and cytokinin in controlling phloem and xylem differentiation. Recent evidence from transgenic plants is also discussed.

P. J. Davies (ed.), Plant Hormones, 531–546.
© 1995 *Kluwer Academic Publishers. Printed in the Netherlands.*

Structure and Development of Vascular Systems

In order to discuss the role of both auxin and cytokinin in vascular differentiation, there is need to present some of the basic features of these tissues.

The first organized vascular system found in lower members of the plant kingdom consists of phloem with no xylem (22). Thus, we find the first developed sieve tubes in members of the brown algae. Much later during the evolution of plants, in the transition from aquatic to terrestrial habitats, the water conducting system developed.

The vascular systems of higher plants are complex tissues, as they consist of several types of cells (37). The conducting elements of the phloem, the sieve-tube elements, lose their nucleus at maturity. Their cytoplasm is connected through the sieve areas to build the sieve tubes. In the xylem, the vessels function as non-living cells after autolysis of their cytoplasm. They are easily detected by their secondary wall thickenings, which keep their shape when they are dead, against the pressure formed by their surrounding cells.

In the embryo, and in the young portions of the shoots and roots, primary phloem and primary xylem are formed from the apical meristem - the *procambium*. Recent studies on the genetic control of the activities of the apical shoot meristem have uncovered genes that are expressed in the procambium (32, 36). These genes are expressed in the pattern of the vascular meristem and are the first molecular markers that may be used to reveal the early stages of the developmental program controlling vascular differentiation (7). During the development of gymnosperms and dicotyledons, a lateral vascular meristem, the *cambium*, differentiates in the older parts of the plant body and produces the secondary phloem and xylem, which increase the width of the woody parts.

Along the plant axis the vascular tissues are organized in longitudinal strands which are also called vascular bundles, or they are arranged in a continuous cylindrical complex. The xylem is formed in the inner side towards the pith and the phloem towards the outside. In some plant families phloem can be found both in the outer and inner side of the xylem (22). Xylem does not differentiate in the absence of phloem. On the other hand, phloem often develops with no xylem. Thus, we find phloem anastomoses (Fig. 1A) between the longitudinal bundles in many plant species (12, 15). The phloem anastomoses may operate as an emergency system which enables the plant to respond to damage by providing alternative pathways for assimilates around the stem (12). Longitudinal bundles of phloem with no concomitant xylem are usually found in the stem of many species. For example, in mature internodes of *Coleus*, on each bundle of phloem with xylem (a collateral bundle), there is another bundle which consists of phloem only (9, 11). In the young internodes there are more phloem- only bundles than collateral bundles (39). In the course of vascular development, phloem

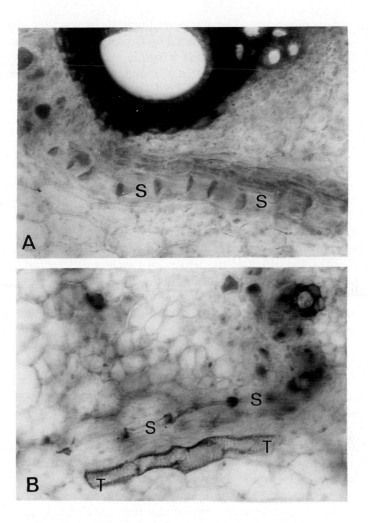

Fig. 1. Transverse sections, taken from internodes of *Luffa cylindrica*, stained with 2% lacmoid in 96% ethanol (1), both x200. A. From an intact untreated plant. The photograph shows a typical phloem anastomosis consisting of sieve elements only. B. From a decapitated stem treated for 10 days with 1.0% naphthaleneacetic acid (NAA) applied in a lanolin paste. The high auxin concentration induced the differentiation of tracheary elements in the anastomosis. S, sieve element; T, tracheary element.

differentiation precedes that of xylem. Therefore, the first elements to differentiate from the procambium are the sieve elements (22). In tissue culture the vascular elements differentiate from parenchyma cells by redifferentiation and have a similar sequential pattern in which the phloem appears before the xylem (2).

Along the plant axis there is a general and gradual increase in the diameter of both sieve-tube and vessel elements from leaves to roots. The narrowest vascular elements are found in the leaves and the largest elements are observed in the roots. The basipetal increase in vascular element size from leaves to roots is often associated with basipetal decrease in vascular element density (4, 16). Thus, for example, vessel density is generally greater in stems where vessels are smaller than in roots where vessels are larger.

THE INDUCTION OF VASCULAR DIFFERENTIATION BY AUXIN AND BY LEAVES

Developing buds and young growing leaves induce vascular differentiation below them. In the spring the young leaves stimulate cambium reactivation and formation of new phloem and xylem which extend downwards from the developing buds toward the root tips (13, 38). The removal of young leaves from stems reduces or prevents vascular differentiation below the excised leaves (26, 30). The effect of the leaves on vascular differentiation is polar: leaves promote vascular differentiation in the direction of the roots (6, 26, 42) but have no effect, or even a slight inhibitory effect on vascular development in the direction of the shoot tip (10). The stimulation effect of the leaves can also be demonstrated by grafting shoot apices with a few leaf primordia onto callus, which results in the formation of both phloem and xylem below the graft in the callus tissue (49).

It is well known that one of the major signals produced by the young leaves is auxin, which moves in a polar fashion towards the roots. Jacobs (26) demonstrated that indole-3-acetic acid is the limiting and controlling factor in xylem regeneration around a wound in *Coleus*. A source of auxin replaces the effect of the young leaves in promoting vascular regeneration (Fig. 2). Auxin alone replaces both qualitatively and quantitatively the effect of the leaves on phloem and xylem regeneration in *Coleus* (30, 48).

In his elegant series of experiments done with pea seedlings, Sachs (42) brings evidence supporting his hypothesis that canalization of auxin flux determines the orderly pattern of vascular differentiation from leaves to roots. According to the '*canalization hypothesis*' (42) the organization of the vascular tissues in orderly bundles is determined by the flow of auxin through the cells prior to vascular differentiation. Auxin movement from an auxin source probably occurs initially by diffusion. As the auxin diffuses through the cells it induces the formation of the polar auxin transport system (see Chapter G3) along a narrow file of cells through which diffusion has been taking place. The continuous polar transport of auxin through the cells induces a further complex series of events which finally result in the formation of a vascular bundle. When the vascular bundle has developed it remains the preferable pathway for auxin transport, as cells possessing the ability to transport auxin are associated with vascular tissues (25, 35).

As the vascular strands are the fastest channels for auxin flow they become the preferable pathways for further auxin transport. Therefore new

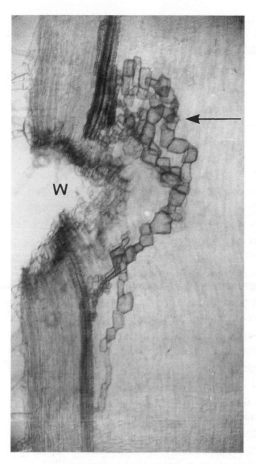

Fig. 2. A longitudinal view of xylem regeneration around a wound induced by IAA in a young internode, with the leaves and buds above it excised. The tissue was taken from a decapitated internode of *Cucumis sativus* treated for 7 days with 0.1% IAA in lanolin, which was applied to the upper side of the internode immediately after the wounding. The tissue was cleared with pure lactic acid and remained unstained, x50. The photograph shows a typical polar pattern of xylem regeneration around a wound which was induced by the basipetal polar flow of auxin. The polar regeneration is characterized by a dense appearance of many regenerative tracheary elements (arrow) immediately above the wound that differentiated close to the wound surface. Below the wound there are a few elements in defined files that connect to the damaged strand at greater distances from the wound. The polar pattern of xylem regeneration reflects the pattern of auxin movement around the wound. As the vascular strands are the preferable pathways for auxin movement, the applied IAA moved basipetally in the damaged strand to where it was interrupted by the wound and was forced to find a new pathway around the obstacle, resulting in a somewhat higher auxin concentration immediately above the wound. This higher auxin concentration resulted in many more regenerative tracheary elements above and close to the wound than below it. W, wound.

streams of auxin from young developing leaves are directed towards the vascular bundles, thus continuing the development of a discrete vascular network whose position is determined by the location of the leaf primordia. For a review on the topic see Sachs (42).

Phloem is induced at low auxin levels while there is a need for higher auxin levels for xylem differentiation in *Coleus* stems (48). Bruck and Paolillo (21) reported that more phloem than xylem differentiated after leaf excision in *Coleus*. They suggested that the phloem-only bundles represent an auxin deficient stage of potentially collateral bundles (consisting of both phloem and xylem), which depends on auxin availability to the bundle for fulfillment. In tissue cultures, phloem differentiates with no xylem at low auxin levels (Fig. 3), whereas both phloem and xylem are induced at higher auxin concentrations (2). Thus, xylem differentiation only takes place at higher auxin levels. As this also fulfills the requirement for phloem differentiation, it explains why xylem does not differentiate in the absence of phloem and always accompanies the pattern of phloem.

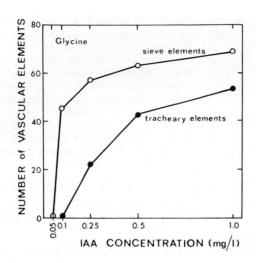

Fig. 3. Effect of IAA on the differentiation of sieve and tracheary elements in tissue culture of *Glycine max* after 21 days. Notice that the low auxin concentration (0.1 mg/L) induced sieve elements with no tracheary elements (2).

I propose that phloem anastomoses, which consist of phloem with no xylem (12, 15), and occur between the longitudinal bundles, are induced by lateral streams of a low auxin level that occur between the longitudinal bundles. It is therefore expected that high auxin levels would induce xylem in these anastomoses. The induction of anastomosis of phloem with xylem by the application of high auxin concentration to decapitated young stems of *Luffa* (Fig. 1B) supports this view.

An additional factor to be considered in the control of vascular differentiation is the capacity of mature vascular tissues to transport auxin. Auxin from mature leaves moves rapidly in a non-polar fashion in the phloem (25, 35). When the phloem is damaged below mature leaves there is a quantitative increase in vascular differentiation below these leaves (20). It is believed that this promoting effect of the wounding is due to an additional source of auxin in the wound region which arrives from the mature phloem. This is an additional mechanism which enables leaves to regulate their supportive vascular system.

The concept that low sucrose levels (1.5-2.5%) induce xylem differentiation with little or no phloem, whereas high sucrose levels (3-4%) favor differentiation of phloem with little or no xylem (49) has been seriously questioned (2, 34, 51). Later studies show that sucrose is needed only as a carbon source and it does not have a specific effect on phloem differentiation or on the relationship between phloem and xylem (2, 34, 51). For a review of this topic see Aloni (4).

The Role of Auxin in Controlling Cell Size in Vascular Tissues

The following *'six-point hypothesis'* (16) has been proposed in order to explain the gradual increase in vessel size and decrease in vessel density (i.e., number of vessels per transverse-sectional area) from leaves to roots. It is based upon the assumption that the stable and steady polar flow of auxin from leaves to roots controls the polar changes in vessel size and density along the plant axis.

(1) Basipetal polar flow of auxin from leaves to roots establishes a gradient of decreasing auxin concentration from leaves to roots.

(2) Local structural or physiological obstruction to auxin flow results in a local increase in auxin concentration.

(3) The distance from the source of auxin to the differentiating cells controls the amount of auxin flowing through the differentiating cells at a given time, thus determining the cells' position in the gradient.

(4) The rate of vessel differentiation is determined by the amount of auxin that the differentiating cell receives; high concentrations cause fast, low concentrations result in slow differentiation. Therefore, the duration of the differentiation process increases from leaves to roots.

(5) The final size of a vessel is determined by the rate of cell differentiation. Since cell expansion is stopped when the secondary wall is deposited, rapid differentiation results in narrow vessels, while slow differentiation permits more cell expansion before secondary wall deposition and therefore results in wide vessels. Decreasing auxin concentration from leaves to roots therefore results in an increase in vessel size in the same direction.

(6) Vessel density is controlled by the auxin concentration; high concentrations induce greater, low concentrations lower densities. Therefore, vessel density decreases from leaves to roots (16).

Experiments with bean (16), *Acer* stems (17) and *Pinus* seedlings (43), support the above '*six-point hypothesis*'. The experiments show that the rate of vessel (16, 17) and tracheid differentiation (43) decreases with increasing distance from the auxin source. The rate of vessel formation in bean was found to be constant at any studied distance from the auxin source (16).

Various auxin concentrations applied to decapitated bean stems induced substantial gradients of increasing vessel diameter (Fig. 4) and decreasing vessel density from the auxin source towards the roots. High auxin concentration yielded numerous vessels that remained small because of their rapid differentiation. Low auxin concentration resulted in slow differentiation and therefore in fewer and larger elements (16). Auxin concentration also influenced the patterns of vessels in the secondary xylem of bean. Immediately below an auxin source the vessels were arranged in layers. Further down along the stem, where lower levels of auxin are expected, the vessels were grouped in bundles (16).

Vascular Adaptation and Wood Formation in Deciduous Hardwoods

Vascular plants grow in different environments, ranging from deserts to rain forests and from alpine regions to the tropics. Comparative anatomical studies reveal similarities in structure of vascular system in plants grown in

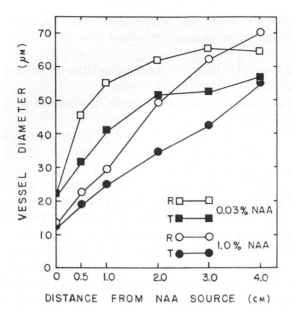

Fig. 4. Effect of low (0.03% NAA; squares) and high (1.0% NAA; circles) auxin concentrations on the radial (R) and tangential (T) diameter of the late-formed vessels along decapitated internodes of *Phaseolus vulgaris*. Both auxin concentrations used induced a substantial gradient of increasing vessel diameter with increasing distance from the site of application. The high auxin concentration yielded the narrowest vessels immediately below the application site (16).

extreme habitats versus ones grown in mesomorphic environments (4). Shrubs and dwarf trees, which occur naturally in extreme environments, show a high density of very narrow vessels. Such vascular systems are typical to extreme habitats and are deemed adaptive safety mechanisms against drought and freezing. Conversely, forest trees and lianas, which are typical in the tropics and rain forests, have vessels of very wide diameter, which affords maximal efficiency of water conduction and is considered to be an adaptation to mesic conditions. Aloni (4) suggested that the environment controls the plant's vascular system through its control of the plant's development, height, and shape. According to the *'vascular adaptation hypothesis'* (4) a limiting growth factor in the plant's environment limits the development and size of the plant and results in a small suppressed shoot, whereas favorable conditions allow the plant to attain its appropriate shape and maximal height. The height of the plant and the degree of its branching determine gradients of auxin along the plant's axis. An increase in the plant's length and a diminution of its branching enhances the gradients of auxin from the young leaves (the sources of auxin) to the lower parts of the stem. In small shrubs, which are typical to extreme habitats, the distances from the young leaves to the roots are very short and no substantial gradient of auxin can be formed. Therefore, the levels of auxin along these small plants are relatively high and result in the differentiation of numerous very small vessels in the greatest densities [as predicted by the *'six-point hypothesis'* (16)]. Conversely, in the large trees and in the long lianas, the very great distances from the young leaves to the roots enable a substantial decrease in auxin levels in the lower parts of the stem and in the roots; this leads to the differentiation of very wide vessels in low density. For experimental evidence supporting this adaptation hypothesis see Aloni (4) and Roberts *et al.* (39).

In temperate deciduous hardwood trees, the size differences of the vessels, or pores, in the early- and latewoods are quite marked and two categories of deciduous trees are determined: diffuse-porous species and ring-porous species. In ring-porous wood, the vessels produced at the beginning of a growth season are significantly wider than those produced at the end of the season, while in diffuse-porous wood, the vessels are more or less uniform (37). Ring-porous trees have originated from diffuse-porous plants (6, 50). Aloni (6) suggested that continuous selective pressures in limiting environments finally resulted in the development of the specialized ring-porous wood that maximized the efficiency of water conduction. Evidence which support the *'limiting-growth hypothesis'* (6) shows that the selection for ring-porous wood has led to a decrease in the intensity of vegetative growth, accompanied by reduced levels of growth regulators. The later was followed by an increase in the sensitivity of the cambium to a relatively low levels of auxin stimulation. These physiological changes created the special internal conditions that enable the differentiation of very wide and long earlywood vessels during spring in the ring-porous trees. It has been recently shown that low-level streams of auxin control the differentiation of wide earlywood vessels in ring-porous trees. While moderate or high auxin levels applied to ring-porous trees at the time of bud break limit the size of the earlywood vessels and result in a diffuse-porous type of wood (6).

Studies on transgenic plants with altered levels of IAA confirm these general relations between auxin level and xylogenesis (28). Auxin-overproducing plants (i.e., overexpressing the *iaaM* gene) contain many more xylem elements than do control plants, and their xylem cells are smaller (29). Conversely, plants with lowered IAA levels (i.e., expressing the *iaaL* gene as an anti-auxin gene) contain fewer xylem elements of generally larger size (40). For additional information on this topic see Chapter E2.

THE INDUCTION OF VASCULAR TISSUES BY CYTOKININ AND BY ROOTS

Roots do not induce vascular differentiation nor must they be present in order to obtain vascular tissues in stems. The roots, however, have two major functions in vascular differentiation, namely: (i) the root orients the pattern of vascular differentiation towards the root tip by acting as a sink for the continuous flow of auxin deriving from young leaves (41), and (ii) the root apices are sources of inductive stimuli that promote vascular development (8). The major developmental signal of the root is cytokinin. Cytokinin, in the presence of an auxin source, controls the early stages of fiber differentiation in stem tissue of *Helianthus* (3, 44) at the stage when many cell divisions occur in the differentiating tissue. Cytokinin also controls the regeneration of vessels and sieve tubes around a wound in *Coleus* internodes (9, 18). It should be emphasized that cytokinin alone, or root apices in the absence of an auxin source, do not induce vascular differentiation in stem tissues.

In spite of the fact that cytokinin is known to influence vascular differentiation in tissue cultures (23, 24, 39), relatively little information has been gained about the role of cytokinin from studies done with intact plants (44). This chapter provides the recent evidence on the role of cytokinin in controlling vascular differentiation in plants.

Vessel differentiation

The role of cytokinin in regenerative differentiation of vessels around a wound was studied in excised internodes of *Coleus* receiving IAA in lanolin at their apical ends and cytokinin in aqueous solution at their base (18). This *in vivo* system enables an experimental study of the role of growth regulators in organized tissues which maintain the natural gradients and polarities inherent in the intact plant. In order to minimize endogenous cytokinin production due to adventitious root formation, 1-mm slices have been removed daily from the basal end of the excised internodes (18). Nevertheless, the control internodes with no cytokinin exhibited a low amount of vessel regeneration (Fig. 5). This low regeneration may have been due to residual amounts of cytokinin in the internodes, or to low cytokinin production by unemerged adventitious root primordia. Experiments with kinetin (Fig. 5), zeatin and 6-benzylaminopurine (BAP) revealed that cytokinin is a controlling factor in the regeneration of vessels around a wound (18). The three cytokinins studied had different effects on vessel regeneration in the *Coleus* internodes. Zeatin displayed its maximum promoting effect

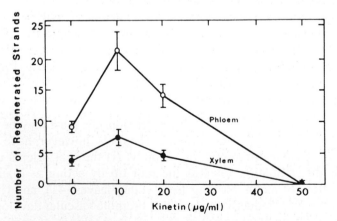

Fig. 5. Effect of cytokinin on the relationship between the regeneration of sieve tubes (○) and vessels (●) in mature internodes of *Coleus blumei* Benth, counted 7 days after wounding. The kinetin was applied in aqueous solution to the bases of the excised internodes treated with low concentration of auxin (0.1% IAA w/w in lanolin) at their apical end. Values are means ± SE (n=10) (9).

under low (0.1% IAA in lanolin) auxin level, kinetin showed its maximum effect under high (1.0%) IAA level, while BAP was the least effective cytokinin at any auxin level tested. Furthermore, kinetin displayed its maximum effects on vessel regeneration in the *Coleus* internodes at 5 and 10 μg ml^{-1} (Fig. 5), while zeatin produced its strongest promoting effect at 20 μg ml^{-1} (18).

There is evidence that along the stem axis of *Coleus* the internodes may show different responses to cytokinin (18). This probably indicates that different sites in the plant body may operate at different levels of cytokinin. In addition, cytokinin may have diverse metabolism in different plant organs (19).

Studies of transgenic plants with altered levels of cytokinin confirm the involvement of cytokinin as a controlling factor in the differentiation of vessels (31, 33). The first study of cytokinin-overproducing plants, i.e., overexpressing the *ipt* gene, shows reduced xylem formation in transgenic plants in comparison with the control (33), while in the second study the vascular tissues are more centralized in the transgenic plants, giving the appearance of a thicker vascular cylinder with more xylem elements than in the control plants (31).

The regeneration of vessels induced by cytokinin in the *Coleus* internodes is polar (Fig. 6). The maturation of regenerated vessels occurred in the acropetal direction (18). In the internodes treated with cytokinin, the regenerated vessel members below the wound were fully differentiated while secondary wall formation on the members at the upper end above the wound was incomplete (Fig. 6). In addition, more regenerated vessel members differentiated below the wound than above it. In extreme cases the regenerated vessel members above the wound were absent, while below the wound there was usually complete regeneration (18).

A major question concerning the role of cytokinin in inducing the acropetal polar pattern of vascular differentiation is the question if root apices can stimulate the development of polar patterns of vessel differentiation similar to those induced by cytokinin. The following evidence (Fig. 7) shows that adventitious roots of *Cucurbita* stimulate differentiation of regenerated vessels in the acropetal direction (8), similar to the pattern induced by cytokinin in the *Coleus* internodes (Fig. 6) (18). The regenerated vessels of the adventitious root develop from the root toward the neighboring strands, usually with an upward tendency towards the leaves (Fig. 7B). In a later developmental stages these regenerated vessels connect the vessel system of the adventitious root with the neighboring vascular strands.

The adventitious roots of *Cucurbita* promote the maturation of regenerated vessels near them (8), similar to the acropetal polar pattern of vessel regeneration induced in the *Coleus* internodes by cytokinin (18). It is therefore suggested that the acropetal polar pattern of regenerated vessel differentiation in *Cucurbita* (Fig. 7) is induced by cytokinin originating in the adventitious roots.

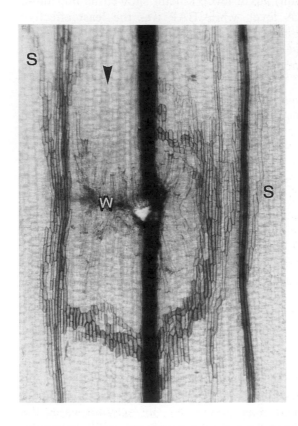

Fig. 6. A longitudinal view of the xylem in a mature internode of *Coleus blumei* Benth, showing a typical polar pattern of xylem regeneration induced by cytokinin around a wound (W), 7 days after wounding, x 32. The internode was treated with 20 μg ml^{-1} zeatin in aqueous solution at its base and with 1.0% IAA (w/w) in lanolin at its apical end. Generally, the cytokinin induced more regenerated vessel members below the wound than above it. There is an absence of regeneration of vessel members immediately above the left side of the wound (arrowhead), demonstrating an interruption to the acropetal movement of zeatin by the wound. The cytokinin induced a novel pattern of xylem regeneration in addition to the familiar one immediately around the wound. The novel pattern of regenerated longitudinal vessels which differentiated away from the wound, was defined as supplementary regeneration (S). The maturation of longitudinal supplementary regenerated vessels occurred in the acropetal direction. It is proposed that the wound interrupts the acropetal movement of zeatin resulting in a local increase of cytokinin concentration immediately below the wound. The high cytokinin concentration increases the responsiveness of the cambium initials located below the wound to the auxin stimulation. The increased sensitivity of the cytokinin-affected cambial initials enables a fast and strong response of these cells to the basipetal polar flow of IAA coming from above the wound (18).

Sieve-tube differentiation

Cytokinin proved to be a controlling factor in sieve-tube regeneration around wounded collateral bundles in excised *Coleus* internodes in which the endogenous cytokinin level had been minimized (9). At appropriate concentrations, both kinetin and zeatin induced a significant increase in sieve-tube regeneration around the wound. Zeatin was most effective with low auxin levels (0.1% IAA in lanolin), while kinetin was most effective with high auxin (1.0% IAA)(9).

There are contradictory results concerning the quantitative relationship between phloem and xylem regeneration in *Coleus* plants. High ratio of phloem/xylem strands (an average of 2.24) was found in regeneration around a wound (11), while low ratios (ranging from 0.07 to 0.37) were reported in grafts (47). At optimal cytokinin levels the ratio of phloem/xylem was the

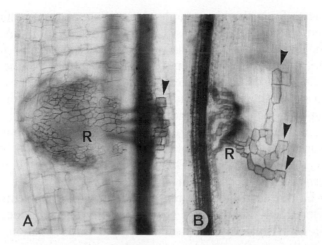

Fig. 7. Longitudinal views of regenerated vessels (arrowheads) originating at the base of adventitious roots (R) in the hypocotyl of *Cucurbita pepo* L, observed 2 days after main root removal. The hypocotyls were cleared in lactic acid and remained unstained, both photographs x 400. The dark longitudinal strand in each photograph is a vascular bundle. **A** shows a side view of an adventitious root, with the first regenerated vessel elements at its base (arrowhead). **B** shows the view from the inside of the hypocotyl and therefore the adventitious root develops away from the eyes of the observer. The photograph shows the maturation pattern of three regenerated vessels (arrowheads) originating from one adventitious root (8).

highest and reached an average of 3.09 (Fig. 5) (9). It seems that the cytokinin-affected cambial initials in *Coleus* internodes become more sensitive to very low levels of auxin, which are known to induce phloem with no xylem (2, 4) and therefore result in high ratios of phloem/xylem.

Recently, we have found that cytokinin and auxin are also involved in controlling callose levels in the sieve tubes. High levels of cytokinin promote callose production on the sieve plates in excised internodes of *Coleus* (9). On the other hand, the first maturing leaves of *Vitis* show the opposite effect, namely that they stimulate the removal of dormancy callose and that this effect can be mimicked by a source of auxin (14). Therefore, it is proposed that production and dissolution of callose in the sieve tubes is controlled by the relative cytokinin/auxin levels. Cytokinin stimulates callose production toward the end of the growing season when the levels of auxin decline. Immediately after the leaves abscise, the effect of cytokinin produced by the roots would predominate, leading to heavy callose production which plugs the sieve tubes for the winter (9).

SUMMARY AND CONCLUSIONS

The polar movement of auxin from leaves to roots induces continuous vascular tissues along the flow of auxin. Auxin, which is the limiting and controlling factor for both phloem and xylem differentiation, induces phloem with no xylem at low auxin levels. Xylem differentiation takes place at higher auxin levels. As this also meets the requirements for phloem differentiation, it explains why xylem does not differentiate in the absence of phloem and always accompanies the pattern of the phloem.

The non-polar transport of auxin in the phloem promotes vascular differentiation in instances of wounding below mature leaves, and serves as an additional mechanism which regulates vascularization in wound regions.

The polar flow of auxin controls the size and density of vascular elements along the plant axis. The general increase in the diameter of vascular elements and decrease in their density from leaves to roots is suggested to be due to a gradient of decreasing auxin concentration from leaves to roots. High auxin levels near the young leaves induce numerous vessels that remain small because of their rapid differentiation, while low auxin concentrations further down result in slow differentiation and therefore in fewer and larger vessels. Vascular adaptation, as well as size and patterns of vessels in hardwood trees are also controlled by auxin.

Cytokinin, which is the major developmental signal of the root, is involved in the differentiation of fibers, vessels and sieve tubes as a limiting and controlling factor. Cytokinin, in the presence of auxin, stimulates early stages of vascular differentiation, at the time when many cell divisions occur in the differentiating tissue. However, late stages of vascular differentiation may occur even in the absence of cytokinin. Cytokinin increases the sensitivity of tissues to auxin stimulation. This increased sensitivity favors sieve-tube differentiation resulting in the highest ratio of phloem/xylem under the optimal level of cytokinin. It is therefore suggested that the differentiation of phloem strands with no xylem in young internodes results from optimal cytokinin levels in the young parts of the shoot.

Cytokinin and auxin influence callose levels in the sieve tubes and are probably involved in controlling dormancy and activity of these phloem conduits.

References

1. Aloni, R. (1979) Role of auxin and gibberellin in differentiation of primary phloem fibers. Plant Physiol. 63, 609-614.
2. Aloni, R. (1980) Role of auxin and sucrose in the differentiation of sieve and tracheary elements in plant tissue cultures. Planta 150, 255-263.
3. Aloni, R. (1982) Role of cytokinin in differentiation of secondary xylem fibers. Plant physiol. 70, 1631-1633.
4. Aloni, R. (1987) Differentiation of vascular tissues. Ann. Rev. Plant Physiol. 38, 179-204.
5. Aloni, R. (1989) Control of xylogenesis within the whole tree. Ann. Sci. For. 46 Suppl., 267-272.
6. Aloni, R. (1991) Wood formation in deciduous hardwood trees. *In*: Physiology of Trees, pp. 175-197, Raghavendra, A.S., ed. Wiley, New York.
7. Aloni, R. (1992) The control of vascular differentiation. Int. J. Plant Sci. 153 Suppl., 90-92.

8. Aloni, R. (1993) The role of cytokinin in organized differentiation of vascular tissues. Aust. J. Plant Physiol. 20, 601-608.
9. Aloni, R., Baum, S.F., Peterson, C.A. (1990) The role of cytokinin in sieve tube regeneration and callose production in wounded *Coleus* internodes. Plant Physiol. 93, 982-989.
10. Aloni, R., W. P. Jacobs, W.P. (1977) Polarity of tracheary regeneration in young internodes of *Coleus* (Labiatae). Amer. J. Bot. 64, 395-403.
11. Aloni, R., Jacobs, W.P. (1977) The time course of sieve tube and vessel regeneration and their relation to phloem anastomoses in mature internodes of *Coleus*. Amer. J. Bot. 64, 615-621.
12. Aloni, R., Peterson, C.A. (1990) The functional significance of phloem anastomoses in stems of *Dahlia pinnata* Cav. Planta 182, 583-590.
13. Aloni, R., Peterson, C.A. (1991) Seasonal changes in callose levels and fluorescein translocation in the phloem of *Vitis vinifera* L. IAWA Bull. n.s. 12, 223-234.
14. Aloni, R., Raviv, A., Peterson, C.A. (1991) The role of auxin in the removal of dormancy callose and resumption of phlocm activity in *Vitis vinifera*. Can. J. Bot. 69, 1825-1832.
15. Aloni, R., Sachs, T. (1973) The three-dimensional structure of primary phloem systems. Planta 113, 345-353.
16. Aloni, R., Zimmermann, M.H. (1983) The control of vessel size and density along the plant axis - a new hypothesis. Differentiation 24, 203-208.
17. Aloni, R., Zimmermann, M.H. (1984) Length, width and pattern of regenerative vessels along strips of vascular tissue. Bot. Gaz. 154, 50-54.
18. Baum, S.F., Aloni, R., Peterson, C.A. (1991) The role of cytokinin in vcsscl regeneration in wounded *Coleus* internodes. Ann. Bot. 67, 543-548.
19. Bayley, A.D., Van Staden, J., Mallett, J.A., Drewes, S.E. (1989) The *in vitro* metabolism of [8-^{14}C]benzyladenine by excised organs of tomato plants. Plant Growth Regul. 8, 193-204.
20. Benayoun, J., Aloni, R., Sachs, T. (1975) Regeneration around wounds and the control of vascular differentiation. Ann. Bot. 39, 447-454.
21. Bruck, D.K., Paolillo, Jr., D.J. (1984) Replacement of leaf primordia with IAA in the induction of vascular differentiation in the stem of *Coleus*. New Phytol. 96, 353-370.
22. Esau, K. (1969) The Phloem. *In*: Encyclopedia of Plant Anatomy, Vol. 5. pt. 2., Zimmermann, W., Ozenda, P., Wulff, H.D., eds. Borntraeger, Berlin.
23. Fosket, D.E., Torrey, J.G. (1969) Hormonal control of cell proliferation and xylem differentiation in cultured tissues of *Glycine max* var Biloxi. Plant Physiol. 44, 781-880.
24. Fukuda, H. (1992) Tracheary element formation as a model system of cell differentiation. Int. Rev. Cytol. 136, 289-332.
25. Goldsmith, M.H.M., Cataldo, D.A., Karn, J., Drenneman, T., Trip, P. (1974) The rapid nonpolar transport of auxin in the phloem of intact *Coleus* plants. Planta 116, 301-317.
26. Jacobs, W.P. (1952) The role of auxin in differentiation of xylem around a wound. Amer. J. Bot. 39, 301-309.
27. Jacobs, W.P. (1970) Regeneration and differentiation of sieve-tube elements. Int. Rev. Cytol. 28, 239-273.
28. Klee, H., Estelle, M. (1991) Molecular genetic approaches to plant hormone biology. Ann. Rev. Plant Physiol. Plant Mol. Biol. 42, 529-551.
29. Klee, H.J., Horsch, R.B., Hinchee, M.A., Hein, M.B., Hoffmann, M.B. (1987) The effects of overproduction of two *Agrobacterium tumefaciens* T-DNA auxin biosynthetic gene products in transgenic petunia plants. Gene & Dev. 1, 86-96.
30. LaMotte, C.E., Jacobs, W.P. (1963) A role of auxin in phloem regeneration in *Coleus* internodes. Dev. Biol. 8, 80-98.
31. Li, Y., Hagen, G., Guilfoyle, T.J. (1992) Altered morphology in transgenic tobacco plants that overproduce cytokinins in specific tissues and organs. Dev. Biol. 153, 386-395.

32. Medford, J. I., Elmer, J.S., Klee, H.J. (1991) Molecular cloning and characterization of genes expressed in shoot apical meristems. Plant Cell 3, 359-370.

33. Medford, J.L., Horgan, R., El-Sawi, Z., Klee, H.J. (1989) Alteration of endogenous cytokinins in transgenic plants using chimeric isopentenyl transferase gene. Plant Cell 1, 403-413.

34. Minocha, S.C., Halperin, W. (1974) Hormones and metabolites which control tracheid differentiation, with or without concomitant effects on growth in cultured tuber tissue of *Helianthus tuberosus* L. Planta 116, 319-331.

35. Morris, D.A., Thomas, A.G. (1978) A microautoradiographic study of auxin transport in the stem of intact pea seedlings (*Pisum sativum* L.). J. Exp. Bot. 29, 147-157.

36. Pri-Hadash, A., Hareven, D., Lifshitz, E. (1992) A meristem-related gene from tomato encodes a dUTPase: analysis of expression in vegetative and floral meristems. Plant Cell 4, 149-159.

37. Raven, P.H., Evert, R.F., Eichhorn, S.E. (1992) Biology of Plants, 5th Ed., Worth Publishers, New York.

38. Reinders-Gouwentak, C.A. (1965) Physiology of the cambium and other secondary meristems of the shoot. *In*: Handbuch der Pflanzenphysiologie XV/1, pp. 1077-1105, Ruhland, W., ed. Springer, Berlin, Gottingen, Heidelberg.

39. Roberts, L.W., Gahan, P.B., Aloni, R. (1988) Vascular Differentiation and Plant Growth Regulators. *In*: Springer Series in Wood Science, Timell, T.E., ed. Springer-Verlag, Berlin, Heidelberg, New York.

40. Romano, C.P., Hein, M.B., Klee, H.J. (1991) Inactivation of auxin in tobacco transformed with the indoleacetic acid-lysine synthetase gene of *Pseudomonas savastanoi*. Genes & Dev. 5, 438-446.

41. Sachs, T. (1968) The role of the root in the induction of xylem differentiation in peas. Ann. Bot. 32, 391-399.

42. Sachs, T. (1981) The control of the patterned differentiation of vascular tissues. Adv. Bot. Res. 9, 152-255.

43. Saks, Y., Aloni, R. (1985) Polar gradients of tracheid number and diameter during primary and secondary xylem development in young seedlings of *Pinus pinea* L. Ann. Bot. 56, 771-778.

44. Saks, Y., Feigenbaum, P., Aloni, R. (1984) Regulatory effect of cytokinin on secondary xylem fiber formation in an *in vivo* system. Plant Physiol. 76, 638-642.

45. Savidge, R.A. (1988) Auxin and ethylene regulation of diameter growth in trees. Tree Physiol. 4, 401-414.

46. Savidge, R.A., Wareing, P.F. (1981) Plant-growth regulators and the differentiation of vascular elements. *In:* Xylem Cell Development, pp. 192-235, Barnett, J.R., ed. Castle House Publications, Kent, UK.

47. Stoddard, F.L., and McCully, M.E. (1980) Effects of excision of stock and scion organs on the formation of the graft union in *Coleus*: a histological study. Bot. Gaz. 141, 401-412.

48. Thompson, N.P., Jacobs, W.P. (1966) Polarity of IAA effect on sieve-tube and xylem regeneration in *Coleus* and tomato stems. Plant Physiol. 41, 673-682.

49. Wetmore, R.H., Reir. J.P. (1963) Experimental induction of vascular tissues in callus of angiosperms. Amer. J. Bot. 50, 418-430.

50. Wheeler, E.A., Baas, P. (1991) A survey of the fossil record for dicotyledonous wood and its significance for evolutionary and ecological wood anatomy. IAWA Bull. n.s. 12, 275-332.

51. Wright, K., Northcote, D.H. (1972) Induced root differentiation in sycamore callus. J. Cell Sci. 2, 319-337.

G5. Hormones and the Orientation of Growth

Peter B. Kaufman, Liu-Lai Wu, Thomas G. Brock and Donghern Kim

Cellular and Molecular Biology Group, Department of Biology, University of Michigan, Ann Arbor, Michigan 48109, USA.

INTRODUCTION

Hormones are chemical messengers which act at target sites to regulate rates and amounts of growth of cells in tissues of roots, stems, leaves, buds, flowers, and fruits. In this chapter, we shall focus on the roles that hormones play in the orientation of growth of plant organs, particularly of roots and shoots. The basic growth-orienting processes that we shall discuss include *phototropism*—the orientation of shoots toward unilateral light sources; *gravitropism*—the orientation of roots downwards and of shoots upwards in response to gravistimulation (placement of plants horizontally); and *thigmotropism*—the change in orientation of growth in stems from one of rapid elongation to one of repressed elongation and promoted lateral expansion as a result of mechanical perturbation.

PHOTOTROPISM

Nature of the Response and Its Adaptive Significance.

It is well-known that unilateral light plays a central role in orienting shoots of plants; that is, it causes them to grow asymmetrically so that shoots become oriented toward sources of higher light intensity, as we have all observed with plants growing in a window. The adaptive significance of this response is quite obvious: it allows the leaves and stems of shoots to capture more radiant energy in the process of photosynthesis.

When shoots grow toward unilateral light sources, we call this a positive phototropic response. Roots are generally negatively phototropic and thus bend away from unilateral light sources. Some vines in the tropics, such as *Monstera deliciosa* and *Philodendron* spp., also exhibit negative phototropic responses in their shoots during seedling and early vegetative stages of development (53). Because of this, their shoots, at these stages, tend to grow toward objects such as tree trunks that cast shadows and are loci where light intensities are lower than in surrounding areas where trees are not growing. The shoots of these lianas (vines) attach themselves by means of aerial roots as they grow up the tree trunk. Later, when the shoots of these plants are

P. J. Davies (ed.), Plant Hormones, 547–571.

older and have reached the canopy of the trees upon which they grow, they exhibit a positive phototropic response. The adaptive significance of both phototropic responses is easily seen.

The Three Primary Components of the Phototropic Response Mechanism in Higher Plants.

The overall mechanism of positive phototropic curvature can be divided into three phases: (a) the *light perception phase*, *(b)* the *transduction phase*, and (c) the *asymmetric growth response phase*. In light reactions such as phototropism, it is inherently assumed that the perception phase involves absorption of particular wavelengths of light by a particular phototropically active pigment. The action spectrum for phototropism in grass seedlings indicates that greatest positive phototropic curvatures occur in blue light (peaks occur at 420, 436, and 475 nanometers) with a second, but lower, peak occurring in the ultraviolet region (peak occurs at 370 nanometers). The nature of the phototropically active pigment has been a matter of controversy, some saying that it is a carotenoid pigment, and others saying that it is a flavin pigment (5, 9, 45, 60); most current evidence supports the idea that a flavin pigment is the primary photoreceptor pigment for the process (11, 12, 45). During the transduction phase, the auxin, indole-3-acetic acid, becomes asymmetrically distributed in most plants—with more accumulating on the shaded side than on the illuminated side. We shall have more to say about the experimental evidence for how this IAA asymmetry may arise in the next section. During the asymmetric growth response phase, cell elongation is inhibited on the illuminated side. As a consequence, the shoot curves toward the light.

How Is Auxin Asymmetry Established in Phototropically Stimulated Shoots?

To answer this question, one can postulate several possible ways by which IAA becomes asymmetrically distributed in shoots of seedlings which are illuminated unilaterally: *(a)* light causes a net transport of IAA to the shaded side; *(b)* light causes destruction of IAA on the illuminated side; (c) less IAA is synthesized on the lighted side; and (d) light causes a decrease in rate of IAA transport on the lighted side.

Two experimental approaches have been used to examine these possibilities (60). One involves the use of diffusates collected in agar blocks which are then assayed by the IAA-specific and sensitive *Avena* coleoptile curvature test. The other involves the use of ^{14}C-IAA applied in donor agar blocks placed on grass coleoptile or dicot seedling shoot tips, and the measurement of radioactivity in both tissue samples and in receiver agar blocks following phototropic stimulation of the seedlings. We shall next examine the results of experiments that have employed each of these protocols.

548

The auxin diffusate—bioassay experiments basically involve excision of *Avena* coleoptile tips, placing them on agar receiver blocks to collect diffusible auxin following various types of light or mechanical treatments, and testing the auxin activity in the diffusates by means of the *Avena* coleoptile curvature bioassay. These experiments were done over the course of several decades, starting in the early 1900s. They showed several important results: *(a)* the coleoptile tip is necessary for phototropic curvature to occur, for if it is absent, no curvature occurs in unilaterally illuminated seedlings; and if the tip is replaced on the cut seedling stump, curvature does occur. Thus, the tip of the grass seedling coleoptile is the primary source of diffusible auxin that mediates the phototropic curvature response. (b) The auxin moves basipetally from the tip region, and when collected from the basal ends of coleoptile tips that have been unilaterally illuminated, one finds about 60% of the diffusible auxin in receptor blocks placed below the shaded side and 40% in blocks placed below the illuminated side. (c) If a mica or glass barrier is placed vertically through coleoptile tips of unilaterally illuminated *Avena* seedlings, no such asymmetry in diffusible auxin in the receptor blocks is seen; the amount of diffusible auxin in essentially the same in receiver blocks placed below shaded and illuminated sides of the coleoptiles. From this series of experiments, it was concluded that light causes a lateral migration of auxin across the coleoptile tip, and because more auxin accumulates on the shaded side, greater cell elongation occurs here, causing the seedling shoot to display a positive phototropic response. This conclusion is the basis for the famous Cholodny-Went hypothesis, first proposed in 1928 (60). It has been invoked to explain both phototropic curvature responses in shoots and gravitropic curvature responses in roots and shoots (see next section), and until recently, has largely been unchallenged.

Studies (5, 11, 12, 56, 60, 62) based on the use of ^{14}C-IAA to follow IAA distribution and rate of movement in tissue to which it is applied in unilaterally illuminated seedlings of both cereal grasses and the dicot, sunflower, are summarized below as follows: When ^{14}C-IAA is applied to the apical portions of excised *Avena* coleoptiles whose tips have been removed (to remove the primary endogenous IAA source), the ratio of radioactivity obtained in basally applied receiver agar blocks is 25:75 for the illuminated and shaded sides, respectively. The ratio of radioactivity in the tissue halves is 35:65 for the illuminated and shaded portions, respectively. These results indicate that for *Avena* seedlings, under these conditions, the IAA that is applied to the coleoptile apex moves in a basipetal direction, and in doing so, becomes asymmetrically distributed in the tissue itself and this leads to a difference in the amounts of IAA that arrive at the receiver agar blocks. Further, under these conditions, the unilateral light treatment, in comparison with dark controls, has no effect on the total amount of radioactivity obtained in the receiver blocks. Thus, one must rule out the possibility that asymmetry in IAA in the tissue comes about by light causing an inactivation of IAA on the illuminated side.

The above results are not always obtained with other plants. A similar experiment with ^{14}C-IAA was performed on sunflower seedlings, and in this case, no asymmetric distribution of radioactive IAA occurs in the stem tissue of unilaterally illuminated seedlings (62). If IAA does not get asymmetrically distributed in such seedlings, it is possible that other hormones, such as native gibberellins, do so, and these could be the effector hormones. It is also possible, that in such seedlings, light has a more direct effect—that of decreasing the rate of cell elongation on the illuminated side.

In studies on the influence of light on the rate of longitudinal transport of ^{14}C-IAA in *Avena* coleoptiles, it was found that light significantly represses the rate of basipetal IAA transport in uniformly illuminated coleoptiles, relative to dark controls (56). The amount of decrease in IAA transport rate is 12 to 17% over a short period following exposure of the coleoptiles to blue light for 15 minutes. Similar results were also obtained for transport of ^{14}C-IAA in maize coleoptiles. It therefore appears that the asymmetry in IAA that develops in tissue of unilaterally illuminated seedlings could be explained by light causing a diminution in the rate of basipetal transport of IAA on the lighted side. How light brings about a decrease in IAA transport rate is unknown and should be explored.

From the above, it is clear that the role of IAA, or of other hormones, in regulating the positive phototropic curvature response needs to be reinvestigated. The Cholodny-Went hypothesis of light inducing IAA transport to the shaded sides of unilaterally illuminated seedlings is open to question, particularly in view of the fact that (a) the rate of IAA transport to the shaded side may be too slow in relation to the kinetics for the bending response, and (b) one could also explain the asymmetry on the basis of light causing a decrease in rate of IAA transport downwards on the illuminated sides of the seedlings. We are further confounded by the fact that no IAA asymmetry has been shown experimentally to occur in some plants, such as the unilaterally illuminated sunflower seedlings. Moreover, all the above plants were first grown in the dark (etiolated), then given unilateral light treatments. In de-etiolated, green seedlings, there is no convincing experimental evidence that such IAA asymmetry does, in fact, occur when the seedling are illuminated unilaterally (62). Here, the positive phototropic curvature response could be explained by light directly causing a diminution in rate of cell elongation on the illuminated sides, as compared with that on the shaded sides of such green seedlings.

A reinvestigation of the role of auxin in the phototropic response of maize coleoptiles has, in fact, been carried out by Briggs and co-workers (3), with the seeming vindication of the Cholodny-Went hypothesis in that system. They have shown that there is a growth rate depression on the irradiated side. There is also a basipetal migration of the growth responses which occurs at the same rate as the growth stimulation caused by exogenous applications of IAA. By applying different concentrations of IAA in lanolin unilaterally or symmetrically to intact coleoptile tips, they demonstrated that the endogenous auxin content was limiting, because applied auxin enhanced growth. The concentration of asymmetrically applied IAA needed to counteract unilateral light was only about twice that which, when applied symmetrically, induced an overall growth rate equivalent to that on the shaded side of a non-auxin-treated, unilaterally illuminated coleoptile. Thus, a two-fold concentration difference between shaded and illuminated sides could explain phototropic curvature of maize coleoptiles. This is within the range of auxin asymmetry commonly found.

GRAVITROPISM

Nature of the Gravitropic Response in Higher Plants and Its Adaptive Significance.

One of the best, and agriculturally important, manifestations of the gravitropic curvature response is seen in shoots of cereal grasses which have become prostrated by the action of wind or rain. This phenomenon is called *lodging*. Once lodged, and provided that the shoots are not too heavy with grain, they begin to manifest an upward bending response. In cereal grasses, this curvature response takes place at swollen localized regions of the leaf sheath bases and, in some grasses (mainly C-4 grasses), also at the bases of the internodes. Such regions are referred to, respectively, as *leaf sheath and internodal pulvini*. It requires almost 60 hours for the shoots of lodged cereal grasses to attain an upright position (90°curvature response) (Fig. 1). At 30°C, they bend upward at the rate of about 1.5°. hour^{-1} [in oat *(Avena)* shoots]. This response is clearly temperature-dependent, as *Avena* shoots, for example, show no such upward bending at 40°C or at 4°C. Those held at the latter temperature "store" growth potential, for if these shoots are now placed

Fig. 1. Gravitropic response in two oat *(Avena)* shoot segments photographed every three hours over a 48-hour period. The segment on the right was held in a normal position with the basal stem portion held firmly to permit the apical portion to bend upwards. The segment on the left has the apical sheath portion held firmly in place, permitting the stem portion to bend upwards. This upward bending response in gravistimulated grass shoots, such as oat*(Avena)*, occurs at the swollen leaf-sheath pulvinus . From (28).

at 30°C, rapid upward bending ensues. On the other hand, the response is very sluggish or nil for shoots held at 40°C for two days, then placed at 30°C.

The upward bending response is even more rapid in prostrated (gravistimulated), dark-grown maize seedlings (2). Upward bending is initiated in the coleoptile within one to three minutes and progresses basipetally to the mesocotyl (first internode) where bending is first seen within five minutes after shoots are placed horizontally. By 60 minutes, the shoots have attained an upright position after a couple of overshoots to the right and left of vertical. Thus, the rate of upward bending is of the order of 1.5 to 2 degrees . minute⁻¹. Time-lapse photographs of this upward bending response in maize seedlings are shown in Figure 2.

What we have just described in older cereal grass shoots and in maize seedlings is termed a *negative gravitropic curvature response*. It is also manifest in setae of mosses and liverworts, shoots of lower vascular plants and ferns, newly developing shoots of conifers (Fig. 3), and shoots of both monocots and dicots at all stages of shoot development. With the exception of grasses, sedges, *Ephedra,* and scouring rushes *(Equisetum* spp.), which have active intercalary growth systems and show the negative gravitropic response at localized sites associated with their nodal regions, most shoots manifest the response over extensive, elongating portions of their internodes. This can be seen in elongating conifer shoots (Fig. 3) and in dicots such as mung bean and sunflower seedlings. The adaptive significance of this negative gravitropic curvature response is that it orients the photosynthetically active shoots toward the sun so that more incident solar radiation is captured in photosynthesis. The upright orientation of shoots involved in sexual reproduction may also be important for pollen dispersal by wind, attracting pollinators (e.g., insects, birds, bats), and in seed dispersal. For humans,

Fig. 2. Time-lapse photograph of a seedling of maize or corn (*Zea mays*) during gravitropic curvature. The initial photograph was taken just as the seedling was placed horizontally (gravistimulated). Successive photographs were taken at 15-minute intervals. The India ink mark at "N" denotes the node between the coleoptile (at left) and the mesocotyl or first internode (at right). From (2)

Fig. 3. Illustration of upward bending, negative gravitropic curvature response in Norway spruce (*Picea abies*) shoots.

upright growth of shoots is essential for proper harvesting of crops such as grains. Plant breeders have developed semi-dwarf cultivars of cereals, such as wheat and rice, which are resistant to lodging during heading (flowering and grain-filling) stages of development. More traditional, taller cultivars are very prone to lodging, and when their shoots do not show upward bending (recovery), yield losses can be very severe.

A positive gravitropic curvature response is displayed by primary roots and most adventitious roots. In prostrated seedlings, it requires about one to two hours for the roots to attain a vertical orientation toward the center of gravity. Lateral roots do not always show such a pronounced curvature response. Some grow outwards at ca. 45°, and others show no downward bending in the normal course of their development. Those roots which show no positive gravitropic response are termed *ageotropic*. Ageotropic growth is also displayed by shoots of 'lazy' mutants (e.g., ageotropica mutant of tomato and lazy mutants of rice, wheat, and maize), stolons or runners (e.g., strawberries), and underground stems or rhizomes (e.g., bamboos, quackgrass, and potato). The adaptive significance of the ageotropic response in stolons and rhizomes is one of facilitating vegetative propagation. The 'lazy' mutants of maize and tomato, of course, have no agricultural value because of their permanently "lodged" growth habit. The positive gravitropic curvature response in primary and adventitious roots facilitates the growth of these roots into the soil and thus provides anchorage for the plant and facilitates the acquisition of water and mineral nutrients.

The Primary Components of the Gravitropic Response Mechanism in Higher Plants.

As with phototropism, we can conveniently subdivide the gravitropic response mechanism into three sequential components: *gravity perception, gravity transduction, and asymmetric growth response.* For convenience, we shall discuss the gravitropic responses in roots and shoots of higher plants in terms of these three components.

Gravity Perception in Roots and Shoots.

In both roots and shoots, gravity is considered to be perceived by starch-containing plastids (amyloplasts or chloroplasts), which we call *statoliths* (1, 18, 21, 23, 41, 60, 62). The cells which contain statoliths are termed *statocytes*. In dicot and monocot roots, the starch statoliths occur in the root-caps, whereas in primary shoots of these plants, they occur in parenchyma cells of cortex, usually in close proximity to vascular bundles of the primary vascular system. When shoots or roots are gravistimulated, the starch statoliths fall to the bottoms of the statocyte cells within less than one minute, usually impinging on the rough endoplasmic reticulum and the plasma membrane. That the starch grains are essential for the gravitropic response to occur in roots is witnessed by the fact that when root-caps are carefully excised (as is easily demonstrated with maize seedlings, which have a meristematic calyptrogen between the root-cap and the root apex, making it easy to separate the cap from the apex), the roots do not show a positive gravitropic response; however, when the root-caps are replaced on the root apices, the gravitropic curvature response is restored (60, 62).

Kiss et al. (1989) and Caspar and Pickard (1989) have observed that a starchless mutant (TC7) of *Arabidopsis thaliana* is gravitropic (8, 32). The latter investigators raised the question as to whether or not starch and amyloplasts play a role in gravity perception. The former researchers, based on their studies of wild type (WT) and TC7 starchless mutant of *Arabidopsis*, concluded that WT roots are more sensitive to gravity than TC7 roots; that is, in the time-course of curvature after gravistimulation, curvatures in TC7 roots is both delayed and reduced in amount compared to WT roots. Further, they indicate that *starch* is not required for gravity perception in TC7 roots, but it is necessary for *full sensitivity*. In sum, then, it is likely that *starch-filled* amyloplasts function as the statoliths in WT *Arabidopsis* root caps, while the *starchless* plastids, which are relatively dense and the most movable component in TC7 mutant *Arabidopsis*-root cap columella cells, may also function as statoliths.

In the leaf-sheath pulvini of cereal grass shoots, the starch statoliths are actually chloroplasts which contain large starch grains. They occur in statocytes located in U-shaped cell clusters on the inner sides of each vascular bundle in the ground parenchyma (30). In *Avena fatua* the acquisition of gravisensitivity correlates with the development of 14 to 16 statocytes in association with each vascular bundle (67). When the shoots are gravistimulated, the starch statoliths in the pulvini cascade to the bottoms of the statocytes within two minutes, and most complete their descent within 15 to 30 seconds. Proof of their essentiality for a negative gravitropic curvature response to occur in these shoots has now been unequivocally established

(58). To determine if starch statoliths do, in fact, act as gravisensors in cereal grass shoots, starch was removed from the starch statoliths by placing 45-days-old intact barley plants (*Hordeum vulgare* cv 'Larker') in the dark at 25°C for 5 days. Evidence from staining with I_2-KI, scanning electron microscopy, and transmission electron microscopy indicated that starch grains were no longer present in plastids in the pulvini of plants placed in the dark for 5 days. Furthermore, gravitropic curvature response in these pulvini was reduced to zero, even though pulvini from vertically oriented plants were still capable of elongating in response to applied auxin plus gibberellic acid. However, when 0.1 M sucrose was fed to dark pretreated, starch statolith-free pulvini during gravistimulation in the dark, they not only reformed starch grains in the starch-depleted plastids in the pulvini, but they also showed an upward bending response. Starch grain reformation appeared to precede reappearance of the graviresponse in these sucrose-fed pulvini. These results strongly support the view that starch statoliths do indeed serve as the gravisensors in cereal grass shoots.

One of the big mysteries in the gravitropic response mechanism is how perception of the gravitropic signal by organelles such as starch statoliths leads to the next phase of the response, namely, transduction. One idea that is extant is that the starch statoliths serve as information carriers (e.g., of enzymes or of Ca^{2+}) (28, 35, 41). As the starch statoliths cascade to the bottoms of the statocyte cells in gravistimulated cereal grass pulvini, they drag down with them tonoplast membranes. These unit tonoplast membranes, as well as the double plastid membranes, could be sites where hormone synthesizing or hormone-deconjugating enzymes are located. Otherwise, the starch statoliths may stimulate the plasma membranes of the statocyte cells to open ion channels. The existence of mechanotransductive ion channels (MCs) has been demonstrated from the measurement of changes in conductivity across the protoplast membrane of tobacco and onion by using the patch clamp technique (15, 50). Opening of these plasma membrane-localized MCs by fallen statoliths may result in the massive ion transport (e.g., Ca^{2+}) across the plasma membrane which, in turn, initiates the cascade of events that occur in the signal transduction pathway. However, the presence of such ion channels in statocyte cells of cereal grass shoot pulvini should be carefully examined.

One novel idea recently put forth by Randy Wayne and colleagues (63) is that gravity perception is not via statoliths, but rather via differential pressure perception at opposite ends of the cell. This is based on studies with giant algal cells of characean algae (e.g., *Chara*), which lack starch. He proposes that (1) proteins in the cell-extracellular matrix (the cell wall) junction may be required for gravisensing, (2) the cell-extracellular matrix junction is, without exception, required for graviperception in statolith-free *Chara* internodal cells, and (3) the function of statoliths is to enhance the overall density of the cytoplasm so that the pressure receptors can function.

Gravity Transduction in Roots and Shoots.

The second component of the gravitropic response mechanism is transduction. The essential component of this process is the development of hormone asymmetry (22, 24, 25, 47, 48, 58, 60, 61, 62). Basic dogma from the Cholodny-Went hypothesis says that the asymmetry comes about as a result of basipetal transport of IAA from the upper to the lower sides of gravistimulated roots and shoots. Cell elongation is then inhibited in the former and stimulated in the latter (where up to a 10,000-fold difference in sensitivity of root cells, as compared to shoot cells, to IAA is the basis for the difference in the cell elongation response) (60). It will become apparent after reading the material which follows (a) that IAA is not the only hormone involved in regulating the upward bending response in gravistimulated shoots; and (b) that non-hormonal messengers, such as Ca^{2+}, may be involved in regulating gravitropic curvature (52, 57). Thus, as with phototropism, the Cholodny-Went hypothesis has been challenged as the sole mechanism for explaining how upward bending occurs in shoots and downward bending occurs in roots. Some investigators (16) have gone so far as to say that hormones may not even be involved in mediating the response. This is highly unlikely in view of what we know about the regulation of gravitropic curvature with exogenously applied hormones such as IAA, GAs, ABA, and ethylene, and endogenous changes which occur in the distribution of one or more of these same hormones in graviresponding organs.

Root Gravitropism

The role of hormones in regulating root gravitropism has been controversial for the past decade (60, 62, 64, 65). IAA has been the primary hormone cited as regulating the process. Many experiments have shown that the root cap produces a growth inhibitor and that this inhibitor causes a reduction of growth on the lower side of the root. The Cholodny-Went hypothesis proposes that this inhibitor is auxin, which moves from the root cap to the lower side of the growing zone of a horizontal root. A primary source of auxin for the root is the shoot, but apparently auxin synthesis can also take place in the root tip itself (55). In addition, IAA applied to the root tip can migrate back to the growing zone (10), and, if placed laterally on the tip, it can induce a bending toward that side. When IAA is applied laterally to the growing zone of a horizontal root, three times more IAA moves downward to the lower side of the root than in the reverse direction (34). This polarity of lateral auxin movement is abolished by the removal of the root cap. If there is a higher level of endogenous IAA in the lower side of a horizontal root (a point not yet conclusively settled), then it would probably inhibit growth via the synthesis of ethylene, because if ethylene synthesis is blocked, then IAA no longer inhibits root growth, but only causes a stimulation (42). According to a model proposed by Hasenstein and Evans (20), IAA synthesized in the shoot is transported to the root cap through the vascular system. In vertically held roots, the statoliths (starch-filled amyloplasts) in root cap cells are at the basal ends of the cells. Under this situation, IAA and calcium ions are symmetrically distributed in the root cap tissue. Then, IAA and Ca^{2+} are transported acropetally to the elongating zone of the root cortex. When roots are placed horizontally, the statoliths settle down to the lateral side of the plasma membranes which face the source of gravity. Somehow, this event initiates the polar transport of Ca^{2+} and IAA to the lower half of the cap. Acropetal transport of asymmetrically redistributed Ca^{2+} and IAA in the root cap results in the accumulation of IAA in the lower side of the root elongation zone. Such a supraoptimal concentration of IAA in the lower side of the elongation zone inhibits cell elongation in this region either by enhanced synthesis of ethylene or by other unknown mechanisms. The IAA role is thus still a viable one. In fact, today, IAA receives most support as being the primary effector hormone that regulates the root gravitropic curvature response. ABA, once thought to be also involved in root gravitropism regulation, has clearly been shown not to be involved because ABA-free mutants still respond to gravity.

When roots are gravistimulated, one of the initial events which takes place is a rapid proton efflux on the upper side of the elongation zone of the root (41, 43). This is easily seen with the proton efflux indicator dye, bromcresol purple, which changes from red at pH 6.5 to yellow at more acid pH's; the shift from red to yellow indicates proton efflux. No such color change occurs on the lower side of the downward bending root. Such a change on the upper side of the root would make the pH in the vicinity of the cell walls lower, and this would favor the enhancement in the activities of one or more cell wall-loosening enzymes. This has not yet been demonstrated experimentally, but it is a likely event. It could explain the rapid increase in cell elongation that occurs on the upper side, in the growing zone, of a gravistimulated root.

We cannot leave the subject of root gravitropism without mentioning an important non-hormonal regulator of the process, namely calcium (35, 36, 41). Exogenously applied ^{45}Ca typically moves to the lower side of tips of gravistimulated roots. When Ca^{2+} is placed on the lower side of the tip, the root will curve downward. EDTA or EGTA, calcium chelators, will block the effect of Ca^{2+} in inhibiting root curvature. Thus, its role in regulating the curvature response is undisputed. One of its primary physiological roles is to activate the enzyme modulator, calmodulin; the calmodulin activates plasma membrane bound Ca-ATPases in plant cells. Ca^{2+} is also essential for the basipetal transport of IAA in both roots and shoots, and it may have important regulatory effects on the cell elongation process itself. It may also be important in gravity perception because the starch statoliths themselves have significant amounts of Ca^{2+} associated with them. The descent of the starch grains in the root cap in gravistimulated roots may provide a mechanism for getting Ca^{2+} to the lower side of the root where it exerts its regulatory effect on cell elongation back of the root apex. The means by which a Ca^{2+} redistribution causes downward bending may be via IAA since the presence of Ca^{2+} on the lower side of the root enhances the downward movement of IAA applied to the upper side (34). In addition Ca^{2+} has been found to be necessary for the growth-inhibitory action of IAA in roots (20). Ca^{2+} redistribution in the root cap may cause auxin to move upward from the root cap. Alternatively, the Ca^{2+} might possibly move to the growing zone and there influence the distribution and inhibitory action of auxin. This auxin could have arrived in the growing zone either by downward movement from the stem, or by upward movement from the root cap. As auxin is only inhibitory in the presence of Ca^{2+}, the gravitropic response could, in fact, be caused by evenly distributed auxin in the presence of an asymmetric distribution of Ca^{2+} (20). The exact mechanism awaits elucidation.

Shoot Gravitropism

In shoots, several hormones are implicated in the regulation of negative gravitropic curvature (28, 43, 46, 47, 54, 58, 60, 61, 62). These include IAA and its *myo*-inositol ester and amide-linked conjugates, GAs and their glucosyl esters and glucoside conjugates, and ethylene. The hormone(s) involved apparently vary with the experimental system (plant) being considered, and as expected, multihormonal control occurs in most of them. We shall first examine hormonal control by IAA in gravistimulated maize seedlings.

As mentioned earlier, the upward bending response in maize seedlings occurs very rapidly, usually within one hour at 30°C (2). It starts within three minutes in the coleoptile and five minutes in the mesocotyl (first internode). The onset of auxin asymmetry is also very rapid. Analyses (2) for free IAA in gravistimulated seedlings indicates that top half / bottom half asymmetry first appears within three minutes in the mesocotyls. This is two minutes before the mesocotyls start to bend upward; thus, IAA could be an important effector hormone in triggering the upward bending response. Full IAA asymmetry is established within 15 minutes after seedlings are oriented horizontally. Fifty-seven percent of the free IAA occurs in the lower halves of the mesocotyls at this time and remains at this level at 30, 60, and 90 minutes after seedlings are first gravistimulated. We can only speculate on how the asymmetry in IAA is established. Further experiments are necessary to determine whether or not basipetal transport is fast enough. It is also possible that the IAA asymmetry arises as a result of release of IAA from its conjugates to a greater extent on the lower side, and/or that more IAA synthesis begins to occur here soon after seedlings are gravistimulated.

In gravistimulated cereal grass shoots at older stages of development (late vegetative and early reproductive stages), the leaf sheath and internodal pulvini are the primary sites for the upward bending response, as we have mentioned earlier. At least two hormones, IAA and GAs, are important in regulating the response during the transduction stage (28). If *Avena* shoots (45 days old) are gravistimulated for 24 hours, they attain an upward curvature that is of the order of 30°. If the leaf sheath pulvini from these plants are partitioned into top and bottom halves to compare with "left" and "right" halves in vertical control plants, and these tissue fractions are analyzed for free IAA, one obtains the results depicted in Figure 4. The vertical control pulvini contain 60 to 70 nanograms of free IAA per gram dry weight of pulvinus tissue. The sum total of free IAA in the gravistimulated pulvini (top plus bottom halves) is 420 nanograms per pulvinus, indicating that enhanced IAA synthesis (a seven-fold increase over that of controls) is induced by gravistimulation in these shoots (28). An average of 120 nanograms IAA occurs in the top halves and 300 nanograms IAA occurs in the lower halves, an asymmetry of ca. 1:2.5 top/bottom. In *Avena fatua* a top/bottom asymmetry of IAA can be detected in as short a time as 15 minutes of gravistimulation, when only 1 to 10 degrees of upward bending has occurred (67). One mechanism for establishing the IAA asymmetry may be the redirecting of the endogenous polar auxin transport system. In upright pulvini, IAA is actively transported basipetally (in a downward direction), from the leaf blade through the pulvinus to the node below (46, 69). A link between this transport system and gravitropism was established by the observation that the appearance of pulvinus gravisensitivity coincides with the development of the polar transport system in *Avena fatua* (67). Another link came with the finding that, if the pulvinus is turned upside-down, then the direction of transport within the tissue reverses, so that IAA is then actively moved in an acropetal, but still downward, direction (68). Similarly, when the pulvinus is placed on its side, transport is reoriented so that IAA moves downward, producing lateral asymmetry (6). This movement of IAA still involves the polar transport system since it is blocked by several inhibitors of that system. Measurable asymmetry of IAA occurs before a detectable growth response following gravistimulation, and the addition of sucrose, which hastens the generation of a growth response, also produces a more rapid redistribution of IAA (6). These results indicate that the reorientation of IAA transport, putting more IAA in the pulvinus bottom than in the top,

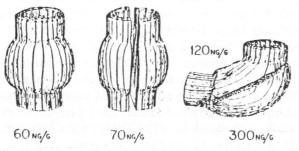

60 NG/G   70 NG/G   120 NG/G   300 NG/G

Fig. 4. Amount of free IAA in ng/g dry weight in *Avena sativa* (oat) leaf-sheath pulvinus tissue. At left, pulvini were also left upright, but were divided into 'left' and 'right' halves. At right, pulvini were gravistimulated for 24 hours so that shoots attained a 30° upward bending response; they were then divided into top and bottom halves for the free IAA analysis.

stimulates the asymmetric growth response. However, inhibitors of the transport system, which completely block lateral IAA transport, only block part of the growth response. Therefore, other processes, such as differential IAA synthesis, occurring more at the bottom than at the top, or the conversion of conjugated IAA to the free (active) form, may also be important in the graviresponse.

Gravistimulation also brings about changes in the levels of free GAs, as well as GA conjugates, in *Avena* leaf sheath pulvini (47). With gravistimulation of shoots to 30° (24 hrs. required), and comparable fractionation of pulvinus tissue into lower and upper halves (compared with "left" and "right" halves of vertical control pulvinus tissue), one finds that GA conjugates predominate in the upper halves, that acidic GA_3, GA_4, and GA_7-like GAs are in greatest abundance in the lower halves, and that total free GAs and GA conjugates decrease slightly as a result of gravistimulation. When the *Avena* shoots are fed ^3H-GA_4, and one "chases" the metabolites formed during gravistimulation, it is clear that the lower halves produce and/or retain more of the free acidic GAs, and the upper halves produce more of the highly water-soluble GA conjugates. Differential GA and/or GA-conjugate synthesis is clearly implicated, but part of the asymmetry could also arise as a result of differential movement of free GAs and of GA conjugates to the respective pulvinus halves.

Changes in the Sensitivity to Auxin of Cereal Grass Shoots During the Gravitropic Response.

Even though it is controversial, the basic concept of the Cholodny-Went theory is widely accepted to represent the involvement of IAA in the regulation of asymmetric growth during tropic responses (61). Asymmetric distribution of IAA, correlated with asymmetric tissue growth, has been detected in different types of plant tissues during the gravitropic response. Asymmetric proton efflux in the elongation zone of graviresponding roots has also been detected (41, 43). These results indicate that 'acid-growth' caused by IAA is a relevant theory to explain the asymmetric growth mechanism during tropic responses. Measuring asymmetric proton efflux, with the pH indicator dye, bromcresol purple in graviresponding cereal grass pulvini is more difficult than for roots. However, by inserting a pH microelectrode into graviresponding *Avena* pulvinus tissue, the pH drops in the elongating portions (bottom) of a pulvinus can be monitored continuously (Fig. 5). Moreover, the negative gravitropic response in pulvini is completely inhibited by pretreating the pulvinus tissues with 0.5 mM vanadate (a potent inhibitor of plasma membrane localized H^+-ATPase) for 2 hrs. These results indicate that the proton efflux induced by H^+-ATPase is an important step in the asymmetric cell elongation in gravistimulated *Avena sativa* (oat) shoot pulvini.

Recent measurements of the effects of exogenously supplied IAA on the gravitropic tissue elongation of soybean hypocotyls (54) and *Avena* shoot

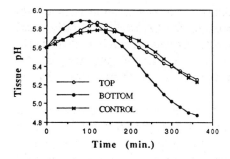

Fig. 5. Changes in tissue pH during the gravitropic response of oat shoot p-1 pulvini. Upward bending of the pulvini starts about 1 hrs after the initiation of gravistimulation. Each point is the average value of three independent replicate experiments.

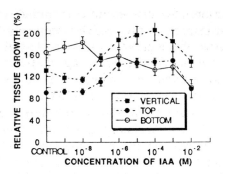

Fig. 6. Relative pulvinus growth induced by IAA with or without gravistimulation.

pulvini clearly indicate that elongating portions of graviresponding pulvini show higher sensitivity to applied auxin. As shown in Figure 6, the maximum tissue elongation of vertical (non-gravistimulated) *Avena* shoot pulvini occurs at 10^{-5}M IAA. Top portions (non-elongating portions) of gravistimulated *Avena* shoot pulvini require the same amount of IAA for the maximum tissue elongation as vertical, non-gravistimulated pulvini. However, only 10^{-8} M IAA is sufficient to cause maximum tissue elongation in bottom portions of gravistimulated *Avena* shoot pulvini. This indicates that the bottom portions of gravistimulated *Avena* shoot pulvini are about 1,000 times more sensitive to IAA in their cell elongation response as compared to that in the top portions. Correlated with this large difference in auxin sensitivity, it has been found that the maximum *in vitro* activation of the plasma membrane-localized proton pump (vanadate sensitive H^{+}-ATPase) in the membranes isolated from top and bottom halves of gravistimulated *Avena* shoot pulvini occurs at 10^{-6} and 10^{-8} M IAA, respectively (Fig.7). These results indicate that the perception of the gravity signal increases auxin sensitivity of bottom portions of *Avena* shoot pulvini by reducing the amount of IAA needed to elicit the maximum proton efflux across the plasma membranes. Since the activation of the proton pump by IAA is mediated by auxin receptor proteins (ARPs), it is possible that the perception of the gravity signal initiates changes in such ARPs, either in their affinity constants, or in their number at the IAA action sites in the bottom portions of graviresponding *Avena* shoot pulvini. However, the presence of such ARPs should be carefully examined in order to evaluate such a hypothesis.

The Role of Ethylene in Shoot Gravitropism.

Ethylene has been implicated as being another one of the regulatory hormones in shoot gravitropism (19, 28, 49, 50). Following gravistimulation of cereal grass shoots, one finds very large increases in ethylene evolution from the tissues where curvature takes place, with more ethylene emanating from the lower halves than from the upper halves. In *Avena* shoots that have been gravistimulated, the burst in ethylene production brought about by this treatment occurs some six hours after shoots are first placed horizontally; this is almost five and one-half hours *after* initiation of upward bending. Thus, one can conclude that ethylene in this system does not play a primary role in the initiation of curvature. Its enhanced production and the greater amount produced on the lower sides of gravistimulated shoots may be brought about, respectively, by the greatly enhanced levels of IAA synthesized in pulvini of gravistimulated shoots and the greater amounts of IAA found in the lower halves. Perhaps ethylene plays a role in the differential expansion of cells that occurs later during the course of upward bending.

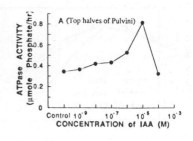

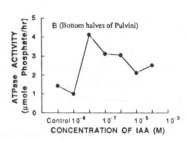

Fig. 7. *In vitro* activation of vanadate-sensitive H^+-ATPase in the presence of various concentrations of IAA.
A: Enzyme preparation from top halves of graviresponding oat pulvini.
B: Enzyme preparation from bottom halves of graviresponding oat pulvini.

In conifer shoots, such as those of *Cupressus arizonica,* ethylene, auxin, and gibberellin all appear to be important in the upward orientation (hyponastic growth response) of elongating lateral shoots (4). In these plants, GA_3, high light intensity, decapitation, and certain levels of IAA increase ethylene evolution and induce upturning of the lateral shoots. The ethylene evolution precedes the initiation of shoot upturning, so it could be important in causing this hyponastic response. Additional evidence in support of this idea comes from exogenous ethylene applications to the shoots; here, such applications cause upward bending. Furthermore, removal of endogenous ethylene from the plants with an external mercuric perchlorate trap causes the branches to grow downwards. Whether ethylene or GAs or IAA or a combination of these hormones initiates the upward bending response in these plants is not yet clear. Studies are needed on the kinetics for ethylene evolution and on changes in levels of endogenous IAA and GAs during the course of hyponastic growth of these lateral branches.

Molecular Biology of the Shoot Gravitropic Response in Pulvinus of Cereal Grass Shoots

Gravity plays an important role in determining plant growth. In recent years, a number of investigators have described the response of plants to gravity, mostly from physiological and biochemical points of view (7, 30, 38, 41, 58).

One component of the gravitropic response mechanism for the upward bending response in pulvini of cereal grass shoots is sucrose hydrolysis, mediated by invertase or β-fructosidase (EC 3.2.1.26) (17, 27, 70, 71). During the gravitropic response in *Avena*, shoots invertase shows a significant increase in its activity from the top half to the bottom half of the upturning pulvinus (17, 71). Invertase is responsible for the hydrolysis of the disaccharide, sucrose, to the two hexoses, D-glucose and D-fructose. These hexoses provide substrate for starch synthesis in the gravisensors and for cell wall biosynthesis in elongating pulvinus cells (17, 38). These hexoses are also responsible for maintaining the turgor pressure in elongating cells of graviresponding pulvini (7, 17, 70, 71).

It has been demonstrated that the activity of invertase the gene(s) is (are) differentially upregulated in top versus bottom halves of graviresponding *Avena* shoot pulvini (70, 71). Invertase activity increases in the pulvini within 3 hrs after gravistimulation treatment (71). The maximal level of enzyme activity is achieved 9-12 h after such treatments. This is followed by a gradual decrease in enzyme activity to the background level characteristic of vertical control pulvini. An asymmetric induction in invertase activity is also observed in graviresponding pulvini. The bottom halves of the pulvini show a four-fold invertase activity greater than that in the top halves of pulvini in response to gravistimulation (Fig. 8).

In an attempt to gain an insight into the molecular mechanism of this gravitropic metabolic response, the regulation of invertase gene(s) expression by gravity was examined in the oat shoot pulvini (70, 71). The focus here is primarily on the transcriptional regulation. The hypothesis is that the upregulation of invertase activity by gravistimulation is most likely due to differential enhancement in invertase mRNA. In order to test this hypothesis, poly(A)⁺RNA was purified from oat pulvini (39). A partial-length invertase cDNA was then synthesized by the polymerase chain reaction (PCR) (31, 33, 59) and used as a probe to investigate the invertase mRNA expressed in response to gravistimulation (70, 71). Using Northern blot analysis, one observes that the level of a 1.90 Kb invertase mRNA is very low at time zero of gravistimulation. But, it is significantly induced 1 h after initiation of gravistimulation treatment, after which time, it decreases (Fig. 9A). This indicates that the induction of invertase mRNA by gravistimulation of oat pulvini is rapid. In order to examine the induction pattern of the

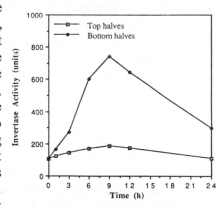

Fig. 8. Time-course changes in invertase activity in top halves and bottom halves of oat shoot pulvini after gravistimulation treatment. Each point is the average of three replicate assays. One unit of enzyme activity is defined as 1 μM sucrose hydrolyzed . gm fresh weight tissue⁻¹ . min⁻¹ at 37°C.

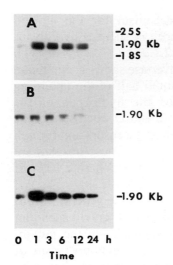

A
−2 5 S
−1.90 Kb
−1 8 S

B
−1.90 Kb

C
−1.90 Kb

0 1 3 6 12 24 h
Time

Fig. 9. Northern blot analyses of oat shoot invertase mRNA isolated from gravistimulated oat shoot pulvini. Partially purified poly(A)⁺RNA (5 μg/well) was electrophoresed in 1% agarose gel, blotted onto a nylon membrane, and hybridized with α-³²PdCTP-labeled invertase cDNA (PCR product) used as a probe. A: mRNA isolated from intact gravistimulated oat shoot pulvini. B: mRNA isolated from top halves of the pulvini. C: mRNA isolated from bottom halves of the pulvini. Lanes 1-6 represent mRNA isolated from oat pulvini during different times of gravistimulation. Lane 1: 0 h is the vertical control. Lane 2: gravistimulated for 1 h. Lane 3: 3 h. Lane 4: 6 h. Lane 5: 12 h. Lane 6: 24 h of gravistimulation, respectively.

gravistimulation-induced invertase mRNA in more detail, mRNA was separately isolated from top and bottom halves of *Avena* shoot pulvini at different times after initiation of gravistimulation treatment. As shown in Figure 9, there is a clear pattern of kinetic changes in invertase mRNA levels during the gravitropic response of the respective halves of oat shoot pulvini. The 1.90 Kb mRNA was detected at a relative low level in the top halves of the pulvini (Fig. 9B). On the other hand, this mRNA occurs at a very high level in the bottom halves of the pulvini (Fig. 9C). The amount of mRNA rapidly increases to a peak at 1 h after initiation of gravistimulation, and then it gradually decreases afterwards. Using a densitometer, the maximum level in these bottom halves of the pulvini that occurs at 1 h represents about a five-fold increase above that of the time zero control (vertical pulvini).

The asymmetric induction of invertase mRNA by gravistimulation (Fig. 9) is consistent with the time-course changes in invertase activities (Fig. 8) in the same tissues. A marked, transient accumulation of invertase transcripts is followed by a transient increase in the enzyme activity. This allows one to speculate that the induction of invertase mRNA level may account for the changes of invertase activity that occur during the graviresponse. The expression of invertase gene(s) in oat pulvini is regulated by gravistimulation at the transcriptional level.

It is important to point out here that IAA redistribution following gravistimulation treatments occurs within 15 minutes in *Avena fatua* (64) and 20 minutes in *Avena sativa* (6). Based on recent studies of Kim, Wu, and Kaufman (unpublished), it is clear that IAA is directly involved in upregulation in the levels of invertase mRNA and subsequent increase in invertase activity in graviresponding oat pulvini.

Auxin-regulated mRNAs Induced During Gravitropism

In addition to the fact that the expression of invertase gene(s) can be rapidly regulated by gravity, a group of small auxin up RNAs (SAURs) genes has been reported to be controlled during the gravitropic response of soybean hypocotyls (40). These auxin-regulated RNAs in vertically oriented seedlings are symmetrically distributed during the elongation of the hypocotyl. However, the RNAs in horizontally oriented seedlings show a rapid asymmetric distribution within 20 min after gravistimulation. The asymmetry of the RNAs for SAUR genes coincides with the onset of rapid bending of the hypocotyl. Therefore, gravity can alter expression of SAUR genes during the development of plants, but we still do not know the functions of such genes in the gravitropic response mechanism, nor do we know what gene products they code for.

In more recent work, Li, Hagen, and Guilfoyle (37) have elegantly shown that an auxin-responsive promoter is differentially induced by auxin gradients during the gravitropic response in shoots of transgenic tobacco plants. Using β-glucuronidase (GUS) as a reporter gene in chimeric gene constructs utilizing the soybean small auxin up RNA (SAUR) promoter and leader sequence, they show that (1) GUS activity expression is responsive to exogenous auxin (IAA) in the range of 10^{-8} to 10^{-3} M; (2) during gravitropic curvature, GUS activity becomes greater on the more rapidly elongating sides of tobacco stems; and (3) auxin transport inhibitors (NPA, TIBA = naphthylphthalamic acid and 2,3,5-triiodobenzoic acid, respectively) block asymmetric distribution of GUS activity in gravistimulated stems. They conclude that their results provide support for the formation of an asymmetric distribution of auxin at sites of action during the gravitropic response. This is very elegant evidence in support of the Cholodny-Went theory of gravitropism. Illustrations of this work and further details relevant to it are provided by Gretchen Hagen in Fig. 7 in Chapter D2.

Basis For the Asymmetric Growth Response after the Transduction Phase of Gravitropic Curvature.

We have already alluded to possible mechanisms that could explain how positive gravitropic curvature might occur in roots. Similar mechanisms may be operating in shoots. These basically involve proton pumping, calcium redistribution, calmodulin synthesis, and activation of calmodulin-binding proteins (e.g., protein kinases, ATPases) (41, 52, 57), and auxin-induced cell wall-loosening and wall synthesis (30). As for the shoot negative gravitropic curvature response, we shall use as our model the grass leaf-sheath pulvinus (30). The scheme in Fig. 10 depicts a cascade of events that could explain the upward bending response mechanism in pulvini of grass shoots. The asymmetric distribution of IAA and of GAs and their conjugates is simply the start of a chain of reactions which leads to asymmetric growth. In this chain of events, protein and cellulose synthesis are essential; likewise, differential

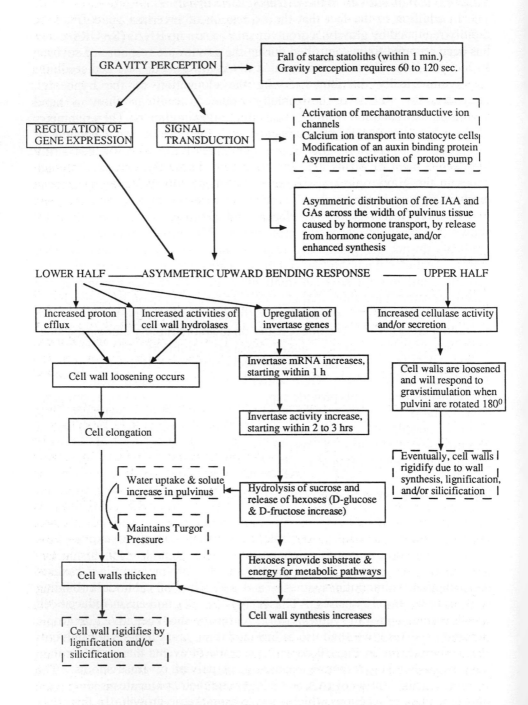

Fig. 10. Cascade of events which could explain the biochemical basis for the upward bending response in gravistimulated cereal grass pulvini. Modified from (30).

566

hormone-stimulated cell wall-loosening is involved. Both processes would occur early in the process of upward bending, the wall loosening starting first, followed by cell wall synthesis. Experimental evidence shows that new proteins are synthesized on both upper and lower sides of upward-bending pulvini, and several have been identified. Cellulase synthesis increases in the upper halves; this could allow the cell walls to become folded under the stress created by upward bending in the pulvinus. Invertase activity increases markedly in the lower halves; this could provide hexose substrate from sucrose for the cell wall synthesis process. Future work will indicate how hormones such as GAs and IAA regulate the expression of genes that synthesize the particular proteins involved in the gravitropic response mechanism. These proteins include H^+-ATPase, cell wall-loosening proteins (enzymatic or non-enzymatic), invertase (sucrose-hydrolyzing enzyme), and cell wall synthesis enzymes, such as glucan synthase.

THIGMOMORPHOGENESIS

Thigmonastic growth responses, or thigmomorphogenetic responses, refer to the effect of mechanical perturbation on plant growth (26). It is manifested in roots which grow away from stones or other barriers in the soil, the contact coiling of tendrils, the closure of leaf traps of *Drosera* (sundew) and *Dionaea* (Venus' fly-trap) upon contact by insects or other contact stimuli, and the diminution in stem length with concomitant thickening caused by mechanical perturbation.

The last-mentioned example, growth retardation by mechanically perturbed plants, has probably been studied most extensively, especially in connection with its regulation by hormones. Several hormones appear to be involved in the response (13). It is well-known that stress induced by various stimuli, such as flooding, or mechanical perturbation (MP), will induce ethylene formation in plants. In beans, MP elicits the production of the ethylene precursor, ACC (l-aminocyclopropane-l-carboxylic acid) from its precursor, SAM (S-adenosyl methionine), peaking at 30 and 90 minutes; ethylene levels then rise, peaking at about two hours and ceasing by four hours. This ethylene production typically causes the internodes to thicken and show reduced linear extension growth. However, the hormonal regulation of this response may be yet more complicated. Recent evidence indicates that MP lowers the levels of endogenous GAs and increases the levels of endogenous IAA and ABA. ^{14}C-IAA basipetal polar transport is also reduced significantly by MP. In this bean system, exogenously applied GA_3 decreases the levels of endogenous IAA and ABA. Most recent evidence, then, suggests that MP of bean internodes induces ethylene synthesis, which in turn, induces the accumulation of high levels of IAA and the production of ABA, both of which contribute to a diminution in internodal extension.

The thickening of the internodes may be due to the effect of ethylene in promoting lateral cell expansion.

Acknowledgments
We thank Richard Pharis of the University of Calgary for his helpful discussions in connection with the writing of this chapter, and the NASA Space Biology Program under the direction of Thora Halstead for financial support under NASA grant, NAG 10-0069.

References

1. Audus, L.J. (1969) *In*: Physiology of Plant Growth and Development, pp. 201-212, Wilkins, M.B., ed. McGraw-Hill Book Co., N.Y.
2. Bandurski, R.S., Schulze, A., Dayanandan, P., Kaufman, P.B. (1984) Response to gravity by *Zea mays* seedlings. Time course of the response. Plant Physiol. 74, 284-288.
3. Baskin, T.I., Briggs, W. R., Iino, M. (1986) Can lateral auxin redistribution account for phototropism of maize coleoptiles? Plant Physiol. 81, 306-309.
4. Blake, T.J., Pharis, R.P., Reid, D.M. (1980) Ethylene, gibberellins, auxin and the apical control of branch angle in a conifer, *Cupressus arizonica*. Planta 148, 64-68.
5. Briggs, W.R. (1963) The phototropic responses of higher plants. Ann. Rev. Plant Physiol. 14, 311 -353.
6. Brock, T.G., Kapen, E.H., Ghosheh, N.S., Kaufman, P.B. (1991) Dynamics of auxin movement in the gravistimulated leaf-sheath pulvinus of oat (*Avena sativa*) J. Plant Physiol. 138, 57-62.
7. Brock, T. G., and P. B Kaufman (1991) Growth regulators: an account of hormones and growth regulation. *In*: Growth and Development (Vol. 10 of Plant Physiology, a Treatise, Steward, F.C., ed) pp. 277-326, Bidwell, R.G.S., ed. Academic Press, New York .
8. Caspar, T. and Pickard, B.G. (1989) Gravitropism by a starchless mutant of *Arabidopsis*: implications for the starch-statolith theory of gravity sensing. Planta 177, 185-197.
9. Curry, G.M., Thimann, K.V. (1961) Phototropism: The nature of the photoreceptor in higher and lower plants. *In*: Progress in photobiology, pp. 127-34, Christensen, B.C., Buchmann B., eds. Elsevier Pub. Co., Amsterdam. Netherlands.
10. Davies, P. J., Doro, J. A., Tarbox, A. W. (1976) The movement and physiological effect of indoleacetic acid following point applications to root tips of *Zea mays*. Physiol. Plant. 36, 333-337.
11. Dennison, D.S. (1979) Phototropism. *In*: Physiology of Movements, Encyclopedia of Plant Physiology, New Series. Vol. 7. pp. 506-566, Haupt, W., Feinleib, M.E., eds. Springer-Verlag, New York.
12. Dennison, D.S. (1984) Phototropism. *In*: Advanced Plant Physiology, pp. 149-162, Wilkins, M.B., ed. Pitman, Marshfield, Mass.
13 Erner, Y., Jaffe, M.J. (1982) Thigmomorphogenesis: the involvement of auxin and abscisic acid in growth retardation due to mechanical perturbation. Plant and Cell Physiol. 23, 935-941.
14. Fahrendorf, T., Beck, E. (1990) Cytosolic and cell-wall-bound acid invertases from leaves of *Urtica dioica* L.: a comparison. Planta 180, 237-244.
15. Falke, L., Edwards, K.L., Pickard, B.G., Misler, S. (1988) A stretch-activated anion channel in cultured tobacco cells. FEBS Lett. 237, 141-144
16. Firn, R.D., Digby, J. (1980) The establishment of tropic curvature in plants. Ann. Rev. Plant Physiol. 31, 31-48.
17. Gibeaut, D.M., Karuppiah, N., Chang, S-R., Brock, T.G., Vadlamudi, B., Kim, D., Ghosheh, N.S., Rayle, D.L., Carpita, N.C., Kaufman, P.B. (1990) Cell wall and enzyme changes during the graviresponse of leaf-sheath pulvinus of oat (*Avena sativa*). Plant Physiol. 94, 411-416.

18. Haberlandt, C. (1928) The statocytes of stems. *In*: Physiological Plant Anatomy, pp. 606-609. Macmillan, N.Y.
19. Harrison, M., Pickard, B.G. (1984) Burst of ethylene upon horizontal placement of tomato seedlings. Plant Physiol. 75, 1167-1169.
20. Hasenstein, K-H., Evans, M.L. (1986) Calcium dependence of rapid auxin action in maize roots. Plant Physiol. 81, 439-443.
21. Heathcote, D.G. (1981) The geotropic reaction and statolith movements following geostimulation of mung bean hypocotyls. Plant Cell Environ. 4, 131-140.
22. Hertel, R., DelaFuente, R.K., Leopold, A.C. (1969) Geotropism and the lateral transport of auxin in the corn mutant amylomaize. Planta 88, 204-214.
23. Iversen, T.H. (1969) Elimination of geotropic responsiveness in roots of cress (*Lepidium sativum*) by removal of statolith starch. Physiol. Plant. 22, 1251-1262.
24. Jacobs, M. (1983) The localization of auxin transport carriers using monoclonal antibodies. What's New In Plant Physiology 14, 17-20.
25. Jacobs, M., Gilbert, S.F. (1983) Basal localization of the Presumptive auxin transport carrier in pea stem cells. Science 220, 297-1300.
26. Jaffe, M.J. (1981) Thigmomorphogenesis and thigmonasty. *In*: McGraw-Hill Yearbook of Science and Technology, pp. 394-395. McGraw-Hill Book Co., New York.
27. Karuppiah, N., B. Vadlamudi, and P. B. Kaufman (1989) Purification and characterization of soluble (cytosolic) and bound (cell wall) isoforms of invertases in barley (*Hordeum vulgare*) elongating stem tissue. Plant Physiol. 91, 993-998.
28. Kaufman, P.B., Dayanandan, P. (1984) Hormonal regulation of the gravitropic response in grass shoots. *In*: Hormonal Regulation of Plant Growth and Development, Vol. 1, pp. 369-386, Purohit, S.S., ed. Agro Botanical Pub., Bikaner, India.
29. Kaufman, P.B., Ghosheh, N., Ikuma, H. (1968) Promotion of growth and invertase activity by gibberellin acid in developing *Avena* internodes. Plant Physiol. 42, 29-34.
30. Kaufman, P.B., Song, I., Pharis, R.P. (1985) Gravity perception and response mechanism in graviresponding cereal grass shoots. *In*: Hormonal Regulation of Plant Growth and Development, Vol. II. pp. 189-200, Purohit, S.S., ed. Agro Botanical Pub., Bikaner, India.
31. Kim, W.T., A. Silverstone, W.K. Yip, J.G. Dong, and S.F. Yang (1992) Induction of 1-aminocyclopropane-1-carboxylate synthase mRNA by auxin in mung bean hypocotyls and cultured apple shoots. Plant Physiol. 98, 465-471.
32. Kiss, J.R., Hertel, R., and Sack, F.D. (1989) Amyloplasts are necessary for full gravitropic sensitivity in roots of *Arabidopsis thaliana* Planta 177, 198-206.
33. Klann, E., Yille, S., Bennett, A. (1992) Tomato fruit acid invertase complementary DNA. Plant Physiol. 99, 351-353 (1992).
34. Lee, J.S., Evans, M.L. (1985) Polar transport of auxin across gravistimulated roots of maize and its enhancement by calcium. Plant Physiol. 77, 824-827.
35. Lee, J.S., Mulkey, T.J., Evans, M.L. (1983) Gravity-induced polar transport of calcium across root tips of maize. Plant Physiol. 73, 874-876.
36. Lee, J.S., Mulkey, T.J., Evans, M.L. (1983) Reversible loss of gravitropic sensitivity in maize roots after tip application of calcium chelators. Science 220, 1375-1376.
37. Li, Y., Hagen, G., Guilfoyle, T.J. (1991) An auxin-responsive promoter is differentially induced by auxin gradients during tropisms. The Plant Cell 3, 1167-1175.
38. Lu, C.R., Kim, D., Kaufman, P.B. (1992) Changes in the ultrastructure of cell walls, cellulose synthesis, and glucan synthase activity from gravistimulated pulvini of oat (*Avena sativa*). Intl. Jour. Plant Sci. 153, 164-170.
39. Maniatis, T., Fritsch, E.F., Sambrook, J. (1989) Molecular Cloning: A Laboratory Manual. Cold Spring Harbor Laboratory Press, Cold Spring Harbor, New York.
40. McClure, B.A., Guilfoyle, T. (1989) Rapid redistribution of auxin-regulated RNAs during gravitropism. Science 24, 91-93.

41. Moore, R., Evans, M.L. (1986) How roots perceive and respond to gravity. Amer. Jour. Bot. 73, 574587.
42. Mulkey, T.J., Kuzmanoff, K.M., Evans, M.L. (1982) Promotion of growth and hydrogen ion efflux by auxin in roots of maize pretreated with ethylene biosynthesis inhibitors. Plant Physiol. 70, 186-188.
43. Mulkey, T.J., Kuzmanoff, K.M., Evans, M.L. (1981) Correlation between proton-efflux patterns and growth patterns during geotropism and phototropism in maize and sunflower. Planta 152, 239-241.
44. Naqvi, S.M., Gordon, S.A. (1966) Auxin transport in *Zea mays* L. Coleoptiles. I. Influence of gravity on the transport of indoleacetic acid-2-^{14}C. Plant Physiol. 41, 1113-1118.
45. Ninnemann, H. (1980) Blue light photoreceptors. BioScience 30, 166-170.
46. Osborne, D.J., Wright, M. (1977) Gravity-induced cell elongation. Proc. Roy. Soc. Lond. B. Biol. Sci. 199, 551-564.
47. Pharis, R.P., Legge, R.L., Noma, M., Kaufman, P.B., Ghosheh, N.S., LaCroix, J.D., Heller, K. (1981) Changes in endogenous gibberellins and the metabolism of GA_4 after geostimulation in shoots of the oat plant *(Avena sativa)*. Plant Physiol. 67, 892-897.
48. Pickard, B.G. (1985) Early events in geotropism of seedling shoots. Ann. Rev. Plant Physiol. 36, 55-75.
49. Pickard, B.G. (1985) Roles of hormones in geotropism. *In*: Hormonal Regulation of Development III. Role of Environmental Factors. Encyclopedia of Plant Physiology, New Series, Vol. 11, pp. 93-281, Pharis, R.P., Reid, D.M., eds. Springer-Verlag, Berlin.
50. Pickard, B.G., Ding, J.P. (1992) Gravity sensing by higher plants. *In*: Advances in Comparative and Environmental Physiology, Vol. 10, pp 81 - 110, Ito, F., ed. Springer-Verlag, Berlin, Heidelberg.
51. Pickard, B.G., Thimann,K.V. (1966) Geotropic response of wheat coleoptiles in absence of amyloplast starch. J. Gen. Physiol. 49, 1065-1086.
52 Poovaiah, B.W., Reddy, A.S.N. (1993) Calcium and signal transduction in plants. Critical Rev. Plant Sci. 12, 185-211.
53. Ray, T. R., Jr. (1979) Slow motion world of plant behavior visible in rain forest. Smithsonian 9 (12), 121-130.
54. Rorabaugh, P.A., Salisbury, F.B. (1989) Gravitropism in higher plant shoots: VI. changing sensitivity to auxin in gravistimulated soybean hypocotyls. Plant Physiol. 91, 1329-1338.
55. Scott, T.K., Matthyse, A. (1984) Function of hormones at the whole plant level of organization. *In*: Hormonal Regulation of Development II. Encyclopedia of Plant Physiology, New Series, Vol. 10, pp. 217-243, Scott, T.K., ed. Springer-Verlag, New York.
56. Shen-Miller, J.,Cooper, P., Gordon, S.A. (1969) Phototropism and photoinhibition of basipetal transport of auxin in oat coleoptiles. Plant Physiol. 44, 491-496.
57. Slocum, R.D., Roux, S.J. (1983) Cellular and subcellular localization of calcium in gravistimulated oat coleoptiles and its possible significance in the establishment of tropic curvature. Planta 157, 481-492.
58. Song, I., Lu, C., Brock, T.G., Kaufman, P.B. (1988) Do starch statoliths act as the gravisensors in cereal grass pulvini? Plant Physiol. 86, 1155-1162.
59. Sturm, A., Chrispeels, M.J. (1990) cDNA cloning of carrot extracellular β-fructosidase and its expression in response to wounding and bacterial infection. The Plant Cell *2,* 1107- 1119.
60. Thimann, K.V. (1977) Hormone Action in the Whole Life of Plants. Univ. of Mass. Press, Amherst, Mass.
61. Trewavas, A.J. (editor). (1992) What remains for the Cholodny-Went theory? Plant, Cell Environ. 15, 759-794.

62. Wareing, P.F., Phillips, I.D.J. (1981) Growth and Differentiation in Plants. 3rd Ed. Pergamon Press, New York.
63. Wayne, R., Staves, M.P., and Leopold, A.C. (1992) The contribution of the extracellular matrix to gravisensing in charcean cells. J. Cell Sci. 101, 611-623.
64. Wilkins, M.B. (1984) Gravitropism. *In*: Advanced Plant Physiology, pp. 163-185, Wilkins, M.B., ed. Pitman Publishing, Inc., Marshfield, Mass.
65. Wilkins, M.B. (1966) Geotropism. Ann. Rev. Plant Physiol. 17, 379-408.
66. Wright, L.Z., Rayle, D.L. (1983) Evidence for a relationship between H^+ excretion and auxin in shoot gravitropism Plant Physiol. 72, 991-9944.
67 Wright, M. (1986) The acquisition of gravisensitivity during development of nodes of *Avena fatua*. J . Plant Growth Reg. 5, 37-47.
68. Wright, M. (1982) The polarity of movement of endogenously produced IAA in relation to a gravity perception mechanism. J. Exp. Bot. 33, 929-934.
69. Wright, M., Mousdale, D.M.A., Osborne, D.J. (1978) Evidence for a gravity-regulated level of endogenous auxin controlling cell elongation and ethylene production during geotropic bending in grass nodes. Biochem. Physiol. Pfl. 172, 581-596.
70. Wu, L.-L., Song, I., Karuppiah, N.B., Kaufman, P.B. (1993) Kinetic induction of oat pulvinus invertase mRNA by gravistimulation and partial cDNA cloning by polymerase chain reaction. Plant Mol. Biol. 21, 1175-1179.
71. Wu, L.-L., Song, I., Kim, D., Kaufman, P.B. (1993) Molecular basis of the increase in invertase activity elicited by gravistimulation of oat-shoot pulvini. J. Plant Physiol. 142, 179-183.

G6. Hormonal Regulation of Apical Dominance

Imre A. Tamas

Biology Department, Ithaca College, Ithaca, NY 14850-7278, USA.

INTRODUCTION

The growing shoot apex is known to regulate a wide range of developmental processes in plants including axillary bud growth, the orientation of laterals, the growth of rhizomes and stolons, leaf abscission, and others (10, 22, 24, 42). These effects are expressions of correlative control, through which the shoot apex exerts a central coordinating influence on plant development. Plant response is affected by environmental variables such as light, soil nutrients and various forms of stress. The correlative signal pathway may involve nutrients and other factors, but plant hormones have a preeminent role. In recent years, the understanding of correlative phenomena in plants has broadened, and thus the developing fruit is now regarded as a potential source of correlative effects which regulate growth in other fruits and in axillary buds.

MANIFESTATIONS OF DOMINANCE IN PLANT DEVELOPMENT

Correlative Regulation of Axillary Bud Growth

Control by the Vegetative Shoot Apex
    Among the correlative effects of the growing shoot apex, the inhibition of axillary bud growth has received most attention, and therefore has become most closely associated with the concept of apical dominance. Axillary buds of developing shoots are generally kept in a state of partial or total inhibition, or "quiescence". The primary source of the bud-inhibiting effect is the growing shoot apex, and the relative strength of the repressive signal is related to the vigor of the apex. Therefore, the axillary buds of rapidly growing shoots are often fully repressed, while those on less vigorous shoots may escape inhibition and develop into lateral branches. The degree of bud growth repression is under genetic control, which explains the predictable extent of branching exhibited by individual plant species.
    The quiescent axillary buds may be regarded as "replacement apices", many of which remain inhibited under unfavorable conditions such as nutrient poor soil, drought or shading. However, the buds may be induced to develop

P. J. Davies (ed.), Plant Hormones, 572–597.
© 1995 *Kluwer Academic Publishers. Printed in the Netherlands.*

into lateral branches in case the plant is exposed to favorable conditions, or the shoot apex is lost (10, 22, 42).

Control by Reproductive Structures

The correlative effect of the shoot apex over axillary buds is most highly expressed early in plant development. Consequently, much of our knowledge of this phenomenon is derived from work with plants in their juvenile state. As plants mature, the emergence of reproductive structures initiates a major rearrangement in the plant's overall morphology. This is likely to shift the correlative relationships among the various organs and could alter the dominant position of the shoot apex. Flower induction in *Perilla* demonstrates this point. After photoperiodic flower induction, an increased number of axillary shoots can be found in these plants (5). In oats, the emergence of the inflorescence releases the lateral buds from inhibition and allows them to develop into tillers (19). Strong correlation can be observed between flower induction and the release of axillary buds in the short-day plant *Chenopodium rubrum*. The photoperiodic conditions required to evoke flowering are similar to those which permit axillary bud outgrowth (30). Flowering, therefore, can decrease the level of dominance expressed by the shoot apex.

Reproductive development at a more advanced stage however, can severely restrict axillary bud growth. When axillary bud growth was investigated in adult *Phaseolus vulgaris* plants, the growth rate was found to be unaffected by the appearance of flowers (60). However, as the growing fruits were approaching full size, axillary bud growth was suddenly terminated. A rapidly developing cultivar, 'Redkloud', showed faster fruit growth and earlier bud inhibition compared to a slower cultivar 'Redkote'. In both cultivars, the cessation of bud growth coincided with the appearance of fully grown fruits. Complete fruit removal allowed the resumption of axillary bud growth in both cultivars and caused up to a two-fold increase in the combined total length of axillary shoots (60). It seems, therefore, that fruits of bean plants are able to exert "reproductive dominance" over the growth of axillary buds. In peas, defloration caused both the continuation of apical growth (which stops otherwise) and the emergence of axillary shoots (15, 34). The effect on lateral growth was most pronounced near the region from which the flowers were detached. It is interesting to note that decapitation, a common method of releasing axillary buds from dominance, had no effect in these plants. Therefore, dominance over bud growth was exerted not by the shoot apex but by the reproductive structures.

Dominance Among Developing Fruits

Interaction among developing fruits is affected by relative age and size. Generally, older fruits assume a position of dominance and repress the development of the younger fruits. Dominance phenomena in fruits have been described in many plants including beans (56, 63), apples (18), tomato (18), and soybeans (23). In *Phaseolus* the proximal fruits on a raceme develop earlier, and therefore grow larger, than the distal ones. In the presence of vigorous older fruits the great majority of the younger fruits fail to develop, and ultimately abort. However, if the older fruits are removed, the incidence of abortion among the younger ones is reduced suggesting that the older fruits are the source of the growth inhibiting signal (63). In aborting *Phaseolus* fruits a sequence of symptoms is noted; first the cessation of seed development, followed by the flattening of pods, loss of green color and abscission of the fruits (63). Cytological analysis of aborting bean fruits shows that they are arrested at the proembryo or the globular stage of embryogeny (48).

Clonal Plant Development

Morphology
Diagravitropic stems such as rhizomes and stolons grow on or below the surface of the soil and produce upright, orthogravitropic shoots from either their terminal or axillary buds. These two types of stems--one horizontal and the other vertical--represent the basic units from which clonal plants are built as modular structures through a reiterative process of development. The resulting system (the genet or clone) consists of a branching diagravitropic structure bearing adventitious roots at the nodes, and supporting a multiplicity of upright, leafy shoots (the ramets).

The growth form of clonal plants varies significantly depending on the length of rhizomes or stolons, the distribution of their axillary buds, the height of shoots, and the rate of axillary bud activation (6). Plants characterized by short internodes and a high rate of lateral bud activation develop in the form of a dense stand or "phalanx". In the so called "guerilla" form, a group of widely spaced ramets are produced by rhizomes or stolons in which the internodes are long and the activation of lateral buds is infrequent (6).

Stems of clonal plants may appear as transitional forms between horizontal and vertical stems. These may represent particular stages in the shoot's ontogeny, or arise in response to environmental effects. In certain species (e.g., iris) the tip of the growing rhizome can be seen at a certain time to turn upright, rise above ground and develop into a leafy shoot. In the aquatic macrophyte *Phragmites communis* a developing underground rhizome may change direction, emerge at a forty-five degree angle and continue to develop as an above-ground stolon. These stolons have been observed to appear generally at the edge of dense stands, traverse rocky or otherwise inhospitable areas, and produce stands of shoots at favorable locations from their axillary buds (Bernard and Tamas, unpublished). Some of the axillary buds may develop as above-ground stolons, or dive below the soil surface and grow as rhizomes.

Shoots of some species regularly change orientation well after their formation has been completed. The current year's shoots of the small woodland sedge, *Carex projecta*, have been noted to fall over in the autumn, overwinter, and then produce adventitious roots at the nodes and axillary shoots from lateral buds (Tamas and Bernard, unpublished). Thus, the heretofore vertically oriented shoots can become organs of vegetative reproduction and facilitate the lateral distribution of the new members of the clone.

The foregoing responses make it clear that the apical and axillary meristems of diagravitropic stems possess an unusual degree of developmental plasticity. Multiple and overlapping regulatory protocols may be operating to coordinate the linear growth of apical buds, or the release of axillary buds, with other responses such as the onset of a new differentiation sequence, or the changing response of the apices to gravity. The evident flexibility of this developmental system contributes much to the ability of the clonal plant to adjust its strategy of growth to its environment, and extract necessary resources efficiently. This point is briefly illustrated below.

Growth Strategy

Foraging. Some clonal plants can exploit resource-rich areas by producing a network of rhizomes or stolons from which a stand of leafy shoots is established. When adverse conditions are encountered, branching in rhizomes and stolons is repressed and their linear growth is increased (13, 24).

Consolidation. Some clonal species in high-resource areas tend to establish dense stands in virtual monoculture. The "consolidation strategy" of these plants requires, first, the colonization of new areas, and second, the consolidation of occupied space. In certain species, rhizomes take different forms according to their developmental role. The longer rhizomes of *Brachypodium pinnatum* undergo intensive branching and thus foster colonization; the shorter ones develop fewer diagravitropic branches but produce--from their numerous buds--groups of closely placed ramets (13).

The Conservative Strategy. Plants adapted to conditions of severe deprivation typically show a slow rate of growth; reduced branching; and the production of short, thin rhizomes and stolons (13).

Gametophore Development in Mosses

The gametophyte generation of *Funaria hygrometrica, Physcomitrella patens*, and other moss species has been used extensively for work on the hormonal and genetic control of development (see Chapter G1 and 8, 11). Mosses provide an excellent system for the study of developmental regulation due to the relative simplicity of the moss structure, well-defined cellular responses to phytohormones, and the ease of identification of developmental mutants in the haploid gametophyte.

Questions of central interest concerning moss gametophyte development have been the regulation of growth in the protonemal filaments, and the initiation of gametophore buds from which the leafy gametophores, or shoots, emerge. The developing protonema produces two morphologically distinct types of filaments. Spore germination first gives rise to chloronema, and after subsequent development, caulonemal filaments appear. While both types of filaments grow by the repeated division of the apical cell, and both develop side branches from the division of subapical cells, only in the caulonemal filaments can subapical cell division produce gametophore buds (11).

The apical cell appears to regulate both the development of the protonemal filament and the initiation of the gametophore bud. The presence of an active apical cell is required for the conversion of chloronema to caulonema, and the preservation of the caulonemal structure. Both are prerequisites for the development of gametophore buds. If the apical cell of a caulonemal filament is detached or injured, the filament reverts to the chloronemal stage (7).

In the developing gametophore of mosses, as in the shoot of seed plants, the growing apex appears to hold a position of dominance over the lateral buds. The gametophore of *Plagiomnium cuspidatum* develops as a leafy shoot with a growing apex but without lateral branches. Close to the axil of each leaf an inhibited lateral bud can be found in the form of a small mound of tissue (40). The removal of the gametophore apex activates mitosis in many of these lateral buds, but in time the uppermost bud imposes dominance over the others and develops into a lateral shoot.

Control of Abscission by the Shoot Apex

Among the numerous correlative effects of the growing shoot apex, the acceleration of leaf abscission has received relatively little attention. Observations on *Phaseolus* (25) and *Gossypium* (36) have revealed that the abscission of petioles induced by deblading is enhanced in the presence of the intact shoot apex. Decapitation of the shoot, on the other hand, causes a delay in the abscission of petioles.

HORMONES IN CORRELATIVE INTERACTIONS

Auxin

The discovery and early characterization of auxin were closely linked to the idea that growing tips of shoots and coleoptiles produce auxin, which may be released to regulate the development of other structures elsewhere in the plant. In their pioneering work on apical dominance, Thimann and Skoog (64) demonstrated that axillary buds were under the correlative control of the growing shoot apex. They found that decapitation of *Vicia* plants caused the outgrowth of the axillary buds, but the treatment of the cut surface with auxin prevented bud growth. These results have since been confirmed in numerous plant species (22). The active substance responsible for the inhibition of axillary buds has been isolated from *Phaseolus* shoot tips (particularly from the young, developing leaves) and identified as indoleacetic acid (IAA) by gas chromatography-mass spectrometry (22).

The expression of dominance by the shoot apex requires basipetal IAA transport in the subapical part of the stem. The application of the IAA transport inhibitor triiodobenzoic acid (TIBA) in lanolin below the shoot apex releases the axillary buds from inhibition (65). Moreover, a comparison of the branching tomato line 'Craigella' (which has weak apical dominance) with the isogenic non-branching line 'Blind' reveals that among the two, only Blind is able to export radiolabeled IAA from the shoot apex (9). This suggests that the branching character is due to the failure of the shoot apex to export IAA. Therefore, the release of IAA from the shoot apex and its subsequent basipetal transport appear to be essential steps in the process of axillary bud growth inhibition.

A similar requirement may exist for IAA participation in reproductive dominance, a condition in which axillary bud growth and the development of younger subordinate fruits are repressed by older dominant fruits. Experiments with *Phaseolus* plants show that the source of the inhibiting signal is the seed of dominant fruits. When older fruits are deseeded, the axillary buds resume growth (57), and the development of younger fruits is released from inhibition (58). However, the replacement of the seeds in the deseeded pods with IAA, or with the synthetic auxin naphthaleneacetic acid (NAA), restores the growth inhibiting effect of the fruits on the axillary buds (57) and the younger subordinate fruits (58). Thus, the dominant effect of fruits can be duplicated by auxin application, suggesting that the correlative signal may be IAA.

High levels of IAA have been found in the developing fruits and seeds of several plant species (3). The ability of fruits to export IAA to neighboring organs was tested in an experiment in which deseeded fruits of *Phaseolus* or *Glycine* were injected with ^{14}C-IAA. Subsequent analysis recovered radiolabeled material from neighboring leaves, fruits and axillary buds (56, 58). Larger fruits exported more applied ^{14}C-IAA than smaller fruits reflecting their relatively greater ability to exert dominance over other organs. The role of a correlative signal for fruit-derived IAA has not been confirmed, but can be inferred from the results of several experiments. When the pedicels of individual fruits on *Phaseolus* plants are treated with the IAA transport inhibitor naphthylphthalamic acid (NPA), the resting axillary buds near the treated fruits resume growth, as do the buds on defruited plants (56). The buds on intact plants remain inhibited. Resumption of bud growth is not the result of decreased competition for nutrients between fruits and axillary buds because on NPA-treated plants fruit development is not inhibited (in fact, fruit growth is substantially enhanced). In *Glycine*, free IAA as well as

IAA-esters have been identified in phloem exudate (20). After the plants are depodded, the IAA-ester level in the exudate decreases to about one-fifth of its previous level, suggesting that the IAA is fruit-derived. Bangerth and his collaborators have shown that the export of endogenous IAA from dominant tomato, apple and other fruits is greater than from the younger inhibited fruits, and that the removal of dominant fruits enhances IAA export from the remaining fruits (18). These results are interpreted to mean that the fruit-derived IAA is involved in the correlative signal that regulates dominance relationships among fruits, and further that the older fruits achieve dominance by repressing IAA export from the other fruits (18).

According to a recent report the seeds of pea plants export 4-Cl-IAA, which regulates gibberellin metabolism in the pod to stimulate pod growth (41). Therefore, auxins other than IAA may also serve as the correlative signal released from seeds.

Cytokinins

Exogenous cytokinin treatment of axillary buds stimulates bud outgrowth in many plant species including apple, *Cuscuta, Macadamia*, oats, peas and soybeans (10). It has been shown also that axillary bud growth is generally well correlated with the level of naturally occurring cytokinins in the buds. Cytochemical localization of cytokinins using immunogold in the apical and axillary buds reveals much less cytokinin in the sideshootless tomato mutant, Craigella Lateral Suppressor (Cls), than in its isogenic parental line Craigella (51). Therefore, axillary bud growth in Cls plants seems to be impaired because of diminished cytokinin biosynthesis or importation into the buds.

Zeatin appears to be the most active among the naturally occurring cytokinins in promoting axillary bud growth. Analysis of extracts from intact *Marsilea drummondii* stolons using high performance liquid chromatography (HPLC) separation and enzyme linked immunosorbent assay (ELISA) shows that the combined level of zeatin and zeatin riboside is highest in the growing apical bud, and the levels in the axillary buds decline with increasing distance of the bud from the apex (43). After decapitation, the youngest axillary bud is the first to resume growth followed by the others in a basipetal sequence. Therefore, the activity of zeatin and its metabolites in the axillary bud is correlated with the bud's growth potential.

Perhaps the dominant organs repress growth elsewhere in the plant by acting as an auxin-activated sink for root-derived cytokinins, and thus preventing cytokinin transport to subordinate structures. In etiolated pea seedlings, decapitation and removal of one cotyledon confers dominance of the cotyledonless bud over the other. The application of the synthetic cytokinin ^{14}C-benzyladenine (^{14}C-BA) to the roots of these plants results in substantially greater accumulation of ^{14}C in the dominant bud compared to the inhibited one (44). When ^{14}C-BA is supplied to the base of decapitated

Solanum andigena cuttings, the treatment of the apical stump with IAA reduces the amount of ^{14}C transported to the axillary buds (67).

Recent experiments by Bangerth suggest that the levels and distribution of naturally occurring cytokinins are under the control of the shoot apex. The results suggest further that this effect of the shoot apex is mediated by auxin. When *Phaseolus vulgaris* and *Pisum sativum* plants are decapitated, the concentration of two predominant cytokinins, zeatin riboside and isopentenyl adenosine, increases dramatically in both species as shown by HPLC and radioimmunoassay (RIA) of the xylem exudate of *Phaseolus* and the stem extracts of *Pisum*. The application of NAA in lanolin to the stump abolishes the effect of decapitation on cytokinin levels (Bangerth, personal communication).

Therefore, either a redirection of the cytokinin supply from the roots may be necessary to initiate axillary bud growth, or the decline in auxin supply from the apex enables the axillary buds to commence cytokinin synthesis. Although added cytokinin in rootless *Solanum* cuttings was able to stimulate axillary bud growth (67), it was subsequently shown that shoots are competent to produce their own cytokinin (66). However, the repressed lateral buds do not do so. Experiments with a purine synthesis inhibitor, hadacidin, suggest that synthesis of endogenous cytokinins within the buds themselves is required for bud growth to occur. Following decapitation, hadacidin treated axillary buds remain inhibited, and the inhibitory effect can be reversed by cytokinin treatment (32).

Work with *Pisum* indicates that the dominant apex regulates not only the transport but the synthesis of active cytokinins as well (27). The conversion of the inactive $N^6(\Delta^2$-isopentenyl) adenine (iP) to the active form, zeatin, in maturing plants is most pronounced at the apex of the stem and declines toward the base, and its rate is directly correlated with the growth potential of the axillary buds. Moreover, zeatin application to the axillary buds enhances their growth in intact plants, but similarly applied iP is effective only when the plants are decapitated. Therefore, decapitation of the shoot seems to enable axillary buds to convert an inactive precursor to an active cytokinin such as zeatin, possibly by altering the concentration of apically derived IAA in or near the buds.

Gibberellins

There is a general lack of information on endogenous gibberellins in growing and repressed axillary buds, and therefore a role of these substances in axillary bud development remains hypothetical. Studies with exogenously applied substances show that gibberellin treatment alone does not release axillary buds from apical dominance, but it can cause rapid elongation in released buds (10).

In peas, gibberellic acid (GA_3) enhances the inhibitory effect of IAA when the two are applied together at the shoot apex. The effect is thought to result from the more efficient release of IAA by the dominant apex because GA₃ stimulates the basipetal transport of radiolabeled IAA in these plants (26).

Abscisic Acid

According to several reports, the inhibition of axillary bud growth is closely correlated with the abscisic acid (ABA) content of the buds. An analysis using gas chromatography-mass spectrometry-selected ion monitoring (GC-MS-SIM) has shown that decapitation of *Phaseolus vulgaris* plants causes both an increase of IAA, and a decrease of ABA levels, in the axillary buds (17). ABA may conceivably act as an inhibitor in the buds because the decline in the ABA level precedes the onset of bud outgrowth by several hours. In the nonbranching Blind and Cls mutants of the tomato, the ABA content of the axillary buds is much higher than in the branching parental line Craigella (65), suggesting a role of ABA in bud growth regulation.

In *Phaseolus*, the presence of older fruits increases the level of endogenous ABA, and inhibits growth, in the axillary buds as well as the subordinate younger fruits (60, 63). The removal or deseeding of the dominant older fruits causes a drop in the ABA level of axillary buds and younger fruits, and allows the resumption of their development (57, 60, 63).

In defruited plants, axillary bud growth inhibition can be reimposed by either direct ABA treatment of the buds or the application of auxin to the pod cavity of the deseeded dominant fruits. However, ABA application to the deseeded pods does not inhibit bud growth suggesting that the bud-suppressing effect of fruits is not caused by ABA release from the fruits. Rather, the ABA that accumulates in inhibited buds probably originates from another source such as the leaves, or is synthesized within the buds under the control of the dominant fruits.

The data indicate, therefore, that ABA is not the correlative signal released by dominant organs such as growing fruits or shoot apices. The evidence suggests instead that ABA acts within the axillary buds--or other subordinate structures--as a second factor whose level can be regulated by the dominant organ. In *Phaseolus,* decapitation lowers the ABA level in the axillary buds, but this is prevented by the application of IAA to the cut stump (29). These results suggest that IAA, moving down from the shoot apex, can maintain a high ABA level in the axillary buds.

Taken together, the foregoing data provide strong evidence that in most of the species under study the dominant structures elicit high endogenous ABA levels in quiescent axillary buds and other subordinate structures, and furthermore that the release of the buds from dominance is correlated with a decrease of ABA concentration in the buds. However, these effects could not be confirmed in some studies (43), and in certain experiments on decapitation the outgrowth of axillary buds preceded--rather than followed--the decline in bud ABA levels (29).

Relatively little is known about the synthesis, transport and cellular effects of ABA in relation to the growth of axillary buds. Progress in these basic questions of ABA physiology must occur before the role of ABA in apical dominance can be adequately assessed.

Ethylene

It has been known for some time that high concentrations of IAA enhance the synthesis of ethylene. A possible consequence of this is that a localized ethylene buildup, triggered by an IAA signal, may cause growth inhibition in specific structures. Indeed, IAA treatment of etiolated pea stem segments was shown to increase the rate of ethylene synthesis in nodal tissue and inhibit axillary bud growth (10). In decapitated *Vicia faba* plants the treatment of the cut surface with IAA repressed axillary bud growth, but the simultaneous application of the ethylene synthesis inhibitor aminoethoxyvinylglycine (AVG) prevented inhibition (46).

Contrary to the foregoing results, several reports have provided evidence that ethylene does not take part in the repression of axillary bud growth. In *Phaseolus vulgaris*, the application of IAA to the cut surface of decapitated plants failed to increase ethylene emanation from the node below (22). Moreover, AVG treatment of axillary buds inhibited rather than stimulated bud outgrowth in decapitated *Phaseolus* plants (22). The introduction of the auxin-overproducing tryptophan monooxygenase transgene into *Arabidopsis* plants resulted in elevated levels of both auxin and ethylene, as well as increased repression of axillary bud growth (45). However, in a cross between these plants and those expressing an ethylene synthesis-inhibiting transgene, strong apical dominance was maintained even though only auxin-- but not ethylene--was overproduced. Similarly, transgenic tobacco plants, in which ethylene production is inhibited through the introduction of the gene for ACC deaminase, looked exactly the same as plants producing ethylene normally (see Chapter E2).

In summary, while some reports have indicated the involvement of ethylene in apical dominance, the evidence (10, 22) shows a general lack of correlation between axillary bud growth inhibition and ethylene levels in bud tissue.

Unrelated to the question of ethylene participation in apical dominance, ethylene appears to stimulate axillary bud growth under conditions of gravity stress. When shoots of *Ipomea nil* are inverted it induces the release of the highest lateral bud adjacent to the bend, and this response is preceded by a sharp increase in ethylene production and the retardation of shoot elongation (10). Treatment of the inverted area with the ethylene antagonist silver nitrate reduces the effect of inversion on shoot inhibition and lateral bud release. Shoot inversion-induced ethylene synthesis and the consequent inhibition of shoot growth also occurs in a number of species including corn, tomato, pea and sunflower (10). The effect of ethylene in releasing axillary buds from inhibition appears to be indirect, either through the repression of the shoot apex (and consequent reduction of its dominance), or the retardation of auxin transport toward the axillary buds. The fact that shoot inversion causes axillary bud outgrowth offers an alternative method to decapitation in experiments on apical dominance, and makes it possible to subject the shoot apex to analysis any time after bud outgrowth has taken place (10).

HORMONES AND CLONAL PLANT DEVELOPMENT

The problem of growth regulation in clonal plants has received relatively little attention, and knowledge is fragmentary on hormonal relations in these plants. The few available studies indicate, however, that general notions about the role of auxins and cytokinins in apical dominance may be applicable to rhizomatous and stoloniferous structures. In the stolons of *Marsilea* for example, the IAA and zeatin levels of apical and axillary buds are closely correlated with the bud's growth potential (43).

Certain morphogenetic effects of the dominant apex are uniquely related to the developmental strategy of clonal plants. In the ancestral potato, *Solanum andigena*, lateral buds develop into leafless, horizontal stolons in the presence of the dominant shoot apex. Decapitation of the main shoot transforms the stolons into orthogravitropic leafy shoots, but only in the presence of roots or applied cytokinins. Stolon production seems to depend on high gibberellin levels, and the effect of the dominant apex can be duplicated by auxin application (22, 42, 67). GA_3 can affect other aspects of clonal plant development such as the stimulation of rhizome branching observed in water hyacinth (10).

Because clonal plants are composed of both horizontal and vertical stems, their development may be particularly influenced by gravimorphic effects (gravity-induced developmental effects). Placing shoots in an horizontal orientation generally releases the axillary buds from inhibition presumably through the restriction of apical growth due to gravity stress-induced ethylene (10). It is not known, however, whether diagravitropic and orthogravitropic stems differ in their sensitivity to, or production of, ethylene. Gravistimulation may also elicit morphogenetic changes in the development of lateral organs. The tying down in a horizontal position of the erect shoots of *Cordyline terminalis* causes the development of lateral buds on the upper side into leafy shoots, and those on the lower side into rhizomes (10). Presumably, this response is caused by the redistribution of auxin between the two sides of the stem.

The great developmental plasticity of many clonal plants depends on the plant's ability to coordinate axillary bud growth and related responses with environmental effects. In their recent review, Hutchings and de Kroon (24) point out that there is remarkably little information on the relationship between environmental variables and hormone activity in plants, and our understanding of environmental control of plagiogravitropic shoot growth is especially poor. The available studies, however, suggest that environmental effects on development are mediated through hormones. The fact that elevated levels of soil nitrogen stimulate cytokinin synthesis, and thereby enhance tiller formation, may be a factor in the consolidation of growth by some clonal plants at nitrogen-rich locations. Shaded plants experience decreased red/far-red ratio in the incident light which increases plant sensitivity to gibberellins, increases internode length, inhibits axillary bud growth and thereby stimulates escape from shaded areas (24). Much of this remains speculative until the mechanism of hormone action in these responses is clarified.

CELLULAR AND GENETIC CONTROL

Cellular Events in Developing Axillary Buds

The development of the axillary bud originates in the growing shoot apex. A small group of cells in the axil of the leaf primordium is detached from the apical meristem and becomes organized into the apex of the axillary bud (16). Continuing growth produces a visible bud. Events in the developing bud are under the control of the shoot apex. In the axillary buds of intact *Pisum* plants mitosis is arrested, but it becomes activated within hours after the shoot is decapitated (54). Some of the genes, known to be involved in the regulation of the cell cycle of many eukaryotes, have been identified recently in *Pisum* (54). The substance that controls the transition from G_1 to the S-phase (and thus initiates mitosis) is the maturation promoting factor (MPF), which consists of a protein kinase ($p34^{cdc2}$ kinase) and cyclin B. The conditions necessary for the initiation of mitosis are high levels of cyclin and the dephosphorylation of the kinase. The *cdc2* gene that encodes the $p34^{cdc2}$ kinase has been isolated recently from peas (54). Another gene, which encodes a plant homologue of the animal mitogen-activated protein kinase (MAP kinase), was identified in the axillary buds of pea plants decapitated 24 hours before analysis (54). MAP kinase is believed to be involved in signal transmission between the cell surface and other parts of the cell. MAP kinase is inactive in nongrowing (G_0) human cells and *Xenopus* oocytes (held at the G_2/M boundary). Many aspects of *Pisum* MAP kinase activity are unknown including the timing of its synthesis, its mode of regulation and the identity of its substrate.

The effect of specific phytohormones on mitosis may relate to their role in apical dominance. In decapitated pea plants the treatment of axillary buds with ABA prevents bud outgrowth, decreases the mitotic index of the meristem and subapical regions of the buds, and delays the entry of the G_{0-1} nuclei of the inhibited buds into the S phase and then into mitosis (39). It is to be noted that both the intact shoot apex and ABA treatment of the axillary buds inhibit the cell cycle in axillary buds at the same stage.

In *Phaseolus vulgaris*, a species with incomplete apical dominance, axillary bud growth proceeds at a slow rate even in the presence of an intact shoot apex (22). Both mitosis and cell expansion can be observed in these buds. In plants with complete apical dominance, such as *Tradescantia*, mitotic activity is inhibited and growth is arrested at an early stage of bud development (38).

If the shoot apex is detached, mitotic activity and growth resume in the lateral buds after some delay. Variation in the length of the lag period may depend on the stage of the cell division cycle at which the inhibited cells are held. In the axillary buds of *Cicer*, mitosis can be observed within an hour after decapitation, followed by the resumption of bud growth and DNA synthesis in that order (22). Presumably the cells have already completed the duplication of DNA at the time of release (i.e., they are at the G_2 stage) and thus are able to undergo mitosis with little delay. On the other hand, *Tradescantia* requires over four days for the resumption of bud growth following shoot decapitation (38). The cells of inhibited *Tradescantia* buds are held at the G_1 stage, and therefore DNA needs to be synthesized before mitosis can occur.

The length of the lag period preceding bud growth may also depend on the degree of inhibition expressed by the plant. The buds of *Phaseolus*, which are incompletely suppressed in the intact plant, can show increased internodal expansion four hours after decapitation of the main shoot (22). This is in contrast to the much longer time required by *Tradescantia*, a species in which axillary bud growth is totally suppressed during the growth of the shoot.

The degree of bud inhibition may also influence the sequence of early events after the removal of the dominant shot apex. In *Tradescantia*, bud outgrowth is preceded by the resumption of mitotic activity in the bud apex (38). In the less suppressed buds of *Phaseolus*, however, the initial growth results entirely from the enlargement of internodal cells of the bud (22). Cell division begins about a day after the onset of growth.

Analysis of axillary buds by Nougarede et al. (39) has shown that changes in nucleolar composition are among the early cellular events caused by decapitation. The cells of the inhibited cotyledonary buds of *Pisum sativum* are held at the G_{0-1} stage (defined as the prolonged G_1 stage) of the cell cycle. Following decapitation the cell cycle is reactivated, and the cells proceed through the S stage, the G_2 stage and mitosis after 6, 12 and 24 hours respectively. In the G_{0-1} nucleolus a dense fibrillar component surrounds the much lighter fibrillar center. Both contain DNA. Twenty-four hours after decapitation the nucleolus becomes enlarged and much granular material appears at its periphery. Staining with an RNase-colloidal gold complex indicates significant accumulation of RNA in the granular component (39).

Gene Expression In Axillary Buds

To characterize the biochemical events associated with axillary bud growth, Stafstrom and Sussex investigated the protein composition of *Pisum sativum* buds at different stages of development (53, 55). They selected four categories of buds for study: dormant (nongrowing) buds on intact plants, growing buds from plants decapitated 24 hours earlier, and those from two transitional stages. Their analysis by two-dimensional polyacrylamide gel electrophoresis (PAGE) reveals that each category of buds synthesizes a unique set of proteins. The pattern of gene expression in the dormant-to-growing transition buds three hours after decapitation (and well before the onset of visible growth at eight hours) is already more similar to that in growing buds than dormant buds, indicating that the changes may be among the early events leading to bud outgrowth.

There is strong similarity also in the pattern of expression between the dormant buds and the ones undergoing transition from the growing to the dormant state. This transition is observed in small, subordinate buds that start growing after the shoot is decapitated, but become inhibited two to three days later under the influence of fast growing larger buds. The excision of the larger bud after five days enables the small bud to resume growth (53, 55). Therefore, individual buds can undergo cycles of growth and dormancy due to changing patterns of dominance relations among neighboring organs.

The pattern of gene expression for the two transition stages differ, and each contain gene products not present at any other stage, suggesting that the molecular events during exit from dormancy are not the same as those that occur during entry into dormancy (53). The results also show that dormant buds incorporate labeled amino acids at the same rate as growing buds. Therefore, the maintenance of bud inhibition appears to be an active process that may involve the activation of dormancy-specific genes or the repression of growth-specific genes (53).

Auxin treatment of the cut surface of decapitated plants stimulates the formation of dormancy-specific proteins in the axillary buds, whereas the application of kinetin to the axillary buds of intact plants results in the synthesis of growth-specific proteins (53). These data support the view that auxins and cytokinins control the development of axillary buds by the activation of genes involved in dormancy and growth.

Some of the genes expressed in *Pisum* axillary buds have proven to be useful developmental or hormone-specific markers (53, 55). RNA gel blot analysis shows that the expression of the ribosomal protein clone pGB8 is very low in the dormant buds of intact plants, but increases sharply within a few hours after decapitation. In the small buds, located at nodes together with larger buds, the increase in gene expression is temporary. A few days after decapitation of the shoot, gene activity declines due to the inhibiting effect of fast growing larger buds.

The treatment of bud-bearing, cultured stem sections with 10 μM IAA (ensuring that the bud itself is not exposed) totally inhibits the expression of pGB8 in the axillary buds (53, 55). Therefore, physiological concentrations of IAA applied to the stem can inhibit both the outgrowth of axillary buds and the activity of growth specific markers in the buds. In contrast, growing axillary buds strongly express the auxin-responsive genes pIAA4/5 and pIAA6 (53) suggesting that IAA concentration increases in growing axillary buds.

Transgenic Plants with Altered Hormone Content or Sensitivity

Plants Transformed by Agrobacterium

Agrobacterium tumefaciens and *A. rhizogenes* are soil-borne plant pathogens that incite crown gall and hairy-root disease respectively in dicotyledonous plants. The cells of the host plant are transformed by the T-DNA fragment of the infectious bacterial plasmid, an event that alters drastically the subsequent developmental events in the host. In *A. tumefaciens* the T-DNA of the Ti plasmid carries, among others, genes for IAA (*iaa*M, *iaa*H) and cytokinin (*ipt*) biosynthesis (50) (see Chapter E1). Tumor-inducing neoplastic growth is stimulated by the incorporation into the host cells of the cytokinin and auxin biosynthetic genes which cause elevated levels of IAA and cytokinin. Among the Ri (root inducing) plasmids in *A. rhizogenes,* certain plasmids harbor two discrete T-regions designated TR-DNA and TL-DNA for the right and left regions respectively. TL-DNA is primarily responsible for Ri plasmid-induced pathogenesis due to the presence of the genes *rolA, rolB* and *rolC*, collectively designated as the root loci *(rol)* genes *(rolABC)*. TR-DNA includes the *iaa*M and *iaa*H genes encoding for IAA synthesis (52), and it plays an ancillary role in tumor induction.

Experiments with hormone overproducing plants have confirmed the long-held view that the development of lateral buds is enhanced by cytokinins and inhibited by auxins. The transfer of the isopentenyl transferase (*ipt*) gene from *A. tumefaciens* to tobacco plants causes a sharp increase in the endogenous cytokinin level, and stimulates axillary bud growth (28). On the other hand, auxin overproducing plants carrying the *iaa*M gene from *A. tumefaciens* show almost total suppression of axillary bud growth. A cross of auxin overproducing plants with cytokinin overproducers results in a decreased expression of apical dominance, suggesting that apical dominance is regulated by the ratio of endogenous auxins and cytokinins (28).

The Rol Genes of A. rhizogenes

Plants regenerated from hairy roots exhibit particular abnormalities attributed to *rol* genes. The symptoms include wrinkled leaves, shortened internodes, reduced apical dominance, and abundant, plagiotropic roots with many laterals (52). Spano et al. (52) have demonstrated that rhizogenesis in transformed plants is not due to elevated auxin levels but to the increased auxin sensitivity of the tissue. Enhanced auxin sensitivity has been expressed also at the cellular level by hairy root protoplasts. Maurel et al. (35) measured the effect of NAA on the transmembrane potential difference, and observed that to induce hyperpolarization in normal protoplasts requires a thousandfold higher NAA concentration than in the transformed ones (35). Unlike the transmembrane potential in protoplasts, cell division is not affected by transformation with regard to its auxin sensitivity. Therefore, different cellular responses to hormone action may be affected selectively by the presence of *rol* genes.

Protoplasts from plants transformed by each of the three *rol* genes individually also display enhanced auxin sensitivity. The effect of NAA on the transmembrane potential difference is greatest for *rolB*, moderate for *rolA*, and lowest for *rolC* (35). The same ranking applies to the root inducing ability of these genes. Therefore, enhanced auxin sensitivity may be a necessary condition for root induction in transformed plants (35).

In contrast to their parallel effect on transmembrane potential, the three *rol* genes have been shown to cause distinct morphological and biochemical changes in transformed plants. Transgenic plants that carry the *rolA* gene have wrinkled leaves and large flowers; the most prominent feature of *rolB* plants is their increased tendency to form adventitious roots; and the plants transgenic for *rolC* are characterized by smaller flowers and decreased expression of apical dominance (49). According to evidence presented by Estruch et al. (14), the *rolB* gene codes for a glucosidase able to hydrolyze indole β-glucosides. The authors suggested that the expression of the *rolB* gene causes the release of active auxins from the inactive β-glucosides (14). It was also indicated that the *rolC* β-glucosidase releases active cytokinins from glucoside conjugates. Therefore, the decrease in the expression of apical dominance observed in *rolC* plants could be accounted for by the higher level of endogenous cytokinins.

Taken together, the foregoing evidence indicates that the *rol* genes harbored by the T-DNA of *A. rhizogenes* may influence plant development by a combination of two effects namely the increased endogenous level of auxins and cytokinins, and the enhanced sensitivity of cells to these phytohormones. Because both auxins and cytokinins are involved in the control of axillary bud growth, plants transformed by *A. rhizogenes* may provide a valuable model system for the study of apical dominance.

Auxin-Regulated Gene Expression Monitored with *Lux* Reporter Genes

The level of IAA in various tissues during development has been studied by Langridge et al. (31) in transgenic tobacco plants using bacterial luciferase *luxA* and *luxB (luxA&B)* as reporter genes. In these plants, the *luxA&B* genes are under the control of the auxin-responsive mannopine synthase *(mas)* dual (P_1, P_2) promoters from *A. tumefaciens*. Therefore, IAA-regulated *mas* promoter activity can be monitored by measuring the amount of light emitted from the luciferase-catalyzed reaction. *Lux* expression is stimulated by applied IAA and other auxins. Video image analysis of luciferase activity reveals enhanced *lux* gene expression in and around the axillary buds within 12 hours after decapitation, suggesting that decapitation causes an increase in the auxin content of the buds (31).

Tamas et al. (59) have investigated the effect of the *rolABC* genes on *lux* expression in the axillary buds of tobacco plants. *RolABC* x *luxA&B* (a genetic cross between *rolABC* and *luxA&B* plants) and *luxA&B* plants were decapitated and luciferase activity was monitored using luminometry in the uppermost axillary bud. While no change occurred in the buds of the *luxA&B* strain in the first 24 hours following decapitation (although activity increased rapidly thereafter), luciferase activity doubled in *rolABC* x *luxA&B* buds in six hours and increased nearly fourfold in 24 hours. These changes in luciferase activity were correlated with the growth response of the two tobacco strains. The buds of the intact *luxA&B* plants were fully suppressed, and started to grow two days after decapitation. In contrast *rolABC* buds, which exhibited a slow rate of growth even in intact plants, began to grow rapidly within a day after decapitation, and grew to twice the size of the buds on the decapitated control plants within a week (59).

These results show, first, that the enhanced expression of *luxA&B* genes in the *rolABC* plants corresponds to the diminished level of apical dominance in these plants; and second, that the increase in *lux* expression within the buds of decapitated plants precedes the onset of bud growth, and thus could indicate important hormonal changes involved in the release of apical dominance. A particularly valuable feature of this system is its high sensitivity. A few milligrams of fresh tissue weight is sufficient for several replicate luciferase assays. Therefore, the *mas* promoter-regulated *lux* reporter gene system provides a simple and sensitive technique to monitor the auxin status of axillary buds during their release from apical dominance.

TISSUE POLARITY AND THE MECHANISM OF CONTROL

Polarity and Auxin Action

Thimann and Skoog (64) observed that the quiescent axillary buds of *Vicia* contain little auxin, but the level increases substantially when bud growth is resumed (this has been confirmed by accurate physicochemical methods [22]). Thus it was recognized that bud growth is directly correlated with auxin production within the bud, even though auxin from the shoot apex represses axillary bud growth. To explain this apparent paradox, it was suggested that the conversion of a precursor to auxin in the axillary bud is inhibited by auxin arriving from the shoot apex (64). Because the latter needs to pass through tissues of the bud without enhancing growth therein, the implication of the suggestion is that the effect of auxin in a tissue may depend on the direction of its movement. Whether this is true in the case of axillary bud growth regulation has not been established, but the potential significance of the idea is illustrated by studies on leaf abscission and xylem differentiation in lateral buds. Each of these processes involves basipetal auxin transport within the respective lateral organ (leaf or bud), and the physiological effect of this auxin is inhibited by the auxin signal from the shoot apex.

Sachs (see Chapter G4 and 47) has made the observation that xylem differentiation in developing stems is oriented toward a source of auxin such as a lateral bud or the site of applied IAA. By making a pattern of incisions, or varying the location of IAA treatment relative to the wound, the direction of differentiation can be altered but it always follows the path of IAA transport. These results are interpreted by Sachs to mean that auxin flux through the tissue determines tissue polarity by inducing the formation of auxin transport channels. Continued auxin flux is required to maintain and reinforce the existing polarity which leads eventually to the differentiation of vascular tissue along the auxin transport channels.

Further analysis of the system reveals that if two polar axes intersect, the one with the stronger polarity can inhibit the formation or development of the other. When auxin is applied laterally on a decapitated stem, it induces the treated area to differentiate xylem tissue which eventually connects to the existing vascular strand. If the latter is provided with a source of auxin from the apical region above (e.g., a young leaf or auxin applied to the stump), the site of lateral IAA application fails to form a vascular connection with the existing vascular bundle (47). These data indicate that a relationship of dominance may develop between the intersecting polar axes of two competing organs based on the relative rate of auxin transport and the resulting strength of the respective polarities. The data suggest further that the expression of dominance by the stronger member inhibits polar auxin transport and retards development in the subordinate organ.

If an analogous mechanism exists for the apical control of axillary bud growth it would presumably meet two assumptions: first, that polar auxin transport in the intact stem represses the formation or maintenance of auxin transport channels in the axillary buds; and second, that the removal of the dominant apex causes the induction or reactivation of auxin transport channels along the bud axis as a precondition for bud outgrowth.

The effect of auxin on leaf abscission depends on the direction of auxin movement in the petiole. When petiole explants are prepared that include the abscission zone at their midpoint, and IAA is applied to the distal end, formation of the abscission layer is prevented. But if IAA application is proximal to the abscission zone, abscission is stimulated (2). These responses seem to be related to the competing effects of the leaf blade and the shoot apex on the cells of the abscission zone. The presence of the leaf blade prevents abscission by virtue of its continuing release of auxin. The shoot apex has the opposite effect, and this too seems to be mediated by auxin. The removal of the shoot apex delays the abscission of petioles (25), and the abscission enhancing effect of the apex can be duplicated by IAA treatment of the stump (2). Thus leaf abscission and axillary bud growth show a fundamental similarity in their response to auxin. First, both are subject to the correlative effect of the shoot apex (which is mediated by auxin), and second the effect of auxin secreted by the shoot apex is opposed by auxin from the appropriate lateral organ (e.g., leaf blade and axillary bud respectively).

The way cells in the abscission zone respond to IAA may in part depend on their sensitivity to, and ability to produce, ethylene. Auxin treatment of the distal end in *Phaseolus* petiole explants shortly after excision prevents abscission, whereas the same treatment given more than 12 hours later accelerates it. During the initial phase (stage 1), the petiole is insensitive to ethylene treatment, but after 12 hours (stage 2) abscission is stimulated by ethylene. Furthermore, auxin action in stage 2 is prevented by the removal of evolved ethylene (1) indicating that the abscission-enhancing effect of auxin can be attributed to increased ethylene production.

Polarity in the Control of Axillary Bud Growth

Recent work using bud-bearing, isolated stem sections of *Phaseolus* has shown that axillary bud growth may be affected differentially depending on the direction of the transport of applied auxin in the stem. When the apical end of the section is inserted into a sterile solid medium containing sucrose and mineral nutrients, the bud at the midpoint of the section (not in direct contact with the medium) resumes growth (62). However, if IAA or NAA is also present in the medium, bud growth is prevented (Fig. 1). If the section is implanted with its basal end, auxin in the medium either has no

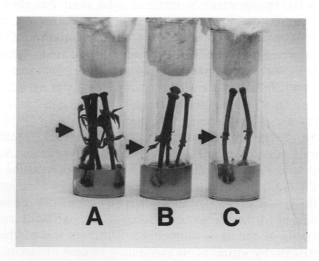

Fig. 1. Cultured *Phaseolus vulgaris* stem segments with the primary node at the segments' midpoint. Shown is the effect of IAA on the growth of axillary buds (arrows) nine days after the segments were excised and inserted with their apical end into the IAA-containing medium. IAA concentrations were 0 (A), 10 (B) and 100 μM (C). From (62).

effect or causes a slight stimulation of bud growth. IAA applied to the apical end also causes bud abscission and prevents the appearance of new lateral buds. Basal application does not have these effects (62). Because basipetal auxin transport is relatively more effective, more auxin may be transported to the bud from the apical end than from the basal end. This, however, is not the explanation of the foregoing data. When ^{14}C-IAA transport is measured in bud-bearing stem sections, the amount of ^{14}C transported to the bud from the apical cut surface does not exceed that transported from the basal end (33). Furthermore, inhibition of lateral bud growth by apically applied IAA is relieved if the stem segment is pretreated with the auxin transport inhibitor N-1-naphthylphthalamic acid (NPA) (62). NPA treatment of the segments also inhibits polar IAA transport within the stem axis but does not substantially alter the amount of IAA entering the axillary bud (33). According to these results, transport of IAA is necessary for the correlative inhibition of axillary buds, but only basipetal transport seems to be effective. Furthermore, there is no difference between basipetal and acropetal IAA transport regarding the amount of IAA accumulation in the bud, yet the two have opposite effects on bud growth. It is concluded that basipetal IAA transport in the stem, rather than IAA accumulation in the bud, is required for bud growth inhibition.

It is not known how the direction of auxin transport can influence developmental responses in the tissue. In an attempt to answer this question regarding the abscission zone, it has been suggested that the relative rates of acropetal and basipetal transport regulate the abscission response by setting up an auxin gradient in the petiole (2). The hypothesis was tested by applying IAA in varying amounts to the two ends of petiole explants. It was found that abscission was delayed if the amount applied distally exceeded the amount applied proximally. The reverse arrangement stimulated abscission. These same results could be obtained even when the total amount of IAA was varied, as long as the ratio of the two applications did not change. The data show that simultaneous IAA fluxes in opposite directions can cancel each other's effect and the outcome (promotion or inhibition) depends on the ratio of the two.

Is there an analogous response at work in axillary bud growth regulation? This question was tested by applying varying amounts of IAA to the two ends of isolated bud-bearing stem segments. It was found that basal IAA treatment relieved the bud inhibiting effect of IAA applied to the apex (62). Furthermore, treatment of the base with nonradioactive IAA reduced the amount of ^{14}C-IAA transported from the apex to all parts of the segment. To explain these results it was suggested that basal IAA application relieves the growth inhibiting effect of apically-derived IAA by altering the ratio of basipetal *versus* acropetal IAA transport.

The mechanism through which polar IAA transport in the stem regulates axillary bud growth is not known. One possibility is that auxin transport in the stem inhibits cell polarization in the axillary buds (62). Decapitation of the dominant shoot apex has been shown to cause increased polar IAA transport *within* the axillary buds of *Phaseolus* (61), and in the subordinate shoot of two-branched *Pisum* plants (37). A similar response occurs among reproductive organs, namely the removal of dominant fruits enhances IAA export from the remaining fruits (18). Due to impaired polar IAA transport along its axis, the correlatively inhibited lateral bud or shoot may fail to develop vascular connections (22), import needed nutrients (42) and growth regulating substances such as root-derived cytokinins (65), or promote cell expansion in its apex (47).

IAA transport is impaired in correlatively inhibited axillary buds (61), subordinate shoots (37) and developing fruits (18), suggesting a common mode of action among different manifestations of correlative control. Perhaps the lack of polar auxin transport in the cells of inhibited shoots results from the randomization of IAA efflux carrier distribution in the plasma membrane (thus eliminating their preponderance at the base of the cell), and the recovery of polar auxin transport, after release from correlative inhibition, involves restoration of efflux carrier asymmetry (37).

The apparently widespread occurrence of IAA transport inhibition in subordinate organs raises the question of what role IAA transport--occurring within the subordinate organ--plays in the correlative relationship. One intriguing possibility is that it is an integral component of the correlative signal pathway. The idea implies that polar IAA transport in the dominant organ impairs the polar transport of IAA within the subordinate organ (e.g., elicits IAA efflux carrier randomization in the cells), and further, that the inhibition of polar IAA transport within the subordinate organ is a necessary condition for growth repression therein. These possible relationships have not been subjected to rigorous analysis.

The Role of Calcium in Hormone Action and Polarity

Calcium ions have been widely implicated in the cellular control of plant growth and development, affecting a broad range of functions including mitosis, cell elongation, gravitropism, polarity, and differentiation (21). An increasing body of evidence suggests that Ca^{2+} is a factor in the intracellular transduction of environmental and hormonal signals. The finding that Ca^{2+} participates in auxin-regulated cell polarization is of special interest (21) because of its possible relevance to apical dominance. It is believed that cell polarization involves the formation of a gradient of Ca^{2+} concentration in the cell resulting from the nonrandom distribution of plasma membrane Ca^{2+} carriers and channels. In tip-growing cells, such as the pollen tube or the rhizoids of the moss *Funaria*, elevated Ca^{2+} levels are generally observed at the growing tip (21). Such polarization of Ca^{2+} distribution is known to be enhanced by IAA transport.

Interaction between IAA and Ca^{2+} occurs in the stems of vascular plants as well. Experiments with *Helianthus* hypocotyls shows that the basipetal transport of IAA is tightly linked to the opposite movement of Ca^{2+}. IAA treatment of the ends of cut segments stimulates the release of Ca^{2+} at the treated end (12). Ca^{2+} extrusion is prevented by the auxin transport inhibitor TIBA, and the absence of Ca^{2+} inhibits IAA transport. The results suggest, first, that the Ca^{2+} status of the cell is directly linked to IAA transport; second, that a change in the direction and rate of IAA movement is likely to alter both the internal concentration and polar distribution of Ca^{2+} in the cell; and third, that a change in the amount of available Ca^{2+} may influence the rate of IAA transport. These changes, in turn, may modify or reverse the course of physiological and developmental events in the cell including those affecting dominance relationships. There is ample evidence showing that the disruption of polarity in Ca^{2+} distribution drastically alters cell structure and development (21). Work by Bangerth and his collaborators have demonstrated that the transport of Ca^{2+} into developing tomato fruits is dependent on the simultaneous export of auxin, and can be inhibited by TIBA and other auxin transport inhibitors (4). The failure of the fruit to import Ca^{2+} causes deficiency disorders indicating that normal fruit development requires the continuing release of auxin from the fruit. When young, developing fruits of tomato plants are subject to correlative inhibition by older, dominant fruits, the export of auxin by the subordinate fruits is impaired by the auxin signal from the dominant fruits (18).

Taken together, the foregoing considerations imply that dominance relations involve Ca^{2+} in at least two ways: first, the calcium ions transported to the dominant organ may assist in the reciprocal release of auxin and thus contribute to the continuing position of dominance for that organ; and second, Ca^{2+} transport to the subordinate organ may diminish because of reduced Ca^{2+} availability or impaired auxin release, and thus cause further weakening of auxin release and induce Ca^{2+} deficiency in the inhibited organ. Thus far, these possible interactions have not been critically tested.

There is evidence indicating that Ca^{2+} and cytokinin interact in the regulation of bud development. In the moss *Funaria*, bud formation in the growing caulonema is induced specifically by cytokinin (7, 8). Bud induction takes place in a subapical 'target' cell at a fixed distance from the caulonema tip. The predictable position of the prospective bud relative to the apex, and the enhancement of growth at the caulonema tip, are two important expressions of the strong polarity that characterizes the caulonema structure. IAA seems to play a key role in the control of polarity because it serves to induce, as well as maintain, the caulonema stage (7, 8). The bud first appears as a localized lateral outgrowth near the apical end of the target cell. After cytokinin treatment, accumulation of membrane-bound calcium is observed in the prospective bud region of the cell (21), followed by cell division that produces the first cell of the bud. If sufficient Ca^{2+} is provided, but its polar distribution is experimentally disrupted, buds can be induced in all target cells even in the absence of cytokinin. On the other hand, the lack of Ca^{2+} prevents the bud-inducing effect of cytokinin. The results indicate that the effect of cytokinin is mediated by Ca^{2+}. The data further suggest that the early events of induction involve the establishment of a new polar axis by the accumulation of Ca^{2+} at the prospective bud site. This raises the interesting question whether axillary bud growth stimulation by cytokinin in higher plants could be viewed as an act of setting a new polar axis. The implication of this idea is that, in addition to stimulating cell division in the bud, cytokinin would also act to reorient IAA transport and cell differentiation in the bud region thus shifting the balance away from the prevailing dominant polarity of the shoot.

CONCLUDING COMMENTS

The basic mechanism of apical dominance remains unresolved even though extensive information is available on specific hormonal effects. Although the correlative signal has not been conclusively identified, IAA is, by all indications, the prospective candidate. All the major classes of growth substances have at least some effect on axillary bud growth, but their interaction is largely undefined. There is strong evidence that cytokinin is a key factor in promoting bud growth. In general, a hormonal regime that enhances the vigor of the apical bud enhances apical dominance, while, when the apical bud is less vigorous, lateral growth may ensue under the influence of growth promotive hormones in the axillary bud.

To understand the mechanism of hormonal control, it will be necessary to explore the fundamental cellular events involved in axillary bud growth inhibition and release. Important progress has been made in recent years regarding the mechanism of auxin transport and its role in tissue polarity. It can be now inferred that auxin-dependent cell polarity is involved in the control of axillary bud growth. How auxin controls polar cellular responses needs to be resolved before the mechanism of apical dominance can be elucidated.

References

1.   Abeles, F.B., Rubinstein, B. (1964) Regulation of ethylene evolution and leaf abscission by auxin. Plant Physiol. 39, 963-969.
2.   Addicott, F.T. (1982) Abscission. Univ. California Press, Berkeley.
3.   Bandurski, R.S., Schulze, A. (1977) Concentration of indole-3-acetic acid and its derivatives in plants. Plant Physiol. 60, 211-213.
4.   Banuelos, G.S., Bangerth, F., Marschner, H. (1987) Relationship between polar basipetal auxin transport and acropetal Ca^{2+} transport into tomato fruits. Physiol. Plant. 71, 321-327.
5.   Beever, J.E., Woolhouse, H.W. (1975) Changes in the growth of roots and shoots when *Perilla frutescens* L. Britt. is induced to flower. J. Exp. Bot. 26, 451-463.
6.   Bernard, J.M. (1990) Life history and vegetative reproduction in *Carex*. Canad. J. Bot. 68, 1441-1448.
7.   Bopp, M. (1983) Developmental physiology of bryophytes. *In*: New manual of bryology, vol. 1, pp. 276-324, Schuster, R.M., ed. The Hattory Botanical Laboratory, Miyazaki.

8. Bopp, M. (1990) Hormones of the moss protonema. *In*: Bryophyte development: physiology and biochemistry, pp. 55-77, Chopra, R.N., Bhatla, S.C., eds. CRC Press, Boca Raton.

9. Brenner, M.L., Wolley, D.J., Sjut, V., Salerno, D. (1987) Analysis of apical dominance in relation to IAA transport. HortSci. 22, 833-835.

10. Cline, M.J. (1991) Apical dominance. Bot. Rev. 57, 318-358.

11. Cove, D.J., Kammerer, W., Knight, C.D., Leech, M.J., Martin, C.R., Wang, T.L. (1991) Developmental genetic studies of the moss, *Physcomitrella patens*. *In*: Molecular biology of plant development, pp. 31-43, Jenkins, G.I., Schuch, W., eds. Symposia of the Society for Experimental Biology, No. XLV. The Company of Biologists Ltd., Cambridge.

12. De Guzman, C.C., Dela Fuente, R.K. (1984) Polar calcium flux in sunflower hypocotyl segments I. The effect of auxin. Plant Physiol. 76, 347-352.

13. De Kroon, H., Schieving, F. (1990) Resource partitioning in relation to clonal growth strategy. *In*: Clonal growth in plants: regulation and function, pp. 113-130, van Groenendael, J., de Kroon, H., eds. SPB Academic Publishing, The Hague.

14. Estruch, J.J., Schell, J., Spena, A. (1991) The protein encoded by the *rolB* plant oncogene hydrolyses indole glucosides. EMBO J. 10, 3125-3128.

15. Garcia-Martinez, J.L., Beltran, J.P. (1992) Interaction between vegetative and reproductive organs during early fruit development in pea. *In*: Progress in plant growth regulation, pp. 401-410, Karssen, C.M., van Loon, L.C., Vreugdenhil, D., eds. Kluwer Academic Publishers, Dordrecht.

16. Garrison, R. (1955) Studies in the development of axillary buds. Am. J. Bot. 42, 257-266.

17. Gocal, G.F.W., Pharis, R.P., Yeung, E.C., Pearce, D. (1991) Changes after decapitation in concentrations of indole-3-acetic acid and abscisic acid in the larger axillary bud of *Phaseolus vulgaris* L. cv Tender Green. Plant Physiol. 95, 344-350.

18. Gruber, J., Bangerth, F. (1990) Diffusible IAA and dominance phenomena in fruits of apple and tomato. Physiol. Plant. 79, 354-358.

19. Harrison, M.A., Kaufman, P.B. (1980) Hormonal regulation of lateral bud (tiller) release in oats (*Avena sativa* L.). Plant Physiol. 66, 1123-1127.

20. Hein, M.B., Brenner, M.L., Brun, W.A. (1984) Effects of pod removal on the transport and accumulation of abscisic acid and indole-3-acetic acid in soybean leaves. Plant Physiol. 76, 955-958.

21. Hepler, P.K., Wayne, R.O. (1985) Calcium and plant development. Ann. Rev. Plant Physiol. 36, 397-439.

22. Hillman, J.R. (1984) Apical dominance. *In*: Advanced plant physiology, pp. 127-148, Wilkins, M.B., ed. Pitman, London.

23. Huff, A., Dybing, C.D. (1980) Factors affecting shedding of flowers in soybean (*Glycine max* L. Merrill). J. Exp. Bot. 31, 751-762.

24. Hutchings, M.J., de Kroon, H. (1994) Foraging in plants: the role of morphological plasticity in resource acquisition. Adv. Ecol. Res. 25, 159-238.

25. Jacobs, W.P. (1955) Studies on abscission: the physiological basis of the abscission-speeding effect of intact leaves. Am. J. Bot. 42, 594-604.

26. Jacobs, W.P., Case, D.B. (1965) Auxin transport, gibberellin and apical dominance. Science 148, 1729-1731.

27. King, R.A., van Staden, J. (1990) The metabolism of $N^6(\Delta^2$-isopentenyl) [^{3}H]adenine by different stem sections of *Pisum sativum*. Plant Growth Regul. 9, 237-246.

28. Klee, H., Estelle, M. (1991) Molecular genetic approaches to plant hormone biology. Annu. Rev. Plant Physiol. Plant Mol. Biol. 42, 529-551.

29. Knox, J.P., Wareing, P.F. (1984) Apical dominance in *Phaseolus vulgaris* L.: The possible roles of abscisic and indole-3-acetic acid. J. Exp. Bot. 35, 239-244.

30. Krekule, J., Machackova, I., Pavlova, L., Seidlova, F. (1989) Hormonal signals in photoperiodic control of flower initiation. *In*: Signals in plant development. Proceedings, 32 and 33 of the 14th Biochemical Congress, pp. 145-162, Krekule, J., Seidlova, F., eds. SPB Academic Publishing, The Hague.

31. Langridge, W.H.R., Fitzgerald, K.J., Koncz, C., Schell, J., Szalay, A.A. (1989) Dual promoter of *Agrobacterium tumefaciens* mannopine synthase genes is regulated by plant growth hormones. Proc. Natl. Acad. Sci., USA 86, 3219-3223.

32. Lee, P.K.-W., Kessler, B., Thimann, K.V. (1974) The effect of hadacidin on bud development and its implications for apical dominance. Physiol. Plant. 31, 11-14.

33. Lim, R., Tamas, I.A. (1989) The transport of radiolabeled indoleacetic acid and its conjugates in nodal stem segments of *Phaseolus vulgaris*. Plant Growth Regul. 8, 151-164.

34. Malik, N.S.A., Berrie, A.M.M. (1975) Correlative effects of fruits and leaves in senescence of pea plants. Planta 124, 169-175.

35. Maurel, C., Barbier-Brygoo, H., Spena, A., Tempe, J., Guern, J. (1991) Single *rol* genes from the *Agrobacterium rhizogenes* TL-DNA alter some of the cellular responses to auxin in *Nicotiana tabacum*. Plant Physiol. 97, 212-216.

36. Morris, D.A. (1993) The role of auxin in the apical regulation of leaf abscission in cotton (*Gossypium hirsutum* L.). J. Exp. Bot. 44, 807-814.

37. Morris, D.A., Johnson, C.F. (1990) The role of auxin efflux carriers in the reversible loss of polar auxin transport in the pea (*Pisum sativum* L.) stem. Planta 181, 117-124.

38. Naylor, J.M. (1958) Control of nuclear processes by auxin in axillary buds of *Tradescantia paludosa*. Can. J. Bot. 36, 221-232.

39. Nougarede, A., Landre, P., Jennane, A. (1990) Intranucleolar visualization of nucleic acids and acidic proteins in inhibited and reactivated pea cotyledonary buds. Protoplasma 156, 183-191.

40. Nyman, L.P., Cutter, E.G. (1981) Auxin-cytokinin interaction in the inhibition, release and morphology of gametophore buds of *Plagiomnium cuspidatum* from apical dominance. Can. J. Bot. 59, 750-762.

41. Ozga, J.A., Reinecke, D.M., Brenner, M.L. (1993) Quantitation of 4-Cl-IAA and IAA in 6DAA pea seeds and pericarp. Plant Physiol. 102, S-7.

42. Phillips, I.D.J. (1975) Apical dominance. Ann. Rev. Plant Physiol. 26, 341-367.

43. Pilate, G., Sossountzov, L., Miginiac, E. (1989) Hormone levels and apical dominance in the aquatic fern *Marsilea drummondii* A. Br. Plant Physiol. 90, 907-912.

44. Prochazka, S., Jacobs, W.P. (1984) Transport of benzyladenine and gibberellin A_1 from roots in relation to the dominance between the axillary buds of pea (*Pisum sativum* L.) cotyledons. Plant Physiol. 76, 224-227.

45. Romano, C.P., Cooper, M.L., Klee, H.J. (1993) Uncoupling auxin and ethylene effects in transgenic tobacco and *Arabidopsis* plants. Plant Cell 5, 181-189.

46. Russell, W., Thimann, K.V. (1990) The second messenger in apical dominance controlled by auxin. *In*: Plant growth substances 1988, pp. 419-427, Pharis, R.P., Rood, S.B., eds. Springer-Verlag, Berlin.

47. Sachs, T. (1991) Pattern formation in plant tissues. Cambridge University Press, Cambridge.

48. Sage, T.L., Webster, B.D. (1987) Flowering and fruiting patterns of *Phaseolus vulgaris* L. Bot. Gaz. 148, 35-41.

49. Schmülling, T., Schell, J., Spena, A. (1988) Single genes from *Agrobacterium rhizogenes* influence plant development. EMBO J. 7, 2621-2629.

50. Schroder, G., Waffenschmidt, S., Weiler, E.W., Schroder, J. (1984) The T-region of Ti plasmids codes for an enzyme synthesizing indole-3-acetic acid. Eur. J. Biochem. 138, 387-391.

51. Sossountzov, L., Maldiney, R., Sotta, B., Sabbagh, I., Habricot, Y., Bonnet, M., Miginiac, E. (1988) Immunocytochemical localization of cytokinins in Craigella tomato and a sideshootless mutant. Planta 175, 291-304.

52. Spano, L., Mariotti, D., Cardarelli, M., Branca, C., Costantino, P. (1988) Morphogenesis and auxin sensitivity of transgenic tobacco with different complements of Ri T-DNA. Plant Physiol. 87, 479-483.

53. Stafstrom, J.P. (1993) Axillary bud development in pea: apical dominance, growth cycles, hormonal regulation and plant architecture. *In:* Cellular communication in plants, pp. 75-86, Amasino, R.M., ed. Plenum Press, New York.

54. Stafstrom, J.P., Altschuler, M., Anderson, D.H. (1993) Molecular cloning and expression of a MAP kinase homologue from pea. Plant Mol. Biol. 22, 83-90.

55. Stafstrom, J.P., Sussex, I.M. (1992) Expression of a ribosomal protein gene in axillary buds of pea seedlings. Plant Physiol. 100, 1494-1502.

56. Tamas, I.A., Davies, P.J., Mazur, B.K., Campbell, L.B. (1985) Correlative effects of fruits on plant development. *In:* World soybean research conference III: Proceedings, pp. 858-865, Shibles, R., ed. Westview Press, Boulder.

57. Tamas, I.A., Engels, C.J., Kaplan, S.L., Ozbun, J.L., Wallace, D.H. (1981) Role of indoleacetic acid and abscisic acid in the correlative control by fruits of axillary bud development and leaf senescence. Plant Physiol. 68, 476-481.

58. Tamas, I.A., Koch, J.L., Mazur, B.K., Davies, P.J. (1986) Auxin effects on the correlative interaction among fruits in *Phaseolus vulgaris* L. *In:* Proceedings, plant growth regulator society of America (PGRSA), pp. 208-215, Cooke, A.R., ed. PGRSA, Lake Alfred, FL.

59. Tamas, I.A., Langridge, W.H.R., Abel, S.D., Crawford, S.W., Randall, J.D., Schell, J., Szalay, A.A. (1992) Hormonal control of apical dominance. Studies in tobacco transformed with bacterial luciferase and *Agrobacterium rol* genes. *In:* Progress in plant growth regulation, pp. 418-430, Karssen, C.M., van Loon, L.C., Vreugdenhil, D., eds. Kluwer Academic Publishers, Dordrecht.

60. Tamas, I.A., Ozbun, J.L., Wallace, D.H., Powell, L.E., Engels, C.J. (1979) Effect of fruits on dormancy and abscisic acid concentration in the axillary buds of *Phaseolus vulgaris* L. Plant Physiol. 64, 615-619.

61. Tamas, I.A., Reimels, A.J. (1989) Increased IAA transport in axillary buds upon release from apical dominance. Plant Physiol., 89, S108.

62. Tamas, I.A., Schlossberg-Jacobs, J.L., Lim, R., Friedman, L., Barone, C.C. (1989) Effect of plant growth substances on the growth of axillary buds in cultured stem segments of *Phaseolus vulgaris* L. Plant Growth Regul. 8, 165-183.

63. Tamas, I.A., Wallace, D.H., Ludford, P.M., Ozbun, J.L. (1979) Effect of older fruits on abortion and abscisic acid concentration of younger fruits in *Phaseolus vulgaris* L. Plant Physiol. 64, 620-622.

64. Thimann, K.V., Skoog, F. (1934) On the inhibition of bud development and other functions of growth substance in *Vicia faba*. Proc. Roy. Soc. B 114, 317-339.

65. Tucker, D.J. (1978) Apical dominance in the tomato: the possible roles of auxin and abscisic acid. Plant Sci. Lett. 12, 273-278.

66. Wang, T.L., Wareing, P.F. (1979) Cytokinins and apical dominance in *Solanum andigena*: lateral shoot growth and endogenous cytokinin levels in the absence of roots. New Phytol. 82, 19-28.

67. Woolley, D.J., Wareing, P.F. (1972) The interaction between growth promoters in apical dominance. I. Hormonal interaction, movement and metabolism of a cytokinin in rootless cuttings. New Phytol. 71, 781-793.

G7. Hormones as Regulators of Water Balance

Terry A. Mansfield and Martin R. McAinsh
Institute of Environmental and Biological Sciences, Division of Biological Sciences, University of Lancaster, Lancaster, Lancashire LA1 4YQ, U.K.

INTRODUCTION

The development of strategies which enable growth to continue without excessive consumption of limited water resources has played a vital part in the evolution of plants which can survive in terrestrial environments. Research over the last two decades has established a clear role for plant hormones in governing the water economy of plants. By influencing stomatal behaviour they can control the expenditure of water, and by regulating the growth and activities of roots, they can exert some control over the uptake of water. Our knowledge of the role of hormones in relation to stomatal functioning is now progressing rapidly and it is appropriate to devote most of this chapter to this topic. Studies of roots have not progressed so rapidly, but nevertheless we have begun to recognise an important role for the roots in regulating activities in the shoot, to provide an integrated strategy for controlling the water balance of the plant.

HORMONES AND STOMATAL BEHAVIOUR

Abscisic Acid

Formation and distribution of abscisic acid in relation to the functioning of stomata.

Wright and Hiron (70) found that the ABA content of wheat leaves increased forty-fold within 30 minutes when they were detached from the plant and subjected to a water deficit severe enough to cause wilting. Subsequent studies have shown that there is a fairly abrupt rise in the ABA content of the leaves of many different species as the water potential falls below -1.0 MPa (= -10.0 bar). The water potential at which there is an accelerated production of ABA is very similar in several different species (Fig. 1) (2). In nutrient deficient plants the production of ABA seems to occur at less negative water potentials (53). Stomatal opening is strongly inhibited by ABA in many different species. Some typical dose-response curves are shown in Figure 2, from which it will be seen that the concentration of K^+ in the medium surrounding the epidermis has a major

P. J. Davies (ed.), Plant Hormones, 598–616.

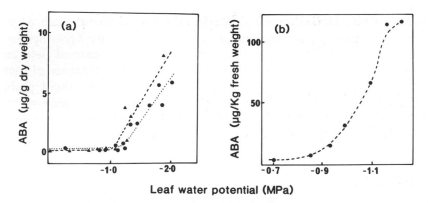

Fig. 1. Relationship between leaf water potential and ABA content. (a) Intact plants of *Ambrosia trifida* (▲) and *Ambrosia artemisifolia* (●). (b) Excised leaves of wheat (ABA produced in 330 min). From (69) and (71).

determining influence. This is not surprising in view of the apparent mechanism of action of ABA on the guard cells (see below). It is clear from these curves that rises in the endogenous levels of ABA in leaves could readily inhibit stomatal opening, and there can be little doubt that individual leaves do possess some capacity to control their own water status by this means. It is, however, clear that mechanisms are present in the plant which ensure that within-leaf production of ABA does not assume an important role until the later, more severe, stages of soil moisture stress.

It is now well established that ABA can be synthesized in roots (10). When roots are in contact with drying soil they produce ABA in increased quantities, which enters the xylem and is transported to the leaves where it inhibits stomatal opening (72). This occurs *before* the shortage of soil moisture causes any measurable change in the water status of the leaves. Thus it is believed that the early stages of soil drying lead to the production of ABA which is transported as a chemical signal to the leaves, where it causes a reduction in transpiration and prevents a decline in water potential or a loss of turgor (13).

These new discoveries make it necessary to reassess the appropriate physiological measurements to determine whether a plant is experiencing water stress. In the past it has been common

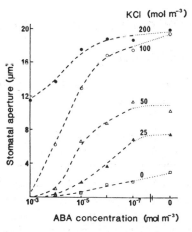

Fig. 2. Effects of ABA on stomata of *Commelina communis* in different concentrations of Kcl. Abaxial epidermis was detached from plants that had been grown carefully to avoid water deficits, i.e., there was a minimum endogenous ABA content. Modified from (61).

599

practice to take leaf samples for determinations of factors such as water potential or relative water content. The data obtained have been supposed to be indicative of the water status of the soil and, in the case of field crops, have been used to indicate irrigation requirements. The existence of root-to-shoot signals which initiate protective mechanisms in the shoot probably means that such measurements provide less accurate information than was thought. The studies of Tardieu *et al.* (66) on field-grown maize have shown that there is a good correlation between stomatal conductance and ABA concentration in the xylem sap (Fig. 3c) and this may be the best above-ground indicator of the water status of the root system. Figure 3 also shows clearly the poor correlation in maize between stomatal conductance and the ABA concentration in the leaf as a whole (3b), leaf water potential (3a), and leaf turgor (3d).

It has been suggested that for the most efficient, long-term exploitation of water in the soil, a plant would benefit from being able to respond to the

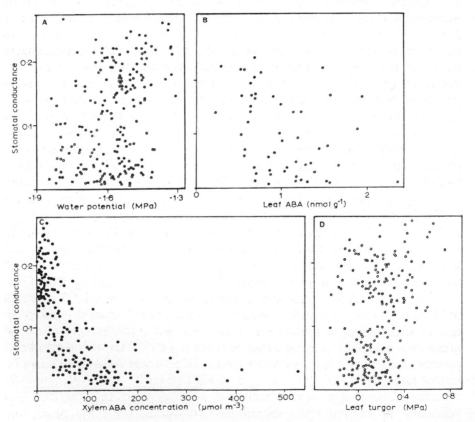

Fig. 3. Relationship between stomatal conductance (in mol m^{-2} s^{-1}) and (a) leaf water potential, (b) bulk ABA concentration ([ABA]) in the lamina, (c) [ABA] in the xylem sap, and (d) leaf turgor (P). Each point represents a coupled potential-conductance or ABA-conductance measurement in the same leaf. Modified from (66).

water potential in different parts of the root system (34). Contrasting types of stomatal behaviour would affect the pattern of water use (35). Four classes of behaviour have been identified: (a) *pessimistic, non-responsive,* in which the stomata open daily in a fixed but restrained manner so as to use initial soil water by the end of the season; (b) *optimistic, non-responsive,* in which the fixed daily opening routine uses water faster than justified from the initial supply in the soil; (c) *pessimistic, responsive,* in which the stomata react to changing conditions to regulate the use of water so that the available supply is used by the end of the season; (d) *initially optimistic, responsive,* in which water consumption begins as in (b) but there is an ability to reduce the daily degree of stomatal opening to prevent very serious water deficits. To understand how these patterns may be determined and controlled we need to explore in detail how hormones and other agents regulate the turgor changes in guard cells.

Mechanisms of Action of Abscisic Acid on Stomata

Stomatal movements result from alterations in the turgor of the pair of guard cells which surround the pore. These changes are driven by fluxes of anions and cations, notably K^+ balanced either by Cl^- or malate, across the plasma membrane and tonoplast (28, 39, 42, 59). Stomatal opening reflects a net accumulation, and stomatal closure reflects a net loss, of K^+. However, the mechanisms of stomatal opening and closure can be regarded as separate processes involving different ion fluxes. Closure involves the stimulation of K^+ efflux and not simply the cessation of the K^+ influx which leads to opening. The ion fluxes involved in the regulation of stomatal aperture occur through specific ion channels, a number of which have been identified in the plasma membrane of stomatal guard cells using whole-cell current-voltage analysis and patch clamp techniques. These comprise both anion channels, including inwardly and outwardly directed K^+ channels, and cation channels (6, 28, 39, 42, 59). ABA stimulates a reduction in stomatal aperture through both the promotion of closure and inhibition of opening (47). Recently whole cell electrical studies have revealed that ABA affects both the influx and efflux of K^+ into guard cells, the inwardly directed K^+ channel being inhibited by ABA whilst the outwardly directed K^+ channel is activated by ABA (39).

External ABA induces stomatal closure as effectively at pH 8.0, at which pH it is not taken up into the guard cells, as at pH 5.0 when it is taken up readily (23). Thus it appears that ABA need not enter the cytosol of the guard cells in order to induce a change in the ion fluxes across the guard cell plasma membrane. This means that the site of action of ABA must either be at the outer surface of the cell, or at a location easily accessible from the outside. There is a considerable body of physical evidence to indicate that ABA can interact with phospholipids (63), indicating that it may be capable of altering the permeability of the plasma membrane. In addition, Hornberg and Weiler (29) have reported the presence of proteins located in the guard cell plasma membrane which bind ABA with a high affinity, although there has been no progress in the purification of the "ABA receptor" since this initial report. Nevertheless, MacRobbie (38) has demonstrated that the ABA response of guard cells exhibits desensitisation, suggesting that there is a specific cellular recognition system for physiologically active ABA.

Ion channels in the plasma membrane of guard cells have been shown to be both voltage-sensitive and/or Ca^{2+}-dependent (6, 28, 39, 42, 59). Of the channels shown to be affected by ABA, the inwardly directed K^+ channel is inhibited by Ca^{2+} whilst the outwardly directed K^+ channel is insensitive to Ca^{2+} (6, 39). Ca^{2+} is known to play an important part in the regulation of many different cellular processes in animals, through its role as a "second messenger", forming an essential link in the pathways through which extracellular signals (e.g. peptide hormones) are transduced into a physiological response (3). Numerous physiological and metabolic processes are also known to be influenced by Ca^{2+} in plants (1, 25, 49) and it is becoming increasingly apparent that Ca^{2+} may have a similar second messenger role to play in plant cells (50). Many of the components of a putative Ca^{2+}-based signal transduction pathway have been identified in plants, including Ca^{2+}-binding proteins such as calmodulin (CaM) (1, 49), Ca^{2+}/CaM-dependent enzymes (1, 26), Ca^{2+} channels (27), Ca^{2+}-ATPases (17), and G-proteins (16) and in the last few years evidence has begun to emerge that Ca^{2+} may act to trigger the intracellular machinery responsible for initiating many of the physiological responses to plant hormones (see Chapter D5).

In the last 10 years the mechanism of action of ABA has been the focus of intense research. Ca^{2+} has long been known to inhibit stomatal opening in some plants (Fig. 4). After finding that at low concentrations of ABA (10^{-9} to 10^{-8} M) the inhibition of stomatal opening is strongly Ca^{2+}-dependent De Silva *et al.* (14) proposed that Ca^{2+} may act as a second messenger during ABA-stimulated stomatal closure, and subsequently it was suggested that there may be an interaction with phospho-inositide metabolism (Fig. 5) (28, 39, 42, 59).

In order for Ca^{2+} to act as a second messenger during the response of stomata to ABA, possibly through the regulation of ion channel activity, it is essential to determine the capacity of the plant hormone to modulate the concentration of cytosolic free Ca^{2+} ($[Ca^{2+}]_{cyt}$) in guard cells. Influx of Ca^{2+} into "isolated" stomatal guard cells has been clearly demonstrated using $[^{45}Ca^{2+}]$ (37). However, until recently attempts to determine whether ABA stimulates a change in $[Ca^{2+}]_{cyt}$

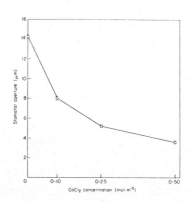

Fig. 4. Inhibition of stomatal opening in *Commelina communis* after incubation of abaxial epidermis for 3 h in a range of concentrations of $CaCl_2$. From (14).

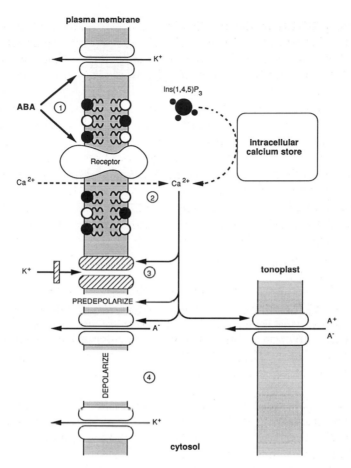

Fig. 5. A possible Ca^{2+}-based ABA signal transduction pathway in stomatal guard cells. (1) *Signal perception.* Binding of ABA to a plasma membrane (PM) receptor (or transport protein) or interaction with membrane lipids. Promotion of stomatal closure through ABA-stimulated activation of K^+-efflux channels in the PM. (2) *Increase in $[Ca^{2+}]_{cyt}$.* Influx of Ca^{2+} or release of Ca^{2+} from intracellular stores through interaction with phosphoinositide metabolism. (3) *Inhibition of stomatal opening.* Ca^{2+}-stimulated inhibition of K^+-influx channels. (4) *Promotion of stomatal closure.* Ca^{2+}-stimulated predepolarisation of the PM, release of anions and cations from the vacuole and activation of PM anion-efflux channels leading to depolarization of the PM and activation of K^+-efflux channels. Modified from (45).

proved inconclusive (37, 39). To resolve this question McAinsh *et al.* (43) used fluorescence ratio techniques to monitor changes in guard cell $[Ca^{2+}]_{cyt}$ in response to ABA. The resting guard cell $[Ca^{2+}]_{cyt}$ recorded using this methodology ranges between 50-350 nM (19, 20, 33, 43, 46, 58). McAinsh *et al.* (43) reported that 10^{-7} M ABA stimulated increases in guard cell $[Ca^{2+}]_{cyt}$ ranging between 2 and 10-fold above the resting level, with peaks up to 1 M (Fig. 6). This observation has subsequently been confirmed in the guard cells of detached epidermis of *Commelina communis* (20, 46) and

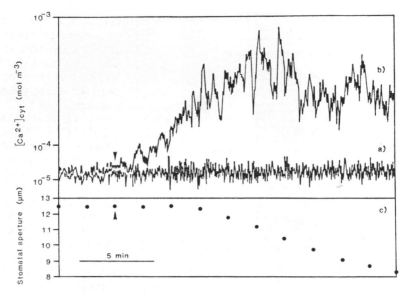

Fig. 6. Changes in guard cell $[Ca^{2+}]_{cyt}$ with time: (a) cells perfused with CO_2-free 10 mol m^{-3} MES, 50 mol m^{-3} Kcl, pH 6.15, at 25°C, and (b) before and after addition of 10^{-4} mol m^{-3} ABA (arrow). (c) Changes in stomatal aperture with time, before and after the addition of 10^{-4} mol m^{-3} ABA (arrow). Modified from (43).

orchid (33), guard cell protoplasts of *Vicia faba* (58), and a number of other cell types (18). This increase precedes stomatal closure by approximately 5 min (33, 43, 46).

The nature of the ABA-stimulated increase in $[Ca^{2+}]_{cyt}$ varies markedly (20, 43, 46), including rapid transient increase in $[Ca^{2+}]_{cyt}$ (Fig. 7) (46, 58). A comparable degree of variability is encountered in certain animal cells (67). Similarly, although all stomata close in response to ABA (20, 43, 46), the proportion of guard cells reported to exhibit an ABA-stimulated increase in $[Ca^{2+}]_{cyt}$ varies from approximately 40% (20) to 68-80% (33, 43, 46). In addition, in *V. faba* the percentage of guard cell protoplasts that exhibit an ABA-stimulated increase in $[Ca^{2+}]_{cyt}$ is equal to the proportion of stomata reported to close in *V. faba* in response to ABA (approximately 37%) (58).

ABA-stimulated increases in guard cell $[Ca^{2+}]_{cyt}$ may be the result of: (a) an influx of apoplastic Ca^{2+}, and/or (b) release of Ca^{2+} from intracellular stores. Studies using guard cell protoplasts of *C. communis* suggest there is no absolute requirement for extracellular Ca^{2+} in the ABA response (60). Similarly, $[^{45}Ca^{2+}]$ flux studies in isolated stomatal guard cells of *C. communis* indicate that there is no sustained ABA-stimulated increase in Ca^{2+} influx at the plasma membrane (37, 39). These techniques, however, are unlikely to detect the release of Ca^{2+} from intracellular stores or short-lived transient alterations in $[Ca^{2+}]_{cyt}$ (46, 58). In contrast, pharmacological studies examining the effects of EGTA and Ca^{2+}-channel blockers on guard cells in detached epidermis of *C. communis* (14, 44) suggest that Ca^{2+} influx and

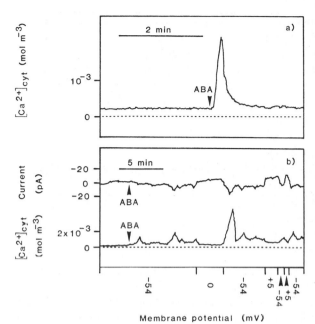

Fig. 7. $[Ca^2]_{cyt}$ and plasma membrane ion currents in guard cell protoplasts of *Vicia faba*. (a) ABA-stimulated $[Ca^{2+}]_{cyt}$ transient. Cells were externally perfused with 10^{-3} mol m^{-3} ABA. The membrane potential of the cell was held at -40 mV. (b) Simultaneous measurement of ABA-stimulated increases in $[Ca^{2+}]_{cyt}$ (lower trace) and inward ion currents (upper trace). Cells were externally perfused with 5×10^{-3} mol m^{-3} ABA. The membrane potential of the cell (Vm) was varied between +5 mV (depolarization) and -54 mV (hyperpolarization). From (58).

Ca^{2+} released from intracellular stores both contribute to ABA-stimulated increases in guard cell $[_{Ca}^{2+}]_{cyt}$. In addition, simultaneous monitoring of ion currents across the plasma membrane of guard cell protoplasts of *V. faba* (using patch clamp techniques) and $[Ca^{2+}]_{cyt}$ (by fluorescence ratio photometry) imply that Ca^{2+} influx, through "ABA-activated" ion channels, contributes to ABA-stimulated increases in $[Ca^{2+}]_{cyt}$ (Fig. 7) (58).

Recently, fluorescence ratio imaging techniques, which allow changes in the spatial distribution of $[Ca^{2+}]_{cyt}$ within a cell to be monitored, have been employed to clarify the origin of ABA-stimulated increases in $[Ca^{2+}]_{cyt}$. Although initial attempts to image ABA-stimulated changes in guard cell $[Ca^{2+}]_{cyt}$ proved unsuccessful (20) subsequent studies indicate that ABA-stimulated increases in $[Ca^{2+}]_{cyt}$ are unevenly distributed across the cytosol of the guard cell (46). Similar heterogeneities have also been observed in animals (67). The distribution of Ca^{2+} varies with time (46). This spatial and temporal localisation of the ABA-stimulated increases in $[Ca^{2+}]_{cyt}$ may reflect both the influx of Ca^{2+} from the apoplast and the release of Ca^{2+} from internal stores. Confocal scanning laser microscopy has also been used to study the distribution and origin of ABA-stimulated changes in $[Ca^{2+}]_{cyt}$, but this technique has so far yielded little additional information (33).

It is apparent that an increase in $[Ca^{2+}]_{cyt}$ is an essential part of the response of stomatal guard cells to ABA (39, 46). However, it must be noted that the outwardly directed K^+ channel in guard cells is insensitive to Ca^{2+} (6, 39). In addition, in flux-tracer studies ABA has been reported to stimulate transient, biphasic increases in K^+ efflux from isolated guard cells of *C. communis*, the initial fast rate of efflux being insensitive to external Ca^{2+} (Fig.

8) (38, 39). This introduces the possibility that the ABA signal transduction pathway may contain Ca^{2+}-independent events. ABA has been shown to stimulate an increase in the cytosolic pH (pH_{cyt}), of between 0.04-0.3 pH units, in guard cells (33). Therefore, pH_{cyt} may be an attractive candidate as an additional putative second messenger in the ABA signal transduction pathway, acting to modify the physiological response to ABA-stimulated changes in $[Ca^{2+}]_{cyt}$ (33, 39).

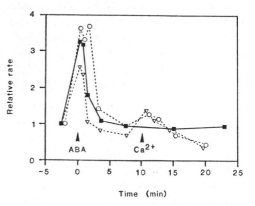

Fig. 8. [$^{86}Rb^+$] efflux transients from isolated guard cells of *Commelina communis* in response to 10^{-2} mol m^{-3} ABA: (■) 10^{-1} mol m^{-3} Ca^{2+} throughout, ($\bigcirc/\triangledown$) 5×10^{-4} mol m^{-3} increased to 0.1 mol m^{-3} after 10 min. Rates are expressed relative to that before ABA was added. Each shows the mean of four strips. Standard errors are not shown. From (38).

These indications of the mode of action of ABA on the guard cells are important because they should eventually lead us to a better understanding of the fine-control mechanisms that enable plants to continue to function under varying conditions of water availability in the field. Such detailed understanding will, however, also depend on a knowledge of the contribution of at least two other groups of hormones, the auxins and cytokinins.

Auxins

Control of Stomatal Behaviour by Auxins

Early studies with IAA failed to reveal clear effects on stomata. For example, Boysen-Jensen in 1936 (7) supplied excised leaves with IAA via their petioles, but could detect no response. Later experiments in the 1940s and 1950s with synthetic auxins did, however, produce positive results and a number of workers found that compounds such as naphth-1-ylacetic acid (NAA) and naphth-2-yloxyacetic acid (NOXA) caused stomatal closure. These discoveries meant that when modern workers looked again for effects of the natural auxin, IAA, their experiments were often designed to show stomatal closure, not opening. Thus IAA was usually applied to stomata that were already open fully, and when there was no response it was concluded that IAA had no effect. We now know that this conclusion was wrong, and that IAA does exert important controls on stomata. Its action is to stimulate opening, and the closing responses to the synthetic auxins bear no resemblance to the effect of the natural auxin.

The first indication of a substantial role for IAA came from the work of Pemadasa (48) on the factors controlling the opening of adaxial and abaxial stomata. In many herbaceous plants there are stomata on both sides of the leaves, but it is common to find fewer stomata and smaller individual apertures on the adaxial surfaces. Pemadasa showed that externally applied IAA caused increased openings of adaxial stomata, but there was very little effect on abaxial stomata under conditions favourable for opening (Fig. 9). However, when apertures of the abaxial stomata were restricted by a reduced supply of K^+ in the medium used for incubating the epidermis, IAA did cause enhanced opening.

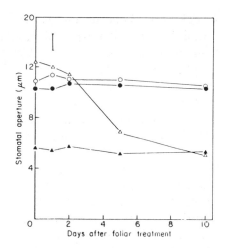

Fig. 9. The effect of application of 1mM IAA on abaxial ($\circ$,$\bullet$) and adaxial ($\triangle$,$\blacktriangle$) stomatal opening in *Commelina communis*. The closed symbols are for the controls untreated with IAA. IAA was applied to the surface of the intact leaves on day 0, and epidermis was then removed at the times shown and incubated for 3 h under conditions favourable for stomatal opening. From (48).

Interactions between IAA and Other Factors

The dose-response relationships between ABA concentration and stomatal closure in epidermal strips are greatly modified by the addition of IAA to the incubation medium (61, 62) (Fig. 10). The inhibitory effect of ABA is virtually absent in the presence of a high concentration of IAA (10^{-4} M). Another important interaction is found with CO_2. Stomata normally close as the CO_2 concentration in the vicinity of the guard cells increases, but the expression of this response is dependent on the concentration of IAA (Fig. 11). High concentrations of IAA drastically reduce the closing reaction to CO_2, but it is restored partially when ABA is supplied along with IAA. These interactions are thought to be very important in the control of the water balance of the plant, and we shall return to them later.

A possible cellular basis for the interaction between IAA and ABA was revealed by the work of Irving *et al* (33) who found that ABA caused alkalinization of the cytosol in guard cells, whereas IAA caused acidification. pH_{cyt} began to change in 2-4 minutes after application of the hormones and it was postulated that, along with Ca^{2+} fluxes, pH is important in determining the activation of K^+ channels in guard cells.

Cytokinins

Since 1977, much evidence has been reported that cytokinins act as important regulators of stomatal movements. When kinetin is applied to isolated epidermis from a grass, *Anthephora pubescens,* it causes stomatal opening

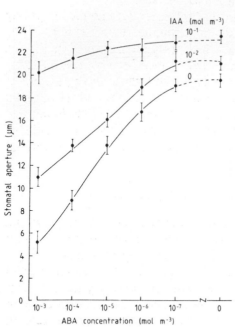

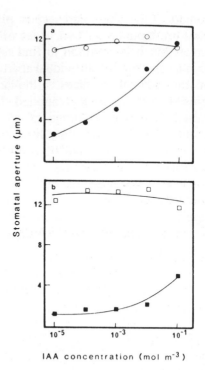

Fig. 10. Influence of IAA on the response of stomata in abaxial epidermis of *Commelina communis* to ABA. A favourable concentration of KCl (100 mol m^{-3}) was present in all treatments. From (61).

Fig. 11. Stomatal opening on detached abaxial epidermis of *Commelina communis* after incubation for 3 h in light in different IAA concentrations. (○) Zero CO_2, (●) 700 μL L^{-1} CO_2. Points are means of 60 measurements. b. as in a. but with 10^{-5} mol m^{-3} ABA in the medium. (□) Zero CO_2, (■) 700 μL L^{-1} CO_2. From (62).

(30). At first it was thought that these responses applied only to the Gramineae, but subsequent work has shown that they are more widespread. Age of the leaves may be a determining factor, for in the *Argenteum* mutant of *Pisum sativum* the stomata show no response to kinetin when the leaves have just reached full expansion, but 15 days later there is a significant stimulation of opening (31). Similarly the stomata of young leaves of *Zea mays* do not respond, but in ageing leaves they open more widely when treated with kinetin (5). Thus it seems that an effect of endogenous cytokinins might be to determine the extent of stomatal opening on leaves of different ages.

The characteristics of stomatal responses to cytokinins have several features in common with the responses to IAA. For example, there are sometimes no effects when cytokinins are supplied to leaves on their own (54). This may be because there is already an adequate supply of endogenous cytokinins and/or because the experiments are conducted under conditions favourable for stomatal opening. It is only when there are factors restricting opening that a clear response to cytokinins can be seen.

Kinetin and zeatin have no distinct effect when they are applied to young leaves of *Zea mays* (4). However, if they are supplied together with ABA, they can overcome the strong inhibitory effect of that hormone (Fig. 12). In *Z. mays* the nature of the CO_2-response is determined both by zeatin and by ABA (5). The data in Fig. 13 can be compared with those in Fig.11, and it will be seen that there are some similarities in the ABA/IAA and ABA/zeatin interactions, though the latter is more complex. Irving *et al.* (33) found that kinetin could influence both pH_{cyt} and the Ca^{2+} status of the guard cells of *Paphiopedilum tonsum*, and this could be the basis of the hormonal interactions.

Hormones and the Fine Control of Stomatal Movements

The discovery that the CO_2 responses of stomata are variable is not new, but recent observations have led to some reassessments of their role in controlling the water relations of the plant (41, 42). It is suggested that a major advantage resulting from the response of stomata to CO_2 is their partial closure as wind speed increases. Wind takes water vapour away from the leaf surface and brings CO_2 towards it. This means that if the stomata close as the amount of CO_2 in their vicinity rises, some control over the rate of transpiration may result. The importance of such control will depend on changes in leaf temperature brought about by the increased air movement: sometimes increasing wind speed can reduce transpiration because cooling reduces the vapour pressure of water in the intercellular spaces. Thus under certain conditions wind itself may cause a reduction in transpiration for purely physical reasons and in these circumstances a response of the stomata may be superfluous. In many situations, however, such as when the input of solar

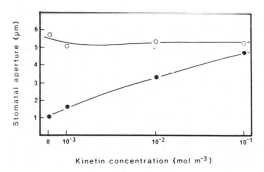

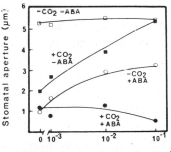

Fig. 12. Stomatal apertures from maize leaf pieces incubated on a range of kinetin concentrations, with (●) or without (○) ABA (10^{-1} mol m^{-3}). Points are means of 60 observations. From (4).

Fig. 13. Stomatal aperture from maize leaf pieces incubated on a range of zeatin concentrations with (+) or without (-) CO_2 (350 µL L^{-1}) and with (+) or without (-) ABA (10^{-1} mol m^{-3}). Points are means of 60 observations. From (5).

radiation is moderate to low, wind does increase transpiration and without a partial closure of the stomata, the water consumption of the plant may rise considerably. It is under such conditions that a closing reaction to CO_2 is seen as an advantage.

Most control mechanisms not only bring gains, but also impose penalties, and that is true in this case. If the stomata remain open when wind increases the CO_2 concentration at the leaf surface, there will be an increase in the intercellular space CO_2 concentration (C_i). This will potentially benefit photosynthesis (depending on irradiance) and such benefit, however small, will be lost if the stomata close as wind brings more CO_2 to the leaf surface. Thus it can be suggested that the stomata should not always close in response to CO_2. If the plant is supplied with sufficient water, it is desirable that C_i should rise with increased wind speed, so that the plant can benefit from a small increase in photosynthesis. It is when the plant is running short of water that stomatal closure in wind will become important. Thus the discovery that three plant hormones, ABA, IAA and cytokinins can become involved in regulating the CO_2 responses of stomata is intriguing (Figs. 11 and 13). This suggests that hormones may provide a much more precise control of stomatal movements than previously recognised. They may contribute to a control system that is extremely elegant - one which changes its characteristics according to the external environment of the plant, especially water supply in the soil. This would mean that hormones are responsible for minute-by-minute control of gaseous diffusion, helping the growing plant to optimise its water consumption relative to the gain of CO_2 for photosynthesis. The existence of a variable CO_2 sensor in the guard cells means that the plant can vary its priorities according to the water supply available (41, 42).

A challenge to plant physiologists in the immediate future will be to determine changes in endogenous levels of cytokinins and auxins in water-stressed plants, to discover how they are related to stomatal movements. ABA has a well-established role in the physiological responses to water stress, and the behaviour of ABA during stress/recovery cycles is fairly well documented. There is little comparable information for cytokinins and auxins, and it will be as difficult to obtain the required precision in analyses (e.g. amounts in the epidermis rather than the leaf as a whole) as has proved to be the case with ABA.

HORMONES AND PHOTOSYNTHESIS

Chemicals which bring about changes in stomatal aperture must also affect rates of net photosynthesis in the intact leaf. Only careful analyses of cause-effect relationships can reveal whether a compound first inhibits mesophyll photosynthesis, and thereafter causes stomatal closure (because of an increased C_i), or whether stomatal closure precedes a reduction in net photosynthesis. In the former case, however, C_i will rise, and in the latter case it will fall. This means that measurements or estimations of C_i can prove valuable in helping to interpret the causes of observed effects.

Cummins *et al* (12) decided that ABA caused a reduction in photosynthesis as a result of stomatal closure, and not because there was a direct action on the photosynthetic capacity of mesophyll chloroplasts. Further studies over the years have appeared to confirm this conclusion, but nevertheless some doubts remain and there is a possibility that ABA has a direct action on mesophyll photosynthesis (55). A possible reason for this is that when ABA is applied to the whole leaf, it may affect the action or distribution of other hormones. IAA has been found to stimulate photosynthetic CO_2 uptake, probably by increasing the coupling between electron transport and phosphorylation (65). ABA often antagonises the effects of IAA in plant tissues (cf Fig. 10).

Hormone-induced changes in the sensitivity of stomata to CO_2 may act to counteract excessive transpiration in windy conditions, as discussed above. Such effects would lead to a temporary interference with photosynthesis. When hormonal changes bring about long-term adjustments of stomatal aperture, however, such as when the plant has suffered a severe water deficit, it may be desirable for the photosynthetic capacity of the mesophyll to be adjusted accordingly. It seems possible, therefore, that changes in IAA in the leaf could be involved both in adjusting the behaviour of stomata to the prevailing conditions, and in altering the activities of the mesophyll to correspond with the modified diffusive conductance of the epidermis. The "down-regulation" in photosynthetic capacity of the mesophyll during water stress may accompany other important changes, such as the accumulation of newly fixed carbon as sucrose which can contribute towards osmotic adjustment (8).

HORMONES AND INTEGRATED RESPONSES OF THE PLANT TO WATER STRESS

We have much more information on the effects of hormones on stomatal conductance than on other activities in the plant which may be relevant to water conservation. This imbalance in the available information partly explains the emphasis given to stomata in this chapter, but such emphasis is probably justified for other reasons because the stomata provide the major point at which a plant can control its rate of water loss. It is, nevertheless, appropriate to make brief mention of some other responses which may assist the plant in its overall adjustment to conditions of water stress.

Roots

ABA increases the permeability of carrot root tissue to water (21), and a similar conclusion was drawn from studies of exudation from isolated roots of maize (9). Effects on exudation rate could be the result of changes in ion flux and/or water permeability (hydraulic conductivity), and a detailed investigation of the effects of ABA on the exudation process in roots of sunflower indicated separate actions on both of these.

A change in root pressure is unlikely to be of much consequence for the bulk flow of water in the xylem of many plants. However, many herbaceous plants are susceptible to cavitation of the sap in their xylem conduits when they are water-stressed, and root pressure at night may be important in refilling the affected vessels. It thus seems possible that ABA formed in shoots under water stress may be transported to the roots where it can stimulate root pressure.

Root growth and development may also be affected by hormonal changes induced by water deficits. There have been indications from studies using inhibitors of carotenoid biosynthesis, and also using mutants deficient in carotenoid biosynthesis, as a means of manipulating the levels of endogenous ABA (56, 57), or using applications of exogenous ABA (11), that ABA limits the growth of the shoot and enhances the growth of the roots at low water potentials. This could explain the reduced shoot:root ratio that is often found in water-stressed plants. Hartung & Davies (22) have proposed a particular role for ABA in roots when they have to penetrate compacted soil. Increased ABA concentrations were found to induce more radial growth immediately behind the root cap, and a greater number of root hairs that could act as an anchor to help penetration through a physical barrier. These morphological effects of ABA could be very important in helping plants to locate new water supplies in compacted soil.

Leaves

ABA inhibits the active efflux of protons thought to be responsible for the co-transport of sucrose in phloem loading in *Ricinus* (40). A reduction in the rate of transport of sugars out of the leaf would enable more solutes to be retained for the maintenance of turgor. Turgor is necessary for the continuation of growth, but under water stress the continued production of a large area of leaves is clearly undesirable. ABA has been found to inhibit the light-stimulated cell enlargement in leaves of *Phaseolus vulgaris* (68). This could provide an important mechanism for limiting expansion of leaves when plants experience water shortage. Convincing evidence is beginning to emerge that leaf growth rate, as well as stomatal conductance, is strongly influenced by an inhibitor (probably ABA) produced by roots in drying soil (22).

Overall Effects of Hormonal Changes

Mutants of several plant species are known which possess different balances of the hormones discussed in this chapter. A deficiency of ABA production produces 'wilty' plants which are unable to maintain turgor under normal conditions (51, 64). Such mutants provide an important practical demonstration of the importance of ABA for the maintenance of normal water relations in the plant. Higher concentrations of auxin-like substances and cytokinins accompany reduced ABA concentrations in some mutants (64). Genetic variation in ABA content in crop plants is now becoming important in the search for genotypes with improved drought resistance and other desirable qualities (52). Studies of the water relations of different cultivars in relation to ABA production have yielded promising results. Varieties of spring wheat with a high capacity for drought-induced ABA formation have been shown to produce higher yields than low-ABA cultivars, probably because of a higher efficiency of water use (32). By contrast, cultivars of sorghum which showed excessive stomatal closure, apparently because of high ABA production, showed a reduced grain yield in response to drought (15). This suggests that the effects of hormonal balance on CO_2 exchange as well as water vapour loss must be fully considered in future studies.

References

1. Allan, E.F., Trewavas, A.J. (1987) The role of calcium in metabolic control. *In*: The Biochemistry of Plants. A Comprehensive Treatise (vol. 12), pp 117-149, Stumpf, P.K., Conn, E.E., eds. Academic Press, New York.
2. Beardsell, M.F., Cohen, D. (1975) Relationship between leaf water status, abscisic acid levels and stomatal resistance in maize and sorghum. Plant Physiol. 56, 208-212.
3. Berridge, M.J., Irvine, R.F. (1989) Inositol phosphates and cell signalling. Nature 341, 197-205.
4. Blackman, P.G., Davies, W.J. (1983) The effects of cytokinins and ABA on stomatal behaviour of maize and *Commelina*. J. Exp. Bot. 34,1619-1626.
5. Blackman, P.G., Davies, W.J. (1984) Modification of the CO_2 responses of maize stomata by abscisic acid and by naturally occurring and synthetic cytokinins. J. Exp. Bot. 35, 174-179.
6. Blatt, M.R., Thiel, G. (1993) Hormonal control of ion channel gating. Annu. Rev. Plant Physiol. Plant Mol. Biol. 44, 543-567.
7. Boysen Jensen, P. (1936) Growth Hormones in Plants. McGraw-Hill Book Co., New York.
8. Chaves, M.M. (1991) Effects of water deficits on carbon assimilation. J. Exp. Bot. 42, 1-16.
9. Collins, J.C., Kerrigan, A.P. (1983) Hormonal control of ion movements in the plant root. *In*: Ion Transport in Plants, pp 589-594, Anderson, W.P., ed. Academic Press, London.
10. Cornish, K., Zeevaart, J.A.D. (1985) Abscisic acid accumulation by roots of *Xanthium strumarium* L. and *Lycopersicon esculentum* Mill in relation to water stress. Plant Physiol. 79, 653-658.
11. Creelman, R.A., Mason, H.S., Bensen, R.J., Boyer, J.S. Mullet, J.E. (1990) Water deficit and abscisic acid cause differential inhibition of shoot versus root growth in soybean seedlings. Analysis of growth, sugar accumulation, and gene expression. Plant Physiol. 92, 205-214.
12. Cummins, W.R., Kende, H., Raschke, K. (1971) Specificity and reversibility of the rapid stomatal response to abscisic acid. Planta 99, 347-351.

13. Davies, W.J., Zhang, J. (1991) Root signals and the regulation of growth and development of plants in drying soil. Annu. Rev. Plant Physiol. Plant Mol. Biol. 42, 55-76.

14. De Silva, D.L.R., Hetherington, A.M., Mansfield, T.A. (1985) Synergism between calcium ions and abscisic acid in preventing stomatal opening. New Phytol. 100, 473-482.

15. Durley, R.C., Kannangara, T., Seetharama, M., Simpson, G.M. (1983) Drought resistance of *Sorghum bicolor*. 5. Genotypic differences in the concentrations of free and conjugated abscisic acid, phaseic acid and indole-3-acetic acid in leaves of field grown drought stressed plants. Can. J. Plant Sci. 63, 131-1 45.

16. Einspahr, K.J., Thompson, G.A., Jr. (1990) Transmembrane signalling via phosphatidylinositol 4,5-bisphosphate hydrolysis in plants. Plant Physiol. 93, 361-366.

17. Evans, D.E., Briars, S-A., Williams, L.A. (1991) Active calcium transport by plant cell membranes. J. Exp. Bot. 42, 285-303.

18. Gehring, C.A., Williams, D.A., Parish, R.W. (1990) Effects of auxin and abscisic acid on cytosolic calcium and pH in plant cells. Proc. Natl. Acad. Sci. USA 87, 9645-9649.

19. Gilroy, S., Read, N.D., Trewavas, A.J. (1990) Elevation of cytoplasmic calcium by caged calcium or caged inositol trisphosphate initiates stomatal closure. Nature 343, 769-771.

20. Gilroy, S., Fricker, M., Read, N.D., Trewavas, A.J. (1991) Role of calcium in signal transduction of *Commelina* guard cells. The Plant Cell 3, 333-344.

21. Glinka, Z., Reinhold, L. (1972) Induced changes in the permeability of plant cells to water. Plant Physiol. 49, 602-606.

22. Gowing, D.J.C., Davies, W.J., Jones, H.G. (1990) A positive root-sourced signal as an indicator of soil drying in apple, *Malus domestica* Borkh. J. Exp. Bot. 41, 1535-1540.

23. Hartung, W. (1983) The site of action of abscisic acid at the guard cell plasmalemma of *Valerianella locusta*. Plant, Cell & Environment 6, 427-428.

24. Hartung, W., Davies, W.J. (1991) Drought-induced changes in physiology and ABA. *In*: Abscisic Acid: Physiology and Biochemistry, pp 63-79, Davies, W.J., Jones, H.G., eds. BIOS Scientific Publishers, Oxford.

25. Hepler, P.K., Wayne, R.O. (1985) Calcium and plant development. Annu. Rev. Plant Physiol. 35, 397-439.

26. Hetherington, A.M., Battey, N.H., Millner, P.A. (1990) Protein kinases. *In*: Methods in Plant Biochemistry, pp 371-384. Lea, P.J. ed. Academic Press, London.

27. Hetherington, A.M., Graziana, A., Mazars, C., Thuleau, P., Ranjeva, R. (1992) The biochemistry and pharmacology of plasma-membrane calcium channels in plants. Phil Trans. R. Soc. Lond. B 338, 91-96.

28. Hetherington, A.M., Quatrano, R.S. (1991) Mechanisms of action of abscisic acid at the cellular level. New Phytol. 119, 9-32.

29. Hornberg, C., Weiler, E.W. (1984) High affinity binding sites for abscisic acid on the plasmalemma of *Vicia faba* guard cells. Nature 310, 321-324.

30. Incoll, L.D., Whitelam, G.C. (1977) The effect of kinetin on stomata of the grass *Anthephora pubescens*. Planta 137, 243-245.

31. Incoll, L.D., Jewer, P.C. (1985) Cytokinins and stomata. *In*: Stomatal Function, Zeiger, E., Farquhar, G.D., Cowan, I.R., eds. Stanford University Press, Palo Alto, CA.

32. Innes, P., Blackwell, R.D., Quarrie, S.A. (1984) Some effects of genetic variation in drought-induced abscisic acid accumulation on the yield and water use of spring wheat. J. Agric. Sci. 102, 341-351.

33. Irving, H.R., Gehring, C.A., Parish, R.W. (1992) Changes in cytosolic pH and calcium of guard cells precede stomatal movements. Proc. Natl. Acad. Sci. USA 89, 1790-1794.

34. Jones, H.G. (1980) Interaction and integration of adaptive responses to water stress: the implications of an unpredictable environment. *In*: Adaptation of Plants to Water and High Temperature Stress, pp. 353-365, Turner, N.C., Kramer, P.J., eds. John Wiley & Sons, London.

35. Jones, H.G. (1983) Plants and Microclimate. Cambridge University Press.
36. MacRobbie, E.A.C. (1981) Effects of ABA in 'isolated' guard cells of *Commelina communis* L. J. Exp. Bot. 32, 563-572.
37. MacRobbie, E.A.C. (1989) Calcium influx at the plasmalemma of isolated guard cells of *Commelina communis*. Effects of abscisic acid. Planta 178, 231-241.
38. MacRobbie, E.A.C. (1990) Calcium-dependent and calcium-independent events in the initiation of stomatal closure by abscisic acid. Proc. R. Soc. Lond. B 241, 214-219.
39. MacRobbie, E.A.C. (1992) Calcium and ABA-induced stomatal closure. Phil. Trans. R. Soc. Lond. B 338, 5-18.
40. Malek, T., Baker, D.A. (1978) Effect of fusicoccin on proton co-transport of sugars in the phloem loading of *Ricinus communis* L. Plant Sci. Lett. 11, 233-39.
41. Mansfield, T.A., Davies, W.J. (1985) Mechanisms for leaf control of gas exchange. BioScience 35, 158-164.
42. Mansfield, T.A., Hetherington, A.M., Atkinson, C.J. (1990) Some current aspects of stomatal physiology. Annu. Rev. Plant Physiol. Plant Mol. Biol. 41, 55-75.
43. McAinsh, M.R., Brownlee, C., Hetherington, A.M. (1990) Abscisic acid-induced elevation of guard cell cytosolic Ca^{2+} precedes stomatal closure. Nature, 343, 186-188.
44. McAinsh, M.R., Brownlee, C., Hetherington, A.M. (1991) Partial inhibition of ABA-induced stomatal closure by calcium channel blockers. Proc. R. Soc. Lond. B 243, 195-201.
45. McAinsh, M.R., Brownlee, C., Sarsag, M., Webb, A.R.R., Hetherington, A.M. (1991) Involvement of second messengers in the action of ABA. *In*: Abscisic acid: Physiology and Biochemistry, pp. 137-152. Davies, W.J., Jones, H.G. eds. BIOS Scientific Publishers, Oxford.
46. McAinsh, M.R., Brownlee, C., Hetherington, A.M. (1992) Visualizing changes in cytosolic-free Ca^{2+} during the response of stomatal guard cells to abscisic acid. The Plant Cell 4, 1113-1122.
47. Mittlehauser, C.G., van Steveninck, R.F.M. (1969) Stomatal closure and inhibition of transpiration induced by RS-abscisic acid. Nature 221, 281-282.
48. Pemadasa, M.A. (1982) Differential abaxial and adaxial stomatal responses to indole-3-acetic acid in *Commelina communis* L. New Phytol. 90, 209-219.
49. Poovaiah, B.W., Reddy, A.S.N. (1987) Calcium messenger systems in plants. CRC Critical Rev. Plant Sci. 6, 47-103.
50. Poovaiah, B.W., Reddy, A.S.N. (1993) Calcium and signal transduction in plants. CRC Critical Rev. Plant Sci. 12, 185-211.
51. Quarrie, S.A. (1982) Droopy: a wilty mutant of potato deficient in abscisic acid. Plant, Cell & Environment 5, 23-26.
52. Quarrie, S.A. (1991) Implications of genetic differences in ABA accumulation for crop production. *In*: Abscisic acid: Physiology and Biochemistry, pp. 137-152, Davies, W.J., Jones, H.G., eds. BIOS Scientific Publishers, Oxford.
53. Radin, J.W. (1984) Stomatal responses to water stress and to abscisic acid in phosphorus-deficient cotton plants. Plant Physiol. 76, 392-394.
54. Radin, J.W., Parker, L.L., Guinn, G. (1982) Water relations of cotton plants under nitrogen deficiency. V. Environmental control of abscisic acid accumulation and stomatal sensitivity to abscisic acid. Plant Physiol. 70, 1066-1070.
55. Raschke, K., Hedrich,R. (1985) Simultaneous and independent effects of abscisic acid on stomata and the photosynthetic apparatus in whole leaves. Planta 163, 105-118.
56. Saab, I.N., Sharp, R.E., Pritchard, J., Voetberg, G.S. (1990) Increased endogenous abscisic acid maintains primary root growth and inhibits shoot growth of maize seedlings at low water potentials. Plant Physiol. 93, 1329-1336.
57. Saab, I.N., Sharp, R.E., Pritchard, J. (1992) Effects of inhibition of abscisic acid accumulation on the spatial distribution of elongation in the primary root and mesocotyl of maize at low water potentials. Plant Physiol. 99, 26-33.

58. Schroeder, J.I., Hagiwara, S. (1990) Repetitive increases in cytosolic Ca^{2+} of guard cells by abscisic acid activation of nonselective Ca^{2+} permeable channels. Proc. Natl. Acad. Sci. USA 87, 9305-9309.
59. Schroeder, J.I., Hedrich, R. (1989) Involvement of ion channels and active transport in osmoregulation and signaling in higher plant cells. Trends Biochem. Sci. 14, 187-192.
60. Smith, G.N., Willmer, C.M. (1988) Effects of calcium and abscisic acid on volume changes of guard cell protoplasts of *Commelina communis*. J. Exp. Bot. 30, 1529-1539.
61. Snaith, P.J., Mansfield, T.A (1982a) Stomatal sensitivity to abscisic acid: can it be defined? Plant, Cell & Environment 5, 309-311.
62. Snaith, P.J., Mansfield, T.A. (1982b) Control of the CO_2 responses of stomata by indol-3-ylacetic acid and abscisic acid. J. Exp. Bot. 33, 360-365.
63. Stillwell, W., Brengle, B., Hester, P., Wassall, S.R. (1989) Interaction of abscisic acid with phospholipid membranes. Biochemistry 28, 2798-2804.
64. Tal, M., Nevo, Y. (1973) Abnormal stomatal behaviour and root resistance, and hormonal imbalance in three wilty mutants of tomato. Biochem. Genet. 8, 291-300.
65. Tamas, I.A., Schwartz, J.M., Hagin, J.W., Simmonds, R. (1974) Hormonal control of photosynthesis in isolated chloroplasts. *In*: Mechanisms of Regulation of Plant Growth, pp. 261-268, Bieleski, R.L., Ferguson, A.R., Cresswell, M.M., eds. Royal Society of New Zealand, Wellington.
66. Tardieu, F., Zhang, J., Katerji, N., Bethenod, O., Palmer, S., Davies, W.J. (1992) Xylem ABA controls the stomatal conductance of field-grown maize subjected to soil compaction or soil drying. Plant, Cell & Environment 15, 193-197.
67. Tsien, R.W., Tsien, R.Y. (1990) Calcium channels, stores, and oscillations. Annu. Rev. Cell Biol. 6, 715-760.
68. Van Volkenburg, E., Davies, W.J. (1983) Inhibition of light-stimulated leaf expansion by abscisic acid. J. Exp. Bot. 34, 835-845.
69. Wright, S.T.C. (1977) The relationship between leaf water potential and the levels of abscisic acid and ethylene in excised wheat leaves. Planta 134, 183-189.
70. Wright, S.T.C., Hiron, R.W.P. (1969) (+) abscisic acid, the growth inhibitor induced in detached wheat leaves by a period of wilting. Nature 224, 719-720.
71. Zabadal, T.J. (1974) A water potential threshold for the increase of abscisic acid in leaves. Plant Physiol. 53, 125-127.
72. Zhang, J., Davies, W.J. (1989) Abscisic acid produced in dehydrating roots may enable the plant to measure the water status of the soil. Plant, Cell & Environment 12, 73-81.

G8. Hormones and Reproductive Development

James D. Metzger
USDA/ARS Biosciences Research Laboratory, Fargo, North Dakota 58105, USA[1].

INTRODUCTION

To farmers, horticulturalists, and others whose livelihood depends on growing plants, it is obvious that the transition from vegetative to reproductive development is a critical phase in the life cycle of higher plants. It is difficult to overestimate the impact the direct products of flowering have on human endeavors. Because they are an integral part of the human diet, production of seeds and fruits form the foundation of all nations' economies. Thus, it becomes equally evident that the ability to manipulate and control flowering with simple treatments has enormous potential both from an economic standpoint as well as increasing food production for an ever growing human population. However, the development of such cultural techniques is predicated on a thorough understanding of the physiological, biochemical, and molecular aspects of reproductive development. Although a great deal of descriptive information exists for many species about the influence of environmental factors on reproductive development, knowledge of the mechanisms by which the floral transition takes place is almost totally lacking.

Flowering, or more precisely, reproductive development, is composed of many independent but highly coordinated processes. To discuss the role of hormones in flowering *per se* would be pointless; it is much more convenient to arbitrarily divide flowering into several temporally related sequences. Flower initiation is the production of flower (or inflorescence) primordia. Evocation, on the other hand, is the term used to describe those processes that occur in the apex prior to, and are required for the formation of flower primordia. Since one cannot precisely determine the exact point at which a flower primordium is formed, it is necessary to wait for the appearance of true floral structures; this is usually referred to as flower formation. Flower development encompasses the processes occurring between flower formation and anthesis.

In many species, the onset of reproductive development is regulated by environmental factors. Many environmental factors such as daylength and

[1] Current address: Department of Horticulture, Ohio State University, Columbus, OH 43210-1096, USA.

P. J. Davies (ed.), Plant Hormones, 617–648.
© 1995 *Kluwer Academic Publishers. Printed in the Netherlands.*

temperature vary regularly during the year. Numerous species use these seasonal variations as cues to coordinate the initiation of reproductive development with the growing season.

Photoperiodism

Plants in which flowering occurs only under certain daylength conditions are said to be photoperiodic. Photoperiodically sensitive plants fall into several response types. Short-day plants (SDP) flower only when the photoperiod is less than some critical length. Actually, it is more accurate to state that SDP flower only when the dark period is greater than a certain critical length, since it is the night length which is measured by the plant. Plants that flower when the daylength is greater than a certain time are called long-day plants (LDP). The requirement can be absolute (obligate) or facultative (quantitative). Plants with a facultative requirement for certain photoperiodic conditions will eventually flower under unfavorable photoperiods, although much later than in inductive conditions. Although LDP and SDP comprise the majority of photoperiodically sensitive plants, three other response types are known. Short-long-day plants (SLDP) are plants that flower only when subjected first to SD followed by LD. Conversely, plants that flower only with the sequence LD then SD are called long-short-day plants (LSDP). Finally, day neutral plants (DNP) are plants with no photoperiodic requirements.

Perception of daylength occurs in the leaves. The primary photochemical event is the absorption of a photon by the chromoprotein phytochrome. The role of the phytochrome in the photoperiodic timing mechanism is discussed in detail in several recent books and reviews (78, 79), and will not be considered further here.

Vernalization

In numerous species from temperate regions, exposure to the low temperatures of winter promotes the initiation of reproductive development upon the return of warmer temperatures the following spring. This phenomenon is known as vernalization, and is unique because the perception and transduction of the environmental cue occurs during the cold period while floral development is manifested later when the plants are under warmer (growth-promoting) temperatures. An exception is brussels sprouts, *Brassica oleracea*, in which flower initiation occurs during vernalization.

Analogous to the various photoperiodic response types, plants can be classified according to their response to low (thermoinductive) temperatures. Summer annuals (or simply annuals), which have no requirement for a thermoinductive treatment, complete their entire life cycle in one growing season. Seeds of winter annuals germinate during late summer or early autumn and overwinter as seedlings. These plants then flower the following growing season. Usually the cold requirement is facultative; that is, flowering will eventually occur without vernalization, but takes considerably longer than plants subjected to thermoinductive temperatures. Furthermore, winter annuals as a group tend to be sensitive to thermoinductive temperatures at most stages of development. Indeed, imbibed seeds of many winter annuals can be vernalized (seed vernalization). Biennials, in contrast, require one full season of vegetative growth, have an obligate requirement for vernalization, and exhibit a period of juvenility in which plants cannot be thermoinduced. Numerous perennials also have a cold requirement for flowering.

Vernalization also promotes flowering in species not normally considered cold requiring. The LDP *Spinacia oleracea* responds to thermoinductive temperatures with a shortening of the critical daylength requirement. In other LDP as well, low night temperatures can compensate for a long dark period (i.e., SD) resulting in flowering under normally noninductive conditions.

In contrast to photoperiodic plants, the site of perception of temperature lies in the apical region of the plant. Often, cold requiring plants also have a requirement for certain photoperiodic conditions following thermoinduction. Most winter annuals and biennials behave as LDP once thermoinduction is complete. On the other hand, the cold requiring perennial *Chrysanthemum morifolium* needs SD after thermoinduction.

Table 1 shows a list of representative plants in each of the various response groups. Both photoperiodism and vernalization have important ecological consequences in allowing plants to fill special niches. Biennials and winter annuals are often the first plants to set seed in spring and early summer. Many LDP flower in midsummer while SDP produce seed toward autumn. Precise timing of flowering also ensures that many individuals from a given population will be flowering at the same time thereby maximizing outcrossing. Although one usually thinks of vernalization and photoperiodism as adaptations that evolved to avoid initiation of reproductive development right before the onset of winter, and possibly not being able to produce seed, they are also important in timing reproductive development to coincide with other seasonal environmental factors. Some tropical plants are strongly photoperiodic (certain species can respond to changes in daylength as little as 15 minutes) and this apparently is to coordinate flowering with seasonal variations in rainfall (i.e., monsoon or rainy season).

Juvenility

In many plants, flowering cannot be induced despite being subjected to the proper inductive conditions until a certain size or age is obtained. This refractory period is known as the juvenile phase. The length of the juvenile period can be as short as a few days or weeks in some herbaceous plants or as long as forty years in some species of trees (Table 2). Juvenility is a serious obstacle in breeding programs for economically important forest trees.

Table 1. Representative plants of each flowering response type

Species (common names)	Family
LONG DAY PLANTS	
Rudbeckia bicolor (cone flower)	Compositae
Anethum graveolens (dill)	Umbelliferae
Spinacia oleracea (spinach)	Chenopodiaceae
Hyoscyamus niger (henbane, annual strain)	Solanaceae
Nicotiana sylvestris (nicotiana)	Solanaceae
Agrostemma githago (corn cockle)	Caryophylaceae
Silene armeria (sweet William campion)	Caryophylaceae
Lolium perrene (perennial ryegrass)	Gramineae
SHORT DAY PLANTS	
Xanthium strumarium (cocklebur)	Compositae
Kalanchoë blossfeldiana (kalanchoë)	Crassulaceae
Perilla crispa (perilla)	Labiatae
Nicotiana tabacum (tobacco, var Maryland Mammoth)	Solanaceae
Euphorbia pulcherrima (poinsettia)	Euphorbiaceae
Chrysanthemum morifolium(Chrysanthemum var. Honeysweet)	Compositae
Pharbitis nil (Japanese morning glory)	Convolvulaceae
Glycine max (soybean var. Biloxi)	Leguminosae
LONG SHORT DAY PLANTS	
Bryophyllum daigremontianum (bryophyllum)	Crassulaceae
Cestrum nocturnum (night jessamine)	Solanaceae
SHORT LONG DAY PLANTS	
Coreopsis grandiflora (tickseed)	Compositae
Echeveria harmsii (Echeveria)	Crassulaceae
DAY NEUTRAL PLANTS	
Zea mays (maize or corn)	Gramineae
Cucumis sativus (cucumber)	Cucurbitaceae
Glycine max (soybean var Wilkins)	Leguminosae
Lycopersicon esculentum (tomato)	Solanaceae
Nicotiana tabacum (tobacco var. Wisconsin 38)	Solanaceae
Gossypium hirsutum (cotton)	Malvaceae
COLD-REQUIRING PLANTS	
Althaea rosea (hollyhocks)	Malvaceae
Thlaspi arvense (field pennycress)	Cruciferae
Brassica oleracea (brussels sprouts)	Cruciferae
Hyoscyamus niger (henbane, biennial strain)	Solanaceae
Triticum aestivum (winter wheat)	Gramineae
Beta vulgaris (beet)	Chenopodiaceae
Daucus carota (carrot)	Umbelliferae
Chrysanthemum morifolium (Chrysanthemum var. Shuokin)	Compositae

Table 2. Comparison of the Duration of the Juvenile Phase in Various Species

Species (common name)	Duration of Juvenile Phase
Chenopodium rubrum (coast blite)	0
Pharbitis nil (Japanese morning glory)	0
Perilla crispa (perilla)	1-2 months
Bryophyllum daigremontianum (bryophyllum)	1-2 years
Malus pumila (apple)	6-8 years
Citrus sinensis (orange)	6-7 years
Citrus paradisi (grapefruit)	6-8 years
Pinus sylvestris (Scotch pine)	5-10 years
Betula pubescens (birch)	5-10 years
Pyrus communis (pear)	8-12 years
Larix decidua (European larch)	10-15 years
Pseudotsuga menziesii (Douglas-fir)	15-20 years
Fraxinus excelsia (ash)	15-20 years
Acer pseudoplatanus (sycamore maple)	15-20 years
Picea abies (Norway spruce)	20-25 years
Abies alba (white fir)	25-30 years
Quercus robur (English oak)	25-30 years
Fagus sylvatica (European beech)	30-40 years

The transition from the juvenile phase to one permissive of flowering (adult or mature phase) is called phase change. It is important to understand that the transition to the adult state does not necessarily mean that the plant has been induced to flower; certain environmental conditions such as photoperiod or thermoinductive temperatures may still be required. Plants that have become mature but have not flowered because of improper conditions are termed ripe-to-flower. At present, there is no good method to distinguish a juvenile plant from a mature one, except by the ability to form flowers.

Another feature exhibited by the two phases is a marked stability through cell division. Reversion to the juvenile phase is not observed in cuttings from flowering (mature) plants. Generally speaking, grafting scions from plants in one state to receptors of the other does not usually alter the phase of either graft partner. Re-juvenation normally occurs only with sexual reproduction.

Often other morphological characteristics differ in juvenile and mature plants. For example, the shape and thickness of leaves, phyllotaxis, or the growth habit of stems may be different. What relationships, if any, these changes have with the ability to form flowers is not known since none of these characteristics are always associated with a transition to maturity. A comparison of correlative morphological characteristics in juvenile and adult plants is shown in Table 3. Rooting ability of cuttings is one of these

Table 3. Juvenile and adult characteristics in selected species. Adapted from (16)

Characteristic	Species	Juvenile form	Adult form
Growth Habit	*Hedera helix*	plagiotropic	orthotropic
	Ficus punula	plagiotropic	orthotropic
	Metrosideros diffusa	plagiotropic	orthotropic
	Euonymus radicans	plagiotropic	orthotropic
Leaf Shape	*Cupressus* spp.	acicular	scale-like
	Acacia spp.	pinnate	phyllodes
	Eucalyptus spp.	oval, sessile	lanceolate with petioles
	Pinus spp.	flat, glaucous	scale- and bract-like
	Hedera helix	palmate	ovate, entire
Phyllotaxis	*Eucalyptus* spp.	opposite	alternate
	Hedera helix	alternate	spiral
Anthocyanin	*Malus pumila*	+	-
Pigmentation in	*Carya illinoisiensis*	+	-
Leaves	*Acer rubrum*	+	-
	Hedera helix	+	-
Thorniness	*Robinia pseudoacacia*	thorns	no thorns
	Malus robusta	thorns	no thorns
	Citrus spp.	thorns	no thorns
Autumn Leaf	*Fagus sylvatica*	keep leaves	abscise
Abscission in	*Quercus* spp.	keep leaves	abscise
Deciduous Trees	*Robinia pseudoacacia*	keep leaves	abscise
	Carpinus spp.	keep leaves	abscise
Rooting Ability	*Hedera helix*	+	-
of Cuttings	*Quercus* spp.	+	-
	Fagus sylvatica	+	-
	Pinus spp.	+	-
	Pyrus malus	+	-

characteristics which has some economic importance. Often cuttings from forest trees lose the ability to root following phase change to the adult state, thus hindering mass introduction of genetically improved clones.

It appears that juvenility provides a mechanism whereby flowering is prevented until the plant is large enough to survive supplying growing reproductive structures and subsequent developing seeds with assimilates. This is particularly critical with perennials which also have demands for those same assimilates from storage organs.

Floral stimulus

In photoperiodically-sensitive plants, the site of perception of daylength is the leaf, whereas the apex is where the morphological change occurs. This suggests that some message is transferred from the leaf to the apex causing the transition to flower formation. This signal is termed the floral stimulus, florigen, or flower hormone. Further evidence for the existence of the floral stimulus are the observations that non-induced plants flower when a leaf from an induced plant is grafted onto them. Successful flowering in the non-induced receptor has also been obtained with several interspecific grafts between different photoperiodic response types. Likewise, receptor plants with a cold requirement can be made to flower without vernalization by grafting flowering donors that are either SDP, LDP or DNP. A summary of successful flower induction in receptor plants under non-inductive conditions by grafting flowering donors is shown in Table 4.

It is tempting to conclude from these experiments that the floral stimulus is very similar or even identical in all plants. However, numerous examples exist in which successful intra- and interspecific graft unions have been made but no apparent transmission of the floral stimulus occurred (77, 91). These negative results have been interpreted as evidence against a unique or ubiquitous floral stimulus (4, 5). This has led a number of investigators to consider the possibility that flower formation is controlled by the known classes of plant hormones acting in concert (5).

Floral inhibitors

When a receptor of the DNP *Nicotiana tabacum* cv. Trapezond was grafted to the LDP *N. sylvestris* and maintained in SD, flowering of the receptor was essentially suppressed. Noninduced *Hyoscyamus niger* (annual strain, LDP) also served as an inhibitory donor (47). This suggests the existence of a graft-transmissible floral inhibitor. Another good example of active (and presumably chemical) inhibition of flowering is in the SDP *Fragaria* x *ananassa*. Subjecting daughter plants to SD while still attached by stolons to the parent plants maintained under non-inductive LD resulted in a great reduction of flowering in the daughter plants. Severing the stolon connection removed the inhibition. Maximum inhibition was observed when assimilate flow from the parents to the daughters was the greatest (31). The most logical interpretation of these results is that a phloem-mobile compound is produced in non-induced leaves, and then transported to the apex where it actively prevents flower initiation.

As with the floral stimulus, the identity of the floral inhibitor(s) remains a mystery. Extractions have not provided any clues to its nature. The floral inhibitor remains a physiological concept.

Active floral inhibitors do not appear to be the basis for all cases of flower inhibition by non-induced leaves. In *Perilla crispa*, the flower promoting ability of induced donor leaf is nullified if non-induced leaves of the receptor lie between the donor leaf and the apex. But this inhibition can be explained entirely by the alteration of assimilate translocation patterns and, hence, movement of the floral stimulus (44). Likewise, non-induced leaves

Table 4. Selected examples of successful transmission of the floral stimulus between graft partners. Adapted from Lang (46) and Zeevaart (90)

DONOR			RECIPIENT			FAMILY
Species	Flowering Type	Photoperiod/ Temperature	Species	Flowering Type	Photoperiod/ Temperature	
Glycine max	SDP	SD	Glycine max	SDP	LD	Leguminosae
Perilla crispa	SDP	SD	Perilla crispa	SDP	LD	Labiatae
Xanthium strumarium	SDP	SD	Xanthium strumarium	SDP	LD	Compositae
Beta vulgaris	LDP	LD	Beta vulgaris	LDP	SD	Chenopodiaceae
Bryophyllum daigremontianum	LSDP	LD→SD	Bryophyllum daigremontianum	LSDP	LD or SD	Crassulaceae
Chrysanthemum morifolium var. Shuokin	CRP[a]	TI[b]→LD	Chrysanthemum morifolium var. Honeysweet	SDP	LD	Compositae
Anethum graveolens	LDP	LD	Daucus carota	CRP	LD	Umbelliferae
Sinapis alba	LDP	LD	Brassica oleracea	CRP	LD	Cruciferae
Brassica nigra	CRP	TI→LD	Brassica oleracea	CRP	LD	Cruciferae
Blitum capitatum B. virgatum	LDP	LD	Chenopodium rubrum	SDP	LD	Chenopodiaceae
Petunia hybrida	LDP	LD	Hyoscyamus niger	CRP	LD	Solanaceae
Nicotiana tabacum var. Maryland Mammoth	SDP	SD	Hyoscyamus niger	CRP	SD	
Nicotiana tabacum var. Maryland Mammoth	SDP	SD	Nicotiana sylvestris	CRP	SD	Solanaceae
Gossypium hirsutum	DNP	LD	Gossypium davidsonii	LDP	LD	Malvaceae
Bryophyllum daigremontianum	LSDP	LD→SD	Echeveria harmsii	SLDP	SD	Crassulaceae
Pharbitis nil	SDP	SD→LD	Pharbitis nil	SDP	LD	Convolvulaceae
Lunaria annua	CRP	TI→LD	Lunaria annua	CRP	LD	Cruciferae
Kleinia articulata	SDP	SD→LD	Kleinia repens	LSDP	LD	Compositae
Silene armeria	LDP	LD→SD	Silene armeria	LDP	SD	Caryophyllaceae

[a]CRP=Cold-requiring plant; [b]TI=Thermoinduced

of the SDP *Xanthium strumarium* apparently do no more than to serve as sources of assimilates devoid of the floral stimulus, thereby diluting the floral stimulus and reducing its effective concentration at the apex (95).

Measurement of flowering

A plant is either flowering or it is not; thus, at first glance, measurement should be only a simple qualitative determination. However, such an analysis is too simplistic. A plant may be induced to flower by two different treatments, but in one treatment it produced only one or a couple of flowers, while many flowers are produced following the other treatment. Alternatively, the time required for the completion of some aspect of reproductive development might be affected by different inductive treatments. Clearly, there are quantitative aspects to flowering.

As in the study of other developmental processes, it is important to quantify flowering in relation to various treatments. The methods used to quantitate flowering are numerous, but they basically fall into five categories depending on the experimental design. The simplest is the percentage of plants that have flowered following a particular treatment. Another way to measure the flowering response is to determine the number of buds, flowers, or flowering nodes on an individual plants. An experimenter might choose to determine the number of leaves produced from the beginning of a treatment until flowering is observed. This technique gives temporal data, with the fewer number of leaves, the faster the rate of development. Similar data could be obtained by determining the time (usually in days or weeks) for flowering to occur. As a final alternative, there are scales with numbered values assigned to different stages of apical development. At various times after the start of a treatment, the apex is examined microscopically and assigned a number, based on an arbitrary scale developed for that particular species. This measure differs from the others in that it is destructive and therefore requires significantly more plants, but it does provide data on flower initiation. The relative advantages and disadvantages of each of these techniques have been discussed in considerable detail (4,46).

HORMONES AND JUVENILITY/PHASE CHANGE

The hormonal role in the control of phase change is not well established, although gibberellins (GAs) may be involved. It does not appear that a specific hormone is required for the maintenance of either the juvenile or adult phase. Reviews of various aspects of juvenility have appeared over the past 30 years (16, 68, 100). The following discussion on juvenility will be limited to a comparison of evidence for the hormonal basis of juvenility in three greatly different types of plants.

Bryophyllum daigremontianum is a LSDP with a juvenile phase lasting until the development of 10 to 12 pairs of leaves. It has been shown that the leaves perceive the transfer from LD to SD, but in juvenile plants, they are incapable of producing the floral stimulus (87). Application of GA_3 to the leaves of juvenile plants will promote flowering only under SD. When GA_3 is applied in LD, the plants will subsequently flower only if transferred to SD. This period of time between the treatment in LD until the transfer to SD when GA_3 is still effective can be remarkably long: up to four months. Since it has been shown that in adult plants flower formation following the transfer from LD to SD requires GA biosynthesis (99), the biochemical basis for juvenility in *Bryophyllum* could reside in an inability for GA biosynthesis following the sequence LD to SD (87).

Fig. 1. Exogenous GAs will cause precocious "flowering" in juvenile conifers. This four-year-old rooted cutting of *Tsuga heterophylla* was subjected to six weekly spray treatments of a solution containing 200 $mg/GA_{4/7}$. Untreated plants of the same age failed to flower. Photograph courtesy of Dr. Stephen Ross.

In conifers, where "flowering" is not normally observed until plants are 10-20 years old, exogenous GAs can induce flowering in 3-12 month old plants (60, 61) (Fig. 1). Polar GAs such as GA_3 are effective in members of Cupressaceae and Taxodiaceae families but not in species from Pinaceae. Less polar GAs, such as GA_9 and $GA_{4/7}$, are necessary for promotion of precocious flowering in Pinaceae species (60, 61).

It has been postulated that in juvenile conifers, GAs are the limiting factor preventing phase change. However, cessation of GA application to juvenile plants results in a decline in the flowering response and concomitant abscission of newly formed cones (60). Thus, exogenous GA does not cause a true phase change since the mature state should be stable through many mitotic cycles. It may be that in conifers, GAs are limiting for flower formation and development, and overcoming this limitation may be one of many alterations resulting from the phase change to the adult state.

In contrast with *Bryophyllum* and conifers, exogenous GA_3 causes reversion of many adult woody plants to the juvenile state (94, 100) (Fig. 2). These results suggest a positive relationship between GAs and the maintenance of the juvenile phase. Consistent with this is the observation that the levels of GA-like substances were higher in extracts of apical buds from juvenile plants of *Hedera helix* than adult apical buds (22). If it is true that high levels of endogenous GAs are required to maintain the juvenile

Fig. 2. In contrast to conifers, exogenous GAs cause rejuvenation of mature *Hedera helix* (English ivy) plants. Left to right: Juvenile control. GA-induced reversion to the juvenile form of a mature shoot treated with 5 nmole of GA_3. ABA (5 μmole) prevented rejuvenation caused by 5 nmole GA_3. Mature control shoot. Note the distinct differences in leaf shape and internode length between juvenile and mature forms. From (65).

phase, one would predict that a decline in GA levels in juvenile plants would lead to the transition to the adult phase. However, attempts to artificially reduce endogenous GA levels with growth retardants (inhibitors of GA biosynthesis) have produced equivocal results. Application of CCC (2-chloroethyltrimethyl ammonium chloride) to juvenile *Hedera* plants resulted in dwarf plants but paradoxically also caused an increase in the levels of endogenous GA-like substances (23).

Under certain conditions, mature *Hedera* plants will revert to plants that are either juvenile or show several juvenile characteristics. This fact has been used to study the possible role of hormones in phase maintenance. Rejuvenation of adult shoots was observed when grafted on juvenile stocks. However, the adult leaves had to be removed (15). Rejuvenation of adult plants has also been reported when adult and juvenile plants were grown together in the same culture solution (21). Together, these results suggest that juvenile plants produce a substance(s) that maintains the apex in the juvenile state. Since exogenous GA_3 can also cause rejuvenation (22), it is possible that GAs are involved. Under conditions of low intensity light, adult *Hedera* plants spontaneously revert to the juvenile state; this reversion can be prevented by inhibitors GA biosynthesis (65).

The above results are consistent with the notion that in *Hedera*, the ability to maintain high GA levels is necessary for the maintenance of the juvenile state, and, conversely, a diminution of this capacity is associated with the phase change to the adult state. Thus, rather than controlling phase change per se, GAs (or lack of GAs) may be involved in phase stabilization (65).

In conclusion, no unified picture for the role of hormones in juvenility and phase change can be presented. Although there is evidence that GAs play a role, it appears to be different in the case of *Bryophyllum* and *Hedera*. This indicates that the physiological and biochemical basis for juvenility and subsequent phase change can be distinct in different species. In the *Bryophyllum*, for example, the physiological basis for the failure of juvenile plants to flower following transfer from LD to SD lies in an inability of the leaves to produce the floral stimulus (87). A similar conclusion was reached for the SDP *Perilla crispa* (83). In other instances, it appears that the apex itself is insensitive to the floral stimulus. Juvenile apices of both *Larix* and *Hedera* failed to flower when grafted to mature stock (15, 64). This interpretation assumes that the floral stimulus is produced in the leaves and then transported to the apex akin to photoperiodically sensitive herbaceous plants. Even if this is not so, the experiments with *Larix* and *Hedera* do demonstrate that phase change is a property of the apex rather than the leaf as was shown for *Bryophyllum* and *Perilla*. Thus, the concept of phase change is in reality only an operational definition, useful to describe the ability of plants to flower under inductive conditions rather than implying common mechanisms.

HORMONES AND FLOWER FORMATION

This section will examine the role of the known classes of plant hormones in flower formation. Each class will be dealt with separately. In general, the pattern of experimentation has been to first apply the hormone to the plant under non-inductive conditions (or inductive conditions if inhibition of flower formation is being scrutinized). Second, endogenous hormone levels are measured in relation to flower formation. Unfortunately, most quantitative work was done before modern analytical techniques using physico-chemical methods were available to physiologists. Instead, relating hormone levels to flower formation was performed with bioassays. Although a good first approximation can be made with bioassays, much of this work must be viewed with caution. The pitfalls of bioassays have been discussed elsewhere (12 and Chapter F1). Wherever possible, I have attempted to use examples in which the investigators have employed physico-chemical techniques for quantitative analysis.

Another useful technique is to use inhibitors of hormone biosynthesis and/or action. In this way, the effect of reducing endogenous hormone levels on flower formation can be assessed. In a similar vein, it has been possible to select hormone-deficient mutants. In certain instances, such mutants have proven extremely valuable in assessing the role of a particular hormone in flower formation.

Auxins

Members of the Bromeliaceae are unique among plants in that they exhibit a strong flowering response following auxin application. This effect is due to auxin-induced ethylene production (92) and will be discussed in greater detail later when the role of ethylene in flower formation is considered. For the most part, however, application of various auxins (either indole acetic acid or naphthalene acetic acid) tend to be inhibitory to flower formation under inductive conditions. This is true for the various response types including: the SDP *Pharbitis nil* and *Chenopodium rubrum*; the LDP *Lolium temulentum* and *Sinapis alba*; and the cold-requiring plants *Lunaria annua* and *Cichorium intybus* (5, 92). These results have been interpreted to mean that the role of auxins in flowering is to prevent flower formation under noninductive conditions. However, quantitative analyses of endogenous auxin levels have been equivocal. In any event, such a proposed role for auxin in the control of flower formation is probably an oversimplification. It now appears these inhibitory effects are also due to auxin-induced ethylene production (92).

In certain instances, application of low amounts of auxin levels promote flower formation (5), but only under conditions of marginal or partial induction. The significance of these results are uncertain. Nevertheless, it is possible that auxins play a role in some of the processes associated with evocation such as a loss in apical dominance and an alteration in phyllotaxis, but this remains to be shown conclusively.

Ethylene

In many plants, exogenous ethylene, applied either as the gas or by the use of ethylene-releasing agents such as ethrel, inhibits or delays flower formation. As yet, there is no evidence that this inhibition is part of the natural regulating mechanism (92).

In contrast to most plants, species from the Bromeliaceae flower in response to exogenous ethylene. A common and economically important horticultural practice is to induce flowering of bromeliads, particularly pineapples, at will with ethylene-releasing agents (5, 92).

Little is known on the role of ethylene in flower formation in bromeliads. In one report, treatments that resulted in flower induction were invariably associated with increased ethylene production. Flower induction by one of the treatments, namely mechanical perturbation, could be suppressed by AVG (aminoethoxyvinyl glycine), an inhibitor of ethylene biosynthesis. This inhibition could be reversed by adding ethylene. Furthermore, the immediate precursor to ethylene biosynthesis, aminocyclopropane carboxylic acid, could also induce flowering (14). These results strongly suggest that ethylene is indeed an important regulator of flower formation in bromeliads.

Cytokinins

Flower formation is promoted in a number of species by exogenous cytokinins (5, 92). In many of these cases, the promotive effect of cytokinins is seen only in induced or marginally induced plants. Furthermore, in the SDP *Pharbitis nil*, the promotion of flowering by cytokinin was shown to be indirect, i.e., due to enhanced translocation of the floral stimulus and assimilates from the induced leaves (59).

Quantitative analyses of cytokinins in relation to flower formation have not clarified the picture either. In the LDP *Sinapis alba*, an increase in the endogenous cytokinin levels in both leaves and phloem exudate was observed 16 hours after the beginning of LD (1, 48). This positive correlation is in sharp contrast to the negative relationship between endogenous cytokinins and photoperiodic induction observed in the SDP *Xanthium strumarium* (37).

Thus, at present, there is no convincing evidence linking cytokinins and flower induction in a cause and effect relationship. However, a single low dose of cytokinin applied to the apex of noninduced *Sinapis* plants mimicked the early stimulation of mitotic activity caused by an inductive LD (3). Other features of this mitotic activity were identical in the two treatments. The cytokinin effects corresponded very well with higher cytokinin-like activity in the leaves as well as increased cytokinin flux in the phloem following transfer to LD, indicating that cytokinins may play a role in regulating evocational processes (1, 3, 48).

Abscisic Acid

As discussed earlier, physiological evidence exists for the presence of a graft-transmissible flower inhibitor. With the discovery of ABA as a potent, naturally occurring growth inhibitor that was possibly involved in the control of bud dormancy, it was reasonable to suspect that ABA might be the flower inhibitor. Indeed, application of ABA to the LDP *Spinacia oleracea* and *Lolium temulentum* repressed LD-induced flower formation (18, 19). However, no causal relationship was observed between endogenous ABA levels and the ability to flower; ABA levels in *Spinacia* were in fact higher under LD than SD (45, 89). The apparent inhibition of flower formation might have been the result of a delay in inflorescence development rather than inhibition of flower initiation, indicating that ABA does not have a role in the regulation of flower formation in LDP as a graft-transmissible inhibitor (92).

Although ABA promotes flowering in several SDP, the positive effect is observed only under partially inductive conditions (17). In other SDP, ABA has no effect or is inhibitory (5, 92). As in LDP, there is no apparent role for ABA in flowering formation in SDP.

Gibberellins

Of all the plant hormones that have been applied to plants under strictly noninductive conditions, only GAs have been shown to effectively cause flower formation in a wide variety of species. In general, LDP and plants with a cold requirement are responsive to exogenous GAs while SDP and DNP are not. GA-sensitive LDP and cold-requiring plants usually grow as rosettes in noninductive conditions. These generalizations do not always hold, however. For example, the LDP *Hieracium aurantiacum* and *Blitum virgatum*, and the cold-requiring plants *Geum urbanum* and *Lunaria annua* do not form flowers under non-inductive conditions in response to exogenous GA$_3$ (94). On the other hand, application of GA$_3$ caused flower formation in the SDP *Impatiens balsamina* and *Zinnia elagans* maintained in LD (5, 94). In still other plants GA inhibits flower formation, particularly woody fruit trees such as cherry, peach, apricot, almond, and lemon (94). These species often exhibit a period of juvenility and as discussed earlier, the basis for GA inhibition of flowering in these species may reside in rejuvenation of adult plants. A comprehensive list of the effects of exogenous GAs on flower formation can be found in (5, 94).

In view of the large number of species in which exogenous GAs cause vegetative plants to flower under noninductive conditions, it is logical to conclude that GAs have a critical role in the regulation of flower formation. As yet, however, such a role cannot be defined. This is partly due to the analytical difficulties encountered when investigating GA physiology. At present there are over 80 different GAs, although only a fraction of these appear to be present in any given species. The endogenous GAs appear to be metabolically related. It is likely that only one is responsible for mediating the physiological process under GA control with the others being either precursors or deactivation products (36). Thus, when attempting to correlate endogenous levels of GAs to certain physiological processes, it is important to know the identity of those GAs. However, the instrumentation and analytical expertise necessary to conclusively identify and quantitate endogenous GAs in vegetative tissues has, until recently, been largely unavailable to plant physiologists. Nevertheless, the picture beginning to emerge indicates that GAs may have different roles in the various response types. In the remainder of this section, specific examples will be used to illustrate this point.

As noted earlier, flower formation can be induced in many rosetted LDP with exogenous GAs suggesting that GAs may be limiting in SD. Consistent with this idea are the numerous observations that transfer to LD results in an increase in the levels of one or more GAs or GA-like substances (10, 52, 74). Associated with LD induction are both increased GA biosynthesis and metabolism (10, 86, 96). Nevertheless, reduction of endogenous GA levels with inhibitors of GA biosynthesis completely suppressed LD-induced stem elongation (bolting) but had no effect on flower formation in *Spinacia oleracea* (86), *Silene armeria* (10) and *Agrostemma githago* (43). However, since the inhibitors failed to totally eliminate endogenous GA production, it is possible that although stem growth was blocked, tissue concentrations of GAs were sufficient for flower initiation. Indeed, a deletion (and therefore presumably non-leaky) mutation of kaurene synthase in *Arabidopsis thaliana* results in a failure to flower in SD, although flower formation does occur in LD (81). This led the authors to conclude that GA is critical for flower initiation under certain conditions. Alternatively, it is also possible that the molecular changes associated with floral induction occur despite SD and a lack of GA. In this scenario, the role of GA may be in the regulation of growth and development of floral organs. A total lack of GA may prevent the formation of microscopic structures even though floral induction has occurred.

In other rosette LDP, application of GA_3 to vegetative plants induces stem elongation but not flowering (77). This may be interpreted that in most rosetted LDP, GAs are not limiting for flower initiation but do mediate the photoperiodic control of stem elongation. An alternative explanation is that there is specificity in the action of individual GAs such that certain GAs stimulate stem elongation while others induce flowering. Thus it is possible there exists special "florigenic" GAs. Evidence for this notion was obtained in studies using the LDP *Lolium temulentum* in which it was found that structural features of exogenous GAs that stimulate stem growth are different from those promoting flower formation (20,62,63). However, no endogenous GAs have yet been isolated from *Lolium* or any other species that appear as likely candidates as special florigenic GAs.

A similar situation has been observed in many rosetted cold requiring plants. Changes in endogenous GA-like substances following thermoinduction appear to be more important for thermoinduced stem elongation than for flower formation per se (94). Growth retardants inhibit thermoinduced stem elongation but not flower formation in *Raphanus sativus* (77), *Daucus carota* (39) and *Thalaspi arvense* (53). Application of GA_3 to nonthermoinduced *Lunaria annua* and *Brassica napus* elicits only stem elongation (70, 82).

Even though it does not appear that GAs are directly involved in the transition to flowering in many LDP and cold-requiring plants, the question still remains as to the mechanism by which GAs induce flowering in vegetative plants. One possible answer is that exogenous GA acts indirectly through the production of the floral stimulus. This is supported by grafting experiments using two lines of the LDP *Silene*: one in which exogenous GA_3 induces flowering and stem elongation and another that responds to exogenous GA_3 only with stem elongation. Flowering in the latter can be induced in SD by grafting it to the other line following induction of flowering with GA_3 (77). This must mean that the exogenous GA induced the production of the floral stimulus. A similar conclusion can be drawn from the cold requiring plant *Chrysanthemum morifolium*. Non-vernalized plants induced to flower with GA_3 can cause flowering in SD (non-cold-requiring) lines maintained under LD. Exogenous GA_3 does not induce flowering in the SD lines when the plants are in LD (33).

The vast majority of SDP do not respond to exogenous GA; this fact leads to the conclusion that GAs are usually not limiting for flower formation in LD. But in an ironic contrast to rosetted LDP and cold-requiring plants, growth retardants prevent flower formation in the SDP *Pharbitis nil* if applied before or during the inductive dark period (71, 84). About 100 times more GA_3 was required to reverse the inhibition of internode elongation than was needed for flower formation (71). Moreover, the growth retardants acted at the shoot apex and not in the cotyledons, the site of perception of photoperiod. This indicates that the reduction of endogenous GAs reduced the ability of the apex to respond to the floral stimulus (71, 84). Thus, in *Pharbitis*, GAs are required for floral initiation, but not for production of the floral stimulus (84).

The opposite conclusion can be made for the LSDP *Bryophyllum daigremontianum*, namely that GAs are necessary for production of the floral stimulus but not initiation. *Bryophyllum* can be induced to flower when GA is applied in SD but not LD, indicating that GA can substitute for the LD portion of the inductive photoperiod (87). GA_{20} has been identified as the major biologically active GA in leaves of *Bryophyllum* although it was 20 times less effective than GA_3 in inducing flower formation under SD (25). The transfer of adult plants from LD to SD caused a dramatic increase in the level of GA_{20}. In contrast, plants maintained permanently in SD had no detectable GA_{20} (90). The growth retardant CCC completely inhibited flower formation in *Bryophyllum* when applied during the inductive SD period. This inhibition could easily be reversed with GA_3 (98).

Taken together these results strongly suggest high levels of endogenous GA (presumably GA_1 formed by the hydroxylation of GA_{20}) are required for flower formation in *Bryophyllum*. The site of GA action in *Bryophyllum* is in the leaves, not the shoot apex where flower formation occurs (88). Zeevaart (98, 99) concluded that the LD part of photoinduction leads to GA production which is absolutely necessary for the production of the floral stimulus in SD. Thus in *Bryophyllum*, GAs regulate the production of the floral stimulus.

To summarize, it was shown that GAs have a variety of roles in reproductive development depending on the species and the response type. In many rosetted LDP and cold requiring plants, GAs are not required for flower formation, but do regulate the closely related phenomenon of bolting. GAs also can play an essential role in the regulation of the production of the floral stimulus as was shown for *Bryophyllum* or in controlling the events associated with the response of the apex to the floral stimulus (i.e., *Pharbitis*).

HORMONES AND THE FLORAL STIMULUS

To date, there is a massive collection of evidence indicating the existence of a stimulus which initiates the transition of an apex from a vegetative state to one committed to reproductive development. From grafting experiments (Table 4), it is logical to conclude that the floral stimulus is very similar or even identical in all response types. Despite many attempts to isolate the floral stimulus, its chemical nature remains as much a mystery today as it was over 55 years ago when its existence was first proposed by Chailakhyan (7). Since its conception, the floral stimulus has been envisioned as a single or at most a few substances different from the known plant hormones (4, 91). The failure to identify the chemical nature of the floral stimulus has led some to question this hypothesis. As noted earlier, while there are numerous reports in which floral induction can be transferred between graft partners of different response types, there have also been many failures. Furthermore, with two exceptions, all successful grafts have been within families. The exceptions are the induction of flowering in the LDP *Silene* (Caryophyllaceae) following grafting to the SDP *Perilla* (Labiatae) and *Xanthium* (Compositae) (77, 80). These experiments were criticized, however, for not having enough non-induced control grafts (91).

Bernier and colleagues have suggested that instead of a unique and specific flower-inducing compound, the stimulus is primarily a certain balance of the known plant hormones and assimilates arriving at the apex in a specific sequence (3, 4, 5, 48). Support for this hypothesis comes from the observations that exogenous hormones can often mimic one or several evocational processes. For example, application of a single low dose of cytokinin to the apex of the LDP *Sinapis alba* under non-inductive conditions caused several evocational events including an increase in the mitotic index and the sub-division of vacuoles into smaller ones (3). Since cytokinin application cannot completely substitute for LD in inducing flowering, only "partial evocation" occurred (3, 4). Based on defoliation experiments, transport of a mitotic stimulus from the leaves began about 16 h after the beginning of the LD treatment (1). This corresponded with higher cytokinin-like activity in the leaves and an increase in cytokinin flux in the phloem (2, 48). Other evocational processes relating to energy metabolism seemed to be regulated by soluble sugars (5).

It is difficult, however, to reconcile this theory with some important experimental observations. First, if flower initiation is the sum of partial evocational events, then one should be able to induce flowering by using a combination of treatments. To date, this has not yet been successfully accomplished. Second, it is unlikely that a specific ratio of the known hormones arriving at the apex is the critical factor in floral initiation. It has been shown that a single leaf under inductive conditions can induce flower formation. In the presence of non-induced leaves, a special ratio of hormones emanating from the induced leaf would almost certainly be altered by the time it reached the apex (91). Third, it is difficult to see how a certain specific sequence of essential compounds could be an overriding factor in flower initiation. Leaves from induced *Perilla* plants continue to export the floral stimulus in non-inductive conditions for up to three months (83). Under these conditions how could the sequential arrival of hormones and intermediary metabolites at different apices be maintained? Fourth, there is always a danger in over-interpreting the results of experiments in which substances are applied to plants and assuming that the observed effect is identical to a certain aspect of evocation. The action of exogenous compounds can be indirect and this should be considered before formulating a hypothesis concerning the role of a hormone in flower formation. A case in point is the instance discussed earlier where application of cytokinins to the cotyledons of suboptimally induced plants of the SDP *Pharbitis nil* promoted flowering (59). It was shown that the action of cytokinin was in the cotyledons, not the apex. Furthermore, it was believed that the enhanced flowering response by cytokinin was due to greater export of assimilates along with a concomitant increase in the amount of floral stimulus moving from the cotyledons to the apex (59). Finally, the results of grafting experiments between different response types are more easily explained on the basis of a unique floral stimulus. The site of action of GA when applied to non-induced *Bryophyllum* plants is in the leaves not the apex (87). In this case, the action of exogenous GA seems to be linked to the production of a stimulus in the leaves, which is different from GA, and is capable of initiating the transition to floral development (87, 98). Further evidence for the uniqueness of the floral stimulus was obtained in grafting experiments with the SLDP *Echeveria harmsii* and *Bryophyllum*. *Echeveria* receptors flowered in SD when grafted onto *Bryophyllum* plants induced with GA_3. Since GA_3 does not induce flowering in *Echeveria* under any photoperiodic condition the inescapable conclusion is that the floral stimulus is different from GA (93). In a similar vein, two lines of the LDP *Silene armeria* have been selected based on their response to exogenous GA_3 in SD. One line will respond to GA_3 with both stem elongation and flower formation, while in the other, only stem elongation results. When the former is induced to flower in SD with exogenous GA_3 and grafted to the other, the receptor flowers (77). Again the most logical explanation is that GA is not a component of the floral stimulus, and that exogenous GA causes flowering indirectly by increasing production of the floral stimulus (77, 94).

The failure to successfully transfer flower induction in some grafting experiments cannot be taken as certain evidence against a ubiquitous floral stimulus. Sensitivity to the floral stimulus may vary among species with some requiring more stimulus than others for flower induction. Still another difference could be the amount of the floral stimulus that is produced and/or transported. *Bryophyllum* was an effective donor for the SDP *Kalanchoë blossfeldiana*. However, a weak flowering response was observed in the reciprocal graft, suggesting that the two species differ either in the amount of floral stimulus produced under inductive conditions, or their sensitivity to the floral stimulus (77).

Another possible explanation for negative results in some grafting experiments could be that the donor reverted to vegetative growth when moved back to noninductive conditions (91). When induced donors of the LDP *Blitum virgatum* were grafted to receptors of the same species in SD, no transfer of flower induction was observed. Yet flowering in a receptor could be induced in two-branched plants with the donor in LD and the receptor in SD (40, 41).

In conclusion, there is no compelling evidence to date that forces abandonment of the hypothesis that the floral stimulus is a single or a few substances unique from the presently known plant hormones and is very similar or even identical among higher plants. This does not mean, however, that plant hormones do not interact with and/or modulate the events initiated by the floral stimulus; indeed this is likely.

THE ROLE OF HORMONES IN OTHER ASPECTS OF REPRODUCTIVE DEVELOPMENT

Following the initiation of flower primordia a series of highly integrated and complex developmental steps occur that culminate in flower opening and fertilization. In the interests of space, only two aspects of this later development will be discussed: sex expression and stem/inflorescence growth in rosette plants. The role of GAs in other areas such as petal growth and pollen development are the subject of a recent review (61); literature encompassing the role of hormones in other aspects of reproductive development has also been reviewed (30,58,68).

Gibberellins and Stem Growth in Rosette Plants

As discussed in the previous section, many LDP and cold-requiring plants grow as rosettes prior to the inductive treatment. Flower formation in these plants is associated with rapid stem elongation (bolting) (Fig. 3). Although closely related temporally, flower formation and bolting are separable developmental processes. Many rosette plants will respond to exogenous GA with rapid stem elongation (Fig. 4). This suggests that GAs are somehow limiting in noninduced plants. Experiments with inhibitors of GA biosynthesis support this idea for both LDP (10, 36, 86) and cold requiring plants (39, 53, 77) (Fig. 5).

The biochemical basis for this limitation in *Spinacia* has been investigated in greater detail by Zeevaart and co-workers. Initially, six endogenous GAs in *Spinacia* (Fig. 6) were identified by combined gas chromatography-mass spectrometry (GC-MS) (51). Quantitative analysis of the endogenous GAs as a function of photoperiodic treatment showed that under SD, the level of GA_{19} was high while the level of GA_{20} was low (Fig.

Fig. 3. The effect of a transfer from SD to LD conditions on the growth habit of *Spinacia oleracea* L. cv. Savoy Hybrid 612. SD consisted of 8 hr of light from fluorescent and incandescent lamps followed by 16 hr of darkness. LD contained the same 8 hr light period as the SD conditions but was followed by 16 hr of low intensity illumination from incandescent lamps. Photograph by the author.

7). Following transfer to LD, the level of GA_{19} progressively declined. The GA_{20} content, on the other hand, increased in proportion to the lowering of the GA_{19} level (Fig. 7). An increase in the level of GA_{29} was also observed following the transfer to LD; however, the start of the increase lagged a day or so behind the increase in GA_{20} levels.

The rise in the levels of both GA_{20} and GA_{29} occurred just prior to the onset of stem elongation (Fig. 7). The other endogenous GAs (GA_{17} and GA_{44})

Fig. 4. Application of GA_3 to *Spinacia* plants under SD causes the plants to assume the LD growth habit (elongated petioles and stem). Ten micrograms of GA_3 were applied to the plant on alternate days for 10 days. Photograph courtesy of Dr. Jan A. D. Zeevaart.

Fig. 5. Inhibitors of GA biosynthesis retard the appearance of morphological characteristics associated with the transfer of *Spinacia* plants to LD. Approximately 17 mg of AMO 1618 [2-isopropyl-4-trimethylammonium chloride)-5-methylphenyl piperidine-1-carboxylate] in 10 ml of water was applied daily to the roots. The inhibition caused by the growth retardant was reversed with exogenous GA$_3$. Application of GA$_3$ was identical to that described in Fig. 4. Photograph courtesy of Dr. Jan A. D. Zeevaart.

remained fairly constant under either photoperiodic regime (52).

The rise in the level of GA$_{20}$ and the concomitant decline in the level of its immediate precursor, GA$_{19}$, suggests that the conversion of GA$_{19}$ to GA$_{20}$ is under photoperiodic control (52). This has subsequently been confirmed. [^{2}H]-GA$_{53}$ was metabolized by *Spinacia* shoots to [^{2}H]-GA$_{44}$ and [^{2}H]-GA$_{19}$ in SD, while in LD, [^{2}H]-GA$_{20}$ was also formed (26).

When plants were allowed to accumulate [^{2}H]-GA$_{44}$ and [^{2}H]-GA$_{19}$ in SD and then transferred to LD, a decline was observed in the levels of both [^{2}H]-GA$_{44}$ and [^{2}H]-GA$_{19}$ while at the same time the amount of [^{2}H]-GA$_{20}$ increased (26). In later work, photoperiodic regulation of the conversion of GA$_{19}$ to GA$_{20}$ was directly demonstrated using [^{14}C]-GA$_{19}$ (27).

Cell-free preparations from *Spinacia* leaves were made which could also catalyze this reaction. Enzyme activity was proportional to the length of the light period that the plants were exposed prior to preparation of the cell-free extract. Once the plants were returned to darkness, extractable enzyme activity declined rapidly (28, 29). An identical pattern for the light activation of the enzyme that catalyzes the conversion of GA$_{53}$ to GA$_{44}$ was also observed. The molecular basis for the light-induced increases in enzyme activity remains unknown (28, 29).

Subsequently, the relationship between photoperiod and endogenous GA levels was investigated in more detail. In addition to the six endogenous GAs previously identified, both GA$_1$ and GA$_8$ were also detected (Fig. 6); LD

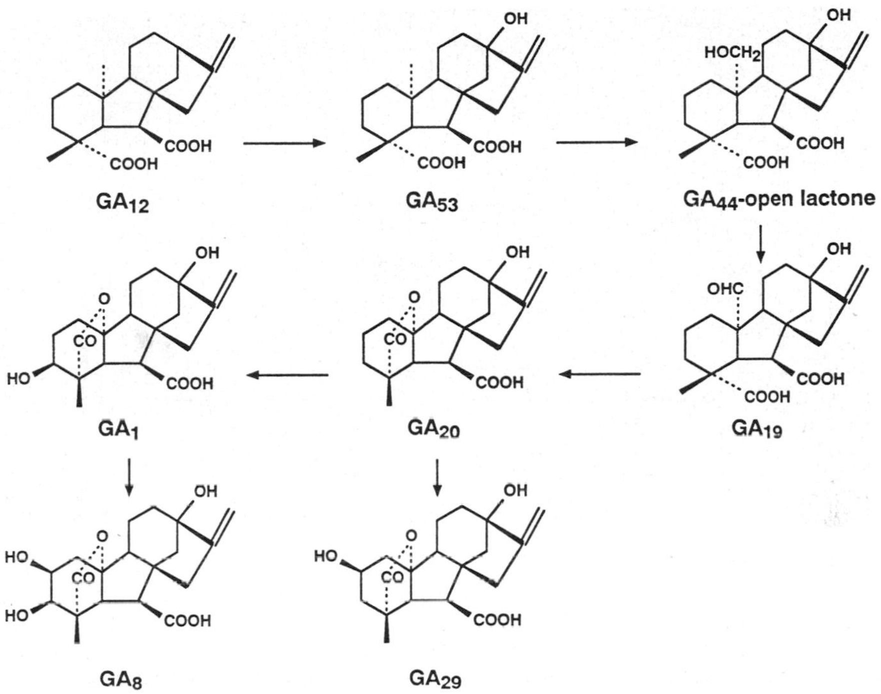

Fig. 6. The structure of the endogenous GAs identified in shoots of *Spinacia oleracea* L. and *Silene armeria* L. by combined gas chromatography-mass spectrometry. The structures are arranged in their proper biosynthetic sequence. In both *Spinacia* and *Silene*, GA_1 is the GA responsible for biological activity. The 2β-hydroxylated GAs GA_8 and GA_{29} represent deactivation products. From (51, 74, 76).

resulted in an increase in the levels of these two GAs (76, 97). Moreover, GA_1 was shown to be the GA responsible for bioactivity and, therefore, the increase in its endogenous level in LD is the primary factor for stem elongation (97). Nevertheless, the conversion of GA_{20} to GA_1 is not under direct photoperiodic control since the ability of exogenous GA_{20} and GA_1 to elicit stem growth in SD is the same (52).

Although less thoroughly investigated, LD-induced bolting in *Silene armeria* also appears to be mediated by an increase in endogenous GA_1 levels. When plants were transferred from SD to LD conditions there was a 10-fold increase in GA_1 levels. In addition, an excellent correlation was observed with the accumulation of GA_1 in the shoot subapical meristem and increased mitotic activity in that tissue which forms the cellular basis for LD-induced stem growth (74, 75). The higher GA_1 levels appear to be the result of enhanced metabolism of GA_{53} (74).

In a situation analogous to LD-induced bolting in *Spinacia*, thermoinduced stem growth in the cold requiring plant *Thlaspi arvense*, a cruciferous winter annual weed, is associated with the activation of specific

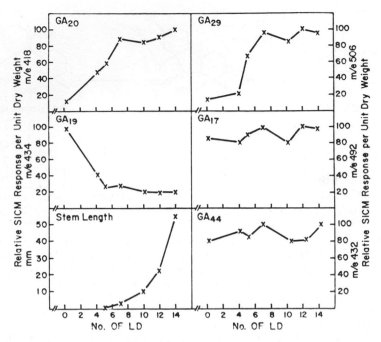

Fig. 7. Changes in the relative levels of five GAs and stem length in *Spinacia* shoots as affected by different durations of LD treatment. GA levels measured by combined gas chromatography selected ion current monitoring. The highest concentration (ion current response/unit dry wt.) of each GA was arbitrarily assigned a value of 100, and the other concentrations were expressed in proportion to this value. The ion current response of the molecular ion was used in all cases except for GA_{19}, in which case the base peak was used. From (51).

steps in GA metabolism. By comparing the abilities of various exogenous GAs and GA precursors in GA-depleted vernalized and non-vernalized plants, it was inferred that the metabolism of *ent*-kaurenoic acid (KA) to GAs was blocked prior to vernalization. This led to the hypothesis that thermoinduced stem growth in *Thlaspi* is the result of increased conversion of KA to GAs induced by vernalization (55). Corroborating evidence for this hypothesis was obtained comparing the metabolism of exogenous [^{2}H]-KA in vernalized and non-vernalized plants. The metabolism of exogenous [^{2}H]-KA was roughly 19 times more rapid in shoot tips of vernalized plants. Although [^{2}H]-labeled GAs were detected only in the induced material, the amount of incorporation of deuterium was extremely low. In contrast, the metabolism of [^{2}H]-KA to GAs in the leaves was unaffected by vernalization (34). In addition, quantitative analysis by GC-MS showed that vernalization resulted in a rapid and dramatic decline in endogenous KA levels in shoot tips, while no effect was observed in the leaves (Fig. 8). These results are consistent with the observation that the site perception of thermoinductive temperatures in *Thlaspi* is the shoot apical meristem (35, 54).

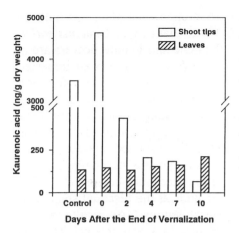

Fig. 8. Changes in the endogenous levels of *ent*-kaurenoic acid in *Thlaspi* shoot tips and leaves when the plants were returned to 21°C following a four week vernalization treatment at 6°C. *ent*-Kaurenoic acid levels determined by combined gas chromatography-chemical ionization mass spectrometry using [²H]-labeled *ent*-kaurenoic acid as an internal standard. Control represents the value obtained from non-vernalized plants. Adapted from (35).

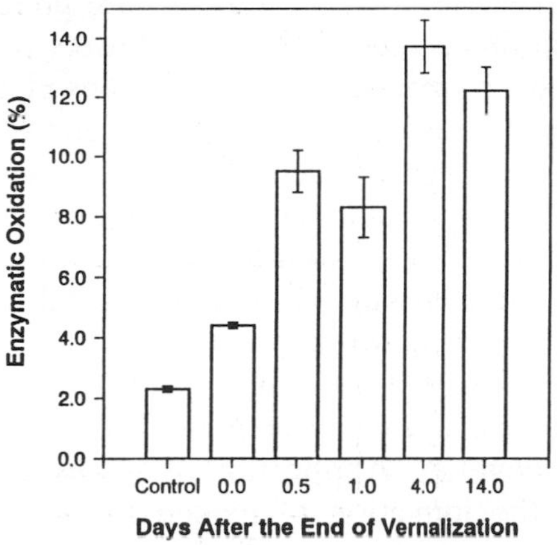

Fig. 9. The kinetics of changes in *ent*-kaurenoic acid hydroxylase activity in *Thlaspi* shoot tips following a four week vernalization treatment at 6°C. Enzyme activity measured by assessing the ability of microsomal preparations to convert [¹⁴C]-*ent*-kaurenoic acid to [¹⁴C]-*ent*-7α-hydroxykaurenoic acid. Data presented as the percentage of total extractable radioactivity that co-chromatographed with authentic *ent*-7α-hydroxykaurenoic acid. Control represents the value obtained from non-vernalized plants. Vertical bars show the standard error of the mean. Adapted from (35).

The biochemical basis for the thermoinduced change in KA metabolism was investigated. A cell free system derived from *Thlaspi* tissues that enzymatically converts KA to *ent*-7αhydroxy kaurenoic acid (KA hydroxylase) was developed (35). This enzyme has properties suggesting that it is a cytochrome P450-linked mixed function oxidase (42). Vernalization resulted in the rapid induction of enzyme activity in the shoot tips shortly after the plants were returned to warm temperatures following a four week cold treatment (Fig. 9). In contrast to the situation in *Spinacia* in which enzyme activity is reversibly modulated by light (28), the thermoinduced increase in KA-hydroxylase activity in *Thlaspi* shoot tips is permanent and occurs after the stimulus is withdrawn (35). Vernalization had no effect on KA-hydroxylase activity in the leaves (35).

Hormones and Sex Expression

There are two basic groups of flowers: they are either perfect, in which the flower contains both stamens and pistils, or they are imperfect, in which case either the pistils or stamens are present. A monoecious plant has both the staminate and the pistillate flowers on the same plant (e.g., *Cucumis sativus* and *Zea mays*) while in dioecious plants, the male and female flowers are on separate plants (e.g., *Spinacia* and *Cannabis sativa*). The sex of imperfect flowers has a genetic basis, but environmental factors such as photoperiod, temperature, and nitrogen status have an influential role (38, 68). Exogenous hormones can modify the sexuality of flowers suggesting that hormones mediate genetic and environmental control of sex expression (8). The role of hormones in sex expression of three species is discussed below.

The role of hormones in sex expression is perhaps best understood in cucurbits. In *Cucumis*, perfect flowers are initiated but one sex organ fails to develop. Application of auxins to flower buds at the bisexual stage leads to the formation of female flowers, while exogenous GAs results in male flower formation (61, 67). It now appears that IAA acts through ethylene (70). Treatments which reduce endogenous ethylene levels promote maleness (6). Inhibitors of GA biosynthesis, on the other hand, cause a tendency towards feminization (32, 61).

Endogenous hormone levels are also consistent with their postulated role in sex expression. Shoots of a hermaphroditic line of *Cucumis* had a higher auxin content than an andromonoecious line (24). Endogenous levels of ethylene and GA-like substances were also correlated with sex expression: high levels of ethylene production were associated with plants containing pistillate flowers (6, 66) and staminate plants contained more GA-like substances than their female counterparts (67). In total, the results strongly suggest that sex expression in *Cucumis* and cucurbits in general is regulated by the internal balance of auxins acting through ethylene and GAs.

Nevertheless, the hormonal balance may not be the sole factor in determining sex expression in *Cucumis*. In a *C. sativus* cultivar in which the number of pistillate flowers increases in response to SD, the endogenous levels of GA-like substances were higher and ethylene production lower in SD than LD, the opposite to what one would predict (73). This suggests that environmental control of sex expression in this species is not necessarily mediated through a balance of GAs and ethylene.

Cannabis sativa is a dioecious plant in which, like *Cucumis*, flower primordia are uncommitted at the time of flower initiation. Application of auxins, ethylene, and GAs affect sex expression in similar fashion as in *Cucumis* (8, 38, 61). Growth retardants feminize *Cannabis* plants, which can be reversed by GA_3 (8). In addition, cytokinins also promote femaleness (8). Following experiments in which plants were defoliated or derooted, it was proposed that leaves play an essential role in sex expression in *Cannabis* by supplying GAs to the flower bud (8). Male plants tend to have higher levels of GA-like substances as well. An increased cytokinin content was associated with female plants (8).

In *Spinacia*, sex expression is under genetic control. There are two sex chromosomes designated as X and Y. Female plants have a genotype of XX and male plants are XY. Photoperiod also influences sex expression. LD increases a tendency towards femaleness despite a genotype of XY. Again, as in many other species, auxins, ethylene, and cytokinins promote female flowers while treatment with GA_3 increases the tendency for the formation of male flowers (8, 13).

The qualitative pattern of GA-like substances was different in extracts from male and female plants. Female plants contained high amounts of a GA-like substance that was probably GA_{19} and low amounts of a GA-like substance that was probably GA_{20}. Male plants showed the opposite pattern (13). Since GA_{20} is the direct precursor of GA_1, the GA responsible for biological activity in *Spinacia* (76, 97), the increased level of GA_{20} in male plants is consistent with a role of GAs in sex expression. However, the trend towards feminization observed when plants are subjected to LD is obviously inconsistent with this hypothesis. In addition, male plants bolt sooner in LD than female plants, but both types flower at about the same time. This might be an indication that the higher levels of GA_{20} in male plants are a reflection of more rapid stem elongation and may not have anything to do with sex expression.

FINAL COMMENTS

In countless experiments, plant hormones have been applied to plants in a pharmacological approach to determine the basis for the control of the transition from a vegetative to the reproductive state. Needless to say, the fruits of thousands of hours of labor do not provide us with any certain mechanisms. Perhaps the overriding impediment is the paucity of information about the floral stimulus. Although some success in isolating active fractions has been reported (9, 49, 50) the results have not been reproduced in other laboratories (91). Until we know the nature of the floral stimulus--whether it is a unique morphogen or a combination of many factors--it is unlikely that much will be learned about the internal mechanisms controlling reproductive development.

Why has it been so difficult to isolate and characterize the floral stimulus? Fundamental to this problem is that there is no suitable assay. Without such an assay the difficulties simply compound themselves into a vicious circle: how can the presence of an active principle be ascertained if there is no sure way of detecting it? On the other hand, perhaps the lack of an assay is more apparent than real because the active stimulus was not present in any of the fractions that were applied. It should be noted that the majority of extraction experiments assumed *a priori* that because many other plant hormones are acidic compounds soluble in organic solvents, the floral stimulus has similar chemical properties. But what if the floral stimulus were a peptide, carbohydrate, or a volatile lactone? Certainly, these possibilities were precluded by the extraction methods employed in the earlier work.

Other uncertainties are lability and method of application to assay plants. While this bewildering maze of obstacles may cause most experimenters to shrug their shoulders and move onto other areas, it would be wise to bear in mind the comment by Zeevaart (76) that with the exception of ABA none of the other hormones were discovered by extracting higher plant tissue.

Flowering is almost certainly the result of selective gene expression--both turning on and switching off of genes. In a number of species, mutants for various aspects of floral development have been isolated, (11, 56). Use of the techniques in molecular biology to analyze the structure and function of genes has been quite useful in determining the molecular basis for these mutations. Already this kind of approach has allowed advances in the understanding how genes control, on a molecular level, floral development and the specification of floral organs (11, 57, 69). While this avenue would not directly lead to the identification of the floral stimulus, it certainly would tell us the molecular details of what the floral stimulus does. Such a vantage point could provide the insights necessary for a breakthrough.

References

1. Bernier, G., Bodson, M., Kinet, J-M., Jacqmard, A., Havelange, A. (1974) The nature of the floral stimulus in mustard. *In*: Plant growth substances 1973, pp. 980-986. Hirokawa, Tokyo.
2. Bernier, G., Kinet, J-M., Claes, A. (1979) The role of cytokinin in floral evocation in *Sinapis alba*. Tenth International Conference on Plant Growth Substances, Madison, Wisconsin, pg. 30 (abstract).
3. Bernier, G., Kinet, J-M., Jacqmard, A., Havelange, A., Bodson, M. (1977) Cytokinin as a possible component of the floral stimulus in *Sinapis alba*. Plant Physiol. 60, 282-285.
4. Bernier, G., Kinet, J-M., Sachs, R.M. (1981) The physiology of flowering, Vol. I, The initiation of flowers. CRC Press, Boca Raton.
5. Bernier, G., Kinet, J-M., Sachs, R.M. (1981) The physiology of flowering, Vol. II, Transition to reproductive growth. CRC Press, Boca Raton.
6. Byers, R.E., Baker, L.R., Sell, H.M., Herner, R.C., Dilley, D.R. (1972) Ethylene: A natural regulator of sex expression in *Cucumis melo* L. Proc. Natl. Acad. Sci. USA 69, 717-720.
7. Chailakhyan, M.Kh. (1936) On the hormonal theory of plant development. Dokl. Acad. Sci. USSR 12, 443-447 (In Russian).
8. Chailakhyan, M.Kh., Khryanin, V.N. (1980) Hormonal regulation of sex expression in plants. *In*: Plant growth substances 1979, pp. 331-344, Skoog, F., ed. Springer-Verlag, New York.
9. Chailakhyan, M.Kh., Lozhnikova, V., Seidlova, F., Krekule, J., Dudko, N., Negretzky, V. (1989) Floral and growth responses in *Chenopodium rubrum* L. to an extract from flowering *Nicotiana tabacum* L. Planta 178, 143-146.
10. Cleland, C.F., Zeevaart, J.A.D. (1970) Gibberellins in relation to flowering and stem elongation in the long day plant *Silene armeria*. Plant Physiol. 46, 392-400.
11. Coen, E.S. (1991) The role of homeotic genes in flower development and evolution. Annu. Rev. Plant Physiol. Plant Mol. Biol. 42, 241-279.
12. Crozier, A., Durley, R.C. (1983) Modern methods of analysis of gibberellins. *In*: The biochemistry and physiology of gibberellins, Vol. II, pp. 485-538, Crozier, A., ed. Praeger, New York.
13. Cúlafic, L., Neskovic, M. (1974) A study of auxins and gibberellins during shoot development in *Spinacia oleracea* L. Arh. Biol. Nauka. Beograd. 26, 19-27.

14. DeProft, M., Van Dijck, R., Philippe, L., DeGreet, J.A. (1985) Hormonal regulation of flowering and apical dominance in bromeliad plants. Twelfth International Conference on Plant Growth Substances, Heidelberg, FDR., pg. 93 (abstract).

15. Doorenbos, J. (1954) "Rejuvenation" of *Hedera helix* in graft combinations. Proc. Koninkl. Ned. Akad. Wetenschap., Ser. C 57, 99-102.

16. Doorenbos, J. (1965) Juvenile and adult phases in woody plants. *In*: Encyclopedia plant physiol, Vol. XV/1, pp. 1222-1235, Ruhland, W., ed. Springer, Berlin.

17. El-Antably, H.M.M., Wareing, P.F. (1966) Stimulation of flowering in certain short-day plants by abscisin. Nature 210, 328.

18. El-Antably, H.M.M., Wareing, P.F., Hillman, J. (1967) Some physiological responses to d.l. abscisin (dormin). Planta 73,74-90.

19. Evans, L.T. (1966) Abscisin. II. Inhibitory effect on flower induction in a long-day plant. Science 151, 107.

20. Evans, L.T., King, R.W., Chu, A., Mander, L.N., Pharis, R.P. (1990) Gibberellin structure and florigenic activity in *Lolium temulentum*, a long-day plant. Planta 182, 97-106.

21. Frank, H., Renner, O. (1956) Uber verjungung bei *Hedera helix* L. Planta 47, 105-14.

22. Frydman, V.M., Wareing, P.F. (1973) Phase change in *Hedera helix* L. I. Gibberellin-like substances in the two growth stages. J. Exp. Bot. 24, 1131-1138.

23. Frydman, V.M., Wareing, P.F. (1974) Phase change in *Hedera helix* L. III. The effects of gibberellins, abscisic acid and growth retardants on juvenile and adult ivy. J. Exp. Bot. 25, 420-429.

24. Galun, E., Izhsar, S., Atsmon, D. (1965) Determination of relative auxin content in hermaphrodite and andromonoecious *Cucumis sativus* L. Plant Physiol. 40, 321-326.

25. Gaskin, P., MacMillan, J., Zeevaart, J.A.D. (1973) Identification of gibberellin A_{20}, abscisic acid, and phaseic acid from flowering *Bryophyllum daigremontianum* by combined gas chromatography-mass spectrometry. Planta 111, 347-352.

26. Gianfagna, T., Zeevaart, J.A.D., Lusk, W.J. (1983) The effect of photoperiod on the metabolism of deuterium-labeled GA_{53} in spinach. Plant Physiol. 72, 86-89.

27. Gilmour, S.J., Zeevaart, J.A.D., Schwenen, L., Graebe, J.E. (1985) The effect of photoperiod on gibberellin metabolism in cell-free extracts from spinach. Plant Physiol. 77(S), 92.

28. Gilmour, S.J., Zeevaart, J.A.D., Schwenen, L., Graebe, J.E. (1986) Gibberellin metabolism in cell-free extracts from spinach leaves in relation to photoperiod. Plant Physiol. 82, 190-195.

29. Gilmour, S.J., Bleecker, A.B., Zeevaart, J.A.D. (1987) Partial purification of gibberellin oxidases from spinach leaves. Plant Physiol. 85, 87-90.

30. Goodwin, P.B. (1978) Phytohormones and fruit growth. *In*: Phytohormones and related compounds: A comprehensive treatise, Vol. II, pp. 175-249, Letham, D.S., Goodwin, P.B., Higgins, T.J.V. eds. Elsevier, Amsterdam.

31. Guttridge, C.G. (1959) Evidence for a flower inhibitor and vegetative growth promoter in the strawberry. Ann. Bot. 23, 351-360.

32. Halevy, A.H., Rudich, J. (1967) Modification of sex expression in muskmelon by treatment with the growth retardant B-995. Physiol. Plant. 20, 1052-1058.

33. Harada, H. (1962) Etude des substances naturelles de croissance en relation avec la floraison-isolement d'une substance de montaison. Rev. Gén. Bot. 69, 201-297.

34. Hazebroek, J.P., Metzger, J.D. (1990) Thermoinductive regulation of gibberellin metabolism in *Thlaspi arvense* L. I. Metabolism of [^{2}H]-*ent*-kaurenoic acid and [^{14}C]gibberellin A_{12}-aldehyde. Plant Physiol. 94, 157-165.

35. Hazebroek, J.P., Metzger, J.D. (1993) Thermoinductive regulation of gibberellin metabolism in *Thlaspi arvense* L. II. Cold induction of enzymes in gibberellin biosynthesis. Plant Physiol. 102, 547-552.

36. Hedden, P., MacMillan, J., Phinney, B.O. (1978) The metabolism of the gibberellins. Annu. Rev. Plant. Physiol. 29, 149-192.

37. Henson, I.E., Wareing, P.F. (1977) Cytokinins in *Xanthium strumarium* L.: Some aspects of the photoperiodic control of endogenous levels. New Phytol. 78, 35-45.

38. Heslop-Harrison, J. (1972) Sexuality of angiosperms. *In*: Plant Physiology: A treatise. Vol. VIC, Physiology of development: From seeds to sexuality, pp. 133-290, Steward, F.C., ed. Academic Press, New York.

39. Hiller, L.K., Kelly, W.C., Powell, L.E. (1979) Temperature interactions with growth regulators and endogenous gibberellin-like activity during seedstalk elongation in carrots. Plant Physiol. 63, 1055-1061.

40. Jacques, M. (1969) Les differents aspects morphologiques florcusion chez les *Blitum capitatum* et *virgatum* en rapport avec les modalités des processus d' induction floral. C.R. Acad. Sci. 268D, 1045-1047.

41. Jacques, M. (1973) Transfert par voie de greffage du stimulus photoperiodique. C.R. Acad. Sci. 276D, 1705-1708.

42. Jennings, J.C., Coolbaugh, R.C., Nakata, D.A., West, C.A. (1993) Characterization and solubilization of kaurenoic acid hydroxylase from *Gibberella fujikuroi*. Plant Physiol. 101, 925-930.

43. Jones, M.G., Zeevaart, J.A.D. (1980) Gibberellins and the photoperiodic control of stem elongation in the long-day plant, *Agrostemma githago* L. Planta 149, 269-273.

44. King, R.W., Zeevaart, J.A.D. (1973) Floral stimulus movement in *Perilla* and flower inhibition caused by noninduced leaves. Plant Physiol. 51, 727-738.

45. King, R.W., Evans, L.T., Firn, R.D. (1977) Abscisic acid and xanthoxin contents in the long-day plant *Lolium temulentum* L. in relation to photoperiod. Aust. J. Plant Physiol. 4, 217-223.

46. Lang, A. (1965) Physiology of flower initiation. *In*: Encyclopedia of plant physiology, pp. 1380-1536, Ruhland, W., ed. Springer-Verlag, Berlin.

47. Lang, A., Chailakhyan, M.Kh., Frolova, I.A. (1977) Promotion and inhibition of flower formation in a dayneutral plant in grafts with a short-day plant and a long-day plant. Proc. Natl. Acad. Sci. USA 74, 2412-2416.

48. Lejeune, P., Kinet, J.-M., Bernier, G. (1988) Cytokinin fluxes during floral induction in the longday plant *Sinapis alba* L. Plant Physiol. 86, 1095-1098.

49. Lincoln, R.G., Cunningham, A., Hamner, K.C. (1964) Evidence for a florigenic acid. Nature 202, 559-561.

50. Lincoln, R.G., Mayfield, D.L., Cunningham, A. (1961) Preparation of a floral initiating extract from *Xanthium*. Science 133, 756.

51. Metzger, J.D., Zeevaart, J.A.D. (1980) Identification of six endogenous gibberellins in spinach shoots. Plant Physiol. 65, 623-626.

52. Metzger, J.D., Zeevaart, J.A.D. (1980) Effect of photoperiod on the levels of endogenous gibberellins in spinach as measured by combined gas chromatography-selected ion current monitoring. Plant Physiol. 66, 844-846.

53. Metzger, J.D., (1985) Role of gibberellins in the environmental control of stem growth in *Thlaspi arvense* L. Plant Physiol 78, 8-13.

54. Metzger, J.D. (1988) Localization of the site of perception of thermoinductive temperatures in *Thlaspi arvense* L. Plant Physiol. 88, 424-428.

55. Metzger, J.D. (1990) Comparison of biological activities of gibberellins and gibberellin-precursors native to *Thlaspi arvense* L. Plant Physiol. 94, 151-156.

56. Meyerowitz, E.M., Smyth, D.R., Bowman, J.L. (1989) Abnormal flowers and pattern formation in floral development. Development 106, 209-217.

57. Meyerowitz, E.M., Bowman, J.L., Brockman, L.L., Drews, G.L., Jack, T., Sieburth, L.E., Weigel, D. (1991) A genetic and molecular model for flower development in *Arabidopsis thaliana*. Development 112 (Suppl. 1), 157-167.

58. Nitsch, J.P. (1965) Physiology of flower and fruit development. *In*: Encyclopedia of plant physiology, Vol. 15, Part 1, pp. 1537-1647, Ruhland, W., ed. Springer-Verlag, Berlin.

59. Ogawa, Y., King, R.W. (1979) Indirect action of benzyladenine and other chemicals on flowering of *Pharbitis nil* Chois. Plant Physiol. 63, 643-649.

60. Pharis, R.P., Morf, W. (1968) Physiology of gibberellin induced flowering in conifers. *In*: Biochemistry and physiology of plant growth substances, pp. 1341-1356, Wightman, F., Setterfield, G., eds. Runge Press, Ottawa, Canada.

61. Pharis, R.P., King, R.W. (1985) Gibberellins and reproductive development in seed plants. Annu. Rev. Plant Physiol 36, 517-568.

62. Pharis, R.P., Evans, L.T., King, R.W., Mander, L.N. (1987) Gibberellins, endogenous and applied, in relation to flower induction in the longday plant *Lolium temulentum*. Plant Physiol. 84, 1132-1138.

63. Pharis, R.P., Evans, L.T., King, R.W., Mander, L.N. (1989) Gibberellins and flowering in higher plants—Differing structures yield highly specific effects. *In*: Plant reproduction: From floral induction to pollination, pp. 29-41, Bernier, G., Lord, E., eds. Am. Soc. Plant Physiol., Rockville, MD.

64. Robinson, L.W., Wareing, P.F. (1969) Experiments on the juvenile-adult phase change in some woody species. New Phytol. 68, 67-78.

65. Rogler, C.E., Hackett, W.P. (1975) Phase change in *Hedera helix*. Stabilization of the mature form with abscisic acid. Physiol. Plant. 34, 148-152.

66. Rudich, J., Halevy, A.H., Kedar, N. (1972) Ethylene evolution from cucumber plants as related to sex expression. Plant Physiol. 49, 998-999.

67. Rudich, J., Halevy, A.H., Kedar, N. (1972) The level of phytohormones in monoecious and gynoecious cucumbers as affected by photoperiod and ethephon. Plant Physiol. 50, 585-590.

68. Schwabe, W.W. (1971) Physiology of vegetative reproduction and flowering. *In*: Plant physiology: A treatise. Vol. VIA, Physiology of development: Plants and their reproduction, pp. 233-411, Steward, F.C., ed. Academic Press, New York.

69. Schwartz-Sommer, Z., Huijser, P., Nacken, W., Saedler, H. (1990) Genetic control of flower development by homeotic genes in *Antirrhinum majus*. Science 250, 931-936.

70. Shannon, S., De LaGuardia, M.D. (1969) Sex expression on the production of ethylene induced by auxin in the cucumber (*Cucumis sativum* L.). Nature 223, 186.

71. Suge, H. (1980) Inhibition of flowering and growth in *Pharbitis nil* by ancymidol. Plant Cell Physiol. 21, 1187-1192.

72. Suge, H., Rappaport, L. (1968) Role of gibberellins in stem elongation and flowering in radish. Plant Physiol. 43, 1208-1214.

73. Takahashi, H., Saito, T., Suge, H. (1983) Separation of the effects of photoperiod and hormones on sex expression in cucumber. Plant and Cell Physiol. 24, 147-154.

74. Talon, M., Zeevaart, J.A.D. (1990) Gibberellins and stem growth as related to photoperiod in *Silene armeria* L. Plant Physiol. 92, 1094-1100.

75. Talon, M., Tadeo, F.R., Zeevaart, J.A.D. (1991) Cellular changes induced by exogenous and endogenous gibberellins in shoot tips of the long-day plant *Silene armeria*. Planta 185, 487-493.

76. Talon, M., Zeevaart, J.A.D., Gage, D.A. (1991) Identification of gibberellins in spinach and the effect of light and darkness on their levels. Plant Physiol. 97, 1521-1526.

77. Van de Pol, P.A. (1972) Floral induction, floral hormones and flowering. Meded. Landbouwhogesch. Wageningen 72, 1-89.

78. Vince-Prue, D. (1975) Photoperiodism in plants. McGraw Hill and Co., London.

79. Vince-Prue, D. (1983) Photomorphogenesis and flowering. *In*: Encyclopedia of plant physiology (N.S.), Vol. 16b, pp. 457-490, Shropshire Jr., W., Mohr, H., eds. Springer-Verlag, Berlin.

80. Wellensiek, S.J. (1970) The floral hormones in *Silene armeria* L. and *Xanthium strumarium* L. Z. Pflanzenphysiol. 63, 25-30.

81. Wilson, R.N. (1992) Gibberellin is required for flowering in *Arabidopsis thaliana* under short days. Plant Physiol. 100, 403-408.
82. Zanewich, K.P. (1993) Vernalization and gibberellin physiology of winter canola. M.Sc. University of Lethbridge, Lethbridge, AB, Canada.
83. Zeevaart, J.A.D. (1968) Flower formation as studied by grafting. Meded. Landbouwhogesch. Wageningen 58, 1-88.
84. Zeevaart, J.A.D. (1964) Effects of the growth retardant CCC on floral initiation and growth in *Pharbitis nil*. Plant Physiol. 39, 402-408.
85. Zeevaart, J.A.D. (1968) Vernalization and gibberellin in *Lunaria annua* L. *In*: Biochemistry and physiology of plant growth substances, pp. 1357-1370, Wightman, F., Setterfield, P., eds. Runge Press, Ottawa.
86. Zeevaart, J.A.D. (1971) Effects of photoperiod on growth rate and endogenous gibberellins in the long-day rosette plant spinach. Plant Physiol. 47, 821-827.
87. Zeevaart, J.A.D. (1969) *Bryophyllum*. *In*: The induction of flowering: Some case histories, pp. 435-456, Evans, L.T., ed. Cornell Univ. Press, Ithaca, New York.
88. Zeevaart, J.A.D. (1969) The leaf as a site of gibberellin action in flower formation in *Bryophyllum daigremontianum*. Planta 84, 339-347.
89. Zeevaart, J.A.D. (1971) (+)-Abscisic acid content of spinach in relation to photoperiod and water stress. Plant Physiol. 49, 86-90.
90. Zeevaart, J.A.D. (1973) Gibberellin GA_{20} content of *Bryophyllum daigremontianum* under different photoperiodic conditions as determined by gas-liquid chromatography. Planta 114, 285-288.
91. Zeevaart, J.A.D. (1976) Physiology of flower formation. Annu. Rev. Plant. Physiol. 27, 321-348.
92. Zeevaart, J.A.D. (1978) Phytohormones and flower formation. *In*: Plant hormones and related compounds, Vol. II, pp. 291-327, Letham, D.S., Goodwin, P.B., Higgins, T.J.V., eds. Elsevier/North Holland, Amsterdam.
93. Zeevaart, J.A.D. (1982) Transmission of the floral stimulus from a long-short-day plant, *Bryophyllum daigremontianum*, to the short-long-day plant *Echeveria harmsii*. Ann. Bot. 49, 549-552.
94. Zeevaart, J.A.D. (1983) Gibberellins and flowering. *In*: The biochemistry and physiology of gibberellins, Vol. 2, pp. 333-374, Crozier, A., ed. Praeger, New York.
95. Zeevaart, J.A.D., Brede, J.M., Cetas, C.B. (1977) Translocation patterns in *Xanthium* in relation to long day inhibition of flowering. Plant Physiol. 60, 747-753.
96. Zeevaart, J.A.D., Gage, D.A. (1993) *ent*-Kaurene biosynthesis is enhanced by long photoperiods in the long-day plants *Spinacia oleracea* L. and *Agrostemma githago* L. Plant Physiol. 101, 25-29.
97. Zeevaart, J.A.D., Gage, D.A., Talon, M. (1993) Gibberellin A_1 is required for stem elongation in spinach. Proc. Natl. Acad. Sci. USA 90, 7401-7405.
98. Zeevaart, J.A.D., Lang, A. (1962) The relationship between gibberellin and floral stimulus in *Bryophyllum daigremontianum*. Planta 53, 531-542.
99. Zeevaart, J.A.D., Lang, A. (1963) Suppression of floral induction in *Bryophyllum daigremontianum* by a growth retardant. Planta 59, 509-517.
100. Zimmerman, R.H., Hackett, W.P., Pharis, R.P. (1985) Hormonal aspects of phase change and precocious flowering. *In*: Encyclopedia of plant physiology (NS), Vol. II, pp. 79-115, Pharis, R.P., Reid, D.M., ed. Springer-Verlag, New York.

G9. The Role of Hormones in Photosynthate Partitioning and Seed Filling

Mark L. Brenner[1] and Nordine Cheikh[2]

[1]Department of Horticultural Science, and [2]Department of Agronomy and Plant Genetics, University of Minnesota, St. Paul, Minnesota 55108, USA.

INTRODUCTION

The movement of photoassimilates from sites of synthesis in leaf tissue (source) to the sites of net accumulation in a different tissue (sink) potentially can be regulated at numerous points. Regulation of the net flow of photoassimilates is an integrated process. It is generally accepted that the concentration gradient of photoassimilates between the source and sink is the primary determinant of the current rate of transport and pattern of partitioning (20, 26, 84). However, close examination of the various components involved in the overall process of partitioning indicates that endogenous plant hormones may serve as modulators of many of the specific rate limiting components. This chapter focuses on the involvement of plant hormones as natural regulators of partitioning of photoassimilates especially to developing seeds.

THE PATHWAY OF PHOTOSYNTHATE PARTITIONING

In simplest terms, regulation of photosynthate partitioning can occur within the leaf, along the transport pathway, or within the seed. For clarity, each of these components will be discussed separately (Fig. 1).

The extent of partitioning within the leaf may be controlled by the availability of recently fixed carbon which is determined by the rate of photosynthesis itself (26). The recently fixed carbon is converted to triose-phosphate (triose-P) which in turn either is converted into starch for storage within the chloroplast or is available for export through the chloroplast envelope to the cytosol. Formation of sucrose from triose-P involves both cytosolic fructose-1,6-bisphosphatase (FBPase), sucrose-phosphate-synthase (SPS), and sucrose phosphatase. The release of inorganic phosphate (Pi) from triose-P during sucrose synthesis stimulates triose-P export by the phosphate transporter ("PT" in Fig. 1) in the chloroplast membrane (73).

649

P. J. Davies (ed.), Plant Hormones, 649–670.

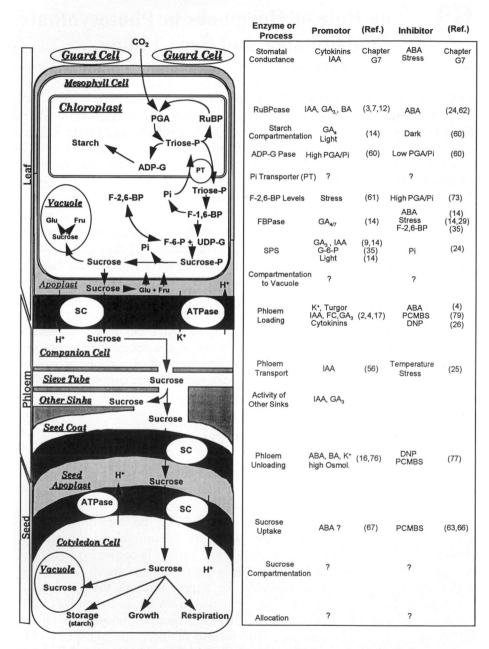

Fig. 1. Schematic representation of the path and possible control points of photosynthate metabolism and partitioning to developing seeds. See text for details including definition of abbreviations. SC=sucrose carrier.

In many plants sucrose is the prime sugar exported from source tissue to sinks. Sucrose produced within mesophyll cells can be partitioned to either the vacuole for temporary storage (20) or released to the apoplast where it subsequently moves to the sieve elements and companion cells where it is loaded for transport in the phloem (26). Once in the phloem, sucrose is transported and partitioned to other parts of the plant such as stems, roots, expanding leaves, axillary buds and reproductive organs (all referred to as sinks) where it is either utilized for growth and maintenance or storage. Sucrose imported by developing seeds differs from that of other parts of the plant in that phloem transport terminates in the maternal tissue and photoassimilates must move apoplastically to the developing zygotic tissue where they are absorbed and utilized (77). The unloading of phloem for seed development occurs in the testa of dicotyledonous species while in monocotyledonous plants unloading occurs either in placental-chalazal tissue of plants like maize (*Zea mays* L.) or in pericarp tissue of plants like wheat (*Triticum aestivum* L.), barley (*Hordeum vulgare* L.) and rice (*Oryza sativa* L.).

CORRELATION OF ENDOGENOUS PLANT HORMONES WITH SOURCE-SINK RELATIONS

Hormone Content and Distribution in Developing Seeds

To consider hormonal regulation of source-sink relations, it is important to first understand where the hormones occur and if possible to understand their respective sites of origin. While the occurrence of plant hormones in specific tissues does not prove hormonal role in source-sink relations, patterns of occurrence of hormones in relation to changes in source-sink processes might indicate the involvement of hormones. In general, under normal conditions developing seeds have higher concentrations of plant hormones than all other plant parts. There are numerous reports of high concentrations of auxins, cytokinins, gibberellins (GAs) and abscisic acid (ABA) in developing seeds (40). As will be described below, several authors have reported that hormone content changes distinctly during seed and fruit development.

In early works, hormonal changes were measured only on a total seed basis. With the possible exception of cytokinins, the maximum amount of hormones generally occurs during the time of rapid dry matter accumulation (rapid filling period). From later work, it has become evident that hormonal content may be vastly different in the various seed tissues. On a whole tissue basis, the greatest quantity of ABA is generally found in the prime storage tissue of seeds or grain such as cotyledons of soybeans (*Glycine max* L.[67]), or embryo and endosperm of maize (38). However, on a concentration basis, a measure presumably of greater physiological significance, ABA is found to be present in higher concentrations in the embryonic axis of soybean (30) and in the embryo of maize (38). ABA is also found in high concentrations in

651

the tissues where phloem unloading occurs, in the seed coats of soybeans (30, 67). Likewise, in maize, ABA is found in high concentrations in the pedicel/placental-chalazal tissue prior to full growth of the embryo (38), the tissue in which unloading occurs in this species. In general the highest concentrations of ABA found in these sites of unloading are observable during the rapid filling stage of seed development.

The occurrence of indoleacetic acid (IAA) in developing soybean seeds on a total seed basis is similar to that for ABA except the maximum amount of IAA is observed several days before the maximum for ABA. IAA is found in highest concentration in the embryonic axis at the time of maximum pod wall elongation, while in the seed coat, IAA reaches a distinct maximum coincident with maximum seed filling (30). In peas (*Pisum sativum*), high concentrations of auxin-like[1] materials are found in the liquid endosperm at the time of maximum pod elongation, and in the embryo (cotyledon and embryonic axis) a small increase in auxin-like material is observed coincident with rapid filling of the seed (23). In cereal grains, most authors have assumed that IAA is generally localized in the endosperm (28). Lur and Setter (47) have reported that IAA concentrations are low in the endosperm of maize kernels early in development, then abruptly increase at about 10 days after pollination (DAP). This increase coincides with an increase in DNA content per nucleus. However, it is not clear yet whether the embryo, aleurone layer and/or the pericarp are also important sites of localization of auxin.

Similar to the occurrence of auxin-like material, gibberellin-like material is highest in liquid endosperm of pea at the time of rapid pod elongation (23). In whole pea seeds, GAs (GA_9, GA_{17}, GA_{20}, GA_{29}, and GA_{29}-catabolite) occur maximally during the period of rapid seed growth compared to very low quantities during early seed development or seed maturation (71). In barley, the levels of GA-like material within the grain parallel the pattern of dry weight accumulation rate increase of the developing grain, with the highest quantities recoverable from the endosperm (51). The maxima observed for "free" GAs is at 9 DAP in the husks and 21 DAP in the endosperm, and the "conjugated" GAs peaks at 32 DAP (51). In general, gibberellins are prevalent in developing seeds, grains, and fruits of several species; with their levels, biochemical composition, and time of occurrence varying. A more detailed information on GAs and developing reproductive organs is reviewed by Pharis (58).

In cereals, peas, and beans, cytokinins are generally found in highest concentrations in the endosperm of developing seeds. Moreover, in maize kernels there is a dramatic increase (as much as 500-fold) and a subsequent decline in cytokinin levels (15, 39, 47, 50). Particularly zeatin and zeatin-riboside reach a maximum amount between 8 and 12 DAP. Developing rice

---

[1] Auxin-like or GA-like is used to designate that analyses of hormones were based on bioassays rather than on physico-chemical determinations. The substance detected responded similarly to a hormone standard (IAA for auxin-like, GA_3 for GA-like).

and wheat grains appear to be similar with their having a transient increase in levels of cytokinins 4 to 5 days post-anthesis (50). In *Phaseolus coccineus*, high amounts of zeatin-like material occur in the suspensor early in seed development, while later in development more polar cytokinins occur (46). Lorenzi et al. (46) propose that the suspensor is the source of cytokinins for the embryo.

Movement of Hormones in Relation to Source-Sink Partitioning

The occurrence of high concentrations of hormones in developing seeds may indicate that 1) they function at that site or in the surrounding tissue; 2) they are produced at that site to be exported and function in some other site such as source tissue; or, 3) they are accumulating at that site thus relieving some other site of excess levels of a given hormone. The latter two require movement of hormones between seeds and source tissue.

Indoleacetic Acid

The classic work of Nitsch on developing achenes providing auxin for strawberry fruit growth led to his suggesting (53) that achenes release auxin that affects processes in leaves. Hein et al. (31) used an EDTA enhanced exudation technique (Fig. 2) to estimate the amount of IAA moving between seeds and source leaves of soybeans. IAA, primarily in the form of ester conjugate(s), was found to be moving acropetally (toward leaf laminae) in petioles (Fig. 3). The highest amount of IAA ester(s) was found in petiole exudate during the mid to late stages of seed filling. Removal of fruits 36 hours prior to exudation reduced the amount of IAA ester recovered in exudate, indicating that fruits were a major source of the IAA conjugate observed in the petiole exudate. A small amount of labeled material that co-chromatographs with IAA can be recovered in source leaves

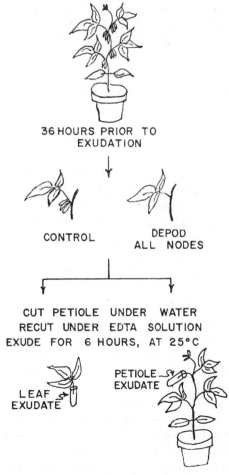

36 HOURS PRIOR TO EXUDATION

CONTROL

DEPOD ALL NODES

CUT PETIOLE UNDER WATER RECUT UNDER EDTA SOLUTION EXUDE FOR 6 HOURS, AT 25°C

LEAF EXUDATE

PETIOLE EXUDATE

Fig. 2. Diagram of the experimental protocol used to obtain EDTA-enhanced exudation of hormones transported in soybean leaf petioles. From (31).

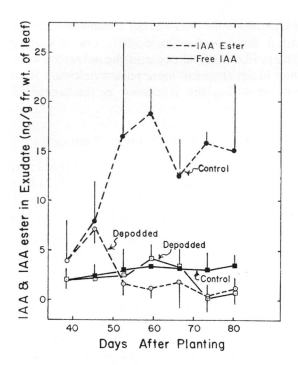

Fig. 3. Amount of free and ester IAA exuded acropetally into 20 mM EDTA from soybean petioles as detailed in Fig. 2. n=6±SE. From (31).

following application of ^{14}C-tryptophan to soybean pods, and thus provides further evidence that seeds may export IAA to leaves (32). Though depodding nearly eliminated exudation of the IAA esters, the level of IAA in leaves was unaffected 36 hours after the depodding (31). It is possible that the IAA transported to the leaf accumulates and functions in specialized cells of the leaf. Sampling the entire leaf would have masked detection of a hormonal change in a localized area. The fact that IAA will promote stomatal opening (Chapter G7) is especially relevant.

Gibberellins

Defruiting grape plants resulted in leaves containing less GA-like material, suggesting that developing fruit also export GAs to source leaves (34). However, further research is required to verify if fruit export GAs to affect processes in the leaves.

Abscisic Acid

ABA movement about the plant is dynamic. Developing sinks may control the level of ABA in leaves indirectly by serving as sites for accumulation of ABA produced within the leaves (69, 70). Based on experiments that followed the movement of trace quantities of radiolabeled ABA, it appears that ABA rapidly moves from mature leaves to all other parts of vegetative plants and will generally accumulate in sink tissues (9). Labeled-ABA can be found in soybean roots within 15 minutes following application to a leaf, and the ABA is recycled via the xylem back to the shoot apex during the next 30 minutes. ABA applied to source leaves of fruiting plants is exported to developing seeds (21, 70); while ABA applied to filling seeds is immobile (21). Setter et al. (69) found that when translocation from soybean leaves is obstructed or the plant is totally depodded, ABA accumulates in the leaves resulting in stomatal closure and depressed photosynthesis within one hour. Thus it seems that sinks can regulate processes in the source by drawing away ABA which may be inhibiting some processes in the source.

In soybean plants the pattern of ABA distribution changes diurnally (14) and with the stage of plant development (unpublished data, Cheikh and Brenner; 31). Developing soybean seeds are the major site for ABA accumulation at the time of mid-pod fill (31), and it is only at this stage of plant development that defruiting the plant will result in ABA accumulation within leaves (Fig. 4). Presumably at this stage of development the filling seeds are the only active sinks, while at other stages there are multiple sinks for ABA accumulation.

Further proof that the maternal plant exports ABA to developing seeds is obtained by defoliating a soybean plant and then examining the seeds 48 hours later for ABA content. Sampling seeds of soybean (cv. Clay) resulted in a one-third reduction of the concentration of ABA recovered in the seeds 48 hours after plants were defoliated (10). However, similar treatment to another soybean genotype had no affect on ABA concentration in the seeds from the defoliated plants (68). Since defoliation reduced ABA only in the Clay cultivar, it seems that the treatment eliminated a source of ABA rather than removing a substrate for ABA biosynthesis in the seeds. Thus we have hypothesized that the maternal contribution of ABA to developing soybean seeds can vary genotypically from a minimal contribution (the seeds appear to be autonomous for ABA) to a substantial one. Additional proof that seeds are capable of synthesizing ABA is demonstrated by the culturing of isolated soybean cotyledons on a range of sucrose concentrations and observing elevated levels of ABA in the tissue presumably due to osmotic stress (8). It appears maize kernels synthesize most of their ABA since culturing kernels *in vitro* in the

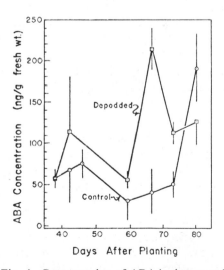

Fig. 4. Concentration of ABA in the second most recently expanded trifoliate leaf of soybean during fruit development. Depodding treatments consisted of removing all reproductive tissue 32 hours before sampling. n=6±SE. From (31).

absence of added hormones contain similar concentrations and distribution patterns of ABA to those kernels removed from field grown ears (38). These studies have also indicated that prior to growth of the embryo, both pedicel/placental-chalazal complex and the endosperm contain high concentrations of ABA. However, levels of ABA in these tissues decline as the concentration in the embryo increases during its development. Finally, isolated wheat ears can also synthesize ABA which accumulates in the grain (42).

In summary, as shown in Fig. 5, it appears that at least for soybeans, ABA moves from leaves to filling pods. ABA also is transported from leaves to roots via the phloem and then recycled back to other sinks in the xylem stream (9). Additional reports have also indicated that roots can be a source for ABA synthesis. Particularly, in cases where ABA is believed to serve as a root-to-shoot signal of root stress, such as dehydrated (6), flooded (36), and salt-stressed roots (82). IAA-ester is transported from pods to the leaves (31, 32). Cytokinins produced in the roots are carried in the xylem stream to the shoot (6, 13). Though there is no direct evidence that these cytokinins accumulate in developing fruit, it is known that they do influence fruit set of soybean (13) and presumably subsequent fruit growth. However, developing seeds (maize kernels) are known to synthesize their own cytokinins; because excised- and *in vitro*-cultured maize kernels produce similar levels of cytokinins as the kernels left attached to the mother plant (15). Since there

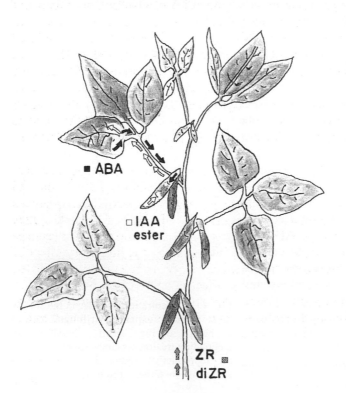

Fig. 5. Diagram of the apparent direction of hormone movement in soybeans in relation to control of source-sink relations.

is minimal evidence of GAs moving between source and sink tissues, GAs were not included in Fig. 5.

HORMONAL REGULATION OF PROCESSES RELATED TO PHOTOSYNTHATE PARTITIONING

Regulation of Processes in the Leaf

Regulation of Photosynthesis

There is evidence that IAA enhances stomatal opening (see Chapter G7). Early reports described IAA promotion of photosynthesis within one hour of application to leaves (7). Follow-up experiments indicated IAA imparted this effect by increasing photophosphorylation and CO_2 fixation when tested on isolated chloroplasts (75). However, when similar experiments were repeated (64) IAA only reduced the aging of isolated chloroplasts but seemed to have no direct effect on photophosphorylation of active intact chloroplasts maintained in the presence of BSA (bovine serum albumen).

Applications of cytokinins promote photosynthetic activity mainly by means of increase in leaf chlorophyll content, accelerating chloroplast development, or modifying other components of photosynthesis; such as, CO_2 assimilation capacity and activity of photosynthesis enzymes (12).

A number of reports indicate that GAs also should be considered as possible promoters of photosynthesis. Application of GA_3 will stimulate leaf photosynthesis (3) if treatments are applied to intact plants at least several hours before measurement of photosynthesis. However, when isolated chloroplasts were treated with GA_3, no effect was observed (64). Fruit removal on grape plants causes decreased leaf photosynthesis (34), and this effect has been associated with a decrease in GA-like material in the leaves of the defruited plants. Since removing seeded grapes also results in ABA accumulation in leaves (B. Loveys, personal communication), further studies are required to determine if the decrease in photosynthesis is related to changes in both ABA and GAs.

It is well known that increases of ABA in leaves leads to stomatal closure (see Chapter G7) resulting in decreased photosynthesis due to depressed intracellular CO_2 levels. However, ABA may also directly depress photosynthesis (62), perhaps by reducing ribulose 1,5-bisphosphate carboxylase activity (24), or increasing ribulose-1,5-bisphosphate oxygenase activity, and photorespiration (59). And interestingly, the inhibitory effect of ABA is only observed when applied to intact leaf tissue rather than to isolated chloroplasts which suggests the primary effect is on stomatal conductance.

Ethylene was also cited to inhibit photosynthesis, but it is believed that this is done mainly through ethylene-induced stomatal closure and increase in ABA (49).

657

Regulation of Sucrose Formation

The activities of cytoplasmic SPS and FBPase enzymes (14, 35) are major determinants of the pool size of sucrose in the leaf. In soybean leaves, both endogenous sucrose concentration and SPS activity positively correlate with assimilate export when measured at midday (35). However, when similar comparisons are made on a diurnal basis, these relationships do not hold up during the dark periods, making it clear that there are additional important factors controlling the export of carbon beside SPS activity and net photosynthesis. Gibberellin applications (10^{-6}M GA_3 and 10^{-5}M $GA_{4/7}$) to intact soybean leaves increase SPS activity and protein levels, but have inconsistent effects on FBPase (14). However, either imposition of water stress or applications of ABA (10^{-6} M) inhibit FBPase activity, but have no significant effect on SPS activity (9, 14, 29). These studies show that plant hormones ABA and GAs play a regulatory role in leaf sucrose metabolism; but, their sites of control are different. Fructose-2,6-bisphosphate (Fru-2,6-BP) is a potent regulator of sucrose formation by inhibition of cytosolic FBPase activity (72). When Fru-2,6-BP levels increase, cytosol FBPase activity is suppressed resulting in less fructose-6-phosphate (F-6-P) synthesis and a buildup of triose-P in chloroplasts. This leads to starch synthesis and a concomitant decrease in sucrose formation in the cytosol (35, 72). Regulation of concentration of Fru-2,6-BP represents another potent point for hormonal regulation of partitioning, but we do not know if hormones affect the level of Fru-2,6-BP in plants as is the case in animal liver cells (33). Additional studies have reported that levels of Fru-2,6-BP in leaves increase in response to environmental stresses, namely water stress (61).

Regulation of Phloem Transport

Regulation of Phloem Loading

The capacity of a tissue to load photoassimilates into phloem tissue for long distance transport functionally establishes that tissue as a source. Factors that determine the extent of phloem loading include the availability of sucrose for export, the distribution of the sucrose within the leaf (intra- and intercellular), transfer of sucrose (generally believed to be in the apoplast) towards the sites of phloem loading, and the rate of phloem loading (4, 26). Loading of phloem involves transfer of sucrose through the sieve-element plasmamembrane from the vascular boundary. Alternatively, a total symplastic pathway from mesophyll cells to the phloem can be argued (4, 79) with loading occurring across the plasmamembrane of mesophyll cells (48).

The transfer process into the sieve element-companion cells of the phloem involves both a high affinity system and a linear system (20, 48). The high affinity system involves a proton/sucrose cotransport directed into the phloem (symport) and is pH dependent. It is seemingly coupled with an outward proton translocating ATPase (4). To compensate for the proton influx, there is a transient efflux of K^+, but there is also a H^+/K^+ exchanging

ATPase (26) which maintains the high K^+ content of the phloem. The linear transport system is considered pH independent (20).

Based on experiments where hormones were applied (4) it appears that IAA or the fungal toxin fusicoccin (FC) will enhance phloem loading. Both of the later compounds promote membrane proton extrusion (see Chapter D1). Treatment with ABA inhibits sucrose loading (4). In addition, BA or kinetin also enhance sucrose loading in stems of *Ricinus*. This is based on experiments where petioles were perfused with cytokinins and exudate from the petioles was sampled for sucrose and K^+. The loading of sucrose into isolated phloem tissue of celery occurs at greater rates when the tissue is placed in a medium adjusted to 200 to 300 m osmolality (with a nonpenetrating solute, PEG 3350 [17]). IAA (0.1-100 μM) or GA_3 (at 10 μM) promote greater rates of uptake in 200 m osmolal medium but not in media at 50 or 400 m osmolal. Application of GA_3 to excised-mature leaves of broad beans enhances sucrose export from source leaves (2). This promotion of phloem loading by GA_3 occurs after a duration of only 10 minutes by a mechanism still unknown. However, this effect could not be explained by changes in the photosynthetic rate or by chemical partitioning of carbon. Furthermore, other investigations (18) indicate the presence of an indirect and long-term (24 h) effect of GA_3 on phloem loading mediated by changes in sink activity. Since prior studies have not examined the potential interaction of hormones and turgor on phloem loading, it appears that this area deserves further examination.

Regulation of Phloem Transport

The rate and magnitude of phloem transport is generally believed to be determined by the hydrostatic pressure gradient between the source end and the sink end of the phloem, which is regulated by the rate of phloem loading in the source and by the rate of phloem unloading in the sink. These basic elements describe the Munch pressure hypothesis (20). This generally accepted theory also states that there is minimal control exerted along the transport pathway. Thus, if the Munch pressure concept is correct, there would not be any role for plant hormones to directly regulate flow along the phloem transport pathway. Indirectly, however, plant hormones may regulate phloem transport by regulating loading and/or unloading.

A refinement of the explanation describing phloem transport invokes turgor regulation (44). This hypothesis envisages a gradient of solute concentration consisting primarily of sucrose and K^+. While sucrose is preferentially accumulated, K^+ concentration is modulated to maintain the turgor pressure gradient down the phloem pathway. A distinct aspect of this premise is pathway control of phloem transport obtained by adjusting the K^+ concentration through radial exchange of K^+ along the pathway. Data of Vreugdenhil (81) provide evidence that such gradients of K^+ do occur in the phloem of both cassava and castor bean. However, more rigorous experiments are required in which source/sink conditions can be manipulated to determine if K^+ gradients correlate with changes in phloem transport.

Recognizing that several hormones are known to effect K^+ transmembrane movement, future studies should consider if hormones modulate K^+ exchange between sieve elements and surrounding tissue. Though it is known that hormones affect K^+ exchange in guard cells (see Chapter G7) existing data on the application of hormones to directly influence phloem transport are inconclusive (20). Perhaps the most interesting observation is by Patrick (56) who indicated that IAA promotes mobilization of photoassimilate to the point of application and the cut stump of decapitated bean plants by acting upon transfer processes along the transport pathway. Since IAA has been implicated as a regulator of membrane ATPases, it is reasonable to consider IAA as a possible regulator of the K^+ concentration in the phloem. This would mean that IAA might be regulating transport by controlling turgor of the phloem through adjustment of K^+ concentration.

Regulation of Phloem Unloading

Withdrawal of photoassimilates from the phloem plays a major role in establishing the osmotic gradient between source and sink tissues, thereby influencing the rate of phloem transport. As mentioned above, photoassimilates are unloaded in the testa of dicotyledonous plants and in the placental-chalazal tissue of monocotyledonous plants. Since there are no vascular connections between these sites of unloading and the developing zygotic tissue, it has been experimentally feasible to surgically remove the developing embryo and sample substances released into a bathing medium added to the empty maternal "cup". Most of these experiments have been performed on developing legume seeds. One approach to examine phloem unloading systems from seed cups is to preload developing seed coats with radiolabeled photoassimilates (via transport from labeled leaves), then excise the seeds, remove the embryos, and sequentially wash the seed coats with various bathing media (16, 84). A second approach is to sample the unloading of substances from attached seed coats. This is accomplished by cutting through the pod wall to surgically remove the distal half of a seed. The remaining embryo tissue is removed and sampling media is added to the seed cup, which is still attached via the funiculus to the pod wall (77).

The addition of a high osmolality solution (400 mmol consisting of sucrose and mannitol) to attached empty seed cups of *Vicia faba* and peas, permits obtaining unloading of radiolabeled sucrose into the trapping medium within the cups at rates equivalent to that which moves into intact seeds (83). Filling empty seed cups with solutions of lower osmotic potential resulting in reduced unloading. It is proposed (83) that solutions of high solute concentration will lower the osmotic potential of the solution in the seed-coat apoplast. Consequently, water flows from the phloem system of the seed coats and reduces turgor pressure in the sink end of the phloem. This reduced phloem turgor is transmitted along the transport pathway to the source tissue, where phloem loading is increased. Alternatively, the high turgor treatment may have imparted its effect by altering the activity of membrane bound ATPase/proton carriers, as seems to be the case with sugar beet taproot tissue (84).

Inclusion of metabolic inhibitors in the sampling medium with these seed cup techniques indicates that the unloading of sucrose requires metabolic energy (77). Addition of FC promotes ^{14}C-photo-assimilate efflux from attached soybean seed coats. As mentioned above, the fungal toxin FC promotes membrane proton extrusion (see Chapter D1). In contrast to FC's effect on empty soybean seed coats, treatment of both excised and attached *Phaseolus vulgaris* seed coats with FC inhibits ^{14}C-photoassimilate efflux even though it does cause acidification of the medium (80). These effects of FC can be reduced by including orthovanadate or ABA in the *Phaseolus vulgaris* seed cups. One model which might explain the above results describes phloem unloading from *Phaseolus vulgaris* seed coats via two independent (and simultaneous) proton carrier pathways located in the plasmalemma (80). One carrier is an outward (into the apoplast) ATPase proton pump electrochemically balanced by K^+ uptake into the phloem. The second carrier is an outward proton/sucrose symporter. It is suggested that FC inhibits sucrose release because it stimulates the first proton carrier which reduces the proton/sucrose symport carrier activity by reducing the number of protons available for symport with sucrose. Additional data indicate that both the synthetic cytokinin, 6-benzylaminopurine (BAP), or ABA added to excised seed cups stimulate photoassimilate unloading from excised bean seed coats (16). The cytokinin effect is almost immediate, while ABA imparts its effect within 12 minutes of application to the seed cup. IAA, NAA, GA_3 and ACC (1-aminocyclopropane-1-carboxylic acid) are inactive in this unloading system. The action of ABA may be through restricting the ATPase driven proton carrier, thereby allowing the sucrose/proton symport carrier activity to be enhanced. More recent reports have implicated cytokinins as stimulators of the plasmamembrane proton pump. However, this cytokinin effect on membrane transport is believed to be unrelated to their hormonal action and more likely it reflects changes in purine metabolism through changes in levels of cytokinins in the cytoplasm (55). The observation that both ABA and cytokinins act in a similar manner seems quite different to their opposing action on guard cell turgor (Chapter G7).

Application of ABA to filling grain of wheat (21) and barley (78) has been found to enhance the mobilization of recently fixed photoassimilates to the filling grain. It has been proposed that the promotive action of ABA is upon the unloading process (76). The promotive effect of applied ABA on increasing the import of assimilates to intact barely grains appears to be inversely related to the endogenous ABA content of the grain. Treatments with ABA were only promotive when applied to young ears (2 weeks after anthesis). High concentrations of ABA (10^{-3}M) inhibited assimilate import when applied to older ears (3 weeks after anthesis) when their endogenous ABA content had increased five fold (78). This might indicate that while

ABA can promote mobilization in cereals, high concentrations might be inhibitory. It should be noted, however, that the promotive effect of ABA on assimilate import to wheat has not been repeatable by other scientists (43). In fact, in developing kernels of winter wheat Borkovec and Prochazka (11) have looked at the interaction between cytokinins and ABA and its effect on transport of ^{14}C-sucrose into the kernels. Applications of ABA (10^{-4} M) during pre- or post-anthesis reduces transport of ^{14}C-sucrose into developing kernels. This ABA-inhibitory effect is reversed by addition of cytokinins, but only when applied at anthesis. Application of cytokinins (10^{-6} M benzyladenine, BA) to developing wheat kernels transport of ^{14}C-sucrose into the grain. However, this increase is observed only when BA is added at pre-anthesis (11).

Active unloading of photoassimilates is not ubiquitous process for all plant species. For example, unloading into empty maize kernels appears to be a passive process (see 77). Currently, there is no evidence for hormonal regulation of phloem unloading in maize.

Regulation of Assimilate Accumulation

For both legumes and temperate cereals, the sucrose that is unloaded into the apoplast is directly accumulated by the respective cotyledons or endosperm tissues as they develop (77). In contrast, sucrose is inverted to glucose and fructose in the apoplast of the maize pedicel-parenchyma tissue and these are accumulated by the basal endosperm tissue (38).

In vitro uptake of sucrose by developing embryos of legumes occurs by both saturable and nonsaturable mechanisms (45). Follow-up experiments (66) on protoplasts prepared from soybean cotyledons revealed three distinct uptake mechanisms for sucrose: a) a saturable (<10 mM) carrier that is energy-dependent and sensitive to the nonpenetrating sulfhydryl inhibitor p-chloromercuribenzene sulfonate (PCMBS); b) a nonsaturable carrier (at least up to 50 mM) that is suppressed when high concentrations of PCMBS are added; c) a simple diffusive mechanism. The sucrose concentration in the interfacial (apoplastic) region between the seed coat and cotyledon is in the range of 150 to 200 mM and appears to be relatively stable during both the day and night (77). This means that uptake of sucrose in peripheral regions of developing cotyledons by the saturable carrier-mediated component may be of little physiological significance. However, if sucrose moves primarily apoplastically through soybean cotyledons, as it appears to do in beans (57), then it is reasonable to suggest that the sucrose concentration in the central portion of soybean cotyledons may be lower than the saturation point of the sucrose carrier. This would be due to substantial withdrawal of sucrose by the peripheral cells first exposed to the unloaded sucrose.

Sucrose appears to be the prime sugar taken up by wheat endosperm tissue. In addition to a diffusional component, the uptake of sucrose also occurs by an energy-dependent carrier that is sensitive to PCMBS inhibition

(63). In contrast, hexoses are the principal sugars taken up by maize endosperm tissue, and their uptake appears to be a nonsaturable passive process (38). Maize embryos take up hexoses in part by a facilitative, metabolically driven process which may be how the embryo can functionally compete with endosperm for available hexose (38).

Exogenous ABA can enhance the accumulation of sucrose by isolated soybean cotyledons (when tested with low sucrose concentrations, 10 mM; [67]), though we have only been able to demonstrate this effect on a specific genotype (cv. Clay) grown in nonstressed environments. Tests with several genotypes grown in the field showed no effect of ABA on *in vitro* sucrose uptake, probably due to the high endogenous ABA in these field derived tissues. However, endogenous ABA content of cotyledons is positively correlated to the sucrose uptake. The concept that ABA may function to enhance sucrose accumulation is further supported by the following manipulative study. Subjecting certain soybean genotypes (cvs. Clay and Evans) to brief periods of drought stress followed by rewatering, results in an accumulation of stress-produced ABA in the developing seeds, presumably from the leaves. The cotyledons of these seeds have significantly greater sucrose uptake compared to cotyledons derived from nonstressed plants. Stressing a different genotype (PI 416.845) does not alter the ABA content of its seeds and also has no effect on subsequent *in vitro* sucrose uptake. We have found that the seeds of this genotype appear to synthesize their own ABA and do not depend on the maternal plant to supply ABA (68). In cereal grains, treatment with ABA of intact plants and isolated kernel of barley, during grain development, increased sucrose accumulation in the endosperm (22). However, conflicting results do exist in this area . When a diverse set of soybean genotypes were investigated, seed growth rate was not limited by endogenous concentrations of ABA in the seed of the various genotypes examined (68). The role of ABA in the regulation of assimilate partitioning to developing seeds is not supported by results obtained from studies using genetic mutants of some species. Investigations using the mutants *aba1* (*Arabidopsis*) [41], *wil* (peas) [19], and *sit^w* (tomato) [27], which are all ABA-deficient mutants (having less than 10% of ABA detected in the wildtype); reveal that when these mutants are fertilized with pollen from either the wild type or mutants (*aba1, wil,* or *sit^w*), they show no significant differences in assimilate uptake. These results lead us to believe that the presence of ABA and its tissue levels do not play an important role in the regulation of assimilate accumulation by developing seeds of these species. However, more studies are still required in this area before ruling out a role of ABA, present in developing reproductive organs, in assimilate uptake and accumulation.

Unlike the promotive effect of ABA on *in vitro* sucrose uptake by soybean embryos (67), neither ABA nor BAP (each promotes unloading from *Phaseolus vulgaris* seed cups) affect *in vitro* sucrose uptake by *Phaseolus vulgaris* embryos (54). A unifying hypothesis proposed by Offler and Patrick

(54) to explain the difference between these two legumes is that regulation of the substrate concentration (osmotic potential) in the interface between seed coat and embryo tissue is a critical factor affecting photosynthate exchanges from seed coats to embryos. They suggest that alteration of osmotic potential in this interfacial zone affects the turgor of these tissues, and turgor is one of the key determinants of the rate of unloading. Consequently, in *Phaseolus vulgaris* unloading from seed coats may be the major determinant for assimilate accumulation, while in soybean, uptake of assimilates by embryos may be more significant component compared to unloading by seed coats.

The observation that ABA increases the accumulation of photoassimilates is not unique to developing seeds. Application of ABA to root tissue discs of sugar beet causes as much as a three-fold increase in sucrose accumulation and the promotive effect is observable within one hour (65). Both IAA and K^+ inhibit the accumulation of sucrose by the sugar beet tissue. ABA also may promote the accumulation of photoassimilate by young developing soybean pods through an increase in invertase activity within the pod tissue (1). Since glucose is superior to sucrose as a substrate for *in vitro* culture of young soybean seeds, it is an interesting idea that ABA may promote the growth of young seeds by increasing the pool size of glucose through increasing invertase activity in the pod tissue.

When pea seeds are grown *in vitro*, GAs do not appear to be required since treatment with GA-biosynthesis inhibitors blocks the accumulation of GAs, but has no effect on seed growth (5). These data indicate that GAs do not seem to act within the seed for seed growth and therefore it seems that GAs are not involved with sucrose uptake by seeds. However, it has been reported that pretreatment of intact barley plants with GA_3 (10^{-4} M) induced translocation of injected ^{14}C-sucrose from the flag leaf to the ear (52). So, it is possible that gibberellins might be regulating sucrose translocation to the sink tissue rather than its unloading or accumulation.

Indirect Effects of Hormones on Assimilate Partitioning

Compartmentation of photoassimilates is a likely mechanism for sink tissue to maintain the steepness of the concentration gradient from the source tissue. In maize and temperate cereals, the number of endosperm cells and starch granules formed in the endosperm is highly correlated with kernel growth rate and size at maturity (37, 38). In cereals, the number of endosperm cells and amyloplasts is established during the early phases of grain development. Since these parameters are not controlled by the levels of available carbohydrates during grain filling (37), it may be speculated that cytokinins might affect the numbers of both endosperm cells and amyloplasts formed within those endosperm cells (39). Cytokinins generally are considered to play a major role promoting cell division, but there is no information on their effect on endosperm cell number. It has been shown that cytokinins can promote chloroplast development (12). Since chloroplasts and amyloplasts are both derived from proplastids, it is possible that amyloplast development also might be altered by cytokinins. As mentioned earlier, cytokinin content is high in developing seeds at this critical period but it is not known if any source-sink manipulations that lead to reduced seed development can be related to reduced cytokinin activity.

Plant hormones may be involved in the regulation of sink potential by regulating cell division and differentiation of developing sinks. This process may involve not only a particular hormone, but possibly a balance between cytokinins, ABA, IAA, and GAs. The balance between cytokinins and ABA may be critical for establishing the tolerance of maize-kernel to heat stress and determines sink potential by regulating cell division and seed set (15). The balance between cytokinin and IAA in developing-maize endosperm at the time of cell differentiation, particularly during DNA endoreduplication, may also be critical (47). At this crucial stage in kernel development there is an increase in IAA and decrease in zeatin and zeatin riboside.

Developing pea seeds that are GA-deficient as a consequence of having a mutation (*lh^i* allele) that reduces gibberellin content in developing pea seeds are smaller and have a higher frequency to abort as compared to the wild type (*LhLh*) [74]. Fertilization of *lh^{i}lh^i* plants with *LhLh* pollen produces *lh^{i}Lh* seeds having higher GA$_1$ and GA$_{20}$ content and greater seed weight compared to *lh^{i}lh^i*.

Hormones, especially cytokinins, also may indirectly affect partitioning by altering the ratio of source to sink tissue by regulating seed set (13). Applications of cytokinins at pollination significantly reduces grain abortion of developing maize kernels (15, Cheikh and Jones, personal communication). Most of our major crop plants produce only a fraction of their potential seeds compared to the number of flowers produced on the respective plant types. The failure of fertilized flowers to produce seeds has been attributed to insufficient photoassimilates at the time of early fruit development or to a hormonal imbalance at this time.

Finally, partitioning may be indirectly altered by regulating the duration of seed fill. This can occur by either delaying leaf senescence or extending the seed-filling period by delaying the onset of maturation of seeds. Though minimal information is available, it is reasonable to peculate that hormones may also play a role in regulating this process.

SUMMARY

As depicted in Fig. 1, there are experimental data indicating that hormones may act at a number of steps involved with the assimilation and transport of photoassimilates to developing seeds. The numerous manipulative studies done to investigate regulation of this overall process have amply demonstrated that overall regulation is extensively integrated. Hormones are logical candidates to play key roles in coordinating the respective processes. For those plants that are limited by source activity, transport of promotive hormones from sinks to source tissue may regulate the processes associated with source activity, namely photosynthesis, production of sucrose, and phloem loading. Auxins and gibberellins exported from sinks are reasonable candidates to serve as the promotive signals (3, 4, 31, 32, 34, 53, 75). Relieving source tissue of excess ABA may be a further mechanism by which sinks can attenuate source leaf activity (24, 31, 62, 69, 70).

Hormones certainly may function to enhance sink activity by increasing the net amount of photoassimilates transferred from the maternal tissue to developing seeds. Action may occur at the sites of phloem unloading (16, 54) or at sites of assimilate accumulation (65, 67). Turgor also seems to be an important factor regulating phloem loading and unloading (17, 44, 77, 83, 84). It is reasonable to consider that a change in hydrostatic pressure within the phloem is all that is necessary to facilitate communication between source and sinks (83, 84). Based on the observations of Daie (17) that both IAA and GA_3 promote phloem loading when osmotic pressure of the bathing medium is adjusted to 200 to 300 m osmolality, it is obvious that considerable attention should be directed at examining the possible interaction of turgor and hormones in regulating partitioning of photosynthates.

Plant hormones could also function by enhancing sink potential via increasing cell number and/or regulating cell differentiation such as plastid biogenesis and DNA amplification, or by modifying the duration or rate of dry mass accumulation of a developing reproductive organ.

Acknowledgements
The authors wish to thank W. A. Brun, R. J. Jones, S. L. Maki and J. R. Schussler for their thoughtful discussions and critical comments during the preparation of this manuscript. Supported in part by the United States Department of Agriculture under grant 84-CRCR-1-1484 from the Competitive Research Grants Office, and by a grant from the Minnesota Soybean Research and Promotion Council. Contribution from the University of Minnesota Agricultural Experiment Station, St. Paul, MN 55108. Paper No. 2045 of the miscellaneous journal series.

References

1. Ackerson, R.C. (1985) Invertase activity and abscisic acid in relation to carbohydrate status in developing soybean reproductive structures. CropSci. 25, 615-618.
2. Aloni, B., Daie, J., Wyse, R.E. (1986) Enhancement of ^{14}C-sucrose export from source leaves of *Vicia faba* by gibberellic acid. Plant Physiol. 82, 962-966.
3. Arteca, R.N., Dong, C.H. (1981) Increased photosynthetic rates following gibberellic acid treatments to the roots of tomato plants. Photosynth. Res. 2, 343-349.
4. Baker, D.A. (1985) Regulation of phloem loading. British Plant Growth Regulator Group, Monograph 12, 163-176.
5. Baldev, B., Lang, A., Agatep, A.O. (1967) Gibberellin production in pea seeds developing in excised pods: Effect of growth retardant AMO-1618. Science 147, 155-156.
6. Bano, A., Dorffling, K., Bettin, D., Hahn, H. (1993) Abscisic acid and cytokinins as possible root-to-shoot signals in xylem sap of rice plants in drying soils. Aust. J. Plant Physiol. 20, 109-115.

7. Bidwell, R.G.S., Levin, W.B., Tamas, I.A. (1968) The effects of auxin on photosynthesis and respiration. *In*: Biochemistry and Physiology of Plant Growth Substances, pp. 361-376, Wightman, F., Setterfield G., eds. The Runge Press, Ottawa.

8. Bray, E.A., Beachy, R.R. (1985) Regulation by ABA of ß-Conglycinin expression in cultured developing soybean cotyledons. Plant Physiol. 79, 746-750.

9. Brenner, M.L., Brun, W.A. Schussler, J., Cheikh, N. (1986) Effects of endogenous and exogenous plant growth substances on development and yield of soybeans. *In*: Plant Growth Substances 1985, pp. 380-386, Bopp M., ed. Springer-Verlag, Berlin, Heidelberg.

10. Brenner, M.L., Hein, M.B., Schussler, J., Daie, J., Brun, W.A. (1982) Coordinate control: The involvement of ABA, its transport and metabolism. *In*: Plant Growth Substances 1982, pp. 343-352, Wareing P.F., ed. Academic Press, New York.

11. Borkovec, V. and Prochazka, S. (1992) Interaction of cytokinins with ABA in the transport of ^{14}C-sucrose in the developing kernels of winter wheat. *In*: Physiology and Biochemistry of cytokinins in Plants, Kaminek, M., Mok, D.W.S., Zazimalova, E., eds. SPB Academic Publishing, The Hague, The Netherlands.

12. Cacrs, M., Vendrig, J.C. (1986) Benzyladenine effects on the development of the photosynthetic apparatus in *Zea mays* : studies on photosynthetic activity, enzymes and (etio) chloroplast ultrastructure. Physiol. Plant. 66, 685-691.

13. Carlson, D.R., Dyer, D.J., Cotterman, C.D., Durley, R.C. (1987) The physiological basis for cytokinin induced increases in pod set in IX93-100 soybeans. Plant Physiol. 84, 233-239.

14. Cheikh, N. and Brenner, M.L. (1992) Regulation of key enzymes of sucrose biosynthesis in soybean leaves. Effect of dark and light conditions and role of gibberellins and abscisic acid. Plant Physiol. 100, 1230-1237.

15. Cheikh, N., Jones, R.J., Gengenbach, B.G. (1993) The effect of heat stress on carbohydrate metabolism and hormonal levels of developing maize kernels. Agronomy Abstracts, pp. 110.

16. Clifford, P.E., Offler, C.E., Patrick, J.W. (1986) Growth regulators have rapid effects on photosynthate unloading from seed coats of *Phaseolus vulgaris* L. Plant Physiol. 80, 635-637.

17. Daie, J. (1987) Interaction of cell turgor and hormones of sucrose uptake in isolated phloem of celery. Plant Physiol. 84, 1033-1037.

18. Daie, J., Watts, M., Aloni, B., Wyse, R.E. (1986) *In vitro* and *in vivo* modification of sugar transport and translocation in celery by phytohormones. Plant Sciences 46, 35-41.

19. de Bruijn, S.M., Vreugdenhil, D. (1992) Abscisic acid and assimilate partitioning to developing seeds. I. Does abscisic acid influence the growth rate of pea seeds? J. Plant Physiol. 140, 201-206.

20. Delrot, S., Bonnemain, J.L. (1985) Mechanism and control of phloem transport. Physiol. Veg. 23, 199-220.

21. Dewdney, S.J., McWha, J.A. (1979) Abscisic acid and the movement of photosynthetic assimilates towards developing wheat (*Triticum aestivum* L.) grains. Z. Pflanzenphysiol. 92, 193-186.

22. Dorffling, K., Tietz, A., Fenner, R., Naumann, R., Dingkuhn, M. (1984) The role of abscisic acid on assimilate transport and on assimilate accumulation. Ber. Deutsche Bot. Ges. 97, 87-99.

23. Eeuwens, C.J., Schwabe, W.W. (1975) Seed and pod wall development in *Pisum sativum* L. in relation to extracted and applied hormones. J. Exp. Bot 26, 1-14.

24. Fisher, E., Stitt, M., Raschke, K. (1986). Effects of abscisic acid on photosynthesis in whole leaves: Changes in CO_2 assimilation, levels of carbon reduction cycle intermediates, and activity of ribulose-1,5-bisphosphate carboxylase. Planta 169, 536-545.

25. Geiger, D.R., Sovonick, S.A. (1975) Effects of temperature, anoxia and other metabolic inhibitors on translocation [Phloem]. Encycl. Plant Physiol. New Series 1975. 1, 256-286.

26. Giaquinta, R.T. (1983) Phloem loading of sucrose. Ann. Rev. Plant Physiol. 34, 347-387.

667

27. Groot, S.P.C., van Yperen, I.I., Karssen, C.M. (1991) Strongly reduced levels of endogenous abscisic acid in developing seeds of tomato mutants *sitiens* do not influence *in vivo* accumulation of dry matter and storage proteins. Physiol. Plant. 81, 73-78.

28. Hall, P.J., Bandurski, R.S. (1978) Movement of indole-3-acetic acid and tryptophan-derived IAA from the endosperm to the shoot of *Zea mays* L. Plant Physiol. 61, 425-429.

29. Harn, C., Daie, J. (1992) Regulation of the cytosolic fructose-1,6-bisphosphatase by post-translational modification and protein level in drought-stressed leaves of sugarbeet. Plant Cell Physiol. 33, 763-770.

30. Hein, M.B., Brenner, M.L., Brun, W.A. (1984) Concentrations of indole-3-acetic acid and abscisic acid in soybean seeds during development. Plant Physiol. 76, 951-954.

31. Hein, M.B., Brenner, M.L., Brun, W.A. (1984) Effect of fruit removal on the transport and accumulation of abscisic acid and indole-3-acetic acid in soybean leaves. Plant Physiol. 76, 955-958.

32. Hein, M.B., Brenner, M.L., Brun, W.A. (1986) Accumulation of ^{14}C-radiolabel in leaves and fruits after injection of ^{14}C-tryptophan into seed of soybean. Plant Physiol. 82, 454-456.

33. Hers, H.G., van Schaftingen, E. (1982) Fructose-2,6-bisphosphate 2 years after its discovery. Biochem J. 206, 1-12.

34. Hoad, G.V., Loveys, B.R., Skenek, G.M. (1977) The effect of fruit-removal on cytokinins and gibberellin-like substance. Planta 136, 25-30.

35. Huber, S.C., Kerr, P.S., Kalt-Torres, W. (1985) Regulation of sucrose formation and movement. *In*: Regulation of Carbon Partitioning in Photosynthetic Tissue, pp. 199-214, Heath, R.L., Preiss, J., eds. Amer. Soc. Plant Physiol., Rockville, MD.

36. Jackson, M.B., Young, S.F., Hall, K.C. (1988) Are roots a source of abscisic acid for the shoots of flooded pea plants? J. Exp. Bot. 39, 1631-1637.

37. Jenner, C.F. (1985) Control of the accumulation of starch and protein in cereal grains. British Plant Growth Regulator Group, Monograph 12, 195-209.

38. Jones, R.J., Brenner, M.L. (1987) Abscisic acid in maize kernel during grain tilling. Plant Physiol. 83, 905-909.

39. Jones, R.J., Schreiber, B.M.N., McNeil, K.J., Brenner, M.L., Faxon, G. (1992) Cytokinin levels and oxidase activity during maize kernel development. *In*: Physiology and Biochemistry of cytokinins in plants, pp. 235-239, Kaminek M., Mok B.W.S., Zazimalova E., eds. SPB Academic Publishing, The Hague, The Netherlands.

40. Karssen, C.M. (1982) The role of endogenous hormones during seed development and the onset of primary dormancy. *In*: Plant Growth Substances 1982 pp. 623-632, Wareing P.F., ed. Academic Press, London.

41. Karssen, C.M., van Loon L.C. (1992) Probing hormone action in developing seeds by ABA-deficient and -insensitive mutants. *In*: Progress in Plant Growth Regulation, pp. 43-53, Karssen, C.M., van Loon, L.C., Vreugdenhil, D., eds. Kluwer Academic Publishers, Dordrecht, The Netherlands.

42. King, R.W. (1979) Abscisic acid synthesis and metabolism in wheat ears. Aust. J. Plant Physiol. 6, 99-108.

43. King, R.W., Patrick, J.W. (1982) Control of assimilate movement in wheat. Is abscisic acid involved? Z. Pflanzenphysiol. 106, 375-380.

44. Lang, A. (1983) Turgor-regulated translocation. Plant, Cell Environ. 6, 683-689.

45. Lichner, F.T., Spanswick, R.M. (1981) Electrogenic sucrose transport in developing soybean cotyledons. Plant Physiol. 67, 869-874.

46. Lorenzi, R., Bennici, A., Cionini, P.G., Alpi, A., D'Amato, F. (1978) Embryo-suspensor relations in *Phaseolus coccineus*: Cytokinins during seed development. Planta 143, 59-62.

47. Lur, H-S., Setter, T.L. (1993) Role of auxin in maize endosperm development: Timing of nuclear DNA endoreduplication, zein expression, and cytokinins. Plant physiol. 103, 273-280.

48. Lucas, W.J. (1985) Phloem-loading: A metaphysical phenomenon? *In*: Regulation of

Carbon Partitioning in Photosynthetic Tissue, pp. 254-271, Heath, R.L., Preiss J., eds. Amer. Soc. Plant Physiol., Rockville, MD.

49. Matsushima, J., Yonemori, K. (1981) Effect of ethylene on fruit drop, growth, photosynthesis, diffusive resistance and abscisic acid of citrus trees. Proc. Int. Soc. Citriculture Vol.1, 304-307.

50. Morris, R.D., Blevins, D.G., Dietrich, J.T., Durely, R.C., Gelvin, S.B., Gray, J., Hommes, N.G., Kaminek, M., Mathews, L.J., Meilan, R., Reinbott, T.M., Sayavedra-Soto, L. (1993) Cytokinins in plant pathogenic bacteria and developing cereal grains. Aust. J. Plant Physiol. 20, 621-637.

51. Mounla, M.A.Kh. (1978) Gibberellin-like substances in parts of developing barley grain. Physiol. Plant 44, 268-272.

52. Nath-kothiala, N., Mishra, S.D. (1991) Plant growth regulator-induced photoassimilate partitioning at different stages of growth in barley (*Hordeum vulgare* L.). Photosynthetica 25, 589-595.

53. Nitsch, J.P. (1959) Auxines et croissance des fruits, II. *In*: Recent Adv. Bot. 2, 1089-1093. Univ. Toronto Press, Toronto.

54. Offler, C.E., Patrick, J.W. (1986) Cellular pathway and hormonal control of short-distance transfer in sink regions. *In*: Phloem Transport, pp. 295-306, Cronshaw, J., Lucas, W.J., Giaquinta, R.T., eds. A.R. Liss, New York.

55. Parsons, A., Sanders, D. (1989) Cytokinin-Stimulation of the plasmamembrane proton pump. Its role in hormonal stimulus transduction. British Plant Growth Regulator Group, Monograph 18, pp.27-39.

56. Patrick, J.W. (1979) Auxin-promoted transport of metabolites in stems of *Phaseolus vulgaris* L. J. Exp. Bot. 30, 1-13.

57. Patrick, J.W.., McDonald, R. (1980) Pathway of carbon transport within developing ovules of *Phaseolus vulgaris* L. Aust. J. Plant Physiol. 7, 671-684.

58. Pharis, R.P. (1985) Gibberellins and reproductive development in seed plants. Ann. Rev. Plant Physiol. 36, 517-568.

59. Popova, L.P., Tsonev, T.D., Vaklinova, S.G. (1987) A possible role of abscisic acid in the regulation of photosynthetic and photorespiratory carbon metabolism in barley leaves. Plant Physiol. 83, 820-824.

60. Preiss, J. (1982) Regulation of the biosynthesis and degradation of starch. Ann. Rev. Plant Physiol. 33, 431-454.

61. Quick, P., Siegle, G., Neuhans, E., Feil, R., Stitt, M. (1989) Short term water stress leads to a stimulation of sucrose-phosphate synthase. Planta 177, 536-546.

62. Raschke, K., Hedrich, R. (1985) Simultaneous and independent effects of abscisic acid on stomata and the photosynthetic apparatus in whole leaves. Planta 163, 105-118.

63. Rijven, A.H.G.C., Gifford, R.M. (1983) Accumulation and conversion of sugars by developing wheat grains. 3. Non-diffusional uptake of sucrose, the substrate preferred by endosperm slices. Plant Cell Environ. 6, 417-425.

64. Robinson, S.P., Wiskich, J.T., Paleg, L.G. (1978) Effects of Indoleacetic acid on CO_2 fixation, electron transport and phosphorylation in isolated chloroplasts. Aust. J. Plant Physiol. 5, 425-431.

65. Saftner, R.A., Wyse, R.E. (1984) Effect of plant hormones on sucrose uptake by sugar beet root tissue discs. Plant Physiol. 74, 951-955.

66. Schmitt, M.R., Hitz, W.D., Lin, W., Giaquinta, R.T. (1984) Sugar transport into protoplasts isolated from developing soybean cotyledons. Plant Physiol. 75, 941-946.

67. Schussler, J.R., Brenner, M.L., Brun, W.A. (1984) Abscisic acid and its relationship to seed filling in soybeans. Plant Physiol. 76, 301-306.

68. Schussler, J.R., Brenner, M.L., Brun, W.A. (1991) Relationship of endogenous abscisic acid to sucrose level and seed growth rate of soybeans. Plant Physiol. 96, 1308-1313.

69. Setter, T.L., Brun, W.A., Brenner, M.L. (1980) Effect of obstructed translocation on leaf abscisic acid, and associated stomatal closure and photosynthesis decline. Plant Physiol.

65, 1111-1115.

70. Setter, T.L., Brun, W.A., Brenner, M.L. (1981) Abscisic acid translocation and metabolism in soybeans following depodding and petiole girdling treatments. Plant Physiol. 67, 774-779.

71. Sponsel, V.M. (1983) The localization, metabolism, and biological activity of gibberellins in maturing and germinating seeds of *Pisum sativum*, cv. Progress No. 9. Planta 159, 774-779.

72. Stitt, M. (1990) Fructose-2,6-bisphosphate a regulatory molecule in plants. Ann. Rev. Plant Physiol. Plant Mol. Biol. 41, 153-185.

73. Stitt, M., Quick, W.P. (1989) Photosynthetic carbon partitioning: Its regulation and possibilities for manipulation. Physiol. Plant. 77, 633-641.

74. Swain, S.M. (1993) Gibberellins and seed development in Pisum. PhD thesis, University of Tasmania, Hobart, Australia.

75. Tamas, I.A., Schwartz, J.W., Breithaupt, B.J., Hagin, J.M., Arnold, P.H. (1973) Effect of indoleacetic acid on photosynthetic reactions in isolated chloroplasts. *In*: Plant Growth Substances pp. 1159-1168, Hirokawa Publ. Co. Tokyo.

76. Tanner, W. (1980) On the possible role of ABA on phloem unloading. Ber. Deutsche Bot. Ges. 93, 349-351.

77. Thorne, J.H. (1985) Phloem unloading of C and N assimilates in developing seeds. Ann. Rev. Plant Phys. 36, 317-343.

78. Tietz, A., Ludwig, M., Dingkuhn, M., Dorffling, K. (1981) Effect of abscisic acid on the transport of assimilates in barley. Plant 152, 557-561.

79. van Bel, A.J.E. (1993) Strategies of phloem loading. Ann. Rev. Plant Physiol. Plant Mol. Biol. 44, 253-281.

80. van Bel, A.J.E., Patrick, J.W. (1984) No direct linkage between proton pumping and photosynthate unloading from seed coats of *Phaseolus vulgaris* L. Plant Growth Regul. 2, 319-336.

81. Vreugdenhil, D. (1985) Source-to-sink gradient of potassium in the phloem. Planta 163, 238-240.

82. Wolf, O., Jeschke, W.D., Hartung, W. (1990) Long distance transport of abscisic acid in Na-Cl-treated plants of *lupinus albus*. J. Exp. Bot. 41, 593-600.

83. Wolswinkel, P., Ammerlaan, A. (1986) Turgor-sensitive transport in developing seeds of legumes: the role of the stage of development and the use of excised vs. attached seed coats. Plant Cell Environ. 9, 133-140.

84. Wyse, R.E. (1986) Sinks as determinants of assimilate partitioning: Possible sites for regulation. *In*: Phloem Transport, pp. 197-209, Cronshaw J., Lucas, W.J., Giaquinta, R.T., eds. A.R. Liss, New York.

G10. The Role of Hormones During Seed Development

Christopher D. Rock and Ralph S. Quatrano

Department of Biology, University of North Carolina, Chapel Hill, North Carolina 27599-3280, U.S.A.

INTRODUCTION

The formation of a seed in the life cycle of higher plants is a unique adaptation. It incorporates embryo development with various physiological processes that insure the survival of the plant in the next generation. These adaptations include the accumulation of nutritive reserves, an arrest of tissue growth and development, and the ability to withstand desiccation, all of which are of considerable agronomic importance (e.g., nutritive value, yield, germination). The extent of these adaptations are quite spectacular. For example, the embryo must acquire the ability to withstand a reduction in water content from about 85% to 10%; in other plant tissues, such a severe desiccation is lethal. To survive long periods of time in this dry state until environmental conditions are favorable to resume development into a seedling, numerous plants have acquired different mechanisms of seed dormancy. The term "dormancy" is not entirely appropriate for many higher plants; this term can be defined as the absence of germination during environmental conditions which otherwise promote germination. Typically some external stimulus such as light or chilling (stratification) is required. However, many angiosperms undergo the developmental program of maturation, developmental arrest, and desiccation without true dormancy.

From a more basic viewpoint, seed development has represented a convenient experimental system for the study of the underlying mechanisms of physiological and molecular regulation of cell and tissue development (21, 35). The stages of seed development and germination involve both spatial and temporal regulation of cell and tissue growth and function. The sequence of events begins with rapid endosperm and embryo growth and differentiation after fertilization, followed by the transition from a state of high metabolic activity and growth to a quiescent state, and finally during germination a switch back to active growth of the embryo to form a seedling (Fig. 1).

Hormones are thought to play an important role in these processes, since the levels and activities of various hormones change dramatically during this developmental sequence (5, 35). Cytokinins (CK), auxins (IAA), gibberellins (GA) and abscisic acid (ABA) are found in relatively high concentrations in extracts from seeds of different developmental stages. In fact, a large part of

P. J. Davies (ed.), Plant Hormones, 671–697.
© 1995 *Kluwer Academic Publishers. Printed in the Netherlands.*

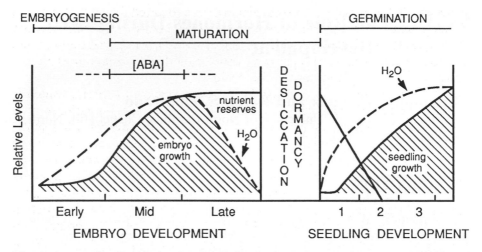

Fig. 1. A generalized graph showing the relative levels of nutrient reserves, water content and growth during the embryogenic, maturation and germination stages of embryo and seedling development. Time periods of embryo development vary with species and are not included. Times stated for seedling development are in days. Desiccation separates the end of maturation from the initiation of seedling growth; dormancy is not found in all species. High ABA levels are temporally correlated with the onset of maturation and prevention of precocious germination during mid-embryo development.

our knowledge of hormone biosynthesis and metabolism has been obtained using young seeds. One must critically ask, however, whether the changes in hormone levels/activity are correlated with changes in embryo growth, and, do they play a causal role in embryo development?

EMBRYO DEVELOPMENT

What is unique about higher plant embryogenesis when compared with embryo development in other plant groups? During embryo maturation, there is a buildup of nutrient reserves, an arrest of tissue growth, development of desiccation tolerance and the acquisition of dormancy mechanisms. All of these events occur in the latter half of embryo development in most higher plants. Following desiccation, activation of the embryo results in initiation of a meristematic-type growth pattern, and the utilization of nutrient reserves to support the development of the seedling. These processes can be referred to as the "seed strategy," *i.e.* mechanisms which have evolved to insure survival of the young embryo. Of course, there are natural exceptions, such as the mangrove (*Rhizophora mangle*) which lives in an aquatic environment and forms seedlings on the plant (vivipary) without any developmental arrest, desiccation or germination (59). Embryos from lower plants complete embryogeny without the intervening processes of maturation and desiccation. Since these processes are unique to seed plants, one can ask: which of the processes that occur late in seed development are required for normal seedling development? A number of experiments with mutants and *in vitro* culture techniques (to artificially provide the embryo with the required nutrient and environmental conditions) support the conclusion that these processes are not obligatory for completion of the life cycle.

Normal seedlings (notwithstanding some somaclonal variation which is more pronounced under certain conditions in some species/varieties than others) can be formed *in vitro* from totipotent callus cultures via somatic embryogenesis or organogenesis; these developmental sequences occur in the absence of developmental arrest. In most higher plant species, isolated embryos are capable of precocious germination into seedlings (after a period of seed development) if cultured on a nutrient medium containing salts, reduced nitrogen (e.g., glutamine) and sucrose (see below). Now the question arises: when seedlings are compared from embryos which have and have not undergone developmental arrest, desiccation, *etc.*, are they identical? What role, if any, do "maturation" processes and desiccation play in subsequent stages in the life cycle of the plant? Are the function(s) of these processes confined to the seed and germination stages or do they have effects later in the next generation? The *viviparous* mutants of maize and *Arabidopsis* form seedlings in the absence of desiccation if rescued from the normal dehydration program of the maturing seed (47a, 50, 51, 54). Thus, although the maturation events and associated specific gene products may be of critical importance for some process, agronomic trait, or survival in nature, they do not appear to be required for completion of the life cycle. They can be thought of as a set of physiological processes that comprise an important but non-obligatory pathway of seed development (Fig. 2, pathway 1). Thus, embryos can proceed directly from embryogeny to germination by bypassing dormancy (Fig. 2, pathway 2). Because of these experimental manipulations and mutants available to investigators, approaches can be taken in an attempt to understand the regulation of this maturation sequence.

As a start we can ask: at which stage is the embryo capable of germinating if removed from the seed and cultured in a nutrient medium? We will divide seed development into three periods: early, middle and late, each representing about a third of the time between fertilization and desiccation. Using this standard, embryos from wheat, maize, rice, rapeseed, soybean, bean, and probably most others, can precociously germinate in culture during the latter two-thirds of early development (28, 34). If embryos are capable of germinating by the start of the middle period, what is present in the seed environment around/in the embryo that normally prevents precocious germination *in vivo*? Cytokinins, IAA, GA and ABA levels have been shown generally to be high during these stages. What do we know about the role of these hormones in seed development and their possible effect on the prevention of precocious germination and the initiation of the maturation pathway? Many of the studies which show changes in hormones levels during seed development and associations of hormones with various

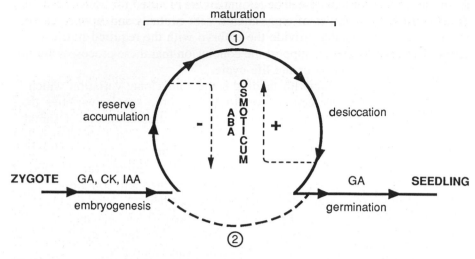

Fig. 2. A generalized representation of embryo development in angiosperms, showing three major stages from zygote to seedling: embryogenesis, maturation, and germination, and the importance of plant hormones in development. The maturation "loop" (1) can be by-passed in culture by precocious germination of embryos (2). This occurs in the absence of ABA and the presence of reduced nitrogen (e.g., glutamine) and carbon (e.g., sucrose), and in some cases by premature desiccation. In the presence of high osmoticum and/or ABA, maturation is promoted and precocious germination inhibited. Upon desiccation these processes are finalized, and normal germination will occur in response to imbibition. In culture, embryos can reversibly enter or leave the maturation loop by application (+) or removal (-), respectively, of high osmoticum or ABA.

developmental events are reviewed elsewhere (5, 35). This chapter is not meant to be an exhaustive review but will focus on recent progress in our understanding of embryo development with special emphasis on the role of hormones in gene expression, specifically ABA.

CHANGES IN LEVELS/ACTIVITY OF HORMONES DURING SEED DEVELOPMENT

Early work on hormones in seeds gave rise to a general scheme to explain dormancy on the basis of changing levels of growth-promoting and -inhibiting hormones. This paradigm has been the basis of all plant hormone physiology, biochemistry and molecular biology to date. However, observed changes in hormone levels cannot explain many hormone phenomena; changes in tissue sensitivity to the hormones are also important (61). Phytohormone action in general has three characteristics that apply to all physiological responses to a hormone: 1) the response occurs only at a specific developmental stage and in a tissue that is competent to respond, 2) the response is specifically and quantitatively correlated with a range of hormone concentrations (the dose-response curve), and 3) the tissue can vary in its sensitivity to the hormone (displacement of the dose-response curve).

Early Embryogenesis

The effects of auxins on growth and morphogenesis of embryos have been studied extensively, and it can be concluded that in general, low concentrations of auxins promote growth of embryos, and high concentrations inhibit their growth. In embryos with well-differentiated embryonic organs (e.g., *Phaseolus vulgaris*), the plumule has a higher auxin concentration optimum than the radicle with respect to promotion and inhibition of growth, respectively.

Cytokinins are found in relatively high concentrations in the liquid endosperm stage of early seed growth, and their presence coincides with the highest rate of mitosis. Studies with isogenic lines of barley that vary in grain weight demonstrate that large-grain lines contain higher amounts of CK at this very early stage of seed development than small-grain lines. Based on results such as these, it has been suggested that CK activity at this stage is responsible for enhanced seed size by increasing cell number, resulting in larger storage capacity.

A number of studies suggest a role of GA and CK in the function of the suspensor during this early stage of embryo growth. Higher concentrations of these hormones are found in the suspensor compared to the embryo, and exogenous GA can substitute for the suspensor in supporting embryo growth in culture. Protein levels are decreased in embryos of *Phaseolus* when embryos are cultured after being detached from the suspensor. The addition of GA at 0.1-1.0 µM restores the protein content to that of freshly excised embryos. In *Pisum* the endosperm and maternal tissues of the fertilized ovule are the source of GAs or other growth factors necessary for fruit set and pod development. Application of GA induces parthenocarpic ovary growth in pea (20). The timing and concentration of GA and CK supplied to the embryo through the suspensor appears to be critical for early embryo growth.

The use of mutants provides an approach to pursue questions dealing with early embryo development and function. For example, defective kernel (*dek*) mutants of corn affect both embryo and endosperm development while the embryo lethal (*emb*) mutants are blocked at specific stages early in corn embryo development but do not effect endosperm development (10). Similar embryo mutants of *Arabidopsis* have been identified (45). Based on the large number of non-allelic embryo-lethal mutants (over 50 in maize and *Arabidopsis*), it is clear there are many genes involved in embryogenesis. Hormones are undoubtedly important in this complex process and may be absolutely essential. In fact, no mutant which is null for accumulation of a hormone has been found in plants; all the characterized hormone mutants are leaky to some extent. If exogenous hormones can be shown to rescue any of the embryo-lethal mutants, then our understanding of the role of hormones in plant development will be greatly advanced.

Mid-embryogenesis

In general, high auxin (IAA) and GA levels have been associated with two phases of reproductive development: active seed growth by cell expansion, and fruit growth (35). The GA's found, however, are often the type that have moderate to very low bioactivity. Both GA and auxins are highest during early to mid-embryo development in a number of plants, at a stage when CK is decreasing rapidly and there is little or no ABA detectable. The timing of these increases in peas and wheat are correlated with increases in pod and grain length. Using the same barley genotypes discussed above which differ in yield, investigators found that high single grain weight correlated with higher IAA content (48).

Whereas early pod/grain growth is correlated with increases in GA and CK, the role of these hormones in the middle and late stages of seed development is more obscure. There does not seem to be a causal relationship between the amount of bioactive GA and late pea seed growth, nor is there an increased degradation of bioactive GA's to inactive ones at this stage. Studies with the GA-deficient mutants of *Arabidopsis* and tomato support the conclusion that high GA levels are not important for seed development (27). During early stages of wheat grain development, treatment of aleurone cells with GA does not induce secretion of endosperm-mobilizing enzymes. Cornford *et al.* (12) showed that the GA insensitivity was manifested at the level of α-amylase mRNA accumulation. This result underscores the importance of tissue sensitivity to hormones. Although GA is present in relatively high concentrations in seeds, its major role seems to be confined to early embryo development and subsequently to germination and seedling growth. The same appears to be true for CK and IAA. There is no evidence that ethylene is involved in seed development, but it is probably involved in germination.

Based on these studies, the roles of CK, GAs and IAA in the mid and late stages of embryogenesis are not clear. Caution should be exercised when interpreting such experiments. The local concentration and/or change in tissue sensitivity to endogenous levels of these hormones may be very critical. The ratios of hormones are also important as controlling factors. For example, in most plants the middle and late periods of embryo growth are generally characterized by decreasing amounts of/sensitivity to GA with a corresponding increase in amounts of/sensitivity to ABA. This phenomenon was elegantly exploited by Koornneef *et al.* (38) to isolate an ABA-deficient mutant of *Arabidopsis*. A non-germinating GA-deficient mutant was mutagenized and revertants of the non-germinating phenotype were selected. This screen identified a gene which affects ABA biosynthesis (see Chapter

676

B5), and demonstrates the importance of ABA/GA ratios in seed germination (27).

Mid to Late Embryogenesis

In a number of different species of monocots and dicots, ABA levels begin to rise and reach their highest levels during the middle and late period of seed development (see Fig. 1), at a time of decreasing GA and IAA levels and the initiation of maturation events. The ABA levels then decrease rapidly to very low levels in the dry seed. In general, ABA reaches a maximum at the same time as does seed growth, and declines sharply after the accumulation of reserves and the beginning of desiccation. In several species there are dual peaks of ABA accumulation that occur during the middle and late periods. In *Arabidopsis* and maize it has been shown that the first peak is associated with a maternal biosynthetic origin, whereas the second has a zygotic origin (28, 33). The levels of ABA that accumulate in seeds are in the physiological range of activity, varying roughly between 1-10 μM. Hence, the levels and timing of ABA accumulation *in vivo* is consistent with ABA having a role in physiological events occurring at this time, *i.e.* the maturation events of embryo development. What evidence, other than a temporal correlation between high ABA levels and the initiation of embryo growth, maturation, and prevention of germination, links ABA with its proposed role as a natural regulator of this developmental pathway? The evidence can be summarized as follows: 1) There is a correlation between low ABA levels and precocious germination of cultured embryos. 2) Exogenous ABA prevents germination and promotes growth of immature embryos of several species cultured *in vitro*. Not only does ABA prevent germination, but in many cases it results in embryo growth and an accumulation of storage reserves (Fig. 3). 3) Sensitivity to ABA is highest when endogenous ABA levels are the highest. 4) ABA-deficient mutants have very low ABA levels and have reduced dormancy or germinate while still in the fruit. Similarly, application of the carotenoid and ABA biosynthesis inhibitor fluridone to developing seeds results in vivipary. 5) The ABA-insensitive mutants of *Arabidopsis* and

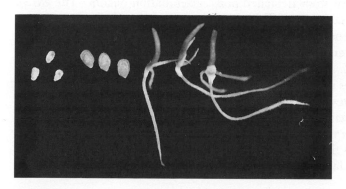

Fig. 3. Stage III wheat embryos (1.5 mm length) at the beginning of incubation (left), after 5 days in culture + 100 μM ABA (middle), and after 5 days in culture - ABA (right). Note embryo growth and scutellum enlargement in + ABA embryos. From Rogers and Quatrano, Am. J. Bot. 70, 308-311 (1983).

maize exhibit vivipary and fail to express a set of ABA-inducible maturation proteins. Most work has focused on isolated embryos in culture and analysis of mutants, and is detailed below.

ABA Effects On Embryos In Culture

As stated above, immature embryos placed in culture at the start of mid-embryo development will precociously germinate. The germination-specific enzyme, carboxypeptidase C, appears in cotton embryos within one day of precocious germination. The same is true of the small subunit of ribulose bisphosphate carboxylase in wheat, and for isocitrate lyase in germinating castor bean embryos (34). Although most embryos switch completely from embryogenic to germination processes, *Brassica* embryos seem to be an exception. They concurrently express characteristics of both, gradually acquiring the seedling traits (17).

In the late 1960s, several investigators added ABA to cultures of immature embryos to determine if it prevented precocious germination. The rationale was that ABA was originally isolated and characterized as an inducer of dormancy and was known for its inhibitory effects on seed germination, and as discussed above, because of its high levels during the mid and late stages of embryo development *in vivo*. In many species, if embryos are removed at the beginning of the middle period and cultured in 0.1-10 μM ABA, precocious germination is inhibited (34; Fig. 3). ABA stimulates growth and protein accumulation in soybean embryos isolated during the early part of mid- embryogenesis. If embryos from the later stages are cultured in ABA, growth is suppressed but the protein increase is not. Thus ABA is involved in the growth and development of the early embryo as well as the later maturation processes. ABA has been implicated in control of endosperm cell division (52a). There is also a correlation between embryo ABA concentration and ability to germinate. Soybean embryos cannot precociously germinate before 21 days (*i.e.*, until the middle stage of embryo development), at which time the ABA concentration is approximately 10 μg/g fresh weight. If the endogenous ABA concentration is experimentally reduced in embryos during mid-maturation by washing-out, or drying slowly within or after removal from detached pods, germination occurs. The extent of germination is correlated with length of washing or drying treatments which reduces endogenous ABA (1). Similar results are seen with application of fluridone, a carotenoid and ABA biosynthesis inhibitor, to developing maize kernels (28).

There is evidence in wheat, maize, soybean and in rapeseed embryos that with declining ABA levels in late development, there is also a corresponding decrease in tissue sensitivity to ABA. This indicates that dehydration, and not ABA, is probably responsible for inhibiting germination at this stage (1, 17, 34). Based on the accumulation kinetics of 12 classes of mRNAs expressed during the late stages of cotton embryo development, both *in vivo* and in excised, ABA-treated embryos, Galau *et al.* (19) have proposed that ABA is not involved in maintenance of the dicot embryo maturation program. They postulate that there are other developmentally regulated factors which control embryo development after abscission of the ovule in late embryogenesis. Desiccation *in situ* is probably the normal trigger, not only to prevent germination of late embryos, but also to switch the developmental program from maturation to germination. Premature drying of *Phaseolus* and castor bean seeds alters the developmental potential such that upon rehydration, they germinate and express germination markers rather than resuming maturation. Although dehydration is not required in culture for most embryos to switch, it may be necessary for certain embryos such as rapeseed that continue to express certain maturation traits during precocious germination (17).

Finally, previous studies showed that osmotic stress (e.g., sucrose or sorbitol) can both inhibit germination and maintain excised embryos of wheat and rapeseed in the maturation pathway (34). It has been assumed that osmotic stress results in increased ABA levels in embryos, similar to its effect on other plant organs such as leaves. Excised cotyledons of soybean cultured separately from the embryo axis increased ABA levels in the presence of 10% sucrose. However, ABA levels in rapeseed embryos did not increase significantly, even though the ABA-inducible storage protein mRNAs accumulate in response to osmoticum (17, 65). The fact that inhibition of germination and maintenance of storage protein synthesis can be uncoupled from high ABA levels suggests that ABA is probably not the primary effector regulating these processes in rapeseed.

ABA Mutants

Mutants with reduced sensitivity to ABA, or altered ABA biosynthesis resulting in reduced tissue levels of ABA have been studied in potato (*droopy*), pea (*wilty*), tomato (*flacca, sitiens, notabilis*), *Arabidopsis* (*abi, aba*), maize (*viviparous*), *Nicotiana* (*ckr1*), and barley (Az34, "cool"). We will focus our discussion on mutants that specifically affect seed, rather than leaf, physiology.

In *Arabidopsis*, five mutant loci (*abi1, abi2, abi3, abi4* and *abi5*) for sensitivity to ABA have been isolated by selecting for germination in the presence of exogenous ABA (57a). Developing seeds of some of these mutants contain higher than normal ABA levels. These five loci fall into two classes based on their responses to altered water balance: mutations in the *ABI1* and *ABI2* genes are semi-dominant and result in plants which transpire excessively and wilt like the ABA-deficient *aba* plants, whereas *abi3* mutant plants resemble wild type in their water relations (16a, 18, 39). ABI3, ABI4 and ABI5 appear to act in the same seed-specific signaling pathway (16a). The *ABI3* gene product is essential for seed development; null alleles of this gene result in seeds which exhibit precocious germination, desiccation intolerance, lack of storage protein accumulation, and incomplete seed development (50, 54). The dose-response curve of germination inhibition by ABA is shifted toward higher ABA concentrations in the *abi* mutants (Fig. 4, 16a). When seeds of *abi* double mutants are assayed, the *abi1/abi3* and *abi2/abi3* double mutants show a much more pronounced ABA insensitivity to germination inhibition than the *abi* single mutants or the *abi1/abi2* double mutant (Fig. 4). This synergistic interaction of *abi*1 and *abi*2 with *abi*3 (as opposed to the epistatic interaction between *abi*1 and *abi*2, Fig. 4) suggests the *abi*1 and *abi*2 genes act via a different ABA response pathway than does *abi*3 (18).

Genetic analysis of *Arabidopsis* ABA-deficient (*aba*) and ABA-insensitive (*abi*) mutants has shown that the maternal and zygotic sources of ABA function in different developmental pathways; the maternal ABA is required for seed development and the zygotic ABA for dormancy (16b, 37). Reciprocal crosses between wild-type *Arabidopsis* and the ABA-deficient (*aba*) genotype show that homozygous *aba/aba* mutant seeds developing on a heterozygous *aba/*+ mother plant are ABA deficient and non-dormant. Thus, the genotype of the embryo, and not the mother plant, determines seed ABA content and dormancy. Similar results are seen with the ABA-deficient mutants of tomato and maize (24, 28). Seeds of a double mutant containing leaky alleles of *aba* and *abi3* were normal when they developed on a mother plant heterozygous for the *aba* allele (37). These genetic studies have shown that there are at least two different effects of ABA in developing seeds; a high sensitivity (or low ABA threshold) response for seed development, and a low sensitivity (high ABA threshold) response for seed dormancy. Other *Arabidopsis* mutants which affect seed development are not impaired in ABA responses, indicating that there are regulatory networks besides ABA which

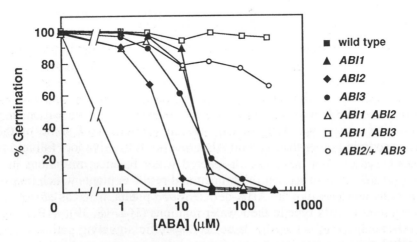

Fig. 4. Effect of ABA on germination of mono- and digenic *abi* mutants of *Arabidopsis*. Germination percentage was scored 4 days after plating on minimal media containing 0, 1, 3, 10, 30, 100, or 300 µM ABA. From (18).

control embryo maturation (47a).

The *viviparous* mutants of corn were the first to provide insight into ABA biosynthesis and action. It has been found that *vp1* embryos have normal amounts of ABA and carotenoids, while *vp2*, *vp5*, *vp7*, *vp8*, and *vp9* all have reduced levels of both (51). The latter class of mutants are all affected in the early steps of carotenoid biosynthesis and are seedling lethal because of photodestruction of chloroplasts in the absence of carotenoids. An inhibitor of carotenoid synthesis, fluridone, produces a phenocopy of this class of mutant when applied to wild type ears at a specific time in embryogeny. The vivipary produced is partially reduced by the application of exogenous ABA (28). This correlation of ABA and carotenoid biosynthetic activities provides strong evidence for the "indirect pathway" of ABA biosynthesis via oxidative cleavage of the epoxy-carotenoids. ABA biosynthesis in fruits and seeds is developmentally controlled, in contrast to vegetative tissues where ABA accumulation is environmentally controlled through changes in cell turgor (69).

Responsiveness to ABA has been measured in *vp* mutants by removing the immature embryos from the seed at different times in development and measuring the effect of various concentrations of ABA on the inhibition of growth in culture. The *vp1* mutation has unique pleiotropic effects on seed development, seed storage proteins, germination enzymes, and anthocyanin pigments. The *Vp1* gene confers the ability to respond to ABA in the seed; mutations in this gene result in a shifting of the ABA dose-response curve towards higher ABA concentrations (51, 57). Mutant *vp1* seedlings survive if rescued from desiccation, and vegetative tissues respond normally to ABA, e.g., the stomata close in response to exogenous ABA (51).

The Function of ABA in Embryo Development

All of the above results strongly link the initiation of the maturation pathway and inhibition of precocious germination with ABA. ABA initiates, but does not accompany desiccation of the seed; the mechanisms of desiccation which paradoxically occur during the period of decreasing water potential (and perhaps surprisingly, decreasing ABA), are unknown. The disappearance of ABA in the mature, dry seed is apparently a developmental adaptation which allows the mature embryo to germinate upon imbibition. It has become clear in recent years that there are unique gene products associated with the development of quiescence, and the expression of these genes can be increased or decreased by ABA. Towards the latter third of seed development in both monocots and dicots, a set of gene products begins to accumulate in embryos and includes a variety of proteins such as a lectin, amylase inhibitor, lipid body membrane protein, storage proteins and a number of functionally uncharacterized proteins that appear at the time embryos acquire the ability to withstand desiccation (13, 14, 34). For example, Williamson and Quatrano (67) showed that a 10 kD soluble protein in wheat embryos, *Em*, accumulates during the latter third of grain development. When wheat embryos are removed from the grain at the end of the first third of grain development and cultured in the presence of ABA, *Em* accumulates. In the absence of ABA, the immature embryos do not accumulate the mRNA or protein product from the *Em* gene and precociously germinate.

A proposed function of ABA in embryos, in addition to promoting embryogenesis and preventing germination, is to regulate the synthesis of proteins involved in desiccation tolerance. During the mid to late stages of seed development, specific mRNAs accumulate in embryos at the time of high endogenous ABA levels. A number of cDNAs encoding LEA (*Late Embryogenesis-Abundant*) genes have been cloned from numerous species, including cotton, rapeseed, barley, rice, wheat, and maize (9a, 13). If embryos are isolated at earlier developmental stages and exposed to exogenous ABA, the LEA and other mRNAs are precociously accumulated. LEA proteins are highly homologous, very soluble, basic, and have a biased amino acid composition high in glycine and lysine and low in hydrophobic residues. These characteristics make the LEA proteins very hydrophilic and stable to boiling. These features led Dure *et al.* (14) to propose that the proteins are not enzymes, but rather function in protecting proteins and membranes from damage due to loss of water in the cytoplasm during desiccation. However, a direct relationship between LEAs and desiccation tolerance has yet to be demonstrated (9a).

When vegetative plant tissues are exposed to water-stress by high osmoticum, NaCl, or desiccation, specific mRNAs are accumulated, some of which are identical to those induced during embryo development. These include the *Em* gene from wheat and *Arabidopsis* (4, 16) and 15-25 kDa proteins from maize (56), tomato (7), rice (51), and barley (11). The ABA- and stress-inducible expression of these genes supports the hypothesis that they are involved in a general mechanism of desiccation tolerance. However, conclusive evidence that ABA has a causal role in embryo development, dormancy and desiccation tolerance will require an understanding of the molecular mechanisms of ABA action. The ABA responsiveness of the *Em* gene appears to be tissue-independent, making it a good candidate to study the basic mechanism responsible for ABA-dependent gene expression. Hence, we will focus most of our subsequent discussion on what is known about the mechanisms of *Em* gene expression.

REGULATION OF GENE EXPRESSION BY ABA

Cis-acting sequences

Plant protoplasts are now widely used to rapidly identify sequences involved in hormone-regulated gene expression (*cis*-acting sequences; 32, 43, 49). The results from these more rapid transient analyses can then serve as a guide for the generation of transgenic plants for detailed *in vivo* analyses. By generating chimeric gene constructs that contain various segments of the 5' flanking sequences of the *Em* gene linked to a reporter gene, which produces an easily assayed enzyme (ß-glucuronidase, GUS), Marcotte *et al.* (43) have demonstrated that a 646 base pair (*bp*) *Em* promoter segment is necessary and sufficient for the ABA response. The response of the *Em* promoter-GUS fusion to ABA is rapid, with enzyme activity being detectable less than one hour after addition of ABA. Induction level is proportional to the concentration of ABA used and occurs at the same concentrations of ABA that had been previously shown to inhibit precocious germination of cultured wheat embryos (63). When present in the opposite orientation, the 646 *bp* promoter of *Em* is totally inactive with or without ABA. In addition, the transcription start site is the same in rice protoplasts transiently expressing either the *Em*-GUS fusion or the wheat *Em* genomic clone and is identical to the transcription start site in wheat embryos (43).

Marcotte *et al.* (44) have demonstrated that a GUS fusion construct containing the *Em* promoter segment from -168 to +92 gives the same approximate 25-fold induction as longer *Em* promoter fragments in response to ABA, but with a reduced level of GUS expression, *i.e.* the response is qualitatively, not quantitatively, similar (Fig. 5). When a slightly shorter *Em* promoter fragment is used (-106 to +92), the ABA response and expression is essentially abolished (Fig. 5). Examination of the sequences between -554 and -168 reveals the presence of three regions, 40 to 70 nucleotides each, which contain at least 84% A plus T (A/T) residues. Similar A/T rich regions are observed in several other plant genes and have been shown to be associated with high levels of gene expression (9). As such, the drop in expression levels in the *Em*-GUS fusions mentioned above has been attributed to the removal of these non-specific transcriptional enhancer sequences (44). Other investigators have found a similar phenomenon using a 5' deletion strategy for genes which respond to different signals (32).

Results of the deletion analysis of the *Em* promoter in the transient assay has identified a 62 *bp* region (-168 to -106 from the transcription start site) that is likely to contain at least part of a specific ABA response element (ABRE). Comparison of sequences from seed and ABA-regulated promoter regions has revealed several conserved sequence motifs in the *Em* promoter region near the ABRE (Em1a, Em1b, and Em2 in 45; 50, 59). One such conserved sequence, Em1, is found in most ABA-regulated promoters for which sequence data are available. To test the functionality of the region of the *Em* promoter containing the sequences Em1a/b and Em2, various oligonucleotides containing these conserved motifs have been synthesized and linked to the 5' end of a truncated 35S(-90)-GUS fusion (Fig. 5). A 20 base pair oligonucleotide which contains only Em1a is sufficient to confer ABA responsiveness and, as such, constitutes a minimal ABRE (Fig. 5). Transient expression analyses demonstrated that mutations at Em1a reduce or eliminate the ABA response (25). The results to date strongly suggest that the sequences necessary for the ABA response in the rice protoplast transient expression system reside in the sequences at Em1a. The conserved sequence

Em PROMOTER

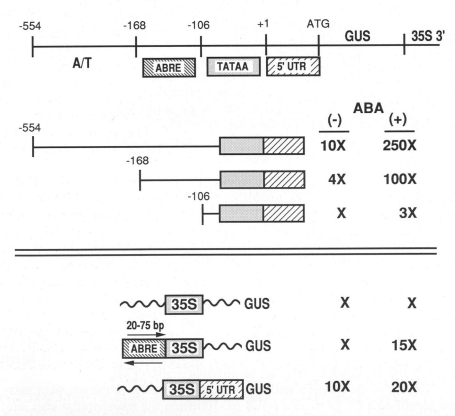

Fig. 5. Diagram of the *Em* chimeric gene construct utilized in the transient assay which includes the *Em* promoter, the reporter gene GUS, and the 3' segment of the CaMV 35S gene. Representative deletions and the corresponding GUS activities in the absence (-) or presence (+) of ABA (100 µM) illustrate the functional dissection of the *Em* promoter into an AT-rich enhancer region, the ABA response element (ABRE), and the 5' untranslated leader (UTR). Removal of the ABRE and 5' UTR from the *Em* promoter and linking them to a minimal CaMV 35S (-90) promoter demonstrates their ability to confer ABA inducibility and enhance expression, respectively.

Em2, found in some ABA-regulated promoters and also in many seed-specific promoters (e.g., the α' subunit of α-conglycinin), appears not to be essential for the ABA response and as such, may be involved in the seed-specific regulation of these genes (21). There are other Em1a-like elements within the ABRE and upstream from Em1a (complex II) which may have a similar function to Em1a (A. Hill and R. S. Quatrano, unpublished). Similar functional analyses of ABA-responsive gene promoters and sequence comparisons have produced an ABRE consensus sequence: CACGTGGC (25, 58). Interestingly, this core sequence, termed the G-box, is present in the upstream regulatory elements of a number of diverse plant genes, including various seed and embryo specific genes, non-ABA-inducible stress-induced genes, and light-regulated genes (see below). Nuclear factors from yeast and animal cells bind to G-box-like *cis* motifs, suggesting that transcriptional regulatory mechanisms may be conserved among eukaryotes (25).

To extend the above observations and to determine if the 646 *bp* promoter fragment from wheat could be properly regulated in embryos of developing tobacco seeds, tobacco leaf disks have been transformed using *Agrobacterium* and regenerated transgenic plants have been analyzed for GUS expression (44). No GUS activity is detected in vegetative tissue nor in young seeds from the *Em*-GUS transformants. However, in mature seeds removed from the *Em*-GUS transformants, GUS expression is very high and is confined to the embryo. In addition, the expression pattern of this fusion correlates with the developmental rise in ABA levels during tobacco seed development.

The accumulation of *Em* transcripts in response to ABA is controlled at least in part at the level of transcription (67); however, the ABA-induced expression of the *Em* gene is also controlled at post-transcriptional and/or translational level(s) (4, 67). In the absence of exogenous ABA, *Em* transcripts are rapidly degraded within the first several hours of imbibition. The requirement for ABA to maintain levels of the *Em* transcript found in mature embryos occurs even in the presence of α-amanitin at concentrations that specifically inhibit RNA synthesis from polymerase II (67). As such, ABA may have an effect on the stability/translatability of the *Em* mRNA. Post-transcriptional regulation of the ABA response is also seen in vegetative tissue. When comparable levels of *Em* mRNA are induced specifically by ABA in embryonic and vegetative tissues, antibodies to the *Em* protein detect lower levels of *Em* in seedlings compared to embryos (4).

To further investigate the question of post-transcriptional regulation, and to determine if sequences in the 5' and/or 3' untranslated regions (UTR) of the *Em* gene are involved in the ABA response, chimeric genes have been constructed which contain various combinations of the Em 5'/3' UTR sequences on either side of GUS. Inclusion of the Em 5' UTR between the (-90) 35S promoter and the translational initiation codon, ATG, increases GUS activity in protoplasts to double the level of the full-length (-338) 35S promoter in the absence of ABA. In the presence of ABA, an additional doubling of GUS activity is observed. Therefore, compared to the truncated 35S promoter alone, inclusion of the Em leader from wheat leads to a 10-fold increase in activity without ABA and a 20-fold increase in the presence of ABA (Fig. 5). It is clear from these results that the +6 to +86 segment of the *Em* promoter has a major effect on GUS expression in rice protoplasts in the absence of ABA, and may indicate a component of post-transcriptional regulation involving ABA (44).

Trans-acting factors

The control of gene transcription is mediated by interactions of DNA sequences (*cis* elements) with proteins (transcription or "*trans*-acting" factors) within the nucleus. The modulation of gene expression levels by hormonal, environmental and developmental signals is accomplished, in part, through changes in these interactions. In more complex examples, the interaction of several transcription factors, as well as modification of these factors by kinases *etc.*, affects the ultimate level of gene expression (8). A number of these transcription factors have a characteristic juxtaposition of conserved amino acid sequence motifs linked in tandem; a conserved basic (b) region next to either a leucine heptad repeat[1] (ZIP), a helix-loop-helix motif (HLH), or a combination of HLH and ZIP. Examples of the bZIP group of transcription factors include Jun and Fos from mammals, GCN4 from yeast, and O2, OHP1, and HBP1 in cereals. The B, R, and C1 transcription factors from maize are representative of bHLH. Other groups of DNA-binding proteins have been identified in plants and implicated in the control of transcription, for example the homeobox, zinc-finger, and MADS box proteins (8). The following sections summarize what is known of the proteins which bind the *Em* promoter and related *cis*-elements from plants.

The binding of nuclear proteins to the wheat *Em* promoter can be investigated by an electrophoretic mobility shift assay (EMSA). This assay exploits the reduction in mobility of a DNA/protein complex upon electrophoresis relative to an unbound DNA probe. Proteins extracted from crude nuclear preparations are incubated with a radiolabeled DNA probe and applied to a gel. After electrophoresis, the positions of the free and bound probe are revealed by autoradiography. The DNA complexed with the bound probe, being larger than probe alone, moves less distance upon electrophoresis. Specificity of the interaction is determined with the addition of unlabeled DNA competitors to the binding reaction. Only a DNA sequence containing the putative binding site should compete for binding, and in excess, should eliminate the autoradiographic signal of the bound complex. A 300 *bp* A/T-rich fragment (300KS) from the 5' region of the *Em* promoter (-559 to -263) forms a specific complex with proteins in both the rice (R) and wheat (W) extracts (Fig. 6, lanes 2,3). An unlabeled plasmid (S)

---

[1] A repetition of leucine residues at every seventh position, which allows factor dimerization via α-helical "leucine zippers", ZIP.

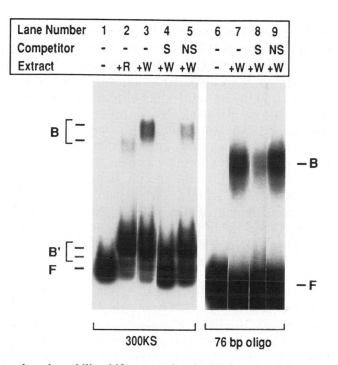

Lane Number	1	2	3	4	5	6	7	8	9
Competitor	-	-	-	S	NS	-	-	S	NS
Extract	-	+R	+W	+W	+W	-	+W	+W	+W

300KS   76 bp oligo

Fig. 6. Electrophoretic mobility shift assays using the 300 bp AT-rich fragment (300 KS) and a 76-bp oligonucleotide fragment (ABRE), both sequences from the *Em* promoter (25, 44). Wheat (W) and rice (R) nuclear extracts demonstrate the presence of protein factors that bind to both the 300 KS and ABRE, resulting in a shift in mobility of the radioactive bound (B) fragment compared to the free probe (F). The binding is eliminated by specific (S) competitions using excess non-radioactive 300 KS (lane 4) and ABRE (lane 8) fragments, but not by a nonspecific (NS) competitor (an *Em* coding region sequence). From Quatrano *et al.*, *in* Control of Plant Gene Expression, pp. 69-90, Verma, D.P.S., ed. CRC, Boca Raton, FL.

incorporating the A/T-rich fragment competes for the binding (lane 4), while non-specific (NS) plasmid at the same concentration does not (lane 5). Previous studies in animal systems have shown that A/T-rich sequence elements are frequently bound by high mobility group chromosomal proteins (HMGs)[2]. Purified wheat HMGs interact specifically with the A/T-rich fragment of the *Em* promoter (55). The finding that the A/T-rich element can confer quantitative enhancement of *Em* gene expression levels (44) is consistent with previous studies showing that HMGs bind preferentially to actively transcribing regions of chromatin (55). Proteins in nuclear extracts from wheat embryos and cultured rice cells can interact with a 76 *bp* oligonucleotide containing the ABRE element responsible for ABA mediated enhancement of gene expression (Fig. 6, lanes 6-9; 25). Additionally, a 2 *bp* mutation of the ABRE (mABRE) eliminates the ability to compete for the binding (25). The 2 base pairs mutated in the mABRE reside within Em1a.

---

[2] Defined operationally as 10-30 kD chromosomal proteins extractable with salt and soluble in 2% trichloroacetic acid.

The binding of rice and wheat nuclear proteins to the 76 *bp* ABRE is more precisely defined using methylation interference footprinting (25). This method involves partially methylating the G residues of a DNA sequence, reacting this methylated DNA probe with a protein extract and separating the bound DNA probe from the free probe by EMSA. The DNA in each fraction is then isolated, cleaved at methylated G residues with piperidine, then electrophoresed on a DNA sequencing gel. Guanine residues within the recognition sequence, which normally interact intimately with a DNA binding protein, interfere with complex formation when methylated. This results in the depletion, or footprinting, of the specific fragments in the bound DNA sample, relative to the free sample. Wheat and rice nuclear proteins are footprinted to a small region of the ABRE spanning the consensus motif Em1a (25). Each G residue in the ABRE is either completely or partially abolished in the bound DNA relative to the free DNA samples. No binding is observed in the Em2 or Em1b boxes.

Guiltinan *et al.* (25) cloned a leucine zipper (bZIP) type DNA-binding transcription factor (EmBP-1) by probing a wheat cDNA lambda phage expression library with a radioactive ABRE probe. The protein appears to be expressed constitutively in seeds and vegetative tissue, and is not induced by ABA (M.J. Guiltinan and R.S. Quatrano, unpublished). The specificity of EmBP-1 binding has been determined by both EMSA and methylation interference footprinting and is identical to that seen with the nuclear extract from wheat embryo and rice tissue culture cells. The EMBP-1 gene has homology to numerous plant bZIP factors which also interact with the CACGTG motif found within the context of a number of plant promoters that are regulated by different signals such as light (3, 36, 53).

Analyses of the response of the *Em* gene to ABA have identified several components that must be included in any general model to explain the tissue and chemical specificity of a phytohormone response: 1) a *cis* response element in the *Em* promoter, composed of at least 20 *bp* (ABRE), which has been shown to confer ABA responsiveness to a non-responsive viral promoter, and, 2) a *trans*-acting protein, structurally similar to the bZIP transcription factor class, which recognizes and binds to the ABRE of the *Em* gene. One of the most interesting observations that has emerged from the analyses of *Em* activation by ABA is that the ABRE and EmBP-1 are not unique to ABA-responsive genes or to plants. How does one achieve the specificity of expression from such different signals as light and ABA, when each signal appears to act through response elements which share a common core (CACGTG) and which bind proteins from the same transcription factor family (bZIP)? Some possibilities are discussed below.

Leucine zipper transcription factors bind DNA as dimers, and are found to comprise multigene families in eukaryotes. Four G-box-binding bZIPs (GBF1-4) have been cloned from *Arabidopsis*, and their protein products can form heterodimers which exhibit differing binding affinities for a G-box target DNA sequence (3, 36). Thus by heterodimerizing, similar bZIPs could generate the complexity necessary to define a specific response pathway by binding to specific *cis* elements, even though the target sites are similar. Conversely, differences in the sequences adjacent to the G-box may encode the specificity for binding to bZIP heterodimers (31).

Another mechanism of transcriptional regulation is phosphorylation of transcription factors by kinases involved in signal transduction. Klimczak *et al.* (36) have shown that the DNA binding activity of a bZIP G-box factor (GBF1) is stimulated by phosphorylation by a kinase from nuclear extracts. Phosphorylation of bZIP factors in vivo modulates their nuclear localization (25a). This observation is especially interesting in light of the recent discoveries of an ABA-inducible gene which has homology to protein kinases (2), and an ABA-inducible gene which binds to nuclear localization signals in a phosphorylation-dependent manner (23a). These results suggest possible mechanisms of ABA-inducible gene expression by phosphorylation of transcription factors, or modulation of their nuclear transport.

The *VP1* locus of maize has been cloned by transposon tagging, which is a powerful approach utilizing classical genetics and molecular biology. Essentially, if a mutable allele (showing an unstable phenotype because of transposon excision) of the gene of interest is available, then the gene can be cloned by using the transposable element DNA as a molecular probe of mutant DNA to select and clone the flanking DNA encoding the inactivated gene. *VP1* encodes a novel seed-specific transcription factor which interacts with the upstream regulatory regions of ABA-inducible genes (47). Expression of the 7S storage globulin, *Em*, *C1* (an anthocyanin biosynthesis regulatory factor of the bHLH class), and numerous other genes is induced by ABA; however, these genes are not expressed in *vp1* mutant kernels, indicating that *VP1* is required for ABA-inducible gene expression. Remarkably, *VP1* also functions to repress the transcription of germination-specific genes that are down-regulated by ABA, e.g., α-amylase and lipase (29). Overexpression of *VP1* (using the 35S CaMV promoter) in a maize protoplast transient expression assay gives a 5-fold increase in the expression of an ABA-inducible reporter construct (Em-GUS of ref. 43) in the absence of exogenous ABA (Fig. 7). Addition of ABA to the expression system further induces Em-GUS expression in a dose-dependent manner (Fig. 7). Thus, *VP1* acts synergistically with ABA at a specific time in embryo development to activate or repress ABA-responsive genes. Because the *vp1* mutant has pleiotropic effects on expression of numerous genes, it may function by specifically interacting with numerous transcription factors, or by interacting with DNA, or both.

Recently a gene has been cloned (*GF14*) the protein product of which is associated with transcriptional complexes that bind to the G-box element found in ABA-responsive and other inducible genes (42). This gene has homology to a mammalian regulatory protein family; it binds calcium, is phosphorylated by an endogenous kinase, and is found in both the cytoplasm and nucleus (42a). The characterization of these protein-protein interactions will undoubtedly shed light on the mechanisms of transcriptional activation in plants and may lead to identification of the factor(s) which confer specificity to the signal transduction pathways.

Interactions with other signals

The *Em* gene is not expressed in response to IAA, CK, salicylic acid, GA, chilling, heat, or UV-light. Environmental signals that are believed to be transduced, at least in part via changes in ABA levels (e.g., desiccation, salinity, wounding) can result in *Em* gene expression. For example, desiccation of rice suspension cultures to 10% of their initial fresh weight is accompanied by an increase in *Em* mRNA that parallels the 2-fold increase in ABA levels (6). Response-saturating concentrations of either ABA (50 μM) or NaCl (0.4 M) rapidly induce the accumulation of *Em* mRNA.

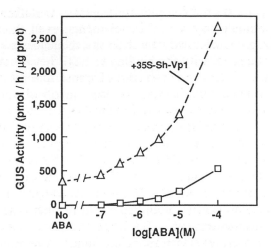

Fig. 7. The over-expressed *VP1* gene product transactivates a GUS gene driven by the ABA-responsive *Em* promoter in transfected maize protoplasts. Em-GUS contains 554 base pairs of the 5' flanking sequence of the *Em* gene fused to the bacterial GUS gene (43). 35S-Sh-Vp1 contains the 35S CaMV promoter and Shrunken1 intron (to increase expression) fused to the *Vp1* coding region. Maize protoplasts (plant cells with the cell wall removed) were cotransformed by electroporation with 20 μg each of effector (35S-Sh1-Vp1) and/or reporter (Em-GUS) plasmid DNA. GUS activity was determined after 2 days of incubation. Squares represent protoplasts electroporated with Em-GUS alone, and triangles represent protoplasts electroporated with Em-GUS together with 35S-Sh-Vp1. *Vp1* overexpression increases the magnitude of the ABA response in the absence of exogenous ABA and has a synergistic effect on ABA-inducible Em-GUS expression. From (47).

At least a part of the salt effect is via changes in endogenous ABA levels; inhibition of ABA biosynthesis by fluridone during NaCl treatment reduces the levels of endogenous ABA by four-fold and *Em* expression by 50 percent. Data from the rice culture studies suggests that salt interacts synergistically with ABA, in part because of the increased sensitivity of rice cells to ABA. Similar results have recently been obtained with osmotic agents such as mannitol and sucrose (J.-L. Magnard and R. S. Quatrano, unpublished). The effect of osmotic stress on *Em* gene expression in rice suspension cells appears to operate through two pathways; one is mediated through increases in the level of ABA, the other is via a unique salt response pathway that includes an intermediate that is common to both the salt and ABA response chains. A similar conclusion has been drawn from studies with barley LEAs (15).

Some ABA-inducible genes, including seed storage genes, are also regulated by jasmonic acid (JA; see Chapter C2A). In fact, the promoter regions necessary for JA-inducible gene expression have been functionally localized to G-boxes. JA may be an intermediate in the ABA signal transduction pathway, or an independent regulator of growth and development which controls expression, via conserved transcriptional mechanisms, of gene sets which overlap with ABA-responsive gene sets.

Secondary messengers act to amplify the information contained in the primary signal through direct regulation of cellular metabolism or the generation of additional signals. ABA has been shown to stimulate transient increases in the cytosolic Ca^{2+} concentration in stomatal guard cells, roots, and corn coleoptiles (26, 46). In roots and coleoptiles, cytosolic pH also increases concomitantly with Ca^{2+}, indicating the possible regulation of the plasmamembrane H^+-ATPase. In barley aleurone protoplasts, GA causes an increase in cytosolic Ca^{2+} (over several hours) which is necessary for induction of germination-specific genes such as α-amylase (22). ABA can antagonize this response by inhibiting the cytosolic Ca^{2+} increase. Furthermore, it has been shown directly that ABA causes a rapid decrease in cytosolic Ca^{2+} and an increase in pH in these cells, and that the cytosolic alkalinization is necessary, but not sufficient, for ABA-inducible gene expression (62). The ionic fluxes in response to ABA in aleurone may play a role in regulation of gene expression, but further work is needed to firmly establish this signal transduction mechanism.

Response pathways

The diverse physiological effects mediated by ABA (e.g., fast responses such as stomatal closure and slower responses such as alteration in patterns of gene expression) indicate that there are multiple mechanisms of ABA action. It is a reasonable and testable hypothesis that ABA may act through response mechanisms similar to those found in animals, including hormone receptors. The direct evidence for ABA receptors, despite a few promising reports (30, 41), is still lacking. This area of research is of tantamount importance to elucidation of a complete ABA signal transduction pathway.

The biological activities of a large number of ABA analogs have been compared. Although structure-activity correlations are complicated by the unknown stability of the analogs in bioassays, the observed large activity differences resulting from minor structural variations, especially chiral enantiomers, argue for the necessity of a precise molecular fit in a receptor. In the response of stomata to ABA, only (+)-ABA is active, while seed germination and α-amylase production are inhibited by both (+)-and (-)-ABA, as well as phaseic acid (52, 64). When ABA-regulated gene expression is assayed with various optically pure ABA analogs, differential induction of various ABA-responsive genes is observed, suggesting there may be more than one ABA receptor or different signal transduction pathways in embryos

(64). Recently, one of these biologically inactive ABA analogs was shown to compete for an ABA perception site (receptor?) and inhibit the ABA response (66).

Sensitivity is an important parameter in signal transduction and can be defined as the efficiency of coupling all the necessary components involved in a particular response (61). Sensitivity could be the result of rapid metabolism of the hormone or signal, differential uptake, the concentration of receptors, or any rate-limiting step in the stimulus-response pathway. There is the open question of whether a change in hormone concentration or hormone sensitivity is the pivotal factor in the ABA response. In rice suspension cultures, both ABA concentration and sensitivity to ABA increase in response to osmotic stress (6). The degree of sensitivity to ABA is seen in differences in the ABA dose-response curves of low and high dormancy cultivars of wheat germinated in the presence of varying concentrations of ABA (63). Seeds of both cultivars contain similar concentrations of ABA, but respond according to their sensitivity to the hormone.

Genetics is a powerful tool for dissecting pathways and has played no small part in our understanding of ABA biosynthesis and action. Comparison of the phenotypic expression of monogenic and digenic *abi* lines of *Arabidopsis* led Finkelstein and Somerville (18) to suggest that ABA responses during seed maturation are regulated by two parallel pathways defined by the gene products of *ABI1* and *ABI3* (see Fig. 4, 57a). This hypothesis is interesting in light of the results of Pla *et al.* (56), who propose an alternative ABA response pathway for expression of a maize ABA-responsive gene (*RAB28*). This gene does not require *VP1* for ABA-induced expression in seeds, unlike the ABA-inducible 7S globulin gene or *Em* (40, 47, 68). Similar results have recently been reported for another maize *LEA* gene (61) and an isoform of catalase (68). The existence of two distinct ABA response pathways in both a monocot and a dicot raises the possibility that the pathways may be homologous and conserved between the two classes of plants. Indeed, the recent cloning of the *ABI3* gene of *Arabidopsis* by chromosome walking has established that the gene sequence is highly homologous to *VP1* of maize (23).

Recently, the research groups of Giraudat and Grill have cloned the *Arabidopsis* gene for *abi1* (ABA insensitivity) by chromosomal walking, and have shown it encodes a novel serine-threonine protein phosphatase with calcium binding domains (41a, 47b). The ABI1 gene product predicts a novel type of signaling enzyme in that the amino-terminal portion displays the structural elements for an EF hand type of calcium binding site. The carboxyl-terminal domain of the protein has 57% similarity (35% identity) with the 2C class of serine-threonine protein phosphatases (PP2C) identified in animals and yeast. This combination of EF hand- and PP2C-domains is novel; the only Ca^{2+}-dependent ser-thr phosphatases known are the calcineurins (or PP2Bs). Based on inhibitor studies, a calcineurin-like protein phosphatase is involved in ABA regulation of stomatal aperture through

Ca^{2+}-induced inactivation of potassium channels (42b). The ABI1 gene product is a likely candidate for this calcineurin-like protein. The molecular characteristics of ABI1, as well as its requirement for most ABA responses so far tested, are tantalizingly suggestive of a central intermediate in the cellular signaling pathways that have been implicated in ABA responses.

Protein phosphorylation is an attractive model to explain several mechanisms by which the ABA signal is transduced to activate transcription: modulation of bZIP DNA binding specificities, nuclear localization of transcription factors, and protein/protein interactions. Although the molecular mechanisms of *ABI1* and *ABI3/(vp1)* function are not yet known, the structures of these genes provide a conceptual framework for further studies. A kinase regulatory cascade, modulated by ABI1, kinases and calcium ions, might act at multiple steps in multiple response pathways; for example, the cell cycle, ion channels, nuclear transport, and transcriptional activation. Because numerous ABA- and other response-insensitive loci remain to be characterized and cloned, there will be exciting discoveries and a wealth of knowledge on plant hormone signaling from *Arabidopsis* research in the future.

Acknowledgements

The authors thank Janice Davis and Alison Hill for critical reading of the manuscript. This work was supported by NIH Postdoctoral Fellowship GM14752 to C.D.R., and by NIH grant GM44288 to R.S.Q.

References

1. Ackerson, R.C. (1984) Abscisic acid and precocious germination in soybeans. J. Exp. Bot. 35, 414-421.
2. Anderberg, R.J., Walker-Simmons, M.K. (1992) Isolation of a wheat cDNA clone for an abscisic acid-inducible transcript with homology to protein kinases. Proc. Natl. Acad. Sci. U.S.A. 89, 10183-10187.
3. Armstrong, G.A. Weisshaar, B. Hahlbrock, K. (1992) Homodimeric and heterodimeric leucine zipper proteins and nuclear factors from parsley recognize diverse promoter elements with ACGT cores. Plant Cell 4, 525-37.
4. Berge, S.K., Bartholomew, D.M., Quatrano, R.S. (1989) Control of the expression of wheat embryo genes by abscisic acid. *In*: Molecular Basis of Plant Development, pp. 193-201, Goldberg, R.L., ed. Liss, New York, NY.
5. Black, M. (1991) Involvement of ABA in the physiology of developing and mature seeds. *In*: Abscisic Acid: Physiology and Biochemistry, pp 90-124. Davies, W.J., Jones, H.G., eds. BIOS, Oxford UK.
6. Bostock, R.M., Quatrano, R.S. (1992) Regulation of *Em* gene expression in rice: interaction between osmotic stress and abscisic acid. Plant Physiol. 98, 1356-1363
7. Bray, E.A. (1991) Regulation of gene expression by endogenous ABA during drought stress. *In*: Abscisic Acid: Physiology and Biochemistry, pp. 81-98, Davies, W.J., Jones, H.G., eds., Bios, Oxford, UK.
8. Brunelle, A.N., Chua, N.-H. (1993) Transcription regulatory proteins in higher plants. Curr. Top. Genet. Devel. 3, 254-258.

9. Bustos, M.M., Guiltinan, M.J., Jordano, J., Begum, D., Kalkan, F.A., Hall, T.C. (1989) Regulation of ß-glucuronidase expression in transgenic tobacco plants by an A/T-rich, *cis*-acting sequence found upstream of a french bean ß-phaseolin gene. Plant Cell 1, 839-853.

9a. Chandler, P.M., Robertson, M. (1994) Gene expression regulated by abscisic acid and its relation to stress tolerance. Ann. Rev. Plant Physiol. Plant Mol. Biol. 45, 113-141.

10. Clark, J.K., Sheridan, W.F. (1991) Isolation and characterization of 51 embryo-specific mutations of maize. Plant Cell 3, 935-951.

11. Close, T.J., Koort, A.A., Chandler, P.M. (1989) A cDNA-based comparison of dehydration-induced proteins (dehydrins) in barley and corn. Plant Mol. Biol. 13, 95-108.

12. Cornford, C.A., Black, M., Chapman, J.M., Baulcombe, D.C. (1986) Expression of α-amylase and other gibberellin-regulated genes in aleurone tissue of developing wheat grains. Planta 169, 420-428.

13. Dure, L. III (1993) The LEA proteins of higher plants. *In*: Control of Plant Gene Expression. pp. 325-335, Verma, D.P.S., ed. CRC, Boca Raton, FL.

14. Dure, L. III, Crouch, M., Harada, J., Ho, T.-H.D., Mundy, J., Quatrano, R., Thomas, T. Sung, Z.R. (1989) Common amino acid sequence domains among the *LEA* proteins of higher plants. Plant Mol. Biol. 12, 475-486.

15. Espelund, M., Sæbøe-Larssen, S., Hughes, D.W., Galau, G.A., Larsen, F., Jakobsen, K.S. (1992) Late embryogenesis-abundant genes encoding proteins with different numbers of hydrophyllic repeats are regulated differentially by abscisic acid and osmotic stress. Plant J. 2, 241-252.

16. Finkelstein, R.R. (1993) Abscisic acid insensitive mutations provide evidence for stage specific signal pathways regulating expression of an *Arabidopsis* late embryogenesis abundant gene. Mol. Gen. Genet. 238, 401-408..

16a. Finkelstein, R.R. (1994) Mutations at two new Arabidopsis ABA response loci are similar to the *abi3* mutations. Plant J. 5, 765-771.

16b. Finkelstein, R.R. (1994) Maternal effects govern variable dominance of two abscisic acid response mutations in *Arabidopsis thaliana*. Plant Physiol. 105, 1203-1208.

17. Finkelstein, R.R., Tenbarge, K.M., Shumway, J.E., Crouch, M.L. (1985) Role of ABA in maturation of rapeseed embryos. Plant Physiol. 78, 630-636.

18. Finkelstein, R.R. Somerville, C.R. (1990) Three classes of abscisic acid (ABA)-insensitive mutations of *Arabidopsis* define genes that control overlapping subsets of ABA responses. Plant Physiol. 94, 1172-1179.

19. Galau, G.A., Jakobsen, K.S., Hughes, D.W. (1991) The controls of late dicot embryogenesis and early germination. Physiol. Plant. 81, 280-288.

20. García-Martínez, J.L., Martí, M., Sabater, T., Maldonado, A., Vercher, Y. (1991) Development of fertilized ovules and their role in the growth of the pea pod. Physiol. Plant. 83, 411-416.

21. Gatehouse, J.A., Shirsat, A.H. (1993) Control of expression of seed storage protein genes. *In*: Control of Plant Gene Expression, pp. 357-375, Verma, D.P.S., ed. CRC, Boca Raton, FL.

22. Gilroy, S., Jones, R.L. (1992) Gibberellic acid and abscisic acid coordinately regulate cytoplasmic calcium and secretory activity in barley aleurone protoplasts. Proc. Natl. Acad. Sci. U.S.A. 89, 3591-3595.

23. Giraudat, J., Hauge, B.M., Valon, C., Smalle, J., Parcy, F. Goodman, H.M. (1992) Isolation of the *Arabidopsis ABI3* gene by positional cloning. Plant Cell 4, 1251-1261.

23a. Goday, A., Jensen, A.B., Culiáñez-Macià, F.A., Albà, M.M. , Figueras, M., Serratosa, J., Torrent, M., Pagès, M. (1994) The maize abscisic acid-responsive protein Rab17 is located in the nucleus and interacts with nuclear localization signals. Plant Cell 6, 351-360.

24. Groot, S.P.C., Karssen, C.M. (1992) Dormancy and germination of abscisic acid-deficient tomato seeds: studies with the sitiens mutant. Plant Physiol. 99, 952-958.

25. Guiltinan, M. J., Marcotte, W. R., Jr., Quatrano, R. S. (1990) A plant leucine zipper protein that recognizes an abscisic acid response element. Science 250, 267-271.

25a. Harter, K., Kircher, S., Frohnmeyer, H., Krenz, M., Nagy, F., Schäfer, E. (1994) Light-regulated modification and nuclear translocation of cytosolic G-box binding factors in parsley. Plant Cell 6, 545-559.

26. Hetherington, A. R. S. Quatrano 1991 Mechanisms of action of abscisic acid at the cellular level. New Phytol. 119, 9-32.

27. Hilhorst, H.W.M., Karssen, C.M. (1992) Seed dormancy and germination: the role of abscisic acid and gibberellins and the importance of hormone mutants. Plant Growth Reg. 11, 225-238.

28. Hole, D.J., Smith, J.D., Cobb, B.G. (1989) Regulation of embryo dormancy by manipulation of abscisic acid in kernels and associated cob tissue of *Zea mays* L. cultured *in vitro.* Plant Physiol. 91, 101-105.

29. Hoecker, U., Rosenkrans, L., Vasil, I.K., McCarty, D.R. (1993) *VP1* is a repressor as well as an activator of transcription in the developing seed. J. Cell. Biochem. 17B, 29.

30. Hornberg, C. Weiler, E.W. (1984) High-affinity binding sites for abscisic acid on the plasmalemma of *Vicia faba* guard cells. Nature 310, 321-324.

31. Izawa, T., Foster, R., Chua, N.-H. (1993) Plant bZIP protein DNA binding specificity. J. Mol. Biol. 230, 1131-1144.

32. Jacobsen, J.V. Close, T.J. (1991) Control of transient expression of chimaeric genes by gibberellic acid and abscisic acid in protoplasts prepared from mature barley aleurone layers. Plant Mol. Biol. 16, 713-724.

33. Karssen, C. M., Brinkhorst-Van Der Swan, D. L. C., Breekland, A. E., Koornneef, M. (1983) Induction of dormancy during seed development by endogenous abscisic acid: studies on abscisic acid deficient genotypes of *Arabidopsis thaliana* L. Heynh. Planta 157, 158-165.

34. Kermode, A.R. (1990) Regulatory mechanisms involved in the transition from seed development to germination. Crit. Rev. Plant Sci. 9, 155-195.,

35 Khan, A.A,(1982) Gibberellins and seed development. *In.* The Physiology and Biochemistry of Seed Development, Dormancy and Germination, pp. 111-135, Khan, A.A., ed. Elsevier Biomedical, Amsterdam.

36. Klimczak, L.J., Schindler, U., Cashmore, A.R., (1992) DNA binding activity of the *Arabidopsis* G-box binding factor GBF1 is stimulated by phosphorylation by casein kinase II from broccoli. Plant Cell 4, 87-98.

37. Koornneef, M., Hanhart, C. J., Hihorst, H. W. M., Karssen, C.M. (1989) *In vivo* inhibition of seed development and reserve protein accumulation in recombinants of abscisic acid biosynthesis and responsiveness mutants in *Arabidopsis thaliana.* Plant Physiol. 90, 463-469.

38. Koornneef, M., Jorna, M.L., Brinkhorst-van der Swan, D.L.C. Karssen, C.M. (1982) The isolation of abscisic acid (ABA) deficient mutants by selection of induced revertants in non-germinating gibberellin sensitive lines of *Arabidopsis thaliana* L. Heynh. Theor. Appl. Genet. 61, 385-393.

39. Koornneef, M., Reuling, G., Karssen, C. M. (1984) The isolation and characterization of abscisic-acid-insensitive mutants of *Arabidopsis thaliana.* Physiol. Plant. 61, 377-383.

40. Kriz, A.R., Wallace, M.S., Paiva, R. (1990) Globulin gene expression in embryos of maize *viviparous* mutants. Plant Physiol. 92, 538-542.

41. Ladyzhenskaya, É.P., Dardzhaniya, L.G., Korableva, N.P. (1989) Kinetic parameters of binding of gibberellic and abscisic acids with a plasmalemma preparation from potato tubers. Sov. Plant Physiol. 36, 820-825.

41a. Leung, J., Bouvier-Durand, M., Morris, P.-C., Guerrier, D., Chefdor, F., Giraudat, J. (1994) *Arabidopsis* ABA response gene ABI1: features of a calcium-modulated protein phosphatase.Science 264, 1448-1452.

42. Lu, G., DeLisle, A.J., de Vetten, N.C., Ferl, R.J. (1992) Brain proteins in plants: an *Arabidopsis* homolog to neurotransmitter pathway activators is part of a DNA binding complex. Proc. Natl. Acad. Sci. U.S.A. 89, 11490-11494.

42a. Lu, G., Sehnke, P.C., Ferl, R.J. (1994) Phosphorylation and calcium binding properties of an *Arabidopsis* GF14 brain protein homolog. Plant Cell 6, 501-510.

42b. Luan, S., Li, W., Rusnak, F., Assmann, S.M., Schreiber, S.L. (1993) Immunosuppressants implicate protein phosphatase regulation of K$^+$ channels in guard cells. Proc. Natl. Acad. Sci. U.S.A. 90, 2202-2206.

43. Marcotte, W.R. Jr., Bayley, C.C., Quatrano, R.S. (1988) Regulation of a wheat promoter by abscisic acid in rice protoplasts. Nature 335, 454-457.

44. Marcotte, W. R., Russell, S.H., Quatrano, R.S. (1989) Abscisic acid-responsive sequences from the *Em* gene of wheat. Plant Cell 1, 969-976.

45. Mayer, U., Büttner, G., Jürgens, G. (1993) Apical-basal pattern formation in the *Arabidopsis* embryo: studies on the role of the *gnom* gene. Development 117, 149-162.

46. McAinsh, M.R., Brownlee, C., Hetherington, A.M. (1992) Visualizing changes in cytosolic-free Ca^{2+} during the response of stomatal guard cells to abscisic acid. Plant Cell 4, 1113-1122.

47. McCarty, D.R., Hattori, T., Carson, C.B., Vasil, V., Lazar, M., Vasil, I.K. (1991) The *viviparous-1* developmental gene of maize encodes a novel transcriptional activator. Cell 66, 895-905.

47a. Meinke, D.W., Franzmann, L.H., Nickle, T.C., Yeung, E.C. (1994) Leafy cotyledon mutants of *Arabidopsis*. Plant Cell 6, 1049-1064.

47b. Meyer, K., Leube, M.P., Grill, E. (1994) A protein phosphatase 2C involved in ABA signal transduction in *Arabidopsis thaliana*. Science 264, 1452-1455.

48. Mounler, M.A.Kh., Bangerth, F., Story, V.(1980) Gibberellin-like substances and indole type auxins in developing grains of normal- and high-lysine genotypes of barley. Physiol. Plant. 48, 568-753.

49. Mundy, J., Yamaguchi-Shinozaki, K., Chua, N.-H. (1990) Nuclear Proteins Bind Conserved Elements in the Abscisic Acid-responsive Promoter of a Rice *rab* gene. Proc. Natl. Acad. Sci. U.S.A. 87, 1406.

50. Nambara, E., Naito, S., McCourt, P. (1992) A mutant of *Arabidopsis* which is defective in seed development and storage protein accumulation is a new *abi3* allele. Plant J. 2, 435-441.

51. Neill, S.J., Horgan, R., Rees, A.F. (1987) Seed development and vivipary in *Zea mays* L. Planta 171, 358-364.

52. Nolan, R.C., Ho, T.-H.D. (1988) Hormonal regulation of α-amylase expression in barley aleurone layers. The effects of gibberellic acid removal and abscisic acid and phaseic acid treatments. Plant Physiol. 88, 588-593.

52a. Ober, E.S., Setter, T.L., Madison, J.T., Thompson, J.F., Shapiro, P.S. (1991) Influence of water deficit on maize endosperm development. Plant Physiol. 97, 154-164.

53. Oeda, K., Salinas, J., Chua, N.-H. (1991) A tobacco bZIP transcription factor (TAF-1) binds to a G-box-like motif conserved in plant genes. EMBO J. 10, 1793-1802.

54. Ooms, J.J.J., Léon-Kloosterziel, K.M., Bartels, D., Koornneef, M., Karssen, C.M. (1993) Acquisition of desiccation tolerance and longevity in seeds of *Arabidopsis thaliana*. Plant Physiol. 102, 1185-1191.

55. Pederson, T.J., Arwood, L.J., Spiker, S., Guiltinan, M.J., Thompson, W.F. (1991) High mobility group chromosomal proteins bind to AT-rich tracts flanking plant genes. Plant Mol. Biol. 16, 95-104.

56. Pla, M., Gómez, J., Goday, A. Pagès, M. (1991) Regulation of the abscisic acid-responsive gene *rab28* in maize *viviparous* mutants. Mol. Gen. Genet. 230, 394-400.

57. Robichaud, C. S., Wong, J., Sussex, I. M. (1980) Control of *in vitro* growth of viviparous embryo mutants of maize by abscisic acid. Develop. Genet. 1, 325-330.

57a. Rock, C.D., Quatrano, R.S. (1994) Plant signals: insensitivity is in the genes. Current Biol., in press.

58. Skriver, K., Mundy, J. (1990) Gene expression in response to abscisic acid and osmotic stress. Plant Cell 2, 503-512.

59. Sussex, I. (1975) Growth and metabolism of the embryo and attached seedling of the viviparous mangrove *Rhizophora mangle*. Am. J. Bot. 62, 948-953.

60. Thomann, E.B., Sollinger, J., White, C., Rivin, C.J. (1992) Accumulation of group 3 late embryogenesis abundant proteins in *Zea mays* embryos. Roles of abscisic acid and the *Viviparous-1* gene product. Plant Physiol. 99, 607-614.

61. Trewavas, A. (1991) How do plant growth substances work? II. Plant Cell Environ. 14, 1-12.

62. van der Veen, R., Heimovaara-Dijkstra, S., Wang, M. (1992) Cytosolic alkalinization mediated by abscisic acid is necessary, but not sufficient, for abscisic acid-induced gene expression in barley aleurone protoplasts. Plant Physiol. 100, 699-705.

63. Walker-Simmons, M. (1988) ABA levels and sensitivity in developing wheat embryos of sprouting resistant and susceptible cultivars. Plant Physiol. 84, 61-66.

64. Walker-Simmons, M.K., Anderberg, R.J., Rose, P.A., Abrams, S.R. (1992) Optically pure abscisic acid analogs-- tools for relating germination inhibition and gene expression in wheat embryos. Plant Physiol. 99, 501-507.

65. Wilen, R.W., Mandel, R.M., Pharis, R.P., Holbrook, L.A., Moloney, M.M. (1990) Effects of abscisic acid and high osmoticum on storage protein gene expression in microspore embryos of *Brassica napus*. Plant Physiol. 87, 875-881.

66. Wilen, R.W., Hays, D.B., Mandel, R.M., Abrams, S.R., Moloney, M.M. (1993) Competitive inhibition of abscisic acid-regulated gene expression by stereoisomeric acetylenic analogs of abscisic acid. Plant Physiol. 101, 469-476.

67. Williamson, J. D., Quatrano, R.S. (1988) ABA-regulation of two classes of embryo-specific sequences in mature wheat embryos. Plant Physiol. 86, 208.

68. Williamson, J.D., Scandalios, J.G. (1992) Differential response of maize catalases to abscisic acid: Vp1 transcriptional activator is not required for abscisic acid-regulated Cat1 expression. Proc. Natl. Acad. Sci. U.S.A. 89, 8842-8846.

69. Zeevaart, J.A.D., Creelman, R.A. (1988) Metabolism and physiology of abscisic acid. Ann. Rev. Plant Physiol. Plant Mol. Biol. 39, 439-473.

G11. The Role of Hormones In Potato (*Solanum Tuberosum* L.) Tuberization

Elmer E. Ewing

Department of Fruit and Vegetable Science, Cornell University, Ithaca New York 14853 USA.

INTRODUCTION

Tuber initiation in the potato plant is accompanied by extensive morphological and biochemical changes above and below ground. Plants that are capable of tuber initiation are said to be "induced" to tuberize. It has long been postulated that the changes leading to induction are mediated hormonally. Before considering the evidence, it is helpful to summarize what is known about the tuberization process, its genetic and environmental control, and the techniques that have been employed to study it. For a more comprehensive treatment of these topics and for literature citations, see the recent review article by Ewing and Struik (8).

Description of Tuberization

A potato tuber is a modified stem, with nodes (the "eyes") and internodes. Its leaves are tiny and scale-like, often dehiscing during harvest and handling. The axis of the tuber is shortened and greatly thickened, and its tissues are packed with starch, though the anatomy of the tuber resembles that of a typical stem. Normally tubers form underground on rhizomes, more commonly called stolons. Tuber initiation begins in the subapical zone of the stolon, and swelling of the tuber occurs acropetally in internodes that were already present at the time of initiation. At about the same time that the swelling occurs, the meristematic activity in the stolon apex ceases. Starch deposition occurs very early in the ontogeny of the tuber.

Tubers form most readily underground. Why this is so is not entirely clear. The most important factor is probably darkness, but the physical resistance of the soil particles may also play a role. Under unusual circumstances tubers form above ground, usually at axillary buds. Such "aerial" tubers are much smaller and contain chlorophyll. Aerial tubers illustrate the point that although the tuber most often forms on a stolon, any bud or shoot apex of the potato is capable of tuberizing, given the proper conditions. For example, axillary buds cut from plants that are tuberizing will usually develop tubers if the buds are buried in the soil, particularly if there is a leaf attached to the bud cutting. This ability of leaf-bud cuttings to tuberize affords a useful tool for the study of tuberization (Fig. 1).

P. J. Davies (ed.), Plant Hormones, 698–724.

Fig. 1. Leaf-bud cuttings, illustrating the typical progression of responses to increasing levels of induction prior to cutting. The bud of each cutting was buried in potting mix. Cuttings were kept in a mist bench for 11 days under long photoperiods. No induction prior to cutting gives dormant buds (not shown) or more commonly growth as a shoot or a stolon (A). Development of a tuber at the tip of the stolon (B) is indicative of moderate induction to tuberize. Formation of a sessile tuber (C) at the buried bud is evidence that the cutting was taken from a plant that was strongly induced to tuberize. From Ewing, American Potato J. 55, 43-53 (1978).

Control of Tuber Initiation

Environment

The degree to which a particular potato plant is induced to tuberize is controlled by many factors. The environmental factor that has been most investigated is the effect of photoperiod. It is well known that long nights favor the induction of tuberization: with respect to tuber formation, potatoes are short day plants. (Although the controlling factor is the length of the dark period, the convention of referring to the daylength will be followed in the rest of this chapter.) Presumably the tuberization response is mediated by phytochrome, since five minutes of red light interrupting the daily dark period will reduce tuberization, and far-red light tends to reverse the effect of the red light.

Temperature also has a pronounced effect on the level of induction to tuberize. Cool temperatures (day temperatures below 30 C and night temperatures below 20 C) favor tuber induction. The effects of photoperiod and temperature on tuber induction depend on the irradiance under which the plant is growing. Effects of long days or high temperatures are exaggerated at low levels of irradiance. A fourth environmental factor affecting tuber induction is the amount of nitrogen available to the plant. Heavy applications of nitrogen fertilizer reduce the level of tuber induction. Under hydroponic conditions, withdrawal of nitrogen from the nutrient solution may produce immediate tuberization on stolons; addition of nitrogen again may cause tubers to revert to stolons.

The mother tuber

Although potato plants can be propagated from seeds, ordinarily they are grown from tubers. The condition of the planted tuber exercises considerable influence over the degree to which the potato plant will respond to the environment. Tubers planted soon after they have passed through their normal dormant period give rise to plants that are less readily induced to tuberize than are plants developing from old tubers. Tubers that are extremely old, whether because of prolonged storage at cold temperatures or at warmer temperatures with repeated sprout removal, will not develop normal sprouts. Instead, new tubers will form directly at the eyes and there will be no production of new stems and leaves--a disorder variously referred to as "sprout tubers" or "little potato."

Genetics

There are great genetic differences among potatoes in their responses to all of the above factors. The potato was brought under cultivation in the Andean highlands, where photoperiods are about 12 hours and temperatures are cool. Varieties adapted to these conditions are classified as *Solanum tuberosum* L. ssp. *andigena*. When grown under the higher temperatures of the lowland tropics or the long summer days of temperate zones, Andigena potatoes will tuberize weakly or not at all. The potatoes adapted to summers in the temperate zone--the potatoes of southern Chile, Europe, and North America--are classified as *S. tuberosum* ssp. *tuberosum*. Tuberosum potatoes give low yields in the highland tropics because tuberization is "turned on" so early and so intensely by the cool temperatures and short days that there is too little shoot growth to support good tuber yields. There are also substantial genetic differences in response to photoperiod within each group.

Most cultivated potatoes are tetraploid, which complicates interpretation of their genetics. When Andigena potatoes were crossed with Tuberosum, inheritance of the ability to tuberize under long days appeared to be quantitative (36). More definitive evidence was obtained by working at the diploid level and by using Restriction Fragment Length Polymorphisms (RFLP's) to map loci that control tuberization. Alleles at 11 distinct loci on eight chromosomes affected tuberization in populations obtained by backcrossing a hybrid of *S. tuberosum* x *S. berthaultii* to the diploid parent species (60). Earliness was favored by a *S. tuberosum* (*tbr*) allele in eight cases, and by an allele from *S. berthaultii* (*ber*) in three cases. Four of the eight *tbr* alleles and two of the three *ber* alleles favoring earliness were at least partly dominant.

Table 1. Relation between the number of loci carrying favorable alleles for tuberization and mean values for tuberization. Data were grouped according to the status of the plants at six loci that affected tuberization. Numbers in the left column indicate the number of loci that possessed alleles favorable for tuberization. (Note that if, for example, there were two loci with favorable alleles, this includes plants with favorable alleles at *any* two of the six loci; i.e., not all plants included in the average had favorable alleles at the *same* two loci.) Rating of cuttings was on a scale of 1 to 9, where 1 indicates no growth of the buried bud, and 9 indicates a sessile tuber at the buried bud. The percentage of tuberized plants was determined 6 to 8 weeks (depending upon replication) after *in vitro* plantlets were transplanted into pots.

No. of loci carrying alleles favorable to tuberization	Rating of cuttings for tuberization	Percentage of plants with tubers
0	3.0	13
1	3.6	43
2	4.3	69
3	4.5	78
4	5.9	96
5	6.4	100
6	7.0	98

The aggregate effects of the loci identified were substantial (60). Table 1 compares the mean responses of plants that contained from zero to six alleles favoring tuberization at six loci. If no favorable allele was present at any of the six loci, the mean rating for tuberization on cuttings was only 3.0 on a scale of 1 to 9, and only 13% of whole plants tuberized by about 6 weeks after planting (60). If any three of the six loci had alleles favoring tuberization, then the mean rating was 4.5, and the early tuberization on whole plants was 41%. The respective figures when all six loci had favorable alleles were 7.0 and 98% (60).

The effects of possessing multiple alleles were not all additive; epistasis was observed between some of the loci (60). The most common form of epistasis was that if an allele favoring tuberization was found at either of two interacting loci, the result was the same as when both loci had favorable alleles. Redundancy among some of the genes that control tuberization would be one possible explanation for this type of epistasis.

Several RFLP markers that were associated with earliness traits were also associated with other morphological characteristics. Three of five alleles for branching, and two of three for leaf size, corresponded to loci for earliness of tuberization. At these "pleiotropic" loci, alleles promoting earliness were correlated with reduced branching and increased leaf size. These relationships are what might be expected based upon plant response to photoperiod-- exposing a given genotype to short days promotes tuberization, increases leaf size, and decreases branching. There were similar pleiotropic effects for loci controlling tuberization and length of tuber dormancy. Tuber dormancy, like tuber initiation, is thought to be controlled by a balance between GA's and inhibitors. Do the various pleiotropic responses all reflect changes in levels of one or more hormones that are controlled at these loci?

Leaf area

A final controlling factor in tuberization is plant size. When genetic and environmental conditions are highly favorable for tuberization, tubers may form on very small plants. For example, Tuberosum plants grown from seeds sometimes tuberize when only one leaf above the cotyledons has developed, especially if photoperiods are short. At the other extreme, Andigena plants may grow for more than six months without tuberizing when exposed to long photoperiods. Thus there is no particular stage of development that necessarily coincides with tuber initiation. However, experiments with cuttings having different leaf areas demonstrate that when other conditions are equal, cuttings with a large leaf area are more likely to develop tubers than are those with a smaller leaf area. Whether a given set of environmental conditions is favorable for tuber induction may therefore depend upon the leaf area of the plant. A small plant may fail to tuberize, yet tubers may form as the plant attains a greater leaf area even though the environment does not change. It is as though every leaflet were contributing its part toward tuber initiation, with initiation occurring only when the sum of the contributions from all of the leaflets on the plant has attained the necessary level. The threshold will be reached at a small leaf area if conditions strongly favor induction, at a correspondingly greater leaf area if conditions are less favorable for induction, and not at all under very unfavorable conditions.

Partitioning of Assimilate and Tuber Induction

Plants grown under conditions that favor tuber induction have a very high percentage of their assimilate directed into tubers. The long term effects of strong induction are a reduction in root growth, shoot growth, flowering, and fruiting. The overall changes in plant morphology accompanying induction raise the question whether tuber induction should be considered as an indirect effect of the restricted plant growth. Before hormonal theories of tuberization became popular it was often assumed that under inducing conditions the amount of assimilate required by shoots and roots is so diminished that the surplus assimilate stimulates tuberization. Short term observations show that this is not the case. Moving plants from long to short photoperiods did not cause a decrease in rate of leaf expansion until well after a strong tuber sink had been formed (30). For the two weeks after daylength was shortened, both leaves and tubers were significant sinks, and total sink strength relative to photosynthetically active radiation was highest for the short day treatment (30).

Carbohydrate Relationships

There is, however, some reason to think that the daytime accumulation of assimilate in the leaf plays a role in induction. *In vitro* tuberization is highly responsive to sucrose concentration, and this is not merely an osmotic effect (13, and many subsequent studies). Neither glucose nor fructose is as effective as sucrose in meeting the sugar requirement, even though either is superior to sucrose in promoting shoot growth *in vitro* (6). Transfer to short photoperiods increases daytime accumulation of leaf starch in potato (as in many other species, e.g. ref. 4), and leaf starch accumulation continues even after a strong tuber sink has been established that makes heavy demands for assimilate (31). Transgenic plants that contained a yeast invertase gene expressed in the leaf apoplast showed only a small reduction in leaf starch during the night, presumably because very little sucrose--the only form in which carbohydrate is transported to the potato tuber (16, 44)--was available for export via the sieve tubes (16). All this presents the possibility that when normal plants are induced, higher levels of sucrose transported from the leaf during the night accumulate in the stolon tip and contribute to tuber initiation.

Although analyses of *S. demissum* stolon tips for sucrose and reducing sugars have not supported the hypothesis that sucrose accumulation in the stolon tip contributes to tuber initiation (64), there are problems in identifying on intact plants which stolon tips are just on the verge of tuberizing. Even under strong induction, not all tips are destined to become tubers, and it is impossible to predict which ones are likely to swell or when they will do so. Obviously, by the time swelling of a given tip is visible it is too late to analyze that tip for changes which occur in the earliest stages of tuberization. To overcome these problems Vreugdenhil and Helder (64) plotted sugar concentrations against starch content, taking the latter value as an indication of tuberization stage. The level of sucrose did not change as starch content of the stolon tip increased, and the levels of glucose and fructose decreased. There appeared to be a greater decrease in fructose than glucose at the same time that starch content of the stolon tip increased above 9% of the dry weight (64).

Genes for the synthesis of patatin and proteinase inhibitor II, two proteins that seem to be intimately associated with tuberization, are "turned on" by exposing potato leaf tissue to high sucrose concentrations (46, 53). However, the hypothesis that the expression of these two storage proteins is induced *in planta* by high concentrations of sucrose was not supported by evidence from a transgenic potato in which tuber starch synthesis was inhibited. This was accomplished by expressing a chimeric gene that encoded antisense mRNA for ADPG pyrophosphorylase (40). The starch content in tubers of these plants was lowered to 2 to 5% of wild-type levels, and sucrose accounted for up to 30% of the tuber dry weight. In spite of the high sucrose content, the synthesis of patatin and proteinase inhibitor II was reduced in the transgenic plants, suggesting that the blocking of starch synthesis directly affected the expression, at either the transcriptional level or the RNA stability level, of genes encoding these proteins.

Other effects brought about by blocking tuber starch synthesis included a substantial increase in tuber number and a decrease in tuber weight, interpreted as a lessening of sink strength (40). Thus although starch deposition is probably essential for development of a strong photosynthate sink, and starch deposition is one of the earliest anatomical changes detected at the site of tuber initiation, tuberization can be initiated in the absence of tuber starch biosynthesis.

In summary, evidence for the role of carbohydrate accumulation in tuber initiation is mixed. An increase in leaf starch accumulation during the day can be measured within 2 d after moving plants to inducing conditions. The starch is broken down at night and exported as sucrose, but analysis of whole plants has not demonstrated a sucrose accumulation in the stolon tip associated with tuber starch accumulation. However, the possibility remains that there is a transitory increase in sucrose concentration *preceding* (and helping to trigger) the first increases in tuber starch accumulation.

Grafting Experiments

Intra-specific grafts

Grafting experiments led Gregory (13) to hypothesize that tuber induction is triggered by a stimulus that is produced in the leaf. If a leaf of a potato plant that has been exposed to short photoperiods is grafted to a plant that has been exposed to long photoperiods, tuberization will occur. When the leaf of a special Andigena clone that tuberizes under all photoperiods was grafted to a stem of an ordinary Andigena plant in long days, tubers were produced at the buried bud (Fig. 2). In the reciprocal graft, no tubers formed. Thus it appears that in the leaf of the special clone a

Fig. 2. Grafted cuttings taken from plants exposed to 20-h photoperiods. The lower leaf on each stock was excised at time of cutting. The portion below the graft union (dotted line) was inserted in potting mix. Cuttings were maintained in a mist chamber under a 20-h photoperiod for 12 days after cutting and grafting. A) Scion taken from a selected clone of Andigena that is able to tuberize under continuous light; stock from ordinary Andigena, requiring short days to tuberize. Three of four stocks tuberized. B) Reciprocal grafts from those shown in A. There was no tuberization. From Ewing and Wareing, Plant Physiol. 61, 348-353 (1978).

stimulus is produced under long photoperiods that can be translocated through a graft union, causing tuberization on a clone which ordinarily would not tuberize under long photoperiods.

Inter-specific grafts

More intriguing results have been produced by grafting other species. The stimulus will pass through leafless eggplant or tomato stem segments inter-grafted between induced potato scions and noninduced potato stocks; but leaves of tomato or eggplant diminish tuberization when grafted to potato even if such leaves are exposed to short photoperiods. On the other hand, it was shown by Nitsch (41) that the Jerusalem artichoke (*Helianthus tuberosus* L.)--the tuberization of which is also favored by short photoperiods--would tuberize when grafted to sunflower leaves that had been exposed to short days, but not when grafted to sunflower leaves exposed to long days. This implied that leaves of the sunflower were able to receive and transmit to the Jerusalem artichoke a tuberization signal. These interesting results and their implications for the potato lay fallow in the literature for many years. Then, more or less simultaneously, workers in France and the USSR independently carried out almost identical experiments stimulated by Nitsch's reports. Inasmuch as flowering of the sunflower is controlled by photoperiod, these researchers decided to find a species the flowering of which would be sensitive to photoperiod and which could be grafted to potato. Both groups chose tobacco. Several species of tobacco were utilized, including plants that flower only under long days, plants that flower only under short days, and ones that flower under either long or short days. Results are summarized in Table 2. It will be seen that a tobacco requiring short days for flowering would induce tuberization on potato only if the tobacco leaves received short days; whereas a tobacco that requires long days for flowering caused tuberization on potato when the tobacco leaves were exposed to long days. The experiments seem to show that the stimulus for flowering of the tobacco is graft-transmissible to the potato, where it induces tuberization.

Site of Production and Movement of the Stimulus

Role of roots

Shoot cuttings of Andigena plants were grown hydroponically with the developing adventitious roots continually removed. Leaf-bud cuttings taken from the de-rooted plants still tuberized, provided the shoots had been given short photoperiods. Therefore roots are not essential for the production of the induced condition.

Leaf age

Other experiments with cuttings indicate that both young leaves and old leaves are effective in producing the induced condition, but younger leaves are more effective per unit of leaf area. Even the apical bud promotes

Table 2. Summary of grafting experiments in which potato or tobacco scions were grafted to potato stocks from which all leaves were excised. Plants were kept under long or short photoperiods. From Chailakhyan et al., Doklady Akademic Nauk S.S.S.R. 257, 1276-1280 (1981) and from Martin et al., C.R. Acad. Sc. Paris 295, 565-568 (1982).

| | Short Days | | Long Days | |
| | Flowers on tobacco scion[y] | Tubers on potato stock[y] | Flowers on tobacco scion | Tubers on potato stock |
Scion				
Mammoth tobacco	+	+	0	0 or +[z]
Xanthi tobacco	+	+	+	+
Trapezond tobacco	+	+	+	+
Sylvestris tobacco	0	0	+	+
Andigena type potato[x]	n.a.	+	n.a.	0
Tuberosum potato	n.a.	+	n.a.	+

[x] Andigena or a hybrid of Tuberosum X Demissum that required short days for tuberization.
[y] + = present, 0 = absent, n.a.= not applicable.
[z] Results were inconsistent between experiments.

tuberization on induced plants, although there is evidence that under noninductive conditions, tuberization is favored by bud excision.

Girdling

When one stem of a two-branched plant was exposed to short days and the other to long days, girdling of either stem blocked the effect of that stem on tuber induction. That is, tuberization was improved if the stem receiving long days was girdled, and tuberization was inhibited if the stem receiving short days was girdled.

Response target

The lowest of several buried buds on a cutting is most likely to tuberize, but this pattern is an effect of neither gravity nor node age. There is some indication that tuberization is expressed most strongly at the bud which is most distant from an illuminated leaf or stem.

THE HORMONAL CONTROL OF TUBERIZATION

Although other explanations cannot be excluded, much of the information already presented in this chapter suggests that tuberization is under hormonal control: the alteration of the total morphology of the plant during the induced state, even if tuberization is surgically prevented; transmissibility of the stimulus across grafts; the effects of girdling; and the contribution of the mother tuber to induction would all seem consistent with a hormonal role. Is there a hormone that is produced in the leaves in response to long nights, that is transmissible across grafts, and that induces changes in the morphology and physiology of the plant including tuberization? If we accept the evidence from tobacco/potato grafts, then it does not seem that the stimulus for tuberization is unique to tuberizing species of *Solanum*. Moreover, the stimulus could be a single compound or a balance in concentrations of a group of compounds--some, but not necessarily all, of them hormonal.

As the number of short days to which a plant is exposed increases, the response at buried buds of cuttings taken from the plant changes from no growth, to a shoot, to a stolon, to a tuber. Is the stimulus for shoot or stolon formation at the underground bud the same stimulus that when present at a higher concentration induces tuberization, or are there two separate stimuli? Tuber induction is also associated with changes in leaf morphology (8). Leaves are larger, thinner, and have a flatter angle to the stem; axillary branching is suppressed; flower buds abort more frequently; and senescence is hastened when tubers are more strongly induced. Any theory that would explain the hormonal control of tuberization should take into account the manifold effects of tuber induction. Let us now consider the evidence for the involvement of particular hormones.

Endogenous Hormones

Gibberellins

There have been many attempts to measure changes in endogenous hormones as correlated with changes in the degree of induction of the potato plant. One of the first changes that was noted was a decrease in GA-like activity (29, 42, 48, 51). As few as two short days produced a decline in the GA-like activity extracted from Andigena leaves that were previously grown under long photoperiods, and even the chloroplast fraction of Andigena leaves showed reductions in GA-like substances as photoperiods shortened (52). Low irradiance, which tends to inhibit tuberization, greatly increased the GA-like substances in leaves exposed to short days, while long-day leaves were high in such substances regardless of irradiance levels (67). Similarly, high temperatures increased GA-like activity of shoots, though in this respect buds appeared to be much more affected by high temperatures than were leaves (Table 3). If potato plants were grown at high temperatures and buds were excised or chemically inhibited, then the deleterious effect of high temperature on tuberization was largely ameliorated (38). Not only do long photoperiods, high temperatures, and low irradiance have similar effects on tuberization, they all produce effects on shoot morphology that are consistent with known effects of GA's (8). Other evidence comes from growing plants in nutrient solution: a continuous supply of N, which inhibits tuberization, causes higher GA activity in shoots than does discontinuing the N supply, which favors tuberization (26). Assays of GA-like activity in stolons and newly initiated tubers (24, 55) indicate that there is a substantially lower activity associated with the conversion from stolons to tubers (Fig. 3).

Table 3. Effect of temperature on GA-like activity in apical buds of potato plants. From Menzel, Ann. of Bot. 52, 5-69 (1983).

Day/Night Temperatures °C	GA activity[x] (ug GA$_3$ equivalent/kg f.wt.)	
	Buds	Leaves
20/15	4.0	1.1
35/30	71.8	4.3
	LSD (P=0.05) 15.17 (log transformation)	

[x] Lettuce hypocotyl bioassay (means of three plants)

The metabolic pathways for GA's are the same in potato as in most other species that have been investigated. By following the metabolic products from applied [^{14}C]GA$_{12}$ and [^{14}C]GA$_{12}$-aldehyde, eight GA's from the early 13-hydroxylation-pathway (GA's 53-aldehyde, 53, 44, 19, 20, 29, 1, and 8) were detected in Andigena potato shoot apices, indicating that this pathway operates in potato shoots and is probably the predominant pathway (61). This was confirmed by time-course studies and by re-feeding with early

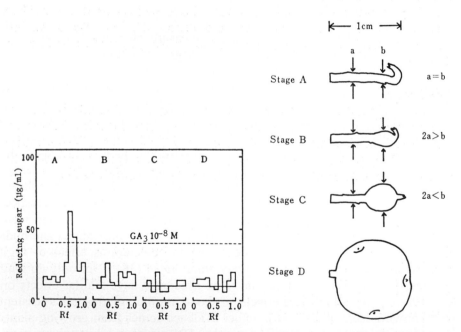

Fig. 3. Changes in the levels of GA-like substances at different stages of tuber development. Stolons at stage A showed no indication of tuberization. At stage B, tuber initiation had started, but the diameter of the swelling was less than twice the diameter of the stolon immediately behind it. Stage C tubers were more than twice the diameter of the stolon. Tubers at stage D were 1 to 2 cm in diameter. Each extract, equivalent to 10 g fresh weight, was chromatographed on paper and assayed by the Avena endosperm test. Broken line represents the amount of reducing sugar liberated from the endosperms by 10^{-8} M GA$_3$. From (24).

metabolites from the pathway (61). Kinetic evidence favored GA_{12} as the main intermediate between GA_{12} and GA_{53} (61). Although the bulk of the radioactivity was recovered from members of the 13-OH-pathway, there was evidence that the non-hydroxylation pathway was also present. GA_{51} was detected (61), and GA_{15} has been reported in potato berries (21).

No differences were found in $[^{14}C]GA_{12}$-aldehyde or $[^{14}C]GA_{12}$ metabolism between plants grown under 16-h or 10-h photoperiods. Photoperiod would thus seem not to act via GA metabolism. It is possible that a photoperiodic control exists prior to GA_{12}-aldehyde (70).

Taken together, there is convincing evidence of a negative correlation between the degree of induction to tuberize and the GA-like activity in shoots and stolons.

Inhibitors

What causes the GA activity to decrease during induction? Is there simply a decline in its net production, or is a GA inhibitor produced during tuber induction? An unidentified inhibitor has been reported (27, 48, 55), but more attention has been paid to the possibility that ABA performs the role of GA inhibition. There is some indication that ABA activity increases under inducing conditions (26), but in most cases differences reported have not been very great or have been in the opposite direction (27, 66). There was little increase in ABA activity of stolons at the earliest stage of tuber initiation (Fig. 4).

Jasmonates

A compound related to jasmonic acid (JA; Fig. 5A; 63) is the most recent prospect for a natural inhibitor that counteracts the effects of GA and induces tuberization (69). The active compound (Fig. 5C) has been identified as 3-oxo-2-(5'-ß-D-glucopyranosyloxy-2'-z-pentenyl)-cyclopentane-1-acetic acid, the glucoside of 12-OH-jasmonic acid (69). The aglycone (Fig. 5B) was

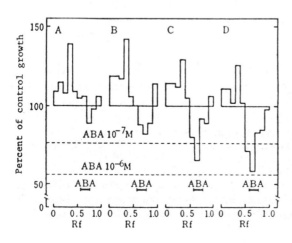

Fig. 4. Changes in the levels of ABA-like substances at different stages of tuber development. Stages and sampling were as explained in Figure 3. Each extract was chromatographed and assayed by inhibition of *Avena* coleoptile straight growth induced by 10^{-7} M IAA. Broken lines represent the percent of control growth obtained by 10^{-7} and 10^{-6} M ABA. Markers in the lower part of figure indicate the position of ABA. From (24).

named "tuberonic acid" (TA). The emphasis was first on the glucoside as the tuber inducing substance; concentrations as low as 3×10^{-8} M added to the agar medium stimulated tuberization *in vitro* (25). The same laboratory reported that JA, its methyl ester, TA, and the glucoside of tuberonic acid (TAG) all showed similar promotion of tuberization in the bioassays utilized (22). Cucurbic acid, which differs from JA by possessing a hydroxy group instead of oxygen at C-3, and the methyl ester of cucurbic acid were active, but required higher concentrations to obtain the same rate of tuberization in the bioassay (22).

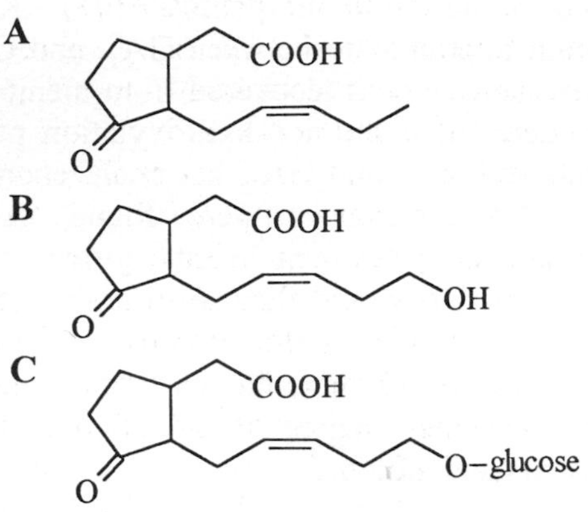

Fig. 5. Chemical structures of A) jasmonic acid, B) tuberonic acid, and C) the glucoside of tuberonic acid. Tuberonic acid is 12-OH jasmonic acid.

Somewhat different conclusions derive from recent studies with *S. demissum* Lindl., which has an absolute requirement of short days for tuberization. Leaves from short-day plants had more 11-OH-jasmonic acid than 12-OH-jasmonic acid (17). Neither compound could be detected in leaves from long-day plants in the same experiments. There may be technical problems with the stability of the hydroxylated compounds and their glycosides during isolation and identification. For this reason it is not clear whether the 11-OH compound was more prevalent than the 12-OH compound because of the species difference or because of variations in analytical methods employed.

Another dissimilarity in the *S. demissum* studies was that no glycosides of 11-OH-jasmonic acid or 12-OH-jasmonic acid were found in either short-day or long-day leaves (17). Again it is difficult to say whether this should be attributed to the difference in species or to the difference in analytical methods. The conclusion that the glycosides were absent from the *S. demissum* leaves was based upon failure to find the aglycones following cellulase digestion of the ethyl acetate fraction derived from partitioning between ethyl acetate and water at pH 3. The aqueous fraction from this partitioning was discarded (17). It is our experience that TAG moves to the aqueous phase rather than to ethyl acetate at this pH, so it may be that no TAG was found simply because the wrong fraction was examined.

Even though Helder et al. (1993) did not find TAG, they pointed out that the glycosylated form would probably be more easily transported out of the leaf and is hence a better candidate to play a role in tuber initiation. Because photoperiod did not affect JA levels in leaves of *S. demissum*, they also suggested that photoperiod affects the activity of one or more enzymes controlling the hydroxylation of JA (17). Combining the two ideas leads to the interesting hypothesis that photoperiod controls hydroxylating enzymes, which convert the jasmonates to compounds that can be glycosylated, thus making them transportable from the leaf to the stolon so that they can initiate tuberization. This does not explain why JA promotes tuberization *in vitro*, but it may be that transportability in the phloem is less of an issue when the JA is bathing the target tissue.

The isolation of TAG derived from findings that crude extracts from potato leaves promoted *in vitro* tuberization better when the extracts were from plants grown under short rather than long days. TAG was isolated from the short day leaf extracts. So far, however, the effects of daylength on levels of TAG in *S. tuberosum* leaves have not been determined. Another unresolved question is the relative importance of TA, TAG and their methyl esters (TAme, TAGme) in tuberization. TAGme is active in the potato bioassay (68) and has been found in potato leaves (Omer, Davies, Koch, Ewing, and van den Berg, unpublished) as well as in leaves of Jerusalem artichoke (68).

Cytokinins

Another group of hormones that reportedly changes under inducing conditions is cytokinins. The major cytokinin in potato leaves has been identified as *cis-zeatin* riboside (33), which seems somewhat unlikely--the *trans-* rather than the *cis-* form of cytokinins is considered to predominate in higher plants (Chapter B3). In any case, the zeatin riboside was found to be 29% higher in extracts from induced tissues (33). Maximal levels of cytokinins in shoots and underground tissues occurred four and six days, respectively, after exposure to inducing conditions (10, 28). The decline in cytokinin activity of shoots after four days of inducing conditions was attributed to transport to the stolon tips, where metabolic sinks were created that led to tuberization (10, 28). If cytokinin is responsible for induction, and if the levels in the shoots decrease after four days of inducing conditions, then one would expect that leaf-bud cuttings taken after inductive periods progressively longer than four days would tuberize less and less. In our experience the contrary pattern applies; the more days of inducing conditions, the stronger the tuberization on cuttings, until all form sessile tubers. We find no evidence that tuberization depletes the supply of tuberization stimulus in the leaf (6).

The association between tuberization and increased cytokinin activity in shoots has also been noted when the nitrogen supply was removed from hydroponically grown potatoes, though a different interpretation was given (54). In this case it was speculated that some factor other than the cytokinins caused tuber initiation and that once tubers were present, the sink effect increased the photosynthetic activity of the leaf, which in turn produced increased cytokinin activity of the shoot. We have, then, two opposing explanations for the association between the increased cytokinin activity of leaves and tuber initiation: 1) the cytokinins in the leaves are translocated to the stolon where they induce tuberization (10, 28); or 2) the initiation of tubers stimulates cytokinin production in the leaves (54). If the first hypothesis is correct, then cytokinins should increase in stolon tips before tubers are initiated. However, cytokinin assays of stolons and small tubers (20, 24) indicated no substantial increase in cytokinin-like activity until tubers were well into their enlargement phase (Fig. 6).

Other hormones

Tuberization in dahlia is associated with changes in evolution of ethylene (2), but similar findings from potato have not been reported. One role hypothesized for ethylene is that it is produced when stolon tips push against soil particles, thereby restricting their extension growth and favoring tuberization (65). Auxin, the other major category of plant hormone, has received little attention with respect to changes in endogenous content during tuber induction, and one review of the topic has suggested that this may be a serious omission (35). An association between tuber growth rate and auxin content has been noted (32), but, as shown in Figure 7, only a relatively small increase in auxin activity was found in the first stage of tuberization (24).

Hormone Applications to Whole Plants

Assays for activities of endogenous hormones have revealed no unequivocal answer as to the role of hormones in the control of tuberization, with the possible exception that tuber initiation is associated with a decline in GA activity. A second approach to the question is through application of exogenous hormones.

Gibberellins

It is abundantly clear from a great number of such experiments that GA applications have a strong inhibitory effect on tuberization of whole plants

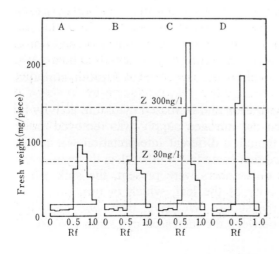

Fig. 6. Changes in the levels of butanol-soluble cytokinins at different stages of tuber development. Stages and sampling were as explained in Figure 3. Each extract was chromatographed and assayed using soybean callus. Broken lines represent the callus yields with 30 and 300 mg/1 zeatin. From (24).

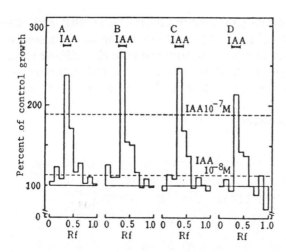

Fig. 7. Changes in the levels of auxin during the course of potato tuberization. Stages and sampling were as explained in Figure 3. Each extract was chromatographed and assayed by Avena coleoptile straight growth test. Broken lines represent the percent of control growth by 10^{-8} and 10^{-7} M IAA. Markers in the upper part of figure indicate the position of authentic IAA. From (24).

(42). GA's 1, 3, 4, 5, 7, and 9 all inhibited tuberization of leafed stem cuttings (59). GA applications also promote shoot growth on whole plants or leaf-bud cuttings (37). In both respects exogenous GA's appear to mimic the effects of noninductive conditions. There is also a biochemical similarity between plants grown under noninductive conditions and plants grown under inductive conditions but treated with GA's. Buried petioles of leaf cuttings taken from either kind of plant had lower levels of the glycoprotein patatin than petioles of cuttings from induced plants (14). Treatment with chlormequat chloride (CCC), which blocks GA synthesis, affords another way of examining the effects of these compounds. The general effect of CCC applications has been an improvement in tuberization, especially where noninductive conditions prevailed (37).

Other hormones

Promotion of tuber initiation on whole plants through the application of other classes of growth substances--including auxin and related compounds, ABA, cytokinins, 2-chloroethylphosphonic acid (ethephon), and a large group of miscellaneous compounds--has been relatively ineffective. In evaluating such experiments it should be noted that almost any chemical applied in dosages that are slightly phytotoxic to the shoot may increase the number of tubers initiated under inductive conditions. The result is typically a larger number of tubers, but smaller average tuber size, producing no increase in total tuber yield. Of much greater interest would be a chemical that could lead to tuber initiation under environmental conditions that would ordinarily prevent all tuberization.

Hormone Applications to Sprout Tubers

The application of the ethylene producing compound, ethephon, to potato tubers that are so old physiologically as to be producing small tubers rather than normal sprouts, produced sprouts that elongated normally and that contained somewhat higher levels of endogenous GA's than were present in the sprout tubers (5). Cytokinin activity was similar in the sprouts and small tubers. It is not clear whether ethylene plays a role in normal sprout growth to prevent tuberization, or whether this is a special effect in very old tubers.

Hormone Applications to Cuttings

Abscisic acid

When single node cuttings were taken from Andigena plants that had received 20 short days, tuberization resulted (66). Removal of the leaf from a cutting at the time it was taken led to an orthotropic, elongated shoot rather than a tuber; but application of ABA or grafting of a leaf back to the cutting gave a tuber. The leaf grafted to the cutting did not need to be from a plant that had been exposed to short days, as long as the bud to which it was grafted came from a plant that had received the short days. In fact, even a tomato leaf produced a tuber when grafted to such a stock. A key point is that the ABA applied to the bud, or the leaf grafted to the cutting, produced a tuber only if the bud had been taken from a plant that had received the short days. Neither the ABA nor the noninduced leaf resulted in tubers when the original cutting came from a plant exposed only to long days (66).

In interpreting the above results, it is important to note that if leafless cuttings of the type mentioned had come from plants that had been exposed to several weeks of strongly inductive photoperiods, then tubers would have formed even without ABA or a grafted leaf. Thus the buds on the cuttings used in the experiments were somewhere near the verge of being able to tuberize alone. The ABA, or whatever substance(s) were provided by the grafted leaves, "tipped" them toward tuberization. The ABA would not have caused buds on leafless cuttings from long day plants to tuberize; thus by itself it cannot be considered to be the tuberization stimulus, although it could be one component of the stimulus.

Cytokinin

A second set of experiments with cuttings deals with dipping strongly induced leaf-bud cuttings in cytokinin (34). Repeated treatment with 6-benzyladenine interfered with normal development of sessile tubers. Tuberization was delayed; and instead of the formation of sessile tubers, thickened stolons preceded tuberization. The reason for the interference is not known; perhaps the cytokinins created a sink in the detached leaves that hindered movement of metabolites to the developing tubers.

Hormone Applications to Tissue *In Vitro*

Procedures

Two papers, which in other respects have had an enormous impact on tuberization research, reported a technique that has been largely ignored.

Gregory (13) and Chapman (3) cut small pieces of stem containing axillary buds from plants that had been strongly induced to tuberize, disinfected them, and placed them on an agar medium in the dark. Within about four days after cutting, tubers formed at the buds. Nodes similarly treated from noninduced shoots produced only leafy shoots until cultures were more than three weeks old. To obtain good tuberization it was necessary to add sucrose to the agar medium, and increasing the sucrose concentration up to 10% gave larger tubers (13). Gregory pointed out that the technique might form the basis of a kind of bioassay for the hypothetical tuberization stimulus. If *in vitro* cuttings from, for example, Andigena plants exposed to long photoperiods tuberized only after addition to the agar medium of a substance isolated from induced plants, that would constitute evidence for the role of the isolated compound in controlling tuberization. Unfortunately, since the original papers there has been relatively little published work utilizing the technique, perhaps because--as Gregory himself mentioned (13)--microbial contamination is often difficult to overcome (6).

Much subsequent work on tuberization (12, 45, 56, 57) has been done utilizing a very different *in vitro* system. This latter technique, instead of taking nodes directly from shoots exposed to varying photoperiods, calls for subculturing of sprouts or stolons in the dark prior to the *in vitro* test for tuberization. In this case there is no obvious way to study the effects of photoperiod--the plant material being investigated is cultured under continuous darkness. Another question that may be raised about the technique is that subculturing might deplete the supply of a particular nutrient normally present in the tissue to the point where it would be difficult to distinguish between substances merely required for normal growth and those that are uniquely associated with tuberization. Still another objection is that very long periods are often required before tuberization takes place, and eventually the control treatments may also develop tubers (23).

There is yet a third form of *in vitro* culture. During the last 15 years seed certification agencies have become very interested in growing potato plantlets *in vitro* as a method of preserving stock cultures free from diseases. Plantlets are subcultured from nodal cuttings under aseptic conditions in agar media containing sugar and a variety of other organic and inorganic nutrients. Various photoperiods are employed, ranging from 12 hours to continuous light. The usual method employed to propagate such cultures is to dissect the plantlets into nodal cuttings every few weeks; but if plantlets are left for longer periods, two months or more, it is common to find tubers under the agar or at the aerial nodes. The small tubers may be convenient for long term storage or as a means of mailing germ plasm to other locations, so attention has been given to factors controlling the tuberization (18, 19). Again, we must be cautious in extrapolating the results of such research to whole plants. The *in vitro* plantlets are relatively insensitive to photoperiod, even when Andigena types are grown. The tubers are forming in the light on plants growing in a highly artificial medium. It sometimes appears that the tuberization is what happens when the growth of the *in vitro* plantlets has been slowed by a limitation of nutrients or by the development of toxic factors. Tuberization as a response to exhaustion may provide a poor model of what happens in the whole plant. With these caveats in mind, let us review some of the results from *in vitro* studies.

Results with subcultured stolons

Subcultured stolons tuberized only if they were supplied cytokinin and sucrose. Tuberization was not complete until about 25 days, considerably later than in the Gregory system (13). Inhibitors of nucleic acid synthesis and protein synthesis did not block the tuberization. GA_3 strongly depressed tuberization (23), and IAA and NAA promoted it weakly (23, 56). There is some disagreement on the effects of ethylene (12), but most studies have shown that it decreases or eliminates the tuberization of subcultured stolons (39). Additions of ABA to agar media containing 2% sucrose caused a slight swelling of the subapical region of the stolon, but no further development to tubers (23). On 8% sucrose media, ABA alone did not increase tuberization, although it partially overcame the deleterious effects of GA_3 (23). Phenolic acids (47) and coumarin also (56) favored tuberization. The enhancement of tuberization by coumarin was blocked by actinomycin D and chloramphenicol (57), a fact which is somewhat surprising considering that these inhibitors of nucleic acid and protein synthesis had little effect on tuberization promoted by cytokinin (45). (The promotive effects of JA and related compounds on *in vitro* tuberization have already been mentioned.)

Polyamines, synthesized by the ornithine decarboxylase (ODC) pathway, appear to be needed for tuberization and for shoot formation of node explants grown *in vitro*; treatment with difluoromethylornithine (DFMO), an inhibitor of ODC, produced a reduction in both responses (49). Polyamines probably are required during cell division, an early event in tuber formation. A gene turned on soon after induction to tuberize is associated with polyamines (58). Spermidine is the principal polyamine in potato tubers, and spermidine applications to potato cuttings were found to promote the transport of K^+($^{86}Rb^+$) to the tuber (9). However, there is currently no evidence that polyamines represent the tuberization stimulus *per se*.

Results with light-grown plantlets

The temperature and photoperiod under which *in vitro* plantlets are grown has been shown to affect their size and morphology (18), but it is unclear to what extent this is an indirect photosynthetic response. Shortening the photoperiod of plantlets grown under long days did not favor tuberization (19). Tuberization tended to be promoted by 6-benzylaminopurine, high sucrose, and CCC. It was inhibited by GA_3 and ethylene. Additions of ABA had little consistent effect (19).

Results with stem pieces

The few experiments employing the *in vitro* system proposed by Gregory (13) have produced little indication that addition of plant hormones to the agar medium can substitute for exposing plants to short photoperiod before bud excision. The best evidence for such a substitution is with kinetin (11), but even in this case results were not entirely clear-cut; the long photoperiod treatment also gave partial tuberization. Other experiments have shown no benefit from kinetin (7). As expected, GA_3 decreases tuberization (15). ABA has no effect (6), in contrast to its effect in promoting tuberization of leafless Andigena cuttings from moderately induced plants (66). Perhaps the difference in results is to be explained by the difference in cultivars used or by the presence of sucrose and ammonium nitrate in one case (6) and not in the other (66). There are no reports that JA or its relatives have been tested in the system proposed by Gregory (13).

Genetic Studies

Many genetic variants of the potato may be related to hormonal differences. For example, the "droopy potato" mutant lacks ABA (50). It nevertheless tuberizes, which further weakens the case for ABA being the tuberization stimulus. Mutants called "giant hills" or "bolters," which tuberize late and have the appearance of high GA content, are fairly common. At the other extreme, dwarf plants (Fig. 8) of Andigena tuberize under much longer photoperiods than do their wild type siblings (1). Metabolic patterns indicated a partial metabolic block between GA_{12} and GA_{53} in the dwarf plants (62). As a result less $[^{14}C]GA_1$ was formed in the dwarfs than in wild types when $[^{14}C]GA_{12}$-aldehyde or $[^{14}C]GA_{12}$ were fed (62).

The "topiary" gene gives a rosette plant totally lacking in apical dominance, with short stolons and strong tuberization (29). A similar morphological response has been achieved by the incorporation into a Tuberosum clone of T-DNA genes from *Agrobacterium tumefaciens* (43). Shoots, which contained a 10-fold increase in cytokinins, were characterized by small leaves, reduced apical dominance, and aerial tubers. Subterranean tubers were smaller and more numerous than on the original clone. The larger number of small tubers and the smaller leaf size are reminiscent of symptoms that accompany foliar applications of cytokinins (or various other growth regulators) to potato plants. In considering whether this should be taken as evidence for the role of cytokinins as the tuberization stimulus, it is worth remembering that photoperiodic induction to tuberize is associated with larger rather than smaller leaf size. Cytokinins may be present at phytotoxic levels whether produced internally or applied externally; in either case, the phytotoxicity may increase the number of tubers initiated.

The difference between the photoperiodic reactions of early and late cultivars of Tuberosum is substantial, and within Andigena populations developed by several plant breeders are clones that cover a broad range of

Fig. 8. Wild type (left) and Dwarf (right) siblings of Andigena grown under long photoperiods. A) Above ground growth. B) Underground growth, with a well developed tuber only on the dwarf plant.

critical photoperiods. Some have an absolute requirement for photoperiods less than 12 hours, even at the most favorable temperatures. Others are able to tuberize under high temperatures and continuous light. Physiologists would do well to consider this wealth of genetic diversity when investigating the role of hormones in tuberization.

SUMMARY AND CONCLUSIONS

Gibberellins

Among the known hormones, the most convincing case for a critical role in control of tuberization is for GA's. 1) The environmental changes that increase induction to tuberize cause decreased GA activity of shoots. 2) GA activity of stolon tips declines at the earliest stage of tuber initiation. 3) High temperatures cause GA content to increase far more in buds than in leaves, and excision of the buds alleviates much of the deleterious effect of high temperature on tuberization. 4) Exogenous GA's are highly effective in reducing or eliminating tuberization, and also mimic other effects of the noninductive environment. 5) The effects of exogenous GA application are ameliorated by the GA inhibitor, CCC, which also shows some ability to promote tuberization in noninductive environmental conditions. 6) A dwarf Andigena that has a block in GA metabolism tuberizes under longer photoperiods than do its wild type siblings.

The problem, of course, is that the effects of GA are all negative in terms of looking for a tuberization stimulus. To argue that tuberization is under the control of GA, we must turn the concept of a tuberization stimulus on its head and consider that potato buds and stolon tips are programmed to tuberize unless their apical meristems are stimulated to develop into shoots or stolons. Short photoperiods would then promote tuberization by decreasing the supply of GA's, permitting the tubers to form. Sprout tubers would form on old tubers after the supply of GA's in the mother tuber was depleted.

Inhibitors

An obvious alternative is to invoke a GA inhibitor as the tuberization stimulus. This would be consistent with the observed changes in endogenous GA's and the effects of exogenous GA's. There have been attempts to isolate such an inhibitor, but to date they have not had much success. Or instead of a specific inhibitor of GA's, a more generalized growth inhibitor has been sought. ABA has attracted considerable attention as a logical candidate in this category, but the balance of the evidence is against it. Various phenolic compounds have been suggested, but no data have been published to show that any particular compounds are actually involved. The case for TAG and other relatives of JA seems promising, but more evidence is needed as to whether concentrations of these compounds actually increase when potato leaves are exposed to short photoperiods and which forms are active.

Cytokinins

There is no doubt that cytokinins must be present in order for tuberization to take place, inasmuch as cell division is one of the early events following tuber initiation. This is not to say, however, that a change in the level of cytokinins in whole plants is the event that triggers tuberization--the switch that shuts off cell division at the apical meristem, and turns on cell enlargement, cell division, and starch deposition in the subapical meristem. Cytokinins increase in shoots as plants are induced to tuberize, but the increases are smaller than might be expected for hormonally controlled processes. The reported decline of cytokinin concentrations in shoots after

719

four days of inductive conditions seems inconsistent with the pattern of tuberization on cuttings taken at various periods after induction begins. Cytokinins increase in stolon tips during tuberization, but the increase is relatively small until well after tubers have been initiated (Fig. 6). There have been numerous attempts to improve tuberization of whole plants by foliar applications of cytokinins. Yield differences have generally been marginal, and not strikingly different from results obtained with a great variety of growth substances similarly applied. There is no report that cytokinin application to a noninduced Andigena plant caused it to tuberize.

The other evidence for cytokinins as the tuberization stimulus comes from *in vitro* experiments. Where these have been performed on subcultured stolons, it is difficult to know whether the cytokinin is the unique stimulus for tuberization, or whether it, perhaps like sugar, is a necessary ingredient for tuber formation once tuber induction has occurred. There have been only a few tests of the ability of cytokinin added *in vitro* to substitute for exposure of plants to short days when buds are excised from stems of plants grown under noninducing conditions. Results were not conclusive.

Considering all the data, it seems premature to identify cytokinin as the sole component of the tuberization stimulus.

Balance of Factors

There is no reason why we must seek a single compound to be the tuberization stimulus, especially when one considers the large number of genes that, according to RFLP data, control tuberization. Inducing conditions might lead to simultaneous changes in the concentrations of a number of compounds, the balance of which may control tuberization and the many associated morphological changes. The preponderance of thinking at present seems to favor the ratio of GA's to cytokinin, or to ABA, or to TAG, as the stimulus. However, there is no justification for ruling out other compounds. Auxin could be a component, especially in controlling the degree of plagiotropism of stolons or in other morphological responses associated with intermediate levels of induction.

Nor should we limit our thinking to hormones. Earlier investigators attributed control of tuberization to carbohydrate levels or to the carbon/nitrogen ratio. Hormonal theories by and large have supplanted such explanations; but it may be that sugar concentrations play a role along with hormone levels.

The hypothetical tuberization stimulus has proved to be as elusive as the analogous flowering stimulus; yet there is good evidence that induction to tuberize produces complex hormonal changes in the potato plant. As more sophisticated and sensitive analytical methods for hormones become available, as gene probes for proteins associated with the first stages of tuber induction become more fully available, and as researchers begin to take better advantage of the vast genetic variation available for physiological studies, we can hope to learn whether there is a unique compound controlling tuberization and all its attendant changes, or whether many compounds operate in concert.

References

1. Bamberg, J.B., Hanneman, R.E., Jr. (1991) Characterization of a new gibberellin related dwarfing locus in potato (*Solanum tuberosum* L.). Am. Potato J. 68, 45-52.
2. Biran, I., Gur, I., Halevy, A.H. (1972) The relationship between exogenous growth inhibitors and endogenous levels of ethylene and tuberization of Dahlias. Physiol. Plant. 27, 226-230.
3. Chapman, H.W. (1958) Tuberization in the potato plant. Physiol. Plant. 11, 215-224.
4. Chatterton, N.J., Silvius, J.E. (1980) Photosynthate partitioning into leaf starch as affected by daily photosynthetic period duration in six species. Physiol. Plant. 49, 141-144.
5. Dimalla, G.G., van Staden, J. (1977) Effect of ethylene on the endogenous cytokinin and gibberellin levels in tuberizing potatoes. Plant. Physiol. 60, 218-221.
6. Ewing, E.E. (1985) Cuttings as simplified models of the potato plant. *In*: Potato Physiology, pp.153-207, Li, P.H., ed. Academic Press, Orlando.
7. Ewing, E.E., Senesac, A.H. (1981) *In vitro* tuberization on leafless stem cuttings. *In*: 8th Triennial Conference of the European Association for Potato Research (Abs.) pp.7-8.
8. Ewing, E.E., Struik, P.C. (1992) Tuber formation in potato: Induction, initiation, and growth. Hortic. Rev. 14, 89-198.
9. Feray, A., Hourmant, A., Beraud, J., Brun, A., Cann-Moisan, C., Caroff, J., Penot, M. (1992) Influence of polyamines on the long distance transport of K (^{86}Rb) in potato cuttings (*Solanum tuberosum* cv. Sirtema)--comparative study with some phytohormones. Jour. Exp. Bot. 43, 403-408.
10. Forsline, P.L., Langille, A.R.(1975) Endogenous cytokinins in *Solanum tuberosum* as influenced by photoperiod and temperature. Physiol. Plant. 34, 75-77.
11. Forsline, P.L., Langille, A.R. (1976) An assessment of the modifying effect of kinetin on *in vitro* tuberization of induced and non-induced tissues of *Solanum tuberosum*. Can. J. Bot. 54, 2513-2516.
12. Garcia-Torres, L., Gomez-Campo, C. (1973) *In vitro* tuberization of potato sprouts as affected by ethrel and gibberellic acid. Potato Res. 16, 73-79.
13. Gregory, L.E. (1956) Some factors for tuberization in the potato. Ann. Bot. 41, 281-288.
14. Hannapel, D.J., Miller, J.C., Jr., Park, W.D. (1985) Regulation of potato tuber protein accumulation by gibberellic acid. Plant Physiol. 78, 700-703.
15. Harmey, M.A., Crowley, M.P., Clinch, P.E.M. (1966) The effect of growth regulators on tuberization of cultured stem pieces of *Solanum tuberosum*. Eur. Potato J. 9, 146-151.
16. Heinecke, D., Sonnewald, U., Büssis, D., Günter, G., Leidreiter, K., Wilke, I., Raschke, K., Willmitzer, L., Heldt, H.W. (1992) Apoplastic expression of yeast-derived invertase in potato. Effects on photosynthesis, leaf solute composition, water relations, and tuber composition. Plant Physiol. 100, 301-308.
17. Helder, H., Miersch, O., Vreugdenhil, D., Sembdner, G. (1993) Occurrence of hydroxylated jasmonic acids in leaflets of *Solanum demissum* plants grown under long- and short-day conditions. Physiol. Plant. 88, 647-653.
18. Hussey, G., Stacey, N.J. (1981) *In vitro* propagation of potato (*Solanum tuberosum* L.). Ann. Bot. 48, 787-796.
19. Hussey, G., Stacey, N.J. (1984) Factors affecting the formation of *in vitro* tubers of potato (*Solanum tuberosum* L.). Ann. Bot. 53, 565-578.
20. Jameson, P.E., McWha, J.A., Haslemore, R.M. (1985) Changes in cytokinins during initiation and development of potato tubers. Physiol. Plant . 63, 53-57.
21. Jones, M.G., Horgan, R., Hall, M.A. (1988) Endogenous gibberellins in the potato (*Solanum tuberosum*). Phytochemistry 27, 7-10.

22. Koda, Y., Kikuta, Y., Tazaki, H., Tsujino, Y., Sakamura, S., Yoshihara, T. (1991) Potato tuber-inducing activities of jasmonic acid and related compounds. Phytochemistry 40, 1435-1438.

23. Koda, Y., Okazawa, Y. (1983a) Influences of environmental, hormonal and nutritional factors on potato tuberization *in vitro*. Japan. Jour. Crop Sci. 52, 582-591.

24. Koda, Y., Okazawa, Y. (1983b) Characteristic changes in the levels of endogenous plant hormones in relation to the onset of potato tuberization. Japan. Jour. Crop Sci. 52, 592-597.

25. Koda, Y., Omer, E.A., Yoshihara, T., Shibata, H., Sakamura, S., Okazawa, Y. (1988) Isolation of a specific potato tuber-inducing substance from potato leaves. Plant Cell Physiol. 29, 969-974.

26. Krauss, A., Marschner, H. (1982) Influence of nitrogen nutrition, daylength and temperature on contents of gibberellic and abscisic acid and on tuberization in potato plants. Potato Res. 25, 13-21.

27. Kumar, D., Wareing, P.F. (1974) Studies on tuberization of *Solanum andigena*. New Phytol. 73, 833-840.

28. Langille, A.R., Forsline, P.L. (1974) Influence of temperature and photoperiod on cytokinin pools in the potato *Solanum tuberosum* L. Plant Sci. Lett. 2, 189-191.

29. Leue, E.F., Peloquin, S.J. (1982) The use of the topiary gene in adapting *Solanum* germ-plasm for potato improvement. Euphytica 31, 65-72.

30. Lorenzen, J.H., Ewing, E.E. (1990) Changes in tuberization and assimilate partitioning in potato (*Solanum tuberosum*) during the first 18 days of photoperiod treatment. Ann. Bot. 66, 457-464.

31. Lorenzen, J.H., Ewing, E.E. (1992) Starch accumulation in leaves of potato (*Solanum tuberosum*) during the first 18 days of photoperiod treatment. Ann. Bot. 69, 481-485.

32. Marschner, H., Sattelmacher, B., Bangerth, F. (1984) Growth rate of potato tubers and endogenous contents of indolylacetic acid and abscisic acid. Physiol. Plant. 60, 16-20.

33. Mauk, C.S., Langille, A.R. (1978) Physiology of tuberization in *Solanum tuberosum* L.:cis-zeatin riboside in the potato plant--its identification and changes in endogenous levels as influenced by temperature and photoperiod. Plant Physiol. 62, 438-442.

34. McGrady, J.J., Struik, P.C., Ewing, E.E. (1986) Effects of exogenous cytokinin on bud development of potato cuttings. Potato Res. 29,191-205.

35. Melis, R.J.M., Van Staden, J. (1984) Tuberization and hormones. Z. Pflanzenphysiol. 133, 271-283.

36. Mendoza, H.A., Haynes, F.L. (1977) Inheritance of tuber initiation in tuber bearing *Solanum* as influenced by photoperiod. Am. Potato J. 54, 243-252.

37. Menzel, B.M. (1980) Tuberization in potato *Solanum tuberosum* cultivar Sebago at high temperatures: responses to gibberellin and growth inhibitors. Ann. Bot. 46, 259-266.

38. Menzel, B.M. (1981) Tuberization in potato at high temperatures:promotion by disbudding. Ann. Bot. 47, 727-733.

39. Mingo-Castel, A.M., Negm, F.B., Smith, O.E. (1974) Effect of carbon dioxide and ethylene on tuberization of isolated potato stolons cultured *in vitro*. Plant Cell Physiol. 53, 798-801.

40. Müller-Röber, B., Sonnewald, U., Willmitzer, L. (1992) Inhibition of the ADP-glucose pyrophosphorylase in transgenic potatoes leads to sugar-storing tubers and influences tuber formation and expression of tuber storage proteins. EMBO Jour. 11, 1229-1238.

41. Nitsch, J.P. (1965) Existence d'un stimulus photopériodique non spécifique capable de provoquer la tubérisation chez *Helianthus tuberosus* L. Bull. Soc. Bot. Fr. 112, 333-340.

42. Okazawa, Y. (1960) Studies on the relation between the tuber formation of potato plant and its natural gibberellin content. Proc. Crop Sci. Soc. Japan 29, 121-124.

43. Ooms, G., Lenton, J.R. (1985) T-DNA genes to study plant development: precocious tuberisation and enhanced cytokinins in *A. tumefaciens* transformed potato. Plant Molecular Biology 5, 205-212.

44. Oparka, K.J. (1986) Phloem unloading in the potato tuber, pathways and sites of ATPase. Protoplasma 131, 201-210.

45. Palmer, C.E., Smith, O.F. (1970) Effects of kinetin on tuber formation on isolated stolons of *Solanum tuberosum* L. cultured *in vitro*. Plant Cell Physiol. 11, 303-314.

46. Park, W.D. (1990) Molecular Approaches to tuberization in potato. *In*: The molecular and cellular biology of the potato, pp. 43-56, Vayda, M.E., Park, W.D., eds. C.A.B. International, Redwood Press Ltd., Melksham, UK.

47. Paupardin, C., Tizio, R. (1970) Action de quelques composés phénoliques sur la tuberisation de la Pomme de terre. Potato Res. 13, 187-198.

48. Pont-Lezica, R.F. (1970) Evolution des substances de type gibbérellines chez la pomme de terre pendant la tubérisation, en relation avec la longueur du jour et la temperature. Potato Res.13, 323-331.

49. Protacio, C.M., Flores, H.E. (1992) The role of polyamines in potato tuber formation. In Vitro Cell. Dev. Biol. 28P, 81-86.

50. Quarrie, S.A. (1982) Droopy: a wilty mutant of potato deficient in abscisic acid. Plant, Cell Environment 5, 23-26.

51. Racca, R.W., Tizio, R. (1968) A preliminary study of changes in the content of gibberellin-like substances in the potato plant in relation to the tuberization mechanism. Eur. Potato J. 11, 213-220.

52. Railton, I.D., Wareing, P.F. (1973) Effects of daylength on endogenous gibberellins in leaves of *Solanum andigena*. I. Changes in levels of free acidic gibberellin-like substances. Physiol. Plant. 28, 88-94.

53. Sanchez-Serrano, J.J., Amati, S., Keil, M., Pena-Cortes, H., Prat, S., Recknagel, C., Willmitzer, L. (1990) Promoter elements and hormonal regulation of proteinase inhibitor II gene expression in potato. *In*: The molecular and cellular biology of the potato, pp. 57-70, Vayda, M.E., Park, W.D., eds. C.A.B. International, Redwood Press Ltd., Melksham, UK.

54. Sattelmacher, B.,Marschner, H.(1978) Relation between nitrogen, cytokinin activity and tuberization in *Solanum tuberosum* L. Physiol. Plant 44, 65-68.

55. Smith, O.E., Rappaport, L. (1969) Gibberellins, inhibitors, and tuber formation in the potato, *Solanum tuberosum*. Am. Potato J. 46, 185-191.

56. Stallknecht, G.F., Farnsworth, S. (1982a) General characteristics of coumarin-induced tuberization of axillary shoots of *Solanum tuberosum* L.cultured *in vitro*. Am. Potato J. 59, 17-32.

57. Stallknecht, G.F., Farnsworth, S. (1982b) The effect of the inhibitors of protein and nucleic acid synthesis on the coumarin-induced tuberization and growth of excised axillary shoots of potato sprouts (*Solanum tuberosum* L.) cultured *in vitro*. Am. Potato J. 59, 69-76.

58. Taylor, M.A., Arif, S.A.M., Kumar, A., Davies, H.V., Scobie, L.A., Pearce, S.R., Flavell, A.J. (1992) Expression and sequence analysis of cDNAs induced during the early stages of tuberization in different organs of the potato plant (*Solanum tuberosum* L). Plant Molecular Biol. 20, 641-651.

59. Tizio, R. (1971) Action et rôle probable de certaines gibbérellines (Al, A3, A4, A5, A7, A9 et A13) sur la croissance des stolons et la tubérisation de la Pomme de terre (*Solanum tuberosum* L.) Potato Research 14, 193-204.

60. Van den Berg, J.H., Bonierbale, M.W., Ewing, E.E., Plaisted, R.L., McMurry, S.E. (1994) Use of RFLP-linkage to study the genetics and physiology of potato (*Solanum tuberosum* and *S. berthaultii*) tuberization. Theor. Appl. Gen. (submitted)

61. Van den Berg, J.H., Davies, P.J., Ewing, E.E., Halinska, A. (1994a) Gibberellin metabolism in *Solanum tuberosum* ssp. *andigena*. Submitted.

62. Van den Berg, J.H., Davies, P.J., Ewing, E.E., Halinska, A. (1994b) Gibberellin metabolism and morphology of dwarf and wild type *Solanum tuberosum* ssp. *andigena* grown under long and short photoperiods. Submitted.

63. Van den Berg, J.H., Ewing, E.E. (1991) Jasmonates and their role in plant growth and development, with special reference to the control of potato tuberization: A review. Am. Potato J. 68, 781-794.
64. Vreugdenhil, D., Helder, H. (1992) Hormonal and metabolic control of tuber formation. *In*: Progress in Plant Growth Regulators, pp. 393-400. Karssen, C.M., Van Loon, L.C., Vreugdenhil, D., eds. Kluwer Academic Publishers, The Netherlands.
65. Vreugdenhil, D., Struik, P.C. (1989) An integrated view of the hormonal regulation of tuber formation in potato (*Solanum tuberosum*). Physiol. Plant. 75, 525-531.
66. Wareing, P.F., Jennings, A.M.V. (1980) The hormonal control of tuberization in potato. *In*: Plant Growth Substances, pp. 293-300. Skoog, F., ed.Springer-Verlag, New York.
67. Woolley, D.J., Wareing, P.F. (1972) Environmental effects on endogenous cytokinins and gibberellin levels in *Solanum tuberosum*. New Phytol. 71, 1015-1025.
68. Yoshihara, T., Matsuura, H., Ichihara, A., Kikuta, Y., Koda, Y. (1992) Tuber forming substances of Jerusalem artichoke (*Helianthus tuberosus* L.) *In*: Progress in Plant Growth Regulators, pp. 286-290. Karssen, C.M., Van Loon, L.C., Vreugdenhil, D., eds. Kluwer Academic Publishers, The Netherlands.
69. Yoshihara, T., Omer, E.A., Koshino, H., Sakamura, S., Kikuta, Y., Koda, Y. (1989) Structure of a tuber-inducing stimulus from potato leaves (*Solanum tuberosum* L.). Agr. Biol. Chem. 53, 2835-2837.
70. Zeevaart, J.A.D., Gage, D.A. (1993) *ent*-Kaurene biosynthesis is enhanced by long photoperiods in the long-day plants *Spinacia oleracea* L. and *Agrostemma githago* L. Plant Physiol. 101, 25-29.

G12.   Postharvest Hormone Changes in Vegetables and Fruit

Pamela M. Ludford

Fruit and Vegetable Science Department, Cornell University, Ithaca, NY 14850, USA

INTRODUCTION

At the time of harvest, there is a large potential for change in physiological processes going on in edible plant tissue. On removal from the parent plant, vegetables are deprived of their normal supply of water, minerals, and organic molecules including hormones, which normally would be supplied by translocation from other parts of the plant.   Although little new photosynthesis is being carried out, there is active transpiration, and tissues can transform many of the constituents already present. While postharvest changes in fresh vegetables cannot be stopped, they can be slowed down within certain limits.

The kind and extent of physiological activity in detached plant parts determine their storage longevity to a large extent. Thus some, such as seeds, tubers, bulbs, and fleshy roots, are morphologically and physiologically adapted to maintain the tissue in a dormant state, both innate and imposed, until environmental conditions are favorable for germination or growth. Metabolic activity is depressed, but not halted, in such organs. Regrowth is triggered in the spring, probably by a change in the hormone balance. The term "dormancy" in this chapter is used as a state in which growth is temporarily suspended due to unfavourable conditions   i.e. imposed dormancy, while innate dormancy (*true dormancy* or *rest*) occurs when growth cannot take place even under favourable conditions due to the condition of the plant material itself. Most vegetables stored over extended periods in the fresh state are biennials that break rest and eventually sprout during storage, hence terminating their usefulness for commercial purposes. The cells of most other plant parts, such as fruit, leaves, stems, petioles, etc. differ in that they are physiologically primed for senescence rather than dormancy.   Fruit ripening is usually associated with the development of optimal eating quality and constitutes the final stages of maturation, and thus could be regarded as a typical senescence phenomenon. Many plant materials classified as fruits botanically are considered to be vegetables for commercial or legal purposes. The cucurbits are consumed in both the immature and the fully mature states as cucumbers or zucchini and melons.

P. J. Davies (ed.), Plant Hormones, 725–750.
© 1995 *Kluwer Academic Publishers. Printed in the Netherlands.*

Most cultivated vegetable parts are removed suddenly from the natural environment, and often held for short-term transportation or long-term storage in stressful environments, including low temperatures, artificial atmospheres, or both in combination, to reduce respiration rates. Storage of fruit and vegetables can be prolonged by ethylene removal using ethylene scrubbers (e.g., Purafil), flushing with nitrogen gas, or by hypobaric storage. Transient ethylene production is also triggered by stress or injury, so wounding during harvest and transport has an obvious effect on storage, added to which many bacterial pathogens have the capacity to synthesise ethylene. Quite apart from ethylene effects in fruit ripening, the effect on leaf abscission has an immediate result on leafy vegetables. The commercial storage of cabbage along with apples can be disastrous.

Research reports available on stored plant materials have suggested that endogenous hormones continue to function and appear to control physiological events. This conclusion is apparent from correlative evidence of hormonal balances in detached plant organs and easily observed physiological events such as rest, dormancy and compulsive regrowth. The application of growth substances to plants or their excised parts has been widely used to extrapolate to endogenous hormonal responses. In many cases there are problems with this approach, often because of difficulties in uptake and distribution into bulky tissue such as fruit or tubers. All the commonly identified endogenous hormones i.e. auxins, gibberellins, cytokinins, abscisic acid, and ethylene, appear to be present, as well as polyamines and jasmonic acid.

Interaction and balance between opposing promotory and inhibitory hormonal factors is the idea behind the control of metabolism in postharvest storage, be it of rest and regrowth in vegetable storage organs or of maturation and ripening in vegetable fruit. The interesting difference is in what constitutes these opposing factors. In fruit, ethylene is one important promoter of ripening, and ABA may be another, while auxins, GAs and cytokinins are possible candidates for the role of ripening inhibitors. The latter are high in young seeds and developing fruit, and may affect the changing sensitivity of maturing climacteric fruit to ethylene. In leafy vegetative tissue, ethylene causes leaf abscission and cytokinins retard senescence. In storage organs, ABA acts more as an inhibitor of regrowth, while auxins and GAs are likely to promote it.

FRUIT RIPENING

Under normal conditions, fruit ripening occurs as an integrated sequence of changes including softening, colour change, and the accumulation of sugars and aromatics, coupled with a decline in organic acids, catalysed by specific enzymes (62). These active metabolic processes are accompanied in many fruits by an increase in respiration termed the *climacteric*. Just prior to the increase in respiration in climacteric fruit there is a pronounced increase in the production of ethylene. In some cases, ripening can be promoted by simple substances such as galactose or N-glycans (51). At least some of the ripening changes can be separated from the respiration climacteric experimentally, including softening and carotenoid synthesis (62), but inhibition of ethylene synthesis or even perception inhibits ripening (31). Ethylene application can initiate the respiration response even in non-climacteric fruit such as citrus, but here continued application is necessary. The main difference between climacteric and non-climacteric fruit is seen in their ability to produce ethylene autocatalytically in response to threshold levels of ethylene.

Commonly referred to as the ripening hormone, ethylene has a cascade effect in climacteric fruit leading to the 'one rotten apple in the barrel' syndrome. It plays a significant role in the changes that occur with the climacteric in fruit ripening, and its production is intimately involved in fruit ripening changes. Exogenous ethylene application, in the form of ethephon or 'liquid ethylene', is registered for a harvest aid to promote ripening in cherry, grape, pepper, blackberry, boysenberry, and pineapple, as well as tomato and apple. The role of ethylene in the ripening of non-climacteric fruit such as strawberry is not thought to be great, and strawberry fruit are relatively unresponsive to exogenous ethylene and insensitive to inhibitors of ethylene synthesis or of ethylene perception (1).

The most desirable fresh state for consumption of the climacteric tomato fruit, *Lycopersicon esculentum* Mill., is at the termination of ripening and the beginning of senescence, the red ripe stage. However, to facilitate shipment for commercial purposes, fruit are frequently harvested at the mature green (MG) stage and then ripened by ethylene application. Tomato ripening starts in the interior with gel formation in the locule, placenta, and then pericarp. A difficulty in commercial harvesting is the determination of the precise MG stage, as some immature fruits respond to exogenous ethylene but do not undergo a normal ripening. This may account for complaints of poor quality in winter-shipped tomatoes. The ripening of green bananas (*Musa acuminata*) with ethylene gas is also a common practice, and again bananas ripen from the inside out, so that pulp ripening precedes peel yellowing. However, externally applied ethylene reverses this process (1).

Cultivars of bell pepper, *Capsicum annuum* L., cannot be ripened to a satisfactory red color if removed from the plant in the immature or green stage. On the other hand, some avocado cultivars *(Persea americana* Mill.) will not undergo ripening or the climacteric rise in respiration while still attached to the plant, with the peduncle suspected of supplying a ripening inhibitor possibly involving indoleacetic acid or polyamines (58). This is an extreme example of the tree factor seen in apple (*Malus sylvestris*) and other fruit, where ripening is accelerated by fruit detachment. Exogenous ethylene is equally effective on both attached and detached apples in the induction of endogenous ethylene synthesis after harvest, so that the effect of detachment on induction does not appear to be related to ethylene-responsive developmental changes (34).

The rate of protein synthesis increases during the early stages of ripening in several climacteric fruit. While this may partly reflect an increase in protein turnover, it is also related to *de novo* synthesis of ripening-specific enzymes (62). These include cell wall degrading enzymes that influence fruit softening. For instance, the enzyme polygalacturonase (PG) is absent from green tomato pericarp tissue, is first detected when fruit begin to colour, and increases progressively during ripening, along with acid invertase. Until the first appearance of ethylene in tomato fruit, neither the respiratory climacteric and appearance of PG nor the increase in polysomes and cytoplasmic mRNA takes place. Ethylene is also involved in the synthesis of cellulase in ripening of avocado fruit (58). Attached apple fruit show a decline in cellulase activity as they go through the respiratory climacteric. However, harvested strawberry fruit (*Fragaria* sp.) show a large increase in cellulase activity during maturation and softening, but ethylene does not increase cellulase levels in this non-climacteric fruit (1).

The role of ethylene as the ripening hormone was brought into question and the investigation of ripening is aided by several non-ripening tomato mutants, for example *Nr* - never ripe; *rin* - ripening inhibitor; and *nor* - non-ripening. Both *rin* and *nor* fruit fail to ripen (with the exception of seed maturation) and do not display a climacteric rise in CO_2 or ethylene evolution. They have reduced lycopene levels, their chlorophyll content remains high, with very low, if any, PG levels. External ethylene applications have little effect in inducing ripening, though it will bring about a temporary stimulation of CO_2 evolution, while by contrast, wounding the fruit causes an increase in both CO_2 and ethylene production, so the capability to produce ethylene is not lacking (54). Furthermore, in wild tomato species, that ripen on the vine but remain green, two species show ethylene production correlated with fruit softening, while in two others external ripening changes are not correlated with ethylene production (27). Ripening may thus be determined by changes in sensitivity to ethylene rather than the amount produced. There is also the mutant *alc* (Alcobaca), which ripens partially and has a long shelf life. Its possible connection with increased polyamine levels is discussed by Picton et al. (Chapter E4).

Recent advances in molecular biology have opened new windows for the understanding of fruit ripening, and a more detailed discussion of this aspect is provided in Chapter E4. The identification and isolation of ripening-related genes, along with the introduction of antisense mRNA and its resulting down-regulation, provide opportunities to control ripening of fruits. The cloning of genes involved in ethylene synthesis, as well as of other genes induced during fruit ripening, have allowed the construction of ripening mutants in tomato using reverse genetics. The production of antisense RNA inhibits the translation and expression of mRNA for the ripening-induced enzyme, PG, but fails to give a strong effect on softening (31). This is particularly the case from mature-green to turning stages, although there is a definite retardation from turning to red, at which time a striking arabinose- and rhamnose-containing uronide solubilization usually begins. Both control and PG antisense fruit also show solubilization of a gal-containing polymer, which may implicate a β-galactosidase (11). Similarly, pectin methylesterase alone will not bring about softening. Thus PG is probably not the only, or even the primary, cause of softening in tomato, confirmed by introduction of the PG gene behind a fruit specific promoter into *rin*, where little induction of softening was achieved despite the production of active PG (31).

The inhibition of ethylene production itself is more effective in inhibiting ripening. This is accomplished either by limiting formation of one of the ethylene synthesis enzymes, ACC synthase or ACC oxidase, or by breaking ACC down to α-ketobutyric acid and ammonia. The latter approach was accomplished by overexpressing the bacterial gene for ACC deaminase from *Pseudomonas* in transgenic tomato plants (33). This resulted in 90-97% inhibition of ethylene production during fruit ripening, and the transgenic plants showed significant delays in ripening of fruit detached at the breaker stage (see Chapter E2).

Fruit from plants with ACC oxidase antisense RNA (from pTOM13) had a 95% reduction in ethylene production, and, although colour change was initiated at the normal time during ripening, the extent of red colour production was reduced, along with carotenoid reduction (8). Fruit from plants with ACC synthase antisense RNA had even more inhibition of ethylene production (99%) and no red color development (33). They could be kept in air or even remain attached on the plant for 91 to 150 days without softening, developing an aroma, or turning red although they eventually became orange, a condition that could be reversed by ethylene treatment.

The transcription of the tomato E8 gene is fruit specific, and is activated at the onset of ripening and in unripe fruit treated with exogenous ethylene. The E8 predicted protein is a dioxygenase related to ACC oxidase, and its reduction in quantity by antisense RNA results in overproduction of ethylene during the ripening of detached tomato fruit. It was suggested that the E8 gene may participate in the feedback inhibition of ethylene production during fruit ripening, or may be part of the ethylene receptor. By a series of deletions incorporated into transformed fruit, it was found that DNA sequences of the E8 gene that confer responsiveness to exogenous ethylene in unripe fruit are distinct from DNA sequences needed for expression during fruit ripening (18).

Similar approaches to investigations of gene expression during fruit ripening have been carried out in avocado, pear, and apple (19), as well as peach (46).

The overall signal transduction pathway involved in the ethylene response has the hormone at one end (suggesting the binding to a receptor molecule) and the up-regulation of transcriptional (or post-transcriptional) activity at the other end of the pathway. Theologis (59) drew three main

729

conclusions about ripening from the tomato mutant results and antisense technology. The ethylene-mediated process requires continuous transcription of the necessary genes, ethylene is indeed autocatalytically regulated, and, rather than acting as a switch, ethylene acts as a rheostat for controlling the ripening process. Further, transcriptional and post-transcriptional regulation of genes involved in lycopene and aroma synthesis, and respiratory metabolism, are dependent on the ethylene signal during ripening, along with the expression of the ACC synthase gene. However, other genes are developmentally regulated, such as those for chlorophyllase, ACC oxidase, and even PG, which are not dependent on the ethylene signal for transcription but probably for translation.

Ethylene

There is a large variation between species in rates of ethylene production during the climacteric, and in apple alone the rise can vary from small to 40-fold. Passion fruit (*Passiflora edulis* Sims.) have such high rates of ethylene production that they were suggested as a commercial source of ethylene (1).

Fruit tissue was used for the establishment of parts of the ethylene synthetic pathway via l-aminocyclopropane-l-carboxylic acid (ACC). Levels of endogenous ACC and ACC synthase are higher in the inside parts of freshly-harvested ripening tomato fruit (septa, pulp, and seeds) than in the outer pericarp. At the preclimacteric MG stage, ACC synthase activity, ACC content, and ACC oxidase activity are low, but increase markedly on ripening following the breaker stage. Addition of exogenous ACC to many vegetative tissues results in greatly increased ethylene production, but this is not the case with preclimacteric fruit of apple and cantaloupe, so that both ACC synthase and ACC oxidase are restricted in these preclimacteric fruit. In preclimacteric MG tomato fruit, ACC slightly enhances the ripening process and ethylene production, and exogenous ethylene treatment increases the capability to convert ACC to ethylene, i.e. increases ACC oxidase activity before ACC synthase activity. This is also true for cantaloupe, *Cucumis melo* L., where wounding increases the activity of both ACC synthase and ACC oxidase. In over-ripe or postclimacteric fruit, ACC can accumulate considerably, possibly because of the inactivation of ACC oxidase in these fruit. Fruit tissue is also capable of conjugating ACC to N-malonyl-ACC. Under normal physiological conditions the malonylation is irreversible, but the reaction is possible, as shown in watercress stems (33).

ACC oxidase, as measured by ethylene production in the presence of a saturating concentration of ACC, is present in most tissues of higher plants with the exception of unripe fruits. Preclimacteric cantaloupe and tomato fruits have low levels of ACC oxidase, whose activity is enhanced following treatment with ethylene (33). During fruit ripening, the level of ACC oxidase increases markedly and effectively regulates ethylene production. Helped by information given by the cDNA clone, pTOM13, ACC oxidase has finally been extracted in a soluble form from pericarp of melon fruit. Its activity is dependent upon the presence of the appropriate cofactors (65). Different signals activate specific ACC synthase and ACC oxidase genes, both of which are encoded by multigene families (33).

The inhibitor of ACC synthesis, aminoethoxyvinylglycine (AVG), has been tried as a preharvest spray with apples to suppress ethylene production in harvested fruit and to provide a possible alternative for controlled atmosphere storage rooms (low O_2 and high CO_2). AVG pre-treated 'Golden Delicious' fruit stored at 3°C do not produce autocatalytic ethylene and have a lower respiration rate than control fruit (28). However, AVG does not affect chlorophyll breakdown, and production of aroma volatiles is reduced and cannot be stimulated by exogenous ethylene after 4 months storage. Also other authors find marginal effects which vary among apple cultivars. Treatment with AVG is effective in inhibiting ethylene synthesis in slices of green tomatoes, but relatively ineffective in pink and red tomato fruit, possibly reflecting relatively high endogenous levels of ACC at these stages since the isolated ACC synthase enzyme from pink and red fruit is sensitive to low levels of AVG (33).

Of the many mutants identified in tomato, two interesting single-gene ethylene mutants are derived from cv. VFN8. These are *dgt* (diageotropica) and *epi* (epinastic). The former is characterized by a horizontal growth habit (70), which can be partially normalized by exposure to very low concentrations of ethylene. The fundamental lesion of this *dgt* mutant may be an insensitivity to auxin, although IAA concentrations themselves do not differ significantly between the two mutants and their parent. The *epi* mutant has a contrasting phenotype, with elevated ethylene levels and greater ACC content. This may not represent the fundamental lesion, since blocking ethylene synthesis or action does not normalize the *epi* phenotype, nor does it result from increased auxin sensitivity (24). Both mutants set fruit which ripen normally and have viable seed.

Some fruit have a chilling requirement as in cv. D'Anjou, a late maturing cultivar of pear (*Pyrus communis* L.), where low temperature stimulates ethylene synthesis and ripening. On the other hand, other fruit can be subjected to stress by low temperature. Ethylene production in a number of chilling-sensitive vegetables is stimulated by chilling temperatures of 0 to 15°C. This may occur in tissue which does not usually produce significant amounts of ethylene, such as preclimacteric fruit. In chilled cucumber (*Cucumis sativus* L.), increased ethylene production is due to increased capacity to make ACC, but the increase in ACC, ACC synthase, and ethylene is not apparent until subsequent warming. An increase in ACC synthase activity during the warming period can be inhibited by cycloheximide treatment but not by cordycepin or α-amanitin, suggesting stimulated production of mRNA coding for ACC synthase during chilling (36), i.e., mRNA is transcribed during the chilling stage but translation is not completed until transfer to warmer temperatures. However, in both tomato fruit (10) and

'Honey Dew' melon (36), ACC accumulates during the chilling period without waiting for subsequent warming, so it seems that not all sensitive fruit respond in the same way to chilling, any more than all fruit are likely to show exactly the same ripening control. Under most commercial conditions, chilling is not likely to exceed four days, and the ethylene generated upon warming probably is responsible for the chlorophyll loss and pitting noted in chilled cucumbers. Prolonged chilling exposure results in a reduction of ACC oxidase activity, as also does high temperature (1).

Abscisic acid

The linkage of abscisic acid (ABA) with abscission is almost limited to young cotton fruit, where the name originated. However, similar relationships are found with fruit drop of the non-climacteric litchee, *Litchi chinensis* (68). Here ABA levels in seeds of abscising fruit are higher than those of persisting fruit and a similar tendency is seen in the peel. While ABA is a germination inhibitor in many seeds (see Chapter G10), in tomato fruit the osmotic environment within the tissues is more important than endogenous ABA in preventing precocious germination of the developing seeds (7).

In a number of fruit (e.g. pear, avocado), the level of free ABA is constant during maturation and increases during ripening, with the rise in ABA seen in both climacteric and non-climacteric (e.g. citrus, grape, cherry) fruit (54). However, this was not the case in kiwi fruit (*Actinidia deliciosa*), but these are also unusual in that the respiratory climacteric and ethylene burst occur very late in ripening (1). In tomato, levels of ABA increase during growth of the fruit reaching a peak at the MG stage, and decline prior to the respiration climacteric and ripening (Fig. l), even in the non-ripening mutant *rin*. The ABA peak comes a little later in the mutant *Nr*, and about 20 days later in *nor*, where ABA levels are lower (45). This increase to a peak before falling is found both in the pericarp and in the seed, though the latter peak is reached earlier. In the tomato mutant *sitiens* (with the *sit*^{w} gene) where ABA synthesis is impaired, neither accumulation nor peak is seen in the seed (7).

Developing sinks serve as sites for accumulation of ABA produced in source leaves, as in developing soybean seeds. However, free ABA accumulates in both attached and detached tomato fruit unless the fruit is detached very early, showing that ABA is also synthesised in the detached fruit and is not dependent only on translocation from the plant. Changes in bound ABA reflect those of free ABA, but at about one-seventh of the levels, similar to the 10:1 ratio of free/bound ABA throughout avocado ripening. Thus the increase in free ABA must represent net synthesis rather than release of the bound form (54).

Endogenous ABA is correlated to sucrose uptake in cotyledons and ABA may function to enhance sucrose accumulation (see Chapter G9). It has been suggested that ABA enhances sucrose unloading from phloem and also

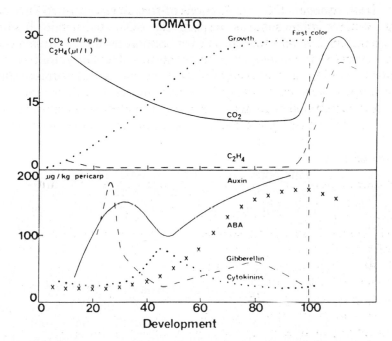

Fig. 1. Trends in free hormone levels in tomato pericarp tissue during development and ripening. From (45).

inhibits phloem loading, or possibly reloading in fruit. Both sugar and ABA levels have been linked to chilling injury effects in fruit. Chilling MG tomatoes at 2°C for 12 days results in a 2 to 3 fold increase in free ABA in the pericarp (39). Sucrose levels also increase during chilling in several cultivars of tomato fruit (16), and reducing-sugar levels in peel are highest in "Marsh" grapefruit (*Citrus paradisi*) when seasonal resistance of the fruit to chilling injury is highest (52).

Many genes that are induced by drought are responsive to ABA, and several genes expressed during drought in vegetative tissues are also expressed during desiccation of developing seeds. Furthermore hybridization of halved tomato fruit blots show one such tomato gene, pLE25, expressed only in seeds of MG fruit, while at the breaker and red stage hybridization is observed in locular tissue (9).

Auxins

It is known that indoleacetic acid (IAA) influences ethylene formation through the induction of ACC synthase. Endogenous auxin concentrations are thought to be highest after pollination during the early stages of fruit development and lowest during maturation. The early work of Nitsch showed control by developing seeds and/or applied auxin over early development of strawberry receptacles. When quantitated during strawberry fruit development, endogenous concentrations of free and conjugated IAA peaked 14 and 8 days after anthesis respectively (43), and auxin binding to strawberry fruit membranes has been demonstrated. Auxin controls stage-specific formation of polypeptides during strawberry fruit receptacle development, and two cDNA clones have been isolated corresponding to auxin-induced mRNAs. Auxin not only induces but also represses the expression of developmental genes. For instance, during colour development in ripening, the removal of strawberry achenes (a source of auxin) induces anthocyanin formation, while externally applied auxin delays it, and ethylene has no effect (43).

Contrasting results obtained by different types of application of auxin (e.g. dipping or spraying fruit vs. vacuum infiltration) can be explained by limited penetration of auxin into the tissue when applied by dipping. Using dipped whole fruit and vacuum-infiltrated cut discs, it was shown that 2,4-D causes a dual effect in tomato fruit tissue, namely an increase in ethylene production which promotes ripening, but also a delay in ripening (1). The last effect prevails, but depends on the uniformity of the auxin distribution and its concentration. Infiltration of green banana slices with 2,4-D inhibits both colour change of the peel and softening of the pulp, but does not prevent ethylene-induced respiration. Studies on locating the signal for fruit abscission in apple suggest that fruit drop is related to production of auxin-like hormone in the seeds (6).

In infiltrated tomato pericarp, IAA is conjugated to yield both IA-glucose and IA-aspartic acid, the latter being an irreversible conjugation (13). In green immature tomatoes IAA is deactivated primarily by conversion to IA-aspartate and further metabolites, while in mature pink fruit there is more IA-glucose, a potential storage product.

Gibberellins

Similarly, endogenous levels of gibberellins (GAs) are thought to be high in very young fruit and they may play a role in retarding senescence. Normal pod or pericarp growth in pea (*Pisum sativum*) requires the presence of seeds, which have high GA levels, and GA can replace the requirement for seeds, indicating that transported seed GAs regulate pericarp growth. It was later shown that seeds promote pericarp growth by maintaining GA synthesis in the pericarp. Pea pericarp has the capacity to metabolize GA_{12} to GA_{19} and on to GA_{20} and GA_1, but the presence of the seed is needed particularly for the conversion of GA_{19} to GA_{20}. The seed factor regulating this conversion was shown to be 4-chloro-IAA (49).

Early exogenous GA treatment of pear fruit, i.e. during flowering, produces misshapen and parthenocarpic fruit, whereas later treatment during petal fall results in mostly seeded, well-shaped fruit in "Agua de Aranjuez" pear (29). This was speculated to be by prolonging embryo sac viability or longevity, and thus increasing the chances of fertilization taking place. Apple skin russeting was thought to have a localised origin probably relating to GA levels in the skin, and in fact GA_4 can reduce the incidence and severity of apple skin russeting, but results in more angular fruit (38). For sweet cherries (*Prunus* sp.), foliar applications of GA_3 about a month before harvest are used in the Pacific Northwest (USA) to delay fruit ripening, reduce surface marking, and increase size and firmness, resulting in an improved canned product (37) and a firmer brined product (23).

Flower and Fruit Colour

Gibberellins play a role in the promotion of anthocyanin synthesis in the corolla of petunia flowers. For instance, GA_3 is required for the activation of transcription of the chalcone synthase gene in an indirect manner (67). Applied GAs have also been found to induce colour changes in fruit. Fruit ripening is associated with the conversion of chlorophyll-containing chloroplasts to carotenoid-containing chromoplasts. Rind of Valencia oranges (*Citrus sinensis*) reaches maximum orange colour during winter months, but tends to regreen during spring and summer as chromoplasts revert back to chloroplasts. This is enhanced by GA and cytokinins, and GA_3 is highly persistent in citrus peel (3). For citrus fruit stored either on the tree or after harvest in cold storage, GA_3 delays fruit rind colour development from green to orange and maintains rind quality. In mandarin orange GA treatment results in a delay of colouration and senescence and promotes peel thickening. Similarly peel resistance to puncture is increased by preharvest GA treatment of grapefruit, along with peel oil concentration (3). This may increase resistance to attack by the Caribbean fruit fly.

Citrus fruit thus often have green peel colour when fully mature and are de-greened for visual quality by exposure to ethylene, which advances the ripening of oranges (1). This is normally done before other packing house operations, since pre-washing fruit inhibits de-greening. In grapefruit this effect of washing was overcome by increasing the ethylene concentration to 250 μl/l, but a similar effect of film-wrapping could not be overcome in this way (2). Although applied cytokinins are not necessarily very effective, kinetin also delays ethylene-induced degreening in peel of banana slices and loss of chlorophyll in other fruit (45).

A GA postharvest dip extends the shelf life of mango fruit (*Mangifera indica* L.), delaying ripening during storage in terms of colour and aroma and retarding chlorophyll degradation, ascorbic acid decrease, and the decline in activity of α-amylase and peroxidase. GA_3 delays appearance of plastid-localized lycopene during ripening of tomatoes, but colour change does not appear to be due to differential transcriptional control during the chloroplast-chromoplast conversion. There are only moderate changes in the stability of plastid transcripts, with post-transcriptional processing and/or translation the most likely areas of specific gene expression control in plastids (44). The molecular role played by GA is therefore even more unclear. Despite the effect of applied GAs on delaying fruit colouration in fruit of several species, no close correlation has yet been found between colour change in fruit ripening and endogenous gibberellin content (45).

Cytokinins

Senescence in leaves can generally be delayed by cytokinin treatment, and, in one aspect, fruit ripening can be regarded as a senescence phenomenon. High levels of endogenous cytokinin can also delay fruit ripening, and levels may decline as ripening proceeds. Cytokinin activity is high for two weeks after anthesis in cherry tomato when cell division is most active (32), and declines from MG to the red ripe stage in standard varieties. The non-ripening mutant *rin* not only shows less decline in cytokinin activity than standard fruit, but it also has high levels of zeatin-O-glucoside activity, which could act as a storage form to ensure continued high zeatin levels. Increased cytokinin levels in seeded tomato fruit can be obtained by reducing the ratio of foliage to fruit and hence lowering sink competition (64), and this delays the rate of ripening after the breaker stage.

There are many instances of parthenocarpic fruit growth induced by auxins and gibberellins, but few involving cytokinins. Application of a diphenylurea-derivative cytokinin, N-(2-chloro-4-pyridyl)-N'-phenylurea, to kiwi fruit before anthesis induces parthenocarpic fruit development. However, parthenocarpic fruit are of little value in commercial kiwi fruit since size of fruit is important, and more than 1000 seeds should be present for large-size fruit, which can entail hand pollination during adverse weather conditions. Spraying of cytokinin on open-pollinated flowers about 3 weeks after anthesis causes stimulation of fruit growth, most effective when seed number is over a threshold number, so that cytokinins alone do not control fruit expansion. Both zeatin and zeatin ribotide are present in flesh and seeds of 20-day old and mature kiwi fruit, but only the former was found at 40 and 60 days after flowering (26).

Thus, IAA, GAs, and possibly cytokinins are implicated in the delay of fruit ripening, while ABA levels may increase until ripening is in progress.

Polyamines

Polyamines are associated with rapid cell division, and senescence in many plant organs is correlated with a decline in polyamines. Free polyamine levels decline during fruit development in avocado, apple, pear, and tomato fruit (55), but the behaviour of individual polyamines varies with species (12), with spermine being the major one in avocado, putrescine in tomato, and spermidine in pear.

In young tomato fruit, as in apple, during the cell division phase the level of polyamines is high, mostly in the form of conjugates (21). Comparatively lower levels are found during cell expansion and fruit ripening, and they are free rather than conjugated. Levels of free putrescine are high at the immature green stage, decline at the MG stage, and remain low through ripening. However, in fruit of the Alcobaca tomato mutant (with the recessive allele *alc*), which have prolonged keeping qualities, putrescine levels rise again after the MG stage and become three times as high as in normal fruit at the ripe stage (53). Similar changes are seen in another longer keeping tomato, cv. Liberty (55). The elevated levels of putrescine in *alc* fruit appear to be age-related and take place whether or not they ripen, which is light dependent in *alc*, and are not due to changes in conjugation or metabolism, but to an increase in arginine decarboxylase (ADC) activity. However, a lack of correlation between increasing ADC activity and ADC mRNA levels, which peak at the breaker stage, in ripening *alc* fruit suggests translational and/or posttranslational regulation of ADC expression in tomato fruit.

Both *alc* and 'Liberty' fruit also show a decrease in climacteric ethylene production. S-adenosylmethionine is a precursor common to both ethylene and the polyamines, spermidine and spermine, and added polyamines appear to inhibit ethylene production in apple fruit discs and protoplasts (22). However, during ripening of detached tomato fruit the onset of synthesis and accumulation of the ethylene precursor, ACC, is not a consequence of a decrease in spermidine synthesis (12). In addition, while treatment of *alc* pericarp discs with polyamines reduces ethylene synthesis, it only does so at concentrations higher than those which delay ripening and overripening (Davies et al, unpublished). Further, during vegetative growth norbornadiene prevents the depletion of putrescine resulting from the citrus exocortis viroid without affecting the enhanced ethylene synthesis, while putrescine levels are unaffected by norbornadiene in the ethylene-overproducing epinastic tomato mutant (*epi*) compared with normal (5).

Putrescine infiltration of mature green fruit of 'Rutgers' or the line 'Alcobaca-red', both of which have normal ripening and the *Alc* allele, increased their storage life in darkness, not by slowing down ripening but by slowing softening (35). Controlled atmosphere storage of apples similarly reduces the rate of postharvest softening and also maintains higher levels of polyamines. When 'McIntosh' and 'Golden Delicious' apples harvested at optimum commercial maturity are infiltrated with polyamines there is an immediate increase in firmness, but the effect on ethylene production is negligible (66). Endogenous putrescine accumulation is correlated with chilling injury in three species of *Citrus* and one of *Capsicum*, and increased levels are found in response to other stress conditions. All this confirms a possible working model involving binding of polyamines to membranes, prevention of lipid peroxidation, and quenching of free radicals (22).

Jasmonates

Jasmonic acid and methyl jasmonate are regarded by some as strong candidates for intracellular or intercellular messengers, and have been identified in many plants, including apple and tomato fruit. However, their location within tissues and cells is not certain.

Externally applied methyl jasmonate stimulates ethylene production, including ACC oxidase activity, in all stages of tomato ripening (57), and increases ACC content, ACC oxidase activity, and ethylene production in preclimacteric apples. However, it has the opposite effect on levels in climacteric and postclimacteric apple fruit, except for ACC oxidase activity. Even the increase in the latter seems to depend on the cultivar, since methyl jasmonate had little or no effect on ACC oxidase activity or was inhibitory in cv. Jonathan (48), possibly due to differing levels of phenolic compounds in the cultivars. The role of jasmonic acid through ethylene in fruit ripening obviously needs further study, especially in nonclimacteric fruit (57).

UNDERGROUND STORAGE ORGANS

Wounding Responses

Ethylene effects on bulky storage organs are of interest because of possible wound ethylene production after harvest. Treatment with ethylene results in a sharp rise in respiration, especially in the presence of O_2 rather than air. These organs can be separated in their respiration response to cyanide (CN). One group yields CN-sensitive fresh slices (eg. potato, turnip, rutabaga, Jerusalem artichoke), while the other gives CN-resistant slices using the alternative path of respiration, as in carrot and parsnip. After ethylene treatment, the former group produces CN-resistant fresh slices, rather than CN-sensitive. Other changes resulting from ethylene treatment include increased numbers of polysomes and changes in gene expression, reflected in induced mRNA levels. However, it was shown for carrot (*Daucus carota*) that the ethylene-stimulated changes in mRNA levels are not necessarily correlated with induced respiration enhancement (47).

Endogenous hormones have been found in all plant storage organs investigated, including potato tuber, the swollen stem of kohlrabi, carrot root, beet, sweet potato, and Jerusalem artichoke tuber. They increase after wounding (see also Chapter E5), as with IAA and cytokinin activity in potato (*Solanum tuberosum*), although this could indicate differences in sensitivity or form (bound vs. free) rather than concentration differences.

The wound reactions and aging of cut slices from storage organs, such as potato tuber, carrot root or sugar beet, include the induction of mRNA synthesis and an increase in nucleolar size and protein synthesis along with RNase activity. Exogenous applications of hormones can amplify endogenous hormone activities and affect the metabolic activity of cut slices. The addition of GA_3 stimulates protein synthesis, RNA synthesis, nucleolar size and RNA polymerase activity still further in wounded potato tissue, although cells of uninjured tubers are not responsive. Inactive ribosomes from resting potato tubers can be activated by added mRNA from polysomes of wounded potato.

A genomic clone isolated from carrot storage roots encodes a proline-rich cell wall protein very similar to other proline-rich protein cDNAs (20). Its expression is developmentally regulated, beginning at the earliest visible stages of carrot storage root growth. This can be induced to a higher level by IAA or other auxin treatment, and is also wound induced. There is no expression of this cDNA in leaves or in vegetative (non-storage) roots, whether wounded or not. It seems that auxin may therefore play a role in the wound-induced expression of this gene in carrot storage roots.

Polyamines inhibit RNase activity in cut potato, and spermidine and spermine also inhibit the rise in betacyanin leakage that normally takes place from cut discs of beet root, *Beta vulgaris*. The presence of more than two amino groups, as in spermine and spermidine, appear necessary to counteract the detrimental effects of ammonium sulphate or ethylene, applied as ethephon, on cell permeability and pigment leakage in beet root and rose petals (50). Free radical scavenging by polyamines is also correlated with the number of amino groups (22). Polyamines may affect wound-induced or senescence-induced destabilization of cell membranes in plant storage organs.

Thus hormones may work together to counteract the catabolic processes caused and to stimulate the synthetic activity necessary to "heal a wound".

Hormonal Changes During Storage

Onion

The onion, *Allium cepa* L., most important of the bulb crops, demonstrates innate dormancy or rest, dormancy, and regrowth, whose morphological changes correlate with changes in endogenous hormonal balances. In common commercial practice, onion bulbs are cured, i.e. air-dried for about 10 days after harvest, then stored at low temperatures for an extended time period, e.g. 2°C for nine months, September through May. Preharvest foliar spraying with maleic hydrazide (MH) delays sprouting of onion (as well as potato and garlic) during storage and is regularly used. While onion storage in the U.S. is usually around 2°C, MH treatment also is effective with high storage temperatures (30°C) found in the Middle East (56).

Hormonal changes which induce the rest period are initiated while the onion is maturing and still in the field (32). Morphological changes consist of a rapid lateral growth of the bulb combined with a steady senescence of the leaves. Auxin activity decreases in both the green foliage and bulb apices as the leaves senesce from lush green to the weakened soft-neck stage. When all foliage has become procumbent on the ground, auxin activity disappears while ABA activity is initiated in the tops. At this developmental stage the crop is usually lifted from the soil (undercut) and set on top of it and then harvested within a few days. Inhibitors appear to be synthesized in the leaf and translocated to the bulb apex, since ABA activity in harvested bulbs is high. Also bulbs whose green tops were prematurely dried by a desiccator spray or removed while green show early sprouting. Premature defoliation should be avoided in practice since movement of inhibitor to the bulb apex appears necessary to establish dormancy during storage (32).

Bioassays of endogenous hormones in bulbs (central plugs containing apical tissue) stored at 2°C in the dark show a hormonal balance high in ABA activity and low in IAA, GA, and cytokinin activities from harvest to mid-winter. This is the rest period and no sprout growth or cell division occurs for several weeks even under the most favorable conditions. In January, the beginning of imposed dormancy is shown by a rapid rise in sprouting and apical cell division on transfer to favourable conditions. There is a decline in ABA activity during dormancy and the following regrowth stage, accompanied by an increase in growth promoter activity (Fig. 2). By April, when some bulbs sprout even in storage and most have well-developed internal leaves (Fig. 3), most hormone activity declines with the balance in favor of growth promoters. Application of exogenous growth substances to excised onion apices shows that only kinetin can break rest in non-temperature induced apices, while ABA prolongs the innate dormancy

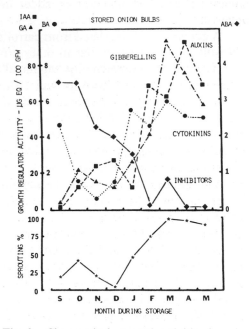

Fig. 2. Changes in hormonal activities in onion bulbs during storage at 5-8°C (3 year average). From Isenberg et al., Proc. XIX Intern. Hort. Congress, Warszawa, Vol. 2, pp. 129-138 (1974).

Fig. 3. Sprouting of onion in storage.

period. Thus cytokinins may be part of the breaking of innate dormancy, and ABA may be part of the inhibitor complex maintaining this state.

Potato

The potato tuber is a swollen underground stem or stolon that can overwinter in the soil or under proper storage conditions. It also shows the biennial postharvest features of innate dormancy, dormancy and regrowth. The investigation of tuberization under inducing conditions has involved much hormone work (see Chapter G11), and tuberonic acid has strong tuber-inducing activities in potato, as well as in yam (*Discorea batatas* Decne.) and Jerusalem artichoke (*Helianthus tuberosus* L.). However, postharvest hormone studies are more concerned with the induction and termination of the innate dormancy period and the possible role of an endogenous inhibitor. The latter, found in potato peel extract, was shown to decline naturally with the termination of rest, and several authors have shown a correlation between a low level of ABA and the breaking of bud dormancy in stored potatoes. Tubers on one-leaf cuttings growing under sprout-promoting conditions ($35°C$ plus excised leaf) have a lower level of ABA compared to controls under cool conditions (63). However, this difference in ABA is preceded by changes in carbohydrate levels, with a conversion of starch to soluble sugars taking place under warm conditions, so it is not certain whether ABA or sugar levels are the growth trigger.

A rise in endogenous GA activity along with a decline in ABA activity seems to be involved in the dormancy break of potato tubers, and the application of GA_3 to promote sprouting in seed potatoes is an approved commercial practice in some countries. Increases in cytokinin activities are found during the transition from rest to dormancy, with conjugation being of possible importance in cytokinin movement from storage tissue to active meristems. The polyamines putrescine, spermidine and spermine are also present in all parts of dormant potato tubers, and their levels increase in apical buds with breaking of dormancy and beginning of sprout growth, along with levels of their biosynthetic enzymes, arginine decarboxylase, ornithine decarboxylase, and SAM decarboxylase. This is in contrast to more constant levels in non-bud tissue or even dormant lateral buds (32). Thus the break of dormancy may also involve changes in polyamine levels, although it is uncertain whether these changes are the cause or the result of the breaking of dormancy.

Root Crops

Low temperatures along with high humidities are generally used for commercially-stored root crops, most of which are biennial plants. Although no rest period has been demonstrated, all show some degree of dormancy and regrowth and, with cold-induction, flowering. Treatment with exogenous GA_3 can cause carrot, among other species, to bolt following non-vernalizing temperatures, and endogenous GA-like activity increases during vernalization at 5°C. However, while GAs may be involved in stem elongation, bolting precedes floral differentiation and endogenous GAs are not implicated in cold-induced flowering in carrot (32).

Roots are also sites of synthesis for cytokinins, which are exported in the xylem. Cambium tissue was found to be the site for cytokinin synthesis in carrot root (14). ABA in root tissue may derive from leaves and build up during vegetative growth, and in sugar beet is preferentially catabolized to phaseic acid and dihydrophaseic acid in the root, although ABA conjugation could take place in both the taproot sink and source leaves (17).

LEAFY CROPS

Cabbage

Within the Cruciferae, white cabbage (*Brassica oleracea* L., Capitata group) is an important fresh vegetable crop that can be stored for up to six months. The traditional method involves harvesting during low field temperatures and storing with ventilation of cold winter air (common storage). This can result in quality loss, typified by loss of chlorophyll and weight loss, to the point of making it unmarketable, a condition to which ethylene may be a contributing factor. Use of technical improvements in storage such as refrigeration, ethylene removal, and controlled atmosphere (CA) has made possible longer storage times. Cabbage can be satisfactorily stored at 2°C under CA conditions (5% CO_2, 2.5% O_2), extending the storage life for several months. There is considerable diversity in keeping quality among cultivars.

The long storing cultivar "Green Winter" apparently has an inhibitor-controlled dormant period for several months after harvest, since ABA activity rises rapidly for about eight weeks after harvest and then declines. The level of IAA activity is slightly higher in heads stored in refrigerated air than in CA-stored heads, along with most GA activity. Cytokinin activity is not affected differently by CA or air conditions, being high at harvest, declining to a low point during early dormancy, and increasing when regrowth begins in the apices. Cytokinin profiles show varietal differences between cv. "Excel" and three cabbage breeding lines of different storage capabilities (32).

High levels (100 ppm) of ethylene in air at 0°C for five weeks storage of fresh cabbage result in 6-fold increases in endogenous auxin activity and 12-fold in GA activity over the air control, while ABA activity is undetectable (32). The external leaves of all heads become bleached, desiccated and abscised from the main stalk (Fig. 4). Even at 1 ppm, ethylene in air has a detrimental effect, magnifying or accelerating changes taking place over the long storage period, whether these are in degreening and leaf abscission, weight loss, sugar loss, changes in organic acid content, or increased respiration rates (30). The increased CO_2 and reduced O_2 of CA

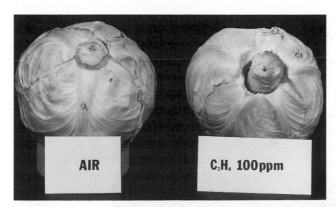

Fig. 4. Cabbage stored in ethylene atmosphere showing abscised leaves.

conditions seem to counteract the ethylene effects seen in air storage.

When decapitated trimmed stems from sampled heads of cabbage are rooted in potting mix and grown in the greenhouse, regrowth of axillary buds takes place with time, sometimes followed by flowering. The rate and type of regrowth varies depending on the state of the cabbage head, possibly reflecting the hormonal state (40). Regrowth is slow for cabbage straight from the field, and also after four weeks of storage. All plants form heads with no flowering (Fig. 5A). This could represent the dormant period with a preponderance of ABA. After four months in refrigerated air storage, regrowth on planting is fast and most plants bolt and flower, with differences between cultivars in the rate of regrowth (Fig. 5B). Cabbage stored under CA conditions show no flowering even after five months. By this time ABA levels may have dropped, as in air storage, but possibly the ratio between IAA, GA and cytokinins is not optimal for flowering. However, after a storage time of ten weeks in air,

Fig. 5. Regrowth of cabbage stem axillary buds. A) Vegetative heads; B) Flowering.

Fig. 6. Regrowth and flowering of stem axillary buds from "Bartola" cabbage heads stored for 10 weeks. A) Air; B) Controlled atmosphere.

some cultivars bolt and flower, while others remain vegetative, indicating a possible borderline hormone balance between either IAA, GA, cytokinins, or ABA and these promoters. Regrowth and flowering in cv. Bartola is speeded up by the presence of ethylene during air storage (Fig. 6A), while 5 ppm ethylene even in CA storage allows flowering at this ten week stage as well as after 5 months (Fig. 6B). Actual measurements made on these heads indicate that ABA still tends to be high at 10 weeks, and IAA levels remain high in the non-flowering CA-stored heads. However, these analyses were done on large "apical samples" which included stem, leaf bases, and axillary buds as well as apices. Axillary buds contain twice as much free ABA for instance as the rest of the "apical" tissue, and chance inclusion of more axillary buds could disproportionately slant the hormonal results (Ludford, unpublished data).

Other Leafy Vegetables

While cabbage cultivars can be stored for months, others in the *B. oleracea* L. species, e.g. Brussels sprout (Gemmifera group), cauliflower (Botrytis group) and broccoli (Italica group) can be stored from only a few days to a few weeks due to their rapid loss of edible quality. With broccoli and Brussels sprouts, this involves desiccation, yellowing, and abscission of leaves, while cauliflower also shows curd growth called "riciness".

Quality loss is accentuated by ethylene. Treatment of broccoli with AVG, the ethylene synthesis inhibitor, reduces ethylene production and respiration and retards yellowing and senescence so that it is still in saleable condition with green colour retention and compactness (40). Ethylene, known to promote loss of chlorophyll in fruit ripening, was used to promote the blanching of celery (*Apium graveolens*) as far back as 1924 (60). Pithiness of the edible celery petiole can be stimulated by flooding and nutrition-deficiency after a prolonged period, but much faster by water stress. Not too surprisingly, the latter is associated with an increase in endogenous free ABA. However, exogenous ABA application also stimulates petiole pithiness of detached celery leaves, and shortens the storage life of broccoli and Brussels sprout (32). Jasmonic acid is even more efficient than ABA in the promotion of leaf senescence (57).

In asparagus (*Asparagus officinalis*), higher concentrations of IAA are found in green than in white spears (grown under black plastic or sawdust mulch), which is consistent with the greater elongation of green spears, although white had higher fresh and dry weights in 5mm tip samples (41). However, higher levels of ABA are also found. In comparison, IAA levels are higher in light-grown seedlings of pea than in much taller etiolated dark-grown seedings (4), so that this is more a light/dark sensitivity than an elongation effect.

Applied cytokinins are often effective in prolonging the shelf life of leafy vegetables by slowing down senescence. Synthetic and natural cytokinins extend the storability of both Brussels sprout and broccoli (32). Good quality in broccoli is maintained by benzyladenine (BA) treatment and low temperatures of 2°C, reflected in sensory evaluations of the cooked broccoli (40). Application of BA as a postharvest dip delays senescence and maintains green colour and fresh appearance in many other leafy vegetables besides crucifers, including endive, escarole, spinach, parsley, green onion, celery, and asparagus. Exogenous cytokinin treatments enhance chlorophyll retention in lettuce (32), but are of little practical value since only outer leaves are affected, and these would be trimmed by the time they reach the consumer because of damage in handling. However, application for the use of BA and other synthetic cytokinins on leafy vegetables has not been approved in the U.S.

It was suggested that these opposing effects of applied BA and ABA could be due to a relationship between senescence and stomatal aperture (60), ABA enhancing closure of stomata in the light, and cytokinins maintaining opening. While this may be understandable in intact plants, harvested leafy vegetables are detached plant parts, and transpiration with the resulting wilting is a major factor in initial deterioration. The older view of cytokinins delaying senescence through maintenance of RNA and protein synthesis is more adaptable, with a resultant delay in protein and chlorophyll degradation and a retardation of the respiration rate. A microsomal enzyme fraction from fresh cauliflower heads hydroxylates applied isopentenyl adenine and isopentenyl adenosine to zeatin and zeatin riboside, again demonstrating that cytokinin biosynthesis or conversion is not limited to roots and seeds (15). Ethylene treatment of the enzyme system reduces the conversion by 28-43%. Part of the senescence-promoting activity of ethylene could thus act through preventing cytokinin synthesis.

Although most reports discuss exogenous hormone treatment, endogenous cytokinin activities decrease during accelerated aging of the outer leaves of Brussels sprouts, while GA and inhibitor activities increase (32). Lateral bud development of Brussels sprout is less inhibited than in most other species, and buds attain a large size as the plant matures. English gardeners remove the terminal buds after the axillary buds or sprouts have begun to form to increase the size of sprouts, since decapitation results in increased auxin activity in the top lateral buds creating new nutrient sinks. However, decapitation at a younger stage results in axillary shoots, and lateral buds of the younger plants contain more total GA-like activity than those of older plants which would not show shoot extension (40). The "riciness" of cauliflower or elongation of the floret peduncles may be GA-controlled, and cauliflower curd growth can be inhibited by treatment with growth inhibitors including chlormequat, an inhibitor of GA synthesis (32). Good quality sprouts would therefore appear to need high cytokinin content along with low GA, with the latter improving cauliflower quality also.

Similarly, in rosette plants such as spinach, *Spinacia oleracea*, low GA levels would be advantageous unless seed stalks are required. The rate of GA synthesis in long-day rosette plants is lower during vegetative growth under short-day conditions than it is under long-day conditions, when stem elongation and flower formation take place. Both Chinese cabbage and oilseed rape bolt in response to GA even when unvernalised, but the flowering that follows is dependent upon photoperiod (42). In spinach, photoperiod has been shown to regulate the conversion of GA_{53} to GA_{44} and of GA_{19} to GA_{20} (25), and recently, by application of the growth retardant, tetcyclacis, to regulate the rate of *ent*-kaurene accumulation (69). Seedlings of lettuce, *Lactuca sativa*, were used for one of the earliest GA bioassays, but little is known about the postharvest endogenous hormone content. Two cytokinins have been isolated from butterhead lettuce, mostly from the innermost developing leaflets (32).

Polyamines provide considerable protection against ozone injury when they are given in solution at low rates to the cut stems of 21-day tomato shoots (22). If this positive effect still takes place in intact plants of other species when applied in a more practical way, such as foliar sprays, it could be a useful protectant.

CONCLUSIONS

Plant hormones obviously play an important role in the postharvest physiology of vegetables. As is usually the case, the effects observed are due more to a hormonal balance than to the activity of any one hormone. However, the same physiological event can sometimes be initiated by a variety of physiologically relevant factors, e.g. minerals, carbohydrate, light, CO_2, temperature, water, indicating that there is a redundancy in signalling, of which growth substances are a part. Sensitivity to growth substances must also be considered as a possible controlling factor in addition to the hormone level itself, along with the characterization of receptors (61). The controversy is over the primacy of the change in concentration of a plant growth substance over the corresponding sensitivity to the rate of plant development.

References

1. Abeles, F.B., Morgan, P.W., Saltveit, M.E. (1992) Fruit ripening, abscission, and postharvest disorders. *In*: Ethylene in Plant Biology, pp. 182-221. Academic Press Inc., San Diego.
2. Ahrens, M.J., Barmore, C.R. (1988) Interactive effects of washing, film wrapping and ethylene concentration on color development in grapefruit flavedo. Sci. Hortic. 34, 275-281.
3. Baldwin, E.A. (1993) Citrus fruit. *In*: Biochemistry of Fruit Ripening, pp. 107-149, Seymour, G.B., Taylor, J.E., Tucker, G.A., eds. Chapman and Hall, London.
4. Behringer, F.J., Davies, P.J., Yang, T., Law, D.M. (1992) The role of indole-3-acetic acid in mediating changes in stem elongation of etiolated *Pisum* seedlings following exposure to light. *In*: Progress in Plant Growth Regulation, pp. 437-445, Karssen, C.M., Van Loon, L.C., Vreugdenhil, D., eds. Kluwer Academic Publishers, Dordrecht.
5. Belles, J.M., Tornero, P., Conejero, V. (1992) Pathogenesis-related proteins and polyamines in a developmental mutant of tomato, *Epinastic*. Plant Physiol. 98, 1502-1505.
6. Beruter, J., Droz, P. (1991) Studies on locating the signal for fruit abscission in the apple tree. Sci. Hortic. 46, 201-214.
7. Black, M. (1991) Involvement of ABA in the physiology of developing and mature seeds. *In*: Abscisic Acid: Physiology and Biochemistry, pp. 99-124, Davies, W.J., Jones, H.G., eds. Bios Scientific Publishers, Oxford.
8. Bouzayen, M., Hamilton, A., Picton, S., Barton, S., Grierson, D. (1992) Identification of genes for the ethylene-forming enzyme and inhibition of ethylene synthesis in transgenic plants using antisense genes. Biochem. Soc. Trans. 20, 76-79.
9. Bray, E.A. (1991) Regulation of gene expression by endogenous ABA during drought stress. *In*: Abscisic Acid: Physiology and Biochemistry, pp. 81-98, Davies, W.J., Jones, H.G., eds. Bios Scientific Publishers, Oxford.
10. Brown, J.W. (1990) Chilling effects on color and 1-aminocyclopropane--1-carboxylic acid content of chilling-sensitive and -tolerant tomato fruit. Ph.D. Thesis, Cornell University.
11. Carrington, C.M.S., Greve, L.C., Labavitch, J.M. (1993) Cell wall metabolism in ripening fruit. Plant Physiol. 103, 429-434.

12. Casas, J.L., Acosta, M., Del Rio, J.A., Sabater, F. (1990) Ethylene evolution during ripening of detached tomato fruit: its relation with polyamine metabolism. Plant Growth Regul. 9, 89-96.

13. Catala, C., Ostin, A., Chamarro, J., Sandberg, G., Crozier, A. (1992) Metabolism of indole-3-acetic acid by pericarp discs from immature and mature tomato. Plant Physiol. 100, 1457-1463.

14. Chen, C-M., Ertl, J.R., Leisner, S.M., Chang, C-C. (1985) Localization of cytokinin biosynthetic sites in pea plants and carrot roots. Plant Physiol. 78, 510-513.

15. Chen, C-M., Leisner, S.M. (1984) Modification of cytokinins by cauliflower microsomal enzymes. Plant Physiol. 75, 442-446.

16. Crooks, J.R. (1985) Chilling injury in tomato fruit cultivars with different sensitivities to low temperature. Ph.D. Thesis, Cornell University.

17. Daie, J., Campbell, W.F., Seeley, S,D. (1981) Temperature-stress-induced production of abscisic acid and dihydrophaseic acid in warm- and cool-season crops. J. Amer. Soc. Hort. Sci. 106, 11-13.

18. Deikman, J., Kline, R., Fischer, R.L.(1992) Organization of ripening and ethylene regulatory regions in a fruit-specific promoter from tomato (*Lycopersicon esculentum*). Plant Physiol. 100, 2013-2017.

19. Dilley, D.R., Wilson, I.D. (1992) Molecular biological investigations of gene expression attending fruit ripening: current status and future prospects. HortTechnology 2, 294-301.

20. Ebener, W., Fowler, T.J., Suzuki, H., Shaver, J., Tierney, M.L. (1993) Expression of DcPRP1 is linked to carrot storage root formation and is induced by wounding and auxin treatment. Plant Physiol. 101, 259-265.

21. Egea-Cortines, M., Cohen, E., Arad, S.M., Bagni, N., Mizrahi, Y. (1993) Polyamine levels in pollinated and auxin-induced fruits of tomato (*Lycopersicon esculentum*) during development. Physiol. Plant. 87, 14-20.

22. Evans, P.T., Malmberg, R.L. (1989) Do polyamines have roles in plant development? Ann. Rev. Plant Physiol. Plant Mol. Biol. 40, 235-269.

23. Facteau, T.J., Chestnut, N.E., Rowe, K.E., Payne, C. (1992) Brine quality of gibberellic acid-treated 'Napoleon' sweet cherries. HortScience 27, 118-122.

24. Fujino, D.W., Nissen, S.J., Jones, A.D., Burger, D.W., Bradford, K.J. (1988) Quantification of indole-3-acetic acid in dark-grown seedlings of the *diageotropica* and *epinastic* mutants of tomato (*Lycopersicon esculentum* Mill.). Plant Physiol. 88, 780-784.

25. Gilmour, S.J., Zeevaart, J.A.D., Schwenen, L., Graebe, J.E. (1986) Gibberellin metabolism in cell-free extracts from spinach leaves in relation to photoperiod. Plant Physiol. 82, 190-195.

26. Given, N.K. (1992) Kiwifruit. *In*: Biochemistry of Fruit Ripening, pp. 235-254, Seymour, G.B., Taylor, J.E., Tucker, G.A., eds. Chapman and Hall, London.

27. Grumet, R., Fobes, J.F., Herner, R.C. (1981) Ripening behavior of wild tomato species. Plant Physiol. 68, 1428-1432.

28. Halder-Doll, H., Bangerth, F. (1987) Inhibition of autocatalytic C_2H_4-biosynthesis by AVG applications and consequences on the physiological behaviour and quality of apple fruits in cool storage. Sci. Hortic. 33, 87-96.

29. Herrero, M. (1989) Fruit shape as a response to time of GA_3 treatment in 'Agua de Aranjuez' pear. Acta Hortic. 256, 127-132.

30. Hicks, J.R., Ludford, P.M. (1981) Effects of low ethylene levels on storage of cabbage. Acta Hortic. 116, 65-73.

31. Hobson, G., Grierson, D. (1993) Tomato. *In*: Biochemistry of Fruit Ripening, pp. 405-442, Seymour, G.B., Taylor, J.E., Tucker, G.A., eds. Chapman and Hall, London.

32. Isenberg, F.M.R., Ludford, P.M., Thomas, T.H. (1987) Hormonal alterations during the postharvest period. *In*: Postharvest Physiology of Vegetables, pp. 45-94, Weichmann, J., ed. Marcel Dekker, Inc., New York, Basel.

33. Kende, H. (1993) Ethylene biosynthesis. Annu. Rev. Plant Physiol. Plant Mol. Biol. 44, 283-307.

34. Knee, M. (1993) Pome fruits. *In*: Biochemistry of Fruit Ripening, pp. 325-346, Seymour, G.B., Taylor, J.E., Tucker, G.A., eds. Chapman and Hall, London.

35. Law, D.M., Davies, P.J., Mutschler, M.A. (1991) Polyamine-induced prolongation of storage in tomato fruits. Plant Growth Regul. 10, 283-290.

36. Lipton W.J., Wang, C.Y. (1987) Chilling exposures and ethylene treatment change the level of ACC in 'Honey Dew' melons. J. Amer. Soc. Hort. Sci. 112, 109-112.

37. Looney, N.E., Lidster, P.D. (1980) Some growth regulator effects on fruit quality, mesocarp composition, and susceptibility to postharvest surface marking of sweet cherries. J. Amer. Soc. Hort. Sci. 105, 130-134.

38. Looney, N.E., Granger, R.L., Chu, C.L., McArtney, S.J., Mander, L.N., Pharis, R.P. (1992) Influences of gibberellins A_4, A_{4+7} and A_4+iso-A_7 on apple fruit quality and tree productivity. I. Effects on fruit russet and tree yield components. J. Hortic. Sci. 67, 613-618.

39. Ludford, P.M., Hillman, L.L. (1990) Abscisic acid content in chilled tomato fruit. HortScience 25, 1265-1267.

40. Ludford, P.M., Isenberg, F.M.R. (1987) Brassica crops. *In*: Postharvest Physiology of Vegetables, pp. 497-522, Weichmann, J., ed. Marcel Dekker, Inc., New York and Basel.

41. Makus, D.J., Guinn, G. (1992) Higher levels of ABA and IAA are found in green than in white asparagus spears. HortScience 27, 1047.

42. Mandel, R.M., Rood, S.B., Pharis, R.P. (1992) Bolting and floral induction in annual and cold-requiring biennial *Brassica* spp.: effects of photoperiod and exogenous gibberellin. *In*: Progress in Plant Growth Regulation, pp. 437-445, Karssen, C.M., Van Loon, L.C., Vreugdenhil, D., eds. Kluwer Academic Publishers, Dordrecht.

43. Manning, K. (1993) Soft fruit. *In*: Biochemistry of Fruit Ripening, pp. 347-377, Seymour, G.B., Taylor, J.E., Tucker, G.A., eds. Chapman and Hall, London.

44. Marano, M.R., Carrillo, N. (1992) Constitutive transcription and stable RNA accumulation in plastids during the conversion of chloroplasts to chromoplasts in ripening tomato fruits. Plant Physiol. 100, 1103-1113.

45. McGlasson, W.B. (1978) Role of hormones in ripening and senescence. *In*: Postharvest Biology and Biotechnology, pp. 77-96, Hultin, H.O., Milner, M., eds. Food and Nutrition Press, Inc., Westport, CT.

46. Morgens, P.H., Callahan, A.M., Wright, P., Nichols, K.A. (1990) Identification of a peach gene whose expression correlates with fruit ripening and rate of ethylene accumulation during wounding. *In*: Polyamines and Ethylene: Biochemistry, Physiology, and Interactions, pp. 319-320, Flores, H.E., Arteca, R.N., Shannon, J.C., eds. Amer. Soc. Plant Physiol., Rockville, MD.

47. Nichols, S.E., Laties, G.G. (1985) Differential control of ethylene-induced gene expression and respiration in carrot roots. Plant Physiol. 77, 753-757.

48. Nowacki, J., Saniewski, M., Lange, E. (1990) The inhibitory effect of methyl jasmonate on ethylene-forming enzyme activity in apple cultivar Jonathan. Fruit Sci. Rep. (Skierniewice) 17, 179-186.

49. Ozga, J.A., Reinecke, D.M., Brenner, M.L. (1993) Quantification of 4-Cl-IAA and IAA in 6 DAA pea seeds and pericarp. Plant Physiol. 102, S7.

50. Parups, E.V. (1984) Effects of ethylene, polyamines and membrane stabilizing compounds on plant cell membrane permeability. Phyton 44, 9-16.

51. Priem, B., Gross, K.C. (1992) Mannosyl- and xylosyl-containing glycans promote tomato (*Lycopersicon esculentum* Mill.) fruit ripening. Plant Physiol. 98, 399-401.

52. Purvis, A.C., Kawada, K., Grierson, W. (1979) Relationship between midseason resistance to chilling injury and reducing sugar level in grapefruit peel. HortScience 14, 227-229.

53.   Rastogi, R., Dulson, J., Rothstein, S.J. (1993) Cloning of tomato (*Lycopersicon esculentum* Mill.) arginine decarboxylase gene and its expression during fruit ripening. Plant Physiol. 103, 829-834.
54.   Rhodes, M.J.C. (1980) The maturation and ripening of fruits. *In*: Senescence in Plants, pp 157-205, Thimann, K.V., ed. CRC Press, Inc., Boca Raton, FL.
55.   Saftner, R.A., Baldi, B.G. (1990) Polyamine levels and tomato fruit development: possible interaction with ethylene. Plant Physiol. 92, 547-550.
56.   Salama, A.M., Hicks, J.R. (1987) Respiration and fresh weight of onion bulbs as affected by storage temperature, humidity and maleic hydrazide. Trop. Sci. 27, 233-238.
57.   Sembdner, G., Parthier, B. (1993) The biochemistry and the physiological and molecular actions of jasmonates. Annu. Rev. Plant Physiol. Plant Mol. Biol. 44, 569-589.
58.   Seymour, G.B., Tucker, G.A. (1993) Avocado. *In*: Biochemistry of Fruit Ripening, pp. 53-81, Seymour, G.B., Taylor, J.E., Tucker, G.A., eds. Chapman and Hall, London.
59.   Theologis, A. (1992) One rotten apple spoils the whole bushel: the role of ethylene in fruit ripening. Cell 70, 181-184.
60.   Thimann, K.V. (1980) The senescence of leaves. *In*: Senescence in Plants, pp. 85-115, Thimann, K.V., ed. CRC Press, Inc., Boca Raton, FL.
61.   Trewavas, A. (1992) Growth substances in context: a decade of sensitivity. Biochem. Soc. Trans. 20, 102-108.
62.   Tucker, G.A., Grierson, D. (1987) Fruit ripening. *In*: The Biochemistry of Plants: a Comprehensive Treatise, Vol. 12 Physiology of Metabolism, pp. 265-318, Davies, D.D., ed. Academic Press, New York.
63.   Van den Berg, J.H., Vreugdenhil, D., Ludford, P.M., Hillman,L.L., Ewing, E.E. (1991) Changes in starch, sugar, and abscisic acid contents associated with second growth in tubers of potato (*Solanum tuberosum* L.) one-leaf cuttings. J. Plant Physiol. 139, 86-89.
64.   Varga, A., Bruinsma, J. (1974) The growth and ripening of tomato fruit at different levels of endogenous cytokinins. J. Hort. Sci. 49, 135-142.
65.   Ververidis, P., John, P. (1991) Complete recovery *in vitro* of ethylene forming enzyme activity. Phytochemistry 30, 725-727.
66.   Wang, C.Y., Kramer, G.F. (1990) Effect of polyamine treatment on ethylene production of apples. *In*: Polyamines and Ethylene: Biochemistry, Physiology, and Interactions, pp. 411-413, Flores, H.E., Arteca, R.N., Shannon, J.C., eds. Amer. Soc. Plant Physiol., Rockville, MD.
67.   Weiss, D., van Blokland, R., Kooter, J.M., Mol, J.M.N., van Tunen, A.J. (1992) Gibberellic acid regulates chalcone synthase gene transcription in the corolla of *Petunia hybrida*. Plant Physiol. 98, 191-197.
68.   Yuan, R., Huang, H. (1988) Litchi fruit abscission: its patterns, effect of shading and relation to endogenous abscisic acid. Sci. Hortic. 36, 281-292.
69.   Zeevaart, J.A.D., Gage, D.A. (1993) *ent*-kaurene biosynthesis is enhanced by long photoperiods in the long-day plants *Spinacia oleracea* L. and *Agrostemma githago* L. Plant Physiol. 101, 25-29.
70.   Zobel, R.W. (1973) Some physiological characteristics of the ethylene-requiring tomato mutant *diageotropica*. Plant Physiol. 52, 385-389.

G13. Natural and Synthetic Growth Regulators and Their Use in Horticultural and Agronomic Crops

Thomas Gianfagna
Plant Science Department, Rutgers University, New Brunswick, NJ 08903, USA

INTRODUCTION

Plant growth regulators have been an important component in agricultural production even prior to the identification of plant hormones. As early as the turn of the century, fires were lit adjacent to fields in order to synchronize flowering in mango and pineapple (50). Gas powered generators were used as the source for the post-harvest heat treatment of lemons that stimulated ripening and degreening (12). The ethylene generated as a result of incomplete combustion, rather than higher temperatures, stimulated flowering and ripening in both cases, although this fact was unknown at the time. Plant growth regulators are now used on over one million hectares worldwide on a diversity of crops each year (61). Most of these applications are, however, confined to high-value horticultural crops rather than field crops, although there are several significant exceptions. Chlormequat chloride (2-chloroethyltrimethylammonium chloride) is used to reduce lodging in wheat. This application is generally limited to Europe, where under conditions of high fertility, lodging in small grains contributes significantly to yield reduction. Glyphosine (N,N,bis(phosphonomethyl)glycine) is used in sugarcane to increase the sucrose content of the cane. This compound acts by diverting carbohydrate into sucrose storage rather than fiber production.

In cotton, defoliants such as DEF (S,S,S-tributlyphosphorotrithioate) are used to remove green leaves that can then fall free of the lint from the open cotton boll. These compounds are, therefore, useful aids in the mechanical harvesting of the crop. Application is limited, however, to those parts of the world where mechanical harvesting is practiced, an estimated one-sixth of the total worldwide acreage.

There have also been several promising results with plant growth regulators for soybean and corn. Triiodobenzoic acid (TIBA) was registered briefly to increase yield in soybean. TIBA reduced plant height and petiole length, and stimulated branching and fruit set. In corn, dinoseb, used primarily as a pre-emergence herbicide, has been found to increase grain yield by 10-15%. Dinoseb treatment stimulated earlier tasseling and improved ear filling. Nevertheless, both compounds have failed to provide consistent yield

P. J. Davies (ed.), Plant Hormones, 751–773.
© 1995 *Kluwer Academic Publishers. Printed in the Netherlands.*

increases, and there was considerable genotype variation in response to treatment. Significant opportunities certainly still exist for the development of plant growth regulators to increase yield in the major crops.

Most of the current uses for plant growth regulators in the high-value horticultural crops are not, however, for compounds that increase crop yield directly, either by increasing the total biological yield or the harvest index. Rather, compounds that provide economic benefit by enhancing crop quality or aid in more efficient crop management are more common. For example, gibberellic acid is used to reduce the incidence of physiological rind disorders in citrus, and daminozide (N,N,dimethylaminosuccinamic acid) application to apple stimulates color development of the fruit. Both treatments increase the value of the crop but not necessarily the yield.

Plant growth regulators that aid in crop management fall into several categories. First, mechanical harvesting of crops can reduce substantially the cost of production. Compounds that reduce the fruit removal force have in some cases, such as sweet cherry production, allowed the development and use of mechanical harvesting equipment, which was not particularly effective alone, because the force required to remove the crop damaged the trees. Second, plant growth regulators are used for manipulation of the harvest date in a wide range of crops. Ethephon (2-chloroethylphosphonic acid) is used to accelerate and concentrate ripening of tomato prior to a single mechanical harvest. Gibberellic acid is used to extend the lemon harvest season by preventing senescence of the rind. This allows a greater percentage of the crop to be sold when demand for fresh lemons is high. Conversely, gibberellic acid accelerates flower bud development in artichoke, significantly advancing maturity and increasing the value of the crop. Third, plant growth regulators can be used as direct replacements for hand labor in crop production practices other than harvesting. In apple, naphthalene acetic acid (NAA) is used to reduce excessive fruit set that would otherwise result in many small fruit, and for some varieties, inhibit flower bud production for the following season's crop. Maleic hydrazide (1,2-dihydro-3,6-pyradazinedione) is used in tobacco to control the growth of lateral buds that reduce leaf quality. In both cases, fruit thinning in apple and sucker control in tobacco, a labor intensive cultural practice can be accomplished efficiently with plant growth regulators.

In most of the cases described above, plant growth regulators are useful because they can in some way modify plant development. This may occur by interfering with the biosynthesis, metabolism, or translocation of plant hormones, or the plant growth regulator may replace or supplement the plant hormones when their endogenous levels are below that needed to change the course of plant development. The plant growth retardants such as chlormequat chloride and ancymidol (α-cyclopropyl-α-(4-methoxyphenyl)-5-pyrimidinemethanol), which are used to control internode elongation in lily, poinsettia, and other floricultural crops, act by inhibiting gibberellin biosynthesis. Ethephon, on the other hand, accelerates ripening and

752

abscission by releasing ethylene at a time when endogenous levels of this hormone are low.

Other compounds may more directly affect plant metabolism, circumventing the hormonal control system. For example, glyphosine is used to increase the sucrose content of sugar cane, and lateral bud development in tobacco can be prevented by inhibiting cell division with maleic hydrazide.

While the biochemical mechanism of action of many plant growth regulators is not well understood, it is still possible to classify compounds on the basis of their similarity in action to the naturally occurring plant hormones. This approach will be taken in the following description of the chemistry, physiology, and use of plant growth regulators in agriculture.

AUXINS

The auxin-type plant growth regulators comprise some of the oldest compounds used in agriculture. Shortly after indole acetic acid (IAA) was identified, it was synthesized and became readily available. IAA was not found in itself to be useful in agriculture because it is rapidly broken down to inactive products by light and microorganisms. Nevertheless, a number of synthetic compounds were found to act similarly to IAA in the auxin bioassay tests. Indolebutyric acid (IBA) and NAA were found to increase root development in the propagation of stem cuttings. 2,4-dichlorophenoxyacetic acid (2,4-D) stimulates excessive, uncontrolled growth in broadleaf plants for which it is used as a herbicide. NAA and naphthalene acetamide (NAAm) are used to reduce the number of fruit that have set in apple, whereas 4-chlorophenoxyacetic acid (4-CPA) is used to increase fruit set in tomato. The auxins 2,4,5-trichlorophenoxypropionic acid (2,4,5-TP) and the dichlorophenoxy analog (2,4-DP) are used to prevent abscission of mature fruit in apple.

Propagation

Rooting of stem cuttings was one of the first uses of auxins. The most common compound used is IBA, which has only weak auxin activity, but is relatively stable and insensitive to the auxin degrading enzyme systems. It is also not readily translocated. Other compounds such as NAA and 2,4-D will also promote root development, however, these compounds are more easily translocated to other parts of the stem cutting where they may have toxic effects (29). The auxins stimulate root development by inducing root initials that differentiate from cells of the young secondary phloem, cambium, and pith tissue.

Stimulation of fruit set

One of the first recorded effects of auxins was the stimulation of fruit set in unpollinated ovaries of solanaceous plants. Gustafson (25) demonstrated that pollen was a rich source of auxin, and that in some species pollination alone was all that was required for fruit set to occur. In tomato, chemical stimulation of fruit set is all that is needed for fruit growth to take place as well. In addition, compounds that block the transport of auxin from the ovary to the pedicel of the flower also stimulate fruit set (4). It seems likely, therefore, that given environmental conditions somewhat inhibitory to fruit set, application of auxin to flowers could promote this process. In California, the early spring crop of tomatoes is treated with 4-CPA at 25-50 ppm to stimulate fruit set at a time of the year when cool night temperatures that inhibit fruit set in tomato are likely. This treatment results in an increase in yield and earlier harvest.

Chemical thinning

Removal of excessive numbers of young fruit from apple and pear trees is a common orchard management practice, although this results in drastic reductions in the total biological yield. There are two main reasons for removing as much as 80% of the flowers: first, to increase the total marketable yield by increasing the size of the remaining fruit, and second, to reduce the phenomenon of biennial bearing in order to maintain production levels from year to year. The effect of fruit thinning on fruit size is probably related to the leaf/fruit ratio. As this ratio is reduced below 30/1, fruit size is reduced as well (38). The time in which fruit thinning is done is as important to fruit size as the amount of fruit thinning. In order to improve flower bud production and fruit size in apple, thinning should take place within 30 days from full bloom. In apple, the period of cell division in the fruit is brief, ending approximately 20 days after full bloom (67). Removing excess fruit during this period can stimulate cell division within the remaining fruit. This early period is also of critical importance for floral initiation, the time when next year's crop will be partially determined. The two auxin-type compounds used in chemical thinning of apple and pear are NAA and NAAm. NAA is applied at 2-5 ppm, 7-20 days after full bloom, whereas NAAm is effective during the same time period, but higher rates (17-34 ppm) are used (68).

Several mechanisms have been proposed to explain the effect of auxins on fruit abscission. Observations that auxin application reduced the early drop of flowers and fruit soon after the bloom period led to the suggestion that auxin first stimulated fruit set, and then due to increased competition between fruits for nutrients and assimilates, a greater percentage of fruit abscised during the June-drop period (60). This early increase in fruit set, however, is not always observed. Luckwill (36) proposed that fruit abscission occurs because NAA and NAAm induce embryo abortion. Without seed growth, fruit senescence and abscission take place prematurely. While the number of viable seeds is often correlated with fruit abscission, this is not always the case, suggesting that embryo abortion may not be the primary factor resulting in abscission. NAA does cause increased ethylene evolution from apple fruit within one day after application (66) (Fig. 1). Ethylene is known to reduce auxin transport from the leaf blade to the petiole (3), and to induce the synthesis of enzymes that degrade the abscission zone (2). Perhaps the induction of fruit abscission by NAA is mediated by ethylene, which stimulates abscission of fruit in a similar manner to its effect on the abscission of leaves.

While not strictly speaking an auxin, the compound carbaryl (1-naphthyl-N-methyl-1-carbamate) is used commonly for chemical thinning in apple and is effective over a longer period of time than NAA. Carbaryl seems to induce abscission by interfering with assimilate or hormonal transport between the developing seeds and the fruit.

Prevention of fruit drop

Frequently, the mature fruit of apple, pear, lemon, and grapefruit will abscise prior to the time of commercial harvest. This obviously reduces the potential crop yield, and may result in the tendency to begin harvesting the crop earlier than is desirable, resulting in lower quality fruit. Under natural conditions, there seems to be an inverse relationship between auxin content of the fruit, and the tendency toward abscission (37). The role of auxin in abscission is complicated. Clearly, application of auxin soon after fruit set results in an acceleration of abscission, however, when auxins such an 2,4,5-TP, NAA, and 2,4-D are applied during the mid-stages of fruit growth, abscission is delayed or prevented (56). In addition, auxin application may decrease the response of fruit abscission zones to exogenously applied ethylene. NAA and 2,4,5-TP are used at 10-20 ppm just prior to the beginning of fruit drop in apple. Repeat applications may be necessary with NAA; 2,4,5-TP prevents fruit drop for a longer period. In citrus, 2,4-D at 25 ppm prevents premature fruit drop and allows an extension of the harvest season into the summer (59).

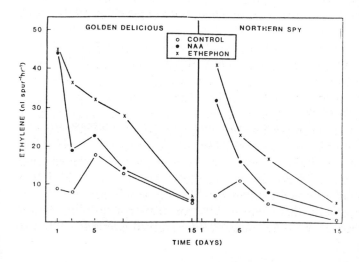

Fig. 1. Ethylene evolution from fruit spurs in response to post-bloom treatments of 'Golden Delicious' and 'Northern Spy' apple trees with NAA and ethephon. From (66).

Herbicidal action

2,4-D and picloram (4-amino-3,5,6-trichloropicolinic acid) are two auxin-type herbicides that at low concentration bring about growth responses in plants similar to IAA. At higher concentrations they are herbicidal. 2,4-D is commonly used to control broadleaf weeds in grasses, and picloram is used for vegetation control on non-crop land because of its high activity and soil persistence. Both compounds cause epinastic bending in leaves, a cessation in growth in length, and increased radial expansion. After several days tumors may form, followed by a softening and collapse of the tissue. Epinastic bending and stem swelling are characteristic of ethylene effects on plants, and auxin induced ethylene biosynthesis may partially account for the effect of these compounds on plant growth. Treatment with inhibitors of ethylene synthesis or action in the presence of 2,4-D, however, do not reverse the herbicidal effects of the auxins (1). Auxin herbicides cause an increase in DNA, RNA, and protein levels in treated tissue. The greatest effect, however, is on RNA levels (7). Specific mRNAs are induced by auxin treatment, and ethylene apparently plays no role in the expression of these mRNAs (64). In addition, in resistant plants the level of RNAase activity is higher than in sensitive plants. One aspect of the herbicidal activity of the synthetic auxins seems clearly to be a disturbance in RNA metabolism of the cell.

GIBBERELLINS

Despite considerable enthusiasm for the potential uses for gibberellic acid in agriculture that existed when this compound was rediscovered by US and British scientists in the 1950's, major GA use remains limited to fruit crops, the malting of barley, and extension of sugarcane growth in certain production regions.

There are about 90 gibberellins found in both higher plants and the *Gibberella* fungus, although only two commercial products are available, GA_3 and a mixture of GA_4 and GA_7. Both are produced by fermentation cultures of the fungus. A formulation of $GA_{4/7}$ and benzyladenine is also available that is being used to induce apple fruit elongation, and to increase the extent of lateral branching in young trees.

Increasing fruit size in grape

GA is used extensively on seedless grape varieties to increase the size and quality of the fruit (8). Pre-bloom sprays of 20 ppm induce the rachis of the fruit cluster to elongate. This creates looser clusters that are less susceptible to disease during the growing season. GA also reduces pollen viability, as well as decreasing ovule fertility in grape. Application of GA at bloom, therefore, results in a decrease in fruit set, which reduces the number of berries per cluster, but increases the weight and length of the remaining fruit. An additional application of GA during the late bloom to early fruit set period will further increase berry size. It has been suggested that this later application of GA increases the mobilization of carbohydrates to the developing fruit (54) (Table 1).

Seeded varieties generally do not respond favorably to GA treatment. However, in Japan, the seeded variety 'Delaware', is cluster-dipped in 100 ppm GA to induce parthenocarpic fruit development and increase berry size.

Table 1. Effects of GA_3 applied at bloom and/or fruit set on fruit growth in 'Thompson Seedless' grapes. From (8)

Treatment	Berry Number	Cluster compactness	Berry Wt. (g)	Berry Length (cm)
Control	91	3.9	2.7	1.8
GA at bloom	61	2.8	3.3	2.3
GA at fruit set	90	4.1	4.0	2.3
GA at bloom+fruit set	64	3.0	4.8	2.6

* Cluster compactness increases with higher values.

Stimulating fruit set

Not all crops respond as positively as the tomato to auxin-induced fruit set. However, a number of deciduous fruit tree species such as apple and pear, as well as some citrus species, can be induced to set fruit with GA, or a combination of GA and auxin. In Europe, poor fruit set in apple and pear can drastically reduce crop yield. Consistently unfavorable weather during the pollination period has led to the development of a hormone mixture to induce parthenocarpic fruit set in apple (33). In pear, spring frost injury to the ovules or style can prevent fertilization and the stimulation for fruit set. Application of GA_3 at 15-30 ppm can induce parthenocarpic fruit and salvage what would have been a lost crop.

In citrus, fruit set of mandarin oranges is often light. Application of GA during full bloom can increase fruit set. Girdling branches will also increase fruit set as well as cause an increase in endogenous GA-like substances in the region above the branch girdle (65).

Effects on fruit ripening

GAs are used to delay fruit ripening in lemon in order to increase the availability of fruit during the months of May-August when demand is high, but production is low. GA is applied in November or December in order to delay the harvest date, and increase storage life of the fruit (10) (Table 2).

Delaying harvest is also important for a number of other citrus species including 'Navel' oranges and grapefruit. While fruit abscission can be controlled by 2,4-D, after maturity is attained, changes associated with the senescence of the rind will decrease the quality of the fruit. This reduces the usefulness of holding the fruit on the tree longer in order to allow harvest to

Table 2. Influence of gibberellic acid on lemon tree harvest patterns one year after treatment. (From *The Citrus Industry* vol.2)

GA Treatment* (ppm)	Harvest Dates (1960-61)					
	March	May	June	August	October	January
	percentage of crop harvested					
0	9.2	46.5	23.7	12.3	5.8	2.5
5	10.0	40.6	24.1	16.9	5.8	2.6
10	10.1	39.4	23.7	17.0	7.2	2.6
20	11.2	37.7	22.7	17.9	8.0	2.5
40	10.3	36.2	22.8	18.6	9.5	2.6

*GA treatment applied on Nov. 17, 1959.

take place during the period of high consumer demand. GA application will reduce the occurrence of physiological rind disorders such as water spot, creasing, rind staining, and softening by delaying senescence of the rind tissue (9). GA_{4+7} is registered for use on 'Golden Delicious' apple to reduce russeting, a physiological disorder that results from abnormal cell division in the epidermal layer of the fruit.

Increasing yield in sugarcane

Sugarcane growth is very sensitive to the reductions in average daily temperature normally experienced during the winter months in many cane producing regions of the world, especially Hawaii. GA application is used to overcome the reduced growth of the 3-5 internodes undergoing expansion during the cooler winter season. GA treatment has resulted in an increase in fresh weight of harvested cane of 10.9 ton/ha, and has increased sucrose yield by 1.1 ton/ha or 2.8% (40).

Malting of barley

GA is used to increase the yield of barley malt and to decrease the time required for this process to occur. Embryo growth and yield of malt extract are competitive processes, by increasing the rate of malting relative to embryo growth, a greater yield of malt extract occurs (45). Application of GA to germinating barley supplements the endogenous content of this hormone and accelerates the production and release of hydrolytic enzymes that degrade the storage proteins and carbohydrates of the endosperm into the sugars and amino acids that comprise the malt extract.

Controlling flower bud production

Spring application of GA is used extensively to accelerate flower bud production in artichoke to allow earlier harvesting dates. If treatment is delayed to coincide with the appearance of flower buds, increases in head size and number have also been reported (55).

Overcoming environmental constraints on growth

GA is used to break dormancy in plants that have not received an adequate chilling period for the resumption of growth to occur. For rhubarb crowns transplanted in the fall for forcing, GA application can substitute for the cold period normally required for bud development and subsequent petiole elongation (62). Potato tuber dormancy can also be broken by application of GA (49). This treatment is of value in the identification and screening of virus infected tubers. In warm climates, where it is possible to plant two crops in a single year, GA treatment can break dormancy of the seed tubers from the first crop in time for a second planting.

In celery production, GA is used to increase petiole elongation under cool weather conditions, where growth is reduced. GA is also being used as a seed treatment in rice to stimulate germination and initial elongation of semi-dwarf cultivars. This allows deeper planting, which improves germination and stand establishment.

Uses in plant breeding

GA can be used to induce precocious cone production in conifers. This may be an especially important aid to genetic improvement in silviculture. Douglas fir for example normally requires 20 years before seed production will occur, with $GA_{4/7}$, 6-year-old trees can be induced to produce seed (47).

GA has also been used to control flower sex expression in cucumbers and squash. GA application tends to promote maleness in these plants. When gynoecious cucumbers are treated with GA, staminate flowers are produced for breeding purposes.

Bolting and seed stalk formation are promoted by GA in many normally biennial vegetables. This facilitates hybrid seed production for commercial purpose as well as accelerating vegetable variety improvement.

ETHYLENE-RELEASING AGENTS

While the biological effects of ethylene on plant growth have been documented for some time, little practical use of ethylene in agricultural was possible due to its gaseous nature. In the early 1970s experimental formulations of compounds became available that decompose on or within a plant to release ethylene. One of the first of these compounds, ethephon, (2-chloroethylphosphonic acid) is stable at pH values of 4 or less, but at higher pH values, the compound decomposes to produce ethylene, chloride and phosphate ions. Since the cytoplasmic pH is greater than 4, once ethephon is absorbed, cleavage to ethylene inside the cell begins. Two other compounds, etacelasil (2-chloroethyl-tris-ethoxymethoxy silane) and 2-chloroethyl-bis-phenylmethoxy silane, also decompose to ethylene, but much more rapidly than ethephon, and are less sensitive to changes in pH.

Increasing latex flow in Hevea

The amount of rubber produced in the form of coagulated latex is a function of the duration of latex flow from the tapping cut that is made in the tree bark. Ethephon is applied to a region near the tapping cut and causes latex flow to increase in duration, resulting in an increase in the volume of latex collected. Rubber yield increases of 50-100% are common. The mechanism for increased flow of latex by ethephon is not well understood. It is believed that lutoids, non-rubber containing bodies within the latex, are disrupted by tapping and cause coagulation or plugging of the latex vessels, as a result of changes in osmotic potential, or shear forces imposed by the high flow rate through the narrow pores of the vessel. Ethephon may stabilize the lutoids making them less susceptible to disruption. Alternatively, it has been proposed that ethephon treatment leads to an increase in cell wall thickening of the vessels making the walls less likely to contract during tapping, and therefore fewer lutoids would be disrupted, all of which will increase latex flow (43).

Promoting abscission

The use of mechanical harvesting devices in cherry production had been limited because the force required to remove the fruit at the time of fruit maturity resulted in damage to the trees. Ethephon may be applied approximately 10 days before anticipated harvest to reduce the fruit removal force to allow mechanical harvesting of the crop without tree injury (5) (Fig. 2).

Walnuts are also harvested mechanically after treatment with ethephon. The edible kernel of the walnut reaches maturity some 3-4 weeks before harvest, due to the time required for hull dehiscence. Ethephon treatment accelerates this process. The quality of the harvested nuts is also increased because they do not remain on the tree for long periods of time after maturation and, therefore, avoid decomposition due to heat and disease (39).

Olive fruit also have a high fruit removal force at maturity. In addition, the fruit is attached to long willowy branches that do not lend themselves to

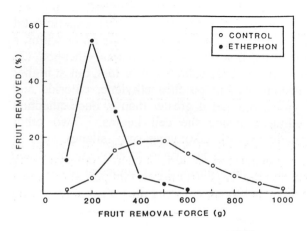

Fig. 2. The effect of ethephon on the fruit removal force in sour cherry 6 days after application. From (5).

mechanical shaking. Ethephon causes fruit abscission, but also excessive leaf abscission reducing flowering the following spring. Etacelasil can be used, however, because it reduces the attachment force without defoliation. This compound releases ethylene at a much faster rate than does ethephon. There may be a requirement for elevated ethylene levels of longer duration to induce leaf abscission than for fruit abscission, making the silyl compounds more useful as fruit abscission agents than ethephon (26).

Ethylene-releasing agents are also being used to remove young fruit from apple and peach trees that have set a potentially excessive fruit crop. In peach, 2-chloroethylmethyl-bis-phenylmethoxy silane has provided acceptable fruit abscission without defoliation in many areas of the southeastern US (14).

Promoting fruit ripening

The ripening process in mature fruit can be accelerated by ethephon application. Presumably the fruit are sensitive to ethylene at this stage of development, but have not produced enough endogenous ethylene to stimulate the ripening process. In apple, ethephon can be used to accelerate fruit softening and advance fruit color production by several weeks, although an additional application of compounds that delay abscission must be made in conjunction with ethephon (17) (Table 3).

In tomato, ethephon is used to accelerate ripening and concentrate maturity of the fruit for mechanical harvesting. Ethephon stimulates the production of lycopene by fruits and therefore can increase total yield of ripe fruit in the production of processing tomatoes, since this crop is harvested at one time only (13).

In grape, ethephon has been found to promote color development and decrease total fruit acidity. In some of the cooler grape growing regions acidity is often excessive for optimum wine quality. Ethephon treatment may also be useful when natural fruit color development is poor.

Table 3. Effects of ethephon, daminozide and NAA on abscission, firmness, and color of 'McIntosh' apples. From (15)

Treatment	Drop (%)	Firmness (kg)	Red color (%)
Control	29	6.6	51
Daminozide	2	7.4	56
Daminozide+ethephon	77	7.0	67
Daminozide + ethephon + NAA	5	7.0	90

Delaying flowering in fruit crops

Application of ethephon in the fall of the year prior to the spring flowering period delays bud expansion and anthesis in cherry and peach (48, 23). Ethephon appears to increase the length of the dormant period of flower buds (15), which results in a delay in bloom, reducing the potential for spring

ETHEPHON
100 PPM
15 OCTOBER 1988

CONTROL
CRESTHAVEN
CREAM RIDGE. N.J.
TWIGS CUT 31 MARCH 1988
TODAYS DATE 4 APRIL 1988

Fig. 3. Effect of ethephon on cold hardiness and bloom date in peach. Note that the ethephon-treated twigs contain more live flower buds and that they are at an earlier stage of development.

frost damage to flowers. In some cases, ethephon also increases mid-winter flower bud cold hardiness by maintaining high levels of the cryoprotectants sorbitol and sucrose in the flower pistil (16) (Fig. 3).

Promoting leaf senescence in tobacco
Ethephon is used to promote leaf yellowing in flue-cured tobacco. Ethephon increases the number of leaves that may be harvested at one time, and decreases the curing time for the leaf (58). This treatment is especially useful in the cooler tobacco growing regions where the uppermost leaves, which are the last to be harvested, may be damaged by frost.

GROWTH RETARDANTS

The growth retardants are a diverse group of synthetic compounds that reduce stem elongation and generally increase the green color of leaves. These compounds inhibit cell division in the subapical meristem of the shoot, but generally have little effect on the production of leaves or on root growth. The physiological effects of the growth retardants can be reversed by application of GA, but generally no other compounds are effective. Incorporation of growth retardants into the media of cultures of *G. fujikuroi* generally inhibit GA production. In higher plants, the activity of the enzymes involved in kaurene synthesis and oxidation are inhibited by growth retardants. Three general classes of compounds have emerged. Compounds such as AMO1618 and phosphon D inhibit the enzymes involved in kaurene synthesis, whereas ancymidol and the triazole analogs inhibit the kaurene oxidation sequence of reactions. The cyclohexanetriones such as prohexadione-Ca (BX-112) inhibit later stages of GA oxidation (41). In all three cases the level of active gibberellin is reduced. A number of the compounds that inhibit GA biosynthesis also inhibit sterol production in plants. There is, however, little evidence that sterols are involved in stem elongation, and it would appear, therefore, that the primary effect of the growth retardant on internode length is due to an inhibition in GA biosynthesis.

Controlling stem growth in greenhouse crops

The application of growth retardants to potted plants results in shorter more rigid stems and darker green foliage, characteristics that increase the value of the crop. In chrysanthemums, daminozide is effective as a foliar spray and ancymidol may be used as both a foliar spray or a soil drench (34). Ancymidol treatment may, however, result in a delay in flowering.

In poinsettia, chlormequat chloride is used extensively for height control since it is less expensive than ancymidol. In Easter lily, ancymidol is used because it is the most effective compound for reducing stem height in this plant (Fig. 4). Paclobutrazol (1-(4 chlorophenyl)4,4-dimethyl-2-(1,2,4-triazol-1-yl)pentan-3-ol) and the triazole fungicide triademefon will also control stem height, but higher concentrations are required in comparison to ancymidol (70). Another triazole, uniconazole (Sumagic), is also registered for use on

Fig. 4. Effect of ancymidol concentration on stem height in Easter Lily. Left to right: no treatment; 0.25 mg; 0.5 mg; 1 mg as a soil drench; 50 ppm; 100 ppm as a foliar spray. (Photo courtesy of J.G. Seeley)

Fig. 5. Effect of uniconazole (Sumagic) and paclobutrazole (Bonzi) on stem height in Poinsettia. (Photo courtesy of G.J. Wulster).

poinsettias. This compound is less persistent than paclobutrazol, but provides good height control in a variety of herbaceous and woody ornamental plants (Fig. 5).

Controlling rank growth in cotton

Under certain conditions of high fertility and favorable environmental conditions excessive vegetative growth of cotton results. Mepiquat chloride (1,1,dimethylpiperidinium chloride) applied at the time of flowering can reduce growth by 20-30% (28). Early yield of cotton is often increased by this treatment presumably due to greater light penetration into the canopy, thus allowing fruit set to take place in flowers produced on the lower nodes of the plant. Reduced vegetative growth also allows greater coverage of insecticides, fungicides, and defoliants, the latter increasing the efficiency of mechanical harvesting.

Lodging control in cereals

Stem lodging is one of the most serious problems in wheat, when this crop is grown under the conditions of high fertility in Europe. The ability to use nitrogen to increase yield is limited by its adverse effect on stem growth. Chlormequat chloride can be used to reduce stem height and increase stem diameter. Yield is increased as a result of reduced stem lodging (31) (Fig. 6). In addition, in some years when lodging is not a problem, yield may still be increased because the growth retardant treatment results in a stimulation of tillering. Other cereals do not respond as well to chlormequat chloride as wheat. However, lodging control has been obtained in barley and rye with a combination of mepiquat chloride and ethephon.

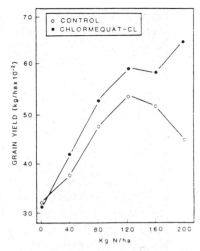

Reducing growth of turfgrass

Chemical control of grass growth especially on sites such as highway dividers, near airfields, on steep slopes that are difficult or dangerous to mow, can be economically attractive alternative (18). Several compounds such as chlorflurenol (methyl-2-chloro-9-hydroxyfluorene-9-carboxylate), mefluidide (N-(2,4-dimethyl-5(trifluoromethyl)sulfonyl)amino)-phenylacetamide), and paclobutrazol have been registered for use as plant growth regulators for grass control. Chlorflurenol

Fig. 6. Effect of chlormequat chloride and nitrogen fertilization on yield of wheat. (Euphytica 1: 215-218).

acts by inhibiting cell division in the shoot meristem, whereas mefluidide inhibits cell elongation. Both compounds have been found to reduce the frequency of mowings required over the course of the growing season. Paclobutrazol reduces mowing frequency to an even greater extent than the other compounds; however, it does not effectively suppress seedhead formation. Combinations of paclobutrazol with either maleic hydrazide, chlorflurenol or mefluidide should provide adequate seedhead suppression and persistence throughout the growing season for almost complete control of grass growth.

Increasing fruit set in grape

Application of chlormequat chloride to vinifera grapes before bloom increases fruit set of seeded berries (11). Cluster fresh weight is increased as a result of treatment. Daminozide is more effective than chlormequat chloride in increasing fruit set of the labrusca varieties (63). In addition to increasing cluster yield, vine growth is reduced by growth retardant treatment. It is not clear whether the increase in fruit set by the growth retardant is due to a direct effect on this process by decreasing GA levels (GA is used for berry thinning) or an indirect effect resulting from decreased vegetative growth. Exceedingly vigorous shoot growth is often associated with poor fruit setting in the field. Moreover, if shoot tips are removed, fruit set in grape can be increased, and the growth retardants are not capable of further increasing fruit set in detopped plants.

Advancing fruit color development

Daminozide may be used to advance anthocyanin production in the fruit skin and flesh of sweet cherry (6). The rate of color development is increased as well as the total amount of pigment synthesized. Other processes associated with fruit ripening such as fruit softening are not affected by daminozide treatment. Daminozide will also increase anthocyanin synthesis in apple, as well as reduce fruit softening in cold storage and pre-harvest fruit drop (17). Physiological disorders that develop at harvest or in storage have also been reported to be less severe after a mid-summer application of daminozide. The mechanism by which daminozide enhances color development in fruit is not clear. In apple, daminozide will inhibit ethylene production by blocking the conversion of methionine to aminocyclopropane-1-carboxylic acid (24), and delay the appearance of the respiratory climacteric (35). This will permit a delay in the harvest date and perhaps allow anthocyanin production to continue for a longer time before harvest. In some cases, however, daminozide will not only accelerate color production, but also stimulate the production of greater amounts of anthocyanin in the apple skin or the flesh of cherry, therefore, suggesting a more direct effect of the compound on pigment synthesis. Faust (19) has shown that anthocyanin production in apple is associated with increasing activity of the pentose phosphate pathway in the catabolism of carbohydrate, whereas See and Foy (53) found that daminozide inhibits succinate dehydrogenase activity in isolated mitochondria. Perhaps by inhibiting Krebs cycle activity, greater carbon flow occurs in the pentose pathway, which forms the essential precursors for anthocyanin. In isolated apple skin discs, however, it was not possible to demonstrate a shift in carbon metabolism to the pentose pathway or increase anthocyanin production in the presence of daminozide (22).

Induction of flower bud formation

Both apple and pear trees do not generally come into full production until the trees are at least 5 years of age. Flowering can be stimulated in young trees by daminozide application. Increasing return bloom of mature trees will also occur after daminozide application in apple, or chlormequat chloride treatment of pear. The growth retardants decrease shoot elongation in fruit trees, and perhaps by inhibiting vegetative growth, flower bud initiation is promoted.

Controlling tree size

Paclobutrazol and other triazole analogs are probably the most effective compounds found to date for controlling shoot elongation in fruit trees (57). Controlling tree size with these compounds will be an effective way of maintaining tree height for maximum spraying and harvesting efficiency in conjunction with modern pruning practices such as summer mowing of the tree canopy. Growth of woody landscape plants may also be effectively controlled using the triazoles, paclobutrazol and uniconazole (32). This practice is particularly useful during nursery container production.

ABSCISIC ACID

There are no practical uses of abscisic acid (ABA) because of the high cost of synthesis and its instability in UV light. However, given the effect of ABA on abscission, dormancy, and transpiration, synthetic ABA analogs might play a significant role in crop production. The ABA analogs LAB 173 711 and LAB 144 143 are acetyleneacetal-type compounds that have been found to reduce water use in crop plants (51), increase cold hardiness (20), and delay flowering in peach (21).

CYTOKININS

Benzyladenine (Pro-Shear) is used on white pine to increase lateral bud formation and subsequent growth and branching, while tetrapyranylbenzyladenine (Accel) is registered for use on carnations and roses for increased lateral branching. Promalin, a mixture of benzyladenine and $GA_{4/7}$, is used to control fruit shape in 'Delicious' apple. High temperature during the bloom period will often reduce the length to diameter ratio resulting in rounder fruit, uncharacteristic of the more normally elongated fruit of this variety that consumers expect. Promalin applied at bloom will increase length to diameter ratio of the fruit (69) (Fig. 7). Increased fruit size may also result from treatment.

Promalin is also being used to increase lateral branching in non-bearing apple trees. Young trees typically have a strong, vigorously growing central leader with a few upright growing branches. For fruit production, this is an undesirable tree shape and mechanical devices are used to force the lateral branches to grow more horizontally. Promalin will stimulate branching and increase the branch angle, as well as increase shoot elongation, all of which aid in the development of a scaffold branching system more suitable for fruit production.

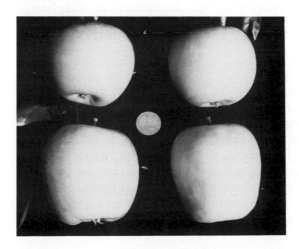

Fig. 7. Effect of Promalin (below) on fruit shape of 'Mutsu' apple. (Photo courtesy of J.A. Hopfinger).

MISCELLANEOUS COMPOUNDS

Maleic hydrazide

Maleic hydrazide has been used since the 1950s for tobacco sucker growth control, the prevention of bud sprouting in onions and potato, and for the control of turfgrass growth. At one time, maleic hydrazide accounted for almost 90% of the sales of plant growth regulators.

In tobacco production, the terminal bud is removed from the plant after a selected number of leaves has been produced. This practice, called topping, increases the size, weight, and quality of the cured leaf. Axillary buds, which develop as a result of topping, will reduce the effect of terminal bud removal on leaf yield and quality. Maleic hydrazide will provide excellent control of axillary bud growth when applied as a foliar spray to the upper two-thirds of the plant, after terminal bud removal (46).

Maleic hydrazide is also used to control storage sprouting of onions and potatoes. The compound is applied as a pre-harvest foliar spray, since it is rapidly translocated to the storage organs. Maleic hydrazide inhibits cell division in a wide range of plants, and the ability of the compound to be translocated to meristematic tissue probably accounts for the effect of the compound on axillary bud growth in tobacco and the sprouting of tubers and bulbs (42). Maleic hydrazide is an analog of uracil and may inhibit cell division by reducing nucleic acid biosynthesis in shoot and root meristems.

Citrus abscission agents

Several abscission agents are being developed for the mechanical harvesting of oranges intended for processing use rather than fresh market. The products, Release (5-chloro-3-methyl-4-nitro-1-pyrazole) and Pik-Off (ethandiol dioxime), induce abscission by causing superficial injury to the rind of the fruit (30). Wound ethylene is synthesized and presumably is the cause of the reduction in fruit removal force. Application of ethephon to trees with mature fruit will also induce abscission, however, significant defoliation often will occur with this chemical.

Sugarcane ripeners

One of the more useful compounds for increasing yield in sugarcane is glyphosine, which in Hawaii, has increased sucrose yield by 10-15%. The herbicide glyphosate, an analog of glyphosine, is also effective, and lower rates of application can be used. Glyphosine will decrease terminal growth of the cane, and other workers have shown that removing the upper leaves of the stalk increases sucrose translocation from lower leaves into the stem and ripening joints (27). In addition, these compounds apparently alter the partitioning of carbohydrate in the sugarcane internode. More carbohydrate goes into sucrose storage at the expense of fiber production (44).

Cotton defoliants

The organophosphate compounds DEF and Folex are used as leaf abscission agents before mechanical harvesting of cotton. Two new compounds are being evaluated for this purpose. Dimethipin (2,3-dihydro-5,6-dimethyl-1,4-dithiin-1,1,4,4-tetraoxide) and thidiazuron (1-phenyl-3 (1,2,3-thiadiazol-5-yl)urea) induce defoliation and provide control of regrowth vegetation after leaf abscission.

FUTURE PROSPECTS

Most of the uses for plant growth regulators increase crop yield or quality by modifying plant development through the hormone system, either by blocking the synthesis or action of a hormone, or by supplementing its supply at a given time. It may be possible, however, to directly affect plant metabolism and increase crop yield by modifying key physiological processes such as photosynthesis, nitrogen fixation, mineral ion uptake and senescence (52).

Increasing photosynthetic efficiency by reducing photorespiration could potentially increase dry matter production. Inhibitors of photorespiration have been identified, unfortunately net photosynthesis has not been increased by these compounds. Altering assimilate partitioning may be another approach to increasing the efficient use of photosynthates. Certainly hormones such as IAA and GA are known to stimulate assimilate translocation to the site of application of these compounds, although no practical use of this phenomenon has been made. Nitrogen fixation requires large amounts of energy and is often limited by photosynthetic capacity. Growth regulators to improve nitrogen fixation efficiency could greatly affect crop yield as would compounds capable of increasing the efficiency of mineral ion uptake by the root system.

Controlling plant senescence also offers potential for yield increase. It may be possible to delay leaf senescence so that more of the growing season could be used for crop growth, in addition to inducing senescence for the remobilization of nutrients and assimilates at times when the potential for seed or fruit growth is high.

Other areas of active research include the search for chemical hybridizing agents, which can induce male sterility in crop plants, facilitating cross pollination and the development of hybrid seeds. Herbicide antidotes have also been discovered. These compounds reduce the toxicity of the herbicide to the crop plants, but not the weed, therefore, allowing increased selective use of currently available compounds in a greater variety of weed crop situations.

Despite considerable advances in our understanding of basic physiological processes that affect crop growth, and the identification of potentially useful experimental compounds, there have been only a few new growth regulators registered for use in the last 5 years, and most of these, such as the triazole growth retardants, have been for non-food crops exclusively. The reasons for sluggish growth in this field are economic and sociological rather than scientific. Research costs are high because many compounds must be screened to obtain a potentially useful product. Development time is long because extensive testing is required to determine efficacy, toxicity, and environmental persistence. In addition, the chemical phobia of the general public has caused the agrichemical industry to become reluctant to invest in new products. The growth regulator daminozide (Alar), for example, was removed from registration for food crops in the US because of unsubstantiated claims of carcinogenicity, and was not reinstated, despite the findings of a National Science Foundation review panel that the compound posed no health risk to the general public. The profitability of growing apples, especially in the eastern part of North America, has been reduced due to increased preharvest drop, rapid fruit softening, and poor color development in varieties such as 'MacIntosh'. Moreover, fruit removed from cold storage is often of lower quality.

References

1. Abeles, F.B. (1969) Herbicide-induced ethylene production: Role of the gas in sublethal doses of 2,4-D. Weed Sci. 16, 498-500.
2. Abeles, F.B., Leather, G.R., Forrence, L.E., Craker L.E. (1971) Abscission: regulation of senescence, protein synthesis and enzyme secretion by ethylene. HortScience 6, 371-3756.
3. Beyer, E.M. (1973) Abscission-support for a role of ethylene modification of auxin transport. Plant Physiol. 52, 1-5.
4. Beyer, E.M., Jr., Quebedeaux B. (1974) Parthenocarpy in cucumber; mechanism of action of auxin transport inhibitors. J. Amer. Soc. Hort. Sci. 99, 385-390.
5. Bukovac, M.J., Zucconi, F., Larsen, R.P., Kesner C.D.. (1970) Chemical promotion of fruit abscission in cherries and plums with special reference to 2-chloroethylphosphonic acid. J. Amer. Soc. Hort. Sci 94, 226-230.
6. Chaplin, M.H., Kenworth A.L. (1970) The influence of succinamic acid 2,2-dimethyl hydrazide on fruit ripening of the 'Windsor' sweet cherry. J. Amer. Soc. Hort. Sci. 95, 532-536.
7. Chen L.G., Switzer, C.M., Fletcher R.A. (1972) Nucleic acid and protein changes induced by auxin-like herbicides. Weed Sci. 20, 53-55.
8. Christodoulou, A.J., Weaver, R.J., Pool, R.M.. (1968) Relation of gibberellin treatment to fruit-set, berry development and cluster compactness in *Vitis vinifera* grapes. Proc. Amer. Soc. Hort. Sci. 92, 301-310.
9. Coggins, C.W., Jr., Hield, H.Z.. (1965) Navel orange fruit response to potassium gibberellate. Proc. Amer. Soc. Hort. Sci. 81, 227-230.
10. Coggins, C.W., Jr., Hield, H.Z.., Boswell, S.B. (1960) The influence of potassium gibberellate on 'Lisbon' lemon trees and fruit. Proc. Amer. Soc. Hort Sci. 76, 199-207.
11. Coombe, B.G. (1965) Increase in fruit set of *Vitis vinifera* by treatment with growth retardants. Nature 205, 305-306.
12. Denny, F.E. (1924) Effect of ethylene upon respiration of lemons. Bot. Gaz. 77, 322-329.
13. Dostal, H.C., Wilcox, G.E. (1971) Chemical regulation of fruit ripening of field-grown tomatoes with (2-chloroethyl) phosphonic acid. J. Amer. Soc. Hort. Sci. 96, 656-660.
14. Dozier, W.A., Jr., Carlton, C.C., Short, K.C., McGuire, J.A. (1984) Thinning 'Loring' peaches with CGA-15281. HortScience 16, 56-57.

15. Durner, E.F., Gianfagna, T.J. (1991) Ethephon prolongs dormancy and delays dehardening in peach flower buds. J. Amer. Soc. Hort. Sci. 116, 500-506.

16. Durner, E.F., Gianfagna, T.J. (1991) Effect of ethephon on peach pistil carbohydrates, moisture content, and growth during controlled deacclimation. J. Amer. Soc. Hort. Sci. 116, 507-511.

17. Edgerton, L.J., Blanpied, G.D. (1970) Interaction of succinic acid 2,2-dimethyl hydrazide, 2-chloroethylphosphonic acid and auxins on maturity, quality, and abscission of apples. J. Amer. Soc. Hort. Sci. 95, 664.

18. Elkins, D.M. (1983) Growth regulating chemicals for turf and other grasses. *In*: Plant Growth Regulating Chemicals Vol. 2, pp. 113-128, Nickel, L.G., ed.

19. Faust, M. (1965) Physiology of anthocyanin development in 'McIntosh' apple. I. Participation of pentose phosphate pathway in anthocyanin development. Proc. Amer. Soc. Hort. Sci. 87, 1-9.

20. Flores, A., Grau, A., Laurich, F., Dorffling, K. (1988) Effect of new terpenoid analogues of ABA on chilling and freezing resistance. J. Plant Physiol. 132, 362-369.

21. Gianfagna, T.J. (1991) The effect of LAB 173711 and ethephon on time of flowering and cold hardiness of peach flower buds. J. Plant Growth Regul. 10, 191-195.

22. Gianfagna, T.J., Berkowitz, G.A. (1986) Glucose catabolism and anthocyanin production in apple fruit. Phytochemistry 25, 607-609.

23. Gianfagna, T.J., Marini, R.P., Rachmiel, S. (1986) Effect of ethephon and GA_3 on time of flowering in peach. HortScience 21, 69-70.

24. Gushman, C.D., Salas, S., Gianfagna, T.J. (1993) Daminozide inhibits ethylene production in apple fruit by blocking the conversion of methionine to aminocyclopropane-1-carboxylic acid (ACC). Plant Growth Regul. 12, 149-154.

25. Gustsfson, F.G. (1937) Parthenocarpy induced by pollen extracts. Amer. J. Bot. 24, 102-107.

26. Hartmann, H.T., Reed, W., Opitz, K. (1976) Promotion of olive fruit abscission with 2-chloroethyl-tris (2-methoxyethoxy)-silane. J. Amer. Soc. Hort. Sci. 101, 278-281.

27. Hartt, C.E., Kortschak, H.P., Burr, G.O. (1964) Effects of defoliation eradication and darkening the blade upon translocation of ^{14}C in sugarcane. Plant Physiol. 39, 15-22.

28. Heilman, M.D. (1981) Interactions of nitrogen with Pix on the growth and yield of cotton. Proc. Beltwide Cotton Prod. Res. Conf. 47.

29. Hitchcock, A.E., Zimmerman, P.W. (1942) Root-inducing activity of phenoxy compounds in relation to their structure. Contrib. Boyce Thomp. Inst. 12, 497-507.

30. Holm, R.E., Wilson, W.C. (1977) Ethylene and fruit looscning from combinations of citrus abscission chemicals. J. Amer. Soc. Hort. Sci. 102, 576-579.

31. Humphries, E.C., Welbank, P.J., Witts, K.J. (1965) Effect of CCC (chlorocholine chloride) on growth and yield of spring wheat in the field. Ann. Appl. Biol. 56, 351-361.

32. Keever, G.J., Foster, W.J., Stephenson, J.C. (1990) Paclobutrazol inhibits growth of woody landscape plants. J. Environ. Hort. 8, 41-47.

33. Kotob, M.A., Schwabe, W.W. (1971) Induction of parthenocarpic fruit in 'Cox's Orange Pippin' apples. J. Hort Sci. 46, 89.

34. Larson, R.A., Kimmins, R.K. (1971) Response of *Chrysanthemum morifolium* Ramat to foliar and soil applications of ancymidol. HortScience. 7, 192-193.

35. Looney, N.E. (1968) Inhibition of apple ripening by succinic acid 2-2- dimethyl hydrazide and its reversal by ethylene. Plant Physiol. 43, 1133- 1137.

36. Luckwill, L.C. (1953) Studies of fruit development in relation to plant hormones. II. The effect of naphthalene acetic acid on fruit set and fruit development in apples. J. Hort. Sci. 28, 25-40.

37. Luckwill, L.C. (1953) Studies of fruit development in relation to plant hormones. I. Hormone production by the developing apple seed in relation to fruit drop. J. Hort. Sci. 28, 14-24.

38. Magness, J.E., Overly, F.L. (1929) Relation of leaf area to size and quality of apples and pears. Proc. Amer. Soc. Hort. Sci. 26, 160-162.

39. Martin, G.C. (1971) 2-chloroethylphosphonic acid as an aid to mechanical harvesting of English walnuts. J. Amer. Soc. Hort. Sci. 96, 434-436.

40. Moore, P.H., Osgood, R.V., Carr, J.B., Ginoza, H.S. (1982) Sugarcane studies with gibberellin. V. Plot harvests vs stalk harvests to assess the effect of applied GA$_3$ on sucrose yield. J. Plant Growth Reg. 1, 205-210.

41. Nakayama, I., Miyazawa, T., Kobayashi, M., Kamiya, Y., Abe H., Sakurai, A. (1990) Effects of a new plant growth regulator prohexadione calcium (BX-112) on shoot elongation caused by exogenously applied gibberellins in rice (*Oryza sativa* L.) seedlings. Plant Cell Physiol. 31, 195-200.

42. Nooden, L.D. (1969) The mode of action of maleic hydrazide: inhibition of growth. Physiol. Plant 22, 260.

43. Osborne, D.J., Sargent, J.A. (1974) A model for the mechanism of stimulation latex flow in *Hevea brasiliensis* by ethylene. Ann. Appl. Biol. 78, 83-88.

44. Osgood, R.V., Moore, P.M., Ginoza, H.S. (1981) Differential dry matter partitioning in sugarcane cultures treated with glyphosate. Proc. Plant Growth Reg. Soc. Amer. 8, 97.

45. Palmer, G.M. (1974) The industrial use of gibberellic acid and its scientific basis - A review. J. Inst. Brewing 80, 13-30.

46. Petersen, E.L. (1952) Controlling tobacco sucker growth with maleic hydrazide. Agron. J. 44, 332.

47. Pharis, R.P., Ross, S.D., McMullan, E. (1980) Promotion of flowering in the Pinaceae by gibberellins. III. Seedlings of Douglas fir. Physiol. Plant. 50, 119-126.

48. Proebsting, E.L., Jr., Mills, H.H. (1972) Bloom delay and frost survival in ethephon treated sweet cherry. HortScience 8, 46-47.

49. Rappaport, L., Timm, H., Lippert, L.F. (1957) Sprouting, plant growth and tuber production as affected by chemical treatment of white potato seed pieces. I. Breaking the rest period with gibberellic acid. Amer. Potato J. 34, 254-260.

50. Rodriguez, A.G. (1932) Influence of smoke and ethylene on fruiting of pineapple (*Ananas sativus* Schult.). J. Agric. Univ. P.R. 15, 5.

51. Schubert, J., Roser, K., Grossmann, K., Sauter, H., Jung, J. (1991) Transpiration-inhibiting ABA analogs. J. Plant Growth. Regul. 10, 27-32.

52. Scott, T.K. (1978) Plant Growth Regulation and World Agriculture. Plenum Press.

53. See, R.M., Foy, C.L. (1982) Effect of butanedioic acid mono (2,2-dimethyl hydrazide) on the activity of membrane bound succinate dehydrogenase. Plant Physiol. 70, 350-352.

54. Sidahmed, D.A., Kliewer, W.M. (1980) Effects of defoliation, gibberellic acid and 4-chlorophenoxy acetic acid on growth and composition of 'Thompson Seedless' grape berries. Amer. J. Enol. Vitic 31, 149.

55. Snyder, M.J., Welch, N.C., Rubatzky, V.E. (1971) Influence of gibberellin on time of bud development in globe artichoke. HortScience 6, 484-485.

56. Southwick, F.W., Demoranville, I.E., Anderson, J.F. (1953) The influence of some growth regulating substances on pre-harvest drop, color, and maturity of apples. Proc. Amer. Soc. Hort. Sci. 59, 155-162.

57. Steffins, G.L. (1988) GA biosynthesis inhibitors: Comparing growth retarding effectiveness on apple. J. Plant Gr. Regul. 7, 27-36.

58. Steffins, G.L., Alphin, J.G., Ford, Z.T. (1970) Ripening tobacco with the ethylene releasing agent 2-chloroethylphosphonic acid. Beitr. Tobakforschung 5, 262.

59. Stewart, W.S., Hield, H.Z. (1949) Effect of 2,4-dichlorophenoxyacetic acid and 2,4,5-trichlorophenoxyacetic acid on fruit drop, fruit production and leaf drop of lemon trees. Proc. Amer. Soc. Hort. Sci. 55, 163-171.

60. Struckmeyer, B.E., Roberts, R.H. (1950) A possible explanation of how naphthalene acetic acid thins apples. Proc. Amer. Soc. Hort. Sci. 56, 76-78

61. Thomas, T.H. (1982) Plant Growth Regulator Potential and Practice. BCPC Publications, Croydon, U.K.

62. Tompkins, D.R. (1966) Rhubarb rest period as influenced by chilling and gibberellin. Proc. Amer. Soc. Hort. Sci. 87, 371-379.

63. Tukey, L.D., Fleming, H.K. (1968) Fruiting and vegetative effects of N-dimethylaminosuccinamic acid on 'Concord' grapes, *Vitis labrusca* L. Proc. Amer. Soc. Hort. Sci. 93, 300-310.

64. Walker, J.C., Legocka, J., Edelman, L., Key, J.L. (1985) Plant Physiol. 77, 847-850.

65. Wallerstein, I., Goren, R., Monselise, S.P. (1973) Seasonal changes in gibberellin-like substances in 'Shamouti' orange (*Citrus sinensis* (L.) Osb.) trees in relation to ringing. J. Hort. Sci. 48, 75-82.

66. Walsh, C.S., Swartz, H.J., Edgerton, L.J. (1979) Ethylene evolution in apple following post-bloom thinning sprays. HortScience 14, 704-706.

67. Westwood, M.N. (1978) Temperate Zone Pomology. W.H. Freeman and Co., San Francisco, CA.

68. Williams, M.W. (1979) Chemical Thinning of Apples. Hort. Rev. 1, 270-300.

69. Williams, M.W., Stahly, E.A. (1969) Effect of cytokinins and gibberellins on shape of 'Delicious' apple fruits. J. Amer. Soc. Hort. Sci. 94, 17-19.

70. Wulster, G. J., Gianfagna, T.J., Clark, B.B. (1987) Using sterol-inhibiting fungicides as growth retardants in Easter lily. HortScience 22, 601-603

G14. Hormones in Tissue Culture and Micropropagation

Abraham D. Krikorian

Department of Biochemistry and Cell Biology, Division of Biological Sciences, State University of New York at Stony Brook, Stony Brook, New York 11794-5215 USA.

INTRODUCTION

Identification and characterization of the role of hormones in plant growth regulation has been closely linked to studies on development *in vitro*. Aseptic culture techniques, and the use of hormones in nutrient media that usually goes with it, are now more widely used than ever in basic and applied research (39). Several categories of hormone are recognized, but search for new growth regulatory molecules continues and inevitably new categories will be erected. In all this, study of the identity, effects and mechanism of action of hormones will rely heavily on *in vitro* techniques.

The first successful *in vitro* cultures were of excised root tips and study of roots in aseptic culture continues, especially to foster specialized biosyntheses (7, 48). Somewhat after the initial successes with root tip cultures, it became possible to grow callus derived from storage organ explants or the cambial region of woody species. Cultures grew slowly and the stimuli to cell division generally entailed use of auxins such as indole-3-acetic acid (IAA) or naphthaleneacetic acid (NAA) in otherwise simple media comprised of mineral salts, a reduced carbon source (usually sucrose) and a few vitamins, especially thiamine (31).

Progress was made when it was shown that one could foster vigorous increase in the level and rate of cell division in carrot storage root explants on addition of natural fluids, e.g., liquid endosperms like coconut water, to the culture medium. This discovery was significant because the explants did not include cambium, and this implied that quiescent cells could be stimulated to divide. In fact, the carrot root plugs were very small and excision was far enough away from the cambium to guarantee that the component cells would have stopped dividing (31).

Tobacco stem explants containing bits of vascular cylinder as well as pith required only auxin as an additive to the culture medium to achieve significant callus growth. When pith explants without vascular tissue were tested, they would not grow well unless complex additives such as yeast or malt extract, or coconut water were supplied in addition to an auxin. This work opened the way to the discovery of kinetin. It soon followed that

P. J. Davies (ed.), Plant Hormones, 774–796.
© 1995 *Kluwer Academic Publishers. Printed in the Netherlands.*

kinetin in the presence of IAA or a synthetic substitute could bring about cell division in tobacco and a number of other species in a completely defined medium (31).

Simultaneous appreciation of the complexity of growth requirements of tissues *in vitro* came about from studies on crown gall tumors. Proliferative growths on stems, such as those of tobacco, sugar beet, kalanchoe and periwinkle could be initiated by inoculation with *Agrobacterium tumefaciens*. The time that the crown gall bacterium was allowed to remain in contact with the tissue prior to its elimination by an appropriate heat (or antibiotic) treatment determined the complexity of the additives needed in a culture medium to sustain the proliferation when explanted. The basis for this is that during tumor induction, the bacterium transfers specific genes into the plant cells where these genes enter the nucleus and integrate into the nuclear DNA. The more fully transformed the plant tissue, the more autotrophic it is. Thus, crown gall tumors can be maintained on a simple medium of mineral salts and a carbon source since they are synthesizing their own growth substances (28).

So-called minimal media without phytohormonal additives only occasionally serve, however, as a vehicle to sustain rapid growth of explants of normal tissue. Growth stimulating substances may be supplied in the form of natural fluids like coconut water which contain auxins, cytokinins, gibberellins and sugar alcohols like *myo*-inositol, or as known chemical compounds (31). In either case, the guiding principle is that *in vitro* (and *in situ*) one requires at least two systems to be operative. One involves auxins and the other cytokinins. The extent to which this classic view now needs to be modified or amplified will emerge in the course of this chapter.

ASEPTIC CULTURE SYSTEMS

"Tissue culture" has evolved to signify the *in vitro* culture of virtually any plant part at any level of organization (39, 50, 68). This includes protoplasts, cells, callus, excised organs and even whole plants. Precise designation of what is being cultured and why, thus becomes very important, for in this way one communicates the strategies a tissue culturist might adopt. Since this determines which (and the way) growth regulators are used, it will be useful to review some culture techniques.

Embryo Culture

Since the early 1920s it has been possible to facilitate growth, otherwise unobtainable or erratic, of certain embryos in aseptic culture. In some cases, embryos with poorly developed food reserves do not germinate because they are dependent on an external nutrient source (43). For instance, orchid seeds contain a very small embryo comprised of a simple cell mass (3). The embryo is dependent upon exogenous sugar, provided in nature by a symbiotic mycorrhizal relationship, to germinate. In the case of the Makapuno mutant of coconut, erroneously regarded in the past as an embryo lethal mutant, lack of galactomannan-degrading enzymes in the endosperm results in deficiency of nutrients and energy for the embryo (57), but embryo rescue permits germination (33, 34). Another example where embryos fail to germinate involves the formation of inhibitors in the seed. Here embryos often germinate only after an appropriate period of dormancy or leaching. In plants like iris one can eliminate both the dormancy requirement and the effect of germination inhibitors that may be present in the seed by excising embryos and rearing them to a size sufficient for independent growth in sterile culture in the absence of the chemical constraints. Embryo rescue has become widely adopted (54). The growth regulators used must foster normal growth if the strategy is to be effective. This means no callus (4).

Meristem, Shoot, or Stem-Tip Culture

Excised shoot apices of the orchid *Cymbidium* when appropriately cut and grown in aseptic culture, produce protuberances which resemble normal protocorms that can grow to plantlets (3). In the 1960s this provided a dramatic impetus for the further development of procedures for multiplying and maintaining a variety of plants in aseptic culture (31). Shoot tip cultures from many other plants are now exploited in the obtaining, maintaining and multiplication of stocks. In some cases, a single plantlet is generated from a cultured shoot tip; in others, multiple shoots can be stimulated by addition of appropriate levels of cytokinin. As long as development of shoots emerging from the proliferated area at the base of a shoot tip explant can be maintained at a rate consistent with their removal by excision, an "open-ended" system is achieved (71). One tries to maintain a hormone balance, frequently higher in auxin, which favors the continued formation of undifferentiated growth that can organize and develop in culture by subsequently adjusting the medium-- usually by increasing cytokinin or lowering auxin. As these shoots (with or without roots) are removed from the proliferating mass and are transferred to an environment or different medium conducive to root production and growth, say auxin-rich (10), new proliferations grow to replace them (68, 71).

A variation on this theme involves the stimulation of precocious axillary shoots in profusion when stimulated by cytokinins. Since axillary shoots can, in turn, produce additional axillary branches as each newly formed shoot is subcultured, the method is a good one for rapid clonal multiplication. It has been applicable to species ranging from herbaceous foliage plants to bulbous monocotyledons and woody species (68, 71).

Other Organ Cultures

When we understand better what developing organs receive in terms of stimuli and nutrients from the rest of the plant, and are able to recreate the environment in which they originate *in situ*, we should be able grow them separately, and, if not from their initiating cells, then from their primordia. For some years now, the achievements of classical organ culture have applied more to roots than other organs. Leaves and fruits only occasionally reach significant size *in vitro* (31).

Anthers and Ovules as Sources of Haploid Cells

In the mid-1960s, cultured anthers of several plant species were shown to be capable of yielding haploid somatic embryos (31, 54). Cultures can be initiated from anthers containing immature pollen grains, i.e., microspores prior to the development of the mature male gametophyte. In tobacco, for instance, the vegetative nucleus divides to give rise to the proembryo while still within the original wall of the pollen grain. The process, androgenesis, has potential for raising haploid plants and presents considerable benefits to plant breeders and geneticists. Homozygous diploids can be raised by the use of colchicine or by taking advantage of the fact that many cultured cells and tissues spontaneously undergo endopolyploidization. Androgenesis generally does not involve use of exogenous hormones to initiate the system, and presumably they produce what they need. It is a matter of catching them at the right stage (4, 25). Gynogenesis, the process whereby haploid somatic embryos are produced *in vitro* by induction of haploid tissues from the female gametophyte, is not so technically advanced as androgenesis, and generally involves more complex media (4, 25).

Protoplast Cultures

Enzymatic isolations, *en masse*, of viable, wall-less cells (protoplasts) have been familiar since 1960 (31). Cellulases and pectinases, generally derived from certain wood-degrading fungi, capable of dissolving the intercellular components and cell wall, are used to produce protoplasts from different plants and organs and from their cultured tissues and cells. With reconstituted walls, they are able to divide, proliferate and in some instances, eventually yield plants (52). In still other cases, fusion of protoplasts may lead to non-heterokaryotic nuclei and the production of reconstituted cells from which new plants can be regenerated. If the protoplasts were from haploid cells (as from anther or ovule culture) new diploid cells could be produced in a sort of artificial syngamy (parasexual or somatic hybridization). In select cases one might exploit the methods in the production of novel plants, even between evolutionarily disparate organisms, by fusion breeding techniques (50). Successful regeneration of plants from protoplasts can also have value in the implementation of genetic engineering/transformation studies (4, 39, 50). The hormonal requirements for regeneration, chemical composition and biochemical activity of wall are not understood (59) and regeneration media are often complex (52). Once walled cells are formed, the cells are grown in the way specific for the plant.

USE OF AUXINS, CYTOKININS, AND OTHER GROWTH REGULATORS FOR CALLUS INDUCTION AND MAINTENANCE

Auxins

Auxins cause cell enlargement and elongation but in excised, cultured systems they may promote cell division. Because of their relative stability and ability to withstand autoclaving, synthetic auxins are extensively employed. The most commonly used are 2,4-dichlorophenoxyacetic acid (2,4-D), 1-naphthaleneacetic acid (NAA) and indole-3-butyric acid (IBA). There are also many compounds which are derivatives of the chloro substituted phenylacetic or phenoxyacetic acids that have found application (4). In some cases, compounds which are not strictly auxins, such as dicamba (3,6-dichloro-*o*-anisic acid) (23) or picloram (4-amino-3,5,6-trichloro-pyridine-2-carboxylic acid) (both of which, are herbicides at higher concentrations), are used as auxin substitutes (24).

The use of auxins in tissue culture is an art and rarely can one specify a particular concentration to be used in any single case. There is an extensive literature on the initiation of cultures of various plants or cultivars (4, 39, 50).

One of the best examples of a nominally permanent change in auxin requirement of a culture is that of "habituation" to auxin. Here, tissues that originally required an exogenous auxin for growth, gradually or suddenly lose this requirement (41). (The same applies to cytokinins (60).

In many instances, addition of any one of the auxins to a basal medium may be enough to initiate and sustain callus growth. Therefore, one ordinarily uses a single auxin at a time but it can be helpful to use more than one auxin simultaneously in recalcitrant cases (32). Tissue culture of monocotyledons, particularly of cereals and palms has been achieved in some cases through rather high levels of synthetic auxins like 2,4-D. The levels used would ordinarily be considered to be toxic, but organized centers of growth in the absence of exogenous cytokinin may be achievable, and development progresses (either production of somatic embryos or adventitious organs) when the auxin is removed or lowered in the medium (33, 34). There is some evidence that growth of cultures can be stimulated or modulated by addition of substances that regulate the level of endogenous IAA. For instance, catecholamines like dopamine which inhibit IAA oxidase activity and thus prevent IAA oxidation can promote growth of tissue and organ cultures (51). Inhibitors of auxin synthesis like 5-hydroxy-nitrobenzyl bromide (HNB) and 7-azaindole stimulate somatic embryogenesis in habituated citrus callus (30). Phenolic anti-auxins or auxin antagonists like *trans*-cinnamic acid, transport inhibitors like 2,3,5-triiodobenzoic acid (TIBA) (4, 20) and N-(1-naphthyl)phthalamic acid (NPA) (58), 2,4,6-trichloro-phenoxy acetic acid (2,4,6-T) (44), and *p*-(chlorophenoxy)-2-methyl propionic acid (CMPA) (17, 66), can affect morphogenesis and embryogenesis in addition to having effects on growth (1, 58).

Cytokinins

Kinetin and zeatin, which is about 10 times more potent and long considered the prototype of the naturally-occurring cytokinins, are both routinely used in tissue culture. Dihydrozeatin, also naturally occurring, is not widely used. The synthetic N^6-benzylaminopurine (BAP) is much more widely used than either kinetin or zeatin (4, 50). N^6-Δ^2-isopentenyl adenine (2iP) is also widely used. Not surprisingly, whatever is active, readily available and least costly is employed. When tested in the carrot root phloem or tobacco pith assay, any of these cytokinins are active, but only in the presence of an auxin. In those cases where an exogenous auxin supply is not needed, it is assumed that the system is synthesizing its own auxin.

Cytokinin-like compounds

The substituted phenylureas comprise another class of substances active as cytokinins. The ability of N,N'-diphenylurea, first isolated from coconut water (although there is a likelihood that it was actually a contaminant picked up during the purification), and related compounds to substitute for cytokinin-active adenine derivatives can be demonstrated in several callus culture bioassay systems, although the levels needed are somewhat high (5, 31). Particular phenylurea derivatives such as N-phenyl-N'-4-pyridylurea are as active as zeatin, or even more so, but because they are not readily available, they are still not extensively used. Thidiazuron (N-phenyl-N'-1,2,3-thiadiazol-5-ylurea), used commercially as a cotton defoliant due in part at least to its ability to stimulate ethylene production, is but one example that shows considerable activity in both callus assays showing cytokinin-dependence and in the fostering of shoot bud initiation and proliferation *in vitro* (15). Thus, although they are more widely used, adenyl cytokinins are but one class of substance active in cell division. One example of many other substances with considerable cytokinin activity that emerge when looked for is dihydroconiferyl alcohol (originally identified from bleeding sap of *Acer pseudoplatanus* and maple syrup and now shown to be widely occurring). It is active in various callus assays, albeit at concentrations considerably higher than those normally viewed as consistent with being classified a cytokinin (46). Another, less satisfying example, is the amino acid *S*-lathyrine, from the legume *Lathyrus*, which although active in the soybean callus growth assay, and even potentiates kinetin, is inactive in other standard callus assays (53). Leucoanthocyanins and other phenolics (perhaps acting as auxin protectants?) provide yet other examples (42).

Liquid endosperms can be pivotal in specific culture systems and in those cases can rarely be substituted for with completely defined media (31). Also, the device of using "conditioned" medium (i.e., in which prior growth has occurred) to facilitate growth of protoplasts, cells and other tissues indicates a role for various other growth promoters, some of which are large molecules and active in small amounts (12, 57). Similarly, the device of using "nurse" cultures and "feeder" layers of cells which are alive but cannot divide (because of X-ray inactivation etc.) to nourish protoplasts and/or cells all emphasize that we do not have an adequate understanding of the full nutritional or regulatory needs (31, 41), which indeed are not limited to small molecules (13).

Gibberellins

Despite the wide ranging physiological effects of gibberellins, their addition, primarily as GA_3, to culture media has only been minimal (18, 45). In some cases like the carrot root bioassay, they result in cell divisions rather than enlargement (31). Anti-gibberellins have attracted more attention as additives (72).

Abscisic Acid (ABA)

ABA slows growth and moderates the effects of cytokinins and auxins. It can bring about a more normal progression of somatic embryo development especially when hormone production has been perturbed and there is too much callus. It also prevents precocious germination in somatic embryos (2, and see Chapter G10).

Ethylene

Ethylene (both normal and stress metabolism-derived, or even from synthetic precursors such as ethephon, [2-chloroethylphosphonic acid; CEPA]), and air pollutants from hydrocarbons (frequently associated with emissions from sterilizer flames or burners), are generally viewed as noxious substances that should be avoided in the *in vitro* environment, be it in culture vessels or laminar flow hoods (4, 68). Use of ethylene inhibitors such as silver nitrate or sulfate, cobaltous or nickel chloride, and aminoethoxyvinylglycine (AVG) or salicylic acid supposedly increase shoot regeneration and somatic embryo production, but results are contradictory (40). Ethylene is all pervasive, and effects such as acceleration of the breakdown of cytokinins and subsequent stimulation of rooting in some systems is but one example of importance to tissue culturists (10). In daylily (*Hemerocallis)* cultures, elevated ethylene in the culture vessel fosters a mature phenotype, whereas lower levels produce juvenile forms (63).

CYTOKININ-AUXIN RATIOS AND ADVENTITIOUS ORGANOGENESIS FROM INTACT ORGANS AND CALLUS CULTURES

Study of the interaction of kinetin and auxin in cell division in the excised tobacco pith system showed that manipulating auxin:cytokinin ratios can effect organogenesis. When the level of auxin relative to that of cytokinin is high, roots form; when cytokinin relative to that of auxin is high, shoots form. When the ratios are about the same, a callus mass is produced (4, 50). The device of adjusting auxin:cytokinin ratios to induce shoots and roots is now a well-established procedure.

In the so-called "indirect organogenesis" route for multiplication *in vitro* (4, 71), the strategy is to induce a callus by whatever means, stimulate shoots on the callus, and then remove the shoots and root them (cf. Table 1). Each of these steps may not be discrete or synchronous and the levels of the needed exogenous hormone are different. For example an auxin, or an auxin and a cytokinin, might be needed to establish a callus; then a higher level of cytokinin relative to auxin might be needed to stimulate shoot buds and then, only an auxin might be used to induce roots on the separated shoot. The

---

Table 1. Strategies for Multiplication of Higher Plants *In Vitro*

- Stimulation of indirect organogenesis
  adventitious shoot and/or root formation on a callus

- Stimulation of direct organogenesis
  adventitious shoot and/or root formation on an organ or tissue explant without an intervening callus

- Use of shoots from terminal, axillary, or lateral buds for precocious branching and multiple bud formation
  shoot apical meristems (no leaf primordia present)
  shoot tips (leaf primordia or young leaves present)
  buds
  nodes
  shoot buds on roots

- Stimulation of somatic embryogenesis
  direct formation on a primary explant
  indirect formation from cells grown in suspension or semi-solid media
- Stimulation of direct plantlet formation via an organ of perennation formed *in vitro*

- Implementation of micrografting

- Ovule culture

- Embryo rescue

- Mega- and microspore culture

- Infection with a crown gall plasmid genetically altered to give teratoma-like tumors

---

same principles apply in "direct organogenesis" (Table 1). The nominal difference is that minimal, if any, callus is involved in the proliferation of organized structures. These form adventitiously "directly" on an explanted organ (4, 71). The shoots are removed and root formation is fostered with auxin. But there are still many instances where morphogenesis cannot be induced despite use of elaborate protocols. Even in *Nicotiana* manipulation of auxin and cytokinin ratios is not equally effective among the many species and varieties. The "model" system for the classic relationship is the Wisconsin 38 strain of the 'Havana' cultivar of *N. tabacum*. Many plants respond to manipulation, however, and commercialization has been possible in a range of horticultural cultivars (31, 71).

Regimens for the regeneration of plants from cells or from protoplasts fit into the principles outlined here. Small cell clusters, obtained either directly from liquid cultures or from small pieces of callus produced first on semi-solid medium, are plated in semi-solid media. These are then provided with a sequence of hormones (or their withdrawal) to induce or foster progression of morphogenesis (4, 31, 39).

HORMONES AND SOMATIC EMBRYOGENESIS

Many tissues comprise cells which can be induced *in vitro* to an embryogenic state by the use of auxins such as 2,4-D. This was first noted in carrot and other umbellifers (36), and extends to an expanding list of species. Cultures of very early stage somatic embryos can proliferate indefinitely without further progression to later stages of embryo development. Frequently, these early stage embryos, variously referred to as proembryos, preglobular stage embryos, embryogenic masses or units etc., release into the medium by budding numerous immature somatic embryos. When these early stage, minimally developed, embryos pass through later stages of embryo ontogeny, they ultimately grow into plantlets that develop and reproduce. Thus, at least some cells of the plant body can be induced to express the full information present in the zygotic nucleus and they can be multiplied and sustained in this embryogenic state through serial subculture. Such cultured cells provide valuable material for the investigation of hormonal stimuli and other factors that modulate development (2, 4, 6, 31).

Carrot somatic embryos were first obtained in suspension cultures supplemented with coconut water (31), but it was later learned that coconut water was not required for the process and that an auxin such as 2,4-D or NAA could be used for induction (36). Most protocols, whether carried out in liquid or on semi-solid medium, now involve the exposure of an explant to a synthetic auxin such as 2,4-D, NAA, picloram or dicamba. This initial stage of culture is critical for right at the outset responsive cells become induced to express their embryogenic capacity. The cell division rate of induced cells may be slow at first, but populations of cells are conserved and their numbers increased. Early stage somatic embryos or preglobular stage somatic embryos are easily recognized by their generally dense cytoplasmic contents and general absence of vacuolization. They can be sustained in this embryogenic state for varying periods (36). The removal or drastic lowering of the level of the exogenous auxin leads to the continued development of preglobular stage embryos and their progression through various stages of embryogenesis. (Addition of cytokinin reduces somatic embryo production (37)). The proembryos can further pass through later stages of development, e.g., in a dicotyledonous species the heart stage, the torpedo stage, the cotyledonary stage. All stages, but especially the cotyledonary forms, may be reared to mature plants. Evidence that normal development in embryogenic cultures is modulated by the presence of a variety of growth factors and environmental parameters is quite strong (2, 36). More striking is the evidence that the type of growth--by cell division or enlargement, normally or abnormally organized, or unorganized--is dependent upon their balanced effects and interaction with the external environment. In some cases the production of somatic embryos within a suspension occurs at high frequency, but in others imperfectly or not at all. Occasionally, when the expression of the developmental program is imperfect (22), one gets abnormal forms resembling embryos but which are not capable of further growth. I call these neomorphs. That the cells of neomorphs may be redirected is indicated by the fact that neomorphs can be treated *in vitro*, and, when the proper inductive and permissive sequence and environment is supplied, normal somatic embryos develop (62).

Long term sub-culture of embryogenic cells is frequently accompanied by a loss of their ability to progress. Similarly, the smaller the unit size of embryogenic cells that are tested for their ability to progress through all stages of embryogenesis, the greater is their requirement for exogenous growth-promoting substances, proper osmotic environment and, in some cases, environmental stimuli such as darkness etc. (6, 36).

There remain a number of species, however, in which theoretically-existent embryogenic (or even organogenic) capacity has not been demonstrated. Even when a system works, there may be a problem of mixed response and yield, for many cells do not develop but merely proliferate. Others do not grow at all. But even in those cases where expression of competence is realized relatively easily, in no case is the growth uniform, much less synchronous, and virtually never is the yield 100%. For example, while many small preglobular stage embryos progress to the globular stage, and these may proceed to the heart shaped stage, torpedo stage and cotyledonary stage and thence to a plantlet, there are variations on this pattern and the earliest stages are especially vulnerable to perturbation. Somatic embryo production has earned an increasingly important place in research (2, 6, 36) in *in vitro* propagation (55).

Some investigators distinguish between an *indirect* route of somatic embryo formation and a *direct* route (the process described in some detail here). Semantics and ambiguity of such terminology aside, the direct route would involve formation of embryos, without involvement of an undifferentiated callus stage (Table 1). In cultures which are already induced by use of auxin etc., the procedure is to remove the auxin, and the early stage somatic embryos continue developing in an essentially basal medium (2, 36). In direct somatic embryogenesis, auxin is added and the embryos develop and may even progress and mature in the presence of the hormone if its level is sufficiently lowered by metabolism etc. (39). The significance of all this for our discussion of hormones is that induction and expression of competence are not readily separable in the laboratory into two discrete phases and the literature is frequently confused on this point. It is generally stated that induction occurs *after* a callus forms; in fact, the so-called "callus" is more often a mass of embryogenic tissue that is prevented from progressing through later stages of embryogenesis! Finally, direct somatic embryogenesis may provide an avenue for true cloning of plants since chances for change in genotype are presumably minimized (4,31,49).

MICROPROPAGATION OR MULTIPLICATION USING PRE-FORMED OR ORGANIZED PROPAGULES

Meristem, shoot- or stem tip, node, axillary, or lateral bud culture

Micropropagation was originally defined as any aseptic culture procedure involving the manipulation of organs, tissues or cells that produces a population of plantlets by bypassing the normal sexual process or non-aseptic vegetative propagation (4, 71). In practice, stem tips and lateral buds are the most commonly used starting point (71). Use of callus, cell cultures or other of the more labor-intensive approaches listed above rarely comes to mind when the word micropropagation is used.

Familiarity with the morphology of any given plant is the first of several prerequisites to successful manipulation of organized propagules *in vitro*. Plant biologists appreciated long before aseptic culture came into laboratory practice that the development of the higher plant body involved the suppression of many, and the development of relatively few, existing or potential primordia. Production of new growing regions, organs and even new plants are held in check by correlative influences and inhibitions (31). IAA or IBA were early used to induce root formation on cuttings (11). The key feature of micropropagation via precocious axillary branching, however, directly arose out of the early work on kinetin. Wickson and Thimann (70) were taken by the possibility that kinetin might well exert an effect on the development of buds, not just their initiation as had been suggested by the work of Skoog and Miller (3, 31). Figure 1 summarizes the observation that cytokinin can antagonize the inhibitory effect of auxin on lateral bud elongation. One is essentially manipulating the variable tendency for a shoot

784

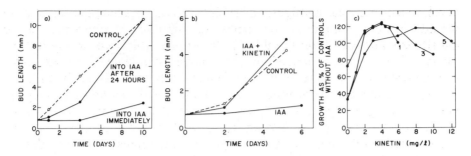

Fig. 1. Interactions of auxin and kinetin in lateral bud growth of pea. a) bud elongation is immediately inhibited by 4mg/l IAA 24 hours after excising them from the parent plant. The elongation is slightly retarded but thereafter growth is the same or slightly faster than the control. This indicates that if auxin supply is interrupted, the laterals begin to grow; b) the inhibition of bud elongation by IAA is completely reversed by 4mg/l kinetin; c) the level of kinetin needed to overcome inhibition of lateral bud growth by auxin increases as the concentration of auxin increases. Note that the level of kinetin needed to overcome inhibition when auxin is provided at 1.3 and 5 mg/l is roughly equal. (From 70).

apex to exhibit apical dominance over the lateral buds.

Some view it as helpful to segregate the sequence of events associated with the multiplication process into stages as follows: I - establishment stage, II - shoot multiplication, III- rooting.

Specific media or culture conditions have become associated with each of these stages for a given species and viewing the problem in this orderly fashion can be useful in developing a strategy or interpretation of data. But one should not assume that these stages are causally associated with a particular hormonal regimen. Also, the stages are not always temporally discrete. Three stages (I, II, III) of *in vitro* propagation were initially conceptualized (71) but a fourth stage (IV), "final transfer to the natural environment", is now an integral part of the process. Moreover, a pre-establishment stage "0", involving selection of the mother plant and adoption of a program of pre-treatment to render any given protocol to be workable in the first place (4, 31, 71), emphasizes micropropagational strategy as a process, not an event.

Procedures, mainly reporting the levels of added hormone and sequences, and reminiscent of "recipes" for *in vitro* regeneration or multiplication, remind one that this aspect of the field continues to be empiric (4, 68). Here again, even when plants can be multiplied by the forcing of precocious axillary branching with cytokinins (Fig. 2), not all plants respond equally well and there are many where an acceptable methodology still needs to be developed (32).

It generally holds that the smaller the primary explant, the greater the difficulty encountered in the establishment of growth and the promotion of shoots. One can, of course, resort to using a large shoot tip explant to minimize constraints brought on by starting with an apical meristem, but there are many situations, such as obtaining of virus-free stock, where the

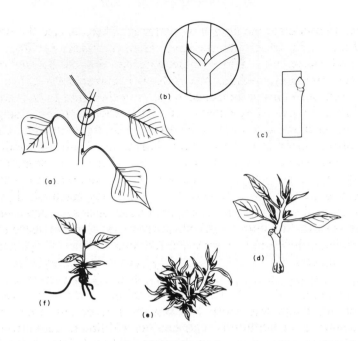

Fig. 2. Diagrammatic scheme of multiplication of *Sapium sebiferum*, a woody member of the Euphorbiaceae. The strategy is similar for many other plants where precocious axillary branching is to be stimulated. a) a young branch with lateral buds from a mature, tested tree; b) close-up of the lateral bud; c) primary explant comprising a single bud; d) precocious branches formed by bud break and; e) a culture with a well established but compact branching system; f) a rooted plantlet. (From Krikorian and Kann (1987), *In*: Tissue Culture in Forestry, 3, 357-369. Martinus Nijhoff, The Hague.)

meristematic dome is the preferred starting material (69).

The points of strategy in terms of hormone use in each of the stages of micropropagation are: Stage 0 involves bringing the donor plant to a vigorous stage of growth. This means selection of material at the right time of year so that endogenous inhibitors provide no obstacles. Breaking of dormancy by growth regulator treatment may be necessary. Increasing light or even etiolation in some cases is useful. Stage I involves design of media to stimulate bud or shoot primordia initiation and development, or to stimulate precocious axillary bud break and development. Here cytokinins and auxins may be supplied, but if the system is synthesizing adequate auxin (or cytokinin) then only a cytokinin (or auxin) may be necessary. Occasionally, a low level of GA_3 may be useful since it will permit some extension growth and allow the system to be managed better. Stage II involves increasing or sustaining the level of bud production or precocious branching. This is generally achieved by increasing the cytokinin level, but it may be preferable to opt for a lower multiplication rate by use of low levels of cytokinin, since too rapid an increase may lead to genetic change (49). Indeed, it is sometimes helpful to pulse a system intermittently with cytokinin between subcultures to modulate the branching patterns. There is also the process of "vitrification" to be taken into account. Here the tissue *in vitro* becomes "glassy" and succulent. While some consider this to be the result of overexposure to high cytokinin, its real basis is unresolved (4, 71). It may be that a compact multiple branched plant is desired and in that case there may be little to be gained by pulsing. In plants where a uniform single stem is preferred, such as forest trees, a system may profit by pulsing, combined with use of a gibberellin to foster stem elongation. High light will tend to keep internodes short. Stage III involves use of auxins to stimulate roots. While this stage is generally done under aseptic conditions, it is possible in many cases to carry this out *ex vitro* (71). The long established principles that apply to rooting of cuttings in conventional propagation procedures (10) apply here as well. Sometimes difficult-to-root species are easier to root after *in vitro* shoot multiplication. Removal or leaching of "excess" cytokinin from shoots is sometimes critical since they generally antagonize rooting. "Flushing" of microcuttings can be carried out in a "blank" medium prior to exposure to auxin (32). Systems known to be hard-to-root because of endogenous inhibitors also profit by such flushing or any other means to destroy or remove the inhibitor(s). Sometimes addition of activated charcoal to the medium to remove excess hormone etc. is helpful but the level of auxin must be increased to account for adsorption (3, 34). Striking or rooting of cuttings is a demanding art in some cases; very easy in others. Some investigators subcategorize Stage III by distinguishing between a Stage III MC (microcutting) and Stage III MS (multiple shooted) (71). In III MC a single microcutting is rooted; in III MS a multiple shooted cutting is rooted and this gives a "bushier" plant. Stage IV involves full establishment of the plant *ex vitro*. This requires skillful management but rarely involves added hormones outside of what may be normally used for a particular crop, e.g., growth retardants to keep plants short.

The horticultural nature of the micropropagation process involving precocious axillary branching is emphasized when seen in the context that plants so multiplied originate from small cuttings (see Fig. 2). Indeed, certain conifers have been manipulated with a combination of traditional and aseptic culture strategies. This involves repetitive cytokinin treatment of the intact tree at very high levels during periods of optimum growth so as to foster buds or shoots (usually from latent axillary bud meristems at the base of a needle cluster or at the apex of the fascicles). These forced shoots are juvenile in morphology and in some cases resemble those of newly germinated seedlings. They may be excised and further forced *in vitro* to produce additional buds which in turn can be separated and rooted (31, 32).

Another example derives from the use of a micropropagational strategy without the aseptic component. For example, leaf bud cuttings from cassava prepared from field grown plants and rooted under mist have been calculated as being able to yield 4,000,000 plants from a stock plant with 500 healthy leaves. As each propagated plant grows and produces additional leaves with buds, new material is available for propagation (31, 33, 35).

MULTIPLICATION VIA ORGANS OF PERENNATION FORMED IN ASEPTIC CULTURE

Some plants form organs of perennation *in vitro*. When this occurs one has the means for multiplication at another level and it may well turn out that direct planting or germ plasm storage of plants can be implemented by this means (35). Potatoes can form miniature tubers, gladioli can form cormlets, bulbils have been encountered in certain lilies, onion, narcissus, hyacinth, *Dioscorea* (35) etc. and, of course, protocorms are produced by orchids (3).

OVERALL DISCUSSION AND COMMENTARY

The problems of growth and development *in vitro* dictate that tissue culturists be bound to an informed empiric approach. Attempts have been made to show that the goals are not modest. One is, in essence, aiming to control growth and development and it is a far cry from merely using the "correct" levels of growth regulator (cf. 10)! Figure 3 indicates some of the points of response in a system and emphasizes why one often resorts to exhaustive media modification. Work on mutants (64), including hormone-resistant ones (38), shows it will be possible to engineer plants to be more amenable to regeneration *in vitro*, but it will take time.

It has been known for years that the growth regulators are only a part of the picture, but their role is very significant even though it may not always be direct (31). Work on embryogenic cultures of carrot shows, for example, that upon withdrawal of 2,4-D, the release of several extracellular proteins occurs; in the presence of 2,4-D the level of protein is reduced or even absent. Phytohormone thus controls the release of the appropriately glycosylated proteins so that somatic embryogenesis can proceed (13). It would be interesting to see if the same proteins are produced in embryogenic carrot systems that are initiated via stress treatments such as insult with heavy metal ions or oxidants like hypochlorite (29), i.e. initiated and maintained in the absence of exogenous hormone.

Still too little attention is paid to the role of components and parameters of *in vitro* systems other than hormones (26). Figure 4 provides some relationships between nutrition and various intra-, inter- and extracellular activities as they relate to cell division, enlargement, and differentiation. Embryogenesis and organogenesis, and their sensitivity relative to nutrition and various other system activities like osmotic values, are key responses to special signals (or an alteration or breakdown in signal processing). These, in turn, derive directly or indirectly from what occurs at a given place at a particular time (spatial and temporal responses). The high degree of

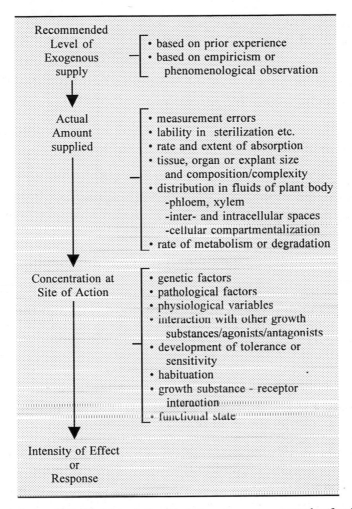

Recommended Level of Exogenous supply
- based on prior experience
- based on empiricism or phenomenological observation

Actual Amount supplied
- measurement errors
- lability in sterilization etc.
- rate and extent of absorption
- tissue, organ or explant size and composition/complexity
- distribution in fluids of plant body
  -phloem, xylem
  -inter- and intracellular spaces
  -cellular compartmentalization
- rate of metabolism or degradation

Concentration at Site of Action
- genetic factors
- pathological factors
- physiological variables
- interaction with other growth substances/agonists/antagonists
- development of tolerance or sensitivity
- habituation
- growth substance - receptor interaction
- functional state

Intensity of Effect or Response

Fig. 3. Some factors that affect the relationship between exogenous supply of a plant growth substance and the response it elicits.

communication between intra- and extracellular functional activities is represented by the dotted lines in Figure 4. The major paths for feedback and regulation are suggested by dashed lines. This scheme emphasizes that there are many points where complex interactions may occur. A major objective for tissue culture workers is to define the points that control progression and to evaluate how they interact and thus understand fundamental mechanisms of action.

Attempts to draw relationships between the nature and level of endogenous growth substances, culturability, and morphogenetic competence have not proven very instructive either theoretically or helpful practically for the very reasons implicit in Figure 4. The multiplicity of effects encountered in growing systems is, in part, a consequence of interactions and various

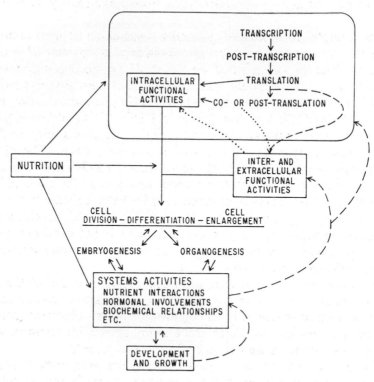

Fig. 4. Skeleton scheme of the relationships between nutrition and various intra-, inter- and extracellular activities as they relate to cell division, differentiation, and enlargement. Embryogenesis and organogenesis, and their sensitivity relative to nutrition and various other systems activities, are key responses to special signals. These in turn derive directly or indirectly from what occurs at a given place at a particular time. The high degree of communication between intra- and extracellular functional activities is represented by the lighter lines. The major paths for feedback and regulation are suggested by darker lines.

levels of sensitivity on the part of essentially heterogeneous cell populations in tissue and organ explants (9, 33). No doubt the prior history of the initial explant is critical. This is tacitly recognized, for instance, when one recognizes the distinct status of Stage 0 of micropropagation. It is also underscored in anther culture work or spore work where little more than a basal medium is normally used. There is no question that physiological state of the initial explant is critical to responsiveness.

Work in my laboratory has shown that external pH is an important experimental variable in the somatic embryogenic process. Preglobular stage embryos can be kept "cycling" and increasing in number provided the pH of the medium is kept below 4.5 (61). If the pH is elevated, the somatic embryos continue their development and proceed through the normally expected stages of embryogenesis---globular, heart, torpedo and cotyledonary (or the equivalent in a monocotyledonous system like daylily, *Hemerocallis*) (62).

The "simple" parameter of pH under these circumstances (where cells have never "seen" exogenous hormone) could arguably be viewed as a second messenger. The near-dogma that the "best" pH for a cell culture medium is such and such (usually around pH 5.6-5.8 or so) clearly now needs to be qualified. If one wants more embryogenic cells or preglobular stage embryos, then the pH should be kept low (below pH 4.5). (It should be emphasized that the low pH does not *confer* embryogenic status or capacity on the cells for they are already embryos, albeit of a very early stage. The pH works only on cells that are already in the embryogenic mode. It is a modulating agent---an important one--not a primary inducing one like 2,4-D (61).

Another factor involves nitrogen supply--be it ammonium or nitrate. Reduced nitrogen (ammonium) will support continued development--i.e. stages beyond preglobular stage embryos; nitrate will not and should therefore not be used alone to support continued somatic embryo development after induction-initiation has begun (36, 43).

No doubt many more controlling factors similar to that of the pH and source of nitrogen will be encountered. Again, Figure 4 draws attention to a few relationships that exist in an organizing culture. Nutrients (nature and concentration) are an important focal point of the interactions, and the hormonal involvements are but part of the entire process.

The wide range of successfully-cultured explants from a single test organism moderates arguments that one tissue source is always necessarily better than another. A rule of thumb among experienced workers is to utilize whatever information may be available as to the normal zones of rapid growth in selecting a primary explant. General analysis of growth substance content whether of auxins, cytokinins, gibberellins or abscisic acid and other inhibitors will be, at best, a rough guide to physiological status. Surely the location of growth substances within a cell or tissue complex is as important as the overall level. There are many interrelationships that can be and are set in motion when explanted tissues and organs are placed in culture. Even the very first step in the process, namely wounding upon excision and its release from positional information, induces many responses and alters the hormone pool. Wound-induced expression (56) of molecules like proteinase inhibitors, the activity of which are dramatically affected by sucrose, auxin or abscisic acid (see Chapter E5), portends that the details of such relationships, and their significance for growth and differentiation of cells *in vitro*, will take time to uncover and understand.

The chemical form, level and sequence of exposure to growth regulators all can play a role. Use of agents such as activated charcoal etc. to modify the rate of release, delivery or absorption of a growth regulator may be explained in some instances from such a perspective (3). These are part of the tissue culturist's armamentarium and can play a decisive role. One can modify response by carefully selecting material to be subjected to *in vitro* culture. Tissues may change in terms of their sensitivity to growth regulator etc. Certain tissues show a seasonal component as to culturability and this in part may be due to accumulation of inhibitors (4, 31, 71). Subtleties such as morphological polarity and position or placement of explant on substratum (e.g., adaxial vs abaxial surface of a leaf explant on agar) may also play a role (21). No doubt some investigators are better than others in getting cultures to grow because of their ability to "diagnose" the requirements.

Hormone receptors in the membranes of cells must perforce be key elements in initiation and those regulatory systems which allow added growth regulators to modify intracellular events and alter function and differentiation of the sorts mentioned. There has long been suggestion of the existence of a multiplicity of plant hormone receptors or binding proteins (see Chapter D4, 47). For instance, combinations of several auxins (or several cytokinins) have dramatically positive effects on culturability even though their combined concentration is no higher on a molar basis than one used singly. Additional evidence for this multiplicity continues to come from many areas including *in vivo* studies of hormone action, *in vitro* bioassays, binding experiments with radioligands, and selective protection and inactivation experiments. One can predict that various hormone receptors will be shown to share structural domains with other receptors which allow, in turn, a convergence of several hormonal or other signals onto common pathways. This would explain getting the same results through different treatments.

A better understanding of the identity and role of signalling molecules is continuously being sought (8), and in this volume several chapters indicate that the concept, first of the kinds of molecules serving as hormones or primary messengers is broadening, and secondly, understanding of the nature and role of secondary messengers and/or cell physiological controls must continue to be expanded.

The role of pH in embryogenic carrot cultures has been mentioned. Another example deals with the demonstrated effects of oligosaccharides in the morphogenesis of cell and protoplast cultures (56). The indication that production of transmembrane signalling molecules via phosphatidylinositol 4,5-bisphosphate (PIP_2) hydrolysis, namely *myo*-inositol-1,4,5-triphosphate (IP_3) and diacylglycerol, which in turn are able to release organelle-bound calcium, are important in cell function via a cascade effect is consistent with tissue culture findings (14, and see Chapter D5)). For example, calcium increases the yield of somatic embryos in embryogenic carrot cultures (27) and glycerol modulates citrus somatic embryogenesis (19). Exogenously added aliphatic polyamines (PAs) like putrescine, spermidine and spermine are yet another group of modulators and they can, among other things, enhance regeneration via organogenesis and somatic embryogenesis, but how they do this is still unresolved (16, 65, and see Chapter C1).

Plants have developed through evolution a number of "strategies" to deal with physiological and environmental constraints. Since higher plants are sessile organisms and cannot "run away from their problems", the adaptive mechanisms for dealing with less than ideal (i.e., stress) conditions are both numerous and redundant. Clearly plants are "overbuilt". Undoubtedly it will be some time before the range of mechanisms put into play as a result of stress sensing are recognized much less understood. Changing gene expression, deriving from changes in availability of growth regulators in the broadest sense of the word "hormone", accounts at least partly for what is known about development *in vitro*. A significant point is that the exquisite plasticity of plants becomes even more exaggerated as the level of spatial and temporal (positional) controls get broken down. *In vitro*, one sees that relatively large organ or tissue explants respond more or less predictably to hormone regimens and treatments since their positional controls have been minimally perturbed and they often behave much as if they were whole plants (31). As the unit size is reduced, i.e., as explanted tissues are broken down into free cells or their protoplasts, and their regulation is disrupted, the vulnerability, plasticity or developmental digressions become more apparent, and hence what needs to be imposed to achieve control becomes more precise.

Identification and characterization of the various hormone receptor subtypes will allow experiments to test the intrinsic specificity of hormone receptor interaction. On this basis, meaningful structure-activity relationships can be established for the domains of ligand binding and of activation of proteins by agonist-liganded receptor. Greater understanding of the cell biological and molecular events should allow the development of new growth regulators free of unwanted effects and selective in their expression of desired physiological responses first identified by empiricism.

References

1. Abe, J., Nakashima, H., Mitsui, K., Mikami, T., Shimamoto, Y. (1991) Tissue culture response of *Beta* germplasm: callus induction and plant regeneration. Plant Cell, Tissue and Organ Culture 27, 123-127.
2. Ammirato, P.V. (1989) Recent progress in somatic embryogenesis. Intl. Assoc. Plant Tiss. Culture Newsletter 57, 2-16.
3. Arditti, J., Ernst, R. (1993) Micropropagation of Orchids. Wiley, New York.
4. Bhojwani, S.S. (ed.) (1990) Plant tissue culture: Applications and limitations. Elsevier, Amsterdam.
5. Burrows, W.J., Leworthy, D.P. (1976) Metabolism of N, N'-diphenylurea by cytokinin-dependent tobacco callus: Identification of the glucoside. Biochem. Biophys. Res. Comm. 70, 1109-1114.
6. Carman, J. (1990) Embryogenic cells in plant tissue cultures: Occurrence and behavior. In Vitro Cell Develop. Biol. 26, 746-753.
7. Charlwood, B.V., Rhodes, M. (eds.) (1990) Secondary products from plant tissue culture. Clarendon Press, Oxford.
8. Chasan, R. (1991) Searching for signals. Plant Cell 3, 848-850.
9. Clowes, F.A.L., Macdonald, M.M. (1987) Cell cycling and the fate of potato buds. Ann. Bot. 59, 141-148.
10. Davis, T.D., Curry, E.A. (1991) Chemical regulation of vegetative growth. Critical Rev. Plant Sci. 10, 151-188.

11. Davis, T.D., Haissig, B., Sankhla, N. (eds.) (1988) Adventitious root formation in cuttings. Dioscorides Press, Portland.

12. De Jong, A.J., Cordewener, J., Lo Schiavo, F., Terzi, M., Vanderkerckhove, J., Van Kammen, A., De Vries, S.C. (1992) A carrot somatic embryo mutant is rescued by chitinase. Plant Cell 4, 425-433.

13. De Vries, S.C., Booij, H., Janssens, R., Vogels, R., Saris, L., LoSchiavo, F., Terzi, M., van Kammen, A. (1988) Carrot somatic embryogenesis depends on the phytohormone-controlled presence of correctly glycosylated extracellular proteins. Genes Develop. 2, 462-476.

14. Einspahr, K.J., Thompson, G.A. (1990) Transmembrane signaling via phosphatidylinositol 4,5-bisphosphate hydrolysis in plants. Plant Physiol. 93, 361-366.

15. Fellman, C.D., Read, P.E., Hosier, M.A. (1987) Effects of thidiazuron and CPPU on meristem formation and shoot proliferation. HortScience 22, 1197-1200.

16. Flores, H.E., Arteca, R.N., Shannon, J.C. (1990) Polyamines and ethylene: biochemistry, physiology, and interactions. Amer. Soc. Plant Physiol. Rockville, MD.

17. Frenkel, C., Haard, N.F. (1973) Initiation of ripening in Bartlett pear with an anti-auxin α(p-chlorophenoxy)isobutyric acid. Plant Physiol. 52, 380-384.

18. Fry, S.C., Street, H.E. (1980) Gibberellin-sensitive cultures. Plant Physiol. 65, 472-477.

19. Gavish, H., Vardi, A., Fluhr, R. (1991) Extracellular proteins and early embryo development in *Citrus* nucellar cell cultures. Physiol. Plantarum 82, 606-616.

20. George, L., Eapen, S., Rao, P.S. (1989) High frequency somatic embryogenesis and plant regeneration from immature inflorescence cultures of two Indian cultivars of sorghum (*Sorghum bicolor* Moench.). Proc. Indian Acad. Sci. (Plant Sci.) 99, 405-410.

21. Goh, H.K.-L., Rao, A.N., Loh, C.-S. (1990) Direct shoot bud formation from leaf explants of seedlings of seedlings and mature mangosteen (*Garcinia mangostana*) L. trees. Plant Science 68, 113-121.

22. Goldberg, R.B., Barker, S.J., Perez-Grau, L. (1989) Regulation of gene expression during plant embryogenesis. Cell 56, 149-160.

23. Gray, DJ.., Conger, B.V. (1985) Influence of dicamba and casein hydrolysate on somatic embryo number and culture quality in cell suspensions of *Dactylis glomerata* (Gramineae). Plant Cell, Tissue Organ Cult. 4, 123-133.

24. Hagen, S.R., Muneta, P. Augustin, J., LeTourneau, D. (1991) Stability and utilization of picloram, vitamins, and sucrose in a tissue culture medium. Plant Cell, Tissue Organ Cult. 25, 45-48.

25. Han, H., Yang, H. (1986) Haploids of higher plants *in vitro*. China Academic Publishers, Beijing; Springer-Verlag, Berlin.

26. Jackson, M.B., Mantell, S.H., Blake, J. (eds.) (1987) Advances in the chemical manipulation of plant tissue cultures. British Plant Growth Regulator Group Monograph No. 16.

27. Jansen, M.A.K., Booij, H., Schel, J.H., de Vries, S.C. (1990) Calcium increases the yield of somatic embryos in carrot embryogenic suspension cultures. Plant Cell Rep. 9, 221-223.

28. Kado, C.I. (1991) Molecular mechanisms of crown gall tumorigenesis. Crit. Rev. Plant Sci. 10, 1-31.

29. Kiyousue, T., Takano, K., Kamada, H., Harada, H. (1990) Induction of somatic embryogenesis in carrot by heavy metal ions. Can. J. Bot. 68, 2301-2303.

30. Kochba, J., Spiegel-Roy, P. (1977) The effects of auxins, cytokinins and inhibitors on embryogenesis in habituated ovular callus of the "Shamouti" orange (*Citrus sinensis*). Z. Pflanzenphysiol. 81, 283-288.

31. Krikorian, A.D. (1982) Cloning higher plants from aseptically cultured tissues and cells. Biol. Rev. 57, 151-218.

32. Krikorian, A.D. (1988) Plant tissue culture: Perceptions and realities. Proc. Indian Acad. Sci. (Plant Sci.) 98, 425-464.

33. Krikorian, A.D. (1989) The context and strategies for tissue culture of date, African oil and coconut palms. *In*: Applications of biotechnology in forestry and horticulture, pp. 119-144, Dhawan, V., ed. Plenum Press, New York.

34. Krikorian, A.D. (1994a) *In vitro* methods for plantation crops. *In*: Plant tissue culture--theory and applications, Vasil, I.K., Thorpe, T.A., eds. Kluwer Academic Publishers, Dordrecht.

35. Krikorian, A.D. (1994b) *In vitro* methods for root and tuber crops. *In*: Plant tissue culture--theory and applications. Vasil, I.K., Thorpe, T.A., eds. Kluwer Academic Publishers, Dordrecht.

36. Krikorian, A.D., Smith, D.L. (1992) Somatic embryogenesis in carrot (*Daucus carota*). *In*: Plant tissue culture manual: fundamentals and applications. Lindsey, K., ed., pp. PTCM-A9 1-32. Kluwer Academic Publishers, Dordrecht.

37. Kysely, W., Jacobsen, H.-J. (1990) Somatic embryogenesis from pea embryos and shoot apices. Plant Cell, Tissue Organ Cult. 20, 7-14.

38. Lincoln, C., Turner, J., Estelle, M. (1992) Hormone-resistant mutants of *Arabidopsis* have an attenuated response to *Agrobacterium* strains. Plant Physiol. 98, 979-983.

39. Lindsey, K. (1992) Plant tissue culture manual. Kluwer Academic Publishers, Dordrecht.

40. Meijer, E.G.M. (1989) Developmental aspects of ethylene biosynthesis during somatic embryogenesis in tissue cultures of *Medicago sativa*. J. Exp. Bot. 40, 479-484.

41. Meins, F. (1989) A biochemical switch model for cell-heritable variation in cytokinin requirement. *In*: The molecular basis of plant development, pp. 13-24, Goldberg, R., ed. Alan R. Liss, New York.

42. Messens, E., Dekeyser, R., Stachel, S.E. (1990) A nontransformable *Triticum monococcum* monocotyledonous culture produces the potent *Agrobacterium* vir-inducing compound ethyl ferulate. Proc. Natl. Acad. Sci. 87, 4368-4372.

43. Murray, D.R. (1988) Nutrition of the angiosperm embryo. John Wiley & Sons, New York.

44. Newcomb, W., Wetherell, D.F. (1970) The effects of 2,4,6-trichlorophenoxyacetic acid on embryogenesis in wild carrot tissue cultures. Bot. Gaz. 131, 242-245.

45. Nissen, P. (1988) Dose responses of gibberellins. Physiol. Plant. 72, 197-203.

46. Orr, J.D., Lynn, D.G. (1992) Biosynthesis of dehydroconiferyl alcohol glucosides: implications for the control of tobacco cell growth. Plant Physiol. 98, 343-352.

47. Palme, K., Hesse, T., Campos, N., Garbers, C., Yanovsky, M., Schell, J. (1992) Molecular analysis of an auxin binding protein gene located on chromosome 4 of *Arabidopsis*. Plant Cell 4, 193-201.

48. Payne, G.F., Bringi, V., Prince, C., Shuler, M.L. (1991) Plant cell and tissue culture in liquid systems. Hanser Publishers, Munich.

49. Peschke, V.N., Phillips, R.L. (1992) Genetic implications of somaclonal variation in plants. Adv. Genetics 30, 41-75.

50. Pollard, J.W., Walker, J.M. (eds.) (1990) Plant cell and tissue culture. Methods in molecular biology Vol. 6. Humana Press, Clifton.

51. Protacio, C.M., Dai, Y.-r., Lewis, E.F., Flores, H.E. (1992) Growth stimulation by catecholamines in plant tissue/organ culture. Plant Physiol. 98, 89-96.

52. Puite, K.J. (ed.) (1988) Progress in plant protoplast research: Proceedings of the 7th International Protoplast Symposium, Wageningen, Netherlands, Dec. 6-11, 1987. Kluwer Academic Publishers, Dordrecht.

53. Purse, J.G., Lee, T.S., Pryce, R.J. (1985) Stimulation of soybean callus growth by S-lathyrine. Phytochemistry 24, 897-900.

54. Raghavan, V. (1986) Embryogenesis in angiosperms: A developmental and experimental study. Cambridge University Press, Cambridge.

55. Redenbaugh, K. (1992) Synseeds. CRC Press, Boca Raton

56. Ryan, C.A., Farmer, E.E. (1991) Oligosaccharides signals in plants: A current assessment. Ann. Rev. Plant Physiol. Mol. Biol. 42, 651-674.

57. Samonte, J.L., Mendoza, E.M.T., Ilag, L.L., De La Cruz, N.B., Ramirez, D.A. (1989) Galactomannan degrading enzymes in maturing normal and makapuno and germinating normal coconut endosperm. Phytochem. 28, 2269-2273.

58. Schiavone, F.M., Cooke, T.J. (1987) Unusual patterns of somatic embryogenesis in the domesticated carrot: developmental effects of exogenous auxins and auxin transport inhibitors. Cell Differentiation 21, 53-62.

59. Showalter, A.M. (1993) Structure and function of plant cell wall proteins. Plant Cell 5, 9-23.

60. Smigocki, A., Owens, L. (1989) Cytokinin-to-auxin ratios and morphology of shoots and tissues transformed by a chimeric isopentenyl transferase gene. Plant Physiol. 91, 808-811.

61. Smith, D.L., Krikorian, A.D. (1990). Low external pH replaces 2,4-D in maintaining and multiplying 2,4-D initiated embryogenic cells of carrot. Physiol. Plant. 80, 329-336.

62. Smith, D.L., Krikorian, A. (1991) Growth and maintenance of an embryogenic cell culture of daylily (*Hemerocallis*) on hormone-free medium. Ann. Bot. 67, 443-449.

63. Smith, D.L., Kelly, K., Krikorian, A.D. (1989). Ethylene-associated phase change from juvenile to mature phenotype of daylily (*Hemerocallis*) *in vitro*. Physiol. Plant. 76, 466-473.

64. Thomas, H., Grierson, D. (Eds.) (1987) Developmental mutants in higher plants. Soc. Exp. Biol. Seminar Series 32. Cambridge University Press, Cambridge.

65. Tiburcio, A.F., Kaur-Sawhney, R., Galston, A.W. (1990) Polyamine metabolism. Biochem. Plants 16, 283-325.

66. Tsai, D-S., Arteca, R.N. (1984) Inhibition of IAA-induced ethylene production in etiolated mung bean hypocotyl segments by 2,3,5-triiodobenzoic acid and 2-(*p*-chlorophenoxy)-2-methyl propionic acid. Physiol. Plant. 62, 448-452.

67. Van Engelen, F.A., De Vries, S.C. (1992) Extracellular proteins in plant embryogenesis. Trends in Genetics 8, 66-70.

68. Vasil, I.K., Thorpe, T.A. (eds.) (1994) Plant tissue culture--theory and applications. Kluwer Academic Publishers, Dordrecht.

69. Wang, P.J. (1990) Regeneration of virus-free plants through tissue culture. Adv. Biochem. Engin. 18, 61-99.

70. Wickson, M., Thimann. K.V. (1958) The antagonism of auxin and kinetin on apical dominance. Physiol Plant. 11, 62-74.

71. Zimmerman, R.H., Debergh, P. (eds.) (1990) Micropropagation. Kluwer Scientific Publishers, Dordrecht.

72. Ziv, M., Ariel, T. (1991) Bud proliferation and plant regeneration in liquid-cultured Philodendron treated with ancymidol and paclobutrazol. J. Plant Growth Regul. 10, 53-57.

INDEX

A23187 312
Abi gene 31, 466, 679, 692-693
Abies alba 621
Abortion 349, 574
ABRE 683, 687, 688
Abscisic acid (ABA) 9-10, **140-155, 255-267, 677-693**
9'-cis-neoxanthin 463
ABA-aldehyde 463
abi gene 31, 466, 679, 692-693
abscission 3, 9, 505, 732
ACC 123
accumulation 732
action **255-267**
activity 152
alcohol 149
aldehyde 149
aleurone 154, 246
α-amylase 10, 153, 254, 261, 266, 291, 689, 691
  gene 30
  transcription 262
analogs 152, 691, 767
antigen 438
apo-carotenoid 463
Arabidopsis 24, 31, 148, 463-467, 680
aspirin 408
assimilate 662-663
ATPase 661
auxin 505
axillary buds 580, 744
barley 154
binding 255, 291-292
bioassay 152
biosynthesis 9, 33, **140-149,** 400, 472, 679, 681
  bean 141
  carotenoid 463
  farnesyl pyrophosphate 463
  inhibitor 677, 678
  fluridone 690
  gene 676
  pathway 141
  seeds 655
  tomato 141
bound 732
buds 141, 583
bulbs 739, 740
Ca²⁺ 299, 301, 304-305, 309, 310-313, 602, 604, 606, 691
calmodulin 312-313
carotenoids 145, 463, 612, 681

cell cycle, axillary buds 583
cell division 678
cell enlargement 612
chloroplasts 145, 149, 154
chromatography of 419
coleoptiles 691
concentration 141, 692
conjugation 742
correlative interactions 580-581
cotyledons 651
defense 10
deficiency 680
deficient mutants 148, 149, 185, 467, 677
desiccation 681, 682, 690
developing seeds 652, 655
disappearance 681
discovery 3
dormancy 3, 9, 140, 247, 399, 678, 680, 740, 741
dose-response curves 692
drought resistance 613
effects 9
efflux of protons 612
Em storage protein 28, 682-685, 688, 690
embryo 651, 652, 656, **677-690**
  cultures 678
  development 677, **681-682**
  maturation 680
  proteins 681
embryogenesis 682
embryonic axis 651
enantiomers 140, 149
  immunoassay 438
endosperm 651, 656
esters 152
ethylene 29, 123, 505, 657
exudation 611, 612
flc gene 462
flower formation 630
free 732
fructose-1,6-bisphosphatase 658
fruit 732-733
GA 10, 452
GA₃ 153, 263, 267
GCMS 425, 462-463
genes 351, 462, 689-690, 692
  expression 260, 674, **682-690**, 691
  transcription 10
germination 247, 255, 465, 467, 677-679, 682, 732
  Ca²⁺ 312

cereal grains 246, 247
  prevention 677
glucose ester 150, 152-153
glucoside 151, 153, 710
1'-glycoside 152
grain development 247
growth 25, 612
guard cells 310-311, 599, 601, 602-606, 691
GUS 30, 683, 690
HPLC 418, 419
hydraulic conductivity 612
hydroxylating enzyme 154
IAA 611
identification 415
immunoassays 437-438, 441
inactivation 152
insensitivity 692
invertase 664
ion flux 612
isomers 691
jasmonate 185 186, 405, 408
K⁺ 601
late embryo abundant (LEA) proteins 682
leaves 580, 600, 654-656
levels 140, 141
maturation 674, 677
metabolism **149-155**
mitotic index 583
molecular ion 425
mRNA 236, 262, 682
mutants 24, 351, 406, 410, 452-453, 455, **462-467,** 472, **679-681**
  biosynthesis 400, 465
  insensitive 467, 677
  water relations **462-467,** 612-613
  wilty 612
myb 267
nature 9
not 462
nuclei 583
osmotic stress 28, 409, 692
perception 256
permeability 611
pH 606, 691
phaseic acid 169
phloem 612, 656
phospho-inositide metabolism 602
phospholipids 601
photoassimilate 661
photorespiration 657
photosynthesis 611, 655, 657
Pin2 gene 399, 405

ABA continued:
plasma membrane 601, 681
pods 656
pools 479
precocious germination 674,
678, 681, 780
promoter 30, 683-684
protein synthesis 9, 261
proteinase inhibitor 10, 399-
405
proton-ATPase 691
quiescence 681
receptor 28, 291-292, 602,
691-692
redistribution 28
regulation 140
response pathway **691-693**
responsive genes 466, 688,
691-692
promoters (element) 30,
266, 683-685, 687-688
responsiveness 28, 30, 681
ribulose 1,5-bisphosphate
carboxylase 657
ribulose-1,5-bisphosphate
oxygenase 657
ripening 726, 732
root 32, 33, 599, 611, 656,
742
growth 9, 25, 455
pressure 612
stress 656
Ca²⁺ 691
salinity 690
seed 9, 140, 141, 655, 732,
733
development 466, 651,
671, 673, 677, 679, 680
dormancy **462-467**, 680
germination 462
maturation 677
desiccation tolerance 467
senescence 501, 745
sensitivity 25, 27, 28, 33,
351, 679, 680, 690, 692
seed development 676
shoot 9, 25, 32, 655
signal transduction 466, 606,
691-693
sink tissues 654
sit gene 462
soil moisture 25
somatic embryo development
780
sucrose phosphate synthetase
658
stomata 9, 27, 33, 311, **598-
606**, 607, 610, 681, 692,
745
Ca²⁺ 691

closure 152, 611, 655, 657,
691
conductance 600
opening 602, 609
storage 153, 742
organs 726
protein 679
structure isomers 140
sucrose 612, 664, 732
synthesis, mutant 400
tissue culture 780
touch 567
trans ABA 427
transpiration 33, 478, 599
transport 9, 25, 32, 33, 153,
599, 656
tuberization 710, 714, 717,
719, 720
turgor 599, 612
uses 767
viviparous mutants 26, 467,
680
VP1 gene 689, 690
water **598-606**, **611-613**
deficit 598
permeability 612
potential 25, 33, 598, 599,
612
relations 679
stress 9, 27, 140-141, 148,
409, 462-463, 611-613
wilting 598
wilty mutants 462, 467
wound(ing) 10, 396, 399-405,
690
xanthoxin 463
xylem 33, 600, 655, 656
Abscisin II 9
Abscission 761
ABA 3, 732
apple 734, 753, 755
auxin 5, 29, 505, 734, 754-
755
axillary bud 591
cherry 760
citrus 755, 768
cytokinins 505
ethephon 752, 760, 761, 768
ethylene 29, 372, 496, 505,
697, 726, 752, 755, 761,
768
binding 292
flowers 349
fruit 503, 754, 755, 757, 768
jasmonate 181
leaf 588, 742
nutrition 495
ripening 726
storage, cabbage 742
zone 495-496, 589

Acacia 622
ACC (1-aminocyclopropane-1-
carboxylic acid) 8, 29,
118-128, 131, 160, 189,
208, 343, 375, 377-378,
488
analogue 378
biosynthesis 122
breakdown 729
chilling 731
daminozide 766
E8 gene 389
ethylene 489, 730
isolation 120
jasmonate 737
level 122
N-malonylation 128
N-malonyltransferase 128
mechanical perturbation 567
metabolism 120, 386
methionine 130
oxidation 127
petals 500
pollen 499-500
polyamines 131,170
production 122
putrescine 131
ripening 729
senescence 497
spermidine 737
stigma 499-500
synthesis, regulation 123
transport 492, 500
wounding 398
yeast 378
ACC deaminase 343, 348,
386-389
apical dominance 346
antisense 390
ethylene 390
gene 344
tomato 729
ACC N-hydroxylase 127
ACC N-Malonyltransferase 128-
129
ACC oxidase 82, 120, **124-128**,
343, 375, **377-380**, **382-
385**, 387-388, 498, 729
amino acid sequence 125, 380
antisense **382-385**, 389, 390
fruit 386
gene 348-349, 379
RNA 377, 382, 729
ripening 384
senescence 384
tomato 382
cantaloupe 730
cDNA 124, 125, 730
chilling 732
cloned 128

ACC oxidase continued:
  cloning 377
  ethylene 489
  genes 730
  identification 377
  isolated 135
  jasmonate 737
  mRNAs 386
  nature 380
  pAE12 apple 125
  pAVOe3 avocado 125
  PCH313 peach 125
  pHTOM5 125
  present 128
  purified 135
  pSR120 carnation 125
  pTOM13 tomato 125
  reaction 128
  ripening 387
  senescence 385, 497
  stigma 499
  stress 128
  TOM 13 378
  tomato 375, 377, 378, 730
  wounding 375, 378, 730
ACC synthase **120-124**, 375, **380-382**, 729
  antibodies 380
  antisense **385-388**, 390
    construct 385, 389
    gene 348
    RNA 729
  Arabidopsis 459
  auxin 123, 733
  brassinolide 208, 210
  cDNA 121, 123, 210, 380
  chilling 731
  cloning 380
  DNA sequences 380
  ethylene 123, 489, 733
  gene 121, 123, 349, 380, 730
  IAA 121, 210
  induction 122
  isoforms 121, 123
  isolated 135
  molecular weight 378
  mRNA 123, 731
  polyamines 170
  purified 135
  ripening 387
  senescence 497,
  stress 121
  tomato 375, 380, 385, 731
  wounding 121, 730
Accel 767
Acer 108, 537, 621, 622, 779
Acer pseudoplatanus 108, 621, 779
Acer rubrum 622
Acetonitrile 418

Acetosyringone 189, 322, 330
Acetylcholine receptor 516
Acetylene 132
Acetyleneacetal 767
Acetylsalicylic acid 188, 406
O-Acetyl serine 111
Acid phosphatase 254
Acid-growth 219-222, 286-289
Acropetal 397, 512, 523, 541, 542, 557, 559, 591
Actin 307
Actinidia deliciosa 732
Action potentials 399
Activated 210
Activator, transcriptional 322
Acylcyclohexanedione 79, 85-88
Adenine 7, 330, 462
Adenine phosphoribosyl transferase 108-109
Adenosine 330, 579
Adenosine kinase 109
Adenosine monophosphate 326
  transferase 327-330, 332
  -Δ2-isopentenyl transferase 105
Adenosine phosphorylase 109
S-Adenosylmethionine 103, **119-121**, 129-131, 160, 170, 343, 375, 390, 489, 737
  mechanical perturbation 567
S-Adenosylmethionine decarboxylase 163, 174, 390
  dormancy 741
  potato 741
S-Adenosylmethionine synthase 131
Adenylate cyclase 274, 275
Adjuvant 433
ADPG pyrophosphorylase 703
Adult 622, 625
Adventitious roots 541, 164
Affinity 13, 14
Ageotropic 553
Aging 497, 505, 657, 738, 746
Agrichemical industry 770
Agrobacterium rhizogenes 24, 287, 288, 320, 359, 361
  auxin 360
  cytokinin 110
  genes 350
  hairy roots 362, 586
  Ri plasmids, root loci genes 586
  rol genes 359, 361
  tumors 362
Agrobacterium tumefaciens
  auxin 318, 360
    biosynthetic genes 324, 345
  crown gall 319-321, 586, 775
  cytokinin 103, 318

  /auxin ratio 21
  biosynthesis 104, 105, 327-328
  prenyl transferases 334
  Em-GUS 685
  genes 325, 327, 350
    expression 189
  genome 320
  hormone biosynthesis 456
    metabolism 341, 456
    regulation 350
  IAA synthesis 48
  ipt 241, 327
  opines 323
  ornithine decarboxylase 175
  plasmid 322
  promoter 238
  rolB 361
  T-DNA 364
    genes for IAA, cytokinin 586
    insertion mutagenesis 450
    ipt 342
    tagging 450
  tms genes 333
  tomato 374, 375
  transformation 374-375, 457
  tumorigenesis 319
  virulence 322
Agrostemma githago 620, 631
ain1 gene 134
Air pollutants 780
D-Alanine 129
Alanyl zeatin 110, 112
Alar 770
Albumin 433
Alcobaca (*alc*) tomato 167, 390, 728, 736
Aleurone 246-249, **250-255**, 256, 260, 264
  α-amylase 248
    isozymes 262
  ABA 153, 246, 247
  auxin 652
  Ca^{2+} 299, 305, 312
  GA 246, 247, 254, 259
    receptor 291
  hydrolases 249, 250
  mRNA levels 261
  nuclei 262
  oat 256
  protoplasts 262
  RNA 259
Alfalfa 172, 182
Alkaline phosphatase 435, 438
Alkaloids 72, 161, 415
Alkanes 132
Allelopathic compounds 188
Allium cepa 739

β-(6-Allylaminopurine-9-yl)
    adenine synthase 110
Almond 111, 631
Alnus glutinosa 206
Alternative oxidase 191, 193-194
Althaea rosea 620
α-Amanitin 685, 731
Amaranthus 100, 417, 441
Ambrosia artemisifolia 599
Ambrosia trifida 599
Amino acid 10
    abbreviations 122
    auxotrophs 452
    sequence 125, 134, 174,
        235-237, 241, 260, 285,
        325, 355, 358, 359, 380,
        504, 686
D-Amino acid malonyltransferase
    129
4-Aminobutyraldehyde 160
4-Aminobutyric acid 160
1-Aminocyclopropane-1-
    carboxylic acid - see ACC
1-Aminocyclopropane-1-carboxy-
    lic synthase - see ACC synthase
Aminoethoxyvinylglycine - see
    AVG
l-Amino-2-ethylcyclopropane-1-
carboxylic acid 124
α-Aminoisobutyric acid 125, 129
Amino(oxyacetic) acid (AOA)
    120, 131, 208, 489
Aminopeptidases 253
Aminopropyl group 131
Aminopropylpyrroline 160
4-Amino-3,5,6-trichloro-pyridine-
    2-carboxylic acid 778
AMO-1618 18, 85, 638, 763
Amp1 gene 457
Amy32b gene 264, 265
α-Amylase **246-267**
    ABA **246-267**, 291, 689, 691
    aleurone 248, 251
    calcium 304, 312
    cDNA 258, 261, 263, 264
    cell-free translation 260, 263
    cis-acting factors 263
    GA 6, 30, **246-267**, 735
    GA₃ 254, 258, 260-266
    genes **257-267**
        ABA 30
        barley 30
        cis-acting factors 266
        expression 265
        promoters 263-266
        trans-activation 267
        transcription 266
        transcripts 261
    germination 246-249
    gibberellin 246, 247

high temperatures 247
    induction 255
    inhibitor 681
    isozymes 252, 255, 257-258,
        260, 263
    mRNA 258, 260-263, 291
    polyamines 169
    promoter 263-266
    reporter 265
    secretion 304
    trans-acting factors 263, 266
β-amylase 252, 253
Amylopectin 252
Amyloplasts 443, 512, 554, 556,
    664
Amylose 252
An1 genes 467
Anaerobiosis 16, 234, 511, 513
Ancymidol 85, 752, 763
Androgenesis 777
Anethum graveolens 620, 624
Animal hormones 274-276
Anise 441
Annuals 618, 619
Anthephora pubescens 607
Anthers 777
Anthesis 617
Anthocyanin 191, 622, 681, 689,
    734, 735, 766
Anthranilate 43, 46
Anti-auxins 778
Anti-gibberellins 780
Antibodies 357, **433-444**
    auxin-binding protein 282,
        284-288
    anti-idiotypic 256, 281, 290-
        291
    defined 433
    ethylene binding protein 291-
        292
    maize, ABP 282, 286-289
    primary 442
    rolB 362
    secondary 442
Antiecdysteroid 210
Antigens **433-444**
Antiporter 302
Antirrhinum majus 378
Antirrihinum 477
Antisense **375-377**
    gene 125, 340
        ACC oxidase 343-344,
            348, 379, **382-385**
        ACC synthase 343, 348,
            385-388
    mRNA 374-376, 703, 729
        ACC oxidase **382-385**, 377
    ethylene **377-387**
    ripening 377
    transgenic plants 375

Antiserum 433
Aphids 190
Apical dominance 509, **572-594**,
    785
    ABA 580
    apical bud 594
    auxin 345, 360
        transport 525
    axillary buds 576, 589
    Ca²⁺ 592
    correlative control 576
    cytokinin 110, 345, 346, 360
    ethylene 345, 581
    flowering 360
    IAA 594
    iaaL gene 456
    luxA&B genes 588
    mitosis 583
    rhizomes 582
    rol plants 362, 588
    shoot apex 576
    stolons 582
    transgenic plants 345
Apical growth 573
Apical meristem 532
Apical senescence 474, 477
Apium graveolens 745
Apo-proteohormone 288
Apple 171, 727, 761
    ABA 33
    abscission 734, 753-755
    ACC 120, 124
        oxidase 125
        synthase 121
    auxin 734
    cytokinin 110
    daminozide 752
    dominant fruits 578
    ELISA 441
    ethephon 755, 761
    ethylene 727, 730-731, 755
    firmness 737
    fruit set 753, 757
    fruit shape 767
    GA 67, 734
    GA₄/₇ 756, 758
    IAA 578
    jasmonate 737
    juvenility 621
    lateral branching 767
    naphthalene acetic acid 752,
        755, 761
    polyamines 736-737
    quality 758
    ripening 372, 729
    russeting 734
    storage 726
    thinning 752, 755
Apricot 631
Aquatic plants 702

Arabidopsis thaliana **449-453**, **463, 465-466, 471**
ABA 31, 148, 186, 465, 692, 693
  -deficient mutant 663, 676
  -insensitive mutants 677
  mutants 465-467, 679, 680
  pools 479
  seeds 677
abi gene 31, 692
ACC synthase 459
adenine phosphoribosyl transferase 109
S-adenosyl methionine 131
apical dominance 346
aux mutants 454
auxin 24, 231
  -binding protein 355-356
  biosynthesis 457
  mutants 453
  overproducing 581
  perception gene 350
  regulated genes 237, 239
CaM 306, 307
cosmid 454
ctr gene 350
cytokinin 108, 455, 457
developmental phenotypes 465
embryo 675, 682
EMS mutagenesis 363
ethylene 24, 133, 581
  binding, 291, 503
  biosynthesis pathway 459
  insensitive mutants 291
  mutants 457, 501, 503-504
  perception 391
  resistant mutants 501,
  response pathway 453
  senescence 501
Etr gene 350
flowering 473, 476-477
G-box-binding bZIPs 688
GA 70, 75, 82-83, 472, 632
  activity 87
  biosynthesis **82-82**, 88, 471
  deficient mutants 27, 676
  dwarf 465
  flowering 478
  genes 236-238, 351, 449-453
genome 449
genomic subtraction 450
germination 465
hormone mutants 344, 449-453
IAA 46
internode length mutants 467
ipt transformed 105
jasmonate 182, 185-186
map-based cloning 454
mutants 24, 364, **449-453**, 478

promoters 237
response mutants 471
SAUR 237
seed 186, 465-466
senescence 712
starchless mutant 554
transgenic plants 344
tryptophan mutants 456
viviparous mutants 673
wilty mutants 463, 465
wounding 410
Arabinofuranosidase 254
Arabinoxylans 252
Arachidonic acid 56, 410
Archaebacteria 359
Arginase 173
Arginine 122, 158, 161, 169, 171, 173, 334, 737, 741
Arginine decarboxylase (ADC) 158, 161-162, 167-174, 737, 741
  activity 171, 737
  cytosol 162
  dormancy 741
  mRNA 737
  polyclonal antibody 174
  potato 741
Aroma 730, 731
Artichoke 356, 705, 712, 738, 741, 752, 758
Arum 11, 201
Asclles 434
Ascorbate 78, 81, 126, 380, 735
Aseptic culture 775
Ash 621
Ashby's law 34
Asparagine 122, 127, 334
Asparagus officinalis 745
Aspirin 188, 406 409
Asplenium 488
Assimilate 5, 210, 475, 623, 624, 658, 661-664, 666, 702, 703, 755, 769
  genes 475
  partitioning 769
  tuberization 702
ATP 130
ATPase 27, 224-225, 254, 286, 565
ATPase/proton carriers 661
Aux gene 364, 453, 454
Auxin - see also Indole-3-acetic acid (IAA) 4-5, **39-61**, **214-225, 228-242, 318-335, 354-367, 534-539**
  abscission 5, 495, 734, 754-755
  ACC 123
    synthase 123, 733
  achenes 653

acid-growth theory **219-222**, 286-289
Agrobacterium 320, 324
aleurone 652
anaerobiosis 16
analogues 287
antagonist 366, 778
anthocyanin 734
antibody 282
apical dominance 5, 345
application 734
assimilate partitioning 5
axi 159, 366
axillary buds 576, 588
basipetal transport 31, 32, 397, 510-512, 523, 534, 535, 537, 542, 549, 550, 556-558, 567, 577 579, 588, 590-592
binding protein (ABP) 24, **277-289, 354-358**
  antibodies 285-288
  classes 278, 281
  fruits 516
  maize 220, 286
  NAA 286
  NPA 278
  photoaffinity labeling 516
  roots 516
  site 356
  stems 516
binding sites 277, 284, 733
bioassays 283
biosynthesis 4, **39-50**
  Agrobacterium 341
  inhibitors 778
  genes 318
  mutants 456
  prokaryotes **318-335**
bound 50
brassinolide 208, 210
buds 784-5
Ca^{2+} 299, 301, 304
callose 543, 544
callus 778, 781
CaM 306
cambial initials 542
carrier 514
caulonemata 462
cDNA 733
cell division 5, 228-229
cell elongation 4, **214-225**, 283
  acid-growth theory **219-222**
  ATP 217
  ATPases 217, 220-221, 224-225
  auxin-binding protein 220

Auxin - cell elongation cont:
  Avena coleoptile 215, 217, 218, 222-225
  Ca^{2+} 221
  cell turgor 217
  chloride channels 221
  coleoptiles 215, 217-218, 222-225
  collenchyma 217
  cycloheximide 220
  cytoplasmic pH 221
  diacylglycerol 221
  dicot stem 215, 217, 229
  epidermis 217
  expansin 223
  G-protein 221
  β-glucans 222
  growth maintenance 223
  H$^+$ 222
  hemicelluloses 222
  inhibitors 216
  Instron stress-strain analysis 218
  IP3 221, 225
  leaf mesophyll protoplasts 220
  monocot 229
  neutral buffers 219
  osmoregulation 223, 225
  phospholipase 221
  phosphorylation 221, 225
  plasma membrane 220-221, 224
  protein kinase 221
  protein synthesis 217
  protons 219-220
  receptor 220, 224
  signal transduction 455
  solutes 223
  sucrose 215
  sugar 216
  time course 215
  tonoplast 221
  turgor 218, 219, 223, 225
cell wall acidification 223
  crosslinks 225
  extension 217
  extensibility 217, 218
  loosening **219-223**
  protein 738
  yield threshold 218
  xyloglucan endotransglycosylase 222
  xyloglucans 222, 225
cellulose microfibrils 223
chemical forms 40-41
coleoptiles 215, 217-218, 222-225, 576
compartmentation 15
concentration 537

conjugates 40, 342, 358-363, 510
  iaaL gene 456
correlative interactions 576-578
cuttings 495, 753, 787
cytokinins 29, 318, 540, 778
deficient mutants 456
developing seeds 651
differentiation 532
diffusible 15
dissociation constant 25
effects 4
efflux 16, 522
  carrier 517-520, 525
elongation, maintenance of 223
embryo 652, 676
embryogenesis 675
ethylene 494, 590, 629, 733
export carriers 282
extractable 15
female flowers 642
flowering 5, 629
flux polarity 589
fruit 733, 757
  abscission 754
  growth 676
  ripening 5, 726, 733-734
  set 753, 754, 757
  thinning 754
GA 29
gametophore 462
GC-MS 422
gene expression **228-242**, 734
  adventitious root 239
  antibodies 242
  antisense constructs 242
  asymmetric growth 239
  auxin-responsive 232
  cadmium 235
  cDNA 230, 232
  DNA 229, 236
  flowers 239
  gene families 237
  gravitropism 239, 521
  growth 238
  heavy metals 235
  in vitro translation 230
  molecular approaches 229
  mRNA 229, 230-232, 234, 235
  phototropism 239
  polyribosome 229
  post-transcriptional events 233
  prokaryotes **318-355**
  promoter 237-238, 242
  protein synthesis inhibitors 233, 234

protein content 229
recombinant DNA 230
regulation 239
reporter genes 237-239
RNA 229
RNA polymerase I 229
roots 239
salicylic acid 235
stress 234
trans-acting factors 242
transcription 230, 233
transgenic plants 239, 240, 242
tropisms 240
germination 454
β-glucuronidase (GUS) 565
gradients 495, 521, 538, 544, 565
gravitropic response 555-561, 565-567
growth 20, 32, **223-224**
habituation 778
herbicides 756
insenstivity 731
intact plants 20
invertase 565
kinetin 785
lateral bud 785
leaf primordium 21
leaves 543
lux genes 587
mechansism of action **354-367**
membrane hyperpolarization 24, 26
metabolism 39, **53-60**
microfibrils 223-224
micropropagation 786
mRNAs 733, 756
mutants 24, 360, 363, **453-457**
nature 4, 40-41
NPA 521, 565
organogenesis 781
polyamine 169
parthenocarpic fruit 736
perception **277-289**
pericarp 652
pericycle 523
phloem 535, 543, 666
photosynthesis 666
phototropism 32, 549-550
phytotropins 524
Pin2 gene 398
plant organ storarage 726
polar transport 509-511, 536, 543
polarity 588-592
pollen 754
precursor, seed 52
prokaryotes, genes 318
proliferation 277

Auxin continued:
  promoter 22, 565
  protectants 779
  protein synthesis 215
  proton translocation 26, 27, 30
  protoplasts 24, 26
  Pseudomonas savastanoi 325
  putrescine 161
  1-pyrenoylbenzoic acid 517
  radiolabeled 510, 512
  receptor 219, **277-289**, 311,
    354, 364, 561
  redistribution 15
  regeneration 277
  regulated channel protein 517
  reporter gene 565
  response mutants 453
  responsive genes **232-241**
    promoter 521
  responsiveness 30
  RNA 565
  root/ing 25, 543, 781
    cuttings 495
    growth 5, 455, 522
    initiation 5, 31, 240
    tissue culture 787
  seed 671, 676, 733
  senescence 5
  sensitivity 24, 30, 516, 522,
    539, 544
  shoot 25, 522, 576, 588, 776
  sieve-tube 542
  signal transduction 232, 236,
    277-289, 455
  slender pea 23
  somatic embryogenesis 782,
    784
  stem growth 4, 20, 31
  stomata 311, 606-607
  stored 48
  synthetic 4, 778
  TIBA 511, 514, 521, 565
  tissue culture 774, 775, **778-
    782**, 792
    polarity 588-592
  transduction **277-289**
  transgenic plants 341, 345,
    360
  transmenbrane potentials 282
  transport 4, 30-32, **51-52**, 208,
    239, **509-530**, 536, 559,
    588
    acropetal, velocity 512
    anaerobiosis 513
    apical dominance 525
    axillary bud 59, 589
    channels 589
    chemiosmotic hypothesis
      513-514
    dominance 589

  ethylene 755
  fruit set 754
  gravitropism 521-522, 525
  hypothesis 513
  inhibitors 355-356, 359,
    511-512, 517-520, 522,
    523-524, 565, 778
  lateral root growth 523
  leaf abscission 589
  light 521, 525
  nodulation genes 525
  NPA 511, 513, 514, 525
  phloem 510, 544
  phototropism 521, 525
  polar 510-514, 519, 589
  polarity 511, 594
  Rhizobium nodule 524
  regulation 524-526
  role 520-524
  roots 522-524
  system 534
  TIBA 511-512, 514
  wounding 512
  xylem 510
tropisms 5, 31
tuberization 712
uptake carrier 514-517
uses **753-756**
vascular bundles 534
  development 21
  differentiation 5, 31, 347,
    531, 534-539
  tissues **531-539**
water stress 610
wounding 398, 512, 536
xylem 535, 543
Auxin-like 39, 49, 285, 323, 613,
  652, 734
Avena 100, 162, 168, 170 172,
  174, 199, 215, 216, 256,
  260, 262-264, 299, 306,
  509, 551, 558-564
  α-amylase 260, 262
  arginine decarboxylase 174
  auxin 216, 549
    translocation 509
  ATPase 307
  Ca²⁺ 299
  coleoptile 549
  ethylene 561
  germination 256
  gravistimulated 561
  IAA 53, 548
  inflorescence 573
  invertase 563
  lateral buds 573
  NPA 356
  polyamines 168-171
  pulvini 560-561
  salicylic acid 200

  seedlings 509
Avena fatua 554, 559, 564
Avena sativa see Avena
AVG 120, 123, 374, 390, 459,
  489, 500, 581, 629, 730,
  780
Avidin-biotin complex 435
Avocado 72, 126, 144, 372, 592,
  727-729, 732, 736
Axi gene 159, 366
Axillary bud **572-573**, **583-588**,
  768
  ABA 579-580, 744
  abscission 591
  apex 583
  auxin 576, 585, 591
  Ca²⁺ 593
  cell differentiation 593
    division 593
    polarization 591
  correlative control 571-572,
    591
  cytokinin 105, 579, 585, 594
  dormancy 585
  gibberellins 579
  growth 576, 589, 591
    inhibition 572
    polar IAA transport 591
    proteins 585
  IAA 585, 591, 593
    inhibition 579
  lux gene 581
  outgrowth 576
  polar axis 593
  quiescence 572
  rol genes 587
  tuberization 698
  zeatin 578
Axillary branching 784
Axillary buds / shoots 573, **583-
  594**
Axr genes 350, 453-455, 364
Az34 gene 679
7-Azaindole 778
Azide 216
5-Azidoindole-3-acetic acid 285
Azospirillum 335
Azotobacter 332, 335

Bacteria 21, 43, 44, 48, 50, 103,
  104, 120, 172, 318-320,
  326, 332, 334, 335, 341,
  344, 358, 360, 457
  auxins 318
  cytokinins 318
Bacteriophage 454
Bactris major 193
Bamboos 553

Bananas 372, 374, 734-735
  2,4-D 734
  ethylene 34, 502, 727
Barley **247-267**, 501, 554, 651-
  652, 661, 663, 756, 765
  α-amylase **246-267**
    gene 30
    isozymes 252
    promoter 263-264
  ABA 153, 679
  aleurone 250-255, 306
  Ca^{2+} 302, 304, 312
  calmodulin 306
  cytokinin 290, 675
  embryo 682
  germination 169, 246, 247
  grain weight 676
  IAA 676
  jasmonate 182, 405
  LEA proteins 690
  malting 756, 758
  mutants 363
Bean 537, 652, 662
  ABA 145
  brassinolides 211
  ELISA 441
  ethylene receptors 133
  GA 88
  germination 673
  IAA 49, 54
  xanthophyll 147
Beech 621
Beet 309, 620, 661, 664, 738,
  739, 742, 775
Benzoic acid 198-200, 356, 358,
  359, 511, 517
  2-hydroxylase 198-199
Benzonitrile 127
Benzyladenine
  (6-benzylaminopurine) 100,
  101, 109-110, 112, 123,
  290, 455, 714
  alanyl conjugates 110
  $GA_{4/7}$ 756
  glucosides 109
  lateral bud 767
  phloem loading 659
  photoassimilate unloading 661
  riboside 109
  senescence 745
  stomata 745
  tissue culture 779
  uses 745
  vegetables 745
  vessels 540
Beta vulgaris 620, 624, 739
Betacyanin 162, 208, 417, 739
Betula pubescens 621
Biennials 619, 725
Binding proteins

abscisic acid 291
auxins 277
Ca^{2+} 303, 305
cytokinin 290
NPA 525-526
nucleic acid 361
Binding sites
  ABA 292
  ethylene 292
Bioassays 17, 100, 101, 109, 111,
  152, 153, 190, 206-208,
  416, 417, 420, 441, 467,
  471, 510, 549, 708, 711,
  712, 716, 753, 779, 780
  auxin 283
  cytokinins 417
  gibberellins 417
Biolistics 263
BiP-related protein 254
Birch 621
Bis-trimethylsilyl acetamide
  (BSA) 422
Bisphosphate carboxylase 678
Bis(phosphonomethyl)glycine 751
Blackberry 727
Blitum capitatum 624
Blitum virgatum 624, 631, 636
Blooming 191, 193
Bolting 6, 18, 631, 633, 636,
  639, 639, 741, 759
  GA 741
  GA_1 639
Bonzi 764
Botrytis 172, 745
Boysenberry 727
Brachypodium pinnatum 575
Bradyrhizobium 189
Branching 572
Brassica 11, 54, 56, 182, 185,
  206, 467, 618, 620, 624,
  632, 678, 742
  brassinosteroids 11
  IAA metabolism 56
  internode length mutants 467
  jasmonate 185
Brassica napus 182, 206, 632
Brassica nigra 624
Brassica oleracea 618, 620, 624,
  742, 745
Brassica rapa 54
Brassinolide 11, 179, 207
Brassinosteroids 3, 11, **206-211**
Brassins-see above
Broccoli 745
Brodiaea 491
Bromeliads 629
Brown algae 532
Brussels sprouts 618, 620, 745,
  746
Bryonia dioica 183

Bryophyllum daigremontianum
  620, 621, 624-626, 633,
  635
Bryophytes 460
Buckwheat 120
Buds 534
  ABA 141
  auxin 784
  axillary 574
  culture 784
  cytokinin 460, 784
  differentiation 166
  kinetin 784
  lateral 785
  spermidine 166
  terminal 574
  tuberization 698, 699, 715,
  719
Bulbs 725, 739
1-Butene 132
BX-112 18-19, 85, 763
bZIP 686, 688, 689, 693

C-4 grasses 551
C19-GAs 66, 67, 72-74, 78, 79,
  81, 82, 86, 88, 470
C20-GAs 66, 79, 81
Cabbage 171, 726, 741-745, 746
Cadaverine 161
Cadmium 235
Calcineurins 692
Calcium **298-313**
  ABA 292, 299, 301, 304, 310,
  312-313, 604, 605, 691
  acropetal transport 557
  aleurone 299, 310, 312-313
  α-amylase 304, 312
  animal hormones 274
  apical dominance 592
  apoplast 605
  ATPase 298, 301, 306, 602
  auxin 221, 288, 299, 301,
  304,593
  axillary bud 593
  binding 298, 689, 692
    protein 254, 305-307
  brassinolide 208
  callose synthase 395
  carriers 302, 592
  cell elongation 557, 592
    polarization 592
    wall 302
  channels 300, 453, 592, 602
  cold shock 299
  cytokinins 299, 301, 311, 593
  cytoplasmic pH 309
  cytosolic 303
  deficiency 593
  differentiation 592
  distribution 558

Calcium continued:
  dominant organ 593
  ED(G)TA 557
  endoplasmic reticulum 304
  enzymes 602
  ethylene 134
  florescence imaging 605
  G-proteins 602
  germination 312
  gibberellin 291, 299, 301,
    304, 312-313
  gradient 592
  gravitropism 52, 299, 556-
    557, 565, 592
  guard cell 299, 310-311,
    603-605
  homeostasis 300, 303-305
  hormones 298, 592, 602
  IAA 557, 592
  inositol phosphates 308
  light 299
  membrane potential 309
  mitosis 592
  morphogenesis 792
  moss 311-312
  NPA 526
  peptide hormone 275
  plasma membrane 300, 304,
    312
  polar transport 557
  polarity 592-594
  polyamines 167
  potassium channels 310, 605,
    692
  protein kinases 306-307
  pumps 302
  root 557
  salinity 299
  second messenger 300, 307
  signal transduction 300, 305-
    307, 602
  statoliths 557
  stomata 299, 310-311, 609
  stores 302-303
  subordinate organ 593
  temperature 304
  touch 299, 304
  transport 300
  wind 299
  yeast 299
Calcium 3,5-dioxo-4-propionyl
  cyclohexanecarboxylic acid
  85
Callitriche platycarpa 491
Callose 395, 543, 544
Callus 181, 534, 775, 778-779,
  781, 784
Calmodulin 254, 274, 300, 305,
  557, 565, 602
  ABA 312-313

animal 305, 274
auxin 306
Ca^{2+}-ATPase 312
chloroplasts 303
GA 306, 312-313
gene 306
germination 254
gravitropism 565
mRNA 306
plant hormones 274
proteins 307
Calorigen 11, 191, 194
Cambium 532, 534, 539, 542
cAMP 275, 308
Canalization hypothesis 534
Cannabis sativa 642
Cantaloupe 125, 128, 730
Capsicum 463, 737
Carbaryl 755
Carbodiimide 437, 438, 443
Carbohydrates, tuberization 703
Carboxylic acids 422
Carboxypeptidase 253, 265, 288,
  678
Carcinogenicity 770
Carex projecta 574
Carnation 26, 29, 121, 126, 131-
  132, 496-499, 767
  senescence 497-498
Carotenoids 70, 142, 678
  ABA 145, 463, 691
  antisense 729
  biosynthesis 377, 380, 384
  biosynthesis inhibitor 677
  ripening 348, 383, 727
Carpinus 622
Carrot 10, 46, 47, 49, 50, 161,
  167, 209, 611, 620, 624,
  632, 738-739, 741-742,
  774, 779, 780, 782, 788,
  792
  bioassay 779, 780
  brassinolides 210
  culture 788
  cultures 49, 792
  cytokinin 742
  embryogenesis 167
  flowering 620, 624, 741-742
  polyamines 10, 161
  somatic embryos 782
  tissue culture 774, 779
Carya illinoisiensis 622
Caryophyllaceae 499
Cassava 659, 787
Castasterone 207
Castor bean 659, 678, 679
Catecholamines 778
Cathepsin 265, 404
Cauliflower 174, 349, 364,
  745-746

Cauliflower mosaic virus
  promoter 174, 364
Caulonemata 460, 462
CCC 627, 633, 714, 716, 719
Cdi gene 406, 407
cDNA 174, 454, 459, 682
  α-amylase 258, 264
  auxin 230-233, 280
    binding protein 285, 288,
      355-356, 359
  axi 159 365
  expression library 230, 688
  maize 285
  pea 232
  Pin2 401
  proteinase inhibitor II 396
  SAUR 234
  tobacco 280
  Zm-p60.1 359
Celery 659, 745, 759
Cell culture 775, 784
Cell cycle 583-584
Cell division 98
  auxin 228, 229, 279
  IAA 775
  kinetin 775
  polyamines 164
  tuberization 716
  lateral buds 584
Cell elongation
  ABA 612
  auxin 4, 214-225, 283
    acid-growth theory 219-
      223
    ATP 217
    ATPase 217, 220-221,
      224-225
    Avena coleoptile 215, 217,
      218, 223
    -binding protein 220
    brassinosteroids 211
    Ca^{2+} 221
    cell turgor 217
    chloride channels 221
    coleoptiles 217, 222-224
    collenchyma 217
    cycloheximide 220
    cytoplasmic pH 221
    diacylglycerol 221
    dicot stem 215, 217
    epidermis 217
    expansin 223
    G-protein 221
    β-glucans 222
    H$^+$ 222
    hemicelluloses 222
    inhibitors 216
    Instron stress-strain
      analysis 218
    IP3 221, 225

Cell elongation - auxin cont:
leaf mesophyll protoplasts
220
neutral buffers 219
osmoregulation 223, 225
peas 216
phospholipase 221
phosphorylation 221, 225
plasma membrane 220-221,
224
polyamines 164
protein kinase 217, 221
protons 219-220
receptor 220, 224
solutes 223
sucrose 215
sugar 216
time course 215
tonoplast 221
turgor 218, 219, 223, 225
wall acidification 223
crosslinks 225
extensibility 217, 218
-loosening factors 219-
220, 222-223
wall yield threshold 218
xyloglucans 222, 225
calcium 592
protein synthesis 215
wall extension 217
Cell permeability 739
Cell polarity 592
Cell turgor 217
Cell wall 222
acidification 223
auxin-binding proteins 281
brassinolides 209
Ca^{2+} 302
extension 214
extensibility 218
loosening 220, 222-224
enzymes 557
protons 219
metabolism 377
Pin2 expression 397
protein 182, 738
regeneration 777
salicylate 198
synthesis 567
wounding 395, 398
vascular differentiation 347
yield threshold 214, 218
Cellulase 281, 496, 567, 728, 777
Cellulose 218, 222-224
Centimorgan 449
Ceratocystis ulmi 183
Cercospora cruenta 143-145
Cercospora rosicola 142-145,
415

Cereal - see also individual
species
aleurone 246-267
abscisic acid 246
α-amylase 246
Ca^{2+} 310
gibberellin 246
brassinolide 211
germination 246-248
gibberellin 246
internodal pulvini 558
leaf sheath 558
gravitropic curvature 551
lodging 551
cytokinin 290
lodging 764
tissue culture 778
transcription factors 686
Cestrum nocturnum 620
Cyclic GMP 308
Chalcone synthase 181, 182, 735
Chara 555
Charcoal 787
Chemiosmotic hypothesis 513,
514
Chenopodium rubrum 442, 573,
621, 624, 629
Cherry 631, 727, 732, 734, 736,
752, 760, 761, 765, 766
Chilling
ACC 731
oxidase 732
Em 690
ethylene 122, 731
fruit 732
injury 171, 732, 737
Chimeric genes 263, 398, 683
Chinese cabbage 171, 746
Chitin 403
Chitinases 194, 395
Chitosan 398
Chlamydomonas 194
Chloramphenicol acetyltransferase
237
Chlorflurenol 765
Chloride channels 221
Chlormequat chloride 713, 746,
751, 752, 763, 764-766
2-Chloroethyl-bis-phenylmethoxy
silane 759
2-Chloroethylphosphonic acid -
see Ethephon
2-Chloroethylmethyl-bis-phenyl
methoxy silane 761
2-Chloroethylphosphonic acid
714, 752, 759, 780
2-Chloroethyltrimethylammonium
chloride 85, 751
2-Chloroethyl-tris-ethoxymethoxy
silane 759

2-Chloro-9-hydroxyfluorene-9-
carboxylic acid 517
4-Chloroindole-3-acetic acid
(4-Cl-IAA) 4, 39, 40, 45,
47
gibberellin metabolism, pod
growth 578
methyl ester 47
parthenocarpic 734
seeds, pea 578
4-Chloroindole-3-acetic acid
aspartate 47
5-Chloro-3-methyl-4-nitro-1-
pyrazole 768
Chloronemata 460
4-Chlorophenoxyacetic acid 753-
754
2-(*p*-Chlorophenoxy)-2-methyl
propionic acid 209, 779
Chlorophenyl)-4,4-dimethyl-2-(1,
2,4,triazol-1-yl)pentan-3-ol
85, 763
5-(4-Chlorophenyl)-3,4,5,9,10-
pentaaza-tetracyclo-5,4,1-
dodeca-3,9-diene 85
Chlorophyll 731, 735
a/b-binding protein 307
cytokinins 657
degradation 348, 735
ethylene 742, 745
ripening 384
Chloroplasts 8, 144, 145, 154,
161, 162, 303, 443, 460,
554, 611, 657, 658, 664,
681, 735
ABA 145, 149, 154
Ca^{2+} 301, 303
CaM 303, 307
cytokinins 657
IAA 657
N-(2-Chloro-4-pyridyl)-N'-phenyl
urea 736
4-Chloro-tryptophan 45
Chlortetracycline 312
5-α-Cholestan 207
Cholodny-Went 521, 549, 556,
560, 565
Chromatin 101, 274, 276, 290,
687
Chromoplasts 735
Chromosomal proteins 687
Chromosome walk 450, 457
Chrysanthemum morifolium
619-621, 624, 632, 692
Chrysanthemums 763
Cicer 584
Cichorium intybus 629
Cider 120
Cinnamic acid 198, 199, 778

Cis-acting elements (sequences) 265, 402-403, 682, 686, 688
  definition 263
Citrus 123, 166, 372, 621, 622, 727, 732, 735, 737, 752, 755, 757, 758, 768, 778, 792
  abscission 755, 768
  callus 778
  embryogenesis 792
  fruit ripening 757
  gibberellic acid 752
  juvenile 622
  ripening 372, 727
Citrus paradisi 621, 732
Citrus sinensis 621, 735
ckr gene 455, 679
cl gene 466
Climacteric 121, 347, 348, 351, 372, 374, 386, 389, 489, 501-503, 726-728, 730, 732, 737, 766
  daminozide 766
  ethylene 489, 726, 737
  fruits 372, 374
  ripening 347, 348
Clonal multiplication 776, 784
Clonal plants 574-575, 582
CO_2 128, 132, 657, 700
Co^{2+} 125, 208, 398
Coast blite 621
Cobalt 378, 780
Cocklebur 129, 208, 620
Coconut water (milk) 98, 774-776, 779, 782
Colchicine 777
Cold
  flower formation 636
  GA 759
  hardening 171
  hardiness 762, 767
  requiring plants 633
  shock 299
Coleoptile 15, 31, 51, 214-216, 217-219, 223-225, 229, 230, 284, 299, 309, 355, 358, 512, 518, 521, 548-552, 558, 710, 714
  auxin 576
  etiolated 511
  IAA 31
  NPA 518
  phototropism 2
  upward bending 552
Coleus 503, 532, 534, 535, 539-543
Collenchyma 217
Colletotrichum lagenarium 196
Colonization 575

Commelina communis 304, 310, 599, 602-604, 606-608
Compartmentation 15, 46, 102, 113, 161, 664
Competence 322, 783, 784, 789
Complementation analysis 462
Conditioned medium 169, 780
Cone 759
Cone flower 620
Conglycinin 684
Conifers 552, 626, 627, 759, 787
  juvenile 787
  micropropagation 787
  tissue culture 787
Coniferyl alcohol 322
Constitutive response mutants 458
Controlled atmosphere storage 132, 171, 374, 731, 737, 742, 744
Copalylpyrophosphate 70, 468, 470
Cordycepin 731
Cordyline terminalis 582
Coreopsis grandiflora 620
Corn - see maize
Corn cockle 620
Correlative effects **572-573, 576-583**, 591-593
Corynebacterium fascians 103
Cosmid 454, 457, 466
Cosmos 711
Cosuppression 340
Cotton 9, 20, 21, 172, 182, 620, 678, 682, 732, 751, 764, 769, 779
  ABA 678
  carboxypeptidase C 678
  embryo proteins 682
  jasmonate 182
  polyamines 172
  precocious germination 678
Cotyledons 651
o-Coumaric acid 198
Coumarin 54, 716
Crassulacean acid metabolism 164
Cress 208
Crop management 752
Crown gall 21, **318-335**, 341, 360, 775
  cytokinin 102-104, 106, 326-335
  culture 775
  IAA 44, 48, 323-326
Cruciferin 182
Cry' gene 472, 477
Ctr gene 134, 350, 391, 459
Cucumber 620, 725, 731, 759
  cell wall protein 223
  GA-binding proteins 290

GAs 87-88
  sex expression 642
  tobacco necrosis virus 196
Cucumis melo 730
Cucumis sativus - see cucumber
Cucurbic acid 181
Cucurbita 541
Cucurbita maxima **79-82**
Cucurbita pepo 514
Culturability 792
Cupressus 562, 622
Cupressus arizonica 562
Cuttings 5, 41, 208, 495, 578, 579, 621, 622, 698, 699, 701, 702, 705, 712-717, 720, 741, 753, 784, 787
  auxin 753, 787
  Solanum 579
  tuberization 698, 699, 701, 704, 715
Cyanide 191, 738
Cyanoalanine 127
Cyanoformic acid 127
Cycads 193
Cyclic AMP 175, 274
Cyclin B 583
Cycloheximide 122, 170, 220, 233, 266, 404, 406, 410, 731
Cyclohexylamine 165
α-Cyclopropyl-α-(4-methoxyphenyl) 5 pyrimidinemethanol 752
Cycocel 85
Cymbidium 776
Cysteine endopeptidases 252
Cytochrome P-450 72
Cytokinin 7-8, **98-113, 318-335, 539-543**
  abbreviations 99
  ACC 123
  action, mode of 8, **100-101**
  activity 111-113
  adventitious roots 541
  aglycones 107
  9-alanyl conjugates 100, 110
  N-alanyl conjugates 111
  amino acid conjugates 107, 110, 111
  analysis **101**, 113
  antigens 439
  apical dominance 7, 110, 345, 346
  Arabidopsis 24, 457
  auxin 540, 544, 778
    genes 318
    ratio 21, 781
    interaction 29
  axillary branching 785
    buds 105, 578, 593
    shoots 776

Cytokinin continued:
  binding proteins 290
  bioassay 100-101, 109, 111,
    417
  biosynthesis 7, **98-106, 326-332**
    Agrobacterium 320, 328-332
    complementation analysis
      462
    genes 318, 320, 342
    mutants 456
  branching 21-22
  brassinosteroids 208-209
  buds 7, 784, 787
  bulbs 740
  Ca^{2+} 299, 301, 311, 593
  callose 543, 544
  callus 778, 781
  cambial initials 542
  caulonemata 462
  cell division 7, 664, 719, 736
    enlargement 7
  chlorophyll 8, 657
  chloroplasts 8, 657, 665
  cis 103
  conjugates 100, **107-111**, 358-363
  correlative interactions 578-579
  definition 98
  differentiation 532
  discovery 2
  DNA 3
  dormancy 740-741
  effects 7
  Em 690
  embryo 359, 653, 675-676
  endosperm 652
  ethylene 123, 780
  etioplasts 8
  fibers 544
  flowers 630, 634, 635
  fruit 735
    growth 656, 676
    ripening 726, 736
  gametophore 462
  GC-MS 424, 441
  genes 7, 237, 328-332
  germination 359
  GLC 421
  glucosidases 110, 362
  glucosides 100, 107, 109-111
  glucosyltransferases 109
  grain abortion 665
  habituation 778
  hadcidin 579
  HPLC 419
  hydrolysis 111
  IAA 362

identification 101
immunoassays 439
immunocytochemistry 442
immunogold 578
ipt gene 541
isomers 103, 104
lateral buds 110
leaf 7, 105
lettuce 746
levels 102
localization 578
metabolism **107-113**
2-methylthio 103
micropropagation 786
mitotic activity 630, 634
morphogenesis 7
moss 460, 462
mutants 24, 453, **455-457**, 457
nature 7
nomenclature **99-100**
nuclease 107
occurrence **100**
organogenesis 781
overproducing 541
oxidase 111, 112
oxidation 107, 110, 112
parthenocarpic fruit 736
phloem 543, 634
  exudate 630
phosphoribosylation 108
photosynthesis 57
plant organ storarage 726
polyamines 169
pool 111
prokaryotes 318
proliferation 277
protein synthesis 101
quantitation 113
radioimmunoassay 441
ratios 782
reduction 107, 111
regeneration 277, 342
5'-ribonucleotidase 111
riboside phosphates 107
ribosides 7, 107, 109, 111
9-ribosyl 100
ribotides 7, 107, 111
RNA synthesis 101
rolC gene 362
rooting 787
roots 7, 22, 105, 539, 540,
  579, 656, 741
seed 102
  development 651, 671, 673
  growth 676
  size 675
senescence 7, 21-22, 348, 726,
  735, 745
sex expression 642
shoot cultures 776

shoots 781
side chain cleavage 112
sieve tubes 540, 542, 544
somatic embryos 783
source:sink relationships 349
sprout tubers 714
stomata 8, 311, 607-609
storage 110, 742
stress 455
structures 99
sucrose 662
suspensor 653, 675
synthesis gene 21
target cell 311
Ti plasmid 322
tissue culture 540, 775, **778-782**, 787, 792
trans 103
transgenic plants 7, 21, 342,
  344-345, 348, 363
transport 7, 578
tRNA 100-104, 318
tuberization 711-712, 714,
  717, 719, 720
tumor 326-328
uptake 108
uses 767
vascular differentiation 531,
  539-543
vessels 540, 544
vitrification 787
water stress 610
wound 540, 738
xylem 541, 656
Cytokinin-like compounds 779
Cytoplasm 532
  pH 309
Cytosolic Ca^{2+} 299, 301

D1-8 genes (GA, maize) 86, 467-469
Dahlia 712
Daisy 320
Daminozide 752, 761-763, 766,
  770
Darwin 509
Daucus carota - see carrot
Day neutral plants 618
Daylength 622
Daylily 780, 791
Death 395
Debranching enzyme 252
Deep-water rice 21
DEF 751, 769
Defense 183, 188, 410
Defloration 573
Defoliants 751, 764, 769, 779
Defoliation 488, 769
Dehydroascorbate 126
Dehydrogenases 129, 145, 147

dek mutants 675
Deletion analysis 30, 365, 683
   mutant 450
1'-Deoxy ABA 143, 144, 415
Desiccation 467, 671-674, 677,
   679, 681, 682, 690, 733,
   745
   ABA 690
   proteins 682
   seeds 733
   tolerance 682
Det gene 473, 476
Detectors 426
Developmental arrest 460,
   671-673
   mutants 478
Dextrinase 252, 253
Di-isopropyl carbodiimide IPC
   443
Diacylglycerol 221, 274, 275,
   308, 792
Diageotropica (*dgt*) 363, 493, 731
Diagravitropic 363, 574
Diamine oxidase 160
1,4-Diaminobutane 158
Diaminopropane 160
Dianthus 497
Diastase 252
Diazocyclopentadiene 132-133,
   489
Diazomethane 419, 422, 437
Dicamba 778, 782
3,6-Dichloro-*o*-anisic acid 778
2,4-Dichlorophenol 54
2,4-Dichlorophenoxyacetic acid
   (2,4-D) 232-234, 235, 282,
   289, 734, 753, 755-756
   abscission 29
   auxin-binding proteins 281
   Ca^{2+} 309
   cell division 228
   fruit abscission 757
   IAA 49
   IAA synthesis 47
   mRNA 236
   polar transport 511
   resistance, *Arabidopsis* 363
   somatic embryogenesis 782
   somatic embryos 782
   tissue culture 49, 778, 788
   791
2,4-Dichlorophenoxy propionic
   acid 753
Dicot 512
Dictyostelium discoideum 105
N,N'-Dicyclohexyl- carbodiimide
   [DCCD] 216
Diethylstilbesterol 216
Differentiation 1, 2, 5, 8, 31, 60,
   61, 98, 164-166, 210, 228,

272, 347, 460, 463, 486,
   491, 493, 509, 531-541,
   543, 544, 575, 588, 589,
   592, 593, 665, 666, 671,
   741, 788, 790-792
   calcium ions 592
   polyamines 164
   vascular tissue polarity 589
α-Difluoromethylarginine 158,
   161-162, 167, 169-170,
   172-173
α-Difluoromethylornithine 158,
   161-164, 166-167, 172-
   173, 716
Dihydroalanyl zeatin 110
Dihydroconiferyl alcohol 779
2,3-Dihydro-5,6-dimethyl-1,4-
   dithiin-1,1,4,4-tetraoxide 769
Dihydrolupinic acid 110
Dihydrophaseic acid 148, 151-
   153, 742
Dihydrophaseic acid glucoside
   151
1,2-Dihydro-3,6-pyradazinedione
   752
ent-6α,7α-dihydroxykaurenoic
   acid 72
Dihydrozeatin 100, 109, 111-112,
   442, 457, 779
Dihydrozeatin-O-glucoside 110
Dill 620
Dimethipin 769
Dimethylallyl pyrophosphate 68,
   326-327
Dimethylallylpyrophosphate:
   5'-AMP transferase 327-330
N,N,dimethylaminosuccinamic
   acid 752
1,1,Dimethylpiperidinium
   chloride 764
N-(2,4-Dimethyl-5(trifluoro
   methyl)sulfonylamino)-
   phenylacetamide 765
2,4-Dinitrophenol 216
Dinoseb 751
Dioecious 5, 6, 8, 642
Dionaea 567
Dioscorea 788
Dioxindoleacetic acid aspartate
   55, 58
Dioxindole-3-acetic acid 39, 54-
   55, 57-58
Dioxygenase 77-78, 82, 83, 85,
   88, 380, 388, 729
Diphenols 54
Diphenylurea 736, 779
Discoloration 488
Discorea batatas 741
Disease resistance 194-197, 200-
   201

Distylium 206
Diterpenes 68, 70
Dm gene 473
Dimethylallyl pyrophosphate:
   AMP transferase 328-329,
   331-332
Dimethylallyl pyrophosphate:
   tRNA prenyl transferase
   327, 331
DNA
   animal hormones 274
   binding 693
   binding protein 31, 688
   biotinylated 450, 451
   hybridization 450
   markers 449
   PCR amplified 471
   protein complex 686
   sequencing 688
DNase footprinting 266
Dne gene 473-475
Dock 208
Dominance 577, 580
Dopamine 778
Dormancy 673, 692, 739
   ABA 3, 140, 247, 678, 680,
   740, 741
   buds 140, 345, 584
   callose 544
   cytokinins 740-741
   elimination 776
   ethylene 505
   GA 741, 759
   low temperature 247
   polyamines 741
   true 10, 725
   tuber 701, 741
Dormin 9
Douglas fir 621, 759
Droopy mutant potato 463, 679
Drosera 567
Drought 28, 122, 171, 172, 398,
   400, 496, 500-501, 538,
   572, 613, 663, 733
   ABA 400
   ethylene 122, 500
   genes 733
   stress 28
   sucrose 663
Dutch Elm disease 183
Dwarf maize bioassay 417
Dwarf mutants 67, 70, 82, 88,
   465, 467-468, 717-718
Dy gene 469

E gene 388-389, 473, 475, 729
Easter lily 763
Echeveria harmsii 620, 624, 635
Echinocystis lobata 154
Eggplant 705

EIA 434, 435, 437
Eicosanoid 183, 406, 410
Eicosatetraynoic acid 409
ein genes 134, 391, 459
Electrical signaling 185
Electrochemical (EC)
    immunoassay 443
Electron capture 426
  spin resonance 514
  transport 611
Electrophoretic mobility shift
    assay 686-687
Elicitor 197, 399
ELISA 435, 438-441, 578
Elongation - see also cell
elongation
  rate 164, 286
Em gene / protein 405, 466, 681-
    688, 690
  chilling 690
  chimeric gene construct 684
  cytokinins 690
  GA 690
  gene expression 687, 690
  GUS 683, 685, 689, 690
  heat 690
  IAA 690
  mRNA 685, 690
  osmotic stress 690
  promoter 683, 685, 687, 688,
    690
  protein 685
  salicylic acid 690
  transcripts 685
  UV-light 690
  wheat 404
emb mutants 675
EmBP-1 gene 688
Embryo 250
  ABA 656
  auxin 652
  axis 652
  culture 775, 782
  cytokinin 290, 359
    binding protein 290
    development 671, **672-674,**
    679
  GA 248
  gene product 681
  germination 249, 260
  lethal mutants 675
  maturation 678
    ABA 680
    desiccation 672
    dormancy 672
    nutrient reserves 672
    tissue growth 672
  proteins 682
  rescue 776
  sac 734

Embryogenesis 166, 167, 182,
    521, **672-682,** 778,
    781-784, 788, 790-792
  auxin 166
  cultures 783
  ethylene 166,
  polyamines 166
Embryoid 164
Emergent growth 16
EMS mutagenesis 363
Endive 745
Endo-(1-3,1-4)-β-glucanases 252
Endo-(1-4)-β-xylanase 252
Endocytosis 281, 289
Endonuclease 322
Endopeptidases 253
Endoplasmic reticulum 154, 250,
    254, 275, 283, 300, 354,
    356, 554
  auxin-binding proteins 283-
    285, 288-289, 354, 356
  Ca^{2+} 300, 302-305, 308, 310
Endopolyploidization 777
Endosperm 31, 46, 47, 49, 55-57,
    67, 79-81, 154, 246-250,
    252, 253, 258, 306, 312,
    359, 441, 443, 651, 652,
    656, 662-665, 671, 675,
    676, 678, 709, 758, 774,
    776
  ABA 651, 656
  CaM 306
  function 247
  GA 248
  germination 312
  hydrolases 249
  IAA 652
  phytohormone conjugates 359
Endoxylanase 252
Enzyme-immunoassays 434
Ephedra 552
epi mutant 737
24-Epibrassinolide 210-211
Epidermis 217, 284, 512
Epifluorescent optics 443
Epinasty 11, 208, 209, 488, 502
Epistasis 24, 453, 459, 701
Epitope 434, 437
Equisetum 552
Erwinia herbicola 48, 319-320,
    325, 332-333
Escarole 745
Escherichia coli 82, 102, 164,
    329, 378, 382
  cytokinins 103
  tms 325
  transformation 326
  glutathione S-transferases 237
Esterase 153
Etacelasil 759, 761

Ethandiol dioxime 768
Ethane 132
Ethephon 233, 398, 487-488,
    496, 501-502, 727, 739,
    752, 755, 759-762, 765,
    768, 780
  abscission 752, 761, 768
  ethylene 487-488
  flower sex 496
  flowering 762
  fruit 761
  grape 761
  Hevea 501
  latex 501, 760
  ripening 727, 752, 761
  senescence 762
  tissue culture 780
  tomato 752
  tuberization 714
L-Ethyl-3 (3-dimethyl-amino-
    propyl) carbodiimide
    hydrochloride 443
Ethylene 8, **118-135,** **372-391,**
    486-505
  ABA 29, 123, 657
  abscission 5, 8, 26, 29, 30,
    372, 486, **495-496,** 505,
    726, 752, 755, 761, 768
    zone 495, 589
  ACC (see also under ACC)
    119-128, 488, 730
    deaminase 386
    -N-malonyltransferase 128-
    129
    oxidase **124-128,** **377-380,**
    382-385, 387, 489, 500
      antisense RNA **382-385,**
      729
  ACC synthase **120-124,** **380-**
    382, **385-386,** 387, 489,
    500
    antisense 385-388
  action **133-134,** 135, 486
    modulators 488-490
  S-adenosyl methionine 119--
    121, 129-131, 489
  aerenchymatous roots 493,
    505
  aminooxyacetic acid (AOA)
    489
  1-aminocyclopropane-1-carbox
    ylic acid - see ACC
  analogues **131-133,** 134-135
  antagonists **131-133,** 134-135
  antisense RNA **375-377**
  apical dominance 345, 581
    hooks 457
  apple 727, 731, 755
  aquatic plants 491
  Arabidopsis 350

Ethylene continued:
  auxin 123, 492, 494, 733
    -induced 497
    transport 754
  Avena 561
  AVG 489
  bananas 34
  binding 132, 133, 292-293,
    457, 458, 503
    protein 133-134, 292-293
    site 22, 489, 504
  biosynthesis 8, **118-131**, 135,
    488, 699
    antisense RNA 382
    brassinosteroids 208, 211
    enzymes **120-131**, 382
    genes 135, 377, 728
    IAA 581
    inhibitors 374, 488-490,
      495
    low temperature 731
    mRNA 26, 375, 377
    pathway 119, 375, 459,
      486
    ripening fruit 343, 373
    salicylic acid 190, 196
    tomato 373
  brassinosteroids 207, 209
  bromeliads 629
  bulbs 491
  cabbage 742-744
  cDNA 498
  cellulase 728
  chilling 122, 731
  chlorophyll 742, 745
  climacteric 501-503, 726, 737
  CO₂ 132, 489
  correlative interactions 581-
    582
  cuttings 495
  cytokinin 123, 746, 780
  daminozide 766
  defoliation 488
  development 488
  diageotropica mutant 493
  diazocyclopentadiene 489
  differentiation 486
  diffusion 487
  discoloration 488
  discovery 3
  dormancy 8, 490-491, 505
  drought 122, 500
  E8 389
  effects 8, 372
  embryogenesis 167
  epinasty 488
  ethephon 487-488
  etiolation 491
  fatty acids 486
  female flowers 642

filaments 497
flooding 29, 567
flower 22, 501
  formation 496, 629
  malformations 490
  opening 496-497
  senescence 486, 497-499
flowering 8, 490, 496, 751
-forming enzyme (EFE) 124,
  375, 377, 612
fruit 22, 486, 502, 503, 505,
  735
  ripening 8, 26, 122,
    347-349, **372-391**,
    486-487, 502-503,
    505, 726, 727-729,
    730-732, 734, 737,
    751
  inhibitor 502
  low temperature 731
  softening 728
GA 29
gas chromatography 422, 486,
  488, 502
genes **372-391**
  protein kinases 453
  protein phosphorylation
    453
genetic engineering 135
germination 122, 457, 490
glycol 134
gravitropism 494, 557, 561
growth 457, 486, 488, 490
  regulators 486
hls mutation 458
hook 458
hyperelongation 21
hypocotyl elongation 457
hyponasty 562
IAA 22, 29, 733
illuminating gas 3
incorporation 134
infection 500-501
inhibitors 374, 488, 780
insensitive mutants 458
insensitivity 457
intercellular spaces 487
jasmonate 737
juvenility 780
latex 505
leaf abscission 500
  senescence 457, 501
light 494
location 34
long distance hormone 487
mechanical perturbation 567
membrane 503
metabolism **134-135**
mobilization, carbohydrates
  490

mode of action **133-134**, 135,
  503-504
molecular biology 505
morphology 22
mRNA 133, 498, 503, 738,
  756
mutants 24, 133, 340, **342-
  344**, **346-349**, **382-388**,
  457-460
2,5-norbornadiene 489
NPA 525
oxidation 133, 134
oxide 133, 134
pathogenesis 30, 398
perception 458
photoaffinity label 489
photosynthesis 657
physical stimuli 494
phytoalexins 501
plant organ storarage 726
pollination 499-500
polyamines 131, 160, 168,
  170, 387, 390, 737
potato sprouting 490
precursor 118, 120
production 26, 123, 696, 699,
  730-731
promoter 729
proteins 133
putrescine 737
receptor 132-134, 292-293,
  389, 457, 503, 729
regulated genes 134
releasing agents 759
resistance mutants 458
respiration 490, 501, 738
response mutants 458
response pathway 453
responsiveness 22, 26, 729
rice, deep water 21 29
root 493
  differentiation 8, 493-494
  formation 29
  growth 8, 455, 493-494
  hairs 494
rooting 495, 780
salicylic acid 780
scrubbers 726
second message 503
secretion 505
seed development 676
  dispersal 503-504
  germination 701
seedlings 457
senescence 8, 122, 349, 372,
  486, 497-499, 501, 503,
  745, 746
  mRNAs 26
sensitivity 13, 22, 29-30, 488,
  505, 726

Ethylene continued:
  sex expression 496
  shoot growth and
    differentiation 8, 491-493
  signal 697
    transduction 350, 391, 729
  silver 5, 504
  soil atmosphere 493
  spermidine 737
  spermine 737
  stems, diagravitropic,
    orthogravitropic 582
  stigma 499
  stomata 657
  storage organs 738, 742
  stress 122, 160, 350, 372,
    486-487, 492, 500 504,
    567, 726
  submerged shoots 491
  tactile stimuli 494
  thidiazuron 779
  tissue culture 780
  tomato 373, 374, 378, 492,
    727, 728, 731
  transduction pathway 134
  transgenic plants 342, 344-
    345, 375
  transport 8
  triple response 457-458, 491
  tuberization 714, 716
  two-component systems 504
  uses **759-762**
  vascular differentiation 531
  water logging 122
  wheat 123
  wilt diseases 501
  wounding 122, 377, 379, 398,
    403, 500-501
  yeast 378
  yellowing 500
etr gene 133, 350, 457-459
Eubacteria 359
Eucalyptus 622
Euonymus radicans 622
Euphorbia pulcherrima 492, 620
Euphorbiaceae 786
Evocation 473, 477, 617, 629,
  634, 635
Exopeptidases 253
Expansin 223
Explants 793

Fabaceae 47
Fagus sylvatica 621, 622
Far red 476
Farnesyl pyrophosphate 68, 142,
  144, 463
Fatty acids 486
FCCP 514
Fe^{2+} 78, 81, 126

Feeder cells 780
Feijoa 713
Female flowers 642-643
Ferns 552
Fiber 539
Fibroblasts 306
Ficus punula 622
Field pennycress 620
Fir 621
Fires 496, 751
Flacca (*flc*) 462, 679
Flame ionisation detector 426
Flavanone-3-hydroxylase 125,
  378
Flavonoids 189, 519, 524
Flexing stress 492
Flooding 29, 491, 504, 567, 745
Floral development 644
  inhibitor 623
  initiation 166, 633
  stimulus 166, 622-626, 628,
    630, 632-636, 643
    assay 643
    cytokinin 635
    extraction 643
    GA 633, 635
Florigen 623
Flower
  bud 758, 766
  colour 735
  development 474
  evocation 473
  formation 617, **628-633**
    ABA 630
    CCC 633
    cold-requiring plants 636
    cytokinins 630
    GAs 631
    hormones 628
    thermoinduction 632
    woody fruit trees 631
  hormone 623
  induction **473-478**, 573, 623
  inhibitor 474, 476, 630
  initiation 164, 475-476, 617,
    634
  longevity 26
  organogenesis 476
  peas 474
  promoter 474, 476
  senescence 26, 497-499
Flowering **617-644**
  apical dominance 360
  auxin 629
  cytokinin 635
  delay 767
  ethephon 762
  ethylene 751
  fruit crops 76, 761-762
  GA 626, 632, 741

gene expression 644
genes 477
genetic control 478
grafts 624
heterochrony 476
hormone 476-478
inhibitors623, 625
measurement 625
mutants 478
peas 473-477
phenolics 190
phenotypes 473
photoperiod 705
physiology 478
polyamines 164-165, 361
precocious, mutation 457
response types 620
salicylic acid 189-190
stimulus 473, 476, **622-625**,
  634-636, 705, 720
tuberization 705
Fluorescein 442
Fluorescence 426, 428, 439, 443,
  603, 605
  detector 426
Fluridone 145, 147, 677, 678,
  681, 690
Folex 769
Footprinting 266, 688
Foraging 575
Fragaria 728
Fragaria x ananassa 623
Fraxinus excelsia 621
Free radicals 737
Freesia 490
Freezing 224, 400
Frost 757, 762
Fructose 1,6-bisphosphatase 163,
  658
Fructose-1,6-bisphosphate 303
Fructose-2,6-bisphosphate 658
Fructose-6-phosphate 658
Fruit **725-747**
  abortion 574
  abscission 574, 503, 726, 754,
    755, 757, 768
  auxin 593
  chilling 731, 732
  climacteric - see Climacteric
  color 735, 761, 765, 770
  correlative inhibition 577, 593
  de-greening 735
  development 388, 577
    correlative effects 572
    cytokinins 656
  dominance 573, 593
  drop 755
  ethephon 761
  ethylene - see under ripening
  below

Fruit continued:
  gibberellins 734
  growth 676
  growth inhibiting signal 574, 577
  hormone changes 725
  IAA 577, 591
  parthenocarpic 734, 736
  polyamines 166, 728, 736
  promoter 729
  quality 757
  reproductive dominance 573
  ripening **372-391**, **726-738**, 757, 761
    ABA 726, 732-733
    abscission 726
    ACC 377-388
    oxidase **377-380**, **382-385**, 387, 729
      antisense 382-385
    synthase **380-382**, **385-388**, 729
      antisense 385-388
    synthesis 123
    antisense **382-388**, 729
      fruit 384
      mRNA 728
    apple 727
    auxin 726, 733-734
    cDNA 374
    chlorophyllase 729
    clones 377
    cytokinins 726, 736
    daminozide 766
    description 726
    E8 389
    enzymes 727
    ethephon 727, 761
    ethylene 122-123, 347-349, **372-391**, 486, 502-503, 505, 726-727, **730-732**, 734, 751
      production 123
      sensitivity 27
    GA 726, 734-736
    genes 348, 388, 728
    hormone 727-728
    IAA 727, 736
    inhibitor 391, 502, 726
    jasmonate 181, 737-738
    molecular biology 728
    non-climacteric 727
    polyamines 164, 168, 727, 737
    polygalacturonase 729
    protein synthesis 727
    putrescine 131
    stages 728
    tomato 374, 727

transgenic plants **347-348**, 351, 375, **382-388**
  set 753-757, 765
  shape 767
  size 754, 756
  softening 728, 766, 770
  subordinate 593
  thinning 754
  trees 766
  wounding 728
Fsd gene 476
Fucose 222
Funaria hygrometrica 299, 311-312, 460, 575, 592-593
Fungal cell walls 398
Fungal elicitors 181, 184, 237
Fungi 43, 142, 165, 172, 181, 183, 777
6-(Furfurylamino) purine 98
Fusicoccin 208, 224, 286, 659, 661

G-box 685, 688, 691
G-protein 221, 275, 308, 453, 517
GA-like 67, 467, 469, 626, 627, 631, 632, 642, 643, 652, 654, 657, 707-709, 741, 746, 757
GA$_1$ 6, 17-21, 28, 29, 33, 34, 66, 70, 78, 80, 82-84, 86-88, **93**, 247, 248, 417, 439, 465, 469-473, 633, 639, 643, 665, 708, 734
  activity 86
  aleurone 248
  bioassay 17
  bolting 18, 639
  cell wall 472
  IAA levels, wall yield threshold 472
  gene 82, 88
  hydroxylation 88
  internode length 18
  location 17
  mutants 470-472
  stem elongation 248, 471
  tallness 17, 21, 33
GA$_3$ 6, 17, 28, 29, 67, 83, 84, 87, 88, **93**, 153, 246-248, 252-255, 258-267, , 299, 304, 306, 439, 453, 465, 467, 468, 472, 555, 560, 562, 567, 579, 582, 626, 627, 631-633, 635, 637, 638, 642, 643, 652, 657-659, 661, 664, 666, 708, 709, 716, 717, 734, 735, 738, 741, 752, 756-758, 780, 786

ABA 153, 263
  gene expression 267
  acid phosphatase 254
  aleurone 254-255, 259
  α-amylase 254, 258, 260-266
    genes 263
    transcription 262
  arabinofuranosidase 254
  artichoke 752
  bolting 741
  CaM 306
  cereal grains 247
  flowering 632
  gene expression 265
  β-glucanase 254
  hydrolases 253
  IAA transport 579
  juvenile plants 626-627
  lycopene 735
  micropropagation 786
  mRNAs 262
  mutants 453
  phloem loading 659
  protease 254
  protein synthesis 255
  rhizome 582
  ribonuclease 254
  ripening 735
  rRNA 262
  senescence 752
  tuberization 716, 717
  xylopyranosidase 254
GA$_4$ 82, 87, 88, **93**, 291
GA$_{4/7}$ 626, 658, 756, 758-759, 767
GA$_5$ 78, 83-85, 88, **93**
GA$_6$ 83, 84, **93**
GA$_7$ 84-87, **93**
GA$_8$ 18, 78, 81, 87-88, **93**, 247, 469-471, 638, 708
GA$_8$-catabolite 78
GA$_9$ 73, 75-77, 79, 82, 84, 85, 87, **93**, 439, 626, 652
GA$_{12}$ 18, 68, 70-76, 79, 81, 82, **93**, 470, 708, 709, 717, 734
GA$_{12}$-aldehyde 18, 68, 70-73, 76, 79, 81, 470, 708, 709, 717
GA$_{13}$ 79, 80, **93**
GA$_{15}$ 73, 76, **93**
GA$_{17}$ 82, 247, **93**
GA$_{19}$ 77, 78, 80, 87, **94**, 247, 636-638, 640, 643, 708, 734, 746
GA$_{20}$ 15, 17-19, 34, 75-78, 80, 82-88, **94**, 247, 417, 424, 439, 441, 469-471, 633, 636-640, 643, 652, 665, 708, 734, 746
  methyl/TMS spectrum 424

GA$_{23}$ 80, **94**
GA$_{24}$ 73, 76, **94**
GA$_{25}$ 73, 82, **94**
GA$_{29}$ 15, 77, 78, 83, 87, **94**, 247, 441, 469-471, 637-640, 652, 708
GA$_{29}$-catabolite 77, 468
GA$_{34}$ 81, **94**, 247
GA$_{38}$ 80, **95**
GA$_{39}$ 79, **95**
GA$_{43}$ 79-81, **95**
GA$_{44}$ 77, 78, 80, 87, **95**, 638, 708, 746
GA$_{48}$ 247, **95**
GA$_{51}$ 76, 77, **95**, 709
GA$_{53}$ 76-80, 87, **95**, 638-639, 708-709, 717, 746
GA$_{53}$-aldehyde 708
Ga genes 70, 465, 469
Gai gene 453, 472
Gaillardia 497
Galactose 40, 49, 726
β-Galactosidase 237, 729
Galls 48, 207, 318-320, 326, 332, 342
Gametophore 460, 462, 575, 576
Garlic 739
Gas-liquid chromatography 415, 417, 420, 424, 426
Gas chromatography/mass spectrometry (GC-MS) 3, 39, 47, 101, 419-424, 428, 429, 431, 440, 441, 443, 462, 477, 580, 636, 641
   abscisic acid 425
   auxins 422
   cytokinins 424
   gibberellins 423
   IAA 422
   zeatin 424
Gel shift assay 30
Gene
   ABA 463
   alternative oxidase 193
   Arabidopsis **449-453**, 463, 465-466, 471
   expression
      α-amylase 250
      ABA 260
      animal hormones 274
      auxin 228-242
      axillary buds 585
      brassinosteroids 210, 211
      GA 260
      protein phosphorylation 459
   flower **473-478**
   internode length **467-473**
   maize **467-469**
   peas 463, **469-477**

promoter 31
   regulatory 479
   response mutants 453
   seed dormancy **462-467**
   structural 479
   tagging 462
   transcription, animal hormones 274
   wilt **462-467**
Genet 574
Genetic engineering 320, 777
Genomic subtraction 450, 471
Genotypes 4
Geranyl pyrophosphate 68
Geranylgeranyl pyrophosphate 68-71, 468, 470
Geranylgeranyl pyrophosphate synthase 69
Germination 673, 674, 759
   α-amylase 246-248
   ABA 247, 465, 677, 679, 682, 732
   auxin 454
   Ca^{2+} 312
   cereal grains 247
   cytokinin 359
   embryo 248
   enzymes 681
   ethylene 122
   genes 689
   gibberellin 27, 247, 465, 759
      barley 246
      -binding proteins 290
      Ca^{2+} 312
      cereal grains 246-248
      production 247
   inhibitors 776
   markers 679
   mRNAs 262
   nucleic acids 253
   osmotic controls 253, 732
   phytohormone conjugates 359
   precocious 400
   putrescine 165
   spermidine 165
   starch 248
   storage protein-hydrolysis 253
Geum urbanum 631
Gi gene 473, 476
Gibberella fujikuroi 66-68, 72, 762
Gibberellane 6, 66
Gibberellin (GA - see also GA numbers above) 6, **66-97**, **246-267**
   ABA 452
   activity 86
   action **255-267**
   aleurone 246-267
   α-amylase 6, **246-267**

gene 30
   mRNA 676
anthocyanin 735
antibodies 256
apical senescence 477
Arabidopsis thaliana **82-83**, 471-472, 477
assimilate translocation 769
axillary buds 579
barley 250, 756
binding protein 255-256, 290-291
bioassay 417, 477, 746
biosynthesis 6, **66-88**, 471, 631
   control of 86
   enzymes 81
   growth retardant 763
   inhibitors **85-86**, 638
      light 27
   pathway 68, 467-469
   seeds 27
   site of 67
bolting 6, 741
brassinosteroids 208
bulbs 740
Ca^{2+} 291, 299, 301, 304, 305, 312-313
CaM 312-313
carboxypeptidase 265
cell division 6
   elongation 6
cereal 246-267
chloroethyltrimethyl ammonium chloride 627
cis-acting elements 265
cold 759
commercial 756
concentration 17
conjugates 67, 77
correlative interactions 579-580
Cucurbita **79-82**
3-deoxy 87
developing seeds 651, 665
developmental mutants 477
discovery 2
dormancy 759
dwarf plants 17, 467-473
effects 6
Em 690
embryo development 675-676
endosperm 652
enzymes 81
ethylene 29
floral stimulus 633, 635
flower bud 758
   initiation 631-632
flowering 6, 477-478, 741
fructose bisphosphatase 658

Gibberellin continued:
fruit 6, 734
  colour 735-736
  development 654
  growth 88, 676
  ripening 726, 734-736, 757
  set 675, 757, 765
  size 756
GC-MS 423, 477, 577, 636
gene expression 260
genes 351, 467-473, 477
germination 6, 27, 255, 465,
  759
  Ca²⁺ 312
  cereal grains 246-248
  endosperm 246
grain development 247
grape 756, 757
growth retardants 85-86, 762
height 6, 17, 21, 33, 467-473
HPLC 418, 419
2/3β-hydroxylase 78, 79, 84,
  88
2/3β-hydroxylation 17, 77-82,
  84, 86-88, 803
13-hydroxylated 18, 79, 82,
  86, 248, 709
IAA 29, 45
immunoassay 436-437, 439
inhibitors **85-86**, 632
internode length **467-473** 478
  genes, maize 467 469
  genes, peas 467, **469-471**
juvenility 625
leaves 654, 658
levels 477
liquid endosperm 652
maize - see *Zea* below
male flower formation 642
malting 756, 758
Marah **83-85**
mechanical perturbation 567
metabolic grid 89
metabolism **66-88**, 477
  control 86
  genes 351
  inhibitors 85
molecular ion 424
mono-oxygenase 72, 79
mutants 351, 452, 465, 477
mutation 257,471, 479
myb 267
nature 6
number 66
ovary growth 675
oxidase 78
20-oxidase 81, 82, 87, 88
  mRNA 81
oxidation 79, 82, 763
polyamines 169

parthenocarpic fruit 736, 757
peas 17, **75-79**, 81, 469-471,
  676, 734
perception 256
petal growth 636
Phaseolus **83-85**
phloem loading 666
photosynthesis 657, 666
physiological disorders 758
plant organ storarage 726
pod development 675
pollen development 636
potato - see tuberization
promoter 264
protein levels 658
  synthesis 246, 261
pumpkin **79-82**
receptor 23, 290-291
response 28
  element 30, 264
  mutants 471, 478
    Arabidopsis 472
  mutations 473
responsiveness 27, 30
ripening - see fruit
roots 68
rosette plants **636-641**
seed 88
  development 671, 673,
    676, 677
  germination 6
  growth 652, 676
  maturation 652
senescence 477, 734
sensitive mutants 471
sensitivity 27, 676
sex expression 636, 759
shoots 88
signal transduction chain 266
spinach 746
sprout tubers 714
stem elongation (growth) 6,
  28, 87, 88, **467-473**, 632,
  636-641, 741
  mutants 478
  rosette plants 636
  tissue culture 787
stereochemistry 66, 85
stomata 311
storage 742
structures 66, **93-97**
sucrose translocation 664
suspensor 675
tallness 6, 17, 21, 33, 467-473
tissue culture 780, 787
trans-acting factor 266
transport 6
tuberization 701, 707-709,
  712-714, 716-720
uses **756-759**

vascular differentiation 531
vernalization 741
woody fruit trees 631
wounding 738
Zea **83-85**, **467-469**
Girdling 757
Gladiolus 490, 788
Globulin 689, 692
β-Glucanase 194, 222, 249, 253-
  255, 265, 395
β-Glucan 222, 249, 252, 253
Glucosamine 398
Glucose 39, 51, 55, 59, 67, 77,
  109, 150, 152, 154, 199,
  252, 359, 563, 662, 664,
  703, 734
β-Glucosidase 110-111, 199-200,
  252, 358-360, 362, 587
  rolC 362
Glucoside - see individual
  hormones
β-Glucuronidase - see GUS
L-Glutamine 130
Glutaraldehyde 439
Glutathione S-transferases 237,
  241
N-Glycans 726
Glycerol 792
Glycine 577
Glycine max 536, 620, 624, 651
Glycoproteins 272
Glycosidase 153
Glycosylation 285
Glyphosate 768
Glyphosine 751, 753, 768
Gossypium 576
Gossypium davidsonii 624
Gossypium hirsutum 620, 624
Graft 33, 534, 621, 623, 624,
  630, 634, 635, 705, 706
Grafting 195, 463, 469, 473, 476,
  534, 621, 623, 624, 632,
  634, 635, 704-706, 715
  flowering 476
Grain 31, 169, 191, 246-250,
  252, 255, 259, 260, 262,
  551, 553, 555, 613, 651,
  652, 656, 661-665, 675,
  676, 681, 751, 777
  filling 664
Grape 654, 657, 727, 732, 756,
  757, 761, 765
  ethephon 761
  fruit set 765
  GA 756
Grapefruit 123, 621, 732, 735,
  755, 757
Grass 222, 548, 549, 551, 552,
  554, 555, 558, 560-562,
  565, 566, 608, 765

Grass continued:
  cell walls 222
Graviresponse 555, 560
Gravisensitivity 554, 559
Gravisensors 555
Gravitropism 5, 22, 52, 239, 241,
    299, 364, 449, 455, 494,
    510, 515, 521-522, 525-
    526, 547, **551-567**, 575,
    581-582, 592
  ABA 557
  adaptive significance 552
  asymmetric growth response
    553
  ATPases 565
  auxin 22, 364, 521-522
    -regulated genes 239
    transport 355, 521
  calcium 52, 299, 557, 565,
    592
  calmodulin 565
  ethylene 494, 558, 582
  GA 558, 560, 56
  gene 567
  glucan synthase 567
  H⁺-ATPase 567
  hormone asymmetry 555
  IAA 15, 52, 558, 564, 567
  invertase 562-563, 56
  molecular biology 562-565
  mutant 364
  negative 552
  perception 553, 555
  positive 553
  protein kinases 565
  proton efflux 557
    pumping 565
  pulvini 562
  root 553, 556-557
  shoot 556, 558 561
  stress 582
  sucrose hydrolysis 562
  transduction 553, 555-561,
    565-567
Greenhouse crops 763
Growth 185, 474
  ageotropic 553
  auxin receptor 357
  auxin-binding 279
  axillary bud 579, 584, 591
  consolidation 582
  inhibiting signal, fruits 574
  intercalary 552
  jasmonate 181
  orientation **547-568**
  polyamines 164
  reproductive 573
  salicylic acid 191
  secondary metabolites 188

Growth regulators - see Plant
    growth regulators
Growth retardants 67, 79, 85, 88,
    762-768, 770
Guanosine di/triphosphate 275
Guard cell 28, 291-292
  ABA 291, 310-311, 599,
    601-605
  auxin 309
  Ca²⁺ 299, 301, 305, 307, 309,
    310-311, 603-605
  Cl⁻ 601
  hormones 601
  ion channels 601, 605
  IP3 308, 311
  K⁺ 601, 311
  malate 601
  patch clamp 601, 605
  plasma membrane 601
  protoplast 605
  signal transduction 310
  tonoplast 601
  turgor 601
  wilty mutants 467
Guerilla form 574
GUS
  ABA 30, 683
  auxin 22, 237, 240, 280, 521
  Em 583, 683, 685, 689, 690
  gravitropism 240
  Pin2 397, 402
Gynoecium 499
Gynogenesis 777
Gypsophila paniculata 48, 320

H⁺-ATPase
  Ca²⁺ 301-302, 307
  IAA 561
  plasalemma 286
H⁺/IAA⁻ symport carrier 514-515
Habituation 778
Hadacidin 579
Hairy roots 361-362, 586
Haploid 396, 460, 575, 777
Hapten 433, 437, 439
Heat 690
Heat shock 105, 211, 236, 348
Heavy isotopes 427-429
Heavy metals 236
Hedera helix 622, 626-628
Helianthus - see Sunflower
Helianthus tuberosus 169, 705,
    741
Helix-loop-helix motif 686
Hemerocallis 780, 791
Hemicellulose 222
Hemocyanin 433
Henbane 620
HEPES 441

Herbicide 72, 509, 751, 753, 756,
    768, 769, 778
Heterostyly 362
Hevea 501, 760
Hieracium aurantiacum 631
High performance liquid
    chromatography: see HPLC
High temperature 732
Histidine kinase 350
Hls mutations 458
Hollyhocks 620
Holoenzyme 312
Holohormone 288
Homeobox 686
Hordeum vulgare - see Barley
Horseradish 53, 54
HPLC 3, 101, 106, 415-417, 426,
    440, 578
  detectors 426
HPLC-MS 3
Hyacinth 582, 788
Hybrid 197, 377, 451, 700, 496,
    759, 769
  seeds 769
Hybridizing agents 769
Hybridoma 434
Hydraulic conductivity 214, 612
Hydrolases 48-50, 248-250, 253,
    254, 325, 326, 331, 341
  GA₃ 253
  genes 250
  mRNA 249
Hydroperoxide 407, 408, 411
  dehydrase 407, 408, 411
N-Hydroxy-ACC 127
α-Hydroxyacetosyringone 189
4-Hydroxybenzoate hydroxylase
    325
Hydroxycinnamoyl acid amides
    164
5-Hydroxy-IAA 359
4'-Hydroxy-α-ionylidene acetic
    acid 143
11/12-Hydroxy-jasmonic acid
    709-711
ent-7α-Hydroxykaurenoic acid
    72, 469-470, 641
3-Hydroxy-3-methyl ABA 151-
    154
6-(4-Hydroxy-3-methylbut-trans-2
    -enylamino) purine - see
    zeatin 98
3-Hydroxy-3-methyl glutaryl
    HM-ABA 152
Hydroxymethylglutaryl-coenzyme
    -A 68
3-Hydroxymethyl-oxindole 54, 57
5-Hydroxy-nitrobenzyl bromide
    778
7-Hydroxyoxindole acetic acid 56

Hydroxy IAA continued:
glucoside 55, 56
2-Hydroxyoxindole-IAAsp 58
Hydroxyperoxide dehydrase 405
13-Hydroxyperoxylinolenic acid
405-408
Hydroxyproline 77, 395, 398
Hygromycin 364
Hygrophorus conicus 55
Hyoscyamus niger 620, 623, 624
Hypersensitive reaction 194-196,
199, 201, 473, 475
Hypobaric storage 374, 726

iaaH gene 325, 326, 333, 341,
360, 586
iaaL gene 326, 333, 344-345,
360-361, 456
iaaM gene 325, 326, 333, 341,
342, 344, 346, 360, 457,
539, 586
Ibuprofen 409
Illuminating gas 3
Imbibition 681
Immunoaffinity 331, 333, 441
Immunoassay 3, 17, 433
Immunocytochemistry 442
Immunofluorescence 251, 284,
442, 444
Immunoglobulin 442
Immunogold 442, 443
Impatiens balsamina 631
In vitro propagation 785
Incipient plasmolysis 218
Indirect organogenesis 781
Indo-1 gene 312
Indole 46, 47, 422, 456, 362
Indole alkaloids 72
Indole-3-acetaldehyde 4, 44
Indole-3-acetaldoxime 45
Indole-3-acetamide 44, 48,
324-326, 341
hydrolase 48, 325, 326, 341
Indole-3-acetic acid (IAA) (see
also Auxin) 4-5, **39-61,**
214-225, 228-242, 318-
335, 354-367, 534-539
ABA 580, 611
abscission zone 589
ACC synthase 733
acid-growth 560
acropetal transport 557
amides 41, 59
amino acid conjugates 41, 59
hydrolases 50
apical dominance 594
assimilate translocation 769
asymmetric distribution 548-
550, 558-560
growth 560

Avena coleoptiles 548-549
axillary bud 581, 591, 593
bacteria 48
benzyladenine 540
basipetal transport 550, 558
binding 357
biosynthesis **41-50,** 559
genes 326, 360
gibberellin 45
indole 46-47
microbial 48, 326
tryptophan 43, 44
brassinosteroids 208
bulbs 740
Ca^{2+} 309
carrier 516
catabolism 53, 57
cell division 775
cell wall protein 738
chloroplasts 17, 657
CO_2 fixation 657
coleoptile 31
concentration 22, 25
conjugates 31, 40, 41, 49, 58,
325, 558, 560, 653, 733
hydrolysis 41, 48-50
synthesis 60
correlative signal 594
crown gall 323-326
cuttings 784
decarboxylation 53, 54, 56, 57
destruction 548
deuterium 45
distribution 52
duration of exposure 25
efflux 16
carrier 526
electrical gradients 514
electron transport 611
Em 690
embryonic axis 652
endogenous 40
endosperm 652
ester 41, 52, 325, 577, 653,
656
ethylene 22, 29, 581, 733
export 591
fluorescence 426
fruit export 591
ripening 736
GC-MS 422
geotropism 15
glucosidase 359
gravitropism 22, 52, 560
growth 60
kinetics 216
rate 20
H^+/IAA^- symport 514-515
HPLC 418, 419
immunoassay 437

immunocytochemistry 443
initial growth response 20, 25
intact plants 19, 20
invertase 564
K^+ concentration 660
leaves 653, 654
mechanical perturbation 567
metabolism 41, 42, **53-60**
methyl ester 422
mobilization 660
morphology 22
mutants 478
overproduction 457
oxidation 53-58, 60, 778
pea growth 32
peptide conjugates 40-41, 49
petioles 653
phloem loading 659
phosphorylation 611, 657
photosynthesis 611, 657
phototropism 15, 23, 31
plasmodesmata 52
polar transport 557
pool 41-43, 53, 60
prolonged growth response 20,
25, 32
proteins 49
proton extrusion 514-515, 659
quercetin 519
redistribution 31
resistance, *Arabidopsis* 363
rolB 362
root 29, 522-523
rooting 41
seed 652-654, 677
sensitivity 516
shoot tips 548
stem elongation 28
growth 20, 32
length 22
segments 19
stomata 606-611, 657
storage 17, 742
sucrose accumulation 664
sugar conjugates 41
tallness 19, 21
tissue culture 41, 774
transgenic plants 21, 22, 344,
360
transport 31, 32, **51-52,** 60,
510, 548, 556, 559
abscission 591
axillary buds 591, 593
Ca2+ 592
chemiosmotic theory 514
coleoptile tip 51
inhibition 592
lateral 515
light 550
seed to shoot 51

IAA transport continued:
 stem 590
  tropisms 52
 tropisms 22, 52, 560
 tryptophan mutants 456
 tuberization 716
 tumor 323-326
 turnover 40
 uptake carrier 514-515
 uses 753
 vacuole 17
 vascular bundle 589
 vessel regeneration 540
 vesicles 514, 516, 520, 522
 wounding 398, 399, 738
 xylogenesis 539
 zeatin 540
Indoleacetic acid-amino acid
 hydrolase 49
Indoleacetic acid-glucose synthase
 51
Indoleacetic acid-glycoprotein 40
Indoleacetic acid oxidase 53, 57-
 58, 124, 778
Indoleacetic acid-phenylalanine
 50
Indole-3-acetonitrile 325, 456
Indoleacetyl-alanine 50
Indoleacetyl-aspartate 50, 58-59,
 734
Indoleacetyl-glucan 40
Indoleacetyl-glucose 41, 58-60,
 734
Indoleacetyl-glucose hydrolase
 50, 61
Indoleacetyl-glucose synthetase
 50, 59, 61
Indoleacetyl-glutamate 59
Indoleacetyl-inositol 31, 41, 48,
 49, 51, 52, 58-59, 61
Indoleacetyl-inositol-arabinose 40,
 49, 59
Indoleacetyl-inositol-galactose 40,
 49, 59
Indoleacetyl-lysine 2, 4, 48, 58,
 323, 534, 577, 727
 gene 60
Indoleacetyl-lysine synthetase
 347,360
 iaaL gene 60, 342, 456
Indole-3-aldehyde 54, 56
Indole-3-butyric acid 39, 40, 47,
 753
 cuttings 784
 tissue culture 778
Indole-3-carboxylic acid 56
Indole-3-ethanol 44
Indole-3-glycerophosphate 43, 46
Indole-β-glucosidase 362
Indole-3-lactic acid 44

Indole-3-methanol 54, 56
Indole-3-methanol glycoside 57
Indole-3-pyruvic acid 44
Indoxyl-O-glucoside 359
Infection 122, 500-501
Inflorescence 573
Inhibition, axillary bud 579
Inhibitors 208, 709
Initial growth response 20
Injury 395
myo-Inositol 39, 49, 59, 359,
 558, 775, 792
 glycosides 49
Inositol-1,4,5-trisphosphate (IP3)
 221, 225, 274-275, 298,
 300-301, 308, 792
 ABA 292
 animal cells 288
 auxin transduction 288
 Ca²⁺ 301, 303, 308, 310
 guard cells 310-311
 vacuoles 308
Inositol phospholipases 453
Insects 122
Insertional mutagenesis 324, 364,
 462
Instron stress-strain analysis 218
Internal standard 427-430, 463,
 641
Internode length genes **467-473**
Invertase 563-565, 567, 664, 703,
 728
Iodine 435
Ion exchange chromatography
 416
α-Ionylidene acetic acid 143-144
Ipomea nil 497, 581
ipt gene 105, 327, 330, 332, 334
 axillary bud growth 586
 apical dominance 345
 cytokinin 586
 leaf senescence 348
 mRNA 105
 senescence 349
 transformed plants 105
Iris 490-491, 574, 776
Iron 77
Isocitrate lyase 678
Isogenic lines 448, 675
Isopentenyl adenine 100, 106,
 111, 112, 327, 330, 332,
 462, 579, 746, 779
Isopentenyl adenosine 100,
 102-104, 106, 107, 111,
 112, 327, 439, 746
Isopentenyl adenosine
 monophosphate 104, 105,
 107, 111, 326-327, 342
Isopentenyl pyrophosphate 68,
 102-104, 342

Isopentenyl transferase 102, 104,
 105, 327, 329, 342, 586
Isopentenyl transferase gene - see
 ipt
2'-isopropyl-4'-(trimethylammon
 ium chloride)-5'-methylphenyl
 piperidine-1-carboxylate 85
Ivy 627
Jasmine 10
Jasmonic acid/Jasmonates 3, 10,
 179-186
 ABA 185-186, 405, 409
 abscission 11
 ACC 737
  oxidase 737
 animal 179, 183
 antifungal agent 183
 Arabidopsis 185
 biosynthesis 10, 179-180, 184
 defense 11, 179, **181-183**
 effects 11
 eicosanoid 179, 183
 ethylene 737
 fruit 11, 737-738
 fungi 181, 183
 gene 181, 182
  induction 184, 691
  products 181
 glucoside 181, 711
 growth 11, 181
 interplant signal 183
 leaf vasculature 184
 lipoxygenase 179 180
 metabolism 179
 methyl ester 10, 180, 710
 mutants 185
 occurrence **179-181**
 pheromone 183
 pigment formation 11
 Pin2 399, 403
 plant organ storarage 726
 promoter 691
 protease inhibitors 181
 proteinase inhibitor 11, 405-
 411
 proteins 182
 root growth 185
 seed germination 11
 senescence 11, 745
 stereoisomers 181
 stress 179, 181, 183
 systemic signal 184-185
 tendrils 11
 tuberization 11, 709-711, 717,
 719
 vegetative storage protein 185
 wound 185, 396, 405-411
Jerusalem artichoke 705, 712,
 738, 741
Jessamine 620

Juvenile(ity) 476, 573, **619, 621-622, 625-628**, 631, 780, 787
  conifers 787
  GA 625-627

Kalanchoë blossfeldiana 320, 323, 361, 620, 635, 775
ent-Kaurenal 71, 469
ent-Kaurene 69-72, 81, 82, 85, 88, 468-470, 746, 762
ent-Kaurene oxidase 72
ent-Kaurene synthetase 70, 85, 468
ent-Kaurenoic acid 71, 72, 310-311, 469-470, 640, 641
Kaurenoids 70
ent-Kaurcnol 71, 469
Kaurenolide 72
α-Ketobutyric acid 343, 344, 386
α-Keto-methylthiobutyric acid 130
Kinase 693, 689
Kinetin 98, 112, 542
  auxin 785
  buds 784-785
  Ca²⁺ 309
  cell division 775
  discovery 774
  organogenesis 781
  phloem loading 659
  senescence 501
  stomata 607-609
  tissue culture 779
  tuberization 717
  vessels 540
Kiwi fruit 377, 502, 732, 736
Kleinia articulata 624
Kleinia repens 624
Kohlrabi 738
Kovats retention index 422, 424

la gene 472, 477
*la cry*ˢ genes 19, 23, 24, 472
LAB-144-143 767
LAB-173-711 767
LAB-198-999 85, 86
Lac promoter 329
Lactone 66, 73, 76, 77, 99, 207, 643
Lactuca sativa 746
Lanthanum 310, 312
Larix decidua (larch) 621, 628
Late embryogenesis abundant (LEA) proteins 182, 404, 466, 682, 692
Lateral branches 572, 767
Lateral bud
  auxin 785
  cell division cycle 584

cell expansion 591
cytokinin 110
elongation 784
IAA transport 591
kinetin 785
nutrients 591
oats 573
vascular connections 591
xylem differentiation 588
Lateral roots 363, 522
Latex 501, 505, 760
S-Lathyrine 779
Lathyrus 779
Laticifers 403
lazy mutants 553
Le gene 17, 18, 34, 86 477
Leᵈ gene 469
LE-ACC2 clone 385
Leaching 776
Leaf 533, 534, 536, 538
  ABA 656
  abscission 191, 500, 588-589
  age 29
  area 105
  auxin transport 589
  bud cuttings 787
  crops **742-747**
  ethylene 500
  IAA 654, 656
  movements 190
  NPA 518
  photosynthesis 657-658
  primordia 534, 535
  senescence 348, 501, 769
  sheath pulvini 554
  shoot apex 589
  tryptophan 654
  vascular system 533-536
  water stress 612
Leaky mutants 451, 469, 470
Lectin 222, 681
Legumes 59, 67, 662, 663
Leguminosae 160, 620, 624
Lemna 441
Lemna gibba 45, 190
Lemnaceae 190
Lemon 631, 751-752, 755, 757, 758
Lettuce 211, 467, 745, 746
Leucine aminopeptidase 182
  heptad 686
  zipper 686, 688
Leucoanthocyanins 779
Leukotrienes 406, 410
Lf gene 473, 476, 477
Lfy gene 476
Lgr gene 472
LH-20 Sephadex 419
Lh gene 469, 470, 477, 479
Light 549, 572

auxin binding 285, 357
Ca²⁺ 299, 303
ethylene 494
PA 167
Lignification 131, 160, 347
Lignin 72, 160, 188, 347, 395
Lily 191-193, 752, 763, 788
Limit dextrinase 252, 253
Linoleic acid 56
Linolenic acid 10, 56, 179, 405-408, 410
Lipase 410, 689
Lipid 261, 681, 737
Lipoxygenase 56, 124, 179, 180, 182-184, 395, 405
  jasmonate 179-180, 184
Liquid endosperm 675, 779
Liquid scintillation counting 427
Litchi chinensis 732
Liverworts 552
lk gene 24, 28-29, 472
Lodging 551, 553, 751, 764
Lolium perrene 620
Lolium temulentum 629, 630, 632
Long-day plants 618-619, 711
ls gene 29, 469, 477
Luciferase 587
Luffa cylindrica 533, 536
Luminal protein 303
Lunaria annua 624, 629, 631, 632
Lupinic acid 110, 112
Lupinus angustifolius 102
Lupinus 110, 112, 150
Lupinus luteus 102, 110, 150
Lutein 145
Luteolin 525
Lux gene 587-588
Lv gene 474
Lycopene 383-385, 388, 728, 730, 735, 761
Lycopersicon esculentum - see tomato
Lysine 39, 58, 60, 122, 161, 334, 341, 342, 347, 360, 361, 456, 682
Lysine oxidase 161
Lysophospholipids 221

Macleaya microcarpa 154
MADS box 686
Maize 42, 46, 51-53, 55, 56, 58, 59, 84, 98, 102, 220, 282, 283, 356, 441, 443, 444, 518-519, 521, 525, 552, 600, 608-609, 612, 620, 642, 651, 652, 663-665, 677
  ABA 33, 267, 463, 467, 655, 677-679

Maize - ABA continued:
mutants 679-680
α-amylase 260
auxin 356-357
coleoptile 222, 229-230
transport 511
auxin-binding protein 282-
289, 355, 358
antibodies 285-289
NPA 518
Ca²⁺ 299, 301, 304, 309
cDNA 285
coleoptiles 222, 229-230, 512,
550, 558
copalyl pyrophosphate 468
cytokinin 441-442, 656
-binding protein 290
oxidase 112-113
dinoseb 751
dwarf mutants 86
embryo-lethal mutants 675
mRNAs 682
proteins 682
ethylene 581
flowering 620
GA **83-85**, 86
biosynthesis genes **467-**
469, 471
geranylgeranylpyrophosphate
468
germination 249, 678
β-glucosidase 358
gravistimulated 558
IAA 22, 49, 443, 558
biosynthesis 46
catabolism 53
conjugates 58-59
distribution 52
glucose synthase 51
oxidation 54-56
transport 51
internode length genes **467-**
469
isopentenyl transferase 102
ent-kaurene 468
kernel development 665
lazy mutants 553
LEA proteins 692
male-sterile 165
mesocotyl 558
mevalonic acid 468
mutants, hormone 449
NPA 518
orange pericarp 46
phototropism 32
plasma membrane ATPase 221
precocious germination 466-
467, 673
protoplasts 690
rab28 gene 30

salicylic acid 191
seed protein 404
sex expression 642
transcription factor 466, 686
tryptohan mutants 456
viviparous mutants 673
ABA 463, 467, 679
wilty mutants 463
Vp1 gene/protein 351 689,
690
Male flower 642-643, 759
sterility 769
Maleic hydrazide 739, 752, 753,
765, 768
Mallotus japonicus 199
Malonyl-ACC 120, 129, 390, 730
Malonyl-CoA 129
Malonyltryptophan 45
Malonylation 128, 129, 730
Malonyltransferase 128, 129
Malt 774, 756
Malus pumila 621, 622
Malus robusta 622
Malus sylvestris 727
Mangifera indica 735
Mango 496, 735, 751
Mangrove 672
Mannich formaldehyde reaction
437
Mannitol 441
MAP kinase 583
Map-based cloning 449, 454,
457, 466
Maple syrup 779
Marah macrocarpus 71, **83-85**,
441
Marsilea drummondii 578, 582
Mas promoter 587
Mass flow 399
Mass spectra 422-425, 429, 431
Mass spectrometer 428-429
Mass spectrometry 3, 39, 101,
146, 420-421, 425, 577,
580, 636, 639, 641
Maturation 505, 583, 621, 674
Mechanical harvesting 751, 752,
760, 761, 764, 769
Mechanical perturbation 547,
567, 629
Mefluidide 765
Melon 125, 372, 380, 502, 725,
730, 731
Membrane
depolarization, salicylic acid
190
hyperpolarization, auxin 24
ion transport, salicylic acid
190
NPA 518
permeability 514

phytotropins 518
polyamines 171
potential 24, 286, 309
wounding 399
Mepiquat chloride 764
Meristem 228, 442, 476, 491,
532, 575, 583, 639, 640,
719, 762, 765, 776, 784,
785
culture 776, 784
MeroCaM 306
Mesocotyl 552
Mesophyll 611
Metaxylem vessels 463
Methionine 8, 103, 118-122,
129-131, 170, 334, 375,
489, 567, 766
cycle 129
Methionine adenosyltransferase
131
3-Methyl-2-butenal 112
Methyl-2-chloro-9-hydroxy
fluorene-9-carboxylate 765
Methyl iodide 422
Methyl jasmonate 10, 179, 183-
184, 737-738
Methyl sulphonyl anion 422
Methyl zeatin riboside 332-333
Methylation 131
Methylation interference
footprinting 687
3-Methylene-oxindole 54, 57
Methylthioadenosine 119, 120,
129
Methylthio isopentenyl adenosine
100
Methylthioribose 119, 129-131
Methylthioribose phosphate 129
Methylthio zeatin riboside 100
Metrosideros diffusa 622
Mevalonate kinase 68
Mevalonate-5-pyrophosphate 68,
102
Mevalonic acid 6, 9, 67, 68, 81,
102, 141, 143-144, 468
Mg²⁺ 81
MGBG 163, 165, 169 170
Mi2 genes 468
MiaA mutants 327, 330, 334
Mice 434
Microautoradiography 512
Microcutting 787
Micropropagation **774-793**
Microsomes 518
Microspores 777
Microtubule 307
Millet 191, 249, 260
Minimal media 680, 775
Mitochondria 303

Mitogen-activated protein kinase 583
Mitosis 583 592
Mixed anhydride reaction 437-438
Mn⁺⁺ 53, 54, 81
Molecular ion 422
Mono-oxygenases 79, 85
Monoclonal antibodies 285, 434, 519
Monocotyledons 776, 778
Monoculture 575
Monoecious 642
Monophenols 53, 54
Monoterpenes 68, 72
Monstera deliciosa 547
Morning glory (*Ipomoea*) 497, 499, 620, 621
Morphactins 517
Morphogenesis 7, 165, 675, 778, 782, 792
Moss 7, 310, 311, 442, 552, **460-462, 575-576**, 592, 593
Mougeotia 299
Multiplication 428, 776, 781, 784-788
Mung bean 121, 123, 125, 128, 129, 169, 207, 208, 210, 231, 236, 237, 278, 494-495, 552
  ACC 123
    oxidase 123
    synthase 121
  auxin 208, 278
  ethylene 208, 292
  brassinolide 208
  SAUR 237
  salicylic acid 191
Musa acuminata 727
Mustard 494
Mutants **340-351, 363-365, 385-388, 448-479**
  ABA 351, 452, 462-467
    deficient 149
    resistant 453
  Arabidopsis 448-453
  auxin 455
    biosynthesis 456
    resistant 356, 453
  biochemical pathways 448
  cytokinin 453, 457, 462
  deletion 450
  epistatic 453
  ethylene 340
  gibberellin 351, 452, 465, 467-473
    insensitive 452-453
  isogenic lines 448
  jasmonate 185
  moss 460-462

response 451, 453
resistant 448, 452
tryptophan biosynthesis 456
zeatin 462
Mutation supressors 471
Myb gene/protein 266-267
Mycorrhizae 775
Myeloma 434
Myosin filaments 307

na gene 34, 468-470, 472, 477, 479
NAD kinase 306
NADPH oxidizing enzyme 163
NADPH-dependent mono-oxygenases 79
NahG gene 351
nana 19, 20, 28, 34
Naphthalene acetamide 325, 753-754
Naphthalene acetic acid (NAA) 235, 323, 753-755, 761
  abscission 29, 755
  auxin-binding protein 281, 284, 286-287
  cell division 25
  β-glucosidase 359
  immunoassay 440
  membrane potential 24
  somatic embryos 782
  stomata 311
  tissue culture 774, 778
  tuberization 716
1-Naphthyl-N-methyl-1-carbamate 755
Naphthylphthalamic acid (NPA) 29, 324, 355-356, 511, 517-518, 525-526, 778
  auxin 565
    binding 357
  binding 277, 518, 522
    antibodies 519
    protein 518-520, 523-526
    vesicles 520
  β-Glucosidase activity 359
  IAA 522-523
  membranes 518
  quercetin 519
  receptor 355-356
  root 523
  stem 590
Napin 182
Narcissus 490, 788
Nematodes 397
Neomorphs 783
Neoxanthin 145, 147-149, 463, 465
Nickel 780
Nicotiana debneyi 197
Nicotiana glutinosa 195, 197

Nicotiana plumbaginifolia 363, 455
Nicotiana sylvestris 620, 623, 624
Nicotiana tabacum - see Tobacco
Nicotine 161, 174
Nifedipine 312
Nigericin 514
Nitellopsis 299
Nitrate 191
Nitrate reductase 191
Nitrenium ion 127
Nitrogen fixation 769, 791
Nitrogen specific detector 426
p-Nitrophenol 435
p-Nitrophenylphosphate 279, 435
Node culture 784
Nodulation genes 525
Nopaline 238, 320, 321, 323, 330
nor gene 728, 732
Norbornadiene 132-133, 374, 700, 737
Norflurazon 145
Norspermidine 172
Norspermine 172
Norway spruce 621
Notabilis (*not*) tomato 462-463, 679
Nr gene 728, 732
NTP kinase 306
Nuclear proteins 687
Nuclease 253
Nuclei 162, 163, 175, 211, 235, 262, 263, 278-280, 291, 293, 443, 583, 777
  Ca2+ 310
Nucleosidase 107, 253
Nucleotidase 107
Nurse cultures 780
Nutrients 349, 577

Oak 621
Oat - see *Avena*
Octadecylsilane 418
Octopine 320, 323-324, 328, 330
Octopine synthase 324
Oleander 320, 332
Olefins 132-133
Oligosaccharides 397, 398, 792
Olive 320, 326, 332, 760
Oncogenesis 330
Onion 555, 739, 740, 745, 768, 788
Ontogeny 574
Opaque-2 265
Opines 323
Orange 170, 383, 390, 621, 729, 735, 757
Orange pericarp, maize 46, 456
Orchid 604, 775, 776, 788
Orchidaceae 499

Organ cultures 775-776, 778
Organogenesis 319, 476, 477,
　　673, 781, 782, 788, 790-
　　792
Organophosphate 769
Ornamental plants 764
Ornithine 158, 164, 169, 173,
　　361, 716, 741
Ornithine decarboxylase 158,
　　161-163, 168, 169, 172,
　　174, 175, 716, 741
orp gene 46
Orthogravitropic shoots 574
Oryza sativa - see rice
Osmoregulation 223, 225
Osmotic environment 732
　　potential 214, 218, 660, 664,
　　760
　　stress
　　　ABA 405, 409, 692
　　　Em gene 690
Osmoticum 171, 217, 674, 679,
　　682
Overripening 737
Ovules 777
Oxindole-3-acetic acid 39, 54-58,
　　60-61
Oxindole-3-carbinol 54
2-Oxoglutarate 77, 78, 81-83, 85-
　　86
2-Oxoglutarate-dependent-di
　　oxygenases 77-79, 85
12-Oxo-phytodienoic acid 405-
　　408
Oxygenase 154
Ozone 746

P-450 mono-oxygenases 85
Paclobutrazol 85, 472, 763-766
Palms 778
Paper and thin layer
　　chromatography 416
Paphiopedilum tonsum 609
Parsley 299, 745
Parsnip 738
Parthenocarpic fruit 757
Particle gun 263
Partitioning 649
Passiflora edulis 730
Passion fruit 730
Patatin 703, 704, 714
Patch clamp techniques 310, 601,
　　605
Pathogen resistance 196
Pathogenesis-related gene 30
Pathogenesis-related proteins 11,
　　194-196, 200 398
Pathogenicity 332
PCR amplified DNA 471
Pel gene 468

Pea 134, **469-477**, 534, 608, 652,
　　660, 664
　　ABA mutants 463, 663, 679
　　ACC 125
　　Alaska 165
　　apical senescence 477
　　auxin 217, 231, 233
　　　epicotyl 229-230
　　　response 20
　　axillary buds 584-585
　　brassinolides 208
　　cDNA clones 232
　　cell cycle 583
　　cotyledon bud 578
　　cytokinin 579
　　defloration 573
　　diamine oxidase 160
　　dwarf 19-20, 34, 86, 469
　　embryos 76-787
　　ethylene 292, 581
　　floral induction 473
　　flowering genes **473-478**
　　fruit growth 676
　　fruit set 675
　　GA 81, 734
　　　binding proteins 290
　　　biosynthesis **75-79**, 88
　　　　genes **469-473**
　　　hydroxylases 88
　　　non responders 28, 472
　　　pathway 86
　　　response mutants 472
　　　seeds 479
　　　shoot 479
　　　slender phenotype 470
　　　internode length 477
　　　pleiotropic 477
　　germination 165
　　grafting 473
　　IAA 20, 32
　　　decarboxylation 56
　　　response 20,
　　　synthesis 28, 45, 46
　　internode length mutants **469-
　　　473**
　　methionine adenosyltransferase
　　　131
　　mutants 449, 467, 469-477
　　nana 19, 20, 28, 34
　　pod growth 48, 675, 734
　　polyamines 169, 170, 175
　　putrescine 165
　　SAUR 237
　　seed GAs 75, 77 676, 734
　　　growth 676
　　slender 19, 20, 23, 470
　　spermidine 165
　　tall 19, 20, 33-34, 469
　　wilty mutants 463

Peach 126, 443, 631, 729, 761,
　　762, 767
Peanut 129
Pear 189, 621, 729, 731, 732,
　　734, 736, 754, 755, 757,
　　766
　　abscission 755
　　ACC 120
　　fruit set 757
　　GA 734
　　polyamines 736
　　ripening 372
Pectin 222, 729
Pectin methylesterase 729
Pectinases 777
Peduncle 477
Pentose phosphate pathway 766
Pepper 211, 727
Peptide hormones 274, 275
Perennials 619, 622
Pericarp 652
Pericycle 523
Perilla crispa 573, 620, 621, 623,
　　624, 628, 634
Periodate reaction 439
Periwinkle 775
Peroxidase 53-58, 124, 282, 283,
　　404, 443, 735
Perry 120
Persea americana 727
PESIGS rules 35
Petiole explants 589
Petunia 26, 27, 344, 346, 347,
　　360, 500, 624, 735
pH 308, 515
Phalanx 574
Phaleonopsis 497
Pharbitis nil 496, 620, 621, 624,
　　629, 630, 633, 635
Phase change 621, 622, 625-628
Phaseic acid 148, 150-154, 691,
　　742
Phaseolus coccineus 653
Phaseolus vulgaris 576, 577 612,
　　661, 664
　　ABA 429, 580
　　apical buds 577
　　auxin 675
　　axillary bud 583
　　cytokinin 110-111, 579
　　　oxidase 112
　　development 583
　　embryos 663, 675
　　ethylene 581
　　　binding protein 292
　　fruits 573-574, 577, 580
　　GA **83-85**, 675
　　　hydroxylases 87
　　IAA 576, 580
　　leaves 577

Phaseolus vulgaris continued:
    naphthylphthalamic acid
        (NPA) 577
    seed drying 679
    xylem exudate 579
Phenanthroline 378
Phenolics 40, 72, 164, 188-190,
        198, 322, 415, 738, 779
    flowering 190
    levels 415
    plant-microbe interactions 189
    roots 188
    tuberization 716
Phenoxyacetic acids 778
Phenylacetamide 325
Phenylacetic acid 4, 40, 509, 519,
        778
D-phenylalanine 129
Phenylalanine ammonia lyase
        181, 182, 409
Phenylcyclopropylamine 127
Phenylureas 769, 779
Pheromone 183, 487
Philodendron 547
Phloem 532-534, **658-662**, 732
    ABA 612, 656
    metabolites 153-154
    anastomoses 532, 535
    auxin 535, 543
        transport 510
    collateral bundles 535
    cytokinin 543
    differentiation 531, 535
    exudate 577
    loading 612, 658-659, 666,
        732
    transport **658-662**
    unloading 660, 666
    seed coats 652
    wound signal 397
Phosphatase 30, 254, 279, 435,
        438, 649, 692
Phosphate transporter 649
Phosphatidylinositol
        4,5-biphosphate (PIP2)
        274-275, 308, 792
Phospholipase 221, 274-275, 308
Phospholipid 254
Phosphon D 763
Phosphorylation 225, 611, 689
Photoaffinity labeling 133 354,
        519
Photoassimilates **649-666**
    unloading 661-662
Photomorphogenesis 472
Photoperiod(ism) 18, 87, 193,
        473-476, **618**-619, 621,
        624, 633, 639, 642, 643,
        699-701, 705, 709, 711,
        716-718, 746

genes 474
mutant 474
polyamines 165
salicylic acid 193
sex expression 643
thermogenesis 193
Photophosphorylation 657
Photorespiration 657, 769
Photosynthate partitioning **649-
        666**, 704
Photosynthesis 181, 299, 310,
        395, 489, 547, 552, **610-
        611**, **649-666**, 725, 769
    ABA 611
    auxins 666
    cytokinins 657
    ethylene 657
    GAs 657
    gibberellins 666
    IAA 657
    jasmonate 181
    stomata 611
Phototropism 2, 15, 239, 299,
        512, 521, 525, **547-551**,
        553, 556
    action spectrum 548
    adaptive significance 548
    auxin 32, 521, 548-549
        asymmetry 551
        -regulated genes 239
        transport 521
    blue light 548
    carotenoid 548
    coleoptile tip 549
    flavinoid 548
    IAA 2, 23, 31
    negative 547
    photoreceptor 548
    positive 547, 548, 550
Phragmites communis 574
pHTOM5 125
Phyllotaxis 621, 622, 629
Physarum polycephalum 163, 175
Physcomitrella patens 460, 462,
        575
Phytate 302
Phytoalexins 72, 181, 184, 188,
        395, 501
Phytochrome 169, 299, 307,
        472-474, 478, 618, 699
    mutants 474, 478
Phytoene 384, 385
Phytotropin 355, 511, 517-520,
        523, 524-525
pIAA1 326
Picea abies 553, 621
Picloram 756, 778, 782
Pigment 181
Pigweed 208
Pik-Off 768

Pin2 gene 396, ,398, 405
Pine 47, 56, 493, 621
Pineapple 496, 727, 751
Pinus 53, 55, 56, 537, 621, 622
Pinus sylvestris 53, 55, 621
Pisonia 488
Pisum sativum - see pea
Plagiomnium cuspidatum 576
Plant
    breeding 759
    culture 775
    defense 181
    immunization 194
    multiplication 781
    regeneration 782
    transgenic 340
Plant growth regulators 2, 60,
        100, 119, 168, 186, 234,
        400, 486, 490, 493, 504,
        531, 539, 540, **751-770**,
        775, 776, 788, 791-793
Plant growth substance 2, 364,
        747, 789
Plant hormone - see also
        individual hormones
    analysis **415-447**
    apical dominance **572-594**
    binding **272-294**
    bioassays 416
    biosynthesis genes / mutants
        318-335, 372-391, 448-479
    calcium **298 313**
    concentration 2, 13, 17-22
    concept **1-2, 13-14**
    correlative interactions **576-
        582**
    definition **1**
    derivitazation 422
    detection 30
    discovery 2-3
    effects 3
    enzyme activities 273
    extraction 416
    flowering **617-644**
    fruit **372-391, 725-747**
    functions **1-11**
    gas chromatography 420-421
    GC-MS 3, 421-425, 428-429
    HPLC 3, 417-420, 426
    HPLC-MS 3
    identification 2-3, **421-422**
    immunoassay **433-444**
    importance 34
    insensitive mutants 451, 452
    interactions 28, 29
    isolation 416
    levels 479
    localization 14
    measurement 14, **426-429**
    micropropagation **774-793**

Plant hormone continued:
  mutants **363-366**, **448-479**
  nature **1-11**
  occurrence **1-11**
  orientation of growth **547-568**
  perception and transduction
    272-294
  photosynthesis 610-611, **649-666**
  potato tuberization **698-721**
  purification 416
  quantitation 2-3, 15-16, **426-430**
  receptors **272-294**, 792-793
  redistribution **31-34**
  reproduction **617-648**
  response genes 453
    mutants 448-449, 451, 478
  -responsive element 31
  responsiveness 13, **22-31**, 479
  role 34
  roots 598
  seed development **671-693**
  sensitive mutants 451
  sensitivity 2, 13, **22-31**, 462,
    586-588
  sex expression 641-642
  signal transduction **272-294**
  signaling pathways 451
  source-sink 651-657
  stomata **598-610**
  target cells 272
  tissue culture **774-793**
  transgenic plants **340-351**
  transduction 30
  transport 1, 2, 13, **31-34**
  vascular tissues **531-544**
  vegetable 725
  water balance **598-613**
  wounding **395-411**
Plasma membrane 221, 225,
  515-520, 526
  auxin receptor 286, 288
    binding proteins 283-284,
      354, 355, 604
  ATPase 220, 221, 224-225
  Ca^{2+} 300, 302, 304, 307, 312,
    604
    ATPase 557
    channels 300-301
  H^+-ATPase 307, 560
  GA receptor 291
  proton pump 661
  vesicles 30, 519-520
Plasmid 48, 104, 105, 110, 238,
  321, 322, 326, 329, 330,
  332, 333, 365, 586, 454,
  686, 690, 781
Plasmodesmata 52, 248, 250

Pleiotropic 272, 364, 449, 467,
  472, 475, 477, 681, 689,
  701, 702
Pod 48, 477, 656
Poinsettia 492, 620, 752, 763-764
Pollen 11, 777
  ACC 499-500
  auxin 754
  dispersal 552
  ethylene 499
Pollination 166, 383, 497, 499-
  500, 652, 665, 733, 736,
  754, 757, 769
  ethylene 499-500
  flower senescence 499
Pollinators 552
Polyacrylamide gel
  electrophoresis 584
Polyamine oxidase 160, 162
Polyamines 3, **158-175**
  abscisic acid 169
  ACC 131
  S-adenosylmethionine 131,
    169
  adventitous shoot formation
    164
  α-amylase 169
  antisenescence activity 160
  auxin 162, 166, 169
  bacteriophage 163
  binding 162
  biosynthesis 10, **159-162**, 168,
    174-175
    enzymes 158, 163
    localization 161
    scheme 160
  bound 161
  calcium 175
  cell division 164, 736
    walls 161
  cellular 164
  chilling injury 171
  chloroplasts 161
  cold hardening 171
  conjugated 164-165, 736
  cytokinins 169
  differentiation **164-168**
  disease control 173
  DNA 162-163
  dormancy 741
  drought 171
  E. coli genes 174
  effects 10
  elongation 164, 169
  embryogenesis ethylene 166
  embryoids 10, 164
  enzyme localization 162
  ethylene 131, 160, 166-168,
    170, 387, 390, 737
  eukaryotes 158

  flowering 164-166, 361
  free 737
  free radical scavenging 739
  fructose-1,6-bisphosphatase
    163
  fruit development 166, 736
    ripening 164, 166-168,
      387, 727, 737
    softening 728
  functions **162-164**
  GA 169
  gene expression 163
  growth **163-169**
  growth regulators 167
  high temperature 171
  hormones 10, 169
  intracellular 161
  K^+ 170
  kinases 175
  light 169
  lipid peroxidation 737
  localization 16 162
  macromolecule synthesis 162-
    163
  meiosis 162
  membrane 162, 171
  metabolism 160, 162
  methylation 130
  MGBG 169
  mitochrondria 161
  mitosis 162
  molecular analysis 174
  morphology 10
  mutants 158, 165
  nucleic acids 162
  organogenesis 792
  ozone 746
  parthenocarpy 169
  pH 161, 162
  phenolic acid conjugates 160
  phosphorylation 163, 175
  photoperiod 165
  phytochrome 169
  phytopathogenic fungi 172
  plant organ storarage 726
  prokaryotes 158
  protein structure 163
  proton-secreting systems 162
  RNase 739
  rolA 361
  root initiation 164
  senescence 131, 164, 166 736
  sexual differentiation 165
  somatic embryogenesis 792
  stress 160, 168, 171-172, 174
  thermophilic bacteria 172
  titer **168-173**
  tomato 160, 736, 746
  transgenic plants 174
  transport 170

Polyamines continued:
  tRNA 163
  tuberization 716
  uptake 170
  vacuole 161
  vascular differentiation 164
  virus 163
  wounding 739
  yeast 163, 174
Polyclonal antibodies 382, 433
Polygalacturonase 386, 390, 398, 728
Polymerase chain reaction 451
Polyphenols 54
Polysaccharides 442
Postharvest hormone change **725-747**
Potassium channels 311, 692
Potassium cyanide 216
Potato 553, 578, 579, 582, **698-721**, 738
  ABA 741
    mutants 404, 463
  andigena 700
  arginine decarboxylase 741
  aseptic culture 788
  bolters 717
  brassinolides 211
  CN-sensitive 738
  cuttings 579
  cytokinin 741, 579
  dormancy 740, 741, 759
  droopy mutant 400, 402, 463, 467, 717
  dwarf 717-718
  GA 708, 717-718, 741
  genetics 700
  giant hills 717
  grafting 706
  in vitro 715
  jasmonate 182, 185, 405, 409, 410
  late embryogenesis genes 404
  maleic hydrazide 739
  ornithine decarboxylase 741
  Pin2 gene 402, 404
  polyamines 739, 741
  proteinase inhibitor 396
  putrescine 741
  S-adenosylmethionine decarboxylase 741
  spermidine 716, 741
  spermine 741
  sprouting 741, 768
  topiary 717
  transgenic 397, 402, 403
  tuberization 181, **698-721**
  tuberosum 700
  water-stressed 404
  wilty mutants 463

wound-induced genes 405, 409
wounding 400, 401, 403, 738
Ppd gene 473-475
Precocious branching 786-787
Precocious germination 467, 671, 673, 674, 678, 681, 732
Prenyl transferase 69, 327-331
Privet 320
Pro-Shear 767
Procambium 532-533
Proembryo 777, 782, 783
Prohexadione-calcium 85, 763
Prokaryotes 158, 318
Proliferation 277
Proline 182, 738
Prolonged growth response 20, 759
Promalin 767
Promoter 48, 232, 237, 239, 240, 256, 263-266, 280, 340, 342-343, 348, 360-363, 402, 417, 587, 657, 683-685, 688, 740, 744, 779
  ABA 404
  α-amylase 263, 266
  auxin-inducible 349, 521
  Ca^{2+}, animals 307
  cauliflower mosaic virus 35S 349, 361, 364
  constitutive 375
  heat shock 348
  iaaII/iaaM 342
  ipt 334
  lux 587
  Pin2 397, 402-403
  ptz 334
  rol 361-362
  tmr 329
Propagules 490, 784
Propyl gallate 407, 409
4 (n-propyl-α-hydroxymethylene) -3,5,-dioxocyclohexanecarb oxylic acid ethyl ester 85
Propylene 132-133, 372, 386
Prostaglandin 60, 406, 410
Protease 133, 167, 168, 181, 184, 185, 254, 281
Protease inhibitors 181, 184
Protein
  binding region 30
  bodies 252, 290
  Ca^{2+}-binding 300
  extracellular 788
  kinase 30, 221, 274, 275, 295, 306, 350, 391, 458-459, 583, 689
  Ca^{2+} 306-307
  CaM 306
  isoforms 306

Pisum 583
  response regulator 457
  animal hormones 274
  gravitropic curvature 565
phosphorylation 30, 134, 453, 693
synthesis 211
  inhibitors 216
Proteinase 10, 11, 182-185, 196, 395, 396, 398, 405, 703, 704, 791
Proteinase inhibitor 182-185, 196, 396, 703, 791
Proteolytic fingerprinting 257
Protocorms 776, 788
Proton
  ATPase 220, 658
  extrusion, IAA 659
  motive force,auxin 513
  -translocating ATPase 658
  /sucrose cotransport 658, 661
Protonema 460, 478, 576
Protoplast(s) 167, 256, 281, 285, 291, 299, 462, 555, 684, 689-690, 777, 780, 782, 792
  cultures 775, 777
  fusion 462, 777
  sucrose uptake 662
  transient expression assay 689
Prunus 501, 734
Pseudomonas 729
  ACC deaminase 386
  IAA conjugation 58
  NahG gene 351
Pseudomonas amygdali 332
Pseudomonas fluorescens 325
Pseudomonas savastanoi 44, 48, 60, 318, 320, 325-326, 342, 456
  auxin 318, 325-326
  cytokinin 318
  galls 325, 342
  genes 324
  iaaH 333
  iaaL 60, 361, 456
  iaaM 333
  tumorigenesis 319
Pseudomonas solanacearum 332, 335
Pseudomonas syringae 196, 360
Pseudotsuga menziesii 621
pTOM13 clone 125, 377-378, 730
  antisense gene 125
ptz gene 332, 334
Pulvinus 551, 555, 558-563, 565-567
  auxin 560-561
  GA 565

Pulvinus continued:
  gravistimulated 558, 560, 562
  IAA 558, 562, 565
  internodal 551
  invertase 564, 567
  leaf sheath 551
  cellulase 567
Pumpkin 79, 81-82 88
Purafil 726
Putrescine 163, 167, 169,
    171-175
  ACC biosynthesis 131
  alkaloids 161
  auxin 161
  chilling injury 737
  conjugation 736
  crassulacean acid metabolism
    164
  dormancy 741
  ethylene 737
  K⁺ 170
  metabolism 736
  NAA 161
  nicotine 161
  organogenesis 792
  phenolics 160
  ripening 131
  somatic embryogenesis 792
  stress 164
  tomato 131, 736
  yeast 164
Polyvinyl polypyrrolidone
    (PVPP) 441
Pyridoxal phosphate 120, 489
Pyrimidine box 264-265
Pyrophosphatases 302
Pyrroline 160
Pyrus communis 621, 731
Pyrus malus 622

Quackgrass 553
Quercetin 519-520
Quercus 621, 622
Quercus robur 621
Quiescent center 442
Quin-2 222
Quinate: NAD+ oxidoreductase
    306

RAB16 gene 404-405
rab28 30
Rabbits 433, 434
Radical scavengers 358
Radioimmunoassay 329, 331,
    333, 434-437, 439-441,
    579
Radish 100, 102, 109, 110, 112,
    207, 208, 211, 524, 632
Raf protein kinase 350, 391, 459
Raffaele Piria 188

Ramets 574
Randomly amplified polymorphic
    DNAs (RAPD) 449
Rapeseed 673, 678, 679, 682, 746
Raphanus sativus - see radish
Rapid defense response 197
Receptivity 13, 14
Receptor 13-14, 272-276
  ABA 291
  acid-growth theory, auxin 286
  affinity 13, 14
  auxin 277-278, 282, 284, 311
  cytokinin 290
  ethylene 292, 389
  gibberellins 255, 290
  IP3 308
  phytohormones 298
  plasamembrane 274
  proton pumps 286
  steroid 279
  tissue culture 792
Red light 209, 284, 356, 699
Red/far-red ratio 582
Regeneration 277, 281, 782
  media 777
"Release" 768
Reporter genes 263, 307
  GUS 397, 402
  lux genes 587
Reproduction 86, 166, 189-191,
    474, 476, 558, 573, 577,
    591, **617-648**, 651, 652,
    655, 663, 666, 676
  development 617
  dominance 573, 577
  hormones 617
  vegetative 575
Reserve 246, 250, 253
Respiration 191, 194, 389, 489-
    490, 501-502, 513, 726,
    727, 731, 732, 734, 738,
    742, 745, 745
Respiratory climacteric 728
Response capacity 14
Responsive element 404
Responsiveness 13, 25
Rest 10, 725, 739
Restriction Fragment Length
    Polymorphisms (RFLP's)
    449, 454, 457, 471, 700-
    701, 720
Reticuloplasmins 303
Reverse phase chromatography
    418
Rhizobium 189, 333, 524
Rhizobium meliloti 525
Rhizomes 553, 572, 574-575, 582
Rhizophora mangle 672
Rhodamin 442, 443

Rhodococcus fascians 319-320,
    332-333
Rhubarb 759
Ri plasmid 110
Ribes rubrum 54
Ribonuclease 253, 254
Ribulose 1,5-bisphosphate
    carboxylase 181, 395, 397,
    657
Rice 651
  α-amylase 260, 263
    genes 257
  ABA 690
    -responsive element
      (ABRE) 687
  deep-water 21
  Em mRNA 690
    promoter 686
  embryo mRNAs 682
    proteins 682
  ethylene 29
    -binding protein 292
  GA 86, 759
  germination 249
  IAA oxidation 54-55
  lazy mutants 553
  lodging 553
  mutants 449
  nuclear proteins 688
  polyamines 168, 169
  precocious germination 673
  RAB 16 gene 404
  Tanginbozu dwarf rice assay
    417
  upward bending 553
  Waito-C 17
Ricinus 612, 659
rin gene 728, 729, 732, 736
Ripening - see Fruit ripening
RNA blot hybridization 232, 235
RNA polymerase 101, 229, 266,
    276, 278, 290, 685, 738

RNA synthesis 278
  antagonists 217
RNase 739, 756
Robinia pseudoacacia 151, 622
Rol genes **361-363**, 586-588
  rolA 27, 361, 363, 586, 587
  rolB 24, 25, 350, 361-363,
    586, 587
  rolC 110, 350, 359, 362-363,
    366, 586, 587
Root(s)
  ABA 599, 611, 656, 742
  adventitious 540, 574
  aerial 547
  ageotropic 553
  auxin 455, 512, 523, 556
    transport 509

Root continued:
  cap 556
  culture 774, 777
  cytokinin 105, 579 656, 741
  dormancy 725
  elongation 523
  growth 181, 209, 455
  hairs 494
  initiation 164
  NPA 518
  pressure 612
  silver 493
  storage 725
  stress 656
  tips 534
  vascular system 533, 536, 538
  water stress 611-612
Root Crops 741
Rooting 41, 456, 621, 753, 780
Rose 142, 189, 197, 422, 497,
    502, 739, 767
Rosette plants 67, 636, 746
Rubber 403, 501, 760
Rudbeckia bicolor 166, 620
Run-on transcription assays 233
Runners 553
Rutabaga 738
Rye 27, 765
Ryegrass 620

Saccharomyces cerevisiae - see
    Yeast
Salicin 188
Salicyl alcohol 188
Salicylic acid 3, 179, **188-202**,
  235
  ABA 11
  alternative oxidase 194
  animals 201
  anthocyanin 191
  Arum 11
  benzoic acid 2-hydroxylase
    199
  biosynthesis **198-199**, **202**
  calorigen 191
  cell wall 198
  cinnamic acid 198
  2,6-dihydroxybenzoic 193
  disease resistance 11, **194-198**,
    201-202
  effects 11
  Em 690
  ethylene 11, 189, 780
    biosynthesis 189, 190, 196
  flowering 189-190
  glucosyl-SA 200
  glucosyl-transferase 199-200
  growth inhibition 191
  history 188
  leaf abscission 191

  membranes 190
  metabolism **199-202**
  nitrate 191
  physical properties 189
  properties **188-191**
  proteinase inhibitors 196
  rice leaves 189
  seed germination 11, 190
  senescence 189
  stomatal closure 191
  thermogenesis 11, 189, **191-**
    194, 195, 201
  tobacco 195
  TMV 195-200
  transgenic plants 351
  transpiration 190
  transport 189
  wound 190-191, 196
  yield 191
Salicylhydroxamic acid 409
Salinity 299, 690
Salix 188
Salt stress 211, 690
Sapium sebiferum 786
Sarcoplasmic reticulum 308
Sauromatum guttatum 191, 193
SAURs (Small Auxin Up-
    regulated RNAs) 232-237,
    239-241
Scarification 495
Schefflera 488
Schiff's base 439
Scotch pine 621
Scutellum 49, 248-250, 253, 255,
    677
Secale 27
Sechium 441
Second messenger 298, 307-308
  Ca^{2+} 300, 307, 313
  cAMP 275
  cytoplasmic pH 309
  inositol triphosphate (IP3)
    275, 307-308
Secondary metabolites 188
Sedges 552
Seed **671-693**
  ABA 140, 141, 655, 733
    desiccation tolerance 467
    mutations 466
  auxin 733
    precursor 52
  cytokinin 102
  coat 652
  desiccation 733
  development 574, **671-693**
    ABA 466, 651-652, 671,
      677, 680
    auxin 651, 671
    cytokinins 651, 671
    GA 651, 671, 677

    IAA 652, 677
    LEA proteins 466
    mRNAs 682
    sucrose 651
  dispersal 503
  dormancy **462-467**, 671
    ABA 465, 466, 680
    sit mutation, tomato 467
  ethylene 490
  fill 649, 655
  germination 185, 467
    ABA 462
    ABA/GA ratios 676
    GA 6
    jasmonate 181
    salicylic acid 190
  growth 676
  IAA 577, 654
  inhibitors 776
  maturation 359, 679
  production 759
  storage genes 690
    proteins 182, 681
  tryptophan 654
  water relations 466
Seedling development 671, 673
Selected ion monitoring 113, 429,
    580
Senescence 185, 725, 762, 769
  ABA 501, 745
  ACC oxidase 131, 385
    antisense 384
  ACC synthase 131
  S-adenosylmethionine synthase
    131
  benzyladenine 745
  brassinosteroids 208
  cDNA 501
  cytokinin 7, 726, 735, 745-
    746
  ethephon 762
  ethylene 122, 349, 372, 486,
    745, 746
  flower 26, 30, 499
  gibberellins 734, 752
  jasmonic acid 181, 745
  kinetin 501
  leaves 501
  mRNA 26, 501
  nutrient mobilization 349
  polyamines 131, 164, 167,
    736
  salicylic acid 189
  sink 349
  stress-induced 501
  transgenic plants 348, 382
  vegetative 349
Sensitivity **13-14**, **22-31**, 747
  ABA 27, 692
    seed development 676

Sensitivity continued:
  developmental stage 26
  differential 25
  ethylene 13, 726
  GA, seed development 676
  genetic regulation 23
  moss 460
  mutants 23
  tissue 26, 792
  transgenic plants 586-587
  regulation 30
Sephadex 419
Serine-threonine protein
      phosphatase 692
  kinases 391
Sesquiterpenes 68, 141
Sex expression 165, 636, 641-
      643, 496, 759
Shading 572
Shasta Daisy 319
Shelf-life 135
Shikimic acid pathway 43
Shoot apex
  auxin 509, 577, 579, 588, 589
  axillary buds, inhibition 577
    basipetal IAA 577
      correlative control 572-
      573, 576, 589
      dominance 577
      GA 579
      leaf abscission 589
      repressive signal 572
  brassinosteroids 209
  culture 776, 784
  inversion 581
Short-day plants 618-619, 711
Shrubs 538
Sieve tubes 531-533, 540, 542
Signal transduction 255, 457, 503
  auxin 232, 236
  calcium 305
Silene armeria 620, 624, 631,
      634-635, 639
Silica 418
Silver ion 5, 132-134, 374, 386,
      389, 398, 489, 497-499,
      504, 581, 780
  abscission of flower petals 704
  carnations 132
  male flower 489
  thiosulfate complex 489
  tissue culture 780
Sinapis alba 166, 624, 629, 630,
      634
Sink 129, 153, 183, 349, 351,
      539, 578, 649, 651, 653,
      654, 656, 657, 659, 660,
      664-666, 702-704, 712,
      715, 736, 742

Sitiens (*sit* gene) 400, 406, 462,
      679, 732
Slender 19, 470, 472
Slime mould 105
sln gene 470, 479
Smoke 3, 490, 496
Sn gene 473-476
Sodium arsenite 236
Soil 572, 582 601
Solanaceae 499
Solanum andigena 579, 582
S. berthaultii 700
S. demissum 703, 710
S. tuberosum - see potato
  ssp. tuberosum 700
  ssp. andigena 700
S. tuberosum x S. berthaulti 700
Solvent partitioning 416
Somaclonal variation 388
Somatic chimeras 472
Somatic embryogenesis 167, 673,
      777-778, 780-784, 788,
      790, 792
Somatic hybridization 462, 777
Sorghum 18, 249, 260, 613
Source:sink relationships 183,
      349, **651-657**, 665
Soybean 652, 654-656, 658,
      661-664
  ABA 678, 679, 732
  auxin 228-229, 231-235, 239
    binding 279
    mRNA 232
  callus bioassay 417
  cDNA 237
  flowering 620
  gene 238
  germination 678
  gravistimulation, hypocotyls
      240
  IAAsp 59
  jasmonate 182, 184
  precocious germination 673
  protein kinase 306
  TIBA 751
  vegetative storage proteins
      405, 409
  wounding 406
Spectrofluorimeter 426
Spermidine 160, 166-170, 172,
      173, 175, 390
  ACC 737
  bud differentiation 166
  cell permeability 739
  DNA 162, 163
  dormancy 741
  ethylene 737
  organogenesis 792
  overproducers 10
  somatic embryogenesis 792

Spermidine synthase 162, 165,
      170, 174
Spermine 160, 165, 167, 169,
      172, 175
  cell permeability 739
  DNA 162, 163
  dormancy 741
  ethylene 162, 737
  organogenesis 792
  somatic embryogenesis 792
Spermine synthase 174
Spinach 619, 630, 636, 638-642,
      745-746
  ABA 154
  flowering 620
  GA 18, 639-640, 643, 746
    biosynthesis inhibitors 18,
      638
    pathway 86
  sex expression 642
  thermoinduced stem growth
      639
Spinacia oleracea - see spinach
Spirodella 190
Spoilage 135
Sprout tubers 713
Spruce 553, 621
Spy gene 472
Squash 121, 122, 171, 759
Squirting cucumber 503
Starch 246, 248-250, 252, 253,
      258, 260, 443, 512,
      554-557, 563, 649, 658,
      664, 698, 703, 704, 719,
      741
  germination 248
  grains 250, 252, 258, 260,
      443, 512, 554, 555, 557
  hydrolysis enzymes 252
  sheath cells 512
  synthesis 658
Statoliths 554-557
Stem
  culture 784
  elongation (growth) 4, 6, 11,
      18-20, 23, 25, 27, 28, 31,
      87, 88, 465, 478, 631, 632,
      635, 636, 637-640, 741,
      746, 762, 763, 764, 787
    auxin 4
    GAs 87, 88, 471, 636, 741
    sterols 763
    thermoinduction 632
  gravitropism 575
  IAA transport 590
  NPA 518, 590
  sections 590
  tip culture 776
Steroid hormone 11, 207, 135,
      255, 276

Steroid contuinued:
    animal 255, 274, 276
Sterols 72, 763
Stigma 499-500
Stolon 553, 572, 574-575, 582,
    623, 698, 699, 703, 704,
    707, 709, 711-713, 716,
    719, 719, 741
    tuberization 716, 719
Stoloniferous structures 582
Stomata
    ABA 311, 404, 598, 607, 610,
    681, 745
    aperture 310
    auxin 311
    behaviour 601
    benzlyadenine 745
    Ca^{2+} 299
    stomatal closure
        ABA 152, 657
        ethylene 657
        salicylic acid 191
    CO_2 607, 610, 611
    cytokinins 311, 610
    GA 311
    IAA 607, 608, 610, 611
    IP3 310
    K^+ 607
    kinetin 608
    naphthylacetic acid (NAA)
        311, 606
    naphth-2-yloxyacetic acid
        (NOXA) 606
    opening 609, 657
    photosynthesis 610, 611
    solar radiation 610
    transpiration 609, 611
    water 609-611
    wind 609, 610
Storage 725-726, **738-742**, 742-
    747
Storage protein 679
Stratification 6, 671
Strawberry 231, 306, 372, 553,
    653, 727, 728, 733, 734
Stress (see also individual
    stresses) 395, 572
    ACC oxidase 128
    auxin 234
    auxin-binding proteins 281
    brassinolide 211
    cytokinin 455
    ethylene 122, 372, 350, 581,
    486, 492, 500
    glutathione S-transferases 241
    inducible genes 682
    jasmonate 181, 183
    light 697
    microorganisms 697

polyamines 160, 168, 170-
    172, 174
    response mutants 455
    second messengers 410
    water 409
    temperature 697
Striga 188
Submergence 491
Subordinate organs 592
Succinate dehydrogenase 766
Sucrose 403, 536, 651, 658, 703,
    732, 768
    accumulation 663
    co-transport 612
Sucrose phosphate synthetase 658
Sugar beet 309, 661, 664, 738,
    742, 775
Sugarcane 751, 753, 756, 758,
    768
Sumagic 763-764
Sundew 567
Sunflower 29, 199, 518, 525,
    539, 549, 550, 552, 581,
    592, 612, 705
Suspension cultures 47, 143, 154,
    172, 199, 229, 282, 690,
    692, 782
Suspensor 675
Sweet potato 738
Sweet William 620
Sycamore maple 621
Syngamy 777
Syngonium 488
Systemic acquired resistance 194-
    197, 351
Systemic signaling 184-185
Systemin 3, 11, 185, 399, 410,
    411

T-DNA 48, 104, 105, 321-324,
    328, 330, 331, 341, 342,
    360, 361, 364-366, 450,
    459, 586, 587, 717
    genes 323
    insertion mutagenesis 450
    tagging 364, 450
Tallness 17, 19, 21, 33, 34
Tanginbozu dwarf rice GA
    bioassay 417
Target cell 273, 274, 291
Td gene 406, 407
Temperature shock 304, 305
Tendril coiling 181, 183
Teratoma 320, 781
Terpenoid 68, 70, 72, 415
Tetcyclacis 85, 746
Tetrapyranylbenzyladenine 767
Theoretical plates 420
Thermogenesis 11, 189, 191,
    194-195, 201

Thermoinduction 618, 619, 621,
    632, 639-641
Thermopolyamines 172
Thermotolerance 211
Thiamine 774
Thidiazuron 769, 779
Thigmomorphogenesis 567
Thigmotropism 307, 547
Thin layer chromatography 416
Thinning 754, 755
Thionin 181-182
Thlaspi arvense 620, 632, 639-
    641
Threonine deaminase 182, 404
Ti plasmid 48, 104, 105, 238,
    321-323, 330, 586
TIBA 209, 355-356, 511, 517,
    565, 593, 751, 778
Tickseed 620
Tillering 764
Tissue 775
    age 26
    plasticity 793
    polarity **588-594**
    printing 235
    responsiveness 25
Tissue culture 5, 7, 163-165, 189,
    210, 228, 232, 360, 388,
    434, 533, 536, 688, **774-
    793**
    2,4-D 791
    auxins 774
    brassinosteroids 210
    carrot 774
    cytokinin 540
    embryos 782
    nitrogen 791
    overall considerations 788
    pH 790, 791
    tobacco 774
tml gene 324
tmr gene 321, 324, 328-331
tms gene 321, 324-325, 333,
    422-424
Tobacco 195, 277, 279, 320, 440,
    456, 620, 623, 624, 777
    ABA mutants 679
    apical dominance 346
    auxin 220, 229, 231, 233,
    278, 281, 282, 309, 356,
    357
        binding protein 283
        conjugate 342
        receptor 286
        resistant mutant 288
    Ca^{2+} 299
    callus bioassay 417
    CaM gene, 307
    cDNA 232, 237, 280
    cytokinin 102

Tobacco - cytokinin cont:
    oxidase 112
    synthesis 327
clones 235, 236
crown gall 328, 775
2,4-D 279
division 775
Em-GUS 685
ethylene 30
    receptors 133
flowering 165, 620
grafting 706
gravistimulation 239
ipt transformed 105
jasmonate 182
leaf yellowing 762
maleic hydrazide 753
morphogenesis 782
NAA 287
ovules 162
polyamines 162, 164-165, 168
PBR-1b gene 30
proliferation 277
protoplasts 24
regeneration 277
rol genes 587
salicylic acid 189, 195, 197-200
spermidine 10
sucker 752, 768
teratoma 319
thermogenesis 194
tissue culture 774, 779
transformed 22, 359
transgenic plants 174, 175,
    305, 344, 361-363, 398,
    400, 403
    callus 365
    ethylene 581
tumors 324
vascular differentiation 347
vesicles 30
wounding 403
Tobacco mosaic virus 195-200
Tobacco necrosis virus 196
TOM13 cDNA 378, 380
Tomato 124, 374, 440, 521, 620,
    727 729, 737
ABA 153, 443, 732, 733
    deficient mutants 463, 663
    mutants 580, 679, 680
ACC 730, 731
    deaminase 348, 386, 729
    oxidase 128, 375, **377-380**,
        382-385, 387, 730
    synthase 121, 122, 343,
        375, **380-382**, **385-387**,
        730, 731
    clone 385
alcobaca (*alc*) 390, 728, 736

antisense fruit 729
apical dominance 577
arginine decarboxylase gene
    174
auxin, fruit 757, 593
    binding protein 285, 355
brassinolides 208, 209
CaM 307
carotenoid biosynthesis 380
cDNA 374, 377
chilling 732
cytokinin
Craigella 577
crown gall 324
cytokinin 107-108, 578, 736
2,4-D 734
dgt mutation 285, 492
dominant fruits 578
embryo mRNAs 682
epinastic 737
ethephon 502, 752, 761
ethylene **372-391**, 502, 727,
    728, 731, 737
    binding protein 292
    receptors 133
    mutants **372-391**, 731
    lateral bud 581
flc mutant 462
flowering 620
fruit 593
    development 166
    ripening 167, **372-391**,
        347-348, 502, 728, 729,
        732
        ethylene sensitivity 27
        mutants 374, 728
        stages 728
        transgenic 351, 388
    set 753, 754
GA 68, 735
    -deficient mutants 676
IAA conjugates 734
internode length mutants 467
jasmonate 182, 185, 405, 410
    737
lazy mutants 553
mutants 363, 449
    ABA 463
    ethylene insensitive 349
    sitiens 400
nor mutant 728
not mutant 462
Nr mutant 728
pericap, hormone levels 732
peroxidase 53
Pin2 gene 406, 410
polyamines 166, 390, 736,
    746
polygalacturonase 728-729
proteinase inhibitor II 396

putrescine 131, 736
rin mutant 728
salicylic acid 191
shoot apex 577
sit mutant 467
softening 729
storage life 737
transformation 375
transgenic 344, 378, 386, 389
    729
    abscission 382
    ethylene 349, 382
    ripening 351, 388
    senscence 382
water-stressed 404
wilty mutants 462, 463
wound 398-400, 403, 408
Tonoplast 221, 283, 300, 555,
    601
Totipotent 673
Touch 299, 301-306, 309
Toxic agents 122
Tracheid 537
Tradescantia 583, 584
Trans-cyclooctene 132
Trans-acting (transcription)
    factors 31, 263, 266, 366,
    686, 688-689
Transcription 123
    activation 689
    auxin-binding 279
    enhancer sequence 364
Transduction chain 30
Transformation
    Agrobacterium 457
    tomato 375
Transgenic plants 4, **340-351**,
    385-388, 586-588
    antisense 376
    apical dominance 345
    Arabidopsis 344
    auxins 345, 360
    cell division / expansion 345
    cytokinin 21, 344-345
    ethylene 340, 344, 345, **385-388**
    fruit ripening 347-348, **385-388**
    hormone content / sensitivity
        586-588
    perception 350
    IAA 21, 344
    polyamines 174
    petunia 344
    potato 703
    rol 362
    tobacco 344
    tomato 344
    vascular differentiation 347
Transglutaminase 163

Transmembrane electrochemical
   potential 190
Transpiration 33, 190, 349, 404,
   462, 478, 510, 599,
   609-611, 725, 745, 767
   ABA 599
   salicylic acid 190
Transport
   auxin 512, 513, 588
   IAA 557, 592
   inhibitors, auxin 565
   transport channels 588
   vascular tissue 588
Transposable element 363
Transposon mutagenesis 324, 689
Tree 188, 206, 403, 492-493,
   501-502, 547, 727, 735,
   757, 758, 760, 761, 766,
   767, 786, 787
Triademefon 763
Triazole 763, 766, 770
S,S,S-Tributlyphosphorotrithioate
   751
2,4,5-Trichlorophenoxyacetic acid
   (2,4,5-T) 233, 235
2,4,6-Trichlorophenoxyacetic acid
   (2,4,6-T) 233, 778
2,4,5-Trichlorophenoxypropionic
   acid 753, 755
3,4,5-Trihydrobenzoic acid propyl
   ester 409
Triiodobenzoic acid - see TIBA
Trimethylsilyl 422-424
Triterpenes 68
Triticum aestivum - see wheat
tRNA 100-104
Tropisms 509
Trp2-1 gene 456
Trypsin inhibitor 182
Tryptophan 43-47, 60, 235, 341,
   358-360, 654
   IAA 323
   biosynthesis
      anthranilate 456
      mutants 456
   monooxygenase 48
      genes 341
   mutants
      Arabidopsis 456
      maize 456
   racemization 45
Tryptophan synthase 46, 456
Tryptophan monooxygenase 325-
   326
Tsuga heterophylla 626
Tuber 725, 759
   dormancy 701
   induction, 707
   initiation 699, 701, 711
   sink 702

sprouting 700
starch 703, 704
Tuberization **698-721**
   ABA 709, 713-714, 717, 719,
      720
   Agrobacterium 717
   alleles 701
   apical dominance 717
   assimilate partitioning 702
   auxin 712-713
   axillary buds 698
   bioassay 715
   buds 698, 699, 715, 719
   carbohydrate 703-704, 720
   CCC 713, 716, 719
   cell division 716
   coumarin 716
   cuttings 698, 699
   cytokinins 711-714, 717, 719,
      720
   environment 699
   epitasis 701
   ethephon 713
   ethylene 713-714
   GA 701, 707-709, 712, 713,
      716-720
   genetics 700, 717-718
   grafts 704-705
   hormonal control **706-721**
   IAA 716
   in vitro 715
   inhibitors 702, 709, 719
   jasmonates 181, 709-711, 717,
      719
   12-OH-jasmonic acid 709-
      711
   kinetin 717
   leaf area 702
   light 718
   loci 701
   NAA 716
   nucleic acid synthesis 716
   ornithine decarboxylase 716
   patatin 703
   phenolics 716, 719
   photoperiods 699, 705, 706,
      717-719
   plant size 702
   polyamines 716
   protein synthesis 716
   proteinase inhibitor 703
   RFLP 720
   signal 705
   starch 703, 704
   stimulus 705-706, 719, 720
   stolons 716, 719
   storage proteins 703
   sucrose 703, 716
   T-DNA 717
   temperature 699, 718

Tuberonic acid 3, 10, 710-711,
   741
   glycoside 711, 719, 720
Tulip 490
Tumor 21, 48, 318-335, 362,
   366, 434, 440, 586, 775
   cytokinin 326-333
   IAA 323-326
Tumor-inducing plasmid - see Ti
   plasmid
Tumorigenesis 319-322
Turfgrass 765, 768
Turgor 214, 218, 223, 225, 599,
   612
Turnip 738
Tyrosine kinase 275, 308
tzs gene 327, 330-334

Ubiquitin 350, 454-455
UDP-glucose:SA-glucosyltrans-
   ferase 199
Uunderground storage organs **738-
   742**
Uniconazole 85, 763-764, 766
Uracil 768
Uromyces phaseoli 173
UV-light 690

Vacuole 153, 308, 310, 354, 377
Valinomycin 514
Vanadate 216
Vascular tissue **531-544**
   auxin 534-539
   Ca²⁺ 307
   cambium 347
   cytokinin 539-543
   development 509
      iaaL gene 456
   differentiation 531, 534
      adventitious roots 541
      auxin 347, 536
      cytokinin 540
      leaves 534-539
      lignified 347
      polar pattern 541
      polyamines 164
      root 539
      transgenic plants 347
      wall 347
   elements 544
   IAA 589
   network 535
   system 538
   leaves 536
   tissue transport 589
veg gene 473, 476
Vegetables **725-747**
Vegetative reversion 476
   propagation 553, 784

Vegetative storage protein 182-83, 405
   genes 184, 186
Venus' fly-trap 567
Verapamil 312
Vernalization 473, 476, **618-619**, 623, 640, 641, 741
   apical bud 476
   GA 640, 741
   ent-kaurenoic acid 640-641
Vesicles 27, 30, 220, 221, 308, 312, 355, 356, 443, 514-520, 522
   IAA 520
   NPA binding 520
Vessel 532-534, 536-541
Vicia faba 604-605, 660
   ABA 144, 153, 311
   AVG 581
   axillary buds 581, 588
   decapitation 576
   ethylene metabolism 134
   guard cell 291, 311
   IAA catabolism 53
      oxidation 55
   IAAsp 58
Victoria regia 193
Vigna 10, 167
Vinca rosea 102, 104, 106, 112, 113
Violaxanthin 141, 145-150
Vir gene 321, 322, 330
Virulence 189, 321, 322, 326, 331
Virus-free stock 785
Vitamins 10
Vitis 543
Vitrification 787
Viviparous mutants 26, 400, 463, 467, 672-673, 677, 680
Voodoo lily 193
Vp1 gene 466, 680, 689, 690

Walnuts 760
Water balance **598-613**
   deficit 404, 598
   logging 122
   potential 33, 599-601
   relations 679
   stress 9, 26, 27, 30, 146-148, 310, 404, 409, 443, 462, 463, **611-613**, 658, 745
      ABA 140-141, 145, 148, 409, 463, 599, 610
      abscisic acid 462
      auxins 610
      cytokinins 610
      ethylene 122-123
      mRNAs 682
      xanthophyll 147

   transport 467
   use 767
Watercress 730
Water lily 193
Wax apple 496
Went, F. W. 2
Wheat 123, 656, 661
   ABA 598, 613, 678, 692
   ABRE 687
   ACC 120
   α-amylase 260, 264
      isozymes 252
   bisphosphate carboxylase 678
   Ca^{2+} 299, 301, 312
   cytokinin 107-109, 290
   dormancy 692
   Em 404, 683
      promoter 404, 686
      GUS 685
   embryo 677, 679
      mRNAs 682
      protein 681
   ethylene 123
   flowering 620
   germ 230
   germination 252
   grain length 676
   internode length mutants 467
   lazy mutants 553
   lodging 553, 751
   nuclear proteins 688
   precocious germination 673
   spring 613
   ubiquitin-activating enzyme 454
   upward bending 553
   winter 662
Willow 201, 441
Wilting 598
Wilty mutants **462-467**, 679
   ABA 467
   Arabidopsis 463
   Capsicum 463
   corn 463
   guard cells 467
   pea 463
   potato 463
   tomato 462
   water transport 467
Wind 299
Wine 761
Witches broom disease 319
Wolffia 190
Wood 539
Woody plants 140, 626
Wound(ing) **395-411**, 542, 544, 738
   ABA 400-405, 690
   ACC oxidase 375, 378, 730
      synthase 730

aspirin 406
auxin 398 536
cell wall protein 738
cytokinin 540
ethylene 122, 377, 379, 500-501, 738
gene 395
hormones 395
jasmonate 182, 184-185, 405-411
polyamines 738
proteinase inhibitor 396-411
responses 190
salicylic acid 190-191, 196
signal transduction 409-411
vegetative storage protein 184

Xanthium strumarium 145, 165, 190, 620, 624, 625, 630, 634
Xanthophyll 141, 143, 145-149
Xanthoxin 141, 142, 145, 147, 149, 463
Xenopus 124, 125, 380
Xylanase 252
Xylem 5-7, 9, 25, 27, 33, 153, 154, 208, 210, 347, 510, 531-537, 539, 541-544, 579, 588, 589, 599, 600, 612, 655, 656, 742, 789
   ABA 656
      metabolites 154
      glucosyl ester 153
   auxin 535
      transport 510
   cavitation 612
   collateral bundles 535
   cytokinins 656
   differentiation 531, 535, 588
   regeneration 542
   sap 612
   water-stressed 612
Xyloglucan endotransglycosylase 222
Xyloglucans 222, 223, 225
Xylopyranosidase 254

YAC 449, 454
Yam 741
Yeast 50, 124, 164, 378, 692
   ACC 378
   artificial chromosomes 449, 454
   Ca^{2+} 299
   CaM 307
   elicitors 299
   ethylene 378
   polyamines 172
   putrescine 164
   transcription factors 686

Yeast continued:
   transformed 378
   ubiquitin-activating enzyme
      454
Yield 191

Zea mays - see Maize
Zeatin 7, 98, 100, 101, 105-107,
     110-112, 327, 330, 332,
     342, 542, 652
   9-alanyl conjugates 110
   axillary bud 578
   cauliflower 746
   GC-MS 424
   identification of 415
   metabolites 108
   molecular ion424

mutant 457
regeneration of vessels 540
stomatal opening 609
tissue culture 779
tomato 736
tumors 328
Zeatin reductase 111
Zeatin riboside 100, 103, 105-
     107, 112, 327, 332, 652
   cis 711
   immunoassays 437
   RIA 436
   transgenic plants 21
   tumors 328
Zeatin ribotide 105-107, 736
Zeatin-O-glucoside 110, 736

Zeatin 7/9-glucoside 109, 112
Zeaxanthin 149, 465
Zinc-finger 686
Zinnia elagans 631
ZIP 686
ZK139817 409
Zm-p60.1 359
Zucchini 121, 122, 285, 343,
     356, 380, 382, 514,
     517-520, 524, 526, 725
   ACC synthase 121, 122, 343,
     382
   auxin 355
   auxin-binding protein 285,
     355
   vesicles 520